ERAU-PRESCOTT LIBRARY

Engineering Design Reliability
HANDBOOK

Engineering Design Reliability
HANDBOOK

Edited by
Efstratios Nikolaidis
Dan M. Ghiocel
Suren Singhal

CRC PRESS

Boca Raton London New York Washington, D.C.

Library of Congress Cataloging-in-Publication Data

Engineering design reliability handbook / edited by Efstratios Nikolaidis, Dan M. Ghiocel, Suren Singhal
    p. cm.
  Includes bibliographical references and index.
  ISBN 0-8493-1180-2 (alk. paper)
   1. Engineering design—Handbooks, manuals, etc. 2. Reliability (Engineering)—Handbooks, manuals, etc. I. Nikolaidis, Efstratios. II. Ghiocel, Dan M. III. Singhal, Suren.

TA174.E544 2004
620'.00452—dc22
                                                                                            2004045850

This book contains information obtained from authentic and highly regarded sources. Reprinted material is quoted with permission, and sources are indicated. A wide variety of references are listed. Reasonable efforts have been made to publish reliable data and information, but the author and the publisher cannot assume responsibility for the validity of all materials or for the consequences of their use.

Neither this book nor any part may be reproduced or transmitted in any form or by any means, electronic or mechanical, including photocopying, microfilming, and recording, or by any information storage or retrieval system, without prior permission in writing from the publisher.

All rights reserved. Authorization to photocopy items for internal or personal use, or the personal or internal use of specific clients, may be granted by CRC Press LLC, provided that $1.50 per page photocopied is paid directly to Copyright Clearance Center, 222 Rosewood Drive, Danvers, MA 01923 USA. The fee code for users of the Transactional Reporting Service is ISBN 0-8493-1180-2/05/$0.00+$1.50. The fee is subject to change without notice. For organizations that have been granted a photocopy license by the CCC, a separate system of payment has been arranged.

The consent of CRC Press LLC does not extend to copying for general distribution, for promotion, for creating new works, or for resale. Specific permission must be obtained in writing from CRC Press LLC for such copying.

Direct all inquiries to CRC Press LLC, 2005 N.W. Corporate Blvd., Boca Raton, Florida 33431.

Trademark Notice: Product or corporate names may be trademarks or registered trademarks, and are used only for identification and explanation, without intent to infringe.

Visit the CRC Press Web site at www.crcpress.com

© 2005 by CRC Press LLC

No claim to original U.S. Government works
International Standard Book Number 0-8493-1180-2
Library of Congress Card Number 2004045850
Printed in the United States of America 1 2 3 4 5 6 7 8 9 0
Printed on acid-free paper

Engineering Design Reliability Handbook

Abstract

Probabilistic and other nondeterministic methods for design under uncertainty are becoming increasingly popular in the aerospace, automotive, and the ocean engineering industries because they help them design more competitive products. This handbook is an integrated document explaining the philosophy of nondeterministic methods and demonstrating their benefits on real life problems.

This handbook is for engineers, managers, and consultants in the aerospace, automotive, civil, industrial, nuclear, and ocean engineering industries who are using nondeterministic methods for product design, or want to understand these methods and assess their potential. Graduate students and professors in engineering departments at U.S. and foreign colleges and universities, who work on nondeterministic methods, will find this book useful too.

This handbook consists of three parts. The first part presents an overview of nondeterministic approaches including their potential, and their status in the industry, the government and the academia. Engineers and managers from companies, such as General Motors, General Electric, United Technologies, and Boeing, government agencies, such as NASA centers and Los Alamos and Sandia National Laboratories, and university professors present their perspectives about the status and the future of nondeterministic methods and propose steps for enabling the full potential of these methods.

In the second part, the handbook presents recent advances in nondeterministic methods. First theories of uncertainty are presented. Then computational tools for assessing the uncertainty in the performance and the safety of a system, and for design decision making under uncertainty are presented. This part concludes with a presentation of methods for reliability certification.

The third part demonstrates the use and potential of nondeterministic methods by presenting real life applications in the aerospace, automotive and ocean engineering industries. In this part, engineers from the industry and university professors present their success stories and quantify the benefits of nondeterministic methods for their organizations.

Dedications

To my parents, George Nikolaidis and Theopisti Nikolaidis

Efstratios Nikolaidis

To my father, Professor Dan Ghiocel
President of the Romanian Academy of Technical Sciences
Civil Engineering Division

Dan M. Ghiocel

To my parents, my wife, Adesh, and children Rashi and Sara,
and to all-too-important mentors, colleagues, and friends

Suren Singhal

Preface

*If one does not reflect, one thinks oneself master of everything;
but when one does reflect, one realizes that one is a master of nothing.*

Voltaire

In today's competitive business environment, decisions about product design involve significant uncertainty. To succeed in this environment, one should replace traditional deterministic approaches for making design decisions with a new risk-based approach that uses rigorous models to quantify uncertainty and assess safety. Probabilistic and other nondeterministic methods for design under uncertainty are becoming increasingly popular in the aerospace, automotive, civil, defense and power industries because they help design safer and cheaper products than traditional deterministic approaches. The term "nondeterministic methods" refers to methods that account explicitly for uncertainties in the operating environment, the material properties and the accuracy of predictive models. The foundation of nondeterministic methods includes theories of probability and statistics, interval arithmetic, fuzzy sets, Dempster-Shafer theory of evidence and Information-Gap theory. Nondeterministic methods have helped companies such as General Electric, United Technologies, General Motors, Ford, DaimlerChrysler, Boeing, Lockheed Martin, and Motorola improve dramatically their competitive positions and save billions of dollars in engineering design and warranty costs. While nondeterministic methods are being implemented in the industry, researchers are making important advances on various fronts including reliability-based design, decision under uncertainty and modeling of uncertainty when data are scarce.

Companies need to educate their designers and managers about the advantages and potential of nondeterministic methods. Professors need to educate their students about nondeterministic methods and increase the awareness of administrators about the importance and potential of these methods. To respond to this need, we put together this handbook, which is an integrated document explaining the philosophy and the implementation of nondeterministic methods and demonstrating quantitatively their benefits.

This handbook is for engineers, technical managers, and consultants in the aerospace, automotive, civil, and ocean engineering industries and in the power industry who want to use, or are already using, nondeterministic methods for product design. Professors and students who work on nondeterministic methods will find this book useful too.

This handbook consists of three parts. The first part presents an overview of nondeterministic approaches including their potential and their status in the industry, academia and national labs. Engineers and managers from companies such as Boeing, United Technologies, General Motors and General Electric, government agencies, such as NASA, Los Alamos and Sandia National Laboratories of the Department of Energy, and university professors present their perspectives about the status and future of nondeterministic methods and propose steps for enabling the full potential of these methods. In the second part, the handbook presents the concepts of nondeterministic methods and recent advances in the area of nondeterministic methods including theories of uncertainty, computational tools for assessing and validating reliability, and methods for design decision-making under uncertainty. The third part demonstrates the use and potential of nondeterministic methods on applications in the aerospace, automotive, defense and ocean engineering industries. In this part, engineers from the industry and university researchers present success stories and quantify the benefits of nondeterministic methods for their organizations.

A fundamental objective of the book is to provide opinions of experts with different background, experience and personality. Therefore, there are several chapters in the book written by experts from academia, national laboratories or the industry with diverse, and occasionally conflicting, views on some issues. We believe that presenting diverse opinions makes the book content richer and more informative and benefits the reader.

We are grateful to all the authors who contributed to this book. We appreciate the considerable amount of time that the authors invested to prepare their chapters and their willingness to share their ideas with the engineering community. The quality of this book depends greatly on the quality of the contributions of all the authors. We also acknowledge CRC Press for taking the initiative to publish such a needed handbook for our engineering community. In particular, we wish to thank Cindy Carelli, Jay Margolis, Helena Redshaw, and Jessica Vakili for their continuous administrative and editorial support.

Efstratios Nikolaidis
The University of Toledo

Dan M. Ghiocel
Ghiocel Predictive Technologies, Inc.

Suren Singhal
NASA Marshall Space Flight Center

About the Editors

Dr. Efstratios Nikolaidis is a professor of mechanical, industrial, and manufacturing engineering at the University of Toledo, Ohio. His research has focused on reliability-based design optimization of aerospace, automotive, and ocean structures, theories of uncertainty, and structural dynamics. He has published three book chapters, and more than 100 journal and conference papers, mostly on nondeterministic approaches.

Dr. Dan M. Ghiocel is the Chief Engineering Scientist of Ghiocel Predictive Technologies Inc., in Rochester, New York that is a small business specialized in nondeterministic modeling for high-complexity engineering problems. Since 2001, Dr. Ghiocel is also Adjunct Professor at Case Western Reserve University, in Cleveland, Ohio. Dr. Ghiocel has accumulated a large experience in probabilistic approaches for civil nuclear facilities and jet engine systems. He is a member with leading responsibilities in several prestigious technical committees and professional working groups, including AIAA, SAE, and ASCE.

Dr. Suren Singhal is a nationally and internationally recognized leader in the field probabilistic methods and technology. He has initiated new committees, panels, and forums in nondeterministic and probabilistic approaches. He has led development of state-of-the-art probabilistic technology and documents dealing with diverse aspects such as cultural barriers and legal issues.

Contributors

Frank Abdi
Alpha STAR Corp
Long Beach, California, U.S.A.

Joshua Altmann
Vipac Engineers & Scientists Ltd.
Port Melbourne, Australia

Andre T. Beck
Universidade Luterana do Brazil
Canoas, Brazil

Yakov Ben-Haim
Technion-Israel Institute of Technology
Haifa, Israel

Jane M. Booker
Los Alamos National Laboratory
Los Alamos, New Mexico, U.S.A.

Jeffrey M. Brown
U.S. Air Force
AFRL Wright-Patterson Air Force Base, Ohio, U.S.A.

Christian Bucher
Bauhaus-University Weimar
Weimar, Germany

John A. Cafeo
General Motors Corporation
Warren, Michigan, U.S.A.

Brice N. Cassenti
Pratt & Whitney
East Hartford, Connecticut, U.S.A.

Tina Castillo
Alpha STAR Corporation
Long Beach, California, U.S.A.

Christos C. Chamis
NASA Glenn Research Center
Cleveland, Ohio, U.SA.

Kyung K. Choi
The University of Iowa
Iowa City, Iowa, U.S.A.

William (Skip) Connon
U.S. Army Aberdeen Test Center
Aberdeen Proving Ground, Maryland, U.S.A.

Thomas A. Cruse
Vanderbilt University
Nashville, Tennessee, U.S.A.

Armen Der Kiureghian
University of California
Berkeley, California, U.S.A.

Xuru Ding
General Motors Corporation
Warren, Michigan, U.S.A.

Joseph A. Donndelinger
General Motors Corporation
Warren, Michigan, U.S.A.

Mike P. Enright
Southwest Research Institute
San Antonio, Texas, U.S.A.

Simeon H. K. Fitch
Mustard Seed Software
San Antonio, Texas, U.S.A.

Eric P. Fox
Veros Software
Scottsdale, Arizona, U.S.A.

Dan M. Frangopol
University of Colorado
Boulder, Colorado, U.S.A.

Jeffrey S. Freeman
The University of Tennessee
Knoxville, Tennessee, U.S.A.

Dan M. Ghiocel
Ghiocel Predictive Technologies
Pittsford, New York, U.S.A.

James A. Griffiths
General Electric Company
Evendale, Ohio, U.S.A.

Mircea Grigoriu
Cornell University
Ithaca, New York, U.S.A.

Jon C. Helton
Sandia National Laboratories
Albuquerque, New Mexico, U.S.A.

Luc J. Huyse
Southwest Research Institute
San Antonio, Texas, U.S.A.

Cliff Joslyn
Los Alamos National Laboratory
Los Alamos, New Mexico, U.S.A.

Lambros S. Katafygiotis
Hong Kong University of Science and Technology
Kowloon, Hong Kong

Dimitri B. Kececioglu
The University of Arizona
Tucson, Arizona, U.S.A.

Artemis Kloess
General Motors Corporation
Warren, Michigan, U.S.A.

Robert V. Lust
General Motors Corporation
Warren, Michigan, U.S.A.

Michael Macke
Bauhaus-University Weimar
Weimar, Germany

Sankaran Mahadevan
Vanderbilt University
Nashville, Tennessee, U.S.A.

Kurt Maute
University of Colorado
Boulder, Colorado, U.S.A.

Laura A. McNamara
Los Alamos National Laboratory
Los Alamos, New Mexico, U.S.A.

David Mease
University of Pennsylvania
Philadelphia, Pennsylvania
U.S.A.

Robert E. Melchers
The University of Newcastle
Callaghan, Australia

Zissimos P. Mourelatos
Oakland University
Rochester, Michigan, U.S.A.

Rafi L. Muhanna
Georgia Institute of Technology
Savannah, Georgia, U.S.A.

Robert L. Mullen
Case Western Reserve University
Cleveland, Ohio, U.S.A.

Prasanth B. Nair
University of Southampton
Southampton, U.K.

Vijayan N. Nair
University of Michigan
Ann Arbor, Michigan, U.S.A.

Raviraj Nayak
General Motors Corporation
Warren, Michigan, U.S.A.

Daniel P. Nicolella
Southwest Research Institute
San Antonio, Texas, U.S.A.

Efstratios Nikolaidis
The University of Toledo
Toledo, Ohio, U.S.A.

Ahmed K. Noor
Old Dominion University
Hampton, Virginia, U.S.A.

William L. Oberkampf
Sandia National Laboratories
Albuquerque, New Mexico, U.S.A.

Shantaram S. Pai
NASA Glenn Research Center
Cleveland, Ohio, U.S.A.

Jeom K. Paik
Pusan National University
Busan, Republic of Korea

Costas Papadimitriou
University of Thessaly
Volos, Greece

Alan L. Peltz
U.S. Army Materiel Systems Analysis Activity
Aberdeen Proving Ground, Maryland, U.S.A.

Chris L. Pettit
U.S. Naval Academy
Annapolis,
Maryland, U.S.A.

Kadambi Rajagopal
The Boeing Company
Canoga Park, California, U.S.A.

David S. Riha
Southwest Research Institute
San Antonio, Texas, U.S.A.

David G. Robinson
Sandia National Laboratories
Albuquerque, Albu Mexico, U.S.A.

Michael J. Scott
University of Illinois at Chicago
Chicago, Illinois, U.S.A.

Youngwon Shin
Applied Research Associates.
Raleigh, North Carolina, U.S.A.

Edward Shroyer
The Boeing Company
Huntington Beach, California, U.S.A.

Thomas J. Stadterman
U.S. Army Materiel Systems Analysis Activity
Aberdeen Proving Ground, Aberdeen, Maryland, U.S.A.

Agus Sudjianto
Ford Motor Company
Dearborn, Michigan, U.S.A.

Robert H. Sues
Applied Research Associates.
Raleigh, North Carolina, U.S.A.

Jun Tang
University of Iowa
Iowa City, Iowa, U.S.A.

Ben H. Thacker
Southwest Research Institute
San Antonio, Texas, U.S.A.

Anil Kumar Thayamballi
Chevron Texaco Shipping Company LLC
San Francisco, California, U.S.A.

Palle Thoft-Christensen
Aalborg University
Aalborg, Denmark

Sviatoslav A. Timashev
Russian Academy of Sciences
Ekaterinburg, Russia

Tony Y. Torng
The Boeing Company
Huntington Beach, California, U.S.A.

Robert Tryon
Vextec Corporation
Brentwood, Tennessee, U.S.A.

Jonathan A. Tschopp
General Electric Company
Evendale, Ohio, U.S.A.

Jian Tu
General Motors Corporation
Warren, Michigan, U.S.A.

Eric J. Tuegel
U.S. Air Force
AFRL Wright-Patterson Air Force Base, Ohio, U.S.A.

Erik Vanmarcke
Princeton University
Princeton, New Jersey, U.S.A.

Chris J. Waldhart
Southwest Research Institute
San Antonio, Texas, U.S.A.

(Justin) Y.-T. Wu
Applied Research Associates.
Raleigh, North Carolina, U.S.A.

Byeng D. Youn
University of Iowa
Iowa City, Iowa, U.S.A.

Contents

Part I: Status and Future of Nondeterministic Approaches (NDAs)

1 Brief Overview of the Handbook
  Efstratios Nikolaidis and Dan M. Ghiocel 1-1

2 Perspectives on Nondeterministic Approaches
  Ahmed K. Noor 2-1

3 Transitioning NDA from Research to Engineering Design
  Thomas A. Cruse and Jeffrey M. Brown 3-1

4 An Industry Perspective on the Role of Nondeterministic Technologies in Mechanical Design
  Kadambi Rajagopal 4-1

5 The Need for Nondeterministic Approaches in Automotive Design: A Business Perspective
  John A. Cafeo, Joseph A. Donndelinger, Robert V. Lust, and Zissimos P. Mourelatos 5-1

6 Research Perspective in Stochastic Mechanics
  Mircea Grigoriu 6-1

7 A Research Perspective
  David G. Robinson 7-1

Part II: Nondeterministic Modeling: Critical Issues and Recent Advances

8 Types of Uncertainty in Design Decision Making
  Efstratios Nikolaidis 8-1

9 Generalized Information Theory for Engineering Modeling and Simulation
Cliff Joslyn and Jane M. Booker 9-1

10 Evidence Theory for Engineering Applications
William L. Oberkampf and Jon C. Helton 10-1

11 Info-Gap Decision Theory for Engineering Design: Or Why "Good" Is Preferable to "Best"
Yakov Ben-Haim 11-1

12 Interval Methods for Reliable Computing
Rafi L. Muhanna and Robert L. Mullen 12-1

13 Expert Knowledge in Reliability Characterization: A Rigorous Approach to Eliciting, Documenting, and Analyzing Expert Knowledge
Jane M. Booker and Laura A. McNamara 13-1

14 First- and Second-Order Reliability Methods
Armen Der Kiureghian 14-1

15 System Reliability
Palle Thoft-Christensen 15-1

16 Quantum Physics-Based Probability Models with Applications to Reliability Analysis
Erik Vanmarcke 16-1

17 Probabilistic Analysis of Dynamic Systems
Efstratios Nikolaidis 17-1

18 Time-Variant Reliability
Robert E. Melchers and Andre T. Beck 18-1

19 Response Surfaces for Reliability Assessment
Christian Bucher and Michael Macke 19-1

20 Stochastic Simulation Methods for Engineering Predictions
Dan M. Ghiocel 20-1

21 Projection Schemes in Stochastic Finite Element Analysis
Prasanth B. Nair 21-1

22 Bayesian Modeling and Updating
    Costas Papadimitriou and Lambros S. Katafygiotis 22-1

23 Utility Methods in Engineering Design
    Michael J. Scott 23-1

24 Reliability-Based Optimization of Civil and Aerospace Structural Systems
    Dan M. Frangopol and Kurt Maute 24-1

25 Accelerated Life Testing for Reliability Validation
    Dimitri B. Kececioglu 25-1

26 The Role of Statistical Testing in NDA
    Eric P. Fox 26-1

27 Reliability Testing and Estimation Using Variance-Reduction Techniques
    David Mease, Vijayan N. Nair, and Agus Sudjianto 27-1

Part III: Applications

28 Reliability Assessment of Aircraft Structure Joints under Corrosion-Fatigue Damage
    Dan M. Ghiocel and Eric J. Tuegel 28-1

29 Uncertainty in Aeroelasticity Analysis, Design, and Testing
    Chris L. Pettit 29-1

30 Selected Topics in Probabilistic Gas Turbine Engine Turbomachinery Design
    James A. Griffiths and Jonathan A. Tschopp 30-1

31 Practical Reliability-Based Design Optimization Strategy for Structural Design
    Tony Y. Torng 31-1

32 Applications of Reliability Assessment
    Ben H. Thacker, Mike P. Enright, Daniel P. Nicolella, David S. Riha, Luc J. Huyse, Chris J. Waldhart, and Simeon H.K. Fitch 32-1

33 Efficient Time-Variant Reliability Methods in Load Space
    Sviatoslav A. Timashev 33-1

34 Applications of Reliability-Based Design Optimization
Robert H. Sues, Youngwon Shin, and (Justin) Y.-T. Wu 34-1

35 Probabilistic Progressive Buckling of Conventional and Adaptive Trusses
Shantaram S. Pai and Christos C. Chamis 35-1

36 Integrated Computer-Aided Engineering Methodology for Various Uncertainties and Multidisciplinary Applications
Kyung K. Choi, Byeng D. Youn, Jun Tang, Jeffrey S. Freeman, Thomas J. Stadterman, Alan L. Peltz, and William (Skip) Connon 36-1

37 A Method for Multiattribute Automotive Design under Uncertainty
Zissimos P. Mourelatos, Artemis Kloess, and Raviraj Nayak 37-1

38 Probabilistic Analysis and Design in Automotive Industry
Zissimos P. Mourelatos, Jian Tu, and Xuru Ding 38-1

39 Reliability Assessment of Ships
Jeom Kee Paik and Anil Kumar Thayamballi 39-1

40 Risk Assessment and Reliability-Based Maintenance for Large Pipelines
Sviatoslav A. Timashev 40-1

41 Nondeterministic Hybrid Architectures for Vehicle Health Management
Joshua Altmann and Dan M. Ghiocel 41-1

42 Using Probabilistic Microstructural Methods to Predict the Fatigue Response of a Simple Laboratory Specimen
Robert Tryon 42-1

43 Weakest-Link Probabilistic Failure
Brice N. Cassenti 43-1

44 Reliability Analysis of Composite Structures and Materials
Sankaran Mahadevan 44-1

45 Risk Management of Composite Structures
Frank Abdi, Tina Castillo, and Edward Shroyer 45-1

Index I-1

Disclaimer

The views and opinions expressed in this book are strictly those of the contributors and the editors and they do not necessarily reflect the views of their companies, their organizations, or the government.

I

Status and Future of Nondeterministic Approaches (NDAs)

1. **Brief Overview of the Handbook** *Efstratios Nikolaidis and Dan M. Ghiocel* 1-1
   Introduction

2. **Perspectives on Nondeterministic Approaches** *Ahmed K. Noor* 2-1
   Introduction • Types of Uncertainties and Uncertainty Measures • Managing Uncertainties • Nondeterministic Analysis Approaches: Categories • Enhancing the Modeling and Simulation Technologies • Verification and Validation of Numerical Simulations • Availability, Reliability, and Performance • Response Surface Methodology (RSM) • Risk Management Process • Robustness • Commercial Software Systems • Key Components of Advanced Simulation and Modeling Environments • Nondeterministic Approaches Research and Learning Network • Conclusion

3. **Transitioning NDA from Research to Engineering Design** *Thomas A. Cruse and Jeffrey M. Brown* 3-1
   Introduction • The Future Nondeterministic Design Environment • Probabilistic Methods in Transition for Systems Design • Other NDA Tools: A Transition Assessment • Transition Issues and Challenges • Conclusion

4. **An Industry Perspective on the Role of Nondeterministic Technologies in Mechanical Design** *Kadambi Rajagopal* 4-1
   Introduction • The Business Case • Strategies for Design Approach for Low-Cost Development • Role of Design Space Exploration Approaches (Deterministic and Nondeterministic) in the Product Development Phase • Sensitivity Analysis • Probabilistic Analysis Approaches • The Need and Role of Multidisciplinary Analysis in Deterministic and Nondeterministic Analysis • Technology Transition and Software Implementation • Needed Technology Advances

5. **The Need for Nondeterministic Approaches in Automotive Design: A Business Perspective** *John A. Cafeo, Joseph A. Donndelinger, Robert V. Lust, and Zissimos P. Mourelatos* 5-1
   Introduction • The Vehicle Development Process • Vehicle Development Process: A Decision-Analytic View • The Decision Analysis Cycle • Concluding Comments and Challenges

6 **Research Perspective in Stochastic Mechanics**   *Mircea Grigoriu* 6-1
   Introduction • Deterministic Systems and Input • Deterministic Systems and Stochastic Input • Stochastic Systems and Deterministic Input • Stochastic Systems and Input • Comments

7 **A Research Perspective**   *David G. Robinson* 7-1
   Background • An Inductive Approach • Information Aggregation • Time Dependent Processes • Value of Information • Sensitivity Analysis • Summary

1
Brief Overview of the Handbook

Efstratios Nikolaidis
The University of Toledo

Dan M. Ghiocel
Ghiocel Predictive Technologies

1.1 Introduction 1-1
Part I: Status and Future of Nondeterministic Approaches • Part II: Nondeterministic Modeling: Critical Issues and Recent Advances • Part III: Applications

1.1 Introduction

This handbook provides a comprehensive review of nondeterministic approaches (NDA) in engineering. The review starts by addressing general issues, such as the status and future of NDA in the industry, government, and academia, in Part I. Then the handbook focuses on concepts, methods, and recent advances in NDA, in Part II. Part III demonstrates successful applications of NDA in the industry.

1.1.1 Part I: Status and Future of Nondeterministic Approaches

Uncertainties due to inherent variability, imperfect knowledge, and errors are important in almost every activity. Managers and engineers should appreciate the great importance of employing rational and systematic approaches for managing uncertainty in design. For this purpose, they need to understand the capabilities of NDA and the critical issues about NDA that still need to be addressed. Managers and engineers should also be aware of successful applications of NDA in product design. Researchers in academia, national labs, and research centers should know what areas of future research on NDA are important to the industry.

The first part of the handbook addresses the above issues with an overview of NDA that includes the capabilities of these approaches and the tools available. This part explains what should be done to enable the full potential of NDA and to transition these approaches from research to engineering design. In Part I, technical managers from the government (Chapters 2 and 3), industry (Chapters 4 and 5), and academia and national laboratories (Chapters 6 and 7) present their perspectives on the potential of NDA technology. We believe that the information presented in this part will help the engineering community understand and appreciate the potential and benefits of the application of NDA to engineering design and help technical managers deploy NDA to improve the competitive positions of their organizations.

1.1.2 Part II: Nondeterministic Modeling: Critical Issues and Recent Advances

The second part of the book presents critical issues that are important to the implementation of NDA to real-life engineering problems, and recent advances toward addressing these issues. These issues include:

1. What are the most important sources of uncertainty in engineering design?
2. What are the most suitable theories of uncertainty and mathematical tools for modeling different types of uncertainty?
3. How do we model uncertainty in quantities that vary with time and space, such as the material properties in a three-dimensional structure, or the particle velocity in a turbulent flow-field?
4. What tools are available for uncertainty propagation through a component or a system, and for reliability assessment? What tools are available for estimating the sensitivity of the reliability of a system to uncertainties in the input variables and for identifying the most important variables?
5. How do we make decisions in the presence of different types of uncertainty?
6. How do we validate the reliability of a design efficiently?

Part II consists of a total of 20 chapters. This part is divided into three groups. The first group studies types of uncertainties and theories and tools for modeling uncertainties. The second group presents computational tools for reliability assessment, propagation of uncertainty through a component or system, and decision making under uncertainty. The third group presents methods for reliability validation using tests.

Following is a summary of the chapters in each group.

1.1.2.1 Uncertainty and Theories of Uncertainty

Chapter 8 studies existing taxonomies of uncertainty in a design decision problem. Then it proposes a new taxonomy of uncertainties that are present in the stages of the solution of a design decision problem. This taxonomy is based on the sources of uncertainty. The stages of a decision include framing of the decision, predicting the outcomes and evaluating the worth of the outcomes according to the decision-maker's preferences. The chapter also examines different types of information that can be used to construct models of uncertainty. Chapter 9 presents a general theory of uncertainty-based information. The term "uncertainty-based information" implies that information is considered a reduction of uncertainty. Theories of uncertainty, such as probability, possibility, and Dempster-Shafer evidence theory, are branches of the general theory of uncertainty-based information. The chapter presents measures for characterizing uncertainty and measures of the amount of uncertainty. Chapter 10 presents Dempster-Shafer evidence theory. The authors of this chapter argue that evidence theory can be more suitable than probability theory for modeling epistemic uncertainty. A principal reason is that evidence theory does not require a user to assume information beyond what is available. Evidence theory is demonstrated and contrasted with probability theory on quantifying uncertainty in an algebraic function of input variables when the evidence about the input variables consists of intervals. Chapters 11 and 12 present theories that are suitable for design in the face of uncertainty when there is a severe deficit in the information about uncertainty. Chapter 11 presents information-gap decision theory, which is based on the theory of robust reliability. Another method is to use an interval for each uncertain variable to characterize uncertainty. Chapter 12 presents an efficient method for computing narrow intervals for the response of a component or system when uncertainty is modeled using intervals. In many design problems, information about uncertainty is obtained mainly from experts. Chapter 13 presents a rigorous approach for eliciting, documenting, and analyzing information from experts.

1.1.2.2 Reliability Assessment and Uncertainty Propagation

The second group of chapters presents tools for propagating uncertainty through a component or system (that is, quantifying the uncertainty in the response given a model of the uncertainty in the input variables), and for assessing the reliability of a component or system. The chapters also present methods for the design of components and systems that account directly for reliability. These chapters focus on methods and tools that are based on probability theory and statistics.

First, two numerical approaches for element reliability analysis are examined in which uncertainties are modeled using random vectors. Chapter 15 presents methods for reliability assessment of systems and illustrates these methods with numerous examples involving civil engineering structures.

First- and second-order reliability methods are employed for this purpose. Chapter 16 presents a class of probability density functions that have a fundamental basis in quantum physics. The chapter explains that these probabilistic models have considerable potential for applications to reliability analysis of components and systems. Chapters 14 and 15 are confined to time-invariant problems. Chapters 17 and 18 study time-variant reliability problems in which the excitation on a structure and/or its strength vary with time. Chapter 17 primarily considers structures whose strength is constant. This chapter presents methods for modeling the excitation, finding the statistics of the response, and the failure probability. Chapter 18 provides a thorough review of methods for calculating the reliability of structures whose strength varies with time.

One way to reduce the computational cost associated with reliability assessment is to replace the deterministic model that is used for predicting the response of a system or component with a response surface. Chapter 19 presents different response surface approximation methods. These methods are widely used in the industry and are effective for stochastic problems with a small number of random variables, e.g., less than ten. Chapter 20 presents Monte Carlo simulation methods. These methods are applicable to a broad range of problems, including both time-invariant and time-variant problems. Uncertainties can be modeled using random vectors, stochastic processes, or random fields (random functions of space variables and/or time). This chapter also includes useful discussions on the use of Monte Carlo simulation in conjunction with hierarchical models for stochastic approximation, application of static, sequential and dynamic sampling, and computation of probabilities for high-dimensional problems. Chapter 21 presents stochastic finite element methods. These methods can be used to analyze the uncertainty in a component or system, when the uncertainty in the input is described using random fields. Thus, the methods allow one to accurately model the spatial variation of a random quantity, such as the modulus of elasticity in a structure. These methods convert a boundary value problem with stochastic partial differential equations into a system of algebraic equations with random coefficients. This system of equations can be solved using different numerical techniques including Monte Carlo simulation, or first- and second-order reliability methods, to compute the probability distribution of the performance parameters of a system or component and its reliability. Bayesian methods are useful for updating models of uncertainty, or deterministic models for predicting the performance of a component or system, using sampling information. Chapter 22 is concerned with statistical modeling and updating of structural models as well as their use for predicting their response and estimating their reliability.

1.1.2.3 Engineering Decision under Uncertainty

The results of nondeterministic approaches are useful for making design decisions under uncertainty. Chapters 23 and 24 present tools for decision making. Chapter 23 introduces and illustrates the concepts of utility theory and tools for assessment of the expected utility. These tools can aid decisions in the presence of uncertainty and/or decisions with multiple objectives. Chapter 24 presents methods for reliability-based design optimization. These methods account explicitly for the reliability of a system as opposed to traditional deterministic design, which accounts for uncertainty indirectly using a safety factor and design values of the uncertain variables. The methods and their advantages over deterministic design optimization methods are illustrated with problems involving civil and aerospace engineering systems and microelectromechanical systems (MEMS).

1.1.2.4 Reliability Certification

Validation of the reliability of a design through tests is an important part of design. Chapter 25 presents accelerated life testing methods for reliability validation. Chapter 26 provides a review of tools for constructing and testing models of uncertainty.

Often, we cannot afford to perform a sufficient number of physical tests to validate the reliability of a design. Chapter 27 presents a rigorous method for reducing dramatically the number of the tests required for reliability validation by testing a small set of nominally identical systems selected so as to maximize the confidence in the estimate of the reliability from the test.

1.1.3 Part III: Applications

After providing a general overview of NDA and explaining the concepts and the tools available, the handbook demonstrates how these approaches work on real-life industrial design problems in Part III. This part reports successful applications of NDA in the following companies: Alpha STAR Corporation; Applied Research Associates, Inc.; Boeing; General Electric; General Motors Corporation; Ghiocel Predictive Technologies, Inc.; Mustard Seed Software; Pratt & Whitney; Shevron Shipping Company, LLC; Vipac Engineers & Scientists LTD; and Vextec Corporation. This part consists of 18 chapters. Success stories of reliability analysis and reliability-based optimization are presented and the benefits from the use of NDA are quantified.

2
Perspectives on Nondeterministic Approaches

2.1	Introduction	2-1
	Definitions of Uncertainty • Brief Historical Account of Uncertainty Modeling • Chapter Outline	
2.2	Types of Uncertainties and Uncertainty Measures	2-4
2.3	Managing Uncertainties	2-5
2.4	Nondeterministic Analysis Approaches: Categories	2-5
	Bounding Uncertainties in Simulation Models • Quality Control and Uncertainty Management in the Modeling and Simulation of Complex Systems	
2.5	Enhancing the Modeling and Simulation Technologies	2-7
	Advanced Virtual Product Development Facilities • Safety Assessment • Uncertainty Quantification	
2.6	Verification and Validation of Numerical Simulations	2-9
2.7	Availability, Reliability, and Performance	2-10
	Availability • Reliability • Performance • Design of Experiments	
2.8	Response Surface Methodology (RSM)	2-11
2.9	Risk Management Process	2-12
2.10	Robustness	2-12
2.11	Commercial Software Systems	2-13
2.12	Key Components of Advanced Simulation and Modeling Environments...	2-13
	Intelligent Tools and Facilities • Nontraditional Methods • Advanced Human–Computer Interfaces	
2.13	Nondeterministic Approaches Research and Learning Network	2-15
2.14	Conclusion	2-17

Ahmed K. Noor
Old Dominion University

2.1 Introduction

Increasingly more complex systems are being built and conceived by high-tech industries. Examples of future complex systems include the biologically inspired aircraft with self-healing wings that flex and react like living organisms, and the integrated human robotic outpost (Figure 2.1). The aircraft is built of multifunctional material with fully integrated sensing and actuation, and unprecedented

FIGURE 2.1 Examples of future complex systems: (a) biologically inspired aircraft and (b) human-robotic outpost.

levels of aerodynamic efficiencies and aircraft control. The robots in the outpost are used to enhance the astronauts' capabilities to do large-scale mapping, detailed exploration of regions of interest, and automated sampling of rocks and soil. They could enhance the safety of the astronauts by alerting them to mistakes before they are made, and letting them know when they are showing signs of fatigue, even if they are not aware of it.

Engineers are asked to design faster, and to insert new technologies into complex systems. Increasing reliance is being made on modeling, simulation and virtual prototyping to find globally optimal designs that take uncertainties and risk into consideration. Conventional computational and design methods are inadequate to handle these tasks. Therefore, intense effort has been devoted in recent years to nontraditional and nondeterministic methods for solving complex problems with system uncertainties.

Although, the number of applications of nondeterministic approaches in industry continues to increase, there is still some confusion—arising from ill-defined and inconsistent terminology, vague language, and misinterpretation—present in the literature.

An attempt is made in this chapter to give broad definitions to the terms and to set the stage for the succeeding chapters.

Perspectives on Nondeterministic Approaches

2.1.1 Definitions of Uncertainty

Uncertainty is an acknowledged phenomenon in the natural and technological worlds. Engineers are continually faced with uncertainties in their designs. However, there is no unique definition of uncertainty. A useful functional definition of uncertainty is: the information/knowledge gap between what is known and what needs to be known for optimal decisions, with minimal risk [1–3].

2.1.2 Brief Historical Account of Uncertainty Modeling

Prior to the twentieth century, uncertainty and other types of imprecision were considered unscientific, and therefore were not addressed. It was not until the beginning of the twentieth century that statistical mechanics emerged and was accepted as a legitimate area of science. It was taken for granted that uncertainty is adequately captured by probability theory. It took 60 years to recognize that the conceptual uncertainty is too deep to be captured by probability theory alone and to initiate studies of nonprobabilistic manifestations of uncertainty, as well as their applications in engineering and science. In the past two decades, significant advances have been made in uncertainty modeling, the level of sophistication has increased, and a number of software systems have been developed. Among the recent developments are the perception-based information processing and methodology of computing with words (Figure 2.2).

2.1.3 Chapter Outline

The overall objective of this chapter is to give a brief overview of the current status and future directions for research in nondeterministic approaches. The number of publications on nondeterministic approaches has been steadily increasing, and a vast amount of literature currently exists on various aspects and applications of nondeterministic approaches. The cited references are selected for illustrating the ideas presented and are not necessarily the only significant contributions to the subject. The discussion in this chapter is kept on a descriptive level; for details, the reader is referred to the cited literature.

The topics of the succeeding sections include types of uncertainties and uncertainty measures; managing uncertainties; nondeterministic analysis approaches; enhancing the modeling and simulation technologies; verification and validation of numerical simulations; risk management process; key

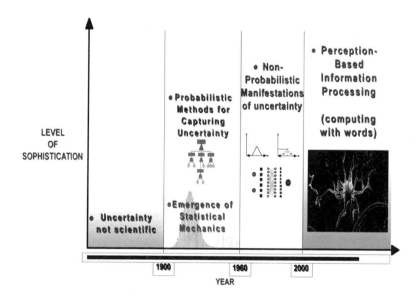

FIGURE 2.2 Evolution of uncertainty modeling.

components of advanced simulation and modeling environments; nondeterministic approaches research and learning networks.

2.2 Types of Uncertainties and Uncertainty Measures

A number of different uncertainty representations and classifications have been proposed. Among these are [1,3–5]:

- The two-type classification — aleatory and epistemic uncertainty in the modeling and simulation process. *Aleatory* uncertainty (originating from the Latin *aleator* or *aleatorius*, meaning dice thrower) is used to describe the inherent spatial and temporal variation associated with the physical system or the environment under consideration as well as the uncertainty associated with the measuring device. Sources of aleatory uncertainty can be represented as randomly distributed quantities. Aleatory uncertainty is also referred to in the literature as variability, irreducible uncertainty, inherent uncertainty, and stochastic uncertainty.

    Epistemic uncertainty (originating from the Greek *episteme*, meaning knowledge) is defined as any lack of knowledge or information in any phase or activity of the modeling process. Examples of sources of epistemic uncertainty are scarcity or unavailability of experimental data for fixed (but unknown) physical parameters, limited understanding of complex physical processes or interactions of processes in the engineering system, and the occurrence of fault sequences or environment conditions not identified for inclusion in the analysis of the system.

    An increase in knowledge or information, and/or improving the quality of the information available can lead to a reduction in the predicted uncertainty of the response of the system. Therefore, epistemic uncertainty is also referred to as reducible uncertainty, subjective uncertainty, and model form uncertainty.
- The two-type classification — uncertainty of information and uncertainty of the reasoning process
- The three-type classification — fundamental (or aleatory), statistical, and model uncertainties. The statistical and model uncertainties can be viewed as special cases of the epistemic uncertainty described previously.
- The three-type classification—variabilities, epistemic uncertainty, and errors.
- The six-type classification with the following six types of uncertainties (Figure 2.3):
    - Probabilistic uncertainty, which arises due to chance or randomness
    - Fuzzy uncertainty due to linguistic imprecision (e.g., set boundaries are not sharply defined)
    - Model uncertainty, which is attributed to a lack of information about the model characteristics
    - Uncertainty due to limited (fragmentary) information available about the system (e.g., in the early stage of the design process)
    - Resolutional uncertainty, which is attributed to limitation of resolution (e.g., sensor resolution)
    - Ambiguity (i.e., one-to-many relations)

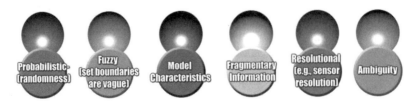

FIGURE 2.3 Six types of uncertainties.

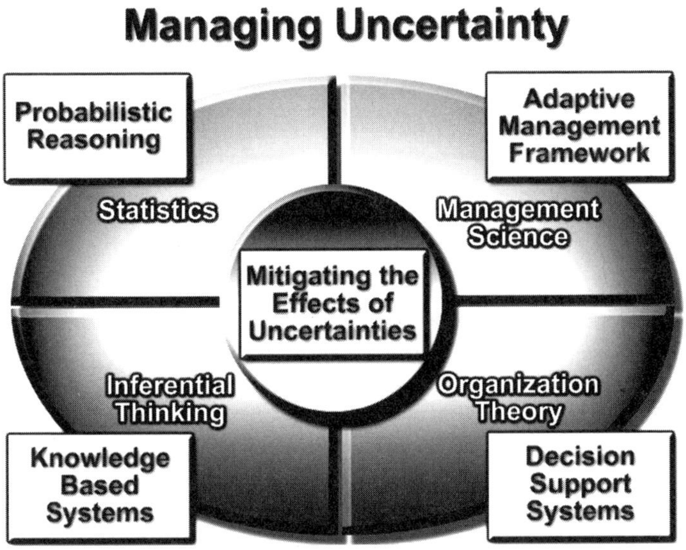

Perspectives on Nondeterministic Approaches 2-5

FIGURE 2.4 Disciplines involved in uncertainty management.

In this classification, the probabilistic uncertainty is the same as the aleatory uncertainty of the first classification; and the remaining five types can be viewed as epistemic uncertainties.

2.3 Managing Uncertainties

While completely eliminating uncertainty in engineering design is not possible, reducing and mitigating its effects have been the objectives of the emerging field of uncertainty management. The field draws from several disciplines, including statistics, management science, organization theory, and inferential thinking (Figure 2.4).

In developing strategies for managing uncertainties, the following items must be considered [1]:

- Developing formal, rigorous models of uncertainty
- Understanding how uncertainty propagates through spatial and temporal processing, and decision making
- Communicating uncertainty to different levels of the design and development team in meaningful ways
- Designing techniques to assess and reduce uncertainty to manageable levels for any given application
- Being able to absorb uncertainty and cope with it in practical engineering systems

2.4 Nondeterministic Analysis Approaches: Categories

Depending on the type of uncertainty and the amount of information available about the system characteristics and the operational environments, three categories of nondeterministic approaches can be identified for handling the uncertainties. The three approaches are [6–12] (Figure 2.5) probabilistic analysis, fuzzy-set approach, and set theoretical, convex (or antioptimization) approach. In probabilistic analysis, the system characteristics and the source variables are assumed to be random variables (or functions), and the joint probability density functions of these variables are selected. The main objective of the analysis is the determination of the reliability of the system. (Herein, reliability refers to the successful operation of the system.)

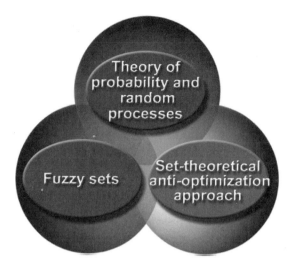

FIGURE 2.5 Three categories of nondeterministic approaches.

If the uncertainty is because of a vaguely defined system and operational characteristics, imprecision of data, and subjectivity of opinion or judgment, fuzzy-set treatment is appropriate. Randomness describes the uncertainty in the occurrence of an event (such as damage or failure).

When the information about the system and operational characteristics is fragmentary (e.g., only a bound on a maximum possible response function is known), then convex modeling is practical. Convex modeling produces the maximum or least favorable response and the minimum or most favorable response of the system under the constraints within the set-theoretic description.

2.4.1 Bounding Uncertainties in Simulation Models

Current synthesis approaches of simulation models involve a sequence of four phases (Figure 2.6):

1. Selection of the models, which includes decisions about modeling approach, level of abstraction, and computational requirements. The complexities arise due to:
   - Multiconstituents, multiscale, and multiphysics material modeling
   - Integration of heterogeneous models

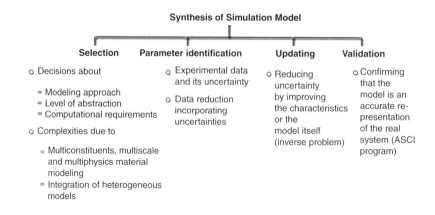

FIGURE 2.6 Four phases involved in the synthesis of simulation models.

2. Parameter identification. Data reduction techniques are used that incorporate uncertainties
3. Model updating, or reducing uncertainty by improving either the model characteristics or the model itself
4. Validation, in the sense of confirming that the model is an accurate representation of the real system

2.4.2 Quality Control and Uncertainty Management in the Modeling and Simulation of Complex Systems

The estimation of total uncertainty in the modeling and simulation of complex systems involves: (1) identification and characterization of the sources of uncertainty, variability, and error; (2) uncertainty propagation and aggregation; and (3) uncertainty quantification. [4,5,13]

Herein, uncertainty is defined as the epistemic uncertainty — a deficiency in any phase of the modeling process due to lack of knowledge (model form or reducible uncertainty), increasing the knowledge base can reduce the uncertainty. The term "variability" is used to describe inherent variation associated with the system or its environment (irreducible or stochastic uncertainty). Variability is quantified by a probability or frequency distribution. An error is defined as a recognizable deficiency that is not due to lack of knowledge. An error can be either acknowledged (e.g., discretization or round-off error) or unacknowledged (e.g., programming error). Uncertainty quantification is discussed in a succeeding subsection.

2.5 Enhancing the Modeling and Simulation Technologies

The synergistic coupling of nondeterministic approaches with a number of key technologies can significantly enhance the modeling and simulation capabilities and meet the needs of future complex systems. The key technologies include virtual product development for simulating the entire life cycle of the engineering system, reliability and risk management, intelligent software agents, knowledge and information, high-performance computing, high-capacity communications, human–computer interfaces, and human performance.

2.5.1 Advanced Virtual Product Development Facilities

Current virtual product development (VPD) systems have embedded simulation capabilities for the entire life cycle of the product. As an example, the top-level system process flow for a space transportation system is shown in Figure 2.7. In each phase, uncertainties are identified and appropriate measures are

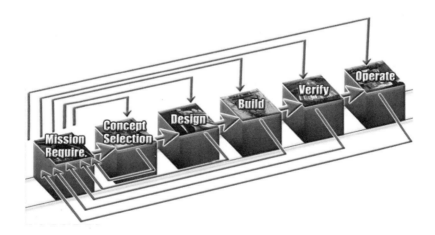

FIGURE 2.7 Top-level system process flow for a space transportation system.

taken to mitigate their effects. Information technology will change the product development from a sequence of distinct phases into a continuous process covering the entire life cycle of the product with full interplay of information from beginning to end and everywhere throughout.

2.5.2 Safety Assessment

Safety assessment encompasses the processes and techniques used to identify risks to engineering systems and missions. The evaluation and identification of risk are based on:

- Probabilistic risk assessment
- Fault tree analysis
- Hazard analysis and failure mode effect analysis coupled with engineering judgment

2.5.3 Uncertainty Quantification

Special facilities are needed for performing uncertainty quantification on large-scale computational models for complex engineering systems. Uncertainty quantification covers three major activities[4,5,13] (Figure 2.8):

1. *Characterization of uncertainty* in system parameters and the external environment. This involves the development of mathematical representations and methods to model the various types of uncertainty. Aleatory uncertainty can generally be represented by a probability or frequency distribution when sufficient information is available to estimate the distribution. Epistemic uncertainty can be represented by Bayesian probability, which takes a subjective view of probability as a measure of belief in a hypothesis. It can also be represented by modern information theories, such as possibility theory, interval analysis, evidence theory (Dempster/Shafer), fuzzy set theory, and imprecise probability theory. These theories are not as well developed as probabilistic inference.
2. *Propagation of uncertainty* through large computational models. The only well-established methodology for uncertainty propagation is based on probability theory and statistics. Use of response surfaces for approximation of the performance of a component or a system can increase the efficiency of this methodology. Nonprobabilistic approaches, which are based on possibility theory,

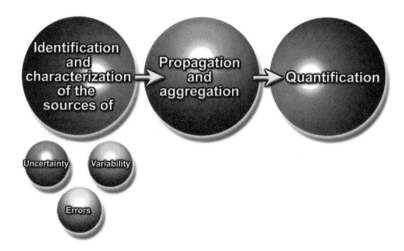

FIGURE 2.8 Major activities involved in uncertainty quantification.

Dempster-Shafer evidence theory, and imprecise probability, are currently investigated. These approaches can be useful when limited data is available about uncertainty.

3. *Verification and validation of the computational models*, and incorporating the uncertainty of the models themselves into the global uncertainty assessment. The verification and validation process is described in the succeeding section.

2.6 Verification and Validation of Numerical Simulations

Quantifying the level of confidence, or reliability and accuracy of numerical simulations has recently received increased levels of attention in research and engineering applications. During the past few years, new technology development concepts and terminology have arisen. Terminology such as virtual prototyping and virtual testing is now being used to describe computer simulation for design, evaluation, and testing of new engineering systems.

The two major phases of modeling and simulation of an engineering system are depicted in Figure 2.9. The first phase involves developing a conceptual and mathematical model of the system. The second phase involves discretization of the mathematical model, computer implementation, numerical solution, and representation or visualization of the solution. In each of these phases there are epistemic uncertainties, variabilities and errors [5,14].

Verification and validation are the primary methods for building and quantifying confidence in numerical simulations. *Verification* is the process of determining that a model implementation, and the solution obtained by this model, represent the conceptual/mathematical model and the solution to the model within specified limits of accuracy. It provides evidence or substantiation that the conceptual model is solved correctly by the discrete mathematics embodied in the computer code. Correct answer is provided by highly accurate solutions. *Validation* is the process of substantiating that a computational model within its domain of applicability possesses a satisfactory range of accuracy in representing the real system, consistent with the intended application of the model. Correct answer is provided by experimental data.

Validation involves computing validation metrics to assess the predictive capability of the computational models, based on distributional predication and available experimental data or other known information.

The development of such metrics is a formidable task, and is the focus of intense efforts at present. Model validation is intimately linked to uncertainty quantification, which provides the machinery to perform the assessment of the validation process.

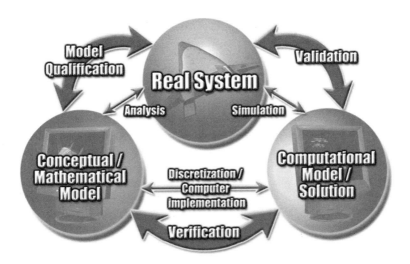

FIGURE 2.9 Verification and validation of numerical simulations.

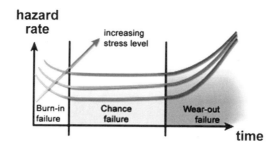

FIGURE 2.10 Failure rate over the lifetime of a component—reliability bathtub curves.

2.7 Availability, Reliability, and Performance

2.7.1 Availability

Availability of a component (or a system) is defined as the probability that the component (or system) will perform its intended functions at a given time, under designated operating conditions, and with a designated support environment. It is an important measure of the component (or system) performance—its ability to quickly become operational following a failure. Mathematically, availability is a measure of the fraction of the total time a component (or a system) is able to perform its intended function.

2.7.2 Reliability

Reliability is defined as the probability that a component (or a system) will perform its intended function without failure for a specified period of time under designated operating conditions. Failure rate or hazard rate is an important function in reliability analysis because it provides a measure of the changes in the probability of failure over the lifetime of a component. In practice, it often exhibits a bathtub shape [15–30] (see Figure 2.10).

Reliability assessment includes selection of a reliability model, analysis of the model, calculation of the reliability performance indices, and evaluation of results, which includes establishment of confidence limits and decision on possible improvements (Figure 2.11).

2.7.3 Performance

Reliability and availability are two of four key performance measures of a component (or a system). The four measures of performance are:

1. Capability: Ability to satisfy functional requirements
2. Efficiency: Ability to effectively and easily realize the objective

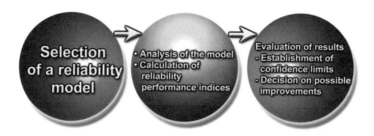

FIGURE 2.11 Reliability assessment.

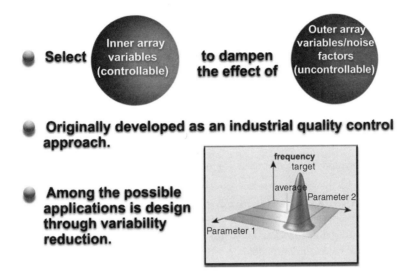

FIGURE 2.12 Taguchi's method.

3. Reliability: Ability to start and continue to operate
4. Availability: Ability to operate satisfactorily when used under specific conditions (in the limit it is the operating time divided by the sum of operating time plus the downtime)

2.7.4 Design of Experiments

These are systematic techniques for investigating (all possible) variations in system performance due to changes in system variables [31].

Two categories of system variables can be identified, namely, (1) inner-array variables, which are controllable; and (2) outer-array variables (also called noise factors), which are functions of environmental conditions and are uncontrollable. Three categories of techniques can be identified: (1) regression analysis, (2) statistical methods, and (3) Taguchi's method [32].

In Taguchi's method, the controllable variables are selected in such a way as to dampen the effect of the noise variables on the system performance. The method was originally developed as an industrial total quality control approach. Subsequently, it has found several other applications, including design optimization through variability reduction (Figure 2.12).

2.8 Response Surface Methodology (RSM)

RSM is a set of mathematical and statistical techniques designed to gain a better understanding of the overall response by designing experiments and subsequent analysis of experimental data. [33] It uses empirical (nonmechanistic) models, in which the response function is replaced by a simple function (often polynomial) that is fitted to data at a set of carefully selected points. RSM is particularly useful for the modeling and analysis of problems in which a response of interest is influenced by several variables and the objective is to optimize the response.

Two types of errors are generated by RSM. The first are random errors or noise, which can be minimized through the use of minimum variance based design. The second is bias or modeling error, which can be minimized using genetic algorithms.

RSM can be employed before, during, or after regression analysis is performed on the data. In the design of experiments it is used before, and in the application of optimization techniques it is used after regression analysis.

FIGURE 2.13 Tasks involved in the risk management process.

2.9 Risk Management Process

Risk is defined as the uncertainty associated with a given design, coupled with its impact on performance, cost, and schedule. Risk management is defined as the systems engineering and program management tools that can provide a means to identify and resolve potential problems [14,18,34,35].

The risk management process includes the following tasks (Figure 2.13):

- Risk planning: development of a strategy for identifying risk drivers
- Risk identification: identifying risk associated with each technical process
- Risk analysis: isolating the cause of each identified risk category and determining the effects

The combination of risk identification and risk analysis is referred to as risk assessment:

- Risk handling: selecting and implementing options to set risk at acceptable levels
- Risk monitoring: systematically tracking and evaluating the performance of risk handling actions

2.10 Robustness

Robustness is defined as the degree of tolerance to variations (in either the components of a system or its environment). A robust ultra-fault-tolerant design of an engineering system is depicted. The performance of the system is relatively insensitive to variations in both the components and the environment. By contrast, a nonrobust design is sensitive to variations in either or both (Figure 2.14).

An example of a robust, ultra-fault-tolerant system is the Teramac computer, which is a one-terahertz, massively parallel experimental computer built at Hewlett-Packard Laboratories to investigate a wide range of computational architectures. It contains 22,000 (3%) hardware defects, any one of which could prove fatal to a more conventional machine. It incorporates a high communication bandwidth that enables it to easily route around defects. It operates 100 times faster than a high-end single processor workstation (for some of its configurations); see Figure 2.15.

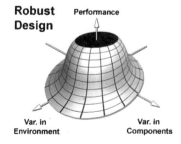 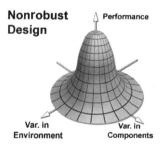

FIGURE 2.14 Robust and nonrobust designs.

- Massively experimental computer built at Hewlett-Packard Labs
- Contains 220,000 hardware defects
- Incorporates a high-communication bandwidth that enables it to easily route around defects
- Operates 100 times faster than a high-end single-processor workstation (for some of its configurations)

FIGURE 2.15 Teramac configurable custom computer.

2.11 Commercial Software Systems

A number of commercial codes have been developed to enable quantifying the variability and uncertainty in the predictions, as well as performing probabilistic design, reliability assessment, and process optimization. Examples of these codes are NESSUS, DARWIN, ProFES, UINPASS, GENOA, DAKOTA, and structural engineering reliability analysis codes such as CalREL, OpenSees, FERUM, and FSG. Also, some of the deterministic analysis codes, like ANSYS and MSC Nastran, are adding probabilistic tools to enable incorporating uncertainties into the analysis.

NESSUS and DARWIN were developed by Southwest Research Institute. NESSUS (Numerical Evaluation of Stochastic Structures Under Stress) integrates reliability methods with finite element and boundary element methods. It uses random variables to model uncertainties in loads, material properties, and geometries [36,37]. DARWIN (Design Assessment of Reliability With Inspection) predicts the probability of fracture of aircraft turbine rotor disks [38,39]. ProFES (Probabilistic Finite Element System) was developed by Applied Research Associates, Inc [40,41]. It performs probabilistic simulations using internal functions. It can also be used as an add-on to the commercial finite element codes or CAD programs. UNIPASS (Unified Probabilistic Assessment Software System) was developed by Prediction Probe, Inc [42,43]. It can model uncertainties, compute probabilities, and predict the most likely outcomes—analyzing risk, identifying key drivers, and performing sensitivity analysis. GENOA was developed by Alpha Star Corporation [44]. It is an integrated hierarchical software package for computationally simulating the thermal and mechanical responses of high-temperature composite materials and structures. It implements a probabilistic approach in which design criteria and objectives are based on quantified reliability targets that are consistent with the inherently stochastic nature of the material and structural properties.

DAKOTA (Design Analysis Kit for Optimization and Terascale Applications) was developed by Sandia National Labs [45]. It implements uncertainty quantification with sampling, analytic reliability, stochastic finite element methods, and sensitivity analysis with design of experiments and parameter study capabilities.

2.12 Key Components of Advanced Simulation and Modeling Environments

The realization of the full potential of nondeterministic approaches in modeling and simulation requires an environment that links diverse teams of scientists, engineers, and technologists. The essential components of the environment can be grouped into three categories: (1) intelligent tools and facilities, (2) nontraditional methods, and (3) advanced interfaces.

2.12.1 Intelligent Tools and Facilities

These include high fidelity – rapid modeling, life-cycle simulation and visualization tools, synthetic immersive environment; automatic and semiautomatic selection of software and hardware platforms, computer simulation of physical experiments, and remote control of these experiments. The visualization tools should explicitly convey the presence, nature, and degree of uncertainty to the users. A number of methods have been developed for visually mapping data and uncertainty together into holistic views. These include the methods based on using error bars, geometric glyphs, and vector glyphs for denoting the degree of statistical uncertainty. There is a need for creating new visual representations of uncertainty and error to enable a better understanding of simulation and experimental data. In all of the aforementioned tools, extensive use should be made of intelligent software agents and information technology (Figure 2.16).

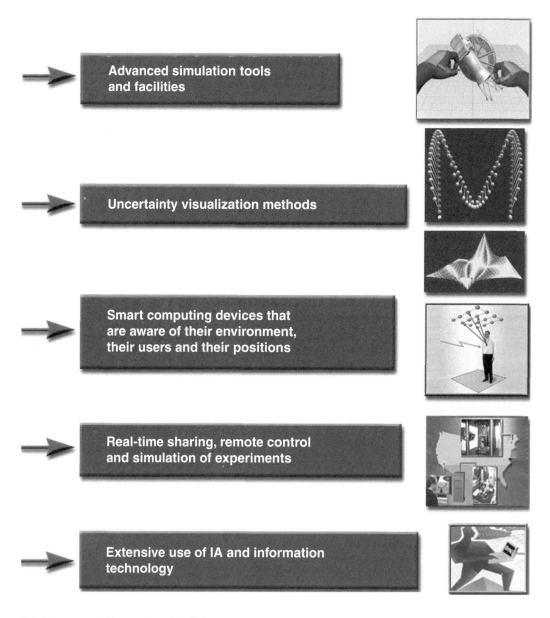

FIGURE 2.16 Intelligent tools and facilities.

2.12.2 Nontraditional Methods

These include multi-scale methods, strategies for highly coupled multi-physical problems, and nondeterministic approaches for handling uncertainty in geometry, material properties, boundary conditions, loading, and operational environments.

2.12.3 Advanced Human–Computer Interfaces

Although the WIMP (windows, icons, menus, pointing devices) paradigm has provided a stable global interface, it will not scale to match the myriad of factors and uses of platforms in the future collaborative distributed environment. Recent work has focused on intelligent multimodal human–computer interfaces, which synthetically combine human senses, enable new sensory-motor control of computer systems, and improve dexterity and naturalness of human–computer interaction [46,47]. An interface is referred to as intelligent if it can:

- Communicate with the user in human language
- Perform intelligent functions
- Adapt to a specific task and user

Intelligent interfaces make the interaction with the computer easier, more intuitive, and more flexible.

Among the interface technologies that have high potential for meeting future needs are perceptual user interfaces (PUIs), and brain–computer (or neural) interfaces (BCIs). PUIs integrate perceptive, multimodal, and multimedia interfaces to bring human capabilities to bear in creating more natural and intuitive interfaces. They enable multiple styles of interactions, such as speech only, speech and gesture, haptic, vision, and synthetic sound, each of which may be appropriate in different applications. These new technologies can enable broad uses of computers as assistants, or agents, that interact in more human-like ways.

A BCI refers to a direct data link between a computer and the human nervous system. It enables biocontrol of the computer—the user can control the activities of the computer directly from nerve or muscle signals. To date, biocontrol systems have utilized two different types of bioelectric signals: (1) electroencephalogram (EEG) and (2) electromyogram (EMG). The EEG measures the brain's electric activity and is composed of four frequency ranges, or states (i.e., alpha, beta, delta, and theta). The EMG is the bioelectric potential associated with muscle movement. BCI technologies can add completely new modes of interaction, which operate in parallel with the conventional modes, thereby increasing the bandwidth of human–computer interactions. They will also allow computers to comprehend and respond to the user's changing emotional states—an important aspect of affective computing, which has received increasing attention in recent years. Among the activities in this area are creating systems that sense human affect signals, recognize patterns of affective expression, and respond in an emotionally aware way to the user; and systems that modify their behavior in response to affective cues. IBM developed some sensors in the mouse that sense physiological attributes (e.g., skin temperature and hand sweatiness), and special software to correlate those factors to previous measurements to gauge the user's emotional state moment to moment as he or she uses the mouse.

Future well-designed multimodal interfaces will integrate complementary input modes to create a synergistic blend, permitting the strengths of each mode to overcome weaknesses in the other modes and to support mutual compensation of recognition errors.

2.13 Nondeterministic Approaches Research and Learning Network

The realization of the full potential of nondeterministic approaches in the design and development of future complex systems requires, among other things, the establishment of research and learning networks. The networks connect diverse, geographically dispersed teams from NASA, other government labs, university consortia, industry, technology providers, and professional societies (Figure 2.17). The

FIGURE 2.17 Nondeterministic approaches research and learning network.

activities of the networks include development of online learning facilities, such as interactive virtual classrooms (Figure 2.18), on nondeterministic approaches.

A European network, SAFERELNET, was established in 2001 to provide safe and cost-efficient solutions for industrial products, systems facilities, and structures across different industrial sectors. The focus of the network is on the use of reliability-based methods for the optimal design of products, production facilities, industrial systems, and structures so as to balance the economic aspects associated with providing preselected safety levels, with the associated costs of maintenance and availability. The approaches used include modeling

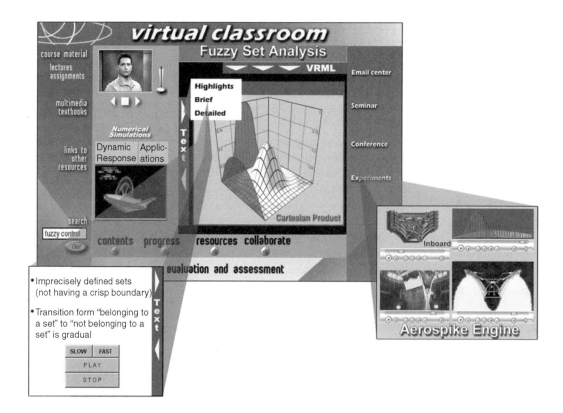

FIGURE 2.18 Interactive virtual classroom.

the reliability of the system throughout its lifetime, and an assessment of the impact of new maintenance and repair schemes on system safety, cycle costs, reliability, serviceability and quality.

2.14 Conclusion

A brief overview is presented of the current status and future directions for research in nondeterministic approaches. Topics covered include definitions of uncertainty, a brief historical account of uncertainty modeling; different representations and classifications of uncertainty; uncertainty management; the three categories of nondeterministic approaches; enhancing the modeling and simulation technologies; verification and validation of numerical simulations; reliability, availability, and performance; response surface methodologies; risk management process; key components of advanced simulation and modeling environments; and nondeterministic approaches research and learning network.

Nondeterministic approaches are currently used as integral tools for reliability, safety, and risk assessments in many industries, including aerospace, nuclear, shipbuilding, and construction.

Future complex engineering systems and information-rich environment will provide a wide range of challenges and new opportunities. The environment will be full of inherent uncertainties, vagueness, noise, as well as conflicts and contradictions. In the next few years, it is expected that the cost effectiveness of nondeterministic approaches tools will be more widely appreciated and, consequently, nondeterministic approaches will become an integral part of engineering analyses and simulations.

References

1. Katzan, H. Jr., *Managing Uncertainty: A Pragmatic Approach*, Van Nostrand Reinhold, New York, 1992.
2. Ben-Haim, Y., *Information-Gap Decision Theory: Decisions Under Severe Uncertainty*, Academic Press, St. Louis, 2001.
3. Natke, H.G. and Ben-Haim, Y., Eds., *Uncertainty: Models and Measures,* John Wiley & Sons, New York, 1998.
4. Oberkampf, W.L., Helton, J.C., and Sentz, K., Mathematical Representation of Uncertainty, presented at Non-Deterministic Approaches Forum, Seattle, WA, April 16–19, 2001, Paper No. 2001-1645, AIAA.
5. Oberkampf, W.L. et al., Error and Uncertainty in Modeling and Simulation, *Reliability Engineering and System Safety*, 75, 333–357, Elsevier Science, New York, 2002.
6. Elishakoff, I. and Tasso, C., Ed., *Whys and Hows in Uncertainty Modeling: Probability, Fuzziness and Anti-optimization,* Springer Verlag, 2000.
7. Soares, C.G., *Probabilistic Methods for Structural Design*, Kluwer Academic Publishers, New York, 1997.
8. Sundararajan, C.R., Ed., *Probabilistic Structural Mechanics Handbook: Theory and Industrial Applications*, Chapman & Hall, New York 1995.
9. Elishakoff, I., *Probabilistic Methods in the Theory of Structures*, John Wiley & Sons, New York, 1999.
10. Elishakoff, I., Possible Limitations of Probabilistic Methods in Engineering, *Applied Mechanics Reviews*, 53, 19–36, 2000.
11. Zadeh, L. and Kacprzyk, J., *Fuzzy Logic for the Management of Uncertainty*, John Wiley & Sons, New York, 1991.
12. Dimitriv, V. and Korotkich, V., Eds., *Fuzzy Logic: A Framework for the New Millennium*, Physica Verlag, New York, 2002.
13. Wojtkiewicz, S.F. et al., Uncertainty Quantification in Large Computational Engineering Models, Paper No. 2001-1455, AIAA.
14. Oberkampf, W.L. and Trucano, T.G., Verification and Validation in Computational Fluid Dynamics, *Progress in Aerospace Sciences*, 38, 209–272, Elsevier Science, New York, 2002.

15. Modarres, M., Kaminskiy, M., and Krivtsov, V., *Reliability Engineering and Risk Analysis: A Practical Guide*, Marcel Dekker, New York, 1999.
16. Ramakumar, R., *Engineering Reliability: Fundamentals and Applications*, Prentice Hall, Englewood Cliffs, NJ, 1993.
17. Knezevic, J., *Reliability, Maintainability and Supportability: A Probabilistic Approach*, McGraw-Hill, New York, 1993.
18. Dhillon, B.S., *Design Reliability: Fundamentals and Applications*, CRC Press, Boca Raton, FL, 1999.
19. Andrews, J.D. and Moss, T.R., *Reliability and Risk Assessment*, ASME Press, New York, 2002.
20. Balakrishnan, N. and Rao, C.R., Eds., *Advances in Reliability, Volume 20*, Elsevier Science and Technology Books, New York, 2001.
21. Birolini, A., *Reliability Engineering: Theory and Practice*, Springer-Verlag Telos, New York, 2001.
22. Blischke, W.R. and Murthy, D.N.P., Eds., *Case Studies in Reliability and Maintenance*, John Wiley & Sons, New York, 2002.
23. Haldar, A. and Mahadevan, S., *Probability, Reliability, and Statistical Methods in Engineering Design*, John Wiley & Sons, New York, 1999.
24. Hoang, P., *Recent Advances in Reliability and Quality Engineering*, World Scientific Publishing, River Edge, 1999.
25. Hobbs, G.K., Ed., *Accelerated Reliability Engineering: Halt and Hass*, John Wiley & Sons, New York, 2000.
26. Kuo, W. et al., *Optimal Reliability Design: Fundamentals and Applications*, Cambridge University Press, New York, 2001.
27. O'Connor, P.D.T., *Practical Reliability Engineering*, Wiley John & Sons, New York, 2002.
28. Wasserman, G.S., *Reliability Verification, Testing, and Analysis in Engineering Design*, Marcel Dekker, New York, 2002.
29. Elsayed, E.A., *Reliability Engineering*, Prentice Hall, Boston, 1996.
30. Hignett, K.C., *Practical Safety and Reliability Assessment*, Routledge, New York, 1996.
31. Coleman, H.W. and Steele, W.G., Jr., *Experimentation and Uncertainty Analysis for Engineers*, John Wiley & Sons, New York, 1999.
32. Roy, R.K., *A Primer on the Taguchi Method,* John Wiley & Sons, New York, 1990.
33. Cornell, J.A., *How to Apply Response Surface Methodology*, ASQ Quality Press, Milwaukee, WI, 1990.
34. Haimes, Y.Y., *Risk Modeling, Assessment, and Management*, John Wiley & Sons, New York, 1998.
35. Bedford, T. and Cooke, R., *Probabilistic Risk Analysis: Foundations and Methods*, Cambridge University Press, Cambridge, MA, 2001.
36. Riha, D.S. et al., Recent Advances of the NESSUS Probabilistic Analysis Software for Engineering Applications, presented at the *Structures, Structural Dynamics and Materials Conference and Exhibit Non-Deterministic Approaches Forum,* Denver, CO, April 22–25, 2002, Paper No. 2002-1268, AIAA.
37. Southwest Research Institute, NESSUS Reference Manual, version 7.5, 2002, (http://www.nessus.swri.org).
38. Wu, Y.T. et al., Probabilistic Methods for Design Assessment of Reliability with Inspection (DARWIN), presented at the *Structures, Structural Dynamics and Materials Conference and Exhibit Non-Deterministic Approaches Forum,* Atlanta, GA, April 3–6, 2000, Paper No. 2000-1510, AIAA.
39. Southwest Research Institute, DARWIN User's Guide, version 3.5, 2002, (http://www.darwin.swri.org).
40. Cesare, M.A. and Sues, R.H., ProFES Probabilistic Finite Element System — Bringing Probabilistic Mechanics to the Desktop, presented at the *Structures, Structural Dynamics and Materials Conference and Exhibit Non-Deterministic Approaches Forum,* St. Louis, MO, April 12–15, 1999, Paper No. 99-1607, AIAA.
41. Aminpour, M.A. et al., A Framework for Reliability-BASED MDO of Aerospace Systems, presented at *Structures, Structural Dynamics and Materials Conference and Exhibit Non-Deterministic Approaches Forum,* Denver, CO, April 22–25, 2002, Paper No. 2002-1476, AIAA.
42. UNIPASS Training Problem Guide, version 4.5, Prediction Probe, Inc., Newport Beach, CA, 2003.

43. UNIPASS Software Specification, version 4.5, Prediction Probe, Inc., Newport Beach, CA, 2003.
44. Abdi, F. and Minnetyan, L., Development of GENOA Progressive Failure Parallel Processing Software Systems, NASA CR No.1999-209404, December 1999.
45. Eldred, M.S. et al., DAKOTA, A Multilevel Parallel Object-Oriented Framework for Design Optimization, Parameter Estimation, Uncertainty Quantification, and Sensitivity Analysis, Version 3.0 Reference Manual, Sandia Technical Report SAND 2001-3515, April 2002.
46. Maybury, M.T. and Wahlster, W., Eds., *Readings in Intelligent User Interfaces*, Morgan Kaufman Publishers, San Francisco, 1998.
47. Carroll, J.M., Ed., Human-Computer Interaction in the New Millennium, ACM Press, New York, 2002.

3
Transitioning NDA from Research to Engineering Design

3.1	Introduction **3**-1
3.2	The Future Nondeterministic Design Environment **3**-2
3.3	Probabilistic Methods in Transition for Systems Design .... **3**-6 Current Nondeterministic Design Technology • Some Transition History • Finite Element Based Probabilistic Analysis
3.4	Other NDA Tools: A Transition Assessment .... **3**-11 Neural Networks [61–66] • Fuzzy Theory [67–80] • Interval Arithmetic [82–85] • Response Surface Methods • Nondeterministic Optimization [79, 92–100] • Design of Experiments [74, 101] • Expert Systems
3.5	Transition Issues and Challenges .... **3**-14 Overview • Verification and Validation of NDA Methods • NDA Error Estimation • Confidence Interval Modeling • Traditional Reliability and Safety Assessment Methods
3.6	Conclusion **3**-19

Thomas A. Cruse
Vanderbilt University

Jeffrey M. Brown
Air Force Research Laboratory, Propulsion Directorate, Components Branch (AFRL/PRTC), Wright-Patterson AFB,

3.1 Introduction

Nondeterministic analysis (NDA) is taken to be the set of technologies that recognize the inherent uncertainties in the model, data, allowables, or any other knowledge that might be applied to predict the reliability of a system or element in the system. Reliability refers to the probability that a design goal can and will be achieved. Given the breadth of NDA, the authors will address a subset of those NDA methods that we deem likely to be most useful in system design.

Within that subset, one can broadly state that there are two classes of NDA methods: (1) those for which a first-order representation (mean and variance) of the uncertain behavior is sufficient and (2) those for which a much higher degree of fidelity is required. The latter are typically needed for the formal processes of certification and apply largely to high-performance systems with human life risk factors that have a high reliability. This review focuses on the latter for their relevance to highly reliable mechanical systems for which accuracy in NDA is needed.

The review begins with a vision for the kind of new design environment in which NDA is central. NDA requires a systems perspective to a much larger degree than do traditional design processes. The new design environment is seen to be more dependent on a comprehensive and consistent interaction of data, modeling, updating, and system model validation than past design practices based on margins to assure reliability. The authors see the need for a new, integrating software framework and a variety of new NDA and NDA-supporting tools operating within that framework to support the full transition of NDA to design. A concept for the framework and possible NDA-supporting tools is presented.

NDA has been used in design practice for systems requiring a high degree of reliability, such as disks in turbine engines for commercial engines through statistically-based design limits for certain failure modes. The review includes a historical review of probabilistic method development that clearly indicates the transition from statistical margins to the current, but still limited use of probabilistic finite elements methods for NDA of structural response. The review then summarizes some pertinent literature on various NDA tools that might be used within this design environment.

The use of NDA in design is clearly still in transition. Some tools and strategies have been successfully developed and are being used in structural design today. Other areas of design from thermal response to aerodynamic performance to material behavior are in early stages of transition. The authors identify some issues and challenges that remain as topics to be worked in order to support the complete transition of NDA to design. Items that are briefly discussed include the design environment software framework, developing more generalized yet rational confidence intervals that have meaning in the context of complex system design, the need for systematic error analysis for the various NDA tools, NDA model verification and validation strategies, linkage of NDA with traditional reliability analysis tools, and the still-important issue of the cost of NDA at the systems level.

3.2 The Future Nondeterministic Design Environment

To a great extent, current NDA methods for system design are not based on physics-of-failure modeling at the system level. The real need for advanced design systems is for comprehensive, multi-physics modeling of system behavior. In only this way can the designer truly trade performance, reliability, quality, and cost metrics as functions of all the design variables.[‡]

Multi-physics modeling with fidelity sufficient for NDA is likely to involve the interactions of multiple physics-based computational algorithms. Given the fact that the design variables and the models themselves have uncertainties a NASA-funded study by the first author concluded that a new design environment was needed to support NDA [1]. Such a design environment must provide for generalized data storage, retrieval, and use by the appropriate physical models. The basic elements of the NDA design process are represented in Figure 3.1. The following paragraphs describe some of the proposed tool sets that might be found in a generalized NDA design environment (see Figure 3.2):

- *Information input.* In the past, data for probabilistic modeling was in the form of statistical data represented by analytical or empirical distribution functions and their parameters. For the future, input will include a much broader range of information. Such NDD information that is subjective or "fuzzy" is envisioned as a critical element in NDA. The elicitation of such fuzzy input is a critical element in the process such that expert bias and prior assumptions do not overly influence the representation of the information.
- *Information fusion.* This critical junction is where one must convert disparate forms of information into a common basis for modeling. A critical element in the technology requirements here is the

---

[‡] Design variables will herein always refer to random variables over which the designer has control. Such random variables may have physical dependencies such as material properties with temperature dependencies. Other variables may have strong statistical dependencies or correlations. The treatment of these conditions is beyond the scope of this chapter but the authors recognize how critical it is that these dependencies be recognized by any NDA methods. In all cases, the analyst is urged to formulate the physics at a primitive enough level to achieve physical and thus statistical independencies of the variables whenever possible.

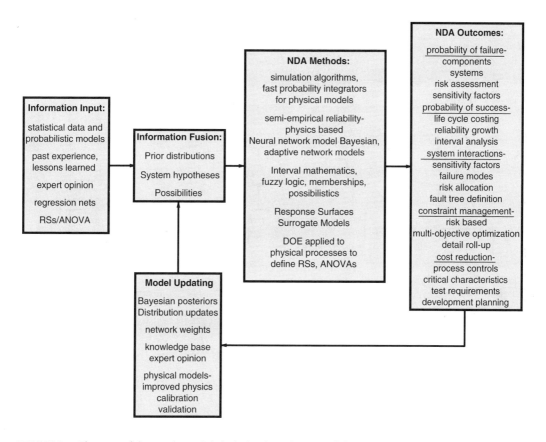

FIGURE 3.1 Elements of the nondeterministic design (NDD) process [1].

translation of properly acquired fuzzy input into sensible probabilistic representations. It is also the place where new information is used to update prior information.

- *NDA methods.* The comments herein address some key current and future NDA elements. Semi-empirical, physics-based modeling of systems using various network algorithms is likely to be important, especially for large systems. Fuzzy logic and its attendant elements of membership functions and possibilistic analysis are not likely to be used for the formal analysis part of the overall design system but are likely to be used for representations of expert-opinion-based input or hyperparametric input models. Response surface (RS) modeling or surrogate (simplified physics) models are likely to be intrinsic to NDD. All probabilistic methods use such models in one way or another. DOE approaches or Taylor series approaches are variations on RS methods. The use of DOE to construct such surfaces from empirical data is seen as an alternative to network models, although the technologies may merge.
- *NDA outcomes.* This is the heart of the design process valuation. A proper NDA design environment in the other entities in this diagram will yield much more useful information for making decisions for design stages from conceptual design to detailed design.
- *Model updating.* A critical element in any design process is updating. Good design operations have evolved means for capturing past experience for new designs. Quantitative risk management programs allow for updating the assumed statistics used in the previous modeling. Bayes' theorem can be used to combine prior distribution models with new data when both the prior model and the new data must be combined. In most data-driven designs, replacing the old dataset with the new one will do the data updating. Model updates will typically also be done by replacement,

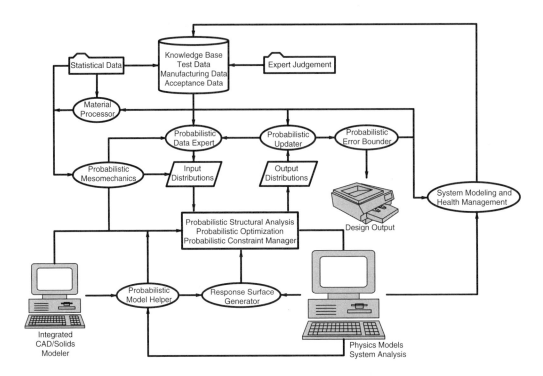

FIGURE 3.2 Integrated nondeterministic design environment.

although there are Bayesian model updating strategies appropriate to certain classes of network models where one cannot validate individual modeling elements. Certainly, the validation process for the physics-based models will lead to model updates. An integrated NDD environment to support these process elements was proposed in [1]. A major challenge in NDA for design is the development of a software framework that supports tool sets with a high degree of interoperability and data transparency for multi-physics modeling of system behavior. The following paragraphs describe some of the proposed toolsets that might be found in a generalized NDA design environment.

- *Probabilistic Data Expert (PDE)*. The PDE provides a systematic means for merging disparate forms of "data" into probabilistic design. The data ranges from subjective expert opinions to raw data such as specimen specific fatigue test data. The PDE relies on nontraditional methods for integrating "fuzzy data" based on qualitative information [2]. The major technology needs here are in creating tools for automating the data acquisition and integration processes.
- *Probabilistic Updater (PU)*. The NDD process for total life-cycle design from concept to field deployment for complex systems generally involves reliability growth processes of one form or another. Early in the design process, the reliability characteristics of the subsystems are either not known or crudely known. Testing during development should focus on those results that are most effective either in demonstrating greater reliability or increasing the assurance (reducing the uncertainty) in the reliability prediction. One can envision a generalized tool that will provide intelligent ways to update the current reliability and assurance intervals (akin to statistical confidence intervals) taking into account new information and data. These intervals are discussed in Reference 3. Bayesian updating of statistical models can be used for certain automated model updates involving the combination of expertise-driven priors with new data or information. However, other updating strategies are likely needed for large system models where the outcomes are used to assess the quality of the model.

- *Probabilistic Error Bounds (PEB)*. Many system design processes have allowed for generalized levels of comfort in product performance—so-called "fuzzy" nondeterministic output. However, both the commercial workplace and the demands of government acquisitions have transformed aerospace vehicle design requirements from nonanalytical reliability bases to specified levels of "demonstrated" reliability that may include both test and analytical bases. A key and, as yet, not addressed and fundamentally critical need is to be able to compute rational reliability error prediction bounds for the various algorithms likely to be used in an advanced aerospace vehicle and propulsion design. Error bounds are needed that span the analysis spectrum from the physics approximations, to the response surface or surrogate models used to represent the physics, to the specific probabilistic algorithms deployed. In fact, it is likely that engineers will need to deploy hybrid combinations of the current probabilistic methods (e.g., RS, fast probability integration (FPI), Monte Carlo (MC) importance sampling) for large system design problems. Error identification, tracking, accountability, and bounds are required for the full process. Such error modeling is seen to be a part of the total model verification and validation process.

- *Response Surface Generator (RSG)*. All current probabilistic algorithms for any but the most trivial problems use a "response surface" or surrogate model to describe the component, subsystem, or system-level behavior as a function of the design variables. Thus, it is a surface in the N-dimensional design space of the problem. In fact, it is increasingly likely to be a union of many such hyper-surfaces that can be used for complex system design. An example of a powerful algebraic system for manipulation of RSs that could be easily extended to nondeterministic design is seen at the Rockwell Science Center link: http://www.rsc.rockwell.com/designsheet/.

- Two methods are currently employed to generate RSs; these are the Taylor series method and the design of experiments (DOE) method [4]. The former is more likely to be employed near a critical design point, while the latter is more often used to span a wider range of the design space. Fast probability methods such as the fast probability integrator (FPI) algorithms [5] or importance sampling (a Monte Carlo simulation strategy) are performed using the RS and do not directly use the actual detailed simulation. As such, the RS methods are highly effective over a range of design levels from conceptual to final. A generalized "tool" for effective development and representation of RSs is needed that will interface both to probabilistics and to the closely related field of multidisciplinary optimization (MDO). Further, the *PEB* tool and the *RSG* tool should be able to work together to adaptively build optimal RS strategies that combine computational efficiency with minimizing the probabilistic approximation errors for large system design problems.

- *Probabilistic Model Helper*. A major limitation in the deployment of NDD analysis methods is the current burden that the analyst be a skilled "probabilistic engineer." Such individuals at this time are probably numbered in the low dozens and are scattered between software firms, universities, government labs, and industry. The field is not one that has received much attention in academic curricula, although the topic is now covered in the most recent undergraduate mechanical design texts. Further, the process of preparing and interpreting a probabilistic design problem involves the inclusion and assessment of many more sources of information and decisions regarding the analysis strategy. Nontraditional methods provide a real basis for the development and deployment of intelligent systems to work as probabilistic "robots" or assistants. The approach will likely require significant use of adaptive networks, software robots, genetic algorithms, expert systems, and other nontraditional methods.

- *Probabilistic Mesomechanics*. Significant progress has been achieved since the early attempts at what was called "level 3" probabilistic material modeling as part of the original NASA Probabilistic Structural Analysis Methods (PSAM) contract.§ Demonstration problems have shown the ability

---

§ Dr. Chris Chamis of NASA/GRC was the driving inspiration for this NASA-funded effort led by the Southwest Research Institute in San Antonio, TX, with partners including the Marc Analysis Research Corporation and Rocketdyne Inc.

to link material processing simulation software with micromechanical models of material behavior at the mesoscale (grain size, grain orientation, flow stress, dislocation density) to predict statistical distributions of material fatigue strength [6, 7]. Such probabilistic modeling advances lend real credibility to the notion of engineered materials that goes well beyond the simpler notions associated with composite material systems. Probabilistic mesomechanics tools that can support the PDE tool or feed "data" directly into the material data distributions are required. Automated, intelligent systems are envisioned with the capability to forward model material processing to define scatter in properties as well as sensitivity links to the independent process and material primitive variables. Further, the inverse problem of optimizing the processing and design of the material microstructure provides the ultimate in an "engineered materials" design capability.

- *System Reliability Interface.* The aerospace system design environment of the future requires the ability to interface multiple models of multiple subsystems in an efficient and accurate manner. Each failure mode of the system has its own probabilistic model in terms of a response surface or surrogate model representation of the physical problem together with the associated nondeterministic design variable descriptions. Current system reliability technology focuses on traditional block-diagram forms for representing system reliability. Each block is typically represented by point estimates of reliability that are not linked to the underlying physics or to the distribution and confidence in the underlying variables. Physics based system modeling provides an information-based opportunity for linking the key design parameters and physics-based models together from the sub-system/component level to the full system level. Intelligent systems can be developed for linking the information and propagating it to the various "top-level" events in order to provide a powerful environment for calculating and managing system level reliability.

- *Health Management System.* Health management (HM) is a philosophy that merges component and system level Health Monitoring concepts, consisting of anomaly detection, diagnostic and prognosis technologies, with consideration to the design and maintenance arenas [8]. Traditionally, system health monitoring design has not been an integral aspect of the design process. This may be partly due to the fact a cost/benefit model of a HM system configuration has not yet been fully realized. Without a doubt, HM technology must "buy" its way into an application. Hence, the need exists to extend the utility of traditional system reliability methods to create a virtual environment in that HM architectures and design trade-offs can be evaluated and optimized from a cost/benefit standpoint. This capability should be present both during the design stage and throughout the life of the system. A new HM design strategy should allow inclusion of sensors and diagnostic/prognostic technologies to be generated in order to produce an enhanced realization of component design reliability requirements at a very early stage. Life-cycle costs can be reduced through implementation of health monitoring technologies, optimal maintenance practices, and continuous design improvement. To date, these areas have not been successfully linked with nondeterministic design methods to achieve cost/benefit optimization at the early design stage.

3.3 Probabilistic Methods in Transition for Systems Design

3.3.1 Current Nondeterministic Design Technology

A simplified overview of nondeterministic design (NDD) must address both empirical and predictive reliability methods. NDA is, formally, a member of the latter. However, one cannot ignore the former because NDA relies at some level on empirical modeling for model input or model validation. NDD also includes system safety and reliability analysis tools such as failure modes and effects analysis (FMEA), criticality analysis (CA), and fault tree analysis (FTA) [9]. One of the major future transition elements for NDA is bringing more analysis into the qualitative reliability and risk assessment processes.

Empirical reliability has primarily focused on determining the fitting parameters for two classical statistical models—the exponential and Weibull distributions [10–12]—although the normal and log normal distributions are also used. The exponential distribution is used to represent the average failure rate for large populations of similar devices or over many hours of operation of a complex system, generally reported as the mean time between failures (MTBF) or the uniform failure rate (MTBF^{-1}). This distribution has most often been applied to modeling the reliability of electronic components that have been through a burn-in usage.

The exponential reliability model does not address the physics of failure. For example, an electrical motor design may fail due to armature fatigue, wear, or other mechanical failure mode. Over a long period of time, the occurrences of these failure modes average out to a single MTBF value. Some effort has been made to convert this purely empirical reliability model to one having some physics effects reflected in the MTBF [13].

The Weibull distribution is generally used to represent mechanical failure modes as it combines the features of wear-out failure rates with single-point (brittle) failure modes [14]. All such empirical models rely on sound statistical methods to infer the proper set of parameters for any reliability model that is to be used along with the supporting data [15, 16].

System safety and reliability analysis [9] is widely used and is closely linked to NDD despite its largely qualitative nature. FMEA is a bottoms-up assessment process based on identifying all of the "important" (perhaps as defined in a criticality analysis [CA] process) failure modes for each subsystem or component and estimating the consequences of subsystem or component failure. FMEA is often used to define corrective actions necessary to eliminate or significantly reduce the failure likelihood or consequences (risk). However, a quantitative extension of the standard FMEA would take advantage of the computational reliability predictions coming from NDA used in design. FMEA can provide more insight than NDA into nonquantitative design issues that can affect product reliability.

Fault tree analysis (FTA) [17] is a top-down reliability assessment based on a specified "top" failure event. FTA then seeks to define all of the causes for the top event by breaking down each cause into its causes and so on until the "bottom" of the fault tree is defined. Point probability estimates can be applied to each event in the tree and these are logically combined using "and" and "or" gates to define the probability of the top event. FTA is also qualitative in nature as far as the tree elements and structures are concerned. Experience with product design and product reliability are important in getting useful and comprehensive FEMA and FTA models for systems. Neither FMEA nor FTA is a physics-of-failure based NDD tool.

Probabilistic risk assessment (PRA) [18] is a reliability tool that was largely developed within the nuclear power industry. PRA typically uses Monte Carlo simulation to combine the statistics of the key, physical variables that combine to cause failure. Often, the reliability problem is posed as a basic stress vs. strength problem such as for fatigue and fracture problems [19]. Following the Challenger accident, NASA made a significant commitment to the use of PRA for the high criticality failure modes for the launch system (e.g., [Reference 20]) and the space station. PRA has also been used for various NDD problems in the turbine engine field for many years [21, 22].

3.3.2 Some Transition History

Mechanical system reliability has evolved in multiple paths owing to the unique elements of mechanical system design, at least as one can see them historically. The two paths can be traced in the design of civil structures such as buildings and bridges, and the design of aircraft and propulsion structures. The first category is driven by issues deriving from the high degree of redundancy in most civil structures and by the attendant use of building codes with specified safety factors. The probabilistic elements of civil structure design have come about largely due to two forms of highly stochastic loadings—earthquakes and sea loads (for offshore structures).

Aircraft and propulsion designs have been driven by the specifics of fatigue and fracture mechanics damage processes. While aircraft are subject to stochastic loading conditions, the large loading conditions are better defined than in civil structures, as the designer controls the stall limits of the airfoil and hence the greatest aerodynamic load that can be seen. Stochastic elements of structural reliability problems for aircraft and propulsion structures are dominated by scatter in the fatigue and crack growth rate characteristics of the materials.

Traditional design for all civil and aerospace structures has been controlled primarily through the use of safety factors or other forms of design margins. Such factors or margins are deterministic in nature but their magnitudes are typically driven by past experience that includes the stated loading or material stochasticity, along with the full range of less important stochastic design variables. There are only two main reasons for changing this, largely successful, design experience. One is economical: design factors and margins are conservative in most cases and the degree of conservancy is not well established. Further, the margins do not reflect the differing degrees of controls on design variations that are a part of the proper design process. All of this costs money and reduces the performance of the system through excess weight or unnecessarily stringent controls.

A key rationale for NDD is its ability to support the design of new systems with new materials or components for that the experience base is lacking, as compared to traditional systems. As material performance is pushed in electronic and mechanical systems, the need for a new design paradigm has become more evident. The past reliance on successful experience and the use of incremental design evolution does not adequately serve the current marketplace. Further, the "customer" in the general sense is no longer willing to pay the price of designs that do not account for intrinsic variability in process or performance.

The following paragraphs seek to provide a limited overview of some of the principal developments that have brought us to the point we are today, where reliability based design is poised to become the new design paradigm for consumer products through advanced aerospace systems.

3.3.2.1 Early Developments

An early interest in the development of reliability was undertaken in the area of machine maintenance. Telephone trunk design problems [23, 24] are cited in a more recent reliability text [25]. Renewal theory was developed as a means of modeling equipment replacement problems by Lotkar [26].

Weibull made a substantial contribution to mechanical system reliability through the development of the extreme value statistic now named after him [27]. His was the first practical tie of the physics of failure to a mathematical model. The model derived from the general extreme value statistical models expounded by Gumbel [28]. The Weibull model contains the concept of brittle failure mechanics that states that any critically sized defect failure causes the entire structure to fail and can be derived as the statistics of failures of the weakest link in a series of links.

3.3.2.2 Use of Weibull Models in Aircraft Engine Field Support

Mechanical failures in gas turbine engine components are driven by specific wear-out modes, including fatigue, creep, crack growth, wear, and interactions of these modes. The components may be bearings, shafts, blades, pressurized cases, or disks. For nearly 30 years, the Weibull reliability model has been successfully used to develop maintenance and field support plans for managing reliability problems defined by field experience. Such methods also have been used successfully in establishing warranty plans for new equipment sales as well as spare parts requirements.

Much of the effectiveness of the developed methods is shrouded in the mist of proprietary data and protectionism. However, it is clear that these specific modes of mechanical system failures are often very well correlated using the Weibull model. The Weibull model has demonstrated the ability to accurately predict subsequent failures from a population based on surprisingly few data points.

The Air Force sponsored some of the more recent work using the Weibull model and the results of the study are in the public domain [14]. This chapter serves as a useful introduction to the use of the Weibull model for various mechanical systems. A more comprehensive examination of the model, along with other models of mechanical reliability is the text by Kapur and Lamberson [10]. There has been considerable growth in the Weibull model applications industry in recent years, as various industries have found ways to use these methods.

A very powerful application of the Weibull method is in probabilistic risk assessment (PRA) for demonstrated mechanical reliability problems [18]. The earliest and probably most extensive use of PRA for large field problems has been in civil transport gas turbine engine field problems [21, 22]. Since about 1978, the Federal Aviation Administration has required the engine manufacturers to use Weibull-based PRA as the basis for setting flight limits on engine components with known field fracture potential. Again, it is the accuracy of the Weibull model to correlate the relevant failure modes on the basis of a few, early failures that allows one to use the model to confidently predict future field failure probabilities. One can then implement the necessary inspection plans that will insure that no actual failures will occur.

Another field application of the Weibull probability model is the prediction of safe operating limits in the field based on successful experimental data. The method is sometimes referred to as a Weibull-Bayesian method. The Weibull-Bayesian method was applied to the problem of safe operating limits for the Space Shuttle auxiliary power unit turbine wheels [29] as part of the return-to-flight studies performed following the Space Shuttle Challenger tragedy.

3.3.2.3 Civil Engineering-Based Reliability Developments

The principal structures that are most closely linked to reliability modeling are those located offshore or in earthquake-prone areas. In both cases, standard design factors do not provide adequate margins nor do they provide sufficient linkage to the actual failure conditions that can occur in specific structural configurations. The notion of a design margin, however, has led to the successful development and use of various probabilistic design methods for these structures.

The stress conditions in complex structures can be computed for various loading conditions, using the extensive computational modeling capabilities that have been available for design modeling since the 1960s. If one can efficiently compute the variability in the stresses due to the stochastic loads, then the design margin between the variable stresses and the variable material strengths can be established. That simple idea underlies much of the developments of reliability design methods for civil structures [30].

The essential feature of these methods is computing the variability of the design response. In simple terms, the variability can be taken to be the statistical "variance" or standard deviation measures. The variance is the second-moment of the statistical distribution of a random variable, while the mean of the distribution is the first moment. The stochastic margin between the stress and the strain was formulated as a reliability index by Cornell [31]. The relation between the reliability index and the safety margin for structural design was then proposed [32].

The original formulation of the reliability index was shown by [33, 34] to give differing reliability indices for the same safety factor, depending only on the algebraic forms of the same relations. The proper reliability index that is invariant to the algebraic form of the safety margin was then given by Hasofer and Lind [35] and by Ang and Cornell [36]. The reliability index for Gaussian functions of stress and strength is directly related to the variance in these two variables, as well as their mean values.

3.3.2.4 Probability Integration Algorithms

Later work in structural reliability has been directed toward computing the reliability index for more complicated structural problems. The failure problem is stated in terms of limit states that separate the safe design from the failed design where the independent variables are the design variables of member size, strength, and loading conditions. The limit states are generally nonlinear functions of the design variables and must be linearized at some design state in order for a reliability index to be computed. The first and enduring algorithm for this problem is that of Rackwitz and Fiessler [37].

One of the major developments in NDA is concerned with the accurate prediction of the reliability of the structural system and not just the variance or reliability index for the design. The earlier NDA design methods are generally referred to as second-moment methods, indicating that the answer is given in terms of the second statistical moment of the problem. To compute an estimated reliability of the performance of the system for low levels of probability, one needs to compute reliability functions for nonGaussian functions. Such computations use approximations that convert the actual probability data for each design variable into "equivalent" Gaussian or normal distributions. The first major improvement to the Rackwitz-Fiessler algorithm was the three-parameter distribution model introduced by Chen and Lind [38]. Wu [39] contributed a further refinement of this algorithm. Applications of the new reliability algorithms have been extensive in the past several years. New and more accurate algorithms than the original Rackwitz-Fiessler algorithm were developed [40] and applied to complex structural systems [41, 42]. These algorithms are now generally referred to as "fast probability integration (FPI)" algorithms.

3.3.3 Finite Element Based Probabilistic Analysis

The principal, general-purpose NDA tool is based on finite element methods (FEM). Two general approaches have been developed for structural mechanics modeling: stochastic finite element methods and probabilistic finite element methods. The former is a second-moment method in which two coupled systems of matrix equations are created, one giving the response and one giving the variance in the response. Stochastic FEM is not suitable for probability calculations in the same way that second-moment methods are not suitable. The response is not a linear function of the random design variables, the variables are typically not normally distributed, and the probability estimates are not suitable for higher performance design problems.

One of the pioneering probabilistic code developments was work sponsored by the NASA (then) Lewis Research Center under the probabilistic structural analysis methods (PSAM) contract effort [43, 44]. The work was done at Southwest Research Institute (SwRI). A number of early papers were generated that document the probabilistic FEM code referred to as NESSUS (Nonlinear Evaluation of Stochastic Structures under Stress).

The NESSUS probabilistic FEM analysis is based on what is known as the advanced mean value (AMV) and AMV+ algorithms. The starting point for these algorithms is the deterministic FEM analysis done at the mean value design point (all random design conditions are set to their mean values). Each RV (again for now assume independence for convenience of the current discussion) is then perturbed a small amount based on its coefficient of variation (COV) to get a linear response surface. [Note: Quadratic and log-mapped RSs can also be used.] Based on the linear RS and design allowable for the performance function of interest, a limit state is generated from which a first-order reliability calculation can be made, as discussed earlier. Nonnormal RVs are mapped according to the Rosenblatt transformation to equivalent normals, using the Wu-FPI algorithm [5].

The Wu-FPI algorithm provides the design point or most probable point (MPP) in terms of the values for each RV. A new deterministic FEM analysis is done at the MPP. While the response changes from the initially approximated value, the AMV algorithm assumes that the probability associated with the selected limit state does not change in this updating step at the probability level for the original MPP. A spreadsheet example of the AMV algorithm is contained in Reference 45. A number of test problems confirmed the viability of the AMV algorithm as a much more accurate result than the usual first-order reliability results [46, 47].

The AMV+ algorithm is an iterative extension of AMV by reinitiating the RS approximation at the current design point; the updating based on the AMV process is continued until a convergence criterion is met. For a deeper discussion of the AMV+ algorithm, see Reference 48. Numerous applications of the NESSUS software and algorithms have been made including static stress [49], rotor dynamics [50], vibration [51], nonlinear response [52], and optimization [53]. NESSUS also provides a wide variety of probability integration algorithms, system reliability, and user-defined analysis models [54].

Other probabilistic FEM codes have been developed over the same time period. A good approach is to combine probabilistic methods with existing deterministic FEM codes. An excellent example of this is the ProFES product [55, 56]. The ProFES system is a three-dimensional graphical analysis environment that allows the analyst to perform a variety of response surface model developments for standard commercial FEM software systems. ProFES provides a variety of response surface simulations including first- and second-order methods and direct Monte Carlo simulation.

CalREL is another shell-type program with a variety of operating modes [57]. This probabilistic structural analysis code can be run in a stand-alone mode or in conjunction with other deterministic FEM codes. The probability algorithms are similar to those in ProFES and NESSUS in those first- and second-order reliability computations are facilitated. First-order reliability calculations can be made for series systems while Monte Carlo simulation is available for more general systems problems.

The two major U.S. turbine engine manufacturers have each developed their own probabilistic FEM approaches to NDA for design. The codes are proprietary and not much has been published on them. However, the general approach in each is to use deterministic FEM codes in support of a design of experiments (DOE) approach for defining a response surface (RS) model of the physical variable or limit state of interest. Monte Carlo simulation is then performed on the RS model using the statistical models for each of the design variables.

The RS approaches generally have been developed to avoid the use of fast probability integration (FPI) algorithms, which some view as unreliable and a source of modeling error [60]. Both of these issues are relevant and a challenge to those developing and using FPI algorithms to provide meaningful convergence assurance and error estimation algorithms in support of these clearly faster algorithms.

The RS methods contain their own modeling error issues that have not been addressed in the reported literature. Largely the companies rely on standard RS fitting statistics to assure a "good" fit. However, such fitting is done in a global sense and likely does not apply at the so-called design point where the limiting failure values of the RV are dominant. Thus, further work on error bounding for RS methods is still needed to support NDA transition to design.

3.4 Other NDA Tools: A Transition Assessment

The previous section addressed some NDA methods that are currently being used to support design. A number of potential NDA technologies were reviewed in Reference 1 as possible contributors to the NDD environment. As such, these methods are in earlier stages of transition to design or are still being assessed by various investigators. The following paragraphs summarize those reviews.

3.4.1 Neural Networks [61–66]

Neural nets (NN) have the ability to represent physical responses somewhat like a Response Surface used in design. A NN model training process is used to establish nodal weights and biases. In general, NN weights do not conform to specific physical variables. A NN would be used to model an RS only if one had datasets and not a physical model. However, the statistical method for constructing RSs is more robust as far as sorting out noise from model behavior and for giving statistics that can be used in constructing assurance intervals for error analysis when applying the RS. As pointed out in [61], NNs rapidly lose their attractiveness for large numbers of NN nodes.

Reference 62 summarizes various uses of networks for data mining. Such applications may be valuable when constructing data models for NDD based on extensive test data. An interesting example of the use of NNs to model a combination of experimental data and varying types of modeling results for airfoil design is given in Reference 63. These authors use the NN as an effective way to solve the inverse problem for finding a RS from disparate data sources. The also report that their strategy for training results in nodal weights that can be tied to specific design variables.

A Bayesian network is one in that the nodal weights represent conditional event probabilities, as discussed in Reference 65. Such modeling appears to be useful for large system reliability problems and provides a direct means for simulating the top event and knowing its probabilistic design variable causes. They state that the Bayesian network method is preferable to Monte Carlo for system simulation for the additional information on conditional probabilities that is gained.

The literature suggests that an important NDA role for NNs is to be found in the proposed Model Helper and Data Expert tools. The application of NNs with a strategy to tie the nodal weights to physical variables appears to be an effective data mining strategy to support probabilistic input. Bayesian networks appear to be useful for system reliability modeling. Finally, NNs have a valid and important role in the data gathering and interpretation role in System Health Management. It does not seem likely that standard NNs will be used in the ND design analysis algorithms.

3.4.2 Fuzzy Theory [67–80]

Fuzzy logic is a potential NDA tool. Reference 80 compares probabilistic and fuzzy modeling for reliability. Fuzzy set theory and the closely related field of possibilistics have been touted for NDA. We do not support that position. Fuzzy systems are based on the mathematics of fuzzy set theory. The above references provide an adequate introduction to these fields. Generally speaking, fuzzy systems are used where so-called "crisp" probabilistic models do not exist, such as in linguistic representations. Characteristic mathematical models are assigned to fuzzy functions that are then combined using fuzzy logic. All input variables to the mathematical-physical model are given fuzzy representations and the physical model is written using the "algebra" of "fuzzy logic" to yield "fuzzy output." As in the paper by Rao [81], one finds that fuzzy physics is violated physics. That is, one loses the constraints of the physical model under fuzzy logic formulations. The only advice offered by that author is that one must impose one's physical intuition to properly understand the results. This is not scalable as a problem-solving strategy for large problems.

Reference 71 applies fuzzy logic to classical reliability engineering. The authors contend that this is the right approach to problems where the input is estimated from engineering experience and judgment. They also contend that this is the right way to represent degraded states of operation, given that the word "degraded" is subjective. However, work at the Los Alamos National Lab [2] has shown that one can take fuzzy input such as these authors refer to and convert that input into probabilistic distributions so that one can perform robust probabilistic calculations on these fuzzy-derived distributions.

Bezdek and Pal [69] are quoted as saying that "fuzzy models belong wherever they can provide collateral or competitively better information about a physical process." Bezdek and Pal give an example case wherein two bottles of water are lying in the desert and are found there by a very thirsty wanderer. The first bottle has a membership in the set of potable waters of 0.91 while the other has a probability of being potable that is 0.91. In the first case, the water shares a high degree of characteristics in common with potable water while the second has a 9% chance of being totally nonpotable (poison!). Which do *you* drink?

Fuzzy input is an important capability for NDA-based design. However, mathematically rigorous probabilistic algorithms are required as the processing element for de-fuzzified input data to support complex system designs. Fuzzy logic might be deployed in program management decision and risk prediction tools.

3.4.3 Interval Arithmetic [82–85]

This topic appears to be much more interesting to nondeterministic design. Interval arithmetic operates under some very precise algebraic rules and does not lead to systems that violate physics. In fact, interval arithmetic was developed as a way of representing floating point arithmetic errors. As cited in Reference 82, interval arithmetic is now released as a new variable type in the Sun Microsystems Fortran compiler.

Elishakoff and co-workers have applied interval methods to some structural analysis problems. Reference 86 combines interval analysis with stochastic finite element methods (second-moment method) to structural frames. Elishakoff applied strict interval analysis methods to computing the range of structural natural frequencies in Reference 87.

In a related technology to interval methods, Elishakoff and co-workers have utilized what they refer to as convex modeling [88] to compute one-sided bounds on structural behaviors such as natural frequencies and buckling loads [89]. The method is based on defining bounds on uncertain parameters as convex sets bounded by ellipsoids. For such cases, convex methods can be used to compute a corresponding convex set of results, or upper bounds.

Interval methods provide a formal mathematical treatment for problems where (1) no input information can be defined beyond upper and lower bounds for design variables and where (2) response ranges are desired for preliminary design purposes. Such methods may be worth considering as part of the required error bounds work proposed herein. Interval arithmetic may offer, especially as a data type, the opportunity to do rapid analysis of the output range for design problems. The major concern with interval methods is that the intervals may be too large to be useful for NDD.

3.4.4 Response Surface Methods

There appears to be some confusion here over terminology as applied to probabilistic design. Probabilistic designers have used Taylor series expansions of the physical response of systems in terms of the design variables to construct locally linear (first-order) and locally quadratic (second-order) polynomial fits to model results. Such representations are updated as the "design point" moves through the design space (so-called advanced mean value [AMV] algorithm). This approach might best be called the local-RS approach.

A second approach, and one currently favored by the major turbine engine companies, derives from classical statistics [90, 91]. These RS methods have been developed as part of experimental statistics wherein the experiment is designed with varying patterns of high-low values for each identified parameter that can be changed. Mathematical methods are then used to define the RS that is most likely that defined by the experiment and the statistics of variance regarding that fit. One also determines the independence or coupling of the variables in the modeled physical response. The use of certain experimental designs leads to first-order surfaces (with interactions) or second- and higher-order surfaces.

While the use of RSs by the turbine engine manufacturers is based on fitting a response surface to deterministic and not to random (i.e., experimental) system responses, the users define their approach as an RS approach consistent with statistical derivations. The reason they use this terminology is that they use a variety of experimental designs (DOE) to select the points in the design space that are used to evaluate the response and from that the first- or second-order surface (with all interaction terms) is fitted. Such an RS method might best be called the DOE-RS method.

The key in all cases is that the RS is a representation of what is probably a model of the physical system. The specific fitting algorithm and probability algorithm define the errors in the nondeterministic results. The two RS approaches have different characteristics in terms of accuracy (local and global) and efficiency.

3.4.5 Nondeterministic Optimization [79, 92–100]

All optimization algorithms can and should be nondeterministically applied. Two design environment references [99, 100] clearly indicate this combination is the future for NDD. Nonprobabilistic elements in optimization can be in the objective function or in the constraints. The issues related to the use of nontraditional optimization strategies (nongradient, for example) such as genetic algorithms and simulated annealing have nothing to do with nondeterministic issues. They have a lot to do with the size, complexity, and the nature of the design environment [96, 97].

3.4.6 Design of Experiments [74, 101]

DOE strategies seek to define an RS by selecting a subset of a full factorial design procedure in order to reduce the experimental cost. The actual patterns are particular to various statisticians and practitioners. The second element is the separation of variables into controlled variables and noise or uncontrolled variables. When trying to target manufacturing processes using Taguchi strategies, one seeks to minimize process variability caused by noise factors. Commercial packages exist that simplify the DOE process (e.g., Reference 102).

3.4.7 Expert Systems

Expert systems are at the heart of an increasing number of applications including the fuzzy logic control systems, as previously discussed. Dedicated expert systems will be important parts of NDA methods deployed to support future design environments and system health monitoring [8, 103].

3.5 Transition Issues and Challenges

3.5.1 Overview

Much of the NDA technology needed to support complex system design for high reliability exists today. Some transition success has been achieved, primarily in areas where reliability is a premium. The following items discussed in this section are not roadblocks to NDA transition but they do represent, at least to these authors, some of the key items that require continuing or new efforts.

The transition issues include the need for a new design environment, verification and validation strategies for NDA models, error estimation models or bounds for NDA methods, linkage of NDA to traditional reliability and safety assessment methods, and the cost of NDA methods for significant (real) design problems. The issue of the design environment was addressed at the outset of this review; essentially, there is a need for a comprehensive information-based operating environment or software framework that supports the comprehensive system design process. Traditional design methods based on margins have worked in the context of individual design environments where design information sharing is very limited. NDA for design is highly dependent on consistent data modeling of common random variables that span the total system problem. The following topics briefly discuss other issues and challenges for NDA transition to design.

3.5.2 Verification and Validation of NDA Methods

Verification and Validation (VV) are critical issues for the modeling and simulation tools used for design—both for deterministic and probabilistic approaches. The Defense Department has published requirements for VV for modeling and simulation [104]. The defined terms for VV are the following.

- *Verification:* the process of determining that a model implementation accurately represents the developer's conceptual description and specifications.
- *Validation:* the process of determining the degree to which a model is an accurate representation of the real world from the perspective of the intended uses of the model.

These are generally accepted definitions within the VV community with variations only in the specific words used [105, 106]. Thus, verification refers to determining if the code does what was intended while validation refers to determining if the code/model represent real physics. Verification is typically done by reference of code output to benchmark solutions with analytical results, while validation is done by reference of code output to experimental results, where the key physics are operative.

We will not address verification issues beyond the need for error estimation requirements for some of the NDA methods. That topic will be addressed in the next subsection. The remainder of this subsection is devoted to the issue of NDA model validation. We propose that NDA model validation will provide a statistical measure of the agreement between the NDA prediction and available test data. We will further suggest that validation for system reliability problems must be inherently modular in nature.

Validation issues for deterministic modeling have been addressed in great detail by the computational mechanics community (e.g., Reference 105) and by the computational fluid mechanics community [107]. The Department of Energy is also active in developing specific modeling strategies to define quantitative measures of model validation [108–110]. This latter set of references demonstrates that validation can result in (1) an efficient use of experimental validation tests and (2) a statistical model for the uncertainty in the deterministic model. The latter can be used as statistical error models in a probabilistic application of the model as alluded to in Reference 3.

To date, no one has defined a comprehensive VV approach to NDA for design. However, we can suggest the elements that need to be included in such an approach. We will not address the issue of validating the deterministic models, *per se*, as we believe such validation needs have been recognized in past design practices.

- *Validate the deterministic models used for nondeterministic system response.*

These models may be based on the full deterministic analysis methods used to support NDA modeling or they may be based on using reduced-order models such as a derived RS [4] or AMV-type of RS [40] that are used as surrogates for the complete computational model. Typically, the issue here that differs significantly from the deterministic models is the need to validate "off-design" or "off-nominal" conditions.

The usual deterministic method validation process involves comparing a deterministic model to experimental results for selected test condition(s). Such testing is most likely to involve design variable conditions quite close to the nominal or expected values of those design variables. The critical reliability condition that drives the design, however, results from off-nominal values of the RVs that dominate the system failure probability.

Thus, a deterministic model may be "validated" at a nominal set of design variables, typically using some calibration factor. In margin-based design practices, it is good practice to assure that the calibration results in a conservative response condition when the reliability issues are considered. Such a calibration approach is unable to determine to what extent the physics of the response are properly modeled for the off-nominal but critical design condition that drives the system reliability.

- *Validate the deterministic model at off-nominal design conditions.*

Ideally, off-nominal validation should be performed at the values of the critical RVs that define the limiting reliability condition for the system design. However, it is likely that such off-nominal testing is impractical. Thus, some compromise must be made and judgment used to extrapolate from the test conditions for the critical RVs to the limiting reliability condition. As discussed in the Confidence Interval subsection, the off-nominal validation process must define the deterministic model errors for both the nominal and off-nominal conditions. The recommended approach is given in the following item.

- *Establish the statistical uncertainties in the deterministic models [110].*

The referenced approach uses probabilistic modeling to propagate uncertainties in the prediction using statistical models for those RVs whose values are not controlled in the testing. The approach also recognizes that the experimental results must themselves have defined uncertainties associated with experimental error. The result is a statistical measure of the probability of having a valid model by a likelihood function comparison of the experimental results with scatter and the NDA model with random factors for the experiment.

Extension of this approach to off-nominal conditions is likely to involve many fewer data points than one might expect for nominal validation testing. This is a practical recognition of the difficulties of performing off-nominal testing. As discussed in a recent Sandia VV report [106], there are practical limits to the cost of VV that dictate sometimes more pragmatic approaches. Nonetheless, it is critical to model the confidence one has in such cost-limited approaches to the experimental validation effort.

Thus, one must define the potential modeling uncertainty using some confidence interval method. That topic is addressed herein in the next subsection. Much further work is required to develop practical deterministic model validation methods that include the off-nominal conditions.

- *Validate the input probabilistic models being used to describe the RVs.*

Typically, such a validation effort will amount to establishing statistical confidence intervals on model parameters. However, it is not yet established how to represent the validation of the probability distribution itself in terms of the probability distributions themselves. Further work is needed on this subject.

- *Validate the total system model.*

The system model validation issue involves two steps. First, one must validate the individual NDA elements based on the individual RVs and statistics used for each. This modular approach would be based on the above validation recommendations. The second element is to perform sufficient system level testing to validate the RV interactions between the system elements. The process must assure that all of the key interaction variables are defined and that valid statistical measures and confidence interval models are represented in the system model. This system-level testing can be done using intermediate system-level testing that addresses blocks of the total system and their interactions.

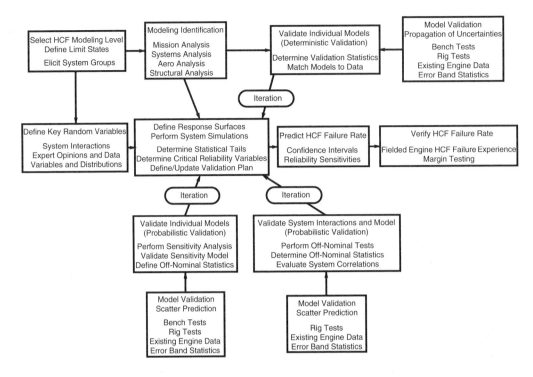

FIGURE 3.3 Notion of probabilistic validation: an HCF design process example.

A suggested logic diagram for validating a probabilistic high cycle fatigue design system is proposed in Figure 3.3. The figure shows the likely iterations that would be needed to update models, especially the error representations. Considerable effort is clearly required to develop a comprehensive validation strategy for each NDA design problem. One can see some of the key elements in the recommended approach used in a recently presented study on probabilistic disk design by Pratt & Whitney [111]. The final validation often must be by reference to a set of actual field reliability data for existing designs.

This approach, used in the reference Pratt & Whitney study combines the modular validation steps for individual NDA elements, with a total system reliability prediction. This is the only practical approach for problems where system level reliability testing is impractical. This is true of most, if not all, real design problems for reliability-critical systems.

3.5.3 NDA Error Estimation

The previous section outlined the VV issues with reference to the usual design analysis tools used in NDA of systems. In addition to these tools, the use of NDA for design must also include statistical measures of algorithm errors in the NDA probabilistic elements. These have been identified in the appropriate sections of this chapter and we will simply summarize them here. In all NDA methods reviewed by these authors, no such error estimations for probabilistic algorithms have been developed. This discussion must be classified as a verification effort where one is comparing actual computational performance of certain NDA elements as compared to benchmark or analytical results.

- *Probability integration algorithms.* The FPI algorithms in existence have varying levels of robustness and accuracy (e.g., References 5, 60). The user must be provided with automated ways of assuring robustness for the particular RV distributions. The final probability of failure result must include a margin for the FPI algorithm error. Monte Carlo simulation errors due to truncation must also be accounted for in RS NDA algorithms.

The use of any FPI algorithm must be supported by established limits on which types of distributions for which the method is robust. Additionally, FPI and MC algorithms must be accompanied by error prediction bounds that can be used in a general approach to confidence interval analysis.

- *RS errors.* The RS approach is suggested in two areas of NDA for design. One is the RS used as a surrogate for the physical or computational model that is too large for application with Monte Carlo methods. As developed by Pratt & Whitney, statistical error measures have been used to validate a RS in a global sense [58] but not at the design point where the probability of failure is being driven.

  The RS method must be implemented in such a way that the error between the surrogate model and the full computational model can be estimated. Often, this amounts to a single deterministic analysis at the most probable point (MPP) at the failure condition. The RS modeling error must be factored into the final reliability prediction in a manner consistent with all modeling errors.

- *Validation errors.* The Sandia National Laboratory validation effort includes statistical measures of the deterministic modeling errors [110]. The Sandia approach includes the important contribution of experimental errors in model validation. More is needed to extend this approach to the off-nominal conditions that are required for probabilistic model validation.

  A statistical model that captures both the experimental and computational modeling error must be defined for each tool and the design regime in which it is being used. This modeling error must be included in the final confidence interval as a statistical distribution.

- *Unknown-unknown errors.* One cannot model what one cannot conceive of as a failure mode for complex systems. The only rational way we can conceive of for reducing the contribution of this type of error is by a full integration of NDA methods that are essentially quantitative with the qualitative methods such as FMEA and FTA [9]. At the same time, it is critical that designers reference comprehensive "lessons learned" data or knowledge bases.

  The nondeterministic design environment must provide for a formal linkage between the FMEA processes and the reliability-based design. Failure modes not considered in the normal NDA design process must all be addressed and resolved. Field failure data records must also be reviewed for any failure modes not considered and such modes resolved for the new design.

3.5.4 Confidence Interval Modeling

The use of NDA for design results in reliability predictions that are the expected value(s) of the system performance. The design data used to generate the expected level of system performance includes probability distributions and their parameters, deterministic models of the physical processes, and system models that tie the variables and the physical process models together. Typically, each of these elements is treated as "truth." Some of the questions one can raise regarding each of these design elements that affect our confidence in the predicted outcome are the following:

- *Incomplete data to fully define the probability distributions being used in the NDA modeling.* Statistically speaking, one can only express the probability of making an error in accepting a certain probability distribution to model individual data sets. The assumption is that the distribution giving the highest statistical evaluation of fitting data is "correct" although any mathematical model is just that—a model. There are many other statistical issues here that go well beyond what can be addressed by this simple discussion.

- *Incomplete data to fully define the parameters used in the probability distributions.* Assuming for now that there are physical grounds for selecting a given distribution, the datasets used to fit the parameters of that distribution are always finite. Our confidence in the values for application of NDA near the center of the distribution is greater than our confidence if the design conditions are near the tails of the data.

NDA for design to a reliability goal typically determines that the limit condition for that reliability condition involves combinations of the RVs such that some of these RVs are in the tails of their distributions. For design problems, there are at least two primary concerns. The design limit condition may be based on applying the probability distribution for that variable beyond the existing data for that variable. Such extrapolations lead to a high risk or low confidence that either the probability distribution or the parameters are accurate at the extrapolated conditions. At this time, there are no formal statistical methods to treat this uncertainty and *ad hoc* methods are needed.

The second problem of incomplete data is the classical statistical problem of empirical or analytical confidence intervals on the distribution given the fitted parameters [112, 113]. Such confidence intervals expand significantly in the tails of the distributions. Strict adherence to these CIs typically imposes significant burdens of the designer to increase the size of the datasets needed to support design for reliability.

The usual basis for NDA for system design is the application of deterministic models of physics but with uncertainty in the modeling parameters, boundary conditions, geometry, etc. This report has identified model verification and validation as required steps in any design problem. The validation process that has been discussed can be used to derive statistical models that link the models to the experimental results recognizing that both have uncertainties associated with them.

Added to the usual validation requirement is the stated requirement that the physical models must be validated in the off-nominal condition, especially weighted toward the conditions that NDA determines control the system reliability. The report also identifies numerical errors in some of the analytical methods that are used to support NDA such as response surface, fast probability integration, and the Monte Carlo simulation algorithms. The uncertainties of each of these must be incorporated in the CI estimation.

The issue of user or operator errors is also a real modeling problem that is linked to confidence in the reliability prediction. We do not address this issue herein beyond indicating the need for the design process to provide quality assurance steps that assure consistency and reliability of the predicted outcomes. Various manufacturers have different approaches to qualifying users of any design tools. These approaches must be expanded to include NDA.

The last area of uncertainty in the NDA prediction of system reliability is the system model itself. This uncertainty is more deterministic in that the system model validation process must be based on sufficient system level testing to assure that the principal RVs and their linkages between the physical modeling modules are correct. We have already cited the Pratt & Whitney probabilistic design system approach to validating the system level model by reference to past performance data [111]. Such system level NDA model validation is the culmination of the validation process for each of the elements in the system model.

Before NDA can fully transition to the design environment for reliability critical applications, the NDA community must address the full set of confidence issues. We strongly believe that the NDA-based design process must be capable of predicting not only the expected system reliability but also the appropriate bounds on the predicted reliability at some confidence level (in the Bayesian sense not the frequentist sense), and the major random variables that govern both. A notional representation of the operation of such an NDA design system with CIs over the design timeframe is shown in Figure 3.4.

The only appropriate formal basis for the definition of CIs for the totality of the defined confidence issues is the Bayesian belief form of confidence [114]. We have developed a generalized application of this general concept in the context of the Markov Chain Monte Carlo algorithm for a representative systems problem [3]. However, work is needed to define rational design approaches to establishing such Bayesian CIs for system problems that combine the uncertainties in distributions, data, models, and the results of validation. Strict statistical approaches simply are not applicable to this critical problem.

3.5.5 Traditional Reliability and Safety Assessment Methods

One of the major opportunities and challenges for NDA in design is achieving a complete closure with standard reliability and safety assessment methods such as FMEA and FTA modeling. The latter already has been addressed in limited ways in various software packages such as NESSUS [54].

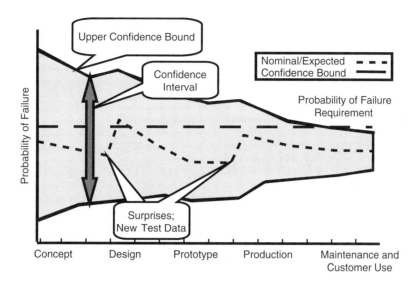

FIGURE 3.4 NDA confidence interval depiction.

A key issue in safety assessment design is the fact that most reliability critical designs begin with a reliability allocation to various subsystems and then to the components themselves. Often, such reliability assessments as are needed to support this kind of top-down reliability allocation approach are based on past hardware data and not on NDA quantitative assessments. Education and communication are required to bring such safety programs into the world of quantitative assessments, with updating methods such as Bayesian methods to achieve consistency between the hardware experience and the quantitative tools coming from NDA technology.

The FMEA process is the most qualitative but provides an important opportunity to address the issues raised by the unknown-unknown reliability problem. That is, the FMEA process, when done properly, has a reasonably high likelihood of identifying a comprehensive set of failure modes—that is its purpose! The FMEA process does not prioritize the failure modes except by reference to operating experience with similar designs. By combining the FMEA and NDA methodologies, one has the ability to be (1) comprehensive in predicting the system reliability, (2) prioritizing by use of the quantitative NDA tools, (3) valid by reference to operating experience, and (4) cost effective by reducing the amount of testing that nonquantitative FMEA results often require.

3.6 Conclusion

The design community is poised on the edge of a major transition from traditional, margin-based design to designs based on NDA. There are many technologies available to support NDA-based design but critical transition issues remain. Some of the transition issues are clearly related to concerns over the cost associated with the need for greater amounts of design data in the appropriate forms for NDA and the greater expense of large NDA models in relation to deterministic design. Nonetheless, there is also a growing understanding that margin-based design is, in many cases, overly conservative such that NDA can result in cost and weight savings in performance critical applications. There is also a growing understanding that NDA methods provide new and critical insights into variable dependencies and interactions that, when properly used, will result in more robust designs.

Acknowledgments

The authors wish to acknowledge the support of the Air Force Research Laboratory, Propulsion Directorate, for the work that underlies this report. The first author also acknowledges the direct support of

Universal Technologies Inc. and Michele Puterbaugh of UTC for her support through the high cycle fatigue program.

References

1. Cruse, T.A., Non-deterministic, Non-traditional Methods (NDNTM), NASA/CR-2001-210976, July 2001.
2. Meyer, M.A. and Booker, J.M., Eliciting and Analyzing Expert Judgment, American Statistical Association and the Society for Industrial and Applied Mathematics, Alexandria, Virginia and Philadelphia, Pennsylvania, 2001.
3. Cruse, T.A. and Brown, J.M., Confidence Intervals for Systems of Random Variables, submitted for publication.
4. Fox, E.P., Issues in Utilizing Response Surface Methodologies for Accurate Probabilistic Design, *Proceedings of the 37th AIAA/ASME/ASCE/AHS/ASC Structures, Structural Dynamics, and Materials Conference,* AIAA-96-1496, April 1996.
5. Wu, Y.-T., Demonstration of a new fast probability integration method for reliability analysis, *J. Engrg. Indus.,* 109, 24–28, 1987.
6. Tryon, R.G., Probabilistic Mesomechanical Fatigue Model, NASA Contractor Report 202342, April 1997.
7. Tryon, R.G. and Cruse, T.A., A Reliability-Based Model to Predict Scatter in Fatigue Crack Nucleation Life, *Fatigue & Fract. Engrg. Matls. & Struct.,* 21, 257–267, 1998.
8. Roemer, M.J. and Kacprzynski, G.J., Advanced Diagnostics and Prognostics for Gas Turbine Engine Risk Assessment ASME/IGTI Turbo-Expo 2000, Munich Germany, 2000-GT-30, May 2000.
9. Lewis, E.E., Introduction to Reliability Engineering, John Wiley & Sons, New York, 1987.
10. Kapur, K.C. and Lamberson, L.R., *Reliability in Engineering Design,* John Wiley & Sons, New York, 1977.
11. Meyer, P.L., *Introductory Probability and Statistical Applications,* Addison-Wesley, Reading, MA, 1970.
12. Cruse, T.A, Ed., *Reliability-Based Mechanical Design,* Marcel Dekker Inc., New York, 1997.
13. Anon., *Handbook of Reliability Prediction Procedures for Mechanical Equipment,* Report Carderock Division NSWC-92/L01, Naval Surface Warfare Center, Bethesda, MD, 1992.
14. Abernethy, R.B., Breneman, J.E., Medlin, C.H., and Reinman, G.L., *Weibull Analysis Handbook,* Air Force Wright Aeronautical Laboratory Report AFWAL-TR-83-2079, 1983.
15. Beck, J.V. and Arnold, K.J., *Parameter Estimation in Engineering and Science,* John Wiley & Sons, New York, 1977.
16. Meeker, W.Q. and Escobar, L.A., *Statistical Methods for Reliability Data,* John Wiley & Sons, Inc., New York, 1998.
17. Kececioglu, D., *Reliability Engineering Handbook,* Vol. 2, Chapter 11: Fault Tree Analysis, Prentice Hall, Englewood Cliffs, NJ, 1991.
18. Kumamoto, H. and Henley, E.J., *Probabilistic Risk Assessment and Management for Engineers and Scientists,* 2nd edition, Chapter 3: Probabilistic Risk Assessment, IEEE Press, 1996.
19. Tryon, R.G., Cruse, T.A., and Mahadevan, S., Development of a reliability-based fatigue life model for gas turbine engine structures, *Engrg. Fract. Mechs.,* 53, 5, pp. 807–828, 1996.
20. Maggio, G. and Fragola, J.R., Combining computational-simulations with probabilistic-risk assessment techniques to analyze launch vehicles, *Proceedings, 1995 Annual R&M Symposium,* pp. 343–348.
21. Cruse, T.A., Engine Components, Practical Applications of Fracture Mechanics, NATA AGARDograph #257, Chapter 2, 1980.
22. Cruse, T.A., Mahadevan, S., and Tryon, R.G., Fatigue Reliability of Gas Turbine Engine Structures, NASA/CR-97-206215.

23. Khintchine, A. Ya., Mathematisches uber die Erwartung von einen offentlicher Schalter, *Matemalliche Sbornik*, 1932.
24. Palm, C., Arbetskraftens Fordelning vid betjaning av automatskiner, *Industritidningen Norden*, 1947.
25. Barlow, R. E. and Proschan, F., *Mathematical Theory of Reliability*, J. Wiley & Sons, New York, p. 1, 1965.
26. Lotka, A.J., A contribution to the theory of self-renewing aggregates with special reference to industrial replacement, *Annual of Mathematical Statistics*, 10, 1–25, 1939.
27. Weibull, W., A statistical theory of the strength of materials, *Proceedings, Royal Swedish Institute of Engineering Research*, 151, Stockholm, Sweden, 1939.
28. Gumbel, E. J., Les valeurs extremes des distributions statistiques, *Annales de l'Institute Henri Poincare*, 4, 2, 1935.
29. Cruse, T. A., McClung, R. C., and Torng, T. Y., NSTS Orbiter Auxiliary Power Unit Turbine Wheel Cracking Risk Assessment, *Journal of Engineering for Gas Turbines and Power*, 114, ASME, pp. 302–308, April 1992.
30. Madsen, H.O., Krenk, S., and Lind, N.C., *Methods of Structural Safety*, Prentice Hall, Englewood Cliffs, NJ, 1986.
31. Cornell, C.A., Bounds on the reliability of structural systems, *J. Structural Division*, 93, American Society of Civil Engineering, pp. 171–200, 1967.
32. Lind, N.C., The design of structural design norms, *J. Structural Mechanics*, 1, 3, pp. 357–370, 1973.
33. Ditlevsen, O., Structural reliability and the invariance problem, Technical Report No. 22, Solid Mechanics Division, University of Waterloo, Ontario, Canada, 1973.
34. Lind, N.C., An invariant second-moment reliability format, Paper No. 113, Solid Mechanics Division, University of Waterloo, Ontario, Canada, 1973.
35. Hasofer, A.M. and Lind, N.C., Exact and invariant second-moment code format, *J. Engineering Mechanics Division*, 100, American Society of Civil Engineers, EM1, pp. 111–121, 1974.
36. Ang, A.H.-S. and Cornell, C.A., Reliability bases of structural safety and design, *J. Structural Division*, 11, American Society of Civil Engineers, pp. 1755–1769, 1974.
37. Rackwitz, R. and Fiessler, B., Structural reliability under combined random load sequences, *J. Engineering Mechanics Division*, 100, American Society of Civil Engineers, pp. 489–494, 1978.
38. Chen, X. and Lind, N.C., Fast probability integration by three-parameter normal tail approximation, *Structural Safety*, 1, pp. 269–276, 1983.
39. Wu, Y.-T., Demonstration of a new, fast probability integration method for reliability analysis, *J. Engineering for Industry*, 109, American Society of Mechanical Engineers, pp. 24–28, 1987.
40. Wu, Y.-T., Millwater, H.R., and Cruse, T.A., Advanced probabilistic structural analysis method for implicit performance functions, *J. of the AIAA*, 28(9), American Institute for Aeronautics and Astronautics, pp. 1663–1669, 1990.
41. Cruse, T.A., Rajagopal, K.R., and Dias, J.B., Probabilistic structural analysis methodology and applications to advanced space propulsion system components, *Computing Systems in Engineering*, 1(2-4), pp. 365–372, 1990.
42. Cruse, T.A., Mahadevan, S., Huang, Q., and Mehta, S., Mechanical system reliability and risk assessment, *AIAA Journal*, 32, 11, pp. 2249–2259, 1994.
43. Cruse, T.A., Burnside, O.H., Wu, Y.–T., Polch, E.Z., and Dias, J.B., Probabilistic structural analysis methods for select space propulsion system structural components (PSAM), *Computers & Structures*, 29, 5, pp. 891–201, 1988.
44. Cruse, T.A., Rajagopal, K.R., and Dias, J.B., Probabilistic structural analysis methodology and applications to advanced space propulsion system components, *Computing Systems in Engineering*, 1,(2–4), pp. 365–372, 1990.
45. Cruse, T.A., Mechanical reliability design variables and models, Chapter 2, in *Reliability-Based Mechanical Design*, T.A. Cruse, Editor, Marcel Dekker, New York, 1997.

46. Wu, Y.-T., Millwater, H.R., and Cruse, T.A., An advanced probabilistic structural analysis method for implicit performance functions, *AIAA Journal,* 28(9), pp. 1663–1669, Sept. 1990.
47. Riha, D.S., Millwater, H.R., and Thacker, B.H., Probabilistic structural analysis using a general purpose finite element program, *Finite Elements in Anal. Des.,* 11, pp. 201–211, 1992.
48. Wu, Y.-T., Computational methods for efficient structural reliability and reliability sensitivity analysis, *AIAA J.,* 32(8), pp. 1717–1723, 1994.
49. Thacker, B.H., Application of advanced probabilistic methods to underground tunnel analysis, *Proceedings of the 10th ASCE Engineering Mechanics Conference,* S. Sture, Editor, Vol. I, pp. 167–170, 1995.
50. Millwater, H.R., Smalley, A.J., Wu, Y.-T., Torng, T.Y., and Evans, B.F., Computational Techniques for Probabilistic Analysis of Turbomachinery, ASME Paper 92-GT-167, June 1992.
51. Cruse, T.A., Unruh, J.F., Wu, Y.-T., and Harren, S.V., Probabilistic structural analysis for advanced space propulsion systems, *J. Eng. Gas Turb. Power,* 112, pp. 251–260, 1990.
52. Millwater, H., Wu, Y., and Fossum, A., Probabilistic Analysis of a Materially Nonlinear Structure, Paper AIAA-90-1099, 31st AIAA/ASME/ASCE/AHS/ASC Structures, Structural Dynamics and Materials Conference, April 1990.
53. Wu, Y.-T. and Wang, W., Efficient probabilistic design by converting reliability constraints to approximately equivalent deterministic constraints, *Trans. Soc. Design Process Sc.,* 2(4), pp. 13–21, 1998.
54. Riha, D.S., Thacker, B.H., Millwater, H.R., and Wu, Y.-T., Probabilistic Engineering Analysis Using the NESSUS Software, Paper AIAA 2000-1512, *Proceedings of the 41st AIAA/ASME/ASCE/ASC Structures, Structural Dynamics, and Materials Conference and Exhibit, AIAA Non-Deterministic Approaches Forum,* Atlanta, Georgia, 3–6 April 2000.
55. Cesare, M.A. and Sues, R.H., ProFES probabilistic finite element system—Bringing probabilistic mechanics to the desktop, *Proceedings of the 40th AIAA/ASME/ASCE/ASC Structures, Structural Dynamics, and Materials Conference and Exhibit, AIAA Non-Deterministic Approaches Forum,* April 1999.
56. Sues, R.H. and Cesare, M., An innovative framework for reliability-based MDO, *Proceedings of the 41st AIAA/ASME/ASCE/ASC Structures, Structural Dynamics, and Materials Conference and Exhibit, AIAA Non-Deterministic Approaches Forum,* Atlanta, Georgia, 3–6 April 2000.
57. Liu, P.-L., Lin, H.-Z., and Der Kiureghian, A., CALREL User Manual, Report No. UCB/SEMM-89/18, Structural Engineering, Mechanics and Materials, Department of Civil Engineering, University of California, Berkeley, CA, 1989.
58. Fox, E.P., The Pratt & Whitney Probabilistic Design System, Paper AIAA-94-1442, *Proceedings of the 35th AIAA/ASME/ASCE/AHS/ASC Structures, Structural Dynamics, and Materials Conference,* 1994.
59. Roth, P.G., Probabilistic Rotor Design System (PRDS), AFRL Contract F33615-90-C-2070 Final Report, General Electric Aircraft Engines, June 1999.
60. Fox, E.P. and Reh, S., On the Accuracy of Various Probabilistic Methods, Paper AIAA-2000-1631, 2000.
61. Wacholder, E., Elias, E., and Merlis, Y., Artificial networks optimization method for radioactive source localization, *Nuclear Technology,* 110, pp. 228–237, May 1995.
62. Harmon, L. and Schlosser, S., CPI plants go data mining, *Chemical Engineering,* pp. 96–103, May 1999.
63. Rai, M.R. and Madavan, N.K., Aerodynamic design using neural networks, *AIAA Journal,* 38(1), January 2000, pp. 173–182.
64. Steppe, J.M. and Bauer, K.W., Jr., Feature saliency measures, *Computers Math. Applic.,* 33(8), pp. 109–126, 1997.
65. Yu, D.C., Nguyen, T.C., and Haddaway, P., Bayesian network model for reliability assessment of power systems, *IEEE Transactions on Power Systems,* 14(2), pp. 426–432, May 1999.
66. Xia, Y. and Wang, J., A general methodology for designing globally convergent optimization neural networks, *IEEE Transactions on Neural Networks,* 9(6), pp. 1331–1343, January-February, 1999.
67. Mendel, J.M., Fuzzy logic systems for engineering: a tutorial, *Proceedings of the IEEE,* Vol. 83, No. 3, 345–377, March 1995.

68. Zadeh, L.A., Outline of a new approach to the analysis of complex systems and decision processes, *IEEE Transactions on Man and Cybernetics*, Vol. SMC-3, No. 1, 28–44, 1973.
69. Bezdek, J. and Pal, S.K., *Fuzzy Models for Pattern Recognition*, IEEE Press, New York, 1992.
70. Jang, J.-S. Roger and Sun, C.-T., Neuro-fuzzy modeling and control, *Proceedings of the IEEE*, 83(3), 378–405, March 1995.
71. Bowles, J.B. and Peláez, C.E., Application of fuzzy logic to reliability engineering, *Proceedings of the IEEE*, 83(3), 435–449, March 1995.
72. Misra, K. B. and Onisawa, T., Use of fuzzy sets theory part II: applications, in *New Trends in System Reliability Evaluation*, Elsevier, Amsterdam, 551–587, 1993.
73. Roy, R., A primer on the Taguchi method, Van Nostrand Reinhold, New York, 1990.
74. Bonissone, P.P., Badami, V., Chiang, K.H., Khedkar, P.S., Marcelle, K.W., and Schutten, M.J., Industrial applications of fuzzy logic at General Electric, *Proceedings of the IEEE*, 83(3), 450–464, March 1995.
75. Procyk, T. and Mamdani, E., A linguistic self-organizing process controller, *Automatica*, 15(1), 15–30, 1979.
76. Zadeh, Lotfi A., Soft Computing and Fuzzy Logic, *IEEE Software*, pp. 48–58, November 1994.
77. Kosko, B., *Neural Networks and Fuzzy Systems: A Dynamical Systems Approach to Machine Language*, Prentice Hall, Englewood Cliffs, NJ, 1991.
78. Karr, C., Genetic algorithms for fuzzy controllers, *AI Expert*, pp. 26–33, Nov. 1991.
79. Mulkay, E.L. and Rao, S.S., Fuzzy heuristics for sequential linear programming, *J. Mechanical Design*, 120, pp. 17–23, March 1998.
80. Chen, S., Nikolaidis, E., and Cudney, H.H., Comparison of probabilistic and fuzzy set methods for designing under uncertainty, Paper AIAA-99-1579, pp. 2660–2674, 1999.
81. Rao, S.S. and Weintraub, P.N., Modeling and analysis of fuzzy systems using the finite element method, Paper AIAA-2000-1633.
82. Walster, G. William, Introduction to interval arithmetic, unpublished note supporting Sun Microsystems compilers for interval programs, May 19, 1997.
83. Alefield, G. and Claudio, D., The basic properties of interval arithmetic, its software realizations and some applications, *Computers and Structures*, 67, pp. 3–8, 1998.
84. Walster, G. William, Interval arithmetic: the new floating-point arithmetic paradigm, unpublished notes from the Fortran Compiler Technology group at Sun Microsystems.
85. Hansen, E.R., *Global Optimization Using Interval Analysis*, Marcel Dekker, Inc., New York, 1992.
86. Köylüoglu, H. Ugur and Elishakoff, I., A comparison of stochastic and interval finite elements applied to shear frames with uncertain stiffness properties, *Computers & Structures*, 67, 91–98, 1998.
87. Qiu, Z.P., Chen, S.H., and Elishakoff, I., Natural frequencies of structures with uncertain but nonrandom parameters, *J. Optimization Theory Appl.*, 86(3), 669–683, 1995.
88. Ben-Haim, Y. and Elishakoff, I., *Convex Models of Uncertainty in Applied Mechanics*, Elsevier Science Publishers, Amsterdam, 1990.
89. Li, Y.W., Elishakoff, I., Starnes, J.H., Jr., and Shinozuka, M., Prediction of natural frequency and buckling load variability due to uncertainty in material properties by convex modeling, *Fields Institute Communications*, 9, 139–154, 1996.
90. Khuri, A. I. and Cornell, J. A., *Response Surfaces*, Marcel Dekker Inc., New York, 1987.
91. Box, G.P. and Draper, N.R., *Empirical Model-Building and Response Surfaces*, J. Wiley & Sons, New York, 1987.
92. Toh, A.T.C., Genetic algorithm search for critical slip surface in multiple-wedge stability analysis, *Can. Geotech. J.*, 36, pp. 382–391, 1999.
93. Xia, Y. and Wang, J., A general methodology for designing globally convergent optimization neural networks, *IEEE Transactions on Neural Networks*, 9(6), pp. 1331–1343, January–February, 1999.
94. Mosher, T., Conceptual spacecraft design using a genetic algorithm trade selection process, *J. of Aircraft*, 36(1), pp. 200–208, January–February, 1999.

95. Alberti, N. and Perrone, G., Multipass machining optimization by using fuzzy possibilistic programming and genetic algorithms, *Proc. Inst. Mech. Engrs.*, 213, Part B, pp. 261–273, 1999.
96. Hajela, P., Nongradient methods in multidisciplinary design optimization—status and potential, *J. Aircraft*, 36(1), pp. 255–265, January 1999.
97. Josephson, J.R., Chandrasekaran, B., Carroll, M., Iyer, N., Wasacz, B., Rizzoni, G., Li, Q., and Erb, D.A., An architecture for exploring large design spaces, *Proceedings of the 1998 AAAI*, http://www.aaai.org/.
98. Fogel, D. B., Fukuda, T., and Guan, L., Scanning the Issue/Technology—special issue on computational intelligence, *Proc. IEEE*, 87(9), pp. 1415–1421, September 1999.
99. Bailey, M.W., Irani, R.K., Finnigan, P.M., Röhl, P.J., and Badhrinath, K., Integrated multidisciplinary design, manuscript submitted to AIAA, 2000.
100. Röhl, P.J., Kolonay, R., Irani, M., Sobolewski, R.K., Kao, M., and Bailey, M.W., A federated intelligent product environment, *AIAA-2000-4902*, 2000.
101. Lochner, R.H. and Matar, J.E., *Design for Quality*, Quality Resources, A Division of The Kraus Organization Ltd., White Plains, NY, 1990.
102. Anon., Design-Expert Software Version 6 User's Guide, Stat-Ease, Inc., Minneapolis, MN, 55413-9827.
103. DePold, H.R. and Gass, F.D., The application of expert systems and neural networks to gas turbine prognostics and diagnostics, *J. Engrg. Gas Turbines Power*, 121, pp. 607–612, October 1999.
104. Anon., DoD Modeling and Simulation (M&S) Verification, Validation, and Accreditation (VV&A), Department of Defense Instruction No. 5000.61, April 29, 1996.
105. Roache, P.J., *Verification and Validation in Computational Science and Engineering*, Hermosa Publishing, Albuquerque, NM, 1998.
106. Oberkampf, W.L., Trucano, T.G., and Hirsch, C., Verification, Validation, and Predictive Capability in Computational Engineering and Physics, Sandia National Laboratory Report SAND2003-3769, February 2003.
107. Anon., Guide for the Verification and Validation of Computational Fluid Dynamics Simulations, AIAA Guide G-077-1998.
108. Pilch, M., Trucano, T., Moya, J., Froehlich, G., Hodges, A., and Peercy, D., Guidelines for Sandia ASCI Verification and Validation Plans—Content and Format: Version 2.0, Sandia Report SAND2000-3101, January 2001.
109. Trucano, T.G., Easterling, R.G., Dowding, K.J., Paez, T.L., Urbina, A., Romero, V.J., Rutherford, B.M., and Hills, R.G., Description of the Sandia Validation Metrics Project, Sandia Report SAND2001-1339, July 2001.
110. Hills, R.G. and Trucano, T.G., Statistical Validation of Engineering and Scientific Models: A Maximum Likelihood Based Metric, Sandia Report SAND2001-1783, January 2002.
111. Adamson, J., Validating the Pratt & Whitney Probabilistic Design System, *Proceedings of the 6th Annual/FAA/AF/NASA/Navy Workshop on the Application of Probabilistic Methods to Gas Turbine Engines*, Solomons Island, MD, March 19, 2003.
112. Mann, N.R., Schafer, R.E., and Singpurwalla, N.D., *Methods for Statistical Analysis of Reliability and Life Data*, John Wiley & Sons, New York, 1974.
113. Meeker, W.Q. and Escobar, L.A., *Statistical Methods for Reliability Data*, John Wiley & Sons, New York, 1998.
114. Martz, H.F. and Waller, R.A., Bayesian Reliability Analysis, Reprint Edition, Krieger Publishing Company, Malabar, FL, 1991.

4
An Industry Perspective on the Role of Nondeterministic Technologies in Mechanical Design

Kadambi Rajagopal
Rocketdyne Propulsion and Power, The Boeing Company

4.1	Introduction	4-1
4.2	The Business Case	4-2
4.3	Strategies for Design Approach for Low-Cost Development	4-4
4.4	Role of Design Space Exploration Approaches (Deterministic and Nondeterministic) in the Product Development Phase	4-5
4.5	Sensitivity Analysis	4-8
4.6	Probabilistic Analysis Approaches	4-9
4.7	The Need and Role of Multidisciplinary Analysis in Deterministic and Nondeterministic Analysis	4-11
4.8	Technology Transition and Software Implementation	4-12
4.9	Needed Technology Advances	4-13

4.1 Introduction

Affordable and reliable access to space is a fundamental and necessary requirement to achieve the loftier goals of expanding space exploration. It also plays an important role in achieving the near-term benefits of space-based technologies for mankind. The development, performance, and reliability of liquid rocket propulsion systems have played and will continue to play a critical role in this mission. However, the cost of development of a typical propulsion system in the past was on the order of a few billion dollars. The design approaches using nondeterministic technologies as well as deterministic design space exploration approaches show promise in significantly reducing the development costs and improving the robustness and reliability of newly developed propulsion systems.

4.2 The Business Case

Traditional ideas of application of structural reliability concepts use the probabilistic information of stress and strength (supply and demand) to compute the probability of failure. The probabilistic approach when applied to all facets of structural analysis (not just random vibration) will be a significant change from past practices. In the conventional factor of safety approach to structural design, the probability of failure is neither explicitly calculated nor stated as a design requirement. However, the potential for reducing the development costs of liquid rocket propulsion systems through the use and application of nondeterministic analysis approaches is of great interest in the liquid rocket propulsion industry.

This is illustrated in Figure 4.1 where the high reliability of liquid rocket engines is achieved by extensive testing of the hardware to flush out all the failure modes. This development cycle is typically referred to as "Test-Fail-Fix" cycle. Similar experiences can be found in jet engine as well as in automobile product development. While this approach has proven itself very successful, the new realities of limited budgets have made such an approach not always practical. Further, in addition to cost, the schedule pressures can preclude extensive testing as an avenue for successful product development. It also must be mentioned that in many space use scenarios, it is impossible to duplicate the space environment on ground. The challenge is then to develop product development approaches to achieve the development cost and schedule profile goals shown in Figure 4.1.

A study of the developmental failures in the liquid rocket engine indicates that the failures can be categorized due to three main causes, which are approximately equal in percentage. They are (1) a lack of knowledge of the loading environment, (2) the absence or lack of accuracy in the physics-based models to explain complex phenomena (e.g., combustion-induced high frequency vibration), and (3) lack of understanding of hardware characteristics as manufactured (e.g., unintended defects). (Figure 4.2). Successful robust and reliable product development under the cost and schedule constraints mentioned above requires a new design approach that puts less emphasis on testing. It is in this role that the nondeterministic approaches that explicitly consider *total uncertainty* can play a

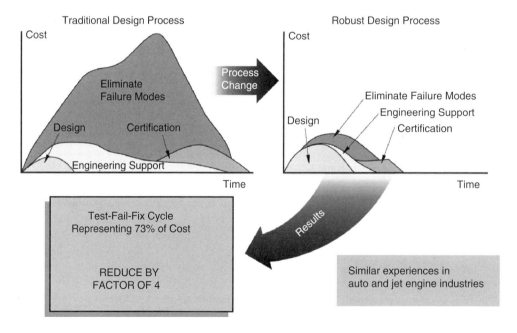

FIGURE 4.1 Historical development cost profile for liquid rocket engine development.

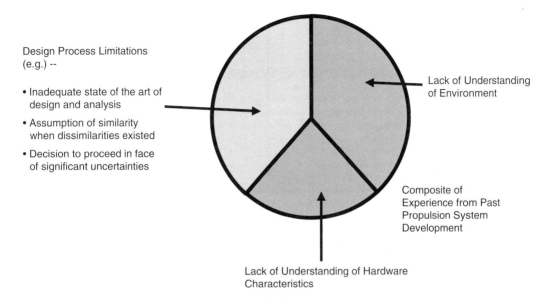

FIGURE 4.2 Categorization of causes of development failures based on historical data.

significant role. The role of epistemic uncertainty (lack of knowledge) resulting in consequences that drive up development cost is significant and comparable to aleatory (system variability) uncertainty based root causes.

Historically, in a typical new engine design, the development testing process works through approximately 25 to 150 failure modes. Most of them are minor but some require major redesign of specific components. The number of failures depends on the technology stretch of the new engine (from past experience), both from engine cycle as well as production process points of view. Considering that the typical engine contains thousands of parts, the majority of the components are designed correctly from the beginning and they perform flawlessly. But the expense of testing is so high, due to cost of fuel, facilities, and expensive hardware, that even this small amount of failures has very high cost consequences. In the population of these failure modes, some of them could be anticipated and prevented by the improved current state-of-the-art analysis technology (e.g., finite element methods, computational fluid dynamics, etc.) leveraging the quantum improvements in the computational speed. Some of the failure modes are the result of complex system interactions, such as fluid structure coupling, high-frequency excitation due to internal turbulent flow, and nozzle side loads generated due to overexpanded nozzle ratios (the nozzle exit pressure is less than the ambient pressure) during ground testing. The current state-of-the-art model based analysis methodology has not advanced enough to predict reliably the loads induced by these phenomena.

It is necessary to make a clear distinction between activities related to automation in design (computer aided design, CAD) that result in significant cost savings and the technologies that reduce the developmental risk because of their ability to operate in the uncertainty space. A successful practical approach to design should leverage advances in both the technologies, as shown in Figure 4.3. The chart should be read by starting at the right quadrant, going up to estimate the number of anticipated failures based on risk anchored by historical records, and moving to the left quadrant to estimate the number of failures and associated development costs. The risk factor as portrayed in Figure 4.3 could be a measure of design risk using rankings on a scale of 0 to 1.0. It could be calculated either by a subjective approach using a scoring method by a panel of experts with experience in very detailed engine categories or using a quantitative approach.

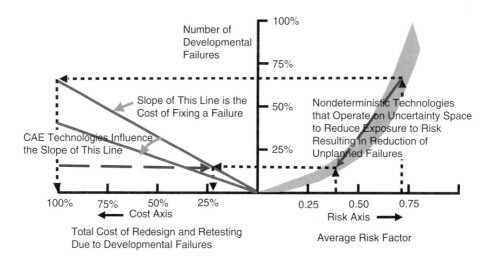

FIGURE 4.3 A scenario that leverages advances in computer aided design and the use of nondeterministic technologies to reduce development risk.

4.3 Strategies for Design Approach for Low-Cost Development

It is clear from the discussion in the previous section that the transient reliability of liquid rocket engines during the development phase has been historically low due to the number of developmental failures. However, it must be emphasized that by the time an engine reaches the certification stage, most of the design errors have been fixed using extensive testing, of course, with the corresponding cost consequence. During the service life, historically the reliability of man rated booster liquid rocket engines has been high enough that there have been no engine-related failures in the manned U.S. Space Program. It is a common occurrence (due to harsh and complex environments) that some of the parts do not meet the life requirements after they have entered into service, thus resulting in expensive inspection and maintenance costs.

So the new tools and approaches on the use of nondeterministic technologies must be tailored to their use in two design phases. During the development stage, the technologies used should provide an approach to reduce the failures using both analysis and testing. They should provide a rigorous quantitative approach for the initial design stages, in which there are significant epistemic and aleatory uncertainties, making the design insensitive, to the extent possible, to these uncertainties. This is referred to as Robust Design. Testing plays an important role during development by contributing to the reduction of epistemic uncertainty. One of the most important challenges that needs to be addressed is in designing a focused test program that maximizes information return with minimal testing. The challenge is a demanding one because the economics of putting a payload in space requires high thrust-to-weight ratios for the engines. This translates into small margins but yet safe operation, and losing an engine even during the development program is not a planned option (e.g., testing until failure occurs).

Once the lack of knowledge has been significantly reduced and the variabilities present in the system due to manufacturing capabilities are estimated and quantified using historical data, the challenge shifts to using a design/analysis approach that further improves the reliability of the engine to very high levels dictated by customer and contractual requirements. This might require redesign of some percentage of parts to meet the high reliability goal. Typically, the very high computed reliability values, as a minimum, will allow for comparative evaluation of competing designs to choose a better design, based on reliability constraints. The goal in this phase is to dramatically reduce service failures.

4.4 Role of Design Space Exploration Approaches (Deterministic and Nondeterministic) in the Product Development Phase

One of the fundamental design approaches to avoiding development failures could be to adopt the philosophy of "anticipate-prevent." This means that engineers must perform numerous "what-if" scenarios to meet the new design philosophy as opposed to "Test-Fail-Fix." The engineers' understanding of the performance of the engine and its components under nominal and off-nominal conditions must be complete. When deterministic technologies are applied to the various "what-if" scenarios, the consequences can be evaluated. Using a nondeterministic approach, for example probability or other uncertainty theories (e.g., Dempster-Shafer theory of evidence), the likelihood of their occurrence can be quantified. In some instances, the uncertainty in the probability statements themselves could be quantified using confidence metrics. The important point is that the effective use of deterministic and nondeterministic technologies can help meet the goals of successful design by understanding the design sensitivities —the word "sensitivity" being used in the most general sense.

The design philosophy is best explained using Figure 4.4 and Figure 4.5. Figure 4.4 describes the design approach in general, and Figure 4.5 goes into the details of the design approach. The design approach does not force the use of all design processes to analyze and design every component. It is common practice that only a subset of available design approaches is used for a large set of components, which inherently have large margins or are not safety critical. However, for safety-critical components with severe failure consequences, the use of all design views to design a robust product is encouraged. This approach has not become universal practice in the industry to the extent one would hope, but the trend in using the design approach has been very encouraging with many instances of success.

The example used in Figure 4.5 is a hypothetical one involving two design variables (X1 and X2) and one response variable. The contours of the response variable are plotted, but in general are not known *a priori*. Each square in the chart represents a particular "view" of the design space using either deterministic or nondeterministic design approaches. It is arranged such that one can proceed from a well-established and widely understood design approach to more recent nondeterministic and robust design approaches. Each circle represents the engineer's knowledge of the behavior of the system under each

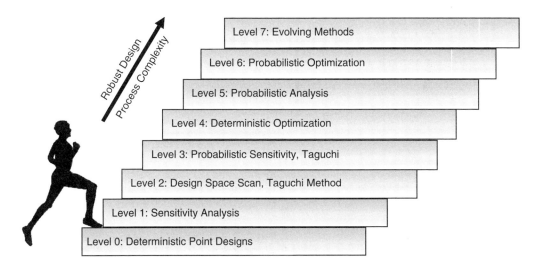

FIGURE 4.4 A design approach use ladder that can systematically improve the understanding of component behavior over the entire design space.

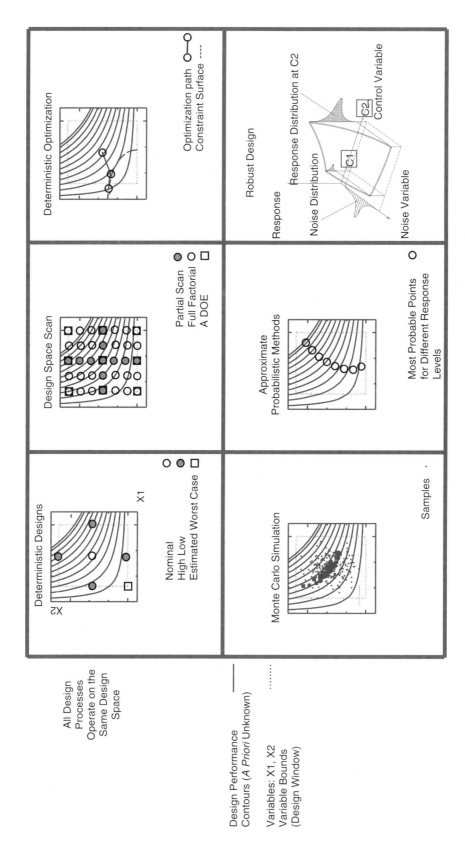

FIGURE 4.5. Multiple views of design space where sampling is performed on the same design space to meet a given "view" objective.

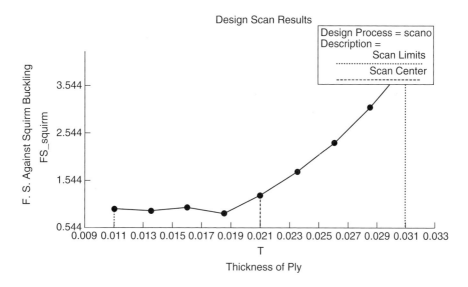

FIGURE 4.6 An example of partial design scan results identifying the nonlinearity in system response.

specific "view." For example, designing a component with conventional, but often heuristically determined worst-case scenario can provide a safe design but it sheds little information on the behavior of the system in nominal or at other "corner" conditions.

The deterministic design space scan is one of the most useful tools in understanding the design space and the potential nonlinearity present in the system. Understanding the extent of the nonlinearity present in the system is very important and in many cases can point to the cause of hardware failures. One can also readily identify nonrobust operating points of the design space from the scan. Frequently, many algorithms in the deterministic optimization or probabilistic analysis fail when significant nonlinearity is present. Even more troubling could be that the algorithms may provide inaccurate answers in the presence of significant nonlinearity and the errors may go unnoticed due to lack of rigorous error bounds with the methods. An example of such a behavior is seen in scan results on metal bellows analysis used in ducting of rocket engines. Figure 4.6 shows the Squirm buckling safety factor as a function of the thickness of each ply for a specific bellow design. Widespread use of such systematic analysis has only been recently made possible due to advances in automation.

While with partial scans one cannot study the interaction effects, one can still obtain valuable information regarding system "sensitivities" and nonlinearity under the umbrella of "Main Effects." When a deterministic scan is performed on noise variables, the results contain the information of tolerance effects or noise effects on performance in the deterministic sense. It is just that a probability measure has not been calculated, which could be done as a second step if sufficient information is available.

Other fractional or full factorial designs can be employed to study the interaction effects, which can be significant. When the dimensionality of the system becomes large, it becomes essential to use more efficient and practical approaches wherein the computational burden can be minimized. The Design of Experiments concept originally developed for physical experiments has been successfully applied to numerical experiments with modifications. The modifications pertain to elimination of the replication present in the original designs, which in the case of numerical experiments adds no new information. The deterministic optimization is widely performed as part of the engineering design process. As practiced by the engineering community, the deterministic optimization process (maximizing performance) does not directly address the question of Robustness (which is the domain of probabilistic optimization), which is defined as performance insensitivity under uncontrollable variations (noise). It has been observed that, in the current economic environment, the customer requirements emphasize equally, if not more, the reliability and robustness over performance.

When low-order statistical moments of responses need to be estimated, Monte Carlo solutions have been used most successfully. The Monte Carlo or other simulation-based methods (such as Latin Hyper Cube) have performed well over a wide class of problems. From several points of view, such as robustness of the algorithm, generality, and simplicity, Monte Carlo simulation is the preferred choice in the industry (including cases that require high reliability numbers) when computational burden is not an issue. When computational expense is prohibitive, approximate probabilistic methods generally referred to as First-Order Reliability Methods (FORMs) have been successfully applied. The word "approximate" is probably controversial as Monte Carlo simulation methods are also approximate. A by-product of FORMs, the most probable point information has the potential for playing an even more important role in defining a focused test matrix, to verify the design. For example, in a focused test scenario, the most probable point information could be used to design tests that can validate the extent of physics model accuracy in the regions of extreme event occurrence by seeding, for example, defects for the purposes of testing.

The Taguchi Analysis-based robust design concepts have been successfully applied in experimentally based design approaches. The use of these concepts in numerical model-based design approaches has been limited. However, the idea of using mathematical optimization techniques to achieve robustness has gained significant attention. Industry and government are making significant research investments in the area of mathematical probabilistic optimization-based technologies and they are discussed later.

4.5 Sensitivity Analysis

The design sensitivity factors have been used extensively to rank the input variables by their importance to performance variables. These factors are finding their way into formal design review meetings. This certainly gives the approving authorities an appreciation of the sensitivity of performance variables to input assumptions and the variable ranking in terms of their importance to satisfying the end requirement. In this context, the sensitivity factors derived from deterministic as well as nondeterministic approaches have been extremely valuable (Figure 4.7). One form of sensitivity factor ranking is based on measuring changes in performance to small perturbations in variables, which is purely deterministic. Because the design space is typically comprised of both input and response variables that span a multidisciplinary domain with a mixture of units, a pure gradient quantity is not of much practical value. Many normalization schemes have been proposed and used but they all have some shortcomings.

Typically, the results are portrayed as a pie or a bar chart with sorted sensitivity factors for each response variable. When this exercise is performed in a multidisciplinary space, they provide sensitivity values across discipline boundaries (e.g., thermal and structures) that have been invaluable. In general, the computed sensitivity factors are a function of the expansion point, which can be varied over the design window to obtain a more realistic assessment of the sensitivity factors over the entire design space. The probabilistic sensitivity factors are measures of the effect of input scatter to output scatter. The measures that have been used are computed such that the probabilistic information is integrated with deterministic sensitivity values (the exact algorithmic detail is a function of definition of the sensitivity measure). Because of the variety of definitions, the sensitivity measures have only been used in a qualitative way to rank the variables. An example of the sensitivity measure obtained for a mildly nonlinear problem with approximately similar coefficients of variation among input design variables is shown in Figure 4.7. When the degree of scatter among the input design variables is markedly different, the differences between deterministic and probabilistic sensitivity measures can be expected to be very high. The use of sensitivity measures in design is a success story as it was used to control variations during an engine development program.

In summary, in the initial stages of design, deterministic as well probabilistic methods that emphasize low-order statistics to screen different concepts are widely used. One practice is to establish safety margins

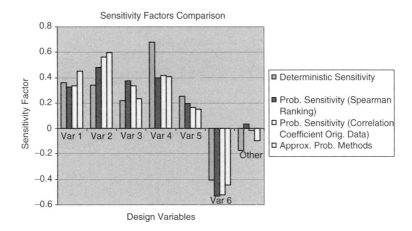

FIGURE 4.7 Comparison of different sensitivity measures for a mildly nonlinear problem with same order of coefficients of variation among input variables.

based on computed standard deviation, with less emphasis on the exact value of the reliability, which is usually designed in to be very high. The computed reliability of the product using probabilistic methods is recomputed, refined, and updated once more reliable data from actual tests is obtained.

4.6 Probabilistic Analysis Approaches

In practice, the use of probabilistic analysis approaches can be categorized based on their use emphasis, either for computing low-order statistics or for computing very high-order reliability quantities. The low-order statistical quantities of choice in the industry are the mean and the standard deviation for the random quantities. This is so because there is a long tradition of defining extreme quantities of design variables as a function of multiples of standard deviation among structural engineers, dynamic loads specialists, material scientists, and thermal analysts.

For the low-order statistics computation, the linear function approximation methods to propagate the "errors" and compute the standard deviation of response quantities have been very successful. However, the industry practice and preference has drifted toward Monte Carlo simulation based approaches to compute the low-order statistics. This is because of the ready availability or ease of writing of Monte Carlo simulation software and the widespread availability of cluster computing. The generality and stability of Monte Carlo simulation methods, the ability to increase the sample size to meet a given error bound, and the availability of high-end cluster computing to obtain quick turnaround are all reasons for its preferred use. It is common practice to perform Monte Carlo simulation on the order of 100 to 5000 sample values (depending on the computational burden in the function evaluation) using the best available knowledge of the input distribution. As an example, this approach has been so successful that detailed uncertainty analysis at the very early stages of engine design are routinely conducted to estimate the uncertainty in engine performance parameters such as thrust, engine mixture ratio, and specific impulse, a measure of engine efficiency (Figure 4.8). When needed for screening purposes, approximate reliability numbers are also computed, either by directly using the simulation results (when enough samples have been run for the required reliability estimate) or by using the computed mean and standard deviation values using the normality assumption for responses.

When more detailed reliability calculations are needed, approximate probabilistic analysis methods have been used. These include the Mean Value First Order (MVFO) method, the First-Order Reliability Method (FORM), the Fast Probability Integration (FPI) method, and many others such as SORM (Second-Order Reliability Method). In the above-mentioned technologies, a form of response function approximation at the region of interest (based on the probability of event occurrence) is fundamental to the algorithmic

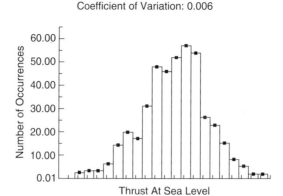

FIGURE 4.8 Example Monte Carlo simulation results from an engine performance model during the conceptual design phase.

approach. On the other hand, explicit computation of a response surface using fractional factorial designs commonly known as Design of Experiments (DOE) has also been widely used. Once the explicit functional form of the approximate response surface is available, it is used in deterministic and probabilistic optimization approaches as well in basic probabilistic analysis using Monte Carlo simulation.

In practice, there are numerous applications in industry where the degree of nonlinearity of the function behavior is moderate. In such cases, the performance of FPI and explicit response surface based applications has been good. An example of the results of probabilistic stress analysis of a turbine blade is shown in Figure 4.9. These methods have been used with success in evaluating the failure risk of products in service with defects or to evaluate the added risk under newly discovered or revised loads after the design is complete.

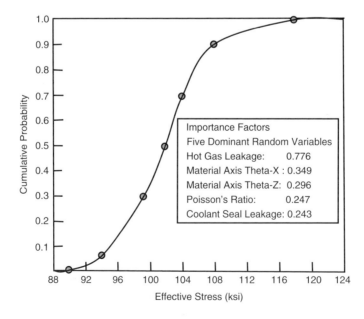

FIGURE 4.9 Example probabilistic stress analysis for a turbo-pump turbine blade using the First-Order Reliability Method.

The use of the above approximate methods for new designs is done on a case-by-case basis where there is prior customer concurrence of the methodology and the input assumptions that will be used in the analysis. The use of the above approximate methods in new designs will become even more widespread if there is a well-established and computationally tractable approach to include the epistemic uncertainty and provide confidence statements or error bounds on the computed reliability values.

4.7 The Need and Role of Multidisciplinary Analysis in Deterministic and Nondeterministic Analysis

All design activities in an industrial setting are multidisciplinary. Traditionally, the design for performance under the purview of each discipline is performed within each discipline, with information and data sharing among disciplines performed manually or in patches of automation. It is increasingly recognized that such an approach results in nonoptimal designs from a systems point of view. Thus, the industry involved in mechanical design in recent years has invested heavily in developing tools, generally component specific, for an effective multidisciplinary analysis. This has cost, schedule, and performance impacts in developing a product in a very positive way.

In the current context, a multidisciplinary analysis is defined as a process wherein consistent data/information is shared across all disciplines for a given realization. That is, as part of the process, a design point set is defined and collected as a specific realization of input design variables and corresponding output response variables spanning multiple disciplines. An example of such an application is illustrated in Figure 4.10, wherein consistent geometry, thermal, structural, and material property data is available for each realization, orchestrated by any design process (see Figure 4.5). The linked sets of models form the basis for deterministic as well as nondeterministic multidisciplinary analysis and optimization. The role of response surface methodology has been effective in system level optimization. In this approach, the response surface models are used in the linked set of models in lieu of computationally intensive models (e.g., computational fluid dynamics) for system optimization.

It is important to recognize the significance of a multidisciplinary linked set of models/codes to an accurate probabilistic analysis. There are uncertainties at all stages of manufacturing and use conditions, as illustrated in Figure 4.11. For an accurate reliability computation, in addition to accounting for those uncertainties, they should be introduced as appropriate numerical models linked to form a multidisciplinary model. This will capture accurately the correlation information through dependency relationships

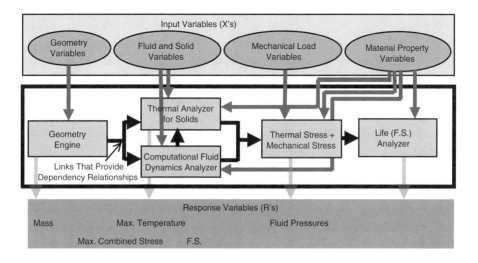

FIGURE 4.10 Example multidisciplinary network of physics-based models connected to form a multidisciplinary design space of input and response variables.

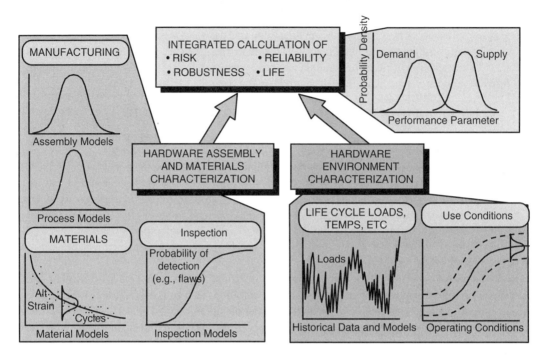

FIGURE 4.11 End structural reliability is affected by uncertainties introduced at all levels of manufacturing and use conditions.

as defined by the multidisciplinary model. The issue of the assumption of independence of random variables in a typical probabilistic analysis is important. Many of the probabilistic tools and codes that are readily available to the industry do not provide effective approaches for treatment of correlated input variables in a probabilistic analysis. In limited instances, tools provide approaches to probabilistic analysis under normality and linear correlation model assumptions. Hence, in the current state of the art, an accurate probabilistic analysis is performed starting with modeling the physical process from a stage that can justify the assumption of using independent input variables. The strategy is then to rely on linked models spanning multiple disciplines to automatically provide the correlation information. When used with Monte Carlo simulation techniques, the use of multidisciplinary models provides the probabilistic information (e.g., variance) for design performance measures across all disciplines for the same computational effort. Further, the approach provides an opportunity to capture accurately the variance information as well as the correlation between intermediate variables.

4.8 Technology Transition and Software Implementation

To use the methodologies described above in an industrial setting, it is essential that commercial-grade software with the current technologies implemented be readily available. Some of the key requirements for efficient software include availability of user-friendly graphic user interfaces for pre- and post-processing of the problem, amenable for nonexpert use, a framework for efficiently linking multidisciplinary models, efficient computational strategies for fast turn around time, implementation of stable algorithms, and a software implementation that allows the user to progress easily from simple to more complex design processes without having to repose the same problem. Satisfying the above requirements is the key to a successful technology transition strategy. A few commercial software programs that satisfy some of these requirements have emerged and others will continue to emerge.

The availability of an integration framework for linking multidisciplinary models has become a reality. Commercial as well as proprietary software programs are now available that can perform over a network

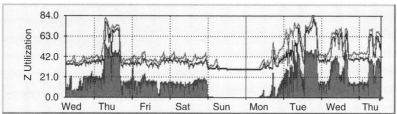

Max LSF Job Util.: 55.0%   Average LSF Job Util.: 12.0%   Current LSF Job Util.: 31.0%
Max CPU Util.: 84.0%   Average CPU Util.: 40.0%   Current CPU Util.: 59.0%

100 Engineering Workstations Cluster
LSF -   Load Sharing Facility (A Job Queuing System)
Green -   Percent of Total Cluster Job Utilization to Nondeterministic Methods
Blue -   Percent of Total Cluster CPU Utilization to Nondeterministic Methods
Dark Green -   Percent Maximal Cluster 1 Minute Job Utilization to Nondeterministic Methods
Magenta   -   Percent Maximal Cluster 1 Minute CPU Utilization to Nondeterministic Methods

FIGURE 4.12 Typical workload of engineering cluster workstations dominated by analysis performed in support of nondeterministic and deterministic design exploration methods.

of heterogeneous computers with different operating systems, chip architectures, and manufacturers. Systematic use of the design processes mentioned above (such as sensitivity analysis, design scans using factorial designs, Response Surface Methodology, Monte Carlo simulation and its variants, and deterministic and probabilistic optimization) all require an order of magnitude more function evaluations than conventional worst-case analysis based design approaches. Hence, for practical use of the above technologies in industry under ubiquitous schedule constraints, leveraging of the information technology revolution in software and hardware is essential. Many of the algorithms implementing the above design processes can effectively perform parallel computations to reduce the wall clock turnaround time of computer runs. It is not uncommon to utilize hundreds of engineering computer workstations to achieve the required computational speed to solve a design problem. An example of such a use scenario is shown in Figure 4.12. In Figure 4.12, software usage for nondeterministic analysis consumed a significant amount of the available CPU cycles. In general, the use of these newer analysis technologies does not cause any additional cost consequence, as they tend to use otherwise idle machine capacity. Without considering analysis jobs supporting newer approaches to design, in large corporations, the CPU average usage of engineering workstations at user desktops is in single-digit percentage points. The network and software reliability are high enough that in a typical year, hundreds of thousands of analyses can be performed in support of deterministic and nondeterministic analysis approaches.

4.9 Needed Technology Advances

The transition of the powerful array of technology and tools described above to a majority of practicing design engineers in a design shop is a challenging one. It is important to recognize that to have a pervasive impact on the design practice, it is necessary to put the above tools in the hands of integrated product design team engineers who perform most of the product design. It is emphasized that this is the group that must be targeted for technology transition and not the advanced engineering groups comprised of specialists. It is granted that a systematic application of the technologies described above will inevitably result in more reliable and robust products. If successful, it will be a quantum improvement over the current practice. However, there are a number of challenges that still need to be adequately addressed and this section attempts to enumerate them.

At first, a far greater recognition among structural reliability researchers of the role epistemic uncertainty is needed. The structural reliability research should make technology advances that treat both the epistemic and aleatory uncertainty together in the reliability assessment process. The theoretical basis for the developed methodology should provide practicing engineers with a tool that is capable of a more rigorous treatment of lack of knowledge, ignorance, in addition to the treatment of variability.

1. Significant new research in developing new design approaches that optimize/effectively combine analysis model results with focused testing is needed. Frequently, current approaches to testing are reactionary to an observed failure mode. In some instances, the testing is considered an independent verification of design. The suggested new approaches should consider both analysis and testing as part of a unified holistic approach to improve confidence in product performance, which either model or testing alone could not provide.
2. Many sub-elements of the broad research topic described above include a strategy for model calibration across scales (e.g., size effect), design of a test matrix that maximizes information return leveraging or giving adequate credit to analysis model results where appropriate, new forms of confidence measures for combined analysis, and focused test data.
3. Advances in stochastic optimization that will allow engineers to solve optimization problems such as "minimize testing subject to target confidence" or "least uncertainty subject to cost constraint on testing."
4. In addition to the fundamental challenge of understanding the "operational definition" for these methods, all the above will present computational challenges and numerical and "software implementation" issues.

Acknowledgments

Many of the ideas and concepts discussed in this chapter evolved over a period of 20 years while performing technology development research and application efforts. The extensive contributions of my colleagues — Drs. Amitabha DebChaudhury, George Orient, and Glenn Havskjold (in the Structures Technology group at Rocketdyne Division of The Boeing Company) — in developing the methodologies and necessary software is gratefully acknowledged.

5
The Need for Nondeterministic Approaches in Automotive Design: A Business Perspective

John A. Cafeo
General Motors Research and Development Center

Joseph A. Donndelinger
General Motors Research and Development Center

Robert V. Lust
General Motors Research and Development Center

Zissimos P. Mourelatos
Oakland University

5.1 Introduction 5-1
5.2 The Vehicle Development Process 5-2
5.3 Vehicle Development Process:
    A Decision-Analytic View... 5-3
5.4 The Decision Analysis Cycle....................................... 5-4
    Illustration: Door Seal Selection • Deterministic
    Phase • Probabilistic Phase • Information Phase
5.5 Concluding Comments and Challenges 5-16

5.1 Introduction

This chapter presents the importance of uncertainty characterization and propagation in the execution of the vehicle development process (VDP). While the VDP may be viewed from many perspectives, we consider it to be a series of decisions. In the absence of uncertainty, this series of decisions can, in principle, be posed as a very complex multidimensional optimization problem. Decisions, however, are actions taken in the present to achieve an outcome in the future. Because it is impossible to predict the outcomes of these decisions with certainty, the characterization and management of uncertainty in engineering design are essential to the decision making that is the core activity of the vehicle development process.

Uncertainties are present throughout the vehicle development process—from the specification of requirements in preliminary design to build variation in manufacturing. Vehicle program managers are continually challenged with the task of integrating uncertain information across a large number of functional areas, assessing program risk relative to business goals, and then making program-level decisions. Engineers struggle to develop design alternatives in this uncertain environment and to provide the program managers with credible, timely, and robust estimates of a multitude of design related vehicle performance attributes. Marketplace pressures to continuously shorten the vehicle development process drive the increasing use of mathematical models (as opposed to physical prototypes) for providing estimates of vehicle performance to support decision-making under uncertainty. For the calculations from the mathematical models to be useful, the decision makers must have confidence in the results. This confidence is formally developed through the model validation process.

In the following sections, we discuss these issues in some depth and illustrate them through a heuristic example. They conclude with a summary and a number of important challenges for embedding nondeterministic approaches in automotive design.

5.2 The Vehicle Development Process

The vehicle development process is the series of actions and choices required to bring a vehicle to market. For domestic (U.S.) vehicle manufacturers, the VDP is structured around a traditional systems engineering approach to product development. The initial phase of the VDP focuses on identifying customer requirements and then translating them into lower-level requirements for various functional activities, including product planning, marketing, styling, manufacturing, finance, and a broad array of engineering disciplines. Work within this phase of the VDP then proceeds in a highly parallel fashion. Engineers design subsystems to satisfy the lower-level requirements; then the subsystems are integrated to analyze the vehicle's conformance to customer requirements and to assess the compatibility of the subsystems. Meanwhile, other functional staffs work to satisfy their own requirements: product planning monitors the progress of the VDP to ensure that the program is proceeding on time and within its budget, marketing ensures that the vehicle design is appropriate to support sales and pricing goals, finance evaluates the vehicle design to ensure that it is consistent with the vehicle's established cost structure, manufacturing assesses the vehicle design to ensure that it is possible to build within the target assembly plant, etc. This is typically the most complex and the most iterative phase of the VDP, as literally thousands of choices and trade-offs are made. Finally, the product development team converges on a compatible set of requirements and a corresponding vehicle design. Engineers then release their parts for production and the vehicle proceeds through a series of preproduction build phases, culminating in the start of production.

Naturally, the VDP is scaled according to vehicle program scope; it is significantly more complex and longer in duration for the design of an all-new family of vehicles than it is for a vehicle freshening with relatively minor changes to feature content and styling cues. At any scale, however, there is an appreciable level of uncertainty at the beginning of the VDP. By the end of the VDP, this level of uncertainty is much lower, and the vehicle is being produced in a manufacturing plant with knowable build variation.

We distinguish between variation and uncertainty as follows. *Variation* is an inherent state of nature. The resulting uncertainty may not be controlled or reduced. Conceptually, variation is easy to incorporate in mathematical models using Monte Carlo simulation. For any number of random variables, the input distributions are sampled and used in the model to calculate an output. The aggregate of the outputs is used to form a statistical description. However, there are challenges in implementation. Monte Carlo simulation requires many evaluations of the mathematical model. If the mathematical models are very complex (e.g., finite element models with hundreds of thousands of elements used for analyzing vehicle structures), it is often prohibitively expensive or impossible to perform a Monte Carlo simulation within the time allotted for the analysis within the VDP.

In contrast to variation, *uncertainty* is a potential deficiency due to a lack of knowledge. In general, this is very difficult to handle. Acquiring and processing additional knowledge, perhaps by conducting experiments or by eliciting information from experts, may reduce uncertainty. The most difficult challenge is that you may not be aware that you do not know some critical piece of information. Thus it is necessary to undertake an iterative process of discovery to reduce uncertainty. It may also be necessary to allocate resources (e.g., people, time, money) to reduce uncertainty, and these allocation decisions are often very difficult given limited resources.

5.3 Vehicle Development Process: A Decision-Analytic View

Uncertainty, then, is our focal point. It is an inherent part of the VDP and we include it here in our decision-analytic view of vehicle development. We begin by discussing the work of Clark and Fujimoto [1] who, alternatively, view the product development process from an information-processing perspective. From this perspective, they identify three key themes:

1. The product development process is a simulation of future production and consumption.
2. Consistency in the details of product development is important.
3. Product integrity is a source of competitive advantage.

"The information-processing perspective focuses on how information is created, communicated, and used and thus highlights critical information linkages within the organization and between the organization and the market [1]."

"Critical information linkages" implies that information must be transferred from people who have it to the decision makers who need it to support critical decisions. Previous studies by Eppinger [2] and Cividanes [3] have focused on information flow within our VDP. Here, we extend this focus beyond the structure of information flow to include the presentation of structured design alternatives to vehicle program decision-makers. That is, rather than focusing on the information linkages, we will focus on the reason for the linkages: making decisions.

At this point it is prudent to discuss the definition of a decision. Although there has been a considerable amount of research into decision-making in engineering design, there is not yet a consensus within the engineering design community as to the definition of a decision. Conceptually, Herrmann and Schmidt [4] view decisions as value-added operations performed on information flowing through a product development team. In contrast, Carey et al. [5] view decisions as strategic considerations that should be addressed by specific functional activities at specific points in the product development process to maximize the market success of the product being developed. While we monitor this work with genuine interest, we subscribe to the explicit and succinct definition from Matheson and Howard [6]:

"A decision is an irrevocable allocation of resources, in the sense that it would take additional resources, perhaps prohibitive in amount, to change the allocation."

Commonly, we understand that a decision is a selection of one from among a set of alternatives after some consideration. This is illustrated well by Hazelrigg [7] in his discussion of the dialogue between Alice and the Cheshire cat in *Alice in Wonderland*. He notes that in every decision, there are alternatives. Corresponding to these alternatives are possible outcomes. The decision maker weighs the possible outcomes and selects the alternative with the outcomes that he or she most prefers. Although apparently simple, this discussion contains several subtle but powerful distinctions. One of these is that the decision is made according to the preferences of the decision maker—not those of the decision maker's stakeholders, or customers, or team members, or for that matter anyone's preferences but the decision maker's. Another is that the decision maker's preferences are applied not to the alternatives, but to the outcomes.

It is clear that decisions are actions taken in the present to achieve a desired outcome in the future. However, the future state cannot be known or predicted with absolute certainty. This is the reason that incorporating nondeterministic methods into decision-making processes (in our case, into decision making in our VDP) is crucial. Because we cannot know future outcomes with certainty, the outcomes resulting from selection of our alternatives must be expressed in terms of possible future states with some corresponding statement of the likelihood of occurrence. This, then, is the core activity of engineers in our VDP. Design engineers formulate sets of subsystem design alternatives. Development engineers assess the performance of these design alternatives in vehicles, considering uncertainty due to manufacturing build variation, mathematical model fidelity, and variations in customer usage.

5.4 The Decision Analysis Cycle

Decision makers are, by definition, people who have the authority to allocate an organization's resources. In a vehicle development program, these people are typically executives or senior managers. They make decisions (knowingly or not) using the Decision Analysis Cycle (Figure 5.1) described briefly below, and described completely in [6].

The discussion of the phases in the Decision Analysis Cycle contains several precisely defined terms from the language of formal decision analysis [6]. The term "value" is used to describe a measure of the desirability of each outcome. For a business, the value is typically expressed in terms of profit. The term "preferences" refers to the decision-maker's attitude toward postponement or uncertainty in the outcomes of his decision. The three phases of the Decision Analysis Cycle that precede the decision are:

1. *Deterministic Phase.* The variables affecting the decision are defined and related, values are assigned, and the importance of the variables is measured without consideration of uncertainty.
2. *Probabilistic Phase.* Probabilities are assigned for the important variables. Associated probabilities are derived for the values. This phase also introduces the assignment of risk preference, which provides the solution in the face of uncertainty.
3. *Informational Phase.* The results of the previous two phases are reviewed to determine the economic value of eliminating uncertainty in each of the important variables of the problem. A comparison of the value of information with its cost determines whether additional information should be collected.

Decisions are made throughout the VDP using the Decision Analysis Cycle. At the outset of the VDP, the decision makers' prior information consists of all their knowledge and experience. This prior information is then supplemented by the information collected and the outcomes of decisions made throughout the course of the VDP. In the Deterministic Phase, engineers provide decision makers with design alternatives and corresponding performance assessments. Other functional staffs, such as finance, marketing, and manufacturing, also provide their assessments of outcomes corresponding to the engineers' design alternatives, such as potential changes in cost and revenue and required changes to manufacturing facilities. In the Probabilistic Phase, the engineers as well as the other functional staffs augment their assessments to comprehend uncertainties; the application of nondeterministic methods is absolutely essential in this phase.

The Informational and Decision Phases are the domain of the decision makers—the senior managers and executives. These decision makers determine whether to gather additional information or to act based on the value of the information and their risk tolerance. If there are profitable further sources of information, then the decision should be made to gather the information rather than to take action. This

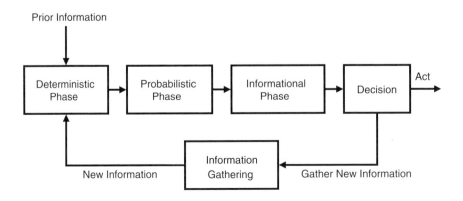

FIGURE 5.1 The Decision Analysis Cycle [6].

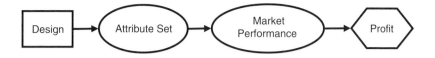

FIGURE 5.2 Abbreviated decision diagram for a design alternative.

cycle can be executed until the value of new analyses and information is less than its cost; then the decision to act will be made.

Another issue that is relevant to incorporating nondeterministic methods into the vehicle development process is the distinction between a good decision and a good outcome.

> "A good decision is based on the information, values and preferences of a decision-maker. A good outcome is one that is favorably regarded by a decision-maker. It is possible to have good decisions produce either good or bad outcomes. Most persons follow logical decision procedures because they believe that these procedures, speaking loosely, produce the best chance of obtaining good outcomes [6]."

This is critical to the idea of a true learning organization. Recriminations because something did not work become pointless. The questions that should be asked include "Was all of the available information used?" or "Was our logic faulty in the decision(s) that we made?" or "Were the preferences of the decision maker properly taken into account in our process?"

We now present a more specific view of decision making in the VDP (Figure 5.2). The rectangular block in this diagram represents the decisions to be made. In the VDP, these are the design alternatives for both the vehicle and its manufacturing system. The ovals in this diagram represent uncertain quantities. In the VDP, we represent the vehicle's Attribute Set and its Market Performance as uncertainties. Finally, the hexagon in this diagram represents the decision maker's value. As would generally be true in business, the value in the VDP is a measure of profit. The arrows in the diagram represent our beliefs about the relationships between our decisions, the relevant uncertainties, and our value. Thus, the interpretation of this diagram is that Profit depends on the vehicle's market performance, which depends on its Attribute Set, which in turn depends on the selection of design alternatives.

The decision diagram shown in Figure 5.2 is applied to the Decision Analysis Cycle shown in Figure 5.1. In the Deterministic Phase, we identify the specific set of design alternatives to be considered and the corresponding set of analytical and experimental methods and tools we will use to relate the design alternatives to their effects on the vehicle's attributes, the vehicle's market performance, and ultimately to the profit generated by selling the vehicle. In the Probabilistic Phase, we characterize the uncertainties in our estimates of the vehicle's attributes and its market performance and use them to examine their effects on profit. In the Information Phase, we examine whether or not it is prudent to allocate additional resources to gather more information (usually through further studies with our analytical tools) to increase our likelihood of realizing greater profit. Finally, in the Decision Phase, the decision maker commits to a course of action developed using the Decision Analysis Cycle.

5.4.1 Illustration: Door Seal Selection

Now let us now begin to illustrate the role of nondeterministic methods in decision making during the vehicle development process. While there are many different types of decisions made during a vehicle development program, here we will begin to discuss decision making in the context of a single vehicle subsystem: a door sealing system. In the following, we introduce the problem. We then continue to discuss this example as we proceed through this chapter.

5.4.1.1 The Door Sealing System

The door seal is typically an elastomer ring that fills the space between the door and the vehicle body. Figure 5.3 illustrates a typical door body and seal cross-section within a cavity. The function of the door seal is to prevent the flow of air, water, and environmental debris into the interior of the vehicle. The door seal is usually larger than the space it must seal. The seal compresses as the door is closed, filling the space and providing some restoring force in the seal to accommodate fluctuations in the size of the door-to-body structure gap due to motion of the door relative to the body. This motion can occur as the result of road loads applied or transmitted to the body and door structures.

Functionally, the door seal must satisfy several customer needs. The first and most obvious is that customers expect the passenger compartment to be sealed against water leaks. Second, customers are annoyed by air leaks and their accompanying wind noise, especially at highway speeds. And third, air leaks can carry dust into the vehicle, causing problems for people with allergies. If this were our only concern, we could fill the space with a very large seal. However, there is a competing objective. The larger the seal, the more force it takes to compress it. We also know that customers like to be able to close the doors with minimum effort. So from this standpoint, we would like the seal to have very little compression when the door is being shut. Thus, the door seal designer must strike a delicate balance when designing the seal and its resultant compression force. If the force is too low, the vehicle will be viewed as poorly designed because of the resulting air and water leaks; if it is too high, the vehicle will be viewed as poorly designed because of the high door-closing effort.

In luxury cars, it is typical to have a door seal system that contains three seals. The primary seal is a continuous bulbous seal, located deep within the door-to-body gap, which completely surrounds the door opening. It can be mounted on either the body or door side of the cavity. The other two seals are usually combined and are located across the top of the door and down the windshield pillar. The upper part of this combined seal is a barrier across the gap between the door and the body while the lower part fills the gap directly below. This provides three barriers to isolate the interior from the exterior environment. In economy cars, it is typical to just use a primary seal which provides one barrier.

The seal cross-section shape and size are dependent on the specific door and body geometry in the cavity when the door is closed. This cross-section is designed to deflect and compress during the door-closing event to seal the cavity while maintaining a good door-closing effort. The material for the seal is

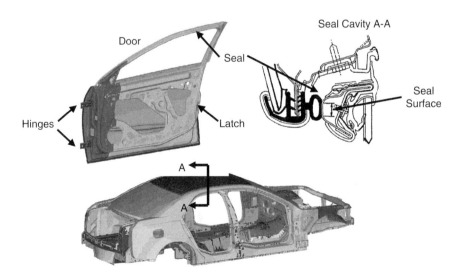

FIGURE 5.3 Finite element representation of a typical door, body, and cavity cross-section showing sealing system.

chosen for both its mechanical and durability properties. Other important design parameters are the location of the hinges and the latching mechanism.

5.4.2 Deterministic Phase

The Deterministic Phase mainly focuses on quantifying the complex relationship between our design decisions (engineering design variables) and our value (profit). We previously presented a simplified view of this relationship in Figure 5.2. However, in reality, it is much more complex. Figure 5.4 provides an expanded and more complete view of decision making in the VDP.

Fundamentally, there are two approaches to managing the complexity of product development decision making. One approach is to decompose the product development system into smaller units; perhaps using product design parameters, product architecture, organizational structure, or tasks in the product development process (as discussed by Browning [8]) as the basis for the decomposition. The enterprise *value* is then decomposed into a set of specific objectives for each of the units so that they can work in parallel to achieve a common enterprise-level goal. This decomposition simplifies decision making in some respects; however, it often leads to sub-optimal results for the enterprise due to insufficient coordination of actions between the units. There is no guarantee that the objectives given to each unit will lead to the greatest benefit for the enterprise. Therefore, if the product development system is to be decomposed, mechanisms must be put in place for rebalancing the objectives allocated to each of the units. Kim et al. [9] have made progress toward developing these rebalancing methods, but this remains a very challenging problem.

The alternative approach is to retain the interconnectivity within the product development system with a fully integrated cross-functional design environment. The benefit of this approach is that decisions are consistently made in the best interest of the enterprise. Its drawback is that it requires a set of engineering design tools that are compatible with one another both in terms of their level of detail and their flow of information. While it can be challenging to conduct these analyses even by assembling a

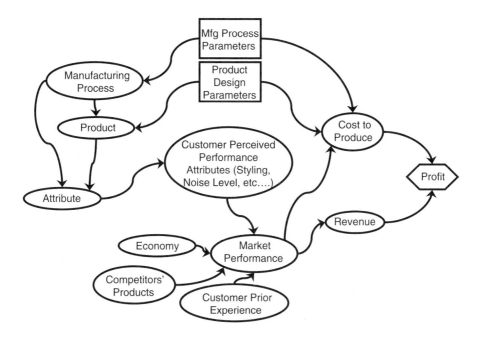

FIGURE 5.4 Detailed decision diagram for a design alternative.

team of cross-functional experts to provide their judgments, recent research suggests that it is now possible to conduct these analyses in a model-based environment using a combination of well-established and emerging technologies.

A number of commercial tools are available for representing the product and the manufacturing process. Mature tools are also available for correlating engineering evaluations of vehicle attributes to customer perceptions, for modeling the effects of the economic climate and of customers' prior experiences on purchase behavior. These tools are usually developed internally by the vehicle manufacturers and are considered proprietary.

More recently, a number of methods have been developed for linking the engineering and business domains by estimating the impacts of engineering designs on cost and revenue. The procedures published by Dello Russo et al. [10] are broadly applicable for cost estimating of new designs and the Technical Cost Modeling process developed by Kirchain [11] is particularly well suited to assessing the cost impacts of implementing new materials and manufacturing processes. Market demand can be estimated using either Discrete Choice Analysis, as discussed by Wassenaar et al. [12] or the S-Model, as demonstrated by Donndelinger and Cook [13]. Competitors' reactions to marketplace changes can, in principle, be simulated using game theoretic methods such as those implemented by Michalek et al. [14] Once market demand and cost are known, revenue and profit can be estimated using conventional financial practices.

Multidisciplinary design frameworks such as those developed by Wassenaar and Chen [15] and Fenyes et al. [16] can be applied to conduct a series of mathematical analyses relating engineering designs to enterprise values with a compatible level of details and information flow. These frameworks are therefore suitable for analysis of design alternatives in the Deterministic Phase of decision making. This type of analysis is necessary but definitely not sufficient for design decision making; the Probabilistic and Information phases of the Decision Analysis Cycle must also be completed before the committed decision maker can reach a decision. We continue with a discussion of the work in subsequent phases of the Decision Analysis Cycle in the next section. But first, let us return to the door seal selection example.

5.4.2.1 Door Seal Selection Continued: Deterministic Phase

In the Deterministic Phase of the Decision Analysis Cycle, we identify the decisions to be made, the value we seek to achieve, and the intermediate quantities that relate our decisions to our value. The decisions that we will study are "Which sealing system and its associated parameters should we choose for a particular new vehicle that we are developing?" and "Are any modifications to our manufacturing facilities required to improve our door attachment process capability?" To begin framing these decisions, we will stipulate that the business plan for the vehicle program has been approved, meaning among other things that the vehicle's architecture and body style have already been chosen and that an assembly plant has already been allocated for production of the vehicle. The strategic decisions under consideration for the door seal at this stage relate to the conceptual design of the door seal system and include the number of seals, the seal's cross-sectional shape, and its material properties. The decisions related to improvement of the door attachment process (hinges and latching system) are also strategic decisions at this stage and will be considered concurrently with the conceptual design of the door seal. The decisions related to detail design of the door seal system are tactical decisions at this stage; however, we may need to consider some of the seal size parameters to be tentative decisions if they are relevant to the selection of the door seal design concept.

The Decision Analysis Cycle begins with the generation of a set of design alternatives to present to the decision maker. There are many methods and theories about the best ways to create sets of alternatives; however, we do not discuss them here. For this discussion, suffice it to say that the set of alternatives will be created by considering a number of door seal designs carried over from similar vehicles and perhaps some innovative new sealing technologies developed by our suppliers or through our internal research and development activities. Then, a manageable list of specific design alternatives can be engineered and analyzed (a coupled and iterative process).

The analyses conducted to relate the design alternatives to their corresponding values follow the VDP decision diagram in Figure 5.4. Decisions about the door seal design concept and the door attachment process capability influence our value of profit through effects on both cost and revenue. The effect on cost can be determined "simply" by conducting a cost estimate. The effect on revenue, however, requires a much more intricate series of analyses.

The first of these analyses covers the synthesis of the vehicle design concept. These analyses are inherently complex because every design engineer on the program is conducting them in parallel. Thus, while one design engineer is developing alternative door seal concepts, another is developing alternative concepts for the inner and outer door panels, and yet another is developing alternative concepts for the door ring on the outside of the body structure. Vehicle concept engineers are faced with the challenging job of integrating each engineer's design alternatives into vehicle-level representations that can then be used to analyze the performance of the various subsystem design concepts at the vehicle level.

Once the vehicle-level design representations are created, all of the relevant vehicle performance attributes can be assessed for each of the design alternatives. For the door seals, these would include assessments of wind noise, water penetration, pollen count, and door-closing effort. Engineers are faced with difficult choices over the construction and use of mathematical models while performing these assessments. In this example, finite element analysis techniques could be applied to model the air and water penetration using seal pressure as a surrogate for air and water leakage. Meanwhile, an energy analysis could be performed to model the door-closing effort. We could construct a variety of models at various levels of complexity, allocating a little or a lot of time and engineering resources in the process. The prudent choice is to construct the model that provides the appropriate level of information and accuracy to the decision maker to support the decision. It may be fruitful to err on the side of simplicity and let the decision maker inform and guide improvement to our modeling in the Information Phase of the Decision Analysis Cycle.

The engineering assessments of the vehicle attributes must be translated into customer terms before they can be used to estimate the market demand for the vehicle. From a technical perspective, the most interesting translation in this example would be for wind noise. The results of the engineering analyses for wind noise are sound pressure levels. These results would then be translated into terms more meaningful to a customer, such as into an Articulation Index (as discussed in Reference [17]) or some company-proprietary metric to reflect the perceived quietness of the vehicle's interior. Most of the results in this example, however, would be mapped into vehicle-level quality measures. Water leaks, high door-closing efforts, and any misalignment of the door relative to the rest of the vehicle will all lead to some warranty claims and will also negatively affect the customer's overall perception of the vehicle's quality. These measures of customer-perceived quality can then be used as inputs to a market demand model that can be used to explore the marketability of each of the design alternatives in terms of changes in the vehicle's market share and in customers' willingness to pay for the vehicles.

At this point, only a few more steps remain. The effects of the economic climate and of customers' prior experiences on the market demand for the vehicle must be considered; however, most decision makers would not consider these to be conditioned upon the performance of the door seal. The discussion of competitive action is more interesting: the changes in the vehicle's market share and in customers' willingness to pay for the vehicle based on the performance of the door seal could be influenced by competitors' actions. If our competitors make significant improvements to the interior quietness or the door-closing efforts in their vehicles, it will likely shift the customers' level of expected performance, meaning that the competitive advantage we would realize by improving our vehicle's interior quietness or door-closing efforts would be decreased. Once these assessments are made, the resulting effects on the vehicle's revenue can be computed. Our value of profit can then be determined as a function of revenue and cost.

5.4.3 Probabilistic Phase

In the Probabilistic Phase, we estimate, encode, and propagate the uncertainties that have been identified as relevant to the design alternative decisions [6]. In the formal decision process discussed in [6], all the uncertainties that have been identified are characterized by the decision maker's subjective probability estimates. This Bayesian approach is consistent with the main objective of the probabilistic phase: to incorporate the element of risk preference into the values corresponding to the design alternatives. Risk is inherent in the decision-making process because of uncertainty, especially the epistemic uncertainty associated with the outcome of future events.

A growing body of formal methodologies is available to aid the decision makers in identifying key uncertainties and in mitigating the risks associated with them. The Design-for-Six-Sigma (DFSS) methodology [18] is frequently applied to this end. DFSS can be applied to identify, prioritize, and monitor both the product attributes critical to satisfying the customers' expectations as well as the uncertainties that pose the greatest challenges in consistently delivering them. Robust Design [19] principles can then be applied to develop subsystem and component designs that are less sensitive to these uncertainties, thereby increasing the likelihood that the designs will satisfy customers' expectations.

The results generated using these methodologies (or, for that matter, the results generated using any mathematical model, whether deterministic or nondeterministic) can be used to augment the decision maker's state of information. However, they are not a direct substitute for the belief-based probability assessments that must be used as the basis for the decision. A decision maker's beliefs about these uncertainties are ultimately based on his general state of information. The extent to which the results of an analytical model are substitutable for a decision maker's beliefs is determined through a process of model validation that is discussed in the next section.

5.4.3.1 Uncertainty Characterization Methods

The previous paragraphs discussed the importance of characterizing and quantifying the uncertainties associated with vehicle attributes. Although probability theory is perhaps the first method that comes to mind when formally characterizing uncertainties, other theories have been proposed and are used in various disciplines. We will overview these here and discuss their relationships. At this point, however, it is still unclear precisely how they fit into the decision analysis framework that we have been discussing. We will discuss this briefly at the end of this section.

Generally, uncertainties can be classified into two general types: (1) aleatory (stochastic or random) and (2) epistemic (subjective) [20–24]. *Aleatory uncertainty* is related to inherent variability. It is irreducible because collecting more information or data cannot decrease it. *Epistemic uncertainty* describes subjectivity, ignorance or lack of information in any phase of the modeling process. It is reducible because it can be decreased with an increased state of knowledge or the collection of more data.

Formal theories for handling uncertainty include evidence theory (or Dempster–Shafer theory) [20,21], possibility theory [25], and probability theory [26]. Evidence theory bounds the true probability of failure with belief and plausibility measures. These measures are mutually dual in the sense that one of them can be uniquely determined from the other. When the plausibility and belief measures are equal, the general evidence theory reduces to the classical probability theory.

Classical probability theory models aleatory uncertainty very efficiently and it is extensively used in engineering when sufficient data is available to construct probability distributions. However, when data is scarce or there is a lack of information, probability theory is not as useful because the needed probability distributions cannot be accurately constructed. For example, during the early stages of product development, the probabilistic quantification of the product's reliability or compliance to performance targets is very difficult due to insufficient statistical data for modeling the uncertainties.

When there is no conflicting evidence or information, we obtain a special subclass of dual plausibility and belief measures called possibility and necessity measures, respectively. As a subclass of the general theory of evidence, the possibility theory can be used to characterize epistemic uncertainty when incomplete data is available. The true probability can be bounded using the possibility theory, based on the

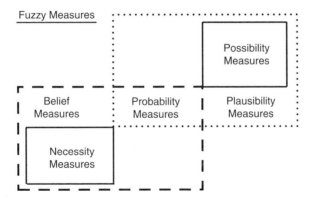

FIGURE 5.5 Fuzzy measures to characterize uncertainty.

fuzzy set approach at various confidence intervals (a-cuts). The advantage of this is that as the design progresses and the confidence level on the input parameter bounds increases, the design need not be reevaluated to obtain the new bounds of the response.

A fuzzy set is an imprecisely defined set that does not have a crisp boundary. It provides, instead, a gradual transition from "belonging" to "not belonging" to the set. The process of quantifying a fuzzy variable is known as *fuzzification*. If any of the input variables is imprecise, it is considered fuzzy and must therefore be fuzzified in order for the uncertainty to be propagated using fuzzy calculus. Fuzzification is done by constructing a possibility distribution, or membership function, for each imprecise (fuzzy) variable. Details can be found in References [27] and [28]. After the fuzzification of the imprecise input variables, the epistemic uncertainty must be propagated through the transfer function in order to calculate the fuzzy response. For that, explicit and implicit formulations are available in the literature [29,31].

In summary, the classical probability theory is a subset of the possibility theory, which in turn is a subset of the evidence theory (Figure 5.5). There is no overlap between the probability and possibility measures, although both are special classes of the plausibility measures. Probability theory is an ideal tool for formalizing uncertainty when sufficient information is available; or equivalently, evidence is based on a sufficiently large body of independent random experiments. When there is insufficient information, possibility theory can be used if there is no conflicting evidence. If there is conflicting evidence, then evidence theory should be used instead. It should be noted that, in practice, it is very common to have conflicting evidence even among "experts." Finally, when evidence theory is used, the belief and plausibility measures can be interpreted as lower and upper probability estimates, respectively.

Now let us return to the issue raised previously. How do these methodologies fit into the formal decision analytic view of vehicle development? Perhaps the most straightforward way is that they can be applied to the models we use within the general guiding framework of the Design-for-Six-Sigma methodology. The results—whether an interval estimate, a probability distribution, or a belief value—are used to inform the decision maker's general state of information. Then, as before, the decision maker assigns subjective probability estimates to the uncertainties and proceeds as outlined in Reference [6].

Alternatively, it might be possible to use the most relevant theory (evidence, possibility, probability) independently for each uncertainty and then combine them in a way where probability is not the overarching framework. We have not explored this in detail and put it forward as a research opportunity.

5.4.3.2 Door Seal Selection Continued: Probabilistic Phase

We stress the importance of the Probabilistic Phase because it extends beyond the identification of a deterministic optimum. We believe that the commonly faced problem of decision makers hesitating to accept the conclusions of deterministic optimization studies is because these studies disregard the uncertainty that is inherent and fundamental to the decision-making process. Robust engineering is gaining popularity because it offers the promise of mitigating risk through management of uncertainty. The decision analytic approach provides further benefits: it includes provisions for the decision maker's estimates of uncertainty based on his general state of information (vs. the estimates generated by models that are limited by the scope of the models and the assumptions made in constructing them) and it incorporates the decision maker's attitude toward risk. We believe that implementing the decision-analytic approach is a means of bridging this gap between presenting the conclusions of engineering analyses to decision makers and enabling them to commit to a course of action and then to execute it.

The decision maker will consider all the uncertainties shown in Figure 5.4. One key consideration in "product uncertainties" is the probabilities assigned to the selection of the various design alternatives of the chosen interfacing systems. For example, the window opening subsystem can have an effect on both the door's stiffness and its mass, which will in turn affect the door-closing effort and the wind noise. However, the product design parameters still need to be chosen to create various alternatives. In this example, the design parameters are the seal material, thickness, and cross-section geometry. These parameters affect the engineering performance attributes of seal deflection and restoring force for a given load. Other design parameters include the geometry of the cavity between the door and body as well as the type and location of both the upper and lower hinges and the latch mechanism location.

Next, we consider "manufacturing uncertainty." For this example, there are two types of hinges available; one is a bolt-on system and one is a weld-on system. The weld-on system requires less labor in the manufacturing process but can lead to a higher variation in door location. This not only affects the size of the interior cavity between the door and the body (and hence closing efforts and wind noise), but also makes visible the gap that can be viewed between the door and the body from the exterior of the vehicle. The visual consistency, size, and flushness of this gap are known to be very important to the customers' perception of vehicle quality. The manufacturing facility and its processes are thus dependent on the various design alternatives. This description shows the close linkages illustrated in Figure 5.4 between the Manufacturing System, the Product, and the Attribute uncertainties.

During engineering design, the wind noise and door-closing attributes are calculated for the various alternatives. Typically, manufacturing variations are considered in these models. They produce a range or distribution for the wind noise and the door-closing efforts. A further consideration is the math model itself. Any model is an approximation to reality and, as such, there is some uncertainty in the actual number calculated. Therefore, an estimate of the model uncertainty must be included as part of the results. This helps the decision maker understand how well the results can be trusted, which in turn has a critical effect on the subjective probabilities that will be assigned to each alternative. Because this is a critical issue, we discuss it in more detail under the name of model validation in Section 5.4.4.

Each of the other uncertainties detailed in Figure 5.4 is important, deserving a lengthy discussion that we will not include here. Some of the key issues are:

- *Translation uncertainties:* noise in the statistical models used to translate attributes from engineering terms to customer terms; also epistemic uncertainty due to imperfect knowledge of how customers perceive vehicles.
- *Economic uncertainties:* those most relevant to the auto industry include unemployment levels to the extent that they drive the total industry volume, interest rates because they affect affordability through financing and availability of capital for R&D, exchange rates because they affect the relative price position of foreign and domestic vehicles, and oil prices because they can affect customers' choices of vehicle types.
- *Competitive uncertainties:* risks due to the unknown extent to which our competitors will improve their products through performance enhancements, attractive new styling, and addition of new technology and feature content.
- *Cost uncertainties:* changes to the purchase price of parts due to design changes over the product's life cycle, amounts of supplier givebacks due to productivity improvements, fluctuations in raw material prices and exchange rates, and the emergence or decline of suppliers.

5.4.4 Information Phase

In the Information Phase, we use the results of the previous two phases to determine the economic value of eliminating uncertainty in each of the important variables of the problem. A comparison of the value of information with its cost determines whether additional information should be collected. If the decision maker is not satisfied with the general state of information, then he will use the diagram shown in Figure 5.4 (and its subsequent analysis) to determine which uncertainty he needs to reduce. Once the dominant uncertainty is determined, it will be clear what needs to be done to improve the information state. If, for example (referring to Figure 5.4), the "Customer Perceived Performance" attribute is the dominant uncertainty, then a marketing research study of potential customers will be conducted.

Often, a major uncertainty is the estimation of the vehicle attributes corresponding to a specific set of design parameters. These attributes are usually calculated using some mathematical model. This model can take many different forms: a statistical (empirical) model based on measured attributes of previous vehicles, a crude physics- (or first principles-) based model of the attribute, or a detailed physics model. Regardless of the exact type of model, the way to reduce uncertainty is to gain more confidence in the results. This is the issue addressed by the model validation process.

5.4.4.1 Validation of Mathematical Models to Inform the Decision Maker

The purpose of the model validation process is to establish confidence that the mathematical model can predict the reality of interest (the attribute). This helps the builders of the models during the model development phase. It enables them to change parameters or modify assumptions to improve the model's predictive power. It also informs the person using the results from the model and helps them estimate their subjective uncertainty during the decision-making process. Because engineering models that calculate the performance attributes of the vehicle are fundamental to the vehicle design process, we will focus on them and on the attribute uncertainty illustrated in Figure 5.4.

During the product development process, the person responsible for a decision will ask and answer the following two basic questions when presented with model results: (1) "Can I trust this result?" and (2) "Even if I can trust it, is it useful (i.e., does it help me make my decision)?" Most times, there is no formal objective measure of trust. Instead, the trust in the model results is equivalent to trust in the modeler, a subjective measure based on previous experiences. The chief engineer will ask the modeler to assess and report his confidence in the results. But this is difficult without a formal model validation process. What the engineer really needs is an objective measure of confidence to present to the decision maker. This can only be obtained as the product of the model validation process.

Previously we described and advocated a formal decision-making process based on the subjective probabilities of the decision maker. Here we advocate the use of a formal model validation process as a logical, complementary, and necessary way to determine the confidence we have in our models. This inherently statistical process determines the degree to which a model is an accurate representation of the real world from the perspective of the intended uses of the model. It is statistical because this (statistics) is the science developed to deal with the idea of uncertainties.

In practice, the processes of mathematical model development and validation most often occur in concert; aspects of validation interact with and feed back to the model development process (e.g., a shortcoming in the model uncovered during the validation process may require change in the mathematical implementation). It is instructive, therefore, to look at some of the current practices to begin to understand the role that nondeterministic analysis should play in this process. We do this now.

5.4.4.2 Model Validation: Current Practice

Model validation often takes the form of what is commonly called a model correlation exercise. This deterministic process involves testing a piece of hardware that is being modeled and running the model at the nominal test conditions. Typically, only a single set of computational results and one set of test measurements are generated. These results (test and computational) are then compared. Often, the results are overlaid in some way (e.g., graphed together in a two-dimensional plot) and an experienced engineer decides whether or not they are "close enough" to declare that the model is correlated (this measure of correlation is often referred to as the "viewgraph norm"). On initial comparison, the degree of correlation may be judged to be insufficient. Then the engineer will adjust some of the model parameters in an attempt to make the model results match the test results more closely. If and when sufficient agreement is obtained, the model is accepted as a useful surrogate for further hardware tests.

During this process, several important issues must be considered; these include:

1. This process, as described, implicitly assumes that the test result is the "correct answer" (i.e., a very accurate estimate of reality). In many cases, this may be a good assumption but experience has shown that testing can be prone to error. Consequently, the test used should have a quantified repeatability. Also, every effort must be made to ensure that the test results are free from systematic error. The use of independent test procedures can help here. For example, the dynamic strain results from a strain gage time history can be used to confirm the results from an accelerometer. If the accuracy of the test results cannot be quantified and assured through procedural controls, then replicated experiments should be run to estimate test uncertainty.
2. Changing model parameters to obtain "better" agreement between test and analysis results (calibration) can lead to the false conclusion that the modified model is better than its original version. Often, many model parameters can be chosen that will allow the same level of agreement to be forced. This nonuniqueness means that the engineer must be very disciplined when choosing the calibration parameters. They should only be chosen when there are significant reasons to believe that they are in error and therefore "need" to be calibrated.
3. Experimental data that has been used to adjust the model should not be the sole source of data for establishing the claim that the model is validated. Additional data for model validation should be obtained from further experiments. Ideally, these experiments should be carefully and specifically planned for model validation and represent the design domain over which the model will be used. However, with great care, data from hardware development and prototype tests can sometimes be used as a surrogate for (or an augmentation to) further validation testing.

These issues make it difficult for the engineer to calculate objective confidence bounds for the results of a mathematical model. To do this, a formal model validation process is necessary. Such a process is discussed next.

5.4.4.3 Model Validation: A Formal Process

A validated model has the following characteristics: (1) an estimate of the attribute (the output of the model), (2) an estimate of the bias (or difference between the computer experiments and the field "measured" experiments), (3) a measure of the confidence (tolerance) in the estimate of the bias, and finally (4) full documentation of the model parameters along with the validation process and the data used in the process. The bias and its tolerance measure results from three causes:

1. The approximation assumptions made during the development of the mathematical model
2. The uncertainty in the model parameters, boundary, and initial conditions used when running the computer model
3. The systematic errors present in the measurement and processing of field data

Figure 5.6 shows a process that is capable of producing a validated model having the characteristics described above. This figure is an adaptation of a similar figure given in Reference [32]. Specific modeling and simulation activities are shown with solid lines; while assessment activities are shown with dashed lines. The reality of interest is the particular performance attribute we are calculating (e.g., stress field or modes of vibration). The mathematical model is the result of the conceptual modeling activity.

The computer program is the realization of the mathematical model for implementing the calculations. It takes inputs, boundary conditions, initial conditions, and parameters (both physical and numerical) and produces simulation outcomes. The associated assessment activities are code and

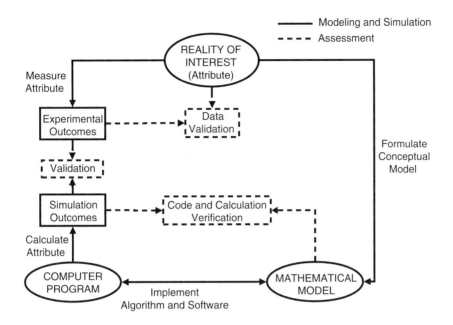

FIGURE 5.6 Block diagram of mathematical model validation and verification outside of the model development context.

calculation verification. Code verification is the process that determines that the computer code is an accurate representation of the discrete mathematical model. Calculation verification is the process that determines the numerical accuracy of code calculations (e.g., numerical error estimation). The main idea here is to make sure that the computer model accurately represents the conceptual mathematical model. A complicating factor from an industrial standpoint is that most models are constructed and exercised with commercial modeling and analysis software packages that are outside the direct control of the engineer.

Finally, the results from carefully designed physical experiments, as well as corresponding computer runs, are gathered and compared. This comparison generates an objective measure of agreement, the bias, and its tolerance bounds. The measure of agreement is what the engineer presents to the decision maker when asked about the validity of the results (see Bayarri et al. [33] for an example).

The decision maker, who has a particular risk tolerance, then decides whether the results are useful for making the decision. This is a critical step. A useful model is one with predictive power that is sufficient for making decisions with respect to the decision maker's risk tolerance. This can only be assessed within the overarching framework of decision making under uncertainty.

Once again, let us return to the door seal selection example.

5.4.4.4 Door Seal Selection Continued: Information Phase

In the Information Phase, the decision maker must decide either to continue collecting information or to commit to a course of action. In this example, the decision would be either to conduct additional analyses and/or experiments in order to better understand the performance and cost of the various door seal design alternatives, or to select a door seal design and authorize any necessary changes to the manufacturing system. "Selection," in the context of this design decision, means that the decision maker irrevocably allocates resources as the result of his or her actions. Thus, the selection of a door seal system would occur in a binding fashion by signing a letter of intent or a contract with a supplier. Likewise "authorization" of changes to the manufacturing system would include signing documents authorizing the exchange of funds from the vehicle program's budget to the manufacturing facility. Until these actions have occurred, no decisions have been made.

In this illustration, we know that the rubber model we use has some high uncertainty associated with it. Two issues that could cause this high uncertainty are the material properties and the nonlinear model used to analyze the potentially complex cross-section seal deflection characteristics. In this case, the decision maker may find the tolerance bounds on the attribute calculations too wide for the risk tolerance he has. He may ask the engineer to estimate the resource requirements to improve the model to a certain tolerance level. The value of information is the cost of conducting additional analyses relative to the changes we will see in the values of the alternatives as a result of conducting the analyses. A question that can be answered from analyzing the decision diagram shown in Figure 5.4 is: "Would a change in the information cause the decision maker to select a different design alternative?" A change in information that does not change the preference order of the design alternatives (i.e., order in terms of our value from highest to lowest; we prefer more profit to less) is not worth conducting.

It is important to note that the need to improve a predictive model should be driven by the decision it supports, *not* by the engineer who is constructing it. For example, if the dominant uncertainty influencing this decision is the customer perception of wind noise for this vehicle, then the decision maker will allocate his resources to reduce that uncertainty, not to improve the door seal model.

5.5 Concluding Comments and Challenges

In this chapter we have discussed why the use of nondeterministic methods is critical from a business perspective. The VDP can be viewed as a series of decisions that are made by humans in an uncertain and creative environment. Our methods for design need to acknowledge and embrace this reality. We advocate a nondeterministic world view in a predominantly deterministically minded world. However, thinking in terms of a number of possible outcomes with varying probabilities of achieving those outcomes is a skill that very few people have developed. The fact that casinos and lotteries are thriving businesses is a powerful testament to this. Getting entire organizations to the point that they have healthy attitudes about decision making under uncertainty is a major challenge. This includes two key elements. First, they must accurately identify possible future states and encode the uncertainties about their potential occurrences, such that they migrate from "gambling" or overly risk-averse behavior to carefully calculated strategies with favorable reward-to-risk ratios. Second, they must accept the aleatory uncertainties that they cannot control and must migrate from the attitude that "they should have known" to "they could not have known with certainty and they made the best decision that they could at the time." These are cultural changes that cannot be made quickly—but unless they are made, a decision-analytic design framework such as the one we have described cannot flourish.

We have discussed the notion of using subjective probability as the overarching framework for decision analysis. We believe that other uncertainty methods fit within this framework. However, turning this notion around to where another uncertainty methodology is used as the overarching framework should be explored and researched.

While we know that the use of validation metrics is essential to help inform the decision maker, the technology for computing metrics like tolerance bounds for calculated attributes is still an area of research. The main challenge lies in the cost and complexity of the data and calculations that need to be assembled to estimate the tolerance bounds.

Uncertainty in calculating performance is intertwined with uncertainty in setting targets. The use of probability-based specifications in a flow-down and roll-up balancing process requires further research.

Finally, in a large enterprise, it is critical to evenly develop analytical capability across all of the product development disciplines. It would be difficult to operate a decision-making framework with varying levels of underlying mathematical modeling across disciplines. Engineers, for example, might think that marketing has sketchy data, has not really done any detailed analysis, and is just guessing. Marketing, meanwhile, might think that engineers are overanalyzing some mathematical minutia and are missing some important elements of the big picture. The challenge is to make both sides recognize that they are both partly right and can learn some valuable lessons from each other.

References

1. Clark, K.B. and Fujimoto, T., Product Development Performance—Strategy, Organization, and Management in the World Auto Industry, Harvard Business School, Boston, MA, 1991.
2. Eppinger, S., Innovation at the speed of information, *Harvard Business Review*, 79 (1), 149, 2001.
3. Cividanes, A., Case Study: A Phase-Gate Product Development Process at General Motors, S.M. thesis, Massachusetts Institute of Technology, 2002.
4. Herrmann, J. and Schmidt, L., Viewing product development as a decision production system, in *Proc. ASME Design Eng. Tech. Conf.*, Montreal, Canada, 2002.
5. Carey, H. et al., Corporate decision-making and part differentiation: a model of customer-driven strategic planning, *Proc. ASME Design Eng. Tech. Conf.*, Montreal, Canada, 2002.
6. Matheson, J.E. and Howard, R.A., An Introduction to Decision Analysis (1968), from Readings on The Principles and Applications of Decision Analysis, Strategic Decisions Group, 1989.
7. Hazelrigg, G.A., The cheshire cat on engineering design, submitted to *ASME Journal of Mech. Design*.
8. Browning, T., Applying the design structure matrix to system decomposition and integration problems: a review and new directions, *IEEE Trans. on Eng. Management*, 48 (3), 292, 2001.
9. Kim, H. et al., Target cascading in optimal system design, *Journal of Mechanical Design*, 125(3), 474–480, 2003.
10. Dello Russo, F., Garvey, P., and Hulkower, N., Cost Analysis, Technical Paper No. MP 98B0000081, MITRE Corporation, Bedford, MA, 1998.
11. Kirchain, R., Cost modeling of materials and manufacturing processes, *Encyclopedia of Materials: Science and Technology*, Elsevier, 2001.
12. Wassenaar, H., Chen, W., Cheng J., and Sudjianto, A., Enhancing Discrete Choice Demand Modeling for Decision-Based Design, *Proc. ASME Design Eng. Tech. Conf.*, Chicago, IL, 2003.
13. Donndelinger, J. and Cook, H., Methods for analyzing the value of automobiles, *Proc. SAE Int. Congress and Exposition*, Detroit, MI, 1997.
14. Michalek, J., Papalambros, P., and Skerlos, S., A study of emission policy effects on optimal vehicle design decisions, *Proc. ASME Design Eng. Tech. Conf.*, Chicago, IL, 2003.
15. Wassenaar, H. and Chen, W., An approach to decision-based design with discrete choice analysis for demand modeling, *Journal of Mechanical Design*, 125 (3), 490–497, 2003.
16. Fenyes, P., Donndelinger, J., and Bourassa, J.-F., A new system for multidisciplinary analysis and optimization of vehicle architectures, *Proc. 9th AIAA/ISSMO Symposium on Multidisciplinary Analysis and Optimization*, Atlanta, GA, 2002.
17. Beranak, L.L. and Vér, I.L., *Noise and Vibration Control Engineering—Principles and Applications*, John Wiley & Sons, New York, 1992.
18. Chowdhury, S., *Design for Six Sigma: The Revolutionary Process for Achieving Extraordinary Profits*, Dearborn Trade, 2002.
19. Phadke, M.S., *Quality Engineering Using Robust Design*, Pearson Education, 1995.
20. Oberkampf, W., Helton, J., and Sentz, K., Mathematical representations of uncertainty, *Proc. AIAA/ASME/ASCE/AHS/ASC Structures, Structural Dynamics and Materials Conf.*, Seattle, WA, 2001.
21. Sentz, K. and Ferson, S., Combination of Evidence in Dempster–Shafer Theory, Sandia National Laboratories Report SAND2002-0835, 2002.
22. Klir, G.J. and Yuan, B., *Fuzzy Sets and Fuzzy Logic: Theory and Applications*, Prentice Hall, New York, 1995.
23. Klir, G.J. and Filger, T.A., *Fuzzy Sets, Uncertainty, and Information*, Prentice Hall, New York, 1988.

24. Yager, R.R., Fedrizzi, M., and Kacprzyk, J., Eds., *Advances in the Dempster–Shafer Theory of Evidence*, John Wiley & Sons, New York, 1994.
25. Dubois, D. and Prade, H., *Possibility Theory*, Plenum Press, New York, 1988.
26. Haldar, A. and Mahadevan, S., *Probability, Reliability and Statistical Methods in Engineering Design*, John Wiley & Sons, New York, 2000.
27. Zadeh, L.A., Fuzzy sets as a basis for a theory of possibility, *Fuzzy Sets and Systems*, 1, 3, 1978.
28. Ross, T.J., *Fuzzy Logic with Eng. Applications*, McGraw-Hill, New York, 1995.
29. Chen, L. and Rao, S.S., Fuzzy finite element approach for the vibration analysis of imprecisely defined systems, *Finite Elements in Analysis and Design*, 27, 69, 1997.
30. Mullen, R.L. and Muhanna, R.L., Bounds of structural response for all possible loading combinations, *ASCE Journal of Structural Eng.*, 125(1), 98–106, 1999.
31. Akpan, U.O., Rushton, P.A., and Koko, T.S., Fuzzy probabilistic assessment of the impact of corrosion on fatigue of aircraft structures, Paper AIAA-2002-1640, 2002.
32. Oberkampf, W.L., Sindir, M., and Conlisk, A.T., Guide for the verification and validation of computational fluid dynamics simulations, AIAA G-077-1998, 1998.
33. Bayarri, M.J., Berger, J.O., Higdon, D., Kennedy M.C., Kottas A., Paulo, R., Sacks, J., Cafeo, J.A., Cavendish, J., Lin, C.H., and Tu, J., A framework for validation of computer models, *Proc. of Foundations 2002—A Workshop on Model and Simulation Verification and Validation for the 21st Century*, Johns Hopkins University Applied Physics Lab, Laurel, MD, 2002.

6
Research Perspective in Stochastic Mechanics

Mircea Grigoriu
Cornell University

6.1 Introduction **6**-1
6.2 Deterministic Systems and Input **6**-2
　　Current Methods • Research Needs and Trends
6.3 Deterministic Systems and Stochastic Input **6**-4
　　Current Methods • Research Needs and Trends
6.4 Stochastic Systems and Deterministic Input **6**-7
　　Current Methods • Research Needs and Trends
6.5 Stochastic Systems and Input **6**-8
　　Current Methods • Research Needs and Trends
6.6 Comments **6**-10

6.1 Introduction

The response and evolution of mechanical, biological, and other systems subjected to some input can be characterized by equations of the form

$$\mathcal{D}[\mathcal{X}(x,t)] = \mathcal{Y}(x,t), \quad t \geq 0, \quad x \in D \subset \mathcal{R}^d, \tag{6.1}$$

where $\mathcal{D}$ can be an algebraic, integral, or differential operator with random or deterministic coefficients characterizing the properties of the system under consideration, $\mathcal{Y}(x,t)$ denotes an $\mathbb{R}^{d'}$-valued random or deterministic function, and $d, d' \geq 1$ are integers. The output $\mathcal{X}$ depends on the properties of $\mathcal{D}$ and $\mathcal{Y}$ and the initial/boundary conditions, which can be deterministic or random.

Four classes of problems, referred to as stochastic problems, can be distinguished, depending on the systems and input properties: (1) deterministic systems and input, (2) deterministic systems and stochastic input, (3) stochastic systems and deterministic input, and (4) stochastic systems and input. The solution of the problems in these classes requires specialized methods that may involve elementary probabilistic considerations or advanced concepts on stochastic processes.

It has been common to develop methods for solving stochastic problems relevant to a field of applications without much regard to similar developments in related fields. For example, reliability indices developed for structural systems began to be used only recently in geotechnical and aerospace engineering because they were perceived as applicable only to buildings, bridges, and other structural systems [1]. Methods for characterizing the output of mechanical, structural, electrical, and other systems subjected to time-dependent uncertain input have been developed independently although the defining equations for the output of these systems coincide [2, 3].

It is anticipated that the emphasis of future research on stochastic problems will be on the development of methods for solving various classes of problems irrespective of the field of applications. This trend is a direct consequence of the interdisciplinary nature of current research. A broad range of stochastic problems relevant to different fields can be described by the same equations. For example, the solution of the random eigenvalue problem can be used to find directions of crack initiation in materials with random properties, characterize the frequency of vibration for uncertain dynamic systems, and develop stability criteria for systems with random properties.

This chapter (1) briefly reviews current methods for solving stochastic problems and (2) explores likely future research trends in stochastic mechanics and related fields. The chapter is organized according to the type of Equation 6.1 rather than the field of application in agreement with the anticipated research trends. It includes four sections corresponding to the type of the operator $\mathcal{D}$ and the input $\mathcal{Y}$ in Equation 6.1.

6.2 Deterministic Systems and Input

Initial and boundary value deterministic problems have been studied extensively in engineering, science, and applied mathematics. Engineers and scientists have focused on numerical results, for example, the calculation of stresses and strains in elastic bodies, flows in hydraulic networks, temperatures in material subjected to heat flux, waves in elastic media, and other topics [4, 5]. Applied mathematicians have focused on conditions for the existence and uniqueness of a broad class of partial differential equations [6]. The difference of objectives has resulted in a limited interaction between mathematicians and engineers/scientists. Also, the engineers and scientists have emphasized numerical solutions for Equation 6.1 because applications usually involve complex equations, which rarely admit analytical solutions.

6.2.1 Current Methods

Numerical solutions to problems in science and engineering are commonplace in today's industry, research laboratories, and academia. Together with analytical and experimental methods, numerical simulations are accepted as an invaluable tool for the understanding of natural phenomena and the design of new materials systems, processes, and products.

The *finite element method* reigns supreme as the method of choice in computational solid mechanics, with the boundary element and finite difference methods running substantially far behind. While these methods are extremely powerful and versatile, they are not optimal for all applications. For example, some of the drawbacks of the finite element method are that it always provides a global solution, needs domain discretization, and involves solution of large linear algebraic systems [7].

Global methods providing values of the stress, displacement, and other response functions at all or a finite number of points of the domain of definition of these functions are generally used to solve mechanics, elasticity, physics, and other engineering problems. These methods can be based on analytical or numerical algorithms. The analytical methods have limited value because few practical problems admit closed form solutions and these solutions may consist of slowly convergent infinite series [5]. The finite element, boundary element, finite difference, and other numerical methods are generally applied to solve practical problems. Some of the possible limitations of these numerical methods include: (1) the computer codes used for solution are relatively complex and can involve extensive preprocessing to formulate a particular problem in the required format, (2) the numerical algorithms may become unstable in some cases, (3) the order of the errors caused by the discretization of the continuum and the numerical integration methods used in analysis cannot always be bounded, and (4) the field solution must be calculated even if the solution is needed at a single point, for example, stress concentration, deformations at particular points, and other problems.

6.2.2 Research Needs and Trends

The analysis of complex multidisciplinary applications requires realistic representations for the operator $\mathcal{D}$ and the input $\mathcal{Y}$ in Equation 6.1, as well as efficient and accurate numerical algorithms for solving this equation. Also, methods for estimating or bounding errors related to numerical calculations and discretization need to be developed and implemented in future algorithms. For example, useful results for the growth of microcracks in aluminum and other materials are possible if the analysis is based on (1) detailed representations of the material microstructure at the grain level and (2) powerful and accurate numerical algorithms for solving discrete versions of Equation 6.1 obtained by finite element or other methods.

Close interaction between engineers, scientists, and applied mathematicians is deemed essential for developing efficient and accurate numerical algorithms for the solution of Equation 6.1 This interaction is expected to deliver numerical algorithms needed to solve complex multidisciplinary applications considered currently in science and engineering.

There are some notable developments in this direction. For example, recent work in applied mathematics is directed toward the estimation of errors generated by various numerical solutions of Equation 6.1. Some of these studies use probabilistic concepts for error quantification [8]. Also, alternative techniques to the traditional numerical methods, referred to as local methods, are currently explored for solving locally some types of partial differential equations. The methods delivers the solution of these partial differential equations at an arbitrary point directly, rather than extracting the response value at this point from the field solution, and are based on probabilistic concepts. The theoretical considerations supporting these solutions are relatively complex. They are based on properties of diffusion processes, Itô's formula for continuous semimartingales, and Monte Carlo simulations [9–12]. However, the resulting numerical algorithms for solution have attractive features. These algorithms are (1) simple to program, (2) always stable, (3) accurate, (4) local, and (5) ideal for parallel computation ([13], Chapter 6).

Consider for illustration the Laplace equation

$$\Delta u(x) = 0, \quad x \in D, \quad u(x) = \xi_r(x), \quad x \in \partial D_r, r = 1,\ldots,m \qquad (6.2)$$

where D is an open bounded subset in $\mathbb{R}^d$ with boundaries ∂D_r, $r = 1,\ldots,m$, and ξ_r denote functions defined on these boundaries. If $m = 1$, then D is a simply connected subset of $\mathbb{R}^d$. Otherwise, D is multiply connected.

Denote by B an $\mathbb{R}^d$-valued Brownian motion starting at $x \in D$. The samples of B will exit D through one of the boundaries ∂D_r of D. Figure 6.1 shows three samples of a Brownian motion B in $\mathbb{R}^2$ and their exit points from a multiply connected set D. Let $T = \inf\{t > 0 : B(t) \notin D\}$ denote the first time B starting at $B(0) = x \in D$ exits D. Then, the local solution of this equation is ([13], Section 6.2.1.3)

$$u(x) = \sum_{r=1}^{m} E^x[\xi_r(B(T)) \mid B(T) \in \partial D_r] p_r(x) \qquad (6.3)$$

where E^x is the expectation operator corresponding to $B(0) = x$ and $p_r(x)$ denotes the probability that B starting at x exits D through ∂D_r.

For example, let $u(x)$ be the steady-state temperature at an arbitrary point $x = (x_1, x_2)$ of an eccentric annulus

$$D = \{(x_1, x_2) : x_1^2 + x_2^2 - 1 < 0, (x_1 - 1/4)^2 + x_2^2 - (1/4)^2 > 0\}$$

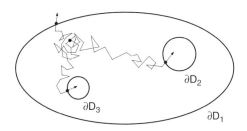

FIGURE 6.1 Local solution of the Laplace equation for a multiply connected domain D in $\mathbb{R}^2$.

in $\mathbb{R}^2$ with boundaries $\partial D_1 = \{(x_1,x_2): x_1^2 + x_2^2 - 1 = 0\}$ and $\partial D_2 = \{(x_1,x_2): (x_1 - 1/4)^2 + x_2^2 - (1/4)^2 = 0\}$ kept at the constant temperatures 50 and 0, respectively. The local solution in Equation 6.3 is $u(x) = 50\, p_1(x)$ for any point $x \in D$. Because $p_1(x)$ cannot be found analytically, it has been calculated from samples of B starting at x. The largest error recorded at $x = (0.7, 0)$, $(0.9, 0)$, $(0, 0.25)$, $(0, 0.5)$, and $(0, 0.75)$ was found to be 2.79% for $n_s = 1{,}000$ independent samples of B with time steps $mit\ \Delta t = 0.001$ or 0.0001. Smaller time steps were used at points x close to the boundary of D. The error can be reduced by decreasing the time step or increasing the sample size. The exact solution can be found in [5] (Example 16.3, p. 296).

6.3 Deterministic Systems and Stochastic Input

Generally, the input to biological, electrical, mechanical, and physical systems is uncertain and can be modeled by stochastic processes. The output of these systems is given by the solution of Equation 6.1 with $\mathcal{D}$ deterministic and $\mathcal{Y}$ random.

Records and physical considerations are used to estimate — at least partially — the probability law of these processes. For example, the symmetry of the seismic ground acceleration process about zero indicates that this process must have mean 0 and that its marginal density must be an even function. It is common to assume that the ground acceleration process is Gaussian although statistics of seismic ground acceleration records do not support this assumption ([14], Section 2.1.4). On the other hand, pressure coefficients recorded in wind tunnels at the eaves of some building models exhibit notable skewness, as demonstrated in Figure 6.2.

The broad range of probability laws of the input processes in these examples show that general Gaussian and nonGaussian models are needed to describe inputs encountered in applications.

6.3.1 Current Methods

Most available methods for finding properties of the output $\mathcal{X}$ of Equation 6.1 have been developed in the framework of random vibration [3, 15]. The methods depend on the system type, which defines the functional form of the operator $\mathcal{D}$ in Equation 6.1, input properties, and required output statistics. For example, the determination of the second-moment output properties for linear systems subjected to white noise involves elementary calculations, and can be based on a heuristic second-moment definition of the driving noise. If the noise is assumed to be Gaussian, then the output $\mathcal{Y}$ is Gaussian so that the probability law of $\mathcal{Y}$ is completely defined by its second-moment properties. On the other hand, there are no general methods for finding the probability law of the output of a linear system subjected to an arbitrary nonGaussian input. Similarly, the probability law of $\mathcal{Y}$ for nonlinear systems can be found only for some special input processes. For example, it can be shown that the marginal distribution of the output of these systems subjected to Gaussian white noise satisfies a partial differential equation, referred to as the Fokker-Planck equation, provided that the white noise input is interpreted as the formal derivative of the Brownian motion or Wiener process ([13], Section 7.3.1.3). Unfortunately, analytical solutions of the Fokker-Planck equation are available only for elementary systems and numerical solutions of this equation are possible

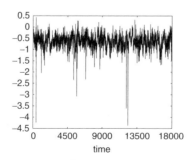

FIGURE 6.2 A record of pressure coefficients.

only for small dimension output processes ([13], Sections 7.3.1.3 and 7.3.1.4). This limitation is the main reason for the development of a variety of heuristic methods for solution; for example, equivalent linearization, moment closure, and other techniques ([13], Sections 7.3.1.1 and 7.3.1.5).

The above exact and approximate methods have been used to model climatic changes, describe the evolution atomic lattice orientation in metals, and perform reliability studies in earthquake engineering and other fields ([13], Sections 7.4.1.1, 7.4.2, 7.4.3, and 7.4.4). Early applications encountered difficulties related to the interpretation of the white noise input as a stationary process with finite variance and correlation time approaching zero. These difficulties have been gradually overcome as differences between physical white noise and mathematical white noise and between the Itô and Stratonovich integrals were understood ([16], Section 5.4.2). A significant contribution to the clarification on these issues has been the available mathematical literature on stochastic differential equations [17].

6.3.2 Research Needs and Trends

There are no general methods for finding the probability law of the output of linear systems subjected to arbitrary nonGaussian input processes and nonlinear systems with Gaussian and nonGaussian input. An additional difficulty relates to the limited number of simple models available for representing non-Gaussian input processes, for example, the process in Figure 6.2. A close interaction between engineers, scientists, and mathematicians is needed to overcome some of these difficulties.

Recent research addresses some of these needs. For example, efficient simple methods have been developed for calculating moments of any order for the state of linear systems driven by a class of non-Gaussian processes defined as polynomials of Gaussian diffusion processes. The analysis is based on properties of diffusion processes and Itô's calculus ([13], Section 7.2.2.3). For example, consider the linear system

$$\ddot{X}(t) + 2\zeta v \dot{X}(t) + v^2 X(t) = \sum_{l=0}^{n} a_l(t) S(t)^l \tag{6.4}$$

where $dS(t) = -\alpha S(t)\,dt + \sigma\sqrt{2\alpha}\,dB(t)$, $\alpha > 0$, σ, $\zeta > 0$, and $v > 0$ are some constants, B denotes a Brownian motion, $n \geq 1$ is an integer, and a_l are continuous functions. The process X represents the displacement of a linear oscillator with natural frequency v and damping ratio ζ driven by a polynomial of the Ornstein-Uhlenbeck process S.

The moments $\mu(p, q, r; t) = E[X(t)^p \dot{X}(t)^q S(t)^r]$ of order $s = p + q + r$ of $Z = (X, \dot{X}, S)$ satisfy the ordinary differential equation

$$\dot{\mu}(p, q, r; t) = p\mu(p-1, q+1, r; t) - qv^2 \mu(p+1, q-1, r; t)$$
$$- (q\beta + r\alpha)\mu(p, q, r; t) + q\sum_{l=0}^{n} a_l(t)\mu(p, q-1, r+l; t)$$
$$+ r(r-1)\alpha\sigma^2 \mu(p, q, r-2; t) \tag{6.5}$$

at each time t, where $\beta = 2\zeta v$ and $\mu(p, q, r; t) = 0$ if at least one of the arguments p, q, or r of $\mu(p, q, r; t)$ is strictly negative. The above equation results by taking the expectation of the Itô formula applied to the function $X(t)^p \dot{X}(t)^q S(t)^r$ ([13], Section 4.6).

If $(X, \dot{X}, S)$ becomes stationary as $t \to \infty$, the stationary moments of this vector process can be obtained from a system of algebraic equations derived from the above moment equation by setting $\dot{\mu}(p, q, r; t) = 0$.

Table 6.1 shows the dependence on damping $\beta = 2\zeta v$ of the coefficients of skewness and kurtosis of the stationary process X for $v^2 = 1.6$, $v^2 = 9.0$, and the input $\sum_{l=0}^{n} a_l(t) S(t)^l$ with $n = 2$, $a_0(t) = a_1(t) = 0$, $a_2(t) = 1$, $\alpha = 0.12$, and $\sigma = 1$. The skewness and the kurtosis coefficients are increasing functions of β. If the linear system has no damping ($\beta = 0$), these coefficients are 0 and 3, respectively, so that they match the values for Gaussian variables. This result is consistent with a theorem stating roughly that the output

TABLE 6.1 Coefficients of Skewness and Kurtosis of the Stationary Displacement X of a Simple Linear Oscillator Driven by the Square of an Ornstein-Uhlenbeck Process

	Skewness		Kurtosis	
β	$v^2 = 1.6$	$v^2 = 9.0$	$v^2 = 1.6$	$v^2 = 9.0$
0	0	0	3	3
0.25	1.87	1.79	9.12	8.45
0.5	2.48	2.34	11.98	11.24
1	2.84	2.72	13.99	13.32
1.5	2.85	2.77	14.51	13.95

of a linear filter with infinite memory to a random input becomes Gaussian as time increases indefinitely ([14], Section 5.2). The linear filter in this example has infinite memory for $\beta = 0$. Similar results are available for linear systems driven by Poisson white noise interpreted as the formal derivative of the compound Poisson process ([13], Section 7.2.2.3). There are some limited research results on the output of nonlinear systems to Gaussian and nonGaussian input. For example, simple tail approximations have been recently found for the output of simple nonlinear systems with additive Lévy noise, interpreted as the formal derivative of α-stable processes [18]. Also, partial differential equations have been developed for the characteristic function of the output of some linear and nonlinear systems subjected to Gaussian white noise ([13], Section 7.3.1.2). These preliminary results need to be extended significantly to be useful in applications.

The Monte Carlo method is general in the sense that it can be used to estimate the probability law of the output for arbitrary systems and input processes. However, its use is limited because the computation time needed to calculate the system output to an input sample is usually excessive in applications. Hence, only a few output samples can be obtained within a typical computation budget so that output properties cannot be estimated accurately. Several methods are currently under investigation for improving the efficiency of current Monte Carlo simulation algorithms. Some of these methods constitute an extension of the classical importance sampling technique for estimating expectations of functions of random variables to the case of stochastic processes, and are based on Girsanov's theorem ([19]). Additional developments are needed to extend the use of these methods from elementary examples to realistic applications. The essentials of the importance sampling method can be illustrated by the following example.

Let $P_s = P(X \in D)$ and $P_f = P(X \in D^c)$ denote the reliability and probability of failure, where $D \subset \mathbb{R}^d$ is a safe set. The probability of failure is

$$P_f = \int_{\mathbb{R}^d} 1_{D^c}(x) f(x) \, dx = E_P[1_{D^c}(X)] \quad \text{or} \quad P_f = \int_{\mathbb{R}^d} \left[1_{D^c}(x) \frac{f(x)}{q(x)} \right] q(x) \, dx = E_Q \left[1_{D^c}(X) \frac{f(X)}{q(X)} \right]$$

where f and q are the densities of X under the probability measures P and Q, respectively. The measure P defines the original reliability problem. The measure Q is selected to recast the original reliability problem in a convenient way. The estimates of P_f by the above two formulas, referred to as direct Monte Carlo and importance sampling, are denoted by $\hat{p}_{f,MC}$ and $\hat{p}_{f,IS}$, respectively.

Let X have independent $N(0, 1)$ coordinates and D be a sphere of radius $r > 0$ centered at the origin of $\mathbb{R}^d$, that is, $D = \{x \in \mathbb{R}^d : \|x\| \leq r\}$, for $d = 0$. The density q corresponds to an $\mathbb{R}^d$-valued Gaussian variable with mean $(r, 0, \ldots, 0)$ and covariance matrix $\sigma^2 i$, where $\sigma > 0$ is a constant and i denotes the identity matrix. The probability of failure P_f can be calculated exactly and is 0.053, 0.8414×10^{-4}, and 0.4073×10^{-6} for $r = 5, 6$, and 7, respectively. Estimates of P_f by direct Monte Carlo simulation based on 10,000 independent samples of X are $\hat{p}_{f,MC} = 0.0053$, 0.0, and 0.0 for $r = 5, 6$, and 7, respectively. On the other hand, importance sampling estimates of P_f based on 10,000 independent samples of X are $\hat{p}_{f,IS} =$ 0.0; 0.0009; 0.0053; 0.0050; 0.0050, $\hat{p}_{f,IS} = 0.0001 \times 10^{-4}$; 0.1028×10^{-4}; 0.5697×10^{-4}; 1.1580×10^{-4}; 1.1350×10^{-4}, and $\hat{p}_{f,IS} = 0.0$; 0.0016×10^{-6}; 0.1223×10^{-6}; 0.6035×10^{-6}; 0.4042×10^{-6} for $r = 5, 6$, and 7, respectively. The values of $\hat{p}_{f,IS}$ for each value of r correspond to $\sigma = 0.5; 1; 2; 3; 4$, respectively. The direct Monte Carlo method is inaccurate for relatively large values of r, that is, small probabilities of failure. The success of the importance sampling method depends on the density q. For example, $\hat{p}_{f,IS}$ is in error for $d = 10$ and $\sigma = 0.5$ but becomes accurate if σ is increased to 3 or 4. Details on these calculations can be found in ([13] Section 5.4.2.1).

6.4 Stochastic Systems and Deterministic Input

The solution of Equation 6.1 with stochastic operator $\mathcal{D}$ and deterministic input $\mathcal{Y}$ is relevant for many problems in science and engineering. For example, this solution can be used to calculate effective properties for random heterogeneous materials; develop criteria for the stochastic stability of mechanical, biological, and other systems; and examine localization phenomena in nearly periodic systems ([13], Section 8.1).

6.4.1 Current Methods

There is no general methodology for the solution of Equation 6.1 with stochastic operator $\mathcal{D}$ and deterministic input $\mathcal{Y}$. The available methods for solution can be divided in several classes, depending on the form of Equation 6.1, the level of uncertainty in $\mathcal{D}$, and the objective of the analysis. If the uncertainty in $\mathcal{D}$ is small, moments and other properties of the solution $\mathcal{X}$ can be obtained approximately by Taylor series, perturbation, Neumann series, and other methods. These methods are based on relatively simple probabilistic concepts, and can also be used to characterize partially the eigenvalues and eigenfunction of the homogeneous version of Equation 6.1 and calculate approximately effective properties for random heterogeneous materials ([13], Sections 8.3.2, 8.4.2, and 8.5.2.4). On the other hand, methods for evaluating the stability of the stationary solution for a class of equations of the type in Equation 6.1 and assessing the occurrence of localization phenomena are based on properties of diffusion processes and Itô calculus ([13], Sections 8.7 and 8.8).

The Monte Carlo simulation method can be used to estimate the probability law of the solution of the version of Equation 6.1 considered here. The method is general but can be impractical in some applications because of excessive computation time. Also, the use of this method requires one to specify completely the probability law of all random parameters in Equation 6.1 even if only a partial characterization of the output is needed. This requirement may have practical implications because the available information is rarely sufficient to specify the probability law of the uncertain parameters uniquely.

6.4.2 Research Needs and Trends

At least two research areas have to be emphasized in future studies: (1) the development of more powerful probabilistic models for the representation of the uncertain coefficients in Equation 6.1 and (2) the development of general and efficient methods for solving this equation. Research in these areas is essential for the formulation and solution of stochastic problems of the type considered in this section, and requires a close collaboration between engineers, scientists, and applied mathematicians.

Recent work seems to address some of these research needs. For example, new probabilistic models are currently being developed for random heterogeneous materials [20]. Generally, these models are calibrated to the available information consisting of estimates of the first two moments of some material properties, so that there exists a collection of random fields consistent with this information. The selection of an optimal model from this collection has not yet been addressed systematically in a general context. Model selection is a relevant research topic because the properties of the output corresponding to models of $\mathcal{D}$ that are consistent with the available information can differ significantly. For example, suppose that Equation 6.1 has the elementary form

$$Y = \frac{1}{X}, \quad X > 0 \text{ a.s.} \tag{6.6}$$

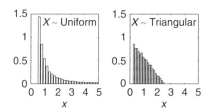

FIGURE 6.3 Histograms of Y for two models of X with the same mean $\mu = 1$, variance $\sigma^2 = a^2/3$, and $a = 0.95$.

where Y can be interpreted as the tip deflection of a cantilever with stiffness X. Suppose that the available information on the uncertain parameter X in $\mathcal{D}$ is limited to the first two moments of X and the requirement that $X > 0$ with probability 1. Let $\mu = E[X]$ and $\sigma^2 = E[(X - \mu)^2]$ denote the mean and variance of X. Hence, any random variable with range in $(0, \infty)$ and first two moments (μ, σ^2) is a valid model for the beam stiffness. Figure 6.3 shows histograms of Y for two models of X with mean $\mu = 1$, variance $\sigma^2 = a^2/3$, and $a = 0.95$. The left plot is for X uniformly distributed in the interval $(1 - a, 1 + a)$, and the right plot is for X assumed to follow a triangular density in the range $(1 - h/3, 1 + 2h/3)$ with the nonzero value at the left end of this interval and $h = a\sqrt{6}$. The histograms of the output Y differ significantly although correspond to models of X that are consistent with the available information.

Similar results can be found in [21], where several random fields are used to represent a two-phase microstructure. The second-moment properties of these fields and the target microstructure nearly coincide. However, other microstructure properties predicted by these fields differ significantly, for example, the density of the diameter of the phases generated by these fields. This lack of uniqueness can have notable practical implications.

There is limited work on the selection of the optimal model from a collection of probabilistic models for $\mathcal{D}$ consistent with the available information.

Also, new methods for finding effective properties of a class of multiphase microstructures described by versions of Equation 6.1 are under development. These methods are based on multiscale analysis, are developed in the context of material science, and involve engineers, scientists, and applied mathematicians [22].

6.5 Stochastic Systems and Input

In most applications, both the operator $\mathcal{D}$ and the input $\mathcal{Y}$ in Equation 6.1 are uncertain. The forms of this equation considered in the previous two sections are relevant for applications characterized by a dominant uncertainty in input or operator. If the uncertainty in both the input and the output is weak, Equation 6.1 can be viewed as deterministic.

6.5.1 Current Methods

Monte Carlo simulation is the only general method for solving Equation 6.1. The difficulties involving the use of this method relate, as stated previously, to the computation time, which may be excessive, and the need for specifying completely the probability law of all random parameters in $\mathcal{D}$ and $\mathcal{Y}$.

Some of the methods frequently applied to find properties of $\mathcal{Y}$ in Equation 6.1 constitute direct extensions of the techniques outlined in the previous two sections. For example, conditional analysis is an attractive alternative when $\mathcal{D}$ depends on a small number of random parameters so that for fixed values of these parameters the methods applied to solve Equation 6.1 with deterministic operator and stochastic input apply. Methods based on Taylor, perturbation, and Neumann series can also be used for finding properties of $\mathcal{X}$ approximately. Other approximate methods for solving Equation 6.1 can be found in ([13] Section 9.2).

Over the past 15 years there have been many studies attempting to extend the finite element and other classical numerical solutions used to solve the deterministic version of Equation 6.1 to the case in which both the operator and the input of this equation are uncertain. Unfortunately, these extensions, referred to as stochastic finite element, finite difference, and boundary element, fell short of expectations. The solutions of Equation 6.1 by these methods are based on (1) approximations of the solution by Taylor, perturbation, or Neumann series for cases in which the uncertainty in $\mathcal{D}$ is small ([13], Section 9.2.6); (2) approximations of the solution by polynomial chaos, that is, sums of Hermite polynomials in Gaussian variables [23]; and (3) formulation of the solution as a reliability problem so that FORM/SORM and other approximate methods for reliability analysis can be used for solution [1].

The Monte Carlo simulation method and many of the above approximate methods have been applied to solve a variety of problems from mechanics, physics, environment, seismology, ecology, and other fields. In addition to these methods, alternative solutions have been used to solve particular applications. For example, properties of diffusion processes and the Itô calculus have been used to characterize noise induced transitions in a randomized version of the Verhulst model for the growth of a biological population ([13], Section 9.4.3) and find properties of the subsurface flow and transport in random heterogeneous soil deposits ([13], Section 9.5.3).

6.5.2 Research Needs and Trends

Methods are needed for solving efficiently and accurately Equation 6.1 for a broad range of stochastic operators $\mathcal{D}$ and input processes $\mathcal{Y}$. It may not be possible to develop a single method for solving an arbitrary form of Equation 6.1. The contribution of applied mathematicians to the development of solutions for Equation 6.1 may be essential for advances in this challenging area.

There are some limited developments that provide approximate solutions to Equation 6.1. For example, equations can be developed for the gradients of the solution $\mathcal{X}$ with respect to the uncertain parameters in $\mathcal{D}$. These gradients, referred to as sensitivity factors, and the uncertainty in the random parameters in $\mathcal{D}$ can be used to identify the most relevant sources of uncertainty and calculate approximately properties of the output $\mathcal{X}$. For example, let $X(t; R)$ be the solution of the differential equation

$$dX(t; R) = -R X(t; R) + \sigma \, dB(t), \quad t \geq 0, \tag{6.7}$$

where the random variable R with mean μ_r and variance σ_r^2 is strictly positive, σ is a constant, and B denotes a Brownian motion. Consider the first order approximation

$$X(t; R) \approx X(t; \mu_r) + Y(t; \mu_r)(R - \mu_r), \tag{6.8}$$

of the solution of the above equation, where the sensitivity factor $Y(t; \mu_r)$ is the partial derivative $\partial X(t; R)/\partial R$ evaluated at $R = \mu_r$. The sensitivity factor $Y(t; \mu_r)$ is the solution of the differential equation

$$dY(t; \mu_r) = -\mu_r Y(t; \mu_r) \, dt - X(t; \mu_r) \, dt \tag{6.9}$$

derived from Equation 6.7 by differentiation with respect to R and setting $R = \mu_r$. The defining equations for $X(t; \mu_r)$ and $Y(t; \mu_r)$ can be considered simultaneously and solved by classical methods of linear random vibration. Figure 6.4 shows the evolution in time of the exact and the approximate variance functions of $X(t; \mu_r)$ and $X(t; R)$, respectively, for $X(0) = 0$, $\sigma = \sqrt{2}$, $\mu_r = 1$, and $\sigma_r = 1.4/\sqrt{12}$. The plots show that the uncertainty in R increases notably the variance of the process X. The use of sensitivity analysis replaces the solution of a differential equation with random coefficients and input with the solution of two differential equations with deterministic coefficients and stochastic input.

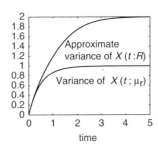

FIGURE 6.4 Exact and approximate variance functions of $X(t; \mu_r)$ and $X(t; R)$.

6.6 Comments

The formulation and solution of the future problems encountered in science and engineering require (1) extensive interaction between experts from different disciplines and (2) methods for solving stochastic problems that focus on the functional form of the defining equations rather than the particular field of application. The contribution of applied mathematicians is considered to be essential for these developments.

Two research directions have been identified as critical. The first relates to the development of probabilistic models capable of representing a broad range of input processes and characteristics of uncertain systems. Moreover, practical methods are needed to select optimal models from a collection of random variables, stochastic processes, and random fields whose members are consistent with the available information. Some useful results have already been obtained in this area.

The second direction relates to the development of efficient and accurate numerical algorithms for solving complex algebraic, differential, and integral equations with random coefficients and input. Significant work is needed in this area.

References

1. Madsen, H.O., Krenk, S., and Lind, N.C., *Methods of Structural Safety*, Prentice Hall, Englewood Cliffs, NJ, 1986.
2. Papoulis, A., *Probability, Random Variables, and Stochastic Processes*, McGraw-Hill, New York, 1965.
3. Soong, T.T. and Grigoriu, M., *Random Vibration of Mechanical and Structural Systems*, Prentice Hall, Englewood Cliffs, NJ, 1993.
4. Crandall, S.H., *Engineering Analysis. A Survey of Numerical Procedures*, McGraw-Hill, New York, 1956.
5. Greenberg, M.D., *Foundations of Applied Mathematics*, Prentice Hall, Englewood Cliffs, NJ, 1978.
6. John, F., *Partial Differential Equations, 4th edition*, Springdf-Verlag, New York, 1982.
7. Brebbia, C.A. and Connor, J.J., *Fundamentals of Finite Element Techniques*, Butterworth, London, 1973.
8. Glimm, J., Uncertainty Quantification for Numerical Simulations, presented at the *SIAM Conference on Computational Science and Engineering*, February 10–13, 2003, San Diego, CA.
9. Chung, K.L. and Williams, R.J., *Introduction to Stochastic Integration*, Birkhäuser, Boston, 1990.
10. Courant, R. and Hilbert, D., *Methods of Mathematical Physics*, Vol. 2, Interscience, New York, 1953.
11. Durrett, R., *Stochastic Calculus. A Practical Introduction*, CRC Press, Boca Raton, FL, 1996.
12. Øksendal, B., *Stochastic Differential Equations*, Springer-Verlag, New York, 1992.
13. Grigoriu, M., *Stochastic Calculus. Applications in Science and Engineering*, Birkhäuser, Boston, 2002.
14. Grigoriu, M., *Applied Non-Gaussian Processes: Examples, Theory, Simulation, Linear Random Vibration, and MATLAB Solutions*, Prentice Hall, Englewood Cliffs, NJ, 1995.

15. Crandall, S.H. and Mark, W.D., *Random Vibration in Mechanical Systems,* Academic Press, New York, 1963.
16. Horsthemke, W. and Lefever, R., Phase transitions induced by external noise, *Physics Letters,* 64A(1), 19–21, 1977.
17. Protteer, P., *Stochastic Integration and Differential Equations,* Springer-Verlag, New York, 1990.
18. Samorodnitsky, G. and Grigoriu, M., Tails of solutions of certain nonlinear stochastic differential equations driven by heavy tailed Lévy motions, *Stochastic Processes and Their Applications,* submitted 2002.
19. Naess, A. and Krenk, S., Eds., *Advances in Nonlinear Stochastic Mechanics,* Kluwer Academic, Boston, 1996.
20. Torquato, S., Thermal conductivity of disordered heterogeneous media form the microstructure, *Reviews in Chemical Engineering,* 4(3 and 4), 151–204, 1987.
21. Roberts, A.P., Statistical reconstruction of three-dimensional porous media from two-dimensional images, *Physical Review E,* 56(3), 3203–3212, 1997.
22. Hornung, U., *Homogenization and Porous Media,* Springer-Verlag, New York, 1997.
23. Ghanem, R.G. and Spanos, P.D., *Stochastic Finite Elements: A Spectral Approach,* Springer-Verlag, New York, 1991.

7
A Research Perspective

7.1	Background..7-1
	Critical Element of the Design Process • The Need for a New Approach
7.2	An Inductive Approach..7-5
	Family of Bayesian Methods
7.3	Information Aggregation 7-7
	Confidence Intervals • Data Congeries
7.4	Time Dependent Processes 7-9
	Distribution or Process Model?
7.5	Value of Information 7-11
7.6	Sensitivity Analysis 7-13
7.7	Summary 7-13

David G. Robinson
Sandia National Laboratories

7.1 Background

As with all probability-related methods, nondeterministic analysis (NDA) methods have gone through numerous cycles of interest to management and as topics of investigation by a wide spectrum of engineering disciplines. In concert with these cycles has been the support for research in NDA methods. As always, research funding is ultimately driven by the perceived benefit that decision makers from management and engineering feel is to be gained by the time and money needed to support the inclusion of nondeterministic methods in their project.

The focus of this chapter is a suggested research direction or focus that I believe will result in nondeterministic methods having more value to the broad spectrum of customers and will, as a result, become more tightly woven in the overall systems design and analysis process.

From a broad perspective, future research in the NDA area needs to begin to focus on being applicable across the entire systems design process, through all phases of initial design conception and across the entire system life cycle. Historically, the NDA field has tackled important pieces of the process and a number of very elegant mathematical techniques have been developed. However, there are areas where there remains much to be done. Inherent with these areas is the critical need for direct or indirect customers to see the value added via the application of NDA methods: How are NDA methods going to permit better system level decisions? Will these methods shorten my time to market? How are NDA methods going to provide information regarding trade-off during design?

By its nature the system engineering process is adrift in a sea of information and data: field data on similar components, handbook data on material properties, laboratory test results, engineering experience, etc. A complete "information science" has grown around the collection, storage, manipulation, and retrieval of information. Perversely, as expansive as this information space appears, the information is incomplete. A critical goal of nondeterministic methods is to characterize or quantify this lack of information.

Historically, research into nondeterministic methods has focused on point estimates and amalgamation of those estimates into a point estimate at the system level. To demonstrate the value of NDA methods, researchers must move outside the "point estimate" box. From a broad perspective, challenges facing researchers in the NDA field include:

- Relating lower-level analyses to system level performance through a more objective integration of information and by identifying those aspects of the system where generation of additional information (e.g., focused testing or simulation) will reduce the uncertainty in system performance. The majority of the discussion in this chapter focuses on this particular challenge.
- Characterization of the uncertainty associated with degradation in system performance throughout the life cycle of the system. The integration of temporal system characteristics and spatial properties result in an extremely complicated analysis problem. Yet with the aging civil infrastructure and the increased dependence by military organizations on long-term storage of complex electronics and materials, characterization of time-dependent system behavior is becoming a critically important part of the decision-making process. However, as Niels Bohr noted, prediction is extremely difficult — especially about the future.
- The logical next step is the combination of the first two challenges: development of techniques to combine information from accelerated laboratory tests, results from periodic testing of system articles, and output from computer simulation models to anticipate failure and better plan maintenance support will be a particularly challenging task.

Information comes at a price — nondeterministic methods can be a powerful tool for placing value on this information. Beyond the above challenges, new directions for research in NDA methods are likely to be in support of information science and, in turn, the systems engineering process. The value of all information lies in the ability to reduce the uncertainty associated with the consequences of the solution chosen. Nondeterministic methods provide the potential for answering many questions, including: *What is the likelihood that the system will perform as needed now and in the future, and where should the limited resources be allocated: in materials testing, complex computer modeling, component testing or system testing to be as confident as possible in how the system will perform?*

7.1.1 Critical Element of the Design Process

Systems engineering is a multidisciplinary approach to problem solution that ensures that system requirements are satisfied throughout the entire system life cycle. This approach has gained popularity because of its proven track record in increasing the probability of system success and reducing costs through the entire life of the system.

Figure 7.1 depicts the systems engineering process used by Lockheed-Martin [1]. This is a common approach: "pushing" information down based on top-level requirements while simultaneously "pulling" information up from lower-level analyses. Figure 7.2 is a similar depiction focusing on the application of nondeterministic methods. Top-level requirements drive subsystem requirements, component requirements, etc. Application of nondeterministic methods are often applied to either amalgamate failure modes or integrated with simulation models (e.g., finite element model, electrical circuit simulation). In either case, these models are often supported in turn by complex materials aging models.

Regardless of the particular perspective on system engineering, the critical point is the recognition that resources expended on each subsequent level of indenture are driven by the impact that analysis at the lower level will have on the characterization of the top-level system. The overriding goal is to assess the impact of various uncertainties associated with internal and external variables on the uncertainty in system performance: an estimate in the probability that the system will perform, in conjunction with the confidence that we have in that estimate.

Clearly, a balance must be obtained between the information gained from testing and modeling at various levels of system indenture, the resources required to perform testing and develop models, and the uncertainties associated with predicting system performance. The ability to logically combine information

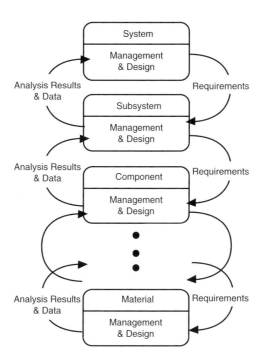

FIGURE 7.1 Systems engineering process.

from these various areas — as well as the organization, characterization, and quantification of this myriad of uncertainties — are critical elements of the system design and analysis processes. Importantly, although the ultimate objective is to obtain accurate predictions, it must be recognized that real merit also exists for simply obtaining more robust solutions than are presently possible (i.e., simply improving the confidence in the estimates of useful life).

7.1.2 The Need for a New Approach

As noted by the Greek philosopher Herodotus in roughly 500 B.C., "A decision was wise, even though it led to disastrous consequences, if the evidence at hand indicated it was the best one to make; and a decision was foolish, even though it led to the happiest possible consequences, if it was unreasonable to expect those consequences." Fundamental to implementing a systems approach will be the need to consider all the evidence at hand through the use of nondeterministic methods, in particular through a Bayesian approach. Classical statistical methods, those based on a frequency approach, permit the combination of point estimates but stumble when confidence in system-level performance must be characterized. In addition, classical Bayesian methods permit the logical combination of test data but do not fully incorporate all available information. Specifically, *classical* Bayesian methods assume that the articles under test are not related in any manner although the articles may be identical. Alternatively, *hierarchical* Bayesian methods permit the relationship between test articles to be explicitly included in the analysis.

In particular, the use of this more modern methodology provides a formal process for synthesizing and "learning" from the data that is in agreement with current information sciences. It permits relevant expert opinion, materials aging, sparse field data, and laboratory failure information to be codified and merged.

In addition to providing an objective means of combining information, Bayesian methods provide a structured tool for robust model development. For example, if laboratory experiments and detailed mathematical simulations are too expensive, empirical models based on data from field measurements and component/subsystem testing can drive model refinement. In addition, this methodology has the potential to address both failures due to aging effects and random failures due to unaccounted-for latent defects.

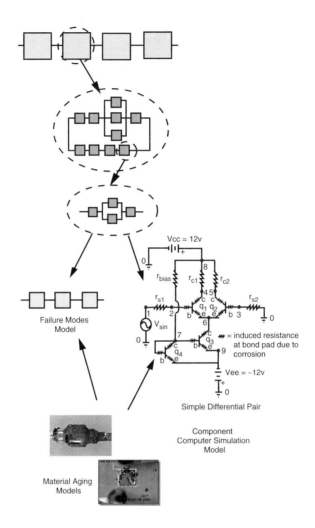

FIGURE 7.2 Systems engineering approach to reliability design.

The foundation for what is referred to today as Bayesian analysis was first formalized by Presbyterian minister Thomas Bayes of Tunbridge Wells, Kent. His article entitled "Essay towards Solving a Problem in the Doctrine of Chances" was published posthumously in 1763 in the *Philosophical Transactions of the Royal Society of London*. However, the work was rediscovered in 1774 by Pierre Laplace and it is his (i.e., Laplace's) formulation that is commonly referred to today as Bayes theorem.

In the early twentieth century, mathematicians became uncomfortable with the concept of only using available information to characterize uncertainty. The concept of long-run relative frequency of an event based on an infinite number of trials became popularized by researchers such as Kolomogorov, Fischer, and Neyman, among others. Practitioners of this latter mathematical philosophy are commonly referred to as Frequentists. However, because it was actually impossible to observe an infinite number of trials, various data transformations (i.e., statistics) were developed. The result has been an array of different statistics for different applications, leading to the cookbook type of statistics commonly practiced today.

At about this same time, Jefferys [2] published one of the first modern books on Bayes theory, uncovering the work by Laplace many years earlier. Unfortunately, the work appeared at the same time that maximum likelihood estimation was being popularized in the literature and much of the potential impact was not realized.

In recent years, the work of Bayes, Laplace, and Jefferys has received increasing attention for a wide variety of related reasons. Information sciences and the rise of decision theory as an aid to management are the most recent additions to the list. For engineers, the high cost of testing has led to a push to investigate some alternative means to extort as much information as possible from existing data.

To appreciate why alternative methods of processing information are important, consider the problem of a bag containing five red balls and seven green balls [3]. On a particular draw we choose a ball, with probability 5/12 and 7/12 of picking a red or a green ball, respectively. If, after the initial selection, the ball is not returned to the bag, then the chance of picking either a green or red ball on the next selection depends on the prior selection. On the other hand, if no information regarding the result of the first selection is available and a green ball is chosen on the second draw, what can be said about the probability of choosing a red or green ball on the first pick? Intuition suggests that the results of the second selection should not influence the probability of choosing a red or green ball on the first draw. However, before answering this, consider the situation where there is only one red ball and one green ball in the bag. Clearly, the information available as a result of the second draw influences the guess as to the first selection. It is this use of information in a *conditional* manner that provides additional insight into problems not otherwise possible and is the key to a Bayesian approach to test plan design and data analysis.

Beyond costly system-level testing, possible information sources include materials testing, subsystem laboratory testing as well as data from similar systems, and finally computer simulations. From the perspective of increasing confidence in the final decision (within budget constraints), it is of immense importance to efficiently use information from all available sources. For example, information from analysis or testing at the subsystem level is commonly used to make system-level assessments because laboratory testing is generally much less expensive than full system testing in an operational environment.

This approach has been particularly appealing for those situations where extensive system testing is impractical for a variety of reasons (e.g., cost of prototype development) or even impossible due to international treaties (e.g., nuclear weapons). While there is no substitute for full-scale testing in a realistic operational environment, it is always difficult to justify not considering data from all relevant sources.

The use of condition-based logic — given this information, then I expect these results — contrasts greatly with the more popular approach — this exact situation has happened many times, so I expect it will happen again — or equivalently — this has never happened before so it will never happen in the future. The application of Bayesian methods, and inductive reasoning in general, permits the analyst to provide answers to a variety of questions with increased confidence. For engineers, Bayesian methods provide a logical, structured approach to assess the likelihood of new events that are outside the current reality and cannot be directly measured.

7.2 An Inductive Approach

From basic probability theory the conditional distribution of variable **Y** given variable **X** is defined as:

$$f(y|x) = \frac{f(y, x)}{f_x(x)}$$

where $f(y, x)$ is the joint density function of (**Y**, **X**) and $f_x(x)$ is the marginal density function for random variable **X**. Because we know that: $f(y, x) = f(y)f(x|y)$, it follows that:

$$f(y|x) = \frac{f(y, x)}{f_x(x)} = \frac{f(y)f(x|y)}{f_x(x)} = \frac{f(y)f(x|y)}{\int_x f(y, x)dy} = \frac{f(y)f(x|y)}{\int_x f(y)f(x|y)dy} \quad (7.1)$$

summarizing the foundation of Bayes theorem. The distribution $f(y)$ characterizes random variable **Y** before information about **X** becomes available and is referred to as the *prior* distribution function.

Similarly, the distribution $f(y|x)$ is referred to as the *posterior* distribution. Note also that because $\int_x f(y)f(x|y)dy$ depends only on the **X** and not on **Y**, the integration is simply a normalization constant and therefore:

$$f(y|x) \propto f(y)f(x|y)$$

The distribution $f(x|y) \equiv l(y|x)$ and is referred to as the likelihood function, incorporating information available about **X** into the characterization of **Y**.

In our context, to contrast Bayes theorem with the traditional Frequentist approach, let us characterize a random variable **Y**, probability density function $f(y)$, with for example a parameter vector $\boldsymbol{\theta} = \{\alpha, \beta\}$. From a Frequentist perspective, the parameters α and β are fixed quantities that can be discovered through repeated observation. However, from a Bayesian point of view, parameter $\boldsymbol{\theta}$ is unknown and *unobservable*. Further, the probability that a sequence of values is observed for **Y** is explicitly conditioned on the value that this parameter assumes that:

$$f(y_1, y_2, \ldots, y_n | \boldsymbol{\theta})$$

7.2.1 Family of Bayesian Methods

As previously alluded to, within the family of Bayesian analysis techniques there are two broad frameworks: (1) empirical Bayes and (2) hierarchical Bayes. The classifications "empirical" and "hierarchical" are unfortunately all too common in the literature; all Bayesian methods are empirical in nature and all can be described as being hierarchical in the fashion in which data is accumulated.

Hierarchical Bayes is the more recent addition to the family, is considerably more efficient than the empirical Bayesian approach, and is suggested as the general direction for future research. The hierarchical Bayes approach is also less sensitive to the choice of the prior distribution parameters, typically the focus of a great deal of emotional discussion.

7.2.1.1 Empirical Bayes

The foundation for empirical Bayes — or more specifically, parametric empirical Bayes — has been in place since von Mises in the 1940s, but really came into prominence in the 1970s with the series of papers by Efron and Morris, (see, e.g., [4]). There have been a number of excellent publications in which the authors have made the effort to explain the theory and logic behind empirical Bayes and its relationship to other statistical techniques [5, 6].

Again, consider the random variable **Y**, probability density function $f(y)$, with for example a parameter vector $\boldsymbol{\theta} = \{\alpha, \beta\}$. From an empirical Bayes perspective, **Y** is explicitly conditioned on the value that this parameter assumes that $f(y_1, y_2, \ldots, y_n | \boldsymbol{\theta})$, where $\boldsymbol{\theta} = \{\alpha, \beta\}$ is unknown, unobservable, and the prior beliefs about $\boldsymbol{\theta}$ are estimated from data collected during analysis.

However, some subtle problems exist with the empirical Bayesian approach. First, point estimates for the prior parameters lack consideration for modeling uncertainty. Second, in general, any available data would have to be used once to form a prior and then again as part of the posterior; that is, data would be used twice and result in an overly conservative estimate of the elements of $\boldsymbol{\theta}$.

In general then, empirical Bayesian methods represent only an approximation to a full Bayesian analysis. They do not represent a true Bayesian analysis of the data because a traditional statistical approach is used to estimate the parameters of the prior distribution. Alternatively, in a hierarchical Bayesian approach, data analysis, and all prior and posterior distribution characteristics are estimated in an integrated fashion.

7.2.1.2 Hierarchical Bayes

The distinguishing feature of this Bayesian technique is the hierarchical nature in which information is accumulated. Using the example from the previous section, it is now assumed that the parameter

$\boldsymbol{\theta} = \{\alpha, \beta\}$ is an *unknown* hyperparameter vector described via a prior distribution $f(\boldsymbol{\theta})$. Now it is assumed that $\boldsymbol{\theta}$ is also a random variable and the uncertainty in the hyperparameters is addressed explicitly. For example, the complete Bayesian analysis for characterizing the density function $f(\alpha | \mathbf{y})$ requires a description of the vector of random variables $\boldsymbol{\theta} = \{\alpha, \beta\}$ with joint prior distribution:

$$f(\alpha, \beta) = f(\alpha | \beta) f(\beta)$$

and joint posterior distribution:

$$\begin{aligned} f(\boldsymbol{\theta} | \mathbf{y}) &\propto f(\alpha, \beta) l(\alpha, \beta | \mathbf{y}) \\ &= f(\alpha, \beta) f(\mathbf{y} | \alpha, \beta) \end{aligned} \quad (7.2)$$

Note that the joint posterior density function $f(\alpha, \beta | \mathbf{y})$ can be written as a product of the hyperprior $f(\boldsymbol{\theta})$, the population distribution $f(\alpha | \beta)$, and the likelihood function $l(\boldsymbol{\theta} | \mathbf{y})$. Under the assumption of an independent and identically distributed set of samples, $\mathbf{y} = \{y_1, \ldots, y_N\}$, an analytical expression for the conditional posterior density of $f(\alpha | \beta, \mathbf{y})$, can be easily constructed as the product of the density functions $f(y_i)$.

In the case of conjugate density functions, a solution is available directly. Once an expression for the joint posterior function is found, the marginal posterior distribution can be found through direct evaluation or via integration:

$$f(\alpha | \mathbf{y}) = \frac{f(\boldsymbol{\theta} | \mathbf{y})}{f(\beta | \alpha, \mathbf{y})} = \int f(\alpha, \beta | \mathbf{y}) d\beta. \quad (7.3)$$

Because it is rare that either direct evaluation or integration is possible, an alternative is the use of simulation to construct the various conditional density functions:

- Generate a sample of the hyperparameter vector β from the marginal distribution function, $f(\beta | \mathbf{y})$.
- Given β, generate a sample parameter vector α from $f(\alpha | \beta, \mathbf{y})$.
- A population sample can be then be generated using the likelihood function $l(\mathbf{y} | \alpha, \beta)$.

(For a more detailed discussion see the summary by Robinson [7].) Generally, these steps will be difficult to accomplish due to the problems associated with generating samples from complex conditional distributions. A simulation technique particularly suited for this task, Markov Chain Monte Carlo simulation, is introduced in the following section. Two recent applications of hierarchical Bayesian methods in the structural reliability area that utilize this approach are Wilson and Taylor [8] and Celeux et al. [9]. Because of the truly objective fashion with which data is integrated together, further efforts in the hierarchical Bayesian area hold the most promise for further research. A key piece of technology that makes the use of hierarchical Bayesian methods possible is Markov Chain Monte Carlo simulation [10].

7.3 Information Aggregation

Before addressing the challenge of aggregating data from various sources and levels of analysis, it is important to understand a key difference (and benefit) between Frequentist and Bayesian analysis techniques: the interpretation of confidence statements. It is the unique nature with which confidence intervals are interpreted in Bayesian methods that suggests their use in a systems engineering framework by permitting the aggregation of information.

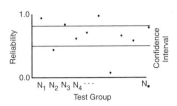

FIGURE 7.3 Confidence intervals: Frequentist.

7.3.1 Confidence Intervals

A Frequentist does not view the system parameters as random variables but rather as *constants* that are unknown. As such, it is not possible to make statements regarding the *probability* that the parameters lie within a particular interval. To characterize confidence intervals from a Frequentist perspective, it is necessary to imagine a very large (potentially infinite) ensemble of tests. All parameters in each group of tests are assumed to be held fixed, with the exception of the parameter of interest (e.g., system reliability). Therefore, the reliability of each group will be different and can be calculated. After an extremely large number of groups are tested and the reliability calculations collected, a fraction of reliability values would fall within a band of reliabilities. This "band" is referred to as a confidence interval for the reliability (see Figure 7.3).

Note that a large number of groups of tests are not actually conducted but are only imagined. It also bears repeating that, because system reliability is not a random variable, it cannot be stated that there is a probability that the true reliability falls within these confidence intervals.

The Bayesian interpretation of confidence intervals is slightly less abstract. Under a Bayesian approach, the reliability of the system is a random variable with density function $f(p)$. The parameters of the density function can be estimated from existing data, as in the case of empirical Bayesian methods, or can also be random variables with associated probability density functions (e.g., hierarchical Bayes).

In any case, because the reliability is a random variable, it is possible to make probability statements regarding the likelihood of the true reliability being in a particular interval as depicted in Figure 7.4. And because the reliability of each subsystem is a random variable, it will be seen that the aggregation of information from various sources and levels of analysis is therefore straightforward.

7.3.2 Data Congeries

A number of authors have suggested methods for combining subsystem and system data into congeries under a Bayesian framework (see, e.g., [11–13]). In general, these papers, along with a number of others, support analysis approaches that permit the collection of test data from a number of levels of system indenture, into an overall estimate of the system reliability. The methods all depend on a combination of analytical techniques for combining test information and inherently depend on assumptions regarding the underlying distribution function.

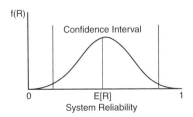

FIGURE 7.4 Confidence intervals: Bayesian.

A Research Perspective

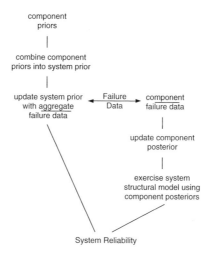

FIGURE 7.5 Data aggregation options.

As noted in the papers above, the aggregation of system and component level data can become involved. In general, the system-level reliability distribution derived from component data is used as a prior for the reliability distribution based on system level data (Figure 7.5). However, when component- and system-level failure and performance information is collected at the same time, aggregation of component data into a system-level analysis may not result in the same reliability prediction as obtained from the system data alone (see, e.g., Azaiez and Bier [14]).

Preliminary efforts in this area are discussed in [7], but considerable work remains. In particular, aggregation issues must be strongly considered as new analysis methods are developed to support time-dependent analysis of material aging and degradation processes.

7.4 Time Dependent Processes

Consideration of potential nondeterministic characteristics of time-dependent processes in concert with a total systems approach is one of the more challenging aspects of future NDA research. There are two critical aspects: (1) the estimation of the current state of the system in the presence of past degradation and, (2) the prediction of future system performance with the expectation of continued aging. The overall goal is to use NDA techniques to improve the ability to *anticipate* the occurrence of a failure, something not possible with deterministic methods (Figure 7.6).

However, before delving into possible research directions, it is important to appreciate the distinction between the failure rate of the underlying time-to-failure distribution and the process failure rate. It is important for the engineer interested in characterizing time-dependent processes (specifically degradation) to understand how each concept applies to the particular problem at hand.

7.4.1 Distribution or Process Model?

An often-confusing aspect of aging/degradation analysis is the distinction between hazard function for the distribution and the failure rate for the underlying degradation process.

Let $N(t)$ denote the number of cumulative failures that occur in the interval $(0, t]$. The set of points on the timeline associated with occurrence of failures form what is referred to as a *point process*. The expected number of failures in the interval $(0, t]$ is defined as $E[N(t)]$. When addressing the situation

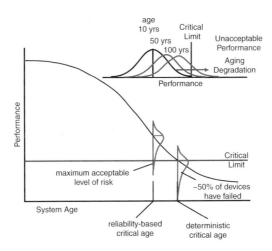

FIGURE 7.6 Anticipation of failure.

that might exist during design, $E[N(t)]$ will be assumed to be a continuous, nondecreasing function and, then, assuming its derivative exists, let

$$\lambda(t) = \frac{dE[N(t)]}{dt}.$$

Therefore, $\lambda(t)$ represents the instantaneous change in the number of failures per unit time; that is, the rate at which failures occur and $\lambda(t)$ is referred to as the *failure rate of the process*. Assuming that simultaneous failures do not occur:

$$\lambda(t) = \lim_{\Delta t \to 0} \frac{P[N(t, t+\Delta t) \geq 1]}{\Delta t} \qquad (7.4)$$

For small Δt, $\lambda(t)\Delta t$ is approximately the probability of a failure occurring in the interval $(t, t+\Delta t]$.

Now consider the hazard function: the probability of a failure per unit time occurring in the interval, given that a failure has not already occurred prior to time t. Let the density and distribution functions of the system lifetime **T** be $f(T)$ and $F(T)$, respectively. Define $1 - F(T) = \overline{F}(T)$. Then:

$$h(t) = \frac{P[t \leq T \leq t + \Delta t \mid t \leq T]}{\Delta t}. \qquad (7.5)$$

Substituting the appropriate expressions and taking the limit yields:

$$h(t) = \lim_{\Delta t \to 0} \frac{\overline{F}(t) - \overline{F}(t + \Delta t)}{\Delta t \overline{F}(t)} = \frac{f(t)}{1 - F(t)} \qquad (7.6)$$

Thus, we have the instantaneous *failure rate of the distribution* of the system lifetimes.

To compare process and distribution failure rates, suppose for example that we have n units with identical hazard rate characteristics as discussed above. Let $N(t)$ represent the number of units that have failed by time t. Thus, $\{N(t)\}$ represents the point process discussed previously.

The random variable $N(t)$ is then binomially distributed:

$$P[N(t) = k] = \binom{n}{k} F^k(t)[1-F(t)]^{n-k} \qquad k = 0, 1, \ldots, n \qquad (7.7)$$

The expected value of $N(t)$ is given by $nF(t)$, and the rate at which failures occur is:

$$\frac{dE[N(t)]}{dt} = nf(t)$$

the failure rate of the process. However, the failure rate of the distribution is given by $h(t)$.

Care must be taken to ensure which changing failure rate is of interest because the difference between a changing failure rate for the process will be significantly different from the failure rate for the density function. The above discussion focused on continuous parameter systems. Similar results hold for discrete parameter systems.

Significant progress has been made in understanding and characterizing those situations where, in the presence of either growth or degradation, the uncertainties in system parameters and system age are separable. In particular, characterization of the improvement in the probability of system performance as the design process progresses is referred to as reliability growth and has received a great deal of attention in the past few decades. The appreciation that reliability is changing during the course of design has led to a significant decrease in the development time of complex systems and has resulted in greatly improved reliability of deployed systems. Unfortunately, until rather recently, little attention has been paid to the "other end" of the system life cycle — the degradation of system performance.

A large number of publications have dealt with time-dependent system performance for those situations where the uncertainties in physical characteristics are separable from the temporal characteristics. This emphasis has resulted from a number of factors, the foremost being (1) a significant number of real problems fall into this category and (2) the mathematics are within reach of most engineers. The benefit of these applications is clear in many areas (the area of fatigue life, for example); however, a number of substantial gaps remain.

A common theme of previous discussion in this chapter has been the need to incorporate historical data to improve the confidence in the analyses. This need is even more critical when dealing with time-dependent processes; the projection of system performance 10, 30, or even 100 years into the future requires substantial confidence in the ability to characterize *current* system performance. One direction of potentially fruitful research lies in the areas of longitudinal and nonlinear Bayesian analysis. Longitudinal analysis involves repeated measurements on a collection of individuals taken over a period of time. A substantial body of work exists in the area of pharmacokinetics and trajectory analyses of disease incidence and mortality [15, 16].

Another area that holds promise for making significant inroads into this problem is with the continued research into those situations where time is inseparable from the uncertainties. The area of stochastic differential equations holds promise for providing insight into the additional complexities that present themselves in these unique problems. Discussion of this area is left for more qualified hands in other chapters.

7.5 Value of Information

Previous discussion focused on tackling problems from a systems perspective and highlighted the need to develop new methods for bringing data together efficiently from possibly disparate sources. On the other side of the discussion is the decision as to where to focus resources to collection information as efficiently as possible. However, there is no such thing as a free lunch; each laboratory experiment, each flight test brings a new piece of the puzzle — but at a price. Even computer simulations are no longer "free" in the usual sense; some simulations can take on the order of days, making the additional effort of considering uncertainty costly from a variety of perspectives, including the time to complete a full analysis.

So how does a decision maker decide where to spend his or her resources: additional computer model fidelity or additional computer simulations? On which variables should the experimental tests be focused, and will specific test equipment have to be developed? Clearly, as depicted in Figure 7.7, the value of new information decreases as the uncertainties decrease, and the focus needs to be on where in the process is the most cost-effective way to reduce this uncertainty.

The expected value of information will be defined as the difference between the benefit or value $u(x, \theta)$ of an optimal decision after the data x is available and the value of an optimal decision if the data had not been collected $u(\theta)$:

FIGURE 7.7 Value of information.

$$EVI = \int_x \left[\int_\upsilon u(x,\theta) f(\theta|x) d\theta \right] f_x(x) dx \\ - \int_\theta u(\theta) f(\theta) d\theta \tag{7.8}$$

The value of information about a parameter θ depends on the prior distribution characterizing the current information $f(\theta)$. The impact of a data x from an experiment on the parameter θ is found through application of Bayes theorem:

$$f(\theta|x) = \frac{f(\theta, x)}{f_x(x)} = \frac{f(\theta) f(x|\theta)}{\int_\theta f(\theta) f(x|\theta) d\theta} \tag{7.9}$$

After data is collected (e.g., computer simulation or laboratory experiment), the expected posterior distribution $f(\theta|x)$:

$$E[f(\theta|x)] = \int_x f(x) f(\theta|x) dx \\ = \int_x f_x(x) \left[\frac{f(\theta) f(x|\theta)}{f_x(x)} \right] dx \tag{7.10} \\ = f(\theta)$$

which indicates that the *expected* uncertainty (as characterized by the posterior) after the data has been collected is fully characterized by the prior distribution. Clearly, information will be gained by the additional data, but the content of that information cannot be anticipated from the information currently available. So what can be discerned about the value of additional information? Consider the well-known situation where $f(\theta)$ is Gaussian and the likelihood function $l(\theta|x) \equiv f(x|\theta)$ is also Gaussian. In this case, any additional data cannot increase uncertainty (i.e., increase posterior variance) and, in general, the expected variance of the posterior distribution cannot exceed the prior variance because:

$$V[f(\theta)] = E[V[f(\theta|x)]] + V[E[f(\theta|x)]] \tag{7.11}$$

Unfortunately, this well-known relationship is true only in expectation and can be misleading when it is used to assess the value of specific information. In particular, it can be disconcerting for those who depend on this as an argument to justify additional data that, in general, *additional data does not guarantee a decrease in uncertainty* [17]. For example, a larger posterior variance results in the situation where both prior and likelihood functions are *t*-distributions. Situations can be constructed where the resulting posterior is actually bi-modal, further complicating the characterization of information value.

Clearly, research into how the value of information is measured is necessary. Some efforts have been made in utilizing information theory (e.g., entropy) as a basis for measuring value of additional data. For example, Shannon's measure of entropy can be used to measure the difference between the prior and posterior distributions (defined over common support Ω):

$$V = \int_\Omega f(\boldsymbol{\theta}|x) \ln\left(\frac{f(\boldsymbol{\theta}|x)}{f(\boldsymbol{\theta})}\right) dx \qquad (7.12)$$

When discussion turns to the value of information, it is generally assumed that the topic is related to experiments of a physical nature. However, as noted above, there is increased reliance on computer simulation to support characterization of uncertainty. A critical area for research is the identification of efficient computer simulations; what are the minimum scenarios that need to be explored to characterize uncertainty to the required degree of confidence? The need is even greater and the research gap even wider when time dependency is required in the analysis. Some significant inroads have been made, for example, in the area of importance sampling; however, much important research remains.

A great deal of research must be successfully accomplished before a useful tool is available to assist decision makers in assessing the value of information, be it from computer simulations or from laboratory experiments. With the increasing scarcity and escalating cost of information, research in this area will become critical and has the potential to position nondeterministic methods as one of the significant design and analysis tools during the life of a system.

7.6 Sensitivity Analysis

An area closely related to value of information is *sensitivity analysis*. There is a vast quantity of literature related to sensitivity analysis typically based on some form of variance decomposition. Because this is a topic discussed in other chapters, the discussion herein is very limited. It is noted however that a drawback of current sensitivity analysis methods is the limited consideration for *global* sensitivity. The strength of the variance decomposition methods common in the literature lies in the assumption that the uncertainty in the response of a system can be characterized through an at least approximate linear transformation of internal and external uncertainties. Unfortunately, the applicability of such linear relationships is becoming increasingly rare as systems being analyzed and associated models have become increasingly complex (and nonlinear) except under very localized conditions within the problem.

7.7 Summary

The previous discussion focused on those areas of research that I believe must be tackled before nondeterministic methods can enter mainstream use for engineering design and analysis. The emphasis has been on specific topics with a central theme: to have a long-term impact, research in nondeterministic methods must support a systems approach to design and analysis. Clearly, a discussion suggesting that particular areas of research deserve particular attention will be biased toward the perceptions and experience of the author. In defense of the above discussion, it should be noted that successful progress in any of the areas would have immediate value.

There have been a couple of notable omissions. In particular, an important area not discussed where increased attention would be of immediate value relates to *verification and validation*. This topic is also left in more knowledgeable hands for discussion in another chapter.

Another area that has received a great deal of recent focus in the literature relates to the classification of uncertainty into different categories: that is, probabilistic, possibilistic, epistemic, aleatory, etc. One important objective of these efforts is to provide engineers and decision makers with insight into the role that each of these variables plays in the analysis, with a goal of trying to better model and control these uncertainties. Clearly, the subject of this chapter has been constrained to probabilistic-type random variables but there is considerable promise in investigating possibilistic or fuzzy types of nondeterministic analysis. One benefit of the fuzzy approach is that the methods are relatively immune to the complexity of the system under scrutiny and, as such, hold promise for having a long-term impact. Unfortunately, they remain viewed by some as "fringe" technologies, much as probabilistic-based methods were a few decades ago.

Classification of probabilistic variables into epistemic and aleatory has also received a great deal of attention from the research community. However, similar to classifying variables as "strength" and "stress" variables, there does not seem to be much to be gained in return for the investment because the classification is somewhat arbitrary and can possibly change during the course of a single analysis iteration.

Finally, the growing volume of literature in the nondeterministic methods area guarantees that a contribution in any one of the research topics might have been overlooked and therefore my apologies in advance for my ignorance; the omission of a particular topic is certainly not intended to diminish any research currently in progress.

References

1. Lockheed Missiles and Space Company, Inc., SSD Systems Engineering Manual, Code 66, Sunnyvale, CA, 1994.
2. Jeffreys, H., *Theory of Probability*, 3rd ed., Oxford University Press, 1939.
3. Jaynes, E.T., Clearing up mysteries—the original goal, in *Maximum Entropy and Bayesian Methods in Applied Statistics*, J.H. Justice, Ed., Cambridge University Press, Cambridge, 1989.
4. Efron, B. and Morris, C., Limiting the risk of Bayes and empirical Bayes estimators. Part II: The empirical Bayes case, *J. Am. Stat. Assoc.*, 67, 130, 1972.
5. Deely, J.J. and Lindley, D.V., Bayes empirical Bayes, *J. Am. Stat. Assoc.*, 76, 833, 1981.
6. Casella, G., An introduction to empirical Bayes data analysis, *Am. Stat.*, 39, 753, 1985.
7. Robinson, D., A Hierarchical Bayes Approach to Systems Reliability Analysis, SAND 2001-3513, Sandia National Laboratories, Albuquerque, NM, 2001.
8. Wilson, S. and Taylor, D., Reliability assessment from fatigue micro-crack data, *IEEE Trans. Rel.*, 46, 165, 1997.
9. Celeux, G., Persoz, M., Wandji, J., and Perrot, F., Using Markov chain Monte Carlo methods to solve full Bayesian modeling of PWR vessel flaw distributions, *Rel. Eng. Sys. Safety*, 66, 43, 1999.
10. Gilks, W.R., Richardson, S., and Spiegelhalter, D.J., Eds., *Markov Chain Monte Carlo in Practice*, Chapman & Hall, New York, 1996.
11. Mastran, D., Incorporating component and system test data into the same assessment: A Bayesian approach, *Oper. Res.*, 24, 491, 1976.
12. Mastran, D.V. and Singpurwalla, N.D., A Bayesian estimation of the reliability of coherent structures, *Oper. Res.*, 26, 663, 1978.
13. Martz, H.F. and Waller, R.A., *Bayesian Reliability Analysis*, John Wiley & Sons, New York, 1982, Chap. 11.

14. Azaiez, M. and Bier, V., Perfect aggregation for a class of general reliability models with Bayesian updating, *Appl. Mathematics and Computation,* 73, 281, 1995.
15. Steimer, J.L., Mallet, A., Golmard, J., and Boisieux, J., Alternative approaches to estimation of population pharmacokinetic parameters: comparison with the nonlinear effect model, *Drug Metab. Rev.,* 15, 265, 1985.
17. Carlin, B., Hierarchical longitundal modeling, in *Markov Chain Monte Carlo in Practice,* Gilks, W.R., Richardson, S., and Spiegelhalter, D.J., Eds., Chapman & Hall, New York, 1996, chap. 17.
18. Hammit, J.K., Can more information increase uncertainty?, *Chance,* 8, 15, 1991.

II

Nondeterministic Modeling: Critical Issues and Recent Advances

8 **Types of Uncertainty in Design Decision Making** *Efstratios Nikolaidis* 8-1
Introduction • Taxonomies of Uncertainty and the Relation between These Taxonomies • Proposed Taxonomy • Chapters in This Book on Methods for Collecting Information and for Constructing Models of Uncertainty • Conclusion

9 **Generalized Information Theory for Engineering Modeling and Simulation**
Cliff Joslyn and Jane M. Booker 9-1
Introduction: Uncertainty-Based Information Theory in Modeling and Simulation • Classical Approaches to Information Theory • Generalized Information Theory • Conclusion and Summary

10 **Evidence Theory for Engineering Applications** *William L. Oberkampf and Jon C. Helton* 10-1
Introduction • Fundamentals of Evidence Theory • Example Problem • Research Topics in the Application of Evidence Theory

11 **Info-Gap Decision Theory for Engineering Design: Or Why "Good" Is Preferable to "Best"** *Yakov Ben-Haim* 11-1
Introduction and Overview • Design of a Cantilever with Uncertain Load • Maneuvering a Vibrating System with Uncertain Dynamics • System Identification • Hybrid Uncertainty: Info-Gap Supervision of a Probabilistic Decision • Why "Good" Is Preferable to "Best" • Conclusion: A Historical Perspective

12 **Interval Methods for Reliable Computing**
Rafi L. Muhanna and Robert L. Mullen 12-1
Introduction • Intervals and Uncertainty Modeling • Interval Methods for Predicting System Response Due to Uncertain Parameters • Interval Methods for Bounding Approximation and Rounding Errors • Future Developments of Interval Methods for Reliable Engineering Computations • Conclusions

13 Expert Knowledge in Reliability Characterization: A Rigorous Approach to Eliciting, Documenting, and Analyzing Expert Knowledge *Jane M. Booker and Laura A. McNamara* 13-1
Introduction • Definitions and Processes in Knowledge Acquisition • Model Population and Expert Judgment • Analyzing Expert Judgment • Conclusion

14 First- and Second-Order Reliability Methods *Armen Der Kiureghian* 14-1
Introduction • Transformation to Standard Normal Space • The First-Order Reliability Method • The Second-Order Reliability Method • Time-Variant Reliability Analysis • Finite Element Reliability Analysis

15 System Reliability *Palle Thoft-Christensen* 15-1
Introduction • Modeling of Structural Systems • Reliability Assessment of Series Systems • Reliability Asessment of Parallel Systems • Identification of Critical Failure Mechanisms • Illustrative Examples

16 Quantum Physics-Based Probability Models with Applications to Reliability Analysis *Erik Vanmarcke* 16-1
Introduction • Background: Geometric Mean and Related Statistics • Probability Distributions of the "Quantum Mass Ratio" and Its Logarithm • Logarithmic Variance and Other Statistics of the "Quantum Mass Ratio" • Probability Distribution of the "Quantum Size Ratio" • Extensions and Applications to Reliability Analysis • Conclusion

17 Probabilistic Analysis of Dynamic Systems *Efstratios Nikolaidis* 17-1
Introduction and Objectives • Probabilistic Analysis of a General Dynamic System • Evaluation of Stochastic Response and Failure Analysis: Linear Systems • Evaluation of the Stochastic Response of Nonlinear Systems • Reliability Assessment of Systems with Uncertain Strength • Conclusion

18 Time-Variant Reliability *Robert E. Melchers and Andre T. Beck* 18-1
Introduction • Loads as Processes: Upcrossings • Multiple Loads: Outcrossings • First Passage Probability • Estimation of Upcrossing Rates • Estimation of the Outcrossing Rate • Strength (or Barrier) Uncertainty • Time-Dependent Structural Reliability • Time-Variant Reliability Estimation Techniques • Load Combinations • Some Remaining Research Problems

19 Response Surfaces for Reliability Assessment *Christian Bucher and Michael Macke* 19-1
Introduction • Response Surface Models • Design of Experiments • Reliability Computation • Application to Reliability Problems • Recommendations

20 Stochastic Simulation Methods for Engineering Predictions *Dan M. Ghiocel* 20-1
Introduction • One-Dimensional Random Variables • Stochastic Vectors with Correlated Components • Stochastic Fields (or Processes) • Simulation in High-Dimensional Stochastic Spaces • Summary

21 Projection Schemes in Stochastic Finite Element Analysis *Prasanth B. Nair* 21-1
Introduction • Finite Element Formulations for Random Media • Polynomial Chaos Expansions • Polynomial Chaos Projection Schemes • The Stochastic Krylov Subspace • Stochastic Reduced Basis Projection Schemes • Post-Processing Techniques • Numerical Examples • Concluding Remarks and Future Directions

22 Bayesian Modeling and Updating *Costas Papadimitriou and Lambros S. Katafygiotis* 22-1
Introduction • Statistical Modeling and Updating • Model Class Selection • Application to Damage Detection • Uncertainty Reduction Using Optimal Sensor Location • Structural Reliability Predictions Based on Data • Conclusion

23 Utility Methods in Engineering Design *Michael J. Scott* 23-1
Introduction • An Engineering Design Example • Utility • Example • Conclusion

24 **Reliability-Based Optimization of Civil and Aerospace Structural Systems**
Dan M. Frangopol and Kurt Maute 24-1
Introduction • Problem Types • Basic Formulations • Sensitivity Analysis in Reliability-Based Optimization • Optimization Methods and Algorithms • Multicriteria and Life-Cycle Cost Reliability-Based Optimization • Examples in Civil Engineering • Examples in Aerospace Engineering • Examples in MEMS • Conclusions

25 **Accelerated Life Testing for Reliability Validation** *Dimitri B. Kececioglu* 25-1
What It Is, and How It Is Applied • Accelerated Reliability Testing Models • Recommendations • Log-Log Stress-Life Model • Overload-Stress Reliability Model • Combined-Stress Percent-Life Model • Deterioration-Monitoring Model • The Step-Stress Accelerated Testing Model

26 **The Role of Statistical Testing in NDA** *Eric P. Fox* 26-1
Introduction/Motivation • Statistical Distributions of Input Variables in Probabilistic Analysis • Output Variables in Probabilistic Analysis • Statistical Testing for Reliability Certification • Philosophical Issues • Conclusions

27 **Reliability Testing and Estimation Using Variance-Reduction Techniques**
David Mease, Vijayan N. Nair, and Agus Sudjianto 27-1
Introduction • Virtual Testing Using Computer Models • Physical Testing • A Review of Some Variance-Reduction Techniques • Use of Variance-Reduction Techniques with Physical Testing • Illustrative Application • Practical Issues on Implementation • Concluding Remarks

8
Types of Uncertainty in Design Decision Making

	8.1	Introduction 8-1
		The Role of Uncertainty in Design Decision Making • Why It Is Important to Study Theories of Uncertainty and Types of Uncertainty • Definition of Uncertainty • Objectives and Outline
	8.2	Taxonomies of Uncertainty and the Relation between These Taxonomies 8-6
		Taxonomies According to Causes of Uncertainty • Taxonomies According to the Nature of Uncertainty
	8.3	Proposed Taxonomy... 8-12
		Types of Uncertainty • Types of Information • Type of Information for Various Types of Uncertainty
Efstratios Nikolaidis	8.4	Chapters in This Book on Methods for Collecting Information and for Constructing Models of Uncertainty.. 8-18
The University of Toledo	8.5	Conclusion 8-19

8.1 Introduction

Uncertainty and types of uncertainty are studied in the context of decision making in this chapter. Concepts from decision theory are used to define a measure of information and the notion of uncertainty. The chapter "The Need for Nondeterministic Approaches in Automotive Design: A Business Perspective," in this book, presents a similar approach to study uncertainty in automotive design.

8.1.1 The Role of Uncertainty in Design Decision Making

Decision is an irrevocable allocation of resources to achieve a desired outcome. Three elements of a decision are alternative courses of action (or choices), outcomes of the actions, and payoffs (Figure 8.1). First, the decision maker defines the decision problem; he or she defines the objectives, the criteria, and the alternative actions. The decision maker tries to predict the outcome of an action using a model of reality. Each outcome is characterized by its attributes. The payoff of an outcome to the decision maker depends on these attributes and the decision maker's preferences. The decision maker wants to select the action with the most desirable outcome.

Decisions involve uncertainty. When framing a decision (defining the context of the decision), the decision maker may not know if the frame captures the decision situation at hand. Also, the decision maker may not know all possible courses of action. Often, a decision maker is uncertain about the

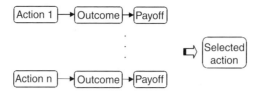

FIGURE 8.1 Decision: Selecting among alternative actions the one with the highest payoff.

outcome of an action, because of a lack of knowledge or inherent randomness. Thus, to the best of the decision maker's knowledge, an action could result in a single outcome, a discrete set of different outcomes, or a continuous range of outcomes (Figure 8.2). Even more challenging are decisions for which the decision maker does not know all possible outcomes of an action. Finally, the decision maker faces uncertainty in assessing the value of the outcomes in a way that is consistent with his or her preferences.

Design of a system or a component, such as a car, is a decision because it involves selection of a design configuration, which is an allocation of resources [1]. The designer is to select the design configuration that will result in the most desirable outcome (e.g., maximize the profit for his or her company).

Example 1: Uncertainty in Design Decisions in Automotive Engineering

The design of an automotive component involves decisions under uncertainty. Figure 8.3a shows two decisions in the design of an automotive component (Figure 8.3b). The objectives of a designer are to (1) find a design with good performance, and (2) minimize the time it takes to find this design. First, a designer determines alternative designs. For simplicity, Figure 8.3a shows only two designs. The designer uses computer-aided engineering (CAE) analysis to predict the performance attributes of these designs (e.g., the stresses or vibration amplitudes at critical locations). The designer must decide whether to select one design or reject both if they fail to meet some minimum acceptable performance targets. In the latter case, he or she must find a new design. In this decision, the designer does not know the true values of the performance attributes of the designs because of:

1. Product-to-product variability
2. Modeling uncertainty and numerical errors in CAE analysis
3. Inability to predict the real-life operating conditions

Because of these uncertainties, the designer builds and tests a prototype and uses the results to decide whether to sign off the design for production or reject it. The test eliminates modeling uncertainty and numerical errors but it introduces measurement errors. Therefore, even after the designer obtains the results of tests, he or she remains uncertain about the true value of the attributes of the design when it will be massively produced. Moreover, when the designer tries to decide whether to sign off or reject the design, he or she is uncertain about the cost and time it will take to find a better, acceptable design.

Decision makers use models of real systems to predict the outcomes of actions. A model is a mathematical expression, $\hat{Y}(X)$, relating one or more quantities of interest, Y (e.g., the stiffness of a structure) to a set of measurable variables, X. We use a model to predict the outcome of a design decision

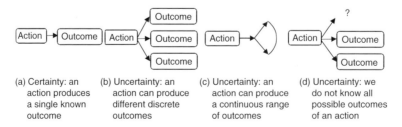

FIGURE 8.2 Decision under certainty (a) and decision under uncertainty (b-d).

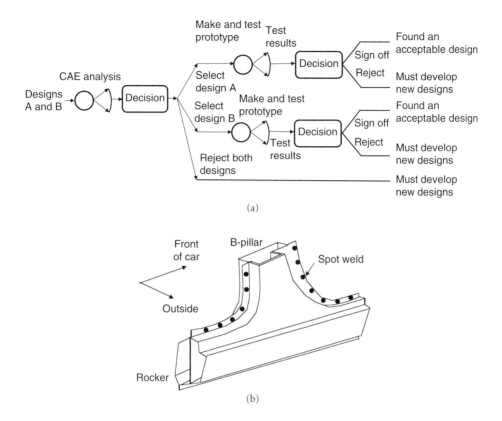

FIGURE 8.3 Uncertainty in design of an automotive component. (a) Two decisions in design of an automotive component. (b) Automotive component in example 1. *Note*: The decision maker is uncertain of the outcome of the test of the prototype (the measured values of the attributes) before the test. The test can produce a continuous range of values of the attributes.

by predicting the attributes of a selected design (e.g., the stiffness). Figure 8.4 shows the procedure for developing an analytical tool for predicting the outcome of an action. First, a conceptual model of a process or a system is built. This is an abstraction of the system. Then a mathematical model, which is a symbolic model consisting of equations, is constructed based on the analytical model. The equations are solved using a numerical technique to estimate the performance. The figure shows various errors in the analytical prediction of the attributes. These are described in Section 8.2.1.

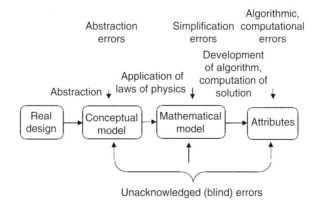

FIGURE 8.4 Analysis of a design to predict its performance.

There is always error in the predictions of analytical tools. Errors result from:

- Use of imperfect model forms (e.g., a linear expression is used, whereas the actual relation between the quantity we want to predict and the variables is nonlinear)
- Variables that affect the outcome of an action but are omitted in the model

The above deficiencies in building models can be due to a poor understanding of the physics of a problem or to intentional simplifications.

The set of variables that affect the outcome of an action can be divided into two subsets: those that are included in a model, $\mathbf{X}$, and those that are missing, $\mathbf{X}'$. Let $\mathbf{Y}(\mathbf{X}, \mathbf{X}')$ be the true relation between the attributes describing the outcome and the uncertain variables. This relation is rarely known; instead, an approximate relation $\hat{\mathbf{Y}}(\mathbf{X})$ is available from an analytical solution and is used as a surrogate of $\mathbf{Y}(\mathbf{X}, \mathbf{X}')$ for making decisions.

Example 2: Developing a Model for Predicting the Response of an Automotive Component

In Example 1, the conceptual model for predicting the performance of a design includes a geometric model, such as a CAD model. The mathematical model is a set of partial differential equations solved by finite element analysis.

Many automotive components consist of thin metal shells fastened by spot welds (Figure 8.3b). An approximate model of a component may replace spot welds with a rigid continuous connection of the flanges of the shells. Then, vector $\mathbf{X}'$ in the previous paragraph includes variables such as the spot weld pitch and diameter, which are missing from the model. Model approximations induce errors, $\varepsilon = \hat{\mathbf{Y}}(\mathbf{X}) - \mathbf{Y}(\mathbf{X}, \mathbf{X}')$, in the response of the component to given loads.

A designer makes decisions when developing a predictive model and solving the model to predict the outcome of an action. The designer has to choose those characteristics of the system or process that should be included in the model, and also choose how to model these characteristics so as to construct a model that is both accurate and affordable. The designer must also choose a numerical algorithm and the parameters of the algorithm (e.g., the step size) so as to compute accurately the attributes of a design at a reasonable cost. Decisions are also involved in solving the model using a numerical algorithm. The payoff of these decisions is a function of (1) the cost of developing and using a predictive model and the solution algorithm, and (2) the information from solving the model to predict the performance. The information could be measured in a simplified manner by the error of the model. A more sophisticated measure of the information a model provides can be the probability that the designer will obtain the most desirable achievable outcome using the model predictions to determine the preferred design [1, p. 341].

8.1.2 Why It Is Important to Study Theories of Uncertainty and Types of Uncertainty

When facing uncertainty, the decision maker cannot select the action that guarantees the most desirable outcome but he or she can select the action with maximum probability to achieve this outcome. To be successful, a decision maker should manage uncertainty effectively, which requires tools to quantify uncertainty. Theories of uncertainty — including probability theory, evidence theory, and generalized information theory — are available for quantifying uncertainty, but there is no consensus as to which theory is suitable in different situations. An important objective of this book is to help readers understand uncertainty and its causes, and the tools that are available for managing uncertainty.

It is important for making good decisions to know all important types of uncertainty and understand their characteristics to construct good models of uncertainty. Decision makers face different types of uncertainty, including uncertainty due to inherent randomness, lack of knowledge, and human intervention. We will see in this chapter that different types of information (e.g., numerical, interval valued or linguistic) are suitable for different types of uncertainty. Understanding different types of uncertainties

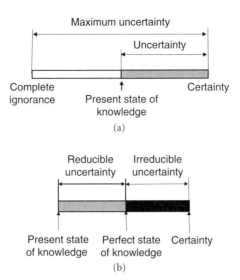

FIGURE 8.5 Illustration of the definition of uncertainty. (a) Uncertainty is the gap between certainty and the present state of knowledge. Information reduces uncertainty. (b) Reducible and irreducible uncertainty.

helps designers allocate resources effectively to collect the most suitable type of information, select the most suitable tools to model uncertainty, and interpret correctly the results of their calculations.

8.1.3 Definition of Uncertainty

We can define uncertainty indirectly based on the definition of certainty. Certainty, in the context of decision theory, is the condition in which a decision maker knows everything needed in order to select the action with the most desirable outcome. This means that, under certainty, the decision maker knows every available action possible and the resulting outcome. Uncertainty is the gap of what the decision maker presently knows and certainty (Figure 8.5a).

When there is complete ignorance, uncertainty is maximum. In this state, the decision maker does not know anything about the likelihood of the outcomes of the alternative actions. Information reduces uncertainty. It helps the decision maker select an action so as to increase the chance of getting the most desirable outcome. When information becomes available to the decision-maker (e.g., sampling information from which the decision maker can estimate the probabilities of alternative actions), uncertainty is reduced. The new state is marked "present state of knowledge" in Figure 8.5a.

Example 3: Uncertainty in Design of the Automotive Component in Example 1

An inexperienced designer knows nothing about the two designs in Figure 8.3. The designer wants to select the best design that meets the targets. But he or she cannot do so because he or she does not know the attributes of the designs. When information from CAE analysis becomes available, some uncertainty is eliminated, which increases the probability that the designer will select the action that will yield the most preferred outcome.

Hazelrigg [1, pp. 340 to 344] defined measures of information and uncertainty by using concepts of decision theory. The amount of information and uncertainty depends on the decision at hand. A measure of the amount of information is the probability that a decision maker's preferred action will lead to the most desired achievable outcome. The preferred action is the one with maximum expected utility, where the utility is determined using the presently known probabilities of the outcomes of the alternative actions. Then Hazelrigg defined the following measure of uncertainty: informational entropy of the decision is the negative natural logarithm of the probability that the preferred choice will *not* result in the most desirable outcome. He showed that this measure of uncertainty has the same form as Shannon's entropy.

Hazelrigg's two measures require that the decision maker know all the possible outcomes of each action and can estimate their probabilities.

8.1.4 Objectives and Outline

The objectives of this chapter are to help the reader understand the concept of uncertainty and its significance in decision making, present different categorizations of uncertainty and the types of information for each type of uncertainty. This helps the reader understand which theories of uncertainty and tools for quantifying uncertainty have been designed for each type of uncertainty. These theories and tools are presented in chapters 9 to 12.

This chapter focuses on uncertainties that a designer faces. These uncertainties include uncertainties in simulation to predict the attributes of designs, including their reliability. We appreciated the importance of uncertainties involved in business decisions, such as uncertainties in predicting the demand for a product and the effect of price and marketing on demand. However, these uncertainties are beyond the scope of this chapter.

The chapter starts with a review of various taxonomies of uncertainty. It illustrates these taxonomies on the example of the design of the automotive component in Example 1. The relation between various taxonomies is studied. Then guidelines for what type of information is suitable for each type of uncertainty are presented.

8.2 Taxonomies of Uncertainty and the Relation between These Taxonomies

Uncertainty can be categorized either based on its causes or its nature. There are two main ways for constructing taxonomies of uncertainty. The first primarily considers three sources of uncertainty: inherent randomness, lack of knowledge, and human error. The second way considers the nature of uncertainty.

8.2.1 Taxonomies According to Causes of Uncertainty

Uncertainty is often categorized into aleatory (random) and epistemic. Aleatory uncertainty is due to variability, which is an intrinsic property of natural phenomena or processes. We cannot reduce variability unless we change the phenomenon or the process itself. For example, there is variability in the thicknesses of metal plates produced in an automotive supplier's factory. We cannot reduce variability by collecting data on plate thicknesses. We can only reduce variability by changing the manufacturing process, for example by using better equipment or more skillful operators.

Epistemic uncertainty is a potential deficiency in selecting the best action in a decision (the action with the highest probability of resulting in the most desirable outcome) due to lack of knowledge. Epistemic uncertainty may or may not result in a deficiency; hence the qualifier "potential" is used in the definition. Epistemic uncertainty is reducible; we can reduce it by collecting data or acquiring knowledge. For example, we have reducible uncertainty in predicting the stress in a structure due to deficiencies in the predictive models. Improving our predictive models can reduce this uncertainty. Reducible uncertainty has a bias component. For example, crude finite element models of a structure tend to underestimate deflections and strains.

Example 4: Aleatory and Epistemic Uncertainty in the Design of an Automotive Component

Suppose that the designer of the automotive component in Examples 1 through 3 does not know the probability distribution of the thicknesses of the plates from the supplier's factory. In this example, assume that the only uncertainty is in the thicknesses of the plates. The designer faces both reducible uncertainty (e.g., does not know the probability distribution of the thickness) and irreducible uncertainty (e.g., even if the designer knew the probability distribution of the thickness, the designer could not know the thickness of a particular plate from the factory). This state is marked as "present state of knowledge" in Figure 8.5b.

The designer starts measuring the thicknesses of plates and estimates the probability distribution of the thickness. As the designer collects more data, the estimated probability distribution of the thickness approaches the true one and the epistemic uncertainty decreases. Eventually, the designer determines the true probability distribution. In this state, which is marked "perfect state of knowledge" in Figure 8.5b, there is no reducible uncertainty because the designer knows all that could be known. However, there is still irreducible (aleatory) uncertainty because the designer cannot know the thickness of a new plate from the factory. The designer can only know that the new plate comes from a known probability distribution.

The best that a decision maker can do in a decision involving only aleatory uncertainty is to determine the true probabilities of the outcomes of the actions. But consider a decision in which there is only epistemic uncertainty. If the decision maker eliminates epistemic uncertainty by collecting information that allows him or her to understand the physics of the problem and construct accurate predictive models, then the decision maker will know the outcomes of the actions.

Several researchers have refined the above taxonomy of uncertainty into aleatory and epistemic types by considering different species within each type and considering other types of uncertainty. Der Kiuregian [2] considered the following four causes of uncertainty in structural analysis and design: (1) inherent variability, (2) estimation error, (3) model imperfection, and (4) human error, as shown in Figure 8.6a. Estimation error is due to incompleteness of sampling information and our inability to estimate accurately the parameters of probabilistic models that describe inherent variability. Model imperfection is due to lack of understanding of physical phenomena (ignorance) and the use of simplified structural models and probabilistic models (errors of simplification). Imperfections in probabilistic models mean errors in the choice of a parameterized probability distribution. Human errors occur in the process of designing, modeling, constructing and operating a system. We can reduce human error by collecting additional data or information, better modeling and estimation, and improved inspection. Cause (1) is irreducible, whereas causes (2) and (3) are reducible. Human error is also reducible. However, human errors tend to occur randomly.

Gardoni et al. [3] developed a methodology for constructing probabilistic models of the capacity of structures, which accounts for both variability and reducible uncertainty. These models consist of three terms: (1) a deterministic model for predicting the capacity of a structure, (2) a correction term for the bias inherent in the model, which is a function of a set of unknown parameters, and (3) a random variable for the scatter in the model. The user of Gardoni's model has to estimate the parameters of the term for the bias and the standard deviation of the term for the scatter using Bayesian methods and using experimental measurements.

Haukass [4] considered inherent variability and measurement error as two components of aleatory uncertainty (Figure 8.6b). He described two types of epistemic uncertainty: modeling and statistical uncertainty. Modeling uncertainty is due to imperfect idealized representations of reality. In Haukass' study, "statistical uncertainty" refers to the uncertainty caused by the inability to predict the parameters of a probability distribution from a sample (for example, to predict the mean value and the standard deviation of the thickness of plates from the factory from a sample of plates). Note that the term "statistical uncertainty" has been also used for aleatory uncertainty by some researchers. Another type of uncertainty is human error.

Haukaas said that the division of aleatory and epistemic uncertainty is not disjunctive; one component of modeling uncertainty is due to missing some random variables in a model and this component is also aleatory uncertainty (Figure 8.6b). Suppose that in the analysis of the automotive component, the designer neglects the spot welds by assuming that the flanges of the shells are rigidly connected. Then the model that he or she uses neglects the variability in the spot weld pitch and diameter as well as the variability due to failed spot welds.

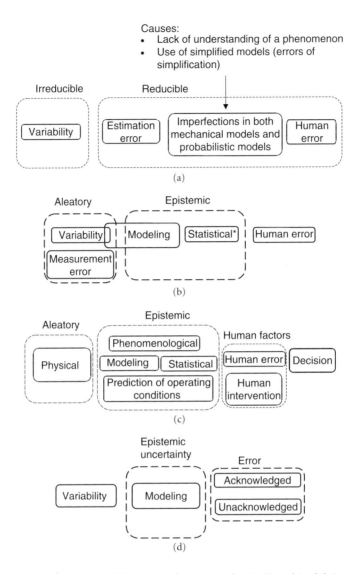

FIGURE 8.6 Taxonomies of uncertainty. (a) Taxonomy of uncertainty by Der Kiureghian [2]. Boxes marked by dashed lines indicate broad categories of uncertainty; solid lines indicate special types within the broad categories. (b) Taxonomy of uncertainty by Haukaas [4]. Modeling uncertainty and variability overlap because modeling uncertainty due to missing random variables in a deterministic model causes variability. *Note:* The term "statistical uncertainty" refers to the uncertainty caused by the inability to predict the parameters of a probability distribution from a sample — it does not refer to variability. (c) Taxonomy of uncertainty by Melchers [5]. (d) Taxonomy by Oberkampf et al. [6]. *Note:* Modeling uncertainty is caused by Vagueness, Nonspecificity and Conflict in the available information.

Melchers [5] considered physical uncertainty as a component of aleatory uncertainty (Figure 8.6c). In addition to physical uncertainty, he described six types of uncertainty:

1. Phenomenological uncertainty (caused by difficulties in understanding and modeling the physics of a novel system)
2. Modeling uncertainty
3. Uncertainty in prediction of the operating conditions, loads, and deterioration of a system during its life
4. Statistical uncertainty

5. Uncertainty due to human factors
6. Decision uncertainty

Melchers divided causes of human error into natural variation in task performance and gross errors. Gross errors occur in tasks within accepted procedures or they are the direct result of ignorance or oversight of the fundamental behavior of systems.

Decision uncertainty is uncertainty in deciding the state of a system, (e.g., deciding if a design whose maximum deflection is given is acceptable).

Oberkampf et al. [6, 7] focused on causes of uncertainty in simulation of physical and chemical processes. He considered (1) variability, (2) epistemic uncertainty, and (3) error (Figure 8.6d). Oberkampf et al. defined epistemic uncertainty as a potential deficiency in any phase of modeling that is due to lack of knowledge. Epistemic uncertainty is caused by incomplete information, which is due to Vagueness, Nonspecificity, and Dissonance (Conflict). These terms will be defined in Section 8.2.2. Epistemic uncertainty was divided into parametric (the analyst does not know the values of the parameters of a system), modeling, and scenario abstraction.

According to Oberkampf et al., error is a recognizable deficiency in modeling that is not due to lack of knowledge. If error is pointed out, it can be corrected or allowed to remain. Error was divided into acknowledged (intentional) and unacknowledged. Discretization errors or round-off errors are examples of acknowledged (intentional) errors. Programming errors, blunders, and mistakes are examples of unacknowledged error.

Oberkampf's definition of error is broader than the definition of human error in other studies presented here. Error includes simplifications of mathematical models and approximations in numerical algorithms. Oberkampf et al. stated that error is not uncertainty because it is not due to lack of knowledge.

Both variability and error cause uncertainty in predicting the outcomes of actions. For example, a designer may be unable to predict if a particular car from an assembly line meets crash requirements because of human errors in the assembly and variability in material properties. The distinction of error and of the effect of epistemic uncertainty is blurred. According to Oberkampf et al., error is a departure from the correct or acceptable modeling procedure. But in many cases there is no consensus of what is an acceptable or correct modeling procedure.

Nowak and Collins [8, pp. 289 to 313] divided causes of uncertainty in building structures into natural and human. Human causes can arise from within the building process and from outside. Examples of the first are errors in calculating loads and load effects and errors in construction, while examples of the second are fires and collisions. Human causes of uncertainty from within the building process occur if one deviates from the acceptable practice, or even if one follows the acceptable practice.

Human errors are defined as deviations from acceptable practice. They can be divided into three types:

1. Conceptual errors, which are departures from the acceptable practice due to lack of knowledge of a particular professional
2. Errors of execution, which are due to carelessness, fatigue, or lack of motivation
3. Errors of intention, (e.g., simplifications in numerical algorithms to reduce computational cost)

Example 5: Different Types of Uncertainty in Automotive Component Design

The following list provides examples illustrating the types of uncertainty presented above.

- *Aleatory uncertainty:* variability in the thicknesses of plates, material properties, spot weld pitch, and diameters of spot welds.
- *Epistemic uncertainty:* uncertainty due to deficiencies in the analytical and mathematical model of a structural component caused by lack of knowledge. Neglecting nonlinearities is an example of a cause of such deficiencies.
- *Blind error:* errors in typing the input data file for finite element analysis, errors in reading stresses and deflections from the output file.
- *Acknowledged error:* a designer assumes that the probability density function of the thickness of a plate is normal without evidence supporting this assumption. The reason is that the designer is under pressure to produce results quickly.

- *Statistical uncertainty:* uncertainty in the mean value and standard deviation of a plate thickness caused by estimating these statistics from a finite sample of measurements. Neglecting the spatial variation in the thickness of a plate is another example of a cause of statistical uncertainty.
- *Measurement error:* errors in measuring the dimensions of plates or spot weld diameters. Errors in measuring the response of a prototype structure to loads.
- *Uncertainty in predicting operating conditions:* uncertainty in predicting the loads on a component during its life.
- *Phenomenological uncertainty:* uncertainty due to the inability to model a new design of the thin wall structure in Figure 8.3b, in which plates are joined by adhesives instead of spot welds.
- *Decision uncertainty:* the component in Examples 1 through 4 was tested under a repeated cyclic loading and a spot weld failed. The designer is uncertain if he or she should consider that the component failed.

8.2.1.1 Importance of Uncertainty Due to Human Errors

Results of surveys show that human error is responsible for the majority of catastrophic failures ([8], p. 292). For example, according to the above reference, 70% of accidents in aviation have been attributed to crew error, and similar results are thought to apply to other industries. Incidents such as Three Mile Island and Chernobyl indicate the importance of human error in the safety of nuclear plants. One cannot obtain an estimate of the failure probability of a system that is indicative of the true failure probability of this system unless one considers uncertainty due to human error. Moreover, if one neglects uncertainty due to human error, one cannot even obtain an estimate of the probability of failure that can be used to compare the safety of designs with different construction types.

8.2.2 Taxonomies According to the Nature of Uncertainty

Studies presented in this subsection consider uncertainty as knowledge deficiency caused by incompleteness in the acquired knowledge. They classify uncertainty according to its cause and the available information.

Zimmermann [9] considered the following causes of uncertainty: lack of information, complexity (which forces one to simplify models of reality), conflicting evidence, ambiguity, measurement error, and subjectivity in forming beliefs. According to Zimmermann, the available information can be numerical, interval-based, linguistic, and symbolic. Numerical information can be nominal (defines objects), ordinal (defines order of objects), and cardinal (in addition to order, it defines differences between ordered objects). An example of nominal numerical information is the plate number of a car, which only defines the particular car. Ordinal information is the ranking of the students in a class based on their performance. This information defines the order of the students in terms of how well they did in the class but does not indicate differences in their performance. Cardinal information is the overall percentage grades of the students in the class — it defines the difference in the performance of the students in addition to their order.

Ayyub [10] presented a comprehensive study of ignorance (lack of knowledge). He divided ignorance into conscious (one is aware of his or her ignorance) and blind ignorance (one does not know that he or she do not know). He considered uncertainty (knowledge incompleteness due deficiencies in acquired knowledge) as a special case of conscious ignorance. He classified uncertainty based on its sources into ambiguity, approximations, and likelihood. Ayyub did not consider conflict as a component of uncertainty. He also explained which theory can handle each type of ignorance.

Klir and Yuan [11] and Klir and Weirman [12] categorized uncertainty on the basis of the type of deficiency in the available information, into vagueness (fuzziness), and ambiguity (Figure 8.7). Ambiguity is a one-to-many relationship. Klir further divided Ambiguity into Conflict and Nonspecificity. We can have Conflict because the same action results in different outcomes (repeated tosses of a die result in different numbers from one to six) or because there is conflicting evidence (three experts provide different estimates of the error in the estimate of the stress in a structure). Nonspecificity occurs when multiple

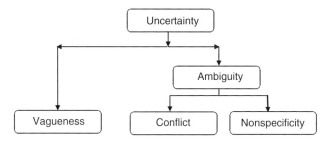

FIGURE 8.7 Taxonomy of uncertainty by Klir and Yuan [11].

outcomes of an action are left unspecified. Fuzziness is the inability to define the boundaries of sets of objects sharing a common property.

Measurements from observations of repeated experiments are often precise but are in conflict. Subjective information from experts is often nonspecific. For example, suppose that we ask an expert for a subjective estimate of the difference between the true value of the stress in a structural component and the predicted value of the stress from a finite element model. It is more likely for the expert to estimate a range for the difference rather than a precise value. The less confident the expert feels about the estimate of the difference, the broader is expected to be the range he or she will provide.

Klir and Weirman have proposed axioms that measures of Conflict and Nonspecificity should satisfy. They also proposed measures for these types as well as a measure of the total uncertainty. These include equations for computing these measures. This is an important contribution because if one disputes the validity of Klir's definitions and measures, then one has to explain which axiom is wrong.

Example 6: Nonspecificity and Vagueness in Error in the Estimate of the Stress in an Automotive Component

The designer estimated the stress in the automotive component described in Examples 1 through 5 using finite element analysis, but does not know the error in the estimate (e.g., the difference between the true value and the estimate of the stress). Figure 8.8a shows a condition called maximum uncertainty, in which the designer does not know anything about the error. The designer asks an expert for an estimate of the error. Often, an expert provides an interval containing the error (Figure 8.8b). This is nonspecific evidence because the exact value of the error is not specified. However, there is no conflict. Finally, the designer conducts an experiment in which he or she measures the stress accurately and determines the exact value of the error for this single experimental validation. Now there is no uncertainty.

Often, an expert will specify the error in the stress in plain English. For example he or she could say that "the error is small." To explain what "small error" means, the expert might specify a few values of the error and the degree to which he considers them small, or construct a plot showing the degree of which the error is small as a function of the value of the error (Figure 8.8c). This is an example of vague or fuzzy evidence. Contrast this case with the case of crisp evidence (the expert provides a sharp boundary between small and not small errors). This latter case is rare because it is difficult to explain why an error of 9.99% is small while and error of 10.01% is not small. Fuzziness in the boundaries of a set is one of the causes of decision uncertainty (see classification by Melchers).

Example 7: Nonspecificity and Conflict in the Thickness of a Plate Selected Randomly from a Batch

A batch of plates is received from the supplier factory. Suppose the designer only knows the range of the thickness based on the factory specifications. Then he or she faces both Nonspecificity and Conflict (Figure 8.9a). There is Nonspecificity because the evidence consists of a range — not a specific value. There is also Conflict because thickness varies from plate to plate in the batch. The designer measures the thicknesses of all the plates in the batch. Now Nonspecificity has been eliminated but Conflict remains for the population that was inspected (Figure 8.9b).

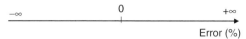

(a) Maximum uncertainty: The error in the stress (e.g., the true value of the stress minus the predicted value) can assume any value. There is only Nonspecificity.

(a)

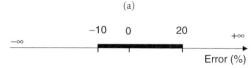

(b) The error is in the range [−10%, 20%]. Nonspecificity is smaller than in case (a).

(b)

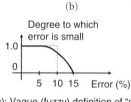

(c): Vague (fuzzy) definition of "small error" (solid curve) vs. crisp definition (dashed line).

(c)

FIGURE 8.8 Illustration of Nonspecificity (a, b), and Vagueness (c).

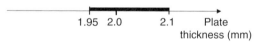

(a) Thicknesses of plates in a batch are known to be in the range [1.95 mm, 2.1 mm]. There is both Nonspecificity and Conflict.

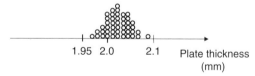

(b) Dot diagram of the measured plate thickness in the batch. There is only Conflict.

FIGURE 8.9 Nonspecificity and Conflict in the thickness of a plate. Initially there is both Nonspecificity and Conflict (a). After measuring all plate thicknesses there is only Conflict (b).

8.3 Proposed Taxonomy

8.3.1 Types of Uncertainty

Uncertainty in three stages of a decision — namely, framing a decision, predicting the outcomes of actions, and evaluating the payoff of outcomes — will be classified according to its causes. For each type of uncertainty, certain types of information are typically available. These are presented in Section 8.3.2.

Types of Uncertainty in Design Decision Making

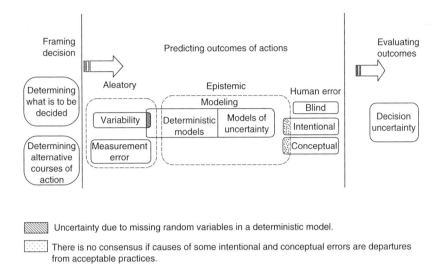

Uncertainty due to missing random variables in a deterministic model.

There is no consensus if causes of some intentional and conceptual errors are departures from acceptable practices.

FIGURE 8.10 Proposed taxonomy of uncertainty based on its causes.

8.3.1.1 Framing a Decision

When framing a decision, the decision maker is uncertain if he or she has correctly determined what is to be decided and if all relevant features of the problem are known. For example, although the performance of the automotive component in Figure 8.3b depends on the design of the adjacent beams, a designer neglects this dependence and designs the automotive component in isolation. Moreover, the decision maker is uncertain if he or she has determined all alternative courses of action available. For example the designer of the automotive component may not know all alternative design configurations, which may result in missing the optimum action (the action with highest probability of resulting in the most desirable outcome). This uncertainty is epistemic and is due to lack of creativity and knowledge.

8.3.1.2 Predicting the Outcome of an Action

Here we have the three basic types of uncertainty presented in Section 8.2.1: aleatory, epistemic, and uncertainty due to human error (Figure 8.10). Aleatory uncertainty is caused by variability and measurement error.

We adopt Oberkampf's [6, 7] definition of epistemic uncertainty; epistemic uncertainty is a potential deficiency in any phase of the modeling process that is due to lack of knowledge. There is epistemic uncertainty in both deterministic models for predicting the attributes of a design and in models of uncertainty. As mentioned in the introduction, uncertainty in deterministic models is due to use of incorrect model forms and missing variables. If a probabilistic model of uncertainty is used, uncertainty in models of uncertainty is caused by selecting the wrong type of probability distribution for a random variable or vector and the inability to estimate accurately the parameters of the distribution from a finite sample (Figure 8.10).

The boxes representing variability and epistemic uncertainty in deterministic models partially overlap, indicating that the division of aleatory and epistemic uncertainty is not disjunctive. Specifically, one component of modeling uncertainty is due to missing some random variables in a model and this component is also aleatory uncertainty [4].

Example 8: Uncertainty in Models of Uncertainty in the Design of an Automotive Component

A designer tries to model the probability distribution of the number of defective welds in a component from the numbers of defective welds in a sample of nominally identical components. The designer faces statistical uncertainty. But the sample could be small or nonexistent so that a designer would rely on first principles-based models, such as a model of the spot welding process, to derive a probability distribution of the number of defective welds in a component. In this case, the designer also faces modeling uncertainty because he or she uses a physics-based model of the spot welding process, which always involves simplifications.

Human error is a departure from accepted procedures. According to Nowak and Collins [8], human error can be divided into execution, intentional, and conceptual errors. The boundary between intentional and conceptual errors and epistemic uncertainty is fuzzy, as mentioned earlier. Sometimes, there is no consensus or dominant view of what acceptable practice is. What one professional considers departure from acceptable practice another professional considers acceptable. For example, many proponents of probability theory consider the use of nonprobabilistic models of uncertainty (e.g., possibility theory or evidence theory) as departure from acceptable practice of modeling uncertainty. Moreover, the definition of acceptable practices changes with time.

8.3.1.3 Evaluating the Payoff of the Outcomes of Actions

Decision makers face uncertainty in assessing the payoffs of outcomes of actions in a way that they are consistent with their preferences. One of the reasons is that decision makers often miss important criteria for determining the payoff of outcomes. Melchers [5] classified this uncertainty as decision uncertainty.

8.3.2 Types of Information

Following are types of information available for quantifying uncertainty:

- *Experimental observations.* These are observed values of the uncertain variables. If the sample size is sufficiently large,[1] then we can construct a reasonable probability distribution using the sample. However, if the sample is small, one must use additional information, such as expert opinion, to model uncertainty.

  When it is impractical to collect data, we could use surrogate data on uncertainties that are similar to the uncertainties we want to model. For example, the probability distribution of the thicknesses of plates from a particular factory can be estimated from measurements on plates from another factory.

  It is impractical to obtain observations for extremely rare or one-time events, such as a catastrophic nuclear accident or a nuclear war. The likelihood of such events can be estimated on the basis of judgment. An alternative approach is to identify all possible failure scenarios that can result in the rare catastrophic event and to estimate the probabilities of these scenarios.

  Observations are not obtainable for future events. Suppose that we want to predict the price of gas in 2020. We cannot rely only on prices from the past to model prices in the future. The reason is that the price of gas as a function of time is a nonstationary, random process.[2] Therefore, in addition to data from the past, we should consider subjective information to construct models of the uncertainty in these events.

- *Intervals.* This information is usually obtained from experts. An expert could provide one interval believed to contain the true value of an uncertain variable or nested confidence intervals containing the true value with certain probabilities. Figure 8.8.b shows an interval that contains the true value of the error in the stress computed from an analytical model.

- *Linguistic information.* This information is obtained from experts. For example, an expert could quantify the error in the stress by saying that "the error is small" (Figure 8.8c).

---

[1] The required size of a sample depends on the problem. Usually, we can estimate the main body of a probability distribution using 30 experimental observations; but if we want to estimate probabilities of rare events, then we need more experimental observations.

[2] The statistical properties of a nonstationary random process change with shifts in the time origin.

- *First principles-based models of uncertainty.* The type of the probability distribution function of a variable can often be derived from first principles of mechanics or chemistry, and probability. Following are examples of standard probability distributions that were derived from first principles:
  - The Weibull distribution. This distribution models the minimum value of a sample of values drawn from a probability distribution bounded from below. The Weibull distribution is often used for the strength of structures or materials. If the load capacity of a system is the minimum of the capacities of the components and the capacities are statistically independent, then the probability distribution of the load capacity of the system converges to the Weibull distribution as the number of components tends to infinity.
  - An extreme value distribution. There are three distributions, called Type I, Type II, and Type III extreme value distributions. These are used to model the maximum and minimum values of a sample from a population. Type I and Type II extreme value distributions are suitable for modeling loads caused by extreme environmental events, such as storms and earthquakes.
  - The Raleigh distribution. The peaks of a narrowband[3] random process, such the elevation of ocean waves, follow the Raleigh distribution, assuming that the peaks are statistically independent. Thus, a designer may be able to determine the type of the probability distribution governing an uncertain variable without collecting data.
- *Models based on principles of uncertainty.* When little information is available, one can select a model of uncertainty based on principles of uncertainty. These include the principles of maximum uncertainty, minimum uncertainty and uncertainty invariance ([11], pp. 269 to 277).

  The maximum uncertainty principle states that a decision maker should use all the information available but not unwittingly add additional information. Therefore, one should select the model with maximum uncertainty consistent with the available evidence.

Example 9: Modeling Uncertainty in the Stress in a Component When We Only Know the Stress Range

A designer knows that the stress in a particular component, resulting from a given load, is between 10 MPa and 50 MPa. He or she should select the model with maximum uncertainty whose support is the above stress range. In probability theory, uncertainty is measured using Shannon's entropy. According to the maximum uncertainty principle, the designer should use the uniform probability distribution for the stress because this distribution has the largest entropy among all distributions supported by the interval [10 MPa, 50 MPa].

This example is only intended to illustrate the maximum entropy principle. One can question the use of a uniform distribution in this example because there is Nonspecificity type of uncertainty in the stress and because probability cannot account for Nonspecificity [11, p. 259]. When the designer uses a uniform probability distribution, he or she assumes that all values of the stress in the specified range are equally likely. But the evidence does not support this assumption — it only indicates that the stress is in the range [10 MPa, 50 MPa].

- *Tools from numerical analysis for numerical error.* Numerical analysis includes methods to quantify errors in numerical algorithms for solving models. These methods determine the effect of the algorithm parameters (e.g., step size, or the size of a panel) on the error in the solution. Convergence studies also show the effect of the finite element mesh size, or the step size on the solution of a mathematical model.
- *Tools from sampling theory.* Sampling theory [13] provides equations for estimating the probability distributions of the statistics of a sample, such as the sample mean, standard deviation and higher moments of samples drawn from a population.

---

[3]The energy of a narrowband process is confined in a narrow frequency range.

TABLE 8.1 Types of Information for Types of Uncertainty

Type of Information Type of Uncertainty or Source	Experimental Observations	Interval and Linguistic	Type of Probability Distribution from First Principles	Type of Probability Distribution from Principles of Uncertainty	Numerical Analysis Tools and Results of Convergence Studies	Sampling Theory
Aleatory	X	X	X	X		
Epistemic (predictive deterministic models, models of uncertainty)	X*	X				
Epistemic (parameters of a probabilistic model)	X	X				X
Execution error	X	X				
Error in numerical algorithms		X			X	

Note: One should be cautious using experimental observations to update predictive models that are used for a broad range of applications because it is often impractical to obtain a sample that is representative of the population.

8.3.3 Type of Information for Various Types of Uncertainty

Here we explain which types of the information reviewed in the previous subsection should be used for various types of uncertainty. Table 8.1 summarizes the recommendations in this subsection.

8.3.3.1 Information for Aleatory Uncertainty

One should seek experimental observations to model aleatory uncertainty. If there is insufficient data to construct a probability distribution, then a first principles-based model for the type of the probability distribution could be chosen. Then the parameters of the distribution could be estimated using the experimental data. The chapter entitled "The Role of Statistical Testing in NDA" in this book presents methods to construct models of aleatory uncertainty.

In some cases, very limited numerical data or no data is available for aleatory uncertainty. Then information from experts, such as intervals or linguistic information, together with principles of uncertainty (e.g., the maximum entropy principle) can be used for modeling uncertainty. A standard probability distribution can also be used. Subjective probabilistic models can be updated using Bayes' rule when additional information becomes available later. In these cases, aleatory and epistemic uncertainties in probabilistic models coexist.

In some problems, such as gambling problems, one can divide the space of all possible events into elementary events. Moreover, in these problems, there is no reason to assume that the events have different probabilities. For example, there is no reason to assume that the six faces of a die have different probabilities to appear in a roll. According to the principle of insufficient reason, these events should be assumed equiprobable. Then one could estimate the probability of an event based on the number of favorable elementary events normalized by the total number of possible elementary events.

Traditionally, the effect of aleatory uncertainty on the performance of a system was quantified with a probability distribution. For example, the Weibull probability distribution was typically used for the strength or fatigue life. In a modern, bottom-up approach, designers quantify aleatory uncertainty by developing probabilistic models of the input variables or drivers, **X**. Then they calculate the probability distributions in the output variables, **Y**, (e.g., the attributes of a product or the response of a system) using probability calculus together with a model relating the output to the input variables. Chapters 14 to 22 present methods that can be used for this purpose for both static and time-invariant problems.

Table 8.1 summarizes the above recommendations for aleatory uncertainty. Experimental information, interval or linguistic information from experts can be used for aleatory uncertainty. We can determine the type of the probability distribution of a variable from principles of uncertainty or we can use a standard probability distribution.

8.3.3.2 Information for Epistemic Uncertainty

One typically relies on expert opinion for epistemic uncertainty in a predictive model. Experts provide nonspecific information (e.g., intervals instead of exact values) or linguistic information about the variables that describe model uncertainty. For example, it is wiser for an expert to say that the error in the stress in a component is between 0 and 20% rather than to pinpoint the exact value of the error.

Often, we represent the modeling uncertainty of a predictive model with random parameters. For example, we can represent modeling uncertainty in a simplistic way as a random correction coefficient with which we multiply the output of a model to estimate the response of a system. Prior probability distributions of the random parameters are estimated on the basis of judgment and later updated using experimental data and using Bayes rule (see, for example, Cardoni et al. [3] and Chapter 22 "Bayesian Modeling and Updating" in this book). We can use a similar approach to represent modeling uncertainty in a probabilistic model of uncertainty — we can consider the distribution parameters random (e.g., the mean value and the standard deviation), estimate their prior probability distributions, and update these distributions using sampling information.

The above approach can develop reasonable probabilistic models for the epistemic uncertainty in a model for a particular application; for example the prediction of the compressive collapse load of a family of plates with similar geometry and materials. Often, we want to quantify modeling uncertainty in a generic model or modeling practice that is used for a broad range of applications. For example, we want to quantify the modeling uncertainty in loads or load effects for design codes for buildings, bridges or offshore platforms. There are concerns about constructing probabilistic models of uncertainty using experimental data and Bayesian updating for these applications because it is often impractical to obtain a sample of experimental measurements that is representative of the population.

One should primarily rely on expert judgment to model modeling uncertainty in this case, in addition to experimental measurements.

Example 10: Experimental Data on Modeling Error in the Activity Coefficient of a Chemical Solvent

Figure 8.11 shows the errors in the predictions of the activity coefficient of ten chemical solvents. The predictive model was originally developed for 2-propanol but it was used for the ten solvents, which are similar to 2-propanol. This figure was created using data reported by Kubic [14].The vertical axis is the subjective degree to which each solvent is similar to 2-propanol. The horizontal axis is the error in the model

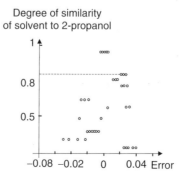

FIGURE 8.11 Errors (discrepancy between predicted and measured values) in activity coefficients for solvents.

predictions. For each solvent, predictions were obtained at different temperatures and the errors are shown as small circles in the figure. The errors for each solvent are on the same horizontal line. It is observed that the errors are different for each solvent and they also vary in a different way as a function of the temperature.

Because the error is not random, one cannot obtain a representative sample of the population of the errors for the ten solvents from the results for a subset of solvents. For example, the errors for the solvent with degree of similarity around 0.8 are clustered around 0.03. None of the results for the errors for the other solvents exhibits similar behavior as the errors in the above solvent. If a designer built a probability distribution of the error using the results in Figure 8.11, then it would be very unlikely that the errors in the predictions for another solvent would fit to the probability distribution. Therefore, the probability distribution for the error developed using the numerical data in the figure would not be useful for constructing a model of the error.

There is epistemic uncertainty in models of uncertainty. We may not know the type of probability distribution of a variable and we may be unable to estimate accurately the parameters of the distribution from a finite sample. One can use expert opinion for the uncertainty in the type of the probability distribution (Table 8.1, third row). Results from theory of sampling [13] are available for modeling the uncertainty in the values of the parameters of a distribution. An alternative approach, which can be used if few experimental measurements are available (e.g., less than 30), is to start with a prior probability distribution of a distribution parameter and update it using experimental measurements and Bayes' rule.

In conclusion, intervals or linguistic information from experts can be used for characterizing epistemic uncertainty (Table 8.1). We can also characterize epistemic uncertainty in predictive models by combining subjective information and experimental measurements for a model that is used for a particular range of applications. Results of theory of sampling are available for modeling uncertainty in the estimates of the parameters of a probabilistic model from a finite sample.

8.3.3.3 Information for Human Errors

We can estimate the frequency of occurrence of execution errors from sampling data. For example, investigations have shown that typical (micro) tasks used in detail structural design have produced an estimate of 0.02 errors per mathematical step on a desktop calculator ([5], p. 40). Sampling data were also used for constructing a histogram of the bending moment capacities of a portal frame computed by different designers.

8.3.3.4 Information for Error in Numerical Solution of Predictive Models

Error in numerical algorithms is easier to quantify than the effect of epistemic uncertainty. Results from numerical analysis and convergence studies help quantify errors due to numerical approximations.

8.4 Chapters in This Book on Methods for Collecting Information and for Constructing Models of Uncertainty

Methods for collecting information, constructing, and testing probabilistic models of uncertainty are presented in Chapter 26 by Fox. Booker's and McNamara's Chapter 13 focuses on elicitation of expert knowledge for quantifying reliability. Theories of uncertainty are presented in Chapter 9 by Joslyn and Booker. Oberkampf and Helton focus on evidence theory in Chapter 10. Two chapters deal with theories and methods for modeling uncertainty when there is a severe information deficit: Ben Haim presents Information Gap theory in Chapter 11, and Muhanna and Mullen present a method using interval arithmetic in Chapter 12. Chapter 22, by Papadimitriou and Katafygiotis, presents Bayesian methods for modeling uncertainty. These methods allow a designer to fuse expert knowledge and experimental data and they are suitable for problems in which limited data is available.

8.5 Conclusion

Uncertainty is important in making design decisions. Because of uncertainty, a decision maker's preferred choice is not guaranteed to result in the most desirable outcome. The best the decision maker can do is to determine the choice that maximizes the odds of getting the most desirable outcome.

To manage uncertainty effectively, a designer should understand all important types of uncertainty. This will enable him or her to allocate resources effectively, to collect the most suitable type of information for each type of uncertainty, and to use the proper theory to quantify each type of uncertainty.

There are three basic types of uncertainty in predicting the outcomes of a decision: aleatory, epistemic, and uncertainty due to human error. The source of aleatory uncertainty is inherent randomness, while the source of epistemic uncertainty is lack of knowledge. There is epistemic uncertainty in deterministic models for predicting the outcome of an action and in models of uncertainty. The latter uncertainty can be in the type of the probability distribution of a random variable and in the parameters of the distribution. Human errors are not due to lack of knowledge and can be corrected if pointed out. Epistemic uncertainty and human error can be reduced by collecting information or by inspection, respectively, whereas aleatory uncertainty cannot be reduced by collecting information.

Aleatory uncertainty is modeled using probability distributions of the driver variables or the response variables of a system.

Sampling data from observations and evidence from experts is usually available for aleatory uncertainty. For epistemic uncertainty, information from experts is suitable. This information is often nonspecific (e.g., consists of intervals or sets of possible outcomes). Sampling information can also be used for epistemic uncertainty in a predictive model for a specific application. Numerical data about the frequency of the effect of gross human errors can be used to construct models of these effects. Errors in numerical solutions of mathematical models can be quantified using tools from numerical analysis and convergence studies.

Acknowledgments

The author wishes to thank Dr. W. Oberkampf and Prof. Z. Mourelatos for reviewing this chapter. The author acknowledges the help of Prof. A. Der Kiureghian in finding publications on types of uncertainty and modeling of epistemic uncertainty using Bayesian updating. The author also wishes to thank Prof. T. Haukass for proving information about human error.

References

1. Hazelrigg, G.A., *Systems Engineering: An Approach to Information-Based Design*, Prentice Hall, Englewood Cliffs, NJ, 1996.
2. Der Kiureghian, A., Measures of Structural Safety under Imperfect States of Knowledge, *J. Structural Eng.*, ASCE, 115(5), 1119–1139, May 1989.
3. Gardoni, P., Der Kiureghian, A., and Mosalam, K.M., Probabilistic Capacity Models and Fragility Estimates for Reinforced Concrete Columns Based on Experimental Observations, *J. Eng. Mechanics*, ASCE, 128(10), 1024–1038, October 2002.
4. Haukass, T., Types of Uncertainties, Elementary Data Analysis, Set Theory, *Reliability and Structural Safety: Lecture Notes*, University of British Columbia, 2003.
5. Melchers, R.E., *Structural Reliability, Analysis and Prediction*, John Wiley & Sons, New York, 1999.
6. Oberkampf, W.L., DeLand, S.M., Rutherford, B.M., Diegert, K.V., and Alvin, K.F., A New Methodology for the Estimation of Total Uncertainty in Computational Simulation, *AIAA Paper No. 99-1612*, 1999.
7. Oberkampf, W.L., DeLand, S.M., Rutherford, B.M., Diegert, K.V., and Alvin, K.F., Error and Uncertainty in Modeling and Simulation, *Reliability Engineering and System Safety*, 75, 333–357, 2002.
8. Nowak, A.S. and Collins, K.R., Uncertainties in the Building Process, in *Reliability of Structures*, McGraw-Hill, New York, 2000.

9. Zimmermann, H.-J., Uncertainty Modelling and Fuzzy Sets, *Uncertainty: Models and Measures,* Proceedings of the International Workshop, Lambrecht, Germany, Akademie Verlag, 1996, 84–100.
10. Ayyub, B., From Dissecting Ignorance to Solving Algebraic Problems, Epistemic Uncertainty Workshop, Sandia National Laboratories, Albuquerque, NM, 2002.
11. Klir, G.J. and Yuan, B., *Fuzzy Sets and Fuzzy Logic*, Prentice Hall, Upper Saddle River, NJ, 1995.
12. Klir, J.G. and Wierman, M.J., *Uncertainty-Based Information*, Physica-Verlag, 1998.
13. Deming, W.E., *Some Theory of Sampling,* Dover, New York, 1966.
14. Kubic, W.,L., "The Effect of Uncertainties in Physical Properties on Chemical Process Design," Ph.D. dissertation, Chemical Engineering Department, Lehigh University, 1987.

9
Generalized Information Theory for Engineering Modeling and Simulation

9.1	Introduction: Uncertainty-Based Information Theory in Modeling and Simulation 9-1
9.2	Classical Approaches to Information Theory 9-3
	Logical and Set Theoretical Approaches • Interval Analysis • Probabilistic Representations
9.3	Generalized Information Theory 9-12
	Historical Development of GIT • GIT Operators • Fuzzy Systems • Monotone and Fuzzy Measures • Random Sets and Dempster-Shafer Evidence Theory • Possibility Theory
9.4	Conclusion and Summary 9-34

Cliff Joslyn
Los Alamos National Laboratory

Jane M. Booker
Los Alamos National Laboratory

9.1 Introduction: Uncertainty-Based Information Theory in Modeling and Simulation

Concepts of information have become increasingly important in all branches of science, and especially in modeling and simulation. In the limit, we can view all of science as a kind of modeling. While models can be physical or scale models, more typically we are referring to mathematical or linguistic models, such as $F = ma$, where we measure quantities for mass m and acceleration a, and try to predict another measured quantity force F. More cogently to the readers of this volume, computer simulation models manifest such mathematical formalisms to produce numerical predictions of some technical systems.

A number of points stand out about all models. To quote George Box, "All models are wrong; some models are useful." More particularly:

- All models are necessarily incomplete, in that there are certain aspects of the world which are represented, and others which are not.
- All models are necessarily somewhat in error, in that there will always be some kind of gap between their numerical output and the measured quantities.
- The system being modeled may have inherent variability or un-measurability in its behavior.

In each case, we wish to be able to measure or quantify these properties, that is, the fidelity and accuracy of our models. We therefore care about the concept of "uncertainty" and all its related concepts: How certain can I be: that I am capturing the properties I'm trying to? that I'm making accurate predictions? that the quantities can be confidently accepted?

We refer to **Uncertainty Quantification** (UQ) as this general task of representing amounts, degrees, and kinds of uncertainty in formal systems. In this context, the concept of uncertainty stands in a dual relation to that of "information." Classically, we understand that when I receive some information, then some question has been answered, and so some uncertainty has been reduced. Thus, this concept of information is that it is a reduction in uncertainty, and we call this **uncertainty-based information**.

Through the 20th century, uncertainty modeling has been dominated by the mathematics of probability, and since Shannon and Weaver [1], information has been defined as a statistical measure of a probability distribution. But also starting in the 1960s, alternative formalisms have arisen. Some of these were intended to stand in contrast to probability theory; others are deeply linked to probability theory but depart from or elaborate on it in various ways. In the intervening time, there has been a proliferation of methodologies, along with concomitant movements to synthesize and generalize them. Together, following Klir [2], we call these **Generalized Information Theory** (GIT).

This chapter surveys some of the most prominent GIT mathematical formalisms in the context of the classical approaches, including probability theory itself. Our emphasis will be primarily on introducing the formal specifications of a range of theories, although we will also take some time to discuss semantics, applications, and implementations.

We begin with the classical approaches, which we can describe as the kinds of mathematics that might be encountered in a typical graduate engineering program. **Logical and set-theoretical** approaches are simply the application of these basic formal descriptions. While we would not normally think of these as a kind of UQ, we will see that in doing so, we gain a great deal of clarity about the other methods to be discussed. We then introduce **interval analysis** and the familiar **probability theory** and related methods.

Following the development of these classical approaches, we move on to consider the GIT proper approaches to UQ. What characterizes a GIT approach is some kind of generalization of or abstraction from a classical approach [3]. **Fuzzy systems theory** was the first and most significant such departure, in which Zadeh generalized the classical, Boolean notions of both set inclusion and truth valuation to representations which are a matter of degree.

A fuzzy set can also be seen as a generalization of a probability distribution or an interval. Similarly, a **monotone** or **fuzzy measure** can be seen as a generalization of a probability measure. A **random set** is a bit different; rather than a generalization, it is an extension of a probability measure to set-valued, rather than point-valued, atomic events. Mathematically, random sets are isomorphic to **Dempster-Shafer bodies of evidence**. Finally, we consider **possibility theory**, which arises as a general alternative to classical information theory based on probability. Possibility measures arise as a different special case of fuzzy measures from probability measures, and are generated in extreme kinds of random sets; similarly, possibility distributions arise as a different special case of fuzzy sets, and generalize classical intervals.

The relations among all the various approaches discussed in this chapter is shown in Figure 9.18. This diagram is somewhat daunting, and so we deliberately show it toward the end of the chapter, after the various subrelations among these components have been explicated. Nonetheless, the intrepid reader might wish to consult this as a reference as the chapter develops.

We also note that our list is not inclusive. Indeed, the field is a dynamic and growing area, with many researchers inventing novel formalisms. Rather, we are trying to capture here the primary classes of GIT theories, albeit necessarily from our perspective. Furthermore, there are a number of significant theoretical components that we will mention in Section 9.4 only in passing, which include:

- **Rough sets** as representations of multi-resolutional structures, and are equivalent to classes of possibility distributions

- Higher-order hybrid structures such as **type II** and **level-II** fuzzy sets and **fuzzified Dempster-Shafer theory**; and finally
- **Choquet capacities** and **imprecise probabilities**, which provide further generalizations of monotone measures.

9.2 Classical Approaches to Information Theory

Throughout this chapter we will assume that we are representing uncertainty claims about some system in the world through reference to a universe of discourse denoted $\Omega = \{\omega\}$. At times we can specify that Ω is finite, countable, or uncountable, depending on the context.

9.2.1 Logical and Set Theoretical Approaches

As mentioned above, some of the most classical mathematical representations can be cast as representations of uncertainty in systems, albeit in a somewhat trivial way. But by beginning this way, we can provide a consistent development of future discussions.

We can begin with a simple proposition A, which may or may not be true of any particular element $\omega \in \Omega$. So if A is true of ω, we can say that the truth value of A for ω is 1: $T_A(\omega) = 1$; and if it is false, that $T_A(\omega) = 0$. Because there are two logical possibilities, 0 and 1, the expression $T_A(\omega)$ expresses the uncertainty, that it might be $T_A(\omega) = 0$, or it might be that $T_A(\omega) = 1$.

Surely the same can be said to be true for any function on Ω. But in this context, it is significant to note the following. First, we *can*, in fact, characterize Boolean logic in this way, characterizing a predicate A as a function $T_A : \Omega \mapsto \{0,1\}$. The properties of this value set $\{0, 1\}$ will be crucial below, and will be elaborated on in many of the theories to be introduced.

Second, we can gather together all the $\omega \in \Omega$ for which T_A is true, as distinguished from all those $\omega \in \Omega$ for which T_A is false, and call this the subset $A \subseteq \Omega$, where $A := \{\omega \in \Omega : T_A(\omega) = 1\}$. It is standard to represent the set A in terms of its characteristic function $\chi_A : \Omega \mapsto \{0, 1\}$, where

$$\chi_A(\omega \in \Omega) = \begin{cases} 1, & \omega \in A \\ 0, & \omega \notin A \end{cases}$$

It is not insignificant that, in fact, $\chi_A \equiv T_A$: the truth value function of the predicate A is equivalent to the characteristic function of the set A. Indeed, there is a mathematical isomorphism between the properties of Boolean logic and those of set theory. For example, the truth table for the logical disjunction ("or") of the two predicates A and B, and the "set disjunction" (union operation) of the two subsets A and B, is shown in Table 9.1. Table 9.2 shows the isomorphic relations among all the primary operations. Graphical representations will be useful below. Letting $\Omega = \{x, y, z, w\}$, Figure 9.1 shows the characteristic function of the subset $A = \{x, z\}$.

TABLE 9.1 Truth Table for Logical Disjunction and Set Disjunction $\cup$

$T_A(\omega)$	$T_B(\omega)$	$T_{A \text{ or } B}(\omega)$	$\chi_A(\omega)$	$\chi_B(\omega)$	$\chi_{A \cup B}(\omega)$
0	0	0	0	0	0
0	1	1	0	1	1
1	0	1	1	0	1
1	1	1	1	1	1

TABLE 9.2 Isomorphisms between Logical and Set Theoretical Operations

Logic		Set Theory	
Negation	$\neg A$	Complement	A^c
Disjunction	A or B	Union	$A \cup B$
Conjunction	A and B	Intersection	$A \cap B$
Implication	$A \to B$	Subset	$A \subseteq B$

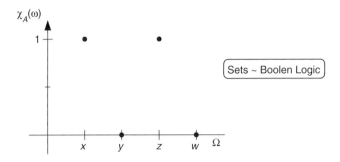

FIGURE 9.1 The crisp set $A = \{x, z\} \subseteq \Omega$.

So far this is quite straightforward, but in so doing we are able to point out the general elements of an uncertainty theory, in particular we can identify:

- The primary objects, in this case sets or propositions, A
- Compound objects as collections of these objects on which the uncertainty can be valued, in this case the power set 2^Ω, that is, the set of all subsets of Ω, so that $2^\Omega = \{A : A \subseteq \Omega\}$
- A range of possible uncertainty quantities for each element $\omega \in \Omega$ with respect to the object $A \in 2^\Omega$, in this case the two binary choices 0 and 1
- Standard operations to combine different objects A and B

The other necessary element is a measure of the total uncertainty or information content $U(A)$ of a particular set or proposition A. For set theory and logic, as well as all the subsequent theories to be presented, such measures are available. Because this is not the primary subject of this chapter, we refer the interested reader elsewhere [4]; nonetheless, for each of the structures we present, we will attempt to identify the community's current best definition for U for that theory. In this case, it is the Hartley measure of information, which is quite simply

$$U_{\text{logic}}(A) = \log_2(|A|), \tag{9.1}$$

where $|\cdot|$ indicates the cardinality of the set. For this and all other uncertainty-based information measures, logarithmic scales are used for their ability to handle addition of two distinct quantities of information, and they are valued in units of bits.

Finally, in each of the cases below, we will identify and observe an **extension principle**, which effectively means that when we generalize from one uncertainty theory to another, then, first, the results from the first must be expressible in terms of special cases of the second, and furthermore the particular properties of the first are recovered exactly for those special cases. However, it is a corollary that in the more general theory, there is typically more than one way to express the concepts that had been previously unequivocal in the more specific theory. This is stated abstractly here, but we will observe a number of particular cases below.

9.2.2 Interval Analysis

As noted, as we move from theory to theory, it may be useful for us to change the properties of the universe of discourse Ω. In particular, in real applications it is common to work with real-valued quantities. Indeed, for many working scientists and engineers, it is always presumed that $\Omega = \mathbb{R}$. The analytical properties of $\mathbb{R}$ are such that further restrictions can be useful. In particular, rather than working with arbitrary subsets of $\mathbb{R}$, it is customary to restrict ourselves to relatively closed sets, specifically closed intervals $I = [I_l, I_u] \subseteq \mathbb{R}$ or half-open intervals $I = [I_l, I_u) \subseteq \mathbb{R}$. Along these lines, it can be valuable to identify

$$\mathcal{D} := \{[a, b) \subseteq \mathcal{R} : a, b \in \mathcal{R}, a \leq b\} \tag{9.2}$$

as the **Borel field** of half-open intervals.

In general, interval-valued quantities represent uncertainty in terms of the upper and lower bounds I_l and I_u. That is, a quantity $x \in \mathbb{R}$ is known to be bounded in this way, such that $x \in [I_l, I_u]$, or $I_l \leq x \leq I_u$. Because I is a subset of $\mathbb{R}$, it has a characteristic function $\chi_I : \mathbb{R} \mapsto \{0, 1\}$, where

$$\chi_I(x) = \begin{cases} 1, & I_l \leq x \leq I_u \\ 0, & \text{otherwise} \end{cases}$$

The use of intervals generally is well-known in many aspects of computer modeling and simulation [5].[1] We can also observe the components of interval analysis necessary to identify it as an uncertainty theory, in particular:

- The basic objects are the numbers $x \in \mathbb{R}$.
- The compound objects are the intervals $I \subseteq \mathbb{R}$.
- Note that while the universe of discourse has changed from the general set Ω with all its subsets to $\mathbb{R}$ with its intervals, the valuation set has remained $\{0, 1\}$.
- The operations are arithmetic manipulations, of the form $I * J$ for two intervals I and J, where $* \in \{+, -, \times, \div, \min, \max\}$, etc. In general, we have

$$I * J := \{x * y : x \in I, y \in J\}. \tag{9.3}$$

For example, we have

$$I + J = [I_l, I_u] + [J_l, J_u] = \{x + y : x \in I, y \in J\} = [I_l + I_u, J_l + J_u].$$

Note, however, that in general for an operator $*$ we usually have

$$I * J \neq [I_l * I_u, J_l * J_u].$$

- Finally, for uncertainty we can simply use the width of the interval $U_{\text{int}}(I) := |I| = |I_u - I_l|$, or its logarithm

$$U_{\text{int}}(I) := \log_2(|I|). \tag{9.4}$$

The interval operation $[1, 2] + [1.5, 3] = [2.5, 5]$ is shown in Figure 9.2. Note that

$$U([2.5, 5]) = \log_2(2.5) = 1.32 > U([1, 2]) + U([1.5, 3]) = \log_2(1) + \log_2(1.5) = .58.$$

Also note that interval analysis as such is a kind of set theory: each interval $I \subseteq \mathbb{R}$ is simply a special kind of subset of the special universe of discourse $\mathbb{R}$. Thus we can see in Figure 9.2 that intervals have effectively the same form as the subsets $A \subseteq \Omega$ discussed immediately above in Section 9.2.1, in that they are shown as characteristic functions valued on $\{0, 1\}$ only.

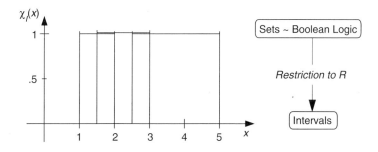

FIGURE 9.2 Interval arithmetic: $[1, 2] + [1.5, 3] = [2.5, 5] \subseteq \mathbb{R}$.

---

[1]For example, see the journal *Reliable Computing*.

Finally, we can observe the extension principle, in that a number $x \in \mathbb{R}$ can be represented as the degenerate interval $[x, x]$. Then, indeed, we do have that $x * y = [x, x] * [y, y]$.

9.2.3 Probabilistic Representations

By far, the largest and most successful uncertainty theory is probability theory. It has a vast and crucial literature base, and forms the primary point of departure for GIT methods.

Probability concepts date back to the 1500s, to the time of Cardano when gamblers recognized that there were rules of probability in games of chance and, more importantly, that avoiding these rules resulted in a sure loss (i.e., the classic coin toss example of "heads you lose, tails I win," referred to as the "Dutch book"). The concepts were still very much in the limelight in 1685, when the Bishop of Wells wrote a paper that discussed a problem in determining the truth of statements made by two witnesses who were both known to be unreliable to the extent that they only tell the truth with probabilities p_1 and p_2, respectively. The Bishop's answer to this was based on his assumption that the two witnesses were independent sources of information [6].

Mathematical probability theory was initially developed in the 18th century in such landmark treatises as Jacob Bernoulli's *Ars Conjectandi* (1713) and Abraham DeMoiver's *Doctrine of Chances* (1718, 2nd edition 1738). Later in that century, articles would appear that provided the foundations of modern interpretations of probability: Thomas Bayes' "An Essay Towards Solving a Problem in the Doctrine of Chances," published in 1763 [7], and Pierre Simon Laplace's formulation of the axioms relating to games of chance, "Memoire sur la Probabilite des Causes par les Evenemens," published in 1774. In 1772, the youthful Laplace began his work in mathematical statistics and provided the roots for modern decision theory.

By the time of Newton, physicists and mathematicians were formulating different theories of probability. The most popular ones remaining today are the relative frequency theory and the subjectivist or personalistic theory. The latter development was initiated by Thomas Bayes [7], who articulated his very powerful theorem, paving the way for the assessment of subjective probabilities. The theorem gave birth to a subjective interpretation of probability theory, through which a human's degree of belief could be subjected to a coherent and measurable mathematical framework within the subjective probability theory.

9.2.3.1 Probability Theory as a Kind of GIT

The mathematical basis of probability theory is well known. Its basics were well established by the early 20th century, when Rescher developed a formal framework for a conditional probability theory and Jan Lukasiewicz developed a multivalued, discrete logic circa 1930. But it took Kolmogorov in the 1950s to provide a truly sound mathematical basis in terms of measure theory [8].

Here we focus only on the basics, with special attention given to casting probability theory in the context of our general development of GIT.

In our discussions on logic and set theory, we relied on a general universe of discourse Ω and all its subsets $A \in 2^\Omega$, and in interval analysis we used $\mathbb{R}$ and the set of all intervals $I \in \mathcal{D}$. In probability theory, we return to a general universe of discourse Ω, but then define a **Boolean field** $\mathcal{E} \subseteq 2^\Omega$ on Ω as a collection of subsets closed under union and intersection:

$$\forall A_1, A_2 \in \mathcal{E}, \quad A_1 \cup A_2 \in \mathcal{E}, \quad A_1 \cap A_2 \in \mathcal{E}.$$

Note that while 2^Ω is a field, there are fields which are not 2^Ω.

We then define a probability measure Pr on $\mathcal{E}$ as a set function $\Pr : \mathcal{E} \mapsto [0, 1]$ where $\Pr(\Omega) = 1$ as a **normalization** condition, and if $A_1, A_2, \ldots$ is a countably infinite sequence of mutually disjoint sets in $\mathcal{E}$ whose union is in $\mathcal{E}$, then

$$\Pr\left(\bigcup_{i=1}^{\infty} A_i\right) = \sum_{i=1}^{\infty} \Pr(A_i).$$

When all of these components are in place, we call the collection $\langle \Omega, \mathcal{E}, \Pr \rangle$ a **probability space**, where we use $\langle \cdot \rangle$ to indicate a general n-tuple, in this case an ordered triple.

We interpret the probability of an event $A \in \mathcal{E}$ as the uncertainty associated with the outcome of A considered as an event. Implicit in this definition are two additional concepts, time, t, and "history" or background information, H, available for contemplating the uncertain events, at t and H. Thus we can also use the revised notation $\Pr(A; H, t)$. Below, we will freely use either notation, depending on the context.

The calculus of probability consists of certain rules (or axioms) denoted by a number determined by $\Pr(A; H, t)$, in which the probability of an event, A, is related to H at time t. When the event A pertains to the ability to perform a certain function (e.g., survive a specified mission time), then $\Pr(A; H, t)$ is known as the product's reliability. This is a traditional definition of reliability, although we must note that treatments outside of the context of probability theory, indeed, outside of the context of any uncertainty-based information theory, are also possible [9].

The quantity $\Pr(A_1 | A_2; H, t)$ is known as the conditional probability of A_1, given A_2. Note that conditional probabilities are in the subjunctive. In other words, the disposition of A_2 at time t, were it to be known, would become a part of the history H at time t. The vertical line between A_1 and A_2 represents a supposition or assumption about the occurrence of A_2.

We can also define a function called a **probability distribution** or **density**, depending on the context, as the probability measure at a particular point $\omega \in \Omega$. Specifically, we have $p: \Omega \mapsto [0, 1]$ where $\forall \omega \in \Omega$, $p(\omega) := \Pr(\{\omega\})$. When Ω is finite, then we tend to call p a discrete distribution, and we have

$$\forall A \subseteq \Omega, \quad \Pr(A) = \sum_{\omega \in A} p(\omega),$$

and as the normalization property we have $\sum_{\omega \in \Omega} p(\omega) = 1$.

When $\Omega = \mathbb{R}$, then we tend to use f, and call it a **probability density function** (pdf). We then have

$$\forall A \subseteq \mathbb{R}, \quad \Pr(A) = \int_A f(x) dx,$$

and for normalization $\int_{-\infty}^{\infty} f(x) dx = 1$. In this case, we can also define the **cumulative distribution** as

$$\forall x \in \mathbb{R}, \quad F(x) := \Pr((-\infty, x]) = \int_{-\infty}^{x} f(x) dx,$$

and for normalization $\lim_{x \to \infty} F(x) = 1$.

To make the terminological problems worse, it is common to refer to the cumulative distribution as simply the "distribution function." These terms, especially "distribution," appear frequently below in different contexts, and we will try to use them clearly.

We have now introduced the basic components of probability theory as a GIT:

- The objects are the points $\omega \in \Omega$.
- The compound objects are the sets in the field $A \in \mathcal{E}$.
- The valuation set has become the unit interval $[0, 1]$.

We are now prepared to introduce the operations on these objects, similar to logic and intervals above. First we exploit the isomorphism between sets and logic by introducing the formulation

$$\Pr(A \text{ or } B; H, t) := \Pr(A \cup B; H, t), \quad \Pr(A \text{ and } B; H, t) := \Pr(A \cap B; H, t).$$

The calculus of probability consists of the following three primary rules:

1. **Convexity:** For any event $A \in \mathcal{E}$, we have $0 \leq \Pr(A; H, t) \leq 1$. Note that this is effectively a restatement of the definition, since $\Pr: 2^\Omega \mapsto [0, 1]$.
2. **Addition:** Assume two events A_1 and A_2 that are mutually exclusive; that is, they cannot simultaneously take place, so that $A_1 \cap A_2 = \emptyset$. Then we have

$$\Pr(A_1 \text{ or } A_2; H, t) = \Pr(A_1; H, t) + \Pr(A_2; H, t).$$

In general, for any two sets $A, B \in \mathcal{E}$, we have

$$\Pr(A_1 \text{ or } A_2; H, t) = \Pr(A_1; H, t) + \Pr(A_2; H, t) - \Pr(A_1 \cap A_2; H, t). \tag{9.5}$$

3. **Multiplication:** Interpreting $\Pr(A_1 | A_2; H, t)$ as a quantification of the uncertainty about an event A_1 supposing that event A_2 has taken place, then we have

$$\Pr(A_1 \text{ and } A_2; H, t) = \Pr(A_1 | A_2; H, t) \Pr(A_2; H, t).$$

Finally, $\Pr(A_1 \text{ and } A_2; H, t)$ also can be written as $\Pr(A_2 | A_1; H, t) \Pr(A_1; H, t)$ because at time t both A_1 and A_2 are uncertain events and one can contemplate the uncertainty about A_1 supposing that A_2 were to be true or vice versa.

To complete the characterization of probability theory as a GIT, we can define the total uncertainty for a discrete as its statistical entropy:

$$U_{\text{prob}}(p) = -\sum_{\omega \in \Omega} p(\omega) \log_2(p(\omega)). \tag{9.6}$$

It is not so straightforward for a continuous pdf, but these concepts are related to variance and other measures of the "spread" or "width" of the density.

9.2.3.2 Interpretations of Probability

The calculus of probability does not tell us how to interpret probability, nor does the theory define what probability means. The theory and calculus simply provide a set of rules by which the uncertainties about two or more events combine or "cohere." Any set of rules for combining uncertainties that are in violation of the rules given above are said to be "incoherent" or inconsistent with respect to the calculus of probability. But it is crucial to note that it is exactly these "inconsistencies" that have spurred much of the work in generalizing the structures reported herein.

Historically speaking, there have been at least 11 different significant interpretations of probability; the most common today are relative frequency theory and personalistic or subjective theory.

Relative frequency theory has its origins dating back to Aristotle, Venn, von Mises, and Reichenbach. In this interpretation, probability is a measure of an empirical, objective, and physical fact of the world, independent of human knowledge, models, and simulations. Von Mises believed probability to be a part of a descriptive model, whereas Reichenbach viewed it as part of the theoretical structure of physics. Because probability is based only on observations, it can be known only *a posteriori* (literally, after observation). The core of this interpretation is in the concept of a random collective, as in the probability of finding an ace in a deck of cards (the collective). In relative frequency theory, $\Pr(A; H, t) = \Pr(A)$; there is no H or t.

Personalistic or subjective interpretation of probability has its origins attributed to Borel, Ramsey, de Finetti, and Savage. According to this interpretation, there is no such thing as a correct probability, an unknown probability, or an objective probability. Probability is defined as a degree of belief, or a willingness to bet: the probability of an event is the amount (say p) the individual is willing to bet, on a two-sided bet, in exchange for $1, should the event take place. By a two-sided bet is meant staking

$(1 - p)$ in exchange for $1, should the event not take place. Probabilities of one-of-a-kind or rare events, such as the probability of intelligent life on other planets, are easily handled with this interpretation.

The personalistic or subjective probability permits the use of all forms of data, knowledge, and information. Therefore, its usefulness in applications where the required relative frequency data are absent or sparse becomes clear. This view of probability also includes Bayes theorem and comes the closest of all the views of probability to the interpretation traditionally used in fuzzy logic. Therefore, this interpretation of probability can be the most appropriate for addressing the uncertainties in complex decisions surrounding modern reliability problems.

9.2.3.3 Bayes Theorem and Likelihood Approaches for Probability

In 1763, the Reverend Thomas Bayes of England made a momentous contribution to probability, describing a relationship among probabilities of events (A_1 and A_2) in terms of conditional probability:

$$\Pr(A_1 | A_2; H) = \frac{\Pr(A_2 | A_1; H) \Pr(A_1; H)}{\Pr(A_2; H)}.$$

Its development stems from the third "multiplication" axiom of probability defining conditional probability. Bayes theorem expresses the probability that event A_1 occurs if we have observed A_2 in terms of the probability of A_2 given that A_1 occurred.

Historical investigation reveals that Laplace may have independently established another form of Bayes Theorem by considering A_1 as a comprised of k sub-events, $A_{11}, A_{12}, \ldots, A_{1k}$. Then the probability of $A_2, \Pr(A_2; H)$ can be rewritten as

$$\Pr(A_2 | A_{11}; H) \Pr(A_{11}; H) + \Pr(A_2 | A_{12}; H) \Pr(A_{12}; H) + \cdots + \Pr(A_2 | A_{1k}; H) \Pr(A_{1k}; H).$$

This relationship is known as the **Law of Total Probability** for two events, A_1 and A_2, and can be rewritten as:

$$\Pr(A_2 = a_2; H) = \sum_j \Pr(A_2 = a_2 | A_1 = a_{1j}; H) \Pr(A_1 = a_{1j}; H),$$

where lower case a values are particular values or realizations of the two events.

The implications of Bayes theorem are considerable in its use, flexibility, and interpretation in that [10]:

- It demonstrates the proportional relationship between the conditional probability $\Pr(A_1 | A_2; H)$ and the product of probabilities $\Pr(A_1; H)$ and $\Pr(A_2 | A_1; H)$.
- It prescribes how to relate the two uncertainties about A_1: one prior to knowing A_2, the other posterior to knowing A_2.
- It specifies how to change the opinion about A_1 were A_2 to be known; this is also called "the mathematics of changing your mind".
- It provides a mathematical way to incorporate additional information.
- It defines a procedure for the assessor, i.e., how to bet on A_1 should A_2 be observed or known. That is, it prescribes the assessor's behavior before actually observing A_2.

Because of these implications, the use of Bayesian methods, from the application of this powerful theorem, have become widespread as an information combination scheme and as an updating tool, combining or updating the prior information with the existing information about events. These methods also provide a mechanism for handling different kinds of uncertainties within a complex problem by linking subjective-based probability theory and fuzzy logic.

The prior about A_1 refers to the knowledge that exists prior to acquisition of information about event A_1. The fundamental Bayesian philosophy is that prior information is valuable, should be used, and can be mathematically combined with new or updating information. With this combination, uncertainties can be reduced.

Bernoulli appears to be the first to prescribe uncertainty about A_1 if one were to observe A_2 (but assuming $A_2 = a_2$ has not yet occurred). By dropping the denominator and noting the proportionality of the remaining terms on the right-hand side, Bayes rule becomes:

$$\Pr(A_1 \mid A_2; H) \propto \Pr(A_2 \mid A_1; H) \Pr(A_1; H).$$

However, $A_2 = a_2$ is actually observed, making the left-hand side written as $\Pr(A_1; a_2, H)$. Therefore,

$$\Pr(A_1; a_2, H) \propto \Pr(A_2 = a_2 \mid A_1 = a_1; H) \Pr(A_1 = a_1; H).$$

However, there is a problem because $\Pr(A_2 = a_2 \mid A_1 = a_1; H)$ is no longer interpreted as a probability. Instead, this term is called the **likelihood** that $A_1 = a_1$ in light of H and the fact that $A_2 = a_2$. This is denoted $L(A_1 = a_1; a_2, H)$. This likelihood is a function of a_1 for a fixed value of a_2. For example, the likelihood of a test resulting in a particular failure rate would be expressed in terms of $L(A_1 = a_1; a_2, H)$.

The concept of a likelihood gives rise to another formulation of Bayes theorem:

$$\Pr(A_1; a_2, H) \propto L(A_1 = a_1; a_2, H) \Pr(A_1 = a_1; H).$$

Here, $\Pr(A_1 = a_1; H)$ is again the *prior probability* of A_1 (i.e., the source for information that exists "prior" to test data (a_2) in the form of expert judgment and other historical information). By definition, the prior represents the possible values and associated probabilities for the quantity of interest, A_1. For example, one decision is to represent the average failure rate of a particular manufactured item. The likelihood $L(A_1 = a_1; a_2, H)$ is formed from data in testing a specified number of items. Test data from a previously made item similar in design forms the prior. $\Pr(A_1; a_2, H)$ is the posterior distribution in the light of a_2 (the data) and H, produced from the prior information and the data.

The likelihood is an intriguing concept but it is not a probability, and therefore does not obey the axioms or calculus of probability. In Bayes theorem, the likelihood is a connecting mechanism between the two probabilities: the prior probability, $\Pr(A_1; H)$, and the posterior probability, $\Pr(A_1; a_2, H)$. The likelihood is a subjective construct that enables the assignment of relative weights to different values of $A_1 = a_1$.

9.2.3.4 Distribution Function Formulation of Bayes Theorem

Bayes theorem has been provided for the discrete form for two random variables representing the uncertain outcomes of two events, A_1 and A_2. For continuous variables X and Y, the probability statements are replaced by pdfs, and the likelihood is replaced by a likelihood function. If Y is a continuous random variable whose probability density function depends on the variable X, then the conditional pdf of Y given X is $f(y \mid x)$. If the prior pdf of X is $g(x)$, then for every y such that $f(y) > 0$ exists, the posterior pdf of X, given $Y = y$ is

$$g(x \mid y; H) = \frac{f(y \mid x; H) g(x; H)}{\int f(y \mid x; H) g(x; H) dx},$$

where the denominator integral is a normalizing factor so that $g(x \mid y; H)$, the posterior distribution, integrates to 1 (as a proper pdf).

Alternatively, utilizing the likelihood notation, we have

$$g(x \mid y; H) \propto L(x \mid y; H) g(x; H),$$

so that the posterior is proportional to the likelihood function times the prior distribution.

In this form, Bayes theorem can be interpreted as a weighting mechanism. The theorem mathematically weights the likelihood function and prior distribution, combining them to form the posterior. If these two distributions overlap to a large extent, this mathematical combination produces a desirable result: the uncertainty (specifically, the variance) of the posterior distribution is smaller than that produced by a simple weighted combination, $w_1 g_{\text{prior}} + w_2 L_{\text{likelihood}}$, for example. The reduction in the uncertainty results from the added information of combining two distributions that contain similar information (overlap).

Contrarily, if the prior and likelihood are widely separated, then the posterior will fall in the gap between the two functions. This is an undesirable outcome because the resulting combination falls in a region unsupported by either the prior or the likelihood. In this situation, one would want to reconsider using Bayesian combination and either seek to resolve the differences between the prior and likelihood or use some other combination method such as a simple weighting scheme; for example, consider $w_1 g_{\text{prior}} + w_2 L_{\text{likelihood}}$.

As noted above, a major advantage of using Bayes theorem to combine distribution functions of different information sources is that the spread (uncertainty) in the posterior distribution is reduced when the information in the prior and likelihood distributions are consistent with each other. That is, the combined information from the prior distribution and the data has less uncertainty because the prior distribution and data are two different information sources that support each other.

Before the days of modern computers and software, calculating Bayes theorem was computationally cumbersome. For those times, it was fortunate that certain choices of pdfs for the prior and likelihood produced easily obtained posterior distributions. For example, a beta prior with a binomial likelihood produces a beta posterior whose parameters are simple functions of the prior beta and binomial parameters, as the following example illustrates. With modern computational methods, these analytical shortcuts, called conjugate priors, are not necessary; however, computation still has its difficulties in how to formulate, sample from, and parameterize the various functions in the theorem. Simulation algorithms such as Metropolis-Hastings and Gibbs sampling provide the how-to, but numerical instabilities and convergence problems can occur with their use. A popular simulation technique for simulation and sampling is Markov Chain Monte Carlo (MCMC). A flexible software package for implementation written in Java is YADAS[2] [11].

9.2.3.5 Binomial/Beta Reliability Example

Suppose we prototype a system, building 20 units, and subject these to a stress test. All 20 units pass the test [10]. The estimate of success/failure rates from test data alone is $n_1 = 20$ tests with $x_1 = 20$ successes.

Using just this information, the success rate is 20/20 = 1, and the failure rate is 0/20 = 0. This fundamental reliability (frequentist interpretation of probability) estimate, based on only 20 units, does not reflect the uncertainty in the reliability for the system and does not account for any previously existing information about the units before the test.

A Bayesian approach can take advantage of prior information and provide an uncertainty estimate on the probability of a success, p. Prior knowledge could exist in many forms: expertise of the designers, relevant data from similar systems or components, design specifications, historical experience with similar designs, etc. that can be used to formulate the prior distribution for p, $g(p)$. The beta distribution is often chosen as a prior for a probability because it ranges from 0 to 1 and can take on many shapes (uniform, "J" shape, "U" shape and Gaussian-like) by adjusting its two parameters, n_0 and x_0. That beta prior is denoted as beta(x_0, n_0), and its pdf is:

$$g(p) = \frac{\Gamma(n_0)}{\Gamma(x_0)\Gamma(n_0 - x_0)} x^{x_0 - 1}(1-x)^{n_0 - x_0 - 1}.$$

---

[2] Yet Another Data Analysis System.

For this example, assume the prior information is in the form of an estimate of the failure rate from the test data done on a similar system that is considered relevant for this new system with $n_0 = 48$ tests on a similar system with $x_0 = 47$ successes.

The new prototype test data forms the likelihood, $L(p; x)$. Because this data represents the number of successes, x_1, in n_1 trials it conforms to the binomial distribution with the parameter of interest for success, p. The beta distribution, $g(p)$, is a conjugate prior when combined with the binomial likelihood, $L(p; x)$, using Bayes theorem. Thus, the resulting, posterior distribution, $g(p \mid x)$ is also a beta distribution with parameters $(x_0 + x_1, n_0 + n_1)$:

$$g(p \mid x) = \frac{\Gamma(n_0 + n_1)}{\Gamma(x_0 + x_1)\Gamma(n_0 + n_1 - x_0 - x_1)} p^{x_0 + x_1 - 1}(1-p)^{n_0 + n_1 - x_0 - x_1 - 1},$$

or

$$g(p \mid x) = \frac{\Gamma(68)}{\Gamma(67)\Gamma(1)} p^{66}(1-p)^0.$$

The mean success rate of the beta posterior is

$$\frac{x_0 + x_1}{n_0 + n_1} = \frac{67}{68} = 0.985,$$

or, in terms of a mean failure rate for the beta posterior, approximately $1 - 0.985 = 0.015$ failure rate. The variance of the beta posterior distribution is:

$$\frac{(x_0 + x_1)[(n_0 + n_1) - (x_0 + x_1)]}{(n_0 + n_1)^2(n_0 + n_1 + 1)} = 0.00021.$$

The engineering reliability community gravitates to the binomial/beta conjugate prior because many of the failures are binomial in nature and the parameters of the prior and posterior can have a reliability-based interpretation: n_0 = number of tests and x_0 = number of successes for the prior parameter interpretation. Similarly, $n_0 + n_1$ = number of pseudo tests and $x_0 + x_1$ = number of pseudo successes for the posterior parameter interpretation, provided these values are greater than 1.

9.3 Generalized Information Theory

We now turn our attention to the sub-fields of GIT proper. Most of these formalisms were developed in the context of probability theory, and are departures, in the sense of generalization from or elaborations, of it. However, many of them are also intricately interlinked with logic, set theory, interval analysis, combinations of these, combinations with probability theory, and combinations with each other. We emphasize again that there is a vast literature on these subjects in general, and different researchers have different views on which theories are the most significant, and how they are related. Our task here is to represent the primary GIT fields and their relations in the context of probability theory and reliability analysis. For more background, see work elsewhere [12–15].

9.3.1 Historical Development of GIT

As mentioned in the introduction, we will describe the GIT sub-fields of fuzzy systems, monotone or fuzzy measures, random sets, and possibility theory. While these GIT sub-fields developed historically in the context of probability theory, each also has progenitors in other parts of mathematics.

In 1965, Lotfi Zadeh introduced his seminal idea in a continuous-valued logic called fuzzy set theory [16, 17]. In doing so, he was recapitulating some earlier ideas in multi-valued logics [13].

Also in the 1960s, Arthur Dempster developed a statistical theory of evidence based on probability distributions propagated through multi-valued maps [18]. In so doing, he introduced mathematical structures that had been identified by Choquet some years earlier, and described as general "capacities." These Choquet capacities [19, 20] are generalizations of probability measures, as we shall describe below. In the 1970s, Glenn Shafer extended Dempster's work to produce a complete theory of evidence dealing with information from more than one source [21]. Since then, the combined sub-field has come to be known as "Dempster-Shafer Evidence Theory" (DS Theory). Meanwhile, the stochastic geometry community was exploring the properties of random variables valued not in $\mathbb{R}$, but in closed, bounded subsets of $\mathbb{R}^n$. The random sets they described [22, 23] turned out to be mathematically isomorphic to DS structures, although again, with somewhat different semantics. This hybrid sub-field involving Dempster's and Shafer's theories and random sets all exist in the context of infinite-order Choquet capacities.

While random sets are defined in general on $\mathbb{R}^n$, for practical purposes, as we have seen, it can be useful to restrict ourselves to closed or half-open intervals of $\mathbb{R}$ and similar structures for $\mathbb{R}^n$. Such structures provide DS correlates to structures familiar to us from probability theory as it is used, for example pdfs and cumulative distributions. It should be noted that Dempster had previously introduced such "random intervals" [24].

In 1972, Sugeno introduced the idea of a "fuzzy measure" [25], which was intended as a direct generalization of probability measures to relax the additivity requirement of additive Equation 9.5. Various classes of fuzzy measures were identified, many of the most useful of which were already available within DS theory. In 1978, Zadeh introduced the special class of fuzzy measures called "possibility measures," and furthermore suggested a close connection to fuzzy sets [26]. It should be noted that there are other interpretations of this relation, and to some extent it is a bit of a terminological oddity that the term "fuzzy" is used in two such contexts [27]. For this reason, researchers are coming to identify fuzzy measures instead as "monotone measures."

The 1980s and 1990s were marked by a period of synthesis and consolidation, as researchers completed some open questions and continued both to explore novel formalisms, but more importantly the relations among these various formalisms. For example, investigators showed a strong relationship between evidence theory, probability theory, and possibility theory with fuzzy measures [28].

One of the most significant developments during this period was the introduction by Walley of an even broader mathematical theory of "imprecise probabilities," which further generalizes fuzzy measures [29].

9.3.2 GIT Operators

The departure of the GIT method from probability theory is most obviously significant in its use of a broader class of mathematical operators. In particular, probability theory operates through the familiar algebraic operators addition + and multiplication ×, as manifested in standard linear algebra. GIT recognizes + as an example of a generalized disjunction, the "or"-type operator, and × as an example of a generalized conjunction, the "and"-type operator, but uses other such operators as well.

In particular, we can define the following operations:

Complement: Let $c: [0, 1] \mapsto [0, 1]$ be a **complement** function when

$$c(0) = 1, \quad c(1) = 0, \quad x \leq y \mapsto c(x) \geq c(y).$$

Norms and Conorms: Assume associative, commutative functions $\sqcap: [0, 1]^2 \mapsto [0, 1]$ and $\sqcup: [0, 1]^2 \mapsto [0, 1]$. Because of associativity, we can use the operator notation $x \sqcap y := \sqcap(x, y)$, $x \sqcup y := \sqcup(x, y)$. Further assume that $\sqcup$ and $\sqcap$ are monotonic, in that

$$\forall x \leq y, z \leq w, \quad x \sqcap z \leq y \sqcap w, \quad x \sqcup z \leq y \sqcap w.$$

Then $\sqcap$ is a **triangular norm** if it has identity 1, with $1 \sqcap x = x \sqcap 1 = x$; and $\sqcup$ is a **triangular conorm** if it has identity 0, with $0 \sqcup x = x \sqcup 0 = x$.

While there are others, the prototypical complement function, and by far the most commonly used, is $c(x) = 1 - x$. Semantically, complement functions are used for logical negation and set complementation.

In general, there are many continuously parameterized classes of norms and conorms [13]. However, we can identify some typical norms and conorms that may be familiar to us from other contexts. Below, use $\wedge, \vee$ for the maximum and minimum operators, let $x, y \in [0, 1]$, and let $\lfloor x \rfloor$ be the greatest integer below $x \in \mathbb{R}$, and similarly $\lceil x \rceil$ the least integer above x. Then we have:

Norms:
- Min: $x \wedge y$
- Times: $x \times y$
- Bounded Difference: $x -_b y := (x + y - 1) \vee 0$
- Extreme Norm: $\lfloor x \rfloor \times \lfloor y \rfloor$

Conorms:
- Max: $x \vee y$
- Probabilistic Sum: $x +_p y := x + y - xy$
- Bounded Sum: $x +_b y := (x + y) \wedge 1$
- Extreme Conorm: $\lfloor x \rfloor \times \lfloor y \rfloor$

In general, $\wedge$ is the greatest and $\lfloor x \rfloor \times \lfloor y \rfloor$ the least norm, and $\vee$ is the least and $\lceil x \rceil \times \lceil y \rceil$ the greatest conorm. The relations are summarized in Table 9.3.

9.3.3 Fuzzy Systems

In 1965, Zadeh published a new set theory that addressed the kind of vague uncertainty that can be associated with classifying an event into a set [16, 17]. The idea suggested that *set membership* is the key to decision making when faced with linguistic and nonrandom uncertainty. Unlike probability theory, based upon crisp sets, which demands that any outcome of an event or experiment belongs to a set A or to its complement, A^c, and not both, fuzzy set theory permits such a joint membership. The degree of membership that an item belongs to any set is specified using a construct of fuzzy set theory, a membership function.

Just as we have seen that classical (crisp) sets are isomorphic to classical (crisp) logic, so there is a fuzzy logic that is isomorphic to fuzzy sets. Together, we can thus describe **fuzzy systems** as systems whose operations and logic are governed by these principles. Indeed, it can be more accurate to think of a *process* of "fuzzification," in which a formalism that has crisp, binary, or Boolean choices are relaxed to admit degrees of gradation. In this way, we can conceive of such ideas as fuzzified arithmetic, fuzzified calculus, etc.

9.3.3.1 Fuzzy Sets

Zadeh's fundamental insight was to relax the definition of set membership. Where crisp sets contain objects that satisfy precise properties of membership, fuzzy sets contain objects that satisfy imprecise properties of membership; that is, membership of an object in a fuzzy set can be approximate or partial.

We now introduce the basic formalism of fuzzy sets. First, we work with a general universe of discourse Ω. We then define a **membership function** very simply as any function $\mu: \Omega \mapsto [0, 1]$. Note, in particular, that the characteristic function χ_A of a subset $A \subseteq \Omega$ is a membership function, simply because $\{0, 1\} \subseteq [0, 1]$.

TABLE 9.3 Prototypical Norms and Conorms

Triangular Norm	$x \sqcap y$:	$x \wedge y$	$\geq$	$x \times y$	$\geq$	$0 \vee (x + y - 1)$	$\geq$	$\lfloor x \rfloor \times \lfloor y \rfloor$
Triangular Conorm	$x \sqcup y$:	$x \vee y$	$\leq$	$x + y - xy$	$\leq$	$1 \wedge (x + y)$	$\leq$	$\lceil x \rceil \times \lceil y \rceil$

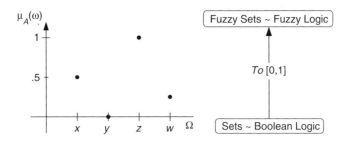

FIGURE 9.3 The fuzzy set $\tilde{A} = \{\langle x, .5\rangle, \langle y, 0\rangle, \langle z, 1\rangle, \langle w, .25\rangle\} \tilde{\subseteq} \Omega$.

Thus, membership functions generalize characteristic functions, and in this way, we can conceive of a **fuzzy subset** of Ω, denoted $\tilde{A} \tilde{\subseteq} \Omega$, as being defined by some particular membership function $\mu_{\tilde{A}}$. For a characteristic function χ_A of a subset A, we interpret $\chi_A(\omega)$ as being 1 if $\omega \in A$, and 0 if $\omega \notin A$. For a membership function $\mu_{\tilde{A}}$ of a fuzzy subset $\tilde{A}$, it is thereby natural to interpret $\mu_{\tilde{A}}(\omega)$ as the *degree* or *extent to which* $\omega \in \tilde{A}$.

Then note especially how the extension principle holds. In particular, consider $\omega_1, \omega_2 \in \Omega$ such that $\mu_{\tilde{A}}(\omega_1) = 0$ and $\mu_{\tilde{A}}(\omega_2) = 1$. In these cases, we can still consider that $\omega_1 \in \tilde{A}$ and $\omega_2 \notin \tilde{A}$ unequivocally.

Below, it will frequently be convenient to denote $\tilde{A}(\omega)$ for $\mu_{\tilde{A}}(\omega)$. Moreover, when $\Omega = \{\omega_1, \omega_2, \ldots, \omega_n\}$ is finite, we can denote a fuzzy subset as a set of ordered pairs:

$$\tilde{A} = \{\langle \omega_1, \tilde{A}(\omega_1)\rangle, \langle \omega_2, \tilde{A}(\omega_2)\rangle, \ldots, \langle \omega_n, \tilde{A}(\omega_n)\rangle\}.$$

Consider the simple example shown in Figure 9.3. For $\Omega = \{x, y, z, w\}$, we might have

$$\tilde{A} = \{\langle x, .5\rangle, \langle y, 0\rangle, \langle z, 1\rangle, \langle w, .25\rangle\},$$

so that z is completely in $\tilde{A}$, y is completely not in $\tilde{A}$, and x and w are in $\tilde{A}$ to the intermediate extents .5 and .25, respectively.

When a membership function actually reaches the line $\mu = 1$, so that $\exists \omega \in \Omega, \tilde{A}(\omega) = 1$, then we call $\tilde{A}$ **normal**. This usage is a bit unfortunate because it may indicate probabilistic additive normalization, so we will try to distinguish this as fuzzy normalization. Fuzzy normalization is also the criterion for $\tilde{A}$ to be a possibility distribution, which we will discuss below in Section 9.3.6.

To continue our characterization of fuzzy sets as a GIT, we need to define the correlates to the basic set operations. Not surprisingly, we will do this through the generalized operators introduced in Section 9.3.2. Below, presume two fuzzy sets $\tilde{A}, \tilde{B} \tilde{\subset} \Omega$. Then we have:

Fuzzy Complement: $\mu_{\tilde{A}^c}(\omega) = c(\tilde{A}(\omega))$

Fuzzy Union: $\mu_{\tilde{A} \cup \tilde{B}}(\omega) = \tilde{A}(\omega) \sqcup \tilde{B}(\omega)$

Fuzzy Intersection: $\mu_{\tilde{A} \cap \tilde{B}}(\omega) = \tilde{A}(\omega) \sqcap \tilde{B}(\omega)$

Fuzzy Set Equivalence: $\tilde{A} = \tilde{B} := \forall \omega \in \Omega, \tilde{A}(\omega) = \tilde{B}(\omega)$

Fuzzy Subsethood: $\tilde{A} \tilde{\subseteq} \tilde{B} := \forall \omega \in \Omega, \tilde{A}(\omega) \leq \tilde{B}(\omega)$

Typically, we use $\sqcup = \vee, \sqcap = \wedge$, and $c(\mu) = 1 - \mu$, although it must always be kept in mind that there are many other possibilities. The extension principle can be observed again, in that for crisp sets, the classical set operations are recovered.

Proposition 7: Consider two crisp subsets $A, B \subseteq \Omega$, and let $\mu_{\tilde{C}} = \chi_A \sqcup \chi_B$, and $\mu_{\tilde{D}} = \chi_A \sqcap \chi_B$ for some general norm $\sqcap$ and conorm $\sqcup$. Then $\mu_{\tilde{C}} = \chi_{A \cup B}$ and $\mu_{\tilde{D}} = \chi_{A \cap B}$.

So again, we have the ideas necessary to cast fuzzy systems as a kind of GIT. As with classical sets, the basic objects are the points $\omega \in \Omega$, but now the compound objects are all the fuzzy subsets $\tilde{A} \tilde{\subseteq} \Omega$, and the valuation is into $[0, 1]$ instead of $\{0, 1\}$. The operations on fuzzy sets are defined above.

So now we can introduce the measure of the information content of a fuzzy set. There are at least two important concepts here. First, we can consider the "size" of a fuzzy set much like that of a crisp set, in terms of its cardinality. In the fuzzy set case, this is simply

$$|\tilde{A}| := \sum_{\omega \in \Omega} \tilde{A}(\omega),$$

noting that in accordance with the extension principle, this fuzzy cardinality of a crisp set is thereby simply its cardinality.

We can also discuss the "fuzziness" of a fuzzy set, intuitively as how much a fuzzy set departs from being a crisp set, or in other words, some sense of "distance" between the fuzzy set and its complement [13]. The larger the distance, the "crisper" the fuzzy set. Using Z to denote this quantity, and recalling our fuzzy complement operator above, we have:

$$Z(\tilde{A}) := \sum_{\omega \in \Omega} |\tilde{A}(\omega) - c(\tilde{A}(\omega))|,$$

which, when $1 - \cdot$ is used for c, becomes:

$$Z(\tilde{A}) = \sum_{\omega \in \Omega} (1 - |2\tilde{A}(\omega) - 1|).$$

Finally, we note the presence of the extension principle everywhere. In particular, all of the classical set operations are recovered in the case of crisp sets, that is, where $\forall \omega \in \Omega, \tilde{A}(\omega) \in \{0, 1\}$.

9.3.3.2 Fuzzy Logic

We saw in Section 9.2.1 that we can interpret the value of a characteristic function of a subset χ_A as the truth value of a proposition T_A, and in this way set theoretical operations are closely coupled to logical operations, to the extent of isomorphism. In classical predicate logic, a proposition A is a linguistic, or declarative, statement contained within the universe of discourse Ω, which can be identified as being a collection of elements in Ω which are strictly true or strictly false.

Thus, it is reasonable to take our concept of a fuzzy set's membership function $\mu_{\tilde{A}}$ and derive an isomorphic fuzzy logic, and indeed, this is what is available. In contrast to the classical case, a fuzzy logic proposition is a statement involving some concept without clearly defined boundaries. Linguistic statements that tend to express subjective ideas and that can be interpreted slightly differently by various individuals typically involve fuzzy propositions. Most natural language is fuzzy, in that it involves vague and imprecise terms. Assessments of people's preferences about colors, menus, or sizes, or expert opinions about the reliability of components, can be used as examples of fuzzy propositions.

So mathematically, we can regard a fuzzy subset $\tilde{A}$ as a fuzzy proposition, and denote $T_{\tilde{A}}(\omega) := \tilde{A}(\omega) \in [0, 1]$ as the extent to which the statement "ω is $\tilde{A}$" is true. In turn, we can invoke the GIT operators analogously to fuzzy set theory to provide our fuzzy logic operators. In particular, for two fuzzy propositions $\tilde{A}$ and $\tilde{B}$ we have:

Negation: $T_{\neg \tilde{A}}(\omega) = c(\tilde{A}(\omega))$
Disjunction: $T_{\tilde{A} \text{ or } \tilde{B}}(\omega) = \tilde{A}(\omega) \sqcup \tilde{B}(\omega)$
Conjunction: $T_{\tilde{A} \text{ and } \tilde{B}}(\omega) = \tilde{A}(\omega) \sqcap \tilde{B}(\omega)$
Implication: There are actually a number of expressions for fuzzy implication available, but the "standard" one one might expect is valid: $T_{\tilde{A} \to \tilde{B}}(\omega) = T_{\neg \tilde{A} \text{ or } \tilde{B}}(\omega) = c(\tilde{A}(\omega)) \vee \tilde{B}(\omega)$

Table 9.4 shows the isomorphic relations among all the primary fuzzy operations. Again, the extension principle holds everywhere for crisp logic. Note, however, that, in keeping with the multi-valued nature of mathematical ideas in the more general theory, the implication operation is only roughly equivalent to the subset relation, and that there are other possibilities. Figure 9.3 shows our simple example again, along with the illustration of the generalization of classical sets and logic provided by fuzzy sets and logic.

TABLE 9.4 Isomorphisms between Fuzzy Logical and Fuzzy Set Theoretical Operations

Fuzzy Logic		Fuzzy Set Theory		GIT Operation
Negation	$\neg \tilde{A}$	Complement	$\tilde{A}^c$	$c(\mu_{\tilde{A}})$
Disjunction	$\tilde{A}$ or $\tilde{B}$	Union	$\tilde{A} \cup \tilde{B}$	$\mu_{\tilde{A}} \sqcup \mu_{\tilde{B}}$
Conjunction	$\tilde{A}$ and $\tilde{B}$	Intersection	$\tilde{A} \cap \tilde{B}$	$\mu_{\tilde{A}} \sqcap \mu_{\tilde{B}}$
Implication	$\tilde{A} \to \tilde{B}$	Subset	$\tilde{A} \subseteq \tilde{B}$	$c(\mu_{\tilde{A}}) \sqcup \mu_{\tilde{B}}$

9.3.3.3 Comparing Fuzzy Systems and Probability

The membership function $\tilde{A}(\omega)$ reflects an assessor's view of the extent to which $\omega \in \tilde{A}$, an epistemic uncertainty stemming from the lack of knowledge about how to classify ω. The subjective or personalistic interpretation of probability, Pr(A), can be interpreted as a two-sided bet, dealing with the uncertainty associated with the outcome of the experiment. While this type of uncertainty is usually labeled as random or aleatory, there is no restriction on applying subjective probability to characterize lack of knowledge or epistemic uncertainty. A common example would be eliciting probability estimates from experts for one-of-a kind or never observed events.

However, just because probabilities and fuzzy quantities can represent epistemic uncertainties does not guarantee interchangeability or even a connection between the two theories. As noted above, their axioms are quite different in how to combine uncertainties represented within each theory. Therefore, the linkage between the two theories is not possible by modifying one set of axioms to match the other. Other fundamental properties also differ. It is not a requirement that the sum over ω of all $\tilde{A}(\omega)$ equals one, as is required for summing over all probabilities. This precludes $\tilde{A}(\omega)$ from being interpreted as a probability in general. Similarly, pdfs are required to sum or integrate to one, but membership functions are not. Therefore, membership functions cannot be equated with pdfs either.

At least one similarity of probability and membership functions is evident. Just as probability theory does not tell how to specify Pr(A), fuzzy set theory does not tell how to specify $\tilde{A}(\omega)$. In addition, specifying membership is a subjective process. Therefore, subjective interpretation is an important common link to both theories.

Noting that $\tilde{A}(\omega)$, as a function of ω, reflects the extent to which $\omega \in \tilde{A}$, it is an indicator of *how likely* it is that $\omega \in \tilde{A}$. One interpretation of $\tilde{A}(\omega)$ is as the likelihood of ω for a fixed (specified) $\tilde{A}$. A likelihood function is not a pdf. In statistical inference, it is the relative degree of support that an observation provides to several hypotheses. Specifying the likelihood is also a subjective process, consistent with membership function definition and the subjective interpretation of probability.

As noted above, likelihoods are mostly commonly found in Bayes theorem. So, Bayes theorem links subjective probability with subjective likelihood. If membership functions can be interpreted as likelihoods, then Bayes theorem provides a valuable link from fuzzy sets back into probability theory. A case is made for this argument [30], providing an important mathematical linkage between probability and fuzzy theories. With two theories linked, it is possible to analyze two different kinds of uncertainties present in the same complex problem. An example application in the use of expert knowledge illustrates how these two theories can work in concert as envisioned by Zadeh [17] can be found in [31]. Additional research is needed to link other GITs so that different kinds of uncertainties can be accommodated within the same problem.

9.3.3.4 Fuzzy Arithmetic

Above we considered the restriction of sets from a general universe Ω to the line $\mathbb{R}$. Doing the same for fuzzy sets recovers some of the most important classes of structures.

In particular, we can define a **fuzzy quantity** as a fuzzy subset $\tilde{I} \subseteq \mathbb{R}$, such that $\mu_{\tilde{I}} : \mathbb{R} \mapsto [0, 1]$. Note that a fuzzy quantity is any arbitrary fuzzy subset of $\mathbb{R}$, and as such may not have any particular useful properties. Also note that every pdf is a special kind of fuzzy quantity. In particular, if $\int_{\mathbb{R}} \mu_{\tilde{I}}(x)dx = 1$, then $\mu_{\tilde{I}}$ is a pdf.

We can discuss another special kind of fuzzy quantity, namely a **possibilistic density** or **distribution function** (again, depending on context), which we will abbreviate as π-df, pronounced "pie-dee-eff." In contrast with a pdf, if $\tilde{I}$ is fuzzy normal, so that $\sup_{x \in \mathbb{R}} \tilde{I}(x) = 1$, then $\tilde{I}$ is a π-df. We note here in passing that where a pdf is a special kind of probability distribution on $\mathbb{R}$, so a π-df is a special kind of possibility distribution on $\mathbb{R}$. This will be discussed in more detail below in Section 9.3.6.

We can also discuss special kinds of π-dfs. When a π-df is convex, so that

$$\forall_{x,y \in \mathbb{R}} \forall_{z \in [x,y]} \tilde{I}(z) \geq \tilde{I}(x) \wedge \tilde{I}(y), \tag{9.8}$$

then $\tilde{I}$ is a **fuzzy interval**. We can also define the **support** of a fuzzy interval as $U(\tilde{I}) := \{x : \tilde{I}(x) > 0\}$, and note that $U(\tilde{I})$ is itself a (possibly open) interval. When a fuzzy interval $\tilde{I}$ is unimodal, so that $\exists ! x \in \mathbb{R}, \tilde{I}(x) = 1$, then $\tilde{I}$ is a **fuzzy number**, where $\exists !$ means "exists uniquely."

These classes of fuzzy quantities are illustrated in Figure 9.4. Note in particular that the fuzzy quantity and π-df illustrations are cartoons: in general, these need not be continuous, connected, or unimodal. Some of the cases shown here will be discussed further in Section 9.3.6.4.

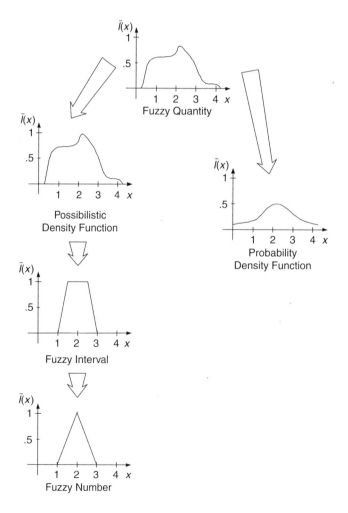

FIGURE 9.4 Kinds of fuzzy quantities.

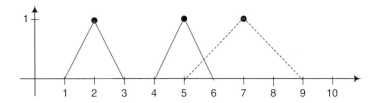

FIGURE 9.5 The fuzzy arithmetic operation [1, 2, 3] + [4, 5, 6] = [5, 7, 9].

Fuzzy intervals and numbers are named deliberately to invoke their extension from intervals and numbers. In particular, if a fuzzy interval $\tilde{I}$ is crisp, so that $\mu_{\tilde{I}}\mathbb{R} \mapsto \{0,1\}$, then $\tilde{I}$ is a crisp interval I with characteristic function $\chi_I = \mu_{\tilde{I}}$. Similarly, if a fuzzy number $\tilde{I}$ is crisp with mode x_0, so that $\tilde{I}(x_0) = 1$ and $\forall x \neq x_0, \tilde{I}(x) = 0$, then $\tilde{I}$ is just the number x_0, also characterized as the crisp interval $[x_0, x_0]$.

So this clears the way for us to define operations on fuzzy intervals, necessary to include it as a branch of GIT. As with crisp intervals, we are concerned with two fuzzy intervals $\tilde{I}, \tilde{J}$, and operations $* \in \{+, -, \times, \div\}$, etc. Then we have $\forall x \in \mathbb{R}$,

$$\mu_{\tilde{I}*\tilde{J}}(x) := \bigsqcup_{x = y * z} \left[\tilde{I}(y) \sqcap \tilde{J}(z) \right], \tag{9.9}$$

for some conorm $\sqcup$ and norm $\sqcap$. Again, the extension principle is adhered to, in that when $\tilde{I}$ and $\tilde{J}$ is crisp, Equation 9.3 is recovered from Equation 9.9.

An example of a fuzzy arithmetic operation is shown in Figure 9.5. We have two fuzzy numbers, each indicated by the triangles on the left. The leftmost, $\tilde{I}$, is unimodal around 2, and the rightmost, $\tilde{J}$, is unimodal around 6. Each is convex and normal, dropping to the x-axis as shown. Thus, $\tilde{I}$ expresses "about 2," and $\tilde{J}$ "about 5," and because they can be characterized by the three quantities of the mode and the x-intercepts, we denote them as

$$\tilde{I} = [1, 2, 3], \quad \tilde{J} = [4, 5, 6].$$

Applying Equation 9.9 for $* = +$ reveals $\tilde{I} + \tilde{J} = [5, 7, 9]$, which is "about 7."

Note how the extension principle is observed for fuzzy arithmetic as a generalization of interval arithmetic, in particular, we have

$$\mathbf{U}(\tilde{I} + \tilde{J}) = [5, 9] = [1, 3] + [4, 6] = \mathbf{U}(\tilde{I}) + \mathbf{U}(\tilde{J}),$$

where the final operation indicates interval arithmetic as in Equation 9.3.

The relations among these classes of fuzzy quantities, along with representative examples of each, is shown in Figure 9.6

9.3.3.5 Interpretations and Applications

Some simple examples can illustrate the uncertainty concept and construction of a fuzzy set, and the corresponding membership function.

First, let Ω be the set of integers between zero and ten, inclusive: $\Omega = \{0, 1, 2, \ldots, 10\}$. Suppose we are interested in a subset of Ω, $\tilde{A}$, where $\tilde{A}$ contains all the *medium* integers of Ω: $\tilde{A} = \{\omega : \omega \in \Omega$ and ω is medium$\}$. To specify $\tilde{A}$, the term "medium integer" must be defined. Most would consider 5 as medium, but what about 7? The uncertainty (or vagueness) about what constitutes a medium integer is what makes $\tilde{A}$ a fuzzy set, and such sets occur in our everyday use (or natural language). The uncertainty of

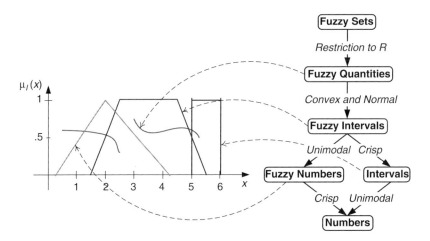

FIGURE 9.6 Fuzzy quantities.

classification arises because the boundaries of $\tilde{A}$ are not crisp. The integer 7 might have some membership (belonging) in $\tilde{A}$ and yet also have some degree of membership in $\tilde{A}^c$. Said another way, the integer 7 might have some membership in $\tilde{A}$ and yet also have some membership in another fuzzy set, $\tilde{B}$, where $\tilde{B}$ is the fuzzy set of large integers in Ω.

For a more meaningful example, assume we have a concept design for a new automotive system, like a fuel injector. Many of its components are also new designs, but may be similar to ones used in the past, implying that partial knowledge exists that is relevant to the new parts, but also implying large uncertainties exist about the performance of these parts and the system. The designer of this system wants to assess its performance based upon whatever information is currently available before building prototypes or implementing expensive test programs. The designer also wants to be assured that the performance is "excellent" with a high confidence. This desire defines a reliability **linguistic variable**. While reliability is traditionally defined as the probability that a system performs its functions for a given period of time and for given specifications, the knowledge about performance (especially new concepts) may only be in the form of linguistic and fuzzy terms.

For example, a component designer may only have access to the information that "if the temperature is too hot, this component won't work very well." The conditions (e.g., "too hot") can be characterized by a fuzzy set, and the performance (e.g., "won't work very well") can also be represented by a fuzzy set. Chapter 11 of Ross, Booker, and Parkinson [15] illustrates how fuzzy sets can be used for linguistic information and then combined with test data, whose uncertainty is probabilistic, to form a traditionally defined reliability.

Combining the probabilistic uncertainty of outcomes of tests and uncertainties of fuzzy classification from linguistic knowledge about performance requires a theoretical development for linking the two theories. Linkage between the probability and fuzzy set theories can be accomplished through the use of Bayes theorem, whose two ingredients are a prior probability distribution function and a likelihood function. As discussed in Section 9.3.3.3, Singpurwalla and Booker [30] relax the convention that the maximum value of $\tilde{A}(\omega)$ is set to 1.0, because that better conforms to the definition of a likelihood. Their theoretical development demonstrates the equivalency of likelihood and membership.

In the example above, if test data exists on a component similar to a new concept design component then probability theory could be used to capture the uncertainties associated with that data set, forming the prior distribution in Bayes theorem. Expert knowledge about the new design in the form of linguistic information about performance could be quantified using fuzzy membership functions, forming the likelihood. The combination of these two through Bayes theorem produces a posterior distribution, providing a probability based interpretation of reliability for the component. See [15] for more details on this kind of approach.

9.3.4 Monotone and Fuzzy Measures

In discussing probability theory in Section 9.2.3, we distinguished the probability measure Pr valued on sets $A \subseteq \Omega$ from the probability distribution p valued on points $\omega \in \Omega$. Then in Section 9.3.3 we characterized the membership functions of fuzzy sets $\tilde{A}$ also as being valued on points $\omega \in \Omega$, and, indeed, that probability distributions and pdfs are, in fact, kinds of fuzzy sets. It is natural to consider classes of measures other than Pr which are also valued on subsets $A \subseteq \Omega$, and perhaps related to other kinds of fuzzy sets.

This is the spirit that inspired Sugeno to define classes of functions he called **fuzzy measures** [25, 32]. Since then, terminological clarity has led us to call these **monotone measures** [33].

Assume for the moment a finite universe of discourse Ω, and then define a monotone measure as a function $v: 2^\Omega \mapsto [0,1]$, where $v(\emptyset) = 0, v(\Omega) = 1$, and

$$A \subseteq B \to v(A) \leq v(B). \tag{9.10}$$

When Ω is uncountably infinite, continuity requirements on v come into play, but this will suffice for us for now.

We can also define the **trace** of a monotone measure as the generalization of the concept of a density or distribution. For any monotone measure v, define its trace as a function $\rho_v : \Omega \mapsto [0,1]$, where $\rho_v(\omega) := v(\{\omega\})$.

In general, measures are much "larger" than traces, in that they are valued on the space of subsets $A \subseteq \Omega$, rather than the space of the points of $\omega \in \Omega$. So, for finite Ω with $n = |\Omega|$, a trace needs to be valued n times, one for each point $\omega \in \Omega$, while a measure needs to be valued 2^n times, one for each subset $A \subseteq \Omega$. Therefore, it is very valuable to know if, for a particular measure, it might be possible not to know all 2^n values of the measure independently, but rather to be able to calculate some of these based on knowledge of the others; in other words, to be able to break the measure into a small number of pieces and then put those pieces back together again. This greatly simplifies calculations, visualization, and elicitation.

When this is the case, we call such a monotone measure **distributional** or **decomposable**. Mathematically, this is the case when there exists a conorm $\sqcup$ such that

$$\forall A, B \subseteq \Omega, \quad v(A \cup B) + v(A \cap B) = v(A) \sqcup v(B). \tag{9.11}$$

It follows that

$$v(A \cup B) = v(A) \sqcup v(B) - v(A \cap B).$$

It also follows that when $A \cap B = \emptyset$, then $v(A \cup B) = v(A) \sqcup v(B)$.

Decomposability expresses the idea that the measure can be broken into pieces. The smallest such pieces are just the values of the trace, and thus decomposability is also called "distributionality," and can be expressed as

$$v(A) = \bigsqcup_{\omega \in A} \rho_v(\omega).$$

Finally, we call a monotone measure **normal** when $v(\Omega) = 1$. When v is both normal and decomposable, it follows that

$$\bigsqcup_{\omega \in \Omega} \rho_v(\omega) = 1.$$

So it is clear that every probability measure Pr is a monotone measure, but not vice versa, and the distribution p of a finite probability measure and the pdf f of a probability measure on $\mathbb{R}$ are both traces of the corresponding measure Pr. Indeed, all of these concepts are familiar to us from probability theory, and are, in fact, direct generalizations of it.

In particular, a probability measure Pr is a normal, monotone measure that is decomposable for the bounded sum conorm $+_b$ (see Section 9.3.2), and whose trace is just the density. Note, however, that because we presume that a probability measure Pr is always normalized, when operating on probability values, the bounded sum conorm $+_b$ becomes equivalent to addition $+$. For example, when $\sum_{\omega \in \Omega} p(\omega) = 1$, then $\forall \omega_1, \omega_2 \in \Omega, p(\omega_1) +_b p(\omega_2) = p(\omega_1) + p(\omega_2)$. In this way we recover the familiar results for probability theory:

$$\Pr(A \cup B) = \Pr(A) + \Pr(B) - \Pr(A \cap B)$$
$$A \cap B = \varnothing \rightarrow \Pr(A \cup B) = \Pr(A) + \Pr(B)$$
$$\Pr(A) + \Pr(A^c) = 1$$
$$\Pr(A) = \sum_{\omega \in A} p(\omega). \tag{9.12}$$

Note that a trace $\rho_\nu : \Omega \mapsto [0, 1]$ is a function to the unit interval, and is thus a fuzzy set. This will be important below, as in many instances it is desirable to interpret the traces of fuzzy measures such as probability or possibility distributions as special kinds of fuzzy sets.

A measure of information content in the context of general fuzzy or monotone measures is an area of active research, and beyond the scope of this chapter (see elsewhere for details [33]). Below we consider some particular cases in the context of random sets and possibility theory.

9.3.5 Random Sets and Dempster-Shafer Evidence Theory

In our historical discussion in Section 9.3.1, we noted that one of the strongest threads in GIT dates back to Dempster's work in probability measures propagated through multi-valued maps, and the subsequent connection to Shafer's theory of evidence and random sets. We detail this in this section, building on the ideas of monotone measures.

9.3.5.1 Dempster-Shafer Evidence Theory

In particular, we can identify **belief** Bel and **plausibility** Pl as dual fuzzy measures with the properties of super- and sub-additivity, respectively:

$$\text{Bel}(A \cup B) \geq \text{Bel}(A) + \text{Bel}(B) - \text{Bel}(A \cap B)$$
$$\text{Pl}(A \cap B) \leq \text{Pl}(A) + \text{Pl}(B) - \text{Pl}(A \cup B).$$

Note the contrast with the additivity of a probability measure shown in Equation 9.12. In particular, it follows that each probability measure Pr is both a belief and a plausibility measure simultaneously. Also, while Pr is always decomposable in $+_b$, Bel and Pl are only decomposable under some special circumstances.

Again, in contrast with probability, we have the following sub- and super-additive properties for Bel and Pl:

$$\text{Bel}(A) + \text{Bel}(A^c) \leq 1, \quad \text{Pl}(A) + \text{Pl}(A^c) \geq 1$$

Also, Bel and Pl are dually related, with:

$$\text{Bel}(A) \leq \text{Pl}(A), \tag{9.13}$$
$$\text{Bel}(A) = 1 - \text{Pl}(A^c), \quad \text{Pl}(A) = 1 - \text{Bel}(A^c). \tag{9.14}$$

So, not only are Bel and Pl co-determining, but each also determines and is determined by another function called a **basic probability assignment** $m: 2^\Omega \mapsto [0,1]$ where $m(\emptyset) = 0$, and

$$\sum_{A \subseteq \Omega} m(A) = 1. \tag{9.15}$$

$m(A)$ is also sometimes called the "mass" of A.

$$\text{Bel}(A) = \sum_{B \subseteq A} m(B), \qquad \text{Pl}(A) = \sum_{B \cap A \neq \emptyset} m(B)$$

We then have the following relations:

$$m(A) = \sum_{B \subseteq A} (-1)^{|A-B|} \text{Bel}(B) = \sum_{B \subseteq A} (-1)^{|A-B|} (1 - \text{Pl}(B^c)), \tag{9.16}$$

where Equation 9.16 expresses what is called a **Möbius inversion**. Thus, given any one of m, Bel, or Pl, the other two are determined accordingly.

Some other important concepts are:

- A **focal element** is a subset $A \subseteq \Omega$ such that $m(A) > 0$. In this chapter, we always presume that there are only a finite number N of such focal elements, and so we use the notation $A_j, 1 \leq j \leq N$ for all such focal elements.

- The **focal set** $\mathcal{F}$ is the collection of all focal elements:

$$\mathcal{F} = \{A_j \subseteq \Omega : m(A_j) > 0\}.$$

- The **support** of the focal set is the global union:

$$\mathrm{U} = \bigcup_{A_j \in \mathcal{F}} A_j.$$

- The **core** of the focal set is the global intersection:

$$\mathrm{C} = \bigcap_{A_j \in \mathcal{F}} A_j.$$

- A **body of evidence** is the combination of the focal set with their masses:

$$\mathcal{S} = \langle \mathcal{F}, m \rangle = \langle \{A_j\}, \{m(A_j)\} \rangle, \qquad 1 \leq j \leq N.$$

- Given two independent bodies of evidence $\mathcal{S}_1 = \langle \mathcal{F}_1, m_1 \rangle$, $\mathcal{S}_2 = \langle \mathcal{F}_2, m_2 \rangle$, then we can use **Dempster combination** to produce a combined body of evidence $\mathcal{S} = \mathcal{S}_1 \oplus \mathcal{S}_2 = \langle \mathcal{F}, m \rangle$, where $\forall A \subseteq \Omega$,

$$m(A) = \frac{\sum_{A_1 \cap A_2 = A} m_1(A_1) m_2(A_2)}{\sum_{A_1 \cap A_2 \neq \emptyset} m_1(A_1) m_2(A_2)}.$$

While Dempster's rule is the most prominent combination rule, there are a number of others available [34].

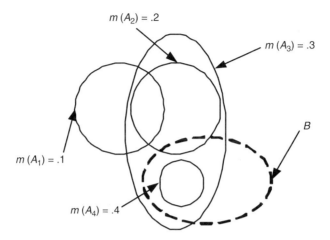

FIGURE 9.7 A Dempster-Shafer body of evidence.

- Assume a body of evidence S drawn from a finite universe of discourse with $\Omega = \{\omega_i\}$ with $1 \leq i \leq n$. Then if $N = n$, so that the number of focal elements is equal to the number of elements of the universe of discourse, then we call S **complete** [35].

An example is shown in Figure 9.7. We have:

$$\mathcal{F} = \{A_1, A_2, A_3, A_4\}$$

$$m(A_1) = .1, \quad m(A_2) = .2, \quad m(A_3) = .3, \quad m(A_4) = .4$$

$$\text{Bel}(B) = \sum_{A_j \subseteq B} m(A_j) = m(A_4) = .4$$

$$\text{Pl}(B) = \sum_{A_j \cap B \neq \emptyset} m(A_j) = m(A_2) + m(A_3) + m(A_4) = .2 + .3 + .4 = .9.$$

9.3.5.2 Random Sets

Above we identified the structure $S = \langle \mathcal{F}, m \rangle = \langle \{A_j\}, \{m(A_j)\} \rangle$ as a body of evidence. When Ω is finite, we can express this body of evidence in an alternative form: instead of a pair of sets ($\mathcal{F}$ and m), now we have a set of pairs, in particular, pairings of focal elements $A_j \in \mathcal{F}$ with their basic probability value $m(A_j)$:

$$\langle \{A_j\}, \{m(A_j)\} \rangle \mapsto \{\langle A_j, m(A_j) \rangle\}.$$

We call this form the **random set** representation of the DS body of evidence.

This alternative formulation is triggered by recalling that $\sum_{A_j \in \mathcal{F}} m(A_j) = 1$, so that m can be taken as a discrete probability distribution or density on the various sets A_j. In other words, we can interpret $m(A_j)$ as the probability that A_j occurs compared to all the other $A \subseteq \Omega$.

Note that despite a superficial similarity, there is a profound difference between a probability measure $\Pr(A)$ and a basic probability assignment $m(A)$. Where it must *always* be the case that for two sets $A, B \subseteq \Omega$, Equation 9.12 must hold, in general there need be *no* relation between $m(A)$ and $m(B)$, other than Equation 9.15, that $\sum_{A \subseteq \Omega} m(A) = 1$.

To explicate this difference, we recall the definition of a **random variable**. Given a probability space $\langle \Omega, \mathcal{E}, \Pr \rangle$, then a function $S: X \mapsto \Omega$ is a random variable if S is "Pr-measurable," so that $\forall \omega \in \Omega$,

$S^{-1}(\omega) \in \mathcal{E}$. S then assigns probabilities to the items $\omega \in \Omega$. Similarly, we can think of a random set simply as a random variable that takes values on *collections* or *sets* of items, rather than points.

General Random Set: $S: X \mapsto 2^{\Omega} - \{\emptyset\}$ is a random subset of Ω if S is Pr-measurable: $\forall \emptyset \neq A \subseteq \Omega$ $S^{-1}(A) \in \Sigma$. m acts as a density of S.

Given this, we can then interpret the DS measures in a very natural way in terms of a random set S:

$$m(A) = \Pr(S = A),$$

$$\mathrm{Bel}(A) = \Pr(S \subseteq A),$$

$$\mathrm{Pl}(A) = \Pr(S \cap A \neq \emptyset).$$

In the remainder of the chapter we will generally refer to random sets, by which we will mean finite random sets, which are isomorphic to finite DS bodies of evidence. Also, for technical reasons (some noted below in Section 9.3.6.2), there is a tendency to work only with the plausibility measure Pl, and to recover the belief Bel simply by the duality relation (Equation 9.14). In particular, for a general random set, we will generally consider its trace specifically as the plausibilistic trace ρ_{Pl}, and thereby have

$$\rho_{\mathrm{Pl}}(\omega_i) = \mathrm{Pl}(\{\omega_i\}) = \sum_{A_j \ni \omega_i} m_j.$$

The components of random sets as a GIT are now apparent. The basic components now are not points $\omega \in \Omega$, but rather subsets $A \subseteq \Omega$, and the compound objects are the random sets S. The valuation set is again $[0, 1]$, and the valuation is in terms of the evidence function m. Finally, operations are in terms of the kinds of combination rules discussed above [34], and operations defined elsewhere, such as inclusion of random sets [36].

9.3.5.3 The Information Content of a Random Set

The final component of random sets as a GIT, namely the measure of the information content of a random set, has been the subject of considerable research. Development of such a measure is complicated by the fact that random sets by their nature incorporate two distinct kinds of uncertainty. First, because they are random variables, they have a probabilistic component best measured by entropies, as in Equation 9.6. But, unlike pure random variables, their fundamental "atomic units" of variation are the focal elements A_j, which differ from each other in size $|A_j|$ and structure, in that some of them might overlap with each other to one extent or another. These aspects are more related to simple sets or intervals, and thus are best measured by measures of "nonspecificity" such as the Hartley measure of Equations 9.1 and 9.4.

Space precludes a discussion of the details of developing information measure for random sets (see [33]). The mathematical development has been long and difficult, but we can describe some of the highlights here.

The first good candidate for a measure of uncertainty in random sets is given by generalizing the nonspecificity of Equations 9.1 and 9.4 to be:

$$U_N(S) := -\sum_{A_j \in \mathcal{F}} m_j \log_2(|A_j|). \tag{9.17}$$

This has a number of interpretations, the simplest being the expectation of the size of a focal element. Thus, both components of uncertainty are captured: the randomness of the probabilistic variable coupled to the variable size of the focal element.

While this nonspecificity measure U_N captures many aspects of uncertainty in random sets, it does not as well capture all the attributes related to conflicting information in the probabilistic component m, which is reflected in probability theory by the entropy of Equation 9.6. A number of measures have been suggested, including **conflict** as the entropy of the singletons:

$$U_S(\mathcal{S}) := -\sum_{\omega_i \in \Omega} m(\{\omega_i\}) \log_2(m(\{\omega_i\})),$$

and **strife** as a measure of entropy focused on individual focal elements:

$$U_S(\mathcal{S}) := -\sum_{A_j \in \mathcal{F}} m_j \log_2 \left[\sum_{k=1}^{N} m_k \frac{|A_j \cap A_k|}{|A_j|} \right].$$

While each of these measures can have significant utility in their own right, and arise as components of a more detailed mathematical theory, in the end, none of the them alone proved completely successful in the context of a rigorous mathematical development. Instead, attention has turned to single measures that attempt to directly integrate both nonspecificity and conflict information. These measures are not characterized by closed algebraic forms, but rather as optimization problems over sets of probability distributions. The simplest expression of these is given as an **aggregate uncertainty**:

$$U_{AU}(\mathcal{S}) := \max_{p: \forall A \subseteq \Omega, \Pr(A) \leq \text{Pl}(A)} U_{\text{prob}}(p), \tag{9.18}$$

recalling that $\Pr(A) = \sum_{\omega_i \in A} p(\omega_i)$ and $U_{\text{prob}}(p)$ is the statistical entropy (Equation 9.6). In English, U_{AU} is the largest entropy of all probability distributions consistent with the random set $\mathcal{S}$.

9.3.5.4 Specific Random Sets and the Extension Principle

The extension principle also holds for random sets, recovering ordinary random variables in a special case. So, given that random sets are set-valued random variables, then we can consider the special case where each focal element is not, in fact, a set at all, but really just a point, in particular a singleton set. We call such a focal set **specific**:

$$\forall A_j \in \mathcal{F}, \quad |A_j| = 1, \quad \exists! \omega_i \in \Omega, A_j = \{\omega_i\}.$$

When a specific random set is also complete, then conversely we have that $\forall \omega_i \in \Omega, \exists! A_j \in \mathcal{F}, A_j = \{\omega_i\}$.

Under these conditions, the gap between $\text{Pl}(A)$ and $\text{Bel}(A)$ noted in Equation 9.13 closes, and this common DS measure is just a probability measure again:

$$\forall A \subseteq \Omega, \quad \text{Pl}(A) = \text{Bel}(A) = \Pr(A). \tag{9.19}$$

And when $\mathcal{S}$ is complete, the plausibilistic trace reverts to a probability distribution, with $p(\omega_i) = m_j$ for that A_j which equals $\{\omega_i\}$.

As an example, consider Figure 9.8, where $\Omega = \{x, y, z, w\}$, and $\mathcal{F} = \{\{x\}, \{y\}, \{z\}, \{w\}\}$ with m as shown. Then, for $B = \{z, w\}, C = \{y, w\}$, we have

$$\text{Bel}(B) = \text{Pl}(B) = \Pr(B) = .3 + .4 = .7, \quad \Pr(B \cup C) = \Pr(B) + \Pr(C) - \Pr(B \cap C) = .9,$$

and, using vector notation, the probability distribution is $p = \langle .1, .2, .3, .4 \rangle$.

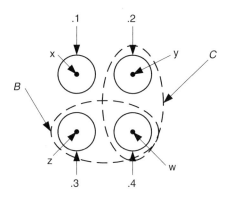

FIGURE 9.8 A specific random set, which induces a probability distribution.

9.3.5.5 Random Intervals and P-Boxes

Above we moved from sets to intervals, and then fuzzy sets to fuzzy intervals. Now we want to similarly move from random sets to **random intervals,** or DS structures on the Borel field $\mathcal{D}$ defined in Equation 9.2. Define a random interval $\mathcal{A}$ as a random set on $\Omega = \mathbb{R}$ for which $\mathcal{F}(\mathcal{A}) \subseteq \mathcal{D}$. Thus, a random interval is a random left-closed interval subset of $\mathbb{R}$. For a random interval, we denote the focal elements as intervals $I_j, 1 \leq j \leq N$, so that $\mathcal{F}(\mathcal{A}) = \{I_j\}$.

An example is shown in Figure 9.9, with $N = 4$,

$$\mathcal{F} = \{[2.5, 4), [1, 2), [3, 4), [2, 3.5)\},$$

support $U(\mathcal{F}(\mathcal{A})) = [1, 4)$, and m is as shown.

Random interval approaches are an emerging technology for engineering reliability analysis [37–41]. Their great advantage is their ability to represent not only randomness via probability theory, but also imprecision and nonspecificity via intervals, in an overall mathematical structure that is close to optimally simple. As such, they are superb ways for engineering modelers to approach the world of GIT.

So, random intervals are examples of DS structures restricted to intervals. This restriction is very important because it cuts down substantially on both the quantity of information and computational and human complexity necessary to use such structures for modeling. Even so, they remain relatively complex structures, and can present challenges to modelers and investigators in their elicitation and interpretation. In particular, interpreting the fundamental structures such as possibly overlapping focal elements and basic probability weights can be a daunting task for the content expert, and it can be desirable to interact with investigators over more familiar mathematical objects. For these reasons, we commonly introduce simpler mathematical structures that approximate the complete random interval by representing a portion of their information. We introduce these now.

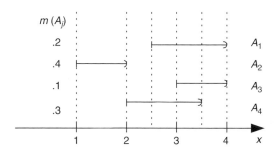

FIGURE 9.9 A random interval.

A probability box, or just a **p-box** [42], is a structure $\mathcal{B} := \langle \underline{B}, \overline{B} \rangle$, where $\underline{B}, \overline{B} : \mathbb{R} \mapsto [0, 1]$,

$$\lim_{x \mapsto -\infty} B(x) \to 0, \quad \lim_{x \mapsto \infty} B(x) \to 1, \quad B \in \mathcal{B},$$

and $\underline{B}, \overline{B}$ are monotonic with $\underline{B} \leq \overline{B}$. $\underline{B}$ and $\overline{B}$ are interpreted as bounds on cumulative distribution functions (CDFs). In other words, given $\mathcal{B} = \langle \underline{B}, \overline{B} \rangle$, we can identify the set of all functions $\{F : \underline{B} \leq F \leq \overline{B}\}$ such that F is the CDF of some probability measures Pr on $\mathbb{R}$. Thus each p-box defines such a class of probability measures.

Given a random interval $\mathcal{A}$, then

$$\mathcal{B}(\mathcal{A}) := \langle \text{BEL}, \text{PL} \rangle \tag{9.20}$$

is a p-box, where BEL and PL are the "cumulative belief and plausibility distributions" PL, BEL: $\mathbb{R} \mapsto [0, 1]$ originally defined by Yager [43]

$$\text{BEL}(x) := \text{Bel}((-\infty, x)), \quad \text{PL}(x) := \text{Pl}((-\infty, x)).$$

Given a random interval $\mathcal{A}$, then it is also valuable to work with its plausibilistic trace (usually just identified as its trace), where $r_{\mathcal{A}}(x) := \text{Pl}(\{x\})$. Given a random interval $\mathcal{A}$, then we also have that $r_{\mathcal{A}} = \text{PL} - \text{BEL}$, so that for a p-box derived from Equation 9.20, we have

$$r_A = \overline{B} - \underline{B}. \tag{9.21}$$

See details elsewhere [44, 45].

The p-box generated from the example random interval is shown in the top of Figure 9.10. Because $\overline{B}$ and $\underline{B}$ partially overlap, the diagram is somewhat ambiguous on its far left and right portions, but note that

$$\overline{B}((-\infty, 1)) = 0, \quad \underline{B}((-\infty, 2,)) = 0,$$

$$\overline{B}([3, \infty)) = 1, \quad \underline{B}([3.5, \infty)) = 1.$$

The trace $r_A = \overline{B} - \underline{B}$ is also shown.

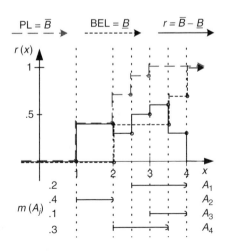

FIGURE 9.10 The probability box derived from a random interval.

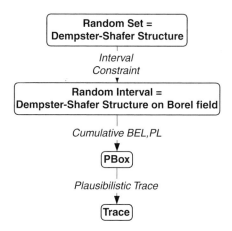

FIGURE 9.11 Relations among random sets, random intervals, p-boxes, and their traces.

So each random interval determines a p-box by Equation 9.20, which in turn determines a trace by Equation 9.21. But conversely, each trace determines an equivalence class of p-boxes, and each p-box an equivalence class of random intervals. In turn, each such equivalence class has a canonical member constructed by a standard mechanism. These relations are diagrammed in Figure 9.11, and see details elsewhere [45].

9.3.6 Possibility Theory

So far, we have discussed classical uncertainty theories in the form of intervals and probability distributions; their generalization to fuzzy sets and intervals; and the corresponding generalization to random sets and intervals. In this section we introduce **possibility theory** as a form of information theory, which in many ways exists as an alternative to and in parallel with probability theory, and which arises in the context of both fuzzy systems and DS theory.

9.3.6.1 Possibility Measures and Distributions

In Section 9.3.4 we identified a probability measure Pr as a normal monotone measure that is decomposable for the bounded sum conorm $+_b$. Similarly, a **possibility measure** Π is a normal, monotone measure that is decomposable for the *maximum* conorm $\vee$. In this way, the familiar results for probability theory shown in Equation 9.12 are replaced by their maximal counterparts for possibility theory. In particular, Equation 9.11 yields

$$\forall A, B \subseteq \Omega, \quad \Pi(A \cup B) \vee \Pi(A \cap B) = \Pi(A) \vee \Pi(B), \qquad (9.22)$$

so that from Equation 9.10 it follows that

$$\Pi(A \cup B) = \Pi(A) \vee \Pi(B), \qquad (9.23)$$

whether A and B are disjoint or not.

The trace of a possibility measure is called a **possibility distribution** $\pi : \Omega \mapsto [0, 1], \pi(\omega) = \Pi(\{\omega\})$. Continuing our development, we have the parallel results from probability theory:

$$\Pi(A) = \bigvee_{\omega \in A} \pi(\omega), \quad \Pi(\Omega) = \bigvee_{\omega \in \Omega} \pi(\omega) = 1.$$

In Section 9.3.5.4 and Figure 9.8 we discussed how the extension principle recovers a "regular" probability measure from a random set when the focal elements are specific, so that the duality of belief and plausibility collapses and we have that Bel = Pl = Pr. In a sense, possibility theory represents the opposite extreme case. In a specific random set, all the focal elements are singletons, and are thus maximally small and disjoint from each other. Possibility theory arises in the alternate case, when the focal elements are maximally large and intersecting.

In particular, we call a focal set **consonant** when $\mathcal{F}$ is a nested class, so that $\forall A_1, A_2 \in \mathcal{F}$, either $A_1 \subseteq A_2$ or $A_2 \subseteq A_1$. We can then arbitrarily order the A_j so that $A_i \subseteq A_{i+1}$, and we assign $A_0 := \emptyset$. If a consonant random set is also complete, then we can use the notation

$$\forall 1 \leq i \leq n, \quad A_i = \{\omega_1, \omega_2, \ldots, \omega_i\} \tag{9.24}$$

and we have that $A_i - A_{i-1} = \{\omega_i\}$.

Whenever a random set is consonant, then the plausibility measure Pl becomes a measure Π. But unlike in a specific random set, where the belief and plausibility become maximally close, in a consonant random set the belief and plausibility become maximally separated, to the point that we call the dual belief measure Bel a **necessity measure**:

$$\eta(A) = 1 - \Pi(A^c).$$

Where a possibility measure Π is characterized by the maximum property by Equation 9.23, a necessity measure η is characterized by

$$\eta(A \cap B) = \eta(A) \wedge \eta(B) \tag{9.25}$$

For the possibility distribution, when $\mathcal{S}$ is complete, and using the notation from Equation 9.24, we have:

$$\pi(\omega_i) = \sum_{j=i}^{n} m_j, \quad m(A_i) = \pi(\omega_i) - \pi(\omega_{i+1}),$$

where $\pi_{n+1} := 0$ by convention.

An example of a consonant random set is shown in Figure 9.12, where again $\Omega = \{x, y, z, w\}$, but now

$$\mathcal{F} = \{\{x\}, \{x, y\}, \{x, y, z\}, \{x, y, z, w\}\}$$

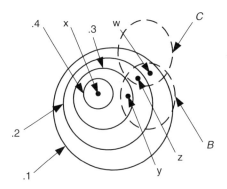

FIGURE 9.12 A consonant, possibilistic random set.

with m as shown. Then, for $B = \{y, z, w\}, C = \{z, w\}$, we have

$$Pl(B) = .1 + .2 + .3 = .6, \quad Pl(C) = .1 + .2 = .3,$$
$$Pl(B \cup C) = .1 + .2 + .3 = .6 = Pl(B) \vee Pl(C),$$

thus characterizing Pl as, in fact, a possibility measure Π. We also have, using vector notation, $\pi = \langle 1, .6, .3, .1 \rangle$.

In considering the information content of a possibility measure, our intuition tells us that it would best be thought of as a kind of nonspecificity such as in Equation 9.17. However, while the nonspecificity of a specific random set vanishes, the strife or conflict of a consonant random set does not. Indeed, it was exactly this observation that drove much of the mathematical development in this area, and thus, technically, the information content of a possibility measure is best captured by an aggregate uncertainty such as Equation 9.18.

However, for our purposes, it is useful to consider Equation 9.17 applied to consonant random sets. Under these conditions, we can express Equation 9.17 in terms of the possibility distribution as:

$$U_N(\pi) := \sum_{i=2}^{n} \pi_i \log_2\left(\frac{i}{i-1}\right) = \sum_{i=1}^{n} (\pi_i - \pi_{i+1}) \log_2(i).$$

9.3.6.2 Crispness, Consistency, and Possibilistic Histograms

In the case of the specific random set discussed in Section 9.3.5.4, not only do the belief measure Bel and plausibility measure Pl collapse together to the decomposable probability measure Pr, but also their traces ρ_{Bel} and ρ_{Pl} collapse to the trace of the probability measure ρ_{Pr}, which is just the probability density p.

But the possibilistic case, which is apparently so parallel, also has some definite differences. First, as we saw, the belief and plausibility measures are distinct as possibility Π and necessity η. Moreover, the necessity measure η is not decomposable, and indeed, the minimum operator $\wedge$ from Equation 9.25 is not a conorm. But moreover, the relation between the measure Pl and its trace ρ_{Pl} is also not so simple.

First consider the relation between intervals and fuzzy intervals discussed in Section 9.3.3.4. In particular, regular (crisp) intervals arise through the extension principle when the characteristic function takes values only in $\{0,1\}$. Similarly, for general possibility distributions, it might be the case that $\pi(\omega_i) \in \{0, 1\}$. In this case, we call π a **crisp possibility distribution**, and otherwise identify a noncrisp possibility distribution as a **proper possibility distribution**. Thus, each fuzzy interval is a general (that is, potentially proper) possibility distribution, while each crisp interval is correspondingly a crisp possibility distribution.

Note that crispness can arise in probability theory only in a degenerate case, because if $\exists \omega_i, p(\omega_i) = 1$, then for all other values, we would have $p(\omega_i) = 0$. We call this case a **certain distribution**, which are the only cases that are both probability and possibility distributions simultaneously.

Now consider random sets where the core is non-empty, which we call **consistent**:

$$C = \bigcap_{A_j \in \mathcal{F}} A_j \neq \emptyset.$$

Note that all consistent random sets are consonant, but not *vice versa*. But in these cases, the trace ρ_{Pl} is a maximal possibility distribution satisfying

$$\bigvee_{\omega \in \Omega} \rho_{Pl}(\omega) = 1,$$

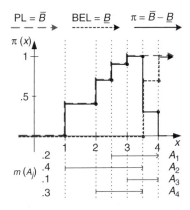

FIGURE 9.13 A consistent random interval with its possibilistic histogram π.

but Pl is not a possibility measure (that is, Equation 9.22 is *not* satisfied). However, it can be shown that for each consistent random set S, there is a unique, well-justified consonant approximation S^* whose plausibilistic trace is equal to that of S [44].

When a consistent random interval is shown as a p-box, it also follows that the trace $\overline{B} - \underline{B} = \pi$ is this same possibility distribution. We have shown [44] that under these conditions, not only is $r = \pi$ a possibility distribution, but moreover is a fuzzy interval as discussed in Section 9.3.3.4. We then call π a **possibilistic histogram**.

An example is shown in Figure 9.13, with m as shown. Note in this case the positive core

$$C = \bigcap_{A_j \in \mathcal{F}} A_j = [3, 3.5),$$

and that over this region, we have $\overline{B} = 1, \underline{B} = 0, r = 1$.

Returning to the domain of general, finite random sets, the relations among these classes is shown in Figure 9.14. Here we use the term "vacuous" to refer to a random set with a single focal element:

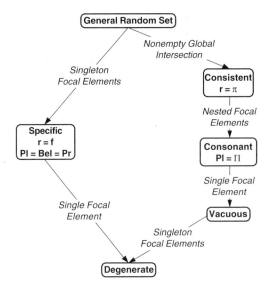

FIGURE 9.14 Relations among classes of random sets.

$\exists A \subseteq \Omega$, $\mathcal{F} = \{A\}$; and "degenerate" to refer to the further case where that single focal element has only one element: $\exists \omega \in \Omega$, $\mathcal{F} = \{\{\omega\}\}$.

9.3.6.3 Interpretations and Applications

Although max-preserving measures have their origins in earlier work, in the context of GIT possibility theory was originally introduced by Zadeh [26] as a kind of information theory strictly related to fuzzy sets. As such, possibility distributions were intended to be measured as and interpreted as linguistic variables.

And, as we have seen, in the context of real-valued fuzzy quantities, it is possible to interpret fuzzy intervals and numbers as possibility distributions. Thus we would hope to use possibility theory as a basis for representing fuzzy arithmetic operations as in Equation 9.8. However, just as there is not a strict symmetry between probabilistic and possibilistic concepts, so there is not a clean generalization here either. In particular, we have shown [45] that the possibilistic properties of fuzzy quantities are not preserved by fuzzy arithmetic convolution operations such as Equation 9.8 outside of their cores and supported.

One of the primary methods for the determination of possibilistic quantities is to take information from a probability distribution and convert it into a possibility distribution. For example, given a discrete probability distribution as a vector $p = \langle p_1, p_2, \ldots, p_n \rangle$, then we can create a possibility distribution $\pi = \langle \pi_1, \pi_2, \ldots, \pi_n \rangle$, where

$$\pi_i = \frac{p_i}{\max p_i}.$$

There are other conversion methods, and an extensive literature, including how the information measure is preserved or not under various transformations [46, 47]. However, one could argue that all such methods are inappropriate: when information is provided in such a way as to be appropriate for a probabilistic approach, then that approach should be used, and *vice versa* [27]. For the purposes of engineering reliability analysis, our belief is that a strong basis for possibilistic interpretations is provided by their grounding in random sets, random intervals, and p-boxes. As we have discussed, this provides a mathematically sound approach for the measurement and interpretation of statistical collections of intervals, which always yield p-boxes, and may or may not yield possibilistic special cases, depending on the circumstances. When they do not, if a possibilistic treatment is still desired, then various normalization procedures are available [48, 49] to transform an inconsistent random set or interval into a consistent or consonant one.

9.3.6.4 Relations between Probabilistic and Possibilistic Concepts

We are now at a point where we can summarize the relations existing between probabilistic and possibilistic concepts in the context of random set theory. To a certain extent, these are complementary, and to another distinct.

Table 9.5 summarizes the relations for general, finite random sets and the special cases of probability and possibility. Columns are shown for the case of general finite random sets, and then the two prominent probabilistic and possibilistic special cases. Note that these describe complete random sets, so that $n = N$ and the indices i on ω_i and j on A_j can be used interchangeably.

In Section 9.3.4 we noted that, formally, all traces of monotone measures are fuzzy sets. Thus, in particular, each probability and possibility distribution is a kind of fuzzy set. In Section 9.3.3.4, we similarly discussed relations among classes of fuzzy quantities, which are fuzzy sets defined on the continuous $\mathbb{R}$. The relations among classes of fuzzy sets defined on a finite space Ω are shown in Figure 9.15, together with examples of each for $\Omega = \{a, b, c, d, e\}$.

Figure 9.16 shows the relations between these distributions or traces as fuzzy sets (quantities defined on the points $\omega \in \Omega$) and the corresponding monotone measures (quantities defined on the subsets $A \subseteq \Omega$). Note that there is not a precise symmetry between probability and possibility. In particular, where probability distributions are symmetric to possibility distributions, probability measures collapse the duality of belief and plausibility, which is exacerbated for possibility.

TABLE 9.5 Summary of Probability and Possibility in the Context of Random Sets

	Random Set	Special Cases: Complete Random Sets, $N=n, i \leftrightarrow j$	
		Probability	Possibility
Focal Sets	Any $A_j \subseteq \Omega$	Singletons: $A_i = \{\omega_i\}$ $\{\omega_i\} = A_i$	Nest: $A_i = \{\omega_1, \ldots, \omega_i\}$ $\{\omega_i\} = A_i - A_{i-1}, A_0 := \emptyset$
Structure	Arbitrary	Finest Partition	Total order
Belief	$\mathrm{Bel}(A) = \sum_{A_j \subseteq A} m_j$	$\mathrm{Pr}(A) := \mathrm{Bel}(A)$	$\eta(A) := \mathrm{Bel}(A)$
Plausibility	$\mathrm{Pl}(A) = \sum_{A_j \cap A \neq \emptyset} m_j$	$\mathrm{Pr}(A) := \mathrm{Pl}(A)$	$\Pi(A) := \mathrm{Pl}(A)$
Relation	$\mathrm{Bel}(A) = 1 - \mathrm{Pl}(A^c)$	$\mathrm{Bel}(A) = \mathrm{Pl}(A) = \mathrm{Pr}(A)$	$\eta(A) = 1 - \Pi(A^c)$
Trace	$\rho_{\mathrm{Pl}}(\omega_i) = \sum_{A_j \ni \omega_i} m_j$	$p(\omega_i) := m(A_j), A_j = \{\omega_i\}$	$\pi(\omega_i) = \sum_{j=i}^{n} m_j$
Measure		$m_i = p_i$ $\mathrm{Pr}(A \cup B) = \mathrm{Pr}(A) + \mathrm{Pr}(B) - \mathrm{Pr}(A \cap B)$	$m_i = \pi_i - \pi_{i+1}$ $\Pi(A \cup B) = \Pi(A) \vee \Pi(B)$
Normalization		$\sum_i p_i = 1$	$\bigvee_i \pi_i = 1$
Operator		$\mathrm{Pr}(A) = \sum_{\omega_i \in A} p_i$	$\Pi(A) = \bigvee_{\omega_i \in A} \pi_i$
Nonspecificity	$U_N(\mathcal{S}) = \sum_{j=1}^{N} m_j \log_2 |A_j|$		$U_N(\pi) = \sum_{i=2}^{n} \pi_i \log_2 \left[\frac{i}{i-1}\right] = \sum_{i=1}^{n} (\pi_i - \pi_{i+1}) \log_2(i)$
Conflict	$U_S(\mathcal{S}) = -\sum_{i=1}^{n} m(\{\omega_i\}) \log_2(m(\{\omega_i\}))$	$U_{\mathrm{prob}}(p) = -\sum_i p_i \log_2(p_i)$	
Strife	$U_S(\mathcal{S}) = -\sum_{j=1}^{N} m_j \log_2 \left[\sum_{k=1}^{N} m_k \frac{|A_j \cap A_k|}{|A_j|}\right]$	$U_{\mathrm{prob}}(p) = -\sum_i p_i \log_2(p_i)$	
Aggregate Uncertainty	$U_{AU}(\mathcal{S}) = \max_{p: \forall A \subseteq \Omega, \mathrm{Pr}(A) \leq \mathrm{Pl}(A)} U_{\mathrm{prob}}(p)$		

9.4 Conclusion and Summary

This chapter described the basic mathematics of the most common and prominent branches of both classical and generalized information theory. In doing so, we have emphasized primarily a perspective drawing from applications primarily in engineering modeling. These principles have use in many other fields, for example data fusion, image processing, and artificial intelligence.

It must be emphasized that there are many different mathematical approaches to GIT. The specific course of development espoused here, and the relations among the components described, is just one among many. There is a very large literature here that the diligent student or researcher can access.

Moreover, there are a number of mathematical components that properly belong to GIT but space precludes a development of here. In particular, research is ongoing concerning a number of additional mathematical subjects, many of which provide even further generalizations of the already generalized

Generalized Information Theory for Engineering Modeling and Simulation

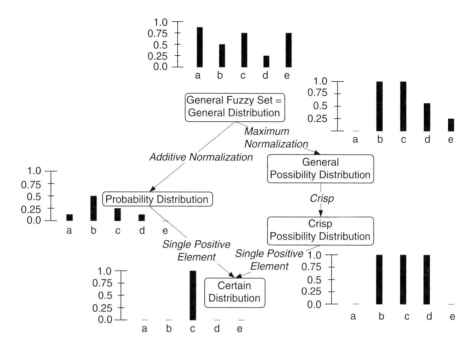

FIGURE 9.15 Relations among classes of general distributions as fuzzy sets.

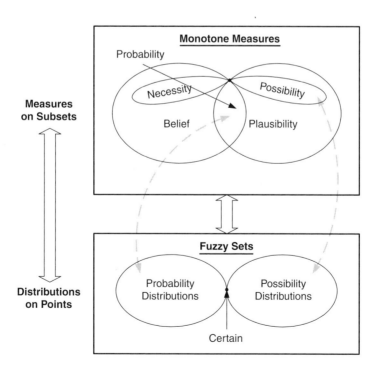

FIGURE 9.16 Relations between distributions on points and measures on subsets.

topics introduced here. These include at least the following:

Rough Sets: Pawlak [50] introduced the structure of a "rough" set, which is yet another way of capturing the uncertainty present in a mathematical system. Let $\mathcal{C} = \{A\} \subseteq 2^\Omega$ be a partition of Ω, and assume a special subset $A_0 \subseteq \Omega$ (not necessarily a member of the partition). Then $\mathbf{R}(A_0) := \{\underline{A}_0, \overline{A}_0\}$ is a **rough set** on Ω, where

$$\underline{A}_0 := \{A \in \mathcal{C} : A \subseteq A_0\}, \quad \overline{A}_0 := \{A \in \mathcal{C} : A \cap A_0 \neq \varnothing\}.$$

Rough sets have been closely related to the GIT literature [51], and are useful in a number of applications [52]. For our purposes, it is sufficient to note that $\overline{A}_0$ effectively specifies the support **U**, and $\underline{A}_0$ the nonempty core **C**, of a number of DS structures, and thus an equivalence class of possibility distributions on Ω [53].

Higher-Order Structures: In Section 9.3.3, we described how Zadeh's original move of generalizing from $\{0, 1\}$-valued characteristic functions to $[0, 1]$-valued membership functions can be thought of as a *process* of "fuzzification." Indeed, this lesson has been taken to heart by the community, and a wide variety of fuzzified structures have been introduced. For example, Type II and Level II fuzzy sets arise when fuzzy weights themselves are given weights, or whole fuzzy sets themselves. Or, fuzzified DS theory arises when focal elements Aj are generalized to fuzzy subsets $\tilde{A}_j \subseteq \Omega$. There is a fuzzified linear algebra; a fuzzified calculus, etc. It is possible to rationalize these generalizations into systems for generating mathematical structures [54], and approach the whole subject from a higher level of mathematical sophistication, for example by using category theory.

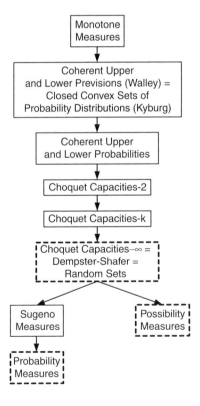

FIGURE 9.17 Relations among classes of monotone measures, imprecise probabilities, and related structures, adapted from [33].

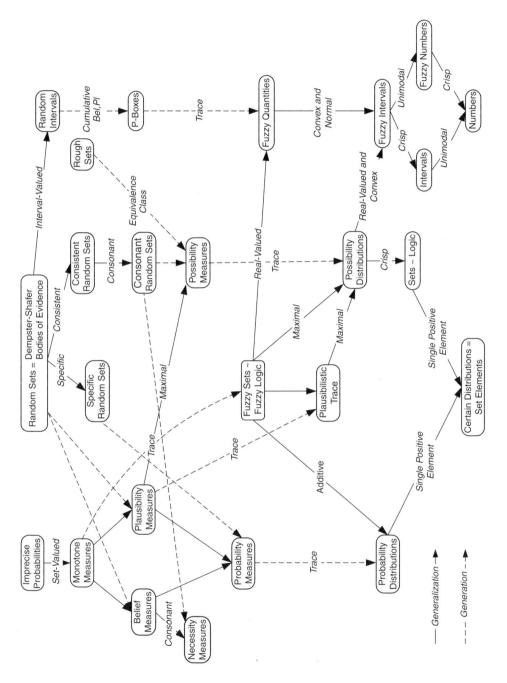

FIGURE 9.18 Map of the various sub-fields of GIT.

Other Monotone Measures, Choquet Capacities, and Imprecise Probabilities: We have focused exclusively on belief and plausibility measures, and their special cases of probability, possibility, and necessity measures. However, there are broad classes of fuzzy or monotone measures outside of these, with various properties worthy of consideration [32]. And while DS theory and random set theory arose in the 1960s and 1970s, their work was presaged by prior work by Choquet in the 1950s that identified many of these classes. Within that context, we have noted above in Section 9.3.1 that belief and plausibility measures stand out as special cases of **infinite order Choquet capacities** [19, 20].

Then, just as monotone measures generalize probability measures by relaxing the additivity property of Equation 9.5, it is also possible to consider relaxing the additivity property of random sets in Equation 9.15, or their range to the unit interval (that is, to consider possibly negative m values), and finally generalizing away from measures on subsets $A \subseteq \Omega$ altogther. Most of the most compelling research ongoing today in the mathematical foundations of GIT concerns these areas, and is described in various ways as **imprecise probabilities** [29] or **convex combinations of probability measures** [33, 55]. Of course, these various generalizations satisfy the extension principle, and thus provide, for example, an alternative basis for the more traditional sub-fields of GIT such as probability theory [56].

Research among all of these more sophisticated classes of generalized measures, and their connections, is active and ongoing. Figure 9.17, adapted from [33], summarizes the current best thought about these relations. Note the appearance in this diagram of certain concepts not explicated here, including previsions, Choquet capacities of finite order, and a class of monotone measures called "Sugeno" measures proper.

We close this chapter with our "grand view" of the relations among most of the structures and classes discussed in Figure 9.18. This diagram is intended to incorporate all of the particular diagrams included earlier in this chapter. Specifically, a solid arrow indicates a mathematical generalization of one theory by another, and thus an instance of where the extension principle should hold. These are labeled with the process by which this specification occurs. A dashed arrow indicates where one kind of structure is generated by another; for example, the trace of a possibility measure yields a possibility distribution.

Acknowledgments

The authors wish to thank the Los Alamos Nuclear Weapons Program for its continued support, especially the Engineering Sciences and Applications Division. This work has also been supported by a research grant from Sandia National Laboratories as part of the Department of Energy Accelerated Strategic Computing Initiative (ASCI).

References

1. Shannon, CE and Weaver, W: (1964) *Mathematical Theory of Communication*, University of Illinois Press, Urbana.
2. Klir, George: (1991) "Generalized Information Theory", *Fuzzy Sets and Systems*, v. **40**, pp. 127–142.
3. Bement, T; Booker, JM; McNulty, S; and N. Singpurwalla: (2003) "Testing the Untestable: Reliaiblity in the 21st Century", *IEEE Transactions in Reliability*, v. **57**, pp. 114–128.
4. Klir, George and Wierman, Mark J: (1997) *Uncertainty-Based Information Elements of Generalized Information Theory*, in: *Lecture Notes in Computer Science*, Creighton University, Omaha.
5. Moore, RM: (1979) *Methods and Applications of Interval Analysis*, in: *SIAM Studies in Applied Mathematics*, SIAM, Philadelphia.
6. Lindley, D: (1987) "Comment: A Tale of Two Wells", *Statistical Science*, v. **2**:1, pp. 38–40.
7. Bayes, T: (1763) "An Essay Towards Solving a Problem in the Doctrine of Chances", *Philosophical Trans. of the Royal Society*, v. **53**, pp. 370–418.

8. Kolmogorov, AN: (1956) *Foundations of the Theory of Probability*, Chelsea, New York.
9. Ben-Haim, Yakov: (2001) *Information-Gap Decision Theory*, Academic Press, London.
10. Sellers, K and Booker, J: (2002) "Bayesian Methods", in: *Fuzzy Logic and Probability Applications: Bridging the*, ed. Ross, Booker, and Parkinson, SIAM, Philadelphia.
11. Graves, T: (2001) "YADAS: An Object-Oriented Framework for Data Analysis Using Markov Chain Monte Carlo", *LANL Technical Report LA-UR-01-4804*.
12. Dubois, Didier and Prade, Henri: (1993) "Fuzzy sets and Probability: Misunderstandings, Bridges and Gaps", in: *Proc. FUZZ-IEEE '93*.
13. Klir, George and Yuan, Bo: (1995) *Fuzzy Sets and Fuzzy Logic*, Prentice-Hall, New York.
14. Klir, George J: (1997) "Uncertainty Theories, Measures, and Principles: An Overview of Personal Views and Contributions", in: *Uncertainty: Models and Measures*, ed. HG Natke and Y Ben-Haim, pp. 27–43, Akademie Verlag, Berlin.
15. Ross, Timothy and Parkinson, J: (2002) "Fuzzy Set Theory, Fuzzy Logic and, Fuzzy Systems", in: *Fuzzy Logic and Probability Applications: Bridging the Gap*, ed. Ross, Booker, and Parkinson, SIAM, Philadelphia.
16. Zadeh, Lotfi A: (1965) "Fuzzy Sets and Systems", in: *Systems Theory*, ed. J. Fox, pp. 29–37, Polytechnic Press, Brooklyn NY.
17. Zadeh, Lotfi A: (1965) "Fuzzy Sets", *Information and Control*, v. **8**, pp. 338–353.
18. Dempster, AP: (1967) "Upper and Lower Probabilities Induced by a Multivalued Mapping", *Annals of Mathematical Statistics*, v. **38**, pp. 325–339.
19. Lamata, MT and Moral, S: (1989) "Classification of Fuzzy Measures", *Fuzzy Sets and Systems*, v. **33**, pp. 243–153.
20. Sims, John R and Wang, Zhenyuan: (1990) "Fuzzy Measures and Fuzzy Integrals: An Overview", *Int. J. of General Systems*, v. **17**:2–3, pp. 157–189.
21. Shafer, Glen: (1976) *Mathematical Theory of Evidence*, Princeton University Press, Princeton, NJ.
22. Kendall, DG: (1974) "Foundations of a Theory of Random Sets", in: *Stochastic Geometry*, ed. EF Harding and DG Kendall, pp. 322–376, Wiley, New York.
23. Matheron, G: (1975) *Random Sets and Integral Geometry*, Wiley, New York.
24. Dempster, AP: (1968) "Upper and Lower Probabilities Generated by a Random Interval", *Annals of Mathematical Statistics*, v. **39**:3, pp. 957–966.
25. Sugeno, Michio: (1972) "Fuzzy Measures and Fuzzy Integrals", *Trans. SICE*, v. **8**:2.
26. Zadeh, Lotfi A: (1978) "Fuzzy Sets as the Basis for a Theory of Possibility", *Fuzzy Sets and Systems*, v. **1**, pp. 3–28.
27. Joslyn, Cliff: (1995) "In Support of an Independent Possibility Theory", in: *Foundations and Applications of Possibility Theory*, ed. G de Cooman et al., pp. 152–164, World Scientific, Singapore.
28. Klir, George and Folger, Tina: (1987) *Fuzzy Sets, Uncertainty, and Information*, Prentice Hall.
29. Walley, P: (1990) *Statistical Reasoning with Imprecise Probabilities*, Chapman & Hall, New York.
30. Singpurwalla, N and Booker, J: (2002) "Membership Functions and Probability Measures of Fuzzy Sets", *Los Alamos Technical Report LAUR 02-0032.*, LANL.
31. Booker, JM and Singpurwalla, N: (2003) "Using Probability and Fuzzy Set Theory for Reliability Assessment", in: *Proc. 21st Int. Modal Analysis Conference (IMAC-XXI)*, pp. 259, Kissimmee FL.
32. Wang, Zhenyuan and Klir, George J: (1992) *Fuzzy Measure Theory*, Plenum Press, New York.
33. Klir, George and Smith, Richard M: (2000) "On Measuring Uncertainty and Uncertainty-Based Information: Recent Developments", *Annals of Mathematics and Artificial Intelligence*, v. **32**:1-4, pp. 5–33.
34. Sentz, Kari and Ferson, Scott: (2002) "Combination Rules in Dempster-Shafer Theory", in: *6th World Multi-Conference on Systemic, Cybernetics, and Informatics*.
35. Joslyn, Cliff: (1996) "Aggregation and Completion of Random Sets with Distributional Fuzzy Measures", *Int. J. of Uncertainty, Fuzziness, and Knowledge-Based Systems*, v. **4**:4, pp. 307–329.
36. Dubois, Didier and Prade, Henri: (1990) "Consonant Approximations of Belief Functions", *Int. J. Approximate Reasoning*, v. **4**, pp. 419–449.

37. Helton, JC and Oberkampf, WL, eds.: (2004) "Special Issue: Alternative Representations of Epistemic Uncertainty", *Reliability Engineering and Systems Safety*, 95:1-3.
38. Joslyn, Cliff and Helton, Jon C: (2002) "Bounds on Plausibility and Belief of Functionally Propagated Random Sets", in: *Proc. Conf. North American Fuzzy Information Processing Society (NAFIPS 2002)*, pp. 412–417.
39. Joslyn, Cliff and Kreinovich, Vladik: (2002) "Convergence Properties of an Interval Probabilistic Approach to System Reliability Estimation", *Int. J. General Systems*, in press.
40. Oberkampf, WL; Helton, JC; Wojtkiewicz, SF, Cliff Joslyn, and Scott Ferson: (2004) "Uncertainty in System Response Given Uncertain Parameters", *Reliability Reliability Engineering and System Safety*, 95:1-3, pp. 11–20.
41. Tonon, Fulvio and Bernardini, Alberto: (1998) "A Random Set Approach to the Optimization of Uncertain Structures", *Computers and Structures*, v. **68**, pp. 583–600.
42. Ferson, Scott; Kreinovich, V; and Ginzburg, L et al.: (2002) "Constructing Probability Boxes and Dempster-Shafer Structures", *SAND Report 2002-4015*, Sandia National Lab, Albuquerque NM, http://www.sandia.gov/epistemic/Reports/SAND2002-4015.pdf.
43. Yager, Ronald R: (1986) "Arithmetic and Other Operations on Dempster-Shafer Structures", *Int. J. Man-Machine Studies*, v. **25**, pp. 357–366.
44. Joslyn, Cliff: (1997) "Measurement of Possibilistic Histograms from Interval Data", *Int. J. General Systems*, v. **26**:1-2, pp. 9–33.
45. Joslyn, Cliff and Ferson, Scott: (2003) "Convolutions of Representations of Random Intervals", in preparation.
46. Joslyn, Cliff: (1997) "Distributional Representations of Random Interval Measurements", in: *Uncertainty Analysis in Engineering and the Sciences*, ed. Bilal Ayyub and Madan Gupta, pp. 37–52, Kluwer.
47. Sudkamp, Thomas: (1992) "On Probability-Possibility Transformations", *Fuzzy Sets and Systems*, v. **51**, pp. 73–81.
48. Joslyn, Cliff: (1997) "Possibilistic Normalization of Inconsistent Random Intervals", *Advances in Systems Science and Applications*, special issue, ed. Wansheng Tang, pp. 44–51, San Marcos, TX, ftp://wwwc3.lanl.gov/pub/users/joslyn/iigss97.pdf.
49. Oussalah, Mourad: (2002) "On the Normalization of Subnormal Possibility Distributions: New Investigations", *Int. J. General Systems*, v. **31**:3, pp. 227–301.
50. Pawlak, Zdzislaw: (1991) *Rough Sets: Theoretical Aspects of Reasoning about Data*, Kluwer, Boston.
51. Dubois, Didier and Prade, Henri: (1990) "Rough Fuzzy Sets and Fuzzy Rough Sets", *Int. J. General Systems*, v. **17**, pp. 191–209.
52. Lin, TY and Cercone, N, eds.: (1997) *Rough Sets and Data Mining: Analysis of Imprecise Data*, Kluwer, Boston.
53. Joslyn, Cliff: (2003) "Multi-Interval Elicitation of Random Intervals for Engineering Reliability Analysis", *Proc. 2003 Int. Symp. on Uncertainty Modeling and Analysis (ISUMA 03)*.
54. Joslyn, Cliff and Rocha, Luis: (1998) "Towards a Formal Taxonomy of Hybrid Uncertainty Representations", *Information Sciences*, v. **110**:3-4, pp. 255–277.
55. Kyburg, HE (1987): "Bayesian and Non-Bayesian Evidential Updating", *Artificial Intelligence*, v. **31**, pp. 271–293.
56. de Cooman, Gert and Aeyels, Dirk: (2000) "A Random Set Description of a Possibility Measure and Its Natural Extension", *IEEE Trans. on Systems, Man and Cybernetics A*, v. **30**:2, pp. 124–130.

10
Evidence Theory for Engineering Applications

10.1 Introduction **10**-1
   Background • Improved Models for Epistemic Uncertainty
   • Chapter Outline

10.2 Fundamentals of Evidence Theory **10**-4
   Belief, Plausibility, and BPA Functions • Cumulative and
   Complementary Cumulative Functions • Input/Output
   Uncertainty Mapping • Simple Conceptual Examples

10.3 Example Problem... **10**-12
   Problem Description • Traditional Analysis Using
   Probability Theory • Analysis Using Evidence
   Theory • Comparison and Interpretation of Results

10.4 Research Topics in the Application of
   Evidence Theory **10**-26

William L. Oberkampf
Sandia National Laboratories

Jon C. Helton
Sandia National Laboratories

10.1 Introduction

10.1.1 Background

Computational analysis of the performance, reliability, and safety of engineered systems is spreading rapidly in industry and government. To many managers, decision makers, and politicians not trained in computational simulation, computer simulations can appear most convincing. Terminology such as "virtual prototyping," "virtual testing," "full physics simulation," and "modeling and simulation-based acquisition" are extremely appealing when budgets are highly constrained; competitors are taking market share; or when political constraints do not allow testing of certain systems. To assess the accuracy and usefulness of computational simulations, three key aspects are needed in the analysis and experimental process: computer code and solution verification; experimental validation of most, if not all, of the mathematical models of the engineered system being simulated; and estimation of the uncertainty associated with analysis inputs, physics models, possible scenarios experienced by the system, and the outputs of interest in the simulation. The topics of verification and validation are not addressed here, but these are covered at length in the literature (see, for example, [1–6]). A number of fields have contributed to the development of uncertainty estimation techniques and procedures, such as nuclear reactor safety, underground storage of radioactive and toxic wastes, and structural dynamics (see, for example, [7–18]).

Uncertainty estimation for engineered systems is sometimes referred to as the simulation of nondeterministic systems. The mathematical model of the system, which includes the influence of the environment on the system, is considered nondeterministic in the sense that: (i) the model can produce nonunique system responses because of the existence of uncertainty in the input data for the model, or (ii) there are multiple alternative mathematical models for the system. The mathematical models, however, are assumed to be deterministic in the sense that when all necessary input data for a designated model is specified, the model produces only one value for every output quantity. To predict the nondeterministic response of the system, it is necessary to evaluate the mathematical model, or alternative mathematical models, of the system multiple times using different input data. This presentation does not consider chaotic systems or systems with hysteresis, that is, mathematical models that map a unique input state to multiple output states.

Many investigators in the risk assessment community segregate uncertainty into *aleatory* uncertainty and *epistemic* uncertainty. Aleatory uncertainty is also referred to as variability, irreducible uncertainty, inherent uncertainty, stochastic uncertainty, and uncertainty due to chance. Epistemic uncertainty is also referred to as reducible uncertainty, subjective uncertainty, and uncertainty due to lack of knowledge. Some of the investigators who have argued for the importance of distinguishing between aleatory uncertainty and epistemic uncertainty are noted in [19–32]. We believe the benefits of distinguishing between aleatory and epistemic uncertainty include improved interpretation of simulation results by decision makers and improved ability to allocate resources to decrease system response uncertainty or risk. Note that in the present work we use the term "risk" to mean a measure of the likelihood and severity of an adverse event occurring [13, 33].

Sources of aleatory uncertainty can commonly be singled out from other contributors to uncertainty by their representation as randomly distributed quantities that take values in an established or known range, but for which the exact value will vary by chance from unit to unit or from time to time. The mathematical representation most commonly used for aleatory uncertainty is a probability distribution. When substantial experimental data is available for estimating a distribution, there is no debate that the correct mathematical model for aleatory uncertainty is a probability distribution. Propagation of these distributions through a modeling and simulation process is well developed and is described in many texts (see, for example, [31, 34–38]).

Epistemic uncertainty derives from some level of ignorance about the system or the environment. For this presentation, *epistemic uncertainty* is defined as any lack of knowledge or information in any phase or activity of the modeling process [39]. The key feature stressed in this definition is that the fundamental source of epistemic uncertainty is incomplete information or incomplete knowledge of some characteristic of the system or the environment. As a result, an increase in knowledge or information can lead to a reduction in the predicted uncertainty of the response of the system, all things being equal. Examples of sources of epistemic uncertainty are: little or no experimental data for a fixed (but unknown) physical parameter, a range of possible values of a physical quantity provided by expert opinion, limited understanding of complex physical processes, and the existence of fault sequences or environmental conditions not identified for inclusion in the analysis of a system. For further discussion of the sources of epistemic uncertainty in engineering systems see, for example, [40, 41].

Epistemic uncertainty has traditionally been represented with a random variable using subjective probability distributions. However, a major concern is that when there is little or no closely related experimental data, a common practice is to simply pick some familiar probability distribution and its associated parameters to represent one's belief in the likelihood of possible values that could occur. Two important weaknesses with this common approach are of critical interest when the assessment of epistemic uncertainty is the focus. First, even small epistemic uncertainty in parameters for continuous probability distributions, such as a normal or Weibull, can cause very large changes in the tails of the distributions. For example, there can be orders-of-magnitude change in the likelihood of rare events when certain distribution parameters are changed by small amounts. Second, when epistemic uncertainty is represented as a probability distribution and when there are multiple parameters treated in this fashion, one can obtain misleading results. For example, suppose there are ten parameters in an analysis that are only *thought* to be within specified intervals; for example, the parameters are estimated from expert opinion, not measurements. Assume each of these parameters is treated as a random variable and assigned the least informative distribution (i.e., a uniform distribution). If extreme system responses correspond to extreme values of these parameters (i.e., values near the ends of the uniform distribution), then their probabilistic combination could predict a very low probability for such extreme system responses. Given that the parameters are only known to occur within intervals, however, this conclusion is grossly inappropriate.

10.1.2 Improved Models for Epistemic Uncertainty

During the past two decades, the information theory and expert systems communities have made significant progress in developing a number of new theories that can be pursued for modeling epistemic uncertainty. Examples of the newer theories include fuzzy set theory [17, 42–46], interval analysis [47, 48], evidence (Dempster-Shafer) theory [49–55], possibility theory [56, 57], and theory of upper and lower previsions [58]. Some of these theories only deal with epistemic uncertainty; most deal with both epistemic and aleatory uncertainty; and some deal with other varieties of uncertainty (e.g., nonclassical logics appropriate for artificial intelligence and data fusion systems [59]).

A recent article summarizes how these theories of uncertainty are related to one another from a hierarchical viewpoint [60]. The article shows that evidence theory is a generalization of classical probability theory. From the perspective of bodies of evidence and their measures, evidence theory can also be considered a generalization of possibility theory. However, in evidence theory and in possibility theory, the mechanics of operations applied to bodies of evidence are completely different [54, 57]. The mathematical foundations of evidence theory are well established and explained in several texts and key journal articles [49–55, 61–64]. However, essentially all of the published applications of the theory are for simple model problems—not actual engineering problems [41, 65–74]. Note that in some of the literature, evidence theory is referred to as the theory of random sets.

In evidence theory there are two complementary measures of uncertainty: belief and plausibility. Together, belief and plausibility can be thought of as defining lower and upper limits of probabilities, respectively, or interval-valued probabilities. That is, given the information or evidence available, a precise (i.e., single) probability distribution cannot be specified. Rather, a range of possible probabilities exists, all of which are consistent with the evidence. Belief and plausibility measures can be based on many types of information or evidence (e.g., experimental data for long-run frequencies of occurrence, scarce experimental data, theoretical evidence, or individual expert opinion or consensus among experts concerning the range of possible values of a parameter or possibility of the occurrence of an event). We believe that evidence theory could be an effective path forward in engineering applications because it can deal with both thoroughly characterized situations (e.g., precisely known probability distributions) and situations of near-total ignorance (e.g., only an interval containing the true value is known).

There are two fundamental differences between the approach of evidence theory and the traditional application of probability theory. First, evidence theory uses two measures—belief and plausibility—to characterize uncertainty; in contrast, probability theory uses only one measure—the probability of an event or value. Belief and plausibility measures are statements about the likelihood related to sets of possible values. There is no need to distribute the evidence to individual values in the set. For example, evidence from experimental data or from expert opinion can be given for a parameter value to be within an interval. Such evidence makes no claim concerning any specific value within the interval or the likelihood of any one value compared with any other value in the interval. In other words, less information can be specified than the least information that is typically specified in applications of probability theory (e.g., the uniform likelihood of all values in the interval).

The second fundamental difference between evidence theory and the traditional application of probability theory is that in evidence theory, the evidential measure for an event and the evidential measure for the negation of an event do not have to sum to unity (i.e., certainty). In probability theory, the measure for an event plus the measure against an event (i.e., its negation) must be unity. This sum to unity implies that the absence in the evidence for an event must be equivalent to the evidence for the negation of the event. In evidence theory, this equivalence is rejected as excessively restrictive; that is, a weak statement of evidence can result in support for an event, but the evidence makes no inference for the support of the negation of the event.

As a final background comment concerning historical perspective of evidence theory, the theory is philosophically related to the approach of Bayesian estimation. Indeed, many of the originators of evidence theory viewed it as an offshoot of Bayesian estimation for the purpose of more properly dealing with subjective probabilities. There are, however, two key differences. First, evidence theory does not assign any prior distributions to a given state of knowledge (i.e., body of evidence) if none are given. The requirement for such a prior distribution is obviated because all possible probability distributions are allowed to describe the body of evidence. Second, evidence theory does not embody the theme of updating probabilities as new evidence becomes available. In Bayesian estimation, a dominant theme is toward continually improving statistical inference as new evidence becomes available; whereas in evidence theory, the emphasis is on precisely stating the present state of knowledge—not updating the statistical evidence.

10.1.3 Chapter Outline

In the following section (Section 10.2), the mathematical framework of evidence theory is explained and contrasted to the traditional application of probability theory. The definitions of belief and plausibility are given, along with the relationships between them. Also discussed is an important quantity called the basic probability assignment (BPA). A few simple examples are given illustrating the interpretation of the belief and plausibility functions and the BPA. Section 10.3 discusses the application of evidence theory to a simple example problem. The simple system is given by an algebraic equation with two uncertain input parameters and one system response variable. The example is analyzed using the traditional application of probability theory and evidence theory. The discussion of each solution approach stresses the mathematical and procedural steps needed to compute uncertainty bounds in the system response, as well as the similarities and differences between each approach. The analyses using traditional probabilistic and evidence theory approaches are compared with regard to their assessment for the system yielding an unsafe response. The presentation concludes in Section 10.4 with a brief discussion of important research and practical issues hindering the widespread application of evidence theory to large-scale engineering systems. For example, the critical issue of propagating BPAs through a "black box" computer code using input/output sampling techniques is discussed.

10.2 Fundamentals of Evidence Theory

10.2.1 Belief, Plausibility, and BPA Functions

Evidence theory provides an alternative to the traditional manner in which probability theory is used to represent uncertainty by allowing less restrictive statements about "likelihood" than is the case with a full probabilistic specification of uncertainty. Evidence theory can be viewed as a generalization of the traditional application of probability theory. By "generalization" we mean that when probability distributions are specified, evidence theory yields the same measures of likelihood as the traditional application of probability theory. Evidence theory involves two specifications of likelihood—a belief and a plausibility—for each subset of the universal set under consideration. Formally, an application of evidence theory involves the specification of a triple $(S, \mathbb{S}, m)$, where (i) S is a set that contains everything that could occur in the particular universe under consideration, typically referred to as the sample space or universal set; (ii) $\mathbb{S}$ is a countable collection of subsets of S, typically referred to as the set of focal elements of S; and (iii) m is a function defined on subsets of S such that

$$m(\mathcal{E}) > 0 \quad \text{if} \quad \mathcal{E} \in \mathbb{S}, m(\mathcal{E}) = 0 \quad \text{if} \quad \mathcal{E} \subset S \text{ and } \mathcal{E} \notin \mathbb{S}, \tag{10.1}$$

and

$$\sum_{\mathcal{E}\in\mathbb{S}} m(\mathcal{E}) = 1. \qquad (10.2)$$

The quantity $m(\mathcal{E})$ is referred to as the basic probability assignment (BPA), or the mass function, associated with the subset $\mathcal{E}$ of $\mathcal{S}$.

The sets $\mathcal{S}$ and $\mathbb{S}$ are similar to probability theory where one has the specification of a triple $(\mathcal{S}, \mathbb{S}, p)$ called a probability space, where (i) $\mathcal{S}$ is a set that contains everything that could occur in the particular universe under consideration, (ii) $\mathbb{S}$ is a suitably restricted set of subsets of $\mathcal{S}$, and (iii) p is the function that defines probability for elements of $\mathbb{S}$ (see [75], Section IV.4). In probability theory, the set $\mathbb{S}$ is required to have the properties that (i) if $\mathcal{E} \in \mathbb{S}$, then $\mathcal{E}^c \in \mathbb{S}$, where $\mathcal{E}^c$ is the complement of $\mathcal{E}$, and (ii) if $\mathcal{E}_1, \mathcal{E}_2, \ldots$ is a sequence of elements of $\mathbb{S}$, then $\cup_i \mathcal{E}_i \in \mathbb{S}$ and $\cap_i \mathcal{E}_i \in \mathbb{S}$. Further, p is required to have the properties that (i) if $\mathcal{E} \in \mathbb{S}$, then $0 \le p(\mathcal{E}) \le 1$, (ii) $p(\mathcal{S}) = 1$, and (iii) if $\mathcal{E}_1, \mathcal{E}_2, \ldots$ is a sequence of disjoint sets from $\mathbb{S}$, then $p(\cup_i \mathcal{E}_i) = \Sigma_i\, p(\mathcal{E}_i)$ (see [75], Section IV.3). In the terminology of probability theory, $\mathcal{S}$ is called the sample space or universal set; elements of $\mathcal{S}$ are called elementary events; subsets of $\mathcal{S}$ contained in $\mathbb{S}$ are called events; the set $\mathbb{S}$ itself has the properties of what is called a σ-algebra (see [75], Section IV.3); and p is called a probability measure.

The sample space $\mathcal{S}$ plays the same role in both probability theory and evidence theory. However, the $\mathbb{S}$ set has a different character in the two theories. In probability theory, $\mathbb{S}$ has special algebraic properties fundamental to the development of probability and contains all subsets of $\mathcal{S}$ for which probability is defined (see [75], Section IV.3). In evidence theory, $\mathbb{S}$ has no special algebraic properties (i.e., $\mathbb{S}$ is not required to be a σ-algebra, as is the case in probability theory) and contains the subsets of $\mathcal{S}$ with nonzero BPAs. In probability theory, the function p actually defines the probabilities for elements of $\mathbb{S}$, with these probabilities being the fundamental measure of likelihood. In evidence theory, the function m is *not* the fundamental measure of likelihood. Rather, there are two measures of likelihood, called belief and plausibility, that are obtained from m as described in the next paragraph. The designation BPA for $m(\mathcal{E})$ is almost universally used, but, unfortunately, m does not define probabilities except under very special circumstances. Given the requirement in Equation 10.2, the set $\mathbb{S}$ of focal elements associated with an evidence space $(\mathcal{S}, \mathbb{S}, m)$ can contain at most a countable number of elements; in contrast, the set $\mathbb{S}$ of events associated with a probability space $(\mathcal{S}, \mathbb{S}, p)$ can, and essentially always does, contain an uncountable number of elements.

The belief, $Bel(\mathcal{E})$, and plausibility, $Pl(\mathcal{E})$, for a subset $\mathcal{E}$ of $\mathcal{S}$ are defined by

$$Bel(\mathcal{E}) = \sum_{\mathcal{U} \subset \mathcal{E}} m(\mathcal{U}) \qquad (10.3)$$

and

$$Pl(\mathcal{E}) = \sum_{\mathcal{U} \cap \mathcal{E} \neq \emptyset} m(\mathcal{U}). \qquad (10.4)$$

Conceptually, $m(\mathcal{U})$ is the amount of likelihood that is associated with a set $\mathcal{U}$ that cannot be further assigned to specific subsets of $\mathcal{U}$. Specifically, no specification is implied concerning how this likelihood is apportioned over $\mathcal{U}$. Given the preceding conceptualization of $m(\mathcal{U})$, the belief $Bel(\mathcal{E})$ can be viewed as the minimum amount of likelihood that *must* be associated with $\mathcal{E}$. Stated differently, $Bel(\mathcal{E})$ is the amount of likelihood that must be associated with $\mathcal{E}$ because the summation in Equation 10.3 involves all $\mathcal{U}$ that satisfy $\mathcal{U} \subset \mathcal{E}$. Similarly, the plausibility $Pl(\mathcal{E})$ can be viewed as the maximum amount of likelihood that *could* be associated with $\mathcal{E}$. Stated differently, $Pl(\mathcal{E})$ is the maximum amount of likelihood that could possibly be associated with $\mathcal{E}$ because the summation in Equation 10.4 involves all $\mathcal{U}$ that intersect $\mathcal{E}$. From the perspective of making informed decisions, the information provided by beliefs and plausibilities is more useful than the information provided by BPAs. This statement is made because a BPA only provides likelihood information that can be attributed to a set, but to none of its subsets. In contrast, a belief provides likelihood information about a set and all its subsets, and a plausibility provides likelihood information about a set and all sets that intersect it.

Belief and plausibility satisfy the equality

$$Bel(\mathcal{E}) + Pl(\mathcal{E}^c) = 1 \qquad (10.5a)$$

for every subset $\mathcal{E}$ of $\mathcal{S}$. In words, the belief in the occurrence of an event (i.e., $Bel(\mathcal{E})$) and the plausibility of the nonoccurrence of an event (i.e., $Pl(\mathcal{E}^c)$) must sum to one. In contrast, probability theory relies on the analogous equation

$$p(\mathcal{E}) + p(\mathcal{E}^c) = 1. \qquad (10.5b)$$

As is well known, Equation 10.5b states that the likelihood in the occurrence of an event and the likelihood of the nonoccurrence of an event must sum to one. Stated differently, in probability theory, the likelihood for an event is the complement of the likelihood against an event, whereas in evidence theory, there is no such assumption of symmetry.

In evidence theory, it can be shown that

$$Bel(\mathcal{E}) + Bel(\mathcal{E}^c) \leq 1. \qquad (10.6)$$

The specification of belief is capable of incorporating a lack of assurance that is manifested in the sum of the beliefs in the occurrence (i.e., $Bel(\mathcal{E})$) and nonoccurrence (i.e., $Bel(\mathcal{E}^c)$) of an event $\mathcal{E}$ being less than one. Stated differently, the belief function is a likelihood measure that allows evidence for and against an event to be inconclusive. For the plausibility function, it can be shown that

$$Pl(\mathcal{E}) + Pl(\mathcal{E}^c) \geq 1. \qquad (10.7)$$

The specification of plausibility is capable of incorporating a recognition of alternatives that is manifested in the sum of the plausibilities in the occurrence (i.e., $Pl(\mathcal{E})$) and nonoccurrence (i.e., $Pl(\mathcal{E}^c)$) of an event $\mathcal{E}$ being greater than one. Stated differently, the plausibility function is a likelihood measure that allows evidence for and against an event to be redundant.

10.2.2 Cumulative and Complementary Cumulative Functions

In probability theory, the cumulative distribution function (CDF) and the complementary cumulative distribution function (CCDF) are commonly used to provide summaries of the information contained in a probability space $(\mathcal{S}, \mathbb{S}, p)$. The CCDF is also referred to as the *exceedance risk* curve in risk assessment analyses. Similarly in evidence theory, cumulative belief functions (CBFs), complementary cumulative belief functions (CCBFs), cumulative plausibility functions (CPFs), and complementary cumulative plausibility functions (CCPFs) can be used to summarize beliefs and plausibilities. Specifically, CBFs, CCBFs, CPFs, and CCPFs are defined by the sets of points

$$\mathcal{CBF} = \left\{ \left[v, Bel\left(\mathcal{S}_v^c\right)\right], v \in \mathcal{S} \right\} \qquad (10.8)$$

$$\mathcal{CCBF} = \left\{ [v, Bel(\mathcal{S}_v)], v \in \mathcal{S} \right\} \qquad (10.9)$$

$$\mathcal{CPF} = \left\{ \left[v, Pl\left(\mathcal{S}_v^c\right)\right], v \in \mathcal{S} \right\} \qquad (10.10)$$

$$\mathcal{CCPF} = \left\{ [v, Pl(\mathcal{S}_v)], v \in \mathcal{S} \right\}, \qquad (10.11)$$

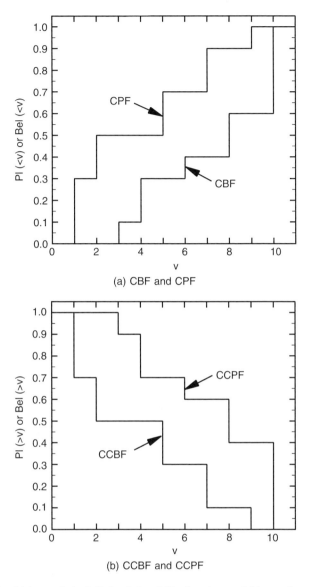

FIGURE 10.1 Example of (a) cumulative belief and plausibility functions and (b) complementary cumulative belief and plausibility functions.

where $\mathcal{S}_v$ is defined as

$$\mathcal{S}_v = \{x : x \in \mathcal{S} \text{ and } x > v\}. \tag{10.12}$$

Plots of the points in the preceding sets produce CBFs, CCBFs, CPFs, and CCPFs (Figure 10.1).

As grouped in Figure 10.1, a CBF and the corresponding CPF occur together naturally as a pair because, for a given value v on the abscissa, (i) the value of the CBF (i.e., $Bel(\mathcal{S}_v^c)$) is the smallest probability for $\mathcal{S}_v^c$ that is consistent with the information characterized by $(\mathcal{S}, \mathbb{S}, m)$ and (ii) the value of the CPF (i.e., $Pl(\mathcal{S}_v^c)$) is the largest probability for $\mathcal{S}_v^c$ that is consistent with the information characterized by $(\mathcal{S}, \mathbb{S}, m)$. A similar interpretation holds for the CCBF and CCPF. Indeed, this bounding relationship occurs for any subset $\mathcal{E}$ of $\mathcal{S}$, and thus $Pl(\mathcal{E})$ and $Bel(\mathcal{E})$ can be thought of as defining upper and lower probabilities for $\mathcal{E}$ [61].

10.2.3 Input/Output Uncertainty Mapping

The primary focus in many, if not most, engineering problems involving uncertainty estimation is on functions

$$y = f(\boldsymbol{x}), \qquad (10.13)$$

where

$$\boldsymbol{x} = [x_1, x_2, \ldots, x_n]$$

and the uncertainty in each x_i is characterized by a probability space $(X_i, \mathbb{X}_i, p_i)$ or an evidence space $(X_i, \mathbb{X}_i, m_i)$. The elements x_i of $\boldsymbol{x}$ are used to construct the input sample space

$$X = \{x : x = [x_1, x_2, \ldots, x_n] \in X_1 \times X_2 \times \ldots \times X_n\}. \qquad (10.14)$$

When the uncertainty in each x_i is characterized by a probability space $(X_i, \mathbb{X}_i, p_i)$, this leads to a probability space $(X, \mathbb{X}, p_X)$ that characterizes the uncertainty in $\boldsymbol{x}$, where $\mathbb{X}$ is developed from the sets contained in

$$\mathbb{C} = \{\mathcal{E} : \mathcal{E} = \mathcal{E}_1 \times \mathcal{E}_2 \times \ldots \times \mathcal{E}_n \in \mathbb{X}_1 \times \mathbb{X}_2 \times \ldots \times \mathbb{X}_n\} \qquad (10.15)$$

(see [75], Section IV.6, and [76], Section 2.6) and p_X is developed from the probability functions (i.e., measures) p_i. Specifically, if the x_i are independent and d_i is the density function associated with p_i, then

$$p_X(\mathcal{E}) = \int_{\mathcal{E}} d(\boldsymbol{x}) \, dV \qquad (10.16)$$

for $\mathcal{E} \in \mathbb{X}$ and

$$d(\boldsymbol{x}) = \prod_{i=1}^{n} d_i(x_i) \qquad (10.17)$$

for $\boldsymbol{x} = [x_1, x_2, \ldots, x_n] \in X$.

In engineering practice, f is often a set of nonlinear partial differential equations that are numerically solved on a computer. The dimensionality of the vector $\boldsymbol{x}$ of inputs can be high in practical problems (e.g., on the order of 50). The analysis outcome y is also often a vector of high dimensionality, but is indicated in Equation 10.13 as being a single system response for notational convenience.

The sample space X constitutes the domain for the function f in Equation 10.13. In turn, the range of f is given by the set

$$\mathcal{Y} = \{y : y = f(\boldsymbol{x}), \boldsymbol{x} \in X\}. \qquad (10.18)$$

The uncertainty in the values of y contained in $\mathcal{Y}$ derives from the probability space $(X, \mathbb{X}, p_X)$ that characterizes the uncertainty in $\boldsymbol{x}$ and from the properties of the function f. In concept, $(X, \mathbb{X}, p_X)$ and f induce a probability space $(\mathcal{Y}, \mathbb{Y}, p_Y)$. The probability p_Y is defined for a subset $\mathcal{E}$ of $\mathcal{Y}$ by

$$p_Y(\mathcal{E}) = p_X(f^{-1}(\mathcal{E})), \qquad (10.19)$$

where

$$f^{-1}(\mathcal{E}) = \{x : x \in \mathcal{X} \text{ and } y = f(x) \in \mathcal{E}\}. \qquad (10.20)$$

The uncertainty in y characterized by the probability space $(\mathcal{Y}, \mathbb{Y}, p_Y)$ is typically presented as a CDF or CCDF.

Equation 10.19 describes the mapping of probabilities from the input space to probabilities in the space of the system outcome y when the uncertainty associated with the domain of the function is characterized by a probability space $(X, \mathbb{X}, p_X)$. When the uncertainty in each x_i is characterized by an evidence space $(X_i, \mathbb{X}_i, m_i)$, the uncertainty in $\boldsymbol{x}$ is characterized by an evidence space $(X, \mathbb{X}, m_X)$, where (i) X is defined by Equation 10.14, (ii) $\mathbb{X}$ is the same as the set $\mathbb{C}$ defined in Equation 10.15, and (iii) under the assumption that the x_i are independent, m_X is defined by

$$m_X(\mathcal{E}) = \begin{cases} \prod_{i=1}^{n} m_i(\mathcal{E}_i) & \text{if } \mathcal{E} = \mathcal{E}_1 \times \mathcal{E}_2 \times \ldots \times \mathcal{E}_n \in \mathbb{X} \\ 0 & \text{otherwise} \end{cases} \quad (10.21)$$

for subsets $\mathcal{E}$ of X.

The development is more complex when the x_i are not independent. For a vector $\boldsymbol{x}$ of the form defined in conjunction with Equation 10.13, the structure of $\mathbb{X}$ for the evidence space $(X, \mathbb{X}, m_X)$ is much simpler than the structure of $\mathbb{X}$ for an analogous probability space $(X, \mathbb{X}, p_X)$. In particular, the set $\mathbb{X}$ for the evidence space $(X, \mathbb{X}, m_X)$ is the same as the set $\mathbb{C}$ in Equation 10.15. In contrast, the set $\mathbb{X}$ for an analogous probability space $(X, \mathbb{X}, p_X)$ is constructed from $\mathbb{C}$ and in general contains an uncountable number of elements rather than the finite number of elements usually contained in $\mathbb{C}$.

For evidence theory, relations analogous to Equation 10.19 define belief and plausibility for system outcomes when the uncertainty associated with the domain of the function is characterized by an evidence space $(X, \mathbb{X}, m_X)$. In concept, $(X, \mathbb{X}, m_X)$ and f induce an evidence space $(\mathcal{Y}, \mathbb{Y}, m_Y)$. In practice, m_Y is not determined. Rather, the belief $Bel_Y(\mathcal{E})$ and plausibility $Pl_Y(\mathcal{E})$ for a subset $\mathcal{E}$ of $\mathcal{Y}$ are determined from the BPA m_X associated with $(X, \mathbb{X}, m_X)$. In particular,

$$Bel_Y(\mathcal{E}) = Bel_X[f^{-1}(\mathcal{E})] = \sum_{\mathcal{U} \subset f^{-1}(\mathcal{E})} m_X(\mathcal{U}) \quad (10.22)$$

and

$$Pl_Y(\mathcal{E}) = Pl_X[f^{-1}(\mathcal{E})] = \sum_{\mathcal{U} \cap f^{-1}(\mathcal{E}) \neq \emptyset} m_X(\mathcal{U}), \quad (10.23)$$

where Bel_X and Pl_X represent belief and plausibility defined for the evidence space $(X, \mathbb{X}, m_X)$.

Similarly to the use of CDFs and CCDFs in probability theory, the uncertainty in y characterized by the evidence space $(\mathcal{Y}, \mathbb{Y}, m_Y)$ can be summarized with CBFs, CCBFs, CPFs, and CCPFs. In particular, the CBF, CCBF, CPF, and CCPF for y are defined by the sets of points

$$\mathcal{CBF} = \left\{[v, Bel_Y(\mathcal{Y}_v^c)], v \in \mathcal{Y}\right\} = \left\{[v, Bel_X(f^{-1}(\mathcal{Y}_v^c))], v \in \mathcal{Y}\right\} \quad (10.24)$$

$$\mathcal{CCBF} = \{[v, Bel_Y(\mathcal{Y}_v)], v \in \mathcal{Y}\} = \{[v, Bel_X(f^{-1}(\mathcal{Y}_v))], v \in \mathcal{Y}\} \quad (10.25)$$

$$\mathcal{CPF} = \left\{[v, Pl_Y(\mathcal{Y}_v^c)], v \in \mathcal{Y}\right\} = \left\{[v, Pl_X(f^{-1}(\mathcal{Y}_v^c))], v \in \mathcal{Y}\right\} \quad (10.26)$$

$$\mathcal{CCPF} = \{[v, Pl_Y(\mathcal{Y}_v)], v \in \mathcal{Y}\} = \{[v, Pl_X(f^{-1}(\mathcal{Y}_v))], v \in \mathcal{Y}\}, \quad (10.27)$$

where

$$\mathcal{Y}_v = \{y : y \in \mathcal{Y} \text{ and } y > v\} \quad (10.28)$$

$$\mathcal{Y}_v^c = \{y : y \in \mathcal{Y} \text{ and } y \leq v\}. \quad (10.29)$$

Plots of the points contained in $\mathcal{CBF}$, $\mathcal{CCBF}$, $\mathcal{CPF}$, and $\mathcal{CCPF}$ produce a figure similar to Figure 10.1 and provide a visual representation of the uncertainty in y in terms of belief and plausibility.

The beliefs and plausibilities appearing in Equation 10.24 through Equation 10.27 are defined by sums of BPAs for elements of $\mathbb{X}$. For notational convenience, let $\mathcal{E}_j$ denote the jth element of $\mathbb{X}$ for the evidence space $(X, \mathbb{X}, m_X)$. Such a numbering is possible because $\mathbb{X}$ is countable due to the constraint imposed by Equation 10.2. For $v \in \mathcal{Y}$, let

$$\mathcal{ICBF}_v = \left\{ j : \mathcal{E}_j \subset f^{-1}\left(\mathcal{Y}_v^c\right) \right\} \tag{10.30}$$

$$\mathcal{ICCBF}_v = \{ j : \mathcal{E}_j \subset f^{-1}(\mathcal{Y}_v) \} \tag{10.31}$$

$$\mathcal{ICPF}_v = \left\{ j : \mathcal{E}_j \cap f^{-1}\left(\mathcal{Y}_v^c\right) \neq \varnothing \right\} \tag{10.32}$$

$$\mathcal{ICCPF}_v = \{ j : \mathcal{E}_j \cap f^{-1}(\mathcal{Y}_v) \neq \varnothing \}. \tag{10.33}$$

In turn, the beliefs and plausibilities in Equation 10.24 through Equation 10.27 are defined by

$$Bel_Y\left(\mathcal{Y}_v^c\right) = Bel_X\left(f^{-1}\left(\mathcal{Y}_v^c\right)\right) = \sum_{j \in \mathcal{ICBF}_v} m_X(\mathcal{E}_j) \tag{10.34}$$

$$Bel_Y(\mathcal{Y}_v) = Bel_X(f^{-1}(\mathcal{Y}_v)) = \sum_{j \in \mathcal{ICCBF}_v} m_X(\mathcal{E}_j) \tag{10.35}$$

$$Pl_Y\left(\mathcal{Y}_v^c\right) = Pl_X\left(f^{-1}\left(\mathcal{Y}_v^c\right)\right) = \sum_{j \in \mathcal{ICPF}_v} m_X(\mathcal{E}_j) \tag{10.36}$$

$$Pl_Y(\mathcal{Y}_v) = Pl_X(f^{-1}(\mathcal{Y}_v)) = \sum_{j \in \mathcal{ICCPF}_v} m_X(\mathcal{E}_j). \tag{10.37}$$

The summations in Equation 10.34 through Equation 10.37 provide formulas by which the CBF, CCBF, CPF, and CCPF defined in Equation 10.24 through Equation 10.27 can be calculated. In practice, determination of the sets in Equation 10.24 through Equation 10.27 can be computationally demanding due to the computational complexity of determining f^{-1}. For example, if f corresponds to the numerical solution of a system of nonlinear PDEs, there will be no closed form representation for f^{-1} and the computation of approximate representations for f^{-1} will require many computationally demanding evaluations of f. Section 10.3.3.4 gives a detailed example of how Equation 10.24 through Equation 10.27 are calculated in a simple example. The issue of convergence of sampling was recently addressed in [77], where it was established that, as the number of samples increases, even with minimal assumptions concerning the nature of f, convergence to the correct belief and plausibility of the system response is ensured.

10.2.4 Simple Conceptual Examples

The theoretical fundamentals given above can be rather impenetrable, even to those well grounded in the theoretical aspects of probability theory. We believe two of the reasons evidence theory is difficult to grasp are the following. First, evidence theory has *three* likelihood measures — belief, plausibility, and BPA functions — any one of which can determine the other two. In probability theory there is only one, the probability measure p. Second, the conversion of data or information into a BPA in evidence theory seems rather nebulous and confusing compared to the construction of probability measures in probability theory. In practice, this conversion and its representation in probability theory is simplified by using probability density functions (PDFs) as surrogates for the corresponding probability measures. For example, in probability theory if one assumes a "noninformative prior" then a uniform PDF is chosen. Or, if one has a histogram of experimental data, it is rather straightforward to construct a PDF. To aid in understanding evidence theory and how it compares with probability theory, the following two simple conceptual examples (similar to those given in [53], p. 75–76) are given.

Example 1

Children are playing with black and white hollow plastic Easter eggs. The children place a chocolate candy in each black Easter egg and put nothing into each white Easter egg. A parent appears and says, "I will secretly put one or more of your various Easter eggs in a paper bag. Tell me what is the probability (these are very precocious children) of you drawing out a black Easter egg from the paper bag (which you cannot see into)?"

The universal set X contains the two possible outcomes: $\{B, W\}$, where B indicates a black egg with chocolate, and W indicates a white egg without chocolate.

A probabilistic solution would traditionally assume that, according to the Principle of Insufficient Reason, there is an equal likelihood of drawing out a black egg and a white egg. Therefore, the answer to the question would be: "There is a 0.5 probability of getting a black egg."

An evidence theory solution would assign the BPA of the universal set a value of 1, which can be written as $m(X) = 1$. Because nothing is known about what this parent might do concerning putting white vs. black eggs in the bag, a BPA of zero is assigned to each possible event: $m(B) = 0$ and $m(W) = 0$. Using Equations 10.3 and 10.4, one computes that $Bel(B) = 0$ and $Pl(B) = 1$ for drawing out a black egg. Therefore, an evidence theory approach would answer the question: "The probability is between 0 and 1 for getting a black egg."

Example 2

The same children are playing with plastic Easter eggs and the same parent shows up. However, this time, the parent paints some of the plastic Easter eggs gray and places chocolates into some of the gray eggs, out of sight of the children. The parent now says, "I have put ten Easter eggs into a paper bag, which you cannot see into. In the bag there are two black eggs (with chocolates), three white eggs (without chocolates), and five gray eggs that may or may not contain a chocolate. Tell me what is the probability of drawing out an egg with a chocolate?"

The universal set X can be written as $\{B, W, G\}$; B and W are as before, and G is a gray egg which may or may not have a chocolate. A probabilistic solution could be based on the probabilities

$$p(B) = 0.2, \; p(W) = 0.3, \; p(G) = 0.5.$$

In addition, according to the Principle of Insufficient Reason, each gray egg containing a chocolate, G_w, has a probability of 0.5, and each gray egg without a chocolate, G_{wo}, has a probability of 0.5. Therefore, they would assign a probability of 0.25 to the likelihood that a gray egg with chocolate would be drawn. The probabilistic answer to the question would then be: "The probability of getting a chocolate from the bag is $p(B) + 0.25 = 0.45$."

The evidence theory solution would assign the following BPAs

$$m(B) = 0.2, \; m(W) = 0.3, \; m(G_w, G_{wo}) = 0.5.$$

With the use of Equations 10.3 and 10.4, the belief for getting a chocolate, $Bel(C)$, and plausibility for getting a chocolate, $Pl(C)$, can be computed:

$$Bel(C) = Bel(B, G_w) = 0.2, \; Pl(C) = Pl(B, G_w) = 0.2 + 0.5 = 0.7.$$

Therefore, an evidence theory answer to the question would be: "There is a probability between 0.2 and 0.7 of getting a chocolate from the bag."

10.2.4.1 Observations

First, on these simple examples, it can be seen, hopefully, that evidence theory accurately represents the range in probabilities that are consistent with the given data; no additional assumptions concerning the given data are imposed. Stated differently, traditional application of probability theory leads one to assign probabilities to all events of the universal set, thereby forcing one to make assumptions that are not supported by the evidence. Evidence theory allows one to assign basic probabilities to *sets* of elements in the universal set, thus avoiding unjustified assumptions. Concern might be expressed that the range of probabilities with evidence theory is so large that little useful information is gained with the approach to aid in the decision-making process. The response to this is that when large epistemic uncertainty is present, the decision maker should be clearly aware of the range of probabilities, rather than having assumptions buried in the analysis disguise the probabilities. If high-consequence decisions are involved instead of chocolates, it is imperative that the decision maker understand the probabilities and resulting risks. If the highest possible risks are unacceptable to the decision maker, then resources must be made available to reduce the epistemic uncertainty.

Second, in these examples the inappropriate use of the Principle of Insufficient Reason, particularly in Example 1, is obvious. However, the assumption, without justification, of a uniform PDF in engineering analyses is very common. Sometimes the assumption is made with the caveat that, "For the first pass through the analysis, a uniform PDF is assumed." If the decision maker acts on the "first pass analysis" and a refined analysis is never conducted, inappropriate risks could be the result. Or, the more common situation might occur: "The risks appear acceptable based on the first pass analysis, and if the funds and schedule permit, we will conduct a more refined analysis in the future." Commonly, the funds and schedule are consumed with "more pressing issues."

10.3 Example Problem

The topics covered in this section are the following. First, an algebraic equation will be given (Section 10.3.1), which is a model for describing the response of a simple nondeterministic system. The nondeterministic character of the system is due to uncertainty in the parameters embodied in the algebraic model of the system. Second, the uncertainty of the response of the system will be estimated using the traditional application of probability theory (Section 10.3.2) and evidence theory (Section 10.3.3). Comparisons will be made (Section 10.3.4) concerning the representation of uncertainty using each approach. Third, in the solution procedure using evidence theory, the steps are described for converting the information concerning the uncertain parameters into input structures usable by evidence theory. A detailed discussion will be given for propagating input uncertainties represented by BPAs through the system response function and computing belief and plausibility measures in the output space.

10.3.1 Problem Description

The following equation is a simple special case of the input/output mapping indicated in Equation 10.13:

$$y = f(a,b) = (a+b)^a. \tag{10.38}$$

The parameters a and b are the analysis inputs and are the only uncertain quantities affecting the system response. Parameters a and b are independent; that is, knowledge about the value of one parameter implies nothing about the value of the other. Multiple expert sources provide information concerning a and b, but the precision of the information is relatively poor. Stated differently, there is scarce and conflicting information concerning a and b, resulting in large epistemic uncertainty. All of the sources for a and b are considered equally credible.

Evidence Theory for Engineering Applications

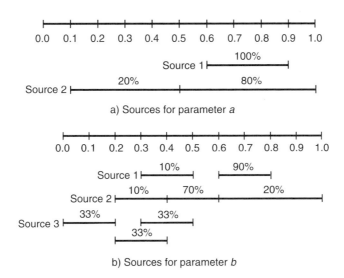

FIGURE 10.2 Information from each source for parameters a (upper) and b (lower). (Originally published in *Investigation of Evidence Theory for Engineering Applications*, Oberkampf, W.L. and Helton, J.C., 4th Nondeterministic Approaches Forum, Denver, AIAA-2002-1569. Copyright © 2002 by the American Institute of Aeronautics and Astronautics, Inc. Reprinted with permission.)

For parameter a, two sources provide information. Source 1 states that he believes that the actual, or true, value lies in the interval [0.6, 0.9]. Source 2 states that, in her opinion, the actual value is in one of two contiguous intervals: in the interval [0.1, 0.5] with a 20% level of subjective belief, or in the interval [0.5, 1.0] with an 80% level of subjective belief. (Note: For clarity, an adjective such as "subjective" or "graded" with the term "belief" is used to designate the common language meaning for "belief." When "belief" is used without such adjectives, reference is being made to the belief function in evidence theory.)

For parameter b, three sources provide information. Source 1 states that the actual value could lie in one of two disjoint intervals: in the interval [0.3, 0.5] with a 10% level of subjective belief, or in the interval [0.6, 0.8] with a 90% level of subjective belief. Source 2 states that the actual value could be in one of three contiguous intervals: [0.2, 0.4] with a 10% level of subjective belief, [0.4, 0.6] with a 70% level of subjective belief, and [0.6, 1.0] with a 20% level of subjective belief. Source 3 states that he believes that three experimental measurements he is familiar with should characterize the actual value of b. The three experimental realizations yielded: 0.1 ± 0.1, 0.3 ± 0.1, and 0.4 ± 0.1. He chooses to characterize these measurements, that is, his input to the uncertainty analysis, as three intervals: [0.0, 0.2], [0.2, 0.4] and [0.3, 0.5], all with equal levels of subjective belief. (See [74] for a description of more complex subjective belief statements.)

The input data for a and b for each source are shown graphically in Figure 10.2a and b, respectively.

To complete the statement of the mathematical model of the system, the system response y is considered to be unsafe for values of y larger than 1.7. It is desired that both the traditional and evidence theory approaches be used to assess what can be said about the occurrence of $y > 1.7$. Although the example problem may appear quite simple to some, the solution is not simple, or even unique, because of the poor information given for the parameters. It will be seen that both methods require additional assumptions to estimate the occurrence of $y > 1.7$. Some of these assumptions will be shown to dominate the estimated safety of the system.

Figure 10.3 presents a three-dimensional representation of $y = f(a, b)$ over the range of possible values of a and b. Several level curves, or response contours, of y are shown (i.e., loci of $[a, b]$ values that produce equal y values). The rectangle defined by $0.1 \le a \le 1$ and $0 \le b \le 1$ is referred to as the input product space.

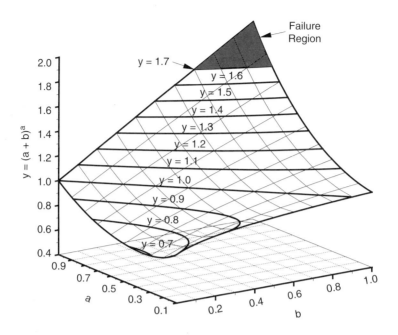

FIGURE 10.3 Three-dimensional representation of $y = (a + b)^a$ on the rectangle defined by $0.1 \leq a \leq 1$ and $0 \leq b \leq 1$. (Originially published in *Investigation of evidence theory for engineering applications*, Oberkampf, W.L. and Helton, J.C., 4th Non-Deterministic Approaches Forum, Denver, AIAA-2002-1569. Copyright © 2002 by the American Institute of Aeronautics and Astronautics, Inc. Reprinted with permission.)

10.3.2 Traditional Analysis Using Probability Theory

10.3.2.1 Combination of Evidence

The information concerning *a* from the two sources is written as

$$\mathcal{A}_1 = \{a : 0.6 \leq a \leq 0.9\} \tag{10.39}$$

$$\mathcal{A}_2 = \begin{cases} a : 0.1 \leq a \leq 0.5 \text{ with } 20\% \text{ belief} \\ a : 0.5 \leq a \leq 1.0 \text{ with } 80\% \text{ belief} \end{cases}. \tag{10.40}$$

Thus, the set

$$\mathcal{A} = \mathcal{A}_1 \cup \mathcal{A}_2 = \{a : 0.1 \leq a \leq 1\} \tag{10.41}$$

contains all specified values for *a*.

Similarly, the information for *b* from the three sources is written as

$$\mathcal{B}_1 = \begin{cases} b : 0.3 \leq b \leq 0.5 \text{ with } 10\% \text{ belief} \\ b : 0.6 \leq b \leq 0.8 \text{ with } 90\% \text{ belief} \end{cases} \tag{10.42}$$

$$\mathcal{B}_2 = \begin{cases} b : 0.2 \leq b \leq 0.4 \text{ with } 10\% \text{ belief} \\ b : 0.4 \leq b \leq 0.6 \text{ with } 70\% \text{ belief} \\ b : 0.6 \leq b \leq 1.0 \text{ with } 20\% \text{ belief} \end{cases} \tag{10.43}$$

$$\mathcal{B}_2 = \begin{cases} b : 0.0 \leq b \leq 0.2 \text{ with } 33\% \text{ belief} \\ b : 0.2 \leq b \leq 0.4 \text{ with } 33\% \text{ belief} \\ b : 0.3 \leq b \leq 0.5 \text{ with } 33\% \text{ belief} \end{cases}. \tag{10.44}$$

Thus, the set

$$\mathcal{B} = \mathcal{B}_1 \cup \mathcal{B}_2 \cup \mathcal{B}_3 = \{b : 0 \le b \le 1\} \tag{10.45}$$

contains all specified values for b.

Given that each of the sources of information specifies only intervals of values, the traditional probabilistic analysis is implemented by assuming that a and b are uniformly distributed over each of their specified intervals. This is a significant assumption beyond what was given for the state of knowledge concerning a and b. The only claim from the expert sources is that the actual value is contained in specified intervals, *not* that all values over each of the intervals are equally likely. Other specifications of probability are possible, e.g., the intervals specified by each source could be divided into a finite number of subintervals, and probability density distributions (PDFs) could be defined for each of the subintervals. However, this is essentially never done. The specification of uniform distributions over each interval is the most common technique used to convert intervals of possible values to PDFs.

Let $[r_i, s_i]$, $i = 1, 2, \ldots n$, denote the specified intervals for a parameter from a given source. Let g_i be the graded level of belief associated with the interval $[r_i, s_i]$, expressed as a decimal. Then the resultant density function d is given by

$$d(v) = \sum_{i=1}^{n} \delta_i(v) g_i / (s_i - r_i), \tag{10.46}$$

where

$$\delta_i(v) = \begin{cases} 1 & \text{for } v \in [r_i, s_i] \\ 0 & \text{otherwise.} \end{cases} \tag{10.47}$$

For parameter a, n has values of 1 and 2 for sources 1 and 2, respectively. For parameter b, n has values of 2, 3, and 3 for sources 1, 2, and 3, respectively.

Because it is given that each source of information is equally credible, the resultant density functions for each source are simply averaged. Thus, if $d_{A1}(a)$ and $d_{A2}(a)$ denote the density functions for sources 1 and 2 for variable a, the resultant combined density function is

$$d_A(a) = \sum_{i=1}^{2} d_{Ai}(a)/2. \tag{10.48}$$

Similarly, the resultant combined density function for b is

$$d_B(b) = \sum_{i=1}^{3} d_{Bi}(b)/3, \tag{10.49}$$

where $d_{B1}(b)$, $d_{B2}(b)$, and $d_{B3}(b)$ denote the density functions from sources 1, 2, and 3, respectively. The density functions $d_A(a)$ and $d_B(b)$ in Equation 10.48 and Equation 10.49 effectively define the probability space used to characterize the uncertainty in a and b, respectively.

10.3.2.2 Construction of Probabilistic Response

Each possible value for y in Equation 10.38 derives from multiple vectors $\mathbf{c} = [a, b]$ of possible values for a and b. For example, each contour line in Figure 10.3 derives from multiple values of $\mathbf{c}$ that produce the same value of y. The set

$$\mathcal{C} = \mathcal{A} \times \mathcal{B} = \{\mathbf{c} = [a, b] : a \in \mathcal{A}, b \in \mathcal{B}\} \tag{10.50}$$

contains all possible values for **c**. Because it is given that a and b are independent, the PDF for $\mathcal{C}$ is given by

$$d_C(c) = d_A(a)d_B(b) \tag{10.51}$$

for $\mathbf{c} = [a, b] \in \mathcal{C}$. For notational convenience, the probability space used to characterize the uncertainty in $\mathbf{c} = [a, b]$ will be represented by $(\mathcal{C}, \mathbb{C}, p_C)$.

Possible values for a and b give rise to possible values for y through the relationship in Equation 10.38. In particular, the set $\mathcal{Y}$ of all possible values for y is given by

$$\mathcal{Y} = \{y : y = f(a, b) = (a+b)^a, [a, b] \in \mathcal{C}\}. \tag{10.52}$$

With a probability-based approach, the uncertainty in y is represented by defining a probability distribution over $\mathcal{Y}$. Ultimately, the probability distribution associated with $\mathcal{Y}$ derives from the nature of the mapping $y = f(a, b)$ and the probability distributions over $\mathcal{A}$ and $\mathcal{B}$.

The probability $p_Y(\mathcal{E})$ of a subset $\mathcal{E}$ of $\mathcal{Y}$ can be formally represented by

$$p_Y(\mathcal{E}) = \int_{\mathcal{E}} d_Y(y) dy = p_C(f^{-1}(\mathcal{E})), \tag{10.53}$$

where d_Y denotes the density function associated with the distribution of y, $p_C(f^{-1}(\mathcal{E}))$ denotes the probability of the set $f^{-1}(\mathcal{E})$ in $\mathcal{C} = \mathcal{A} \times \mathcal{B}$, and

$$f^{-1}(\mathcal{E}) = \{\mathbf{c} : \mathbf{c} = [a, b] \in \mathcal{C} = \mathcal{A} \times \mathcal{B} \text{ and } y = f(a, b) \in \mathcal{E}\}. \tag{10.54}$$

A closed-form representation for the density function d_Y can be derived from f, d_A and d_B. However, in real analysis problems, this is rarely done due to the complexity of the function and distributions involved. The relations in Equation 10.53 are presented to emphasize that probabilities for subsets of $\mathcal{Y}$ result from probabilities for subsets of $\mathcal{C}$.

An alternative representation for $p_Y(\mathcal{E})$ is given by

$$\begin{aligned} p_Y(\mathcal{E}) &= \int_{\mathcal{Y}} \delta_E(y) d_Y(y) dy \\ &= \int_{\mathcal{C}} \delta_E[f(a,b)] d_A(a) d_B(b) da\, db, \end{aligned} \tag{10.55}$$

where

$$\delta_E(y) = \begin{cases} 1 & \text{if } y \in \mathcal{E} \\ 0 & \text{otherwise}. \end{cases} \tag{10.56}$$

The preceding representation underlies the calculations carried out when a Monte Carlo or Latin Hypercube sampling procedure is used to estimate $p_Y(\mathcal{E})$ [78–80].

As indicated in Section 10.2.2, CCDFs provide a standard way to summarize probability distributions and provide an answer to the following commonly asked question: "How likely is y to be this large or larger?" The defining equation for a CCDF is analogous to Equation 10.25 and Equation 10.27 for CCBFs and CCPFs, respectively, in evidence theory. Specifically, the CCDF for y is defined by the set of points

$$\mathcal{CCDF} = \{[v, p_Y(\mathcal{Y}_v)], v \in \mathcal{Y}\} = \{[v, p_C(f^{-1}(\mathcal{Y}_v))], v \in \mathcal{Y}\} \tag{10.57}$$

for $\mathcal{Y}_v$ defined in Equation 10.28. The CCDF for the function $y = f(a, b)$ defined in Equation 10.38 and the probability space $(\mathcal{C}, \mathbb{C}, p_C)$ introduced in this section is presented in Section 10.3.4.

10.3.3 Analysis Using Evidence Theory

10.3.3.1 Construction of Basic Probability Assignments for Individual Inputs

The BPAs for $\mathcal{A}$ and $\mathcal{B}$ are obtained by first defining BPAs for $\mathcal{A}_1$ and $\mathcal{A}_2$, and $\mathcal{B}_1$, $\mathcal{B}_2$, and $\mathcal{B}_3$. This is done using the information for a and b from each source (see Equation 10.39, 10.40 and Equation 10.42 through Equation 10.44). A convenient representational device for BPAs associated with interval data can be obtained with the use of lower triangular matrices. This representation, an extension of that used in [81], is constructed as follows. For each uncertain parameter, an interval is identified that contains the range of the parameter when all of the sources of information for that parameter are combined. The range of each parameter is divided into as many contiguous subintervals as needed to describe the interval value information from each of the sources; that is, any specified interval of values for the parameter is equal to the union of a subset of these contiguous intervals. The columns of the lower triangular matrix are indexed by taking the lower value of each subinterval, and the rows are indexed by taking the upper value of each subinterval.

To represent this lower triangular matrix of potential nonzero BPAs, let $l_1, l_2, \ldots l_n$ be the lower values for the n subintervals, where $l_1 \leq l_2 \ldots \leq l_n$. Let $u_1, u_2, \ldots u_n$ be the upper values of the subintervals, where $u_1 \leq u_2 \ldots \leq u_n$. Consistent with the contiguous assumption, the intervals can be expressed as $[l_1, l_2]$ $[l_2, l_3], \ldots [l_n, u_n]$ with $l_n \leq u_n$, or equivalently as $[l_1, u_1], [u_1, u_2], \ldots [u_{n-1}, u_n]$ with $l_1 \leq u_1$. Let $m([l_i, u_j])$ be the BPA for the subinterval $[l_i, u_j]$. The $n \times n$ lower triangular matrix can then be written as

$$
\begin{array}{c|ccccc}
 & l_1 & l_2 & l_3 & \cdots & l_n \\
u_1 & m([l_1, u_1]) & & & & \\
u_2 & m([l_1, u_2]) & m([l_2, u_2]) & & & \\
u_3 & m([l_1, u_3]) & m([l_2, u_3]) & m([l_3, u_3]) & & \\
\vdots & \vdots & \vdots & \vdots & \ddots & \\
u_n & m([l_1, u_n]) & m([l_2, u_n]) & m([l_3, u_n]) & \cdots & m([l_n, u_n]).
\end{array}
\tag{10.58}
$$

Specifically, m in the preceding matrix defines a BPA for $\mathcal{S} = \{x : l_1 \leq x \leq u_n\}$ provided (i) the values for m in the matrix are nonnegative and sum to 1, and (ii) $m(\mathcal{E}) = 0$ if $\mathcal{E} \subset \mathcal{S}$ and $\mathcal{E}$ does not correspond to one of the intervals with a BPA in the matrix. In essence, this representation provides a way to define a BPA over an interval when all noncontiguous subintervals are given a BPA of zero. When only the diagonal elements of the matrix are nonzero, the resultant BPA assignment is equivalent to the specification of a discrete probability space. For this space, the set $\mathcal{S}$ contains the null set and all sets that can be generated by forming unions of the intervals $[l_1, l_2), [l_2, l_3), \ldots [l_n, u_n]$. Note that half-open intervals are assumed so that the intersection of any two intervals will be the null set. Conversely, from the structure of the matrix it can be seen that the precision of the information decreases as the distance from the diagonal increases. For example, the least precise statement of information appears in the lower-left element of the matrix with the definition of the BPA $m[l_1, u_n]$.

In the present example, the sets $\mathcal{A}$ and $\mathcal{B}$ correspond to the intervals $[0.1, 1]$ and $[0, 1]$, respectively. The corresponding matrices for $\mathcal{A}$ and $\mathcal{B}$ are

$$
\begin{array}{c|cccc}
 & 0.1 & 0.5 & 0.6 & 0.9 \\
0.5 & m_A([0.1, 0.5]) & & & \\
0.6 & m_A([0.1, 0.6]) & m_A([0.5, 0.6]) & & \\
0.9 & m_A([0.1, 0.9]) & m_A([0.5, 0.9]) & m_A([0.6, 0.9]) & \\
1.0 & m_A([0.1, 1.0]) & m_A([0.5, 1.0]) & m_A([0.6, 1.0]) & m_A([0.9, 1.0])
\end{array}
\tag{10.59}
$$

and

$$
\begin{array}{ccccccc}
0.0 & 0.2 & 0.3 & 0.4 & 0.5 & 0.6 & 0.8
\end{array}
$$

0.2 $m_B([0.0, 0.2])$
0.3 $m_B([0.0, 0.3])$ $m_B([0.2, 0.3])$
0.4 $m_B([0.0, 0.4])$ $m_B([0.2, 0.4])$ $m_B([0.3, 0.4])$
0.5 $m_B([0.0, 0.5])$ $m_B([0.2, 0.5])$ $m_B([0.3, 0.5])$ $m_B([0.4, 0.5])$
0.6 $m_B([0.0, 0.6])$ $m_B([0.2, 0.6])$ $m_B([0.3, 0.6])$ $m_B([0.4, 0.6])$ $m_B([0.5, 0.6])$
0.8 $m_B([0.0, 0.8])$ $m_B([0.2, 0.8])$ $m_B([0.3, 0.8])$ $m_B([0.4, 0.8])$ $m_B([0.5, 0.8])$ $m_B([0.6, 0.8])$
1.0 $m_B([0.0, 1.0])$ $m_B([0.2, 1.0])$ $m_B([0.3, 1.0])$ $m_B([0.4, 1.0])$ $m_B([0.5, 1.0])$ $m_B([0.6, 1.0])$ $m_B([0.8, 1.0])$,
(10.60)

respectively.

The information for a provided by the two sources (see Equation 10.39 and Equation 10.40) can be summarized by the following matrices $\mathbf{A}_1$ and $\mathbf{A}_2$ of the form shown in Equation 10.59:

$$
\mathbf{A}_1 = \begin{bmatrix} 0 & & & \\ 0 & 0 & & \\ 0 & 0 & 1 & \\ 0 & 0 & 0 & 0 \end{bmatrix} \quad \text{and} \quad \mathbf{A}_2 = \begin{bmatrix} 0.2 & & & \\ 0 & 0 & & \\ 0 & 0 & 0 & \\ 0 & 0.8 & 0 & 0 \end{bmatrix}. \quad (10.61)
$$

Similarly, the information for b provided by the three sources (see Equation 10.42 through Equation 10.44) can be summarized by the following matrices $\mathbf{B}_1$, $\mathbf{B}_2$, and $\mathbf{B}_3$ of the form shown in Equation 10.60:

$$
\mathbf{B}_1 = \begin{bmatrix} 0 & & & & & & \\ 0 & 0 & & & & & \\ 0 & 0 & 0 & & & & \\ 0 & 0 & 0.1 & 0 & & & \\ 0 & 0 & 0 & 0 & 0 & & \\ 0 & 0 & 0 & 0 & 0 & 0.9 & \\ 0 & 0 & 0 & 0 & 0 & 0 & 0 \end{bmatrix}, \quad \mathbf{B}_2 = \begin{bmatrix} 0 & & & & & & \\ 0 & 0 & & & & & \\ 0 & 0.1 & 0 & & & & \\ 0 & 0 & 0 & 0 & & & \\ 0 & 0 & 0 & 0.7 & 0 & & \\ 0 & 0 & 0 & 0 & 0 & 0 & \\ 0 & 0 & 0 & 0 & 0 & 0.2 & 0 \end{bmatrix},
$$

$$
\text{and} \quad \mathbf{B}_3 = \begin{bmatrix} 0.333 & & & & & & \\ 0 & 0 & & & & & \\ 0 & 0.333 & 0 & & & & \\ 0 & 0 & 0.333 & 0 & & & \\ 0 & 0 & 0 & 0 & 0 & & \\ 0 & 0 & 0 & 0 & 0 & 0 & \\ 0 & 0 & 0 & 0 & 0 & 0 & 0 \end{bmatrix}. \quad (10.62)
$$

The BPAs for all unspecified subsets of $\mathcal{A}$ and $\mathcal{B}$ are zero.

10.3.3.2 Combination of Evidence

Combining the BPAs for the $\mathcal{A}_i$'s to obtain a BPA for $\mathcal{A}$ and combining the BPAs for the $\mathcal{B}_i$'s to obtain a BPA for $\mathcal{B}$ is now considered. The issue of combination of evidence, also referred to as *aggregation of evidence* or *data fusion*, is a topic closely related to evidence theory itself. After the introduction of evidence theory by Dempster in 1967 [61], many researchers believed evidence theory and the issue of combination of evidence were essentially the same topic. This view was reinforced by the close relationship between evidence theory and Dempster's rule of combination of evidence. In fact, much of the criticism of evidence theory has actually been directed at Dempster's rule of combination. It is now recognized that combination of evidence is a separate topic of growing importance in many fields [50, 52, 64, 82–89]. Combination of evidence takes on even more importance in newer theories of uncertainty, such as evidence theory, because many of the newer theories can deal more directly with large epistemic uncertainty than the traditional application of probability theory. The manner in which conflicting evidence is combined can have a large impact on the results of an uncertainty analysis, particularly when evidence theory is used.

In this presentation, the emphasis is on comparing uncertainty estimation results from the traditional application of probability theory and evidence theory. As a result, evidence from the various sources is combined in the same manner as was done for the solution using probability theory. The equivalent formulation to Equation 10.48 and Equation 10.49 for the matrices $\mathbf{A}_i$ and $\mathbf{B}_i$ is

$$\mathbf{A} = \sum_{i=1}^{2} \mathbf{A}_i/2 \quad \text{and} \quad \mathbf{B} = \sum_{i=1}^{3} \mathbf{B}_i/3. \tag{10.63}$$

Applying these equations to Equation 10.61 and Equation 10.62, respectively, we have

$$\mathbf{A} = \begin{bmatrix} 0.1 & & & \\ 0 & 0 & & \\ 0 & 0 & 0.5 & \\ 0 & 0.4 & 0 & 0 \end{bmatrix} \tag{10.64}$$

$$\text{and } \mathbf{B} = \begin{bmatrix} 0.111 & & & & & & \\ 0 & 0 & & & & & \\ 0 & 0.144 & 0 & & & & \\ 0 & 0 & 0.144 & 0 & & & \\ 0 & 0 & 0 & 0.233 & 0 & & \\ 0 & 0 & 0 & 0 & 0 & 0.3 & \\ 0 & 0 & 0 & 0 & 0 & 0.067 & 0 \end{bmatrix}, \tag{10.65}$$

with the additional specification that $m_A(\mathcal{E}) = 0$ if $\mathcal{E}$ is a subset of $\mathcal{A}$ without an assigned BPA in $\mathbf{A}$ and $m_B(\mathcal{E}) = 0$ if $\mathcal{E}$ is a subset of $\mathcal{B}$ without an assigned BPA in $\mathbf{B}$.

10.3.3.3 Construction of Basic Probability Assignments for the Product Space

The variables a and b are specified as being independent. With the use of Equation 10.21, the BPA $m_C(\mathcal{E})$ defined on $\mathcal{C} = \mathcal{A} \times \mathcal{B}$ is given by

$$m_C(\mathcal{E}) = \begin{cases} m_A(\mathcal{E}_A) m_B(\mathcal{E}_B) & \text{if } \mathcal{E}_A \subset \mathcal{A}, \mathcal{E}_B \subset \mathcal{B} \text{ and } \mathcal{E} = \mathcal{E}_A \times \mathcal{E}_B \\ 0 & \text{otherwise} \end{cases} \tag{10.66}$$

for $\mathcal{E} \subset \mathcal{C}$. The resultant nonzero BPAs and associated subsets of $\mathcal{C}$ are summarized in Table 10.1.

Subset 11 is used to illustrate the entries in Table 10.1. This subset corresponds to the interval [0.5, 1.0] for a, and the interval [0.4, 0.6] for b. From Equation 10.66,

$$m_C(\mathcal{E}) = m_A([0.5, 1.0]) m_B([0.4, 0.6]), \tag{10.67}$$

TABLE 10.1 Summary of the Nonzero Values of the BPA m_C for $C = A \times B$

	$m_A([0.1, 0.5]) = 0.1$	$m_A([0.5, 1.0]) = 0.4$	$m_A([0.6, 0.9]) = 0.5$
$m_B([0., 0.2]) = 0.111$	0.0111 (subset 1)	0.0444 (subset 2)	0.0555 (subset 3)
$m_B([0.2, 0.4]) = 0.144$	0.0144 (subset 4)	0.0576 (subset 5)	0.0720 (subset 6)
$m_B([0.3, 0.5]) = 0.144$	0.0144 (subset 7)	0.0576 (subset 8)	0.0720 (subset 9)
$m_B([0.4, 0.6]) = 0.233$	0.0233 (subset 10)	0.0932 (subset 11)	0.1165 (subset 12)
$m_B([0.6, 0.8]) = 0.300$	0.0300 (subset 13)	0.1200 (subset 14)	0.1500 (subset 15)
$m_B([0.6, 1.0]) = 0.067$	0.0067 (subset 16)	0.0268 (subset 17)	0.0335 (subset 18)

where the values for $m_A([0.5, 1.0])$ and $m_B([0.4, 0.6])$ appear in the column and row designators associated with subset 11 (i.e., $\mathcal{E} = [0.5, 1.0] \times [0.4, 0.6]$) and are defined by entries in the matrices **A** and **B** in Equation 10.64 and Equation 10.65, respectively. This yields

$$m_C(\mathcal{E}) = (0.4)(0.233) = 0.0932. \tag{10.68}$$

The magnitude of m_C shown for subsets of C in Table 10.1 indicates the likelihood that can be assigned to a given set, but *not* to any proper subset of that set. As required in the definition of a BPA, the BPAs in Table 10.1 sum to unity.

10.3.3.4 Construction of Belief and Plausibility for the System Response

The belief and plausibility measures for the system outcome y can now be computed. The goal is to obtain an assessment of the likelihood, in the context of evidence theory, that y will be in the failure region. As previously indicated, the threshold of the failure region is $v = 1.7$. Thus, the system failure question is: "How likely is it that y will have a value in the set $\mathcal{Y}_{1.7}$ defined in Equation 10.28?" As indicated in Equation 10.35 and Equation 10.37,

$$Bel_Y(\mathcal{Y}_v) = \sum_{j \in \mathcal{ICCBF}_v} m_C(\mathcal{E}_j) \tag{10.69}$$

and

$$Pl_Y(\mathcal{Y}_v) = \sum_{j \in \mathcal{ICCPF}_v} m_C(\mathcal{E}_j), \tag{10.70}$$

where $\mathcal{E}_j$, $j = 1, 2, \ldots 18$, are the subsets of C with nonzero BPAs given in Table 10.1. The sets $\mathcal{ICCBF}_v$ and $\mathcal{ICCPF}_v$ are defined in Equation 10.31 and Equation 10.33. In turn, the sets

$$\mathcal{CCBF} = \{[v, Bel_Y(\mathcal{Y}_v)] : v \in \mathcal{Y}\} \tag{10.71}$$

and

$$\mathcal{CCPF} = \{[v, Pl_Y(\mathcal{Y}_v)] : v \in \mathcal{Y}\} \tag{10.72}$$

define the CCBF and the CCPF for y.

The quantities $\mathcal{ICCBF}_v$, $\mathcal{ICCPF}_v$, $Bel_Y(\mathcal{Y}_v)$, and $Pl_Y(\mathcal{Y}_v)$ for the example problem are summarized in Table 10.2. The contour lines of $y = (a+b)^a$ are shown in Figure 10.4a to aid in understanding where the jumps in the CCBF and CCPF occur. Along the edges of Figure 10.4a are the values of y at equal increments

TABLE 10.2 Determination of the CCBF and CCPF for $y = f(a, b)$ with Equation 10.67 and Equation 10.68 and the BPAs in Table 10.1.

v	$\mathcal{ICCBF}_v$	$\mathcal{ICCPF}_v$	$Bel_Y(\mathcal{Y}_v)$	$Pl_Y(\mathcal{Y}_v)$
0.69220	1,2,3,4,5,6,7,8,9,10,11,12,13,14,15,16,17,18	1,2,3,4,5,6,7,8,9,10,11,12,13,14,15,16,17,18	1.00000	1.00000
[0.69220, 0.70711]	2,3,4,5,6,7,8,9,10,11,12,13,14,15,16,17,18	1,2,3,4,5,6,7,8,9,10,11,12,13,14,15,16,17,18	0.98889	1.00000
[0.70711, 0.73602]	3,4,5,6,7,8,9,10,11,12,13,14,15,16,17,18	1,2,3,4,5,6,7,8,9,10,11,12,13,14,15,16,17,18	0.94444	1.00000
[0.73602, 0.81096]	4,5,6,7,8,9,10,11,12,13,14,15,16,17,18	1,2,3,4,5,6,7,8,9,10,11,12,13,14,15,16,17,18	0.88889	1.00000
[0.81096, 0.83666]	5,6,7,8,9,10,11,12,13,14,15,16,17,18	1,2,3,4,5,6,7,8,9,10,11,12,13,14,15,16,17,18	0.87444	1.00000
[0.83666, 0.85790]	6,7,8,9,10,11,12,13,14,15,16,17,18	1,2,3,4,5,6,7,8,9,10,11,12,13,14,15,16,17,18	0.81667	1.00000
[0.85790, 0.87469]	6,8,9,10,11,12,13,14,15,16,17,18	1,2,3,4,5,6,7,8,9,10,11,12,13,14,15,16,17,18	0.80222	1.00000
[0.87469, 0.88657]	8,9,10,11,12,13,14,15,16,17,18	1,2,3,4,5,6,7,8,9,10,11,12,13,14,15,16,17,18	0.73000	1.00000
[0.88657, 0.89443]	8,9,10,11,12,13,14,15,16,17,18	2,3,4,5,6,7,8,9,10,11,12,13,14,15,16,17,18	0.73000	0.98889
[0.89443, 0.89751]	9,10,11,12,13,14,15,16,17,18	2,3,4,5,6,7,8,9,10,11,12,13,14,15,16,17,18	0.67222	0.98889
[0.89751, 0.93874]	9,11,12,13,14,15,16,17,18	2,3,4,5,6,7,8,9,10,11,12,13,14,15,16,17,18	0.64889	0.98889
[0.93874, 0.94868]	11,12,13,14,15,16,17,18	2,3,4,5,6,7,8,9,10,11,12,13,14,15,16,17,18	0.57667	0.98889
[0.94868, 0.95620]	12,13,14,15,16,17,18	2,3,5,6,7,8,9,10,11,12,13,14,15,16,17,18	0.48333	0.97444
[0.95620, 1.00000]	12,14,15,17,18	2,3,5,6,7,8,9,10,11,12,13,14,15,16,17,18	0.44667	0.97444
[1.00000, 1.04881]	14,15,17,18	2,3,5,6,8,9,10,11,12,13,14,15,16,17,18	0.33000	0.96000
[1.04881, 1.08957]	15,18	2,3,5,6,8,9,11,12,13,14,15,16,17,18	0.18333	0.93667
[1.08957, 1.11560]	15,18	2,5,6,8,9,11,12,13,14,15,16,17,18	0.18333	0.88111
[1.11560, 1.14018]		2,5,6,8,9,11,12,13,14,15,16,17,18	0.00000	0.88111
[1.14018, 1.20000]		2,5,6,8,9,11,12,14,15,16,17,18	0.00000	0.85111
[1.20000, 1.22474]		5,6,8,9,11,12,14,15,16,17,18	0.00000	0.80667
[1.22474, 1.26634]		5,6,8,9,11,12,14,15,17,18	0.00000	0.80000
[1.26634, 1.35368]		5,8,9,11,12,14,15,17,18	0.00000	0.72778
[1.35368, 1.40000]		5,8,11,12,14,15,17,18	0.00000	0.65556
[1.40000, 1.44040]		8,11,12,14,15,17,18	0.00000	0.59778
[1.44040, 1.50000]		8,11,14,15,17,18	0.00000	0.48111
[1.50000, 1.60000]		11,14,15,17,18	0.00000	0.42333
[1.60000, 1.61214]		14,15,17,18	0.00000	0.33000
[1.61214, 1.78188]		14,17,18	0.00000	0.18000
[1.78188, 1.80000]		14,17	0.00000	0.14667
[1.80000, 2.00000]		17	0.00000	0.02667
2.00000			0.00000	0.00000

of a and b along the boundary of their domain. Also note that these y values are only given to two significant figures. These contour lines approximately define the sets $f^{-1}(\mathcal{Y}_v)$ for selected values of v. For example, $f^{-1}(\mathcal{Y}_{1.04})$ is indicated in Figure 10.4b through Figure 10.4d. The sets $\mathcal{ICCBF}_v$ and $\mathcal{ICCPF}_v$ are obtained by determining the j for which $\mathcal{E}_j \subset f^{-1}(\mathcal{Y}_v)$ and $\mathcal{E}_j \cap f^{-1}(\mathcal{Y}_v) \neq \emptyset$, respectively. Thus, as an examination of Figure 10.4b through Figure 10.4d shows,

$$\mathcal{ICCBF}_{1.04} = \{14, 15, 17, 18\} \tag{10.73}$$

and

$$\mathcal{ICCPF}_{1.04} = \{2, 3, 5, 6, 8, 9, 10, 11, 12, 13, 14, 15, 16, 17, 18\}, \tag{10.74}$$

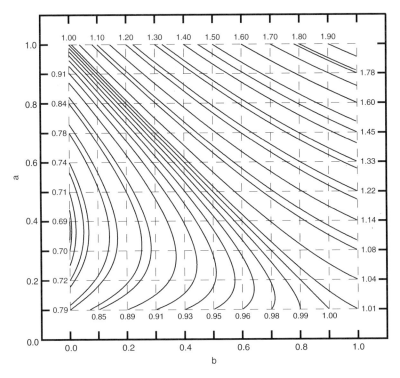

(a) Contour Lines of $y = (a + b)^a$

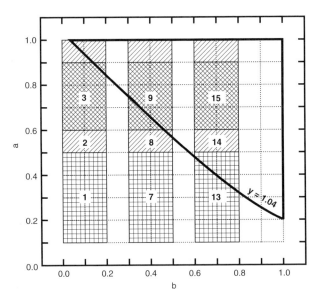

(b) Subsets 1, 2, 3, 7, 8, 9, 13, 14, and 15

FIGURE 10.4 Contour lines of $y = (a + b)^a$ and subsets 1, 2,..., 18 of $\mathcal{C} = \mathcal{A} \times \mathcal{B}$ with nonzero BPAs indicated in Table 10.1: (a) contour lines of $y = (a + b)^a$; (b) subsets 1, 2, 3, 7, 8, 9, 13, 14, and 15; (c) subsets 4, 5, 6, 16, 17, and 18; and (d) subsets 10, 11, and 12. (Originially published in *Investigation of Evidence Theory for Engineering Applications*, Oberkampf, W.L. and Helton, J.C., 4th Nondeterministic Approaches Forum, Denver, AIAA-2002-1569. Copyright © 2002 by the American Institute of Aeronautics and Astronautics, Inc. Reprinted with permission.)

Evidence Theory for Engineering Applications

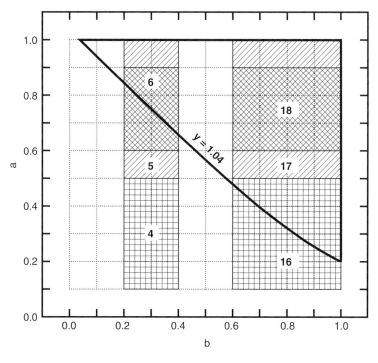

(c) Subsets 4, 5, 6, 16, 17, and 18

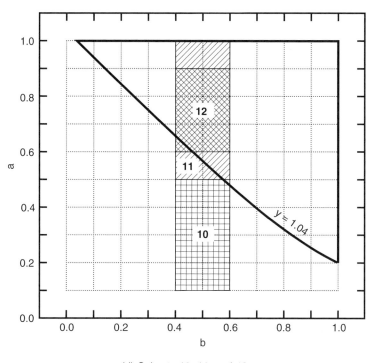

(d) Subsets 10, 11, and 12

FIGURE 10.4 (*Continued*)

which correspond to the values indicated in Table 10.2 for $\mathcal{ICCBF}_{1.04}$ and $\mathcal{ICCPF}_{1.04}$. In turn,

$$\begin{aligned} Bel_{1.04}(\mathcal{Y}_v) &= \sum_{j \in \mathcal{ICCBF}_{1.04}} m_C(\mathcal{E}_j) \\ &= m_C(\mathcal{E}_{14}) + m_C(\mathcal{E}_{15}) + m_C(\mathcal{E}_{17}) + m_C(\mathcal{E}_{18}) \\ &= 0.1200 + 0.1500 + 0.0268 + 0.0335 \\ &= 0.3303 \end{aligned} \tag{10.75}$$

and

$$\begin{aligned} Pl_Y(\mathcal{Y}_{1.04}) &= \sum_{j \in \mathcal{ICCPF}_{1.04}} m_C(\mathcal{E}_j) \\ &= m_C(\mathcal{E}_2) + m_C(\mathcal{E}_3) + m_C(\mathcal{E}_5) + m_C(\mathcal{E}_6) + m_C(\mathcal{E}_8) + \ldots + m_C(\mathcal{E}_{18}) \\ &= 0.9591. \end{aligned} \tag{10.76}$$

The resultant CCBF and CCPF are provided by plots of the points contained in the sets $\mathcal{CCBF}$ and $\mathcal{CCPF}$ defined in Equation 10.71 and Equation 10.72 and are shown in Figure 10.5. We stress that the jumps, or discontinuities, in the CCBF and CCPF shown in Figure 10.5 are accurate and consistent with evidence theory. Using Monte Carlo sampling for traditional probability theory, one occasionally sees jumps or stair-steps in the CCDF. However, these jumps are typically numerical artifacts, that is, numerical approximations due to use of a finite number of samples to compute the CCDF. In evidence theory, the jumps are *not* numerical artifacts but are due to discontinuous assignments of BPAs across the boundaries of subsets.

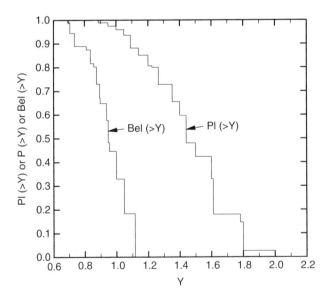

FIGURE 10.5 CCBF and CCPF for example problem $y = (a + b)^a$. (Originially published in *Investigation of Evidence Theory for Engineering Applications*, Oberkampf, W.L. and Helton, J.C., 4th Nondeterministic Approaches Forum, Denver, AIAA-2002-1569. Copyright © 2002 by the American Institute of Aeronautics and Astronautics, Inc. Reprinted with permission.)

10.3.4 Comparison and Interpretation of Results

For the present example, the solution using probability theory is obtained by numerically evaluating Equation 10.55 using Monte Carlo sampling. A random sample of size of one million was used so that an accurate representation of the CCDF could be obtained for probabilities as low as 10^{-4}. The CCDF for y is shown in Figure 10.6. In essence, the CCDF is constructed by plotting the pairs $[v, p_Y(\mathcal{Y}_v)]$ for an increasing sequence of values for v. The resultant CCDF indicates that the probability of the unsafe region, $y > 1.7$, is 0.00646.

Also shown in Figure 10.6 is the CCPF and the CCBF from evidence theory for the example problem. It can be seen in Figure 10.6, and also in Table 10.2, that the highest and lowest probabilities for the unsafe region using evidence theory are 0.18 and 0.0, respectively. That is, the absolutely highest probability that is consistent with the interval data for a and b is 0.18, and the absolutely lowest probability that is consistent with the interval data is 0.0. Stated differently, given the large epistemic uncertainty in the values for a and b, the probability of the unsafe region can *only* be bounded by 0.0 and 0.18. By comparing these interval valued probabilities with the traditional probabilistic result, it is seen that evidence theory states that the likelihood of the unsafe region could be 28 times higher than that indicated by the traditional analysis, or possibly as low as zero.

The key difference between the results obtained with probability theory and evidence theory is that in the probability-based analysis, it was assumed that all values in each specified interval for a and b were equally likely. Stated differently, the probability-based analysis assumed that the probability density function (PDF) was given by a piecewise uniform distribution; whereas in the evidence theory analysis, no additional assumptions were made beyond the uncertainty information supplied by the original sources. In essence, evidence theory permits the specification of partial (i.e., not completely defined)

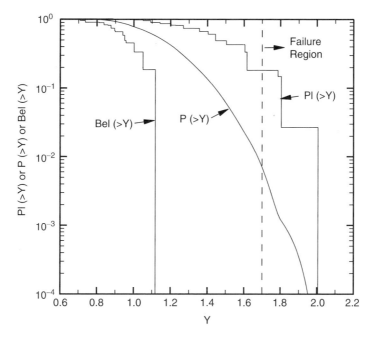

FIGURE 10.6 CCDF, CCPF, and CCBF for example problem $y = (a + b)^a$. (Originially published in *Investigation of Evidence Theory for Engineering Applications*, Oberkampf, W.L. and Helton, J.C., 4th Nondeterministic Approaches Forum, Denver, AIAA-2002-1569. Copyright © 2002 by the American Institute of Aeronautics and Astronautics, Inc. Reprinted with permission.)

probability distributions for *a* and *b*. Thus, evidence theory can be viewed as allowing the propagation of partially specified PDFs through the model of the system, in this case $(a + b)^a$, resulting in a range of likelihoods for *y*. The structure of evidence theory allows the incomplete specification of probability distributions, or more precisely, the complete determination of all possible probability distributions consistent with the input data.

As a final comment, both solution approaches relied on one additional common assumption: namely, that the information from each of the sources of data for *a* and *b* could be combined by a simple averaging procedure. As mentioned, a variety of methods have been developed for combining evidence. The results of an uncertainty analysis can strongly depend on which combination method is chosen for use. The selection of an appropriate combination method is an important open issue in uncertainty estimation. Research is needed to provide guidance for the appropriate selection of a combination method given the particular characteristics of a specific analysis and the type of information. Whatever method of combination is chosen in a given situation, the choice should primarily depend on the nature of the information to be combined. For a recent review of methods for combination of evidence, see [89, 90].

10.4 Research Topics in the Application of Evidence Theory

Although evidence theory possesses some advantages compared to the traditional application of probability theory, there are several prominent open issues that must be investigated and resolved before evidence theory can be confidently and productively used in large-scale engineering analyses. First, consider the use sampling techniques to propagate basic probability assignments (BPAs) through "black-box" computational models. An important practical issue arises: what is the convergence rate of approximations to belief and plausibility in the output space as a function of the number of samples? Our preliminary experience using traditional sampling techniques, such as Monte Carlo or Latin Hypercube, indicates that these techniques are reliable for evidence theory, but they are also expensive from the standpoint of function evaluations. Possibly faster convergence techniques could be developed for a wide range of black-box functions. Second, how can sensitivity analyses be conducted in the context of evidence theory? Rarely is a nondeterministic analysis conducted simply for the purpose of estimating the uncertainty in specified system response variables. A more common question is: what are the primary contributors to the uncertainty in the system response? Thus, sensitivity analysis procedures must be available for use in conjunction with uncertainty propagation procedures. Third, for situations of pure epistemic uncertainty (i.e., no aleatory uncertainty) in input parameters, such as the present example problem with interval data, how does one properly interpret belief and plausibility in the output space as minimum and maximum probabilities, respectively? A closely related question is: how does the averaging technique for aggregating conflicting expert opinion, which is the common technique in classical probabilistic thinking, affect the interpretation of interval valued probabilities in the output space?

The issue of combination of evidence from multiple sources is a separate issue from evidence theory itself. Evidence theory has been criticized in the past because there is no unique method for combining multiple sources of evidence. However, combination of evidence in probability theory is also nonunique and open to question. Further research is needed into methods of combining different types of evidence, particularly highly conflicting evidence from different sources. We believe that the method of combination of evidence chosen in a given situation should be context dependent. Stated differently, there is no single method appropriate for combining all types of evidence in all situations dealing with epistemic uncertainty.

Finally, evidence theory may have advantages over the traditional application of probability theory with regard to representing model form uncertainty (i.e., uncertainty due to lack of knowledge of the physical process being mathematically modeled). Because model form uncertainty is just a special case of epistemic uncertainty, any benefits that evidence theory has in the representation of epistemic uncertainty in general should also apply to model form uncertainty. Although this topic was not specifically addressed in this presentation, evidence theory has the capability to leave unspecified the probability assigned to any given mathematical model among alternative candidate models.

Acknowledgments

We thank Jay Johnson of ProStat, Inc., for computation of the results and generation of the figures. This work was performed at Sandia National Laboratories, which is a multiprogram laboratory operated by Sandia Corporation, a Lockheed Martin Company, for the United States Department of Energy's National Nuclear Security Administration under contract DE-AC04-AL85000.

References

1. AIAA, Guide for the verification and validation of computational fluid dynamics simulations, American Institute of Aeronautics and Astronautics, Reston, VA, AIAA-G-077-1998, 1998.
2. Kleijnen, J.P.C., Verification and validation of simulation models, *European Journal of Operational Research*, 82, 145–162, 1995.
3. Oberkampf, W.L. and Trucano, T.G., Verification and validation in computational fluid dynamics, *Progress in Aerospace Sciences*, 38, 209–272, 2002.
4. Oberkampf, W.L. and Trucano, T.G., Verification, validation, and predictive capability in computational engineering and physics, Sandia National Laboratories, Albuquerque, NM, SAND2003-3769, 2003.
5. Knupp, P. and Salari, K., *Verification of Computer Codes in Computational Science and Engineering*, Chapman & Hall/CRC, Boca Raton, FL, 2002.
6. Roache, P.J., *Verification and Validation in Computational Science and Engineering*, Hermosa Publishers, Albuquerque, NM, 1998.
7. Hora, S.C. and Iman, R.L., Expert opinion in risk analysis: the NUREG-1150 methodology, *Nuclear Science and Engineering*, 102, 323–331, 1989.
8. Hauptmanns, U. and Werner, W., *Engineering Risks Evaluation and Valuation*, 1st ed., Springer-Verlag, Berlin, 1991.
9. Beckjord, E.S., Cunningham, M.A., and Murphy, J.A., Probabilistic safety assessment development in the United States 1972–1990, *Reliability Engineering and System Safety*, 39, 159–170, 1993.
10. Modarres, M., *What Every Engineer Should Know about Reliability and Risk Analysis*, Marcel Dekker, New York, 1993.
11. Kafka, P., Important issues using PSA technology for design of new systems and plants, *Reliability Engineering and System Safety*, 45, 205–213, 1994.
12. Breeding, R.J., Helton, J.C., Murfin, W.B., Smith, L.N., Johnson, J.D., Jow, H.-N., and Shiver, A.W., The NUREG-1150 probabilistic risk assessment for the Surry Nuclear Power Station, *Nuclear Engineering and Design*, 135, 29–59, 1992.
13. Kumamoto, H. and Henley, E.J., *Probabilistic Risk Assessment and Management for Engineers and Scientists*, 2nd ed., IEEE Press, New York, 1996.
14. Helton, J.C., Uncertainty and sensitivity analysis in performance assessment for the Waste Isolation Pilot Plant, *Computer Physics Communications*, 117, 156–180, 1999.
15. Paté-Cornell, M.E., Conditional uncertainty analysis and implications for decision making: the case of WIPP, *Risk Analysis*, 19, 1003–1016, 1999.
16. Helton, J.C. and Breeding, R.J., Calculation of reactor accident safety goals, *Reliability Engineering and System Safety*, 39, 129–158, 1993.
17. Ross, T.J., *Fuzzy Logic with Engineering Applications*, McGraw-Hill, New York, 1995.
18. Melchers, R.E., *Structural Reliability Analysis and Prediction*, 2nd ed., John Wiley & Sons, New York, 1999.
19. Hoffman, F.O. and Hammonds, J.S., Propagation of uncertainty in risk assessments: the need to distinguish between uncertainty due to lack of knowledge and uncertainty due to variability, *Risk Analysis*, 14, 707–712, 1994.
20. Rowe, W.D., Understanding uncertainty, *Risk Analysis*, 14, 743–750, 1994.
21. Helton, J.C., Treatment of uncertainty in performance assessments for complex systems, *Risk Analysis*, 14, 483–511, 1994.

22. Ayyub, B.M., The Nature of Uncertainty in Structural Engineering, in *Uncertainty Modelling and Analysis: Theory and Applications*, B. M. Ayyub and M. M. Gupta, Eds., 1st ed., Elsevier, New York, 1994, 195–210.
23. Hora, S.C., Aleatory and epistemic uncertainty in probability elicitation with an example from hazardous waste management, *Reliability Engineering and System Safety*, 54, 217–223, 1996.
24. Frey, H.C. and Rhodes, D.S., Characterizing, simulating, and analyzing variability and uncertainty: an illustration of methods using an air toxics emissions example, *Human and Ecological Risk Assessment*, 2, 762–797, 1996.
25. Ferson, S. and Ginzburg, L.R., Different methods are needed to propagate ignorance and variability, *Reliability Engineering and System Safety*, 54, 133–144, 1996.
26. Ferson, S., What Monte Carlo methods cannot do, *Human and Ecological Risk Assessment*, 2, 990–1007, 1996.
27. Rai, S.N., Krewski, D., and Bartlett, S., A general framework for the analysis of uncertainty and variability in risk assessment, *Human and Ecological Risk Assessment*, 2, 972–989, 1996.
28. Parry, G.W., The characterization of uncertainty in probabilistic risk assessments of complex systems, *Reliability Engineering and System Safety*, 54, 119–126, 1996.
29. Helton, J.C., Uncertainty and sensitivity analysis in the presence of stochastic and subjective uncertainty, *Journal of Statistical Computation and Simulation*, 57, 3–76, 1997.
30. Paté-Cornell, M.E., Uncertainties in risk analysis: six levels of treatment, *Reliability Engineering and System Safety*, 54, 95–111, 1996.
31. Cullen, A.C. and Frey, H.C., *Probabilistic Techniques in Exposure Assessment: A Handbook for Dealing with Variability and Uncertainty in Models and Inputs*, Plenum Press, New York, 1999.
32. Frank, M.V., Treatment of uncertainties in space nuclear risk assessment with examples from Cassini Mission applications, *Reliability Engineering and System Safety*, 66, 203–221, 1999.
33. Haimes, Y.Y., *Risk Modeling, Assessment, and Management*, John Wiley & Sons, New York, 1998.
34. Ang, A.H.S. and Tang, W.H., *Probability Concepts in Engineering Planning and Design: Vol. I Basic Principles*, 1st ed., John Wiley & Sons, New York, 1975.
35. Ditlevsen, O., *Uncertainty Modeling with Applications to Multidimensional Civil Engineering Systems*, 1st ed., McGraw-Hill, New York, 1981.
36. Ang, A.H.S. and Tang, W.H., *Probability Concepts in Engineering Planning and Design: Vol. II Decision, Risk, and Reliability*, John Wiley & Sons, New York, 1984.
37. Neelamkavil, F., *Computer Simulation and Modelling*, 1st ed., John Wiley & Sons, New York, 1987.
38. Haldar, A. and Mahadevan, S., *Probability, Reliability, and Statistical Methods in Engineering Design*, John Wiley & Sons, New York, 2000.
39. Oberkampf, W.L., DeLand, S.M., Rutherford, B.M., Diegert, K.V., and Alvin, K.F., Error and uncertainty in modeling and simulation, *Reliability Engineering and System Safety*, 75, 333–357, 2002.
40. Oberkampf, W.L., DeLand, S.M., Rutherford, B.M., Diegert, K.V., and Alvin, K.F., Estimation of total uncertainty in computational simulation, Sandia National Laboratories, Albuquerque, NM, SAND2000-0824, 2000.
41. Oberkampf, W.L., Helton, J.C., and Sentz, K., Mathematical representation of uncertainty, *3rd Nondeterministic Approaches Forum*, Seattle, WA, AIAA-2001-1645, 2001.
42. Zadeh, L.A., Fuzzy sets as a basis for a theory of possibility, *Fuzzy Sets and Systems*, 1, 3–28, 1978.
43. Manton, K.G., Woodbury, M.A., and Tolley, H.D., *Statistical Applications Using Fuzzy Sets*, John Wiley & Sons, New York, 1994.
44. Onisawa, T. and Kacprzyk, J., Eds., *Reliability and Safety Analyses under Fuzziness*, Physica-Verlag, Heidelberg, 1995.
45. Klir, G.J., St. Clair, U., and Yuan, B., *Fuzzy Set Theory: Foundations and Applications*, Prentice Hall PTR, Upper Saddle River, NJ, 1997.
46. Dubois, D. and Prade, H., *Fundamentals of Fuzzy Sets*, Kluwer Academic, Boston, 2000.
47. Moore, R.E., *Methods and Applications of Interval Analysis*, SIAM, Philadelphia, PA, 1979.

48. Kearfott, R.B. and Kreinovich, V., Eds., *Applications of Interval Computations*, Kluwer Academic, Boston, 1996.
49. Shafer, G., *A Mathematical Theory of Evidence*, Princeton University Press, Princeton, NJ, 1976.
50. Guan, J. and Bell, D.A., *Evidence Theory and Its Applications*, Vol. I., North Holland, Amsterdam, 1991.
51. Krause, P. and Clark, D., *Representing Uncertain Knowledge: An Artificial Intelligence Approach*, Kluwer Academic Publishers, Dordrecht, The Netherlands, 1993.
52. Kohlas, J. and Monney, P.-A., *A Mathematical Theory of Hints — An Approach to the Dempster-Shafer Theory of Evidence*, Springer, Berlin, 1995.
53. Almond, R.G., *Graphical Belief Modeling*, 1st ed., Chapman & Hall, London, 1995.
54. Klir, G.J. and Wierman, M.J., *Uncertainty-Based Information: Elements of Generalized Information Theory*, Physica-Verlag, Heidelberg, 1998.
55. Kramosil, I., *Probabilistic Analysis of Belief Functions*, Kluwer, New York, 2001.
56. Dubois, D. and Prade, H., *Possibility Theory: An Approach to Computerized Processing of Uncertainty*, Plenum Press, New York, 1988.
57. de Cooman, G., Ruan, D., and Kerre, E.E., *Foundations and Applications of Possibility Theory*, World Scientific Publishing Co., Singapore, 1995.
58. Walley, P., *Statistical Reasoning with Imprecise Probabilities*, Chapman & Hall, London, 1991.
59. Klir, G.J. and Yuan, B., *Fuzzy Sets and Fuzzy Logic*, Prentice Hall, Saddle River, NJ, 1995.
60. Klir, G.J. and Smith, R.M., On measuring uncertainty and uncertainty-based information: recent developments, *Annals of Mathematics and Artificial Intelligence*, 32, 5–33, 2001.
61. Dempster, A.P., Upper and lower probabilities induced by a multivalued mapping, *Annals of Mathematical Statistics*, 38, 325–339, 1967.
62. Wasserman, L.A., Belief functions and statistical inference, *The Canadian Journal of Statistics*, 18, 183–196, 1990.
63. Halpern, J.Y. and Fagin, R., Two views of belief: belief as generalized probability and belief as evidence, *Artifical Intelligence*, 54, 275–317, 1992.
64. Yager, R.R., Kacprzyk, J., and Fedrizzi, M., Eds., *Advances in Dempster-Shafer Theory of Evidence*, John Wiley & Sons, New York, 1994.
65. Dong, W.-M. and Wong, F.S., From uncertainty to approximate reasoning. 1. Conceptual models and engineering interpretations, *Civil Engineering Systems*, 3, 143–154, 1986.
66. Dong, W.-M. and Wong, F.S., From uncertainty to approximate reasoning. 3. Reasoning with conditional rules, *Civil Engineering Systems*, 4, 45–53, 1987.
67. Lai, K.-L. and Ayyub, B.M., Generalized uncertainty in structural reliability assessment, *Civil Engineering Systems*, 11, 81–110, 1994.
68. Tonon, F. and Bernardini, A., A random set approach to the optimization of uncertain structures, *Computers & Structures*, 68, 583–600, 1998.
69. Tanaka, K. and Klir, G.J., A design condition for incorporating human judgement into monitoring systems, *Reliability Engineering and System Safety*, 65, 251–258, 1999.
70. Tonon, F., Bernardini, A., and Elishakoff, I., Concept of random sets as applied to the design of structures and analysis of expert opinions for aircraft crash, *Chaos, Solitons, & Fractals*, 10, 1855–1868, 1999.
71. Rakar, A., Juricic, D., and Ball, P., Transferable belief model in fault diagnosis, *Engineering Applications of Artificial Intelligence*, 12, 555–567, 1999.
72. Fetz, T., Oberguggenberger, M., and Pittschmann, S., Applications of possibility and evidence theory in civil engineering, *International Journal of Uncertainty*, 8, 295–309, 2000.
73. Oberkampf, W.L. and Helton, J.C., Investigation of evidence theory for engineering applications, *4th Nondeterministic Approaches Forum*, Denver, CO, AIAA-2002-1569, 2002.
74. Helton, J.C., Oberkampf, W.L., and Johnson, J.D., Competing failure risk analysis using evidence theory, *5th Nondeterministic Approaches Forum*, Norfolk, VA, AIAA-2003-1911, 2003.
75. Feller, W., *An Introduction to Probability Theory and Its Applications*, Vol. 2, John Wiley & Sons, New York, 1971.

76. Ash, R.B. and Doléans-Dade, C.A., *Probability and Measure Theory*, 2nd ed., Harcourt/Academic Press, New York, 2000.
77. Joslyn, C. and Kreinovich, V., Convergence properties of an interval probabilistic approach to system reliability estimation, *International Journal of General Systems*, in press.
78. McKay, M.D., Beckman, R.J., and Conover, W.J., A comparison of three methods for selecting values of input variables in the analysis of output from a computer code, *Technometrics*, 21, 239–245, 1979.
79. Iman, R.L., Uncertainty and sensitivity analysis for computer modeling applications, *Reliability Technology*, 28, 153–168, 1992.
80. Helton, J.C. and Davis, F.J., Latin hypercube sampling and the propagation of uncertainty in analyses of complex systems, *Reliability Engineering and System Safety*, 81, 23–69, 2003.
81. Luo, W.B. and Caselton, W., Using Dempster-Shafer theory to represent climate change uncertainties, *Journal of Environmental Management*, 49, 73–93, 1997.
82. Kacprzyk, J. and Fedrizzi, M., *Multiperson Decision Making Models Using Fuzzy Sets and Possibility Theory*, Kluwer Academic Publishers, Boston, 1990.
83. Abidi, M.A. and Gonzalez, R.C., *Data Fusion in Robotics and Machine Intelligence*, Academic Press, San Diego, 1992.
84. Goodman, I.R., Mahler, R.P.S., and Nguyen, H.T., *Mathematics of Data Fusion*, Kluwer Academic Publishers, Boston, 1997.
85. Goutsias, J., Mahler, R.P.S., and Nguyen, H.T., *Random Sets, Theory and Applications*, Springer, New York, 1997.
86. Slowinski, R., *Fuzzy Sets in Decision Analysis, Operations Research and Statistics*, Kluwer Academic Publishers, Boston, 1998.
87. Bouchon-Meunier, B., Aggregation and fusion of imperfect information, in *Studies in Fuzziness and Soft Computing*, J. Kacprzyk, Ed., Springer-Verlag, New York, 1998.
88. Ferson, S., Kreinovich, V., Ginzburg, L., Myers, D.S., and Sentz, K., Constructing probability boxes and Dempster-Shafer structures, Sandia National Laboratories, Albuquerque, NM, SAND2003-4015, 2003.
89. Sentz, K. and Ferson, S., Combination of evidence in Dempster-Shafer theory, Sandia National Laboratories, Albuquerque, NM, SAND2002-0835, 2002.

11
Info-Gap Decision Theory for Engineering Design: Or Why "Good" Is Preferable to "Best"

Yakov Ben-Haim
Technion–Israel Institute of Technology

11.1 Introduction and Overview.. 11-1
11.2 Design of a Cantilever with Uncertain Load............... 11-2
 Performance Optimization • Robustness to Uncertain Load • Info-Gap Robust-Optimal Design: Clash with Performance-Optimal Design • Resolving the Clash • Opportunity from Uncertain Load
11.3 Maneuvering a Vibrating System with Uncertain Dynamics... 11-11
 Model Uncertainty • Performance Optimization with the Best Model • Robustness Function • Example
11.4 System Identification 11-17
 Optimal Identification • Uncertainty and Robustness • Example
11.5 Hybrid Uncertainty: Info-Gap Supervision of a Probabilistic Decision 11-22
 Info-Gap Robustness as a Decision Monitor • Nonlinear Spring
11.6 Why "Good" Is Preferable to "Best" 11-25
 The Basic Lemma • Optimal-Performance vs. Optimal Robustness: The Theorem • Information-Gap Models of Uncertainty • Proofs
11.7 Conclusion: A Historical Perspective............................ 11-29

11.1 Introduction and Overview

We call ourselves *Homo sapiens,* in part because we value our ability to optimize, but our sapience is not limited to the persistent pursuit of unattainable goals. Rather, what eons have taught the species is the lesson of balancing goals against the constraints of resources, knowledge, and ability. Indeed, the conjunction of reliability analysis and system design is motivated precisely by the need to balance idealized goals and realistic constraints.

The reliability analyst/system designer seeks to optimize the design, so the question is: What constitutes feasible optimization? First we must recognize that even our best models are wrong in ways we perhaps cannot even imagine. In addition, our most extensive data is incomplete and especially lacks evidence about surprises—catastrophes as well as windfalls—that impact the success and survival of the system. These model and data deficiencies are information gaps (info gaps), or epistemic uncertainties. We will show that optimization of performance is always accompanied by minimization of robustness to epistemic uncertainty. That is, performance and robustness are antagonistic attributes and one must be traded off against the other. A performance-maximizing option will have less robustness against unmodeled information gaps than some suboptimal option, when both are evaluated against the same aspiration for performance. The conclusion is that the performance-suboptimal design is preferable over the performance-optimal design.

The principles just mentioned are explained with four examples in Sections 11.2 through 11.5. These sections can be read independently, although the examples supplement one another by emphasizing different applications, aspects of the problem, and methods of analysis. A theoretical framework is provided in Section 11.6.

Section 11.2 considers the design of the profile of a cantilever that is subjected to uncertain static loads. We first formulate a traditional design analysis, in the absence of uncertainty, which leads to a family of performance-optimal designs. These designs are Pareto-efficient trade-offs between minimizing the stress and minimizing the weight of the beam. We then show that these Pareto-efficient designs in fact have no immunity to info gaps in the load. Robustness to load uncertainty is obtained only by moving off the Pareto-optimal design surface. We also consider the windfall gains that can be garnered from load uncertainty, and examine the relation between robust and opportune designs.

Section 11.3 examines the maneuvering of a vibrating system whose impulse response function is incompletely known. The emphasis is not on control technology, but rather on modeling and managing unstructured info gaps in the design-base model of the system. We formulate a traditional performance optimization of the control input based on the best-available model. We demonstrate that this performance optimization has no immunity to the info gaps that plague the design-base dynamic model. This leads to the analysis of performance-suboptimal designs, which magnify the immunity to uncertainty in the dynamic behavior of the system. A simple numerical example shows that quite large robustness can be achieved with suboptimal designs, while satisfying the performance (making the performance good enough) at levels not too much less than the performance optimum.

Section 11.4 differs from the previous engineering design examples and considers the process of updating the parameters of a system model, based on data, when the basic structure of the model is inaccurate. The case examined is the impact of unmodeled quadratic nonlinearities. We begin by formulating a standard model updating procedure based on maximizing the fidelity between the model and the data. We then show that the result of this procedure has no robustness to the structural deficiencies of the model. We demonstrate, through a simple numerical example, that fidelity-suboptimal models can achieve substantial robustness to model structure errors, while satisfying the fidelity at levels not too far below the fidelity optimum.

Section 11.5 considers hybrid uncertainty: a combination of epistemic info-gaps and explicit (although imprecise) probability densities. A go/no-go decision is to be made based on the evaluation of the probability of failure. This evaluation is based on the best available probability density function. However, this probability density is recognized as imperfect, which constitutes the info-gap that beleaguers the go/no-go decision. An info-gap analysis is used to address the question: How reliable is the probabilistic go/no-go decision, with respect to the unknown error in the probability density? In short, the info-gap robustness analysis supervises the probabilistic decision.

The examples in Sections 11.2 through 11.4 illustrate the assertion that performance optimization will lead inevitably to minimization of immunity to information gaps. This suggests that performance should be satisficed—made adequate but not optimal—and that robustness should be optimized. Section 11.6 provides a rigorous theoretical basis for these conclusions.

11.2 Design of a Cantilever with Uncertain Load

In this section we formulate a simple design problem and solve it by finding the design that optimizes a performance criterion. We then show that this solution has no robustness to uncertainty: infinitesimal deviations (of the load, in this example) can cause violation of the design criterion. This will illustrate the

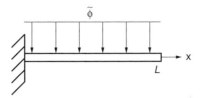

FIGURE 11.1 Cantilever with uniform load.

general conclusion (to be proven later) that optimizing the performance in fact minimizes the robustness to uncertainty. Stated differently, we observe that a designer's aspiration for high performance must be accompanied by the designer's acceptance of low robustness to failure. Conversely, feasible solutions will entail suboptimal performance. Again stated differently, the designer faces an irrevocable trade-off between performance and robustness to failure: demand for high performance is vulnerable to uncertainty; modest performance requirements are more immune to uncertainty.

11.2.1 Performance Optimization

Consider a uniform cantilever of length L [m] subject to a continuous uniform load density $\tilde{\phi}$ [N/m] applied in a single plane perpendicular to the beam axis, as in Figure 11.1. The beam is rectangular in cross section. The beam width, w [m], is uniform along the length and determined by prior constraints, but the thickness in the load plane, $T(x)$ [m], may be chosen by the designer to vary along the beam. The beam is homogeneous and its density is known. The designer wishes to choose the thickness profile to minimize the mass of the beam and also to minimize the maximum absolute bending stress in the beam. These two optimization criteria for selecting the thickness profile $T(x)$ are:

$$\min_{T(x)>0} \int_0^L T(x)\,dx \tag{11.1}$$

$$\min_{T(x)>0} \max_{0\le x\le L} |\sigma_T(x)| \tag{11.2}$$

where $\sigma_T(x)$ is the maximum bending stress in the beam section at x. The positivity constraint on $T(x)$ arises because the thickness must be positive at every point, otherwise the "beam" is not a beam.

These two design criteria are in conflict, so a trade-off between mass and stress minimization will be needed. To handle this, we will solve the stress minimization with the mass constrained to a fixed value. We then vary the beam mass. Consider the following set of thickness profiles corresponding to fixed mass:

$$\Theta(\theta) = \left\{ T(x): \int_0^L T(x)\,dx = \theta \right\} \tag{11.3}$$

For any given value of θ, which determines the beam mass, we choose the thickness profile from $\Theta(\theta)$ to minimize the maximum stress. The performance criterion by which a design proposal $T(x)$ is evaluated is:

$$R(T) = \max_{0\le x\le L} |\sigma_T(x)| \tag{11.4}$$

The design that optimizes the performance from among the beams in $\Theta(\theta)$, which we denote $\hat{T}_\theta(x)$, is implicitly defined by:

$$R(\hat{T}_\theta) = \min_{T(x)\in\Theta(\theta)} \max_{0\le x\le L} |\sigma_T(x)| \tag{11.5}$$

From the small-deflection static analysis of a beam with thickness profile $T(x)$, one finds the magnitude of the maximum absolute bending stress at section x to be:

$$|\sigma_T(x)| = \frac{3\tilde{\phi}(L-x)^2}{wT^2(x)} \text{ [Pa]} \tag{11.6}$$

where $x = 0$ at the clamped end of the beam.

In light of the integral constraint on the thickness profile, $T(x) \in \Theta(\theta)$ in Equation 11.3, and of the demand for optimal performance, Equation 11.5, we find that the optimal design makes the stress uniform along the beam and as small as possible. The optimal profile is a linear taper:

$$\hat{T}_\theta(x) = \frac{2\theta(L-x)}{L^2} \text{ [m]} \tag{11.7}$$

The performance obtained by this design is:

$$R(\hat{T}_\theta) = \frac{3\tilde{\phi}L^4}{4w\theta^2} \text{ [Pa]} \tag{11.8}$$

To understand Equation 11.7, we note from Equation 11.6 that the linear taper is the only thickness profile that achieves the same maximum stress at all sections along the beam. From Equation 11.6 we know that we could reduce the stress in some regions of the beam by increasing the thickness profile in those regions. However, the mass-constraint, Equation 11.3, would force a lower thickness elsewhere, and in those other regions the stress would be augmented. Because the performance requirement is to minimize the maximal stress along the beam, the uniform stress profile is the stress-minimizing solution at this beam mass. Equation 11.8, the performance obtained by this optimal design, is the value of the stress in Equation 11.6 with the optimal taper of Equation 11.7.

We can think of Equation 11.8 as a curve, $R(\hat{T}_\theta)$ vs. θ, representing the trade-off between minimal stress and minimal mass, as shown in Figure 11.2. As the beam mass, θ, is reduced, the least possible maximum stress, $R(\hat{T}_\theta)$, increases. Every point along this curve is optimal in the Pareto sense that either of the design criteria—mass or stress minimization—can be improved only by detracting from the other criterion.

Consider a point P on this optimal design curve, corresponding to the min-max stress σ_1 of a beam of mass θ_1. That is, $\sigma_1 = R(\hat{T}_{\theta_1})$. Let Q be a point to the right of P. Q represents beams whose mass is $\theta_2 > \theta_1$ and whose min-max stress is still only σ_1. These beams are suboptimal: at this min-max stress, they have excessive mass. Alternatively, consider the point R lying above P, which represents beams of

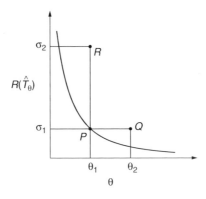

FIGURE 11.2 Optimal (min-max) stress $R(\hat{T}_\theta)$ vs. mass-parameter θ, Equation 11.8.

mass θ_1 whose min-max stress exceeds the optimum for this mass: $\sigma_2 > R(\hat{T}_{\theta_1})$. This again is suboptimal. We can interpret any suboptimal beam as either mass excessive for its min-max stress, or stress excessive for its mass. Finally, because the curve is a Pareto frontier, there are no beams corresponding to points below the curve.

11.2.2 Robustness to Uncertain Load

Now we depart from the performance optimization analysis described above. Any designer wants better performance rather than worse, but aspirations are tempered by the need for feasibility, the need for reliable design. We now consider the very common situation in which the load profile is uncertain and, in response, we will develop a *robust satisficing* design strategy. We are particularly interested in the relation between the optimal design under this strategy, and the performance-maximizing design described in Section 11.2.1.

The designer will choose a thickness profile, $T(x)$, which *satisfices* the aspiration for good performance, that is, which attempts to guarantee that the maximum stress is no greater than a specified level, for a given beam mass. Satisficing is not an optimization, so an element of design freedom still remains. The designer then uses this degree of freedom to maximize the immunity to error in the design-base load. The design specification is *satisfied*, while the robustness to failure is *maximized*.

To implement this we first define an info-gap model of uncertainty and then define the robustness function.

Let $\phi(x)$ [N/m] represent the unknown actual load-density profile, and let $\tilde{\phi}(x)$ denote the designer's best estimate of $\phi(x)$. For example, the nominal estimate may be the constant load density used in Section 11.2.1. Let $\mathcal{U}(\alpha,\tilde{\phi})$ be a set of load profiles $\phi(x)$, containing the nominal estimate $\tilde{\phi}(x)$. An *info-gap model* for the designer's uncertainty about $\phi(x)$ is a *family of nested sets* $\mathcal{U}(\alpha,\tilde{\phi})$, $\alpha \geq 0$. As α grows, the sets become more inclusive:

$$\alpha \leq \alpha' \quad \text{implies} \quad \mathcal{U}(\alpha,\tilde{\phi}) \subseteq \mathcal{U}(\alpha',\tilde{\phi}) \tag{11.9}$$

Also, the nominal load belongs to all the sets in the family:

$$\tilde{\phi}(x) \in \mathcal{U}(\alpha,\tilde{\phi}) \quad \text{for all} \quad \alpha \geq 0 \tag{11.10}$$

The nesting of the uncertainty-sets imbues α with its meaning as a **horizon of uncertainty.** A large α entails great variability of the potential load profiles $\phi(x)$ around the nominal estimate $\tilde{\phi}(x)$. Because α is unbounded ($\alpha \geq 0$), the family of uncertainty sets is likewise unbounded. This means that we cannot identify a worst case, and the subsequent analysis is *not* a worst-case analysis in the ordinary sense, and does *not* entail a min-max as in Equation 11.5. We will see an example of an info-gap model of uncertainty shortly. Info-gap models may obey additional axioms as well [1].

Now we define the *info-gap robustness function*. The designer's aspiration (or requirement) for performance is that the bending stress not exceed the critical value σ_c anywhere along the beam of specified mass. That is, the condition $\sigma(x) \leq \sigma_c$ is needed for "survival"; better performance ($\sigma(x) \ll \sigma_c$) is desirable but is not a design requirement. The designer will choose the critical stress σ_c as small as necessary, but no smaller than needed. The designer attempts to satisfy the design specification with the choice of the thickness profile $T(x)$ from the set $\Theta(\theta)$ in Equation 11.3, but because the actual load profile $\phi(x)$ is unknown when $T(x)$ is chosen, the maximum bending stress is also unknown. The *robustness* of thickness profile $T(x)$ is the greatest horizon of uncertainty, α, at which the maximum stress is guaranteed to be no greater than the design requirement:

$$\hat{\alpha}(T,\sigma_c) = \max\left\{\alpha : \max_{\phi \in \mathcal{U}(\alpha,\tilde{\phi})} \rho(T,\phi) \leq \sigma_c\right\} \tag{11.11}$$

where:

$$\rho(T, \phi) = \max_{0 \leq x \leq L} |\sigma_{\phi,T}(x)| \qquad (11.12)$$

which is the analog of Equation 11.4 for the current case of unknown load profile. $\sigma_{\phi,T}(x)$ denotes the maximum stress in the beam section at x, given load profile $\phi(x)$ and thickness profile $T(x)$.

We can 'read' Equation 11.11 from left to right: The robustness $\hat{\alpha}(T, \sigma_c)$ of thickness profile $T(x)$, given design specification (or aspiration) σ_c, is the maximum horizon of uncertainty α such that the worst performance $\rho(T(x), \phi(x))$, for any realization $\phi(x)$ of the actual load profile up to α is no greater than σ_c. This is a worst-case-up-to-α analysis, but because α is unknown, what we are doing is determining the greatest α that does not allow failure.

More robustness to failure is better than less, provided the design requirements are satisfied. An info-gap robust-optimal design is an allowed thickness profile, $\hat{T}_\theta(x) \in \Theta(\theta)$, which maximizes the robustness while also satisfying the performance:

$$\hat{\alpha}(\hat{T}_\theta, \sigma_c) = \max_{T \in \Theta(\theta)} \hat{\alpha}(T, \sigma_c) \qquad (11.13)$$

This is the info-gap analog of the optimal design criterion in Equation 11.5. Note that, unlike Equation 11.5, we are not minimizing the maximum stress. Rather, we are maximizing the robustness to uncertainty; the stress-requirement is satisfied to σ_c by the robustness function $\hat{\alpha}(T, \sigma_c)$.

Let us consider a concrete example. Suppose that the known nominal load density is the constant non-negative value $\tilde{\phi}$, and that we are also aware that the actual load profile $\phi(x)$ may deviate from $\tilde{\phi}$, but we have no information about this deviation. One representation of this load uncertainty is the envelope-bound info-gap model, which is the following family of nested sets of load profiles:

$$\mathcal{U}(\alpha, \tilde{\phi}) = \{\phi(x): |\phi(x) - \tilde{\phi}| \leq \alpha\}, \quad \alpha \geq 0 \qquad (11.14)$$

$\mathcal{U}(\alpha, \tilde{\phi})$ is the set of load profiles whose deviation from the nominal profile is bounded by α. Because the horizon of uncertainty, α, is unbounded, we have an unbounded family of nested sets of load profiles. Note that $\mathcal{U}(\alpha, \tilde{\phi})$ satisfies the nesting and inclusion properties of Equation 11.9 and Equation 11.10.

The maximum absolute stress in the beam section at x, $|\sigma_{\phi,T}(x)|$, given load profile $\phi(x)$ and thickness profile $T(x)$, is found to be:

$$|\sigma_{\phi,T}(x)| = \frac{6}{wT^2(x)} \int_x^L (v-x)\phi(v)dv \qquad (11.15)$$

Employing this relation with Equation 11.12 in Equation 11.11 yields, after some manipulation, the following expression for the robustness of thickness profile $T(x)$ with design specification σ_c:

$$\hat{\alpha}(T, \sigma_c) = \frac{w\sigma_c/3}{\max_{0 \leq x \leq L}\left(\dfrac{L-x}{T(x)}\right)^2} - \tilde{\phi} \qquad (11.16)$$

provided this expression is positive. A negative value arises if the maximum stress in response to the nominal load exceeds the design requirement, σ_c. A negative value means that, even without uncertainty, the design requirement cannot be achieved. In this case, the robustness to uncertainty vanishes and we define $\hat{\alpha}(T, \sigma_c) = 0$.

11.2.3 Info-Gap Robust-Optimal Design: Clash with Performance-Optimal Design

We now discuss the robustness-maximizing design, $\hat{T}_\theta(x)$ in Equation 11.13. We will find the beam-shape that maximizes the robustness for any choice of the stress-requirement σ_c. Significantly, this beam-shape will be the linear taper that maximizes the performance. However, we will find that when σ_c is chosen on the Pareto-optimal curve, the robustness of this linear taper is precisely zero. That is, performance optimization entails robustness minimization. This motivates the choice of performance suboptimal designs as the only way to obtain positive robustness to uncertainty.

Consider beams of mass θ whose maximum bending stress is no greater than σ_c. We are considering any σ_c, so (θ, σ_c) is not necessarily Pareto-optimal and does not necessarily fall on the optimal-design curve of Figure 11.2. From examination of Equation 11.16 we find that the thickness profile in the allowed-mass set $\Theta(\theta)$ that maximizes the robustness and satisfices the stress (at stress requirement σ_c) is precisely the profile which maximizes the performance: the linear taper $\hat{T}_\theta(x)$ in Equation 11.7. With this robustness-maximizing thickness profile, the robustness in Equation 11.16 becomes:

$$\hat{\alpha}(\hat{T}_\theta, \sigma_c) = \frac{4w\theta^2 \sigma_c}{3L^4} - \tilde{\phi} \tag{11.17}$$

Figure 11.3 illustrates this optimal robustness vs. the maximum-stress design requirement, σ_c.

Figure 11.3 demonstrates one of the most important universal properties of the robustness function: robustness decreases monotonically as the performance requirement becomes more stringent. A small value of σ_c is a demanding specification, while a large value of σ_c is more lenient. A modest stress requirement will be quite robust, while a demanding design will be prone to failure. The value of σ_c at which the robustness becomes zero is denoted in Figure 11.3 by σ^*. This is such an exacting requirement that even infinitesimal deviations of the actual load profile from the nominal profile may entail violation of the design requirement. Clearly, choosing the requirement $\sigma_c = \sigma^*$ is infeasible and unrealistic because σ_c is defined as a stress level that must not be exceeded.

The value of σ^* is obtained by equating $\hat{\alpha}(\hat{T}_\theta, \sigma_c)$, in Equation 11.17, to zero and solving for σ_c. The result is:

$$\sigma^* = \frac{3\tilde{\phi}L^4}{4w\theta^2}\,[\text{Pa}] \tag{11.18}$$

which is precisely the minimal stress obtained by the performance-optimizing design, Equation 11.8. We see here an instance of another general phenomenon of great importance: a design that *optimizes the*

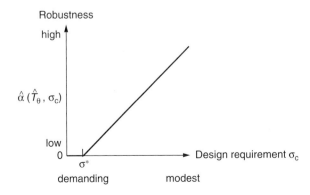

FIGURE 11.3 Optimal robustness curve, $\hat{\alpha}(\hat{T}_\theta, \sigma_c)$, vs. the maximum-stress design requirement σ_c, Equation 11.17.

performance (as in Section 11.2.1) also *minimizes the robustness*. That is, the locus of (θ,σ) values on the optimal-design curve of Figure 11.2 coincides with the zero-robustness points $(\sigma^*,0)$ in Figure 11.3. A point with positive robustness on the curve in Figure 11.3 ($\hat{\alpha}(\hat{T}_\theta,\sigma_2)>0$) corresponds to a point *above* the Pareto-optimal curve like R on Figure 11.2 for which $\sigma_2 > \sigma(\theta)$. Performance optimization leads to the least feasible of all realizable designs. The designer is therefore strongly motivated to *satisfice the performance and maximize the robustness*, as we have done in this section. We explore this further in the next subsection.

11.2.4 Resolving the Clash

We will seek a thickness profile, $T^*(x)$, that has two properties:

1. The design has positive robustness to load-uncertainty, so that $\hat{\alpha}(T^*,\sigma_c)$ in Equation 11.16 is positive.
2. The design has suboptimal performance, so it is not on the optimal trade-off curve between mass and stress, Equation 11.8. This is necessary in order to enable positive robustness.

If the beam mass is constrained to $\Theta(\theta)$, then the min-max stress is given by Equation 11.8. Let us adopt this value, σ^* in Equation 11.18, as the design requirement. We know from our analysis in Section 11.2.3 that we must accept a beam-mass in excess of θ in order to satisfice this stress requirement with positive robustness. That is, following our discussion of Equation 11.8 and Figure 11.2, we must choose a design point to the right of the optimum-performance design curve, such as point Q in Figure 11.2.

From Equation 11.16, the condition for positive robustness is:

$$\frac{w\sigma^*/3}{\max_{0 \le x \le L}\left(\frac{L-x}{T(x)}\right)^2} > \tilde{\phi} \tag{11.19}$$

When $T(x)$ is the performance-optimal linear taper, $\hat{T}_\theta(x)$ of Equation 11.7, and σ^* is given by Equation 11.18, we obtain equality in Equation 11.19 and hence zero robustness, so we must choose the thickness profile so that:

$$T(x) \ge \frac{2\theta(L-x)}{L^2} = \hat{T}_\theta(x) \tag{11.20}$$

with strict inequality over at least part of the beam. There are many available solutions; we consider one simple class of solutions:

$$T^*(x) = \hat{T}_\theta(x) + \gamma \tag{11.21}$$

where $\gamma \ge 0$. From Equation 11.16, the robustness of this profile is:

$$\hat{\alpha}(T^*,\sigma^*) = \frac{w\sigma^*}{3L^2}\left(\frac{2\theta}{L}+\gamma\right)^2 - \tilde{\phi} \tag{11.22}$$

With σ^* from Equation 11.18, this robustness becomes:

$$\hat{\alpha}(T^*,\sigma^*) = \frac{\tilde{\phi}L^2}{4\theta^2}\left(\frac{2\theta}{L}+\gamma\right)^2 - \tilde{\phi} \tag{11.23}$$

whose positivity is controlled by γ, which also controls the extent of deviation of T^* from the performance-optimum solution $\hat{T}_\theta$. When γ is zero, the beam shape lies on the performance-optimal curve,

Equation 11.8: it has minimal mass for the stress requirement σ_c. However, at $\gamma = 0$, the robustness to load uncertainty is zero. As γ becomes larger, the robustness increases but the beam also becomes more mass-excessive.

In Equation 11.20 through Equation 11.23 we have derived a performance-suboptimal design that has positive robustness. This design is formulated as a point Q to the right of the performance-optimal curve in Figure 11.2: we have increased the mass while holding the stress requirement fixed. Another approach to defining designs that are performance-suboptimal and yet have positive robustness is to seek points R that are above the curve. Positive robustness is obtained if the inequality in Equation 11.19 is satisfied. This can be obtained with the linear beam shape of Equation 11.7, and with a stress requirement σ_c in excess of the minimal (optimal) stress for this beam mass given by σ^* in Equation 11.18 (or equivalently by $R(\hat{T}_\theta)$ in Equation 11.8). One way to understand this alternative approach is in terms of the relation between points P and R in Figure 11.2. The mass-optimal linear taper is used, Equation 11.7 with θ_1, but the aspiration for stress performance is weakened: rather than adopting the optimal-stress requirement (σ_1 in Figure 11.2), the designer adopts a less demanding stress requirement (σ_2 in Figure 11.2).

The added beam thickness, γ in Equation 11.21, can be thought of as a safety factor.[1] The designer who proceeded according to the performance-optimization procedure of Section 11.2.1 may add the thickness γ as an ad hoc protection. However, Equation 11.23 enables one to evaluate this deviation from the optimum-performance design in terms of the robustness to uncertainty that it entails. The robustness, $\hat{\alpha}(T^*, \sigma^*)$ in Equation 11.23, is the greatest value of the uncertainty parameter, α in the info-gap model of Equation 11.14, which does not allow the maximum bending stress to exceed $\sigma^*(\theta)$ of Equation 11.18. $\hat{\alpha}$ is the interval of load-amplitude within which the actual load profile, $\phi(x)$, may deviate from the design-base nominal load profile, $\tilde{\phi}$, without exceeding the stress requirement. Suppose the designer desires that the robustness, $\hat{\alpha}$, be equal to a fraction f of the nominal load: $\hat{\alpha} = f\tilde{\phi}$. The thickness safety factor that is needed is obtained by inverting Equation 11.23 to obtain:

$$\gamma = \frac{2\theta}{L}(\sqrt{1+f} - 1) \tag{11.24}$$

$f = 0$ means that no robustness is needed, and this causes $\gamma = 0$, meaning that the optimal performance design, $\hat{T}_\theta$, is obtained. f is increased to represent greater demanded robustness to uncertainty, causing the safety factor, γ, to increase from zero.

11.2.5 Opportunity from Uncertain Load

In Section 11.2.2 we defined the robustness function: $\hat{\alpha}(T, \sigma_c)$ is the greatest horizon of uncertainty that design $T(x)$ can tolerate without failure, when σ_c is the maximum allowed stress. The robustness function addresses the adverse aspect of load uncertainty. It evaluates the immunity to failure, which is why a large value of robustness is preferred to a small value, when the stress limit is fixed.

In this section we explore the idea that uncertainty may be propitious: unknown contingencies may be favorable. The "opportunity function" that we will formulate is also an immunity function: it assesses the immunity against highly desirable windfall outcomes. Because the opportunity function is the immunity against sweeping success, a small value is preferred over a large value. Each immunity function — robustness and opportunity—generates its own preference ranking on the set of available designs. We will see that, in general, these rankings may or may not agree.

As before, σ_c is the greatest acceptable bending stress. Let σ_w be a smaller stress that, if not exceeded at any point along the beam, would be a desirable *windfall* outcome. It is not necessary that the stress be as small as σ_w, but this would be viewed very favorably. The nominal load produces a maximum bending stress that is greater than σ_w. However, favorable fluctuations of the load could produce a

---

[1] The author is indebted to Prof. Eli Altus, of the Technion, for suggesting this interpretation.

maximum bending stress as low as σ_w. The *opportunity function* is the lowest horizon of uncertainty at which the maximum bending stress at any section of the beam can be as low as σ_w:

$$\hat{\beta}(T, \sigma_w) = \min \left\{ \alpha : \min_{\phi \in \mathcal{U}(\alpha, \tilde{\phi})} \rho(T, \phi) \leq \sigma_w \right\} \quad (11.25)$$

where $\rho(T, \phi)$ is the maximum absolute bending stress occurring in the beam, specified in Equation 11.12. $\hat{\beta}(T, \sigma_w)$ is the lowest horizon of uncertainty that must be accepted in order to enable maximum stress as low as σ_w. $\hat{\beta}(T, \sigma_w)$ is the immunity to windfall: a small value implies that windfall performance is possible (although not guaranteed) even at low levels of uncertainty. The opportunity function is the dual of the robustness function in Equation 11.11.

Employing Equation 11.12 we can write the opportunity function more explicitly as:

$$\hat{\beta}(T, \sigma_w) = \min \left\{ \alpha : \min_{\phi \in \mathcal{U}(\alpha, \tilde{\phi})} \max_{0 \leq x \leq L} |\sigma_{\phi, T}(x)| \leq \sigma_w \right\} \quad (11.26)$$

The evaluation of the opportunity function requires a bit of caution because, in general, the order of the inner "min" and "max" operators cannot be reversed. In the current example, however, a simplification occurs.

Let us define a constant load profile, $\phi^* = \tilde{\phi} - \alpha$, that belongs to $\mathcal{U}(\alpha, \tilde{\phi})$ for all $\alpha \geq 0$. One can readily show that this load profile minimizes the maximum stress at all sections, x. That is:

$$|\sigma_{\phi^*, T}(x)| = \min_{\phi \in \mathcal{U}(\alpha, \tilde{\phi})} |\sigma_{\phi, T}(x)| \quad (11.27)$$

Because this minimizing load profile, ϕ^*, is the same for all positions x, we can reverse the order of the operators in Equation 11.26 as:

$$\min_{\phi \in \mathcal{U}(\alpha, \tilde{\phi})} \max_{0 \leq x \leq L} |\sigma_{\phi, T}(x)| = \max_{0 \leq x \leq L} \min_{\phi \in \mathcal{U}(\alpha, \tilde{\phi})} |\sigma_{\phi, T}(x)| \quad (11.28)$$

$$= \max_{0 \leq x \leq L} |\sigma_{\phi^*, T}(x)| \quad (11.29)$$

$$= \max_{0 \leq x \leq L} \frac{3(\tilde{\phi} - \alpha)(L - x)^2}{w T^2(x)} \quad (11.30)$$

The opportunity function is the smallest horizon of uncertainty, α, for which the right-hand side of Equation 11.30 is no greater than σ_w. Equating this expression to σ_w and solving for α yields the opportunity for design $T(x)$ with windfall aspiration σ_w:

$$\hat{\beta}(T, \sigma_w) = \tilde{\phi} - \frac{w \sigma_w / 3}{\max_{0 \leq x \leq L} \left(\frac{L - x}{T(x)} \right)^2} \quad (11.31)$$

This expression is nonnegative unless the right-hand side of Equation 11.30 is less than σ_w at $\alpha = 0$, which occurs if and only if the nominal load entails maximal stress less than σ_w; in this case we define $\hat{\beta}(T, \sigma_w) = 0$.

We have already mentioned that each immunity function—robustness and opportunity—generates its own preference ranking of available designs. Because "bigger is better" for the robustness function, we will prefer T over T' if the former design is more robust than the latter. Concisely:

$$T \succ_r T' \quad \text{if} \quad \hat{\alpha}(T, \sigma_c) > \hat{\alpha}(T', \sigma_c) \quad (11.32)$$

The opportunity function is the immunity against windfall performance, so "big is bad." This means that we will prefer T over T' if the former design is more opportune than the latter:

$$T \succ_o T' \quad \text{if} \quad \hat{\beta}(T, \sigma_w) < \hat{\beta}(T', \sigma_w) \tag{11.33}$$

The generic definitions of the immunity functions do not imply that the preference-rankings in Equation 11.32 and Equation 11.33 agree. The immunity functions are said to be *sympathetic* when their preference rankings agree; they are *antagonistic* otherwise. Both situations are possible. In the present example, the immunities are sympathetic, as we see by combining Equation 11.16 and Equation 11.31 as:

$$\hat{\beta}(T, \sigma_w) = \underbrace{-\frac{\sigma_w}{\sigma_c} \hat{\alpha}(T, \sigma_c)}_{A} + \underbrace{\left(1 - \frac{\sigma_w}{\sigma_c}\right) \tilde{\phi}}_{B} \tag{11.34}$$

Expression "B" does not depend upon the design, $T(x)$, and expression "A" is nonnegative. Consequently, any change in the design that causes $\hat{\alpha}$ to increase (that is, robustness improves) causes $\hat{\beta}$ to decrease (which improves opportunity). Likewise, robustness and opportunity deteriorate together. These immunity functions are sympathetic for any possible design change, although they do not necessarily improve at the same rate; marginal changes may be greater for one than for the other.

In general, robustness and opportunity functions are not necessarily sympathetic. Their sympathy in the current example is guaranteed because B in Equation 11.34 is independent of the design. This need not be the case. If B increases due to a design-change which causes $\hat{\alpha}$ to increase, the net effect may be an increase in $\hat{\beta}$, which constitutes a decrease in opportunity. For an example, see [1, p. 52].

11.3 Maneuvering a Vibrating System with Uncertain Dynamics

In Section 11.2 we considered the design and reliability analysis of a static system subject to uncertain loads. We now consider the analysis and control of a simple vibrating system whose dynamic equations are uncertain. That is, the best model is known to be wrong or incomplete in some poorly understood way, and an info-gap model represents the uncertainty in this system model. Despite the uncertainty in the system model, the designer must choose a driving function that efficiently "propels" the system as far as possible.

11.3.1 Model Uncertainty

Consider a one-dimensional linear system whose displacement $x(t)$ resulting from forcing function $q(t)$ is described by Duhamel's relation:

$$x(t; q, h) = \int_0^t q(\tau) h(t - \tau) d\tau \tag{11.35}$$

where $h(t)$ is the impulse response function (IRF).

The best available model for the IRF is denoted $\tilde{h}(t)$, which may differ substantially from $h(t)$ due to incomplete or inaccurate representation of pertinent mechanisms. For example, for the undamped linear harmonic oscillator:

$$\tilde{h}(t) = \frac{1}{m\omega} \sin \omega t \tag{11.36}$$

where m is the mass and ω is the natural frequency. This IRF is seriously deficient in the presence of damping, which is a complicated and incompletely understood phenomenon.

Let $\mathcal{U}(\alpha, \tilde{h})$ be an info-gap model for uncertainty in the IRF. That is, $\mathcal{U}(\alpha, \tilde{h})$, $\alpha \geq 0$, is a family of nested sets of IRFs, all containing the nominal best model, $\tilde{h}(t)$. That is:

$$\alpha < \alpha' \quad \text{implies} \quad \mathcal{U}(\alpha, \tilde{h}) \subset \mathcal{U}(\alpha', \tilde{h}) \qquad (11.37)$$

and

$$\tilde{h}(t) \in \mathcal{U}(\alpha, \tilde{h}) \quad \text{for all} \quad \alpha \geq 0 \qquad (11.38)$$

As an example we now construct a Fourier ellipsoid-bound info-gap model of uncertainty in the system dynamics. Actual IRFs are related to the nominal function by:

$$h(t) = \tilde{h}(t) + \sum_i c_i \sigma_i(t) \qquad (11.39)$$

where the $\sigma_i(t)$ are known expansion functions (e.g., cosines, sines, polynomials, etc.) and the c_i are unknown expansion coefficients. Let c and $\sigma(t)$ denote the vectors of expansion coefficients and expansion functions, respectively, so that Equation 11.39 becomes:

$$h(t) = \tilde{h}(t) + c^T \sigma(t) \qquad (11.40)$$

A Fourier ellipsoid-bound info-gap model for uncertainty in the IRF is a family of nested ellipsoids of coefficient vectors:

$$\mathcal{U}(\alpha, \tilde{h}) = \{h(t) = \tilde{h}(t) + c^T \sigma(t) : c^T V c \leq \alpha^2\}, \quad \alpha \geq 0 \qquad (11.41)$$

where V is a known, real, symmetric, positive definite matrix that determines the shape of the ellipsoids of c-vectors. V is based on fragmentary information about the dispersion of the expansion coefficients. The size of each ellipsoid is determined by the (unknown) horizon-of-uncertainty parameter α.

11.3.2 Performance Optimization with the Best Model

We now consider the performance-optimal design of the driving function $q(t)$ based on the model $\tilde{h}(t)$ of Equation 11.36, which is, for the purpose of this example, the best-known IRF of the system. The goal of the design is to choose the forcing function $q(t)$ to achieve large displacement $x(T; q, \tilde{h})$ at specified time T with low control effort $\int_0^T q^2(t) dt$. Specifically, we would like to select $q(t)$ so as to achieve an optimal balance between the following two conflicting objectives:

$$\max_{q(t)} x(T; q, \tilde{h}) \qquad (11.42)$$

$$\min_{q(t)} \int_0^T q^2(t) dt \qquad (11.43)$$

Let $\mathcal{Q}(E)$ denote the set of all control functions $q(t)$ whose control effort equals E:

$$\mathcal{Q}(E) = \left\{ q(t) : E = \int_0^T q^2(t) dt \right\} \qquad (11.44)$$

Using the Schwarz inequality, one can readily show that the q-function in $\mathcal{Q}(E)$ that maximizes the displacement $x(T; q, \tilde{h})$ at time T is:

$$q_E^*(t) = \frac{\sqrt{f_0 E}}{m\omega} \sin \omega(T - t) \qquad (11.45)$$

Info-Gap Decision Theory for Engineering Design

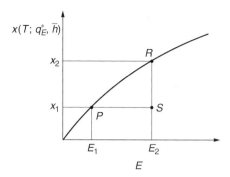

FIGURE 11.4 Maximal displacement $x(T; q_E^*, \tilde{h})$ vs. control effort E, Equation 11.47.

where:

$$f_0 = \frac{4m^2\omega^3}{2\omega T - \sin 2\omega T} \tag{11.46}$$

From this one finds that the greatest displacement at time T, obtainable with any control function in $Q(E)$, is:

$$x(T; q_E^*, \tilde{h}) = \sqrt{\frac{E}{f_0}} \tag{11.47}$$

Equation 11.47 expresses the trade-off between control effort E and maximal displacement x: large displacement is obtained only at the expense of large effort, as shown in Figure 11.4. Like Figure 11.2, this relationship expresses the Pareto-optimal design options: any improvement in control effort (making E smaller) is obtained only by relinquishing displacement (making x smaller).

Points above the curve in Figure 11.4 are inaccessible: no design can realize those (E, x) combinations. Points on the curve are Pareto-optimal and points below the curve are suboptimal designs. For example, point P on the curve is Pareto-optimal: E_1 is the lowest control effort that can achieve displacement as large as x_1. Point S is suboptimal and represents excessive control effort ($E_2 > E_1$) to achieve displacement x_1. Likewise, point R is Pareto-optimal: x_2 is the greatest displacement that can be attained with control effort E_2. So again, S is suboptimal: greater displacement ($x_2 > x_1$) could be achieved with effort E_2.

11.3.3 Robustness Function

We now develop an expression for the robustness, of the displacement $x(T; q, h)$, to uncertainty in the system dynamics $h(t)$.

The first design goal, Equation 11.42, implies that a large value of displacement is needed. The second design goal, Equation 11.43, conflicts with the first and calls for small control effort. In Section 11.3.2 we found that performance optimization leads to a Pareto trade-off between these two criteria, expressed in Equation 11.47 and Figure 11.4. In this section, in light of the uncertainty in the IRF, we take a different approach. For any given control function $q(t)$, we "satisfice" the displacement by requiring that the displacement be at least as large as some specified and satisfactory value x_c. Because $x(T; q, h)$ depends on the unknown IRF, we cannot guarantee that the displacement will be satisfactory. However, we can answer the following question: for given forcing function $q(t)$, by how much can the best model $\tilde{h}(t)$ err without jeopardizing the achievement of adequate displacement? More specifically, given $q(t)$, what is the greatest horizon of uncertainty, α, up to which every model $h(t)$ causes the displacement to be at least as large as x_c? The answer to this question is the robustness function:

$$\hat{\alpha}(q, x_c) = \max\left\{\alpha: \min_{h \in \mathcal{U}(\alpha, \tilde{h})} x(T; q, h) \geq x_c\right\} \tag{11.48}$$

We can "read" this relation from left to right: the robustness $\hat{\alpha}(q, x_c)$ of control function $q(t)$ with displacement-aspiration x_c is the greatest horizon of uncertainty α such that every system-model $h(t)$ in $\mathcal{U}(\alpha, \tilde{h})$ causes the displacement $x(T; q, h)$ to be no less than x_c. If $\hat{\alpha}(q, x_c)$ is large, then the system is robust to model uncertainty and $q(t)$ can be relied upon to bring the system to at least x_c at time T. If $\hat{\alpha}(q, x_c)$ is small, then this driving function cannot be relied upon and the system is vulnerable to uncertainty in the dynamics.

Using Lagrange optimization, one can readily show that the smallest displacement, for any system model $h(t)$ up to uncertainty α, is:

$$\min_{h \in \mathcal{U}(\alpha, \tilde{h})} x(T; q, h) = x(T; q, \tilde{h}) - \alpha \sqrt{b^T V b} \qquad (11.49)$$

where we have defined the following vector:

$$b = \int_0^T q(t) \sigma(T - t) \, dt \qquad (11.50)$$

Equation 11.49 asserts that the least displacement, up to uncertainty α, is the nominal, best-model displacement $x(T; q, \tilde{h})$, decremented by the uncertainty term $\alpha \sqrt{b^T V b}$. If the nominal displacement falls short of the demanded displacement x_c, then uncertainty only makes things worse and the robustness to uncertainty is zero. If $x(T; q, \tilde{h})$ exceeds x_c, then the robustness is found by equating the right-hand side of Equation 11.49 to x_c and solving for α. That is, the robustness of driving function $q(t)$ is:

$$\hat{\alpha}(q, x_c) = \begin{cases} 0 & \text{if} \quad x(T; q, \tilde{h}) \leq x_c \\ \dfrac{x(T; q, \tilde{h}) - x_c}{\sqrt{b^T V b}} & \text{else} \end{cases} \qquad (11.51)$$

Equation 11.51 documents the trade-off between robustness, $\hat{\alpha}(q, x_c)$, and aspiration for performance, x_c, as shown in Figure 11.5. A large and demanding value of x_c is accompanied by a low value of immunity to model uncertainty, meaning that aspirations for large displacements are unreliable and infeasible. Modest requirements (small values of x_c) are more feasible because they have greater immunity to uncertainty. The value of x_c at which the robustness vanishes, x^* in the figure, is precisely the displacement predicted by the best model, $x(T; q, \tilde{h})$. That is:

$$\hat{\alpha}\big(q, x(T; q, \tilde{h})\big) = 0 \qquad (11.52)$$

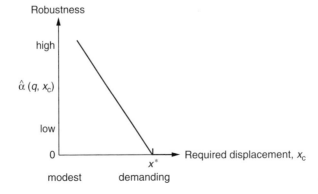

FIGURE 11.5 Robustness $\hat{\alpha}(q, x_c)$ vs. the demanded displacement x_c, Equation 11.51.

This means that, for *any* driving function $q(t)$, the displacement predicted by the best available model $\tilde{h}(t)$ cannot be relied upon to occur. Shortfall of the displacement may occur due to an infinitesimally small error of the model. Because this is true for any $q(t)$, it is also true for the performance-optimum control function $q_E^*(t)$ in Equation 11.45:

$$\hat{\alpha}(q_E^*, x(T; q_E^*, \tilde{h})) = 0 \tag{11.53}$$

While $q_E^*(t)$ is, according to $\tilde{h}(t)$, the most effective driving function of energy E, and while $x(T; q_E^*, \tilde{h})$ is, again according to $\tilde{h}(t)$, the resulting displacement, Equation 11.53 shows that this prediction has no immunity to modeling errors.

It is important to recognize that ordered pairs such as $(E, x(T; q_E^*, \tilde{h}))$ correspond to points such as P and R on the Pareto-optimal design surface in Figure 11.4. That is, E and $x(T; q_E^*, \tilde{h})$ are related by Equation 11.47. Hence, Equation 11.53 shows that all of the performance-optimal designs on the Pareto surface have no immunity to errors in the design-base model of the system. These Pareto-efficient designs are not feasible or reliable predictions of the system performance.

The conclusion from Equations 11.52 and 11.53 is likely to be that, because $x(T; q_E^*, \tilde{h})$ cannot be relied upon to occur, one must moderate one's aspirations and accept a lower value of displacement. The designer might "travel" up and to the left on the robustness curve in Figure 11.5 until finding a value of $x_c < x^*$ at which the robustness is satisfactorily large. But then the question arises: what driving function $q(t)$ *maximizes* the robustness at this selected aspiration for displacement? The optimization studied in Section 11.3.2 was optimization of *performance* (displacement and control effort). We now consider satisficing these quantities and optimizing the *robustness*. Specifically, for any displacement-aspiration x_c, the *robust-optimal* control function $\hat{q}_E(t)$ of energy E maximizes the robustness function:

$$\hat{\alpha}(\hat{q}_E, x_c) = \max_{q(t) \in Q(E)} \hat{\alpha}(q, x_c) \tag{11.54}$$

This robust optimum may not always exist, or it may be inaccessible for practical reasons. In any case, one will tend to prefer more robust over less robust solutions. More specifically, if $q_1(t)$ is more robust than $q_2(t)$, while satisfying the performance at the same level x_c, then $q_1(t)$ is preferred over $q_2(t)$:

$$q_1(t) \succ q_2(t) \quad \text{if} \quad \hat{\alpha}(q_1, x_c) > \hat{\alpha}(q_2, x_c) \tag{11.55}$$

Relating this to our earlier discussion, suppose that E is an accessible control effort and that x_c is a satisfactory level of performance. For x_c to be feasible, it must be less than the best performance obtainable with effort E, namely, $x_c < x(t; q_E^*, \tilde{h})$. This assures that the robustness of the performance-optimal control function, $q_E^*(t)$, will be positive: $\hat{\alpha}(q_E^*, x_c) > 0$. However, we might well ask if there is some other control function in $Q(E)$ whose robustness is even greater. This will often be the case, as we illustrate in the next subsection.

11.3.4 Example

To keep things simple, suppose that $\sigma(t)$ in the unknown part of the IRF in Equation 11.40 is a single, linearly decreasing function:

$$\sigma(t) = \eta(T - t) \tag{11.56}$$

where η is a positive constant. Thus, c is a scalar and the shape-matrix in the info-gap model of Equation 11.41 is simply $V = 1$.

The performance-optimal control function of effort E is $q_E^*(t)$ in Equation 11.45, which is a sine function at the natural frequency of the nominal IRF, $\tilde{h}(t)$. From Equation 11.51, the robustness of this control function is:

$$\hat{\alpha}(q_E^*, x_c) = \frac{\frac{1}{m\omega}\int_0^T q_E^*(t)\sin\omega(T-t)dt - x_c}{\left|\eta\int_0^T q_E^*(t)(T-t)dt\right|} \quad (11.57)$$

Similarly, the robustness of any arbitrary control function $q(t)$ is:

$$\hat{\alpha}(q, x_c) = \frac{\frac{1}{m\omega}\int_0^T q(t)\sin\omega(T-t)dt - x_c}{\left|\eta\int_0^T q(t)(T-t)dt\right|} \quad (11.58)$$

(Presuming the numerator is positive.) In light of our discussion of Equation 11.55 we would like to find a control function $q(t)$ in $\mathcal{Q}(E)$ whose robustness is substantially greater than the robustness of $q_E^*(t)$.

We will illustrate that very substantial robustness benefits can be achieved by abandoning the performance-optimal function $q_E^*(t)$. We will not consider the general maximization of $\hat{\alpha}(q, x_c)$, but only a parametric case. Consider functions of the form:

$$q_\mu(t) = A\sin\mu(T-t) \quad (11.59)$$

where $0 < \mu < \omega$ and A is chosen to guarantee that $q(t)$ belongs to $\mathcal{Q}(E)$ (which was defined in Equation 11.44):

$$A = \sqrt{\frac{4\mu E}{2\mu T - \sin 2\mu T}} \quad (11.60)$$

For the special case that $\omega T = \pi$ the robustness functions of Equations 11.57 and 11.58 become:

$$\hat{\alpha}(q_E^*, x_c) = \frac{\pi\sqrt{\omega E} - \sqrt{2\pi}\, m\omega^2 x_c}{2m\eta\sqrt{\omega E}} \quad (11.61)$$

$$\hat{\alpha}(q_\mu, x_c) = \frac{A\mu^2\left(\frac{\sin[\pi(\mu-\omega)/\omega]}{2(\mu-\omega)} - \frac{\sin[\pi(\mu+\omega)/\omega]}{2(\mu+\omega)}\right) - m\omega\mu^2 x_c}{\eta m A[\omega\sin(\pi\mu/\omega) - \pi\mu\cos(\pi\mu/\omega)]} \quad (11.62)$$

Figure 11.6 shows the ratio of the robustnesses of the suboptimal to the performance-optimal control functions, vs. the frequency of the control function. The robustness at control frequencies μ much less than the nominal natural frequency ω, is substantially greater than the robustness of the performance-maximizing function. For example, at point P, $\mu = 0.2$ and the robustness ratio is $\hat{\alpha}(q_\mu, x_c)/\hat{\alpha}(q_E^*, x_c) = 4.0$, meaning that $q_\mu(t)$ can tolerate a horizon of model-uncertainty four times greater than the uncertainty that is tolerable for $q_E^*(t)$, when satisficing the displacement at x_c. $q_E^*(t)$ maximizes the displacement according to the best IRF, $h(t)$. However, this displacement optimization of the control function leaves little residual immunity to uncertainty in the IRF at this value of x_c. A control function such as $q_\mu(t)$, for $\mu < \omega$, belongs to $\mathcal{Q}(E)$, as does $q_E^*(t)$, but $q_\mu(t)$ is suboptimal with respect to displacement. That is, $x(T; q_\mu, \tilde{h}) < x(T; q_E^*, \tilde{h})$. However, because $q_\mu(t)$ is suboptimal, there are many functions with control effort E that cause displacement as large as $x(T; q_\mu, \tilde{h})$. In other words, there is additional design freedom with which to amplify the immunity to uncertainty. What Figure 11.6 shows is that large robustness-amplification can be achieved. This answers, by way of illustration, the question raised at the end of Section 11.3.3.

Info-Gap Decision Theory for Engineering Design

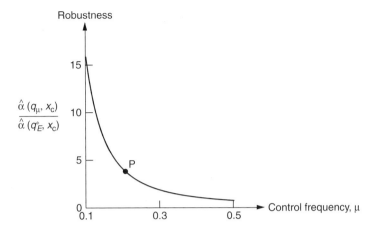

FIGURE 11.6 Ratio of the robustnesses of the suboptimal to performance-optimal control functions, vs. frequency of control function. $x_c = 0.5$, $\omega = E = \eta = m = 1$. $\hat{\alpha}(q_E^*, x_c) = 0.94$.

11.4 System Identification

A common task encountered by engineering analysts is the updating of a system-model, based on measurements. The question we consider in this section is, given that the structure of the model is imperfect, what constitutes optimal estimation of the parameters? More precisely, is it sound procedure to maximize the fidelity between the model and the measurements if the model structure is wrong (in unknown ways, of course)?

11.4.1 Optimal Identification

We begin by formulating a fairly typical framework for optimal identification of a model for predicting the behavior of a system. We then consider an example.

Let y_i be a vector of measurements of the system at time or state i, for $i = 1, \ldots, N$. Let $f_i(q)$ denote the model-prediction of the system in state i, which should match the measurements if the model is good. The vector q, containing real and linguistic variables, denotes the parameters and properties of the model that can be modified to bring the model into agreement with the measurements. We will denote the set of measurements by $Y = \{y_1, \ldots, y_N\}$ and the set of corresponding model-predictions by $F(q) = \{f_1(q), \ldots, f_N(q)\}$.

The overall performance of the predictor is assessed by a function $R[Y, F(q)]$. For example, this might be a mean-squared prediction error:

$$R[Y, F(q)] = \frac{1}{N} \sum_{i=1}^{N} \|f_i(q) - y_i\|^2 \tag{11.63}$$

A performance-optimal model, q^*, minimizes the performance-measure:

$$R[Y, F(q^*)] = \min_q R[Y, F(q)] \tag{11.64}$$

11.4.2 Uncertainty and Robustness

The model $f_i(q)$ is undoubtedly wrong, perhaps fundamentally flawed in its structure. There may be basic mechanisms that act on the system but which are not represented by $f_i(q)$. Let us denote more general models, some of which may be more correct, by:

$$\phi_i = f_i(q) + u_i \tag{11.65}$$

where u_i represents the unknown corrections to the original model, $f_i(q)$. We have very little knowledge about u_i; if we had knowledge of u_i we would most likely include it in $f_i(q)$. So, let us use an info-gap model of uncertainty to represent the unknown variation of possible models:

$$\phi_i \in \mathcal{U}(\alpha, f_i(q)), \quad \alpha \geq 0 \tag{11.66}$$

The centerpoint of the info-gap model, $f_i(q)$, is the known model, parameterized by q. The horizon of uncertainty, α, is unknown. This info-gap model is a family of nested sets of models. These sets of models become ever more inclusive as the horizon of uncertainty increases. That is:

$$\alpha \leq \alpha' \quad \text{implies} \quad \mathcal{U}(\alpha, f_i(q)) \subset (\alpha', f_i(q)) \tag{11.67}$$

In addition, the update model is included in all of the uncertainty sets:

$$f_i(q) \in \mathcal{U}(\alpha, f_i(q)), \quad \text{for all} \quad \alpha \geq 0 \tag{11.68}$$

As before, the model prediction of the system output in state i is $f_i(q)$, and the set of model predictions is denoted $F(q) = \{f_1(q), \ldots, f_N(q)\}$. More generally, the set of model predictions with unknown terms $u_1, \ldots, u_N$ is denoted $F_u(q) = \{f_1(q) + u_1, \ldots, f_N(q) + u_N\}$.

We wish to choose a model, $f_i(q)$ for which the performance index, $R[Y, F_u(q)]$, is small. Let r_c represent an acceptably small value of this index. We would be willing, even delighted, if the prediction-error is smaller, but an error larger than r_c would be unacceptable.

The *robustness to model uncertainty*, of model q with error-aspiration r_c, is the greatest horizon of uncertainty, α, within which all models provide prediction error no greater than r_c:

$$\hat{\alpha}(q, r_c) = \max\left\{\alpha : \max_{\substack{\phi_i \in \mathcal{U}(\alpha, f_i(q)) \\ i=1,\ldots,N}} R[Y, F_u(q)] \leq r_c\right\} \tag{11.69}$$

When $\hat{\alpha}(q, r_c)$ is large, the model $f_i(q)$ may err fundamentally to a great degree, without jeopardizing the accuracy of its predictions; the model is robust to info gaps in its formulation. When $\hat{\alpha}(q, r_c)$ is small, then even small errors in the model result in unacceptably large prediction errors.

Let q^* be an optimal model, which minimizes the prediction error as defined in Equation 11.64, and let r_c^* be the corresponding optimal prediction error: $r_c^* = R(Y, F(q^*))$. Using model q^*, we can achieve prediction error as small as r_c^*, and no value of q can produce a model $f_i(q)$ that performs better. However, the robustness to model uncertainty, of this optimal model, is zero:

$$\hat{\alpha}(q^*, r_c^*) = 0 \tag{11.70}$$

This is a special case of the theorem to be discussed in Section 11.6 that, by optimizing the performance, one minimizes the robustness to info gaps. By optimizing the performance of the model predictor, $f_i(q)$, we make this predictor maximally sensitive to errors in the basic formulation of the model.

In fact, Equation 11.70 is a special case of the following proposition. For any q, let $r_c = R[Y, F(q)]$ be the prediction-error of model $f_i(q)$. The preliminary lemma in Section 11.6 shows that:

$$\hat{\alpha}(q, r_c) = 0 \tag{11.71}$$

That is, the robustness of *any* model, $f_i(q)$, to uncertainty in the structure of that model, is precisely equal to zero, if the error aspiration r_c equals the value of the performance function of that model. No model can be relied upon to perform at the level indicated by its performance function, if that model is subject to errors in its structure or formulation. $R[Y, F(q)]$ is an unrealistically optimistic assessment of model $f_i(q)$, unless we have reason to believe that no auxiliary uncertainties lurk in the mist of our ignorance.

11.4.3 Example

A simple example will illustrate the previous general discussion.

We begin by formulating a *mean-squared-error estimator* for a one-dimensional linear model. The measurements y_i are scalars, and the model to be estimated is:

$$f_i(q) = iq \tag{11.72}$$

The performance function is the mean-squared error between model and measurements, Equation 11.63, which becomes:

$$R[Y, F(q)] = \frac{1}{N} \sum_{i=1}^{N} (iq - y_i)^2 \tag{11.73}$$

$$= \underbrace{\frac{1}{N} \sum_{i=1}^{N} y_i^2}_{\eta_2} - 2q \underbrace{\frac{1}{N} \sum_{i=1}^{N} i y_i}_{\eta_1} + q^2 \underbrace{\frac{1}{N} \sum_{i=1}^{N} i^2}_{\eta_0} \tag{11.74}$$

which defines the quantities η_0, η_1, and η_2. The performance-optimal model defined in Equation 11.64, which minimizes the mean-squared error, is:

$$q^* = \frac{\eta_1}{\eta_0} \tag{11.75}$$

Now we introduce *uncertainty into the model*. The model that is being estimated is linear in the "time" or "sequence" index i: $f_i = iq$. How robust is the performance of our estimator, to modification of the structure of this model? That is, how much can the model err in its basic structure without jeopardizing its predictive power?

Suppose that the linear model of Equation 11.72 errs by lacking a quadratic term:

$$\phi_i = iq + i^2 u \tag{11.76}$$

where the value of u is unknown. The uncertainty in the quadratic model is represented by an interval-bound info-gap model, which is the following unbounded family of nested intervals:

$$\mathcal{U}(\alpha, iq) = \{\phi_i = iq + i^2 u : |u| \leq \alpha\}, \quad \alpha \geq 0 \tag{11.77}$$

The robustness of nominal model $f_i(q)$, with performance-aspiration r_c, is the greatest value of the horizon of uncertainty α at which the mean-squared error of the prediction is no greater than r_c for any model in $\mathcal{U}(\alpha, iq)$:

$$\hat{\alpha}(q, r_c) = \max \left\{ \alpha : \max_{|u| \leq \alpha} R[Y, F_u(q)] \leq r_c \right\} \tag{11.78}$$

The mean-squared error of a model with nonlinear term $i^2 u$ is:

$$R[Y, F_u(q)] = \frac{1}{N} \sum_{i=1}^{N} (iq + i^2 u - y_i)^2 \qquad (11.79)$$

$$= \underbrace{\frac{1}{N} \sum_{i=1}^{N} (iq - y_i)^2}_{\xi_2} + 2u \underbrace{\frac{1}{N} \sum_{i=1}^{N} i^2 (iq - y_i)}_{\xi_1} + u^2 \underbrace{\sum_{i=1}^{N} i^4}_{\xi_0} \qquad (11.80)$$

which defines ξ_0, ξ_1 and ξ_2.

Some manipulations show that the maximum mean-squared error, for all quadratic models ϕ_i up to horizon of uncertainty α, is:

$$\max_{|u| \leq \alpha} R[Y, F_u(q)] = \xi_2 + 2\alpha |\xi_1| + \alpha^2 \xi_0 \qquad (11.81)$$

Referring to Equation 11.78, the robustness to an unknown quadratic nonlinearity $i^2 u$, of the linear model $f_i(q)$, is the greatest value of α at which this maximum error is no greater than r_c.

First we note that the robustness is zero if r_c is small:

$$\hat{\alpha}(q, r_c) = 0, \quad r_c \leq \xi_2 \qquad (11.82)$$

This is because, if $r_c \leq \xi_2$, then max R in Equation 11.81 exceeds r_c for any positive value of α. One implication of Equation 11.82 is that some nonlinear models have prediction errors in excess of ξ_2. If it is required that the fidelity between model and measurement be as good as or better than ξ_2, then no modeling errors of the quadratic type represented by the info-gap model of Equation 11.77 can be tolerated. Recall that ξ_2 is the mean-squared error of the nominal linear predictor, $f_i(q)$. Equation 11.82 means that there is no robustness to model uncertainty if the performance-aspiration r_c is stricter or more exacting than the performance of the nominal, linear model.

For $r_c \geq \xi_2$, the robustness is obtained by equating the right-hand side of Equation 11.81 to r_c and solving for α, resulting in:

$$\hat{\alpha}(q, r_c) = \frac{|\xi_1|}{\xi_0} \left(-1 + \sqrt{1 + \frac{\xi_0 (r_c - \xi_2)}{\xi_1^2}} \right), \quad \xi_2 \leq r_c \qquad (11.83)$$

Relation 11.83 is plotted in Figure 11.7 for synthetic data[2] y_i and for two values of the model parameter q. The figure shows the robustness to model uncertainty, $\hat{\alpha}(q, r_c)$, against the aspiration for prediction error, r_c. The robustness increases as greater error is tolerated. Two curves are shown, one for the optimal linear model, $q^* = 2.50$ in Equation 11.75, whose mean-squared error $r_c^* = R[Y, F(q^*)] = 4.82$ is the lowest obtainable with any linear model. The other model, $q' = 2.60$, has a greater mean-squared error $r_c' = R[Y, F(q')] = 4.94$, so $r_c' > r_c^*$. However, the best performance (the smallest r_c-value) with each of these models, q' and q^*, has no robustness to model uncertainty: $0 = \hat{\alpha}(q^*, r_c^*) = \hat{\alpha}(q', r_c')$.

More importantly, the robustness curves cross at a higher value of r_c (corresponding to lower aspiration for prediction fidelity), as seen in Figure 11.7. If prediction-error $r_c^\circ = 5.14$ is tolerable, then the suboptimal model q' is more robust than, and hence preferable over, the mean-squared optimal model q^*, at the same performance-aspiration. In other words, because $0 = \hat{\alpha}(q^*, r_c^*)$, the analyst recognizes that performance as good as r_c^* is not reliable or feasible with the optimal linear model q^*, and some larger r_c value (representing poorer fidelity between model and measurement) must be accepted; the analyst is motivated

---

[2] $N = 5$ and $y_1, \ldots, y_5 = 1.4, 2.6, 5.6, 8.6, 15.9$.

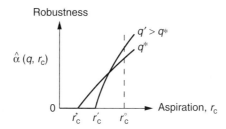

FIGURE 11.7 Robustness vs. prediction-aspiration, Equation 11.83; q fixed.

to "move up" along the q^*-robustness curve. If r_c° is an acceptable level of fidelity, then the suboptimal model q' achieves this performance with greater robustness than the optimal model q^*. In particular, $\hat{\alpha}(q', r_c^\circ) = 0.021$ which is small but still twice as large as $\hat{\alpha}(q^*, r_c^\circ) = 0.011$. In this case, "good" (that is, q') is preferable to "best" (q^*).

We see the robustness preference for a suboptimal model explicitly in Figure 11.8, which shows the robustness vs. the linear model parameter q, for fixed aspiration $r_c = 5.5$ (which is larger than the r_c-values in Figure 11.7, so q^* has positive robustness). The least-squares optimal parameter, $q^* = 2.50$, minimizes the mean-squared error $R[Y, F(q)]$, while the robust-optimal parameter, $\hat{q}_c = 2.65$, maximizes the robustness function $\hat{\alpha}(q, r_c)$. q^* has lower robustness than $\hat{q}_c$, at the same level of model-data fidelity, r_c. Specifically, $\hat{\alpha}(q^*, r_c) = 0.033$ is substantially less than $\hat{\alpha}(\hat{q}_c, r_c) = 0.045$. The mean-squared error of $\hat{q}_c$ is $R[Y, F(\hat{q}_c)] = 5.09$, which is only modestly worse than the least-squares optimum of $R[Y, F(q^*)] = 4.82$. In short, the performance-suboptimal model has only moderately poorer fidelity to the data than the least-squares optimal model $q*$, while the robustness to model uncertainty of $\hat{q}_c$ is appreciably greater than the robustness of q^*.

In summary, we have established the following conclusions from this example.

First, the performance-optimal model, $f_i(q^*)$, has no immunity to error in the basic structure of the model. The model $f_i(q^*)$, which minimizes the mean-squared discrepancy between measurement and prediction, has zero robustness to modeling errors at its nominal prediction fidelity, r_c^*.

Second, this is actually true of *any* model, $f_i(q')$. The value of its mean-squared error is r_c' which, as in Figure 11.7, has zero robustness.

Third, the robustness curves of alternative linear models can cross, as in Figure 11.7. This shows that a suboptimal model such as $f_i(q')$ can be more robust to model uncertainty than the mean-squared optimal model $f_i(q^*)$, when these models are compared at the same aspiration for fidelity between model and measurement, r_c° in the figure.

Fourth, the model that maximizes the robustness can be substantially more robust than the optimal model q^*, which minimizes the least-squared error function, as shown in Figure 11.8. This robustness curve is evaluated at a fixed value of the performance-satisficing parameter r_c.

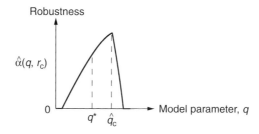

FIGURE 11.8 Robustness vs. model parameter, Equation 11.83; r_c fixed.

11.5 Hybrid Uncertainty: Info-Gap Supervision of a Probabilistic Decision

11.5.1 Info-Gap Robustness as a Decision Monitor

In Sections 11.2 and 11.3 we considered the reliability of technological systems. We now consider the reliability of a decision algorithm itself. Many decisions are based on probabilistic considerations. A foremost class of examples entails acceptance tests based on the evaluation of a probability of failure. Paradigmatically, a go/no-go decision hinges on whether the probability of failure is below or above a critical threshold:

$$P_f(p) \underset{\text{no-go}}{\overset{\text{go}}{\lessgtr}} P_c \tag{11.84}$$

where p is a probability density function (PDF) from which the probability of failure, $P_f(p)$, is evaluated.

This is a valid and meaningful decision procedure when the PDF is well known and when the probability of failure can be assessed with accurate system models. However, a decision algorithm such as Equation 11.84 will be unreliable if the PDF is uncertain (which will often be the case, especially regarding the extreme tails of the distribution) and if the critical probability of failure P_c is small (which is typically the case with critical components). When the PDF is imprecisely known, the reliability of the probabilistic decision can be assessed using the info-gap robustness function.

Let $\tilde{p}$ be the best estimate of the PDF, which is recognized to be wrong to some unknown extent. For the sake of argument, let us suppose that, with $\tilde{p}$, the probability of failure is acceptably small:

$$P_f(\tilde{p}) \leq P_c \tag{11.85}$$

That is, the nominal PDF implies "all systems go." However, because $\tilde{p}$ is suspect, we would like to know how immune this decision is to imperfection of the PDF.

Let $\mathcal{U}(\alpha, \tilde{p})$, $\alpha \geq 0$, be an info-gap model for the uncertain variation of the actual PDF with respect to the nominal, best estimate, $\tilde{p}$. (We will encounter an example shortly.) The robustness, to uncertainty in the PDF, of decision algorithm Equation 11.84, is the greatest horizon of uncertainty up to which all PDFs lead to the same decision:

$$\hat{\alpha}(P_c) = \max\left\{\alpha: \max_{p \in \mathcal{U}(\alpha, \tilde{p})} P_f(p) \leq P_c\right\} \tag{11.86}$$

$\hat{\alpha}(P_c)$ is the greatest horizon of uncertainty in the PDF, up to which all densities p in $\mathcal{U}(\alpha, \tilde{p})$ yield the same decision as $\tilde{p}$. If $\hat{\alpha}(P_c)$ is large, then the decision based on $\tilde{p}$ is immune to uncertainty in the PDF and hence reliable. Alternatively, if $\hat{\alpha}(P_c)$ is small, then a decision based on $\tilde{p}$ is of questionable validity. We see that the robustness function $\hat{\alpha}(P_c)$ is a decision evaluator: it supports the higher-level judgment (how reliable is the probabilistic algorithm?) that hovers over and supervises the ground-level go/no-go decision.

If the inequality in Equation 11.85 were reversed and $\tilde{p}$ implied "no-go," then we would modify Equation 11.86 to:

$$\hat{\alpha}(P_c) = \max\left\{\alpha: \min_{p \in \mathcal{U}(\alpha, \tilde{p})} P_f(p) \geq P_c\right\} \tag{11.87}$$

The meaning of the robustness function as a decision monitor would remain unchanged. $\hat{\alpha}(P_c)$ is still the greatest horizon of uncertainty up to which the decision remains constant.

We can formulate the robustness slightly differently as the greatest horizon of uncertainty at which the probability of failure does not differ more than π_c:

$$\hat{\alpha}(\pi_c) = \max \{\alpha : \max_{p \in \mathcal{U}(\alpha, \tilde{p})} |P_f(p) - P_f(\tilde{p})| \leq \pi_c\} \quad (11.88)$$

Other variations are also possible [2], but we now proceed to a simple example of the use of the info-gap robustness function in the supervision of a probabilistic decision with an uncertain PDF.

11.5.2 Nonlinear Spring

Consider a spring with the following nonlinear relationship between displacement x and force f:

$$f = k_1 x + k_2 x^2 \quad (11.89)$$

The spring fails if the magnitude of the displacement exceeds x_c and we require the probability of failure not to exceed P_c.

The loading force f is nonnegative but uncertain, and the best available PDF is a uniform density:

$$\tilde{p}(f) = \begin{cases} 1/F & \text{if } 0 \leq f \leq F \\ 0 & \text{if } F \leq f \end{cases} \quad (11.90)$$

where the value of F is known. However, it is recognized that forces greater than F may occur. The probability of such excursions, although small, is unknown, as is the distribution of this high-tail probability. That is, the true PDF, shown schematically in Figure 11.9, is:

$$p(f) = \begin{cases} \text{constant} & \text{if } 0 \leq f \leq F \\ \text{variable} & \text{if } F \leq f \end{cases} \quad (11.91)$$

The first question we must consider is how to model the uncertainty in the PDF of the force f. What we **do know** is that f is nonnegative, that $p(f)$ is constant for $0 \leq f \leq F$, and the value of F. What we **do not know** is the actual constant value of $p(f)$ for $0 \leq f \leq F$ and the behavior of $p(f)$ for $f > F$. We face an info gap.

Let $\mathcal{P}$ denote the set of all nonnegative and normalized PDFs on the interval $[0, \infty)$. Whatever form $p(f)$ takes, it must belong to $\mathcal{P}$. An info-gap uncertainty model that captures the information as well as the info gaps about the PDF is:

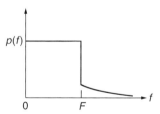

FIGURE 11.9 Uncertain probability density function of the load, Equation 11.91.

$$\mathcal{U}(\alpha, \tilde{p}) = \left\{ p(f) : \ p(f) \in \mathcal{P}; \ \int_F^\infty p(f) df \leq \alpha; \right.$$

$$\left. p(f) = \frac{1}{F}\left(1 - \int_F^\infty p(f) df\right), \ 0 \leq f \leq F \right\}, \quad \alpha \geq 0 \quad (11.92)$$

The first line of Equation 11.92 states that $p(f)$ is a normalized PDF whose tail above F has weight no greater than α. The second line asserts that $p(f)$ is constant over the interval $[0, F]$ and the weight in this interval is the complement of the weight on the tail.

The spring fails if x exceeds the critical displacement x_c. This occurs if the force f exceeds the critical load f_c, which is:

$$f_c = k_1 x_c + k_2 x_c^2 \tag{11.93}$$

With PDF $p(f)$, the probability of failure is:

$$P_f(p) = \text{Prob}(f \geq f_c \mid p) \tag{11.94}$$

We require that the failure probability not exceed the critical probability threshold:

$$P_f(p) \leq P_c \tag{11.95}$$

The robustness of the determination of this threshold exceedence, based on the nominal PDF, is $\hat{\alpha}(P_c)$ given by Equation 11.86. The value of this robustness depends on the values of F and f_c. After some algebra, one finds:

$$\hat{\alpha}(P_c) = \begin{cases} 0 & \text{if } f_c \leq (1-P_c)F \\ 1 - \dfrac{1-P_c}{f_c/F} & \text{if } (1-P_c)F < f_c \leq F \\ P_c & \text{if } F < f_c \end{cases} \tag{11.96}$$

The first line of Equation 11.96 arises when the critical force, f_c, is small enough so that $P_f(p)$ can exceed P_c even when the nominal PDF, $\tilde{p}$, is correct. The third line arises when f_c is so large that only the tail could account for failure. The second line covers the intermediate case. $\hat{\alpha}(P_c)$ is plotted schematically in Figure 11.10.

As we explained in Section 11.5.1, the value of the robustness $\hat{\alpha}(P_c)$ indicates whether the go/no-go threshold decision, based on the best-available PDF, is reliable or not. A large robustness implies that the decision is insensitive to uncertainty in the PDF, while a small value of $\hat{\alpha}(P_c)$ means that the decision can err as a result of small error in $\tilde{p}$. From Equation 11.96 and Figure 11.10 we see that the greatest value that $\hat{\alpha}(P_c)$ can take is P_c itself, the critical threshold value of failure probability. In fact, $\hat{\alpha}(P_c)$ may be much less, depending on the critical force f_c. From Equation 11.86 we learn that $\hat{\alpha}(P_c)$ and α have the

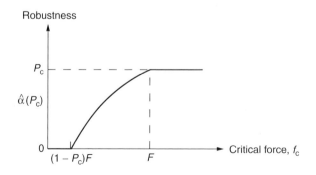

FIGURE 11.10 Robustness vs. critical force, Equation 11.96.

same units. The first line of Equation 11.92 indicates that α is a probability: the statistical weight of the upper tail. Consequently, $\hat{\alpha}(P_c)$ is the greatest tolerable statistical weight of the unmodeled upper tail of $p(f)$. $\hat{\alpha}(P_c)$ must be "large" in order to warrant the go/no-go decision; the judgment whether $\hat{\alpha}(P_c)$ is "small" or "large" depends on a judgment of how wrong $\tilde{p}(f)$ could be. As the tolerable probability of failure, P_c, becomes smaller, the tolerance against probability "leakage" into the upper tail becomes lower as well. $\hat{\alpha}(P_c)$ establishes a quantitative connection between the critical force f_c, the nominally maximum force F, the critical probability P_c, and the reliability of the go/no-go decision.

An additional use of the robustness function is in choosing technical modifications of the system itself that enhance the reliability of the go/no-go decision in the face of load uncertainty. Examination of Equation 11.96 reveals that $\hat{\alpha}(P_c)$ is improved by increasing f_c (if $f_c < F$). Let us consider the choice of the two stiffness coefficients, k_1 and k_2. From Equation 11.93 we note that:

$$\frac{\partial f_c}{\partial k_1} = x_c > 0 \tag{11.97}$$

$$\frac{\partial f_c}{\partial k_2} = x_c^2 > 0 \tag{11.98}$$

Thus, f_c is increased and thereby $\hat{\alpha}(P_c)$ is improved, by increasing either k_1 or k_2 or both. Equation 11.96 quantifies the robustness-enhancement from a design change in k_1 or k_2.

Equation 11.94 implies that increasing f_c causes a reduction in the probability of failure, $P_f(p)$, regardless of how the load is distributed. Thus, a change in the system that reduces the probability of failure also makes the prediction of the failure probability more reliable. There is a sympathy between *system reliability* and *prediction reliability*. This is, in fact, not just a favorable quirk of this particular example. It is evident from the definition of robustness in Equation 11.94 that any system modification which decreases $P_f(p)$ will likewise increase (or at least not decrease) the robustness $\hat{\alpha}(P_c)$.

11.6 Why "Good" Is Preferable to "Best"

The examples in Sections 11.2 through 11.4 illustrated the general proposition that optimization of performance is associated with minimization of immunity to uncertainty. This led to the conclusion that performance should be satisficed—made adequate but not optimal—and that robustness should be optimized. In the present section, we put this conflict between performance and robustness on a rigorous footing.

11.6.1 The Basic Lemma

The designer must choose values for a range of variables. These variables may represent materials, geometrical dimensions, devices or components, design concepts, operational choices such as "go" or "no-go," etc. Some of these variables are expressible numerically, some linguistically. We will represent the collection of the designer's decisions by the decision vector q.

In contrast to q, which is under the designer's control, the designer faces uncontrollable uncertainties of many sorts. These may be uncertain material coefficients, unknown and unmodeled properties such as nonlinearities in the design-base models, unknown external loads or ambient conditions, uncertain tails of a probability distribution, and so on. The uncertainties are all represented as vectors or functions (which may be vector-valued). We represent the uncertain quantities by the uncertain vector u. The uncertainties associated with u are represented by an info-gap model $\mathcal{U}(\alpha, \tilde{u})$, $\alpha \geq 0$. The centerpoint of the info-gap model is the known vector, which is the nominal value of the uncertain quantity u.

Info-gap models are suitable for representing ignorance of u for both practical and fundamental reasons. Practically, probability models defined on multidimensional function spaces tend to be cumbersome and informationally intensive. More fundamentally, info-gap models entail no measure functions, while measure-theoretic representation of ignorance can lead to contradictions [3, Chapter 4].

Many design specifications can be expressed as a collection of inequalities on scalar valued functions. For example, the mechanical deflection must not exceed a given value, while each of the three lowest natural frequencies must be no greater than various thresholds. For a design choice q, and for a specific realization of the uncertainty u, the performance of the system is expressed by the real-valued *performance functions* $R_i(q, u)$, $i = 1, \ldots, N$, where the design specification is the following set of inequalities:

$$R_i(q,u) \leq r_{c,i} \quad \text{for all} \quad i = 1, \ldots, N \tag{11.99}$$

The $r_{c,i}$ are called *critical thresholds*, which are represented collectively by the vector r_c. These thresholds may be chosen either small or large, to express either demanding or moderate aspirations, respectively.

Definition 11.1 A performance function $R_i(q, u)$ is *upper unsatiated* at design q if its maximum, up to horizon of uncertainty α, increases strictly as α increases:

$$\alpha < \alpha' \Rightarrow \max_{u \in \mathcal{U}(\alpha,\tilde{u})} R_i(q,u) < \max_{u \in \mathcal{U}(\alpha',\tilde{u})} R_i(q,u) \tag{11.100}$$

Upper unsatiation is a type of monotonicity: the maximum of the performance function strictly increases as the horizon of uncertainty increases. This monotonicity in α does not imply monotonicity of $R(q,u)$ in either q or u. Upper unsatiation results from the nesting of the sets in the info-gap model $\mathcal{U}(\alpha, \tilde{u})$.

By definition, the robustness of design q, with performance requirements r_c, is the greatest horizon of uncertainty at which all the performance functions satisfy their critical thresholds:

$$\hat{\alpha}(q, r_c) = \max \{\alpha: \max_{u \in \mathcal{U}(\alpha,\tilde{u})} R_i(q,u) \leq r_{c,i}, \quad \text{for all} \quad i = 1, \ldots, N\} \tag{11.101}$$

We now assert the following basic lemma. (Proofs appear in Section 11.6.4.)

Lemma 11.1 Given:

- An info-gap model $\mathcal{U}(\alpha, \tilde{u})$, $\alpha \geq 0$.
- Performance functions $R_i(q, u)$, $i = 1, \ldots, N$, which are all upper unsatiated at q.
- Critical thresholds equaling the performance functions evaluated at the centerpoint of the info-gap model:

$$r_{c,i} = R_i(q, \tilde{u}), \quad i = 1, \ldots, N \tag{11.102}$$

Then the robustness-to-uncertainty of design q vanishes:

$$\hat{\alpha}(q, r_c) = 0 \tag{11.103}$$

$\tilde{u}$ is the centerpoint of the info-gap model: the known, nominal, "best estimate" of the uncertainties accompanying the problem. If $\tilde{u}$ precisely represents the values of these auxiliary variables, then the performance aspirations in Equation 11.102 will be achieved by decision q. However, Equation 11.103 asserts that this level of performance has no immunity to unknown variations in the data and models upon which this decision is based. Any unmodeled factors, such as the higher-order terms in Equation 11.76, jeopardize the performance level vouched for in Equation 11.102. Probabilistically, one would say that things could be worse than the expected outcome. However, our assertion is stronger, because we are considering not only random uncertainties, but rather the info gaps in the entire epistemic infrastructure of the decision, which may include info gaps in model structures and probability densities.

This result is particularly significant when we consider performance-optimization, to which we now turn.

11.6.2 Optimal-Performance vs. Optimal Robustness: The Theorem

We are particularly interested in the application of lemma 11.1 to optimal-performance design. The lemma will show that a design that optimizes the performance will have zero robustness to uncertainty. This means that high aspirations for performance are infeasible in the sense that these aspirations can fail to materialize due to infinitesimal deviations of the uncertain vector from its nominal value. It is true that failure to achieve an ultimate aspiration may entail only a slight reduction below optimal performance. Nonetheless, the gist of the theorem is that zenithal performance cannot be relied upon to occur; a design specification corresponding to an extreme level of performance has no robustness to uncertainty. The designer cannot "sign-off" on a performance-optimizing specification; at most, one can hope that the shortfall will not be greatly below the maximum performance.

Let Q represent the set of available designs from which the designer must choose a design q. Let $\tilde{u}$ denote the nominal, typical, or design-base value of the uncertain vector u. What is an optimal-performance design, from the allowed set Q, and with respect to the design-base value $\tilde{u}$?

If there is only one design specification, so $N = 1$ in Equation 11.99, then an optimal-performance design q^* minimizes the performance function:

$$R(q^*, \tilde{u}) = \min_{q \in Q} R(q, \tilde{u}) \qquad (11.104)$$

If there are multiple design specifications ($N > 1$), then such a minimum may not hold simultaneously for all the performance functions. One natural extension of Equation 11.104 employs the idea of *Pareto efficiency*. Pareto efficiency is a "short blanket" concept: if you pull up your bed covers to warm your nose, then your toes will get cold. A design q^* is Pareto efficient if any other design q' that improves (reduces) one of the performance functions detracts from (increases) another:

$$\text{If:} \quad R_i(q', \tilde{u}) < R_i(q^*, \tilde{u}) \quad \text{for some} \quad i \qquad (11.105)$$

$$\text{Then:} \quad R_j(q', \tilde{u}) > R_j(q^*, \tilde{u}) \quad \text{for some} \quad j \neq i \qquad (11.106)$$

A Pareto-efficient design does not have to be unique, so let us denote the set of all Pareto efficient designs by Q^*. For the case of a single design specification, let Q^* denote all the designs that minimize the performance function, as in Equation 11.104. The following theorem, which is derived directly from lemma 11.1, asserts that any design that is performance-optimal (Equation 11.104) or Pareto efficient (Equations 11.105 and 11.106) has no robustness to uncertain deviation from the design-base value $\tilde{u}$.

Theorem 11.1 Given:
- A set of Pareto-efficient or performance-optimal designs, with respect to the design-base value $\tilde{u}$.
- An info-gap model $\mathcal{U}(\alpha, \tilde{u})$, $\alpha \geq 0$, whose centerpoint is the nominal or design-base value $\tilde{u}$.
- Performance functions $R_i(q^*, u)$, $i = 1, \ldots, N$, which are all upper unsatiated at some $q^* \in Q^*$.
- Critical thresholds equaling the performance functions evaluated at the centerpoint of the info-gap model and at this q^*:

$$r_{c,i} = R_i(q^*, \tilde{u}), \quad i = 1, \ldots, N \qquad (11.107)$$

Then the robustness to uncertainty of this performance-optimal design q^* vanishes:

$$\hat{\alpha}(q^*, r_c) = 0 \qquad (11.108)$$

Theorem 11.1 is really just a special case of lemma 11.1. We know from lemma 11.1 that whenever the critical thresholds, $r_{c,i}$, are chosen at the nominal values of the performance-functions, $R_i(q, \tilde{u})$, and when that nominal value, $\tilde{u}$, is the centerpoint of the info-gap model, $\mathcal{U}(\alpha,\tilde{u})$, then any design has zero robustness. Theorem 11.1 just specializes this to the case where q is Pareto-efficient or performance-optimal.

We can understand the special significance of this result in the following way. The functions $R_i(q^*,\tilde{u})$ represent the designer's best-available representation of how design q^* will perform. $R_i(q^*,\tilde{u})$ is based on the best-available models, and is the best estimate of all residual (and possibly recalcitrant) uncertain factors or terms. We know from lemma 11.1 that any performance-optimal design we choose will have zero robustness to the vagaries of those residual uncertainties. However, the lesson to learn is **not** to choose the design whose performance is optimal; theorem 11.1 makes explicit that this design also cannot be depended upon to fulfill our expectations. To state it harshly, choosing the performance optimum is just wishful thinking, unless we are convinced that no uncertainties lurk behind our models. The lesson to learn, if we seek a design whose performance can be reliably known in advance, is to move off the surface of Pareto-efficient or performance-optimal solutions. For example, referring again to Figure 11.2, we must move off the optimal-performance curve to a point Q or R. We can evaluate the robustness of these suboptimal designs with the robustness function, and we can choose the design to satisfice the performance and to maximize the robustness.

11.6.3 Information-Gap Models of Uncertainty

We have used info-gap models throughout this chapter. In this section we present a succinct formal definition, in preparation for the proof of lemma 11.1 in Section 11.6.4.

An info-gap model of uncertainty is a family of nested sets; Equations 11.14, 11.41, 11.77, and 11.92 are examples. An info-gap model entails no measure functions (probability densities or membership functions). Instead, the limited knowledge about the uncertain entity is invested in the structure of the nested sets of events.

Mathematically, an info-gap model is a set-valued function. Let S be the space whose elements represent uncertain events. S may be a vector space or a function space. Let $\Re$ denote the set of nonnegative real numbers. An info-gap model $\mathcal{U}(\alpha,u)$ is a function from $\Re \times S$ into the class of subsets of S. That is, each ordered pair (α,u), where $\alpha \geq 0$ and $u \in S$, is mapped to a set $\mathcal{U}(\alpha,u)$, which is a subset of S.

Two axioms are central to the definition of info-gap models of uncertainty:

Axiom 11.1 Nesting. An info-gap model is a family of nested sets:

$$\alpha \leq \alpha' \Rightarrow \mathcal{U}(\alpha,u) \subseteq (\alpha',u) \tag{11.109}$$

We have already encountered this property of info-gap models in Equations 11.9, 11.37, and 11.67 where we noted that this inclusion means that the uncertainty sets $\mathcal{U}(\alpha,u)$ become more inclusive as the "uncertainty parameter" α becomes larger. Relation 11.109 means that α is a *horizon of uncertainty*. At any particular value of α, the corresponding uncertainty set defines the range of variation at that horizon of uncertainty. The value of α is unknown, so the family of nested sets is typically unbounded, and there is no "worst case."

All info-gap models of uncertainty share an additional fundamental property.

Axiom 11.2 Contraction. The info-gap set, at zero horizon of uncertainty, contains only the centerpoint:

$$\mathcal{U}(0,u) = \{u\} \tag{11.110}$$

For example, the info-gap model in Equation 11.14 is a family of nested intervals, and the centerpoint is the nominal load, $\tilde{\phi}$, which is the only element of $\mathcal{U}(0,\tilde{\phi})$ and which belongs to the intervals at all positive values of α.

Combining axioms 11.1 and 11.2 we see that u belongs to the zero-horizon set, $\mathcal{U}(0,u)$, and to all "larger" sets in the family.

Additional axioms are often used to define more specific structural features of the info-gap model, such as linear [1, 4], or nonlinear [5] expansion of the sets as the horizon of uncertainty grows. We will not need these more specific axioms. Info-gap models of uncertainty are discussed extensively elsewhere [1, 6].

11.6.4 Proofs

Proof of lemma 11.1. By the contraction axiom of info-gap models, Equation 11.110, and from the choice of the value of $r_{c,i}$:

$$\max_{u \in \mathcal{U}(0,\tilde{u})} R_i(q,u) = r_{c,i} \tag{11.111}$$

Hence, 0 belongs to the set of α-values in Equation 11.101 whose least upper bound equals the robustness, so:

$$\hat{\alpha}(q, r_c) \geq 0 \tag{11.112}$$

Now consider a positive horizon of uncertainty: $\alpha > 0$. Because $R_i(q, r_c)$ is upper unsatiated at q:

$$\max_{u \in \mathcal{U}(0,\tilde{u})} R_i(q,u) < \max_{u \in \mathcal{U}(\alpha,\tilde{u})} R_i(q,u) \tag{11.113}$$

Together with Equation 11.111, this implies that this positive value of α does not belong to the set of α-values in Equation 11.101. Hence:

$$\alpha > \hat{\alpha}(q, r_c) \tag{11.114}$$

Combining Equations 11.112 and 11.114 completes the proof.

Proof of theorem 11.1. Special case of lemma 11.1

11.7 Conclusion: A Historical Perspective

We have focused on the epistemic limitations—information gaps—that confront a designer in the search for reliable performance. The central idea has been that the functional performance of a system must be traded off against the immunity of that system to info gaps in the models and data underlying the system's design. A system that is designed for maximal performance will have no immunity to errors in the models and data underlying the design. Robustness can be obtained only by reducing performance aspirations.

Our discussion is motivated by the recognition that the designer's understanding of the relevant processes is deficient, that the models representing those processes lack pertinent components, and that the available data is incomplete and inaccurate. This very broad conception of uncertainty—including model structure as well as more conventional data "noise"—has received extensive attention in some areas of engineering, most notably in robust control [7]. This scope of uncertainty is, however, a substantial deviation from the tradition of probabilistic analysis that dominates much contemporary thinking.

Least-squares estimation is a central paradigm of traditional uncertainty analysis. The least-squares method was developed around 1800, independently by Gauss (1794–1795) and Legendre (1805–1808), for estimation of celestial orbits [8, 9]. Newtonian mechanics, applied to the heavenly bodies, had established irrevocably that celestial orbits are elliptical, as Kepler had concluded experimentally. However, the data was noisy so it was necessary to extract the precise ellipse that was hidden under the noisy measurements. The data was corrupted, while the model—elliptical orbits—was unchallenged.

Least-squares estimation obtained deep theoretical grounding with the proof of the central limit theorem by Laplace (1812), which established the least-squares estimate as the maximum-likelihood estimate of a normal distribution. The least-squares idea continues to play a major role in modern uncertainty analysis in such prevalent and powerful tools as Kalman filtering, Luenberger estimation [10] and the Taguchi method [11].

What is characteristic of the least-squares method is the localization of uncertainty exclusively on data that is exogenous to the model of the underlying process. The model is unblemished (Newtonian truth in the case of celestial orbits); only the measurements are corrupted. But the innovative designer, using new materials and exploiting newly discovered physical phenomena, stretches models and data to the limits of their validity. The designer faces a serious info gap between partial, sometimes tentative, insights that guide much high-paced modern design, and solid complete knowledge. The present work is part of the growing trend to widen the range of uncertainty analysis to include the analyst's imperfect conceptions and representations. For all our sapience, we are after all only human!

Acknowledgments

The author acknowledges with pleasure the support of the Samuel Neaman Institute for Advanced Studies in Science and Technology. The author is indebted to Professor Eli Altus and Mr. Yakov Saraf of the Technion for useful comments.

References

1. Ben-Haim, Y., *Information-Gap Decision Theory: Decisions Under Severe Uncertainty,* Academic Press, London, 2001.
2. Ben-Haim, Y., Cogan, S., and Sanseigne, L., Usability of Mathematical Models in Mechanical Decision Processes, *Mechanical Systems and Signal Processing.* 12, 121–134, 1998.
3. Keynes, J.M., *Treatise on Probability,* Macmillan & Co. Ltd., London, 1921.
4. Ben-Haim, Y., Set-models of information-gap uncertainty: axioms and an inference scheme, *Journal of the Franklin Institute,* 336, 1093–1117, 1999.
5. Ben-Haim, Y., Robustness of model-based fault diagnosis: decisions with information-gap models of uncertainty, *International Journal of Systems Science,* 31, 1511–1518, 2000.
6. Ben-Haim, Y., *Robust Reliability in the Mechanical Sciences,* Springer-Verlag, Berlin, 1996.
7. Qu, Z., *Robust Control of Nonlinear Uncertain Systems,* Wiley, New York, 1998.
8. Hazewinkel, M., Managing Editor, *Encyclopaedia of Mathematics: An Updated and Annotated Translation of The Soviet "Mathematical Encyclopaedia,"* Kluwer Academic Publishers, Dordrecht, Holland, 1988.
9. Stigler, S.M., *The History of Statistics: The Measurement of Uncertainty before 1900.* The Belknap Press of Harvard University Press, Boston, 1986.
10. Luenberger, D.G., *Optimization by Vector Space Methods,* Wiley, New York, 1969.
11. Taguchi, G., *Introduction to Quality Engineering: Designing Quality into Products and Processes,* translated and published by the Asian Productivity Organization, Tokyo, 1986.

12
Interval Methods for Reliable Computing

12.1	Introduction	12-1
12.2	Intervals and Uncertainty Modeling	12-2
	Definitions, Applications, and Scope • Dependency • Vectors and Matrices • Linear Interval Equations	
12.3	Interval Methods for Predicting System Response Due to Uncertain Parameters	12-8
	Sensitivity Analysis • Interval Finite Element Methods	
12.4	Interval Methods for Bounding Approximation and Rounding Errors	12-17
	Approximation Errors • Rounding-off Errors	
12.5	Future Developments of Interval Methods for Reliable Engineering Computations	12-19
12.6	Conclusions	12-21
Appendix 12.1		12-23

Rafi L. Muhanna
Georgia Institute of Technology

Robert L. Mullen
Case Western Reserve University

12.1 Introduction

This chapter explores the use of interval numbers to account for uncertainties in engineering models. We concentrate on models based on finite element discretization of partial differential equations. Results of such models are used ubiquitously in engineering design. Uncertainties in the results of finite element models can be attributed to the following sources:

1. The appropriateness of the partial differential equation
2. Errors associated with discretization of the partial differential equation
3. Uncertainties in the parameters of the engineering model
4. Errors associated with floating point operations on a digital computer

While the appropriateness of a partial differential equation to a given physical problem is beyond the scope of this chapter, the remaining three sources of errors and uncertainties can be bounded using the concept of interval representation and interval numbers. Quantifying the uncertainties in the results of an engineering calculation is essential to provide a reliable computing environment.

Section 12.2 provides a review of interval representation and the mathematics of interval numbers. Section 12.3 contains a description of interval finite element methods that account for uncertain model parameters. Section 12.4 describes methods for bounding errors associated with floating point operations, while Section 12.5 contains information on bounding discretization errors.

12.2 Intervals and Uncertainty Modeling

12.2.1 Definitions, Applications, and Scope

By definition, an interval number is a closed set in R that includes the possible range of an unknown real number, where R denotes the set of real numbers. Therefore, a real interval is a set of the form

$$x \equiv [x^l, x^u] := \{ \tilde{x} \in R \mid x^l \leq \tilde{x} \leq x^u \} \qquad (12.1)$$

where x^l and x^u are the lower and upper bounds of the interval number x, respectively, and the bounds are elements of R with $x^l \leq x^u$. Degenerated or thin interval numbers are of the form $[x^l, x^l]$ or $[x^u, x^u]$; they are equivalent to the real numbers x^l and x^u respectively.

Based on the above-mentioned definitions, interval arithmetic is defined on sets of intervals, rather than on sets of real numbers, and interval mathematics can be considered as a generalization of real numbers mathematics. The definition of real intervals and operations with intervals can be found in a number of references [1–4]. However, the main interval arithmetic operations are presented in Appendix 12.1 at the end of this chapter.

Early use of interval representation is associated with the treatment of rounding errors in mathematical computations. The idea was to provide upper and lower bounds on rounding errors. For example, in a computational system with three decimal digit accuracy, the number 3.113 would be represented as a member of the interval [3.11, 3.12]. Intervals can also be used to represent rational bounds on irrational numbers. Archimedes used a "two-sided approximation" to calculate the constant π. He considered inscribed and circumscribed polygons of a circle and obtained an increasing sequence of lower bounds and a decreasing sequence of the upper bounds at the same time. Therefore, stopping the process with polygons each of n sides, he obtained an interval containing the desired result, (i.e., the number π.) By choosing n large enough, an interval of arbitrary small width can be found to contain the number π. Inspired by this method, Moore [2], instead of computing a numerical approximation using limited-precision arithmetic, proceeded to construct intervals known in advance to contain the desired exact results. Several authors independently had the idea of bounding rounding errors using intervals [5, 6]; however, Moore extended the use of interval analysis to bound the effect of errors from different sources, including approximation errors and errors in data (see [7]).

In this section, we try to cover the most significant applications of intervals in the two major areas of *mathematics-computations* and *scientific and engineering modeling*. In addition, we introduce a short review of interval definitions and operations.

In the **mathematics-computations** field, there has been increasing interest in solving complex mathematical problems with the aid of digital computers. In general, the solution of these problems is too complicated to be represented in finite terms (floating point representation) or even impossible. But frequently, one is only interested in the existence of a solution within a certain domain or error bounds for the solution (intervals). By the nature of the problems, numerical approximations are insufficient. That motivated the development of interval methods capable of handling the finite representation deficiency. These interval methods are known by different names; for example, self-validating (SV) verification methods or automatic result verification [8]. A detailed introduction of these methods can be found in [4, 8–11]. The goal of self-validating methods, as introduced by Rohn, Rump, and Yamamot [12], is to deliver correct results on digital computers—correct in a mathematical sense, covering all errors such as representation, discretization, rounding errors and others, and more precisely are:

1. To deliver rigorous results
2. In a computing time not too far from a pure numerical algorithm
3. Including the proof of existence (and possibly uniqueness) of a solution

Handling rounding and truncation errors in interval arithmetic is discussed in Section 12.3.

In the area of *scientific and engineering modeling*, intervals represent a simple, elegant, and computationally efficient tool to handle uncertainty. Recently, the scientific and engineering community has begun to recognize the utility of defining multiple types of uncertainty [13]. While the latest computational advances significantly increased the analysis capabilities, researchers started to encounter the limitations of applying only one mathematical framework (traditional probability theory) to present all types of uncertainty. Traditional probability theory cannot handle situations with incomplete or little information on which to evaluate a probability, or when that information is nonspecific, ambiguous, or conflicting. Because of these reasons, many theories of generalized uncertainty-based information have been developed. There are five major frameworks that use interval-based representation of uncertainty: imprecise probabilities, possibility theory, the Dempster-Shafer theory of evidence, fuzzy set theory, and convex set modeling.

For example, "imprecise probability" is a generic term for the many mathematical models that measure chance or uncertainty without sharp numerical probabilities, represented as an interval included in [0, 1]. The Dempster-Shafer theory offers an alternative to traditional probabilistic theory for the mathematical representation of uncertainty. The significant innovation of this framework is that it allows for the allocation of a probability mass to sets or intervals. To illustrate how intervals are used in the case of possibility theory and fuzzy set theory, despite the differences in their interpretation, let us start with the difference between an ordinary subset A and a fuzzy subset $\boldsymbol{A}$ [14]. Let E be a referential set in R. An ordinary subset A of this referential set is defined by its characteristic function.

$$\forall x \in E$$

$$\mu_A(x) \in \{0, 1\}, \tag{12.2}$$

which shows that an element of E belongs to or does not belong to A, according to the value of the characteristic function (1 or 0).

For the same referential set E, a fuzzy subset $\boldsymbol{A}$ will be defined by its characteristic function, called the membership function, which takes its values in the interval [0, 1] instead of in the binary set $\{0, 1\}$.

$$\forall x \in E:$$

$$\mu_A(x) \in [0, 1], \tag{12.3}$$

that is, that the elements of E belong to $\boldsymbol{A}$ with a level of certainty in the closed set [0, 1].

The concept of fuzzy numbers can be presented in many ways. One way is to construct the membership function in terms of an interval of confidence at several levels of presumption or confidence. The maximum level of presumption is considered to be 1 and the minimum of presumption to be at level 0. The level of presumption α, $\alpha \in [0, 1]$ gives an interval of confidence $A_\alpha = [a_1^{(\alpha)}, a_2^{(\alpha)}]$, which is a monotonic decreasing of α; that is,

$$(\alpha_1 < \alpha_2) \Rightarrow \left(A_{\alpha_2} \subset A_{\alpha_1} \right) \tag{12.4a}$$

or

$$(\alpha_1 < \alpha_2) \Rightarrow \left(\left[a_1^{(\alpha_2)}, a_2^{(\alpha_2)} \right] \subset \left[a_1^{(\alpha_1)}, a_2^{(\alpha_1)} \right] \right) \tag{12.4b}$$

for every $\alpha_1, \alpha_2 \in [0,1]$. Such a situation is shown in Figure 12.1. When a fuzzy membership function is expressed in terms of intervals of confidences, the arithmetic of fuzzy numbers can be constructed from interval operations and the fuzzy number can be considered as nested intervals.

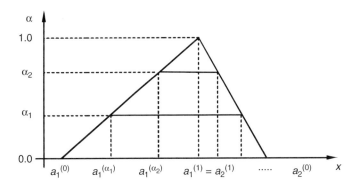

FIGURE 12.1 Definition of fuzzy numbers.

Intervals play a significant role in convex models of uncertainty. To illustrate that, we introduce some basic concepts of the convex sets [15]. A given region is called *convex* if the line segment joining any two points in the region is entirely in the region, and an algebraic formulation of convexity can be presented as follows. Let S be a set of points in n-dimensional Euclidean space, E^n. Then S is convex if all averages of points in S also belong to S. That is, S is convex if, for any points, $p \in S$ and $q \in S$, and any number $0 \leq \alpha \leq 1$

$$\gamma p + (1 - \gamma) q \in S \tag{12.5}$$

The interval models of uncertainty can be seen as an equivalent or a special case of the convex models; for example, within the context of information gap theory [16], the envelope-bound models constrain the uncertain deviations to an expandable envelope. For scalar functions, we can have:

$$\mathcal{U}(\alpha, \tilde{u}) = \{u(t): \; |u(t) - \tilde{u}(t)| \leq \alpha \psi(t)\}, \quad \alpha \geq 0 \tag{12.6}$$

where $\mathcal{U}(\alpha, \tilde{u})$ is the set of all functions whose deviation from the nominal function $\tilde{u}(t)$ is no greater than $\alpha \psi(t)$, $\psi(t)$ is a known function determining the shape of the envelope, and the uncertainty parameter α determines the size. However, envelope-bound models for n-vectors can be formulated by constraining each element of uncertain vector to lie within an expanding envelope:

$$\mathcal{U}(\alpha, \tilde{u}) = \{u_n(t): \; |u_n(t) - \tilde{u}_n(t)| \leq \alpha \psi_n(t), \; n = 1, \ldots, n\}, \quad \alpha \geq 0 \tag{12.7}$$

in which the interval values of each element of vector u result in a family of n–dimensional expanding boxes.

Interval arithmetic has been developed as an effective tool to obtain bounds on rounding and approximation errors (discussed in the next section). The question now is to what extent interval arithmetic could be useful and effective when the range of a number is due to physical uncertainties (i.e., the range in a material's yield stress or modulus of elasticity), rather than rounding errors. In fact, overestimation is a major drawback in interval computations (overestimation handling discussed in Section 12.3). One reason for such overestimation is that only some of the algebraic laws valid for real numbers remain valid for intervals; other laws only hold in a weaker form [4, pp. 19–21]. There are two general rules for the algebraic properties of interval operations:

1. Two arithmetical expressions that are equivalent in real arithmetic are equivalent in interval arithmetic when a variable occurs only once on each side. In this case, both sides yield the range of the expression. Consequently, the laws of commutativity, associativity, and neutral elements are valid in interval arithmetic.
2. If f and g are two arithmetical expressions that are equivalent in real arithmetic, then the inclusion $f(x) \subseteq g(x)$ holds if every variable occurs only once in f.

In fact, the left-hand side yields the range and the right-hand side encloses the range, thus entailing the invalidity of distributive and cancellation laws in interval arithmetic. However, this property implies a weak form of the corresponding laws from real arithmetic. If a, b, and c are interval numbers, then:

$$a(b \pm c) \subseteq ab \pm ac; \quad (a \pm b)c \subseteq ac \pm bc, \quad \text{(subdistributivity)}$$

$$a - b \subseteq (a + c) - (b + c); \quad a/b \subseteq (ac)/(bc);$$

$$0 \in a - a; \quad 1 \in a/a \quad \text{(subcancellation)}$$

If we set

$$a = [-2, 2]; \quad b = [1, 2]; \quad c = [-2, 1],$$

we get

$$a(b + c) = [-2, 2]([1, 2] + [-2, 1]) = [-2, 2][-1, 3] = [-6, 6]$$

However,

$$ab + ac = [-2, 2][1, 2] + [-2, 2][-2, 1] = [-4, 4] + [-4, 4] = [-8, 8]$$

It can be noticed that the interval number $a(b + c) = [-6, 6]$ is contained in the interval number $ab + ac = [-8, 8]$ but they are not equal. Furthermore, in the case of subcancellation;

$$0 \in b - b = [1, 2] - [1, 2] = [-1, 1]$$

and

$$1 \in b/b = [1, 2]/[1, 2] = [1/2, 2]$$

In the subtraction case, the interval number $[-1, 1]$ is enclosing but not equal to zero; and in the division case, the interval $[1/2, 2]$ is enclosing but not equal to the number 1. The failure of the distributive law is a frequent source of overestimation, and appropriate bracketing is advisable. Clearly, sharp determination for the range of an interval solution requires that extreme care be used to identify interval values that represent a single quantity and prevent the expansion of intervals. Another important source of overestimation in interval computation is dependency.

12.2.2 Dependency

The dependency problem arises when one or several variables occur more than once in an interval expression. Dependency may lead to catastrophic overestimation in interval computations. For example, if we subtract the interval $x = [a, b] = [1, 2]$ from itself, as if we are evaluating the function $f = x - x$, we obtain $[a - b, b - a] = [-1, 1]$ as a result. The result is not the interval $[0, 0]$ as we might expect. Actually, interval arithmetic cannot recognize the multiple occurrence of the same variable x, but it evaluates f as a function of two independent variables: namely, $f(x, y) = \{f(\tilde{x}, \tilde{y}) = \tilde{x} - \tilde{y} \mid \tilde{x} \in x, \tilde{y} \in y\}$ instead of a function of one variable, that is, $\{f(\tilde{x}) = \tilde{x} - \tilde{x} \mid \tilde{x} \in \tilde{x}\}$. Thus, interval arithmetic is treating $x - x$ as if evaluating $x - y$ with y equal to but independent of x. Evidently, one should always be aware of this phenomenon and take the necessary precautions to reduce or, if possible, to eliminate its effect. For example, the dependency problem could be avoided in evaluating $x - x$ if it is presented in the form $x(1 - 1)$ instead. In addition, overestimation can occur in evaluating a function $f(x, y)$ of the form $(x - y)/(x + y)$, but not if it is rewritten as $1 - 2/(1 + x/y)$. If $f(x, y)$ is evaluated in the latter form, the resulting interval is the exact range of $f(\tilde{x}, \tilde{y})$ for $\tilde{x} \in x$ and $\tilde{y} \in y$ [7, 11]. The reason behind obtaining the exact range is that the multiple occurrences of variables x and y are avoided in the latter form of the function.

12.2.3 Vectors and Matrices

An interval vector is a vector whose components are interval numbers. An interval matrix is a matrix whose elements are interval numbers. The set of $m \times n$ interval matrices is denoted by $IR^{m \times n}$. An interval matrix $A = (A_{ij})$ is interpreted as a set of real $m \times n$ matrices by the convention [4]

$$A = \{\tilde{A} \in R^{m \times n} \mid \tilde{A}_{ij} \in A_{ij} \text{ for } i = 1, ..., m; \ j = 1, ..., n\} \tag{12.8}$$

In other words, an interval matrix contains all real matrices whose elements are obtained from all possible values between the lower and upper bound of its interval elements. One important type of matrix in mechanics is the symmetric matrix. A symmetric interval matrix contains only those real symmetric matrices whose elements are obtained from all possible values between the lower and upper bound of its interval element. This definition can be presented in the following form:

$$A_{sym} = \{\tilde{A} \in R^{n \times n} \mid \tilde{A}_{ij} \in A_{ij} \text{ with } \tilde{A}^T = \tilde{A}\} \tag{12.9}$$

An interval vector is an $n \times 1$ interval matrix. The set of real points in an interval vector form an n–dimensional parallelepiped with sides parallel to the coordinate axes. Usually, an interval vector is referred to as a *box* [7]. For more information about algebraic properties of interval matrix operations, the reader may refer to [4, 17, 18].

12.2.4 Linear Interval Equations

Systems of linear interval equations engage considerable attention in engineering applications. It is worthwhile to introduce some main features of their solution. A linear interval equation with coefficient matrix $A \in IR^{n \times n}$ and right-hand side $b \in IR^n$ is defined as the family of linear equations

$$\tilde{A}\tilde{x} = \tilde{b} \quad (\tilde{A} \in A, \tilde{b} \in b) \tag{12.10}$$

Therefore, a linear interval equation represents systems of equations in which the coefficients are unknown numbers ranging in certain intervals. The solution of interest is the enclosure for the solution set of Equation 12.10, given by

$$S(A, b) := \{\tilde{x} \in R^n \mid \tilde{A}\tilde{x} = \tilde{b} \text{ for some } \tilde{A} \in A, \ \tilde{b} \in b\}$$

The solution set $S(A, b)$ usually is not an interval vector, and does not need even to be convex; in general, $S(A, b)$ has a very complicated structure. To guarantee that the solution set $S(A, b)$ is bounded, it is required that the matrix A be regular; that is, that every matrix $\tilde{A} \in A$ has rank n. For the solution of a linear interval equation, we usually seek the interval vector x containing the solution set $S(A, b)$ that has the narrowest possible interval components, or what is called the hull of the solution set which can be denoted as

$$A^H b := \Diamond S(A, b) \tag{12.11}$$

where

$$A^H b = \Diamond \{\tilde{A}^{-1} b \mid \tilde{A} \in A, \ \tilde{b} \in b\} \quad \text{for } b \in IR^n \tag{12.12}$$

The expression 12.12 defines a mapping $A^H: IR^n \to IR^n$ that is called the hull inverse of A. The matrix inverse of a regular interval square matrix $A \in IR^{n \times n}$ is defined by

$$A^{-1} := \Diamond\{\tilde{A}^{-1} \mid \tilde{A} \in A\} \qquad (12.13)$$

and it is proven that

$$A^H b \subseteq A^{-1} b, \text{ with equality if } A \text{ is thin} \qquad (12.14)$$

Another case of equality in 12.14 occurs when A is a regular diagonal matrix. To illustrate some of the linear interval equation main properties, let us have the following example (adopted from [4, pp. 93]):

Given

$$A := \begin{pmatrix} 2 & [-1, 0] \\ [-1, 0] & 2 \end{pmatrix}, \quad b := \begin{pmatrix} 1.2 \\ -1.2 \end{pmatrix}$$

Then $\tilde{A} \in A$ iff

$$\tilde{A} := \begin{pmatrix} 2 & -\alpha \\ -\beta & 2 \end{pmatrix} \quad \text{with } \alpha, \beta \in [0, 1]$$

Solving for all possible combinations of α and β, we get the following exact vertices for the solution set:

$$(0.3, -0.6); \; (0.6, -0.6); \; (0.6, -0.3) \quad \text{and} \quad (0.4, -0.4)$$

The solution set is shown in Figure 12.2. Therefore, from Figure 12.2 it can be seen that the hull of the solution set is

$$A^H b = \Diamond S(A, b) = \begin{pmatrix} [0.3, 0.6] \\ [-0.6, -0.3] \end{pmatrix}$$

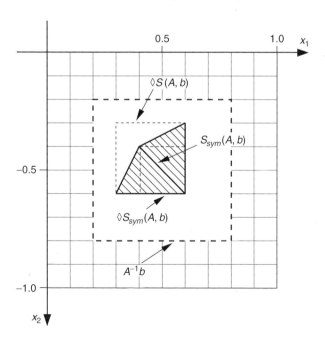

FIGURE 12.2 Solution sets for linear interval equation.

On the other hand,

$$A^{-1} = \begin{pmatrix} [1/2,\ 2/3] & [0,\ 1/3] \\ [0,\ 1/3] & [1/2,\ 2/3] \end{pmatrix}$$

so that

$$A^{-1}b = \begin{pmatrix} [0.2,\ 0.8] \\ [-0.8,\ -0.2] \end{pmatrix} \neq \text{ but } \supset A^H b$$

It is clear that the numerically obtained solution is wider than the actual hull of the solution set $S(A, b)$. In fact, obtaining the exact hull of the solution set for the general case is not known to be achievable. However, obtaining an exact hull of the solution set for the case of interval right-hand side with dependencies is achievable [19].

If the symmetry in A is considered, in that case A stands for the following unknown symmetric matrix:

$$\tilde{A} = \begin{pmatrix} 2 & -\alpha \\ -\alpha & 2 \end{pmatrix}$$

and the solution set will be given by

$$S_{sym}(A,b) = \{\tilde{x} \in R^n \mid \tilde{A}\tilde{x} = \tilde{b} \text{ for some } \tilde{A} \in A, \tilde{b} \in b \text{ with } \tilde{A}^T = \tilde{A}\}$$

The exact solution set is (0.6, –0.6) and (0.4, –0.4), which is smaller than the previous nonsymmetric case, and indicated by a heavy line in Figure 12.2. The hull of the solution set is:

$$\Diamond S_{sym}(A,b) = \begin{pmatrix} [0.4,\ 0.6] \\ [-0.6,\ -0.4] \end{pmatrix} \neq A^H b$$

which is narrower than the hull of the previous general case.

It is clear that care in the application of interval arithmetic to maintain sharp estimate of the interval solution is very important, especially in cases such as bounding the effect of uncertainties on the solution of engineering problems. In general, the sharpest results are obtained when proper bracketing is used, dependency is avoided, and the physical nature of the problem is considered.

12.3 Interval Methods for Predicting System Response Due to Uncertain Parameters

Interval methods are a simple, elegant, and computationally efficient way to include uncertainty when modeling engineering systems with uncertain parameters. As defined earlier, an interval number is the closed set of all real numbers between and including the interval's lower and upper bound [2]. Real-life engineering practice deals with parameters of unknown values: modulus of elasticity, yield stress, geometrical dimensions, density, and gravitational acceleration are some examples. In application to mechanics problems, an interval representation lends itself to the treatment of tolerance specifications on machined parts, components that are inspected, and whose properties will fall within a finite range. For example, the diameter of a rod is given as $D \pm \delta$ or in an interval form as $[D - \delta, D + \delta] = [D^l, D^u]$. In the case of live loads, the value can often be bounded by two extreme values; that is, the live load acting on an office building might be given to be between 1.8 and 2.0 kN/m^2 or in an interval form as [1.8, 2.0] kN/m^2. Experimental data, measurements, statistical analysis, and expert knowledge represent information for defining bounds on the possible ranges of such quantities. Thus, considering the values of unknown parameters to be defined within intervals that possess known bounds might be a realistic or natural way of representing uncertainty in engineering problems.

Tools that provide sharp predictions of system response including uncertainty in an interval form will provide the infrastructure for wide inclusion of uncertainties in engineering models. However, this will only occur if the sharp results corresponding to the physical problem can be obtained.

12.3.1 Sensitivity Analysis

Sensitivity analysis represents a need when the prediction of system response is required due to variations in the system's parameters. For example [4], if x is a vector of approximate parameters and Δx is a vector containing the bounds for the error in components of x, intervals are a powerful tool to obtain the influence of these parameters' variations on the response function $F(x)$; that is, one is interested in finding a vector ΔF such that, for given Δx,

$$|F(\tilde{x}) - F(x)| \leq \Delta F \quad \text{for} \quad |\tilde{x} - x| \leq \Delta x \qquad (12.15)$$

Sometimes the dependence of ΔF on Δx is also sought. Stated in the form

$$F(\tilde{x}) \in [F(x) - \Delta F, F(x) + \Delta F] \quad \text{for} \quad \tilde{x} \in [x - \Delta x, x + \Delta x] \qquad (12.16)$$

Another interesting example is sensitivity analysis in optimization problems [7]. Consider an unperturbed optimization problem in which the objective function and constraints depend on a vector c of parameters. To emphasize the dependence on c, the problem is introduced as

$$\text{Minimize} \quad f(x,c)$$

$$\text{subject to} \quad p_i(x,c) \leq 0 \quad \text{for} \quad i = 1,\ldots,m$$

$$q_j(x,c) = 0 \quad \text{for} \quad j = 1,\ldots,r$$

To show the dependence of the solution on c, the solution value is given as $f^*(c)$ and the solution point(s) as $x^*(c)$, where $f(x^*(c), c) = f^*(c)$.

In the perturbed case, c will be allowed to vary over interval vector C. As c varies over the range of C, the obtained set of solution values is

$$f^*(C) = \{f^*(c): \quad c \in C\}$$

and a set of solution points

$$x^*(C) = \{x^*(c): \quad c \in C\}$$

in this case, the width of the interval $f^*(C)$ represents a measure of the sensitivity of the problem to variation of c over the range of C.

12.3.2 Interval Finite Element Methods

Our focus in this section is the introduction of the principal concept for formulation of finite element methods (FEM) using intervals as a way to handle uncertain parameters. Our goal is to capture the behavior of a physical system. Ad-hoc replacement of scalar quantities by intervals will not provide sharp (or correct) measures of the uncertainty in a physical system. We will present various *intervalizations* of finite element methods, and relate these methods to the corresponding implicit assumption about physical parameter uncertainty that each method represents.

Finite element methods have found wide appeal in engineering practice (i.e., [20–22]). They have been used extensively in many engineering applications such as solids, structures, field problems, and fluid flows. Formulation of the Interval Finite Element Method (IFEM) follows the conventional formulation of FEM except that the values of the variables and the parameters of the physical system are assigned interval values. We then proceed with certain assumptions to idealize it into a mathematical model that leads to a set of governing *interval* algebraic equations.

During the past decade, researchers from different engineering disciplines have been contributing significantly to the development of Interval Finite Element Methods. A review of the major IFEM developments can be found in the authors' works [19, 23]. Additional contributions to IFEM include the work of Kupla, Pownuk, and Skalna [24], in which they use intervals to model uncertainty in linear mechanical systems. Also, Pownuk [25] has introduced an interval global optimization method in an attempt to find all stationary global solutions. Akpan et al. [26] developed a fuzzy interval-base finite element formulation and used the response surface method to predict the system response. The system response is calculated for various sets of input values to develop a quadratic functional approximation between the input variables and the response quantities. Combinatorial optimization is performed on the approximate function to determine the binary combinations of the interval variables that result in extreme responses. Such an approximation might give satisfactory results in the case of interval load; however, the general case requires higher-order functions and the work does not provide any evaluation for the possibility of obtaining the right combinations of the input variables. In one of the examples, 13 finite element runs were required compared to 65 runs for all possible combinations. Chen S. et al. [27] introduced an interval formulation for static problems based on a perturbed stiffness neglecting the higher-order parts in the unknown variable approximation. The introduced results are not compared with the exact solutions of the respective problems. A fuzzy structural reliability analysis is introduced in the work of Savoia [28]; this formulation is based on the interval α-cut concept to compute membership functions of the response variables and is not introduced within the context of FEM. The work of Dessombz et al. [29] introduces an interval application, where free vibrations of linear mechanical systems under uncertainty were studied. Special attention has been devoted in this study to the dependency problem. The fixed-point theorem has been used to achieve a conservative enclosure to the system response in the case of narrow intervals. In the case of wide intervals, an interval partition approach is employed that requires costly repetitive runs. A comparison of different methodologies for uncertainty treatment in oil basin modeling has been conducted in the work of Pereira [30], and a successful alternative interval element-by-element finite element formulation for heat conduction problems has been proposed. In addition, as intervals represent a subset of the convex models, Ganzerli and Pantelides [31] developed a load and resistance convex model for optimum design. They [32] also conducted a comparison between convex and interval models within the context of finite elements, restricted to uncertain load. Both works were based on Cartesian product of the convex sets of the uncertain parameters and the load convex model superposition. Although sharp system response can be calculated by a single run of the response convex model, costly computations are required to obtain the response convex model itself. The present review highlights the general effort in the development of interval finite element, and shows that the main application of intervals to mechanical systems using formulations of IFEM has relied on implicit assumptions on the independence of interval quantities and the assumed width of an interval quantity.

However, for IFEM to gain acceptance as an engineering tool, two fundamental questions in the formulation and behavior of interval finite element must be resolved:

1. Does the replacement of the deterministic parameters by corresponding interval parameters lead to a violation of the underlying laws of physics?
2. What physical uncertainty is implied by the introduction of interval parameters?

We will illustrate the finite element *intervalization* using a simple two-element finite element approximation. We will explore the use of interval numbers to represent uncertainties in the values of Young's modulus, E, and the loading function q, (or boundary conditions τ).

Interval Methods for Reliable Computing

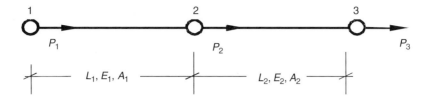

FIGURE 12.3 Two connected linear truss elements.

Figure 12.3 shows the discretization of a bar into two linear truss elements that share node 2. In this example, we assume that the values of E and A are constant over each element. In general, E and A are always positive; thus, as long as the intra-element variation of E is smooth enough and bounded, an interval representation of the integrated element stiffness can be constructed, Dubois, [33].

The resulting noninterval finite element equations are given by

$$\begin{pmatrix} \frac{E_1 A_1}{L_1} & -\frac{E_1 A_1}{L_1} & 0 \\ -\frac{E_1 A_1}{L_1} & \frac{E_1 A_1}{L_1} + \frac{E_2 A_2}{L_2} & -\frac{E_2 A_2}{L_2} \\ 0 & -\frac{E_2 A_2}{L_2} & \frac{E_2 A_2}{L_2} \end{pmatrix} \begin{pmatrix} u_1 \\ u_2 \\ u_3 \end{pmatrix} = \begin{pmatrix} P_1 \\ P_2 \\ P_3 \end{pmatrix} \tag{12.17}$$

Let us consider the case where the values of Young's modulus E and load vector $\boldsymbol{P}$ are uncertain but the bounds of the possible values are known. It is natural to express E and $\boldsymbol{P}$ as an interval quantity and interval vector, respectively. An interval x is a closed set in R defined by:

$$x \equiv [x^l, x^u] := \{\tilde{x} \in R \mid x^l \leq \tilde{x} \leq x^u\} \tag{12.18}$$

where superscripts l and u denote lower and upper bounds, respectively.

Replacement of the scalar E by an interval quantity and vector $\boldsymbol{P}$ by interval vector lead to the following straightforward (**but physically inconsistent**) interval system of equations:

$$\begin{pmatrix} \frac{[E_1^l, E_1^u] A_1}{L_1} & -\frac{[E_1^l, E_1^u] A_1}{L_1} & 0 \\ -\frac{[E_1^l, E_1^u] A_1}{L_1} & \frac{[E_1^l, E_1^u] A_1}{L_1} + \frac{[E_2^l, E_2^u] A_2}{L_2} & -\frac{[E_2^l, E_2^u] A_2}{L_2} \\ 0 & -\frac{[E_2^l, E_2^u] A_2}{L_2} & \frac{[E_2^l, E_2^u] A_2}{L_2} \end{pmatrix} \begin{pmatrix} u_1 \\ u_2 \\ u_3 \end{pmatrix} = \begin{pmatrix} [P_1^l, P_1^u] \\ [P_2^l, P_2^u] \\ [P_3^l, P_3^u] \end{pmatrix} \tag{12.19a}$$

or

$$KU = P \tag{12.19b}$$

with the interval stiffness matrix $K \in IR^{n \times n}$ and the interval load vector $P \in IR^n$. The vector of unknowns U is the interval displacement vector. This linear interval equation is defined as the family of linear equations

$$\tilde{K}\tilde{U} = \tilde{P} \qquad (\tilde{K} \in K, \tilde{P} \in P) \tag{12.20}$$

Thus, in this case we have linear systems of equations in which the coefficients are unknown numbers contained within given intervals. If we do not consider the **underlying physics** of the given system, we will be interested in the enclosure for the solution set of 12.19, given by

$$S(K,P) := \{\tilde{U} \in R^n \mid \tilde{K}\tilde{U} = \tilde{P} \text{ for some } \tilde{K} \in K, \tilde{P} \in P\} \tag{12.21}$$

The exact enclosure for the solution set of 12.19 can be obtained by solving for all possible combinations of the upper and lower bounds of the stiffness matrix coefficients and the right-hand side components.

The stiffness matrix is usually symmetric, and in such a case, the stiffness matrix will be a symmetric *interval* matrix and not an interval matrix in the general sense. The definition of symmetric interval matrix can be found in (Jansson, [34])

$$K_{sym} := \{\tilde{K} \in K, \ \tilde{K}^T = \tilde{K}\} \tag{12.22}$$

and Equation 12.19b takes the form

$$K_{sym} U = P \tag{12.23}$$

We will be interested in the enclosure for the solution set of 12.23, given by

$$S_{sym}(K_{sym}, P) := \{\tilde{U} \in R^n \mid \tilde{K}\tilde{U} = \tilde{P} \text{ for some } \tilde{K} \in K, \tilde{P} \in P \text{ with } \tilde{K}^T = \tilde{K}\} \tag{12.24}$$

The exact enclosure for the solution set of Equation 12.23 can be obtained by solving for all possible combinations of the upper and lower bounds of the stiffness matrix coefficients with the condition $k_{ij} = k_{ji}$ for $i, j = 1, ..., n, i \neq j$ and the right-hand side components.

The solution for both above-mentioned cases (i.e., the *general* and the *symmetric* case) does not represent a correct solution for the given physical problem, even if the solution is based on all possible combinations. Both solutions can overestimate the interval width of the physical problem. This is due to the multiple occurrences of stiffness coefficients in the stiffness matrix. The same physical quantity cannot have two different values (upper and lower bounds) at the same time, and interval arithmetic treats this physical quantity as multiple independent quantities of the same value. In other words, the physical problem has a unique but unknown value of E. Equation 12.19 allows different values of E to exist simultaneously. As will be demonstrated, even exact enclosures of solutions of finite element equations in this form produce physically unacceptable results. Historically, this loss of dependency combined with the well-known computational issues in computing sharp bounds for the solution of interval system of equations encouraged a number of earlier researchers to abandon the application of interval methods to finite element problems.

An alternative method for intervalization of finite element equations is to assume that the same value of Young's modulus E exists in the entire domain. This assumption results in the following interval equation for the two-element problem of Figure 12.2

$$([E^l, E^u]) \begin{pmatrix} \dfrac{A_1}{L_1} & -\dfrac{A_1}{L_1} & 0 \\ -\dfrac{A_1}{L_1} & \dfrac{A_1}{L_1} + \dfrac{A_2}{L_2} & -\dfrac{A_2}{L_2} \\ 0 & -\dfrac{A_2}{L_2} & \dfrac{A_2}{L_2} \end{pmatrix} \begin{pmatrix} u_1 \\ u_2 \\ u_3 \end{pmatrix} = \begin{pmatrix} [P_1^l, P_1^u] \\ [P_2^l, P_2^u] \\ [P_3^l, P_3^u] \end{pmatrix} \tag{12.25}$$

Such a formulation leads to an exact solution for the interval system. However, the material property is limited to a constant interval parameter (E) for the entire domain and is not attractive to engineering applications that require solving realistic problems, where the same parameter might vary within the domain itself.

The exact solutions to the interval finite element problems can only be obtained if the *underlying laws of physics are not violated*. Thus, for the considered case, to obtain the exact *physical* enclosure for the solution set of Equation 12.19 we have to trace back the sources of dependency in the formulation of interval finite elements. To illustrate the concept, let us use again, as an example, the two-bar truss of Figure 12.3.

The element interval stiffness matrices can be introduced in the following form:

$$K_1 = \chi_1 \begin{pmatrix} k_1 & -k_1 \\ -k_1 & k_1 \end{pmatrix}, \quad K_2 = \chi_2 \begin{pmatrix} k_2 & -k_2 \\ -k_2 & k_2 \end{pmatrix} \tag{12.26}$$

$$k_i = \frac{E_i A_i}{l_i} \tag{12.27}$$

where K_i = stiffness matrix of ith finite element (for i = 1, 2); k_i = stiffness coefficient of the ith element; E_i = modulus of elasticity of the ith element; A_i = cross-sectional area of the ith element; l_i = length of the ith element; and χ_i = the interval multiplier of the ith finite element obtained due to uncertainty in E_i, A_i, and l_i. The presented example represents a simple case of structural problems where the stiffness coefficients are the same for each element and are obtained only from axial stress contribution. However, the general formulation of interval finite element is more involved, where usually the stiffness coefficients are not equal for each element and include contributions from all components of the general stress state. The assemblage of global stiffness matrix can be given by

$$K = \sum_e \left\{ L_e^T \left(\sum_i [\chi_i(k_i)] \right) L_e \right\} \tag{12.28}$$

where L_e is the element Boolean connectivity matrix with the dimension (number of degrees of freedom per element × total number of degrees of freedom). In this example, the element Boolean connectivity matrix has the following structure

$$L_1 = \begin{pmatrix} 1 & 0 & 0 \\ 0 & 1 & 0 \end{pmatrix}, \quad \text{and} \quad L_2 = \begin{pmatrix} 0 & 1 & 0 \\ 0 & 0 & 1 \end{pmatrix} \tag{12.29}$$

Finally, the global stiffness matrix takes the following form:

$$K_{phys} = \begin{pmatrix} \chi_1 k_1 & -\chi_1 k_1 & 0 \\ -\chi_1 k_1 & \chi_1 k_1 + \chi_2 k_2 & -\chi_2 k_2 \\ 0 & -\chi_2 k_2 & \chi_2 k_2 \end{pmatrix} \tag{12.30}$$

By imposing the boundary conditions, displacement of the first node equals zero in this case, we obtain:

$$K_{phys} = \begin{pmatrix} \chi_1 k_1 + \chi_2 k_2 & -\chi_2 k_2 \\ -\chi_2 k_2 & \chi_2 k_2 \end{pmatrix} \tag{12.31}$$

The interval load vector (right-hand side) can be introduced in the following form:

$$P_{phys} = \begin{pmatrix} \lambda_1 P_1 \\ \lambda_2 P_2 \\ \lambda_3 P_3 \end{pmatrix} \qquad (12.32)$$

where

$$P_{phys} = P_c + P_b + P_t \qquad (12.33)$$

and P_c = the nodal load vector, P_b = the body force vector, and P_t = traction vector. The body force vector and the traction vector are given by:

$$P_b = \sum_e \left\{ L_e^T \left(\sum_i [\lambda_{ib}(p_{ib})] \right) L_e \right\} \qquad (12.34a)$$

$$P_t = \sum_e \left\{ L_e^T \left(\sum_i [\lambda_{it}(p_{it})] \right) L_e \right\} \qquad (12.34b)$$

where p_{ib} is the body force contribution at the ith integration point, and p_{it} is the traction contributions at the ith integration point acting on the surface Γ, and λ is the interval multiplier for each load.

After imposing the boundary conditions, Equation 12.32 takes the following form:

$$P_{phys} = \begin{pmatrix} \lambda_2 P_2 \\ \lambda_3 P_3 \end{pmatrix} \qquad (12.35)$$

and the interval form of Equation 12.17 is

$$K_{phys} U = P_{phys} \qquad (12.36)$$

The exact enclosure for the solution set of *physical* interval finite Equation 12.36 will be given by

$$S_{phys}(K_{phys}, P_{phys}) := \{ \tilde{U} \in R^n \mid \tilde{K}_{phys}(\chi)\tilde{U} = \tilde{P}_{phys}(\lambda) \text{ for } \tilde{\chi}_i \in \chi_i, \tilde{\lambda}_i \in \lambda_i \} \qquad (12.37)$$

The presented physical solution means that each uncertain parameter throughout the system, regardless of its multiple occurrences, is restricted to take only one unique value within the range of its interval.

The exact enclosure for the solution set of *physical* interval finite equation can be obtained by solving for all possible combinations of lower and upper bounds of uncertain parameters. Such a solution is possible for small size problems (few uncertain parameters); however, engineering applications might require solutions of huge systems where the "all possible combinations" approach is computationally prohibited and such problems are NP-hard problems.

For the solution of interval finite element (IFEM) problems, Muhanna and Mullen [23] introduced an element-by-element interval finite element formulation, in which a guaranteed enclosure for the solution of interval linear system of equations was achieved. A very sharp enclosure for the solution set due to loading, material and geometric uncertainty in solid mechanics problems was obtained. Element matrices were formulated based on the physics, and the Lagrange multiplier method was applied to impose the necessary constraints for compatibility and equilibrium. Other methods for imposing the constraints (such as the penalty method) might also be used. Most sources of overestimation were eliminated. In this formulation, Equation 12.19b can be introduced in the following form:

$$\begin{pmatrix} K & C^T \\ C & 0 \end{pmatrix} \begin{pmatrix} U \\ \lambda \end{pmatrix} = \begin{pmatrix} P \\ 0 \end{pmatrix} \quad (12.38)$$

Here, λ is the vector of Lagrange multipliers, $C = (0 \;\; 1 \;\; -1 \;\; 0)$, $CU = 0$, and K is introduced as

$$DS = \begin{pmatrix} \chi_1 & 0 & 0 & 0 \\ 0 & \chi_1 & 0 & 0 \\ 0 & 0 & \chi_2 & 0 \\ 0 & 0 & 0 & \chi_2 \end{pmatrix} \begin{pmatrix} \frac{E_1 A_1}{L_1} & -\frac{E_1 A_1}{L_1} & 0 & 0 \\ -\frac{E_1 A_1}{L_1} & \frac{E_1 A_1}{L_1} & 0 & 0 \\ 0 & 0 & \frac{E_2 A_2}{L_2} & -\frac{E_2 A_2}{L_2} \\ 0 & 0 & -\frac{E_2 A_2}{L_2} & \frac{E_2 A_2}{L_2} \end{pmatrix} \quad (12.39)$$

where χ_i = the interval multiplier of the ith finite element obtained due to uncertainty in E_i, A_i, and l_i. Such a form (i.e., $D\tilde{S}$) allows factoring out the interval multiplier resulting with an exact inverse for ($D\tilde{S}$).

Equation 12.38 can be introduced in the following equivalent form

$$KU + \tilde{C}^T \lambda = P \quad (12.40a)$$

$$\tilde{C}U = 0 \quad (12.40b)$$

If we express K ($n \times n$) in the form $D\tilde{S}$ and substitute in equation (12.40a)

$$D\tilde{S}U = P - \tilde{C}^T \lambda \quad (12.41)$$

where $D(n \times n)$ is interval diagonal matrix, its diagonal entries are the positive interval multipliers associated with each element, and n = degrees of freedom per element × number of elements in the structure. $\tilde{S}(n \times n)$ is a deterministic singular matrix (fixed point matrix). If we multiply Equation 12.40b by $D\tilde{C}^T$ and add the result to Equation 12.41, we get

$$D(\tilde{S}U + \tilde{C}^T CU) = (P - \tilde{C}^T \lambda) \quad (12.42)$$

or

$$\begin{aligned} D(\tilde{S}U + \tilde{Q}U) &= (P - \tilde{C}^T \lambda) \\ D(\tilde{S} + \tilde{Q})U &= (P - \tilde{C}^T \lambda) \\ D\tilde{R}U &= (P - \tilde{C}^T \lambda) \end{aligned} \quad (12.43)$$

where $\tilde{R}$ is a deterministic positive definite matrix, and the displacement vector U can be obtained from Equation 12.43 in the following form

$$U = \tilde{R}^{-1} D^{-1} (P - \tilde{C}^T \lambda) \quad (12.44)$$

where $\tilde{R}^{-1}D^{-1}$ is an exact inverse of the interval matrix $(D\tilde{R})$. Equation 12.44 can be presented in the form

$$U = \tilde{R}^{-1} M \delta \qquad (12.45)$$

Matrix M has the dimensions ($n \times$ number of elements) and its derivation has been discussed in the previous works of Mullen and Muhanna [19]. The vector δ is an interval vector that has the dimension of (number of elements $\times$ 1): its elements are the diagonal entries of D^{-1} with the difference that every interval value associated with an element is occurring only once. If the interval vector λ can be determined exactly, the solution of Equation 12.45 will represent an exact hull for the solution set of the general interval FE equilibrium equation

$$KU = P \qquad (12.46)$$

As a matter of fact, this is the case in statically determinate structures. An exact hull for the solution set can be achieved because vector λ, in the present formulation, represents the vector of internal forces, and in statically determinate structures the internal forces are independent of the structural stiffness. Consequently, using the deterministic value of λ (mid point of λ, the midpoint of an interval a is defined as $a_c = \text{mid}(a) = \frac{a^u + a^l}{2}$) in Equation 12.44 results in an exact hull for the solution set of statically determinate structures with uncertain stiffness. But in the case of statically indeterminate structures, values of λ depend of the structural stiffness and the use of deterministic value of λ results in a very good estimate for the solution, but of course a narrower one. The details for handling the general case is discussed in [23].

This formulation does not allow any violation to the underlying laws of physics, and results in very sharp interval solutions for linear mechanics problems and in exact solutions for one group of these problems (statically determinate problems).

We will illustrate this discussion by numerical results of the following example.

The example is a triangular truss shown in Figure 12.4. The structure is loaded by a deterministic horizontal load $30KN$ and an interval vertical load $[50, 100]$ KN, at the top node. The truss has the following data: cross-sectional area $A = 0.001$ m^2, Young's modulus $E = 200$ GPa, and 10% uncertainty in E; that is, $E = [190, 200]$ GPa was assumed for each element. The results for displacement of selected nodes are given in Table 12.1.

The results show that the general interval solution is wider than the symmetric interval solution. The symmetric interval solution overestimates the physical diameter by 127.2 to 182.5%, while the general interval solution overestimates the exact diameter by 130 to 186.6%.

The above-mentioned discussion illustrates how interval methods can be used for predicting system response due to uncertain parameters. However, different intervalizations of finite element imply different physical uncertainty: the *general interval approach*, which is a naïve replacement of the deterministic

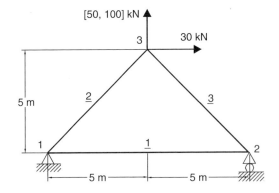

FIGURE 12.4 Triangular truss.

TABLE 12.1 Nodal Displacements of Triangular Truss

	node 2	node 3	
	U	U	V
Node displacements	$U \times 10^{-3}$ m	$U \times 10^{-3}$ m	$V \times 10^{-3}$ m
Comb. (physical)	[-1.8421053, -0.4761905]	[-0.0349539, 0.9138331]	[1.9216828, 4.6426673]
Symmetric interval	[-2.7022107, -0.2451745]	[-0.6531157, 1.0781411]	[1.7879859, 5.24493035]
General interval	[-2.7022107, -0.1784592]	[-0.6531157, 1.1173116]	[1.7111129, 5.24493035]
D_{ph}[a]	1.36591	0.94879	2.72098
D_s	2.445704	1.73126	3.46132
D_i	2.52375	1.77043	3.53819
D_s/D_{ph}	1.799	1.825	1.272
D_i/D_e	1.848	1.866	1.300

[a] D_{ph}, D_s and D_i are the diameter of the exact solution, symmetric interval and general interval solution, respectively.

parameters with interval ones, leads to a violation of the physics and even the exact solutions lead to a catastrophic overestimation that does not reflect any realistic evaluation of the system response to the introduced uncertainty. The *symmetric interval approach*, which accounts for symmetry in the system matrix, treats the diagonal interval coefficients as independent ones and, consequently, leads to results of the same nature of the general interval approach. Within the category of what we call physical formulation, two approaches were discussed; the *first* assumes the same interval value of a given parameter exists in the entire domain. Such an assumption limits the variability of parameters within the considered domain to a constant value, and consequently it might not be considered attractive to engineering applications that require solutions that are more realistic. The *second* introduced approach is the element-by-element formulation, where the interval parameters are allowed to vary within the considered domain including different level of variability between elements. This formulation leads to very sharp results and to a realistic propagation of uncertainty in the system with no violation of the physics.

12.4 Interval Methods for Bounding Approximation and Rounding Errors

12.4.1 Approximation Errors

Intervals and ranges of functions over intervals arise naturally in large number of situations [4, 7, 11]. Frequently, approximate mathematical procedures are used in place of the exact ones, using the mean value theorem, an approximate value for the derivative of a real continuously differentiable function $f(x)$ for $x_1 \neq x_2$, is

$$f'(\xi) = \frac{f(x_2) - f(x_1)}{x_2 - x_1} \tag{12.47}$$

for some ξ between x_1 and x_2. To make more than a theoretical or qualitative use of this formula, interval arithmetic can compute or at least bound the range of f'. Another example for interval methods in bounding approximation errors is the use of Taylor's series. It is well known that Taylor's series provides a means to predict a function value at one point in terms of the function value and its derivatives at another point. In particular, the theorem states that any smooth function can be approximated as a polynomial in the following form

$$f(x_{i+1}) = f(x_i) + f'(x_i)h + \frac{f''(x_i)}{2!}h^2 + \cdots + \frac{f^{(n)}(x_i)}{n!}h^n + \frac{f^{(n+1)}(\xi)}{(n+1)!}h^{n+1} \tag{12.48}$$

where $h = (x_{i+1} - x_i)$ and ξ is a value of x between x_i and x_{i+1}. Once again, interval arithmetic can determine the approximation error by computing an enclosure of $f^{(n+1)}(\xi)$, where ξ ranges over the interval $[x_i, x_{i+1}]$.

12.4.2 Rounding-off Errors

Interval arithmetic is a powerful tool to handle rounding-off errors resulting from the finite floating-point representation of numbers on computers. Using floating-point representation, a fractional number is expressed as

$$m \cdot b^e \tag{12.49}$$

where m is the mantissa, b is the base of the number system used, and e is the exponent. For example, the number 254.67 could be represented as 0.25467×10^3 in a floating-point base-10 system. This results in the fact that there is only a finite set $\mathbb{M} \subset \mathbb{R}$ of machine-representable numbers. Given any real number x lying between two consecutive machine numbers $x_1, x_2 \in \mathbb{M}$, the interval between these exactly represented numbers Δx will be proportional to the magnitude of the number x being represented. For normalized floating-point numbers (the mantissa has no leading zero digit), this proportionality can be expressed, for cases where chopping is employed, as

$$\frac{|\Delta x|}{|x|} \leq \varepsilon \tag{12.50}$$

and, for cases where rounding is employed, as

$$\frac{|\Delta x|}{|x|} \leq \frac{\varepsilon}{2} \tag{12.51}$$

where ε is referred to as the machine epsilon, which can be computed as

$$\varepsilon = b^{1-L} \tag{12.52}$$

where b is the number base and L is the number of significant digits in the mantissa (mantissa length in bits). The inequalities in Equations 12.50 and 12.51 signify that these are error bounds. For example, computers that use the IEEE format allow 24 bits to be used for the mantissa, which translates into about seven significant base-10 digits of precision; that is

$$\varepsilon = b^{1-24} \approx 10^{-7} \tag{12.53}$$

with range of about 10^{-38} to 10^{39}. However, if the number of computer words used to store floating-point numbers is doubled, such computers can provide about 15 to 16 decimal digits of precision and a range of approximately 10^{-308} to 10^{-308}. Despite the provided high precision, rounding-off errors still represent a crucial problem for achieving numerical solutions for a series of problems, such as root finding, function evaluation, optimization, and the formulation of self-validating methods. One might argue that if computed results, using single and double precision, agree to some number of digits, then those digits are correct. The following example of Rump [35] shows that such an argument is not valid. The evaluated function is

$$f(x, y) = 333.75 y^6 + x^2(11 x^2 y^2 - y^6 - 121 y^4 - 2) + 5.5 y^8 + \frac{x}{2y} \tag{12.54}$$

for $x = 77617$ and $y = 33096$. The evaluation was done on an S/370 computer. All input data is exactly representable on the computer, so the only rounding-off error occurs during evaluation.

Single, double, and extended precision were used, which were the equivalent of approximately 6, 17, and 34 decimal digit arithmetic. The following values of f were obtained

Single precision: $f = 1.172603...$
Double precision: $f = 1.1726039400531...$
Extended precision: $f = 1.172603940053178...$

All three results agree in the first seven decimal digits. However, they are all completely incorrect. They do not even have the correct sign. The correct value is

$$f = -0.8273960599468213$$

with an error of at most one unit in the last digit.

Interval arithmetic represents an efficient tool to overcome such a numerical difficulty by using what is referred to as rounding-off error control. For an interval arithmetic with floating point endpoints and floating point operations, we need outward rounding. This can be achieved by multiplying the result by some $1 \pm \varepsilon$, ε denoting the relative rounding error unit. Since the establishment of the IEEE 754 arithmetic standard, optimal floating-point operations in a specifiable rounding mode are available. The important rounding modes are ∇ toward $-\infty$, and Δ towards $+\infty$.

Frequently, a processor may be switched into such a rounding mode. This means that subsequent operations are performed in that mode until the next switch. Interval arithmetic uses the notation that an arithmetic expression in parentheses preceded by a rounding symbol implies that all operations are performed in floating-point in the specified rounding mode. Then for a set $\mathbb{M}$ of floating point numbers (e.g., single or double precision including $\pm\infty$), IEEE 754 defines

$$\begin{aligned} &\forall a, b \in \mathbb{M}, \; \forall o \in \{+, -, \cdot, /\} \\ &\nabla(a \circ b) = \max\{x \in \mathbb{M} : x \leq a \circ b\} \\ &\Delta(a \circ b) = \min\{x \in \mathbb{M} : a \circ b \leq x\} \end{aligned} \qquad (12.55)$$

Thus, rounding is correct and best possible. Note that this is true for all floating-point operands and operations.

12.5 Future Developments of Interval Methods for Reliable Engineering Computations

In the previous sections, interval methods have been demonstrated to be an efficient method for the accounting of uncertainty in parameters in engineering models. Interval methods also account for truncation errors from the computer implementation of engineering models. There is also the possibility that intervals can be used to quantify the impact of other sources of errors in engineering models. One such error is the discretization error associated with the generation of approximations to partial differential equations. Discretization errors have been extensively studied; early work on bounding the discretization error was reported by Aziz and Babuska [36]. Following this work, the convergence behavior of most finite element formulations has been determined. Knowledge of the behavior of discretization errors is key to the development of adaptive solutions strategies [37]. We believe that interval representation can be used to incorporate bounds on the solution associated with discretization of a partial differential equation in combination with the other uncertainties identified above.

We will first illustrate this capability using the one-dimensional truss element as describe above. While the truss element is an exact solution to the PDF when loads are restricted to the endpoints of each element, the element is also appropriate for one-dimensional problems with arbitrary loading along the length of the element. In this case, the finite element solution is not exact and one generates discretization errors. Early work by Babuska and Aziz has shown that the global finite element discretization error is bounded by the interpolation error. Thus, one needs only to examine how close the finite element shape functions can come in representing an exact solution to develop *a priori* error bounds. For example, one can show from simple Taylor theorem calculations [38], the one-dimensional linear element described in Section 12.3, has a discretization error of

$$\|u^h - u\| \leq \frac{h^2}{8} \left\| \frac{\partial^2 u}{\partial x^2} \right\| \quad (12.56)$$

Here, u^h is the finite element solution, u the exact solution to the PDE, and h is the length of the element. From the original differential equation, with uniform values of E and A, the bounds can be written.

$$\|u^h - u\| \leq \frac{h^2}{8} \left\| \frac{q}{AE} \right\| \quad (12.57)$$

Consider the problem of a rod hanging under its own weight, (i.e., q is the weight per unit length), the exact displacement is given by

$$u(x) = \frac{q}{AE}\left(lx - \frac{x^2}{2}\right) \quad (12.58)$$

On the other hand, a single element FEA model predicts

$$u(x) = \frac{q}{AE}\left(\frac{lx}{2}\right) \quad (12.59)$$

giving a maximum error of $ql^2/8AE$ located at the center of the element. The maximum error as a function of element size is given in Table 12.2.

As can be seen, the predicted errors and the observed errors in this problem are identical. While this example provides insight into the potential of bounding finite element errors, it has several limitations: the problem is one-dimensional, a uniform mesh was used, and the nature of the problem was self-similar — inducing the same error pattern in each element independent of size. The self-similar nature of the error in this problem avoids an important difficulty in bounding discretization errors. Most error estimates are global. They provide bounds on measures such as the error in energy in the solution. Application of global discretization errors to each point in the solution results in loss of sharpness in the resulting bounds, often leading to useless results. Sharper bounds on discretization error can be obtained from *a posteriori* errors [39]. Sharp bounds for the energy norm of the discretization error have been developed by Babuska, Strouboulis, and Gangaraj [40]. They employ an iterative evaluation of the element–residuals and provide *a posteriori* sharp bounds on the reliability interval in the energy norm.

Local error bounds have been developed in terms of an additive combination of "pollution errors" and local element residual errors. *A posteriori* error bounds are calculated on a small patch of elements using local element residual errors. The pollution errors, which represent sources of error outside the small patch of elements, are added to provide sharp local error bounds [41, 42].

TABLE 12.2 Finite Element Errors for a Bar under Its Own Weight

Number of Elements	Max FEA Error	Location(s)	Predicted Error
1	0.125 ql/EA	$l/2$	0.125 ql/EA
2	0.03125 ql/EA	$l/4$ and $3l/4$	0.03125 ql/EA
3	0.013888 ql/EA	$l/6$, $l/2$ and $5l/6$	0.013888 ql/EA

Another method of bounding local errors is a concept of attributing all discretization errors to errors in constitutive models [43]. In this method, the governing PDE are considered in terms of separate equilibrium, constitutive, and continuity equations. The displacement finite element solution exactly satisfies the continuity equation. The finite element stress field is transformed into a field that stratifies the equilibrium equation. The changes in the stress field calculated from the FEM and a field that satisfies equilibrium are accommodated by a perturbed constitutive equation. This results in an exact solution to a perturbed problem. All discretization errors are represented by interval bounds on material parameters. In some respects, this method is the inverse problem of IFEM presented in Section 12.3.2.

12.6 Conclusions

Reliable engineering computing is perhaps the only way to ensure a guaranteed solution for engineering problems. A guaranteed solution can be achieved only if all sources of errors and uncertainty are properly accounted for. Errors and uncertainty can be bounded using the concept of interval representation and interval numbers.

Five major frameworks that use interval-based representation of uncertainty have been discussed in this chapter, namely: imprecise probability, possibility theory, the Dempster-Shafer theory of evidence, fuzzy set theory, and convex set modeling. Interval methods for predicting system response under uncertainty were introduced. Sensitivity analysis using intervals has been illustrated; the formulation of Interval Finite Element Method (IFEM) has been discussed, with a focus on dependency problems using Element-By-Element approach. Interval methods for bounding approximation and rounding errors have been introduced. Despite the availability of high-precision computers, rounding-off errors still represent a crucial problem for achieving numerical solutions for a series of problems, such as root finding, function evaluation, optimization, and the formulation of self-validating methods. Intervals show superiority in handling rounding-off errors. There is also the possibility that intervals can be used to quantify the impact of other sources of errors in engineering models. One such error is the discretization errors associated with the generation of approximations to partial differential equations.

Interval representation allows the characterization of uncertainties in model parameters and errors in floating-point representations in digital computers, as well as the potential for representing errors from the discretization of engineering models. This common representation allows for all sources of errors and uncertainty to be represented in a consistent manner. Reliable engineering computations require all sources of uncertainty to be quantified and incorporated into the calculated solutions. Using interval representation may be the only way to accomplish this goal.

References

1. Hansen, E., Interval arithmetic in matrix computation, *J.S.I.A.M., series B, Numerical Analysis*, Part I, 2, 308–320, 1965.
2. Moore, R.E., *Interval Analysis*, Prentice Hall, Englewood Cliffs, NJ, 1966.
3. Alefeld, G. and Herzberger, J., *Introduction to Interval Computations*, Academic Press, New York, 1983.
4. Neumaier, A., *Interval Methods for Systems of Equations*, Cambridge University Press, 1990.
5. Dwyer, P.S., *Computation with Approximate Numbers, Linear Computations*, P.S. Dwyer, Ed., Wiley, New York, 1951, 11–34.
6. Sunaga, T., Theory of interval algebra and its application to numerical analysis, *RAAG Memoirs 3*, 29–46, 1958.
7. Hansen, E., *Global Optimization Using Interval Analysis*, Marcel Dekker, New York, 1992.
8. Rump S.M., Self-validating methods, *Journal of Linear Algebra and Its Application*, 324, 3–13, 2001.

9. Alefeld, G. and Herzberger, J., *Introduction to Interval Computations*, Academic Press, New York, 1983.
10. Hansen, E.R., Global optimization using interval analysis — the one-dimensional case, *J. Optim. Theory*, Appl 29, 331–334, 1979.
11. Moore, R.E., *Methods and Applications of Interval Analysis*, SIAM, Philadelphia, 1979.
12. Rohn, Rump, and Yamamot, Preface, *Linear Algebra and Its Applications*, 324, 1–2, 2001.
13. Ferson and Sentz, Combination of Evidence in Dempster-Shafer Theory, *SAND 2002-0835*, 2002.
14. Kaufman and Gupta, *Introduction to Fuzzy Arithmetic, Theory and Applications*, Van Nostrand Reinhold, 1991.
15. Ben-Haim and Elishakoff, *Convex Models of Uncertainty in Applied Mechanics*, Elsevier Science, New York, 1990.
16. Ben-Haim, Yakov, *Information-Gap Decision Theory*, Academic Press, New York, 2001.
17. Apostolatos, N. and Kulisch, U., Grundzüge einer Intervallrechtung für Matrizen und einige Anwendungen, *Elektron. Rechenanlagen*, 10, 73–83, 1968.
18. Mayer, O., Algebraische und metrische Strukturen in der Intervallrechung und eingine Anwendungen, *Computing*, 5, 144–162, 1970.
19. Mullen, R.L. and Muhanna, R.L., Bounds of structural response for all possible loadings, *Journal of Structural Engineering, ASCE*, 125(1), 98–106, 1999.
20. Bathe, K., *Finite Element Procedures*, Prentice Hall, Upper Saddle River, NJ, 1996.
21. Hughes, T.J.R., *The Finite Element Method*, Prentice Hall, Englewood Cliffs, NJ, 1987.
22. Zienkiewicz, O.C. and Taylor, R.L., *The Finite Element Method*, fourth edition, McGraw-Hill, New York, 1991.
23. Muhanna, R.L. and Mullen, R.L., Uncertainty in mechanics problems — interval-based approach, *Journal of Engineering Mechanics, ASCE*, 127(6), 557–566, 2001.
24. Kulpa, Z., Pownuk, A., and Skalna, I., Analysis of linear mechanical structures with uncertainties by means of interval methods, *Computer Assisted Mechanics and Engineering Sciences*, Vol. 5, Polska Akademia Nauk., 1998.
25. Pownuk, A., New Inclusion Functions in Interval Global Optimization of Engineering Structures, *ECCM-2001 European Conference on Computational Mechanics*, June 26–29, Cracow, Poland, 2001.
26. Akpan, U.O., Koko, T.S., Orisamolu, I.R., and Gallant, B.K., Practical fuzzy finite element analysis of structures, *Finite Elements in Analysis and Design*, 38(2), 93–111, 2001.
27. Chen, S.H., Lian, H.D., and Yang, X.W., Interval static displacement analysis for structures with interval parameters, *International Journal for Numerical Methods in Engineering*, 53(2), 393–407, 2002.
28. Savoia, M., Structural reliability analysis through fuzzy number approach, with application to stability, *Computers & Structures*, 80(12), 1087–1102, 2002.
29. Dessombz, O., Thouverez, F., Laîné, J.P., and Jézéquel, L., Analysis of mechanical systems using interval computations applied to finite elements methods, *Journal of Sound and Vibration*, 239(5), 949–968, 2001.
30. Pereira, S.C.A., *Dealing with Uncertainty in Basin Modeling*, thesis, Department of Civil Engineering, University of Rio De Janeiro, Brazil, 2002.
31. Ganzerli, S. and Pandelides, C.P., Load and resistance convex models for optimum design, *Structural Optimization*, 17, 259–268, 1999.
32. Pantelides, C.P. and Ganzerli, S., Comparison of fuzzy set and convex model theories in structural design, *Mechanical Systems and Signal Processing*, 15(3), 499–511, 2001.
33. Dubios, D. and Prade, H., Towards Fuzzy Differential Calculus Part 2: Intergration on Fuzzy Intervals, *Fuzzy Sets and Systems*, Vol. 8, 105–116, 1982.
34. Jansson, C., Interval Linear System with Symmetric Matricies, Skew-Symmetric Matricies, and Dependencies in the Right Hand Side, *Computing*, 46, 265–274, 1991.
35. Rump, S. M., *Algorithm for Verified Inclusions — Theory and Practice*, Moore, 1988, 109–126.

36. Aziz, A.K. and Babuska, I., Part I, Survey lectures on the mathematical foundations of the finite element method, *The Mathematical Foundations of the Finite Element Methods with Applications to Partial Differential Equations,* Academic Press, New York, 1972, 1–362.
37. Zienkiewicz, O.C., Kelly, D.W., Cago, J.P. de S.R., and Babuska, I., The hierarchical finite element approaches, error estimates and adaptive refinement, J.R. Whiteman, Ed., *The Mathematics of Finite Elements and Applications,* Volume IV, Mafelap, 1982, 313–346.
38. Strang, G. and Fix, G., *An Analysis of the Finite Element Method,* Prentice Hall, Englewood Cliffs, NJ, 1972.
39. Babuska, I. and Rheinboldt, W.C., *A posteriori* error estimate for the finite element method, *International J. for Numerical Methods in Engineering,* 12, 1597–1615, 1978.
40. Babuska, I., Strouboulis, T., and Gangaraj, S.K., Guaranteed computable bounds for the exact error in the finite element solution, *Computer Methods in Applied Mechanics and Engineering,* 176, 51–79, 1999.
41. Oden, J.T. and Feng, Y., Local and pollution error estimates for finite element approximations of elliptical boundary value problems, *Journal of Applied and Computations Mathematics,* 74, 245–293, 1996.
42. Liao, X. and Nochetto, R.H., Local *a posteriori* error estimates and adaptive control of pollution effect, *Numerical Methods Partial Differential Equations,* 19, 421–442, 2003.
43. Ladeveze, P., Rougeot, Ph., Blanchard, P., and Moreau, J.P., Local error estimators for finite element linear analysis, *Computer Methods in Applied Mechanics and Engineering,* 176, 231–246, 1999.

Appendix 12.1   Interval Arithmetic Operations

The main arithmetic operations — addition, subtraction, multiplication, and division — are defined as follows [4]:

Suppose that we have the interval numbers: $x = [a,b]$, $y = [c,d]$.

1. Addition:

$$x + y = [a + c, b + d]$$

2. Subtraction:

$$x - y = [a - d, b - c]$$

3. Multiplication: $(x \times y)$

(1)	if $c \geq 0$ and $d \geq 0$ (2)	if $c < 0 < d$ (3)	if $c \leq 0$ and $d \leq 0$ (4)
if $a \geq 0$ and $b \geq 0$	$[ac, bd]$	$[bc, bd]$	$[bc, ad]$
if $a < 0 < b$	$[ad, bd]$	$[\min(ad, bc), \max(ac, bd)]$	$[bc, ac]$
if $a \leq 0$ and $b \leq 0$	$[ad, bc]$	$[ad, ac]$	$[bd, ac]$

4. Division: (x/y)

(1)	if $c > 0$ and $d > 0$ (2)	if $c < 0$ and $d < 0$ (3)
if $a \geq 0$ and $b \geq 0$	$[a/d, b/c]$	$[b/d, a/c]$
if $a < 0 < b$	$[a/c, b/c]$	$[b/d, a/d]$
if $a \leq 0$ and $b \leq 0$	$[a/c, b/d]$	$[b/c, a/d]$

On the other hand, only some of the algebraic laws valid for real numbers remain valid for intervals; other laws only hold in a weaker form. The following laws hold for intervals $x, y,$ and z:

1. Commutativity:

$$x + y = y + x$$
$$x \times y = y \times x$$

2. Associativity:

$$(x + y) \pm z = x + (y \pm z)$$
$$(x \times y) \times z = x + (y \pm z)$$

3. Neutral elements:

$$x + 0 = 0 + x = x, \quad 1 \times x = x \times 1 = x,$$

$$x - y = x + (-y) = -y + x, \quad x/y = x \times y^{-1} = y^{-1} \times x,$$

$$-(x - y) = y - x, \quad x \times (-y) = (-x) \times y = -(x \times y),$$

$$x - (y \pm z) + (x - y) \mp z, \quad (-x)(-y) = x \times y.$$

and the following laws represent weak forms of several laws from the real arithmetic:

1. Subdistributivity:

$$x \times (y \pm z) \subseteq x \times y \pm x \times z, \quad (x \pm y) \times z \subseteq x \times z \pm y \times z$$

2. Subcancellation:.

$$x - y \subseteq (x + z) - (y + z), \quad x/y \subseteq (x \times z)/(y \times z)$$
$$0 \in x - x, \quad 1 \in x/x$$

13
Expert Knowledge in Reliability Characterization: A Rigorous Approach to Eliciting, Documenting, and Analyzing Expert Knowledge

Jane M. Booker
Los Alamos National Laboratory

Laura A. McNamara
Los Alamos National Laboratory

13.1 Introduction 13-1
13.2 Definitions and Processes in Knowledge Acquisition 13-3
   Expert Knowledge Approach: Philosophy, Definitions, Roles, and Stages • Problem Identification: Deciding When to Use an Expert Knowledge Approach • Problem Definition • Eliciting Expertise for Model Development
13.3 Model Population and Expert Judgment 13-14
   Expert Judgment Elicitation • A Modified Delphi for Reliability Analysis
13.4 Analyzing Expert Judgment 13-23
   Characterizing Uncertainties • Information Integration for Reliability
13.5 Conclusion 13-29

13.1 Introduction

Expert knowledge is technical information provided by qualified people responding to formal elicitation questions that address complex, often under-characterized technical problems or phenomena [1]. Expressions of expert knowledge are of interest to scholars in a wide range of fields: economics, medicine and epidemiology, computational sciences, risk and reliability assessments, engineering, statistics, and others. In its formal expression, expert knowledge is most frequently associated with *expert judgment,* often a significant data source for mathematical and scientific models that attempt to predict or characterize complex, highly uncertain, and rare phenomena—for example, the probability of a nuclear reactor core melt accident. Informally expressed, however, expert knowledge is always a major factor in scientific research. This form of expert knowledge, which in this chapter is referred to as *expertise*, is articulated in the decisions that scientists and engineers make in selecting phenomena for study, identifying appropriate data sources, designing experiments, and choosing computational models. In short, expert knowledge is never absent from any research process, regardless of how formally it is elicited or expressed.

As such, expert knowledge — whether in the form of expert judgment or as expertise — is a valuable and often untapped source of information for analysts attempting to model complex phenomena. This is particularly true when the analyst is not a "native" speaker of the scientific or engineering community in which he or she is working, which is often the case for statisticians and other consulting analysts. However, formally eliciting, representing, and documenting the conceptual models and judgments that experts use in solving complex technical problems can benefit most analyses, particularly when the phenomenon under study is complex, poorly characterized, multidisciplinary, and involves multiple researchers and research sites — a description that can be applied to an increasing number of engineering projects.

Reliability is defined as the probability or likelihood that a system or product will perform its intended functions within specifications for a specified time. While traditional reliability is strictly defined as a probability, we will allow a more general definition such that other mathematical theories of uncertainty (e.g., possibility theory) could be implemented. As in past efforts [1], the goal of this chapter is to provide a brief but practical overview for researchers who are considering the use of expert knowledge in their reliability analyses. As we discuss below, we have incorporated expert knowledge in evaluating the reliability and performance of complex physical systems, as well as problems related to product manufacturing. These problems are comprised of both parts and processes: missiles in ballistic missile defense, concept designs for automotive systems, aging processes in nuclear weapons, and high cycle fatigue in turbine jet engines.

This chapter draws from our experience as well as from the enormous body of literature concerning the elicitation, documentation, and use of expert knowledge. We present a philosophy and a set of methods for the elicitation of expert knowledge, emphasizing the importance of understanding the "native" conceptual structures that experts use to solve problems and the merging of these methods with the analysis or use of this knowledge in solving complex problems where test or observational data is sparse. The elicitation techniques described herein are derived from ethnography and knowledge representation. They provide guidance for the elicitation and documentation of expert knowledge, as well as for merging information in analyses that involve uncertainty representation using probabilistic or General Information Theories (GITs). (See also Chapter 9 by Joslyn and Booker.)

Section 13.2 of this chapter describes the main definitions and processes involved in knowledge acquisition: identification of an adviser-expert, problem definition, elicitation techniques, representational forms, and model development. These stages typically do not take place in a sequential fashion. Rather, the knowledge acquisition process is iterative, with the analyst's questions becoming more sophisticated as his or her understanding of the problem and the domain increases, and the model itself emerging and being refined throughout.

Section 13.3 discusses expert judgment elicitation, including identification of experts, the development of tools, issues of biases and heuristics, and the documentation and refining of expert assessment. The reader will find that we treat expert judgment elicitation as an extension of the knowledge acquisition (KA) process. We find that expert assessments are best gathered by analysts who have built a working fluency with both the domain and the specific problem under study. Again, the process of eliciting expert judgment is iterative, often bringing about changes in the model structure well after the KA process appears to be complete. We also present an expert-oriented approach [2–5] for eliciting and refining expert assessments.

Section 13.4 focuses on analysis, primarily based on statistical methods, for handling expert judgment. Of particular importance is the expression of uncertainty associated with expert information. Also important is aggregation of expert judgment, which can be problematic when multiple experts give varying opinions or estimates. Moreover, the analyst must be able to combine expert opinion with other forms of data, such as experimental results, computer model outputs, historical performance and reliability data, and other available information sources.

We remind the reader that the field of expert knowledge elicitation, representation, and use is extensive, multidisciplinary, and constantly developing and changing. Throughout the chapter, we discuss some of the newer work in this area, including techniques for teaching experts to perform self-elicitation, knowledge reuse, conceptualization of knowledge, and knowledge management. Although this chapter attempts to cover many of the critical issues in the elicitation, representation, quantification, and aggregation of expert judgment, the reader is encouraged to treat it as a starting point for developing an expert elicitation approach to reliability problems. Suggestions for further reading are provided throughout.

13.2 Definitions and Processes in Knowledge Acquisition

13.2.1 Expert Knowledge Approach: Philosophy, Definitions, Roles, and Stages

13.2.1.1 Philosophy of Local Knowledge

As humans interact with the world in our daily lives, we must make sense of an unrelenting deluge of sensory input: tastes, smells, feelings, written information, the spoken word, nonverbal cues, events, and experiences. Humans are very adept at acquiring, structuring, and applying very large amounts of heterogeneous information in relation to complex and emerging problem situations [6]. Experts in cognitive psychology and artificial intelligence have posited that the human brain processes and applies this deluge of information using mental models, or representational abstractions, that simplify reality into a set of phenomenological categories. In the context of a design team or a research project, these categories enable people to organize and structure emergent knowledge and shared cognitive activities so that they can make appropriate judgments and decisions when approaching complex problems.

Documenting these conceptual structures, and using them to provide guidance in quantitative model development and in the elicitation of expert judgment, constitute the core activities in an expert knowledge approach. Even when expert judgment elicitation is not an explicit goal in the project, cognitive models can still provide the analyst with a great deal of information about structuring and modeling a complex decision area. However, despite the fact that conceptual models play an important role in enabling the integration of communally held knowledge, most communities do not bother to represent these models explicitly. Instead, they tend to take their problem-solving process for granted, as something that newcomers just "learn" through working with more experienced experts. Moreover, cognitive models are locally contingent; that is, the form they take depends on the social context in which activity takes place and the objects around which the community is focused. Hence, these models are constantly being redefined as the community's problem-solving process evolves.

The core problem of knowledge elicitation, then, is to develop explicit representations that capture the tacit, shared conceptual models that structure a community's approach to a problem at a particular point in time. Because this process involves transforming the tacit into something explicit, these representations can themselves impact the problem-solving process, insofar as they establish a focal point around which individuals can discuss and debate the problem. Moreover, the process of building these representations and eliciting formal expressions of expert judgment is necessarily collaborative. The analyst often becomes a full member of the research team, and when this occurs, the model she or he develops can impact how community members engage with each other, with their data, and with new problems.

13.2.1.2 Knowledge Concepts

In this chapter we refer to the following important knowledge concepts and definitions:

- *Domains* are bounded areas of human cognitive activity that are constituted by a mixture of real-world referents (i.e., an object such as a missile motor or a fuel injection system) and the experts' cognitive structures (i.e., the conceptual models that an expert uses when engaging in the problem of designing a fuel injection system) [7].

- The social locus of the domain is the *community of practice:* a group of people who organize their problem-solving activity in pursuit of a shared goal or interest. Communities of practice can be synonymous with an institution, although they can also be located across one or more institutions—for example, a long-established research team composed of contractors from two or more corporations [8–11].
- *Problems* are epistemic challenges through which communities of practice identify limits to domain knowledge. In the realm of reliability, these problems are technical in nature, relating to the performance of a product or system, or its domain. They are critical for the survival of the community as well as the domain, insofar as they provide points of focus for shared human activity. As communities of practice resolve domain-related problems, they extend the skills, understandings, tools, and products that define the domain—and often identify new problems in the process. That is, domains are living entities whose boundaries and focus areas are constantly redefined as individuals within communities of practice engage in the activity of defining and solving problems [8].
- *Example:* Aeronautical engineering is a large knowledge domain instantiated in multiple communities of practice, including government (e.g., FAA, Navy, Air Force), academic (e.g., aerospace engineering departments), and industrial research and production facilities (e.g., engine companies). Individuals within and across these communities of practice identify problems in aeronautical engineering, such as high-cycle fatigue (HCF) in turbine engines. The epistemic challenge arises from the lack of first principles knowledge about HCF, insofar as HCF is not simply an extrapolation of the better known low-cycle fatigue. The process of studying high-cycle fatigue will generate new tools for handling the epistemic uncertainties, techniques for comparing computational models to experiments, and understandings for the engineers involved, thereby extending the domain of aeronautical engineering into areas such as computation, statistics, information theory, materials science, structural dynamics, and expert elicitation, to name a few.

13.2.1.3 Roles

An expert knowledge problem exists within a community of practice, so it is critical to understand how experts "operationalize" key domain concepts in solving problems. This information is then used to structure the statistical analysis or engineering modeling. Collaborators will take on the following important roles:

- *Analyst.* An individual or a team of individuals working to develop a model to support decision making. For readers of this chapter, this model will most likely estimate system reliability. In our experience, the team of analysts consists of a knowledge modeler and one or more statisticians. The knowledge modeler, usually a social scientist, elicits expert knowledge and expert judgment, creates appropriate representations, and develops a qualitative model framework for the problem. The statisticians perform the bulk of the quantitative work, including quantifying the model framework as well as data aggregation and analysis.
- *Adviser-expert.* The analysis team collaborates closely with one or two members of the community of practice. These "native" collaborators are referred to as *adviser-experts.* Such individuals should be well-established members of the community. They will act as guides and hosts, providing access to experts as well as data and information sources. This role is described in greater detail below [1].
- *Decision maker.* The adviser-expert may also be the *decision maker*: the individual who requires the output from the analysis to support a course of action, for example, allocating test resources, making design decisions, timing release of a product.
- *Experts.* Finally, the adviser-expert and the analyst will work closely with the *experts*: individuals within the community who own pieces or areas of the problem and who have substantial experience working in the domain. Depending on the domain, the analyst may consider requesting assistance from experts who are not directly involved in the problem, and even experts who are not immediate members of the community under study. For example, estimating nuclear reactor reliability is a process that often draws on experts from throughout the nuclear energy community, and not just those at a specific plant.

Expert Knowledge in Reliability Characterization

13.2.1.4 Stages in the Approach: An Overview

The decision to use an expert knowledge approach cannot be made on-the-fly or taken lightly. Because the elicitation process must be developed in conjunction with other elements of the analysis, the decision to elicit and use expert judgment in any analysis must be made early in the project timeline, well before the model structure has been chosen and developed. Experts must be courted and trained; questions developed and refined; representation, documentation, and storage techniques and tools developed; the model structure and the data vetted by the experts; in addition to any sensitivity and "what-if" analyses. If the process seems time consuming and labor intensive, it is—but it also opens the door to complex and exciting problems that might otherwise seem intractable.

Briefly, the stages in an expert knowledge problem approach include the following:

- *Problem identification.* Can this problem benefit from an expert knowledge approach?
- *Problem definition.* What is the community trying to achieve? How will the reliability analysis support those goals? How will the project be executed?
- *Model development.* What are the core concepts that structure this problem? How are they related to each other? What class of quantitative models best fits the problem as currently structured?
- *Expert judgment elicitation.* Is expert judgment a viable data source for this problem? What is the best way to elicit expert judgment in a given community?
- *Analysis of expert judgment.* What are the uncertainties associated with judgments? How can judgments from multiple experts be combined?
- *Information integration and analysis.* What additional data sources can be used to characterize this problem? What can the populated model tell us about the system?

Each stage in the approach calls for a particular set of methods. Generally speaking, problem identification, definition, and model development are exploratory efforts that rely at first on open-ended, relatively unstructured interview questions. As the model coheres, the methods become increasingly more structured and quantitative. We provide examples of questions and tools for each stage below.

13.2.2 Problem Identification: Deciding When to Use an Expert Knowledge Approach

The problem identification stage is typically characterized by survey conversations between the analyst, one or more decision makers who are interested in pursuing a collaboration, and perhaps one or more members of the engineering design team.

Example: The analyst meets with a section leader in a large automotive parts corporation. The section leader is concerned about high warranty costs the company has recently incurred on some of its products and wants to develop a way to reduce those costs.

13.2.2.1 Criteria for an Expert Elicitation Approach

Not all problems call for a full-scale expert knowledge approach, which requires a great deal of time and forethought. Indeed, some reliability problems require minimal elicitation of expert knowledge for defining the problem and corresponding structure. If the product to be analyzed has a straightforward reliability structure for its parts and processes, and there is a great deal of available and easily accessible data, then a formal elicitation for expert judgments may be unnecessary.

Example: The reliability of a well-established product, such as an automobile fuel injector, quite naturally lends itself to standard model and representation techniques. An experienced fuel injector design team may be able to construct a reliability block diagram for their system in a relatively short time. The diagram can then be populated with test data and a classic model for reliability, such as the Weibull [12], can propagate part and process reliabilities in each block to get an estimate for the entire system.

However, neatly defined problems with easily available data are increasingly rare as multidisciplinary project teams become the norm in science and engineering. Moreover, we have seen an increasing demand for predictive analyses to support decision making in experimental and design engineering projects. Predictive analyses are those that estimate system reliability before designs are finalized and tests are performed. Such analyses are particularly useful in design projects where the data is sparse, difficult to access, only indirectly related to the system being designed, or nonexistent. For problems like these, a standard model such as a reliability block diagram might be an inadequate representation of the problem space. Even trickier are complex technical problems where one subsection of the problem is cleanly defined and well characterized, but the larger problem itself is less neat.

Complex problems tend to have one or more of the following characteristics:

- A poorly defined or understood system or process, such as high cycle fatigue effects on a turbine engine
- A process characterized by multiple exogenous factors whose contributions are not fully understood, such as properties of exotic materials
- Any engineered system in the very early stages of design, such as a new concept design for a fuel cell
- Any system, process, or problem that involves experts from different disciplinary backgrounds, who work in different geographical locations, and whose problem-solving tools vary widely, such as the reliability of a manned mission to Mars
- Any problem that brings together new groups of experts in novel configurations for its solution, such as detection of biological agents in war

Any time that multiple experts are involved, their conceptualizations of the problems can differ widely, depending on the tools they possess, their training, and the community of practice to which they belong. A unified model of an emergent problem, then, can be a difficult thing to develop. Moreover, complex problem domains often have tight timelines—for example, when issues of public or consumer safety are involved.

Even a well-understood fuel injector can carry a high degree of complexity. Although the system's parts may be well defined, its operation relies on a series of less clearly defined processes that impact the injector's functioning and hence its reliability. The assembly process and associated quality control and inspection regimes can impact the system's performance. Moreover, as the system is operating, physical processes such as forces, chemical reactions, thermodynamics, and stresses will affect its reliability.

In the past, processes have often been ignored in estimating reliability because they are difficult to specify and incorporate into the problem structure, much less populate with data. In cases where processes contribute significantly to system reliability, an expert elicitation methodology can specify important processes and provide guidance for locating them in the problem structure. These areas can then be populated with expert judgment and other available data sources. Chapter 30 by Griffiths and Tschopp identifies a likely situation where the methods presented in Chapter 13 for eliciting expert judgment can help engineers construct probability distributions for engine excitation variation.

13.2.3 Problem Definition

Once a problem has been identified as one that might benefit from an expert knowledge approach, the next stage is problem definition. During this stage, the analyst works with the community to identify an analytical approach that will support the community's efforts in a particular problem area.

Example: A group of decision makers in the Department of Defense have identified an experimental missile development program as one that could benefit from a predictive reliability analysis. Before the analysis can proceed, the analyst must make contacts with the design team so that he or she can understand how to align the statistical analysis with the design team's goals.

13.2.3.1 Identifying the Adviser-Expert

Developing an expert knowledge approach requires a productive rapport with one or more subject matter experts who can serve as hosts, advisers, and points of entree into the community for the analyst. Therefore, the first and perhaps most important stage in defining the problem is the identification of an *adviser-expert*: a domain native who can act as a guide to the problem as well as the experts, the sources of information, and data related to the issues under study. Sometimes, the decision maker also plays the adviser-expert role, particularly in smaller corporate or industrial settings. In larger bureaucracies or corporations, the decision maker advocating the analysis may not be a member of the design team. In such cases, the decision maker must suggest one or more candidates within the design team to act as adviser-experts.

There are qualifications for a good adviser-expert, the most important of which is a strong motivation to support the analyst in the elicitation project. A good adviser-expert understands the relationship between the analysis and the problem-solving goals of the community of practice. She or he values the analyst's contributions and is willing to devote time and effort to answering the analyst's questions, assisting the analyst in tool development, locating data sources, reviewing model structure and question formation, identifying reputable experts, and cultivating the community's support for the project. Low motivation on the part of the adviser-expert can make the elicitation process difficult, even impossible.

Experience is another important qualification for the adviser-expert. Because she or he will serve as the analyst's guide to the domain, the adviser-expert must have broad, but not necessarily deep, familiarity with the problem at hand. It goes without saying that a novice will probably not make a good adviser-expert. Rather, the adviser-expert should have a "journeyman's" level of experience within the domain [13]. Ideally, an adviser-expert has been a member of the community for roughly a decade, has direct experience working with problems of the type under study, is well acquainted with the experts assigned to the problem, understands how tasks are distributed among different experts (the social organization of activity), and is respected by other members of the community.

13.2.3.2 Defining a Collaborative Strategy

The first task for the analyst and the adviser-expert is to define the nature and the boundaries of the problem to be addressed. It is important at this stage to differentiate between the goals of the *community* and the *decision maker*, on one hand, and the goals of the *analysis*. For example, a group of engineers working on a new design for a missile might list goals related to the system's reliability and performance: ease of ignition, accuracy of guidance systems, burn time, fuel efficiency, shelf life. The goal of the analysis might be development of a predictive model that enables the engineers to draw on past design experiences to support decision making and resource allocation during the process of designing the new missile. The analyst's goals should be shaped by the community's objectives, so that the two are closely coupled.

Below we offer some questions that we typically use in the problem definition and success/failure stages of elicitation. In Figure 13.1, note that these questions are open-ended interview questions that

PROBLEM DEFINITION QUESTIONS
- What kind of system/process is being developed? How is this similar to past projects? What is this system/process designed to do?
- Who are the customers for this project?
- Who will judge this project's outcomes? What will they use to judge those outcomes?
- What are the motivations in the community for a successful problem outcome (e.g., cash bonus, patent opportunity)?
- Where is this project being conducted? (Single or multiple locations) How many different groups of experts are involved in this project?
- What is the timeline for completing this problem? (Include all milestones.) What constraints does this group face in working this problem (time, financial, human)?
- What is the complexity of the problem space? What parameters must be directly estimated/assessed by the experts? Do the experts have experience with these parameters?
- Do data sources exist? How accessible are existing data sources?
- How much time will people in the organization be willing to devote to this project?

FIGURE 13.1 Sample problem definition questions.

allow the adviser-expert to direct the conversation. At this stage, it is important for the analyst to learn as much as possible from the adviser-expert without making assumptions that could limit the analyst's problem and domain understanding.

Problem definition is not always a straightforward task. If the problem is newly identified or resists definition (we have found this to be the case for many nuclear weapons problems), it is possible that none of the experts will be able to articulate more than a few high-level goals. In this case, it is also likely that the problem definition will change as the project's outlines become clearer. For this reason, we strongly recommend the analyst and the adviser-expert carefully document their problem definition discussions and revisit the topic frequently to assess how changes in the community's problem might impact the goals of the analysis.

13.2.4 Eliciting Expertise for Model Development

The elicitation of expertise and the development of a knowledge model marks the beginning of intensive problem-focused work on the part of the analyst. During this phase, the analyst will focus on developing a qualitative map of the problem area to guide inquiry in subsequent stages of the analysis.

Example: Two design teams in a large aeronautical engineering company are collaborating with a team consisting of one social scientist and two statisticians to develop an expert knowledge approach for making design modifications in their jet engines. Although the engineers share a great deal of information, neither design team is overly familiar with the others' work processes because one team has a parts-oriented design focus, while the other has a manufacturing-oriented design focus. However, the statisticians require a unified representation of the problem, as well as some understanding of the processes the teams use to generate and interpret data. It is also a beneficial learning experience for the two teams with different orientations to learn how the other team views the problem.

13.2.4.1 What Is a Knowledge Model?

A *knowledge model* is a type of knowledge representation designed to capture the key concepts and relationships that make up the community's conceptualization of the problem. As defined by Davis, Schrobe, and Slzolovits [14], a *knowledge representation* is a formally specified medium of human expression that uses symbols to represent a real-world domain. A knowledge representation captures all the categories within a domain, as well as the relationships that define their behavior and interactions, and expresses these in a computationally efficient format. In other words, a representation is a mediating tool that bridges the gap between the as-experienced, real-world domain and a computed version of that domain. In computer science, formally specified knowledge representations play a critical role in the development of databases, query languages, and other common applications. Knowledge representations can also be very useful in developing and specifying statistical models, particularly when the problem under study is very complicated.

Developing a knowledge model is not the same enterprise as developing a statistical model for quantitative analysis, although it provides a foundation for later statistical work. Instead, the knowledge model provides a way to ensure that the primary concepts that comprise the problem space are mapped and documented in a way that all parties involved in the project can come to agreement on what the primary focus areas are, how best to address them, what analytical techniques will be most appropriate, and where to collect data and information [15]. Indeed, developing the knowledge model can be seen as a further iteration of problem definition, albeit a much more detailed one.

The purpose of developing a knowledge model for problem definition is to find out how the community of practice works for solving a problem such as the design of a system or product. The graphical representation should represent the community's knowledge about areas critical to development of a formal model, including the following:

- What level of granularity (detail, resolution) is appropriate for the model?
- Who is responsible for what areas of work?

- Where do data sources exist?
- How do parts in the system relate to produce functions?
- What is the design process?
- How does information flow through the organization?

Developing a knowledge model requires the elicitation of *expertise*. Expertise consists of the information, practices, and background experience that qualified individuals carry to new problems—the implicit ways experts think, make decisions, and solve problems within their community of practice. Expertise is context dependent. It is the distinctive manner and ways that members of a technical community act, make things, perform tasks, and interpret their experience. In other words, expertise is the form of explicit and implicit knowledge that enables expert communities to combine data and information during their problem-solving process. Properly documented, expertise can provide valuable input into model development.

The elicitation and formal representation of expertise has been a key element of software engineering and development since the 1950s. The field of *knowledge acquisition* (KA) emerged as researchers in artificial intelligence and expert systems attempted to elicit and capture experts' problem-solving processes and model them in software applications, such as expert systems. As researchers in these fields quickly realized, the task of eliciting expert knowledge was far more difficult than originally anticipated: experts have a difficult time consciously articulating skills and thought processes that have become automatic. Moreover, many of the bias issues that plague the elicitation of expert judgment can also affect the elicitation of expertise (see discussion below).

Because of these challenges, experts in the field of knowledge acquisition have developed an extensive portfolio of approaches to the elicitation of expert knowledge. In this chapter we provide a general outline of the process below with an emphasis on open-ended and semistructured interview techniques, since this approach can be used with success in most settings. There is an extensive suite of elicitation tools and representational techniques available for capturing expert knowledge. This suite includes methods from fields as diverse as knowledge acquisition and artificial intelligence, statistics, risk analysis, decision analysis, human communication, cognitive science, anthropology and sociology, and other disciplines that study human learning and knowing. Ultimately, the selection of methods is the decision of the analyst, who can draw on many well-established knowledge elicitation techniques (for comprehensive reviews of elicitation techniques, see especially [1, 13, 16]).

13.2.4.2 Process for Eliciting Expertise

The elicitation and representation of expertise is a qualitative, iterative process involving background research on the part of the analyst, and the development of open-ended and semistructured interview schedules. One outcome of the elicitation process is a set of graphical representations that capture the important elements of the knowledge domain (see below). However, the real benefit of the knowledge elicitation process is the opportunity it provides for the analyst and the consulted experts to arrive at an explicit and common understanding of the problem structure and key elements. Briefly, the stages for eliciting expert knowledge include the following:

- Identify key areas and points of contact within the project.
- Gather background materials for project areas.
- Conduct open-ended interviews with experts.
- Develop graphical representations of experts' domain.

As with all elicitation, the analyst should rely on the adviser-expert's assistance in selecting and arranging interviews with experts on the project. A useful tool is to begin with a simple graphical map, similar to an organization chart, that delineates which engineering teams are responsible for what areas of the project. (Note that organizational charts often do not capture project-specific information.) Within each sub-field, the analyst and the adviser-expert should work together to identify points of contact who will provide information and assist in arranging meetings and interviews with specific engineering

> **SAMPLE QUESTIONS FOR DEVELOPING A KNOWLEDGE MODEL:**
> **A Missile System Example**
> - If you had to divide this missile into major subsections or components, what would those be?
> - Provide a list of clear, measurable goals for missile performance. What are the relative states of success and failure for this system?
> - Who are the major groups working on this missile? Are these groups congruous with the missile subsections? Or are they organized differently?
> - What information is required to begin the design process in each group?
> - Where does that information come from? To whom does it flow?
> - How do individual engineers use background information in the design process?
> - How is the system assembled? Where? In what order?
> - At what level does testing take place (material, component, subsystem)? What kind of data do different tests generate?
> - How do you convince yourselves that a part, subsystem, material will work?
> - How do you generate appropriate levels of physical processes (temperature, vibration, shock) for your test process?

FIGURE 13.2 Sample questions for a knowledge model.

experts. Before conducting interviews with the experts, the analyst works with the adviser-expert to learn the basic elements of the problem at hand. Background study is important because, as with all human knowledge, expertise is context dependent and its elicitation requires some familiarity with the community of practice. The adviser-expert can provide the analyst with written material about the project, including statements of work, engineering drawings, project management diagrams, organizational charts, material from presentations, internal reports, journal articles, and other written descriptions of the problem, which would include local jargon and terminology. The analyst should study this material to familiarize herself or himself with the domain before developing questions and interviewing the experts. The adviser-expert's role during this time is to support the analyst's learning process and to assist in the development of appropriately worded questions for the experts. Figure 13.2 provides a set of sample questions used in defining the knowledge domain for a missile system.

Once the analyst and adviser-expert have developed a schedule, identified points of contact, and practiced conducting interviews for eliciting the problem structure from the experts, the elicitation process of problem definition can begin. The expert-advisor's role as a conduit into the experts' community comes into play at this time, providing motivation for the experts to participate in the elicitation.

It is particularly important at this stage to define the appropriate level of detail or *granularity* for the problem, which must be related to the goals that are defined for the analysis. For example, a decision maker may only need a rough comparison of a new design to the old to answer the question, "Is the new one at least as good as the old one?" For a problem like this, it may not be necessary to structure the two systems down to the smallest level of physical quantities to make this determination. Also, the extent of information availability (including data and experts) can dictate how levels of detail are identified and chosen for inclusion in the model. If knowledge is completely lacking at a sub-component level, the problem may be defined and structured at the component level. Different levels of focus are possible within a problem; however, very complex structures pose information integration difficulties when combining all levels for the full system analysis.

During the initial phases of knowledge elicitation, the answers provided by the experts may generate further questions for the analyst. Documentation can be done in the form of simple star-graphs, which can leverage the emergence of more formal knowledge representations (see below).

13.2.4.3 Knowledge Representation Techniques for Modeling Expertise

Although computer scientists use formal logic to develop knowledge representations, simple node-and-arc models provide a tractable way to represent expert knowledge. Because most people find it easy to understand nodes, arcs, and arrows, graphical representations of complex relationships can be used to

Expert Knowledge in Reliability Characterization

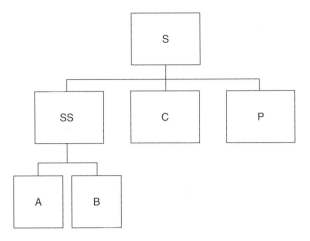

FIGURE 13.3 Reliability block diagram of a simple system.

mediate the negotiation of model structure across communities of experts. Indeed, engineers often communicate with each other using graphical representations and models of their systems, and graphical models are quite common in the field of reliability analysis.

Example: A reliability block diagram uses boxes and connecting lines to represent the intricate relationships in a complex mechanical system. In Figure 13.3, the system, S, is composed of a series of three items: subsystem, SS, component, C, and manufacturing process, P. Subsystem SS is, in turn, a series combination of two other components, A and B.

Some problems can be easily represented in a reliability block diagram. However, others are more complex and require representations that include more information than just parts. The elicitation of knowledge for complex problems should aim at identifying the key concepts and relationships in the knowledge domain and understanding the relationships that tie these concepts together. Even the most complex engineering problems share a core set of concepts and relationships. Some of these concepts will be very easy for the engineers to specify: *component, subsystem, system* may be readily available in design drawings or from tests or vendor documentation. Entities like these, which are explicitly labeled in the community's vocabulary, can be far easier to elicit than entities that are more tacit. For example, *functions* describe what a part or set of parts do as the system is working, but engineering communities rarely specify functions on engineering diagrams and may even lack a formal vocabulary for referring to them. *Processes* are more likely part of the manufacturing community, rather than the design community's everyday language. Processes include the operations required to assemble a system and monitor its quality through inspection or quality control. However, understanding what function a set of parts performs can be critical in developing a system-wide reliability diagram. Similarly, failure modes can be difficult to retrieve from memory.

Simple line graphs are very useful in the initial "brain dump" stages of the elicitation process, as the analyst is working with the adviser-expert to define the problem domain and catalog all the concepts of interest. For example, a *star graph* has a central node to represent the domain and an unlimited number of arcs that radiate outward [17]. Each arc is connected to one of the domain's concepts. Elicitation at this stage is very open-ended: the analyst provides the expert with the concept under study and simply asks the expert to describe it. For example, if the domain under study is high cycle fatigue, a basic star graph that describes the domain might look like Figure 13.4.

Initially, none of the outer nodes are grouped into higher-order categories, nor do additional edges display connections among the outer nodes. Further specifications of the star graph, however, will bring the domain structure into clearer focus. For example, in Figure 13.5, the nodes labeled *blade disk* and *airfoil* might be grouped into a larger category called "parts," while *rotational load, bending, torsion,* and *vibration* can be grouped into a category "stress."

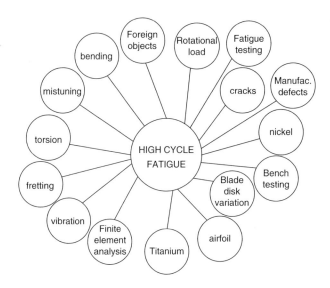

FIGURE 13.4 Star graph for high cycle fatigue.

As the graphical representation is further developed in Figure 13.6, a relationship between the categories "Parts" and "Stress" in the domain "High Cycle Fatigue" can be formed.

A categorical relationship like that specified above can be developed for all areas of the knowledge domain, and will form the structure for the knowledge model. This model, also known as an *ontology* in the field of knowledge representation, displays the basic categories in a domain and the relationships among them. Figure 13.7 is an example of an ontology for a missile system. Note that the lowest level of granularity is "Parts," which are ultimately related to "Events" through "Functions."

As the analyst works through the problem area, the ontology is used to structure later queries that specify instances of each category. For example, the analyst can use the categories to develop a list of all parts, the types of stresses to which they are vulnerable, and potential failure states related to those stresses. Other categories that emerge during the elicitation period, such as "Test" and "Data," can also

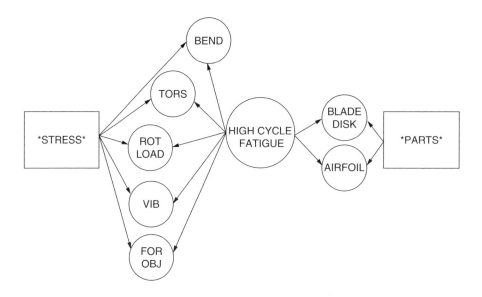

FIGURE 13.5 Grouped representation.

FIGURE 13.6 Fundamental relationship representation.

be represented formally and used to structure further inquiry. The analyst can use these categories to develop a better understanding of how different test processes and data sources are used within the community to generate confidence that parts will not fail under different kinds of stress.

13.2.4.4 Success and Failure for the Problem Area

As part of the knowledge model, it is important to elicit a clear set of definitions for performance—"success" and "failure" of the design team's problem or project, and to have these definitions expressed in terms of clear, discrete, measurable goals. This information, in turn, will guide the analyst in developing an appropriate suite of methods for the consultation and may provide guidance as to what the analysis will be predicting. One tool that we have found very helpful in doing so is a simple *stoplight diagram*: a horizontal chart divided into red, yellow, and green areas that acts as a heuristic for eliciting states of success and failure. The image of a stoplight is commonly used in project management tools and will be familiar to many engineers.

To specify the contents of a stoplight chart, the analyst develops a blank chart and uses it as an elicitation tool with the adviser-expert and the decision maker to identify specific, measurable goals for the project. The event states can be described as catastrophic failure (red), imperfect performance with salvageable outcomes (yellow), or success (green), as in Figure 13.8.

For example, in the missile development ontology, it is possible to identify all the events that the missile designers want their system to perform. The designers can identify red, yellow, and green states elicited for each of these events, which can then be mapped onto the chart. Later, the adviser-expert can

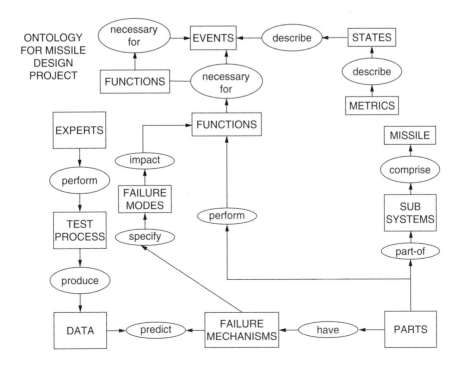

FIGURE 13.7 Ontology for missile design.

FIGURE 13.8 Stoplight chart.

identify members of the community from whom the analyst can elicit various combinations of failures or functions in the system that would lead to the events and each of the possible states for those events, as described on the stoplight chart.

13.3 Model Population and Expert Judgment

13.3.1 Expert Judgment Elicitation

Eliciting problem structure from experts is usually only the first part of the elicitation process. Extracting the specific knowledge that experts have about the various aspects of the structured problem involves elicitation of expert judgment. We refer to knowledge gathered for this type of elicitation as *populating the structured problem* or *model population*. Although formal quantitative expressions are often the goal of expert judgment elicitation, the elicitation process itself is really a *qualitative* and *inductive* one, insofar as elicitation requires the analyst to understand how experts structure the problem, how they weigh and combine different sources of information (for example, computer models vs. experimental data) in making decisions, and how they conceptualize uncertainty. Ideally, a great deal of this knowledge emerges from the KA process described above, so that the analyst can develop questions and instruments that maximize the experts problem-solving abilities.

For studies that use expert judgment to be accepted as valid, they must be conducted in a formal and rigorous manner, with clear documentation throughout the project. In the remainder of this chapter, we describe the methods we have used for designing, performing, and documenting expert elicitation.

13.3.1.1 What Is Expert Judgment?

In this chapter, we treat expert judgment (also referred to as *expert opinion*) as a very specific form of expert knowledge. It is qualitative or quantitative information that reflects an individual's experience, understanding, and assessment of particular technical issues. It is perhaps best thought of as a snapshot of the expert's knowledge about a particular problem at a particular time [18, 19]. And as the term implies, expert judgment refers to the subjective assessments of individuals who are recognized by their peers as having a great deal of experience and fluency with the phenomenon under study.

Expert judgment can be viewed as a subjective probability—a quantitative statement that reflects an individual's degree of belief in the likelihood of a future and uncertain event, based on the knowledge and experience that the individual holds about similar past events. Because this opinion is subjective, and rooted in the individual's experience within the community of practice, it is critical to elicit information in a way that facilitates the experts' reasoning process.

Example: Three experts are selected to provide their opinions about the reliability of a new circuit design. The adviser-expert and the analyst work together to develop a question format that makes sense to the individuals being consulted. The analyst has developed a set of questions that asks about failures over the course of a year. The adviser-expert explains that engineers in this company commonly think of failures per million hours of operation, information that the analyst uses to revise his or her questions.

We recognize the utility and advocate the use of expert judgment in research problems where data sources are scarce or difficult to access. It can provide a valuable source of information for the analyst. For some classes of events—never observed, one time, or very rare events, and phenomena that do not lend themselves to convenient study—expert predictions may be the only source of data available.

Indeed, expert opinion shares many characteristics with data: it can be quantified, its quality is affected by the process used to gather it, and it has uncertainty that can be characterized and subsequently analyzed. Moreover, expert opinion is always part of any analysis, whether in the form of judgments made about appropriate data sources, model structure, or analytical techniques. Rarely are these tacit choices fully documented (or even acknowledged); instead, they are treated as a "natural" part of the research process. For example, model choice is a realization or form of expert judgment, as is the process of populating the model with quantified expressions of expert estimates from judgments.

It is perhaps ironic, then, that studies that make explicit use of expert opinion as a data source are often criticized for being too "soft." Specifically, expert elicitation projects have lacked formality, rigor, and documentation [19], while analysts have failed to carefully consider and document their choice of "experts" for elicitation on a particular subject area [20]. The strongest critique, however, comes from "human fallacy" research that details the heuristics that humans use in making judgment. *Heuristics* are informal mental rules for assessing information and making predictions. These heuristics give rise to predictable biases, which in turn lead to poor coherency and calibration of expert opinion [21].

Such difficulties are not insurmountable. In the following section, we present our philosophy of expert judgment, as well as a set of general principles and tenets for eliciting, documenting, and refining expert opinion. The basic principles of our approach to elicitation can be summarized with the following key words, each described in this section:

- Bias minimization
- Reliance on advisor-expert
- Pilot test
- Verbal protocol
- Verbal probe
- Decomposition
- Conditional information
- Feedback
- Document

These principles are compared to the Delphi technique for eliciting knowledge from experts in Section 13.3.2.

13.3.1.2 What to Elicit?

One of the most significant questions to consider in designing the elicitation instrument is the design of the model. As a general rule of thumb, expert judgment must be elicited so that the expert assessments fit with the model structure.

Example: Consider a model structure in which a Bayesian hierarchical structure of conditional probabilities is used to assess the performance of a simple mechanical system. When populating such a model with expert judgment, the analyst is interested in knowing the probability that any *child* variable (A) will be in a given range of states $(a_1, a_2, \ldots, a_n)$, given the possible range of states for each and all of its *parent* variables (B and C). It makes little sense to elicit probabilities for the states of variable A without taking into consideration the entire range of states for variables B and C, because the model structure specified that the state of A is *conditioned upon* the states of B and C.

As noted in the example above, dependencies among aspects of the model or the structure must be anticipated and included as part of the elicitation process. Dependencies expressed by conditional probabilities can be elicited using a *causal structure:* If event B occurs, then event A will occur with probability p (expressed as $P(a|b) = p$, where a and b are the specific values of A and B, respectively). Conditional probabilities can also be elicited using a *diagnostic structure:* Given that event A has occurred, then event B has occurred with probability p (expressed as $P(b|a) = p$). Bayesian networks [22] are complex structures of such *parent/child* hierarchies that are connected through Bayes theorem.

13.3.1.3 Identifying the Experts

An *expert* can be defined as someone recognized by his peers as having training and experience in his technical field. In our experience, the selection of experts depends on the type of problem facing the analyst, the novelty of the project, and the social organization of tasks in the research setting. The pool of experts may be limited to members of the immediate research or engineering team; for example, in projects that involve a great deal of industrial proprietary or classified information. In cases like these, the analyst works exclusively with all the team members to elicit their expert knowledge.

However, it is often possible to work within a corporation, laboratory, or other research organization to gain information about a system under development from individuals who are not themselves members of the research team, but who may be familiar with one or more of its goals. For example, some corporations have engineers whose job it is to assess the reliability of all electronic parts across programs, and who may be able to provide some prior estimates of the reliability of prototype components being developed within their organization. In some cases, it may even be possible to draw on experts from outside the organization to provide an opinion that can serve as a prior for the model, although again this is not advisable in situations where proprietary information is involved.

All this, however, begs the question of expertise: who is an expert, and what is the best way to assess expertise? Unfortunately, there is no formal test to determine expertise, which, in any case, is problem specific—and therefore not amenable to general assessment tools. One simple way of defining expertise is to assess educational and professional experience: the greater the educational attainment and level of training, the higher the level of expertise. However, some individuals may work for years in a particular field and never be considered true experts within their peer group.

More useful are the definitions given in Hoffman et al. [13]. They draw on the terminology of medieval European guilds to develop a typology of expertise, ranging from *naivette*—one who is totally ignorant of a domain's existence—to master, which they define as an expert whose judgments "set regulations, standards or ideals" for the rest of the community. Between these two extremes lie several intermediate stages where most individuals will reside: initiate, apprentice, journeyman, and expert. The last two—journeyman and expert—are excellent candidates for elicitation because they are members of the community who have achieved an acceptable level of competence so that their peers are comfortable allowing them to work without supervision. More specifically, an expert is one who is called upon to address "tough" cases [13].

Ayyub [23] offers another useful way to think about expertise; an expert, he argues, is an individual who is conscious of ignorance, one who realizes areas of deficiency in her or his own knowledge or that of the community, and strives to amend those gaps. On the other hand, naivettes, initiates, and apprentices tend to be unfamiliar with the knowledge domain and cannot therefore recognize areas of epistemological deficiency. They often solve problems from first principles rather than draw creatively on experience, simply because they lack the domain familiarity of a journeyman, expert, or master.

It is also acceptable to rely on the peer group's opinion. One excellent method is to interview project participants to discover which members of the organization are most frequently tapped for assistance in solving difficult technical problems. Oftentimes, novice or journeyman individuals are excellent sources of information about expertise within an organization because they are often paired with more senior people in mentoring relationships and can identify which individuals are most helpful with difficult problems.

It is also possible to use the ontology to break the problem itself down into a set of subject areas. For example, a missile system might be broken into power subsystems, pneumatic subsystems, software, and other areas. After identifying subject matter experts, the analyst and the adviser-expert can use Hoffman et al.'s typology or a similar tool to identify journeyman-, expert-, and master-level individuals within each topical area. Another alternative is to work with the adviser-expert to determine which individuals are known for making progress on particularly challenging or novel problems—this being a marker of "conscious ignorance" and hence expertise. We have found that it is less helpful to ask individuals to rate their *own* level of expertise, as they frequently tend to downplay or underrate themselves. However, this is another option for the analyst.

No matter what method the analyst chooses in selecting experts for expertise and judgment elicitation, we emphasize again the importance of documenting the rationale and method for doing so. Because expert knowledge projects are often loosely documented, informal, and implicit, they are often criticized for being "soft" or "biased." Careful choice and documentation of the project's methods, rationale, and progress is important throughout, but may become extremely so once the analysis is complete.

13.3.1.4 Single or Multiple Experts?

One of the first considerations in any project is whether to elicit information from single or multiple experts. Statistical sampling theory implies that a greater amount of information can lead to a better understanding of the problem, with corresponding reductions in uncertainty. We are not suggesting that a certain number of experts can represent a statistically valid sample for making inferences about the unknown underlying population. However, logic says that consulting more experts will provide a more diverse perspective on the current state of knowledge.

There are other good reasons to consult multiple experts. While this approach can pose problems of *motivational* bias, it does seem to have some advantage in minimizing *cognitive* bias (see discussion in next section). O'Leary [24] found evidence that judgments acquired from groups of experts are more likely to result in expressions that are consistent with Bayes' theorem [12], and with the axioms of probability, than are judgments elicited from single individuals. One explanation for this phenomenon is that groups of individuals have a wider range of cognitive tools and representations for structuring the problem, enabling individuals in the group to reason more consistently and coherently.

Of course, the answer to the single-vs.-multiple experts may be context dependent. In projects where one or two engineers are the only individuals familiar with the problem or system under discussion, the decision is already made. In others, there may be a great many experts who could proffer opinions, but the project itself may limit participation. For example, in projects that involve classified or proprietary data, there may be little possibility of eliciting information from more than a few experts. Other times, the research setting may offer the chance to gather formal opinions from several experts. In this case, one must decide whether to conduct the elicitation as a focus group, or to interview the experts individually.

Once the decision has been made to conduct elicitation with multiple experts, the next step is to decide if goal of the problem is to seek a *consensus* or a *diversity* of views. Group elicitations can be designed for either situation. However, the analyst must consider the possibility that the experts will have incommensurable views, as is often the case in newly emergent fields. In this case, the challenge of aggregating varying estimates from multiple experts then falls to the analyst. Other issues to consider in expert judgment elicitation are the levels of expertise that individuals bring to the problem. In addition, there may be some sources of inter-expert correlation, so that one must treat them as dependent rather than independent sources of knowledge. These and other issues involved with aggregation are discussed below.

13.3.1.5 Expert Cognition and the Problem of Bias

Perhaps the greatest issue with using expert judgment for predictive modeling is the question of *bias*. Simply put, bias can degrade the quality of elicited data and call into question the validity of using expert judgment as a source of data.

What is bias? Bias in scientific instrumentation means a deviation or drift from the normal or nominal operating conditions. In statistics, bias is understood as a parameter estimation that is off target from its expected value. Similarly, in cognitive science, bias refers to skewing from a reference point or standard. This definition implies the existence of "real" knowledge in the expert's brain, from which the expert deviates due to cognitive or ecological factors.

Since the mid-1950s, researchers in a variety of fields—including statistics, business, and cognitive psychology, among others—have studied human judgment in both experimental and naturalistic settings to assess how individuals make assessments of unknown events with minimal information. Almost without exception, this area of study has benchmarked the quality and utility of expert judgment by comparing human cognitive assessments to the quantitative predictions of probabilistic statistical models.

Set against this standard, early research clearly indicated that human cognition is not consistently logical; instead, it is vulnerable to sources of bias and error that can severely compromise its utility in probabilistic statements. Comparing the accuracy of human clinical assessments to statistical models demonstrated the predictive superiority of the latter, as well as the cognitive inconsistencies of the former (for discussion, see [25]).

Human fallacy research (see discussion in [24]) has shown fairly conclusively that human beings do not reason in accordance with the rules that ground statistical models. On one hand, predictive models must follow the rules of probability theory; however, human cognitive processes for making predictions under uncertainty often violate these rules. Individuals tend not to adequately revise their opinions when provided with new information, which constitutes a failure to reason in accordance with Bayes theorem. Nor do individuals rely instinctively on the axioms of probability. More recently, some researchers [25] have suggested that cognitive studies themselves introduce bias into the results of research on human cognition.

Biases can be broken down into two general categories, relating to their sources and origins. *Cognitive biases* relate to biases of thought processes and problem solving. *Motivational biases* relate to human behavior from circumstances and personal agendas. Both are discussed in detail.

13.3.1.5.1 Cognitive Bias

Some deficiencies in human judgment can be attributed to inherent inconsistencies in cognition. Such cognitive biases become particularly salient when human experts are asked to make judgments under conditions of uncertainty or with incomplete knowledge—a situation that is not uncommon in many decision environments.

In attempting to identify and classify sources of bias in human judgment, Tversky and Kahneman conducted a series of experiments in which they compared predictions made using probabilistic constraints to subjective judgments made by individuals about the same phenomena. In doing so, they identified several significant and consistent types of bias that seem to degrade the quality of human predictive reasoning. They concluded that "Whether deliberate or not, individuals [rely] on natural assessments [as opposed to probabilistic rules] to produce an estimation or a prediction" [26]. Human problem solving is intuitive, not logically extensional; hence, errors in judgment, they argued, were an unavoidable characteristic of the cognitive processes that humans use to integrate information and make predictive assessments.

In defining sources for bias, Tversky and Kahneman identified three primary intuitive heuristics, or cognitive rules of thumb: *representative, availability,* and *adjustment* and *anchoring*.

1. *Representativeness.* Representativeness refers to the tendency to ignore statistically defined probabilities in favor of experientially based, often narrow stereotypes to classify individual phenomena into categories. An expert evaluates the probability of an uncertain event using its most salient properties, leading to misclassification of the event.
    - *Example:* NASA engineers assumed that the Hubble satellite was reliable because it was manufactured by Perkins-Elmer, which had an excellent track record with other national security satellites. In reality, the Hubble faced scientific demands that other satellites did not, making it a far more complicated system—and less reliable [27]. Small samples are particularly vulnerable to representativeness-based biases.
2. *Availability.* Tversky and Kahneman used this term to describe the ease with which humans draw upon certain types of information—recent events, or strikingly memorable examples of some class of events—and over-incorporate them into cognitive assessments without taking into account the relative value of this information. Humans tend to overestimate the likelihood of rare events, particularly when instances of the class of event in the class are catastrophic, traumatic, and recent.
    - *Example:* An individual who has recently been in an automobile accident may overestimate the probability of being in another. Similarly, experts may assign higher likelihoods to the probability of catastrophic events, such as reactor failure, than actual rates of occurrence warrant.

3. *Anchoring.* Anchoring and its counterpart, adjustment, describe the tendency for individuals to undervalue new information when combining it with existing data to make judgments about the likelihood of a particular event. This occurs when experts have a difficult time moving their thinking from preconceptions or baseline experience. For example, an engineer who has never observed aerodynamic stress beyond a particular value may find it difficult to imagine a situation in which higher load levels might occur. Anchoring also appears as an artifact of the elicitation situation, when an expert draws automatically on assumptions or data used in a previous question to answer a new one, regardless of whether or not that data is truly applicable to the new question.
    - *Example:* When asking experts to estimate failure rates for different classes of high explosives, an expert may unconsciously anchor his or her estimate of failure rates for explosive B to the assumptions or data she or he used for explosive A.

In expert judgment situations, these additional heuristics can generate the following forms of bias:

1. *Conjunction fallacy.* Tversky and Kahneman are also credited with pointing out the existence of the *conjunction fallacy*. This term is used to describe situations in which individuals predicted that the co-occurrence of two unrelated phenomena (A and B) was more likely than the occurrence of either one alone — a prediction that directly contradicts a basic axiom of probability, which states the P(A and B) ≤ P(A).
    - *Example:* A component fails if it is under high pressure and under the stress from an external load. An expert estimates the probability of a component being under pressure and being under loading as 0.25, but estimated the probability of it being under pressure as 0.20.
2. *Conservativism/base rate neglect.* The discovery of the conjunction fallacy led to the identification of another heuristic, *conservativism*. This heuristic emerges when individuals are asked to use limited information to place a sample in the appropriate population. When provided with new information that should lead to the updating of one's beliefs (for example, statistical base rates), individuals tend to undervalue the new information in relation to the old, and therefore fail to revise their estimates appropriately. *Conservativism* is also referred to as *base-rate neglect,* and it directly affects the expert's ability to properly apply Bayes theorem — a probability theorem for updating new data in light of prior information.
    - *Example:* An expert would not alter his or her reliability estimate of new component, A, when given additional information, B, directly affecting A's performance. That is, P(A) = P(A|B), which implies the reliability of A (i.e., P(A)) is independent of B.
3. *Inconsistency.* This very common source of bias is both ecological and cognitive in nature. Confusion, fatigue, or memory problems usually lead to inconsistency. As we discuss below, there are some simple steps that the elicitor can take to minimize inconsistency.
    - *Example:* As an interview begins, the expert may make an assumption about initial conditions. However, after an hour of intensive questioning, the expert may forget or change this assumption, resulting in a different value for a temperature than the original assumption would warrant.
4. *Underestimation of uncertainty.* Humans typically believe we know more than we really do and know it with more precision and accuracy.
    - *Example:* A classic and deadly example of overconfidence was that the Titanic could not sink.

13.3.1.5.2 Motivational Bias

Some deficiencies in human judgment can be attributed to environmental factors. Examples of such *motivational* biases might include fatigue or a desire to appear confident in front of a senior expert. Briefly, some common motivational biases include:

1. *Groupthink.* Group or peer social pressure can slant responses or cause silent acquiescence to what is acceptable to that group.
    - *Example:* The classic and deadly example is President Kennedy's decision on the Bay of Pigs. All his advisors told him what they believed he wanted to hear, rather than providing him with their best judgments and estimates.

2. *Misinterpretation.* Misinterpretation is the inadequate translation of elicited knowledge into a response.
    - *Example:* An interviewer records the pressure for vessel burst as 2000 psi when the expert answered the question as an interval from 1500 to 2500 psi.
3. *Wishful thinking.* Experts' hopes influence their judgments and responses. The old adage, *wishing makes it so,* applies here.
    - *Example:* If an expert is defending his or her system design against competitors, she or he may hope and hence state that it will perform beyond requirements.
4. *Impression management.* Experts are motivated to respond according to politically correct interpretations, convincing themselves that this response is correct.
    - *Example:* A manager may verbally choose one design because it is perceived as more environmentally friendly, when she or he knows that this design has a larger cost-to-benefit ratio.

13.3.1.6 Bias Minimization

Like all data sources, the quality of expert judgment is affected by the process used to gather it. Analysts are trained to understand that measurement data must be calibrated to certain standards in order to eliminate instrumentation biases. Similarly, analysts can and should be meticulous in adopting bias minimization techniques to improve the quality of expert judgment. The quality of expert expressions is conditioned on various factors such as question phrasing, information considered by experts, and experts' methods of problem solving. In short, just as experimental design must be considered when attempting to record data, so too does the elicitation process require deliberate planning.

The cognitive and motivational biases affecting how expert knowledge is gathered are addressed by different bias minimization techniques other than physical *calibration*. Studies have shown [21] that for an expert to be calibrated like an instrument, feedback must be frequent, immediate, and relevant to the technical topic. Very few situations, such as in weather prediction, permit calibration. If sufficient data and information exist to calibrate experts, then their expertise is no longer necessary [1].

In minimizing the effects of cognitive bias, awareness is the first step. Anticipating which biases are likely to occur is the next step. Minimizing bias can also be accomplished by informing the experts of what biases are likely to be exhibited in the elicitation. The advisor-expert can help determine which these are and also help in making the other experts aware.

- *Example:* To mitigate impression management:
    - Assess how likely this bias is given the situation (e.g., will experts benefit or suffer respectively from giving an estimate that sounds good/bad).
    - As a general rule, do not allow experts' bosses during the interview because bosses might pressure the expert to adhere to the party line.
    - Consider making judgments nonattributable to a person, maintaining anonymity.
    - Require substantial explanation of reasoning behind the expert's answer, making it more difficult for them to give the party line.

The analyst should constantly monitor for the occurrence of bias. Adjustments can be made during the elicitation if a particular bias is encountered.

Example: If inconsistencies are beginning to appear in a long interview, then fatigue may be the cause. A break can be taken or another session scheduled to mitigate this bias. If a bias is encountered but cannot be mitigated, it should at least be documented.

The use of the formal elicitation principles in this chapter is an important step in bias minimization. It may also be possible to analyze the elicited judgments to determine if a bias is present; however, this usually requires a very controlled elicitation and multiple expert judgments [1].

13.3.1.7 Question Phrasing

The analyst should *pilot test* the interview using the advisor-expert as practice. This exercise helps ensure that the issues are properly specified, background information is sufficient, and definitions are properly phrased. It also provides an estimate of the length of the interview, which should be kept to a reasonable time (such as an hour) to prevent fatigue and to minimize the impact on the experts' schedules.

Care in question phrasing and asking is an obvious, but important, aspect of implementation. Questions must be nonleading and formulated using the terms and jargon of the expert's field. Likewise, responses must be understood and recorded in as original form as possible.

While the advisor-expert can help formulate the appropriate topics and specific questions, the entire elicitation session should be *pilot tested*—practiced prior to scheduling the first expert. The advisor-expert and representatives from the design team can be used as the subject for this pilot test. It is during this pilot testing that question wording, order, *response mode* (see below), and interview flow can be examined and revised if necessary. Not infrequently, the analyst will have to refine terminology, question structure, and even locate additional background information. The pilot test should also be timed, so that the experts can be told how much time to set aside for completing the elicitation.

Common difficulties in formulating questions include the following: program structure is too vague or goals are too general to permit concise questions; subject areas covered are too broad; too many questions covering too many areas for the amount of time available; and a cumbersome or inappropriate response mode is asked.

The *response mode* refers to the form used to ask experts to provide judgments. Some numeric mechanisms include asking experts to specify probabilities, likelihoods, odds, odds ratios, intervals, ratings, rankings or pairwise comparisons. Qualitative responses include verbal or written descriptions, rules, classifications, categorizations, and preferences. It is important to make sure that the information extracted from experts is not distorted. Asking an expert to express a judgment in unfamiliar terms (e.g., probability) or forcing a response beyond the current state of knowledge can open any number of pitfalls that lead to bias in his or her expressions.

Example: An analyst would ask the advisor-expert if experts were comfortable providing ranges of values to represent uncertainty in the reliability of a new component. The expert-advisor may advise that numerical answers might not be possible given the new design. Then descriptors such as *poor, moderate, good,* or *excellent* could be used as the response mode.

One method to minimize potential distortion is to provide feed back to the experts, demonstrating what was done with their supplied judgments and how these were used.

Example: If the analyst has gathered ranges of values to construct a probability density function (PDF) representing the uncertainty of a particular parameter for a stress mechanics finite element model, the analyst shows the expert the final form of this PDF and explains its interpretation. Explanation may be involved if the expert is unfamiliar with PDFs.

It is possible to train experts to provide responses in a given manner, although this can be a time-consuming undertaking. *Training* in probabilistic reasoning can offset some of the mismatch between human cognition and probability theory. Training experts in probability provides them with a new set of perceptual models for structuring their judgments. Also, groups of individuals can make use of a wider universe of perceptual models in making judgments.

Conducting the elicitation using bias minimization techniques is best done employing *verbal probe* and *verbal protocol*. Verbal protocol analysis or verbal report is a technique from educational psychology requiring the expert to think aloud. It is best used in face-to-face interview situations because the elicitor may have to continually remind the experts to verbalize their thoughts. Some experts are more able to accommodate this request. Verbal probe refers to repeated questioning, utilizing the words spoken by an expert, drilling down into details of a subject. Verbal probe is another educational psychology technique for understanding the expert's problem-solving processes. These two methods help minimize biases by avoiding leading questions, by checking for consistency through detailed questioning, and by understanding the expert's problem solving. An example would be using additional questions to expand the expert's thinking in avoiding anchoring or under-estimation of uncertainty.

Drilling for details on a given topic serves another principle of elicitation: *decomposition*. Studies have shown [28] that experts can better tackle difficult technical issues if that issue is broken down into more manageable pieces.

Regardless of what level of detail or complexity of the technical problem, all information is conditional. Probing for and documenting these conditions is an important task for the elicitor and provides a traceable record for the experts. Said another way, the universe is a complex tapestry of multiple variables, relationships among those variables, and dependent information. While it may not be possible to unravel all the complex issues and their relationships, uncovering their existence should be the goal of every question/answer session. Expert knowledge is often conditioned upon the assumptions used, the information seen, the way problems are solved, and experience [29]. Understanding basic conditional and dependent relationships is often the key to resolving differences in experts' knowledge and for combining expert knowledge with other sources of information (see Section 13.4.2).

In our experience it is best to allow the experts time to reflect, research, model, calculate, or just think about the problem and the questions asked. Situations where real-time answers are required do exist, but hastened or pressured responses are subject to bias.

13.3.1.7 Documentation

The old adage that the three secrets to a good business are location, location, location applies to the three secrets for successful elicitation: document, document, document. Information and knowledge are constantly changing. Documentation provides the means for understanding and updating these changes. Documentation not only includes recording everything during the elicitation, but also recording the preparations and pilot study experiences. It includes the way the analysis was done and feedback to the experts. The primary goal of eliciting and utilizing expert knowledge is to capture the current state of knowledge — accomplished by formal elicitation and thorough documentation.

13.3.2 A Modified Delphi for Reliability Analysis

In this section, we describe the Delphi approach for eliciting expert reliability assessments in complex technical problems.

Example: As part of a larger reliability study, an analyst is working with a group of aerospace engineers to determine the reliability of a new control system for a communication satellite. The design team is working to identify a test series that will provide high-quality reliability data at minimal cost to the project. The company has a number of senior engineers who are not working on this particular project but who are willing to assist in the reliability characterization. Two of these engineers are considered senior experts and are known for having very strong opinions about design issues. The analyst and the adviser-experts are concerned that their opinions may unduly sway those of the design team members.

The Delphi technique is often associated with the elicitation of knowledge from experts who are asked to characterize uncertain trends, events or patterns. It is perhaps most accurately characterized as a technique for "structuring a group communication process so that the process is effective in allowing a group of individuals, as a whole, to deal with a complex problem" [5]. The method was developed at the RAND Corporation after World War II to elicit expert forecasts of technology trends for the United States military. In the past 50 years, researchers in a wide range of fields have adopted and modified the technique as a way of eliciting judgments in situations where experts are geographically dispersed, do not typically communicate with each other, and face-to-face meetings are difficult to arrange. The response mode is usually qualitative.

The Delphi procedure begins by identifying experts and assessing their willingness to work on the project. The analyst then provides the participating experts with a series of questions, usually focused on one or two issues (e.g., physical variables). This can be accomplished in several ways:

- Through a website where the experts can access the questionnaire and record responses
- By providing the experts with electronic copies of a questionnaire to be filled out and returned electronically to the analyst
- With a standard paper questionnaire that the experts will complete and return through the post

Once the expert responses are received, the analyst removes identifying information and compiles the expert opinion into a single document. The analyst then sends this revised document to the experts, who are asked to take their peers' answers into account when considering and revising their initial judgments. This procedure can be repeated until an acceptable level of consensus is reached among the experts.

While Delphi has many advantages — including potential cost savings, its ability to minimize groupthink, and its applicability to emergent, poorly-defined problems — there are many potential pitfalls associated with pursuing a Delphi approach. The method has been criticized not so much for its philosophical basis, but for the ways in which it has been applied [5]. As the attached bibliography indicates, analysts interested in learning more about the history and applications of Delphi elicitation will find a great deal of literature and guidance in journals, on the Internet, and in the library. In this discussion, however, we are interested in exploring the application of a Delphi-style approach for eliciting expressions of expert judgment. This includes the following issues:

- Using an expert-oriented elicitation to estimate reliability for parts and processes
- Designing a pilot-tested questionnaire
- Training experts to respond so that the occurrence of bias is minimized

Delphi can be modified to address the criticisms leveled at it, and the approach can provide the analyst with a valuable tool for eliciting expert assessments. As with all expert opinion, our approach, like Delphi is grounded in a fundamental philosophy that formal methods for eliciting and documenting expert knowledge add rigor and provide defensibility. Moreover, formal elicitation augments the analyst's ability to update information as the state of knowledge within the community changes.

This is particularly true in situations when busy schedules make it impossible to gather a group of experts for a group elicitation, or when complicated social dynamics among experts have the potential to limit the quality and quantity of data collected. Examples of the latter include situations when a single expert tends to dominate other members of the organization, influencing the group's consensus. In other organizations, experts simply may not get along with one another. When grouped around a single table, argumentative experts can easily get "stuck" on a fine point of disagreement while failing to address other aspects of the elicitation. Because of this, Delphi requires that experts *not* be in contact during the elicitation. Hence, it can be quite effective in eliciting opinions from individual experts under conditions of anonymity.

13.4 Analyzing Expert Judgment

13.4.1 Characterizing Uncertainties

The main purpose of eliciting expertise and expert judgments is to establish and gather all that is currently known about a problem. However, this knowledge can be in different forms (qualitative or quantitative), dependent upon the current state of what is known. Where knowledge is vague or weak, expert judgments may only be in the form of natural language statements or rules regarding the performance of a product. For example, an expert may say "*if the temperature is too hot, this component will not work very well.*" While this statement is nonnumeric, it contains valuable information, and epistemic uncertainties are inherent in the use of linguistic terms such as *too hot* and *not very well*. Membership functions from Zadeh's fuzzy set theory [30] can be used to characterize the uncertainty associated with these language terms. For example, input membership functions (Figure 13.9) can be constructed to represent the uncertainty in classifying temperatures as *too hot* and output membership functions (Figure 13.10) constructed to classify performance as *not work very well*. *If-then* rules are used to map the input temperatures into the performance descriptions.

Where knowledge is stronger, experts may be able to provide quantitative answers for the physical quantities of interest or for reliability directly. In such cases, quantitative answers for the corresponding uncertainties associated with these quantities may also be elicited. These include *probability, ratings,*

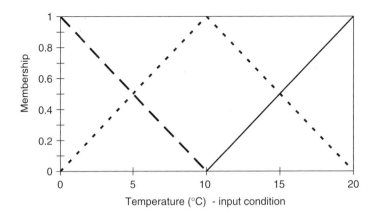

FIGURE 13.9 Input membership functions (— — warm, - - - hot, —— too hot).

rankings, odds ratios, log odds, weighting factors, ranges, and *intervals*. A convenient, useful, and bias-minimizing tool for guiding experts to provide any of the above quantitative estimates is to have them mark their estimates on a drawn real number line. While humans tend to think in linear terms, directing the use of a simple linear *number line*, some physical phenomena may be on a logarithmic scale, directing a log line use (such as for eliciting *log odds*).

The *number line* elicitation begins with experts providing the scale for the line by specifying its endpoints. Then the expert is asked to mark his best estimate on the line with an *x*. To capture uncertainties, he is then asked to provide the extreme high and low values on the line. It is necessary to carefully guide the expert in specifying extreme values to overcome anchoring and under-estimation of uncertainty biases, without leading him. If the expert wants to shade in regions on the number line (rather than marking specific points) for the best and extreme estimates, permit this because he is expressing uncertainties about the values. At this point the analyst has intervals that could be analyzed using *interval* analysis or *random intervals*. (See Chapter 9 on by Joslyn and Booker on General Information Theories.)

If the expert has sufficient knowledge to specify relative frequencies of occurrence for various values on the number line, an entire uncertainty distribution (such as a probability density function) could be elicited [1, 20]. These distributions could also be formulated indirectly by eliciting information from experts about parameters and uncertainties on those parameters for a chosen model. Such an exercise is described by Booker et al. [31], where the parameters of a Weibull reliability model were elicited, because

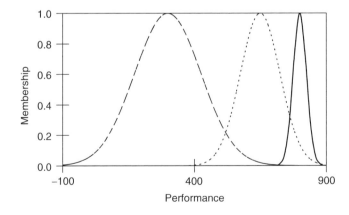

FIGURE 13.10 Output membership functions (— — bad, - - - not well, —— not very well).

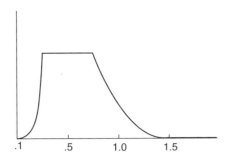

FIGURE 13.11 PDF drawn by hand, expressing an expert's uncertainty about a variable.

the experts customarily thought in terms of failure mechanisms and rates that defined those parameters. It should be noted that elicited PDFs do not necessarily have to follow a particular family or distribution type (e.g., a normal). Empirically based or expert supplied distributions reflect the way experts interpret the likelihood of values associated with the quantity of interest — the variable. Figure 13.11 demonstrates a probability distribution function drawn free-hand by an expert.

Rankings refer to specifying preferences x out of a finite list of n options, alternatives, or similar items. An implicit assumption in using ranks is that the numeric answers are equidistant (i.e., a rank of 4 is twice the rank of 2). If properly directed, experts can use a nonlinear scale (e.g., this rank is "5" out of "50" but "4" through "7" are closer ranked than "1" through "3" or above "8"). Such specifications aid in determining uncertainty.

Rating scales can be used to map words into numbers, percentages, ranks, or ratings. Examples include *Saaty's pairwise comparisons*, which utilize a specified scale description for comparing the degree of comparison between items, two at a time [32]. A scale from Sherman and Kent establishes equivalency between verbal descriptions of likelihood, chances, and percentages. For example, *8 out of 10, 8/10, or 80%* falls between *probable* and *nearly certain* or between *likely* and *highly likely*. Here again, the vagueness of the verbal description imposes a source of epistemic uncertainty.

Odds ratio takes advantage of the concept of betting and chance. Our experience confirms the results of others, indicating that humans are naturally comfortable thinking in these terms.

Uncertainties can be estimated with rankings, ratings, and odds by eliciting ranges rather than single-valued results. Again, GITs can be used to mathematically represent the uncertainties from these ranges. Possibility distributions, random intervals, upper and lower probabilities, and membership functions are among the alternative forms for PDFs. These theories also offer axioms of how to combine their respective distributions of uncertainty within the operations of each theory. However, only one linkage has been formalized between two of the theories: probability and fuzzy membership functions [33].

Twice now the subject of turning linguistic terms into numbers has emerged in the characterization of uncertainties. As demonstrated in Figure 13.4 and Figure 13.5, membership functions from fuzzy sets can be useful in eliciting natural language information and then quantifying the uncertainty associated with those linguistic terms. Fuzzy sets and their corresponding measure of uncertainty, membership, are designed to best capture uncertainties due to ambiguity of classification. For example, what does it mean to be *too hot*? (See Chapter 9 by Joslyn and Booker on General Information Theories, section on fuzzy sets and membership functions.) In contrast to this type of uncertainty, probability is best designed to capture the uncertainty associated with the outcome of an event or experiment.

As a quantification tool for reliability, membership functions map the condition of a component or system into its performance through the use of knowledge contained in *if-then* rules. The previous example holds: *if the temperature is too hot* (input fuzzy membership function), *then this component will not work very well* (output fuzzy membership function). Membership functions can be used to characterize the conditions (x) and *if-then* rules map those conditions into performance (y) membership functions. A reliability example of how to combine linguistic information from a component supplier with probability expert judgments from designers can be found in Kerscher et al. [34].

13.4.1.1 Probability Theory: The Good and the Bad

Probability theory has become a fundamental theory for characterizing *aleatoric uncertainty* — uncertainty associated with phenomena such as random noise, measurement error, and uncontrollable variation. With aleatoric uncertainty, the common conception is that uncertainty cannot be further reduced or eliminated by additional information (data or knowledge). In communities involved in KA and computational sciences, *epistemic uncertainty* refers to an absence of complete knowledge — uncertainty that can be reduced or eliminated by increasing knowledge or sample size. In following the formal elicitation principles, such as preserving the original form of elicited information, uncertainty characterization should be consistent with the way experts think and hence verbalize. Because experts (even those in science and engineering fields) may not readily think in terms of probability and may not be able to characterize their uncertainties with PDFs, probability may not be adequate for handling either form of uncertainty. Studies [35, 36] have also shown that many humans who are comfortable with probabilistic thinking are not able to think consistently within its axioms. Other GITs are examined below in their applicability for elicited judgments.

Yet probability theory has a rich and long legacy of use in science and engineering and in uncertainty analysis and quantification. To those involved with complex decision problems like Probabilistic Risk Assessment (PRA), the uncertainty embodies both: *aleatoric* (random, irreducible uncertainty) and *epistemic* (lack of knowledge, reducible with more information) uncertainties. This interpretation also encompasses uncertainty caused by errors, mistakes, and miscalculations. The absence of a distinction between aleatoric and epistemic uncertainty is also subscribed to by modern subjective probabilists, or Bayesians, and applies to modern, complex reliability analysis. Probability theory (see Chapter 9 by Joslyn and Booker on GITs) provides a calculus for the uncertainty associated with the outcome of an event or experiment (E), designated as P(E). The theory and its basic axioms do not specify how to determine the value of probability nor how to interpret it. Therefore, numerous historical interpretations are of equal value, including the subjective, personalistic or Bayesian interpretation, which says there is no such thing as a correct probability, or an objective probability. This interpretation is best suited for the way expert probabilities are elicited and used. Data is typically sparse for rare or one-of-a-kind events — the perfect use of subjective probability.

13.4.1.2 General Information Theories: The Good and the Bad

While the mathematical theory of many GITs is well developed, the practical implementation of these to real problems, such as reliability and expert elicitation, is lacking. As has already been noted, these theories can offer alternatives for characterizing uncertainties consistent with information elicited from experts. For example, if the extent of knowledge only permits intervals as answers, and those intervals have imprecise limits, then random intervals could provide a consistent analysis tool. Another example is linguistic information and classification uncertainty that can be best represented by fuzzy membership functions.

However, in any complex reliability problem, available information can be in diverse forms, data, models, expert judgment, historical information, practices, and experience. Accompanying uncertainties may also be best formulated using a variety of GITs, including probability. All this information must be combined to assess the reliability of the entire system, so the uncertainty theories must be linked for combination. Unfortunately, the theory and application of the GITs are insufficient to handle cross linkages. This failing is so severe that analysts are often forced to specify a particular GIT for the entire system and work only within that theory.

13.4.1.3 Aggregation of Expert Judgments

A large body of literature specifies various methods for combining expert judgments from multiple experts. These methods range from the use of elicitation to reach consensus or resolve differences to exact analytical methods [37]. Booker and McNamara [38] detail such a scenario, utilizing the tenets of elicitation. To summarize these, differences between experts can often be resolved by thorough examination of their assumptions, sources of information used, heuristics, and problem-solving processes. For example, experts with different assumption sets may be solving slightly different problems. Once common assumptions and problem definition are established, differences can be mitigated. Remaining differences among experts after resolution attempts may be representative of the existing (and perhaps large) state of uncertainty.

Aggregation can be achieved by analytical methods, with combination techniques implemented by an analyst, a decision maker, or by the experts themselves. The maximum entropy solution (when no other information is available to determine weighting factors) is to equally weight the experts. The aggregation (like many others) assumes that all experts are independent, in much the same way as a statistician would consider repeated runs of an experiment as independent draws from a probability distribution representing the population of all possible outcomes. However, this classical inference argument has been disputed when applied to experts who have dependencies due to common experience, access to the same data, and similar education and training. The elicitor, advisor expert, decision maker, or analyst may be able to quantify dependence among experts [1]. From this book, studies have shown that experts are quite capable of determining weights for each other and also that expert judgments are well correlated to how experts solve problems, making precise extraction and documentation of problem solving important.

The many analytical expressions for combining expert judgments can be summarized into three basic formulas for aggregating uncertainty distribution functions, f. Each expert contributes to the construction of these functions by either providing information relating to the uncertainty of the random variable, x, or the parameters, θ, of the functions. Common practice in probability theory is to specify these f's as probability density functions; however, f could also refer to any uncertainty function such as a possibility distribution. Rules for combining these f's within their respective theories will alter the basic combination schemes below (which refer to probability theory):

$$f_c(x, \theta) = w_1 \cdot f_1(x, \theta) + w_2 \cdot f_2(x, \theta) + \ldots \qquad (13.1)$$

$$f_c(x, \theta) = f([w_1 \cdot x_1 + w_2 \cdot x_2 + \ldots], \theta) \qquad (13.2)$$

$$f_c(x, \theta) = f([x_1, x_2, \ldots], \theta) \qquad (13.3)$$

Equation 13.1 illustrates a scheme that weights each expert's function for x and θ in a linear combination. Equation 13.2 weights the random variables from each expert, and that weighted combination is mapped through an overall function. That is, in Equation 13.2, the function on the left hand side is the probability density function of the weighted sum of random variables. Equation 13.3 combines the random variables for a given θ using the concept of a joint distribution function.

Example: Two experts each supply PDF's, f_1 and f_2 as uncertainty distributions on the number of failure incidences per 1000 vehicles (IPTV) manufactured for a new electronic circuit, shown in Figure 13.12. Using Equation 13.1 with equal weights, the combined result is shown in Figure 13.13. Combining expert uncertainty functions is an analytical exercise not unlike combining the diverse sources of information using information integration tools.

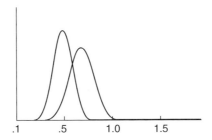

FIGURE 13.12 Two experts' IPTV PDFs.

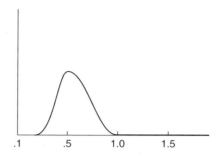

FIGURE 13.13 Equal combination.

13.4.2 Information Integration for Reliability

13.4.2.1 How Expert Knowledge Combines with Other Information

Reliability is typically defined as a probability; specifically, it is the probability that the system will function up to a specified time, t, for given conditions or specifications. This reliability can be denoted as $R(t|\theta)$, where the parameters θ could denote mean time to failure or could refer to two parameters in a Weibull model. In classical (frequentist interpretation of probability) estimation of θ, data is considered as random and θ parameters are considered fixed. In Bayesian (subjective interpretation of probability) estimation, the data is fixed, and θ are random. Either interpretation is valid under the axioms of probability, and in both cases, the process of estimation relies on having *data*. In cases where data are sparse or nonexistent, the estimation process relies on whatever forms of information are available, including expert judgment.

Example: A team of design engineers wants to evaluate the performance of three design ideas for a new actuator on a valve. Design 1 uses a new explosive; design 2 uses a new firing set; and design 3 uses both. Large amounts of data exist on the old explosive and firing sets, making a statistical estimate of reliability possible by calculating successes/tests. No data is available for the new components, but the team estimates a subjective reliability for the new explosive and new firing set. Both the calculated reliability and the experts' subjective reliability values are valid interpretations of probabilities. Both can be used to estimate the performances of the three designs.

Additional information sources could also include historical information about the system, data from similar components, and computer model outputs of physical processes. Systems are composed of more than components and subsystems. Processes are equally important representations of system performance and functioning. If parts are considered the *nouns* of the system, then processes are its *verbs*. Processes include the actions involved in assembly, testing, quality control, inspections, and physical dynamics, to name a few.

Complex or newly designed systems may have one or more parts or processes that have no test or experimental data. Preprototype, preproduction concept design systems may have no relevant test data, resulting in reliance on other information sources to estimate reliability. Information integration techniques have been developed [39] to take advantage of all available information to assess the performance of new, complex, or aging systems for decisions regarding their production, continued use, and maintenance, respectively. Expertise and expert judgment play important roles in these decisions, and expert judgment may be the sole source of information in some cases. A reliability example using this technology can be found in [31].

Information integration techniques are also useful for updating reliability assessments. The state of knowledge is constantly changing, and mathematical updating mechanisms such as Bayes theorem are useful for updating the existing information (under uncertainty) with the new information. Documentation is therefore necessary for all information sources, not just expert judgments, to ensure proper updating with these or any other techniques.

Example: In designing a new fuel pump, the design team of experts used their experience with other fuel pumps to estimate the reliability of one of its new components. They supplied a range of values for the failure

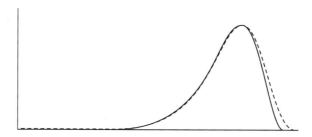

FIGURE 13.14 Experts' prior reliability estimate (solid) and combined estimate with 40 *what-if* test results (dashed) [31].

rate of that component, λ, to be used in a Weibull model for estimating the reliability, R, at time $t = 1$ year:

$$R(t|\lambda) = exp(-\lambda t^{0.8}) \tag{13.4}$$

where the failure rate parameter, λ, decays according to the second parameter in the time exponent with a value of 0.8. To improve reliability, the experts asked *what if* we built 40 prototypes that all passed the testing program; how much would reliability improve? Using Bayes theorem, their initial (prior) reliability distribution (Figure 13.14, solid curve) was combined with a binomial distribution for 40 tests, 0 failures. The resulting posterior for the new combined reliability is the dashed curve (Figure 13.14). Based on the minimal improvement gained from the costly prototype and test program, the decision maker decided not to pursue this program.

13.5 Conclusion

Today's complex reliability problems demand more predictability with less expensive test programs. Awareness of uncertainties and the simple fact that the state of knowledge is constantly changing add to the complexity of these problems. Why not take advantage of an organization's most valuable source of information — the knowledge and expertise of its technical staff. Use of the formal elicitation methods in this chapter provides a traceable, updateable, defensible way to capture this knowledge.

Elicitation methods are used for gathering knowledge and understanding about the structure of a complex problem and for populating that structure with the qualitative and quantitative judgments of the experts. Information integration methods exist to provide a formalism for combining expert knowledge, and its associated uncertainties, with other data or information.

The keys to successful elicitation include the following: Construct an expert-oriented elicitation and analysis that captures the current state of knowledge and permits combination of this knowledge with other data/information. The formal methods outlined in this chapter emphasize bias minimization, capturing knowledge consistent with the way experts think and problem solve, utilizing feedback, and documentation. Finally, the analyst should never compromise the experts' trust.

References

1. Meyer, M.A. and Booker, J.M., *Eliciting and Analyzing Expert Judgment: A Practical Guide*. Philadelphia, PA: Society for Industrial and Applied Mathematics/American Statistical Association, 2001.
2. Helmer, O. and Rescher, N., "On the Epistemology of the Inexact Sciences," *Management Science*, 5 (June), 25–52, 1959.
3. Dalkey, N.C., "An Experimental Study of Group Opinion: The Delphi Method," *Science*, 243, 1668–1673, 1969.
4. Dalkey, N.C. and Helmer, O., "An Experimental Application of the Delphi Method to the User of Experts," *Management Science*, 9(3), 458–67, 1963.

5. Linstone, H.A. and Turoff, M., Eds., *The Delphi Method. Techniques and Applications.* Reading, MA, 1975.
6. Paton, R.C., Nwana, H.S., Shave, M.J.R., and Bench-Capon, T.J.M., "From Real World Problems to Domain Characterisations", *Proceedings of 5th European Knowledge Acquisition Workshop*, Crieff, May, 1991. Reprinted in Linster, M., Ed., *Proceedings of 5th European Knowledge Acquisition Workshop*, GMD-Studien Nr. 211, GMD, Sankt-Augustin, Germany, 1992, 235–256.
7. Paton, R.C., Lynch, S., Jones, D., Nwana, H.S., Bench-Capon, T.J.M., and Shave, M.J.R., "Domain Characterisation for Knowledge Based Systems," *Proceedings of A.I. 94 — Fourteenth International Avignon Conference*, Volume 1, 41–54, 1994.
8. Wenger, E., *Communities of Practice: Learning, Meaning and Identity.* Cambridge, UK: Cambridge University Press, 1998.
9. Chaiklin, S. and Lave, J., Eds., *Understanding Practice: Perspectives on Activity and Context.* Cambridge, UK and New York: Cambridge University Press, 1996.
10. Lave, J. and Wenger, E., *Situated Learning: Legitimate Peripheral Participation.* Cambridge, UK: Cambridge University Press, 1991.
11. Lave, J., *Cognition in Practice.* Cambridge, UK: Cambridge University Press, 1988.
12. Martz, H.F. and Waller, R.A., *Bayesian Reliability Analysis*, New York: John Wiley & Sons, 1982.
13. Hoffman, R.R., Shadbolt, N.R., Burton, A.M., and Klein, G., "Eliciting Knowledge from Experts: A Methodological Analysis," *Organizational Behavior and Decision Management*, 62(2), 129–158, 1995.
14. Davis, R., Shrobe, H., and Szolovits, P., "What Is a Knowledge Representation?" *AI Magazine*, 14(1), 17–33, 1993.
15. Leishman, D. and McNamara, L., "Interlopers, Translators, Scribes and Seers: Anthropology, Knowledge Representation and Bayesian Statistics for Predictive Modeling in Multidisciplinary Science and Engineering Projects," to appear in *Multidisciplinary Studies of Visual Representations and Interpretations.* Malcolm, G. and Paton, R., Eds., Amsterdam: Elsevier Science, 2004.
16. Cooke, N.J., "Varieties of Knowledge Elicitation Techniques," *International Journal of Human-Computer Studies*, 41, 801–849, 1994.
17. Paton, R.C., "Process, Structure and Context in Relation to Integrative Biology," *BioSystems*, 64, 63–72, 2002.
18. Keeney, R.L. and von Winterfeldt, D., "On the Uses of Expert Judgment on Complex Technical Problems," *IEEE Transactions on Engineering Management*, 36, 83–86, 1989.
19. Keeney, R.L. and von Winterfeldt, D., "Eliciting Probabilities from Experts in Complex Technical Problems," *IEEE Transactions on Engineering Management*, 38(3), 191–201, 1991.
20. Kadane, J.B. and Wolfson, L.J., "Experiences in Elicitation," *The Statistician*, 47, 1–20, 1998.
21. Lichtenstein, S., Fischhoff, B., and Phillips, L.D., "Calibration of Probabilities: The State of the Art to 1980," in D. Kahneman, P. Slovic, and A. Tversky, Eds., *Judgment under Uncertainty: Heuristics and Biases*, Cambridge, MA: Cambridge University Press, 1982, 306–334.
22. Jensen, F.V., *An Introduction to Bayesian Networks.* New York: Springer, 1996.
23. Ayyub, B., *Elicitation of Expert Opinions for Uncertainty and Risks.* Boca Raton, FL: CRC Press, 2001.
24. O'Leary, D., "Knowledge Acquisition from Multiple Experts: An Empirical Study," *Management Science*, 44(8), 1049–1058, 1998.
25. Ayton, P. and Pascoe. E., "Bias in Human Judgment under Uncertainty?" *The Knowledge Engineering Review*, 10(1), 21–41, 1995.
26. Tversky, A. and Kahneman, D., "Extensional versus Intuitive Reasoning: The Conjunction Fallacy in Probability Judgment." *Psychological Review*, 90(4), 293–315, 1983.
27. Nelson, M., "The Hubble Space Telescope Program: Decision-Making Gone Awry," Unpublished Report, Nelson Associates, Del Mar, CA, Available at URL http://home.att.net/~maxoccupancy/, 1992.
28. Armstrong, J.S., *Long-Range Forecasting: From Crystal Ball to Computer.* New York: Wiley-Interscience, 1981.
29. Booker, J.M. and Meyer, M.A., "Sources and Effects of Interexpert Correlation: An Empirical Study," *IEEE Transactions on Systems, Man, and Cybernetics*, 18, 135–142, 1988.

30. Zadeh, L., "Fuzzy Sets," *Information and Control*, 8, 338–353, 1965.
31. Booker, J.M., Bement, T.R., Meyer, M.A., and Kerscher, W.J., "PREDICT: A New Approach to Product Development and Lifetime Assessment Using Information Integration Technology," *Handbook of Statistics: Statistics in Industry*, Volume 22 (Rao and Khattree, Editors), Amsterdam, The Netherlands: Elsevier Science, 2003, Chap. 11.
32. Saaty, T.L., *The Analytic Hierarchy Process: Planning, Priority Setting, and Resource Allocation*, New York: McGraw-Hill, 1980.
33. Singpurwalla, N.D. and Booker, J.M., "Membership Functions and Probability Measures of Fuzzy Sets," *Journal of the American Statistical Association*, September 2004 Volume 99 No. 467, 867–877.
34. Kerscher, W.J., Booker, J.M., Meyer, M.A., and Smith, R.E., "PREDICT: A Case Study Using Fuzzy Logic," *Proceedings of International Symposium on Product Quality & Integrity*, Tampa, FL, January 27–30, 2003, and Los Alamos National Laboratory report, LA-UR-02-0732, 2003.
35. Kahneman, D. and Tversky, A., "Subjective Probability: A Judgment of Representativeness," In Kahneman, D., Slovic, P., and Tversky, A., Eds., *Judgment under Uncertainty: Heuristics and Biases*, 32–47. Cambridge, MA: Cambridge University Press, 1982.
36. Hogarth, R., Cognitive processes and the assessment of subjective probability distributions. *Journal of the American Statistical Association*, 70, 271–291, 1975.
37. Lindley, D.V. and Singpurwalla, N.D., "Reliability and Fault Tree Analysis Using Expert Opinions," George Washington University Report GWU/IRRA/TR-84/10, Washington, D.C., 1984.
38. Booker, J.M. and McNamara, L.A., "Solving the Challenge Problems Using Expert Knowledge Theory & Methods," submitted to *Reliability Engineering and System Safety*, Special Issue: Epistemic Uncertainty Workshop and Los Alamos National Laboratory Report, LA-UR-02-6299 (2002).
39. Meyer, M.A., Booker, J.M., and Bement, T.R., "PREDICT-A New Approach to Product Development," *R&D Magazine*, 41, 161, and Los Alamos National Laboratory document, LALP-99-184, 1999.

14
First- and Second-Order Reliability Methods

14.1	Introduction	14-1
14.2	Transformation to Standard Normal Space	14-3
	Statistically Independent Random Variables • Dependent Normal Random Variables • Random Variables with Nataf Distribution • Dependent Nonnormal Random Variables	
14.3	The First-Order Reliability Method	14-6
	Component Reliability by FORM • System Reliability by FORM • FORM Importance and Sensitivity Measures	
14.4	The Second-Order Reliability Method	14-15
	Example: Reliability Analysis of a Column by SORM	
14.5	Time-Variant Reliability Analysis	14-19
	Example: Mean Out-Crossing Rate of a Column under Stochastic Loads	
14.6	Finite Element Reliability Analysis	14-22

Armen Der Kiureghian
University of California, Berkeley

14.1 Introduction

Within the broader field of reliability theory, the class of time-invariant structural reliability problems is characterized by an n-vector of basic (directly observable) random variables $\mathbf{x} = \{x_1,\ldots,x_n\}^\mathrm{T}$ and a subset Ω of their outcome space, which defines the "failure" event. The probability of failure, $p_f = P(\mathbf{x} \in \Omega)$, is given by an n-fold integral

$$p_f = \int_\Omega f(\mathbf{x})d\mathbf{x} \qquad (14.1)$$

where $f(\mathbf{x})$ is the joint probability density function (PDF) of $\mathbf{x}$. This problem is challenging because for most nontrivial selections of $f(\mathbf{x})$ and Ω, no closed form solution of the integral exists. Furthermore, straightforward numerical integration is impractical when the number of random variables, n, is greater than 2 or 3. Over the past two decades, a number of methods have been developed to compute this probability integral. This chapter introduces two of the most widely used methods: the first-order reliability method, FORM, and the second-order reliability method, SORM. Extensions of the above formulation to time- or space-variant problems and to applications involving finite element analysis are also described in this chapter.

In general, the failure domain Ω is described in terms of continuous and differentiable *limit-state* functions that define its boundary within the outcome space of $\mathbf{x}$. Depending on the nature of the problem, the following definitions apply:

Component problem:

$$\Omega \equiv \{g(\mathbf{x}) \leq 0\} \quad (14.2a)$$

Series system problem:

$$\Omega \equiv \bigcup_k \{g_k(\mathbf{x}) \leq 0\} \quad (14.2b)$$

Parallel system problem:

$$\Omega \equiv \bigcap_k \{g_k(\mathbf{x}) \leq 0\} \quad (14.2c)$$

General system problem:

$$\Omega \equiv \bigcup_k \bigcap_{j \in c_k} \{g_j(\mathbf{x}) \leq 0\} \quad (14.2d)$$

The formulation in 14.2a applies to component reliability problems, which are defined in terms of a single limit-state function, $g(\mathbf{x})$. Any outcome $\mathbf{x}$ of the random variables, for which the limit-state function is nonpositive, constitutes the failure of the component. The simplest example for this class of problems is the limit-state function $g(x_1, x_2) = x_1 - x_2$, where x_1 denotes a capacity quantity and x_2 denotes the corresponding demand. More generally, the limit-state function is defined by considering the mechanical conditions under which the failure may occur. For example, the failure due to yielding of a two-dimensional continuum according to the von Mises yield criterion is defined by the limit-state function $g(\mathbf{x}) = \sigma_{yld}^2 - (\sigma_x^2 + \sigma_y^2 - \sigma_x \sigma_y + 3\tau_{xy}^2)$, where σ_x, σ_y and τ_{xy} are the components of stress and σ_{yld} is the yield stress, all being functions of some basic random variables $\mathbf{x}$ describing observable quantities, such as material property constants, structural dimensions and loads. The relationship between the stress components and the basic variables, of course, can be complicated and computable only through an algorithmic procedure, such as a finite element code. This aspect of reliability analysis is briefly described in Section 14.6.

The series system problem in 14.2b applies when the failure domain is the union of componential failure events. A structure having multiple failure modes, each defined in terms of a continuous and differentiable limit-state function $g_k(\mathbf{x})$, belongs to this category of problems. Note that the random variables $\mathbf{x}$ are shared by all the components. As a result, the componential failure events in general are statistically dependent. A good example for a series system is a ductile structural framework having multiple failure mechanisms. Such an example is described later in this chapter.

The parallel system problem in 14.2c applies when the failure domain is the intersection of componential failure events. A redundant structure requiring failure of several components, each defined in terms of a continuous and differentiable limit-state function $g_k(\mathbf{x})$, belongs to this category of problems. A good example is a bundle of brittle wires with random strengths and subjected to a random tension load.

Most structural systems are neither series nor parallel. Rather, the system fails when certain combinations of components fail. We define a cut set as any set of components whose joint failure constitutes failure of the system. Let c_k denote the index set of components in the k-th cut set. The intersection event in 14.2d then defines the failure of all components in that cut set. The union operation is over all the cut sets of the system. For an accurate estimation of the failure probability, one must identify all the minimal cut sets. These are cut sets that contain minimum numbers of component indices; that is, if any component index is removed, what remains is not a cut set. Exclusion of any minimum cut set from consideration will result in underestimation of the failure probability. Thus, the cut-set formulation provides an estimate of the failure probability from below. It is noted that this formulation essentially represents the system as a series of parallel subsystems, each subsystem representing a cut set.

An alternate formulation for a general system defines the complement of Ω in terms of link sets. A link set is a set of components whose joint survival constitutes the survival of the system. Letting l_k define the set of indices in the k-th link set, the complement of the failure event, the survival event, is defined as

$$\overline{\Omega} \equiv \bigcup_k \bigcap_{j \in l_k} \{g_j(\mathbf{x}) > 0\} \tag{14.2e}$$

In this formulation, one only needs to include minimal link sets, that is, link sets that do not include superfluous components. The advantage of this formulation is that exclusion of any link set produces a conservative estimate of the failure probability. However, working with the intersection of survival events as in 14.2e is computationally more difficult. For this reason, the formulation in 14.2d is more commonly used.

FORM, SORM, and several other computational reliability methods take advantage of the special properties of the standard normal space. For this reason, these methods involve a transformation of the random variables $\mathbf{x}$ into standard normal variables $\mathbf{u}$ having the joint PDF $\varphi_n(\mathbf{u}) = (2\pi)^{-n/2} \exp(-\|\mathbf{u}\|^2/2)$, where $\|\cdot\|$ denotes the Euclidean norm, as a first step of the reliability analysis. Useful properties of the standard normal space include rotational symmetry, exponentially decaying probability density in the radial and tangential directions, and the availability of formulas for the probability contents of specific sets, including the half space, parabolic sets, and polyhedral sets. The applicable forms of this transformation are described in the following section.

14.2 Transformation to Standard Normal Space

Let $\mathbf{x}$ denote an n-vector of random variables with a prescribed joint PDF $f(\mathbf{x})$ and the corresponding joint cumulative distribution function (CDF) $F(\mathbf{x}) = \int_{-\infty}^{x_n} \cdots \int_{-\infty}^{x_1} f(\mathbf{x}) d\mathbf{x}$. We wish to construct a one-to-one transformation $\mathbf{u} = \mathbf{T}(\mathbf{x})$, such that $\mathbf{u}$ is an n-vector of standard normal variables. As described below, such a transformation exists, although it may not be unique, as long as the joint CDF of $\mathbf{x}$ is continuous and strictly increasing in each argument. For computational purposes in FORM and SORM, we also need the inverse transform $\mathbf{x} = \mathbf{T}^{-1}(\mathbf{u})$ and the Jacobian of the $\mathbf{x} \to \mathbf{u}$ transformation, $\mathbf{J}_{\mathbf{u},\mathbf{x}}$. These are defined in the following subsections for four distinct cases that occur in practice. For this analysis, let $\mathbf{M} = \{\mu_1 \cdots \mu_n\}^T$ denote the mean vector of $\mathbf{x}$ and $\boldsymbol{\Sigma} = \mathbf{DRD}$ denote its covariance matrix, where $\mathbf{D} = \mathrm{diag}[\sigma_i]$ is the $n \times n$ diagonal matrix of standard deviations and $\mathbf{R} = [\rho_{ij}]$ is the correlation matrix having the elements $\rho_{ij}, i,j = 1,\ldots,n$.

14.2.1 Statistically Independent Random Variables

Suppose the random variables $\mathbf{x}$ are statistically independent such that $f(\mathbf{x}) = f_1(x_1)f_2(x_2)\cdots f_n(x_n)$, where $f_i(x_i)$ denotes the marginal PDF of x_i. Let $F_i(x_i) = \int_{-\infty}^{x_i} f_i(x_i)dx_i$ denote the CDF of x_i. The needed transformation in this case is diagonal (each variable is transformed independently of other variables) and has the form

$$u_i = \Phi^{-1}[F_i(x_i)] \quad i = 1, 2, \ldots, n \tag{14.3}$$

where $\Phi[\cdot]$ denotes the standard normal CDF and the superposed -1 indicates its inverse. Figure 14.1 shows a graphical representation of this transformation. Each point (x_i, u_i) on the curve is obtained by equating the cumulative probabilities $F_i(x_i)$ and $\Phi(u_i)$. Note that this solution of the needed transformation is not unique. For example, an alternative solution is obtained if $F_i(x_i)$ in Equation 14.3 is replaced by $1 - F_i(x_i)$.

The inverse of the transformation in 14.3 is

$$x_i = F_i^{-1}[\Phi(u_i)] \quad i = 1, 2, \ldots, n \tag{14.4}$$

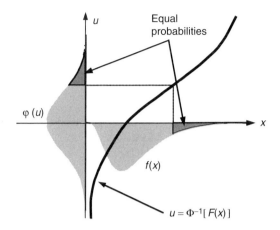

FIGURE 14.1 Transformation to the standard normal space for a single random variable.

If x_i is nonnormal, both transformations 14.3 and 14.4 are nonlinear and may require a numerical root finding algorithm to solve. The Jacobian of the transformation is a diagonal matrix $\mathbf{J}_{u,x} = \text{diag}[J_{ii}]$ having the elements

$$J_{ii} = \frac{f_i(x_i)}{\varphi(u_i)} \qquad i = 1, 2, \ldots, n \tag{14.5}$$

where $\varphi(u) = (2\pi)^{-1/2} \exp(-u^2/2)$ is the univariate standard normal PDF.

14.2.2 Dependent Normal Random Variables

Suppose random variables $\mathbf{x}$ are normally distributed with mean vector $\mathbf{M}$ and covariance matrix Σ. In this case, a convenient form of the transformation to the standard normal space is

$$\mathbf{u} = \mathbf{L}^{-1}\mathbf{D}^{-1}(\mathbf{x} - \mathbf{M}) \tag{14.6}$$

where $\mathbf{D}$ is the diagonal matrix of standard deviations defined earlier and $\mathbf{L}$ is a lower-triangular matrix obtained by Choleski decomposition of the correlation matrix such that $\mathbf{R} = \mathbf{LL}^T$. This decomposition is possible provided the covariance matrix is positive definite, which is the case as long as the random variables $\mathbf{x}$ are not linearly dependent. (If a linear dependence exists, one can eliminate it by reducing the number of random variables.) As can be seen, the transformation in this case is linear. The inverse transform is

$$\mathbf{x} = \mathbf{M} + \mathbf{DLu} \tag{14.7}$$

and the Jacobian is

$$\mathbf{J}_{u,x} = \mathbf{L}^{-1}\mathbf{D}^{-1} \tag{14.8}$$

It is noted that $\mathbf{L}$ being triangular, its inverse is easy to compute.

14.2.3 Random Variables with Nataf Distribution

A set of statistically dependent random variables x_i, $i = 1, \ldots, n$, with prescribed marginal CDFs $F_i(x_i)$ and correlation coefficients ρ_{ij}, $i, j = 1, \ldots, n$, are said to be Nataf-distributed if the marginally transformed random variables

$$z_i = \Phi^{-1}[F_i(x_i)] \qquad i = 1, 2, \ldots, n \tag{14.9}$$

First- and Second-Order Reliability Methods

are jointly normal. Liu and Der Kiureghian [1] have shown that the correlation coefficients of the two sets of random variables are related through the identity

$$\rho_{ij} = \int_{-\infty}^{\infty}\int_{-\infty}^{\infty} \left(\frac{x_i - \mu_i}{\sigma_i}\right)\left(\frac{x_j - \mu_j}{\sigma_j}\right) \varphi_2(z_i, z_j, \rho_{0,ij}) dz_i dz_j \tag{14.10}$$

where $\rho_{0,ij}$ is the correlation coefficient between z_i and z_j and $\varphi_2(z_i, z_j, \rho_{0,ij})$ is their bivariate normal PDF. For given continuous and strictly increasing marginal CDFs $F_i(x_i)$ and a positive-definite correlation matrix $\mathbf{R} = [\rho_{ij}]$, the Nataf distribution is valid as long as $\mathbf{R}_0 = [\rho_{0,ij}]$ is a valid correlation matrix.

Among joint distribution models that are consistent with a set of prescribed marginal distributions and correlation matrix, the Nataf distribution is particularly convenient for reliability applications for two reasons: (1) it can accommodate a wide range of correlations between the random variables $\mathbf{x}$ (see [1] for analysis of the limits on ρ_{ij} for which the Nataf distribution is valid), and (2) the transformation to the standard normal space is simple and independent of the ordering of the random variables. The required transformation is given by

$$\mathbf{u} = \mathbf{L}_0^{-1} \begin{Bmatrix} \Phi^{-1}[F_1(x_1)] \\ \vdots \\ \Phi^{-1}[F_n(x_n)] \end{Bmatrix} \tag{14.11}$$

where $\mathbf{L}_0$ is the Choleski decomposition of the correlation matrix $\mathbf{R}_0$, i.e., $\mathbf{R}_0 = \mathbf{L}_0 \mathbf{L}_0^T$. It can be seen that the above transformation is a synthesis of transformation 14.3 for independent nonnormal random variables and transformation 14.6 for correlated normal random variables. The inverse transform consists in first finding the intermediate variables $\mathbf{z} = \mathbf{L}\mathbf{u}$ and then using $x_i = F_i^{-1}[\Phi(z_i)]$, $i = 1, \ldots, n$. Furthermore, the Jacobian of the transformation is given by

$$\mathbf{J}_{u,x} = \mathbf{L}_0^{-1} \text{diag}[J_{ii}] \tag{14.12}$$

where J_{ii} is as in Equation 14.5.

14.2.4 Dependent Nonnormal Random Variables

For dependent nonnormal random variables other than those with the Nataf distribution, a different transformation must be used. By sequentially conditioning, the joint PDF of the set of random variables x_i, $i = 1, \ldots, n$, can be written in the form

$$f(\mathbf{x}) = f_n(x_n | x_1, \ldots, x_{n-1}) \cdots f_2(x_2 | x_1) f_1(x_1) \tag{14.13}$$

where $f_i(x_i | x_1, \ldots, x_{i-1})$ denotes the conditional PDF of x_i for given values of $x_1, \ldots, x_{i-1}$. Let

$$F_i(x_i | x_1, \ldots, x_{i-1}) = \int_{-\infty}^{x_i} f_i(x_i | x_1, \ldots, x_{i-1}) dx_i \tag{14.14}$$

denote the conditional CDF of x_i for the given values of $x_1, \ldots, x_{i-1}$. The so-called Rosenblatt transformation [2] is defined by

$$\begin{aligned} u_1 &= \Phi^{-1}[F_1(x_1)] \\ u_2 &= \Phi^{-1}[F_2(x_2 | x_1)] \\ &\cdots \\ u_n &= \Phi^{-1}[F_n(x_n | x_1, \ldots, x_{n-1})] \end{aligned} \tag{14.15}$$

Note that this is a triangular transformation; that is, u_i is dependent only on x_1 to x_i. Because of this property, the inverse of this transform is easily obtained by working from top to bottom. Specifically, for a given $\mathbf{u} = \{u_1 \cdots u_n\}^T$, the first equation is solved for x_1 in terms of u_1, the second equation is solved for x_2 in terms of u_2 and x_1, etc. Because of the nonlinearity of the transformation, an iterative root-finding scheme must be used to solve each of these equations. The Jacobian of the transformation, $\mathbf{J}_{u,x}$, is a lower-triangular matrix having the elements

$$J_{11} = \frac{f_1(x_1)}{\varphi(u_1)}$$

$$J_{ij} = 0 \quad \text{for } i < j$$

$$= \frac{f_i(x_i \mid x_1, \ldots, x_{i-1})}{\varphi(u_i)} \quad \text{for } i = j > 1 \quad (14.16)$$

$$= \frac{1}{\varphi(u_i)} \frac{\partial F_i(x_i \mid x_1, \ldots, x_{i-1})}{\partial x_j} \quad \text{for } i > j$$

It is noted that the above transformation is not unique. For example, any reordering of the random variables will produce an alternative transformation to the standard normal space.

In the reliability community, Equation 14.16 is known as the Rosenblatt transformation. This is due to an early work by Rosenblatt [3], which was used by Hohenbichler and Rackwitz in their pioneering use of this transformation in reliability analysis. It must be noted, however, that the above transformation was used much earlier by Segal [4], among others.

14.3 The First-Order Reliability Method

In the first-order reliability method (FORM), an approximation to the probability integral in Equation 14.1 is obtained by linearizing each limit-state function in the standard normal space at an optimal point. The fundamental assumption is that the limit-state functions are continuous and differentiable, at least in the neighborhood of the optimal point. The following subsections describe this method for component and system reliability problems. Simple examples demonstrate the methodology.

14.3.1 Component Reliability by FORM

Consider the component reliability problem defined by Equations 14.1 and 14.2a, which is characterized by the limit-state function $g(\mathbf{x})$ and the joint PDF $f(\mathbf{x})$. Transforming the variables into the standard normal space, the failure probability integral is written as

$$p_f = \int_{g(\mathbf{x}) \leq 0} f(\mathbf{x}) d\mathbf{x} = \int_{G(\mathbf{u}) \leq 0} \varphi_n(\mathbf{u}) d\mathbf{u} \quad (14.17)$$

where $G(\mathbf{u}) \equiv g(\mathbf{T}^{-1}(\mathbf{u}))$ is the limit-state function in the standard normal space. The FORM approximation is obtained by linearizing the function $G(\mathbf{u})$ at a point $\mathbf{u}^*$ defined by the constrained optimization problem

$$\mathbf{u}^* = \arg\min\{\|\mathbf{u}\| \mid G(\mathbf{u}) = 0\} \quad (14.18)$$

where "arg min" denotes the argument of the minimum of a function. It is seen that $\mathbf{u}^*$ is located on the limit-state surface, $G(\mathbf{u}) = 0$, and has minimum distance from the origin in the standard normal space. Because equal probability density contours in the standard normal space are concentric circles centered at the origin, $\mathbf{u}^*$ has the highest probability density among all realizations in the failure domain $G(\mathbf{u}) \leq 0$.

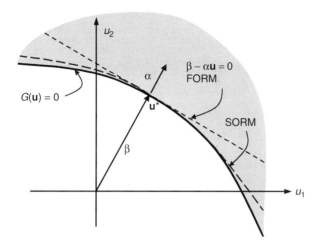

FIGURE 14.2 FORM and SORM approximations for a component problem.

It follows that the neighborhood of this point makes a dominant contribution to the last integral in Equation 14.17. In this sense, $\mathbf{u}^*$ is an optimal point for the linearization of the limit-state function. Another important advantage of this point is that it is invariant of the formulation of the limit-state function, although it may be dependent on the selected form of the transformation from $\mathbf{x}$ to $\mathbf{u}$ space. In the reliability community, this point is commonly known as the *design point*, but other names such as *most probable point* (MPP) and *beta point* are also used.

Noting that $G(\mathbf{u}^*) = 0$, the linearized limit-state function is written as

$$G(\mathbf{u}) \cong G_1(\mathbf{u}) = \nabla G(\mathbf{u}^*)(\mathbf{u} - \mathbf{u}^*) = \|\nabla G(\mathbf{u}^*)\|(\beta - \alpha\mathbf{u}) \tag{14.19}$$

where $\nabla G(\mathbf{u}) = [\partial G/\partial u_1, \ldots, \partial G/\partial u_n]$ denotes the gradient row vector[1], $\alpha = -\nabla G(\mathbf{u}^*)/\|\nabla G(\mathbf{u}^*)\|$ is the normalized negative gradient row vector at the design point (a unit vector normal to the limit-state surface at the design point and pointing toward the failure domain), and $\beta = \alpha \mathbf{u}^*$ is the *reliability index*. In essence, the linearization replaces the failure domain $G(\mathbf{u}) \leq 0$ by the half space $\beta - \alpha\mathbf{u} \leq 0$; see Figure 14.2. The first-order approximation of the failure probability is given by the probability content of the half space in the standard normal space, which is completely defined by the distance β; that is,

$$p_f \cong p_{f1} = \Phi(-\beta) \tag{14.20}$$

where the subscript 1 is used to indicate a first-order approximation. This approximation normally works well because, as mentioned earlier, the neighborhood of the design point makes the dominant contribution to the probability integral 14.17. There are two conditions under which this approximation may not work well: (1) the surface $G(\mathbf{u}) = 0$ is strongly nonflat, and (2) the optimization problem in Equation 14.18 has multiple local or global solutions. A recourse for the first condition is to use a higher-order approximation, such as SORM (see Section 14.4), or a corrective sampling method, such as importance sampling or sampling on the orthogonal plane [5]. All these methods make use of the design point, so they need the FORM solution as a first step. The second condition is quite rare, but it does occur, particularly when dealing with dynamic problems. If the local solutions of Equation 14.18 are

---

[1] To be consistent with the definition of a Jacobian matrix, the gradient of a function is defined as a row vector. Note that for a set of functions $f_i(x_1, \ldots, x_n)$, $i = 1, \ldots, m$, the Jacobian matrix has the elements $J_{ij} = \partial f_i/\partial x_j$, $i = 1, \ldots, m$, $j = 1, \ldots, n$. The gradient ∇f_i is the i-th row of this matrix. The normalized gradient α is taken as a row vector for the same reason.

significant, i.e., the corresponding distances from the origin are not much greater than β, then one recourse is to use multiple linearizations and a series system formulation to obtain an improved FORM approximation (see, e.g., [6]).

It should be clear by now that solving Equation 14.18 is the main computational effort in FORM. Several well-established iterative algorithms are available for this purpose [7]. Starting from an initial point $\mathbf{u}_1 = \mathbf{T}(\mathbf{x}_1)$, one typically computes a sequence of points using the recursive formula

$$\mathbf{u}_{i+1} = \mathbf{u}_i + \lambda_i \mathbf{d}_i \qquad i = 1, 2, \ldots \tag{14.21}$$

where $\mathbf{d}_i$ is a search direction vector and λ_i is a step size. Algorithms differ in their selections of $\mathbf{d}_i$ and λ_i. One simple algorithm that is especially designed for the objective function in Equation 14.18 uses the search direction

$$\mathbf{d}_i = \left[\frac{G(\mathbf{u}_i)}{\|\nabla G(\mathbf{u}_i)\|} + \alpha_i \mathbf{u}_i\right]\alpha_i^{\mathrm{T}} - \mathbf{u}_i \tag{14.22}$$

where $\alpha_i = -\nabla G(\mathbf{u}_i)/\|\nabla G(\mathbf{u}_i)\|$ is the normalized negative gradient row vector. For $\lambda_i = 1$, this algorithm is identical to one originally suggested by Hasofer and Lind [8] and later generalized for nonnormal variables by Rackwitz and Fissler [9]. However, with a unit step size this algorithm does not converge when $1 \leq |\beta \kappa_i|$, where κ_i is a principal curvature of the limit-state surface at the design point. The appropriate step size can be selected by monitoring a merit function, $m(\mathbf{u})$. This is any continuous and differentiable function of $\mathbf{u}$, whose minimum occurs at the solution of Equation 14.18 and for which $\mathbf{d}_i$ is a descent direction at $\mathbf{u}_i$ (i.e., the value of the function $m(\mathbf{u})$ decreases as we move a small distance in the direction $\mathbf{d}_i$ starting from $\mathbf{u}_i$). Zhang and Der Kiureghian [10] have shown that a merit function that satisfies these conditions is

$$m(\mathbf{u}) = \frac{1}{2}\|\mathbf{u}\|^2 + c|G(\mathbf{u})| \tag{14.23}$$

where the penalty parameter c should be selected at each step to satisfy the condition $c_i > \|\mathbf{u}_i\|/\|\nabla G(\mathbf{u}_i)\|$. Using the merit function, the best step size is obtained as

$$\lambda_i = \arg\min\{m(\mathbf{u}_i + \lambda \mathbf{d}_i)\} \tag{14.24}$$

However, strict solution of the above minimization is costly. In practice, it is sufficient to select $\lambda_i \in (0,1]$ such that $m(\mathbf{u}_i + \lambda \mathbf{d}_i) < m(\mathbf{u}_i)$. A popular rule for this purpose is the Armijo rule [11].

The algorithm described by Equation 14.21 through Equation 14.24 is known as the Improved HL-RF algorithm [10]. For most problems, this algorithm converges in just a few steps. To assure that convergence to the global solution is obtained, it is a good practice to repeat the solution starting from a different initial point $\mathbf{x}_1$. In practice, it is very rare that convergence to a local solution occurs, and when this does occur, usually one can detect it from the context of the problem. Nevertheless, one should keep in mind that optimization algorithms do not guarantee convergence to the global solution and caution should be exercised in interpreting their results.

As we have seen, the FORM solution essentially requires repeated evaluation of the limit-state function $G(\mathbf{u})$ and its gradient $\nabla G(\mathbf{u})$ at selected points $\mathbf{u}_i$, $i = 1, 2, \ldots$. Because the limit-state function is defined in terms of the original random variables $\mathbf{x}$, it is necessary to carry out these calculations in that space. For this purpose, for any point $\mathbf{u}_i$ selected in accordance with the optimization algorithm, the inverse transform $\mathbf{x}_i = \mathbf{T}^{-1}(\mathbf{u}_i)$ is used to compute the corresponding point in the original space. Then, $G(\mathbf{u}_i) = g(\mathbf{x}_i)$ and $\nabla G(\mathbf{u}_i) = \nabla g(\mathbf{x}_i)\mathbf{J}_{\mathbf{u},\mathbf{x}}^{-1}(\mathbf{x}_i)$, where $\mathbf{J}_{\mathbf{u},\mathbf{x}}$ is the Jacobian of the transformation, as described in Section 14.2. Note that $\mathbf{J}_{\mathbf{u},\mathbf{x}}$ being a triangular matrix, its inverse is easy to compute.

TABLE 14.1 Description of Random Variables for Example in Section 14.3.1.1

Variable	Marginal Distribution	Mean	c.o.v.	Correlation Coefficient			
				m_1	m_2	p	y
m_1, kNm	Normal	250	0.3	1.0			
m_2, kNm	Normal	125	0.3	0.5	1.0		
p, kN	Gumbel	2,500	0.2	0.3	0.3	1.0	
y, MPa	Weibull	40	0.1	0.0	0.0	0.0	1.0

14.3.1.1 Example: Reliability Analysis of a Column by FORM

Consider a short column subjected to biaxial bending moments m_1 and m_2 and axial force p. Assuming an elastic perfectly plastic material with yield stress y, the failure of the column is defined by the limit-state function

$$g(\mathbf{x}) = 1 - \frac{m_1}{S_1 y} - \frac{m_2}{S_2 y} - \left(\frac{p}{Ay}\right)^\theta \tag{14.25}$$

where $\mathbf{x} = \{m_1, m_2, p, y\}^T$ denotes the vector of random variables, $\theta = 2$ is a limit-state function parameter, $A = 0.190$ m² is the cross-sectional area, and $S_1 = 0.030$ m³ and $S_2 = 0.015$ m³ are the flexural moduli of the fully plastic column section. Assume m_1, m_2, p, and y have the Nataf distribution with the second moments and marginal distributions listed in Table 14.1. Starting from the mean point, the Improved HL-RF algorithm converges in nine steps with the results $\mathbf{u}^* = \{1.21, 0.699, 0.941, -1.80\}^T$, $\mathbf{x}^* = \{341, 170, 3223, 31.8\}^T$, $\alpha = [0.491, 0.283, 0.381, -0.731]$, $\beta = 2.47$, and $p_{f1} = 0.00682$. The "exact" estimate of the failure probability, obtained with 120,000 Monte Carlo simulations and a coefficient of variation of 3%, is $p_f \cong 0.00931$.

14.3.2 System Reliability by FORM

Consider a series or parallel system reliability problem defined by a set of limit-state functions $g_k(\mathbf{x})$, $k = 1, 2, \ldots, m$, and the failure domain as in Equation 14.2b or 14.2c. Let $G_k(\mathbf{u})$, $k = 1, 2, \ldots, m$, denote the corresponding limit-state functions in the standard normal space. A first-order approximation to the system failure probability is obtained by linearizing each limit-state function $G_k(\mathbf{u})$ at a point $\mathbf{u}_k^*$, $k = 1, 2, \ldots, m$, such that the surface is approximated by the tangent hyperplane

$$\beta_k - \alpha_k \mathbf{u} = 0 \tag{14.26}$$

where $\alpha_k = -\nabla G_k(\mathbf{u}_k^*)/\|\nabla G_k(\mathbf{u}_k^*)\|$ is the unit normal to the hyperplane and $\beta_k = \alpha_k \mathbf{u}_k^*$ is the distance from the origin to the hyperplane. An easy choice for the linearization points $\mathbf{u}_k^*$ is the minimum-distance points from the origin, as defined in Equation 14.18. While this is a good choice for series systems, for parallel systems a better choice is the so-called *joint design point*

$$\mathbf{u}^* = \arg\min\{\|\mathbf{u}\| \,|\, G_k(\mathbf{u}) \leq 0, k = 1, \ldots, m\} \tag{14.27}$$

The above is an optimization problem with multiple inequality constraints, for which standard algorithms are available. Figure 14.3 illustrates the above choices for linearization. It is clear that the linearization according to Equation 14.27 will provide a better approximation of the failure domain for parallel systems. Nevertheless, the linearization point according to Equation 14.18 is often used for all system problems because it is much easier to obtain.

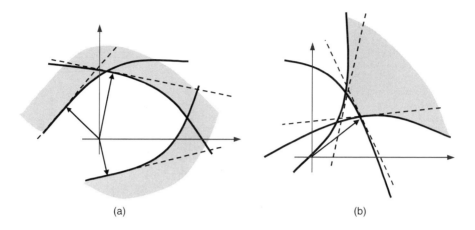

FIGURE 14.3 Linearization points for series and parallel systems: (a) linearization according to Equation 14.18; (b) linearization according to Equation 14.27.

With the limit-state surfaces linearized, the system failure domain is approximated by a hyper-polygon. The corresponding approximations of the failure probability are derived as follows: Let $v_k = \alpha_k \mathbf{u}$, $k = 1, 2, \ldots, m$. It is easy to see that $\mathbf{v} = \{v_1, \ldots, v_m\}^T$ are normal random variables with zero means, unit variances, and correlation coefficients $\rho_{kl} = \alpha_k \alpha_l^T$, $k, l = 1, 2, \ldots, m$. For a series system, one can write [12]:

$$p_{f1,\text{series}} = P\left[\bigcup_{k=1}^{m}(\beta_k \leq v_k)\right]$$
$$= 1 - P\left[\bigcap_{k=1}^{m}(v_k < \beta_k)\right] \qquad (14.28)$$
$$= 1 - \Phi_m(\mathbf{B}, \mathbf{R})$$

where $\Phi_m(\mathbf{B}, \mathbf{R})$ is the m-variate standard normal CDF with correlation matrix $\mathbf{R} = [\rho_{kl}]$ and evaluated at the thresholds $\mathbf{B} = (\beta_1, \ldots, \beta_m)$. For a parallel system, one can write [12]:

$$p_{f1,\text{parallel}} = P\left[\bigcap_{k=1}^{m}(\beta_k \leq v_k)\right]$$
$$= P\left[\bigcap_{k=1}^{m}(v_k < -\beta_k)\right] \qquad (14.29)$$
$$= \Phi_m(-\mathbf{B}, \mathbf{R})$$

where use has been made of the symmetry of the standard normal space. Note that for a single component ($m = 1$), the above relations 14.28 and 14.29 both reduce to 14.20.

It is clear from the above analysis that computing the multi-normal probability function is essential for FORM solution of series and parallel systems. For $m = 2$, the bivariate normal CDF can be computed using the single-fold integral

$$\Phi_2(\beta_1, \beta_2, \rho) = \Phi(\beta_1)\Phi(\beta_2) + \int_0^\rho \varphi_2(\beta_1, \beta_2, r)\,dr \qquad (14.30)$$

where $\varphi_2(\ldots r)$ denotes the bivariate standard normal PDF with correlation coefficient r. For higher dimensions, numerical techniques for computing the multi-normal probability have been developed (see, e.g., [13–15]).

As described in the introduction, a general system can be represented as a series system of parallel subsystems, with each parallel subsystem representing a cut set. Let

$$C_k = \left\{ \bigcap_{j \in c_k} G_j(\mathbf{u}) \leq 0 \right\} \tag{14.31}$$

denote the k-th cut-set event, that is, the event that all components within the k-th cut set have failed. The probability of failure of a system with m cut sets can be written as

$$p_{f,\text{general system}} = P\left(\bigcup_{k=1}^{m} C_k \right) \tag{14.32}$$

Several options are available for computing the above probability. If the cut sets are disjoint (i.e., no two cut sets can simultaneously occur), we can write

$$p_{f,\text{general system}} = \sum_{k=1}^{m} P(C_k) \tag{14.33}$$

where each term $P(C_k)$ is a parallel system problem and can be solved as described above. The difficulty in this approach is that the disjoint cut sets can be numerous and may contain large numbers of components. A more compact system formulation is obtained in terms of the minimum cut sets, that is, cut sets that do not contain superfluous components. However, such cut sets usually are neither mutually exclusive nor statistically independent. If the number of minimum cut sets is not too large, the inclusion-exclusion rule of probability can be used to write Equation 14.32 as

$$p_{f,\text{general system}} = \sum_{k=1}^{m} P(C_k) - \sum_{k=1}^{m-1} \sum_{l=k+1}^{m} P(C_k C_l) + \cdots + (-1)^{m-1} P(C_1 C_2 \cdots C_m) \tag{14.34}$$

Each term in the above expression represents a parallel system reliability problem and can be solved as described earlier. Note, however, that the number of parallel system problems to be solved and the number of components in them will rapidly increase with the number of cut sets, m.

To avoid solving parallel systems with large number of components, bounding formulas for system failure probability have been developed that rely on low-order probabilities. For example, the Kounias-Hunter-Ditlevsen bounds, progressively developed in [16–18], are

$$p_{f,\text{general system}} \geq P(C_1) + \sum_{k=2}^{m} \max\left[0, P(C_k) - \sum_{l=1}^{k-1} P(C_k C_l) \right] \tag{14.35a}$$

$$p_{f,\text{general system}} \leq P(C_1) + \sum_{k=2}^{m} \left[P(C_k) - \max_{l<k} P(C_k C_l) \right] \tag{14.35b}$$

where only joint probabilities of pairs of cut sets are required. Similar formulas involving joint probabilities of three or more cut sets have been developed by Zhang [19]. Recently, Song and Der Kiureghian [20] developed a linear programming algorithm for computing bounds on general system probability for any given information on the marginal and joint component probabilities. Their approach is guaranteed to produce the narrowest possible bounds for any given information.

In the structural reliability literature, the bounds in Equation 14.35 are often considered in connection with series systems, where each component represents a cut set. The above formulation shows that these bounds are equally applicable to general systems represented in terms of multi-component cut sets.

14.3.2.1 Example: Series System Reliability Analysis of a Frame by FORM

Consider the one-bay frame in Figure 14.4a, which has ductile members and is subjected to random horizontal and vertical loads h and v. The frame has random plastic moment capacities m_i, $i = 1, \ldots, 5$, at the joints shown in the figure. Under the applied loads, this frame may fail in any of the three mechanisms shown in Figure 14.4b. Using the principle of virtual work, these mechanisms are described by the limit-state functions

$$g_1(\mathbf{x}) = m_1 + m_2 + m_4 + m_5 - 5h \tag{14.36a}$$

$$g_2(\mathbf{x}) = m_2 + 2m_3 + m_4 - 5v \tag{14.36b}$$

$$g_3(\mathbf{x}) = m_1 + 2m_3 + 2m_4 + m_5 - 5h - 5v \tag{14.36c}$$

where $\mathbf{x} = \{m_1, \ldots, m_5, h, v\}^T$ is the vector of random variables. Table 14.2 shows the assumed distributions and second moments of the random variables.

The reliability of the frame against the formation of a mechanism represents a series system reliability problem with the limit-state functions shown in 14.36. Using the FORM approximation, we obtain the following $\beta_k \alpha_k$ values for the three components:

$$\beta_1 \alpha_1 = 2.29[-0.238 - 0.174 - 0.044 - 0.131 - 0.112 \; 0.939 \; 0.000] \tag{14.37a}$$

$$\beta_2 \alpha_2 = 2.87[-0.263 - 0.356 - 0.425 - 0.204 \; 0.000 \; 0.000 \; 0.763] \tag{14.37b}$$

$$\beta_3 \alpha_3 = 2.00[-0.313 - 0.137 - 0.291 - 0.240 - 0.113 \; 0.792 \; 0.317] \tag{14.37c}$$

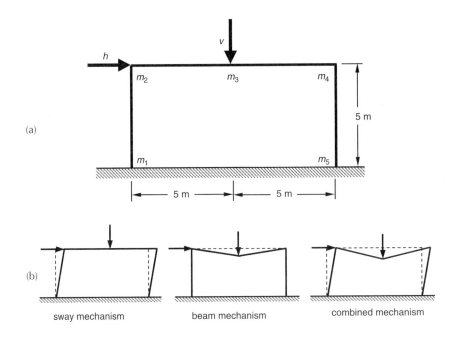

FIGURE 14.4 Ductile frame and its failure mechanisms.

First- and Second-Order Reliability Methods

TABLE 14.2 Description of Random Variables for Example in Section 14.3.2.1

Variable	Distribution	Mean	c.o.v.	Correlation
m_i, $i=1,\ldots,5$, kNm	Joint lognormal	150	0.2	$\rho_{m_i m_j} = 0.3$, $i \neq j$
h, kN	Gumbel	50	0.4	Independent
v, kN	Gamma	60	0.2	Independent

These values are used in Equation 14.28 to obtain the FORM approximation of the failure probability, $p_{f1} = 0.02644$. If the bounds in Equation 14.35 are used, the result is $0.02639 \leq p_{f1} \leq 0.02647$.

14.3.2.2 Example: Reliability Updating of a Frame after Proof Test

Suppose the frame in the example in Section 14.3.2.1 has been proof-tested under a horizontal load of $h_0 = 70$ kN and a vertical load of $v_0 = 72$ kN. Because the frame has survived under these loads, we have observed

$$g_4(\mathbf{x}) = -(m_1 + m_2 + m_4 + m_5 - 5h_0) \leq 0 \tag{14.38a}$$

$$g_5(\mathbf{x}) = -(m_2 + 2m_3 + m_4 - 5v_0) \leq 0 \tag{14.38b}$$

$$g_6(\mathbf{x}) = -(m_1 + 2m_3 + 2m_4 + m_5 - 5h_0 - 5v_0) \leq 0 \tag{14.38c}$$

where h_0 and v_0 are deterministic values, as given above. The updated probability of failure of the frame in light of the proof-test result is the conditional probability

$$p_{f|\text{proof test}} = \frac{P(C_1 \cup C_2 \cup C_3)}{P(C_4)} \tag{14.39}$$

where the index sets for the cut sets C_k, $k=1,\ldots,4$, are $c_1 = (1,4,5,6)$, $c_2 = (2,4,5,6)$, $c_3 = (3,4,5,6)$, and $c_4 = (4,5,6)$, respectively. It is seen that the numerator represents a general system problem, whereas the denominator is a parallel system problem. First-order approximation of the probabilities in Equation 14.39 yields the updated probability $p_{f|\text{proof test}} = 0.0189$. In light of the positive proof-test observation, this result is smaller than the unconditional failure probability estimated earlier. This example demonstrates the use of general system reliability analysis. It also shows how reliability can be updated in light of real-world observations.

14.3.3 FORM Importance and Sensitivity Measures

An important by-product of FORM is a set of importance and sensitivity measures that provide information as to the order of importance of the random variables and the sensitivities of the reliability index or the first-order approximation of the failure probability with respect to parameters in the probability distribution or limit-state models. This section briefly introduces these measures.

Let $G_1(\mathbf{u}) = \|\nabla G\|(\beta - \alpha \mathbf{u})$ denote the linearized limit-state function as in Equation 14.19. Noting that the mean of $\mathbf{u}$ is zero and its covariance is the identity matrix, we obtain the mean and variance of $G_1(\mathbf{u})$ as

$$\mu_{G_1} = \|\nabla G\| \beta \tag{14.40}$$

$$\sigma_{G_1}^2 = \|\nabla G\|^2 \left(\alpha_1^2 + \alpha_2^2 + \cdots + \alpha_n^2\right) = \|\nabla G\|^2 \tag{14.41}$$

where use has been made of the fact that α is a unit vector. The result verifies that $\beta = \mu_{G_1}/\sigma_{G_1}$ is indeed the reliability index for the linearized problem. More importantly, Equation 14.41 shows that α_i^2 is proportional to the contribution of random variable u_i to the total variance of the linearized limit-state function. Clearly, the larger this contribution, the more important random variable u_i is. Hence, the elements of α provide relative measures of importance of the standard normal variables u_i, $i=1,\ldots,n$.

Furthermore, considering the expanded expression $G_1(\mathbf{u}) = \|\nabla G\|(\beta - \alpha_1 u_1 - \cdots - \alpha_n u_n)$, it is clear that a positive (negative) value of α_i is an indication that random variable u_i is of load (capacity) type.

When the basic random variables of a reliability problem are statistically independent, there is a one-to-one correspondence between the original random variables x_i and the standard normal random variables u_i. The order of importance and nature (load or capacity) of random variables x_i then are similar to the corresponding u_i and can be determined in terms of the α vector. However, when the random variables $\mathbf{x}$ are statistically dependent, there is no such one-to-one correspondence. In that case, α does not provide information about the relative importance of the random variables $\mathbf{x}$ in the original space. To derive measures of importance for the basic random variables, consider linearizing the transformation $\mathbf{u} = \mathbf{T}(\mathbf{x})$ at the design point $\mathbf{u}^*$:

$$\mathbf{u} \cong \mathbf{u}^* + \mathbf{J}_{\mathbf{u},\mathbf{x}}(\mathbf{x} - \mathbf{x}^*) \tag{14.42}$$

Replacing the approximation by an equality, we can write

$$\mathbf{u} = \mathbf{u}^* + \mathbf{J}_{\mathbf{u},\mathbf{x}}(\hat{\mathbf{x}} - \mathbf{x}^*) \tag{14.43}$$

where $\hat{\mathbf{x}}$ is slightly different from $\mathbf{x}$. Because $\hat{\mathbf{x}}$ is a linear function of $\mathbf{u}$, it must have the joint normal distribution. Its covariance matrix is

$$\hat{\Sigma} = \mathbf{J}_{\mathbf{u},\mathbf{x}}^{-1}\left(\mathbf{J}_{\mathbf{u},\mathbf{x}}^{-1}\right)^T \tag{14.44}$$

The random variables $\hat{\mathbf{x}}$ are considered as "equivalent normals" of $\mathbf{x}$ at the design point. The covariance matrix $\hat{\Sigma}$ in general depends on the design point and is slightly different from the actual covariance matrix Σ of $\mathbf{x}$. The magnitude of the difference depends on the degree of nonnormality of $\mathbf{x}$. Now using Equation 14.43 in the expression for $G_1(\mathbf{u})$ and noting that $\beta = \alpha \mathbf{u}^*$, we obtain $G_1(\mathbf{u}) = -\|\nabla G\| \alpha \mathbf{J}_{\mathbf{u},\mathbf{x}}(\hat{\mathbf{x}} - \mathbf{x}^*)$. The variance of $G_1(\mathbf{u})$ can now be written as

$$\sigma_{G_1}^2 = \|\nabla G\|^2 \left(\alpha \mathbf{J}_{\mathbf{u},\mathbf{x}} \hat{\Sigma} \mathbf{J}_{\mathbf{u},\mathbf{x}}^T \alpha^T\right) = \|\nabla G\|^2 \left(\|\alpha \mathbf{J}_{\mathbf{u},\mathbf{x}} \hat{\mathbf{D}}\|^2 + \alpha \mathbf{J}_{\mathbf{u},\mathbf{x}}(\hat{\Sigma} - \hat{\mathbf{D}}\hat{\mathbf{D}}) \mathbf{J}_{\mathbf{u},\mathbf{x}}^T \alpha^T\right) \tag{14.45}$$

where $\hat{\mathbf{D}} = \mathrm{diag}[\hat{\sigma}_i]$ is the diagonal matrix of standard deviations of $\hat{\mathbf{x}}$. The first term in the above expression contains the contributions to the variance of $G_1(\mathbf{u})$ arising from the individual variances of the elements of $\hat{\mathbf{x}}$, whereas the second term represents the contributions arising from the covariances of pairs of the random variables. Hence, the elements of the vector $\alpha \mathbf{J}_{\mathbf{u},\mathbf{x}} \hat{\mathbf{D}}$ can be considered to provide relative measures of importance of the elements of $\hat{\mathbf{x}}$, or approximately of $\mathbf{x}$. Normalizing this vector, we define

$$\gamma = \frac{\alpha \mathbf{J}_{\mathbf{u},\mathbf{x}} \hat{\mathbf{D}}}{\|\alpha \mathbf{J}_{\mathbf{u},\mathbf{x}} \hat{\mathbf{D}}\|} \tag{14.46}$$

as the unit row vector defining the relative importance of the original random variables $\mathbf{x}$. A positive (negative) value for the element γ_i of this vector indicates that x_i is of load (capacity) type. It is easy to show that when the random variables are statistically independent, γ reverts back to α.

We now turn our attention to reliability sensitivity measures. Let $f(\mathbf{x}, \theta_f)$ denote the joint PDF of $\mathbf{x}$, where θ_f is a set of distribution parameters, and $g(\mathbf{x}, \theta_g)$ denote the limit-state function, where θ_g is a set of limit-state parameters. It can be shown (see [21] and [22]) that the gradients of β with respect to these parameters are

$$\nabla_{\theta_f}\beta = \alpha \mathbf{J}_{\mathbf{u},\theta_f}(\mathbf{x}^*, \theta_f) \tag{14.47}$$

$$\nabla_{\theta_g}\beta = \frac{1}{\|\nabla G\|} \nabla_{\theta_g} g(\mathbf{x}^*, \theta_g) \tag{14.48}$$

First- and Second-Order Reliability Methods

where $\mathbf{J}_{\mathbf{u},\theta_f}(\mathbf{x}^*,\theta_f)$ is the Jacobian of the probability transformation $\mathbf{u} = \mathbf{T}(\mathbf{x},\theta_f)$ with respect to the parameters θ_f, evaluated at the design point. Using Equation 14.20, the corresponding gradients of the first-order probability approximation are obtained from

$$\nabla_\theta p_{f1} = -\varphi(\beta)\nabla_\theta \beta \qquad (14.49)$$

where $\theta = \theta_f$ or θ_g.

14.3.3.1 Example: Importance and Sensitivity Measures for a Column

Reconsider the example in Section 14.3.1.1. Using Equation 14.46, the importance vector for the basic random variables $\mathbf{x} = \{m_1, m_2, p, y\}^T$ is $\gamma = [0.269, 0.269, 0.451, -0.808]$. From the signs of the elements of the importance vector, it is seen that the first three random variables are load types, whereas the fourth variable, y, is a capacity variable. This finding is intuitively obvious for this problem. However, in a more complex problem, the nature of the random variables may not be obvious and the information provided by the importance vector can be valuable. Of the four basic random variables, y is the most important variable, followed by p, m_1, and m_2. This is in spite of the fact that y has the smallest coefficient of variation among the four.

Using Equation 14.47, the gradient vector of β with respect to the mean values of the random variables (with the standard deviations fixed) is obtained as $\nabla_{\text{mean}}\beta = 10^{-3}[-3.24, -6.48, -0.546, 0.124]$ and the gradient with respect to the standard deviations (with the means fixed) is obtained as $\nabla_{\text{stdev}}\beta = 10^{-3}[-3.92, -7.84, -0.790, -0.245]$. Furthermore, using Equation 14.48, the sensitivity of the reliability index with respect to the limit-state function parameter θ is obtained as $\partial\beta/\partial\theta = 0.552$. The corresponding sensitivities of the first-order failure probability are obtained from Equation 14.49 by scaling the above sensitivity values by $-\varphi(2.47) = -0.0190$.

14.4 The Second-Order Reliability Method

As its name implies, the second-order reliability method, SORM, involves a second-order approximation of the limit-state function. Consider a Taylor series expansion of the component limit-state function $G(\mathbf{u})$ at the design point $\mathbf{u}^*$,

$$\begin{aligned} G(\mathbf{u}) &\cong \nabla G(\mathbf{u}^*)(\mathbf{u}-\mathbf{u}^*) + \frac{1}{2}(\mathbf{u}-\mathbf{u}^*)^T \mathbf{H}(\mathbf{u}-\mathbf{u}^*) \\ &= \|\nabla G(\mathbf{u}^*)\| \left[(\beta - \alpha\mathbf{u}) + \frac{1}{2\|\nabla G(\mathbf{u}^*)\|}(\mathbf{u}-\mathbf{u}^*)^T \mathbf{H}(\mathbf{u}-\mathbf{u}^*) \right] \end{aligned} \qquad (14.50)$$

where α and β are as defined earlier and $\mathbf{H}$ is the second-derivative matrix at the design point having the elements $H_{ij} = \partial^2 G(\mathbf{u}^*)/(\partial u_i \partial u_j)$, $i,j = 1,\ldots,n$. Now consider a rotation of the axes $\mathbf{u}' = \mathbf{Pu}$, where $\mathbf{P}$ is an orthonormal matrix with α as its last row. Such a matrix can be constructed by, for example, the well-known Gram-Schmidt algorithm. This rotation positions the design point on the u'_n axis, such that $\mathbf{u}'^* = [0 \cdots 0 \ \beta]^T$. Because $\mathbf{u} = \mathbf{P}^T \mathbf{u}'$, defining $G'(\mathbf{u}') = G(\mathbf{P}^T\mathbf{u}')/\|\nabla G(\mathbf{u}^*)\|$, we have

$$G'(\mathbf{u}') \cong \beta - \alpha\mathbf{P}^T\mathbf{u}' + \frac{1}{2\|\nabla G(\mathbf{u}^*)\|}(\mathbf{u}'-\mathbf{u}'^*)^T \mathbf{P}\mathbf{H}\mathbf{P}^T(\mathbf{u}'-\mathbf{u}'^*) \qquad (14.51)$$

Noting that $\alpha\mathbf{P}^T\mathbf{u}' = u'_n$ and letting $\mathbf{A} = \mathbf{P}\mathbf{H}\mathbf{P}^T/\|\nabla G(\mathbf{u}^*)\|$, we have

$$G'(\mathbf{u}') \cong \beta - u'_n + \frac{1}{2}(\mathbf{u}'-\mathbf{u}'^*)^T \mathbf{A}(\mathbf{u}'-\mathbf{u}'^*) \qquad (14.52)$$

where $\mathbf{A}$ is a symmetric matrix. Now, consider the partitioning $\mathbf{u}' = \{\mathbf{u}_1'^T \; u_n'\}^T$, where $\mathbf{u}_1' = \{u_1', \ldots, u_{n-1}'\}^T$, and the corresponding partition of $\mathbf{A}$, $\mathbf{A} = \begin{bmatrix} \mathbf{A}_{11} & \mathbf{A}_{12} \\ \mathbf{A}_{12}^T & a_{nn} \end{bmatrix}$, where $\mathbf{A}_{11}$ is the $(n-1)\times(n-1)$ matrix formed by the first $n-1$ rows and columns of $\mathbf{A}$. Expanding the matrix product in Equation 14.52 yields

$$G'(\mathbf{u}') \cong \beta - u_n' + \frac{1}{2}\left[\mathbf{u}_1'^T \mathbf{A}_{11} \mathbf{u}_1' + 2(u_n' - \beta)\mathbf{A}_{12}^T \mathbf{u}_1' + a_{nn}(u_n' - \beta)^2\right] \quad (14.53)$$

Because $u_n' - \beta = 0$ is the tangent plane at the design point, for points on the limit-state surface in the neighborhood of the design point, the last two terms inside the square brackets in Equation 14.53 are of smaller order than the first term. Neglecting these terms, we arrive at

$$G'(\mathbf{u}') \cong \beta - u_n' + \frac{1}{2}\mathbf{u}_1'^T \mathbf{A}_{11} \mathbf{u}_1' \quad (14.54)$$

This is the equation of a paraboloid with its apex at the design point. Now consider a rotation of the axes around u_n' defined by the transformation $\mathbf{u}_1'' = \mathbf{Q}\mathbf{u}_1'$, where $\mathbf{Q}$ is an $(n-1)\times(n-1)$ orthonormal matrix. Since $\mathbf{u}_1' = \mathbf{Q}^T \mathbf{u}_1''$, we can write $\mathbf{u}_1'^T \mathbf{A}_{11} \mathbf{u}_1' = \mathbf{u}_1''^T \mathbf{Q} \mathbf{A}_{11} \mathbf{Q}^T \mathbf{u}_1''$. It follows that by selecting $\mathbf{Q}^T$ as the eigenmatrix of $\mathbf{A}_{11}$, the product $\mathbf{Q}\mathbf{A}_{11}\mathbf{Q}^T$ diagonalizes and Equation 14.54 reduces to

$$G'(\mathbf{u}') = G'(\mathbf{Q}^T \mathbf{u}_1'', u_n') \cong \beta - u_n' + \frac{1}{2}\sum_{i=1}^{n-1} \kappa_i u_i''^2 \quad (14.55)$$

where κ_i are the eigenvalues of $\mathbf{A}_{11}$. The preceding expression defines a paraboloid through its principal axes with $\kappa_i, i = 1, \ldots, n-1$, denoting its principal curvatures. This paraboloid is tangent to the limit-state surface at the design point and its principal curvatures match those of the limit-state surface at the design point. Note that, for $\beta > 0$, a positive curvature denotes a surface that curves away from the origin. Because the apex is at the design point, which is the point nearest to the origin, it follows that the inequality $-1 < \beta \kappa_i$ must hold for each principal curvature.

In SORM, the probability of failure is approximated by the probability content of the above-defined paraboloid. Because the standard normal space is rotationally symmetric, this probability, which we denote as p_{f2}, is completely defined by β and the set of curvatures $\kappa_i, i = 1, \ldots, n-1$. Hence, we can write the SORM approximation as

$$p_f \cong p_{f2}(\beta, \kappa_1, \ldots, \kappa_{n-1}) \quad (14.56)$$

Tvedt [23] has derived an exact expression for p_{f2} under the condition $-1 < \beta \kappa_i$. Expressed in the form of a single-fold integral in the complex plane, Tvedt's formula is

$$p_{f2} = \varphi(\beta)\mathrm{Re}\left\{ \mathrm{i}\sqrt{\frac{2}{\pi}} \int_0^{i\infty} \frac{1}{s} \exp\left[\frac{(s+\beta)^2}{2}\right] \prod_{i=1}^{n-1} \frac{1}{\sqrt{1+\kappa_i s}} ds \right\} \quad (14.57)$$

where $\mathrm{i} = \sqrt{-1}$. A simpler formula based on asymptotic approximations derived earlier by Breitung [24] is

$$p_{f2} \cong \Phi(-\beta) \prod_{i=1}^{n-1} \frac{1}{\sqrt{1+\psi(\beta)\kappa_i}} \quad (14.58)$$

where $\psi(\beta) = \beta$. This formula works well for large values of β. Hohenbichler and Rackwitz [25] have shown that improved results are obtained by using $\psi(\beta) = \varphi(\beta)/\Phi(\beta)$. Note that in the above formula, each term $1/\sqrt{1+\psi(\beta)\kappa_i}$ acts as a correction factor on the FORM approximation to account for the curvature of the limit-state surface in the principal direction u_i''.

To summarize, the SORM approximation according to the above formulation involves the following steps after the design point has been found: (1) Construct the orthonormal matrix $\mathbf{P}$ with α as its last row. (2) Compute the second-derivative matrix $\mathbf{H} = [\partial^2 G/(\partial u_i \partial u_j)]$ at the design point. Usually, a finite-difference scheme in the $\mathbf{u}$ space is used, with the individual function values computed at the corresponding points in the $\mathbf{x}$ space. (3) Compute the matrix $\mathbf{A} = \mathbf{PHP}^{\mathrm{T}}/\|\nabla G(\mathbf{u}^*)\|$ and form $\mathbf{A}_{11}$ by deleting the last row and column of $\mathbf{A}$. (4) Compute the eigenvalues of $\mathbf{A}_{11}$, κ_i, $i = 1, \ldots, n-1$. (5) Use either of the formulas in Equation 14.57 or Equation 14.58 to obtain the SORM approximation. The most difficult part of this calculation is the evaluation of the second-derivative matrix. This is particularly the case if the limit-state function involves numerical algorithms, such as finite element calculations. Two issues arise in such applications. One is the cost of computing the full second-derivative matrix by finite differences when the number of random variables is large. The second is calculation noise in the limit-state function due to truncation errors, which could produce erroneous estimates of the curvatures. Two alternative SORM methods that attempt to circumvent these problems are described below.

Der Kiureghian and De Stefano [26] have shown that certain algorithms for finding the design point, including the Improved HL-RF algorithm described earlier, have the property that the trajectory of trial points $\mathbf{u}_i$, $i = 1, 2, \ldots$, asymptotically converges on the major principal axis of the limit-state surface, that is, the axis u_i'' having the maximum $|\kappa_i|$ value. Using this property, the major principal curvature can be approximately computed in terms of the quantities that are available in the last two iterations while finding the design point. Suppose convergence to the design point is achieved after r iterations with $\mathbf{u}_{r-1}$ and $\mathbf{u}_r$ denoting the last two trial points and α_{r-1} and α_r denoting the corresponding unit normal vectors. Using the geometry in Figure 14.5, the major principal curvature is approximately computed as

$$\kappa_i \cong \frac{\mathrm{sgn}[\alpha_r(\mathbf{u}_r - \mathbf{u}_{r-1})]\cos^{-1}(\alpha_r \alpha_{r-1}^{\mathrm{T}})}{\sqrt{\|\mathbf{u}_r - \mathbf{u}_{r-1}\|^2 - \|\alpha_r(\mathbf{u}_r - \mathbf{u}_{r-1})\|^2}} \tag{14.59}$$

To obtain the second major principal curvature, one repeats the search process (from a randomly selected initial point) in a subspace orthogonal to the major principal axis. The result, using Equation 14.59, is an approximation of the principal curvature having the second-largest absolute value. Next, the search is repeated in a subspace orthogonal to the first two principal axes. The result is the third major principal curvature. This process is continued until all principal curvatures of significant magnitude have been

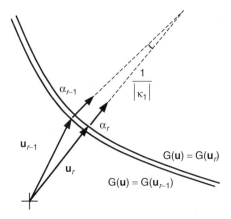

FIGURE 14.5 Geometry for computing the principal curvature.

obtained. The advantages of this approach are twofold: (1) one does not need to compute second derivatives, and (2) one computes the principal curvatures in the order of their importance and can stop the process when the curvature magnitudes are sufficiently small. The latter property is particularly important in problems with large n, where often only a few principal curvatures are significant; that is, the limit-state surface is significantly curved in only a few directions. The reader should consult Der Kiureghian and De Stefano [26] for further details on the implementation of this method.

The SORM method described above constructs the approximating paraboloid by fitting to the principal curvatures at the design point. For this reason, this method is known as the *curvature-fitting* SORM method. As mentioned, there are situations where the computation of the limit-state function involves noise due to truncation errors and computing high-order derivatives is problematic. To circumvent this problem, Der Kiureghian et al. [27] developed a *point-fitting* SORM method, where a piecewise paraboloid surface is defined by fitting to points selected on the limit-state surface on either side of the design point along each axis $u'_i, i=1,\ldots,n-1$. As shown in Figure 14.6 for a slightly modified version of the method, the points for axis u'_i are selected by searching along a path consisting of lines $u'_i = \pm b$ and a semicircle of radius b centered at the design point. The parameter b is selected according to the rule $b = 1$ if $|\beta|<1$, $b=|\beta|$ if $1 \leq |\beta| \leq 3$ and $b=3$ if $3<|\beta|$. This rule assures that the fitting points are neither too close nor too far from the design point. Let (u'^-_i, u'^-_n) and (u'^+_i, u'^+_n) denote the coordinates of the fitting points along axis u'_i, where the superscript signs indicate the negative and positive sides of the u'_i axis. Through each fitting point, a semiparabola is defined that is tangent at the design point (see Figure 14.6). The curvature of the semiparabola at the design point is given by

$$a_i^{\text{sgn}(u'_i)} = \frac{2\left(u'^{\text{sgn}(u'_i)}_n - \beta\right)}{\left(u'^{\text{sgn}(u'_i)}_i\right)^2} \tag{14.60}$$

where $\text{sgn}(u'_i)$ denotes the sign of the coordinate on the u'_i axis. The approximating limit-state function is now defined as

$$G'(\mathbf{u}') \cong \beta - u'_n + \frac{1}{2}\sum_{i=1}^{n-1} a_i^{\text{sgn}(u'_i)} u'^2_i \tag{14.61}$$

When set equal to zero, the above expression defines a piecewise paraboloid surface that is tangent to the limit-state surface at the design point and is coincident with each of the fitting points. Interestingly, this function is continuous and twice differentiable despite the fact that the coefficients $a_i^{\text{sgn}(u'_i)}$ are

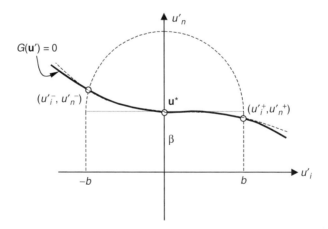

FIGURE 14.6 Definition of fitting points in point-fitting SORM method.

discontinuous. This is because this discontinuity occurs at points where u'_i takes on the zero value. The advantages of this approach are that it is insensitive to noise in the calculation of the limit-state function and it does not require derivative calculations. On the other hand, it requires finding $2(n-1)$ fitting points on the surface. Furthermore, because the fitting points are obtained in the $\mathbf{u}'$ space, the solution may depend on the selected transformation matrix $\mathbf{P}$, which of course is not unique. Der Kiureghian et al. [27] have shown that the maximum error resulting from the worst choice of $\mathbf{P}$ is much smaller than the error in the FORM approximation. That is, even in the worst case, this method improves the FORM approximation.

The above SORM methods are applicable to component reliability problems. For series system reliability problems, a SORM approximation may be obtained by replacing each component reliability index β_k by $\beta'_k = \Phi^{-1}(1 - p_{f2})$, while keeping the unit vector α_k unchanged. The formula in Equation 14.28 is then used to obtain the series system probability. This procedure essentially adjusts the distances to the componential hyper-planes such that the half-space probability for each component is equal to the SORM approximation for the component. Unfortunately, a similar approach may not provide an improved result for nonseries systems.

14.4.1 Example: Reliability Analysis of a Column by SORM

For the example in Section 14.3.1.1, a SORM analysis reveals the principal curvatures $\kappa_1 = -0.155$ and $\kappa_2 = -0.0399$, with κ_3 being practically zero. The curvature-fitting SORM estimate of the failure probability according to Tvetd's formula (Equation 14.57) is $p_{f2} = 0.00936$, whereas the result based on Breitung's formula (Equation 14.58) is $p_{f2} \cong 0.00960$. The result based on the point-fitting SORM method is $p_{f2} \cong 0.00913$. All these results closely match the "exact" result given in the example in Section 14.3.1.1.

14.5 Time-Variant Reliability Analysis

Many problems in engineering involve random quantities that vary in time or space. Such quantities are properly modeled as stochastic processes or random fields. If the failure event of interest is defined over a temporal or spatial domain, then the reliability problem is said to be time- or space-variant, respectively. In this section, we only discuss time-variant problems. One-dimensional space-variant reliability problems have similar features. However, multidimensional space-variant problems require more advanced tools (see, e.g., [28]). Our discussion in this section addresses time-variant problems only in the context of FORM analysis. The more general topic of time-variant reliability analysis is, of course, much broader and involves such topics as stochastic differential equations and random vibrations.

Consider a component reliability problem defined by the limit-state function $g[\mathbf{x}, \mathbf{y}(t)]$, where $\mathbf{x}$ is a vector of random variables and $\mathbf{y}(t)$ is a vector of random processes. According to our definition, $\{g[\mathbf{x}, \mathbf{y}(t)] \leq 0\}$ describes the failure event at time t. Because, for a given t, $\mathbf{y}(t)$ is a vector of random variables, the *instantaneous* failure probability $p_f(t) = P\{g[\mathbf{x}, \mathbf{y}(t)] \leq 0\}$ can be computed by the methods described in this chapter, provided the joint distribution of $\mathbf{x}$ and $\mathbf{y}(t)$ is available.

A more challenging problem results when the failure domain is defined as

$$\Omega = \left\{ \min_{t \in T} g[\mathbf{x}, \mathbf{y}(t)] \leq 0 \right\} \tag{14.62}$$

where T denotes an interval of time, say $T = \{t \mid t_1 < t \leq t_m\}$. This is the well-known first-passage problem: the failure event occurs when the process $g[\mathbf{x}, \mathbf{y}(t)]$ first down-crosses the zero level. One can easily verify that

$$\max_{t \in T} p_f(t) \leq P\left\{ \min_{t \in T} g[\mathbf{x}, \mathbf{y}(t)] \leq 0 \right\} \tag{14.63}$$

That is, the maximum of the instantaneous failure probability over an interval only provides a lower bound to the probability of failure over the interval. This is because the failure event can occur at times other than the time at which $p_f(t)$ assumes its maximum. In the following, we present two approaches for approximately solving the above time-variant reliability problem.

A simple way to solve the problem is to "discretize" the time interval and use a series system approximation. Let $t_k, k = 1, \ldots, m$, be a set of time points selected within the interval T. Then,

$$P\left\{\bigcup_{k=1}^{m} g[\mathbf{x}, \mathbf{y}(t_k)] \leq 0\right\} \leq P\left\{\min_{t \in T} g[\mathbf{x}, \mathbf{y}(t)] \leq 0\right\} \tag{14.64}$$

The problem on the left is a series-system reliability problem with m "time-point" components. It can be solved by the FORM methods described in Section 14.3. This involves transforming the entire vector of random variables, $[\mathbf{x}, \mathbf{y}(t_1), \ldots, \mathbf{y}(t_m)]$, to the standard normal space $\mathbf{u}$, finding the design point for each time-point component, and linearizing the corresponding limit-state surfaces such that $g[\mathbf{x}, \mathbf{y}(t_k)] \propto \beta_k - \alpha_k \mathbf{u}$, $k = 1, \ldots, m$, where β_k and α_k are the distance and the unit normal vector defining the plane tangent to the k-th limit-state surface. The quantities β_k and α_k are used in Equation 14.28 to compute the FORM approximation to the probability on the left side of Equation 14.64. The result is an approximate lower bound on the time-variant failure probability. This lower bound can be improved by increasing the number of discrete time points. When $\mathbf{y}(t)$ and, therefore, $g[\mathbf{x}, \mathbf{y}(t)]$ are nonstationary, advantage can be gained by selecting time points at which the instantaneous failure probability, $p_f(t)$, is high. However, selecting time points that are too closely spaced is not necessary because the failure events associated with such points are strongly correlated (the α_k vectors are nearly coincident) and, therefore, they would not significantly add to the series-system probability. One can increase the number of time points gradually until no appreciable increase in the system probability is observed. The result, then, provides a narrow lower bound approximation to the time-variant failure probability. Obviously, this method can become cumbersome when the process $g[\mathbf{x}, \mathbf{y}(t)]$ has a short correlation length relative to the interval T, because in that case a large number of points are necessary to obtain a good approximation. The main computational effort is in finding the design points associated with the discrete time steps t_k, $k = 1, \ldots, m$, by use of an iterative algorithm, such as the one described in Section 14.3. In this analysis, significant saving is achieved by finding the design points in sequence and using each preceding solution as the initial trial point in the search for the next design point.

An entirely different approach for computing the time-variant failure probability uses the mean rate of down-crossings of the process $g[\mathbf{x}, \mathbf{y}(t)]$ below the zero level. Consider a small time interval $[t, t + \delta t)$ and let $v(t)\delta t$ denote the mean number of times that the process $g[\mathbf{x}, \mathbf{y}(t)]$ down-crosses the level zero during this interval. Let $p(k)$ denote the probability that k down-crossings occur during the interval. We can write

$$v(t)\delta t = 0 \times p(0) + 1 \times p(1) + 2 \times p(2) + \cdots \tag{14.65}$$

For a sufficiently small δt, assuming the process $g[\mathbf{x}, \mathbf{y}(t)]$ has a smoothly varying correlation structure, the probability of more than one down-crossing can be considered negligible in relation to the probability of a single down-crossing. Thus, the second- and higher-order terms in the above expression can be dropped. The probability of a single down-crossing is computed by noting that this event will occur if $0 < g[\mathbf{x}, \mathbf{y}(t)]$ and $g[\mathbf{x}, \mathbf{y}(t + \delta t)] \leq 0$. Hence [29],

$$v(t) = \lim_{\delta t \to 0} \frac{P\left\{-g[\mathbf{x}, \mathbf{y}(t)] < 0 \bigcap g[\mathbf{x}, \mathbf{y}(t + \delta t)] \leq 0\right\}}{\delta t} \tag{14.66}$$

The probability in the numerator represents a parallel-system reliability problem with two components. This problem can be solved by FORM. The solution essentially requires finding the design points for the instantaneous limit-state functions $-g[\mathbf{x}, \mathbf{y}(t)]$ and $g[\mathbf{x}, \mathbf{y}(t+\delta t)]$ and the associated reliability indices $\beta(t)$ and $\beta(t+\delta t)$ and unit normals $\alpha(t)$ and $\alpha(t+\delta t)$, respectively. To avoid dealing with the highly correlated random vectors $\mathbf{y}(t)$ and $\mathbf{y}(t+\delta t)$, one can use the approximation $\mathbf{y}(t+\delta t) \cong \mathbf{y}(t) + \delta t\, \dot{\mathbf{y}}(t)$, where $\dot{\mathbf{y}}(t)$ denotes the derivative vector process. Furthermore, for small δt, we have $\beta(t+\delta t) \cong -\beta(t) = b$ and the inner product of $\alpha(t)$ and $\alpha(t+\delta t)$, which defines the correlation coefficient between the two component events, is nearly -1. For this limiting case, one can show (see [30]) that Equation 14.30 reduces to

$$\Phi_2(-b,b,\rho) \cong \frac{1}{4}\exp\left(-\frac{b^2}{2}\right)\left(1+\frac{\sin^{-1}\rho}{\pi/2}\right) \tag{14.67}$$

This procedure is repeated for a grid of time points to compute $v(t)$ as a function of time. Normally, $v(t)$ is a smooth function and a coarse grid of time points is sufficient.

With the mean down-crossing rate computed, two options are available for estimating the time-variant failure probability. One is the well known upper bound defined by

$$P\left\{\min_{t\in T}[\mathbf{x},\mathbf{y}(t)]\leq 0\right\} \leq P\left\{g[\mathbf{x},\mathbf{y}(t_0)]\leq 0\right\} + \int_{t\in T} v(t)dt \tag{14.68}$$

where $P\{g[\mathbf{x},\mathbf{y}(t_0)]\leq 0\}$ is the instantaneous probability of failure at the start of the interval, a time-invariant problem. This expression usually provides a good approximation of the time-variant failure probability when the process $g[\mathbf{x},\mathbf{y}(t)]$ is not narrow band and the mean rate $v(t)$ is small. The second approach is based on the assumption that the down-crossing events are Poisson, which implies statistical independence between these events. This approximation works only if the limit-state process is ergodic.[2] Thus, for this approximation to work, the random variables $\mathbf{x}$ should not be present and $\mathbf{y}(t)$ must be ergodic. The Poisson-based approximation then is

$$P\left\{\min_{t\in T} g[\mathbf{y}(t)]\leq 0\right\} \cong \exp\left[-\int_{t\in T} v(t)dt\right] \tag{14.69}$$

Provided ergodicity holds, this approximation normally works well when the mean down-crossing rate is small and the limit-state process is not narrow band.

14.5.1 Example: Mean Out-Crossing Rate of a Column under Stochastic Loads

Reconsider the example in Section 14.3.1.1. Assume the applied loads $\mathbf{y}(t) = \{m_1(t), m_2(t), p(t)\}^T$ represent a vector of stationary Gaussian processes with the second moments described in Table 14.1. Also assume the zero-mean stationary Gaussian vector process $\dot{\mathbf{y}}(t)$ has the root-mean square values 300π kNm/s, 150π kNm/s and $2,500\pi$ kN/s, and the correlation coefficients $\rho_{1,2}=0.4$, $\rho_{1,3}=\rho_{2,3}=0.2$. Assuming the yield stress is a deterministic value $y = 40$ MPa, using Equation 14.66, the mean out-crossing rate is obtained as $v = 0.00476$ s^{-1}, which is of course independent of time due to the stationarity of the process. If y is considered to be random as in Table 14.1, the estimate of the mean out-crossing rate is $v = 0.00592$ s^{-1}. This value is slightly greater due to the added uncertainty of the yield stress.

---

[2] A random process is ergodic if its ensemble averages equal its corresponding temporal averages.

14.6 Finite Element Reliability Analysis

The finite element (FE) method is widely used in engineering practice to solve boundary-value problems governed by partial differential equations. Essentially, the method provides an algorithmic relation between a set of input variables **x** and processes **y**(t) and a set of output responses **s**, that is, an implicit relation of the form $\mathbf{s} = \mathbf{s}(\mathbf{x}, \mathbf{y}(t))$. When **x** and **y**(t) are uncertain, **s** is also uncertain. Reliability methods can then be used to compute the probability associated with any failure event defined in terms of the response vector **s**. This may include time- or space-variant problems, in which the failure event is defined over a temporal or spatial domain.

In using FORM or SORM to solve FE reliability problems, one essentially needs to repeatedly solve for the response and the response gradients for a sequence of selected realizations of the basic random variables. In FORM, the realizations of the random variables are selected in accordance to the algorithm for finding the design point (see Section 14.3). The number of such repeated FE solutions usually is not large (e.g., of the order of several tens); furthermore, this number is independent of the magnitude of the failure probability and the number of random variables. In SORM, the number of repeated FE solutions is directly governed by the number of random variables. In either case, if the failure probability is small, the amount of needed computations is usually much smaller than that needed in a straightforward Monte Carlo simulation approach.

An important issue in FORM solution of FE reliability problems is the computation of the response gradients, $\nabla_\mathbf{x}\mathbf{s}$ and $\nabla_\mathbf{y}\mathbf{s}$, which are needed in computing the gradient of the limit-state function. A finite difference approach for computing these gradients is often costly and unreliable due to the presence of computational noise arising from truncation errors. For this reason, direct differentiation methods (DDM) have been developed [31, 32]. These methods compute the response gradients directly, using the derivatives of the governing equations. Of course, this implies coding all the derivative equations in the FE code, for example, the derivatives of the constitutive laws, of the element stiffness and mass matrices, etc. A few modern finite element codes possess this capability. In particular, the OpenSees code developed by the Pacific Earthquake Engineering Research (PEER) Center is so enabled. Furthermore, it embodies the necessary routines for defining probability distributions for input variables or processes and for reliability analysis by FORM and two sampling methods [33]. This code is freely downloadable at http://opensees.berkeley.edu/.

FE reliability analysis by FORM and SORM entails a number of challenging problems, including convergence issues having to do with the discontinuity of gradients for certain nonlinear problems, computational efficiency, processing of large amounts of data, definition of failure events, and solution of time- and space-variant problems. A fuller description of these issues is beyond the scope of this chapter. The reader is referred to [28, 33, 34, 35].

References

1. Liu, P.-L. and Der Kiureghian, A. (1986). Multivariate distribution models with prescribed marginals and covariances. *Probabilistic Engineering Mechanics*, 1(2),105–112.
2. Hohenbichler, M. and Rackwitz, R. (1981). Nonnormal dependent vectors in structural safety. *J. Engineering Mechanics*, ASCE, 107(6), 1227–1238.
3. Rosenblatt, M. (1952). Remarks on a multivariate transformation. *Annals Math. Stat.*, 23, 470–472.
4. Segal, I.E. (1938). Fiducial distribution of several parameters with applications to a normal system. *Proceedings of Cambridge Philosophical Society*, 34, 41–47.
5. Engelund, S. and Rackwitz, R. (1993). A benchmark study on importance sampling techniques in structural reliability. *Structural Safety*, 12(4), 255–276.
6. Der Kiureghian, A. and Dakessian, T. (1998). Multiple design points in first- and second-order reliability. *Structural Safety*, 20(1), 37–49.
7. Liu, P.-L. and Der Kiureghian, A. (1990). Optimization algorithms for structural reliability. *Structural Safety*, 9(3), 161–177.

8. Hasofer, A.M. and Lind, N.C. (1974). Exact and invariant second-moment code format. *J. Engineering Mechanics Division*, ASCE, 100(1), 111–121.
9. Rackwitz, R. and Fiessler, B. (1978). Structural reliability under combined load sequences. *Computers & Structures*, 9, 489–494.
10. Zhang, Y. and Der Kiureghian, A. (1995). Two improved algorithms for reliability analysis. In *Reliability and Optimization of Structural Systems*, Proceedings of the 6th IFIP WG 7.5 Working Conference on Reliability and Optimization of Structural Systems, 1994, (R. Rackwitz, G. Augusti, and A. Borri, Eds.), 297–304.
11. Luenberger, D.G. (1986). *Introduction to Linear and Nonlinear Programming*, Addison-Wesley, Reading, MA.
12. Hohenbichler, M. and Rackwitz, R. (1983). First-order concepts in system reliability. *Structural Safety*, 1(3), 177–188.
13. Gollwitzer, S. and Rackwitz, R. (1988). An efficient numerical solution to the multinormal integral. *Probabilistic Engineering Mechanics*, 3(2), 98–101.
14. Ambartzumian, R.V., Der Kiureghian, A., Oganian, V.K., and Sukiasian, H.S. (1998). Multinormal probability by sequential conditioned importance sampling: theory and application. *Probab. Engrg. Mech.*, 13(4), 299–308.
15. Pandey, M.D. (1998). An effective approximation to evaluate multinormal integrals. *Structural Safety*, 20, 51–67.
16. Kounias, E.G. (1968). Bounds for the probability of a union, with applications. *Am. Math. Stat.*, 39(6), 2154–2158.
17. Hunter, D. (1976). An upper bound for the probability of a union. *J. Applied Probability*, 13, 597–603.
18. Ditlevsen, O. (1979). Narrow reliability bounds for structural systems. *J. Structural Mechanics*, 7(4), 453–472.
19. Zhang, Y.C. (1993). High-order reliability bounds for series systems and application to structural systems. *Computers & Structures*, 46(2), 381–386.
20. Song, J. and Der Kiureghian, A. (2003). Bounds on system reliability by linear programming. *Engineering Mechanics*, ASCE, 129(6), 627–636.
21. Hohenbichler, M. and Rackwitz, R. (1986). Sensitivity and importance measures in structural reliability. *Civil Engineering Systems*, 3, 203–210.
22. Bjerager, P. and Krenk, S. (1989). Parametric sensitivity in first-order reliability theory. *J. Engineering Mechanics*, ASCE, 115(7), 1577–1582.
23. Tvedt, L. (1990). Distribution of quadratic forms in normal space, application to structural reliability. *J. Engineering Mechanics*, ASCE, 116(6), 1183–1197.
24. Breitung, K. (1984). Asymptotic approximations for multinormal integrals. *J. Engineering Mechanics*, ASCE, 110(3), 357–366.
25. Hohenbichler, M. and Rackwitz, R. (1988). Improvement of second-order reliability estimates by importance sampling. *J. Engineering Mechanics*, ASCE, 114(12), 2195–2199.
26. Der Kiureghian, A. and DeStefano, M. (1991). Efficient algorithm for second-order reliability analysis. *J. Engineering Mechanics*, ASCE, 117(12), 2904–2923.
27. Der Kiureghian, A., Lin, H.S., and Hwang, S.-J. (1987). Second-order reliability approximations. *J. Engineering Mechanics*, ASCE, 113(8), 1208–1225.
28. Der Kiureghian, A. and Zhang, Y. (1999). Space-variant finite element reliability analysis. *Comp. Methods Appl. Mech. Engrg.*, 168(1–4), 173–183.
29. Hagen, O. and Tvedt, L. (1991). Vector process out-crossing as parallel system sensitivity measure. *J. Engineering Mechanics*, ASCE, 117(10), 2201–2220.
30. Koo, H. and Der Kiureghian, A. (2003). FORM, SORM and Simulation Techniques for Nonlinear Random Vibrations. *Report No. UCB/SEMM-2003/01*, Department of Civil & Environmental Engineering, University of California, Berkeley, CA, February.
31. Zhang, Y. and Der Kiureghian, A. (1993). Dynamic response sensitivity of inelastic structures. *Computer Methods in Applied Mechanics and Engineering*, 108(1), 23–36.

32. Kleiber, M., Antunez, H., Hien, T.D., and Kowalczyk, P. (1997). *Parameter Sensitivity in Nonlinear Mechanics*. John Wiley & Sons, West Sussex, U.K.
33. Haukaas, T. and Der Kiureghian A. (2004). Finite element reliability and sensitivity methods for performance-based earthquake engineering. PEER Report 2003/4, Pacific Earthquake Engineering Research Center, University of California, Berkeley, CA, April.
34. Liu, P.L. and Der Kiureghian, A. (1991). Finite-element reliability of geometrically nonlinear uncertain structures. *J. Engineering Mechanics*, ASCE, 117(8), 1806–1825.
35. Sudret, B. and Der Kiureghian, A. (2002). Comparison of finite element reliability methods. *Probabilistic Engineering Mechanics*, 17(4), 337–348.

15
System Reliability

15.1 Introduction 15-1
15.2 Modeling of Structural Systems 15-2
Introduction • Fundamental Systems • Modeling of Systems at Level N • Modeling of Systems at Mechanism Level • Formal Representation of Systems
15.3 Reliability Assessment of Series Systems 15-10
Introduction • Assessment of the Probability of Failure of Series Systems • Reliability Bounds for Series Systems • Series Systems with Equally Correlated Elements • Series Systems with Unequally Correlated Elements
15.4 Reliability Assessment of Parallel Systems 15-18
Introduction • Assessment of the Probability of Failure of Parallel Systems • Reliability Bounds for Parallel Systems • Equivalent Linear Safety Margins for Parallel Systems • Parallel Systems with Equally Correlated Elements • Parallel Systems with Unequally Correlated Elements
15.5 Identification of Critical Failure Mechanisms 15-26
Introduction • Assessment of System Reliability at Level 1 • Assessment of System Reliability at Level 2 • Assessment of System Reliability at Level N > 2 • Assessment of System Reliability at Mechanism Level • Examples
15.6 Illustrative Examples 15-38
Reliability of a Tubular Joint • Reliability-Based Optimal Design of a Steel-Jacket Offshore Structure • Reliability-Based Optimal Maintenance Strategies

Palle Thoft-Christensen
Aalborg University, Aalborg, Denmark

15.1 Introduction

A fully satisfactory estimate of the reliability of a structure is based on a system approach. In some situations, it may be sufficient to estimate the reliability of the individual structural members of a structural system. This is the case for statically determinate structures where failure in any member will result in failure of the total system. However, failure of a single element in a structural system will generally not result in failure of the total system, because the remaining elements may be able to sustain the external load by redistribution of the internal load effects. This is typically the case of statically indeterminate (redundant) structures, where failure of the structural system always requires that more than one element fail. A structural system will usually have a large number of failure modes, and the most significant failure modes must be taken into account in an estimate of the reliability of the structure.

From an application point of view, the reliability of structural systems is a relatively new area. However, extensive research has been conducted in the past few decades and a number of effective methods have been developed. Some of these methods have a limited scope and some are more general. One might argue that this area is still in a phase of development and therefore not yet sufficiently clarified for practical application. However, a number of real practical applications have achieved success. This chapter does not try to cover all aspects of structural system reliability. No attempt is made to include all methods that can be used in estimating the reliability of structural systems. Only the β-unzipping method is described in detail here, because the author has extensive experience from using that method.

This section is to some degree based on the books by Thoft-Christensen and Baker [1] and Thoft-Christensen and Murotsu [2].

15.2 Modeling of Structural Systems

15.2.1 Introduction

A real structural system is so complex that direct exact calculation of the probability of failure is completely impossible. The number of possible different failure modes is so large that they cannot all be taken into account; and even if they could all be included in the analysis, exact probabilities of failure cannot be calculated. It is therefore necessary to idealize the structure so that the estimate of the reliability becomes manageable. Not only the structure itself, but also the loading must be idealized. Because of these idealizations it is important to bear in mind that the estimates of, for example, probabilities of failure are related to the idealized system (the model) and not directly to the structural system. The main objective of a structural reliability design is to be able to design a structure so that the probability of failure is minimized in some sense. Therefore, the model must be chosen carefully so that the most important failure modes for the real structures are reflected in the model.

It is assumed that the total reliability of the structural system can be estimated by considering a finite number of failure modes and combining them in complex reliability systems. The majority of structural failures are caused by human errors. Human errors are usually defined as serious mistakes in design, analysis, construction, maintenance, or use of the structure, and they cannot be included in the reliability modeling presented in this chapter. However, it should be stated that the probabilities of failure calculated by the methods presented in this book are much smaller than those observed in practice due to gross errors. The failures included in structural reliability theory are caused by random fluctuations in the basic variables, such as extremely low strength capacities or extremely high loads.

Only truss and frame structures are considered, although the methods used can be extended to a broader class of structures. Two-dimensional (plane) as well as three-dimensional (spatial) structures are treated. It is assumed that the structures consist of a finite number of bars and beams, and that these structural elements are connected by a finite number of joints. In the model of the structural system, the failure elements are connected to the structural elements (bars, beams, and joints); see Section 15.2.3. For each of the structural elements, a number of different failure modes exist. Each failure mode results in element failure, but systems failure will, in general, only occur when a number of simultaneous element failures occur. A more precise definition of system failure is one of the main objects of this chapter. It is assumed that the reliability of a structural system can be estimated on the basis of a series system modeling, where the elements are failure modes. The failure modes are modeled by parallel systems.

15.2.2 Fundamental Systems

Consider a statically determinate (nonredundant) structure with n structural elements and assume that each structural element has only one failure element; see Figure 15.1. The total number of failure elements is therefore also n. For such a structure, the total structural system fails as soon as any structural element fails. This is symbolized by a series system.

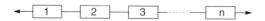

FIGURE 15.1 Series system with n elements.

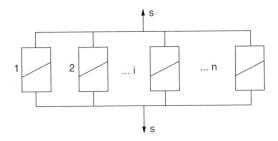

FIGURE 15.2 Parallel system with n elements.

For a statically indeterminate (redundant) structure, failure in a single structural element will not always result in failure of the total system. The reason for this is that the remaining structural elements will be able to sustain the external loading by redistribution of the load effects. For statically indeterminate structures, total failure will usually require that failure takes place in more than one structural element. It is necessary to define what is understood by total failure of a structural system. This problem will be addressed in more detail, but formation of a mechanism is the most frequently used definition. If this definition is used here, failure in a set of failure elements forming a mechanism is called a failure mode. Formation of a failure mode will therefore require simultaneous failure in a number of failure elements. This is symbolized by a parallel system, see Figure 15.2.

In this section it is assumed that all basic variables (load variables and strength variables) are normally distributed. All geometrical quantities and elasticity coefficients are assumed deterministic. This assumption significantly facilitates the estimation of the failure probability, but, in general, basic variables cannot be modeled by normally distributed variables at a satisfactory degree of accuracy. To overcome this problem, a number of different transformation methods have been suggested. The most well-known method was suggested by Rackwitz and Fiessler [3]. One drawback to these methods is that they increase the computational work considerably due to the fact that they are iterative methods. A simpler (but also less accurate) method, called the multiplication factor method, has been proposed by Thoft-Christensen [4]. The multiplication factor method does not increase the computational work. Without loss of generality, it is assumed that all basic variables are standardized; that is, the mean value is 0 and the variance 1.

The β-unzipping method is only used for trussed and framed structures but it can easily be modified to other classes of structures. The structure is considered at a fixed point in time, so that only static behavior has been addressed. It is assumed that failure in a structural element (section) is either pure tension/compression or failure in bending. Combined failure criteria have also been used in connection with the β-unzipping method, but only little experience has been obtained until now.

Let the vector $\bar{X} = (X_1, \ldots, X_n)$ be the vector of the standardized normally distributed basic variables with the joint probability density function φ_n and let failure of a failure element be determined by a failure function $f: \omega \to R$, where ω is the n-dimensional basic variable space. Let f be defined in such a way that the space ω is divided into a failure region $\omega_f = \{\bar{x} : f(\bar{x}) \leq 0\}$ and a safe region $\omega_s = \{\bar{x} : f(\bar{x}) > 0\}$ by the failure surface (limit state) $\partial \omega = \{\bar{x} : f(\bar{x}) > 0\}$, where the vector $\bar{x}$ is a realization of the random vector $\bar{X}$. Then the probability of failure P_f for the failure element in question is given by

$$P_f = P(f(\bar{X}) \leq 0) = \int_{\omega_f} \varphi_n(\bar{x}) d\bar{x}. \qquad (15.1)$$

If the function f is linearized at the so-called design point at the distance β to the origin of the coordinate system, then an approximate value for P_f is given by

$$P_f \approx P(\alpha_1 X_1 + \ldots + \alpha_n X_n + \beta \leq 0) = P(\alpha_1 X_1 + \ldots + \alpha_n X_n \leq -\beta) = \Phi(-\beta) \qquad (15.2)$$

where $\bar{\alpha} = (\alpha_1, \ldots, \alpha_n)$ is the vector of directional cosines of the linearized failure surface. β is the Hasofer-Lind reliability index, and Φ is the standardized normal distribution function. The random variable

$$M = \alpha_1 X_1 + \ldots + \alpha_n X_n + \beta \tag{15.3}$$

is the *linearized safety margin* for the failure element.

Next consider a *series system* with k elements. An estimate of the failure probability P_f^s of this series system can be obtained on the basis of the linearized safety margin of the form in Equation 15.3 for the k elements

$$P_f^s = P\left(\bigcup_{i=1}^{k}(\bar{\alpha}_i \bar{X} + \beta_i \leq 0)\right) = P\left(\bigcup_{i=1}^{k}(\bar{\alpha}_i \bar{X} \leq -\beta_i)\right) = 1 - P\left(\bigcap_{i=1}^{k}(\bar{\alpha}_i \bar{X} > -\beta_i)\right)$$
$$= 1 - P\left(\bigcap_{i=1}^{k}(-\bar{\alpha}_i \bar{X} < \beta_i)\right) = 1 - \Phi_k(\bar{\beta}; \bar{\bar{\rho}}) \tag{15.4}$$

where $\bar{\alpha}_i$ and β_i are the directional cosines and the reliability index for failure element i, $i = 1, \ldots, k$, respectively, and where $\bar{\beta} = (\beta_1, \ldots, \beta_k)$. $\bar{\bar{\rho}} = \{\rho_{ij}\}$ is the correlation coefficient matrix given by $\rho_{ij} = \bar{\alpha}_i^T \bar{\alpha}_j$ for all $i \neq j$. Φ_k is the standardized k-dimensional normal distribution function. Series systems will be treated in more detail in Section 15.3.

For a *parallel system* with k elements, an estimate of the failure probability P_f^p can be obtained in the following way

$$P_f^p = P\left(\bigcap_{i=1}^{k}(\bar{\alpha}_i \bar{X} + \beta_i \leq 0)\right) = P\left(\bigcap_{i=1}^{k}(-\bar{\alpha}_i \bar{X} < -\beta_i)\right) = \Phi_k(\bar{\beta}; \bar{\bar{\rho}}) \tag{15.5}$$

where the same notations as above are used. Parallel systems will be treated in more detail in Section 15.4.

It is important to note the approximation behind Equation 15.4 and Equation 15.5; namely, the linearization of the general nonlinear failure surfaces at the distinct design points for the failure elements. The main problem in connection with application of Equation 15.4 and Equation 15.5 is numerical calculation of the n-dimensional normal distribution function Φ_n for $n \geq 3$. This problem is addressed later in this chapter where a number of methods to get approximate values for Φ_n are mentioned.

15.2.3 Modeling of Systems at Level N

Clearly, the definition of failure modes for a structural system is of great importance in estimating the reliability of the structural system. In this section, failure modes are classified in a systematic way convenient for the subsequent reliability estimate. A very simple estimate of the reliability of a structural system is based on failure of a single failure element, namely the failure element with the lowest reliability index (highest failure probability) of all failure elements. Failure elements are structural elements or cross-sections where failure can take place. The number of failure elements will usually be considerably higher than the number of structural elements. Such a reliability analysis is, in fact, not a system reliability analysis, but from a classification point of view it is convenient to call it *system reliability analysis at level 0*. Let a structure consist of n failure elements and let the reliability index (see, e.g., Thoft-Christensen and Baker [1]) for failure element i be β_i then the system reliability index β_S^0 at level 0 is

$$\beta_S^0 = \min_{i=1,\ldots,n} \beta_i. \tag{15.6}$$

Clearly, such an estimate of the system reliability is too optimistic. A more satisfactory estimate is obtained by taking into account the possibility of failure of any failure element by modeling the structural

FIGURE 15.3 System modeling at level 1.

system as a series system with the failure elements as elements of the system (see Figure 15.3). The probability of failure for this series system is then estimated on the basis of the reliability indices β_i, $i = 1, 2, \ldots, n$, and the correlation between the safety margins for the failure elements. This reliability analysis is called *system reliability analysis at level* 1. In general, it is only necessary to include some of the failure elements in the series system (namely, those with the smallest β-indices) to get a good estimate of the system failure probability P_f^1 and the corresponding generalized reliability index β_S^1, where

$$\beta_S^1 = -\Phi^{-1}\left(P_f^1\right) \tag{15.7}$$

and where Φ is the standardized normal distribution function. The failure elements included in the reliability analysis are called *critical failure elements*.

The modeling of the system at level 1 is natural for a statically determinate structure, but failure in a single failure element in a structural system will not always result in failure of the total system, because the remaining elements may be able to sustain the external loads due to redistribution of the load effects. This situation is characteristic of statically indeterminate structures.

For such structures, *system reliability analysis at level* 2 or higher levels may be reasonable. At level 2, the system reliability is estimated on the basis of a series system where the elements are parallel systems each with two failure elements — so-called *critical pairs of failure elements* (see Figure 15.4). These critical pairs of failure elements are obtained by modifying the structure by assuming, in turn, failure in the critical failure elements and adding fictitious loads corresponding to the load-carrying capacity of the elements in failure. If, for example, element i is a critical failure element, then the structure is modified by assuming failure in element i and the load-carrying capacity of the failure element is added as fictitious loads if the element is ductile. If the failure element is brittle, no fictitious loads are added. The modified structure is then analyzed elastically and new β-values are calculated for all the remaining failure elements. Failure elements with low β-values are then combined with failure element i so that a number of critical pairs of failure elements are defined.

Analyzing modified structures where failure is assumed in critical pairs of failure elements now continues the procedure sketched above. In this way, *critical triples of failure elements* are identified and a *reliability analysis at level* 3 can be made on the basis of a series system, where the elements are parallel systems each with three failure elements (see Figure 15.5). By continuing in the same way, reliability estimates at levels 4, 5, etc. can be performed, but, in general, analysis beyond level 3 is of minor interest.

FIGURE 15.4 System modeling at level 2.

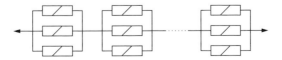

FIGURE 15.5 System modeling at level 3.

15.2.4 Modeling of Systems at Mechanism Level

Many recent investigations in structural system theory concern structures that can be modeled as elastic-plastic structures. In such cases, failure of the structure is usually defined as formation of a mechanism. When this failure definition is used, it is of great importance to be able to identify the most significant failure modes because the total number of mechanisms is usually much too high to be included in the reliability analysis. The β-unzipping method can be used for this purpose, simply by continuing the procedure described above until a mechanism has been found. However, this will be very expensive due to the great number of reanalyzes needed. It turns out to be much better to base the unzipping on reliability indices for fundamental mechanisms and linear combination of fundamental mechanisms.

When system failure is defined as formation of a mechanism, the probability of failure of the structural system is estimated by modeling the structural system as a series system with the significant mechanisms as elements (see Figure 15.6). Reliability analysis based on the mechanism failure definition is called *system reliability analysis at mechanism level*.

For real structures, a mechanism will often involve a relatively large number of yield hinges and the deflections at the moment of formation of a mechanism can usually not be neglected. Therefore, the failure definition must be combined with some kind of deflection failure definition.

15.2.5 Formal Representation of Systems

This section gives a brief introduction to a new promising area within the reliability theory of structural systems called *mathematical theory of system reliability*. Only in the past decade has this method been applied to structural systems, but it has been applied successfully within other reliability areas. However, a lot of research in this area is being conducted and it can be expected that this mathematical theory will also be useful also for structural systems in the future. The presentation here corresponds, to some extent, to the presentation given by Kaufmann, Grouchko, and Cruon [5].

Consider a structural system S with n failure elements $E_1, \ldots, E_n$. Each failure element E_i, $i = 1, \ldots, n$, is assumed to be either in a "state of failure" or in a "state of nonfailure". Therefore, a so-called Boolean variable (indicator function) e_i defined by

$$e_i = \begin{cases} 1 & \text{if the failure element is in a nonfailure state} \\ 0 & \text{if the failure element is in a failure state} \end{cases} \tag{15.8}$$

is associated with each failure element E_i, $i = 1, \ldots, n$.

The state of the system S is therefore determined by the *element state vector*

$$\bar{e} = (e_1, \ldots, e_n). \tag{15.9}$$

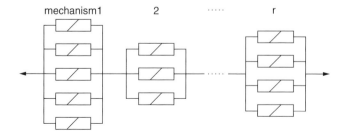

FIGURE 15.6 System modeling at mechanism level.

FIGURE 15.7 Series system with n elements.

The system S is also assumed to be either in a "state of failure" or in a "state of nonfailure". Therefore, a Boolean variable s, defined by

$$s = \begin{cases} 1 & \text{if the system is in a nonfailure state} \\ 0 & \text{if the system is in a failure state} \end{cases} \quad (15.10)$$

is associated with the system S. Because the state of the system S is determined solely by the vector $\bar{e}$, there is a function called the *systems structure function*, $\varphi:\bar{e} \to s$; that is,

$$s = \varphi(\bar{e}). \quad (15.11)$$

—□—

Example 15.1

Consider a series system with n elements as shown in Figure 15.7. This series system is in a safe (non-failure) state if and only if all elements are in a nonfailure state. Therefore, the structural function s_S for a series system is given by

$$s_S = \varphi_S(\bar{e}) = \prod_{i=1}^{n} e_i \quad (15.12)$$

where $\bar{e}$ is given by Equation 15.9. Note that s_S can also be written

$$s_S = \min(e_1, e_2, \ldots, e_n). \quad (15.13)$$

Example 15.2

Consider a *parallel system* with n elements, as shown in Figure 15.8. This parallel system is in a safe state (nonfailure state) only if at least one of its elements is in a nonfailure state. Therefore, the structural function s_P for a parallel system is given by

$$s_P = \varphi_P(\bar{e}) = 1 - \prod_{i=1}^{n} (1 - e_i) \quad (15.14)$$

where $\bar{e}$ is given by Equation 15.9. Note that s_P can also be written

$$s_P = \max(e_1, e_2, \ldots, e_n). \quad (15.15)$$

—□—

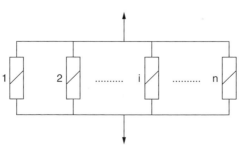

FIGURE 15.8 Parallel system with n elements.

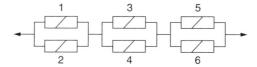

FIGURE 15.9 A level 2 system modeling.

It is easy to combine the structural functions of the system s_S and s_P shown in Equation 15.12 and Equation 15.14 so that the structural function for a more complicated system can be obtained. As an example, consider the system used in system modeling at *level 2* and shown in Figure 15.9. It is a series system with three elements and each of these elements is a parallel system with two failure elements. With the numbering shown in Figure 15.9, the structural function becomes

$$s_S = s_{P_1} s_{P_2} s_{P_3} = [1-(1-e_1)(1-e_2)][1-(1-e_3)(1-e_4)][1-(1-e_5)(1-e_6)] \tag{15.16}$$

where s_{P_i}, $i = 1, 2, 3$, is the structural function for the parallel system i and where e_i, $i = 1, \ldots, 6$, is the Boolean variable for element i.

Consider a structure S with a set of n failure elements $E = \{E_1, \ldots, E_n\}$ and let the structural function of the system s be given by

$$s = \varphi(\bar{e}) = \varphi(e_1, \ldots, e_n) \tag{15.17}$$

where $\bar{e} = (e_1, \ldots, e_n)$ is the element state vector. A subset, $A = \{E_i \mid i \in I\}$, $I \subset \{1, 2, \ldots, n\}$, of E is called a *path set* (or a *link set*) if

$$\left.\begin{array}{l} e_i = 1, i \in I \\ e_i = 0, i \notin I \end{array}\right\} \Rightarrow s = 1. \tag{15.18}$$

According to the definition, a subset $A \subset E$ is a path set if the structure is in a nonfailure state and all elements in $E \setminus A$ are in a failure state.

–□–

Example 15.3

Consider the system shown in Figure 15.10. Clearly, the following subsets of $E = \{E_1, \ldots, E_6\}$ are all path sets: $A_1 = \{E_1, E_2, E_4\}$, $A_2 = \{E_1, E_2, E_5\}$, $A_3 = \{E_1, E_2, E_6\}$, $A_4 = \{E_1, E_3, E_4\}$, $A_5 = \{E_1, E_3, E_5\}$, and $A_6 = \{E_1, E_3, E_6\}$. The path set A_1 is illustrated in Figure 15.11.

–□–

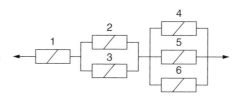

FIGURE 15.10 Example 15.3 system.

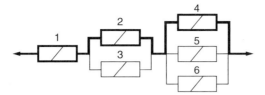

FIGURE 15.11 Path set A_1.

If a path set $A \subset E$ has the property that a subset of A, which is also a path set, does not exist then A is called a *minimal path set*. In other words, a path set is a minimal path set if failure of any failure element in A results in system failure.

Another useful concept is the *cut set* concept. Consider again a structure S defined by Equation 15.17 and let $A \subset E$ be defined by

$$A = \{E_i \mid i \in I\}, \quad I \subset \{1, 2, \ldots, n\}. \tag{15.19}$$

A is then called a cut set, if

$$\left.\begin{array}{l} e_i = 0, i \in I \\ e_i = 1, i \notin I \end{array}\right\} \Rightarrow s = 0. \tag{15.20}$$

According to the definition (Equation 15.20) a subset $A \subset E$ is a cut set if the structure is in a failure state when all failure elements in A are in a failure state and all elements in $E \setminus A$ are in a nonfailure state.

—□—

Example 15.4

Consider again the structure shown in Figure 15.10. Clearly, the following subsets of $E = \{E_1, \ldots, E_6\}$ are all cut sets: $A_1 = \{E_1\}$, $A_2 = \{E_2, E_3\}$, and $A_3 = \{E_4, E_5, E_6\}$. The cut set A_2 is illustrated in Figure 15.12.

—□—

A cut set $A \subset E$ is called a *minimal cut set*, if it has the property that a subset of A is also a cut set. That is a cut set A is a minimal cut set if nonfailure of any failure element in A results in system nonfailure.

It is interesting to note that one can easily prove that any n-tuple $(e_1, \ldots, e_n)$ with $e_i = 0$ or 1, $i = 1, \ldots, n$, corresponds to either a path set or a cut set.

For many structural systems, it is convenient to describe the state of the system by the state of the failure elements on the basis of the system function as described in this section. The next step is then to

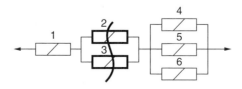

FIGURE 15.12 Cut set A_2.

estimate the reliability of the system when the reliabilities of the failure elements are known. The reliability R_i of failure element E_i is given by

$$R_i = P(e_i = 1) = 1 \times P(e_i = 1) + 0 \times P(e_i = 0) = E[e_i] \quad (15.21)$$

where e_i, the Boolean variable for failure element E_i, is considered a random variable and where $E[e_i]$ is the expected value of e_i.

Similarly, the reliability R_S of the system S is

$$R_S = P(s = 1) = 1 \times P(s = 1) + 0 \times P(s = 0) = E[s] = E[\varphi(\bar{e})] \quad (15.22)$$

where the element state vector $\bar{e}$ is considered a random vector and where φ is the structural function of the system.

Unfortunately, an estimate of $E[\varphi(\bar{e})]$ is only simple when the failure elements are uncorrelated and when the system is simple, for example, a series system. In civil engineering, failure elements will often be correlated. Therefore, the presentation above is only useful for very simple structures.

15.3 Reliability Assessment of Series Systems

15.3.1 Introduction

To illustrate the problems involved in estimating the reliability of systems, consider a structural element or structural system with two potential failure modes defined by safety margins $M_1 = f_1(X_1, X_2)$ and $M_2 = f_2(X_1, X_2)$, where X_1 and X_2 are standardized normally distributed basic variables. The corresponding failure surface and reliability indices β_1 and β_2 are shown in Figure 15.13.

Realizations (x_1, x_2) in the dotted area ω_f will result in failure, and the probability of failure P_f is equal to

$$P_f = \int_{\omega_f} \varphi_{X_1, X_2}(x_1, x_2; 0) d\bar{x} \quad (15.23)$$

where φ_{X_1, X_2} is the bivariate normal density function for the random vector $\bar{X} = (X_1, X_2)$. Let $\beta_2 < \beta_1$ as shown in Figure 15.13, and assume that the reliability index β for the considered structural element or

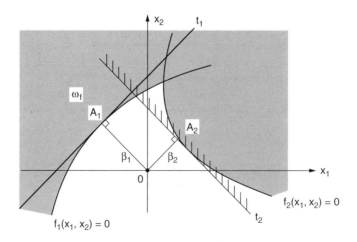

FIGURE 15.13 Illustration of series system with two failure elements.

structural system is equal to the shortest distance from the origin 0 to the failure surface. That is, $\beta = \beta_2$. Estimating the probability of failure by the formula

$$P_f \approx \Phi(-\beta) = \Phi(-\beta_2) \tag{15.24}$$

will then correspond to integrating over the hatched area (to the right of the tangent t_2). Clearly, the approximation (15.2) will in many cases be very different from the exact P_f calculated by (15.1). It is therefore of great interest to find a better approximation of P_f and then define a reliability index β by

$$\beta = -\Phi^{-1}(P_f). \tag{15.25}$$

Let the two failure modes be defined by the safety margins $M_1 = f_1(X_1, X_2)$ and $M_1 = f_2(X_1, X_2)$ and let $F_i = \{M_i \leq 0\}$, $i = 1, 2$. Then, the probability of failure P_f of the structural system is

$$P_f = P(F_1 \cup F_2) \tag{15.26}$$

corresponding to evaluating the probability of failure of a series system with two elements. An approximation of P_f can be obtained by assuming that the safety margins M_1 and M_2 are linearized at their respective design points A_1 and A_2

$$M_1 = a_1 X_1 + a_2 X_2 + \beta_1 \tag{15.27}$$

$$M_2 = b_1 X_1 + b_2 X_2 + \beta_2 \tag{15.28}$$

where β_1 and β_2 are the corresponding reliability indices when $\bar{a} = (a_1, a_2)$ and $\bar{b} = (b_1, b_2)$ are chosen as unit vectors. Then, an approximation of P_f is

$$\begin{aligned} P_f &\approx P\bigl(\{\bar{a}^T \bar{X} + \beta_1 \leq 0\} \cup \{\bar{b}^T \bar{X} + \beta_2 \leq 0\}\bigr) = P\bigl(\{\bar{a}^T \bar{X} \leq -\beta_1\} \cup \{\bar{b}^T \bar{X} \leq -\beta_2\}\bigr) \\ &= 1 - P\bigl(\{\bar{a}^T \bar{X} > -\beta_1\} \cap \{\bar{b}^T \bar{X} > -\beta_2\}\bigr) = 1 - P\bigl(\{-\bar{a}^T \bar{X} < \beta_1\} \cap \{-\bar{b}^T \bar{X} < \beta_2\}\bigr) \\ &= 1 - \Phi_2(\beta_1, \beta_2; \rho) \end{aligned} \tag{15.29}$$

where $\bar{X} = (X_1, X_2)$ is an independent standard normal vector, and ρ is the correlation coefficient given by

$$\rho = \bar{a}^T \bar{b} = a_1 b_1 + a_2 b_2. \tag{15.30}$$

Φ_2 is the bivariate normal distribution function defined by

$$\Phi_2(x_1, x_2; \rho) = \int_{-\infty}^{x_1} \int_{-\infty}^{x_2} \varphi_2(t_1, t_2; \rho) dt_1 dt_2 \tag{15.31}$$

where the bivariate normal density function with zero mean φ_2 is given by

$$\varphi_2(t_1, t_2; \rho) = \frac{1}{2\pi \sqrt{1-\rho^2}} \exp\left(-\frac{1}{2(1-\rho^2)}(t_1^2 + t_2^2 - 2\rho t_1 t_2)\right). \tag{15.32}$$

A formal reliability index β for the system can then be defined by

$$\beta = -\Phi^{-1}(P_f) \approx -\Phi^{-1}\bigl(1 - \Phi_2(\beta_1, \beta_2; \rho)\bigr). \tag{15.33}$$

The reliability of a structural system can be estimated based on modeling by a series system where the elements are parallel systems. It is therefore of great importance to have accurate methods by which the reliability of series systems can be evaluated. In this section it will be shown how the approximation of P_f by Equation 15.29 can easily be extended to a series system with n elements. Further, some bounding and approximate methods are presented.

15.3.2 Assessment of the Probability of Failure of Series Systems

Consider a series system with n elements as shown in Figure 15.3 and let the safety margin for element i be given by

$$M_i = g_i(\overline{X}), \quad i = 1, 2, \ldots, n \tag{15.34}$$

where $\overline{X} = (X_1, \ldots, X_k)$ are the basic variables and where g_i, $i = 1, 2, \ldots, n$ are nonlinear functions.

The probability of failure P_f of *element i* can then be estimated in the following way. Assume that there is a transformation $\overline{Z} = \overline{T}(\overline{X})$ by which the basic variables $\overline{X} = (\overline{X}_1, \ldots, \overline{X}_k)$ are transformed into independent standard normal variables $\overline{Z} = (Z_1, \ldots, Z_k)$ so that

$$P_{f,i} = P(M_i \leq 0) = P(f_i(\overline{X}) \leq 0) = P(f_i(\overline{T}^{-1}(\overline{Z})) \leq 0) = P(h_i(\overline{Z}) \leq 0) \tag{15.35}$$

where h_i is defined by Equation 15.35. An approximation of $P_{f,i}$ can then be obtained by linearization of h_i at the design point

$$P_{f,i} = P(h_i(\overline{Z}) \leq 0) \approx P(\overline{\alpha}_i^T \overline{Z} + \beta_i \leq 0) \tag{15.36}$$

where $\overline{\alpha}_i$ is the unit normal vector at the design point and β_i the Hasofer-Lind reliability index.

The approximation in Equation 15.36 can be written

$$P_{f,i} \approx P(\overline{\alpha}_i^T \overline{Z} + \beta_i \leq 0) = P(\alpha_i^{-T} \overline{Z} \leq -\beta_i) = \Phi(-\beta_i) \tag{15.37}$$

where Φ is the standard normal distribution function.

Return to the *series system* shown in Figure 15.3. An approximation of the probability of failure P_{fs} of this system can then be obtained using the same transformation $\overline{T}$ as for the single elements and by linearization of

$$h_i(\overline{Z}) = g_i(\overline{T}^{-1}\overline{Z})), \quad i = 1, 2, \ldots, n \tag{15.38}$$

at the design points for each element. Then (see, e.g., Hohenbichler and Rackwitz [6])

$$P_{fs} = P\left(\bigcup_{i=1}^{n}\{M_i \leq 0\}\right) = P\left(\bigcup_{i=1}^{n}\{f_i(\overline{X}) \leq 0\}\right) = P\left(\bigcup_{i=1}^{n}\{f_i(\overline{T}^{-1}(\overline{Z})) \leq 0\}\right)$$

$$= P\left(\bigcup_{i=1}^{n}\{h_i(\overline{Z}) \leq 0\}\right) \approx P\left(\bigcup_{i=1}^{n}\{\overline{\alpha}_i^T \overline{Z} + \beta_i \leq 0\}\right) = P\left(\bigcup_{i=1}^{n}\{\overline{\alpha}_i^T \overline{Z} \leq -\beta_i\}\right) \tag{15.39}$$

$$= 1 - P\left(\bigcap_{i=1}^{n}\{\overline{\alpha}_i^T \overline{Z} \leq -\beta_i\}\right) = 1 - P\left(\bigcap_{i=1}^{n}\{\overline{\alpha}_i^T \overline{Z} < \beta_i\}\right) = 1 - \Phi_n(\overline{\beta}; \overline{\overline{\rho}})$$

where $\bar{\beta} = (\beta_1, \ldots, \beta_n)$ and where $\bar{\bar{\rho}} = [\rho_{ij}]$ is the correlation matrix for the linearized safety margins; that is, $\rho_{ij} = \bar{\alpha}_i^T \bar{\alpha}_j$. Φ_n is the n-dimensional standardized normal distribution function.

Using Equation 15.39, the calculation of the probability of failure of a series system with linear and normally distributed safety margins is reduced to calculation of a value of Φ_n.

15.3.3 Reliability Bounds for Series Systems

It has been emphasized several times in this chapter that numerical calculation of the multi-normal distribution function Φ_n is extremely time consuming or perhaps even impossible for values of n greater than, say, four. Therefore, approximate techniques or bounding techniques must be used. In this section, the so-called *simple bounds* and *Ditlevsen bounds* are derived.

15.3.3.1 Simple Bounds

First, the simple bounds are derived. For this purpose, it is convenient to use the Boolean variables introduced in Section 15.2.5. Consider a series system S with n failure elements $E_1, \ldots, E_i, E_{i+1}, E_n$. For each failure element E_i, $i = 1, \ldots, n$, a Boolean variable e_i is defined by (see Equation 15.8)

$$e_i = \begin{cases} 1 & \text{if the failure element is in a nonfailure state} \\ 0 & \text{if the failure element is in a failure state.} \end{cases} \tag{15.40}$$

Then the probability of failure P_{fs} of the series system is

$$P_{fs} = 1 - P\left(\bigcap_{i=1}^{n} e_i = 1\right) = 1 - P(e_1 = 1) \frac{P(e_1 = 1 \cap e_2 = 1)}{P(e_1 = 1)} \cdots \frac{P(e_1 = 1 \cap \ldots \cap e_n = 1)}{P(e_1 = 1 \cap \ldots \cap e_{n-1} = 1)}$$

$$\leq 1 - \prod_{i=1}^{n} P(e_i = 1) = 1 - \prod_{i=1}^{n} (1 - P(e_i = 0)). \tag{15.41}$$

If

$$P(e_1 = 1 \cap e_2 = 1) \geq P(e_1 = 1) P(e_2 = 1) \tag{15.42}$$

etc., or in general,

$$P\left(\bigcap_{j=1}^{i+1} e_j = 1\right) \geq P\left(\bigcap_{j=1}^{i} e_j = 1\right) P(e_{i+1} = 1) \tag{15.43}$$

for all $1 \leq i \leq n-1$. It can be shown that Equation 15.43 is satisfied when the safety margins for E_i, $i = 1, \ldots, n$, are normally distributed and positively correlated. When Equation 15.43 is satisfied, an upper bound of P_{fs} is given by Equation 15.41. A simple lower bound is clearly the maximum probability of failure of any failure element E_i, $i = 1, \ldots, n$. Therefore, the following simple bounds exist when Equation 15.43 is satisfied

$$\max_{i=1}^{n} P(e_i = 0) \leq P_{fs} \leq 1 - \prod_{i=1}^{n} (1 - P(e_i = 0)). \tag{15.44}$$

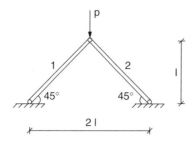

FIGURE 15.14 Structural system for Example 15.5.

The lower bound in Equation 15.44 is equal to the exact value of P_{fs} if there is full dependence between all elements ($\rho_{ij} = 1$ for all i and j) and the upper bound in Equation 15.44 corresponds to no dependence between any pair of elements ($\rho_{ij} = 0$, $i \neq j$).

Example 15.5

Consider the structural system shown in Figure 15.14 loaded by a single concentrated load p. Assume that system failure is failure in compression in element 1 or in element 2. Let the load-carrying capacity in the elements 1 and 2 be $1.5 \times n_F$ and n_F, respectively, and assume that p and n_F are realizations of independent normally distributed random variables P and N_F with

$$\mu_P = 4 kN \quad \sigma_P = 0.8 kN$$

$$\mu_{N_F} = 4 kN \quad \sigma_{N_F} = 0.8 kN.$$

Safety margins for elements 1 and 2 are

$$M_1 = \frac{3}{2} N_F - \frac{\sqrt{2}}{2} P, \quad M_2 = N_F - \frac{\sqrt{2}}{2} P$$

or expressed by standardized random variables $X_1 = (N_F - 4)/0.4$ and $X_2 = (P - 4)/0.8$, where the coefficients of X_1 and X_2 are chosen so that they are components of unit vectors. The reliability indices are $\beta_1 = 3.85$ and $\beta_2 = 1.69$ and the correlation coefficient between the safety margins is $\rho = 0.728 \times 0.577 + 0.686 \times 0.816 = 0.98$.

Therefore, the probability of failure of the system is

$$P_{fs} = 1 - \Phi_2(3.85, 1.69; 0.98).$$

The probabilities of failure of the failure elements E_i, $i = 1, 2$, are

$$P_{f1} = P(e_1 = 0) = \Phi(-3.85) = 0.00006, \quad P_{f2} = P(e_2 = 0) = \Phi(-1.65) = 0.04947.$$

Bounds for the probability of failure P_{fs} are then according to Equation 15.44

$$0.04947 \leq P_{fs} \leq 1 - (1 - 0.00006)(1 - 0.04947) = 0.04953.$$

For this series system, $\rho = 0.98$. Therefore, the lower bound can be expected to be close to P_{fs}.

15.3.3.2 Ditlevsen Bounds

For small probabilities of element failure, the upper bound in Equation 15.44 is very close to the sum of the probabilities of failure of the single elements. Therefore, when the probability of failure of one failure element is predominant in relation to the other failure element, then the probability of failure of series systems is approximately equal to the predominant probability of failure, and the gap between the upper and lower bounds in Equation 15.44 is narrow. However, when the probabilities of failure of the failure elements are of the same order, then the simple bounds in Equation 15.44 are of very little use and there is a need for narrower bounds.

Consider again the above-mentioned series system S shown in Figure 15.7 and define the Boolean variable s by Equation 15.37. Then it follows from Equation 15.12 that

$$s = e_1 \times e_2 \times \ldots \times e_n = e_1 \times e_2 \times \ldots \times e_{n-1} - e_1 \times e_2 \times \ldots \times e_{n-1}(1-e_n) \\ = e_1 - e_1(1-e_2) - e_1 \times e_2(1-e_3) - \ldots - e_1 \times e_2 \times \ldots \times e_{n-1}(1-e_n). \tag{15.45}$$

Hence, in accordance with Equation 15.22,

$$P_{fs} = 1 - R_S = 1 - E[s] \\ = E[1-e_1] + E[e_1(1-e_2)] + E[e_1 e_2(1-e_3)] + \ldots + E[e_1 e_2 \ldots e_{n-1}(1-e_n)]. \tag{15.46}$$

It is easy to see that

$$1 - ((1-e_1)(1-e_2) + \ldots + (1-e_n)) \leq e_1 e_2 \ldots e_n \leq e_i \quad \text{for} \quad i = 1, 2, \ldots, n. \tag{15.47}$$

It is then seen from Equation 15.46

$$P_{fs} \leq \sum_{i=1}^{n} P(e_i = 0) - \sum_{i=2}^{n} \max_{j<i} P(e_i = 0 \cap e_j = 0) \tag{15.48}$$

and

$$P_{fs} \geq P(e_1 = 0) + \sum_{i=2}^{n} \max \left\{ P(e_i = 0) - \sum_{j=1}^{i-1} P(e_i = 0 \cap e_j = 0, 0) \right\}. \tag{15.49}$$

These bounds have been suggested in a slightly different form by Kounias [7]. In structural reliability, they are called *Ditlevsen bounds* [8]. The numbering of the failure elements may influence the bounds in Equation 15.48 and Equation 15.49. However, experience suggests that it is a good choice to arrange the failure elements so that $P(e_1 = 0) \geq P(e_2 = 0) \geq \ldots \geq P(e_n = 0)$, that is, according to decreasing probability of element failure.

The gap between the Ditlevsen bounds of Equation 15.48 and Equation 15.49 is usually much smaller than the gap between the simple bounds in Equation 15.44. However, the bounds of Equation 15.48 and Equation 15.49 require calculation of the joint probabilities $P(e_i = 0 \cap e_j = 0)$ and these calculations are not trivial. Usually, a numerical technique must be used.

15.3.4 Series Systems with Equally Correlated Elements

The n-dimensional standardized normal distribution function $\Phi_n(\bar{x}; \bar{\bar{\rho}})$ can be easily evaluated when $\rho_{ij} = \rho > 0, i = 1, \ldots, n, j = 1, \ldots, n, i \neq j$, that is, when

$$\bar{\bar{\rho}} = \begin{bmatrix} 1 & \rho & \cdots & \rho \\ \rho & 1 & \cdots & \rho \\ \vdots & \vdots & \ddots & \vdots \\ \rho & \rho & \cdots & 1 \end{bmatrix}. \tag{15.50}$$

By the correlation matrix in Equation 15.50, it has been shown by Dunnett and Sobel [9] that

$$\Phi_n(\overline{x}:\overline{\overline{\rho}}) = \int_{-\infty}^{\infty} \varphi(t) \prod_{i=1}^{n} \Phi\left(\frac{x_i - \sqrt{\rho}\,t}{\sqrt{1-\rho}}\right) dt. \quad (15.51)$$

Equation 15.50 can be generalized to the case where $\rho_{ij} = \lambda_i \lambda_j$, $i \neq j$, $|\lambda_i| \leq 1$, $|\lambda_j| \leq 1$, that is,

$$\overline{\overline{\rho}} = \begin{bmatrix} 1 & \lambda_1 \lambda_2 & \cdots & \lambda_1 \lambda_n \\ \lambda_2 \lambda_1 & 1 & \cdots & \lambda_2 \lambda_n \\ \vdots & \vdots & \ddots & \vdots \\ \lambda_n \lambda_1 & \lambda_n \lambda_2 & \cdots & 1 \end{bmatrix}. \quad (15.52)$$

For such correlation matrices, Dunnett and Sobel [9] have shown that

$$\Phi_n(\overline{x}:\overline{\overline{\rho}}) = \int_{-\infty}^{\infty} \varphi(t) \prod_{i=1}^{n} \Phi\left(\frac{x_i - \lambda_i t}{\sqrt{1-\lambda_i^2}}\right) dt. \quad (15.53)$$

For series systems with equally correlated failure elements, the probability of failure P_{fs} can then be written (see Equation 15.39)

$$P_{fs} = 1 - \Phi_n(\overline{\beta}; \overline{\overline{\rho}}) = 1 - \int_{-\infty}^{\infty} \varphi(t) \prod_{i=1}^{n} \Phi\left(\frac{x_i - \sqrt{\rho}\,t}{\sqrt{1-\rho}}\right) dt \quad (15.54)$$

where $\overline{\beta} = (\beta_1, \ldots, \beta_n)$ are the reliability indices for the single failure elements and ρ is the common correlation coefficient between any pair of safety margins M_i and M_j, $i \neq j$.

A further specialization is the case where all failure elements have the same reliability index β_e, that is, $\beta_i = \beta_e$ for $i = 1, \ldots, n$. Then

$$P_{fs} = 1 - \int_{-\infty}^{\infty} \varphi(t) \left[\Phi\left(\frac{\beta_e - \sqrt{\rho}\,t}{\sqrt{1-\rho}}\right)\right]^n dt. \quad (15.55)$$

–⊔–

Example 15.6

Consider a series system with $n = 10$ failure elements, common element reliability index β_e and common correlation coefficient ρ. The probability of failure P_{fs} of this series system as a function of ρ is illustrated in Figure 15.15 for $\rho = 2.50$ and 3.00. Note that, as expected, the probability of failure P_{fs} decreases with ρ.

–⊔–

To summarize for a series system with equally correlated failure elements where the safety margins are linear and normally distributed, the probability of failure P_{fs} can be calculated by Equation 15.54. A formal reliability index β_S for the series system can then be calculated by

$$P_{fs} = \Phi(-\beta_S) \Leftrightarrow \beta_S = -\Phi^{-1}(P_{fs}). \quad (15.56)$$

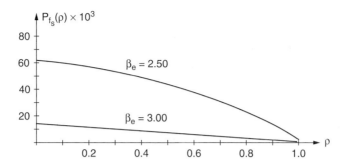

FIGURE 15.15 Probability of failure of series system with ten elements as a function of the correlation coefficient.

15.3.5 Series Systems with Unequally Correlated Elements

It has been investigated by Thoft-Christensen and Sørensen [10] whether an equivalent correlation coefficient can be used in Equation 15.55 with satisfactory accuracy when the failure elements are unequally correlated (or Equation 15.54 if the reliability indices are not equal).

Define the *average correlation coefficient* $\bar{\rho}$ by

$$\bar{\rho} = \frac{1}{n(n-1)} \sum_{\substack{i,j=1 \\ i \neq j}}^{n} \rho_{ij} \tag{15.57}$$

$\bar{\rho}$ is the average of all ρ_{ij}, $i \neq j$. Using $\bar{\rho}$ corresponds to the approximation

$$\Phi_n(\vec{\beta}; \bar{\rho}) \approx \Phi_n(\vec{\beta}; [\bar{\rho}]) \tag{15.58}$$

where the correlation matrix $[\bar{\rho}]$ is given by

$$[\bar{\rho}] = \begin{bmatrix} 1 & \bar{\rho} & \cdots & \bar{\rho} \\ \bar{\rho} & 1 & \cdots & \bar{\rho} \\ \vdots & \vdots & \ddots & \vdots \\ \bar{\rho} & \bar{\rho} & \cdots & 1 \end{bmatrix}. \tag{15.59}$$

Thoft-Christensen and Sørensen [10] have shown by extensive simulation that in many situations,

$$\Phi_n(\vec{\beta}; [\rho]) \leq \Phi_n(\vec{\beta}; [\bar{\rho}]) \tag{15.60}$$

so that an estimate of the probability of failure P_{f_s} using the average correlation coefficient will in such cases be conservative. Ditlevsen [11] has investigated this more closely by a Taylor expansion for the special case $\beta_i = \beta_e$, $i = 1, \ldots, n$ with the conclusion that Equation 15.60 holds for most cases, when $\beta_e >$ 3, $n < 100$, $\bar{\rho} < 0.4$.

Thoft-Christensen and Sørensen [10] have shown that a better approximation can be obtained by

$$\Phi_n(\vec{\beta}; \bar{\rho}) \approx \Phi_n(\vec{\beta}; [\bar{\rho}]) + \Phi_2(\beta_e, \beta_e; \rho_{\max}) - \Phi_2(\beta_e, \beta_e; \bar{\rho}) \tag{15.61}$$

where $\vec{\beta} = (\beta_e, \ldots, \beta_e)$ and where

$$\rho_{\max} = \max_{\substack{i,j=1 \\ i \neq j}}^{n} \rho_{ij}. \tag{15.62}$$

By this equation, the probability of failure P_{fs} is approximated by

$$P_{fs} \approx P_{fs}([\rho_{i,j}]) = P_{fs}([\bar{\rho}]) + P_{fs}(n=2, \rho = \rho_{max}) - P_{fs}(n=2, \rho = \bar{\rho}) \tag{15.63}$$

—□—

Example 15.7

Consider a series system with five failure elements and common reliability index $\beta_e = 3.50$ and the correlation matrix

$$\bar{\bar{\rho}} = \begin{bmatrix} 1 & 0.8 & 0.6 & 0 & 0 \\ 0.8 & 1 & 0.4 & 0 & 0 \\ 0.6 & 0.4 & 1 & 0.1 & 0.2 \\ 0 & 0 & 0.1 & 1 & 0.7 \\ 0 & 0 & 0.2 & 0.7 & 1 \end{bmatrix}.$$

Using Equation 15.58 with $\bar{\rho} = 0.28$ then gives $P_{fs} \approx 0.00115$. It can be shown that the Ditlevsen bounds are $0.00107 \leq P_{fs} \leq 0.00107$. The approximation in Equation 15.63 gives $P_{fs} \approx 0.00111$.

—□—

15.4 Reliability Assessment of Parallel Systems

15.4.1 Introduction

As mentioned earlier, the reliability of a structural system may be modeled by a series system of parallel systems. Each parallel system corresponds to a failure mode, and this modeling is called system modeling at level N, $N = 1, 2, \ldots$ if all parallel systems have the same number N of failure elements. In Section 15.5 it is shown how the most significant failure modes (parallel systems) can be identified by the β-unzipping method. After identification of significant (critical) failure modes (parallel systems), the next step is an estimate of the probability of failure P_{fp} for each parallel system and the correlation between the parallel systems. The final step is the estimate of the probability of failure P_f of the series system of parallel systems by the methods discussed in Section 15.3.

Consider a parallel system with only two failure elements and let the safety margins be $M_1 = f_1(X_1, X_2)$ and $M_2 = f_2(X_1, X_2)$, where X_1 and X_2 are independent standard normally distributed random variables. If $F_i = \{M_i \leq 0\}$, $i = 1, 2$, then the probability of failure P_{fp} of the parallel system is

$$P_{fp} = P(F_1 \cap F_2). \tag{15.64}$$

An approximation of P_{fp} is obtained by assuming that the safety margins M_1 and M_2 are linearized at their respective design points A_1 and A_2 (see Figure 15.16)

$$M_1 = a_1 X_1 + a_2 X_2 + \beta_1 \tag{15.65}$$

$$M_2 = b_1 X_1 + b_2 X_2 + \beta_2 \tag{15.66}$$

where β_1 and β_2 are the corresponding reliability indices when $\bar{a} = (a_1, a_2)$ and $\bar{b} = (b_1, b_2)$ are chosen as unit vectors. Then an approximation of P_{fp} is

$$P_{fp} \approx P\left(\{\bar{a}^T \bar{X} + \beta_1 \leq 0\} \cap \{\bar{b}^T \bar{X} + \beta_2 \leq 0\}\right)$$
$$= P\left(\{\bar{a}^T \bar{X} \leq -\beta_1\} \cap \{\bar{b}^T \bar{X} \leq -\beta_2\}\right) = \Phi_2(-\beta_1, -\beta_2; \rho) \tag{15.67}$$

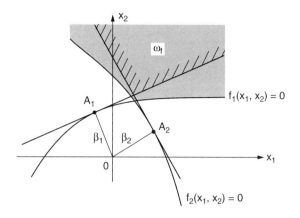

FIGURE 15.16 Illustration of parallel system with two failure elements.

where $\overline{X} = (X_1, X_2)$ and where ρ is the correlation coefficient given by $\rho = \overline{a}^T \overline{b} = a_1 b_1 + a_2 b_2$. Φ_2 is the bivariate normal distribution function defined by Equation 15.32. A formal reliability index β_p for the parallel system can then be defined by

$$\beta_p = -\Phi^{-1}(P_{fp}) \approx -\Phi^{-1}(\Phi_2(-\beta_1, -\beta_2; \rho)). \tag{15.68}$$

Equation 15.67, which gives an approximate value for the probability of failure of a parallel system with two failure elements, will be generalized in Section 15.4.2 to the general case where the parallel system has n failure elements and where the number of basic variables is k.

15.4.2 Assessment of the Probability of Failure of Parallel Systems

Consider a parallel system with n elements as shown in Figure 15.17 and let the safety margin for element i be given by

$$M_i = g_i(\overline{X}), \quad i = 1, \ldots, n \tag{15.69}$$

where $X = (X_1, \ldots, X_k)$ are basic variables and where g_i, $i = 1, 2, \ldots, n$ are nonlinear functions.

The probability of failure P_{fi} of *element i* can then be estimated in the similar way as shown on page 15-12. An approximation of the probability of failure P_{fp} of the parallel system in Figure 15.17 can then be obtained by using the same transformation as for the single elements and by linearization of

$$h_i(\overline{Z}) = g_i(\overline{T}^{-1}(\overline{Z})), \quad i = 1, \ldots, n \tag{15.70}$$

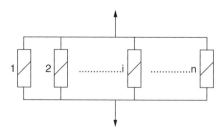

FIGURE 15.17 Parallel system with *n* elements.

at the design points for each element. Then (see, e.g., Hohenbichler and Rackwitz [6])

$$P_{fp} = P\left(\bigcap_{i=1}^{n}\{M_i \leq 0\}\right) = P\left(\bigcap_{i=1}^{n}\{g_i(\bar{X}) \leq 0\}\right) = P\left(\bigcap_{i=1}^{n}\{g_i(\bar{T}^{-1}(\bar{Z})) \leq 0\}\right)$$

$$= P\left(\bigcap_{i=1}^{n}\{h_i(\bar{Z}) \leq 0\}\right) \approx P\left(\bigcap_{i=1}^{n}\{\bar{\alpha}_i^T \bar{Z} + \beta_i \leq 0\}\right) = P\left(\bigcap_{i=1}^{n}\{\bar{\alpha}_i^T \bar{Z} \leq -\beta_i\}\right) \quad (15.71)$$

$$= \Phi_n(-\bar{\beta}; \bar{\bar{\rho}})$$

where $\bar{\beta} = (\beta_1, \ldots, \beta_n)$, and where $\bar{\bar{\rho}} = [\rho_{ij}]$ is the correlation matrix for the linearized safety margins; that is, $\rho_{ij} = \bar{\alpha}_i^T \bar{\alpha}_j$, Φ_n is the n-dimensional standardized normal distribution function.

From Equation 15.71, the calculation of the probability of failure of a parallel system with linear and normally distributed safety margins is reduced to calculation of a value of Φ_n.

15.4.3 Reliability Bounds for Parallel Systems

In general, numerical calculation of the multi-normal distribution function Φ_n is very time consuming. Therefore, approximate techniques or bounding techniques must be used.

Simple bounds for the probability of failure P_{fp} analogous to the simple bounds for series systems Equation 15.44 can easily be derived for parallel systems. Consider a parallel system P with n failure elements $E_1, \ldots, E_n$. For each failure element E_i, $i = 1, \ldots, n$, a Boolean variable e_i is defined by (see Equation 15.8)

$$e_i = \begin{cases} 1 & \text{if the failure element is in a nonfailure state} \\ 0 & \text{if the failure element is in a failure state.} \end{cases} \quad (15.72)$$

Then the probability of failure P_{fp} of the parallel system is

$$P_{fp} = P\left(\bigcap_{i=1}^{n} e_i = 0\right) = P(e_1 = 0)\frac{P(e_1 = 0 \cap e_2 = 0)}{P(e_1 = 0)} \cdots \frac{P(e_1 = 0 \cap \ldots \cap e_n = 0)}{P(e_1 = 0 \cap \ldots \cap e_{n-1} = 0)}$$

$$\geq \prod_{i=1}^{n} P(e_i = 0)$$

(15.73)

if

$$P(e_1 = 0 \cap e_2 = 0) \geq P(e_1 = 0)P(e_2 = 0) \quad (15.74)$$

etc., or in general,

$$P\left(\bigcap_{j=1}^{i+1} e_j = 1\right) \geq P\left(\bigcap_{j=1}^{i} e_j = 1\right) P(e_{i+1} = 1) \quad (15.75)$$

for all $1 \leq i \leq n-1$. A simple upper bound is clearly the maximum probability of failure of the minimum probability of failure of any failure element E_i, $i = 1, \ldots, n$. Therefore, the following simple bounds exist when Equation 15.75 is satisfied:

$$\prod_{i=1}^{n} P(e_i = 0) \leq P_{fp} \leq \min_{i=1}^{n} P(e_i = 0). \quad (15.76)$$

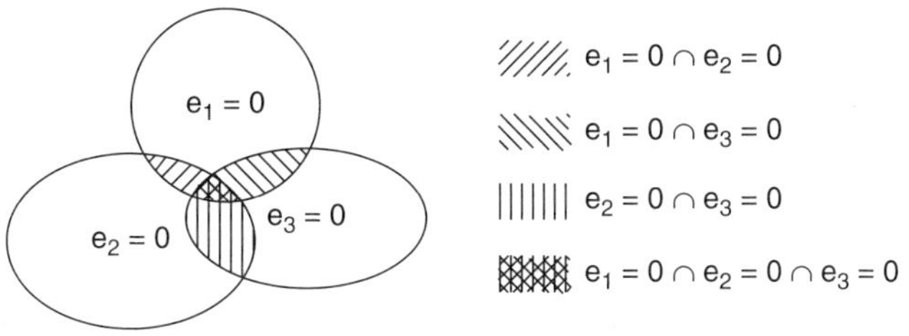

System Reliability

FIGURE 15.18 Derivation of Murotsu's upper bound.

The lower bound in Equation 15.76 is equal to the exact value of P_{fs} if there is no dependence between any pair of elements ($\rho_{ij} = 0$, $i \neq j$), and the upper bound in Equation 15.76 corresponds to full dependence between all elements ($\rho_{ij} = 1$ for all i and j).

The simple bounds in Equation 15.76 will in most cases be so wide that they are of very little use. A better upper bound of P_{fs} has been suggested by Murotsu et al. [12].

$$P_{fp} \leq \min_{i,j=1}^{n} [P(e_i = 0) \cap P(e_j = 0)]. \tag{15.77}$$

The derivation of Equation 15.77 is very simple. For the case $n = 3$, Equation 15.77 is illustrated in Figure 15.18.

—□—

Example 15.8

Consider a parallel system with the reliability indices $\overline{\beta} = (3.57, 3.41, 4.24, 5.48)$ and the correlation matrix:

$$\overline{\overline{\rho}} = \begin{bmatrix} 1.00 & 0.62 & 0.91 & 0.62 \\ 0.62 & 1.00 & 0.58 & 0.58 \\ 0.91 & 0.58 & 1.00 & 0.55 \\ 0.62 & 0.58 & 0.55 & 1.00 \end{bmatrix}.$$

Further, assume that the safety margins for the four elements are linear and normally distributed. The probability of failure of the parallel system then is $P_{fp} = \Phi_4(-3.57, -3.41, -4.24, -5,48; \overline{\overline{\rho}})$.

The simple bounds in Equation 15.76 are

$$(1.79 \times 10^{-4}) \times (3.24 \times 10^{-4}) \times (0.11 \times 10^{-4}) \times (0.21 \times 10^{-7}) \leq P_{fp} \leq 0.21 \times 10^{-7}$$

or $0 \leq P_{fp} \leq 0.21 \times 10^{-7}$. The corresponding bounds of the formal reliability index β_p are $5.48 \leq \beta_p \leq \infty$. With the ordering of the four elements shown above, the probabilities of the intersections of e_1 and e_2 can be shown as

$$[P(e_i = 0 \cap e_j = 0)] = \begin{bmatrix} - & 1.71 \times 10^{-5} & 9.60 \times 10^{-6} & 9.93 \times 10^{-9} \\ 1.71 \times 10^{-5} & - & 1.76 \times 10^{-6} & 9.27 \times 10^{-9} \\ 9.60 \times 10^{-6} & 1.76 \times 10^{-6} & - & 1.90 \times 10^{-9} \\ 9.93 \times 10^{-9} & 9.27 \times 10^{-9} & 1.90 \times 10^{-9} & - \end{bmatrix}.$$

Therefore, from Equation 15.77: $\beta_p \geq -\Phi^{-1}(1.90 \times 10^{-9}) = 5.89$.

—⊔—

15.4.4 Equivalent Linear Safety Margins for Parallel Systems

In Section 15.4.2 it was shown how the probability of failure of a parallel system can be evaluated in a simple way when the safety margin for each failure element is *linear* and normally distributed. Consider a parallel system with n such failure elements. Then, the probability of failure P_{fp} of the parallel system is (see Equation 15.71):

$$P_{fp} = \Phi_n(-\overline{\beta}; \overline{\overline{\rho}}) \tag{15.78}$$

where $\overline{\beta} = (\beta_1, \ldots, \beta_n)$ is a vector whose components are the reliability indices of the failure elements, and where $\overline{\overline{\rho}}$ is the correlation matrix for the *linear* and normally distributed safety margins of the failure elements.

When the reliability of a structural system is modeled by a series system of parallel systems (failure modes), the reliability is evaluated by the following steps:

- Evaluate the probability of failure of each parallel system by Equation 15.78.
- Evaluate the correlation between the parallel systems.
- Evaluate the probability of failure of the series system by Equation 15.39.

Evaluation of the correlation between a pair of parallel systems can easily be performed if the safety margins for the parallel systems are linear. However, in general, this will clearly not be the case. It is therefore natural to investigate the possibility of introducing an *equivalent linear safety margin* for each parallel system. In this section, an equivalent linear safety margin suggested by Gollwitzer and Rackwitz [13] is described.

Consider a parallel system with n elements as shown in Figure 15.17 and let the safety margin for element i, $i = 1, 2, \ldots, n$ be linear

$$M_i = \alpha_{i1} Z_1 + \ldots + \alpha_{ik} Z_k + \beta_i = \sum_{j=1}^{k} \alpha_{ij} Z_j + \beta_i \tag{15.79}$$

where the basic variables Z_i, $i = 1, \ldots, k$ are independent standard normally distributed variables where $\overline{\alpha}_i = (\alpha_{i1}, \ldots, \alpha_{ik})$ is a unit vector, and where β_i is the Hasofer-Lind reliability index. A formal (generalized) reliability index β_P for the parallel system is then given by

$$\beta_P = -\Phi^{-1}(\Phi_n(-\overline{\beta}; \overline{\overline{\rho}})) \tag{15.80}$$

where $\overline{\beta} = (\beta_1, \ldots, \beta_n)$ and $\overline{\overline{\rho}} = [\rho_{ij}] = [\overline{\alpha}_i^T \overline{\alpha}_j]$.

The equivalent linear safety margin M^e is then defined in such a way that the corresponding reliability index β^e is equal to β_P and so that it has the same sensitivity as the parallel system against changes in the basic variables Z_i, $i = 1, 2, \ldots, k$.

Let the vector $\overline{Z}$ of basic variables be increased by a (small) vector $\overline{\varepsilon} = (\varepsilon_1, \ldots, \varepsilon_k)$. Then the corresponding reliability index $\beta_P(\overline{\varepsilon})$ for the parallel system is

$$\beta_P(\overline{\varepsilon}) = -\Phi^{-1}\left(P\left(\bigcap_{i=1}^{n}\left\{\sum_{j=1}^{k} \alpha_{ij}(Z_j + \varepsilon_j) + \beta_i \leq 0\right\}\right)\right) \tag{15.81}$$

$$= -\Phi^{-1}(\Phi_n(-\overline{\beta} - \overline{\overline{\alpha}}\,\overline{\varepsilon}; \overline{\overline{\rho}}))$$

where $\overline{\overline{\alpha}} = [\alpha_{ij}]$.

System Reliability

Let the equivalent linear safety margin M^e be given by

$$M^e = \alpha_1^e Z_1 + \ldots + \alpha_k^e Z_k + \beta^e = \sum_{j=1}^{k} \alpha_j^e Z_j + \beta^e \tag{15.82}$$

where $\bar{\alpha}^e = (\alpha_1^e, \ldots, \alpha_k^e)$ is a unit vector and where $\beta^e = \beta_P$. With the same increase $\bar{\varepsilon}$ of the basic variables, the reliability index $\beta^e(\bar{\varepsilon})$ is

$$\beta_P(\bar{\varepsilon}) = -\Phi^{-1}(\Phi(-\beta^e - \bar{\alpha}^{eT}\bar{\varepsilon})) = \beta^e + \alpha_1^e \varepsilon_1 + \ldots + \alpha_k^e \varepsilon_k. \tag{15.83}$$

It is seen from Equation 15.83 and by putting $\beta_P(\bar{0}) = \beta^e(\bar{0})$ that

$$\alpha_i^e = \frac{\left.\dfrac{\partial \beta_P}{\partial \varepsilon_i}\right|_{\bar{\varepsilon}=\bar{0}}}{\sqrt{\sum_{j=1}^{n}\left(\left.\dfrac{\partial \beta_P}{\partial \varepsilon_i}\right|_{\bar{\varepsilon}=\bar{0}}\right)^2}}, \quad i = 1, \ldots, k. \tag{15.84}$$

An approximate value of α_i^e, $i = 1, \ldots, k$ can easily be obtained by numerical differentiation as shown in Example 15.9.

—□—

Example 15.9

Consider a parallel system with two failure elements and let the safety margin for the failure elements be

$$M_1 = 0.8 Z_1 - 0.6 Z_2 + 3.0 \quad \text{and} \quad M_2 = 0.1 Z_1 - 0.995 Z_2 + 3.5$$

where Z_1 and Z_2 are independent standard normally distributed variables. The correlation ρ between the safety margins is

$$\rho = 0.8 \times 0.1 + 0.6 \times 0.9950 = 0.68.$$

Then, the reliability index β_P of the parallel system is

$$\beta_P = -\Phi^{-1}(\Phi_2(-3.0, -3.5; 0.68)) = 3.83.$$

To obtain the equivalent linear safety margin, Z_1 and Z_2 are in turn given an increment $\varepsilon_i = 0.1$, $i = 1, 2$. With $\bar{\varepsilon} = (0.1, 0)$, one gets

$$-\bar{\beta} - \bar{\bar{\alpha}}\bar{\varepsilon} = \begin{bmatrix} -3.0 \\ -3.5 \end{bmatrix} - \begin{bmatrix} 0.8 & -0.6 \\ 0.1 & -0.995 \end{bmatrix}\begin{bmatrix} 0.1 \\ 0 \end{bmatrix} = \begin{bmatrix} -3.08 \\ -3.51 \end{bmatrix}$$

and by Equation 15.81 $\beta_P = -\Phi^{-1}(\Phi_2(-3.08, -3.51; 0.68)) = 3.87$.

Therefore,

$$\left.\frac{\partial \beta_P}{\partial \varepsilon_1}\right|_{\bar{\varepsilon}=\bar{0}} \approx \frac{3.87 - 3.83}{0.1} = 0.40.$$

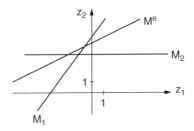

FIGURE 15.19 Safety margins.

Likewise with $\bar{\varepsilon} = (0, 0.1)$,

$$P_{fp} = \int_{-\infty}^{\infty} \varphi(t) \left[\Phi\left(\frac{-\beta_e - \sqrt{\rho}t}{\sqrt{1-\rho}} \right) \right]^n dt$$

and $\beta_p(\bar{\varepsilon}) = -\Phi^{-1}(\Phi_2(-2.94, -3.40; 0.68)) = 3.74$.

Therefore,

$$\left. \frac{\partial \beta_p}{\partial \varepsilon_2} \right|_{\bar{\varepsilon}=\bar{0}} \approx \frac{3.74 - 3.83}{0.1} = -0.90.$$

By normalizing $\bar{\alpha}^e = (\alpha_1^e, \alpha_2^e) = (0.4061, -0.9138)$ and the equivalent safety margin is

$$M^e = 0.4061 \, Z_1 - 0.9138 \, Z_2 + 3.83.$$

The safety margins M_1, M_2, and M^e are shown in Figure 15.19.

—⊔—

15.4.5 Parallel Systems with Equally Correlated Elements

It was shown in Section 15.4.2 that the probability of failure P_{fp} of a parallel system with n failure elements is equal to $P_{fp} \approx \Phi_n(-\bar{\beta}; \bar{\bar{\rho}})$ when the safety margins are linear and normally distributed.

Dunnett and Sobel [9] have shown for the special case where the failure elements are equally correlated with the correlation coefficient ρ; that is,

$$\bar{\bar{\rho}} = \begin{bmatrix} 1 & \rho & \cdots & \rho \\ \rho & 1 & \cdots & \rho \\ \vdots & \vdots & \ddots & \vdots \\ \rho & \rho & \cdots & 1 \end{bmatrix} \quad (15.85)$$

that (see Equation 15.51)

$$\Phi_n(\bar{x}: \bar{\bar{\rho}}) = \int_{-\infty}^{\infty} \varphi(t) \prod_{i=1}^{n} \Phi\left(\frac{x_i - \sqrt{\rho}t}{\sqrt{1-\rho}} \right) dt. \quad (15.86)$$

If all failure elements have the same reliability index β_e, then

$$P_{fp} = \int_{-\infty}^{\infty} \varphi(t) \left[\Phi\left(\frac{-\beta_e - \sqrt{\rho}t}{\sqrt{1-\rho}} \right) \right]^n dt. \quad (15.87)$$

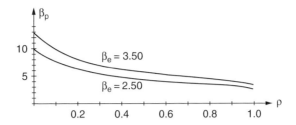

FIGURE 15.20 Safety index of parallel system as a function of correlation coefficient.

A formal reliability index β_p for the parallel system can then be calculated by $\beta_p = -\Phi^{-1}(P_{fp})$.

–⊔–

Example 15.10

Consider a parallel system with $n = 10$ failure elements, and common element reliability index β_e, and common correlation coefficient ρ. The formal reliability index β_p for this parallel system as a function of ρ is illustrated in Figure 15.20 for $\beta_e = 2.50$ and 3.50. Note that, as expected, the reliability index β_p decreases with ρ.

–⊔–

The strength of a *fiber bundle* with n ductile fibers can be modeled by a parallel system as shown in Figure 15.21. The strength R of the fiber bundle is

$$R = \sum_{i=1}^{n} R_i \tag{15.88}$$

where the random variable R_i is the strength of fiber i, $i = 1, 2, \ldots, n$. Let R_i be identically and normally distributed $N(\mu, \sigma)$ with common correlation coefficient ρ. The strength R is then normally distributed $N(\mu_R, \sigma_R)$ where $\mu_R = n\mu$ and $\sigma_R^2 = n\sigma^2 + n(n-1)\rho\sigma^2$.

Assume that the fiber bundle is loaded by a deterministic and time-independent load $S = nS_e$, where S = constant is the load of fiber i, $i = 1, 2, \ldots, n$. The reliability indices of the fibers are the same for all fibers and equal to

$$\beta_e = \frac{\mu - S_e}{\sigma}. \tag{15.89}$$

Therefore, $S = nS_e = n(\mu - \beta_e \sigma)$ and the reliability index β_F for the fiber bundle (the parallel system) is

$$\beta_F = \frac{\mu_R - S}{\sigma_R} = \frac{n\mu - n(\mu - \beta_e \sigma)}{(n\sigma^2 + n(n-1)\sigma^2 \rho)^{1/2}} = \beta_e \sqrt{\frac{n}{1 + \rho(n-1)}}. \tag{15.90}$$

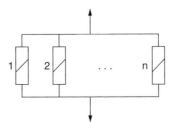

FIGURE 15.21 Modeling of a fiber bundle.

Equation 15.90 has been derived by Grigoriu and Turkstra [14]. In Section 15.4.6 it is shown that Equation 15.90 can easily be modified so that the assumption of the common correlation coefficient ρ can be removed.

15.4.6 Parallel Systems with Unequally Correlated Elements

It was shown earlier that an approximation of the probability of failure of a parallel system with n failure elements with normally distributed safety margins is $P_{fp} \approx \Phi_n(-\bar{\beta}; \bar{\bar{\rho}})$ where $\bar{\beta} = (\beta_1, \ldots, \beta_n)$ is the reliability indices of the failure elements and $\bar{\bar{\rho}} = [\rho_{ij}]$ is the correlation matrix. In Section 15.4.5 it is shown that the reliability index β_F for a *fiber bundle* with n ductile fibers can easily be calculated if the following assumptions are fulfilled:

1. The load S of the fiber bundle is deterministic and constant with time,
2. The strength of the fibers is identically normally distributed $N(\mu, \sigma)$,
3. The fibers have a common reliability index β_e,
4. Common correlation coefficient ρ between the strengths of any pair of fibers. Under these assumptions it was shown that (see Equation 15.90)

$$\beta_F = \beta_e \sqrt{\frac{n}{1 + \rho(n-1)}}. \tag{15.91}$$

Now the assumption 4 above will be relaxed. Let the correlation coefficient between fiber i and fiber j be denoted ρ_{ij}. The reliability index β_F for such a fiber bundle with unequally correlated fibers can then be calculated similarly as used in deriving Equation 15.91 in Section 15.4.5 (see Thoft-Christensen and Sørensen [15])

$$\beta_F = \frac{\mu_R - S}{\sigma_R} = \left(n\mu - (n\mu - n\beta_e \sigma)\right) \left(n\sigma^2 + \sigma^2 \sum_{i,j=1, i\neq j}^{n} \rho_{ij}\right)^{-\frac{1}{2}} = \beta_e \sqrt{\frac{n}{1 + \bar{\rho}(n-1)}} \tag{15.92}$$

where

$$\bar{\rho} = \frac{1}{n(n-1)} \sum_{i,j=1, i\neq j}^{n} \rho_{ij}. \tag{15.93}$$

By comparing Equation 15.91 and Equation 15.92, it is seen that for systems with unequal correlation coefficients, the reliability index β_{F_F} can be calculated by the simple expression in Equation 15.91 by inserting for ρ the average correlation coefficient $\bar{\rho}$ defined by Equation 15.93. $\bar{\rho}$ is the average of all $\rho_{ij}, i \neq j$.

15.5 Identification of Critical Failure Mechanisms

15.5.1 Introduction

A number of different methods to identify critical failure modes have been suggested (see, e.g., Ferregut-Avila [16], Moses [17], Gorman [18], Ma and Ang [19], Klingmüller [20], Murotsu et al. [21], and Kappler [22]). In this section the β-unzipping method [4], [23]–[25] is used.

The β-unzipping method is a method by which the reliability of structures can be estimated at a number of different levels. The aim has been to develop a method that is at the same time simple to use and reasonably accurate. The β-unzipping method is quite general in the sense that it can be used for

System Reliability

two-dimensional and three-dimensional framed and trussed structures, for structures with ductile or brittle elements, and also in relation to a number of different failure mode definitions.

An estimate of the reliability of a structural system on the basis of failure of a single structural element (namely, the element with the lowest reliability index of all elements) is called system reliability at *level* 0. At level 0, the reliability of a structural system is equal to the reliability of this single element. Therefore, such a reliability analysis is, in fact, not a system reliability analysis, but rather an element reliability analysis. At level 0, each element is considered isolated from the other elements, and the interaction between the elements is not taken into account in estimating the reliability. Let a structure consist of n failure elements (i.e., elements or points where failure can take place) and let the reliability index for failure element i be denoted β_i. Then, at level 0, the system reliability index β_0 is simply given as $\beta_s = \min \beta_i$.

15.5.2 Assessment of System Reliability at Level 1

At *level* 1, the system reliability is defined as the reliability of a series system with n elements — the n failure elements, see Figure 15.1. Therefore, the first step is to calculate β-values for all failure elements and then use Equation 15.39. As mentioned earlier, Equation 15.39 can in general not be used directly. However, upper and lower bounds exist for series systems as shown in Section 15.3.3.

Usually for a structure with n failure elements, the estimate of the probability of failure of the series system with n elements can be calculated with sufficient accuracy by only including some of the failure elements, namely those with the smallest reliability indices. One way of selecting is to include only failure elements with β-values in an interval $[\beta_{\min}, \beta_{\min} + \Delta\beta_1]$, where $\beta_{\min}$ is the smallest reliability index of all failure element indices and where $\Delta\beta_1$ is a prescribed positive number. The failure elements chosen to be included in the system reliability analysis at level 1 are called *critical failure elements*. If two or more critical failure elements are perfectly correlated, then only one of them is included in the series system of critical failure elements.

15.5.3 Assessment of System Reliability at Level 2

At *level* 2, the system reliability is estimated as the reliability of a series system where the elements are parallel systems each with two failure elements (see Figure 15.4) — so-called *critical pairs of failure elements*. Let the structure be modeled by n failure elements and let the number of critical failure elements at level 1 be n_1. Let the critical failure element l have the lowest reliability index β of all critical failure elements. Failure is then assumed in failure element l and the structure is modified by removing the corresponding failure element and adding a pair of so-called fictitious loads F_l (axial forces or moments). If the removed failure element is brittle, no fictitious loads are added. However, if the removed failure element l is ductile, the fictitious load F_l is a stochastic load given by $F_l = \gamma_l R_l$, where R_l is the load-carrying capacity of failure element l and where $0 < \gamma_l \leq 1$.

The modified structure with the loads $P_1, \ldots, P_k$ and the fictitious load F_l (axial force or moment) is then reanalyzed and influence coefficients a_{ij} with respect to $P_1, \ldots, P_k$ and a'_{il} with respect to F_l are calculated. The load effect (force or moment) in the remaining failure elements is then described by a stochastic variable. The load effect in failure element i is called $S_{i|l}$ (load effect in failure element i given failure in failure element l) and

$$S_{i|l} = \sum_{j=1}^{k} a_{ij} P_j + a'_{il} F_l. \tag{15.94}$$

The corresponding safety margin $M_{i|l}$ then is

$$M_{i|l} = \min(R_l^+ - S_{i|l}, R_i^- + S_{i|l}) \tag{15.95}$$

where R_i^+ and R_i^- are the stochastic variables describing the (yield) strength capacity in "tension" and "compression" for failure element i. In the following, $M_{i|l}$ will be approximated by either $R_i^+ - S_{i|l}$ or $R_i^+ + S_{i|l}$ depending on the corresponding reliability indices. The reliability index for failure element i, given failure in failure element l, is

$$\beta_{i|l} = \mu_{M_{i|l}} / \sigma_{M_{i|l}}. \tag{15.96}$$

In this way, new reliability indices are calculated for all failure elements (except the one where failure is assumed) and the smallest β-value is called $\beta_{\min}$. The failure elements with β-values in the interval $[\beta_{\min}, \beta_{\min} + \Delta\beta_2]$, where $\Delta\beta_2$ is a prescribed positive number, are then in turn combined with failure element l to form a number of parallel systems.

The next step is then to evaluate the probability of failure for each critical pair of failure elements. Consider a parallel system with failure elements l and r. During the reliability analysis at level 1, the safety margin M_l for failure element l is determined and the safety margin $M_{r|l}$ for failure element r has the form in Equation 15.11. From these safety margins, the reliability indices $\beta_1 = \beta_l$ and $\beta_2 = \beta_{r|l}$ and the correlation coefficient $\rho = \rho_{l,r|l}$ can easily be calculated. The probability of failure for the parallel system then is

$$P_f = \Phi_2(-\beta_1, -\beta_2; \rho). \tag{15.97}$$

The same procedure is then, in turn, used for all critical failure elements and further critical pairs of failure elements are identified. In this way, the total series system used in the reliability analysis at level 2 is determined (see Figure 15.4). The next step is then to estimate the probability of failure for each critical pair of failure elements (see Equation 15.97) and also to determine a safety margin for each critical pair of failure elements. When this is done, generalized reliability indices for all parallel systems in Figure 15.4 and correlation coefficients between any pair of parallel systems are calculated. Finally, the probability of failure P_f for the series system (Figure 15.2) is estimated. The so-called equivalent linear safety margin introduced in Section 15.4.4 is used as an approximation for safety margins for the parallel systems.

An important property by the β-unzipping method is the possibility of using the method when brittle failure elements occur in the structure. When failure occurs in a brittle failure element, then the β-unzipping method is used in exactly the same way as presented above, the only difference being that no fictitious loads are introduced. If, for example, brittle failure occurs in a tensile bar in a trussed structure, then the bar is simply removed without adding fictitious tensile loads. Likewise, if brittle failure occurs in bending, then a yield hinge is introduced, but no (yielding) fictitious bending moments are added.

15.5.4 Assessment of System Reliability at Level N > 2

The method presented above can easily be generalized to higher levels $N > 2$. *At level* 3, the estimate of the system reliability is based on so-called *critical triples of failure elements*, that is, a set of three failure elements. The critical triples of failure elements are identified by the β-unzipping method and each triple forms a parallel system with three failure elements. These parallel systems are then elements in a series system (see Figure 15.5). Finally, the estimate of the reliability of the structural system at level 3 is defined as the reliability of this series system.

Assume that the critical pair of failure elements (l, m) has the lowest reliability index $\beta_{l,m}$ of all critical pairs of failure elements. Failure is then assumed in the failure elements l and m adding for each of them a pair of fictitious loads F_l and F_m (axial forces or moments).

The modified structure with the loads $P_1, \ldots, P_k$ and the fictitious loads F_l and F_m are then reanalyzed and influence coefficients with respect to $P_1, \ldots, P_k$ and F_l and F_m are calculated. The load effect in each of the remaining failure elements is then described by a stochastic variable $S_{i|l,m}$ (load effect in failure element i given failure in failure elements l and m) and

$$S_{i|l,m} = \sum_{j=1}^{k} a_{ij} P_j + a'_{il} F_l + a'_{im} F_m. \tag{15.98}$$

The corresponding safety margin $M_{i|l,m}$ then is

$$M_{i|l,m} = \min(R_i^+ - S_{i|l,m}, R_i^- + S_{i|l,m}) \tag{15.99}$$

where R_i^+ and R_i^- are the stochastic variables describing the load-carrying capacity in "tension" and "compression" for failure element i. In the following, $M_{i|l,m}$ will be approximated by either $R_i^- + S_{i|l,m}$ or $R_i^+ - S_{i|l,m}$ depending on the corresponding reliability indices. The reliability index for failure element i, given failure in failure elements l and m, is then given by

$$\beta_{i|l,m} = \mu_{M_{i|l,m}} / \sigma_{M_{i|l,m}}. \tag{15.100}$$

In this way, new reliability indices are calculated for all failure elements (except l and m) and the smallest β-value is called $\beta_{\min}$. These failure elements with β-values in the interval $[\beta_{\min}, \beta_{\min} + \Delta\beta_3]$, where $\Delta\beta_3$ is a prescribed positive number, are then in turn combined with failure elements l and m to form a number of parallel systems.

The next step is then to evaluate the probability of failure for each of the critical triples of failure elements. Consider the parallel system with failure elements l, m, and r. During the reliability analysis at level 1, the safety margin M_l for failure element l is determined; and during the reliability analysis at level 2, the safety margin $M_{m|l}$ for the failure element m is determined. The safety margin $M_{r|l,m}$ for safety element r has the form of Equation 15.15. From these safety margins the reliability indices $\beta_1 = \beta_l$, $\beta_2 = \beta_{m|l}$, and $\beta_3 = \beta_{r|l,m}$ and the correlation matrix $\overline{\overline{\rho}}$ can easily be calculated. The probability of failure for the parallel system then is

$$P_f = \Phi_3(-\beta_1, -\beta_2, -\beta_3; \overline{\overline{\rho}}). \tag{15.101}$$

An equivalent safety margin $M_{i,j,k}$ can be determined by the procedure mentioned above. When the equivalent safety margins are determined for all critical triples of failure elements, the correlation between all pairs of safety margins can easily be calculated. The final step is then to arrange all the critical triples as elements in a series system (see Figure 15.5) and estimate the probability of failure P_f and the generalized reliability index β_S for the series system.

The β-unzipping method can be used in exactly the same way as described in the preceding text to estimate the system reliability at levels $N > 3$. However, a definition of failure modes based on a fixed number of failure elements greater than 3 will hardly be of practical interest.

15.5.5 Assessment of System Reliability at Mechanism Level

The application of the β-unzipping method presented above can also be used when failure is defined as formation of a mechanism. However, it is much more efficient to use the β-unzipping method in connection with *fundamental mechanisms*. Experience has shown that such a procedure is less computer time consuming than unzipping based on failure elements.

If unzipping is based on failure elements, then formation of a mechanism can be unveiled by the fact that the corresponding stiffness matrix is singular. Therefore, the unzipping is simply continued until the determinant of the stiffness matrix is zero. By this procedure, a number of mechanisms with different numbers of failure elements will be identified. The number of failure elements in a mechanism will often be quite high so that several reanalyses of the structure are necessary.

As emphasized above, it is more efficient to use the β-unzipping method in connection with fundamental mechanisms. Consider an elasto-plastic structure and let the number of potential failure elements (e.g., yield hinges) be n. It is then known from the theory of plasticity that the number of fundamental mechanisms is $m = n - r$, where r is the degree of redundancy. All other mechanisms can then be formed by linear combinations of the fundamental mechanisms. Some of the fundamental mechanisms are so-called joint mechanisms. They are important in the formation of new mechanisms by linear combinations of fundamental mechanisms, but they are not real failure mechanisms. Real failure mechanisms are, by definition, mechanisms that are not joint mechanisms.

Let the number of loads be k. The safety margin for fundamental mechanism i can then be written

$$M_i = \sum_{j=1}^{n} |a_{ij}| R_j - \sum_{j=1}^{k} b_{ij} P_j \qquad (15.102)$$

where a_{ij} and b_{ij} are the influence coefficients. R_j is the yield strength of failure element j and P_j is load number j. a_{ij} is the rotation of yield hinge j corresponding to the yield mechanism i, and b_{ij} is the corresponding displacement of load j. The numerical value of a_{ij} is used in the first summation at the right-hand side of Equation 15.102 to make sure that all terms in this summation are nonnegative.

The total number of mechanisms for a structure is usually too high to include all possible mechanisms in the estimate of the system reliability. It is also unnecessary to include all mechanisms because the majority of them will in general have a relatively small probability of occurrence. Only the most critical or most significant failure modes should be included. The problem is then how the most significant mechanisms (failure modes) can be identified. In this section, it is shown how the β-unzipping method can be used for this purpose. It is not possible to prove that the β-unzipping method identifies all significant mechanisms, but experience with structures where all mechanisms can be taken into account seems to confirm that the β-unzipping method gives reasonably good results. Note that because some mechanisms are excluded, the estimate of the probability of failure by the β-unzipping method is a lower bound for the correct probability of failure. The corresponding generalized reliability index determined by the β-unzipping method is therefore an upper bound of the correct generalized reliability index. However, the difference between these two indices is usually negligible.

The first step is to identify all fundamental mechanisms and calculate the corresponding reliability indices. Fundamental mechanisms can be automatically generated by a method suggested by Watwood [26]; but when the structure is not too complicated, the fundamental mechanisms can be identified manually.

The next step is then to select a number of fundamental mechanisms as starting points for the unzipping. By the β-unzipping method this is done on the basis of the reliability index β_{min} for the real fundamental mechanism that has the smallest reliability index and on the basis of a preselected constant ε_1 (e.g., $\varepsilon_1 = 0.50$). Only real fundamental mechanisms with β-indices in the interval $[\beta_{min}; \beta_{min} + \varepsilon_1]$ are used as starting mechanisms in the β-unzipping method. Let $\beta_1 \leq \beta_2 \leq \ldots \beta_f$ be an ordered set of reliability indices for f real fundamental mechanisms 1, 2, ..., f, selected by this simple procedure. The f fundamental mechanisms selected as described above are now in turn combined linearly with all m (real and joint) mechanisms to form new mechanisms. First, the fundamental mechanism 1 is combined with the fundamental mechanisms 2, 3, ..., m and reliability indices ($\beta_{1,2}, \ldots, \beta_{1,m}$) for the new mechanisms are calculated. The smallest reliability index is determined, and the new mechanisms with reliability indices within a distance ε_2 from the smallest reliability index are selected for further investigation. The same procedure is then used on the basis of the fundamental mechanisms 2, ..., f and a failure tree like the one shown in Figure 15.22 is constructed. Let the safety

System Reliability

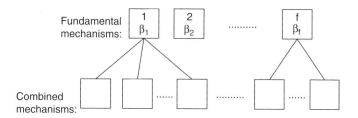

FIGURE 15.22 Construction of new mechanisms.

margins M_i and M_j of two fundamental mechanisms i and j combined as described above (see Equation 15.102) be

$$M_i = \sum_{r=1}^{n} |a_{ir}|R_r - \sum_{s=1}^{k} b_{is}P_s \qquad (15.103)$$

$$M_j = \sum_{r=1}^{n} |a_{jr}|R_r - \sum_{s=1}^{k} b_{js}P_s. \qquad (15.104)$$

The combined mechanism $i \pm j$ then has the safety margin

$$M_{i \pm j} = \sum_{r=1}^{n} |a_{ir} \pm a_{jr}|R_r - \sum_{s=1}^{k} (b_{is} \pm b_{js})P_s \qquad (15.105)$$

where + or − is chosen dependent on which sign will result in the smallest reliability index. From the linear safety margin Equation 15.105, the reliability index $\beta_{i \pm j}$ for the combined mechanism can easily be calculated.

More mechanisms can be identified on the basis of the combined mechanisms in the second row of the failure tree in Figure 15.22 by adding or subtracting fundamental mechanisms. Note that in some cases it is necessary to improve the technique by modifying Equation 15.105, namely when a new mechanism requires not only a combination with $1 \times$ but a combination with $k \times$ a new fundamental mechanism. The modified version is

$$M_{i \pm + k} = \sum_{r=1}^{n} |a_{ir} \pm ka_{jr}|R_r - \sum_{s=1}^{k} (b_{is} \pm kb_{js})P_s \qquad (15.106)$$

where k is chosen equal to, for example, −1, +1, −2, +2, −3 or +3, depending on which value of k will result in the smallest reliability index. From Equation 15.106 it is easy to calculate the reliability index $\beta_{i \pm + k}$ for the combined mechanism $i + kj$.

By repeating this simple procedure, the failure tree for the structure in question can be constructed. The maximum number of rows in the failure tree must be chosen and can typically be $m + 2$, where m is the number of fundamental mechanisms. A satisfactory estimate of the system reliability index can usually be obtained by using the same ε_2-value for all rows in the failure tree.

During the identification of new mechanisms, it will often occur that a mechanism already identified will turn up again. If this is the case, then the corresponding branch of the failure tree is terminated just one step earlier, so that the same mechanism does not occur more than once in the failure tree.

The final step in the application of the β-unzipping method in evaluating the reliability of an elasto-plastic structure at mechanism level is to select the significant mechanisms from the mechanisms identified in the failure tree. This selection can, in accordance with the selection criteria used in making the failure tree, for example, be made by first identifying the smallest β-value β_{min} of all mechanisms in the failure tree and then selecting a constant ε_3. The significant mechanisms are then, by definition, those with β-values in the interval $[\beta_{min}; \beta_{min} + \varepsilon_3]$. The probability of failure of the structure is then estimated by modeling the structural system as a series system with the significant mechanisms as elements (see Figure 15.6).

15.5.6 Examples

15.5.6.1 Two-Story Braced Frame with Ductile Elements

Consider the two-story braced frame in Figure 15.23. The geometry and the loading are shown in the figure. This example is taken from [27] where all detailed calculations are shown. The area A and the moment of inertia I for each structural member are shown in the figure. In the same figure, the expected values of the yield moment M and the tensile strength capacity R for all structural members are also stated. The compression strength capacity of one structural member is assumed to be one half of the tensile strength capacity. The expected values of the loading are $E[P_1] = 100$ kN and $E[P_2] = 350$ kN. For the sake of simplicity, the coefficient of variation for any load or strength is assumed to be $V[\cdot] = 0.1$. All elements are assumed to be perfectly ductile and made of a material having the same modulus of elasticity $E = 0.21 \times 10^9$ kN/m².

The failure elements are shown in Figure 15.23. × indicates a potential yield hinge and | indicates failure in tension/compression. The total number of failure elements is 22, namely $2 \times 6 = 12$ yield hinges in six beams and ten tension/compression possibilities of failure in the ten structural elements. The following pairs of failure elements — (1, 3), (4, 6), (7, 9), (10, 12), (13, 15), and (18, 20) — are assumed fully correlated. All other pairs of failure elements are uncorrelated. Further, the loads P_1 and P_2 are uncorrelated.

The β-values for all failure elements are shown in Table 15.1. Failure element 14 has the lowest reliability index $\beta_{14} = 1.80$ of all failure elements. Therefore, at level 0, the system reliability index is $\beta_S^0 = 1.80$.

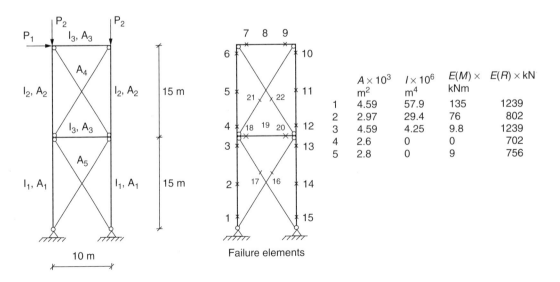

FIGURE 15.23 Two-story braced frame.

TABLE 15.1 Reliability Indices at Level 0 (and Level 1)

Failure element	2	3	4	5	6	7	8	9	10	11
β-value	8.13	9.96	9.92	4.97	9.98	9.84	9.79	9.86	9.89	1.81
Failure element	12	13	14	16	17	18	19	20	21	22
β-value	9.94	9.97	1.80	9.16	4.67	9.91	8.91	9.91	8.04	3.34

Let $\Delta\beta_1 = 0$. It then follows from Table 15.1 that the critical failure elements are 14, 11, 22, and 17. The corresponding correlation matrix (between the safety margins in the same order) is

$$\bar{\bar{\rho}} = \begin{bmatrix} 1.00 & 0.24 & 0.20 & 0.17 \\ 0.24 & 1.00 & 0.21 & 0.16 \\ 0.20 & 0.21 & 1.00 & 0.14 \\ 0.17 & 0.16 & 0.14 & 1.00 \end{bmatrix}. \quad (15.107)$$

The Ditlevsen bounds for the system probability of failure P_f^1 (see Figure 15.24) are

$$0.06843 \leq P_f^1 \leq 0.06849.$$

Therefore, a (good) estimate for the system reliability index at level 1 is $\beta_S^1 = 1.49$. It follows from Equation 15.107 that the coefficients of correlation are rather small. Therefore, the simple upper bound in Equation 15.44 can be expected to give a good approximation. One gets $\beta_S^1 = 1.48$. A third estimate can be obtained by Equation 15.54 and Equation 15.57. The result is $\beta_S^1 = 1.49$.

At level 2, it is initially assumed that the ductile failure element 14 fails (in compression) and a fictitious load equal to $0.5 \times R_{14}$ is added (see Figure 15.25). This modified structure is then analyzed elastically

FIGURE 15.24 Series system used in estimating the system reliability at level 1.

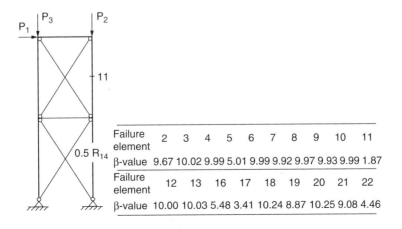

Failure element	2	3	4	5	6	7	8	9	10	11
β-value	9.67	10.02	9.99	5.01	9.99	9.92	9.97	9.93	9.99	1.87
Failure element	12	13	16	17	18	19	20	21	22	
β-value	10.00	10.03	5.48	3.41	10.24	8.87	10.25	9.08	4.46	

FIGURE 15.25 Modified structure when failure takes place in failure element 14.

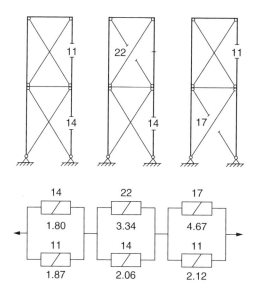

FIGURE 15.26 Failure modes used to estimate the reliability at level 2.

and new reliability indices are calculated for all the remaining failure elements (see Figure 15.25). Failure element 11 has the lowest β-value, 1.87. With $\Delta\beta_2 = 1.00$, failure element 11 is the only failure element with a β-value in the interval $[1.87, 1.87 + \Delta\beta_2]$. Therefore, in this case only one critical pair of failure elements is obtained by initiating the unzipping with failure element 14.

Based on the safety margin M_{14} for failure element 14 and the safety margin $M_{11,14}$ for failure element 11, given failure in failure element 14, the correlation coefficient can be calculated as $\rho = 0.28$. Therefore, the probability of failure for this parallel system is $P_f = \Phi_2(-1.80, -1.87; 0.28) = 0.00347$, and the corresponding generalized index $\beta_{14,11} = 2.70$. The same procedure can be performed with the three other critical failure elements 11, 22, and 17. The corresponding failure modes are shown in Figure 15.26.

Generalized reliability indices and approximate equivalent safety margins for each parallel system in Figure 15.26 are calculated. Then the correlation matrix $\overline{\overline{\rho}}$ can be calculated

$$\overline{\overline{\rho}} = \begin{bmatrix} 1.00 & 0.56 & 0.45 \\ 0.56 & 1.00 & 0.26 \\ 0.45 & 0.26 & 1.00 \end{bmatrix}.$$

The Ditlevsen bounds for the probability of failure of the series system in Figure 15.26 are $0.3488 \times 10^{-2} \le P_f^2 \le 0.3488 \times 10^{-2}$. Therefore, an estimate of the system reliability at level 2 is $\beta_S^2 = 2.70$. At level 3 (with $\Delta\beta_3 = 1.00$), four critical triples of failure elements are identified (see Figure 15.27) and an estimate of the system reliability at level 3 is $\beta_S^3 = 3.30$. It is of interest to note that the estimates of the system reliability index at levels 1, 2, and 3 are very different: $\beta_S^1 = 1.49$, $\beta_S^2 = 2.70$, and $\beta_S^3 = 3.30$.

15.5.6.2 Two-Story Braced Frame with Ductile and Brittle Elements

Consider the same structure as in Section 15.5.5.1, but now the failure elements 2, 5, 11, and 14 are assumed brittle. All other data remains unchanged. By a linear elastic analysis, the same reliability indices for all (brittle and ductile) failure elements as in Table 15.1 are calculated. Therefore, the critical failure elements are 14 and 11, and the estimate of the system reliability at level 1 is unchanged, (i.e., $\beta_S^1 = 1.49$).

The next step is to assume brittle failure in failure element 14 and remove the corresponding part of the structure without adding fictitious loads (see Figure 15.28, left). The modified structure is then linear

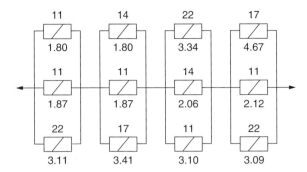

FIGURE 15.27 Reliability modeling at level 3 of the two-story braced frame.

elastically analyzed and reliability indices are calculated for all remaining failure elements. Failure element 17 now has the lowest reliability index, namely the negative value $\beta_{17|4} = -6.01$. This very low negative value indicates that failure takes place in failure element 17 instantly after failure in failure element 14. The failure mode identified in this way is a mechanism and it is the only one when $\Delta\beta_2 = 1.00$. It can be mentioned that $\beta_{16|4} = -3.74$ so that failure element 16 also fails instantly after failure element 14.

Again, by assuming brittle failure in failure element 11 (see Figure 15.28, right), only one critical pair of failure elements is identified, namely the pair of failure elements 11 and 22, where $\beta_{22|11} = -4.33$. This failure mode is not a mechanism. The series system used in calculating an estimate of the system reliability at level 2 is shown in Figure 15.29. Due to the small reliability indices, the strength variables 17 and 22 do not significantly affect the safety margins for the two parallel systems in Figure 15.29 significantly. Therefore, the reliability index at level 2 is unchanged from level 1, namely $\beta_S^2 = 1.49$.

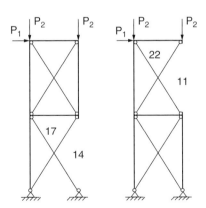

FIGURE 15.28 Modified structures.

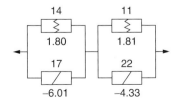

FIGURE 15.29 Modeling at level 2.

As expected, this value is much lower than the value 2.70 (see Section 15.5.5.1) calculated for the structure with only ductile failure elements. This fact stresses the importance of the reliability modeling of the structure.

It is of interest to note that the β-unzipping method was capable of disclosing that the structure cannot survive failure in failure element 14. Therefore, when brittle failure occurs, it is often reasonable to define failure of the structure as failure of just one failure element. This is equivalent to estimating the reliability of the structure at level 1.

15.5.6.3 Elastic-Plastic Framed Structure

In this example, it is shown how the system reliability at mechanism level can be estimated in an efficient way. Consider the simple framed structure in Figure 15.30 with corresponding expected values and coefficients of variation for the basic variables.

The load variables are P_i, $i = 1, \ldots, 4$ and the yield moments are R, $i = 1, \ldots, 19$. Yield moments in the same line are considered fully correlated and the yield moments in different lines are mutually independent. The number of potential yield hinges is $n = 19$ and the degree of redundancy is $r = 9$. Therefore, the number of fundamental mechanisms is $n - r = 10$.

One possible set of fundamental mechanisms is shown in Figure 15.31. The safety margins M_i for the fundamental mechanisms can be written

$$M_i = \sum_{j=1}^{19} |a_{ij}| R_j - \sum_{i=1}^{4} b_{ij} P_j, \quad i = 1, \ldots, 10 \quad (15.108)$$

where the influence coefficients a_{ij} and b_{ij} are determined by considering the mechanisms in the deformed state.

The reliability indices β_i, $i = 1, \ldots, 10$ for the ten fundamental mechanisms can be calculated from the safety margins, taking into account the correlation between the yield moments. The result is shown in Table 15.2.

With $\varepsilon_1 = 0.50$, the fundamental mechanisms 1, 2, 3, and 4 are selected as starting mechanisms in the β-unzipping and combined, in turn, with the remaining fundamental mechanisms. As an example, consider the combination $1 + 6$ of mechanisms 1 and 6. The linear safety margin M_{1+6} is obtained from the linear safety margins M_1 and M_6 by addition, taking into account the signs of the coefficients. The corresponding reliability index is $\beta_{1+6} = 3.74$.

With $\varepsilon_2 = 1.20$, the following new mechanisms $1 + 6$, $1 + 10$, $2 + 6$, $3 + 7$, and $4 - 10$ are identified by this procedure. The failure tree at this stage is shown in Figure 15.16. It contains $4 + 5 = 9$ mechanisms. The reliability indices and the fundamental mechanisms involved are shown in the same figure.

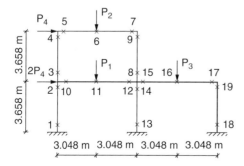

Variables	Expected values	Coefficients of variation
P_1	169 kN	0.15
P_2	89 kN	0.25
P_3	116 kN	0.25
P_4	31 kN	0.25
$R_1, R_2, R_{13}, R_{14}, R_{18}, R_{19}$	95 kNm	0.15
R_3, R_4, R_8, R_9	95 kNm	0.15
R_5, R_6, R_7	122 kNm	0.15
R_{10}, R_{11}, R_{12}	204 kNm	0.15
R_{15}, R_{16}, R_{17}	163 kNm	0.15

FIGURE 15.30 Geometry, loading, and potential yield hinges (×).

System Reliability

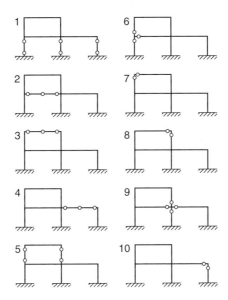

FIGURE 15.31 Set of fundamental mechanisms.

This procedure is now continued as explained earlier by adding or subtracting fundamental mechanisms. If the procedure is continued eight times (up to ten fundamental mechanisms in one mechanism) and if the significant mechanisms are selected by $\varepsilon_3 = 0.31$, then the system modeling at mechanism level will be a series system where the elements are 12 parallel systems. These 12 parallel systems (significant mechanisms) and corresponding reliability indices are shown in Table 15.3. The correlation matrix is

$$\bar{\bar{\rho}} = \begin{bmatrix} 1.00 & 0.65 & 0.89 & 0.44 & 0.04 & 0.91 & 0.59 & 0.97 & 0.87 & 0.81 & 0.86 & 0.92 \\ 0.65 & 1.00 & 0.55 & 0.00 & 0.09 & 0.67 & 0.00 & 0.63 & 0.54 & 0.58 & 0.00 & 0.71 \\ 0.89 & 0.55 & 1.00 & 0.35 & 0.45 & 0.83 & 0.61 & 0.88 & 0.98 & 0.95 & 0.31 & 0.83 \\ 0.44 & 0.00 & 0.35 & 1.00 & 0.00 & 0.03 & 0.00 & 0.45 & 0.37 & 0.04 & 0.89 & 0.08 \\ 0.04 & 0.09 & 0.45 & 0.00 & 1.00 & 0.05 & 0.00 & 0.04 & 0.44 & 0.48 & 0.00 & 0.05 \\ 0.91 & 0.67 & 0.83 & 0.03 & 0.05 & 1.00 & 0.73 & 0.87 & 0.80 & 0.88 & 0.00 & 0.98 \\ 0.59 & 0.00 & 0.61 & 0.00 & 0.00 & 0.73 & 1.00 & 0.58 & 0.60 & 0.65 & 0.00 & 0.64 \\ 0.97 & 0.63 & 0.88 & 0.45 & 0.04 & 0.87 & 0.58 & 1.00 & 0.90 & 0.77 & 0.48 & 0.87 \\ 0.87 & 0.54 & 0.98 & 0.37 & 0.44 & 0.80 & 0.60 & 0.90 & 1.00 & 0.90 & 0.41 & 0.78 \\ 0.81 & 0.58 & 0.95 & 0.04 & 0.48 & 0.88 & 0.65 & 0.77 & 0.90 & 1.00 & 0.00 & 0.88 \\ 0.36 & 0.00 & 0.31 & 0.89 & 0.00 & 0.00 & 0.00 & 0.48 & 0.41 & 0.00 & 1.00 & 0.00 \\ 0.92 & 0.71 & 0.83 & 0.08 & 0.05 & 0.98 & 0.64 & 0.87 & 0.78 & 0.88 & 0.00 & 1.00 \end{bmatrix}.$$

TABLE 15.2 Reliability Indices for the Fundamental Mechanisms

i	1	2	3	4	5	6	7	8	9	10
β_i	1.91	2.08	2.17	2.26	4.19	10.75	9.36	9.36	12.65	9.12

TABLE 15.3 Correlation Coefficients and Safety Indices of 12 Systems at Mechanism Level

No.	Significant mechanisms	β
1	1 + 6 + 2 + 5 + 7 + 3 − 8	1.88
2	1	1.91
3	1 + 6 + 2 + 5 + 7 + 3 + 4 − 8 − 10	1.94
4	3 + 7 − 8	1.98
5	4 − 10	1.99
6	1 + 6 + 2	1.99
7	2	2.08
8	1 + 6 + 2 + 5 + 7 + 3 + 8	2.09
9	1 + 6 + 2 + 5 + 7 + 3 + 8 + 9 + 4 − 10	2.11
10	1 + 6 + 2 + 5 + 9 + 4 − 10	2.17
11	3	2.17
12	1 + 6 + 2 + 5	2.18

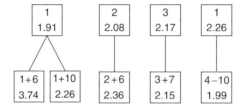

FIGURE 15.32 The first two rows in the failure tree.

The probability of failure P_f for the series system with the 12 significant mechanisms as elements can then be estimated by the usual techniques. The Ditlevsen bounds are $0.08646 \leq P_f \leq 0.1277$. If the average value of the lower and upper bounds is used, the estimate of the reliability index β_s at mechanism level is $\beta_s = 1.25$. Hohenbichler [28] has derived an approximate method to calculate estimates for the system probability of failure P_f (and the corresponding reliability index β_s). The estimate of β_s is $\beta_s = 1.21$. It can finally be noted that Monte Carlo simulation gives $\beta_s = 1.20$.

15.6 Illustrative Examples

15.6.1 Reliability of a Tubular Joint

The single tubular beams in a steel jacket structure will in general at least have three important failure elements; namely, failure in yielding under combined bending moments and axial tension/compression at points close to the ends of the beam and failure in buckling (instability). A tubular joint will likewise have a number of potential failure elements. Consider the tubular K-joint shown in Figure 15.33.

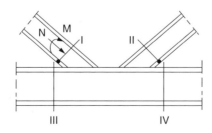

FIGURE 15.33 Tubular K-joint: I through IV are the critical sections and • indicates the hot spots.

This tubular joint is analyzed in more detail in a paper by Thoft-Christensen and Sørensen [29]. It is assumed that the joint has four critical sections as indicated in Figure 15.33 by I, II, III, and IV. The load effects in each critical section are an axial force N and a bending moment M. The structure is assumed to be linear elastic. For this K-joint, 12 failure elements are considered, namely:

- Failure in yielding in the four critical sections I, II, III, and IV
- Punching failure in the braces (cross-sections I and II)
- Buckling failure of the four tubular members (cross-sections I, II, III, and IV)
- Fatigue failure in the critical sections I and II (hot spots are indicated in Figure 15.33)

For these failure elements, safety margins have been formulated by Thoft-Christensen and Sørensen [29]. For failure in yielding the safety margins are of the form

$$M^Y = Z^Y - \left(\left| \frac{M}{M_F} \right| - \cos\left(\frac{\pi}{2} \frac{N}{N_F} \right) \right) \tag{15.109}$$

where M_F and N_F are the yield capacities in pure bending and pure axial loading, and where Z^Y is a model uncertainty variable. For punching failure, the following safety margin is used

$$M^P = Z^P - \left(\left| \frac{N}{Z^N N_U} \right| + \left(\frac{|M|}{Z^M M_U} \right)^{1.2} \right). \tag{15.110}$$

N_U and M_U are ultimate punching capacities in pure axial loading and pure bending, respectively. In this case, three model uncertainty variables Z^P, Z^N, and Z^M are included. The following safety margin is used for buckling failure

$$M^B = Z^B - \left(\frac{N}{N_B} + \frac{M}{M_B} \right) \tag{15.111}$$

where N_B and M_B are functions of the geometry and the yield stress and where Z^B is a model uncertainty variable. Finally, with regard to fatigue failure, the safety margin for each of the two hot points, shown in Figure 15.33, has the form

$$M^F = Z^F - (Z^L)^m K^{-1} (\max(t/32, 1)^{M_1}) g \tag{15.112}$$

where m ($= 3$) and K are constants in the S-N relation used in Miner's rule. K is modeled as a random variable, t is the wall thickness, M_1 is a random variable, and g is a constant. Z^F and Z^L are model uncertainty variables.

For this particular K-joint in a plane model of a tubular steel jacket structure analyzed by Thoft-Christensen and Sørensen [29], only four failure elements are significant:, namely (referring to Figure 15.33) fatigue in cross-section II ($\beta = 3.62$), punching in the same cross-section II ($\beta = 4.10$), buckling in cross-section II ($\beta = 4.58$), and fatigue in cross-section I ($\beta = 4.69$).

The correlation coefficient matrix of the linearized safety margins of the four significant failure elements is

$$\bar{\bar{\rho}} = \begin{bmatrix} 1 & 0 & 0 & 0.86 \\ 0 & 1 & 0.82 & 0 \\ 0 & 0.82 & 1 & 0 \\ 0.86 & 0 & 0 & 1 \end{bmatrix}$$

and the probability of failure of the joint

$$P_f \approx 1 - \Phi_4(\bar{\beta}, \bar{\bar{\rho}}) = 1.718 \times 10^{-4}$$

The corresponding reliability index for the joint is

$$\beta = -\Phi^{-1}(P_f) = 3.58.$$

It is important to note that in this estimate of the reliability of the K-joint, the interaction between the different significant failure elements in, for example, cross-section II is not taken into account. Each failure element (failure mode) is considered independent of the other, although such interaction will influence the reliability of the joint.

15.6.2 Reliability-Based Optimal Design of a Steel-Jacket Offshore Structure

In Thoft-Christensen and Murotsu [2], an extensive number of references to reliability-based optimal design can be found. In this section the optimal design problem is briefly stated and illustrated with an example taken from Thoft-Christensen and Sørensen [30]. In reliability-based optimization, the objective function is often chosen as the weight F of the structure. The constraints can either be related to the reliability of the single elements or to the reliability of the structural system. In the last-mentioned case, the optimization problem for a structure with h elements may be written as

$$\min \quad F(\bar{y}) = \sum_{i=1}^{h} \varphi_i l_i A_i(\bar{y})$$

$$\text{s.t.} \quad \beta^s(\bar{y}) \geq \beta_0^s \tag{15.113}$$

$$y_i^l \leq y_i \leq y_1^u, \quad i = 1, \ldots, n$$

where A_i, l_i and φ_i are the cross-sectional area, the length, and the density of element no. i; $\bar{y} = (y_1, \ldots, y_n)$ are the design variables; β_0^s is the target system reliability index; and y_i^l and y_i^u are lower and upper bounds, respectively, for the design variable y_i, $i = 1, \ldots, n$.

Consider the three-dimensional truss model of the steel-jacket offshore structure shown in Figure 15.34. The load and the geometry are described in detail by Thoft-Christensen and Sørensen [30] and by Sørensen, Thoft-Christensen, and Sigurdsson [31]. The load is modeled by two random variables, and the yield capacities of the 48 truss elements are modeled as random variables with expected values 270×10^6 Nm^{-2} and coefficients of variation equal to 0.15. The correlation structure of the normally distributed variables is described in [31]. The design variables y_i, $i = 1, \ldots, 7$ are the cross-sectional areas (m²) of the seven groups of structural elements (see Figure 15.34). The optimization problem is

$$\min \quad F(\bar{y}) = 125 y_1 + 100 y_2 + 80 y_3 + 384 y_4 + 399 y_6 + 255 y_7 (m^3)$$

$$\text{s.t.} \quad \beta^s(\bar{y}) \geq \beta_0^s = 3.00 \tag{15.114}$$

$$0 = y_i^l \leq y_i \leq y_1^u = 1, \quad i = 1, \ldots, 7.$$

β^s is the system reliability index at level 1. The solution is $\bar{y} = (0.01, 0.001, 0.073, 0.575, 0.010, 0.009, 0.011)$ m² and $F(\bar{y}) = 215$ m³. The iteration history for the weight function $F(\bar{y})$ and the system reliability index β^s is shown in Figure 15.35.

System Reliability

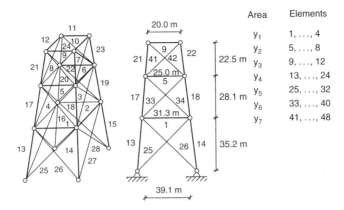

FIGURE 15.34 Space truss tower with design variables.

15.6.3 Reliability-Based Optimal Maintenance Strategies

An interesting application of optimization methods is related to deriving optimal strategies for inspection and repair of structural systems. In a paper by Thoft-Christensen and Sørensen [32], such a strategy was presented with the intention of minimizing the cost of inspection and repair of a structure in its lifetime T under the constraint that the structure has an acceptable reliability. In another paper by Sørensen and Thoft-Christensen [33], this work was extended by including not only inspection costs and repair costs, but also the production (initial) cost of the structure in the objective function.

The purpose of a simple optimal strategy for inspection and repair of civil engineering structures is to minimize the expenses of inspection and repair of a given structure so that the structure in its expected service life has an acceptable reliability. The strategy is illustrated in Figure 15.36, where T is the lifetime of the structure and β is a measure of the reliability of the structure. The reliability β is assumed to be a nonincreasing function with time t. T_i, $i = 1, 2, \ldots, n$ are the inspection times, and $\beta^{\min}$ is the minimum acceptable reliability of the structure in its lifetime.

Let $t_i = T_i - T_{i-1}$, $i = 1, \ldots, n$ and let the quality of inspection at time T_i be q_i, $i = 1, \ldots, n$. Then, with a given number of inspection times n, the design variables are t_i, $i = 1, \ldots, n$ and q_i, $i = 1, \ldots, n$. As an illustration, consider the maintenance strategy shown in Figure 15.36. At time T_1, the solution of the optimization problem is $q_1 = 0$; that is, no inspection takes place at that time. Inspection takes place at time T_2; and according to the result of the inspection, it is decided whether or not repair should be performed. If repair is performed, the reliability is improved. If no repair is performed, the reliability is

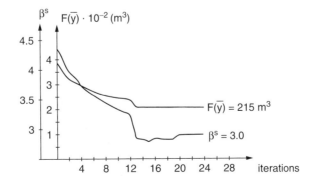

FIGURE 15.35 Iteration history.

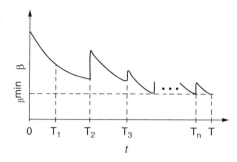

FIGURE 15.36 Maintenance strategy.

also improved because then updating of the strengths of the structural elements takes place. Therefore, the variation of the reliability of the structure with time will be as shown in Figure 15.36. The shape of the curves between inspection times will depend on the relevant types of deterioration, for example, whether corrosion or fatigue is considered.

A very brief description of the application of this strategy is given here, based on [33]. The design variables are cross-sectional parameters $z_1, \ldots, z_m$, inspection qualities $q_1, \ldots, q_N$, and time between inspections $t_1, \ldots, t_N$, where m is the number of cross-sections to be designed and N is the number of inspections (and repairs). The optimization for a structural system modeled by s failure elements can then be formulated in the following way:

$$\min_{\bar{q}=(q_1,\ldots,q_N), \bar{t}=(t_1,\ldots,t_N)} C(\bar{z}, \bar{q}, \bar{t}) = C_I(\bar{z}) + \sum_{i=1}^{N}\sum_{j=1}^{s} C_{IN,j}(q_i)e^{-rT_i} + \sum_{i=1}^{N}\sum_{j=1}^{s} C_{R,j}(\bar{z})E[R_{ij}(\bar{z},\bar{q},\bar{t})]e^{-rT_i} \quad (15.115)$$

$$\text{s.t.} \quad \beta^s(T_i) \geq \beta^{\min}, \quad i=1,\ldots,N, N+1$$

where the trivial bounds on the design variables are omitted; C_I is the initial cost of the structure; $C_{IN,j}(q_i)$ is the cost of an inspection of element j with the inspection quality q_i; $C_{R,j}$ is the cost of a repair of element j; r is the discount rate; and $E[R_{ij}(\bar{z},\bar{q},\bar{t})]$ is the expected number of repairs at the time T_i in element j. $\beta^s(T_i)$ is the system reliability index at level 1 at the inspection time T_i and $\beta^{\min}$ is the lowest acceptable system reliability index.

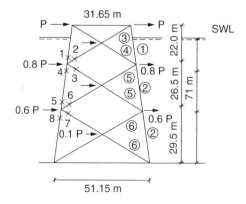

FIGURE 15.37 Plan model of a steel jacket platform.

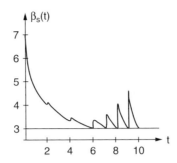

FIGURE 15.38 β^s as a function of the time t for optimal design and inspection variables and for $N = 6$.

Consider the plane model of a steel jacket platform shown in Figure 15.37 (see [33], where all details are described). Due to symmetry, only the eight fatigue failure elements indicated by × in Figure 15.37 are considered. Design variables are the tubular thicknesses of the six groups of elements indicated by ○ in Figure 15.37. Using $\beta^{\min} = 3.00$, $T = 10$ years, $r = 0$, and $N = 6$, the following optimal solution is determined:

$$\bar{t} = (2, 2, 2, 1.19, 0.994, 0.925) \text{ years}$$

$$\bar{q} = (0.1, 0.133, 0.261, 0.332, 0.385, 0.423)$$

$$\bar{z} = (68.9, 62.7, 30.0, 30.0, 50.4, 32.0) \text{ mm}$$

as shown in Figure 15.38.

References

1. Thoft-Christensen, P. and Baker, M. J., *Structural Reliability and Its Applications.* Springer-Verlag, Berlin-Heidelberg-New York, 1982.
2. Thoft-Christensen, P. and Murotsu, Y., *Application of Structural Systems Reliability Theory.* Springer-Verlag, Berlin-Heidelberg-New York, 1986.
3. Rackwitz, R. and Fiessler, B., *An Algorithm for Calculation of Structural Reliability under Combined Loading.* Berichte zur Sicherheitstheorie der Bauwerke, Lab. f. Konstr. Ingb., Tech. Univ. München, München, 1977.
4. Thoft-Christensen, P., *Reliability Analysis of Structural Systems by the β-Unzipping Method.* Institute of Building Technology and Structural Engineering, Aalborg University Centre, Aalborg, Report 8401, March 1984.
5. Kaufmann, A., Grouchko, D., and Cruon R., *Mathematical Models for the Study of the Reliability of Systems.* Academic Press, New York, San Francisco, London, 1977.
6. Hohenbichler, M. and Rackwitz, R., First-Order Concepts in System Reliability. *Structural Safety,* Vol. 1, 1983, pp. 1, 177–188.
7. Kounias, E., Bounds for the Probability of a Union, with Applications. *Annals of Mathematical Statistics,* Vol. 39, 1968, pp. 2154–2158.
8. Ditlevsen, O., Narrow Reliability Bounds for Structural Systems. *J. Struct. Mech.,* Vol. 7(No. 4), 1979, pp. 453–472.
9. Dunnett, C.W. and Sobel, M., Approximations to the Probability Integral and Certain Percentage Points of Multivariate Analogue of Students' t-Distribution. *Biometrika,* Vol. 42, 1955, pp. 258–260.

10. Thoft-Christensen, P. and Sørensen, J. D., Reliability of Structural Systems with Correlated Elements. *Applied Mathematical Modelling*, Vol. 6, 1982, pp. 171–178.
11. Ditlevsen, O., Taylor Expansion of Series System Reliability. *Journal of the Engineering Mechanics*, ASCE, Vol. 110(No. 2), 1984, pp. 293–307.
12. Murotsu, Y., Okada, M., Yonezawa, M., and Taguchi, K., Reliability Assessment of Redundant Structures. *Structural Safety and Reliability, ICOSSAR 81*, Elsevier Scientific, Publishing Company, Amsterdam, 1981, pp. 315–329.
13. Gollwitzer, S. and Rackwitz, R., Equivalent Components in First-Order System Reliability. *Reliability Engineering*, Vol. 5, 1983, pp. 99–115.
14. Grigoriu, M. and Turkstra, C., Safety of Structural Systems with Correlated Resistances. *Applied Mathematical Modelling*, Vol. 3, 1979, pp. 130–136.
15. Thoft-Christensen, P. and Sørensen, J. D., Reliability of Structural Systems with Correlated Elements. *Applied Mathematical Modelling*, Vol. 6, 1982, pp. 171–178.
16. Ferregut-Avila, C. M., *Reliability Analysis of Elasto-Plastic Structures*. NATO Advanced Study Institute, P. Thoft-Christensen, Ed. (ed.), Martinus Nijhoff, The Netherlands, 1983, pp. 445–451.
17. Moses, F., Structural System Reliability and Optimization. *Computers & Structures*, Vol. 7, 1977, pp. 283–290.
18. Gorman, M. R., Reliability of Structural Systems. Report No. 79-2, Case Western Reserve University, Ohio, Ph.D. Report, 1979.
19. Ma, H.-F. and Ang, A. H.-S., Reliability Analysis of Ductile Structural Systems. Civil Eng. Studies, Structural Research Series No. 494, University of Illinois, Urbana, August 1981.
20. Klingmüller, O., Anwendung der Traglastberechnung für die Beurteilung der Sicherheit von Konstruktionen. *Forschungsberichte aus dem Fachbereich Bauwesen*. Gesamthochschule Essen, Heft 9, Sept. 1979.
21. Murotsu, Y., Okada, H., Yonezawa, M., and Kishi, M., Identification of Stochastically Dominant Failure Modes in Frame Structures. *Proc. Int. Conf. on Appl. Stat. and Prob. in Soil and Struct. Eng.*, Universita di Firenze, Italia. Pitagora Editrice 1983, pp. 1325–1338.
22. Kappler, H., Beitrag zur Zuverlässigkeitstheorie von Tragwerken unter Berücksichtigung nichtlinearen Verhaltens. Dissertation, Technische Universität München, 1980.
23. Thoft-Christensen, P., The β-Unzipping Method. Institute of Building Technology and Structural Engineering, Aalborg University Centre, Aalborg, Report 8207, 1982 (not published).
24. Thoft-Christensen, P. and Sørensen, J.D., Calculation of Failure Probabilities of Ductile Structures by the β-Unzipping Method. Institute of Building Technology and Structural Engineering, Aalborg University Centre, Aalborg, Report 8208, 1982.
25. Thoft-Christensen, P. and Sørensen, J.D., Reliability Analysis of Elasto-Plastic Structures. *Proc. 11. IFIP Conf. "System Modelling and Optimization,"* Copenhagen, July 1983. Springer-Verlag, 1984, pp. 556–566.
26. Watwood, V. B., Mechanism Generation for Limit Analysis of Frames. ASCE, *Journal of the Structural Division*, Vol. 105(No. STl), Jan. 1979, pp. 1–15.
27. Thoft-Christensen, P., Reliability of Structural Systems. *Advanced Seminar on Structural Reliability*, Ispra, Italy, 1984. University of Aalborg, Denmark, Report R8401, March 1984.
28. Hohenbichler, M., An Approximation to the Multivariate Normal Distribution Function. *Proc. 155th Euromech on Reliability of Structural Eng. Systems*, Lyngby, Denmark, June 1982, pp. 79–110.
29. Thoft-Christensen, P. and Sørensen, J.D., Reliability Analysis of Tubular Joints in Offshore Structures. *Reliability Engineering*, Vol. 19, 1987, pp. 171–184.
30. Thoft-Christensen, P. and Sørensen, J.D., Recent Advances in Optimal Design of Structures from a Reliability Point of View. *Quality and Reliability Management*, 4, Vol. 4 1987, pp. 19–31.
31. Sørensen, J.D., Thoft-Christensen, P., and Sigurdsson, G., Development of Applicable Methods for Evaluating the Safety of Offshore Structures, Part 2. Institute of Building Technology and Structural Engineering. The University of Aalborg, Paper No. 11, 1985.

32. Thoft-Christensen, P. and Sørensen, J.D., Optimal Strategy for Inspection and Repair of Structural Systems. Civil Engineeringng Systems., Vol. 4, 1987, pp. 94–100.
33. Thoft-Christensen, P., Application of Optimization Methods in Structural Systems Reliability Theory. *Proc. IFIP Conference on "System Modelling and Optimization,"* Tokyo, August/September 1987. Springer-Verlag, 1988, pp. 484–497.

16
Quantum Physics-Based Probability Models with Applications to Reliability Analysis

16.1 Introduction 16-1
16.2 Background: Geometric Mean and Related Statistics 16-2
16.3 Probability Distributions of the "Quantum Mass Ratio" and Its Logarithm 16-2
16.4 Logarithmic Variance and Other Statistics of the "Quantum Mass Ratio" 16-4
16.5 Probability Distribution of the "Quantum Size Ratio" 16-9
16.6 Extensions and Applications to Reliability Analysis 16-11
16.7 Conclusion 16-12

Erik Vanmarcke
Princeton University

16.1 Introduction

We present herein a class of probability models with considerable potential for useful applications to reliability analysis. The new families of probability density functions (PDFs) have a fundamental basis in quantum physics; they describe the inherent uncertainty of properties of single energy quanta emitted by a "perfect radiator" or "blackbody" with known temperature. The starting point for the derivation of the various PDFs is Planck's blackbody radiation spectrum [1], which expresses how the energy absorbed or emitted in blackbody radiation is distributed among quanta of radiation (photons) with different energies (or frequencies or wavelengths), characterizing the aggregate behavior of very many energy quanta in local thermal equilibrium (LTE).

Section 16.2, providing background, reviews the relationship between the geometric and arithmetic means of any nonnegative random variable. Section 16.3, derives the PDFs of various functionally related random properties of single energy quanta when the local-thermal-equilibrium (LTE) temperature T is known or prescribed. The results apply for any LTE temperature between absolute zero ($T = 0$) and the upper limit on the validity of quantum mechanics and general relativity theory, namely the "Planck temperature" ($T_{pl} \sim 10^{32} K$). Section 16.4 deals with how to express the intrinsic uncertainty of the "quantum mass ratio," defined as the quotient of the mass (energy/c^2) of a random quantum of energy and the geometric-mean mass of quanta *given* the LTE-temperature *T*. In Section 16.5, we obtain the probability distribution of the "quantum size ratio," defined as the reciprocal of the quantum mass ratio, representing the relative "sizes" (wavelengths) of single energy quanta sampled from a collection of quanta in (local) thermal equilibrium. Section 16.6 looks into various applications of the probability models to single-mode and system reliability analysis.

16.2 Background: Geometric Mean and Related Statistics

Various random quantities expressed as *products* of random variables can be described efficiently in terms of *geometric* means and related statistics. By definition, the geometric mean $\overline{Y}$ of a nonnegative random variable Y is

$$\overline{Y} \equiv \exp\langle \log Y \rangle = \exp\left\{ \int_0^\infty \log y \, dF_Y(y) \right\}, \tag{16.1}$$

where $\langle \cdot \rangle$ refers to a random quantity's *arithmetic* mean; "log" denotes the natural logarithm; "exp" is the exponential function; and $F_Y(y)$ is the cumulative distribution function (CDF) of Y. The definition in Equation 16.1 implies that the geometric mean of a product of random variables, $Y = Y_1 \ldots Y_n$, equals the product of their geometric means, $\overline{Y} = \overline{Y}_1 \ldots \overline{Y}_n$, and hence that the geometric mean of a power of Y equals $\overline{Y}$ raised to that power. The variability of Y can be quantified in term of its "characteristic value," defined as

$$\widetilde{Y} \equiv \overline{Y} \exp\left\{ \frac{1}{2} Var[\log Y] \right\}, \tag{16.2}$$

which depends on the "logarithmic variance," or the variance of the logarithm, of Y:

$$Var[\log Y] \equiv \langle (\log Y - \langle \log Y \rangle)^2 \rangle = \langle (\log Y)^2 \rangle - \langle \log Y \rangle^2 = 2\log(\widetilde{Y}/\overline{Y}), \tag{16.3}$$

where the term on the right-hand side is just the inverse of Equation 16.2. Both the logarithmic variance and the "stochasticity factor" ($\widetilde{Y}/\overline{Y}$), the quotient of the characteristic value $\widetilde{Y}$ and the geometric mean $\overline{Y}$, measure the spread of the PDF $f_Y(y)$ of any nonnegative random variable Y; and in case Y is deterministic, $Var[\log Y] = 0$ and $\widetilde{Y}/\overline{Y} = 1$. Common measures of uncertainty based on arithmetic means are the variance $Var[Y] \equiv \langle \{Y - \langle Y \rangle\}^2 \rangle = \langle Y^2 \rangle - \langle Y \rangle^2$, its square root, the standard deviation $\sigma_Y \equiv (Var[Y])^{1/2}$, and the ratio of standard deviation to mean, or the "coefficient of variation," $V_Y \equiv \sigma_Y/\langle Y \rangle$. Note that in case Y is lognormally distributed, so the logarithm of Y is normal or Gaussian, the following relations hold: the geometric mean $\overline{Y}$ equals the median (or the fiftieth percentile) of Y, the characteristic value $\widetilde{Y}$ equals the arithmetic mean $\langle Y \rangle$, the stochasticity factor $\widetilde{Y}/\overline{Y}$ is given by $1 + V_Y^2$, and the coefficient of skewness of Y, defined by $\chi_Y \equiv \langle (Y - \langle Y \rangle)^3 \rangle / \sigma_Y^3$, satisfies the relation: $\chi_Y = 3V_Y + V_Y^3$.

16.3 Probability Distributions of the "Quantum Mass Ratio" and Its Logarithm

One of the standard ways of expressing the Planck radiation function [1, 2], characterizing a large number of quanta of radiation (photons) in thermal equilibrium, is in terms of the mean number of quanta per unit volume, dN_λ, having wavelengths within the infinitesimal range between λ and $\lambda + d\lambda$,

$$dN_\lambda = \frac{8\pi}{\lambda^4} d\lambda \bigg/ \left[\exp\left\{ \frac{h_p c}{k_B T \lambda} \right\} - 1 \right], \tag{16.4}$$

in which T is the radiation temperature in degrees Kelvin, $c \approx 2.998 \times 10^{10}\,cm/sec$ is the speed of light, $k_B \approx 8.617 \times 10^{-5}\,eV$ (electron-volt) per degree Kelvin is Boltzmann's constant, and $h_p \approx 4.14 \times 10^{-15}\,eV\,sec$ is Planck's constant, the "quantum of action." Integrating over the wavelength intervals $d\lambda$ for values of λ ranging from zero to infinity yields

$$N_{Total} = \int_0^\infty dN_\lambda = 16\pi\zeta(3)\left(\frac{k_B T}{h_p c}\right)^3 \approx 20.288\,T^3\,cm^{-3} \tag{16.5}$$

for the mean total number of energy quanta per unit volume as a function of temperature, in which $\zeta(u) = 1 + 2^{-u} + 3^{-u} + \ldots$ is Riemann's Zeta function [2].

Our interest henceforth focuses on interpreting the Planck radiation spectrum as a description of the random properties of single energy quanta or particles "sampled" from a collection of quanta or particles in local thermal equilibrium with temperature T. Each quantum possesses a set of functionally related random properties: its energy E or mass $m = E/c^2$ (Einstein's famous equation), wavelength $\lambda = h_p c/E = h_p/(mc)$, frequency $\nu = E/h_p = c/\lambda$, and momentum $p = mc = E/c$. Any of these attributes can be expressed, free of units, in terms of the random "quantum mass ratio"

$$W \equiv \frac{E}{\overline{E}} = \frac{m}{\overline{m}} = \frac{\overline{\lambda}}{\lambda}, \tag{16.6}$$

where $\overline{E} \equiv \exp\{\langle \log E \rangle\} = \overline{m}c^2 = h_p c/\overline{\lambda}$, the geometric mean (see Equation 16.1) of the quantum energy, equals the product of $k_B T$ and the constant

$$c_0 \equiv \frac{\overline{E}}{k_B T} = \left(\frac{h_p c}{k_B T}\right)\frac{1}{\overline{\lambda}} \approx 2.134. \tag{16.7}$$

The natural logarithm of W, or the "quantum-mass-ratio logarithm," is

$$L \equiv \log W = \log E - \log \overline{E} = \log\{E/(k_B T)\} - \ell_0, \tag{16.8}$$

where $\ell_0 \equiv \langle \log\{\overline{E}/(k_B T)\} \rangle = \log c_0 \approx 0.758$. (In general, we denote random variables herein by capital letters and their realizations by the lower case; but the rule is broken when symbols have established definitions or serve overriding demands for clarity of presentation, as is the case for the "random wavelength" λ and the "random mass" m in Equations 16.6 and 16.7.)

Combining Equation 16.4 and Equation 16.6 to Equation 16.8 yields the probability density functions of the quantum mass ratio W and its natural logarithm, $L \equiv \log W$, respectively [3],

$$f_W(w) = \frac{c_0}{2\zeta(3)}\frac{(c_0 w)^2}{e^{c_0 w} - 1}, \quad w \geq 0, \tag{16.9}$$

and

$$f_L(\ell) = \frac{1}{2\zeta(3)}\frac{e^{3(\ell_0 + \ell)}}{e^{e^{(\ell_0 + \ell)}} - 1}, \quad -\infty \leq \ell \leq \infty, \tag{16.10}$$

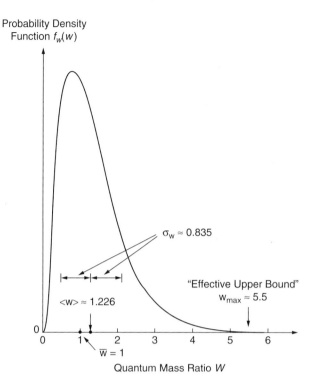

FIGURE 16.1 Probability density function of the "quantum mass ratio" W.

where $c_0[2\zeta(3)]^{-1} \approx 0.8877$ and $[2\zeta(3)]^{-1} \approx 0.416$. The PDFs of W and L are plotted in Figure 16.1 and Figure 16.2, respectively. The expectation of $e^{\iota u L}$, where $\iota = \sqrt{-1}$, defines the "characteristic function" of L; its logarithm $K_L(u) \equiv \log\langle e^{\iota u L}\rangle$ is known as the cumulant function [2], the cumulants of L being the coefficients ($K_{1,L}$, $K_{2,L}$, etc.) in the series expansion of $K_L(u)$:

$$K_L(u) \equiv \log\langle e^{\iota u L}\rangle = (\iota u)K_{1,L} + \frac{(\iota u)^2}{2!}K_{2,L} + \frac{(\iota u)^3}{3!}K_{3,L} + \ldots \tag{16.11}$$

The first three cumulants are the mean, the variance, and the third central moment of L, respectively. The principal property of cumulants is that each cumulant of a sum of independent random variables is itself a sum of component cumulants, so that a sum's cumulant function is also the sum of the component cumulant functions.

16.4 Logarithmic Variance and Other Statistics of the "Quantum Mass Ratio"

In problems involving *products* of (i.i.d.) random variables W, a useful and attractive attribute of the random mass ratio W is that its geometric mean equals one,

$$\overline{W} \equiv \exp\langle \log W\rangle = 1, \tag{16.12}$$

implying $\langle \log W\rangle = \langle L\rangle = 0$. A key measure of uncertainty of the properties of single energy quanta in thermal equilibrium is the logarithmic variance of the quantum mass ratio, or the variance of the

Quantum Physics-Based Probability Models

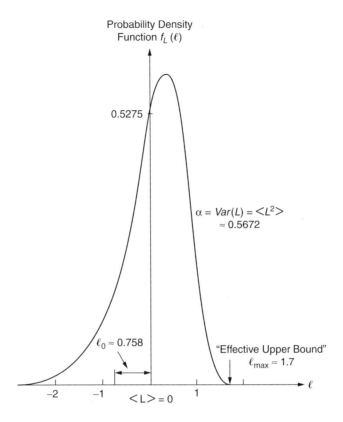

FIGURE 16.2 Probability density function of L, the quantum-mass-ratio logarithm.

mass-ratio logarithm, which we denote by α. We can write:

$$\alpha \equiv Var[\log W] = Var[L] = \sigma_L^2 = \langle L^2 \rangle = \int_0^\infty \ell^2 f_L(\ell) d\ell \approx 0.5672. \tag{16.13}$$

Measuring the variability of the energy (or wavelength or momentum) of single quanta "sampled" from a population of quanta in thermal equilibrium, for any temperature $T \leq T_{Planck} \approx 10^{32} K$, the quantity $\alpha \equiv Var[L] \approx 0.5672$ is a basic constant, expressing the Uncertainty Principle in a way that complements, yet differs from, Heisenberg's (1926) formulation; the latter, proposed in the context of atomic physics, is an *inequality* that limits the precision of joint measurements of *two* attributes of a particle, such as position and momentum, and involves Planck's constant, a tiny number. The functions $\ell f_L(\ell)$ and $\ell^2 f_L(\ell)$ are plotted in Figure 16.3 and Figure 16.4, respectively; their complete integrals, from $-\infty$ to $+\infty$, are, respectively, the mean $\langle L \rangle = 0$ and the variance $\alpha \equiv \langle L^2 \rangle \approx 0.5672$ of the quantum-mass-ratio logarithm.

The statistical moments of the quantum mass ratio W, for any real value $s > -2$, not necessarily integers, are given by

$$\langle W^s \rangle = \int_0^\infty w^s f_W(w) dw = \frac{\Gamma(3+s)\zeta(3+s)}{2(c_0)^s \zeta(3)}, \quad s > -2, \tag{16.14}$$

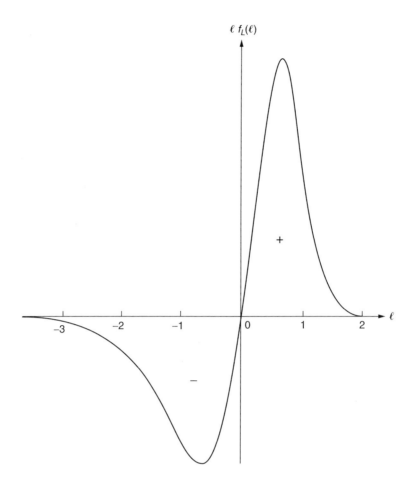

FIGURE 16.3 The function $\ell f_L(\ell)$, whose integral is the (zero) mean of L.

where $\Gamma(\cdot)$ denotes the Gamma function (which for positive integers obeys $\Gamma(m+1)=m!$). The quantities $\langle W^s \rangle$ and its logarithm are plotted in Figure 16.5. The case $s=1$ yields the arithmetic mean $\langle W \rangle \approx 1.226$, while the moment of order $s=-2$ is infinite because $\zeta(1) = \infty$. Other statistics of the mass ratio are its mean square $\langle W^2 \rangle \approx 2.31$, its variance $Var[W] = \langle W^2 \rangle - \langle W \rangle^2 \approx 0.697$, its standard deviation $\sigma_W \equiv (Var[W])^{1/2} \approx 0.835$, and its coefficient of variation $V_W \equiv \sigma_W / \langle W \rangle \approx 0.66$. Also, the quantum-mass-ratio PDF $f_W(w)$ is positively skewed, with coefficient of skewness $\chi_W \approx 1.18$.

The mean energy per particle — in electron-volts (eV) — can be expressed as $\langle E \rangle = c_0 k_B \langle W \rangle T \approx 2.33 \times 10^{-4} T$, where T is the temperature in degrees Kelvin. The mean radiation density per unit volume (in eV/cm^3) is the product of $\langle E \rangle$ and the mean number of quanta per unit volume (given by Equation 16.5), yielding the Stefan-Boltzmann law (expressing mean energy density vs. temperature), namely:

$$u = N_{Total} k_B T c_0 \langle W \rangle = 16\pi \zeta(3) c_0 \langle W \rangle \frac{(k_B T)^4}{(h_p c)^3} \approx 4.723 \times 10^{-3} T^4. \tag{16.15}$$

The function $wf_W(w)$, plotted in Figure 16.6, indicating how energy tends to be distributed among quanta with different w values, is proportional to Planck's energy distribution. (The partial moments of W, such as the partial integral of $wf_W(w)$, are proportional to the Debye functions arising in heat analysis [2].)

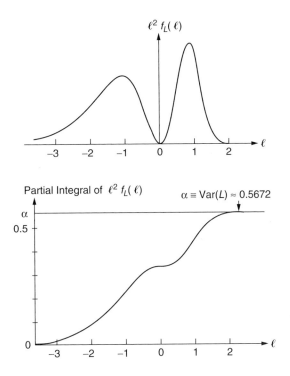

FIGURE 16.4 The function $\ell^2 f_L(\ell)$, whose integral is $\alpha \approx 0.5672$, the variance of L.

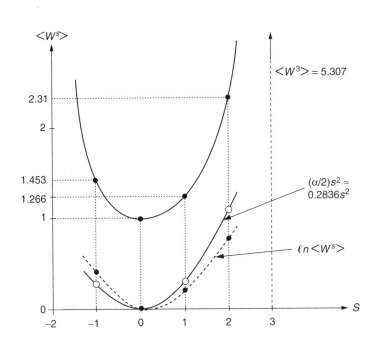

FIGURE 16.5 The moment function of W and its logarithm.

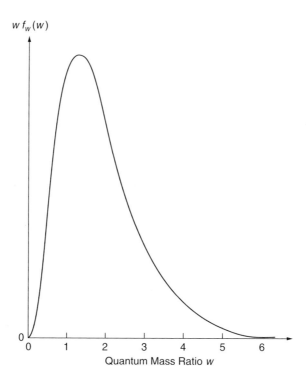

FIGURE 16.6 The function $wf_W(w)$, which indicates how energy tends to be distributed among quanta with different w-values, is proportional to Planck's energy distribution.

The quantum mass ratio W possesses the following geometric statistics: geometric-mean mass ratio $\overline{W} = 1$, logarithmic variance $Var[\log W] \equiv \alpha \approx 0.5672$, and characteristic value (equal to the stochasticity factor) $\widetilde{W} = e^{\alpha/2} \approx 1.328$. The *lognormal* formula for the skewness coefficient, $3V_W + V_W^3 \approx 2.27$, is almost twice the true value, $\chi_W \approx 1.18$, the difference reflecting the fast decay of the PDF of W (or the Planck radiation function) in the high-energy range compared to a lognormal PDF with the same geometric mean ($\overline{W} = 1$) and logarithmic variance ($\alpha \approx 0.5672$). The PDF of W is positively skewed ($\chi_W \approx 1.18$) and that of $L = \log W$ negatively, with skewness coefficient $\chi_L \approx -0.879$. Note in Figure 16.1 and Figure 16.2 the lack of symmetry in the tails of either distribution. The fast decay in the high-energy ("Wien") range of the Planck spectrum means that very-high-energy quanta tend to occur much less frequently than very-low-energy ("Jeans-Rayleigh") quanta. The probability that the mass ratio W (of a randomly chosen quantum of energy, *given* the LTE temperature T) exceeds the value w drops very rapidly for w-values exceeding $w \approx 5.5$, corresponding to $\ell = \log 5.5 \approx 1.70$ (see Figure 16.1 and Figure 16.2); these values may be thought of as crude "apparent upper bounds" on the mass ratio W and its logarithm L, respectively. For similar reasons, the Gaussian approximation for $f_L(\ell)$, with mean $\langle L \rangle = 0$ and variance $\langle L^2 \rangle = \alpha$, significantly overestimates the probability that L is larger than ℓ when the latter exceeds its "apparent upper bound" $\ell_{max} \equiv \log(w_{max}) = \log(5.5) \approx 1.7 \approx 2.26\sigma_L$.

From the definition of the (quantum-mass-ratio-logarithm) cumulant function $K_L(u) = \log\langle e^{\iota uL} \rangle$, where $\iota = \sqrt{-1}$, taking into account $W = e^L$, we can infer that Equation 16.14 in effect expresses $\langle W^{\iota u} \rangle = \exp\{K_L(u)\}$, with ιu now substituting for the exponent s. The cumulant function of L is therefore given by

$$K_L(u) = \log\langle W^{\iota u} \rangle = \log\{\Gamma(3 + \iota u)\} + \log\{\zeta(3 + \iota u)\} - \ell_0 \iota u - \log[2\zeta(3)], \qquad (16.16)$$

where $\ell_0 \equiv \log c_0 \approx 0.758$ and $\log[2\zeta(3)] \approx 0.877$; also, $K_L(0) = \log\langle W^0 \rangle = 0$. Because the first three cumulants of L are known, $K_{1,L} = \langle L \rangle = 0$, $K_{2,L} = Var[L] = \alpha \approx 0.5672$ and $K_{3,L} = \chi_L \alpha^{3/2} \approx -0.3755$ (where

$\chi_L \approx -0.879$ is the skewness coefficient), the series expansion (Equation 16.11) for the cumulant function starts off as follows:

$$K_L(u) = -\alpha \frac{u^2}{2} + \chi_L \alpha^{3/2} \frac{(tu)^3}{3!} + \ldots \approx -0.5672 \frac{u^2}{2} - 0.3755 \frac{(tu)^3}{3!} + \ldots, \qquad (16.17)$$

If L actually had a normal (Gaussian) distribution with mean zero and variance α, its cumulant function would obey $K_L(u) = -\alpha u^2/2$, so the higher-order terms in Equation 16.17 represent physically significant deviations from symmetry and normality. The first-term approximation, $\log\langle W^s\rangle = K_L(s/t) \approx \alpha s^2/2 \approx 0.2836 s^2/2$, is shown, along with exact values, in Figure 16.5.

16.5 Probability Distribution of the "Quantum Size Ratio"

The wavelengths of individual energy quanta can be thought of as representing the sizes (radii) of elementary "cells." Energetic quanta have relatively small wavelengths or "cell radii" λ and short "lifetimes" $\theta = \lambda/c$. The relative sizes of these energy-quantum-specific cells, *given* the (local-thermal-equilibrium) temperature T, are measured by the "quantum size ratio," defined as the reciprocal of the "quantum mass ratio," $D \equiv \lambda/\bar\lambda = 1/W$. The probability density function of the quantum size ratio D is given by $f_D(d) = (1/d^2) f_W(1/d), d \geq 0$, or

$$f_D(d) = \frac{(c_0)^3}{2\zeta(3)} \frac{1}{d^4(e^{c_0/d} - 1)}, \quad d \geq 0, \qquad (16.18)$$

matching the *shape* of the Planck spectrum in Equation 16.4. The probability density function of D is plotted in Figure 16.7.

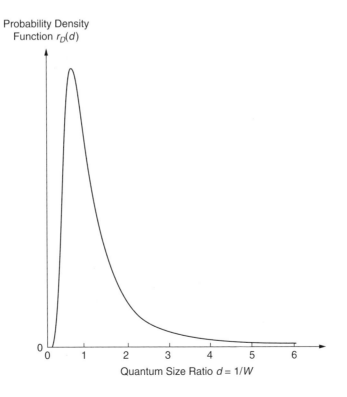

FIGURE 16.7 The probability density function of D, the reciprocal of the quantum mass ration W.

The probability density function $f_{L'}(\ell')$ of the *logarithm* of the quantum size ratio, $L' \equiv \log D = -\log W = -L$, is similar to that of the quantum-mass-ratio logarithm in Equation 16.10. We can write:

$$f_{L'}(\ell') = f_L(-\ell') = \frac{1}{2\zeta(3)} \frac{e^{3(\ell_0 - \ell')}}{e^{e^{(\ell_0 - \ell')}} - 1}, \quad -\infty \leq \ell' \leq \infty. \tag{16.19}$$

The PDF of L' is as shown in Figure 16.2 *provided* the abscissa's direction is reversed, and the two cumulant functions are similarly related: $K_{L'}(u) = K_L(-u)$. The means of L and L' are both zero, their variances are equal ($\alpha \approx 0.5672$), and their coefficients of skewness have the same absolute value but a different sign: $\chi_{L'} = -\chi_L \approx +0.879$.

The "geometric statistics" of $D = e^{-L}$ are the same as those of the quantum mass ratio ($W = e^L$), namely: geometric mean $\overline{D} = (\overline{W})^{-1} = 1$, characteristic value $\tilde{D} \approx 1.328$, and logarithmic variance $Var[\log D] \equiv 2\log(\tilde{D}/\overline{D}) = \alpha \approx 0.5672$. The quantum size ratio's arithmetic mean, $\langle D \rangle = \langle W^{-1} \rangle \approx 1.453$, is the integral of the function $d f_D(d)$. The second moment of the quantum size ratio, $\langle D^2 \rangle = \langle W^{-2} \rangle$, and its variance, $Var[D] = \langle D^2 \rangle - \langle D \rangle^2$, are both *infinite* (corresponding to $s = -2$ in Figure 16.5) due to the slow decay of the upper tail of $f_D(d)$. Closer examination of the asymptotic expression,

$$f_D(d) \to \frac{c_0^2}{2\zeta(3)} \frac{1}{d^3} \approx \frac{1.895}{d^3}, \quad d \gg 1, \tag{16.20}$$

shows that the "truncated" or "conditional" mean square, $\langle D^2 | D < d_0 \rangle$, defined as the integral from 0 to d_0 of the function $d^2 f_D(d)$, shown in Figure 16.8, possesses a term proportional to $\log d_0$, indicating how the "unconditional" mean square $\langle D^2 \rangle$ and variance $Var[D]$ of the quantum size ratio approach infinity. A further consequence is that the conditional or "sample" variance, $Var[D|D < d_0]$, where d_0 represents the size of any actual sampling volume, is predicted to be large but finite. (The predicted sensitivity of size-ratio statistics to the size of sampling regions proves testable in the context of a stochastic model of the so-called inflation phase of the big bang [3]; the model predicts the distribution of radii of "great

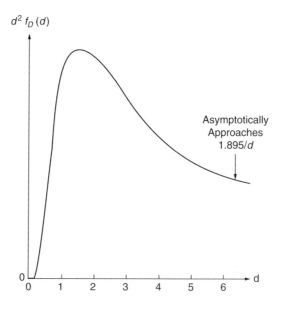

FIGURE 16.8 The function $d^2 f_D(d)$ whose integral, the mean square of D is infinite.

voids" and related geometrical anomalies of the large-scale cosmic structure in terms of basic quantum-size-ratio statistics.)

16.6 Extensions and Applications to Reliability Analysis

Considering a set of quantum-physics-based random variables — a set of W values or L values, for example — that are statistically independent and identically distributed (i.i.d.), one can readily obtain various families of derived probability distributions (and related statistics) for sums or products or extreme values, with the number of i.i.d. random variables (in the sum or product or extremum) serving as a model parameter. Consider, in particular, a set of v independent and identically distributed (i.i.d.) quantum-mass-ratio logarithms, their common probability density function $f_L(\ell)$ given by Equation 16.10. The probability distribution of the sum, $L_v = L_{01} + L_{12} + \ldots + L_{v-1,v}$, can be obtained by iterative convolution, starting with the PDF of $L_2 = L_{01} + L_{12}$, next that of $L_3 = L_2 + L_{23}$, and so on. The PDF of the corresponding product of i.i.d. quantum mass ratios, $W_v \equiv \exp\{L_v\} = W_{01}W_{12}\ldots W_{v-1,v}$, where $W_{i-1,i} = \exp\{L_{i-1,i}\}$ for $i = 1, 2, \ldots, v$, can then be found by a single monotonic (one-to-one) transformation. Let us denote $f_{L_v}(\ell)$ by $f_{L_v}(\ell)$ the PDF of the sum of v independent quantum-mass-ratio logarithms, and by $f_{W_v}(w)$ the corresponding PDF of the product of v quantum mass ratios. In the special case $v = 1$, we have $f_{L_1}(\ell) \equiv f_L(\ell)$ and $f_{W_1}(w) \equiv f_W(w)$, as given by Equation 16.9 and Equation 16.10, respectively.

An elegant alternative to convolution is to express the cumulant function of L_v as a sum of cumulant functions, $K_{L_v}(u) = vK_L(u)$, where $K_L(u)$ denotes the cumulant function of the quantum-mass-ratio logarithm L (given by Equation 16.17), and then compute the inverse-transform of $K_{L_v}(u)$ to obtain $f_{L_v}(\ell)$. Illustrating a typical set of results, Figure 16.9 shows the PDF of the sum of $v = 25$ quantum-mass-ratio logarithms; the dashed line shows the Gaussian probability density function with mean zero and

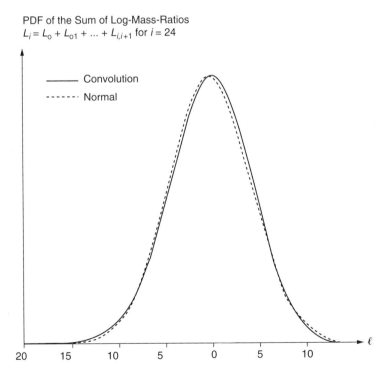

FIGURE 16.9 Probability density function of the sum of $v = 25$ quantum-mass-ratio logarithms; the dashed line shows the Gaussian probability density function with matching mean zero and standard deviation $(25\alpha)^{1/2} \approx 3.766$.

standard deviation $(25\alpha)^{1/2} \approx 3.766$. (The subscripts on L are denoted a bit differently in Figure 16.9, without in any way impacting the results.) Note that perceptible differences remain in the tails of the two distributions (displayed in Figure 16.9), which would be highly significant in applications to reliability analysis. Scaling $L_v \equiv \log W_v$ with respect to its standard deviation, $(v\alpha)^{1/2}$, yields a "standardized" random variable, $U^{(v)} = L_v/(v\alpha)^{1/2}$, whose probability distribution, as v grows, will tend toward the "standard normal" (Gaussian) distribution in its central range. Likewise, the PDF of the derived random variable $V^{(v)} \equiv \exp\{U^{(v)}\} = \exp\{L_v/(v\alpha)^{1/2}\}$ should approach the "standard lognormal" PDF in its central range. The coefficient of skewness of L_v can be expressed in terms of its "cumulants": $\chi_{L_v} \equiv vK_{3,L}/(vK_{2,L})^{3/2} = \chi_L/\sqrt{v} \approx -0.879/\sqrt{v}$, indicating how the skewness decreases as v grows (consistent with the Central Limit Theorem), but also the tendency for the negative skewness to persist.

A generalized set of probabilistic models can be obtained by transforming the "standardized" random variable (introduced above), $U^{(v)} = L_v/(v\alpha)^{1/2}$, modifying its zero mean (by adding a constant) and unit variance (by multiplying by another constant); the result is a three-parameter (quantum-physics-based) family of distributions characterized by their mean, variance, and coefficient of skewness, the latter dependent only on v. A similar transformation of the random variable $V^{(v)} \equiv \exp\{U^{(v)}\}$ results in a generalized set of PDFs of which $f_{W_v}(w)$ are members. These probability models can capture very different types of behavior of random variables (characterized in part by their variability and skewness); the PDFs tend toward the normal (for sums) and lognormal (for products) as the number v of i.i.d. basic random variables grows. Skewness persists, however, and the tails of the probability distributions, on both the low and high ends, continue to differ greatly from the normal and lognormal distributions, even when v becomes large. The PDF of the largest (or smallest) value in a group of i.i.d. values of L can also be expressed straightforwardly in terms of the PDF of L and the group size v.

There appear to be many opportunities for application to system reliability analysis (see, e.g., [4–6]). In single-mode reliability analysis, the focus is often on the *safety margin*, defined as the difference between the random load and the random resistance, or the *safety factor*, defined as the quotient of resistance to load. (A parallel formulation, in economic applications, may be in terms of "supply" and "demand," or "capacity" and "demand".) The two random quantities (i.e., load and resistance) are typically assumed to be statistically independent. The new probability models could be adopted for load and resistance separately, or be applied directly to the safety margin or the safety factor. The safety margin, for example, might be characterized by its mean (above zero) and coefficient of variation; its skewness could also strongly affect the (single-mode) failure probability (i.e., the probability that the actual safety margin is negative). This readily leads to a new set of relationships between, for example, the mean safety margin and the probability of failure (and corresponding reliability indices that depend on the coefficients of variation and skewness).

System reliability concerns the behavior of groups of components or failure modes and the logical relationships between component failures and system failure. In these problem situations as well, the new probabilistic models can be adopted for (or based on data, fitted to) the random "modal" safety margins or safety factors, or the mode-specific loads and resistances, enabling the system reliability to be estimated or bounded (see, e.g., [7]) These bounds are related to the degrees of statistical dependence between the loads and resistances associated with the different failure modes. Here too, the proposed models provide new analytical tools to express statistical dependence. In particular, statistical dependence between, say, component resistances can be realized or simulated by assuming shared (identical) values for some of the basic random variables being summed or multiplied in the process of modeling each failure mode's resistance.

16.7 Conclusion

The probability density functions of a number of functionally related properties of single energy quanta — particles like photons — in thermal equilibrium have been derived from the Planck radiation spectrum. The "quantum mass ratio" W, its reciprocal, the "quantum size ratio" $D = W^{-1}$, and the logarithm $L = \log W = -\log D$ are coupled random properties of individual energy quanta. The ratios W and D have a unit geometric mean, $\overline{W} = \overline{D} = 1$, while the arithmetic mean of their logarithm is zero, $\langle L \rangle = \langle \log W \rangle = \langle \log D \rangle = 0$. The variance of the mass ratio logarithm, $\alpha \equiv Var[L] \approx 0.5672$, measures the intrinsic uncertainty of properties of single energy quanta in thermal equilibrium, for any LTE temperature in the range $0 < T < T_{pl} \approx 10^{32} K$.

The idea is that any quantum-producing "engine" simultaneously gives rise to quantum-physical uncertainty. Wherever individual energy quanta are copiously created, the *least* amount of variability realizable is that associated with a single, unique Planck spectrum, with a constant temperature $T = T^*$, characterized by PDFs and variability as measured by $\alpha = Var[L] = Var[\log W] \approx 0.5672$.

Heisenberg's Uncertainty Principle expresses a fundamental constraint on the precision of paired quantum-scale observations: the product of "errors" in joint measurements of, say, a particle's position and momentum must be greater than or equal to the quotient of Planck's constant hp and 2π. The measure of inherent uncertainty of the properties of energy quanta in thermal equilibrium, $\alpha = Var[L] = Var[\log W] \approx 0.5672$, differs from Heisenberg's classical formulation in that it (1) expresses an equality instead of an inequality; (2) involves only a single attribute (of a quantum of radiation), or one attribute at a time, instead of two; and (3) does not depend on Planck's constant, a tiny number. A further advantage is that the description of quantum-physical uncertainty in terms of $f_L(\ell)$ or $f_W(w)$ or D extends beyond second-order statistics, thereby enabling one to evaluate the relative likelihood of extreme values, high or low, of quantum properties. We have shown that the new probability models have considerable potential for practical application to single-mode and system reliability analysis.

Acknowledgment

The author is grateful to Ian Buck and Ernesto Heredia-Zavoni for their assistance with the computations needed to generate the figures.

References

1. Planck, M. (1900). *Verh. Deutsche Phys. Ges.*, 2, 237; and in [2] Abramowitz, M. and Stegun, I.A. (1965). *Handbook of Mathematical Functions*, New York: Dover.
2. Abramowitz, M. and Stegun, I.A. (1965). *Handbook of Mathematical Functions*, New York: Dover.
3. Vanmarcke, E.H. (1997), *Quantum Origins of Cosmic Structure*, Rotterdam, Holland and Brookfield, VT: Balkema.
4. Barlow, R.E. and Proschan, F, (1975). *Statistical Theory of Reliability and Life Testing*, New York: Holt, Rinehart and Winston.
5. Freudenthal, A.M. (1947). "The Safety of Structures," *Trans. ASCE*, Vol. 112.
6. Madsen, H.O., Krenk, S., and Lind, N.C. (1986). *Methods of Structural Safety*, Englewood Cliffs, NJ: Prentice Hall.
7. Cornell, C.A. (1967). "Bounds on the Reliability of Structural Systems," *Journal of the Structural Division*, ASCE, 93, 171–200.

17
Probabilistic Analysis of Dynamic Systems

17.1	Introduction and Objectives 17-1
17.2	Probabilistic Analysis of a General Dynamic System 17-2
	Modeling Random Processes • Calculation of the Response • Failure Analysis
17.3	Evaluation of Stochastic Response and Failure Analysis: Linear Systems 17-10
	Evaluation of Stochastic Response • Failure Analysis
17.4	Evaluation of the Stochastic Response of Nonlinear Systems 17-17
17.5	Reliability Assessment of Systems with Uncertain Strength 17-18
17.6	Conclusion 17-18

Efstratios Nikolaidis
The University of Toledo

17.1 Introduction and Objectives

Many engineering systems are subjected to loads that are random processes, e.g., random functions. The following are some examples: the wings of an airplane under gust loads, the hull of a ship under wave loads, and an offshore platform under wave loads. Moreover, the properties of many systems, such as their strength (stress at failure) are random and vary in time. The objectives of a random vibration study are to determine the statistics of the response of a system and the probability of failure due to the response. Tools for time invariant reliability problems, such as FORM and SORM, are not directly applicable to random vibration analysis.

Probabilistic analysis of a dynamic system under loads that are random processes involves the following steps:

1. Construct probabilistic models of the excitations. Here we model the excitations using data obtained from measurements and experience. Usually, the excitations are characterized using their means, autocorrelations, and cross-correlations.
2. Calculate the statistics of the response.
3. Calculate of the probability of failure. Failure can occur if the response exceeds a certain level (first excursion failure) or due to accumulation of fatigue damage (fatigue failure). An example of first excursion failure is the collapse of a structure because the stress exceeds a threshold.

This chapter first reviews methods for completing the above three steps, for a general dynamic system with deterministic strength under time varying random excitation. Then the chapter considers special cases where the system of interest is linear or nonlinear. Finally, this chapter explains how to calculate the reliability of a dynamic system when its strength is uncertain.

This chapter studies methods for analysis of systems in which uncertainties are represented as random processes. Chapter 20, "Stochastic Simulation Methods for Engineering Predictions," and Chapter 21, "Projection Schemes for Stochastic Finite Element Analysis," consider random fields.

17.2 Probabilistic Analysis of a General Dynamic System

This section reviews the theory for representing the excitation of dynamic systems using random processes, finding the response of these systems, and determining their probability of failure.

17.2.1 Modeling Random Processes

17.2.1.1 Random Processes and Random Fields

Random process is a random function, $X(t)$, of one parameter, t. Parameter t is called the *index parameter* of the process and can assume discrete values $\{t_1, \ldots, t_i, \ldots\}$ or continuous values in a range, $t \in [0, \infty]$. In the latter case, the random process is called *continuously parametered* or *continuously indexed*. Parameter t usually represents time, but it can represent some other quantity, such as the distance from a reference point. The value of $X(t)$ at a particular value of parameter $t = t_0$, $X(t_0)$, is a random variable.

We can view a random process as a set of possible *time histories*; each time history (or *sample path*) shows the variation of the process with t for a particular realization of the process.

Example 17.1

Figure 17.1 shows four realizations of the elevation, $X(s)$, of a road covered with concrete slabs as a function of the distance, s, from a reference point. The slabs have fixed length and random height. This set of time histories is called *sample* or *ensemble* of time histories.

The concept of *random fields* is a generalization of the concept of random processes. A random field is a random function of two or more parameters. The modulus of elasticity at a point in a solid, $E(x, y, z)$, varies with the location of the point (x, y, z) and is random. In this case, the modulus of elasticity is a

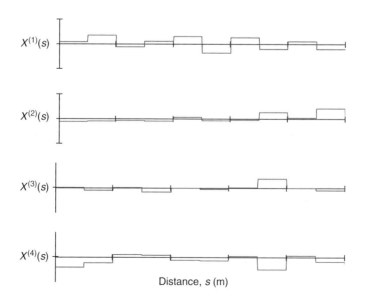

FIGURE 17.1 Four time histories of the elevation of a road covered with concrete labs of fixed length and random height.

random field because it is a random function of the three coordinates of the point of interest. The velocity field in a turbulent flow is also a random field. This is described by the velocity vector $\mathbf{V}(x, y, z, t)$, where (x, y, z) are the coordinates of the point of interest and t is time.

17.2.1.2 Probabilistic Models of Random Processes

A complete probabilistic model of a random process involves the joint probability distribution of the values of the random process for all possible values of the independent parameter. This information is impractical to obtain in most cases. We usually calculate only the first- and second-order statistics of the responses (i.e., their means, autocorrelations, and cross-correlations) because:

- It is impractical to collect sufficient data to determine the higher-order statistics.
- In many problems, we assume that the random processes involved are jointly Gaussian. Then, the first- and second-order statistics define completely the processes.
- Even if the processes involved are not Gaussian, the first two moments contain important information. For example, we can determine an upper bound for the probability that the value of the process at an instant can be outside a given range centered about the mean value using Chebyshev inequality [1, 2].

The first two moments of a single random process include the mean value, $E(X(t))$, and the *autocorrelation* function:

$$R_{XX}(t_1, t_2) = E(X(t_1)X(t_2))$$

where t_1 and t_2 are two time instances at which the value of the process is observed. We can also define the autocorrelation function in terms of parameter $t = t_2$, and the elapsed time between the two observations, $\tau = t_1 - t_2$.

The *autocovariance* of a random process is:

$$C_{XX}(t_1, t_2) = E[(X(t_1) - E(X(t_1)))(X(t_2) - E(X(t_2)))].$$

For two random processes, we also need to define their *cross-correlation*. This function shows the degree to which the two processes tend to move together. The cross-correlation function of two random processes, $X_1(t)$ and $X_2(t)$, is defined as follows:

$$R_{X_1 X_2}(t_1, t_2) = E(X_1(t_1)X_2(t_2)).$$

The *cross-covariance* of two random processes is defined as follows:

$$C_{X_1 X_2}(t_1, t_2) = E[(X_1(t_1) - E(X_1(t_1)))(X_2(t_2) - E(X_2(t_2)))].$$

In the remainder of this chapter, we will consider only random processes with zero mean values. Therefore, their autocorrelation and cross-correlation are equal to their autocovariance and cross-covariance, respectively. If a random process has nonzero mean, then we transform it into a process with zero mean by subtracting the mean from this process.

Example 17.2

Consider a vehicle traveling with constant speed, $V = 100$ km/hr, on the road made of concrete slabs in Figure 17.1. The length of a slab is 10 m. Assume that the wheels are always in contact with the road. Then, the vertical displacement of a wheel is $X(t)$, where $t = s/V$ and $X(.)$ is the road elevation. Figure 17.2 shows the autocorrelation function of the displacement, $R_{XX}(t + \tau, t)$. The autocorrelation is constant

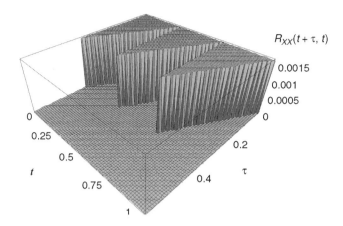

FIGURE 17.2 Autocorrelation of the displacement of the wheel of a car traveling with constant speed on the road made of concrete slabs in Fig. 1.

and equal to the variance of the displacement when the wheel is on the same slab at both instances t and $t + \tau$. Otherwise, it is zero.

A random process can be *stationary* or *nonstationary*. A stationary process is one whose statistics are invariant under time shifts. Otherwise, the process is nonstationary.

The concept of stationarity in random processes corresponds to the concept of steady-state in deterministic functions of time. The wave elevation and the wind speed at a given point in the Atlantic Ocean can be considered stationary during short periods during which weather conditions do not change significantly. The force applied to the landing gear of an airplane during landing is a nonstationary process because this force changes rapidly in time.

A random process, $X(t)$, is called *strictly stationary* if the joint probability density function (pdf) of the process at all possible time instances is invariant under time shifts. Therefore,

$$\begin{aligned}
f_{X(t+\tau)}(x) &= f_{X(t)}(x) \\
f_{X(t_1+\tau),X(t_2+\tau)}(x_1, x_2) &= f_{X(t_1),X(t_2)}(x_1, x_2) \\
&\vdots \\
f_{X(t_1+\tau),X(t_2+\tau),\ldots,X(t_n+\tau)}(x_1, x_2, \ldots, x_n) &= f_{X(t_1),X(t_2),\ldots X(t_n)}(x_1, x_2, \ldots, x_n)
\end{aligned} \qquad (17.1)$$

where $f_{X(t_1),X(t_2),\ldots X(t_n)}(x_1, x_2, \ldots, x_n)$ is the joint PDF of the random process at $t_1, t_2, \ldots,$ and t_n, respectively. In this case, the moments of $X(t)$ are also invariant under time shifts.

Example 17.3

The elevation of the road consisting of concrete slabs in Figure 17.1 is a nonstationary random process. The reason is that the joint pdf of this process at two locations, s_1 and s_2, changes under a distance shift. This is the reason for which the autocorrelation function of the displacement of the wheel in Figure 17.2 varies with t.

Random processes are *jointly stationary* in the strict sense if the joint pdfs of any combinations of values of these processes are invariant under time shifts.

A random process is *stationary in the wide sense* (or *weakly stationary*) if its mean value and autocorrelation function are invariant under shifts of time:

$$E(X(t)) = \text{constant} \qquad (17.2)$$

$$R_{XX}(t_1, t_2) = R_{XX}(\tau) \qquad (17.3)$$

where $\tau = t_1 - t_2$. Note that the autocorrelation function is an even function of τ.

It is often convenient to work with the *power spectral density* (PSD) function of a weakly stationary process, $X(t)$, instead of the autocorrelation. The reason is that it is easy to find the PSD of the response from that of the input for linear systems. The PSD, $S_{XX}(\omega)$, is the Fourier transform of the autocorrelation:

$$S_{XX}(\omega) = \int_{-\infty}^{+\infty} R_{XX}(\tau) e^{-i\omega\tau} d\tau. \tag{17.4a}$$

The PSD shows the distribution of the energy of a random process over the frequency spectrum. Note that the PSD is a real positive number because it is the Fourier transform of a real, even function. Moreover, the PSD is an even function of the frequency. The autocorrelation function is the inverse Fourier transform of the PSD function:

$$R_{XX}(\tau) = \frac{1}{2\pi} \int_{-\infty}^{+\infty} S_{XX}(\omega) e^{-i\omega\tau} d\omega. \tag{17.4b}$$

Random processes are *jointly stationary in the wide sense* if each of them is stationary and their cross-correlations are functions of τ only. The *cross spectral* density function of two jointly stationary processes is calculated using an equation similar to Equation 17.4b. The *spectral matrix* of a vector of random processes contains the power spectral densities and cross spectral densities of these random processes.

If a random process is stationary in the strict sense, then it is also stationary in the wide sense.

Example 17.4

The elevation of a rough road, $X(s)$, is modeled as a stationary Gaussian random process with autocorelation function $R_{XX}(d) = \sigma^2 \exp(|d|/0.5)$, where d is the distance between two locations, s and $s + d$, in meters (m); and σ is the standard deviation of the road elevation. The standard deviation, σ, is 0.02 m. A car travels on the road with constant speed $V = 16$ m/sec. Find the PSD of the wheel of the car, assuming that it remains always in contact with the ground.

The distance that a wheel travels is $s = Vt$. Therefore, the autocorrelation of the elevation of the wheel at two time instances, $t + \tau$ and t, is equal to the autocorrelation function of the road elevation at two locations separated by distance $V\tau$:

$$R_{XX}(\tau) = \sigma^2 e^{-\frac{|V\tau|}{0.5}} = \sigma^2 e^{-32|\tau|}.$$

The units of the autocorrelation are m².

The PSD is the Fourier transform of the autocorrelation function. Using Equation 17.4a we find that the PSD is:

$$S_{XX}(\omega) = \sigma^2 \frac{64}{\omega^2 + 32^2}.$$

Figure 17.3 shows the PSD for positive values of frequency.

17.2.2 Calculation of the Response

In the analysis of a dynamic system we construct a mathematical model of the system. For this purpose, we construct a *conceptual model* that simulates the behavior of the actual system. A conceptual model can be an assembly of discrete elements (such as masses, springs, and dampers) or continuous elements (such as beams and plates). By applying the laws of mechanics, we obtain a deterministic mathematical model of the system relating the responses to the excitations. This is usually a set of differential equations relating the response (output) vector, $\mathbf{Y}(t)$, to the excitation (input) vector, $\mathbf{X}(t)$.

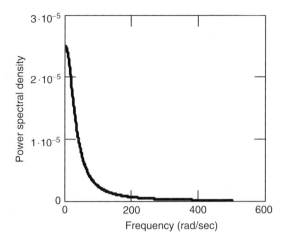

FIGURE 17.3 Power spectral density of road elevation in example 4.

A dynamic system can by thought as a multi-input, multi-output system, such as the one in Figure 17.4. Inputs and outputs are the elements of the excitation vector and response vector, respectively. The following equation relates the response to the excitation:

$$\mathbf{D}(\mathbf{p}, t)[\mathbf{Y}(t)] = \mathbf{X}(t) \qquad (17.5)$$

where $\mathbf{D}(\mathbf{p}, t)[.]$ is a matrix of differential operators applied to the elements of the response. The size of $\mathbf{D}(\mathbf{p}, t)[.]$ is $m \times n$, where m is the number of inputs and n is the number of outputs. $\mathbf{p}$ is a set of parameters associated with the system. We can solve Equation 17.5 analytically or numerically to compute the response.

Many single-input, single-output systems can be analyzed using a *Volterra series* expansion:

$$Y(t) = \sum_{i=1}^{\infty} \int_{-\infty}^{t} \cdots \int_{-\infty}^{t} h_i(t, t_1, \ldots, t_i) \prod_{j=1}^{i} X(t_j) dt_j. \qquad (17.6)$$

Functions $h_i(t, t_1, \ldots, t_i)$, $i = 1, \ldots, n$ are called *Volterra kernels* of the system.

Systems can be categorized as linear and nonlinear, depending if operator $\mathbf{D}(\mathbf{p}, t)$ is linear or nonlinear. For a linear system, the superposition principle holds and it allows us to simplify greatly the equation

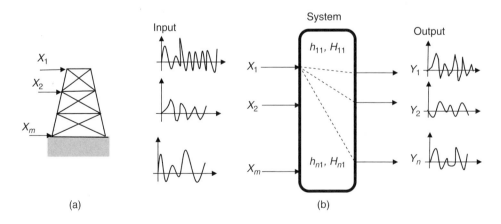

FIGURE 17.4 A structure subjected to excitations $X_1, \ldots, X_m$, and its representation as a multi-input, multi-output system.

for finding the response of the system. Specifically, we can break down the excitation into a series of impulses, find the response to each impulse separately, and then superimpose the responses to find the response of the system. The response can be found as follows:

$$Y(t) = \int_{-\infty}^{t} h(t, t_1) X(t_1) dt_1 \tag{17.7}$$

$h(t, t_1)$ is called *impulse response function*, or *Green's function*. This function represents the response of a system at time t due to a unit impulse that was applied at time t_1. Note that Equation 17.6 reduces to the *convolution integral* in Equation 17.7 if we set $h_i(t, t_1, ..., t_i) = 0$ for $i \geq 2$.

Dynamic systems can be also classified under the following categories:

- *Time invariant systems whose parameters are deterministic.* An example is a system that consists of a spring, a mass, and a viscous damper of which the rate of the spring, the mass, and the damping coefficient are deterministic constants. The response of a time invariant system at time t depends only on the elapsed time between the instant at which the excitation occurred and the current time t. Therefore, $h_i(t, t_1, ..., t_i) = h_i(t - t_1, ..., t - t_i)$. If a time invariant system is also linear, the response is:

$$Y(t) = \int_{-\infty}^{t} h(t - t_1) X(t_1) dt_1. \tag{17.8}$$

- *Time invariant systems whose parameters,* **p**, *are random and time invariant.* In this case parameters, **p**, are random variables.
- *Time variant systems whose parameters are deterministic and are functions of time.* An example is a rocket whose mass decreases as fuel is burned.
- *Time variant systems whose parameters are random.* The parameters of these systems can be modeled as random processes.

Clearly, Equation 17.6 through Equation 17.8 cannot be used directly to assess the safety of a dynamic system because the response is random, which means that it is not enough to know the response due to a given output. We need a relation of the statistical properties of the response (such as the mean and autocorrelation function) to the statistical properties of the excitation. We can calculate the statistical properties of the response using Monte Carlo simulation methods or analytical methods. Analytical methods are described in the following two sections for both linear and nonlinear systems.

Example 17.5

Consider a linear single-degree-of-freedom model of the car traveling on the road in Example 17.4. The model consists of a mass, m, representing the sprung mass, a spring, k, and a damper, c, in parallel representing the suspension and tire (Figure, 17.5). Input is the displacement of the wheel, $x(t)$, and output is the displacement, $y(t)$, of mass, m. The system is underdamped; that is $\zeta = c/(2\sqrt{km}) < 1$. Find the impulse response function.

The equation of motion is $m\ddot{y} + c\dot{y} + ky = kx + c\dot{x}$. Therefore, the transfer function is:

$$H(s) = \frac{Y(s)}{X(s)} = \frac{\omega_n^2 + 2\zeta\omega_n s}{s^2 + 2\zeta\omega_n s + \omega_n^2}$$

where $\omega_n = \sqrt{\frac{k}{m}}$ is the natural frequency. $Y(s)$ and $X(s)$ are the Laplace transforms of the displacements of mass m and the wheel, respectively.

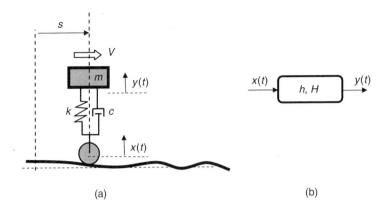

FIGURE 17.5 Model of a vehicle traveling on a rough road and its representation as a single-input, single-output system.

The impulse response function is the inverse Laplace transform of the transfer function:

$$h(\tau) = \frac{\omega_n}{\sqrt{1-\zeta^2}} e^{-\zeta\omega_n \tau} \sin(\omega_d t) - \frac{2\zeta\omega_n}{\sqrt{1-\zeta^2}} e^{-\zeta\omega_n \tau} \sin(\omega_d t - \phi)$$

where, $\phi = \tan^{-1} \frac{\sqrt{1-\zeta^2}}{\zeta}$, and $\omega_d = \omega_n \sqrt{1-\zeta^2}$.

17.2.3 Failure Analysis

A structure subjected to a random, time-varying excitation can fail in two ways: (1) the response exceeds a certain level (e.g., the wing of an aircraft flying in gusty weather can break if the stress exceeds a maximum allowable value), and (2) the excitation causes fatigue failure. In this section, we only consider failure due to exceedance of a level. This type of failure is called *first excursion failure* (Figure 17.6). The objective of this section is to review methods to calculate the probability of first excursion failure during a given period. Fatigue failure is analyzed in the next section.

In general, the problem of finding the probability of first excursion failure during a period is equivalent to the problem of finding the probability distribution of the maximum of a random vector that contains infinite random variables. These variables are the values of the process at all time instants within the period. No closed form expression exceeds for the first excursion probability for a general random process.

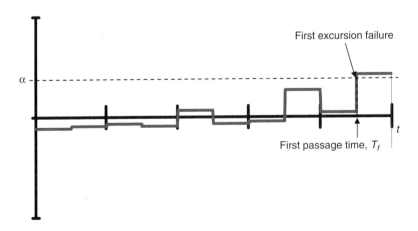

FIGURE 17.6 Illustration of first excursion failure.

Approximate results are available for some special cases. In this section, we review methods to estimate the following quantities for a random process:

- The number of crossings of a level α over a period
- The pdf of the peaks
- The pdf of the first passage time, T_f, which is the time at which the process upcrosses a certain level for the first time (Figure 17.6)

Nigam [1] presents an overview of the methods to estimate the probability of first excursion failure.

17.2.3.1 Number of Crossings of a Level during a Period

Most of the results that we will review have been presented in two seminal papers by Rice [3].

In general, the *average upcrossing rate* (average number of upcrossings of a level α per unit time) at time t is:

$$v^+(\alpha, t) = \int_0^\infty \dot{x} f_{X(t),\dot{X}(t)}(\alpha, \dot{x}, t) d\dot{x} \qquad (17.9)$$

where $v^+(\alpha, t)$ is the average number of upcrossing, and $f_{X(t),\dot{X}(t)}(x, \dot{x}, t)$ is the joint pdf of the process and its velocity.

For a stationary process, this joint density is independent of time: $f_{X(t),\dot{X}(t)}(x, \dot{x}, t) = f_{X(t),\dot{X}(t)}(x, \dot{x})$. Therefore, the average upcrossing rate is also independent of time. The average number of upcrossings over a period T is $v^+(\alpha)T$.

17.2.3.2 Probability Density Function of the Peaks

A random process can have several peaks over a time interval T (Figure 17.7). Let $M(\alpha, t)$ be the expected number of peaks above level α per unit time. Then:

$$M(\alpha, t) = \int_\alpha^\infty \int_{-\infty}^0 |\ddot{x}| f_{X(t),\dot{X}(t),\ddot{X}(t)}(\alpha, 0, \ddot{x}) d\ddot{x}\, dx \qquad (17.10)$$

where $f_{X(t),\dot{X}(t),\ddot{X}(t)}(x, \dot{x}, \ddot{x})$ is the joint pdf of $X(t)$ and its first two derivatives at t.

The expected number of peaks per unit time regardless of the level, $MT(t)$, is obtained by integrating Equation 17.10 with respect to x from $-\infty$ to $+\infty$:

$$MT(t) = \int_{-\infty}^0 |\ddot{x}| f_{\dot{X}(t),\ddot{X}(t)}(0, \ddot{x}) d\ddot{x}. \qquad (17.11)$$

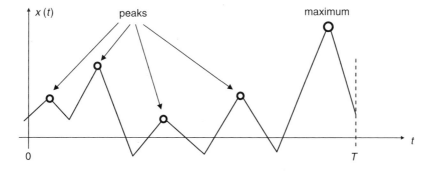

FIGURE 17.7 Local peaks and maximum value of random process, $X(t)$, during period $[0, T]$.

We can find the cumulative probability distribution of a peak at a level by normalizing the expected number of peaks below that level by the expected number of peaks regardless of the level. The pdf of a peak is the derivative of the cumulative distribution of a peak with respect to the level. It can be shown that this pdf of a peak at level α is:

$$f_{peak}(\alpha, t) = \frac{1}{MT(t)} \int_{-\infty}^{0} |\ddot{x}| f_{X(t), \dot{X}(t), \ddot{X}(t)}(\alpha, 0, \ddot{x}) d\ddot{x}. \tag{17.12}$$

17.2.3.3 First Passage Time

Here, the objective is to find the probability distribution of the first passage time, T_f (Figure 17.6). First, we will find the probability distribution of the first passage time for stationary processes and then for nonstationary processes.

Consider a stationary process. If the level is high (e.g., larger than the mean value plus six standard deviations of the process), the upcrossing process can be assumed to be a Poisson process with stationary increments and with interarrival rate $v^+(\alpha)$. Then the first passage time, T_f, follows the exponential distribution:

$$F_{T_f}(t) = 1 - e^{-v^+(\alpha)t}. \tag{17.13}$$

The average crossing rate can be calculated using Equation 17.9. Both the mean value and standard deviation of the first passage time are equal to the inverse of the average crossing rate.

The probability of first excursion failure, $P(F)$, is the probability that the process will exceed α during the time interval $(0, T)$. This is equal to the probability of the first passage time being less than T:

$$P(F) = 1 - e^{-v^+(\alpha)T}. \tag{17.14}$$

The above approach assumes that the peaks of the process are statistically independent. However, subsequent peaks are usually positively correlated, especially if the process is narrowband (e.g., the energy of the process is confined to a narrow frequency range). Yang and Shinozuka [4, 5] proposed a method to account for correlation between subsequent peaks of a process assuming that these peaks are subjected to a Markov chain condition.

If a random process is nonstationary, the average crossing rate is a function of time, $v^+(\alpha, t)$. Then the probability of first excursion failure during an interval $[0, T]$ becomes:

$$P(F) = 1 - e^{-\int_0^T v^+(\alpha, t) dt}. \tag{17.15}$$

The value of the cumulative probability distribution of the first passage time, T_f, is equal to the probability of failure during the period $[0, t]$:

$$F_{T_f}(t) = P(T_f \leq t) = P(F) = 1 - e^{-\int_0^t v^+(\alpha, \tau) d\tau}. \tag{17.16}$$

17.3 Evaluation of Stochastic Response and Failure Analysis: Linear Systems

Closed form, analytical solutions for the response of a linear system to random excitations are available and will be presented here. Results for analysis of failure due to first excursion and fatigue will also be presented.

17.3.1 Evaluation of Stochastic Response

The following two subsections present a method for finding the response of a dynamic system to stationary and nonstationary inputs, respectively.

17.3.1.1 Response to Stationary Inputs

We calculate the autocorrelation and cross-correlations of the responses in three steps:

1. Measure or calculate the PSD of the excitation or the spectral matrix of the excitations. The PSD can be measured directly or it can be derived from the autocorrelation function of the excitation. Similarly, the cross spectral density can be estimated directly or it can be found by calculating the Fourier transform of the cross-correlation.
2. Derive the PSD or the spectral matrix of the response. Consider a linear system with m inputs and n outputs, such as a structure subjected to m loads, for which n stress components are to be calculated. We can consider the structure as a multi-input, multi-output system as shown in Figure 17.4. The second-order statistics of the output are characterized by the spectral matrix of the output, $\mathbf{S}_{YY}(\omega)$, which is calculated as follows:

$$\mathbf{S}_{YY}(\omega) = \mathbf{H}(\omega)\mathbf{S}_{XX}(\omega)\mathbf{H}^{T^*}(\omega). \qquad (17.17)$$

$\mathbf{H}(\omega)$ is an $n \times m$ matrix, whose entry at the i,j position is the sinusoidal transfer function, $H_{ij}(\omega)$, corresponding to the j-th input and i-th output. Sinusoidal transfer function is the Fourier transform of the impulse response function. Superscript T denotes the transpose of a matrix and $*$ complex conjugate. For a single-input, single-output system, the above equation reduces to:

$$S_{YY}(\omega) = |H(\omega)|^2 S_{XX}(\omega) \qquad (17.18)$$

where $S_{XX}(\omega)$ is the PSD of the input, $S_{YY}(\omega)$, is the PSD of the output, and $H(\omega)$ is the sinusoidal transfer function of the system.

2. The autocorrelation and cross-correlation functions of the outputs are the inverse Fourier transforms of the PSDs and cross spectral densities of the corresponding components of the outputs. Using this information we can derive the variances, covariances, and their derivatives with respect to time, which are useful in analysis of failure due to first excursion.

Example 17.6

Find the PSD of the displacement of the vehicle, $Y(t)$, in the Example 17.5. The mass of the vehicle is $m = 1,300$ kg, the rate of the spring is $k = 11,433$ m/N, and the viscous damping coefficient is $c = 600$ N sec/m.

The undamped natural frequency of the system is $\omega_n = 2.96$ rad/sec or 0.472 cycles/sec. The damping ratio is 0.31. The sinusoidal transfer function of this system is:

$$H(\omega) = \frac{k + i\omega c}{-m\omega^2 + k + i\omega c}.$$

Using Equation 17.18 we find the PSD of the displacement $Y(t)$. The PSD is shown in Figure 17.8. Note that the energy of vibration is confined to a narrow range around 3 rad/sec. The reason is that the system acts as a narrowband filter filtering out all frequencies outside a narrow range around 3 rad/sec.

17.3.1.2 Response to Nonstationary Inputs

The inputs are nonstationary in some practical problems. In these problems, the responses are nonstationary.

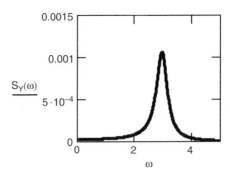

FIGURE 17.8 PSD of the displacement of the mass in Fig. 5.

The *generalized spectral density* of a nonstationary process is defined as the generalized Fourier transform of the covariance, $C_{XX}(t_1, t_2)$, [2]:

$$S_{XX}(\omega_1, \omega_2) = \int_{-\infty}^{+\infty}\int_{-\infty}^{+\infty} C_{XX}(t_1, t_2) e^{-i(\omega_1 t_1 - \omega_2 t_2)} dt_1 dt_2. \tag{17.19}$$

The generalized spectral density of the output of a single-input, single-output linear system can be calculated from the generalized spectral density of the input as follows:

$$S_{YY}(\omega_1, \omega_2) = S_{XX}(\omega_1, \omega_2) H(\omega_1) H^*(\omega_2). \tag{17.20}$$

Note that Equation 17.20 is analogous to Equation 17.17 for the stationary case.

Following are some studies of the response of structures to nonstationary excitation:

- Caughey and Stumpf [6]: study of the transient response of a single-degree-of-freedom system under white noise excitation
- Fung [7]: analysis of stresses in the landing gear of an aircraft
- Amin [8]: analysis of the response of buildings to earthquakes
- Nikolaidis, Perakis, and Parsons [9]: analysis of torsional vibratory stresses in diesel engine shafting systems under a special class of nonstationary processes whose statistics vary periodically with shifts of time

17.3.2 Failure Analysis

17.3.2.1 First Excursion Failure

The average crossing rate for a zero mean, Gaussian, stationary process is:

$$v^+(\alpha) = v^+(0) e^{-\frac{\eta^2}{2}} \tag{17.21}$$

where $v^+(0)$ is the expected rate of zero upcrossings, obtained by the following equation:

$$v^+(0) = \frac{1}{2\pi} \cdot \frac{\sigma_{\dot{X}}}{\sigma_X} \tag{17.22}$$

and η = normalized level = $\frac{\alpha}{\sigma_X}$. σ_X^2 is the variance of the process and $\sigma_{\dot{X}}^2$ is the variance of the derivative of the process:

$$\sigma_{\dot{X}}^2 = \frac{1}{2\pi} \int_{-\infty}^{+\infty} S_{\dot{X}\dot{X}}(\omega) d\omega = \frac{1}{2\pi} \int_{-\infty}^{+\infty} \omega^2 S_{XX}(\omega) d\omega.$$

Once we have calculated the average upcrossing rate, we can estimate the probability of first excursion failure from Equation 17.14.

The peaks (both positive and negative) of a stationary Gaussian process with zero mean follow the *Rice pdf*. This is a weighted average of a Gaussian pdf and a Rayleigh pdf:

$$f_{peak}(\alpha) = (1-\xi^2)^{1/2} \frac{1}{(2\pi)^{1/2}\sigma_X} e^{-\frac{\alpha^2}{2\sigma_X^2(1-\xi^2)}} + \xi\Phi\left[\frac{\xi\alpha}{\sigma_X(1-\xi^2)^{1/2}}\right]\frac{\alpha}{\sigma_X^2}e^{-\frac{\alpha^2}{2\sigma_X^2}} \qquad (17.23a)$$

where $\Phi(.)$ is the cumulative probability distribution function of a standard Gaussian random variable (zero mean, unit standard deviation). Parameter ξ is the ratio of the expected zero upcrossing rate to the expected rate of peaks: $\xi = v^+(0)/MT = \sigma_{\dot{X}}^2/(\sigma_X\sigma_{\ddot{X}})$. This parameter is called *bandwidth parameter*, and it expresses the degree to which the energy of the random process is confined to a narrow frequency range. The derivation of the Rice pdf can be found in the book by Lutes and Sarkani [10, pp. 491–492]. $\sigma_{\ddot{X}}^2$ is the variance of the acceleration:

$$\sigma_{\ddot{X}}^2 = \frac{1}{2\pi}\int_{-\infty}^{+\infty} S_{\ddot{X}\ddot{X}}(\omega)d\omega = \frac{1}{2\pi}\int_{-\infty}^{+\infty} \omega^4 S_{XX}(\omega)d\omega.$$

The pdf of the normalized level, $\eta = \alpha/\sigma$, is:

$$f_H(\eta) = f_{peak}(\eta\sigma)\sigma. \qquad (17.23b)$$

A narrowband process is a process whose energy is confined to a narrow frequency range. The number of peaks of a narrowband process is almost equal to the number of zero upcrossings. Therefore, the bandwidth parameter is close to one ($\xi \cong 1$). If a stationary Gaussian process is also narrowband, then the peaks follow the Rayleigh pdf:

$$f_{peak}(\alpha) = \frac{\alpha}{\sigma_X^2}e^{-\frac{\alpha^2}{2\sigma_X^2}} \cdot 1(\alpha) \qquad (17.24a)$$

where $1(\alpha)$ is a unit step function (e.g., a function that is one for positive values of the independent variable and zero otherwise).

At the other extreme, a broadband process has bandwidth parameter equal to zero ($\xi = 0$). Then the pdf of the peaks reduces to a Gaussian density:

$$f_{peak}(\alpha) = \frac{1}{(2\pi)^{1/2}\sigma_X}e^{-\frac{\alpha^2}{2\sigma_X^2}}. \qquad (17.24b)$$

Shinozuka and Yang [11], and Yang and Liu [12] estimated the probability distribution of the peaks of a nonstationary process. If the time period is much longer than the expected period of the process, then the cumulative probability distribution of the peaks at a level α can be approximated using the following equation:

$$F_{peak}(\alpha,t) \cong 1 - \frac{\int_0^t v^+(\alpha,\tau)d\tau}{\int_0^t v^+(0,\tau)d\tau}. \qquad (17.25)$$

The ratio on the right-hand side of Equation 17.25 is the expected number of upcrossings of level α divided by the number of zero upcrossings. Shinozuka and Yang [11] showed that the distribution of the peaks fits very closely the Weibull distribution:

$$F_{peak}(\alpha, T) = 1 - e^{-\left(\frac{\alpha}{\sigma}\right)^{\beta}} \qquad (17.26)$$

where parameters β and σ are the shape and scale parameters of the distribution, respectively. These parameters depend on the characteristics of the process and duration T.

Example 17.7

Consider the displacement of mass, m, of the system in Examples 17.4 through 17.6. Find:

1. The expected rate of zero upcrossings
2. The expected upcrossing rate of a level equal to 5 standard deviations
3. The probability of first excursion failure of the level in question 2 for an exposure period of 50 seconds.

The displacement of the mass is a Gaussian process because the excitation is Gaussian and the system is linear. Therefore, we can use Equations 17.21 and 17.22 to find the expected upcrossing rate. To find the expected number of zero upcrossings, we calculate the standard deviations of the displacement and the derivative of the displacement:

$$\sigma_X = \left(\frac{1}{2\pi}\int_{-\infty}^{+\infty} S_{XX}(\omega)d\omega\right)^{0.5} = 0.016 \text{ m}$$

$$\sigma_{\dot{X}} = \left(\frac{1}{2\pi}\int_{-\infty}^{+\infty} \omega^2 S_{XX}(\omega)d\omega\right)^{0.5} = 0.046 \text{ m}.$$

The zero upcrossing rate is found to be 0.474 upcrossings/sec from Equation 17.22. This is close to the undamped natural frequency of the system, which is 0.472 cycles/sec. The upcrossing rate of the normalized level of five is 1.768×10^{-6} upcrossings/sec.

The probability of first excursion of the level can be found using Equation 17.14:

$$P(F) = 1 - e^{-v^+(\alpha)T} = 8.84 \times 10^{-5}.$$

Example 17.8

Consider that the PSD of the displacement of mass, m, is modified so that it is zero for frequencies greater than 20 rad/sec. This modification helps avoid numerical difficulties in calculating the standard deviation of the acceleration. Find the pdf of the peaks of the displacement and the probability of a peak exceeding a level equal to five standard deviations of the process.

The standard deviations of the displacement and the velocity remain practically the same as those in Example 17.7. The standard deviation of the acceleration is:

$$\sigma_{\ddot{X}} = \left(\frac{1}{2\pi}\int_{-\infty}^{+\infty} \omega^4 S_{XX}(\omega)d\omega\right)^{0.5} = 0.181.$$

The peaks of the displacement follow the Rice pdf in Equations 17.23a and 17.23a. This equation involves the bandwidth parameter of the process, which is found to be 0.751. This indicates that most

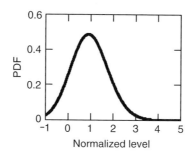

FIGURE 17.9 Probability density function (PDF) of a peak of the displacement normalized by the standard deviation. Since the displacement is a narrowband process, its PDF is close to the Rayleigh PDF.

of the energy of the process is confined to a narrow region (see Figure 17.8). Figure 17.9 shows the pdf of the peaks normalized by the standard deviation of the process. The pdf is a weighted average of the Rayleigh and Gaussian pdfs.

The probability of one peak upcrossing the normalized level, $\eta = 5$, is found to be 2.8×10^{-6} by integrating the pdf of the normalized peaks from 5 to infinity.

17.3.2.2 Fatigue Failure

Fatigue failure is another important failure mode for a vibrating system. This mode is due to accumulation of damage that is inflicted to the system by the oscillating stresses. For example, a spot weld in a car body may fail due to accumulation of damage caused by vibrating stresses. There are two approaches to the problem of fatigue failure: (1) a cumulative damage approach that is usually based on Miner's rule or (2) a fracture mechanics approach. The first approach is popular because it yields reasonably accurate estimates of the lifetime of a component and it is simple. Fracture mechanics-based approaches are more complex. In the following, we use Miner's rule to analyze fatigue failure of a dynamic system under a random, time varying excitation.

Traditionally, the S-N curve is used to characterize the behavior of materials in fatigue. This relates the number of stress cycles, which a material can withstand before failing due to fatigue, to the vibration amplitude of these stress cycles:

$$NS^b = c \tag{17.27}$$

where N is the number of stress cycles, S is the amplitude, and b and c are constants that depend on the material type. Exponent b ranges from 3 to 6. Note that the mean stress is assumed zero and the amplitude is assumed constant. Experimental results have shown that the above equation and the values of b and c are independent of the exact shape of the stress time history.

Suppose that a structure is subjected to n cycles of amplitude S, where n is less than N. Then we can define the damage inflicted to the structure as the ratio of n over N:

$$\text{Damage} = D = \frac{n}{N}. \tag{17.28}$$

If the structure is subjected to n_1 cycles of amplitude S_1, n_2 cycles of amplitude S_2, ..., n_n, cycles of amplitude S_n, then we can calculate the damage inflicted by each set of stress cycles and then add them to get the total damage using Miner's rule:

$$D = \sum_i \frac{n_i}{N_i} \tag{17.29}$$

where N_i is the number of cycles of amplitude S_i to failure. According to Miner's rule, failure occurs when the damage exceeds 1. According to Miner's rule, D is independent of the order of cycles. Although experimental evidence has shown that order is important, and several rules that account for the order have been proposed, none of these rules correlates with experimental results better than Miner's rule.

If the stress is a random variable, so is D. Then the problem of calculating the probability of fatigue failure is equivalent to that of calculating the probability that D exceeds 1. Miles [13] estimated the mean value of the damage for a stationary, narrowband process over a period T:

$$E(D(T)) = v^+(0)T \int_0^\infty \frac{f_{peak}(\alpha)}{N(\alpha)} da \qquad (17.30)$$

where $N(\alpha)$ is the number of stress cycles of amplitude α to failure.

If the random process is also Gaussian, then:

$$E(D(T)) = \frac{v^+(\alpha)T}{c}(\sqrt{2}\sigma_X)^b \Gamma\left(1+\frac{b}{2}\right). \qquad (17.31)$$

Mark [14] derived an expression for the variance of the damage in case of a second-order mechanical system subjected to a stationary Gaussian process.

If the applied stress is wideband, we can scale the damage obtained from Equation 17.31 by a correction factor, λ:

$$E(D(t)) = \lambda \frac{v^+(\alpha)T}{c}(\sqrt{2}\sigma_X)^b \Gamma\left(1+\frac{b}{2}\right). \qquad (17.32)$$

Estimates of this factor can be found in Wirsching and Light [15].

17.3.2.3 Practical Estimation of Damage Due to Cyclic Loading and Probability of Fatigue Failure

Most engineering structures, such as airplanes, ships, and offshore platforms, are subjected to non-stationary excitations that are due to changes in weather, loading, and operating conditions. To calculate fatigue damage in these structures, we divide the spectrum of loading and operating conditions into cells. In each cell, these conditions are assumed constant. The time of exposure to the conditions in each cell is estimated using data from previous designs. Then we calculate the fatigue damage corresponding to operation in each cell and add up the damage over all the cells. An alternative approach finds an equivalent stationary load that causes damage equal to the damage due to operations in all the cells and the damage due to the equivalent load. Wirsching and Chen [16] reviewed methods for estimating damage in ocean structures.

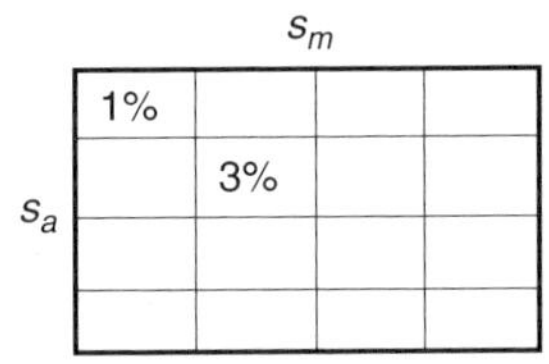

FIGURE 17.10 Load spectrum for alternating and mean stresses.

Consider that the load spectrum (percentage of lifetime during which a structure is subjected to a given mean and alternating load) is given (Figure 17.10). From this spectrum, we can estimate the joint pdf of the mean and alternating stresses, $f_{S_a S_m}(s_a, s_m)$.

For every cell in Figure 17.10, we can find an equivalent alternating stress that causes the same damage as the alternating and mean stresses when they are applied together. This equivalent stress is $(s_{a_i})/[1-(s_{m_j}/S_Y)]$ where s_{a_i} and s_{m_j} are the alternating and mean stresses corresponding to the i, j cell, and S_Y is the yield stress. The above expression is obtained using the Goodman diagram. Suppose n_{ij} is the number of cycles corresponding to that cell. Then, the damage due to the portion of the load spectrum corresponding to the i, j cell is $D_{ij} = \frac{n_{ij}}{N_{ij}}$. N_{ij} is the number of cycles at failure when s_{a_i} and s_{m_j} are applied. From the S-N curve we can find the number of cycles:

$$N_{ij}\left[\frac{s_{a_i}}{1 - \frac{s_{m_j}}{S_Y}}\right]^b = c \Rightarrow N_{ij} = \frac{c\left(1 - \frac{s_{m_j}}{S_Y}\right)^b}{s_{a_i}^b}. \tag{17.33}$$

Therefore, the damage corresponding to the ij cell is:

$$D_{ij} = \frac{n_{ij}\, s_{a_i}^b}{c\left(1 - \frac{s_{m_j}}{S_Y}\right)^b}. \tag{17.34}$$

The total damage is:

$$D = \sum_i \sum_j D_{ij} = \sum_i \sum_j \frac{n_{ij}\, s_{a_i}^b}{c\left(1 - \frac{s_{m_j}}{S_Y}\right)^b} = \sum_i \sum_j \frac{p_{ij}\, s_{a_i}^b}{c\left(1 - \frac{s_{m_j}}{S_Y}\right)^b} \tag{17.35}$$

where p_{ij} is the percentage of the number of cycles corresponding to load cell i, j. This percentage is obtained from information on the load spectrum (for example, from Figure 17.10).

Sources of uncertainty in the damage include:

- The random nature of the environment and the material properties (e.g., uncertainties in parameters b and c of the S-N curve)
- Epistemic (modeling) uncertainties, which are due to idealizations in the models of the environment and the structure

Nikolaidis and Kaplan [17] showed that, for ocean structures (ships and offshore platforms), uncertainty in fatigue life is almost entirely due to epistemic (modeling) uncertainties.

17.4 Evaluation of the Stochastic Response of Nonlinear Systems

Analysis of nonlinear systems is considerably more difficult than that of linear systems. A principal reason is that the superposition principle does not hold for nonlinear systems. Closed form solutions have been found for few simple nonlinear systems.

Methods for analysis of nonlinear systems include:

- Methods using the Fokker-Plank equation, which shows how the pdf of the response evolves in time
- Statistical linearization methods [18]
- Methods using state-space and cumulant equations

The above methods are presented in the book by Lutes and Sarkani [10].

17.5 Reliability Assessment of Systems with Uncertain Strength

So far, we have assumed that the strength of a system is deterministic. The probability of failure in this case can be calculated using Equation 17.14 if the random process representing the load is stationary or using Equation 17.15 if the process is nonstationary. Resistance can also be uncertain and in this case the above equations give the conditional probability of failure given that the resistance is $A = \alpha$. Here, an introduction of methods for reliability assessment of systems whose resistance is random is presented. Chapter 18, "Time-Variant Reliability," in this book presents these methods in detail.

Consider the case where the load process is stationary. Using the total probability theorem we find that the probability of failure over a period [0, T] is the average probability of exceedance over the entire range of values of resistance A:

$$P(F) = \int (1 - e^{-v^+(\alpha)T}) f_A(\alpha) d\alpha. \quad (17.36)$$

Several random variables can affect the resistance. In such a case, we need to calculate the probability of failure using an equation similar to Equation 17.36 involving nested integrals. This is impractical if there are more than three random variables. Wen [19] recommended two approaches to calculate the probability of failure: (1) the ensemble upcrossing rate method and (2) first- and second-order methods.

In the ensemble method, we calculate an average upcrossing rate considering that α is a random variable and then we use the calculated upcrossing rate in Equation 17.14 to find the probability of failure. This approach tacitly assumes that the crossing events form a Poisson process. This is not true because even if the load process is Poisson, if the resistance is high when a crossing occurs, then it will also be high in the next upcrossing. However, Wen demonstrated using an example that if the crossing rate is low (i.e., the product of the upcrossing rate and the length of the exposure period is considerably less than 1), then this approach can give accurate results.

The second method assumes that we can calculate the probability of failure as a function of the resistance and formulates the problem as a time invariant problem. Specifically, the problem of finding the average probability of failure in Equation 17.36 can be recast into the following problem:

$$P(F) = P(g(\mathbf{A}, U) \leq 0)$$

where the performance function $g(\mathbf{A}, U)$ is:

$$g(\mathbf{A}, U) = U - \Phi^{-1}(P(F/\mathbf{A})). \quad (17.37)$$

$\mathbf{A}$ is the vector of uncertain resistance variables, $P(F/\mathbf{A})$ is the conditional probability of failure given the values of the uncertain resistance variables, and U is a standard Gaussian random variable.

This problem can be solved using a first- or second-order method.

17.6 Conclusion

Many systems are subjected to excitations that should be modeled as random processes (random functions) or random fields.

Probabilistic analysis of dynamic systems involves three main steps: (1) modeling of the excitations, (2) calculation of the statistics of the response, and (3) calculation of the probability of failure. To develop a probabilistic model of a random process, we need to specify the joint probability distribution of the values of a random process for all possible values of the index parameter (e.g., the time). This is often impractical. Therefore, often we determine the second-order statistics of the random process, that is, the mean value, and the autocorrelation function. We have analytical tools, in the form of closed form expressions, for the second-order statistics of linear systems. However, few analytical solutions are available for nonlinear systems.

A system may fail due to first excursion failure, in which the response exceeds a level, and fatigue failure due to damage accumulation. Tools are available for estimating the probability of failure under these two modes. For first excursion failure, we have equations for approximating the average upcrossing rate of a level, the pdf of the local peaks of the response and the pdf of the time to failure. For fatigue failure, we have equations for the cumulative damage.

References

1. Nigam, N.C., 1983, *Introduction to Random Vibrations,* MIT Press, Cambridge, MA.
2. Lin, Y.K., 1967, *Probabilistic Theory of Structural Dynamics,* McGraw-Hill, New York.
3. Rice, S.O., 1944, 1945, "Mathematical Analysis of Random Noise," *Bell Systems Technical Journal,* 23:282–332; 24:46–156. Reprinted in *Selected Papers on Noise and Stochastic Processes,* N. Wax, Ed., 1954. Dover, New York, 133–249.
4. Yang, J.N. and Shinozuka, M., 1971, "On the First Excursion Failure Probability of in Stationary Narrow-Band Random Vibration," *ASME Journal of Applied Mechanics,* 38(4), 1017–1022.
5. Yang, J.N. and Shinozuka, M., 1972, "On the First Excursion Failure Probability in Stationary Narrow-Band Random Vibration — II," *ASME Journal of Applied Mechanics,* 39(4), 733–738.
6. Caughey, T.K. and Stumpf, H.J., 1961, "Transient Response of Dynamic Systems under Random Excitation," *Journal of Applied Mechanics,* 28(4), 563.
7. Fung, Y.C., 1955, "The Analysis of Dynamic Stresses in Aircraft Structures during Landing as Nonstationary Random Processes," *Journal of Applied Mechanics,* 22, 449–457.
8. Amin, M., 1966, "Nonstationary Stochastic Model for Strong Motion Earthquakes," Ph.D. thesis, University of Illinois.
9. Nikolaidis, E., Perakis, A.N., and Parsons, M.G., 1987, "Probabilistic Torsional Vibration Analysis of a Marine Diesel Engine Shafting System: The Input-Output Problem," *Journal of Ship Research,* 31(1), 41–52.
10. Lutes, L.D. and Sarkani, S., 2004, *Random Vibrations,* Elsevier Butterworth-Heinemann, Burlington, MA.
11. Shinozuka, M. and Yang, J.N., 1971, Peak Structural Response to Nonstationary Random Excitations, *Journal of Sound and Vibration,* 14(4), 505–517.
12. Yang, J.N. and Liu, S.C., 1980, "Statistical Interpretation and Application of Response Spectra," *Proceedings 7th WCEE,* 6, 657–664.
13. Miles, J.W., 1954, "On Structural Fatigue under Random Loading," *Journal of Aeronautical Science,* 21, 753–762.
14. Mark, W.D., 1961, "The Inherent Variation in Fatigue Damage Resulting from Random Vibration", Ph.D. thesis, Department of Mechanical Engineering, MIT.
15. Wirsching, P.H. and Light, M.C., 1980, "Fatigue under Wide Band Random Stresses," *Journal of Structural Design,* ASCE, 106(ST7), pp. 1593–1607.
16. Wirsching, P.H. and Chen, Y.-N., 1988, "Considerations of Probability-Based Fatigue Design for Marine Structures," *Marine Structures,* 1, 23–45.
17. Nikolaidis, E. and Kaplan, P., 1992, "Uncertainties in Stress Analyses on Marine Structures, Parts I and II," *International Shipbuilding Progress,* 39(417), 19–53; 39(418), 99–133.
18. Roberts, J.B. and Spanos, P.D., 1990, *Random Vibration and Statistical Linearization,* Wiley, New York.
19. Wen, Y.-K., 1990, *Structural Load Modeling and Combination for Performance and Safety Evaluation,* Elsevier, Amsterdam.

18
Time-Variant Reliability

Robert E. Melchers
The University of Newcastle

Andre T. Beck
Universidade Luterana do Brasil

18.1	Introduction	**18-1**
18.2	Loads as Processes: Upcrossings	**18-2**
18.3	Multiple Loads: Outcrossings	**18-3**
18.4	First Passage Probability	**18-3**
18.5	Estimation of Upcrossing Rates	**18-4**
18.6	Estimation of the Outcrossing Rate	**18-5**
18.7	Strength (or Barrier) Uncertainty	**18-6**
18.8	Time-Dependent Structural Reliability	**18-6**
18.9	Time-Variant Reliability Estimation Techniques	**18-7**
	Fast Probability Integration (FPI) • Gaussian Processes and Linear Limit State Functions • Monte Carlo with Conditional Expectation (and Importance Sampling) • Directional Simulation • Ensembled Upcrossing Rate (EUR) Approach • Estimation of the EUR Error • Summary of Solution Methods	
18.10	Load Combinations	**18-11**
18.11	Some Remaining Research Problems	**18-13**

18.1 Introduction

As discussed already in previous chapters, typically the loads acting on structures vary with time. They may be increasing slowly on average, such as with the gradual increase over many years in vehicular loads on bridges, or more quickly, such as wind loads generating a dynamic structural response. In all cases, the loads are of uncertain magnitude and the actual magnitude at any point in time will be uncertain. Moreover, the structural strength may vary with time — typically, it will decrease as a result of fatigue, corrosion, or similar deterioration mechanism. How to deal with these types of situations for structural reliability estimation is the topic for this chapter.

To be sure, it is not always necessary to use time-variant reliability approaches. For example, if just one load acts on a structure, the load may be represented by its extreme (e.g., its maximum) value and its uncertain nature by the relevant extreme value distribution. It represents the extreme value the load would be expected to occur during a defined time interval (usually the design life of the structure). The probability of failure over a defined lifetime can then be estimated employing the usual *time-invariant* methods such as FOSM, FORM and their developments, as well as in most Monte Carlo work. This has been termed the "time-integrated" approach because it transfers the time dependence effect into the way the load acting on the structure is modeled [1].

A further possibility is to consider not the lifetime of the structure, but some shorter, critical periods, such as the occurrence of a storm event and model the load during that event by an extreme value distribution. By considering the probability of various storm events and using the return period notion or probability bounds similar to those for series systems, the lifetime probability can then be estimated [1].

We will not discuss these simplified methods herein but rather concentrate on introducing concepts from the more general (and rich) stochastic process theory.

The time-integrated simplifications are not valid, in general, when more than one load is applied to a structure. In simple terms, it should be evident that it is very unlikely that the maximum of one load will occur at exactly the same time as the maximum of the second. A proper reliability analysis needs to account for this. One way of doing this is to apply the well-known *load combination rules* found in design codes. However, their usual justification is empirical. As discussed elsewhere, they can be derived, approximately, from time-variant structural reliability theory. A better approach is to use time-variant reliability theory directly.

This chapter provides an overview of time-variant reliability theory. Of necessity, it is brief. Moreover, because of space limitations, we have had to be selective in our selection of material. References must be consulted for more details.

18.2 Loads as Processes: Upcrossings

Examples of loads best modeled as processes include earthquake, wind, snow, temperature, and wave loads. Typically, the process fluctuates about a mean value. For wind or earthquake loading, the mean can be very low and there may be long periods of zero activity. This does not change the essential argument, however. Figure 18.1 shows a typical "realization" (that is, a trace) of a stochastic process.

In probability terms, a process can be described by an *instantaneous* (or average-point-in-time) cumulative probability distribution function $F_X(x,t)$. As usual, it represents the probability that the random process $X(t)$ will take a value lower than (or equal to) x. Evidently, this will depend, in general, on time t. The equivalent probability density function (pdf) is $f_X(x,t)$. Figure 18.1 shows a realization of the random process $X(t)$ and various properties. The instantaneous (average-point-in-time) pdf is shown on the left. Various properties may be derived from $f_X(x,t)$ including the process mean $\mu_X(t) = \int_{-\infty}^{\infty} x f_X(x,t) dx$ and the autocorrelation function $R_{XX}(t_1, t_2) = E[X(t_1)X(t_2)] = \int_{-\infty}^{\infty} \int_{-\infty}^{\infty} x_1 x_2 f_{XX}(x_1, x_2; t_1, t_2) dx_1 dx_2$ expressing the correlation relationship between two points in time (t_1, t_2). Here, f_{XX} is the joint probability density function. For (weakly) stationary processes, these two functions are constant in time and R_{XX} becomes dependent on relative time only. In principle, a stationary process cannot start or stop—this aspect usually can be ignored in practical applications provided the times involved are relatively long. Details are available in various texts.

Figure 18.1 also shows a barrier $R = r$. If the process $X(t)$ represents a load process $Q(t)$, say, the barrier can be thought of as the capacity of the structure to resist the load $Q(t)$ at any time t. This is the case for one loading and for a simple time-independent structure. When an event $Q > r$ occurs, the structure "fails."

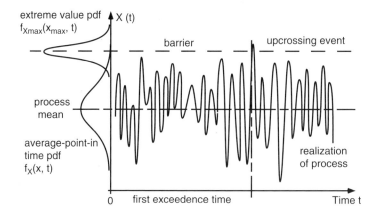

FIGURE 18.1 Realization of a continuous random process (such as windloading) showing an exceedence event and time to first exceedence.

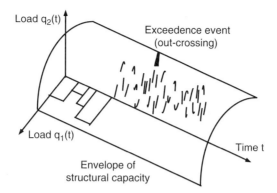

FIGURE 18.2 Envelope of structural strength (resistance) showing a realization of the vector load process and an outcrossing by process $Q_2(t)$.

For a process, this is termed an "upcrossing" or an "exceedence" event. For acceptable structural safety, the time t_1 to the first occurrence of an upcrossing should be sufficiently long. Evidently, t_1 will be a random variable. Estimation of t_1 or its equivalent, the probability of structural failure in a defined time span (the lifetime of the structure), is particularly important for structural reliability estimation, as will be discussed in more detail below. We note that the definition of the failure criterion is a matter for decision by those performing the reliability estimation.

18.3 Multiple Loads: Outcrossings

Most structures are subject to multiple loads, acting individually or in combination, independently or correlated in some way. Moreover, they need not all be modeled as continuous processes such as in Figure 18.1. For example, live loads due to floor loads, car park loads, etc. usually are better modeled as a series of pulses. Each pulse corresponds to a short-term, high-intensity loading event such as an office party or a meeting. Analogous to continuous loading, any one pulse can cause the capacity of the structure to be exceeded.

When several loads act, they can be represented by a vector process $\mathbf{X}(t)$ with corresponding probability density function $f_{\mathbf{X}}(\mathbf{x})$. For Figure 18.2, with loads only as random processes, the corresponding pdf is the vector of load processes $\mathbf{Q}(t)$. Evidently, any one load or any combination of the loads can exceed the structural capacity and produce an exceedence event. This is shown schematically in Figure 18.2 with the outcrossing there being due to the floor load $q_2(t)$. The envelope represents the structural capacity, the region under the envelope is the *safe* domain and the region outside is the *failure* domain. As before, the first exceedence time t_1 is the time until such an exceedence event occurs for the first time, that is, when one of the load processes, or the combined action of the two processes, "outcrosses" the envelope of structural capacity.

18.4 First Passage Probability

The above concepts can be formulated relatively simply using well-established theory of stochastic processes [2, 3]. Let $[0, t_L]$ denote the *design life* for the structure. Then, the probability that the structure will fail during $[0, t_L]$ is the sum of probability that the structure will fail when it is first loaded and the probability that it will fail subsequently, given that it has not failed earlier, or:

$$p_f(t) \leq p_f(0, t_L) + [1 - p_f(0, t_L)] \cdot [1 - e^{-vt}] \tag{18.1}$$

where v is the *outcrossing rate*. Already in the expression $[1 - e^{-vt}]$ in the second term it has been assumed (not unreasonable) that structural failure events are rare and that such events therefore can be represented

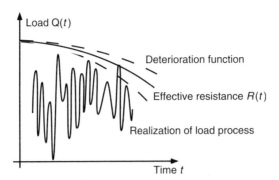

FIGURE 18.3 Exceedence event when there is structural deterioration with time (i.e., when $R = R(t)$).

reasonably closely by a Poisson distribution. Theoretically, this assumption would produce an (asymptotic) upper bound on the actual final term — asymptotic because the bound becomes closer to the true upcrossing probability as the barrier level increases. Despite this, it is has been found that expression 18.1 provides a reasonable estimate for structural failure estimation [4]. However, it would not be appropriate for, say, serviceability failures as these would not, normally, be rare events.

Simulation results tend to confirm that Equation 18.1 is an upper bound. They also show that outcrossings for some types of processes can occur in *clumps*. These are mainly the so-called narrow-band processes because these high values of the peaks of the processes tend to occur together; that is, the high peaks usually are not independent. Various schemes have been proposed for attempting to deal with this problem for different types of process, including being more explicit about individual outcrossings [5].

Equation 18.1 can be applied where the outcrossing rate v is not constant. This can arise, for example, when there is gradual deterioration (or enhancement) of the structural capacity with time (Figure 18.3). As a result, $p_f(0, t_L)$ and v become time dependent $v = v(t)$. Thus, in Equation 18.1, the term vt must be replaced with the time-average upcrossing rate $\int_0^t v^+(\tau)d\tau$.

18.5 Estimation of Upcrossing Rates

When the process is a discrete pulse process, such as Borges, Poisson Counting, Filtered Poisson, and Poisson Square Wave and for Renewal processes more generally, it has been possible to directly apply Equation 18.1 to estimate the first passage probability or the level (up)crossing rate for the process acting individually. The results are available in the literature.

For a continuous random process, the outcrossing rate must be estimated using stochastic process theory. The theory is available in standard texts. It is assumed that the process continues indefinitely in time and is stationary. Moreover, it produces an average value of the upcrossing rate over all possible realizations (i.e., the ensemble) of the random process $X(t)$ at any time t. Hence, it is known as the ensemble average value.

Only if the process is ergodic will the estimated upcrossing rate also be the time average frequency of upcrossings. The latter is a useful property because it allows the upcrossing rate to be estimated from a long record of observations of a single process. It is often assumed to hold unless there is evidence to the contrary.

In the special but important case when the random process $X(t)$ is a stationary normal process $N(\mu_X, \sigma_X^2)$, the upcrossing rate for a barrier level a can be

$$v_a^+ = \frac{1}{2\pi} \frac{\sigma_{\dot{X}}}{\sigma_X} \left[-\frac{(a-\mu_X)^2}{2\sigma_X^2} \right] = \frac{\sigma_{\dot{X}}}{(2\pi)^{1/2}} f_X \left(\frac{1}{\sigma_X} \phi \left[\frac{(a-\mu_X)}{\sigma_X} \right] \right) \tag{18.2}$$

where $\phi(\ )$ is the standard normal distribution function and $\dot{X}(t)$ is the derivative process distributed as $N(0,\sigma_{\dot{X}}^2)$. Details of the derivation and evaluation of these parameters can be found in standard texts. Suffice to note here simply that

$$\sigma_X^2 = R_{XX}(\tau = 0) - \mu_X^2 \tag{18.3}$$

$$\sigma_{\dot{X}}^2 = -\frac{\partial^2 R_{XX}(0)}{\partial \tau^2} \tag{18.4}$$

where R_{XX} is the autocorrelation function. It can be estimated from observations of the process.

For nonnormal processes there are very few closed solutions. Sometimes, Equation 18.2 is used as an approximation but this can be seriously in error. Numerical solutions often are used.

18.6 Estimation of the Outcrossing Rate

For several discrete pulse processes acting in combination, some results are available. One such application is considered later when dealing with load combinations.

For both discrete and continuous processes, the outcrossing situation is as shown in Figure 18.4, with the outcrossing occurring at point A for a vector of (in this case, two) random processes $\mathbf{X} = \mathbf{X}(t)$ with mean located in the safe domain D_S. The time axis is the third axis in Figure 18.4 and has been compressed onto the page. Closed form solutions are available only for a limited range of cases, dealing mainly with two-dimensional normal processes and open and closed square and circular domains [6]. It has been shown that for an m-dimensional vector process that the outcrossing rate can be estimated from:

$$v_D^+ = \int_{D_S} E(\dot{X}_n | \mathbf{X} = \mathbf{x})^+ f_{\mathbf{X}}(\mathbf{x}) d\mathbf{x} \tag{18.5}$$

In Equation 18.5, the term $(\ )^+$ denotes the (positive) component of $\mathbf{X}$ that crosses out of the safe domain (the other components cross back in and are of no interest). The subscript n denotes the outward normal component of the vector process at the domain boundary and is there for mathematical completeness. Finally, the term $f_{\mathbf{X}}(\mathbf{x})$ represents the probability that the process is actually at the boundary between the safe and unsafe domain (i.e., at the limit state). Evidently, if the process is not at this boundary, it cannot cross out of the safe domain.

There are only a few solutions of Equation 18.5 available for continuous processes, mainly for time-invariant domains; and there is an approximate bounding approach [7, 8]. For n discrete Poisson square

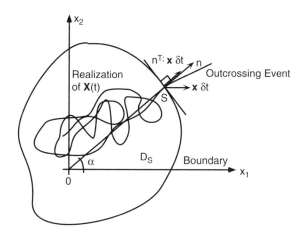

FIGURE 18.4 Vector process realization showing notation for outcrossing and for directional simulation.

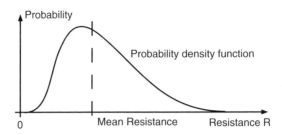

FIGURE 18.5 Schematic probability density function for material strength.

wave processes $\mathbf{X}(t)$ each having pulses of magnitude Y_i normally distributed and having mean arrival rate v_i the mathematics can be rearranged to [9]:

$$v_D^+ = \sum_{i=1}^{n} v_i \int_{-\infty}^{\infty} P(\text{outcrossing due to } Y_i) f_{\mathbf{X}^*}(\mathbf{x}) d\mathbf{x} \qquad (18.6)$$

where $P(\ ) = P[(Y_i, \mathbf{x}^*) \in D_S] \cdot P[(Y_i, \mathbf{x}^*) \notin D_S]$ denotes the probability of a pulse being initially inside the safe domain and then it being outside (i.e., and outcrossing occurs). These are independent events and $\mathbf{X}^*$ denotes the vector without the i-th component that is outcrossing. For various special cases, solutions have been developed for Equation 18.6. These include (1) hyperplanes, (2) hypercubical domains, and (3) linear additions of different processes.

When the safe domain is not convex or cannot be expressed as a linear combination of linear functions, the determination of the outcrossing rate becomes more problematic. Bounding results could be invoked but these tend to be conservative [10]. Nonnormal processes further complicate the problem. Numerical estimation of outcrossing rates must then be used. Figure 18.4 suggests that a procedure using a polar coordinate system is appropriate and indeed this has been developed [11, 12] using simulation (Monte Carlo) methods.

18.7 Strength (or Barrier) Uncertainty

Thus far, attention has focused on estimating the rate at which one or more processes (loads) crosses out of the safe domain — that is, when the loads exceed the capacity of the structure. As previously discussed, actual structures are not of known precise strength or capacity. There are variations in interpretation of design codes, in calculation procedures, and errors of various types can be committed. Moreover, there are variations introduced during construction and materials have variable properties. Hence, when a structure is finally completed, its actual strength R usually is not known precisely.

As a result, the location of the line $R = r$ in Figure 18.1 and Figure 18.3 should be represented as a probabilistic estimate such as shown in Figure 18.5. The line $R = r$ shown in Figure 18.1 and Figure 18.3 is just one "realization" of many possible resistance or strength or capacity outcomes. Because the actual strength outcome is uncertain, all reasonable possibilities must be considered in a time-variant reliability analysis. Moreover, for many realistic situations, the resistance is a function of time $R = R(t)$. This applies, for example, to strength affected by corrosion (of steel, say) or to concrete strength increase with time. Fatigue can be a function of time or, more accurately, of load applications and stress ranges.

18.8 Time-Dependent Structural Reliability

The above concepts can be generalized for random "barriers" and hence applied to estimation of the time dependent reliability of structures. As before, the discussion is limited to relatively high reliability systems such that the outcrossing approach (Equation 18.1) remains valid asymptotically.

The probability of structural failure during a defined structural (nominal) lifetime $[0, t_L]$ can be stated as

$$p_f(t) = P[R(t) \leq S(t)] \quad \forall (t \in [0, t_L]) \tag{18.7}$$

Here, $S(t)$ is the load effect (internal action) process derived from the loads. Equation 18.7 can be rewritten by conditioning on the random variable resistance

$$p_f = \int_{\mathbf{r}} p_f(t_L|\mathbf{r}) f_{\mathbf{R}}(\mathbf{r}) d\mathbf{r} \tag{18.8}$$

where the conditional failure probability $p_f(t_L|\mathbf{r})$ is a function of the vector of load processes $\mathbf{Q}(t)$ or the vector of load effects $\mathbf{S}(t)$ and $f_{\mathbf{R}}(\mathbf{r})$ is the probability density function for the resistance random vector $\mathbf{R}$. The key requirement then is the evaluation of the conditional failure probability $p_f(t_L|\mathbf{r})$. In general, no simple results exist, although Equation 18.1 can be used to provide an upper bound.

In using Equation 18.1 in Equation 18.8, two matters are of interest for evaluation. The first term in Equation 18.1 is $p_f(0)$, the probability of structural failure on the structure first being loaded. This can be substituted on its own directly into Equation 18.8. The outcome is a standard structural reliability estimation problem that can be solved by any of the time-invariant methods (FOSM, FORM, Monte Carlo, etc.).

The second matter of interest is the evaluation of the outcrossisng rate v_D^+. This can be obtained, in principle, from Equation 18.5 or Equation 18.6 for a given realization $\mathbf{R} = \mathbf{r}$ of the structural strength. What then remains is the integration as required in Equation 18.8. The techniques available for these two operations are discussed in the section to follow.

18.9 Time-Variant Reliability Estimation Techniques

18.9.1 Fast Probability Integration (FPI)

For a single normal random process in a problem otherwise described only by Gaussian random variables, the upcrossing rate estimation can be obviated by introducing an "auxiliary" random variable, in standard Gaussian (normal) space, defined as $u_{n+1} = \Phi^{-1}[p_f(t|\mathbf{r})]$. The limit state function then is augmented to [13, 14]:

$$g(u_{n+1}, u_n) = u_{n+1} - \Phi^{-1}[p_f(t|\mathbf{T}^{-1}(u_n))] \tag{18.9}$$

where $\mathbf{u} = \mathbf{T}(\mathbf{r})$ is the transformation to the standard normal space of the resistance random variables $\mathbf{R}$.

Equation 18.9 is exact when $p_f(t_L|\mathbf{r})$ and $f_{\mathbf{R}}(\mathbf{r})$ in Equation 18.8 are independent. However, some error is introduced by approximating the (augmented) limit state function by a linear function in FORM (or a quadratic one in SORM). Applications experience shows that when the conditional failure probability is small (as in most structural reliability problems), this technique may present convergence problems [15].

18.9.2 Gaussian Processes and Linear Limit State Functions

For the special case of fixed, known barriers given as linear limit state functions (i.e., hyperplanes in n-dimensional space) and Gaussian continuous random processes and Gaussian random variables, it is possible to use the result of Equation 18.2 to estimate the outcrossing rate for each hyperplane. Allowance must be made for the area each plane contributes and for the requirement that the process must be "at" the limit state for an outcrossing to occur. The latter is represented by the integral over the hyperplane of $f_X(x)$. For k (i.e., multiple) time-invariant hyperplanes each with surface area ΔS_i the result is then [7]

$$v_D^+ = \sum_{i=1}^{k} v_k^+ = \sum_{i=1}^{k} \left[\frac{\sigma_{\dot{Z}_i} \phi(\beta_i)}{(2\pi)^{1/2} \sigma_{Z_i}} \right] \left[\int_{\Delta S_i} f_X(\mathbf{x}) d\mathbf{x} \right] \quad (18.10)$$

where $\phi(\beta_i)$ represents the usual *safety index* for the i-th hyperplane relative to the origin in standard Gaussian space. The result (Equation 18.10) can be used in Equation 18.1 for fixed (known) barriers. Example calculations are available [17]. The main problem now remaining is the integration over the resistance random variables $\mathbf{R}$. The FPI approach could be used, in principle, but for many problems it appears that integration by Monte Carlo simulation is the only practical approach [18]. Of course, $p_f(0, t_L)$ can be estimated directly by the standard FOSM or FORM methods.

18.9.3   Monte Carlo with Conditional Expectation (and Importance Sampling)

The integration in Equation 18.8 can be carried out using Monte Carlo simulation with conditional expectation [19]. The result is obtained by starting with the usual expression for multi-parameter time-invariant failure probability

$$p_f = \int \cdots \int_{D_F} I[G(\mathbf{x}) \leq 0] f_X(\mathbf{x}) d\mathbf{x} \quad (18.11)$$

where D_F represents the failure domain. With some manipulation, this can be expressed as

$$p_f = \int \cdots \int_{D_1} p_{f|X_1=x_1} f_{X_1}(\mathbf{x}_1) d\mathbf{x}_1 \quad (18.12)$$

where $\mathbf{X}_1$ is a subset of the random vector $\mathbf{X}$ with corresponding sampling space D_1 and $p_{f|X_1=x_1}$ represents a conditional probability of failure for the subset space $D_2 = D - D_1$. If the latter conditional term can be evaluated analytically or numerically for each sampling of vector $\mathbf{x}_1$, variance reduction can be achieved and the Monte Carlo process made more efficient. The integration over D_1 will be performed by Monte Carlo simulation (with importance sampling).

The use of Equation 18.12 lies in recognizing that it has the same form as Equation 18.8, with the components of the random vector $\mathbf{X}$ in domain D_1 in Equation 18.12 corresponding to the vector of resistance random variables $\mathbf{R}$ in Equation 18.8. When importance sampling is to be applied, Equation 18.12 is rewritten in the standard form as

$$p_f = \int \cdots \int_{D_1} \frac{p_{f|X_1=x_1} f_{X_1}(\mathbf{x}_1)}{h_V(\mathbf{x}_1)} h_V(\mathbf{x}_1) d\mathbf{x}_1 \quad (18.13)$$

Some simpler applications are available [19].

18.9.4   Directional Simulation

Directional simulation has been shown to be a reasonably efficient approach for Monte Carlo estimation of the time-invariant probabilities of failure, that is, for random variable problems [12, 20]. When stochastic processes are involved, a more intuitive approach is to consider directional simulation applied in the space of the load processes $\mathbf{Q}(t)$ only.

This way of considering the problem follows naturally from Figure 18.4 with the barrier in that figure now interpreted as one realization of the structural strength or capacity $\mathbf{R}$. It follows that Equation 18.5 now becomes an estimate conditional on the realization $\mathbf{R} = \mathbf{r}$ of the structural resistance; that is, Equation 18.5 estimates $p_f(t_L|\mathbf{r})$ in Equation 18.8. The structural resistance will be represented in this formulation in the space of $\mathbf{Q}(t)$ and will be a function of random variables describing individual member capacities, material strength or stiffness values, dimensions, etc. Thus, $\mathbf{R} = \mathbf{R}(\mathbf{X})$. The main challenge is to derive the probabilistic properties $f_R(\mathbf{r})$ for $\mathbf{R}$. In directional simulation terms, this is written, for each direction $\mathbf{A} = \mathbf{a}$ (see Figure 18.4) as $f_{S|A}(s|\mathbf{a})$ where S is the radial (scalar) distance for measuring the structural strength along the corresponding ray. Several approaches have been proposed [21–23].

Once $f_{S|A}(s|\mathbf{a})$ is available, $p_f(t_L|\mathbf{r})$ in Equation 18.8 can be estimated from Equation 18.1, now written in direction simulation terms as

$$p_f(s|a) \leq p_f(0, s|a) + \{1 - \exp[-v_D^+(s|a)t]\} \tag{18.14}$$

where, as earlier, $p_f(0,)$ represents the failure probability at time $t = 0$ and v_D^+ is the outcrossing rate of the vector stochastic process $\mathbf{Q}(t)$ out of the safe domain D. To support directional simulation, Equation 18.8 has been modified to

$$p_f = \int_{\text{unit sphere}} f_\mathbf{A}(\mathbf{a}) \left[\int_S p_f(s|a) f_{S|A}(s|\mathbf{a}) ds \right] d\mathbf{a} \tag{18.15}$$

where $f_\mathbf{A}(\mathbf{a})$ is the pdf for the sampling direction $\mathbf{A}$. The limitation to this rewriting of Equation 18.8 is that each directional simulation direction is assumed as an independent one-dimensional probability integration. It is a reasonable approximation for high barrier levels, but generally will overestimate the failure probability. Techniques to deal with the integration directly through Equation 18.8 are under investigation.

In common with most other techniques, it is also implicit in this formulation that the limit state functions are not dependent on the history of the load process realizations; that is, the results are assumed *load-path independent*.

18.9.5 Ensembled Upcrossing Rate (EUR) Approach

In the ensembled upcrossing rate approach, the upcrossing rate of a random process through a resistance barrier is averaged over the probability distribution of the resistance:

$$v_{ED}^+(t) = \int_R v^+(r,t) f_R(r,t) dr \tag{18.16}$$

which is then substituted into Equation 18.1 to estimate the failure probability. This is not quite the same as the failure probability estimated using Equation 18.8. Taking the mean over the resistance introduces some level of dependency in the upcrossing rate, thus making the Poisson assumption in Equation 18.1 less appropriate. The EUR approximation consists of approximating the arrival rate of upcrossings through a random barrier by the ensemble average of upcrossings [24]. Typically, this leads to an overestimation of the failure probability [25]. In summary, it would be expected that the EUR approximation has the following characteristics:

1. Appropriate for even higher barrier levels and even lower failure probabilities as compared to a deterministic barrier problem
2. More appropriate for situations with time-dependent random variable or random process barriers, as variations in the barrier tend to reduce the dependency error
3. A function of the relative magnitudes of the resistance and the load process variances

The EUR approach is easily generalized to a vector of random variable resistances **R** and multiple processes. In general, an ensemble outcrossing rate is obtained when the integration boundary (limit state) is considered random in the outcrossing rate calculation (e.g., Equation 18.5 and Equation 18.10). This can be done, for example, through directional simulation in the load-space, by writing Equation 18.15 in terms of outcrossing rates rather than failure probabilities. Ensemble outcrossing rates are also obtained when outcrossing rates are computed as a parallel system sensitivity measure [26] with random resistance parameters included in the solution [27].

The EUR approximation significantly simplifies the solution of time-variant reliability problems when resistance degradation is considered because it changes the order of integrations. In the approaches described earlier, the outcrossing rate is integrated over time for the failure probability (Equation 18.1) and then failure probabilities are integrated over random resistance parameters (Equation 18.8). When resistance degradation is considered, or when load processes are nonstationary, the outcrossing rate becomes time dependent. Hence, outcrossing rate evaluation through Equation 18.5, Equation 18.6, and Equation 18.10 must be repeated over time, for each integration point of Equation 18.8. In the EUR solution, the outcrossing rate is first integrated over random resistance parameters (Equation 18.16) and then over time (Equation 18.1). The ensemble outcrossing rate calculation must also be repeated over time, but this calculation generally represents little additional effort to the computation of conditional outcrossing rates (Equation 18.5). The outer integration for resistance parameters is hence avoided. The integration over time is straightforward, as long as ensemble outcrossing rates can be considered independent.

18.9.6 Estimation of the EUR Error

So far, the EUR approximation has received little attention in structural reliability theory. It is important to understand the error involved in this approximation. This problem was considered in [28]. In summary, Monte Carlo simulation was used to estimate the arrival rate of the first crossing over the random barrier. This is assumed to be an "independent" ensemble crossing rate and is termed v_{EI}^+ in the sequel. By comparing v_{ED}^+ with v_{EI}^+, it was found possible to estimate the error involved using modest numbers of simulations and to extrapolate the results in time.

Figure 18.6 shows a typical comparison of v_{EI}^+ and v_{ED}^+ for narrowband (NB) and broadband (BB) standard Gaussian load processes with time-invariant Gaussian barriers. It is seen that there is a very rapid reduction in v_{EI}^+ with time (cycles), which is nonexistent in v_{ED}^+ (because the barrier is time-invariant) and that the reduction is greater for barriers with higher standard deviations. This is evident also in Figure 18.7, which shows the "order of magnitude" error between v_{EI}^+ and v_{EL}^+. Similar results were obtained for time variant barriers with deterioration characteristics.

The results show that the error involved in the conventional EUR approach can be significant for both narrow and broad band processes and that it becomes greater as the variance of the barrier increases. The study also shows that the EUR error is governed by the parameter:

$$E_P = \frac{1}{\sigma_S} \sqrt{\frac{\sigma_S(\sigma_R^2 + \sigma_S^2)}{(\mu_R - \mu_S)}} \tag{18.17}$$

where indexes R and S are used for resistance and load parameters, respectively. This result is strictly valid only for scalar problems involving Gaussian load processes and Gaussian barriers. It allows establishing practical guidelines for the use of the EUR approximation. For example, the EUR error is known to be smaller than one order of magnitude when $E_P \leq 0.65$. The extension of these results to multidimensional problems and to nonGaussian barriers and processes is under investigation.

18.9.7 Summary of Solution Methods

It should be clear that the estimation of the outcrossing rate for general systems with unrestricted probability descriptions of the random variables and of the stochastic processes involved remains a challenging problem. Some solution techniques are available, however, for special cases.

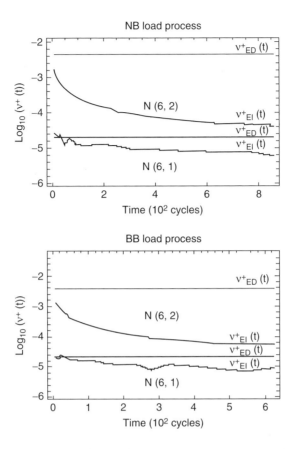

FIGURE 18.6 Ensemble upcrossing rates v_{ED}^+ and v_{EI}^+ for Gaussian random barriers with parameters $N(\mu,\sigma)$ [28].

Only a few analytical solutions exist for estimating the outcrossing rate. The usual approach is through Monte Carlo simulation, including employment of methods of variance reduction [12, 19, 22]. However, it is fair to say that the theoretically accurate results have a high computational demand. Proposals to reduce this all involve various simplifications, often but not always conservative ones.

18.10 Load Combinations

A special but important application of time-variant reliability analysis concerns the possibility of giving the load combination rules used in design codes some degree of theoretical rationality. At the simplest level, the rules provide an equivalent load to represent the combination of two stochastic loads $Q_1(t)$ and $Q_2(t)$, say. As before, these loads are very unlikely to have peak values at the same time.

The stress resultants (internal actions) $S_1(t)$ and $S_2(t)$ from the loads can be taken as additive. Thus, interest lies in the probability that the sum $S_1(t) + S_2(t)$ exceeds some barrier level a. If the two processes each have Gaussian amplitudes and are stationary, the sum will be Gaussian also and Equation 18.10 applies directly.

More generally, Equation 18.5 can be applied for the additive combination of two processes with instantaneous (i.e., arbitrary-point-in-time) pdfs given by $f_{Q_1}(q_1)$ and $f_{Q_2}(q_2)$, respectively. However, this requires joint pdf information that is seldom available. Nevertheless, a reasonably good upper bound is [16]:

$$v_Q^+(a) \leq \int_{-\infty}^{\infty} v_1(u) f_{Q_2}(a-u) du + \int_{-\infty}^{\infty} v_2(u) f_{Q_1}(a-u) du \qquad (18.18)$$

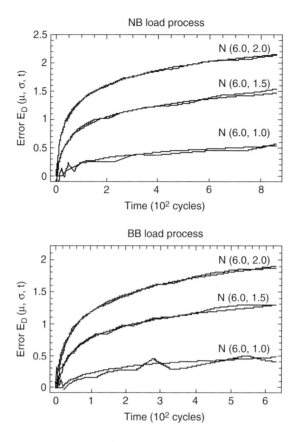

FIGURE 18.7 Ensemble upcrossing rate error as a function of time, numerical result, and analytical approximation.

where $v_i(u)$ is the upcrossing rate for the process $Q_i(t)$ ($i = 1, 2$), evaluated using the theory for single processes. Equation 18.18 is known as a point-crossing formula, and it is exact, mainly for cases where one or both processes has a discrete distribution [29]. For these, the individual process upcrossing rates have a relatively simple form. For example, for the sum of two nonnegative rectangular renewal processes with a probability $p_i (i = 1, 2)$ of zero load value (so-called mixed processes) (see Figure 18.8), the

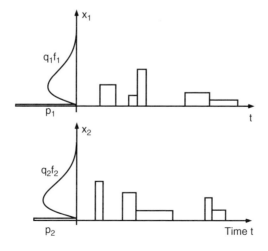

FIGURE 18.8 Realizations of two "mixed" rectangular renewal processes having the shown pdfs.

upcrossing rate is:

$$v_Q^+(a) = v_i[p_i + q_i F_i(a)][q_i(1 - F_i(a))] \tag{18.19}$$

where $v_i q_i$ and $F_i()$ are the mean arrival rate and the cdf of the i-th process, respectively. In this case, the arbitrary-point-in-time probability density function (pdf) for the i-th process is given by $f_{Q_i}(q_i) = p_i \delta(q_i) + q_i f_i(q_i)$, where $\delta()$ is the Dirac-delta function and $q_i = 1 - p_i$ (see Figure 18.8).

Combining Equation 18.18 and Equation 18.19 and making the reasonable assumption that all pulses return to zero after application and that the high pulses are of relatively short duration relative to the life of the structure leads to the approximate result [30]

$$v_Q^+(a) \approx v_1 q_1 G_1(a) + v_2 q_2 G_2(a) + (v_1 q_1)(v_2 q_2) G_{12}(a) \tag{18.20}$$

where $G_i() = 1 - F_i()$ and $F_{12}()$ is the cdf for the total of the two pulses. In Equation 18.20, the first term represents upcrossings of the first process, the second term those of the second process, and the third term upcrossings involving both processes acting at the same time. Typically, when the pulses are of short duration, the joint mean arrival rate $(v_1 q_1)(v_2 q_2)$ is very low and the third term can be neglected. Under this assumption, Equation 18.20 can be interpreted as one example of the conventional design code load combination formula although it is phrased in terms of the upcrossing rate.

Noting that the extreme of a process in a time interval $[0, t_L]$ can be represented approximately by $F_{max}(a) \approx e^{-vt_L} \approx 1 - vt_L$, where $F_{max}() = P(X_{max} < a)$ is the cdf for the maximum of the process in $[0, t_L]$, it follows that

$$G_{max X}(a) = 1 - F_{max X}(a) \approx v_X^+(a) \cdot t_L \tag{18.21}$$

which shows that the upcrossing rate is related directly to the maximum value of a process in a given time interval $[0, t_L]$. Thus, Equation 18.20 can be recast in terms of maximum loads $Q_{i\,max}$ (e.g., greater than the barrier level) or their combinations and the average point-in-time values $\overline{Q}_i$

$$Q_{max} \approx \max[(Q_{1\,max} + \overline{Q}_2); (Q_{2\,max} + \overline{Q}_1)] \tag{18.22}$$

This is known as Turkstra's rule [31]. Although it is of limited validity for accurate structural reliability estimation, it provides an approximate theoretical justification for the intuitive load combination rules in structural design codes.

18.11 Some Remaining Research Problems

It should be clear from the above that the estimation of time-variant structural reliability is a much more complex matter than for time-invariant problems. Much of the effort has gone into the use of classical stochastic process theory and the asymptotic outcrossing estimates available through it. Monte Carlo solutions have been built on this theory but the computational effort involved is significant. It is possible, of course, just to use brute-force Monte Carlo by simulating very many realizations of the processes in a problem and noting the number of outcrossings obtained for a given time period. This requires very extensive computational power but may be attractive in some applications.

For certain discrete pulse processes, there are exact solutions and reasonably good bounds available but the situation is less satisfactory for continuous processes. For these, upper bound solutions based on the asymptotic approximation are available, for fixed barriers particularly. For normal processes but for random variable barriers, integration over the relevant PDF is still required. This includes efforts to recast the problem to more directly in processes-only spaces.

Progress in this area is considered to require a better understanding of the errors involved in the various methods developed so far, improvements in dealing with integration over the resistance random variables and the possibility of employing response surface methodologies to circumvent the high computational requirements.

For nonstationary processes and deteriorating systems, the form and uncertainty characterization of the time-dependence function has only recently begun to receive serious attention. Almost certainly, the conventional design assumptions about material deterioration and about load (and other) process non-stationarity will not be sufficient for high-quality structural reliability analyses [32]. Much remains to be done in this area also.

References

1. Melchers, R.E. (1999). *Structural Reliability Analysis and Prediction*, second edition, John Wiley & Sons, Chichester (reprinted June 2001).
2. Rice, S.O. (1944). Mathematical analysis of random noise, *Bell System Tech. J.*, **23**, 282–332; (1945), **24**, 46–156. Reprinted in Wax, N. (1954) *Selected Papers on Noise and Stochastic Processes*, Dover Publications.
3. Vanmarcke, E.H. (1975). On the distribution of the first-passage time for normal stationary processes, *J. Applied Mech.*, ASME, **42**, 215–220.
4. Engelund, S., Rackwitz, R., and Lange, C. (1995). Approximations of first-passage times for differentiable processes based on higher-order threshold crossings, *Probabilistic Engineering Mechanics*, **10**, 53–60.
5. Kordzakhia, N., Melchers, R.E., and Novikov, A. (1999). First passage analysis of a 'Square Wave' filtered Poisson process, in *Proc. Int. Conf. Applications of Statistics and Probability,* Melchers, R.E. and Stewart, M.G., Eds., Balkema, Rotterdam, 35–43.
6. Hasofer, A.M. (1974). The upcrossing rate of a class of stochastic processes, *Studies in Probability and Statistics*, Williams, E.J., Ed., North-Holland, Amsterdam, 151–159.
7. Veneziano, D., Grigoriu, M., and Cornell, C.A. (1977), Vector-process models for system reliability, *J. Engineering Mechanics Div.*, ASCE, **103**, (EM3) 441–460.
8. Hohenbichler, M. and Rackwitz, R. (1986). Asymptotic outcrossing rate of Gaussian vector process into intersection of failure domains, *Prob. Engineering Mechanics*, **1**(3), 177–179.
9. Breitung, K. and Rackwitz, R. (1982). Nonlinear combination load processes, *J. Structural Mechanics*, **10**(2), 145–166.
10. Rackwitz, R. (1984). Failure rates for general systems including structural components, *Reliab. Engg.*, **9**, 229–242.
11. Deák, I. (1980). Fast procedures for generating stationary normal vectors, *J. Stat. Comput. Simul.*, **10**, 225–242.
12. Ditlevsen, O., Olesen, R., and Mohr, G. (1987). Solution of a class of load combination problems by directional simulation, *Structural Safety*, **4**, 95–109.
13. Madsen, H.O. and Zadeh, M. (1987). Reliability of plates under combined loading, *Proc. Marine Struct. Rel. Symp., SNAME*, Arlington, VA, 185–191.
14. Wen, Y.-K. and Chen, H.C. (1987). On fast integration for time variant structural reliability, *Prob. Engineering Mechanics,* **2**(3), 156–162.
15. Marley, M.J. and Moan, T. (1994). *Approximate Time Variant Analysis for Fatigue, Structural Safety and Reliability,* Schueller, Shinozuka and Yao, Eds.
16. Ditlevsen, O. and Madsen, H.O. (1983). Transient load modeling: clipped normal processes, *J. Engineering Mechanics Div.*, ASCE, **109**(2), 495–515.
17. Ditlevsen, O. (1983). Gaussian outcrossings from safe convex polyhedrons, *J. Engineering Mechanics Div.*, ASCE, **109**(1), 127–148.
18. Rackwitz, R. (1993). On the combination of nonstationary rectangular wave renewal processes, *Structural Safety*, **13**(1+2) 21–28.

19. Mori, Y. and Ellingwood, B.R. (1993). Time-dependent system reliability analysis by adaptive importance sampling, *Structural Safety*, **12**(1) 59–73.
20. Ditlevsen, O., Melchers, R.E., and Gluver, H. (1990). General multi-dimensional probability integration by directional simulation, *Computers & Structures*, **36**(2), 355–368.
21. Melchers, R.E. (1991). Simulation in time-invariant and time-variant reliability problems, *Proc. 4th IFIP Conference on Reliability and Optimization of Structural Systems*, Rackwitz, R. and Thoft-Christensen P., Eds, Springer, Berlin, 39–82.
22. Melchers, R.E. (1995). Load space reliability formulation for Poisson pulse processes, *J. Engineering Mechanics*, ASCE, **121**(7), 779–784.
23. Guan, X.L. and Melchers, R.E. (1999). A load space formulation for probabilistic finite element analysis of structural reliability, *Prob. Engrg. Mech.*, **14**, 73–81.
24. Pearce, H.T. and Wen, Y.K. (1984). Stochastic combination of load effects, *Journal of Structural Engineering*, ASCE, **110**(7), 1613–1629.
25. Wen, Y.K. and Chen, H.C. (1989). System reliability under time-varying loads: I, *Journal of Engineering Mechanics*, ASCE, **115**(4), 808–823.
26. Hagen, O. and Tvedt, L. (1991). Vector process out-crossing as a parallel system sensitivity measure, *Journal of Engineering Mechanics*, ASCE, **117**(10), 2201–2220.
27. Sudret, B., Defaux, G., Lemaire, M., and Andrieu, C. (2002). Comparison of methods for computing the probability of failure in time-variant reliability using the out-crossing approach, *Fourth International Conference on Computational Stochastic Mechanics*, Kerkyra (Corfu), Greece, June 2002.
28. Beck, A.T. and Melchers, R.E. (2002). On the ensemble up-crossing rate approach to time variant reliability analysis of uncertain structures, *Fourth International Conference on Computational Stochastic Mechanics*, Kerkyra (Corfu), Greece, June.
29. Larrabee, R.D. and Cornell, C.A. (1981). Combination of various load processes, *J. Structural Div.*, ASCE, **107**(ST1). 223–239.
30. Wen, Y.-K. (1977) Probability of extreme load combination, *J. Structural Div.*, ASCE, **104**(ST10) 1675–1676.
31. Turkstra, C.J. (1970). *Theory of Structural Design Decisions Study No. 2*, Solid Mechanics Division, University of Waterloo, Waterloo, Ontario.
32. Melchers, R.E. (2001). Assessment of existing structures — some approaches and research needs, *J. Struct. Engrg.*, ASCE, **127**(4), 406–411.

19
Response Surfaces for Reliability Assessment

19.1	Introduction	**19**-1
19.2	Response Surface Models	**19**-3
	Basic Formulation • Linear Models and Regression • Analysis of Variance • First- and Second-Order Polynomials • Exponential Relationships • Polyhedral Models	
19.3	Design of Experiments	**19**-10
	Transformations • Saturated Designs • Redundant Designs • Comparison	
19.4	Reliability Computation	**19**-14
	Choice of Method • Error Checking and Adaptation of the Response Surface	
19.5	Application to Reliability Problems	**19**-15
	Linear Response Surface • Nonlinear Response Surface • Nonlinear Finite Element Structure	
19.6	Recommendations	**19**-21

Christian Bucher
Bauhaus-University Weimar

Michael Macke
Bauhaus-University Weimar

19.1 Introduction

Structural reliability assessment requires one to estimate probabilities of failure that, in general, are of rather small magnitude. Moreover, structural failure is most typically assessed by means of nonlinear, possibly time-variant analyses of complex structural models. In such cases, the computational cost incurred for one single analysis — to decide whether or not a structure is safe — may become quite demanding. Consequently, the application of direct (or even advanced) Monte Carlo simulation — being the most versatile solution technique available — is quite often not feasible. To reduce computational costs, therefore, it has been suggested to utilize the response surface method for structural reliability assessment [1].

Let us assume that the reliability assessment problem under consideration is governed by a vector $\mathbf{X}$ of n basic random variables X_i ($i = 1, 2, \ldots, n$), that is,

$$\mathbf{X} = (X_1, X_2, \ldots, X_n)' \tag{19.1}$$

where $(\cdot)'$ is transpose. Assuming, furthermore, that the random variables $\mathbf{X}$ have a joint probability density function $f(\mathbf{x})$, then the probability of failure $P(F)$ — that is the probability that a limit state will be reached — is defined by

$$P(F) = \int \ldots \int_{g(\mathbf{x}) \leq 0} f(\mathbf{x}) d\mathbf{x} \tag{19.2}$$

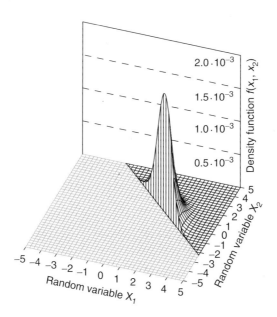

FIGURE 19.1 Integrand for calculating the probability of failure for $g(x_1, x_2) = 3 - x_1 - x_2$.

whereby $g(\mathbf{x})$ is the limit state function that divides the n-dimensional probability space into a failure domain $F = \{\mathbf{x} : g(\mathbf{x}) \leq 0\}$ and a safe domain $S = \{\mathbf{x} : g(\mathbf{x}) > 0\}$. As already mentioned, the computational challenge in determining the integral of Equation 19.2 lies in evaluating the limit state function $g(\mathbf{x})$, which for nonlinear systems usually requires an incremental/iterative numerical approach. The basic idea in utilizing the response surface method is to replace the true limit state function $g(\mathbf{x})$ by an approximation $\eta(\mathbf{x})$, the so-called response surface, whose function values can be computed more easily.

In this context it is important to realize that the limit state function $g(\mathbf{x})$ serves the sole purpose of defining the bounds of integration in Equation 19.2. As such, it is quite important that the function $\eta(\mathbf{x})$ approximates this boundary sufficiently well, in particular in the region that contributes most to the failure probability $P(F)$. As an example, consider a two-dimensional problem with standard normal random variables X_1 and X_2, and a limit state function $g(x_1, x_2) = 3 - x_1 - x_2$. In Figure 19.1, the integrand of Equation 19.2 in the failure domain is displayed. It is clearly visible that only a very narrow region around the so-called design point $\mathbf{x}^*$ really contributes to the value of the integral (i.e., the probability of failure $P(F)$). Even a relatively small deviation of the response surface $\eta(\mathbf{x})$ from the true limit state function $g(\mathbf{x})$ in this region may, therefore, lead to significantly erroneous estimates of the probability of failure. To avoid this type of error, it must be ensured that the important region is sufficiently well covered when designing the response surface.

The response surface method has been a topic of extensive research in many different application areas since the influential paper by Box and Wilson in 1951 [2]. Whereas in the formative years the general interest was on experimental designs for polynomial models (see, e.g., [2, 3]), in the following years nonlinear models, optimal design plans, robust designs, and multi-response experiments — to name just a few — came into focus. A fairly complete review on existing techniques and research directions of the response surface methodology can be found in [4–7]. However, traditionally, the application area of the response surface method is not structural engineering, but, for example, chemical or industrial engineering. Consequently, the above-mentioned special requirement for structural reliability analysis — that is, the high degree of accuracy required in a very narrow region — is usually not reflected upon in the standard literature on the response surface method [8–10].

One of the earliest suggestions to utilize the response surface method for structural reliability assessment was made in [1]. Therein, Lagrangian interpolation surfaces and second-order polynomials are rated as useful response surfaces. Moreover, the importance of reducing the number of basic variables and error checking is emphasized. Support points for estimating the parameters of the response surface are determined by spherical design. In [11], first-order polynomials with interaction terms are utilized as response surfaces to analyze the reliability of soil slopes. The design plan for the support points is saturated — either by full or by fractional factorial design. Another analysis with a saturated design scheme is given in [12], where quadratic polynomials without interaction terms are utilized to solve problems from structural engineering. Polynomials of different order in combination with regression analysis are proposed in [13], whereby fractional factorial designs are utilized to obtain a sufficient number of support points. The validation of the chosen response surface model is done by means of analysis of variance.

In [14] it has been pointed out that for reliability analysis it is most important to obtain support points for the response surface very close to or exactly at the limit state $g(\mathbf{x}) = 0$. This finding has been further extended in [15, 16]. In [17], the response surface concept has been applied to problems involving random fields and nonlinear structural dynamics. It has been shown that a preliminary sensitivity analysis using simplified structural analysis models may help to significantly reduce the computational effort. Because random field problems are typically characterized by a large number of mutually highly correlated variables, a spectral decomposition of the covariance matrix [18] can also considerably reduce the number of random variables required for an appropriate representation of the uncertainties.

In addition to polynomials of different order, piecewise continuous functions such as hyperplanes or simplices can also be utilized as response surface models. For the class of reliability problems defined by a convex safe domain, secantial hyperplane approximations such as presented in [19, 20] yield conservative estimates for the probability of failure. Several numerical studies indicate, however, that in these cases the interpolation method converges slowly from above to the exact result with increasing number of support points. The effort required for this approach is thereby comparable to Monte Carlo simulation based on directional sampling [21].

19.2 Response Surface Models

19.2.1 Basic Formulation

Response surface models are more or less simple mathematical models designed to describe the possible experimental outcome (e.g., the structural response in terms of displacements, stresses, etc.) of a more or less complex structural system as a function of quantitatively variable factors (e.g., loads or system conditions), which can be controlled by an experimenter. Obviously, the chosen response surface model should give the best possible fit to any collected data. In general, we can distinguish two different types of response surface models: regression models (e.g., polynomials of varying degree or nonlinear functions such as exponentials) and interpolation models (e.g., polyhedra).

Let us denote the response of any structural system to a vector $\mathbf{x}$ of n experimental factors or input variables x_i ($i = 1, 2, \ldots, n$) (i.e., $\mathbf{x} = (x_1, x_2, \ldots, x_n)'$), by $z(\mathbf{x})$. In most realistic cases, it is quite likely that the *exact* response function will not be known. Therefore, it must be replaced by a flexible function $q(\cdot)$ that will express satisfactorily the relation between the response z and the input variables x. Taking into account a (random) error term ε, the response can be written over the region of experimentation as

$$z = q(\theta_1, \theta_2, \ldots, \theta_p; x_1, x_2, \ldots, x_n) + \varepsilon \tag{19.3}$$

where θ_j ($j = 1, 2, \ldots, p$) are the parameters of the approximating function $q(\cdot)$. Taking now expectations, that is,

$$\eta = \mathrm{E}[z] \tag{19.4}$$

then the surface represented by

$$\eta = q(\theta_1, \theta_2, \ldots, \theta_p; x_1, x_2, \ldots, x_n) = q(\boldsymbol{\theta}; \mathbf{x}) \tag{19.5}$$

is called a *response surface*. The vector of parameters $\boldsymbol{\theta} = (\theta_1, \theta_2, \ldots, \theta_p)'$ must be estimated from the experimental data in such a way that Equation 19.4 is fulfilled. In the following we will investigate the most common response surface models, methods to estimate their respective parameters, and, most significant, techniques for determining whether a chosen response surface model is suitable or should be replaced by a more appropriate one.

19.2.2 Linear Models and Regression

Let us assume that an appropriate response surface model $q(\cdot)$ has been chosen to represent the experimental data. Then, for estimating the values of the parameters $\boldsymbol{\theta}$ in the model, the method of maximum likelihood can be utilized. Under the assumptions of a Gaussian distribution of the random error terms ε, the method of maximum likelihood can be replaced by the more common method of least squares [8]. In the latter case, the parameters $\boldsymbol{\theta}$ are determined in such a way that the sum of squares of the differences between the value of the response surface $q(\boldsymbol{\theta}; \mathbf{x}^{(k)})$ and the measured response $z^{(k)}$ at the m points of experiment

$$\mathbf{x}^{(k)} = \left(x_1^{(k)}, \ldots, x_n^{(k)}\right)', \quad k = 1, 2, \ldots, m \tag{19.6}$$

becomes as small as possible. That is, the sum of squares function

$$s(\boldsymbol{\theta}) = \sum_{k=1}^{m} \left(z^{(k)} - q(\boldsymbol{\theta}; \mathbf{x}^{(k)})\right)^2 \tag{19.7}$$

must be minimized. This corresponds to a minimization of the variance of the random error terms ε. The minimizing choice of $\boldsymbol{\theta}$ is called a *least-squares estimate* and is denoted by $\hat{\boldsymbol{\theta}}$.

The above regression problem becomes more simple to deal with when the response surface model is linear in its parameters $\boldsymbol{\theta}$. Let us assume that the response surface is given by

$$\eta = \theta_1 q_1(\mathbf{x}) + \theta_2 q_2(\mathbf{x}) + \cdots + \theta_p q_p(\mathbf{x}) \tag{19.8}$$

The observations $z^{(k)}$ made at the points of experiment $\mathbf{x}^{(k)}$ can be represented by this response surface model as

$$\mathbf{z} = \begin{bmatrix} z^{(1)} \\ z^{(2)} \\ \vdots \\ z^{(m)} \end{bmatrix} = \begin{bmatrix} q_1(\mathbf{x}^{(1)}) & q_2(\mathbf{x}^{(1)}) & \cdots & q_p(\mathbf{x}^{(1)}) \\ q_1(\mathbf{x}^{(2)}) & q_2(\mathbf{x}^{(2)}) & \cdots & q_p(\mathbf{x}^{(2)}) \\ \vdots & \vdots & & \vdots \\ q_1(\mathbf{x}^{(m)}) & q_2(\mathbf{x}^{(m)}) & \cdots & q_p(\mathbf{x}^{(m)}) \end{bmatrix} \begin{bmatrix} \theta_1 \\ \theta_2 \\ \vdots \\ \theta_p \end{bmatrix} + \begin{bmatrix} \varepsilon^{(1)} \\ \varepsilon^{(2)} \\ \vdots \\ \varepsilon^{(m)} \end{bmatrix} = \mathbf{Q}\boldsymbol{\theta} + \boldsymbol{\varepsilon} \tag{19.9}$$

with ε as a vector of random error terms. Assuming that the random error terms are normally distributed and statistically independent with constant variance σ^2, i.e.,

$$E[\varepsilon^{(k)}] = 0, \quad \text{Var}[\varepsilon^{(k)}] = \sigma^2 \quad \text{and} \quad \text{Cov}[\varepsilon^{(k)}, \varepsilon^{(l)}] = 0 \quad \text{for} \quad k \neq l \tag{19.10}$$

then the covariance matrix of the observations $\mathbf{z}$ is

$$\text{Cov}[\mathbf{z}] = E[(\mathbf{z} - E[\mathbf{z}])(\mathbf{z} - E[\mathbf{z}])'] = \sigma^2 \mathbf{I} \tag{19.11}$$

with $\mathbf{I}$ as an identity matrix.

The least square estimates $\hat{\boldsymbol{\theta}} = (\hat{\theta}_1, \ldots, \hat{\theta}_p)'$ of the parameter vector $\boldsymbol{\theta}$ are determined such that

$$L = (\mathbf{z} - \mathbf{Q}\hat{\boldsymbol{\theta}})'(\mathbf{z} - \mathbf{Q}\hat{\boldsymbol{\theta}}) = \mathbf{z}'\mathbf{z} - 2\mathbf{z}'\mathbf{Q}\hat{\boldsymbol{\theta}} + (\mathbf{Q}\hat{\boldsymbol{\theta}})'\mathbf{Q}\hat{\boldsymbol{\theta}} \tag{19.12}$$

becomes minimal. A necessary condition is that

$$\frac{\partial L}{\partial \hat{\theta}} = -2\mathbf{z}'\mathbf{Q} + 2(\mathbf{Q}\hat{\theta})'\mathbf{Q} = 0 \tag{19.13}$$

From this follows that

$$\mathbf{Q}'\mathbf{Q}\hat{\theta} = \mathbf{Q}'\mathbf{z} \tag{19.14}$$

The fitted regression model is consequently

$$\hat{\mathbf{z}} = \mathbf{Q}\hat{\theta} \tag{19.15}$$

If the matrix $\mathbf{Q}'\mathbf{Q}$ is not rank deficit (i.e., rank $(\mathbf{Q}'\mathbf{Q}) = p \leq m$), then there exists a unique solution to the above system of equations. The estimated parameter vector is given by

$$\hat{\theta} = (\mathbf{Q}'\mathbf{Q})^{-1}\mathbf{Q}'\mathbf{z} \tag{19.16}$$

This estimator is unbiased; that is,

$$E[\hat{\theta}] = \theta \tag{19.17}$$

with covariance

$$\text{Cov}[\hat{\theta}] = \sigma^2 (\mathbf{Q}'\mathbf{Q})^{-1} \tag{19.18}$$

If the above made assumptions with respect to the random error terms ε do not hold—for example, the error terms are correlated or nonnormally distributed—then a different minimizing function L than given in Equation 19.12 has to be utilized. Typical examples thereof are given in [8].

19.2.3 Analysis of Variance

Because a response surface is only an approximation of the functional relationship between the structural response and the basic variables, it should be evident that, in general, there is always some *lack of fit* present. Therefore, a crucial point when utilizing response surfaces for reliability assessment is to check whether the achieved fit of the response surface model to the experimental data suffices or if the response surface model must be replaced by a more appropriate one. Therefore, different measures have been proposed in the past for testing different aspects of response surface models. In the following, a short overview of the most common measures is given. The basic principle of these measures is to analyze the variation of the response data in comparison to the variation that can be reproduced by the chosen response surface model—that is why this kind of response surface testing is also referred to as *analysis of variance*. Further and more advanced measures or checking procedures can be found in [8–10, 22].

Let us start with a measure of overall variability in a set of experimental data, the *total sum of squares* s_t. It is defined as the sum of squared differences $(z^{(k)} - \bar{z})$ between the observed experimental data $z^{(k)}$ ($k = 1, 2, \ldots, m$) and its average value

$$\bar{z} = \frac{1}{m}\sum_{k=1}^{m} z^{(k)} \tag{19.19}$$

that is,

$$s_t = \mathbf{z}'\mathbf{z} - \frac{1}{m}(\mathbf{1}'\mathbf{z})^2 \tag{19.20}$$

where **1** is a vector of ones. If we divide s_t by the appropriate number of degrees of freedom, that is, $(m-1)$, we obtain the sample variance of the z's, which is a standard measure of variability.

The total sum of squares can be partitioned into two parts, the *regression sum of squares* s_r, which is the sum of squares explained by the utilized response surface model, and the *error sum of squares* s_e, which represents the sum of squares unaccounted for by the fitted model. The regression sum of squares is defined as the sum of squared differences $(\hat{z}^{(k)} - \bar{z})$ between the value $\hat{z}^{(k)}$ predicted by the response surface and the average value $\bar{z}$ of the observed data; that is,

$$s_r = (\mathbf{Q}\hat{\boldsymbol{\theta}})'\boldsymbol{\zeta} - \frac{1}{m}(\mathbf{1}'\boldsymbol{\zeta})^2 \tag{19.21}$$

If the response surface model has p parameters, then the number of degrees of freedom associated with the measure s_r is $(p-1)$.

The sum of squares unaccounted for in the model — called error sum of squares or, sometimes also, residual sum of squares — is defined as the squared difference $(z^{(k)} - \hat{z}^{(k)})$ between the observed experimental data $z^{(k)}$ and the value $\hat{z}^{(k)}$ predicted by the response surface, that is,

$$s_e = \mathbf{z}'\mathbf{z} - (\mathbf{Q}\hat{\boldsymbol{\theta}})'\mathbf{z} \tag{19.22}$$

Obviously, the error sum of square s_e is the difference between the total sum of squares s_t and the regression sum of squares s_r; that is, $s_e = s_t - s_r$. Consequently, the degrees of freedom associated with the measure s_e are $(m - p) = (m - 1) - (p - 1)$. Moreover, it can be shown that

$$E[s_e] = \sigma^2(m - p) \tag{19.23}$$

Thus, an unbiased estimator of σ^2 is given by

$$\hat{\sigma}^2 = \frac{s_e}{m - p} \tag{19.24}$$

From the above defined sums of squares, different kinds of statistics can be constructed that measure certain aspects of the utilized response surface. The first of such measures is the *coefficient of (multiple) determination*

$$r^2 = \frac{s_r}{s_t} = 1 - \frac{s_e}{s_t} \tag{19.25}$$

which measures the portion of the total variation of the values $z^{(k)}$ about the mean $\bar{z}$ which can be explained by the fitted response surface model. We can easily see that $0 \leq r^2 \leq 1$. A large value of r^2 is supposed to indicate that the regression model is a good one. Unfortunately, adding an additional variable to an existing response surface model will always increase r^2 — independent of its relevancy to the model [10]. Therefore, an adjusted r^2-statistic has been proposed, defined as

$$r_A^2 = 1 - \frac{E[s_e]}{E[s_t]} = 1 - \frac{s_e}{s_t}\frac{(m-1)}{(m-p)} \tag{19.26}$$

As has been pointed out in [10], in general the measure r_A^2 does not increase when terms are added to the model, but in fact decreases often if these additional terms are unnecessary.

A different measure, which allows one to test the significance of the fitted regression equation, is the ratio of the mean regression sum of squares and the mean error sum of squares; that is,

$$F_0 = \frac{\mathrm{E}[s_r]}{\mathrm{E}[s_e]} = \frac{s_r}{s_e} \frac{(m-p)}{(p-1)} \qquad (19.27)$$

the so-called *F*-statistic, which follows an *F*-distribution. The *F*-statistic allows to test the null hypothesis

$$H_0: \quad \theta_1 = \theta_2 = \cdots = \theta_p = 0 \qquad (19.28)$$

against the alternative hypothesis

$$H_1: \quad \theta_j \neq 0 \text{ for at least one value of } \theta_j \ (j=1, 2, \ldots, p) \qquad (19.29)$$

For a specified level of significance α, the hypothesis H_0 is rejected if

$$F_0 > F_{\alpha, p-1, m-p} \qquad (19.30)$$

(here, $p-1$ represents the degrees of freedom numerator and $m-p$ represents the degrees of freedom denominator) and we can conclude that at least one or more of the terms of the response surface model are able to reproduce a large extent of the variation observed in the experimental data. Or, if the hypothesis H_0 is not rejected, a more adequate model has to be selected, because none of the terms in the model seem to be of indispensable nature.

In addition to the test with the above-mentioned *F*-statistic, which tests all parameters at once, we are also often interested in the question if an *individual* parameter θ_j is significant to our model. That is, we would like to test if the model can be improved by adding an additional term, or if we can delete one or more variables to make our model more effective. As has been mentioned already above, when one parameter is added to a response surface model, the error sum of squares always decreases. However, adding an unimportant parameter to the model can increase the mean square error, thereby reducing the usefulness of the response surface model [10]. It is therefore an important task to decide if a certain parameter should be used in a response surface model — or better be rejected.

The hypothesis testing concerning individual parameters is performed by comparing the parameter estimates $\hat{\theta}_j$ in the fitted response surface model to their respective estimated variances $\mathrm{Var}[\hat{\theta}_j]$ — these are the diagonal elements of the covariance matrix $\mathrm{Cov}[\hat{\boldsymbol{\theta}}]$ in Equation 19.18. Testing of the null hypothesis

$$H_0: \quad \theta_j = 0 \qquad (19.31)$$

is performed by determining the value of the *t*-statistic

$$t_0 = \frac{\hat{\theta}_j}{(\mathrm{Var}[\hat{\theta}_j])^{1/2}} \qquad (19.32)$$

and compare it with a percentile value α of the *t*-distribution with $(m-p)$ degrees of freedom, which corresponds to the number of degrees of freedom of the unbiased estimator of $\hat{\sigma}^2$ in Equation 19.24. Depending on the alternative hypothesis H_1, that is, if we utilize a two-sided test with

$$H_1: \quad \theta_j \neq 0 \qquad (19.33)$$

or a one-sided-test with either

$$H_1: \theta_j < 0 \quad \text{or} \quad H_1: \theta_j > 0 \tag{19.34}$$

the null hypothesis H_0 is rejected at a confidence level α if, respectively,

$$|t_0| > t_{\alpha/2, m-p} \tag{19.35}$$

or

$$|t_0| > t_{\alpha, m-p} \tag{19.36}$$

If H_0 is rejected, then this indicates that the corresponding term can be dismissed from the model. It should be noted that this is a partial test, because the value of the tested parameter depends on all other parameters in the model. Further tests in a similar vein can be found in [8–10].

We conclude this section with a reminder that the estimated parameters $\hat{\boldsymbol{\theta}}$ are least square estimates; that is, there is a certain likelihood that the true parameter $\boldsymbol{\theta}$ has a different value than the estimated one. Therefore, it is sometimes quite advisable to determine confidence intervals for the parameters. The $(1-\alpha)$ confidence interval for an individual regression coefficient θ_j is given by [10]

$$\hat{\theta}_j - (\text{Var}[\hat{\theta}_j])^{1/2} t_{\alpha/2, m-p} \leq \theta_j \leq \hat{\theta}_j + (\text{Var}[\hat{\theta}_j])^{1/2} t_{\alpha/2, m-p} \tag{19.37}$$

(Joint confidence regions for several regression coefficients are given, e.g., in [10].) Consequently, when utilizing response surfaces in reliability assessment, we should be aware that all predictions — may it be the structural response for a certain design or a reliability measure — show respective prediction intervals.

19.2.4 First- and Second-Order Polynomials

As already mentioned, response surfaces are designed such that a complex functional relation between the structural response and the basic variables is described by an appropriate, but — preferably — as simple as possible mathematical model. The term "simple" means, in the context of response surfaces, that the model should be continuous in the basic variables and should have a small number of terms, whose coefficients can be easily estimated. Polynomial models of low order fulfill such demands. Therefore, in the area of reliability assessment, the most common response surface models are first- and second-order polynomials (see [1, 12, 15, 16]).

The general form of a first-order model of a response surface η, which is linear in its n basic variables x_i, is

$$\eta = \theta_0 + \sum_{i=1}^{n} \theta_i x_i \tag{19.38}$$

with θ_i ($i = 0, 1, \ldots, n$) as the unknown parameters to be estimated from the experimental data. The parameter θ_0 is the value of the response surface at the origin or the center of the experimental design, whereas the coefficients θ_i represent the gradients of the response surface in the direction of the respective basic variables x_i. As can be seen from Equation 19.38, the first-order model is not able to represent even the simplest interaction between the input variables.

If it becomes evident that the experimental data cannot be represented by a model whose basic variables are mutually independent, then the first-order model can be enriched with (simple) interaction terms, such that

$$\eta = \theta_0 + \sum_{i=1}^{n} \theta_i x_i + \sum_{i=1}^{n-1} \sum_{j=i+1}^{n} \theta_{ij} x_i x_j \tag{19.39}$$

The total number of parameters to be estimated is given by $1 + n(n + 1)/2$. In the response surface model of Equation 19.39, there is some curvature present, but only from the twisting of the planes of the respective input variables. If a substantial curvature is required as well, then the above model can be further enriched by n quadratic terms to a complete second-order model of the form

$$\eta = \theta_0 + \sum_{i=1}^{n} \theta_i x_i + \sum_{i=1}^{n} \sum_{j=i}^{n} \theta_{ij} x_i x_j \tag{19.40}$$

The total number of parameters to be estimated is, therewith, given by $1 + n + n(n + 1)/2$. In most common cases, either the first-order or the complete second-order model is utilized as the response surface function.

19.2.5 Exponential Relationships

Although polynomial approximations of the response are most common in reliability assessment, there is also some interest in other forms of approximating functions—most notable in exponential relationships (see, e.g., [13, 23]). There may exist mechanical or mathematical justifications for utilizing exponential relationships (e.g., when approximating solutions of differential equations or probability functions); nevertheless, in most cases, the choice is clearly dominated by an engineer's habit to take the logarithm of the input variables or the response to achieve a better fit of a linear model when the fit so far has not been satisfactory.

Most exponential relationships have been treated in a completely different way as compared to polynomial relationships; nevertheless, they often can be included in the polynomial family. Take, for example, two of the most popular exponential relationships of the form

$$\eta = \theta_0 + \sum_{i=1}^{n} \theta_i \exp(x_i) \tag{19.41}$$

and

$$\eta = \theta_0 \prod_{i=1}^{n} \exp(\theta_i x_i) + \theta_{n+1} \tag{19.42}$$

The model of Equation 19.41 is linear in its parameters. Moreover, by taking the logarithm of the input variables, that is,

$$\eta = \theta_0 + \sum_{i=1}^{n} \theta_i \exp(\log(\tilde{x}_i)) = \theta_0 + \sum_{i=1}^{n} \theta_i \tilde{x}_i \tag{19.43}$$

where $\tilde{x}_i$ are the transformed input variables, the model can be reduced to the first-order polynomial type. In case of Equation 19.42, the same linear form can be obtained by utilizing a logarithmic transformation for the response surface η, that is,

$$\tilde{\eta} = \log(\eta - \theta_{n+1}) = \log(\theta_0) + \sum_{i=1}^{n} \theta_i x_i = \tilde{\theta}_0 + \sum_{i=1}^{n} \theta_i x_i \tag{19.44}$$

where $\tilde{\eta}$ is the transformed response and $\tilde{\theta}_0$ is the transformed parameter θ_0.

When utilizing transformations for a reparametrization of a chosen response surface model, we should take into account that, in general, also the random error terms ε are affected, (i.e., transformed).

Consequently, the assumptions made for the nontransformed model in the regression analysis or the analysis of variance must be revised for the transformed model. Further examples of exponential relationships or other nonlinear models can be found (e.g., in [5, 9]).

19.2.6 Polyhedral Models

In addition to the above-mentioned regression models, there are also interpolation models available for describing response surfaces. A special type of such models are polyhedra, that is, an assemblage of continuous functions. The given rationale for utilizing such models is that they are more flexible to local approximations and should, therefore, indeed converge in the long run to the exact limit state function [19, 20]. This flexibility is an important feature in reliability assessment, where the estimated reliability measure depends quite considerably on a sufficiently accurate representation of the true limit state function in the vicinity of the design point. Therefore, these models utilize only points on the limit state surface $g(\mathbf{x}) = 0$, which separates the failure domain from the safe domain.

Typical examples of such polyhedral models are: *tangential planes* [19] defined by a point on the limit state surface $g(\mathbf{x}) = 0$ and the gradients with respect to the basic variables at this point; *normal planes* [20] defined by one point on the limit state surface $g(\mathbf{x}) = 0$ and the normal vector whose direction is defined by the point on the limit state surface and the origin in standard normal space; and *simplices* [20] defined in n-dimensional random variable space by n nondegenerated points on the limit state surface $g(\mathbf{x}) = 0$. All these polyhedral models allow, by including additional points on the limit state surface, an adaptive refinement of the approximating response surface and, consequently, an improvement of the estimator of the failure probability. However, this refinement is often marred by an excessive need for additional points on the limit state surface.

19.3 Design of Experiments

19.3.1 Transformations

Having chosen an appropriate response surface model, support points $\mathbf{x}^{(k)}$ ($k = 1, 2, \ldots, m$) have to be selected to estimate in a sufficient way the unknown parameters of the response surface. Thereto, a set of samples of the basic variables is generated. In general, this is done by applying predefined schemes, so-called *designs of experiments*. The schemes shown in the following are saturated designs for first- and second-order polynomials, as well as full factorial and central composite designs. As is quite well known from experiments regarding physical phenomena, it is most helpful to set up the experimental scheme in a space of dimensionless variables. The schemes as described in the following perform experimental designs in a space of dimension n, where n is equal to the number of relevant basic variables.

The selected design of experiments provides a grid of points defined by the dimensionless vectors $\boldsymbol{\xi}^{(k)} = (\xi_1^{(k)}, \xi_2^{(k)}, \ldots, \xi_n^{(k)})'$. This grid must be centered around a vector $\mathbf{c} = (c_1, c_2, \ldots, c_n)'$. In the absence of further knowledge, this center point can be chosen equal to the vector of mean values $\boldsymbol{\mu} = (\mu_1, \mu_2, \ldots, \mu_n)'$ of the basic random variables X_i ($i = 1, 2, \ldots, n$). As will be shown below, when further knowledge about the reliability problem becomes available, other choices of the center point are, in general, more appropriate. The distances from the center are controlled by the scaling vector $\mathbf{s} = (s_1, s_2, \ldots, s_n)$. In many cases, it is useful to choose the elements s_i of this scaling vector equal to the standard deviations σ_i of the random variables X_i. So, in general, a support point $\mathbf{x}^{(k)}$ ($k = 1, 2, \ldots, m$) is defined as

$$\mathbf{x}^{(k)} = \begin{bmatrix} x_1^{(k)} \\ x_2^{(k)} \\ \vdots \\ x_n^{(k)} \end{bmatrix} = \begin{bmatrix} c_1 + \xi_1^{(k)} s_1 \\ c_2 + \xi_2^{(k)} s_2 \\ \vdots \\ c_n + \xi_n^{(k)} s_n \end{bmatrix} \tag{19.45}$$

The number m of generated support points depends on the selected method.

Depending on the type of probability functions $F(x_i)$ associated with the random variables X_i, this scheme may generate support points outside the range for which the random variables are defined (such as, e.g., negative values of coordinates for log-normally distributed random variables). To circumvent this type of problem, it may be helpful to transform first all random variables X_i into standardized normal random variables U_i by means of

$$u_i = \Phi^{-1}[F(x_i)] \tag{19.46}$$

whereby $\Phi^{-1}(\cdot)$ is the inverse of the normal distribution function

$$\Phi(u) = \frac{1}{\sqrt{2\pi}} \int_{-\infty}^{u} \exp\left(-\frac{y^2}{2}\right) dy \tag{19.47}$$

Equation 19.46 assumes that the random variables X_i are mutually independent. In normal random variable space, the design would be carried out using, for example, $u_i^{(k)} = \xi_i^{(k)}$ and the corresponding values of $x_i^{(k)}$ can be computed from the inverse of Equation 19.46; that is,

$$x_i^{(k)} = F^{-1}\left[\Phi\left(u_i^{(k)}\right)\right] \tag{19.48}$$

Note that both approaches yield the same support points if the random variables are independent and normally distributed. Nevertheless, for continuous distribution functions, there always exist suitable transformations $T(\cdot)$ that allow one to transform a vector of random variables $\mathbf{X}$ into a vector of mutually independent standard normal variables $\mathbf{U} = T(\mathbf{X})$ [24].

19.3.2 Saturated Designs

Saturated designs provide a number of support points just sufficient to represent a certain class of response functions exactly. Hence, for a linear saturated design, a linear function will be uniquely defined. Obviously, $m = n + 1$ samples are required for this purpose (see Figure 19.2). The factors $\xi_i^{(k)}$ for $n = 3$ are given by

$$\boldsymbol{\xi} = (\boldsymbol{\xi}^{(1)}, \boldsymbol{\xi}^{(2)}, \boldsymbol{\xi}^{(3)}, \boldsymbol{\xi}^{(4)}) = \begin{bmatrix} 0 & +1 & 0 & 0 \\ 0 & 0 & +1 & 0 \\ 0 & 0 & 0 & +1 \end{bmatrix} \tag{19.49}$$

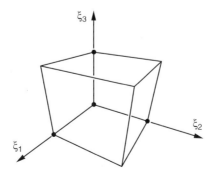

FIGURE 19.2 Saturated linear experimental scheme.

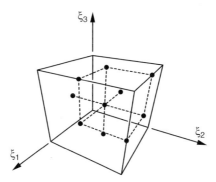

FIGURE 19.3 Saturated quadratic design scheme for $n = 3$.

Here, each column represents one support point. Of course, any variation of the factors above in which some or all values of (+1) were replaced by (−1) would also constitute a valid linear saturated design. Obviously, there is some arbitrariness in the design scheme that can usually be resolved only by introducing additional knowledge about the system behavior.

A saturated quadratic design (Figure 19.3) generates $m = n(n+1)/2 + n + 1$ support points $\mathbf{x}^{(k)}$. The factors $\xi_i^{(k)}$ for $n = 3$ are given by

$$\xi = (\xi^{(1)}, \xi^{(2)}, \ldots, \xi^{(10)}) = \begin{bmatrix} 0 & +1 & 0 & 0 & -1 & 0 & 0 & +1 & +1 & 0 \\ 0 & 0 & +1 & 0 & 0 & -1 & 0 & +1 & 0 & +1 \\ 0 & 0 & 0 & +1 & 0 & 0 & -1 & 0 & +1 & +1 \end{bmatrix} \quad (19.50)$$

Again, each column represents one support point. As mentioned, any change of sign in the pairwise combination would also lead to a saturated design, so that the final choice is somewhat arbitrary and should be based on additional problem-specific information.

19.3.3 Redundant Designs

Redundant experimental design methods provide more support points than required to define the response surface, and thus enable error checking procedures as outlined in the preceding section. Typically, regression is used to determine the coefficients of the basis function. Here also, linear and quadratic functions are being utilized.

The *full factorial* method (Figure 19.4) generates q sample values for each coordinate, thus producing a total of $m = q^n$ support points $\mathbf{x}^{(k)}$ ($k = 1, 2, \ldots, m$). Note that even for moderate values of q and n,

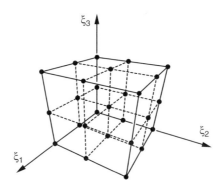

FIGURE 19.4 Full factorial desing scheme for $q = 3$ and $n = 3$.

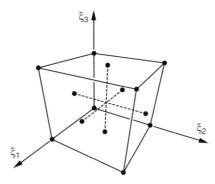

FIGURE 19.5 Central composite design scheme for $n = 3$.

this may become prohibitively expensive. Therefore frequently subsets are chosen that lead to *fractional factorial* designs.

The *central composite* design method (Figure 19.5) superimposes a full factorial design with $q = 2$ and a collection of all center points on the faces of an n-dimensional hypercube. Thus, it generates $m = (2^n + 2n)$ support points $\mathbf{x}^{(k)}$. The factors for $n = 3$ are given by

$$\xi = \begin{bmatrix} +1 & +1 & +1 & +1 & -1 & -1 & -1 & -1 & +1 & 0 & 0 & -1 & 0 & 0 \\ +1 & +1 & -1 & -1 & +1 & +1 & -1 & -1 & 0 & +1 & 0 & 0 & -1 & 0 \\ +1 & -1 & +1 & -1 & +1 & -1 & +1 & -1 & 0 & 0 & +1 & 0 & 0 & -1 \end{bmatrix} \quad (19.51)$$

D-optimal designs attempt to maximize the information content if only a small subset of the otherwise preferable full factorial design can be utilized, for example, due to restrictions on computer capacity. Given a set of candidate factors $\xi^{(k)}$, a subset of size m' is chosen in order to maximize the following function

$$D = \det(\mathbf{Q}'\mathbf{Q}) \quad (19.52)$$

In this equation, $\mathbf{Q}$ denotes a matrix containing values of the basis functions for the response surface evaluated at the selected support points (see Equation 19.9). Typically, the number m' is chosen to be 1.5-times the corresponding number of a saturated design.

19.3.4 Comparison

Table 19.1 shows the number of support points as a function of the number of variables for different experimental schemes. It is quite clear that factorial schemes—especially the full factorial—can become quite unattractive due to the exponential growth of the number of support points with increasing dimension of the problem. On the other hand, the more economical saturated designs do not allow for sufficient redundancy, which is needed to provide error control on the basis of the analysis of variance as outlined in a previous section. Therefore, moderately redundant schemes (such as D-optimal design) may provide the most appropriate solution.

TABLE 19.1 Numbers of Support Points for Different Experimental Designs

n	Linear	Quadratic	Full Factorial $q = 2$	Central Composite	Full Factorial $q = 3$
1	2	3	2	4	3
2	3	6	4	8	9
3	4	10	8	14	27
4	5	15	16	24	81
5	6	21	32	42	243

19.4 Reliability Computation

19.4.1 Choice of Method

Based on a chosen type of response surface and an appropriate design of experiments, an approximation for the limit state function is obtained. The subsequent reliability analysis usually can be performed with any available solution technique, such as first- and second-order method or Monte Carlo simulation. Because the cost of computation for one sample is extremely low, no special consideration needs to be given to achieve high efficiency. It should be carefully noted, however, that in the case of quadratic functions, unexpected phenomena might appear. A quadratic response surface $\eta(\mathbf{x})$ can generally be written in the form of

$$\eta(\mathbf{x}) = \boldsymbol{\theta}_0 + \boldsymbol{\theta}'\mathbf{x} + \mathbf{x}'\boldsymbol{\Theta}\mathbf{x} \qquad (19.53)$$

Here, the scalar coefficient $\boldsymbol{\theta}_0$, the vector $\boldsymbol{\theta}$, and the matrix $\boldsymbol{\Theta}$ have been obtained as outlined in the preceding sections. Unfortunately, there is no guarantee that the matrix $\boldsymbol{\Theta}$ is positive definite. If it turns out to be indefinite (i.e., it has both positive and negative eigenvalues), then the approximated limit state function is a hyperbolic surface. Hyperbolic surfaces are not simply connected, so they consist of several disjoint parts. Such surfaces may cause severe problems for optimization procedures, which are typically used in conjunction with the first-order reliability method. Generally, Monte Carlo-based methods are rather insensitive to the specific shape of the response surface and can be recommended. Adaptive sampling [25] provides a convenient importance sampling method that does not depend on the success of the first-order reliability method approach.

19.4.2 Error Checking and Adaptation of the Response Surface

In view of the statements made regarding the accuracy of the response surface near the region of the largest contribution to $P(F)$ (see Figure 19.1), it is mandatory to make sure that this region has been covered by the design of experiment scheme. Adaptive sampling utilizes the mean value vector $\boldsymbol{\mu}_F$ of the generated samples conditional on the failure domain F; that is,

$$\boldsymbol{\mu}_F = E[\mathbf{X}|\mathbf{X} \in F] \qquad (19.54)$$

If this vector lies outside the region covered by the experimental design scheme, it is advisable to repeat the experimental design around $\boldsymbol{\mu}_F$. This leads to an adaptive scheme in which the response surfaces are updated based on the results of the reliability analysis. A simple concept in this vein has been suggested in [12]. This is sketched in Figure 19.6. As indicated in Figure 19.6, the experimental design shifts all support points closer to the limit state $g(\mathbf{x}) = 0$. Ideally, some support points should lie on the limit state. A repeated calculation of $\boldsymbol{\mu}_F$ can be utilized as an indicator for convergence of the procedure.

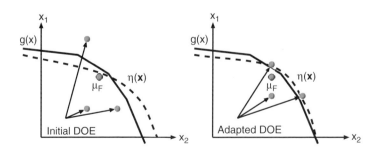

FIGURE 19.6 Simple adaptive response surface scheme.

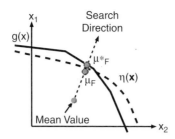

FIGURE 19.7 Simple error check for response surface.

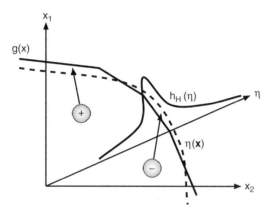

FIGURE 19.8 Correction to response surface using conditional sampling.

If computational resources are extremely limited so that a repetition of the design of experiment scheme is not feasible, it is recommended to at least check that the conditional mean $\boldsymbol{\mu}_F$ (which lies close to $\eta(\mathbf{x}) = 0$) is really close to the exact limit state $g(\mathbf{x}) = 0$. One possibility for that is to perform one single incremental analysis along the direction from the mean values of $\mathbf{X}$ to $\boldsymbol{\mu}_F$ resulting in a point $\boldsymbol{\mu}_F^*$ on the limit state (see Figure 19.7). If the distance between $\boldsymbol{\mu}_F$ and $\boldsymbol{\mu}_F^*$ is small, it may be concluded that the response surface is acceptable.

A somewhat different concept has been presented in [26]. Here, the concept of conditional sampling as introduced by [14] is utilized to generate random samples close to the approximate limit state $\eta(\mathbf{x}) = 0$. For this purpose, an importance sampling function $h(\eta)$ is introduced that allows the generation of suitable samples $\mathbf{x}^{(k)}$. For each of these samples, the exact limit state condition is checked. These samples are utilized to improve the estimate for the failure probability as obtained from the response surface approach. If the sample lies in the failure domain, the estimate for the failure probability is increased, if the sample lies in the safe domain, the estimate is decreased (see the positive and negative symbols in Figure 19.8).

19.5 Application to Reliability Problems

19.5.1 Linear Response Surface

As a first example, a truss-type structure under random static loads $F_1 = X_8$ and $F_2 = X_9$ is investigated (see Figure 19.9). The loads are assumed to be independent, whereby their distribution function is of the Gumbel type with mean $\mu = 2 \cdot 10^7$ N and standard deviation $\sigma = 2 \cdot 10^6$ N. The characteristic length of the structure is $l = 1$ m. Whereas the elasticity modulus E of each truss is assumed to be a deterministic quantity with $E = 2 \cdot 10^{11}$ N/m², the respective areas A_1 to A_7 are mutually independent random quantities that are log-normally distributed with mean $\mu = 0.01$ m² and standard deviation $\sigma = 0.001$ m² ($A_i = X_i$, $i = 1, 2, \ldots, 7$).

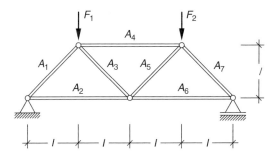

FIGURE 19.9 Truss-type structure.

In the following we are interested in the probability that the midspan deflection exceeds a value of 0.1 m. As the response surface we utilize a first-order polynomial of the form

$$\eta(\mathbf{u}) = \theta_0 + \sum_{i=1}^{9} \theta_i u_i \quad (19.55)$$

whereas u_i are the random variables x_i transformed to standard normal space, that is, $u_i = \Phi^{-1}[F(x_i)]$. As design of experiments we choose a 2_{III}^{9-5} fractional factorial design [27] in standard normal space, that is,

$$\mathbf{u}' = \begin{bmatrix} -1 & -1 & -1 & -1 & -1 & -1 & +1 & -1 & -1 \\ -1 & -1 & +1 & -1 & +1 & +1 & -1 & +1 & -1 \\ -1 & -1 & +1 & +1 & -1 & +1 & -1 & -1 & +1 \\ -1 & -1 & -1 & +1 & +1 & -1 & +1 & +1 & +1 \\ +1 & -1 & +1 & +1 & +1 & -1 & -1 & -1 & -1 \\ +1 & -1 & -1 & +1 & -1 & +1 & +1 & +1 & -1 \\ +1 & -1 & -1 & -1 & +1 & +1 & +1 & -1 & +1 \\ +1 & -1 & +1 & -1 & -1 & -1 & -1 & +1 & +1 \\ -1 & +1 & -1 & +1 & +1 & +1 & -1 & -1 & -1 \\ -1 & +1 & +1 & +1 & -1 & -1 & +1 & +1 & -1 \\ -1 & +1 & +1 & -1 & +1 & -1 & +1 & -1 & +1 \\ -1 & +1 & -1 & -1 & -1 & +1 & -1 & +1 & +1 \\ +1 & +1 & +1 & -1 & -1 & +1 & +1 & -1 & -1 \\ +1 & +1 & -1 & -1 & +1 & -1 & -1 & +1 & -1 \\ +1 & +1 & -1 & +1 & -1 & -1 & -1 & -1 & +1 \\ +1 & +1 & +1 & +1 & +1 & +1 & +1 & +1 & +1 \end{bmatrix} \quad (19.56)$$

An adaptation of the response surface is performed by determining in each step the design-point $\mathbf{u}^* = (u_1^*, u_2^*, \ldots, u_9^*)'$ for the approximating response surface $\eta(\mathbf{u})$ and repeat the experimental design centered at this design-point. As the start point we utilize the origin in standard normal space.

In Table 19.2 the first two steps of the adaptive response surface method as well as the converged result are displayed and compared with the result from first-order reliability method. As can be seen, the response surface provides a very good approximation of the true limit state function already in the second step (i.e., after the first adaptation). To verify that the chosen response surface model is appropriate, we

TABLE 19.2 Analysis of Variance (Truss-Type Structure)

	First-Order Reliability method	Response Surface		
		1. Step	2. Step	Last Step
s_t	—	$4.91 \cdot 10^{-4}$	$4.43 \cdot 10^{-3}$	$2.34 \cdot 0^{-3}$
s_r	—	$4.90 \cdot 10^{-4}$	$4.42 \cdot 10^{-3}$	$2.34 \cdot 10^{-3}$
s_e	—	$8.69 \cdot 10^{-7}$	$9.70 \cdot 10^{-6}$	$5.16 \cdot 10^{-6}$
r^2	—	0.99	0.99	0.99
r_A^2	—	0.99	0.99	0.99
F_0	—	375.83	304.02	302.23
u_1^*	−0.64	−1.44	−0.60	−0.62
u_2^*	−0.44	−1.02	−0.41	−0.43
u_3^*	0.00	0.00	0.00	0.00
u_4^*	−0.93	−2.07	−0.93	−0.93
u_5^*	0.00	−0.10	−0.08	−0.08
u_6^*	−0.44	−1.02	−0.41	−0.43
u_7^*	−0.64	−1.50	−0.66	−0.67
u_8^*	2.55	3.22	2.56	2.37
u_9^*	2.55	3.22	2.56	2.37
β	3.89	5.61	3.89	3.65

determine in each step the coefficient r^2, its adjusted form r_A^2, and the F-statistic. Both r^2 and r_A^2 show values approaching 1, clearly indicating that the response surface model is capable of reproducing the variation in the experimental data almost to its entirety. This is also supported by the F-statistic, because when choosing a confidence level of $\alpha = 1\%$, we can clearly reject the hypothesis, because always $F_0 > F_{0.01,9,6} = 7.98$.

19.5.2 Nonlinear Response Surface

Given is a tension bar with known load $T = 1$, random diameter X_1 and yield strength X_2. The dimensionless limit state function is given as

$$g(x_1, x_2) = \frac{\pi x_1^2}{4} x_2 - T = 0 \qquad (19.57)$$

The random variables X_1 and X_2 are independent. X_1 is log-normally distributed with distribution function $F(x) = \Phi[(\ln(x) - 1)/2]$, $x > 0$. X_2 obeys the Rayleigh distribution function $F(x) = 1 - \exp[-\pi x^2/(4\mu^2)]$, $x \geq 0$, with mean $\mu = 200$. The response surface model we want to use in the following is a first-order polynomial of the form

$$\eta(\mathbf{u}) = \theta_0 + \theta_1 u_1 + \theta_2 u_2 \qquad (19.58)$$

whereas u_1 and u_2, respectively, are the random variables x_1 and x_2 transformed to standard normal space (i.e., $u_i = \Phi^{-1}[F(x_i)]$). As design of experiments we utilize a 2^2-factorial design with one center run in standard normal space, that is,

$$\mathbf{u} = \begin{bmatrix} -1 & +1 & -1 & +1 & 0 \\ -1 & -1 & +1 & +1 & 0 \end{bmatrix} \qquad (19.59)$$

As the method for adaptation, we again determine in each step the design-point $\mathbf{u}^* = (u_1^*, u_2^*)'$ for the approximating response surface $\eta(\mathbf{u})$ and repeat the experimental design centered at this design-point. As the start point we utilize the origin in standard normal space.

When applying this scheme we perform again an accompanying analysis of variance; that is, we determine in each step the coefficient r^2, its adjusted form r_A^2, and the F-statistic. As can be seen from Table 19.3, the

TABLE 19.3 Analysis of Variance (Tension Bar)

	Original Scale			Transformed Scale		
	1. Step	2. Step	Last Step	1. Step	2. Step	Last Step
s_t	$6.60 \cdot 10^9$	$3.19 \cdot 10^8$	16.95	17.41	19.40	22.90
s_r	$5.04 \cdot 10^9$	$2.38 \cdot 10^8$	10.95	17.40	19.38	22.86
s_e	$1.56 \cdot 10^9$	$0.81 \cdot 10^8$	6.00	0.01	0.02	0.04
r^2	0.76	0.74	0.64	0.99	0.99	0.99
r_A^2	0.53	0.49	0.29	0.99	0.99	0.99
F_0	3.23	2.92	1.82	2044.73	857.02	631.24
u_1^*	−0.68	−1.30	−3.54	−4.10	−3.75	−3.41
u_2^*	−0.37	−0.79	−3.51	−1.21	−1.73	−2.24
β	0.77	1.52	4.98	4.27	4.13	4.08

above-mentioned measures indicate a nonsatisfactory performance of the chosen response surface model. Not only do the values of r^2 and r_A^2 differ considerably from each other, but they are also clearly different from the optimal value of 1. This indicates that the model is not able to reproduce appropriately the variation in the experimental data. Furthermore, when testing the null hypothesis H_0 of Equation 19.28 at a confidence level of $\alpha = 1\%$, we cannot reject the hypothesis, because clearly $F_0 < F_{0.01,2,2} = 99$. In other words, the chosen response surface model is not appropriate for the experimentally gained data. This can also be noticed when investigating Figure 19.10, which displays the limit state function $g(\mathbf{u})$ in standard normal space for different values of T. As can be seen, the limit state function is highly nonlinear.

Nevertheless, when taking the logarithm of the response values, a linear response surface would represent a good approximation of the true limit state function. Therefore, a more adequate response surface model would be of the form

$$\eta(\mathbf{u}) = \theta_0 \exp(\theta_1 u_1 + \theta_2 u_2) - 1 \tag{19.60}$$

which can be transformed to its linear form by

$$\tilde{\eta}(\mathbf{u}) = \ln(\eta + 1) = \ln(\theta_0) + \theta_1 u_1 + \theta_2 u_2 = \hat{\theta}_0 + \theta_1 u_1 + \theta_2 u_2 \tag{19.61}$$

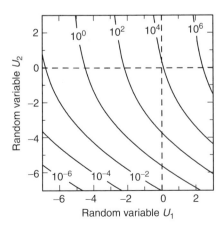

FIGURE 19.10 Limit state function $g(\mathbf{u})$ in standard normal random variable space for different values of T.

When using this response surface model, all measures displayed in Table 19.3 show satisfactory values. Moreover, already in the second step (i.e., after the first adaptation), a reasonable approximation of the exact result can be achieved. The response surface method converges to the point $\mathbf{u}^* = (-3.41, -2.24)'$, which relates to a reliability index $\beta = 4.08$. These results are an excellent approximation of the true design-point $\mathbf{u}^* = (-3.46, -2.29)'$ and the true reliability index $\beta = 4.15$. It should also be noted that if we would have dismissed the indicators in the original scaling as being not relevant, the response surface analysis would not provide a satisfactory result, as can be seen from Table 19.3.

19.5.3 Nonlinear Finite Element Structure

A simple three-dimensional steel frame subjected to three random loadings is considered as shown in Figure 19.11. The three-dimensional frame is modeled by 24 physically nonlinear beam elements (linear elastic-ideally plastic material law, elasticity modulus $E = 2.1 \cdot 10^{11}$ N/m², yield stress $\sigma_Y = 2.4 \cdot 10^8$ N/m²). The cross section for the girder is a box (width 0.2 m, height 0.15 m, wall thickness 0.005 m) and the columns are I-sections (flange width 0.2 m, web height 0.2 m, thickness 0.005 m). The columns are fully clamped at the supports. The static loads acting on the system p_z, F_x, and F_y are assumed to be random variables. Their respective properties are given in Table 19.4. The failure condition is given by total or partial collapse of the structure. Numerically, this is checked either by tracking the smallest eigenvalue of the global tangent stiffness matrix (it becomes 0 at the collapse load) or by failure in the global Newton iteration indicating loss of equilibrium. Because this type of collapse analysis is typically based on a discontinuous function (convergence vs. nonconvergence), it is imperative that the support points for the response surface be located exactly at the limit state. A bisection procedure is utilized to determine collapse loads with high precision (to the accuracy of 1% of the respective standard deviation).

The geometry of the limit state separating the safe from the failure domain is shown in Figure 19.12. The limit points were obtained from directional sampling using 500 samples. The size of the dots indicates the visual distance from the viewer. It can be easily seen that there is considerable interaction between the random variables F_x and F_y at certain levels. The region of most importance for the probability of failure (this is where most of the points from the directional sampling are located) is essentially flat, and mainly governed by the value of F_y. This is clearly seen in Figure 19.13. The box in this figure indicates a space of ±5 standard deviations around the mean values. The probability of failure obtained from directional sampling is $P(F) = 2.3 \cdot 10^{-4}$, with an estimation error of 20%.

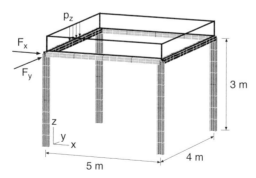

FIGURE 19.11 Three-dimensional steel frame structure.

TABLE 19.4 Random Variables Used in Three-Dimensional Frame Analysis

Random Variable	Mean Value	Standard Deviation	Distribution Type
p_z [kN/m]	12.0	0.8	Gumbel
F_x [kN]	30.0	2.4	Gumbel
F_y [kN]	40.0	3.2	Gumbel

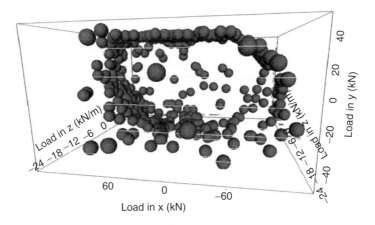

FIGURE 19.12 Visualization of limit state function $g(\mathbf{x})$.

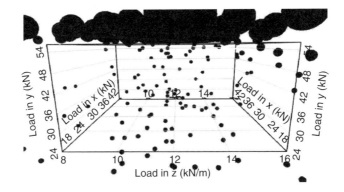

FIGURE 19.13 Visualization of limit state function $g(\mathbf{x})$, details near mean.

A saturated quadratic scheme including pairwise interactions is utilized for the initial layout of the experimental design scheme. The support points thus generated are interpreted as direction vectors along which all loads are incremented. Starting from the mean values, and incrementing along this direction, lead to a set of nine support points on the limit state function. These support points (see Table 19.5) have a function value of $g(\mathbf{x}) = 0$. By adding the mean value as first support point with a function value of $g(\mathbf{x}) = 1$, a quadratic response surface can be defined. Considering lines 2 through 7 in Table 19.5, it was decided to consider combination terms in which all variables are incremented up from the mean. This leads to the final three support points given in lines 7 through 10 of Table 19.5.

TABLE 19.5 Support Points for Response Surface

i-th Support Point $\mathbf{x}^{(i)}$	p_z [kN/m]	F_x [kN]	F_y [kN]	$g(\mathbf{x}^{(i)})$
1	12.000	30.000	40.000	1
2	21.513	30.000	40.000	0
3	−21.516	30.000	40.000	0
4	12.000	113.180	40.000	0
5	12.000	−81.094	40.000	0
6	12.000	30.000	59.082	0
7	12.000	30.000	−59.087	0
8	19.979	109.790	40.000	0
9	13.527	30.000	59.084	0
10	12.000	45.275	59.094	0

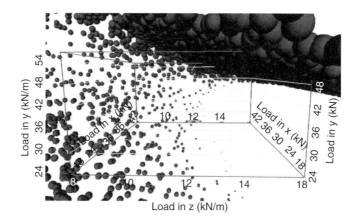

FIGURE 19.14 Visualization of response surface $\eta(\mathbf{x})$, details near mean.

A Monte Carlo simulation based on this quadratic surface is carried out. The sampling scheme utilized is adaptive sampling [25], which iteratively updates the importance sampling density. The resulting probability of failure from three consecutive runs with 3000 samples each was found to be $P(F) = 1.7 \cdot 10^{-4}$, with an estimation error of 2%. Alternatively, directional sampling with 1000 samples was carried out, yielding the same value for $P(F)$. As a by-product, simple visualizations of the response surface can easily be obtained from the directional sampling procedure because this procedure calculates the location of a failure point in a randomly simulated direction. Figure 19.14 shows that the response surface is almost flat in the important region. Consequently, the resulting failure probability matches the exact result very well. The conditional mean vector of the samples in the failure domain is found to be

$$\boldsymbol{\mu}_F = (13.457, 30.160, 59.977)' \tag{19.62}$$

and an incremental search in the direction from the mean values to $\boldsymbol{\mu}_F$ yields a point at the exact limit state

$$\boldsymbol{\mu}_F^* = (13.391, 30.153, 59.080)' \tag{19.63}$$

so that the distance δ between these two points (measured in multiples of the standard deviations) is $\delta = 0.29$. This can be considered reasonably small.

19.6 Recommendations

Based on the examples shown above, the following guidelines and recommendations for the successful application of response surface methods in structural reliability analyses can be given:

- Utilize all available knowledge about the structural behavior, especially sensitivity analysis to eliminate unimportant or unnecessary random variables. In this context, the following hints may be helpful:
  - Use engineering judgment to eliminate random variables whose coefficient of variation is significantly smaller compared to others.
  - Use simple parameter variations, one variable at a time in the magnitude of 1 to 2 standard deviations to assess the sensitivity.
  - Reduce the structural model to include only relevant failure mechanisms.

- Choose an experimental design scheme that is compatible with the type of response surface you wish to utilize.

- Make sure that the experimental design scheme yields support points well within the range of definition of the random variables.
- In general, highly redundant experimental designs do not improve the results significantly over saturated or moderately redundant designs.
- Utilize a simple experimental and a simple response surface with the possibility of adaptation, rather than a complex design of experiments and a complex response surface.
- Make sure that the response surface obtained is "well-behaved" in the sense that its mathematical properties do not interfere with reliability computation. This is of particular importance when using a calculation technique based on the first order reliability method.
- Allow for sufficient computational power to perform some adaptation of the response surface or at least simple error checking.

References

1. Rackwitz, R., *Response Surfaces in Structural Reliability*, Berichte zur Zuverlässigkeitstheorie der Bauwerke, Heft 67, München, 1982.
2. Box, G.E.P. and Wilson, K.B., On the experimental attainment of optimum conditions, *Journal of the Royal Statistical Society, Series B*, 13, 1–45, 1951.
3. Box, G.E.P. and Draper, N.R., A basis for the selection of a response surface design, *Journal of the American Statistical Association*, 54, 622–654, 1959.
4. Hill, W.J. and Hunter, W.G., A review of response surface methodology: a literature survey, *Technometrics*, 8, 571–590, 1966.
5. Mead, R. and Pike, D.J., A review of response surface methodology from a biometric viewpoint, *Biometrics*, 31, 803–851, 1975.
6. Myers, R.H., Khuri, A.I., and Carter, Jr., W.H., Response surface methodology: 1966–1988, *Technometrics*, 31, 137–157, 1989.
7. Myers, R.H., Response surface methodology—Current status and future directions, *Journal of Quality Technology*, 31, 30–44, 1999.
8. Box, G.E.P. and Draper, N.R., *Empirical Model-Building and Response Surfaces*, Wiley, New York, 1987.
9. Khuri, A.I. and. Cornell, J.A., *Response Surfaces: Designs and Analyses*, Dekker, New York, 1996.
10. Myers, R.H. and Montgomery, D.C., *Response Surface Methodology: Process and Product Optimization Using Designed Experiments*, Wiley, New York, 2002.
11. Wong, F.S., Slope reliability and response surface method, *Journal of Geotechnical Engineering*, 111, 32–53, 1985.
12. Bucher, C.G. and Bourgund, U., A fast and efficient response surface approach for structural reliability problems, *Structural Safety*, 7, 57–66, 1990.
13. Faravelli, L., Response-surface approach for reliability analysis, *Journal of Engineering Mechanics, ASCE*, 115, 2763–2781, 1989.
14. Ouypornprasert, W., Bucher, C.G., and Schuëller, G.I., On the application of conditional integration in structural reliability analysis, in *Proc. 5th Int. Conf. on Structural Safety and Reliability*, San Francisco, Ang, A.H.-S., Shinozuka, M., and Schuëller, G.I., Eds., ASCE, New York, 1989, 1683–1689.
15. Kim, S.-H. and Na, S.-W., Response surface method using vector projected sampling points, *Structural Safety*, 19, 3–19, 1997.
16. Zheng, Y. and Das, P.K., Improved response surface method and its application to stiffened plate reliability analysis, *Engineering Structures*, 22, 544–551, 2000.
17. Brenner C.E. and Bucher, C., A contribution to the SFE-based reliability assessment of nonlinear structures under dynamic loading, *Probabilistic Engineering Mechanics*, 10, 265–273, 1995.
18. Ghanem, R. and Spanos, P.D., *Stochastic Finite Elements — A Spectral Approach*, Springer, New York, 1991.

19. Guan, X.L. and Melchers, R.E., Multitangent-plane surface method for reliability calculation, *Journal of Engineering Mechanics, ASCE*, 123, 996–1002, 1997.
20. Roos, D., Bucher, C., and Bayer, V., Polyhedra response surfaces for structural reliability assessment, in *Proc. 8th Int. Conf. on Applications of Statistics and Probability*, Sydney, Melchers, R.E. and Stewart, M.G., Eds., Balkema, Rotterdam, 2000, 109–115.
21. Bjerager, P., Probability integration by directional simulation, *Journal of Engineering Mechanics, ASCE*, 114, 1285–1302, 1988.
22. Böhm, F. and Brückner-Foit, A., On criteria for accepting a response surface model, *Probabilistic Engineering Mechanics*, 7, 183–190, 1992.
23. Yao, T.H.-J. and Wen, Y.K., Response surface method for time-variant reliability analysis, *Journal of Structural Engineering, ASCE*, 122, 193–201, 1996.
24. Hohenbichler, M. and Rackwitz, R., Nonnormal dependent vectors in structural safety, *Journal of the Engineering Mechanics Division, ASCE*, 107, 1227–1238, 1981.
25. Bucher, C.G., Adaptive sampling — An iterative fast Monte Carlo procedure, *Structural Safety*, 5, 119–126, 1988.
26. Schuëller, G.I. and Bucher, C.G., Computational stochastic structural analysis — A contribution to the software development for the reliability assessment of structures under dynamic loading, *Probabilistic Engineering Mechanics*, 6, 134–138, 1991.
27. Montgomery, D.C., *Design and Analysis of Experiments*, Wiley, New York, 1997.

20
Stochastic Simulation Methods for Engineering Predictions

Dan M. Ghiocel
Ghiocel Predictive Technologies, Inc.

20.1 Introduction 20-1
20.2 One-Dimensional Random Variables 20-3
   Uniform Random Numbers • Inverse Probability Transformation • Sampling Acceptance-Rejection • Bivariate Transformation • Density Decomposition
20.3 Stochastic Vectors with Correlated Components 20-5
   Gaussian Vectors • NonGaussian Vectors
20.4 Stochastic Fields (or Processes) 20-8
   One-Level Hierarchical Simulation Models • Two-Level Hierarchical Simulation Models
20.5 Simulation in High-Dimensional Stochastic Spaces 20-21
   Sequential Importance Sampling (SIS) • Dynamic Monte Carlo (DMC) • Computing Tail Distribution Probabilities • Employing Stochastic Linear PDE Solutions • Incorporating Modeling Uncertainties
20.6 Summary 20-31

20.1 Introduction

The term "simulation" comes from the Latin word *simulatio*, which means to imitate a phenomenon or a generic process. In the language of mathematics, the term "simulation" was used for the first time during the World War II period at the Los Alamos National Laboratory by the renowned mathematicians Von Neumann, Ulam, Metropolis and physicist Fermi, in the context of their nuclear physics research for the atomic bomb. About the same time they also introduced an exotic term in mathematics, namely "Monte Carlo" methods. These were defined as numerical methods that use artificially generated statistical selections for reproducing complex random phenomena and for solving multidimensional integral problems. The name "Monte Carlo" was inspired by the famous Monte Carlo casino roulette, in France, that was the best available generator of uniform random numbers.

In scientific applications, stochastic simulation methods based on random sampling algorithms, or Monte Carlo methods, are used to solve two types of problems: (1) to generate random samples that belong to a given stochastic model, or (2) to compute expectations (integrals) with respect to a given distribution. The expectations might be probabilities, or discrete vectors or continuous fields of probabilities or in other words, probability distributions. It should be noted that if the problem is solved, so that random samples are available, then the solution to the problem, to compute expectations, becomes trivial because expectations can be approximated by the statistical averaging of the random samples.

The power of Monte Carlo methods manifests visibly for multidimensional probability integration problems, for which the typical deterministic integration algorithms are extremely inefficient. For very low-dimensional problems, in one dimension or two, Monte Carlo methods are too slowly convergent and therefore there is no practical interest in employing them. For a given number of solution points N, the Monte Carlo estimator converges to the exact solution with rate of the order $O(N^{-1/2})$ in comparison with some deterministic methods that converge much faster with rates up to the order $O(N^{-4})$ or even $O(\exp(-N))$. Unfortunately, for multidimensional problems, the classical deterministic integration schemes based on regular discretization grids fail fatally. The key difference between stochastic and deterministic methods is that the Monte Carlo methods are, by their nature, meshless methods, while the deterministic integration methods are regular grid-based methods. The problem is that the classical grid-based integration methods scale very poorly with the space dimensionality.

The fact that the Monte Carlo estimator convergence rate is independent of the input space dimensionality makes the standard Monte Carlo method very popular for scientific computing applications. Although the statistical convergence rate of the Monte Carlo method is independent of the input space dimensionality, there are still two difficulties to address that are dependent on the input space dimensionality: (1) how to generate uniformly distributed random samples in the prescribed stochastic input domain, and (2) how to control the variance of Monte Carlo estimator for highly "nonuniform" stochastic variations of the integrand in the prescribed input domain. These two difficulties are becoming increasingly important as the input space dimensionality increases. One can imagine a Monte Carlo integration scheme as a sort of a random fractional factorial sampling scheme or random quadrature that uses a refined cartesian grid of the discretized multidimensional input domain with many unfilled nodes with data, vs. the deterministic integration schemes or deterministic quadratures that can be viewed as complete factorial sampling schemes using cartesian grids with fully filled nodes with data. Thus, a very important aspect to get good results while using Monte Carlo is to ensure as much as possible a *uniform filling* of the input space domain grid with statistically independent solution points so that the potential local subspaces that can be important are not missed. The hurting problem of the standard Monte Carlo method is that for high-dimensional spaces, it is difficult to get uniform filling of the input space. The consequence of a nonuniform filling can be a sharp increase in the variance of the Monte Carlo estimator, which severely reduces the attraction for the method.

The classical way to improve the application of Monte Carlo methods to high-dimensional problems is to partition the stochastic input space in subdomains with uniform statistical properties. Instead of dealing with the entire stochastic space, we deal with subdomains. This is like defining a much coarser grid in space to work with. The space decomposition in subdomains may accelerate substantially the statistical convergence for multidimensional problems. Stratified sampling and importance sampling schemes are different sampling weighting schemes based on stochastic space decomposition in partitions. Using space decomposition we are able to control the variation of the estimator variance over the input domain by focusing on the most important stochastic space regions (mostly contributing to the integrand estimate) and adding more random data points in those regions.

Another useful way to deal with the high dimensionality of stochastic space is to simulate correlated random samples, instead of statistically independent samples, that describe a random path or walk in stochastic space. If the random walk has a higher attraction toward the regions with larger probability masses, then the time spent in a region vs. the time (number of steps) spent in another region is proportional to the probability mass distributions in those regions. By withdrawing samples at equal time intervals (number of steps), we can build an approximation of the prescribed joint probability distribution. There are two distinct classes of Monte Carlo methods for simulating random samples from a prescribed multivariate probability distribution: (1) static Monte Carlo (SMC), which is based on the generation of independent samples from univariate marginal densities or conditional densities; and (2) dynamic Monte Carlo (DMC), which is based on spatially correlated samples that describe the random evolutions of a *fictitious* stochastic dynamic system that has a stationary probability distribution identical with the prescribed joint probability distribution. Both classes of Monte Carlo methods are discussed herein.

This chapter focuses on the application of Monte Carlo methods to simulate random samples of a variety of stochastic models — from the simplest stochastic models described by one-dimensional random variable models, to the most complex stochastic models described by hierarchical stochastic network-like models. The chapter also includes a solid section on high-dimensional simulation problems that touches key numerical issues including sequential sampling, dynamic sampling, computation of expectations and probabilities, modeling uncertainties, and sampling using stochastic partial-differential equation solutions. The goal of this chapter is to provide readers with the conceptual understanding of different simulation methods without trying to overwhelm them with all the implementation details that can be found elsewhere as cited in the text.

20.2 One-Dimensional Random Variables

There are a large number of simulation methods available to generate random variables with various probability distributions. Only the most popular of these methods are reviewed in this chapter: (1) inverse probability transformation, (2) sampling acceptance-rejection, (3) bivariate functional transformation, and (4) density decomposition. The description of these methods can be found in many textbooks on Monte Carlo simulations [1–5]. Herein, the intention is to discuss the basic concepts of these methods and very briefly describe their numerical implementations. In addition to these classical simulation methods, there is a myriad of many other methods, most of them very specific to particular types of problems, that have been developed over the past decades but for obvious reasons are not included.

20.2.1 Uniform Random Numbers

The starting point for any stochastic simulation technique is the construction of a reliable uniform random number generator. Typically, the random generators produce random numbers that are uniformly distributed in the interval [0, 1]. The most common uniform random generators are based on the linear congruential method [1–5].

The quality of the uniform random number generator reflects on the quality of the simulation results. It is one of the key factors that needs special attention. The quality of a random generator is measured by (1) the uniformity in filling with samples the prescribed interval and (2) its period, which is defined by the number of samples after which the random generator restarts the same sequence of independent random numbers all over again. Uniformity is a crucial aspect for producing accurate simulations, while the generator periodicity is important only when a large volume of samples is needed. If the volume of samples exceeds the periodicity of the random number generator, then the sequence of numbers generated after this is perfectly correlated with initial sequence of numbers, instead of being statistically independent. There are also other serial correlation types that can be produced by different number generators [4]. If we need a volume of samples for simulating a rare random event that is several times larger than the generator periodicity, then the expected simulation results will be wrong due to the serial correlation of the generated random numbers.

An efficient way to improve the quality of the generated sequence of uniform numbers and break their serial correlation is to combine the outputs of few generators into a single uniform random number that will have a much larger sequence periodicity and expectedly better space uniformity. As general advice to the reader, it is always a good idea to use statistical testing to check the quality of the random number generator.

20.2.2 Inverse Probability Transformation

Inverse probability transformation (IPT) is based on a lemma that says that for any random variable x with the probability distribution F, if we define a new random variable y with the probability distribution F^{-1} (i.e., the inverse probability transformation of F), then $y = F^{-1}(u)$ has a probability distribution F.

The argument *u* of the inverse probability transformation stands for a uniformly distributed random variable.

The general algorithm of the IPT method for any type of probability distribution is as follows:

Step 1: Initialize the uniform random number generator.
Step 2: Implement the algorithm for computing F^{-1}.
Step 3: Generate a uniform random number *u* in the interval [0, 1].
Step 4: Compute the generated deviate *x* with the distribution *F* by computing $x = F^{-1}(u)$.

Because the probability or cumulative distribution functions (CDFS) are monotonic increasing functions, the IPT method can be applied to any type of probability distribution, continuous or discrete, analytically or numerically defined, with any probability density function (PDF) shapes, from symmetric to extremely skewed, from bell-shaped to multimodal-shaped.

Very importantly, IPT can be also used to transform a sequence of Gaussian random deviates into a sequence of nonGaussian random deviates for any arbitrarily shaped probability distribution. A generalization of the IPT method, called the transformation method, is provided by Press et al. [4].

20.2.3 Sampling Acceptance-Rejection

Sampling acceptance-rejection (SAR) uses an auxiliary, trial, or proposal density to generate a random variable with a prescribed distribution. SAR is based on the following lemma: If *x* is a random variable with the PDF *f*(*x*) and *y* is another variable with the PDF *g*(*y*) and the probability of *f*(*y*) = 0, then the variable *x* can generated by first generating variable *y* and assuming that there is a positive constant α such as $\alpha \geq f(x)/g(x)$. If *u* is a uniform random variable defined on [0, 1], then the PDF of *y* conditioned on relationship $0 \leq u \leq f(y)/\alpha g(y)$ is identical with *f*(*x*).

A typical numerical implementation of the SAR method for simulating a variable *x* with the PDF *f*(*x*) is as follows:

Step 1: Initialize the uniform random number generator and select constant α.
Step 2: Generate a uniform number *u* on [0, 1] and a random sample of *y*.
Step 3: If $u > f(y)/\alpha g(y)$, then reject the pair (*u*, *y*) and go back to the previous step.
Step 4: Otherwise, accept the pair (*u*, *y*) and compute *x* = *y*.

It should be noted that SAR can be easily extended to generate random vectors and matrices with independent components.

The technical literature includes a variety of numerical implementations based on the principle of SAR [6]. Some of the newer implementations are the weighted resampling and sampling importance resampling (SIR) schemes [6, 7] and other adaptive rejection sampling [8].

20.2.4 Bivariate Transformation

Bivariate transformation (BT) can be viewed as a generalization of the IPT method for bivariate probability distributions. A well-known application of BT is the Box-Muller method for simulating standard Gaussian random variables [9]. Details are provided in [4–6].

20.2.5 Density Decomposition

Density decomposition (DD) exploits the fact that an arbitrary PDF, *f*(*x*), can be closely approximated in a linear combination elementary density function as follows:

$$f(x) = \sum_{i=1}^{N} p_i g_i(x) \qquad (20.1)$$

in which $0 \leq p_i < 1$, $1 \leq i \leq N$, and $\sum_{i=1}^{N} p_i = 1$.

The typical implementation of DD for a discrete variable x is as follows:

Step 1: Initialize for the generation of a uniform random number and a variable z_i with the PDF $g_i(x)$. This implies that $x = z_i$ with the probability p_i.
Step 2: Compute the values $G_i(x) = \sum_{j=1}^{i} p_j$ for all i values, $i = 1, N$.
Step 3: Generate a uniform random variable u on $[0, 1]$.
Step 4: Loop $j = 1, N$ and check if $u < G_i(x)$. If not, go to the next step.
Step 5: Generate z_j and assign $x = z_j$.

It should be noted that the speed of the above algorithm is highest when the p_i values are ordered in descending order. It should be noted that DD can be combined with SAR. An example of such a combination is the Butcher method [10] for simulating normal deviates.

20.3 Stochastic Vectors with Correlated Components

In this section the numerical simulation of stochastic vectors with independent and correlated components is discussed. Both Gaussian and nonGaussian vectors are considered.

20.3.1 Gaussian Vectors

A multivariate Gaussian stochastic vector $\mathbf{x}$ of size m with a mean vector $\boldsymbol{\mu}$ and a covariance matrix $\boldsymbol{\Sigma}$ is completely described by the following multidimensional joint probability density function (JPDF):

$$f_x(\boldsymbol{\mu}, \boldsymbol{\Sigma}) = \frac{1}{(2\pi)^{\frac{m}{2}} (\det \boldsymbol{\Sigma})^{\frac{1}{2}}} \exp\left[-\frac{1}{2}(\mathbf{x}-\boldsymbol{\mu})^T \boldsymbol{\Sigma}^{-1} (\mathbf{x}-\boldsymbol{\mu})\right] \quad (20.2)$$

The probability distribution of vector $\mathbf{x}$ is usually denoted as $N(\boldsymbol{\mu}, \boldsymbol{\Sigma})$. If the Gaussian vector is a standard Gaussian vector, $\mathbf{z}$, with a zero mean, $\mathbf{0}$, and a covariance matrix equal to an identity matrix, $\mathbf{I}$, then its probability distribution is $N(\mathbf{0}, \mathbf{I})$. The JPDF of vector $\mathbf{z}$ is defined by:

$$f_z(\mathbf{0}, \mathbf{I}) = \frac{1}{(2\pi)^{\frac{m}{2}} (\det \boldsymbol{\Sigma})^{\frac{1}{2}}} \exp\left(-\frac{1}{2}\mathbf{z}^T \mathbf{z}\right) = \prod_{i=1}^{m} \frac{1}{\sqrt{2\pi}} \exp\left(-\frac{1}{2}z_i^2\right) \quad (20.3)$$

To simulate a random sample of a Gaussian stochastic vector $\mathbf{x}$ (with correlated components) that has the probability distribution $N(\boldsymbol{\mu}, \boldsymbol{\Sigma})$, a two computational step procedure is needed:

Step 1: Simulate a standard Gaussian vector $\mathbf{z}$ (with independent random components) having a probability distribution $N(\mathbf{0}, \mathbf{I})$. This step can be achieved using standard routines for simulating independent normal random variables.
Step 2: To compute the vector $\mathbf{x}$ use the linear matrix equation

$$\mathbf{x} = \boldsymbol{\mu} + \mathbf{S}\mathbf{z} \quad (20.4)$$

where matrix $\mathbf{S}$ is called the *square-root matrix* of the positively definite covariance matrix $\boldsymbol{\Sigma}$. The matrix $\mathbf{S}$ is a lower triangular matrix that is computed using the Choleski decomposition that is defined by the equation $\mathbf{SS}^T = \boldsymbol{\Sigma}$.

Another situation of practical interest is to generate a Gaussian vector $\mathbf{y}$ with a probability distribution $N(\boldsymbol{\mu}_y, \boldsymbol{\Sigma}_{yy})$ that is conditioned on an input Gaussian vector $\mathbf{x}$ with a probability distribution $N(\boldsymbol{\mu}_x, \boldsymbol{\Sigma}_{xx})$. Assuming that the augmented vector $[\mathbf{x}, \mathbf{y}]^T$ has the mean $\boldsymbol{\mu}$ and the covariance matrix $\boldsymbol{\Sigma}$ defined by

$$\boldsymbol{\mu} = \begin{pmatrix} \boldsymbol{\mu}_y \\ \boldsymbol{\mu}_x \end{pmatrix} \quad \boldsymbol{\Sigma} = \begin{pmatrix} \boldsymbol{\Sigma}_{yy} & \boldsymbol{\Sigma}_{yx} \\ \boldsymbol{\Sigma}_{xy} & \boldsymbol{\Sigma}_{xx} \end{pmatrix} \quad (20.5)$$

then the statistics of the conditional vector **y** are computed by the matrix equations:

$$\mu_y = \mu_y + \Sigma_{yx} \cdot \Sigma_{xx}^{-1}(\mathbf{x} - \mu_x)$$
$$\Sigma_{yy} = \Sigma_{yy} - \Sigma_{yx}\Sigma_{xx}^{-1}\Sigma_{xy} \quad (20.6)$$

It is interesting to note that the above relationships define a stochastic condensation procedure of reducing the size of a stochastic vector from the size of the total vector $[\mathbf{y}, \mathbf{x}]^T$ with probability distribution $N(\mu, \Sigma)$ to a size of the reduced vector **y** with a probability distribution $N(\mu_y, \Sigma_{yy})$. The above simulation procedures employed for Gaussian vectors can be extended to nonGaussian vectors, as shown in the next subsection.

An alternate technique to the Choleski decomposition for simulating stochastic vectors with correlated components is the popular principal component analysis. Principal component analysis (PCA) is based on eigen decomposition of the covariance matrix. If PCA is used, then the matrix equation in the Step 2 shown above is replaced by the following matrix equation

$$\mathbf{x} = \mu + \Phi\lambda^{1/2}\mathbf{z} \quad (20.7)$$

where λ and Φ are the eigenvalue and eigenvector matrices, respectively, of the covariance matrix.

20.3.2 NonGaussian Vectors

A nonGaussian stochastic vector is completely defined by its JPDF. However, most often in practice, a nonGaussian vector is only partially defined by its second-order moments (i.e., the mean vector and the covariance matrix) and its marginal probability distribution (MCDF) vector. This definition loses information about the high-order statistical moments that are not defined.

These partially defined nonGaussian vectors form a special class of nonGaussian vectors called translation vectors. Thus, the translation vectors are nonGaussian vectors defined by their second-order moments and their marginal distributions. Although the translation vectors are not completely defined stochastic vectors, they are of great practicality. First, because in practice, most often we have statistical information limited to second-order moments and marginal probability distributions. Second, because they capture the most critical nonGaussian aspects that are most significantly reflected in the marginal distributions. Of great practical benefit is that translation vectors can be easily mapped into Gaussian vectors, and therefore can be easily handled and programmed.

To generate a nonGaussian (translation) stochastic vector **y** with a given covariance matrix Σ_{yy} and a marginal distribution vector **F**, first a Gaussian image vector **x** with a covariance matrix Σ_{xx} and marginal distribution vector $\Phi(\mathbf{x})$ is simulated. Then, the original nonGaussian vector **y** is simulated by applying IPT to the MCDF vector **F** as follows:

$$\mathbf{y} = \mathbf{F}^{-1}\Phi(\mathbf{x}) = \mathbf{g}(\mathbf{x}) \quad (20.8)$$

However, for simulating the Gaussian image vector we need to define its covariance matrix Σ_{xx} as a transform of the covariance matrix Σ_{yy} of the original nonGaussian vector. Between the elements of the scaled covariance matrix or correlation coefficient matrix of the original nonGaussian vector, $\rho_{yi,yj}$, and the elements of the scaled covariance or correlation coefficient of the Gaussian image vector, $\rho_{xi,xj}$, there is the following relation:

$$\rho_{yi,yj} = \frac{1}{\sigma_{yi}\sigma_{yj}} \int_{-\infty}^{\infty}\int_{-\infty}^{\infty} \left[F_i^{-1}\Phi(x_i) - \mu_{yi}\right]\left[F_j^{-1}\Phi(x_j) - \mu_{yj}\right]\phi(x_i, x_j)\,dx_i dx_j \quad (20.9)$$

where the bivariate Gaussian probability density is defined by:

$$\phi(x_i, x_j) = \frac{1}{2\pi(1-\rho_{xi,xj}^2)^{0.5} \sigma_{xi}\sigma_{xj}} \exp\left(-\frac{1}{2}\frac{(x_i-\mu_{xi})^2/\sigma_{xi}^2 - 2\rho_{xi,xj}/\sigma_{xi}\sigma_{xj} + (x_j-\mu_{xj})^2/\sigma_{xj}^2}{(1-\rho_{xi,xj}^2)}\right) \quad (20.10)$$

Two probability transformation options are attractive: (1) the components of the Gaussian image stochastic vector have means and variances equal to those of the original nonGaussian vector (i.e., $\mu_y = \mu_x$ and $\sigma_y = \sigma_x$), or (2) the Gaussian image vector is standard Gaussian vector (i.e., $\mu_x = 0$ and $\sigma_x = 1$). Depending on the selected option, the above equations take a simpler form.

Thus, problem of generating nonGaussian vectors is a four-step procedure for which the two steps are identical with those used for generating a Gaussian vector:

Step 1: Compute the covariance matrix of Gaussian vector **x** using Equation 20.9.
Step 2: Generate a standard Gaussian vector **z**.
Step 3: Generate a Gaussian vector **x** using Equation 20.4 or Equation 20.7.
Step 4: Simulate a nonGaussian vector **y** using the following inverse probability transformation of the MCDF, $\mathbf{y} = \mathbf{F}^{-1}\Phi(\mathbf{x})$.

Sometimes in practice, the covariance matrix transformation is neglected, assuming that the correlation coefficient matrix of the original nonGaussian vector and that of the image Gaussian vector are identical. This assumption is wrong and may produce a violation of the physics of the problem. It should be noted that calculations of correlation coefficients have indicated that if $\rho_{xi,xj}$ is equal to 0 or 1, then $\rho_{yi,yj}$ is also equal to 0 or 1, respectively. However, when $\rho_{xi,xj}$ is −1, then $\rho_{yi,yj}$ is not necessarily equal to −1. For significant negative correlations between vector components, the effects of covariance matrix transformation are becoming significant, especially for nonGaussian vectors that have skewed marginal PDF shapes. Unfortunately, this is not always appreciated in the engineering literature. There are a number of journal articles on probabilistic applications for which the significance of covariance matrix transformation is underestimated; for example, the application of the popular Nataf probability model to nonGaussian vectors with correlated components neglects the covariance matrix transformation. Grigoriu [11] showed a simple example of a bivariate lognormal distribution for which the lowest value of the correlation coefficient is −0.65, and not −1.00, which is the lowest value correlation coefficient for a bivariate Gaussian distribution.

For a particular situation, of a nonGaussian component y_i and a Gaussian component x_j, the correlation coefficient $\rho_{yi,xj}$ can be computed by

$$\rho_{yi,xj} = \frac{1}{\sigma_{yi}\sigma_{xj}} \int_{-\infty}^{\infty}\int_{-\infty}^{\infty} [F^{-1}\Phi(x_i) - \mu_{yi}](x_j - \mu_{xj})\phi(x_i, x_j)dx_i dx_j \quad (20.11)$$

As shown in the next section, the above probability transformation simulation procedure can be extended from nonGaussian vectors to multivariate nonGaussian stochastic fields. The nonGaussian stochastic fields that are partially defined by their mean and covariance functions and marginal distributions are called translation stochastic fields.

Another way to generate nonGaussian stochastic vectors is based on the application of the SAR method described for one-dimensional random variables to random vectors with correlated components. This application of SAR to multivariate cases is the basis of the Metropolis-Hastings algorithm that is extremely popular in the Bayesian statistics community. Importantly, the Metropolis-Hastings algorithm [12] is not limited to translation vectors. The basic idea is to sample directly from the JPDF of the nonGaussian vector using an adaptive and important sampling strategy based on the application of SAR. At each simulation step, random samples are generated from a simple proposal or trial JPDF that is different than the target JPDF and then weighted in accordance with the important ratio. This produces dependent samples that represent a Markov chain random walk in the input space. The Metropolis-Hastings algorithm is the cornerstone of the Markov chain Monte Carlo (MCMC) simulation discussed in Section 20.5 of this chapter.

To move from vector sample or state $\mathbf{x}^t$ to state $\mathbf{x}^{t+1}$, the following steps are applied:

Step 1: Sample from the trial density, $\mathbf{y} \approx q(\mathbf{x}^t, \mathbf{y})$, that is initially assumed to be identical to the conditional density $q(\mathbf{y}|\mathbf{x}^t)$.

Step 2: Compute acceptance probability

$$\alpha(\mathbf{x}^t, \mathbf{y}) = \min\left(1, \frac{\pi(\mathbf{y})q(\mathbf{y}, \mathbf{x}^t)}{\pi(\mathbf{x}^t)q(\mathbf{x}^t, \mathbf{y})}\right) \tag{20.12}$$

where π is the target JPDF.

Step 3: Compute $\mathbf{x}^{t+1} = \mathbf{y}$ for sampling acceptance, otherwise $\mathbf{x}^{t+1} = \mathbf{x}^t$ for sampling rejection.

Based on the ergodicity of the simulated Markov chain, the random samples of the target JPDF are simulated by executing repeated draws at an equal number of steps from the Markov chain movement in the stochastic space. There is also the possibility to use multiple Markov chains simultaneously.

A competing algorithm with the Metropolis-Hastings algorithm is the so-called Gibbs sampler [13]. The Gibbs sampler assumes that the probability of sampling rejection is zero; that is, all samples are accepted. This makes it simpler but less flexible when compared with the Metropolis-Hastings algorithm. One step of the Gibbs sampler that moves the chain from state $\mathbf{x}^t$ to state $\mathbf{x}^{t+1}$ involves the following simulation steps by breaking the vector $\mathbf{x}$ in its k components:

$$\begin{aligned}
&\text{Step 1: Sample } x_1^{t+1} \approx \pi\left(x_1 | x_2^t, \ldots, x_k^t\right). \\
&\text{Step 2: Sample } x_2^{t+1} \approx \pi\left(x_2 | x_1^{t+1}, x_3^t, \ldots, x_k^t\right). \\
&\vdots \\
&\text{Step j: Sample } x_j^{t+1} \approx \pi\left(x_j | x_1^{t+1}, \ldots, x_{j-1}^{t+1}, x_{j+1}^t, \ldots, x_k^t\right). \\
&\vdots \\
&\text{Step k: Sample } x_k^{t+1} \approx \pi\left(x_k | x_1^{t+1}, \ldots, x_{k-1}^{t+1}\right).
\end{aligned} \tag{20.13}$$

To ensure an accurate vector simulation for both the Metropolis-Hastings and the Gibbs algorithms, it is important to check the ergodicity of the generated chain, especially if they can be stuck in a local energy minimum. Gibbs sampler is the most susceptible to getting stuck in different parameter space regions (metastable states).

Most MCMC researchers are currently using derivative of the Metropolis-Hastings algorithms. When implementing the Gibbs sampler or Metropolis-Hastings algorithms, key questions arise: (1) How do we need to block components to account for the correlation structure and dimensionality of the target distribution? (2) How do we choose the updating scanning strategy: deterministic or random? (3) How do we devise the proposal densities? (4) How do we carry out convergence and its diagnostics on one or more realizations of the Markov chain?

20.4 Stochastic Fields (or Processes)

A stochastic process or field is completely defined by its JPDF. Typically, the term "stochastic process" is used particularly in conjunction with the time evolution of a dynamic random phenomenon, while the term "stochastic field" is used in conjunction with the spatial variation of a stochastic surface. A space-time stochastic process is a stochastic function having time and space as independent arguments. The term "space-time stochastic process" is synonymous with the term "time-varying stochastic field." More generally, a stochastic function is the output of a complex physical stochastic system. Because a stochastic output can be described by a stochastic surface in terms of input parameters (for given ranges of variability), it appears that the term "stochastic field" is a more appropriate term for stochastic function approximation. Thus, *stochastic field* fits well with stochastic boundary value problems. *Stochastic process* fits well with stochastic dynamic, phenomena, random vibration, especially for stochastic stationary (steady state) problems that assume an infinite time axis. The term "stochastic field" is used hereafter.

Usually, in advanced engineering applications, simplistic stochastic models are used for idealizing component stochastic loading, material properties, manufacturing geometry and assembly deviations. To simplify the stochastic modeling, it is often assumed that the *shape* of spatial random variations is deterministic. Thus, the spatial variability is reduced to a single random variable problem, specifically to a random scale factor applied to a deterministic spatial shape. Another simplified stochastic model that has been extensively used in practice is the traditional response surface method based on quadratic regression and experimental design rules (such as circumscribed central composite design, CCCD, or Box-Benken design, BBD). The response surface method imposes a global quadratic trend surface for approximating stochastic spatial variations that might violate the physics of the problem. However, the traditional response surface method is practical for mildly nonlinear stochastic problems with a reduced number of random parameters.

From the point of view of a design engineer, the simplification of stochastic modeling is highly desired. A design engineer would like to keep his stochastic modeling as simple as possible so he can understand it and simulate it with a good confidence level (obviously, this confidence is subjective and depends on the analyst's background and experience). Therefore, the key question of the design engineer is: Do I need to use stochastic field models for random variations, or can I use simpler models, random variable models? The answer is yes and no on a case-by-case basis. Obviously, if by simplifying the stochastic modeling the design engineer significantly violates the physics behind the stochastic variability, then he has no choice; he has to use refined stochastic field models. For example, stochastic field modeling is important for turbine vibration applications due to the fact that blade mode-localization and flutter phenomena can occur. These blade vibration-related phenomena are extremely sensitive to small spatial variations in blade properties or geometry produced by the manufacturing or assembly process. Another example of the need for using a refined stochastic field modeling is the seismic analysis of large-span bridges. For large-span bridges, the effects of the nonsynchronicity and spatial variation of incident seismic waves on structural stresses can be very significant. To capture these effects, we need to simulate the earthquake ground motion as a dynamic stochastic field or, equivalently, by a space-time stochastic process as described later in this section.

A stochastic field can be homogeneous or nonhomogeneous, isotropic or anisotropic, depending on whether its statistics are invariant or variant to the axis translation and, respectively, invariant or variant to the axis rotation in the physical parameter space. Depending on the complexity of the physics described by the stochastic field, the stochastic modeling assumptions can affect negligibly or severely the simulated solutions. Also, for complex problems, if the entire set of stochastic input and stochastic system parameters is considered, then the dimensionality of the stochastic space spanned by the stochastic field model can be extremely large.

This chapter describes two important classes of stochastic simulation techniques of continuous multivariate stochastic fields (or stochastic functionals) that can be successfully used in advanced engineering applications. Both classes of simulation techniques are based on the decomposition of the stochastic field in a set of elementary uncorrelated stochastic functions or variables. The most desirable situation from an engineering perspective is to be able to simulate the original stochastic field using a reduced number of elementary stochastic functions or variables. The dimensionality reduction of the stochastic input is extremely beneficial because it also reduces the overall dimensionality of the engineering reliability problem.

This chapter focuses on stochastic field simulation models that use a limited number of elementary stochastic functions, also called *stochastic reduced-order models*. Many other popular stochastic simulation techniques, used especially in conjunction with random signal processing — such as discrete autoregressive process models AR, moving-average models MA, or combined ARMA or ARIMA models, Gabor transform models, wavelet transform models, and many others — are not included due to space limitation. This is not to shadow their merit.

In this chapter two important types of stochastic simulation models are described:

1. *One-level hierarchical stochastic field (or stochastic functional) model.* This simulation model is based on an *explicit* representation of a stochastic field. This representation is based on a statistical function (causal relationship) approximation by nonlinear regression. Thus, the stochastic field is approximated by a stochastic hypersurface **u** that is conditioned on the stochastic input **x**. The typical explicit representation of a stochastic field has the following form:

$$\mathbf{u}|\mathbf{x} = \mathbf{u}(\mathbf{x}) = \boldsymbol{\mu}_{u|x} + [\mathbf{u}(\mathbf{x}) - \boldsymbol{\mu}_{u|x}] \qquad (20.14)$$

In the above equation, first the conditional mean term, $\boldsymbol{\mu}_{u|x}$, is computed by minimizing the global mean-square error over the sample space. Then, the randomly fluctuating term $[\mathbf{u}(\mathbf{x}) - \boldsymbol{\mu}_{u|x}]$ is treated as a zero-mean decomposable stochastic field that can be factorized using a Wiener-Fourier series representation. This type of stochastic approximation, based on regression, is limited to a convergence in mean-square sense.

In the traditional response surface method, the series expansion term is limited to a single term defined by a stochastic residual vector defined directly in the original stochastic space (the vector components are the differences between exact and mean values determined at selected sampling points via experimental design rules).

2. *Two-level hierarchical stochastic field (or stochastic functional) model.* This simulation model is based on an *implicit* representation of a stochastic field. This representation is based on the Joint Probability Density Function (JPDF) (non-causal relationship) estimation. Thus, the stochastic field **u** is described by the JPDF of an augmented stochastic system $[\mathbf{x}, \mathbf{u}]^T$ that includes both the stochastic input **x** and the stochastic field **u**. The augmented stochastic system is completely defined by its JPDF $f(\mathbf{x}, \mathbf{u})$. This JPDF defines implicitly the stochastic field correlation structure of the field. Then, the conditional PDF of the stochastic field $f(\mathbf{u}|\mathbf{x})$ can be computed using the JPDF of augmented system and the JPDF of the stochastic input $f(\mathbf{x})$ as follows:

$$f(\mathbf{u}|\mathbf{x}) = f(\mathbf{x}, \mathbf{u})/f(\mathbf{x}) \qquad (20.15)$$

The JPDF of the augmented stochastic system can be conveniently computed using its projections onto a stochastic space defined by a set of locally defined, overlapping JPDF models. The set of local or conditional JPDF describe completely the local structure of the stochastic field. This type of stochastic approximation is based on the joint density estimation convergences in probability sense.

There are few key aspects that differentiate the two stochastic field simulation models. The one-level hierarchical model is based on statistical function estimation that is a more restrictive approximation problem than the density estimation employed by the two-level hierarchical model. Statistical function estimation based on regression can fail when the stochastic field projection on the input space is not convex. Density estimation is a much more general estimation problem than a statistical function estimation problem. Density estimation is always well-conditioned because there is no causal relationship implied.

The two-level hierarchical model uses the local density functions to approximate a stochastic field. The optimal solution rests between the use of a large number of small-sized isotropic-structure density functions (with no correlation structure) and a reduced number of large-sized anisotropic-structure density functions (with strong correlation structure). The preference is for a reduced number of local density functions. Cross-validation or Bayesian inference techniques can be used to select the optimal stochastic models based either on error minimization or likelihood maximization.

20.4.1 One-Level Hierarchical Simulation Models

To simulate complex pattern nonGaussian stochastic fields (functionals), it is advantageous to represent them using a Wiener-Fourier type series [14–16]:

$$\mathbf{u}(\mathbf{x}, \theta) = \sum_{i=0}^{\infty} \mathbf{u}_i(\mathbf{x}) f_i(\theta) = \sum_{i=0}^{\infty} \mathbf{u}_i(\mathbf{x}) f_i(\mathbf{z}(\theta)) \qquad (20.16)$$

where argument $\mathbf{z}$ is a set of independent standard Gaussian random variables and $\mathbf{f}$ is a set of orthogonal basis functions (that can be expressed in terms of a random variable set $\mathbf{z}$). A simple choice is to take the set of stochastic basis functions $\mathbf{f}$ equal to the random variables set $\mathbf{z}$. There are two main disadvantages when using the Wiener-Fourier series approximations. First, the stochastic orthogonal basis functions f_i are multidimensional functions that require intensive computations, especially for high-dimensional stochastic problems for which a large number of coupling terms need to be included to achieve adequate convergence of the series.

Generally, under certain integrability conditions, a stochastic function can be decomposed in an orthogonal structure in a probability measure space. Assuming that the uncorrelated stochastic variables $z_i, i = 1, 2, \ldots, m$ are defined on a probability space and that $u(z_1, \ldots, z_m)$ is a square-integrable function with respect to the probability measure and if $\{p_{i,k}(z_i)\}, i = 1, 2, \ldots, m$ are complete sets of square-integrable stochastic functions orthogonal with respect to the probability density $P_i(dz_i)/dz_i = f_{z_i}(z_i)$, so that $E[p_k(z_i)p_l(z_i)] = 0$ for all $k \neq l = 0, 1, \ldots$ and $i = 1, \ldots m$, then, the function $u(z_1, \ldots, z_m)$ can be expanded in a generalized Wiener-Fourier series:

$$u(z_1, \ldots z_m) = \sum_{k_1=0}^{\infty} \cdots \sum_{k_m=0}^{\infty} u_{k_1 \ldots k_m} p_{k_1}(z_1) \cdots p_{k_m}(z_m) \qquad (20.17)$$

where $\{p_{k_1}(z_1), \ldots, p_{k_m}(z_m)\}, k_1, \ldots, k_m = 0, 1, \ldots$, are complete sets of stochastic orthogonal (uncorrelated) functions. The generalized Wiener-Fourier series coefficients can be computed by solving the integral

$$u_{k_1 \ldots k_m} = \int \cdots \int u(z_1, \ldots, z_m) p_{k_1}(z_1) \cdots p_{k_m}(z_m)\, u_1(z_1) \cdots u_m(z_m) dz_1 \cdots dz_m \qquad (20.18)$$

The coefficients $u_{k_1 \ldots k_m}$ have a key minimizing property, specifically the integral difference

$$D = \int \cdots \int \left[u(z_1, \ldots, z_m) - \sum_{k_1=0}^{M_1} \cdots \sum_{k_m=0}^{M_m} g_{k_1 \ldots k_m} p_{k_m}(z_m) \right]^2 f_z(z_1) \cdots f_z(z_m) dz_1 \cdots dz_m \qquad (20.19)$$

reaches its minimum only for $g_{k_1 \ldots k_m} p_{k_m} = u_{k_1 \ldots k_m} p_{k_m}$.

Several factorization techniques can be used for simulation of complex pattern stochastic fields. An example is the use of the Pearson differential equation for defining different types of stochastic series representations based on orthogonal Hermite, Legendre, Laguerre, and Cebyshev polynomials. These polynomial expansions are usually called Askey chaos series [16]. A major application of stochastic field decomposition theory is the spectral representation of stochastic fields using covariance kernel factorization. These covariance-based techniques have a large potential for engineering applications because they can be applied to any complex, static, or dynamic nonGaussian stochastic field. Herein, in addition to covariance-based factorization techniques, an Askey polynomial chaos series model based on Wiener-Hermite stochastic polynomials is presented. This polynomial chaos model based on Wiener-Hermite series has been extensively used by many researchers over the past decade [15–17].

20.4.1.1 Covariance-Based Simulation Models

Basically, there are two competing simulation techniques using the covariance kernel factorization: (1) the Choleski decomposition technique (Equation 20.4), and (2) the Karhunen-Loeve (KL) expansion (Equation 20.7). They can be employed to simulate both static and dynamic stochastic fields. A notable property of these two simulation techniques is that they can handle both real-valued and complex-valued covariance kernels. For simulating space-time processes (or dynamic stochastic fields), the two covariance-based techniques can be employed either in the time-space domain by decomposing the cross-covariance kernel or in the complex frequency-wavelength domain by decomposing the complex cross-spectral density kernel. For real-valued covariance kernels, the application of the KL expansion technique is equivalent to the application of the Proper Orthogonal Decomposition (POD expansion) and Principal Component Analysis (PCA expansion) techniques [17, 18].

More generally, the Choleski decomposition and the KL expansion can be applied to any arbitrary square-integrable, complex-valued stochastic field, $u(\mathbf{x}, \theta)$. Because the covariance kernel of the complex-valued stochastic field $\text{Cov}[u(\mathbf{x}, \theta), u(\mathbf{x}', \theta)]$ is a Hermitian kernel, it can be factorized using either Choleski or KL decomposition.

If the KL expansion is used, the covariance function is expanded in the following eigenseries:

$$\text{Cov}[u(\mathbf{x}, \theta), u(\mathbf{x}', \theta)] = \sum_{n=0}^{\infty} \lambda_n \Phi_n(\mathbf{x}) \Phi_n(\mathbf{x}') \tag{20.20}$$

where λ_n and $\Phi_n(\mathbf{x})$ are the eigenvalue and the eigenvector, respectively, of the covariance kernel computed by solving the integral equation (based on Mercer's theorem) [19]:

$$\int \text{Cov}[u(\mathbf{x}, \theta), u(\mathbf{x}', \theta)] \Phi_n(\mathbf{x}) d\mathbf{x} = \Phi_n(\mathbf{x}') \tag{20.21}$$

As a result of covariance function being Hermitian, all its eigenvalues are real and the associated complex eigenfunctions that correspond to distinct eigenvalues are mutually orthogonal. Thus, they form a complete set spanning the stochastic space that contains the field u. It can be shown that if this deterministic function set is used to represent the stochastic field, then the stochastic coefficients used in the expansion are also mutually orthogonal (uncorrelated).

The KL series expansion has the general form

$$u(\mathbf{x}, \theta) = \sum_{i=0}^{n} \sqrt{\lambda_i} \Phi_i(\mathbf{x}) z_i(\theta) \tag{20.22}$$

where set $\{z_i\}$ represents the set of uncorrelated random variables that are computed by solving the stochastic integral:

$$z_i(\theta) = \frac{1}{\sqrt{\lambda_i}} \int_D \Phi_n(\mathbf{x}) u(\mathbf{x}, \theta) d\mathbf{x} \tag{20.23}$$

The KL expansion is an optimal spectral representation with respect to the second-order statistics of the stochastic field. Equation 20.23 indicates that the KL expansion can be applied also to nonGaussian stochastic fields if sample data from it are available. For many engineering applications on continuum mechanics, the KL expansion is fast mean-square convergent; that is, only a few expansion terms need to be included.

An important practicality aspect of the above covariance-based simulation techniques is that they can be easily applied in conjunction with the marginal probability transformation (Equation 20.8 and Equation 20.9) to simulate nonGaussian (translation) stochastic fields, either static or dynamic. For nonGaussian (translation) stochastic fields, two simulation models can be employed:

1. *Original space expansion.* Perform the simulation in the original nonGaussian space using the covariance-based expansion with a set $\{z_i\}$ of uncorrelated nonGaussian variables that can be computed as generalized Fourier coefficients by the integral

$$z_i(\theta) = \int_D u_i(\mathbf{x})u(\mathbf{x},\theta)\,d\mathbf{x} \quad (20.24)$$

In particular, for the KL expansion, the Equation 20.23 is used.

2. *Transformed space expansion.* Perform the simulation in the transformed Gaussian space using the covariance-based expansion with a set $\{z_i\}$ of standard Gaussian variables and then transform the Gaussian field to nonGaussian using the marginal probability transformation (Equation 20.8 and Equation 20.9).

In the engineering literature there are many examples of the application of covariance-based expansions for simulating either static or dynamic stochastic fields. In the civil engineering field, Choleski decomposition and KL expansion were used by several researchers, including Yamazaki and Shinozuka [20], Deodatis [21], Deodatis and Shinozuka [22], and Ghiocel [23] and Ghiocel and Ghanem [24], to simulate the random spatial variation of soil properties and earthquake ground motions. Shinozuka [25] and Ghiocel and Trandafir [26] used Choleski decomposition to simulate stochastic spatial variation of wind velocity during storms. Ghiocel and Ghiocel [27, 28] employed the KL expansion to simulate the stochastic wind fluctuating pressure field on large-diameter cooling towers using a nonhomogeneous, anisotropic dynamic stochastic field model. In the aerospace engineering field, Romanovski [29], and Thomas, Dowell, and Hall [30] used the KL expansion (or POD) to simulate the unsteady pressure field on aircraft jet engine blades. Ghiocel [31, 32] used the KL expansion as a stochastic classifier for jet engine vibration-based fault diagnostics.

An important aerospace industry application is the stochastic simulation of the engine blade geometry deviations due to manufacturing and assembly processes. These random manufacturing deviations have a complex stochastic variation pattern [33]. The blade thickness can significantly affect the forced response of rotating bladed disks [34–36]. Due to the cyclic symmetry geometry configuration of engine bladed disks, small manufacturing deviations in blade geometries produce a mode-localization phenomenon, specifically called mistuning, that can increase the airfoil vibratory stresses up to a few times. Blair and Annis [37] used the Choleski decomposition to simulate blade thickness variations, assuming that the thickness variation is a homogeneous and isotropic field. Other researchers, including Ghiocel [38, 39], Griffiths and Tschopp [40], Cassenti [41], and Brown and Grandhi [35], have used the KL expansion (or POD, PCA) to simulate more complex stochastic blade thickness variations due to manufacturing. Ghiocel [36] applied the KL expansion, both in the original, nonGaussian stochastic space and transformed Gaussian space. It should be noted that the blade thickness variation fields are highly nonhomogeneous and exhibit multiple, localized anisotropic directions due to the manufacturing process constraints (these stochastic variations are also technology dependent). If different blade manufacturing technology datasets are included in the same in a single database, then the resultant stochastic blade thickness variations could be highly nonGaussian, with multimodal, leptoqurtique, and platiqurtique marginal PDF shapes. More generally, for modeling the blade geometry variations in space, Ghiocel suggested a 3V-3D stochastic field model (three variables $\Delta x, \Delta y, \Delta z$, in three dimensions x, y, z). For only blade thickness variation, a 1V-2D stochastic field (one variable, thickness, in two dimensions, blade surface grid) is sufficient. It should be noted that for multivariate-multidimensional stochastic fields composed of several elementary one-dimensional component fields, the KL expansion must be applied to the entire covariance matrix of the stochastic field set that includes the coupling of all the component fields.

One key advantage of the KL expansion over the Choleski decomposition, that makes the KL expansion (or POD, PCA) more attractive for practice, is that the transformed stochastic space obtained using the

KL expansion typically has a highly reduced dimensionality when compared with the original stochastic space. In contrast, the Choleski decomposition preserves the original stochastic space dimensionality. For this reason, some researchers [29, 30] consider the KL expansion (or POD, PCA) a *stochastic reduced-order model* for simulating complex stochastic patterns. The KL expansion, in addition to space dimensionality reduction, also provides great insight into the stochastic field structure. The eigenvectors of the covariance matrix play, in stochastic modeling, a role similar to the vibration eigenvectors in structural dynamics; complex spatial variation patterns are decomposed in just a few dominant spatial variation *mode shapes*. For example, blade thickness variation *mode shapes* provide great insight into the effects of technological process variability. These insights are very valuable for improving blade manufacturing technology.

The remaining part of this subsection illustrates the application of covariance-based stochastic simulation techniques to generate dynamic stochastic fields. Specifically, the covariance-based methods are used to simulate the stochastic earthquake ground surface motion at a given site.

To simulate the earthquake ground surface motion at a site, a nonstationary, nonhomogeneous stochastic vector process model is required. For illustrative purposes, it is assumed that the stochastic process is a nonhomogeneous, nonstationary, 1V-2D space-time Gaussian process (one variable, acceleration in an arbitrary horizontal direction, and two dimensions for the horizontal ground surface). The space-time process at any time moment is completely defined by its evolutionary cross-spectral density kernel. For a seismic acceleration field $u(\mathbf{x}, t)$, the cross-spectral density for two motion locations i and k is:

$$S_{ui,uk}(\omega, t) = [S_{ui,ui}(\omega, t) S_{uk,uk}(\omega, t)]^{1/2} Coh_{ui,uk}(\omega, t) \quad \exp[-i\omega(X_{D,i} - X_{D,k})/V_D(t)] \quad (20.25)$$

where $S_{uj,uk}(\omega)$ is the cross-spectral density function for point motions u_j and u_k, and $S_{uj,uj}(\omega)$, $j = i, k$ is the auto-spectral density for location point j. The function $Coh_{ui,uk}(\omega, t)$ is the stationary or "lagged" coherence function for locations i and k. The "lagged" coherence is a measure of the similarity of the two point motions including only the amplitude spatial variation. Herein, it is assumed that the frequency-dependent spatial correlation structure of the stochastic process is time-invariant. The exponential factor $\exp[i\omega(D_1 - D_2)/V_D(t)]$ represents the wave passage effect in the direction D expressed in the frequency domain by a phase angle due to two motion delays at two locations X_i and X_j. The parameter $V_D(t)$ is the apparent horizontal wave velocity in the D direction. Most often in practice, the nonstationary stochastic models of ground motions are based on the assumption that the "lagged" coherence and the apparent directional velocity are independent of time. However, in real earthquakes, the coherence and directional wave velocity are varying during the earthquake duration, depending on the time arrivals of different seismic wave packages hitting the site from various directions.

Because the cross-spectral density is Hermitian, either the Choleski decomposition or the KL expansion can be applied. If the Choleski decomposition is applied, then

$$\mathbf{S}(\omega, t) = \mathbf{C}(\omega, t)\mathbf{C}^*(\omega, t) \quad (20.26)$$

where the matrix $\mathbf{C}(\omega, t)$ is a complex-valued lower triangular matrix. Then the space-time nonstationary stochastic process can be simulated using the trigonometric series as the number of frequency components $NF \to \infty$ [21,22]:

$$u_i(t) = 2 \sum_{k=1}^{NL} \sum_{j=1}^{NF} |C_{i,k}(\omega_j, t)| \sqrt{\Delta\omega} \cos[\omega_j t - \theta_{i,k}(\omega_j, t) + \Phi_{k,j}], \quad \text{for } i = 1, 2, \ldots NL \quad (20.27)$$

In the above equation, NL is the number of space locations describing the spatial variation of the motion. The first phase angle term in Equation 20.27 is computed by

$$\theta_{i,k}(\omega, t) = \tan^{-1}\left\{\frac{\text{Im}|C_{i,k}(\omega, t)|}{\text{Re}|C_{i,k}(\omega, t)|}\right\} \quad (20.28)$$

and the second phase angle term $\mathbf{\Phi}_{k,j}$ is a random phase angle uniformly distributed in [0, 2π] (uniform random distribution is consistent with Gaussian assumption). The above procedure can be used for any space-time stochastic process. For nonGaussian processes, the procedure can be applied in conjunction with the inverse probability equation transformation (Equation 20.8 and Equation 20.9).

If the complex-valued coherence function (including wave passage effects) is used, then its eigenfunctions are complex functions. Calculations have shown that typically one to five coherence function modes are needed to get an accurate simulation of the process [23, 24]. The number of needed coherence function modes depends mainly on the soil layering stiffness and the frequency range of interest; for high-frequency components, a larger number of coherence modes are needed.

20.4.1.2 Polynomial Chaos Series-Based Simulation Models

Ghanem and Spanos [15] discussed theoretical aspects and key details of the application Polynomial Chaos to various problems in their monograph on spectral representation of stochastic functionals.

Polynomial chaos expansion models can be formally expressed as a nonlinear functional of a set of standard Gaussian variables or, in other words, expanded in a set of stochastic orthogonal polynomial functions. The most popular polynomial chaos series model is that proposed by Ghanem and Spanos [15] using a Wiener-Hermite polynomial series:

$$u(\mathbf{x}, t, \theta) = a_0(\mathbf{x}, t)\Gamma_0 + \sum_{i_1=1}^{\infty} a_{i_1}(\mathbf{x}, t)\Gamma_1(z_{i_1}(\theta)) + \sum_{i_1=1}^{\infty}\sum_{i_2=1}^{i_1} a_{i_1 i_2}(\mathbf{x}, t)\Gamma_2(z_{i_1}(\theta), z_{i_2}(\theta)) + \ldots \quad (20.29)$$

The symbol $\Gamma_n(z_{i_1}, \ldots, z_{i_n})$ denotes the polynomial chaoses of order n in the variables $(z_{i_1}, \ldots, z_{i_n})$. Introducing a one-to-one mapping to a set with ordered indices denoted by $\{\psi_i(\theta)\}$ and truncating the polynomial chaos expansion after the pth term, Equation 20.29 can be rewritten

$$u(\mathbf{x}, t, \theta) = \sum_{j=0}^{p} u_j(\mathbf{x}, t)\psi_j(\theta) \quad (20.30)$$

The polynomial expansion functions are orthogonal in L_2 sense that is, their inner product with respect to the Gaussian measure that defines their statistical correlation, $E[\psi_j \psi_k]$, is zero. A given truncated series can be refined along the random dimension either by adding more random variables to the set $\{z_i\}$ or by increasing the maximum order of polynomials included in the stochastic expansion. The first refinement takes into account higher frequency random fluctuations of the underlying stochastic process, while the second refinement captures strong nonlinear dependence of the solution process on this underlying process. Using the orthogonality property of polynomial chaoses, the coefficients of the stochastic expansion solution can be computed by

$$u_k = \frac{E[\psi_k u]}{E[\psi_k^2]} \quad \text{for } k = 1, \ldots, K \quad (20.31)$$

A method for constructing polynomial chaoses of order n is by generating the corresponding multi-dimensional Wiener-Hermite polynomials. These polynomials can be generated using the partial differential recurrence rule defined by

$$\frac{\partial}{\partial z_{ij}}\Gamma_n(z_{i1}(\theta), \ldots, z_{in}(\theta)) = n\Gamma_{n-1}(z_{i1}(\theta), \ldots, z_{in}(\theta)) \quad (20.32)$$

The orthogonality of the polynomial chaoses is expressed by the inner product in L_2 sense with respect to Gaussian measure:

$$\int_{-\infty}^{\infty} \Gamma_n(z_{i1}, \ldots, z_{in})\Gamma_m(z_{i1}, \ldots, z_{in})\exp\left(-\frac{1}{2}\mathbf{z}^T\mathbf{z}\right)dz = n!\sqrt{2\pi}\delta_{mn} \quad (20.33)$$

Although popular in the engineering community, the polynomial chaos series may not necessarily be an efficient computational tool to approximate multivariate nonGaussian stochastic fields. The major problem is the stochastic dimensionality issue. Sometimes, the dimensionality of the stochastic transformed space defined by a polynomial chaos basis can be even larger than the dimensionality of the original stochastic space. Nair [42] discussed this stochastic dimensionality aspect related to polynomial chaos series application. Grigoriu [43] showed that the indiscriminate use of polynomial chaos approximations for stochastic simulation can result in inaccurate reliability estimates. To improve its convergence, the polynomial chaos series can be applied in conjunction with the inverse marginal probability transformation (Equation 20.8 and Equation 20.9) as suggested by Ghiocel and Ghanem [24].

20.4.2 Two-Level Hierarchical Simulation Models

The two-level hierarchical simulation model is based on the decomposition of the JPDF of the implicit input-output stochastic system in local or conditional JPDF models. The resulting stochastic local JPDF expansion can be used to describe, in detail, very complex nonstationary, multivariate-multidimensional nonGaussian stochastic fields. It should be noted that the local JPDF expansion is convergent in a probability sense, in contrast with the Wiener-Fourier expansions, which are convergent only in a mean-square sense.

The stochastic local JPDF basis expansion, or briefly, the local density expansion, can be also viewed as a Wiener-Fourier series of a composite type that provides both a global approximation and a local functional description of the stochastic field. In contrast to the classical Wiener-Fourier series that provides a global representation of a stochastic field by employing global basis functions, as the covariance kernel eigenfunctions in KL expansion or the polynomial chaoses [16], the stochastic local density expansion provides both a global and local representation of a stochastic field.

Stochastic local density expansion can be implemented as a two-layer stochastic neural-network, while Wiener-Fourier series can be implemented as a one-layer stochastic neural-network. It should be noticed that stochastic neural-network models train much faster than the usual multilayer preceptor (MLP) neural-networks. Figure 20.1 describes in pictorial format the analogy between stochastic local JPDF expansion and a two-layer neural-network model.

In the local density expansion, the overall JPDF of the stochastic model is obtained by integration over the local JPDF model space:

$$g(\mathbf{u}) = \int f(\mathbf{u}|\alpha) dp(\alpha) \tag{20.34}$$

where $p(\alpha)$ is a continuous distribution that plays the role of the probability weighting function over the local model space. In a discrete form, the weighting function can be expressed for a number N of local JPDF models by

$$p(\alpha) = \sum_{i=1}^{N} P(\alpha_i) \delta(\alpha - \alpha_i) \tag{20.35}$$

in which $\delta(\alpha - \alpha_i)$ is the Kronecker delta operator. Typically, the parameters α_i are assumed or known, and the discrete weighting parameters $P(\alpha_i)$ are the unknowns. The overall JPDF of the stochastic model can be computed in the discrete form by

$$g(\mathbf{u}) = \sum_{i=1}^{N} g(\mathbf{u}|\alpha_i) P(\alpha_i) \tag{20.36}$$

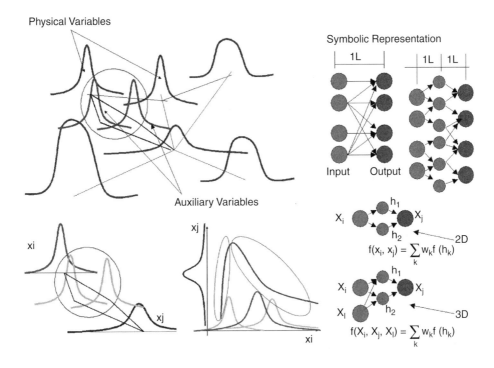

FIGURE 20.1 Analogy betweeen the local JPDF expansion and a two-layer stochastic neural network model.

The parameters α_i can be represented by the second-order statistics (computed by local averaging) of the local JPDF models. Thus, the overall JPDF expression can be rewritten as

$$g(\mathbf{u}) = \sum_{i=1}^{N} f(\mathbf{u} \mid i, \bar{\mathbf{u}}_i, \Sigma_i) P_i \qquad (20.37)$$

where $\bar{\mathbf{u}}_i, \Sigma_i$ are the mean vector and covariance matrix, respectively, of the local JPDF model i. Also, $P_i = n_i/N$, $\sum_{i=1}^{N} P_i = 1$, and $P_i > 0$, for $i = 1, N$. Typically, the types of the local JPDF are assumed and the probability weights are computed from the sample datasets. Often, it is assumed that the local JPDFs are multivariate Gaussian models. This assumption implies that the stochastic field is described locally by a second-order stochastic field model. Thus, nonGaussianity is assumed only globally. It is possible to include nonGaussianity locally using the marginal probability transformation of the local JPDF.

The local JPDF models are typically defined on partially overlapping partitions within stochastic spaces called *soft* partitions. Each local JPDF contributes to the overall JPDF estimate in a localized convex region of the input space. The complexity of the stochastic field model is associated with the number of local JPDF models and the number of stochastic inputs that define the stochastic space dimensionality. Thus, the stochastic model complexity is defined by the number of local JPDFs times the number of input variables. For most applications, the model implementations that are based on a large number of highly localized JPDFs are not good choices because these models have high complexity. For high complexities, we need very informative data to build the locally refined stochastic models. Simpler models with a lesser number of JPDF models have a much faster statistical convergence and are more robust. Thus, the desire is to reduce model complexity at an optimal level.

The complexity reduction of the stochastic field model can be accomplished either by reducing the number of local JPDFs or by reducing their dimensionality. Complexity reduction can be achieved using cross-validation and then removing the local JPDF that produces the minimum variation of the error estimate or the model likelihood estimate. We can also use penalty functions (evidence functions) that

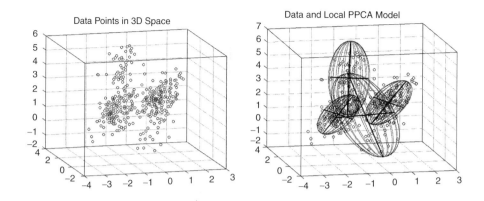

FIGURE 20.2 Sample data and local PPCA models for a trivariate stochastic field.

reduce the model complexity by selecting the most plausible stochastic models based on the evidence coming from sample datasets [44].

Alternatively, we can reduce the model complexity by reducing the dimensionality of local JPDF models using local factor analyzers. Local factor analysis reduces the number of local models by replacing the local variables in the original space by a reduced number of new local variables defined in local transformed spaces using a stochastic linear space projection. If deterministic projection instead of stochastic projection is applied, the factor analysis coincides with PCA. However, in practice, especially for high-dimensional problems, it is useful to consider a stochastic space projection by adding some random noise to the deterministic space projection. Adding noise is beneficial because instead of solving a multidimensional eigen-problem for decomposing the covariance kernel, we can implement an extremely fast convergent iterative algorithm for computing the principal covariance directions. The combination of the local JPDF expansion combined with the local probabilistic PCA decomposition is called herein the local PPCA expansion. Figure 20.2 shows the simulated sample data and the local PPCA models for a complex pattern trivariate stochastic field.

It should be noted that if the correlation structure of local JPDF models is ignored, these local JPDFs lose their functionality to a significant extent. If the correlation is neglected, a loss of information occurs, as shown in Figure 20.3. The radial basis functions, such standard Gaussians, or other kernels ignore the local correlation structure due to their shape constraint. By this they lose accuracy for

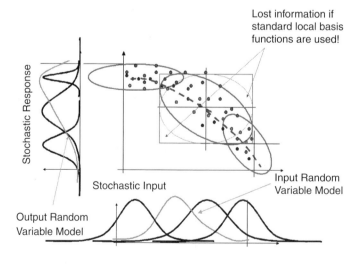

FIGURE 20.3 Loss of information if the local correlation structure is ignored.

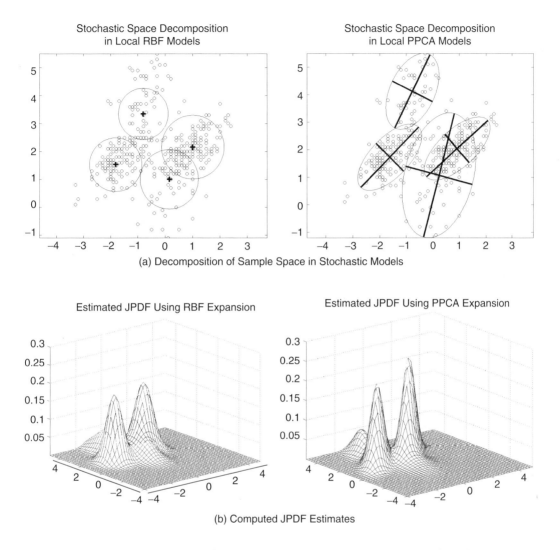

FIGURE 20.4 Local PPCA expansion model versus RBF network model a) decomposition of sample space in local stochastic models b) computed JPDF estimates.

modeling a complex stochastic field. One way to see radial basis function expansions is to look at them as degenerated local JPDF models. These radial functions are commonly used in a variety of practical applications, often in the form of the radial basis function network (RBFN), fuzzy basis function network (FBFN), and adaptive network-based fuzzy inference systems (ANFIS). Figure 20.4 compares the RBFN model with the local PPCA expansion model. The local PPCA expansion model is clearly more accurate and better captures all the data samples, including those that are remote in the tails of the JPDF. For a given accuracy, the local PPCA expansion needs a much smaller number of local models than the radial basis function model. As shown in the figure, the local PPCA models can elongate along long data point clouds in any arbitrary direction. The radial basis functions do not have this flexibility. Thus, to model a long data point cloud, a large number of overlapping radial basis functions are needed. And even so, the numerical accuracy of radial basis function expansions may still not be as good as the accuracy of local PPCA expansions with just few arbitrarily oriented local point cloud models. It should be noted that many other popular statistical approximation models, such as Bayesian Belief Nets (BNN), Specht's Probability Neural Network (PNN), in addition to ANFIS, RBF, and FBF networks, and even the two-layer MLP networks, have some visible conceptual similarities with the more refined local PPCA expansion.

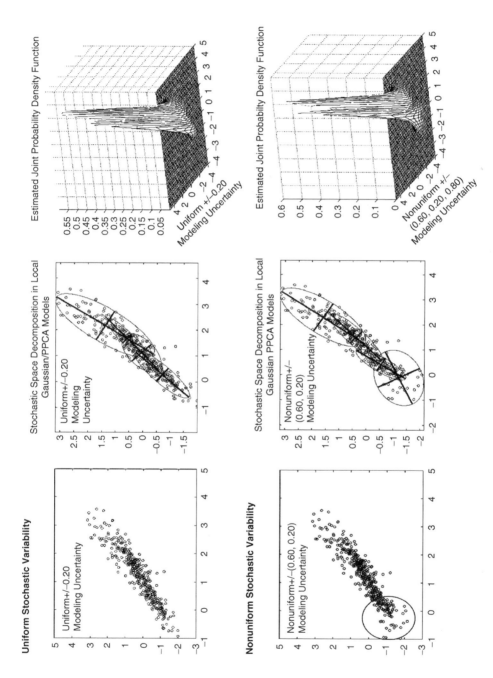

FIGURE 20.5 Local PPCA expansion model captures the nonuniform stochastic variability; a) Initial dataset, uniform data, b) Locally perturbed dataset, nonuniform data.

The stochastic local PPCA expansion is also capable of including accurately local, nonuniform stochastic variations. Figure 20.5 illustrates this fact by comparing the local PPCA expansions for two sample datasets that are identical except for a local domain shown at the bottom left of the plots. For the first dataset (left plots), the local variation was much smaller than for the second dataset (right plots). As seen by comparing the plots, the local PPCA expansion captures well this highly local variability of probabilistic response.

The decomposition of stochastic parameter space in a set of local models defined on soft partitions offers several possibilities for simulating high-dimensional complex stochastic fields. This stochastic sample space decomposition in a reduced number of stochastic local models, also called *states*, opens up a unique avenue for developing powerful stochastic simulation techniques based on stochastic process or field theory, such as dynamic Monte Carlo. Dynamic Monte Carlo techniques are discussed in the next section.

20.5 Simulation in High-Dimensional Stochastic Spaces

In simulating random samples from a joint probability density $g(\mathbf{x})$, in principle there are always ways to do it because we can relate the multidimensional joint density $g(\mathbf{x})$ to univariate probability densities defined by full conditional densities $g(x_j | x_M, \ldots, x_{j+1}, x_{j-1}, \ldots, x_1)$:

$$g(\mathbf{x}) = g(x_1, \ldots, x_K) = \prod_{j=1}^{M} g(x_j | x_{j-1}, \ldots, x_1) \qquad (20.38)$$

Unfortunately, in real-life applications, these (univariate) full conditional densities are seldom available in an explicit form that is suitable to direct sampling. Thus, to simulate a random sample from $g(\mathbf{x})$, we have few choices: (1) use independent random samples drawn from univariate marginal densities (static, independent sampling, SMC) to build the full conditionals via some predictive models; or (2) use independent samples from trial univariate marginal and conditional densities (sequential importance sampling, SIS), defined by the probability chain rule decomposition of the joint density, and then adaptively weight them using recursion schemes; or (3) use independent samples drawn from the univariate full conditional densities to produce spatially dependent samples (trajectories) of the joint density using recursion schemes that reflect the stochastic system dynamics (dynamic, correlated sampling, DMC).

For multidimensional stochastic fields, the standard SMC sampling techniques can fail to provide good estimators because their variances can increase sharply with the input space dimensionality. This variance increase depends on stochastic simulation model complexity. For this reason, variation techniques are required. However, classical techniques for variance reductions such as stratified sampling, importance sampling, directional sampling, control variable, antithetic variable, and Rao-Blackwellization methods are difficult to apply in high dimensions.

For simulating complex pattern, high-dimensional stochastic fields, the most adequate techniques are the adaptive importance sampling techniques, such as the sequential importance sampling with independent samples, SIS, or with correlated samples, DMC. Currently, in the technical literature there exists a myriad of recently developed adaptive, sequential, single sample-based or population-based evolutionary algorithms for high-dimensional simulations [6, 45]. Many of these techniques are based on constructing a Markov chain structure of the stochastic field. The fundamental Markov assumption of one-step memory significantly reduces the chain structures of the full or partial conditional densities, thus making the computations affordable. In addition, Markov process theory is well-developed so that reliable convergence criteria can be implemented.

20.5.1 Sequential Importance Sampling (SIS)

SIS is a class of variance reduction techniques based on adaptive important sampling schemes [45, 46]. A useful strategy to simulate a sample of joint distribution is to build up a trial density sequentially until it converges to the target distribution. Assuming a trial density constructed as $g(\mathbf{x}) = g_1(x_1) g_2(x_2 | x_1) \ldots g_N(x_N | x_1, \ldots x_{N-1})$

and using the same decomposition for the target density $\pi(\mathbf{x}) = \pi(x_1)\pi(x_2|x_1) \ldots \pi(x_N|x_1, \ldots x_{N-1})$, the importance sampling weight can be computed recursively by

$$w_t(\mathbf{x}_t) = w_{t-1}(\mathbf{x}_{t-1}) \frac{\pi(\mathbf{x}_t|\mathbf{x}_{t-1})}{g_t(\mathbf{x}_t|\mathbf{x}_{t-1})} \quad (20.39)$$

An efficient SIS algorithm is based on the following recursive computation steps, using an improved adaptive sampling weight [6, 46]:

Step 1: Simulate $\mathbf{x}_t$ from $g_t(\mathbf{x}_t|\mathbf{x}_{t-1})$, let $\mathbf{x}_t = (\mathbf{x}_{t-1}, \mathbf{x}_t)$
Step 2: Compute the adaptive sampling weight by

$$w_t = w_{t-1} \frac{\pi_t(\mathbf{x}_t)}{\pi_{t-1}(\mathbf{x}_{t-1}) g_t(\mathbf{x}_t|\mathbf{x}_{t-1})} \quad (20.40)$$

The entire sample of the input $\mathbf{x}$ is adaptively weighted until the convergence of the trial density to the target density is reached.

An important application of the SIS simulation technique is the *nonlinear filtering* algorithm based on the linear state-space model that is very popular in the system dynamics and control engineering community. In this algorithm, the system dynamics consist of two major coupled parts: (1) the state equation, typically represented by a Markov process; and (2) the observation equations, which are usually written as

$$\begin{aligned} x_t &\sim q_t(.|x_{t-1}, \theta) \\ y_t &\sim f_t(.|x_t, \phi) \end{aligned} \quad (20.41)$$

where y_t are observed state variables and x_t are unobserved state variables. The distribution of x_t are computed using the recursion:

$$\pi_t(x_t) = \int q_t(x_t|x_{t-1}) f_t(y_t|x_t) \pi_{t-1}(x_{t-1}) dx_{t-1} \quad (20.42)$$

In applications, the discrete version of this model is called the Hidden Markov model (HMM). If the conditional distributions q_t and f_t are Gaussian, the resulting model is the called the *linear dynamic model* and the solution can be obtained analytically via recursion and coincides with the popular Kalman filter solution.

To improve the statistical convergence of SIS, we can combine it with resampling techniques, sometimes called SIR, or with rejection control and marginalization techniques.

20.5.2 Dynamic Monte Carlo (DMC)

DMC simulates realizations of a stationary stochastic field (or process) that has a unique stationary probability density identical to the prescribed JPDF. Instead of sampling directly in the original cartesian space from marginal distributions, as SMC does, DMC generates a stochastic trajectory of spatially dependent samples, or a random walk in the state-space (or local model space). Figure 20.6 shows the basic conceptual differences between standard SMC and DMC. The Gibbs sampler and Metropolis-Hastings algorithms presented in Section 20.3.2 are the basic ingredients of dynamic Monte Carlo techniques.

The crucial part of the DMC is how to invent the best ergodic stochastic evolution for the underlying system that converges to the desired probability distribution. In practice, the most used DMC simulation

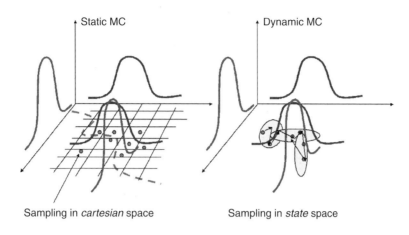

FIGURE 20.6 Conceptual differences between static Monte Carlo and dynamic Monte Carlo.

models are based on irreducible, aperiodic, reversible stationary Markov process models. The DMC simulation based on Markov process models is called Markov Chain Monte Carlo simulation, or briefly MCMC. Higher-order Markov chains or nonMarkovian are more difficult to implement. When using MCMC, a key modeling aspect is how to ensure that the samples of the chain are ergodic, especially when there is a chance that they can be stuck in a local energy minimum. The principal drawback of MCMC is that it has only local moves along its random path. This leads to a very slow convergence to equilibrium or, worse, to a stop in a local state.

A good strategy for improving MCMC movement in space is to combine the local powerful properties of MCMC with the global updating properties of molecular dynamics simulation using a Hamiltonian formulation of system dynamics. This combination is based on the Hamiltonian Markov Chain Monte Carlo (H-MCMC) algorithm [47]. The Hamiltonian MCMC is also called the Hybrid Monte Carlo (HMC). As an alternate to H-MCMC, an MCMC with random jumps has been developed [48].

Herein we describe briefly the H-MCMC algorithm. This algorithm is the preferred MCMC algorithm for high-dimensional stochastic simulation problems. The H-MCMC simulation algorithm can be efficiently implemented in conjunction with conditional PPCA expansion models. Hanson [47] claims in his research paper that "the efficiency of Hamiltonian method does not drop much with increasing dimensionality." Hanson has used H-MCMC for stochastic problems with a parameter space size up to 128 dimensions.

The H-MCMC algorithm is used to simulate a random sample of a canonical (or Gibbs, or Boltzmann) JPDF of the form:

$$\pi(\mathbf{x}) = c \times \exp(-H(\mathbf{x})) \tag{20.43}$$

This distribution is specific to Gibbs fields or equivalently to Markov fields [49]. In the above equation, $H(\mathbf{x})$ is the Hamiltonian of the dynamic system.

In the Hamiltonian formulation, the evolution of the dynamic system is completely known once the total energy or the Hamiltonian function in generalized coordinates, $H(\mathbf{q}, \mathbf{p})$, is defined:

$$H(\mathbf{q}, \mathbf{p}) = \sum_{i=1}^{n} \frac{p_i^2}{2m_i} + E(\mathbf{q}) \tag{20.44}$$

where **q** and **p** are the vectors of the generalized coordinates and generalized momenta (in the phase space), m_i are the masses associated with the individual particles of the system, and E(**q**) is the potential energy that ensures a coupling between the particles. System dynamics is represented by a trajectory in the phase space. For any arbitrary evolutionary function f(**q**, **p**), its rate of change is expressed by its time derivative:

$$\dot{f}(\mathbf{q}, \mathbf{p}) = \sum_{i=1}^{n} \left(\frac{\partial f}{\partial q_i} \dot{q}_i + \frac{\partial f}{\partial p_i} \dot{p}_i \right) = \sum_{i=1}^{n} \left(\frac{\partial f}{\partial q_i} \frac{\partial H}{\partial p_i} - \frac{\partial f}{\partial p_i} \frac{\partial H}{\partial q_i} \right) \quad (20.45)$$

For stationary stochastic dynamics, the Hamiltonian conserves. Now, using the molecular dynamics (MD) approach, we can discretize the Hamiltonian function evolution using finite difference equations over the time chopped in small increments. The most used implementation is the leap-frog scheme [47]:

$$p_i\left(t + \frac{1}{2}\Delta t\right) = p_i(t) - \frac{1}{2}\Delta t \frac{\partial E(q)}{\partial q_i(t)} \quad q_i(t + \Delta t) = q_i(t) + \Delta t p_i\left(t + \frac{1}{2}\Delta t\right) \quad p_i(t + \Delta t)$$

$$= p_i\left(t + \frac{1}{2}\Delta t\right) - \frac{1}{2}\Delta t \frac{\partial E(\mathbf{q})}{\partial q_i(t + \Delta t)} \quad (20.48)$$

The leap-frog scheme is time reversible and preserves the phase space volume. The resulting Hamiltonian-MCMC algorithm is capable of having trajectory jumps between different iso-energy surfaces and converges to a target canonical (or Gibbs, Boltzmann) distribution.

Figure 20.7 and Figure 20.8 show the application of H-MCMC to a simple bivariate stochastic field model composed of three overlapping local Gaussians. Figure 20.7 shows a simulated trajectory using the H-MCMC. The random samples were obtained by drawing sample data at each random walk step. Figure 20.8 compares the target JPDF and the estimated JPDF. The estimated JPDF is computed from the 250 random samples drawn at each step of the Markov chain evolution. It can be observed that there is a visible spatial correlation between the generated random samples. In practice, the random samples should be drawn periodically after a large number of random walk steps, so that the spatial correlation is lost. This improves considerably the convergence in probability of the H-MCMC simulation.

An extremely useful application of the DMC simulation is the stochastic interpolation for stochastic fields with missing data, especially for heavy-tailed nonGaussianity. DMC can be also used to simulate conditional-mean surfaces (response surfaces). Figure 20.9 illustrates the use of H-MCMC in conjunction with the local PPCA expansion for a univariate highly nonGaussian stochastic field with unknown mean. It should be noted that the quality of estimation depends on the data density in the sample space. Figure 20.10 shows another simple example of application of H-MCMC for simulating a univariate stochastic field with known mean using different persistence parameters (that define the magnitude of fictitious generalized momenta). The figure shows the H-MCMC samples and the exact (given) and estimated mean curves. By comparing the two plots in Figure 20.10, it should be noted that larger persistence parameters (larger generalized momenta) improve the simulation results for the mean curve in remote regions where data density is very low. Larger fictitious momenta produce larger system dynamics and, as a result, increase the variance and reduce the bias of the mean estimates. It appears that larger momenta in H-MCMC simulations are appropriate for probability estimations in the distribution tails of the joint distributions. Thus, DMC can be also employed to establish confidence bounds of stochastically interpolated surfaces for different probability levels. From the figure we can note that larger momenta provide less biased confidence interval estimates in the remote area of the distribution tails.

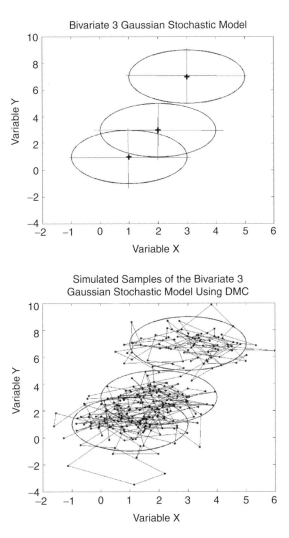

FIGURE 20.7 The prescribed stochastic model and the H-MCMC simulated random samples (drawn point data and random walk).

Other techniques to accelerate the MCMC convergence are simulated tempering and annealing. Simulated tempering and simulated annealing are popular MCMC applications of stochastic simulation to optimization under uncertainty problems. These techniques, which are described in many textbooks, are not discussed herein. This is not to shadow their merit.

20.5.3 Computing Tail Distribution Probabilities

As discussed in the introductory section of this chapter, stochastic simulation can be applied to either (1) simulate random quantities or (2) compute expectations. For probabilistic engineering analyses, both applications of stochastic simulation are of practical interest. For reliability and risk assessment analyses, the second application of simulation is used.

This section addresses the use of stochastic simulation to compute expectations. Only selective aspects are reviewed herein. The focus is on solving high-complexity stochastic problems.

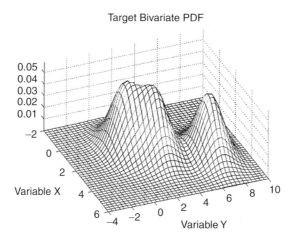

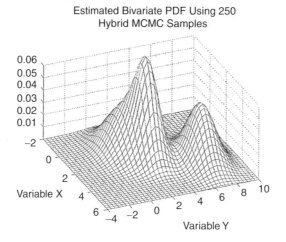

FIGURE 20.8 Target JPDF versus estimated JPDF (using 250 samples drawn at each step).

Expectations can include probability integrals for computing marginal distributions, evidences (normalization constant for the product of likelihood and prior in Bayes theorem), statistical moments, and expectation of functions. Thus, in most stochastic problems, the numerical aspect consists of computing the expected value of a function of interest f(**x**) with respect to a target probability density g(**x**), where **x** is the stochastic input. Assuming that $x_1, x_2, \ldots, x_N$ are an independent identically distributed sample set from g(**x**), then the Monte Carlo estimator $\hat{f}$ is computed by simple statistical averaging over the entire sample set:

$$\hat{f} = \frac{1}{N} \sum_{i=1}^{n} f(x_i) \tag{20.49}$$

The estimator $\hat{f}$ is unbiased and has the variance

$$\mathrm{var}(\hat{f}) = \frac{1}{N} \int [f(\mathbf{x}) - E_g(f)]^2 g(\mathbf{x}) \, d\mathbf{x} \tag{20.50}$$

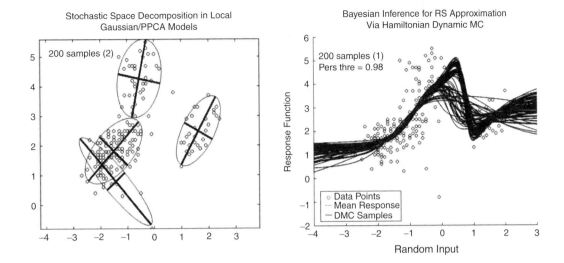

FIGURE 20.9 Stochastic Interpolation Using A Local PPCA Expansion Model via H-MCMC Simulation.

implying that the sampling error, (i.e., standard deviation) of $\hat{f}$ is $O(N^{-1/2})$ independent of stochastic space dimensionality. Thus, the MC simulation estimator always converges with a constant rate that is proportional to the square-root of the number of samples. This is a key property of the standard Monte Carlo estimator that makes the SMC simulation method so popular in all engineering fields.

Many textbooks provide suggestions on how to select the required number of samples to estimate the probability of failure for a given percentage error independent of an assumed probability estimate. If the mean probability estimate $\hat{P}_f$ has a normal distribution, the number of required samples for the confidence $1 - \alpha/2$ is given by

$$n = t^2_{1-\alpha/2, n-1} \hat{\bar{P}}_f (1 - \hat{\bar{P}}_f) \tag{20.51}$$

where t is the T or Student distribution with $n - 1$ degrees of freedom.

Thus, it appears that if we respect the above equation and select a small error, then we have nothing to worry about. This is not necessarily true, especially for high-dimensional problems that involve non-Gaussian stochastic fields (or responses). The delicate problem is that for standard stochastic simulation based on SMC, although the convergence rate is independent of space dimensionality, the probability estimate variance is not independent of space dimensionality. In fact, it can increase dramatically with dimensionality increase. The probability estimate variance increase affects most severely the tail probability estimates (low and high fractiles) that typically are dependent on localized probability mass distributions in some remote areas of the high-dimensional space.

Figure 20.11 shows a simple example of the increase of probability estimate variance with space dimensionality. Consider the computation of failure probability integral, I, defined by a volume that is internal to a multivariate standard Gaussian function:

$$I = \int_0^{1/4} \cdots \int_0^{1/4} \exp\left(\frac{1}{2} x_1^2 + x_2^2 \ldots + x_n^2\right) dx_1 dx_2 \ldots dx_n \tag{20.52}$$

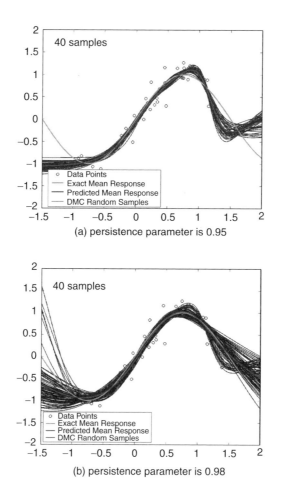

FIGURE 20.10 Effects of persistance (fictitious momenta) on the MCMC stochastic simulations a) persistence parameter is 0.95, b) persistence parameter is 0.98.

Figure 20.11 shows the plot of the coefficient of variation (c.o.v.) and the 90% (normalized) confidence interval (measure of the sampling error) as a function of input space dimension N that varies up to 10. For a number of 10,000 random samples, the c.o.v. of the failure probability estimate increases from about 3% for one dimension to about 87% for ten dimensions.

Thus, for the (random) point estimates of failure probabilities that we are typically getting from our reliability analyses, the use of Equation 20.51 for determining the number of required random samples can be erroneous, and possibly unconservative, as sketched in Figure 20.12. The required number of samples increases with the complexity of the stochastic simulation model, namely with its dimensionality and its nonGaussianity (or nonlinearity of stochastic dependencies).

20.5.4 Employing Stochastic Linear PDE Solutions

A special category of stochastic fields comprises the *physics-based* stochastic fields. These physics-based fields are defined by the output of computational stochastic mechanics analyses. The physics-based stochastic fields, or stochastic functionals, can be efficiently decomposed using *stochastic reduced-order models* (ROM) based on the underlying physics of the problem. This physics is typically described by stochastic partial-differential equations. Stochastic ROMs are computed by projecting

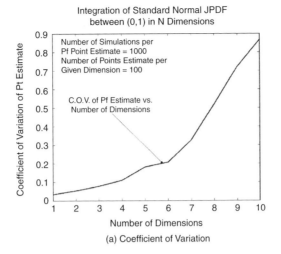

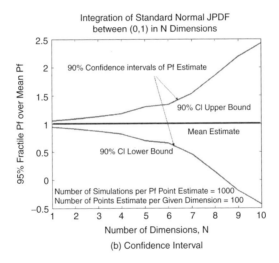

FIGURE 20.11 Effects of stochastic space dimensionality on the computed probability estimates; a) Coefficient of variation, b) Confidence interval.

the original stochastic physical field on reduced-size stochastic subspaces. Iterative algorithms using stochastic preconditioning can be employed in conjunction with physics-based stochastic ROM models.

The most used physics-based stochastic ROM models are those based on stochastic projections in the preconditioned eigen subspace, Taylor subspace, and Krylov subspace. These stochastic subspace projections are accurate for linear stochastic systems (functionals, fields) [42]. Nonlinear variations can be included by stochastically preconditioning Newton-type iterative algorithms. Both approximate and exact solvers can be used as preconditioners in conjunction with either stochastic ROM or full models. The key ingredients of the stochastic ROM implementations are (1) fast-convergent stochastic subspace expansions, (2) stochastic domain decomposition, (3) fast iterative solvers using stochastic preconditioners, and (4) automatic differentiation to compute function derivatives.

A useful industry application example of the stochastic ROM based on a eigen subspace projection is the Subset of Nominal Method (SNM) developed by Griffin and co-workers [50] for computing the mistuned (mode-localization) responses of jet engine bladed-disks due to random manufacturing deviations. The SNM approach is *exact* for proportional small variations of the linear dynamic system mass

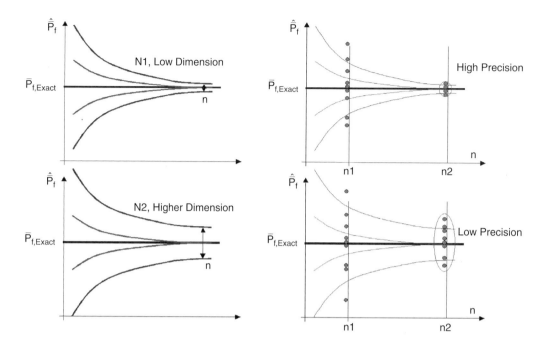

FIGURE 20.12 Space dimensionality effects on the accuracy of the computed probability estimates.

and stiffness matrices. Ghiocel [36] suggested combining the SNM approach with the KL expansion for simulating mistuned responses more realistically by including the effects of nonproportional blade variations due to manufacturing. The main idea is to use the KL expansion to reduce the stochastic input dimensionality and then use SNM iteratively to reduce the stochastic dynamic system dimensionality.

The stochastic field projections in the Taylor and Krylov subspaces can be very efficient for both static and dynamic stochastic applications. Figure 20.13 illustrates the application of the Krylov subspace expansion for computing the vibratory mistuned response of a 72-blade compressor engine blisk system. The system stochasticity is produced by the random variations in airfoil geometry due to manufacturing. Figure 20.13 shows the blade tip stochastic responses in frequency domain (transfer functions) for a given engine order excitation [34, 51]. The stochastic Krylov expansion requires only 14 basis vectors to converge. This means that only a reduced-size system of 14 differential equations must be solved at each stochastic simulation step.

The application of physics-based stochastic ROMs to stochastic simulation is limited to stochastic fields that can be described by *known* linear partial differential equations. It is obvious that the application of these stochastic ROM to computational stochastic mechanics problems is straightforward, as presented by Nair [42]. However, for stochastic fields that have an *unknown* functional structure, the physics-based stochastic ROMs are not directly applicable.

20.5.5 Incorporating Modeling Uncertainties

Engineering stochastic simulation procedures should be capable of computing confidence or variation bounds of probability estimates. The variation of probability estimates is generated by the presence of epistemic or modeling uncertainties due to (1) a lack of sufficient collection of data (small sample size issue); (2) nonrepresentative collection of statistical data with respect to the entire statistical population characteristics or stochastic system physical behavior (nonrepresentative data issue); (3) a lack of fitting the stochastic model with respect to a given statistical dataset, i.e., a bias is typically introduced due to smoothing (model statistical-fitting issue); and (4) a lack of accuracy of the prediction model with respect to real system physical behavior for given input data points, i.e., a bias is introduced at each predicted data point due to prediction inaccuracy (model lack-of-accuracy issue).

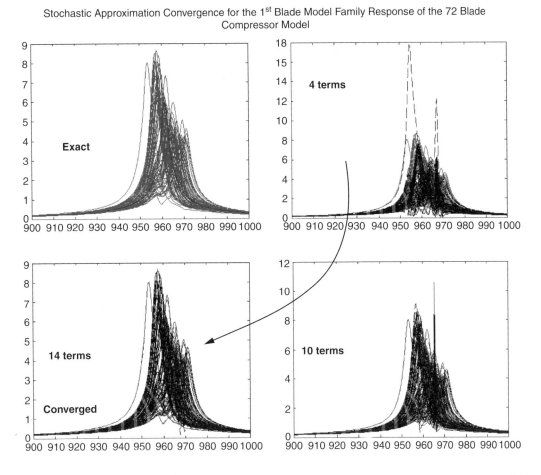

FIGURE 20.13 Convergence of stochastic Krylov subspace expansion for the random mistuned response of the blades compressor engine blisk; solid line is the full-model solution and dashed line is the reduced-model solution.

The first three modeling uncertainty categories are all associated with statistical modeling aspects. The fourth modeling uncertainty category is different. It addresses the prediction uncertainty due to the accuracy limitation of our computational models. In addition to these four uncertainty categories containing information that are objective in nature, there is also an uncertainty category related to the subjective information coming from engineering experience and judgment. Engineering judgment is an extremely useful source of information for situations where only scarce data or no data is available.

The above categories of uncertainties are described in detail in other chapters of the book. The purpose of mentioning these categories is to highlight the necessity of including them in the stochastic simulation algorithms. Unfortunately, this also increases the overall dimensionality of the stochastic simulation problem. Strategies to reduce the overall stochastic space dimensionality are needed [52].

20.6 Summary

Stochastic simulation is a powerful engineering computational tool that allows accurate predictions for complex nondeterministic problems. The most important aspect when employing simulation is to ensure that the underlying stochastic model is compatible with the physics of the problem. By oversimplifying stochastic modeling, one can significantly violate the physics behind stochastic variabilities.

Because the stochastic simulation is an almost endless subject, the focus of this chapter is on the conceptual understanding of these techniques, and not on the implementation details. This chapter reviews only the stochastic simulation techniques that in the author's opinion are the most significant for engineering design applications.

The chapter describes various stochastic simulation models — from simple, one-dimensional random variable models to complex stochastic network-based models. Significant attention is given to the decomposition of stochastic fields (or functionals) in constituent basis functions. Advances of stochastic functional analysis for developing efficient simulation techniques based on stochastic reduced-order models are included.

Another topic of a special attention is the stochastic simulation in high-dimensional spaces. The chapter reviews recent developments in stochastic simulation in high-dimensional spaces based on sequential importance sampling and dynamic simulation via a Hamiltonian formulation. Other key topics that are only briefly discussed include the computation of tail probabilities and the incorporation of epistemic or modeling uncertainties in simulation.

References

1. Hammersley, J.M. and Handscomb, D.C., *Monte Carlo Methods*, Methuen and Co. Ltd., London, 1964.
2. Knuth, D.C., *The Art of Computer Programming*, Addison-Wesley, Reading, MA, 1969.
3. Vaduva, I., *Simulation Models for Computer Applications*, The Series Mathematical Basis of Operational Research, Technical Editure, Bucharest, 1977.
4. Press, W.H., Teukolsky, S.A., Vetterling, W.T., and Flannery, B.P., *Numerical Recipes in FORTRAN: The Art of Scientific Computing*, Cambridge University Press, New York, 1992.
5. Law, A.M. and Kelton, W.D., *Simulation Modeling and Analysis*, McGraw-Hill, New York, 1991.
6. Gamerman, D., "Markov Chain Monte Carlo: Stochastic Simulation for Bayesian Inference," *Text in Statistical Science*, Chapman & Hall, New York, 1997.
7. Rubin, D.B., Using SIR Algorithm to Simulate Posterior Distributions (with discussion), in *Bayesian Statistics, Vol. 3*, Oxford University Press, Oxford, England, 395–402, 1987.
8. Gilks, W.R. and Wild, P., Adaptive Rejection Sampling for Gibbs Sampling, *Applied Statistics*, 41, 337–348, 1992.
9. Box, G.E. and Muller M.E., A Note of the Generation of Random Normal Deviates, *Annals of Mathematical Statistics*, 29, 610–611, 1958.
10. Butcher, J.C., Random Sampling from the Normal Distributions, *Computing Journal*, 3, 1961.
11. Grigoriu M., *Applied Non-Gaussian Processes*, Prentice Hall, Englewood Cliffs, NJ, 1995.
12. Metropolis, N., Rosenbluth, A.W., Rosenbluth, M.N., Teller, A.H., and Teller, E., Equations of State Calculations by Fast Computing Machines, *Journal of Chemical Physics*, 21(6), 1087–1091.
13. Geman, S. and Geman, D., Stochastic Relaxation, Gibbs Distributions and the Bayesian Restoration of Images, *IEEE Transactions on Pattern Analysis and Machine Intelligence*, 6, 721–741, 1984.
14. Ghiocel, D.M., Stochastic Field Models for Advanced Engineering Applications, *42nd AIAA/ASME/ASCE/AHS/ASC Structures, Structural Dynamics, and Materials Conference, AIAA/ASCE/ASME Non-Deterministic Approaches Forum*, Seattle, WA, April 16–19, 2001.
15. Ghanem, R.G. and Spanos, P.D., *Stochastic Finite Elements: A Spectral Approach*, Springer-Verlag, Heidelberg, 1991.
16. Wiener, N., The Homogenous Chaos, *American Journal of Mathematics*, 60, 897–936, 1938.
17. Pearson, K., On Lines and Planes of Closest Fit to Systems of Points in Space, *Philosophical Magazine*, 2, 559–572, 1958.
18. Everitt, B.S. and Dunn, G., *Applied Multivariate Data Analysis*, Arnold, Oxford University Press, London, 2001.
19. Grigoriu, M., *Stochastic Calculus*, Birkhauser, Boston, MA, 2002.

20. Yamazaki, K. and Shinozuka, M., Safety Evaluation of Stochastic Systems by Monte Carlo Simulation *9th SMiRT Conference,* Volume M, Lausanne, August 17–21, 1987.
21. Deodatis, G., Non-Stationary Stochastic Vector Processes: Seismic Ground Motion Applications, *Probabilistic Engineering Mechanics,* 11, 149–168, 1996.
22. Shinozuka, M. and Deodatis, G., Stochastic Process Models for Earthquake Ground Motion, *Journal of Probabilistic Engineering Mechanics,* 3, 191–204, 1988.
23. Ghiocel, D.M., Uncertainties of Seismic Soil-Structure Interaction Analysis: Significance, Modeling and Examples, Invited Paper at the *US–Japan Workshop on Seismic SSI, Organized by the NSF/USGS,* Menlo Park, CA, September 22–24, 1998.
24. Ghiocel, D.M. and Ghanem, R.G., Stochastic Finite Element Analysis for Soil-Structure Interaction, *Journal of Engineering Mechanics,* ASCE, 128(1), January 2002.
25. Shinozuka, M., Stochastic Fields and Their Digital Simulation, in *Stochastic Methods of Structural Dynamics,* Schueller, G.I. and Shinozuka, M., Eds., Martinus Nijhoff, Dordrecht, 1987, 93–133.
26. Ghiocel, D.M. and Trandafir, R., Application of Space-Time Stochastic Processes in Load Modeling for Structural Design of Special Facilities, *6th National Symposium on Informatics in Civil Engineering,* Timisoara, June 1988 (in Romanian).
27. Ghiocel, D.M. and Ghiocel, D., Structural Behavior of Large Cooling Towers on Wind Load, *Scientific Bulletin of Civil Engineering,* Vol. 1, ICB, Bucharest, 1988 (in Romanian).
28. Ghiocel, D.M. and Ghiocel, D., Stochastic Modeling and Simulation of Wind Fluctuating Pressure Field on Large Diameter Cooling Tower Structures, *ASCE Probabilistic Structural Mechanics Conference,* Sandia National Laboratories, Alburquerqe, NM, July 24–26, 2004 (submitted).
29. Romanowski, M., Reduced Order Unsteady Aerodynamic and Aeroelastic Models Using Karhunen-Loeve Eigenmodes, *AIAA Journal,* Paper AIAA 96–3981, 1996.
30. Thomas, J.P., Dowell, E.H., and Hall, K.C., Three Dimensional Transonic Aeroelasticity Using Proper Orthogonal Decomposition-Based Reduced Order Models, *Journal of Aircraft,* 40(3), 544–551, 2003.
31. Ghiocel, D.M., Refined Stochastic Field Models for Jet Engine Vibration and Fault Diagnostics, *Proceedings of International Gas Turbine & Aeroengine Congress,* 2000 TURBO EXPO, Paper 2000-GT-654, Munich, Germany, May 8–11, 2000.
32. Ghiocel, D.M., A New Perspective on Health Management Using Stochastic Fault Diagnostic and Prognostic Models, *International Journal of Advanced Manufacturing Systems, Special Issue on "Non-Deterministic Methods for Design and Manufacturing Under Uncertainties,"* Vol. 4, Issue 1, October 2001.
33. Hilbert, G. and Blair, A., Accuracy of Forced Response Prediction of Gas Turbine Rotor Blades, *4th National Turbine Engine HCF Conference,* Monterey, CA, March 6–9, 1999.
34. Ghiocel, D.M., Mistuning Analysis of Bladed-Disks, in *STI BLADE-GT Engineering Research Report for AFRL/PRT at Wright-Patterson Air Force Base,* Proprietary Information, USAF Contract F33615-96-C-2678, Dayton, OH, June 2002.
35. Brown J. and Grandhi, R.V., Probabilistic Analysis of Geometric Uncertainty Effects on Blade-Alone Forced Response, *Proceedings of the 2004 ASME Turbo Expo,* IGTI 2004, Vienna, June 14–17, 2004.
36. Ghiocel, D.M., Critical Probabilistic Modeling and Implementation Issues for Geometry Mistuning in Engine Blisks, presentation at the *6th Annual FAA/AF/NASA/Navy Workshop on the Application of Probabilistic Methods to Gas Turbine Engines,* Solomon Island, MD, March 18–20, 2003.
37. Blair A. and Annis, C., Development of Probabilistic Methods for Campbell Diagram Frequency Placement, presentation at the *7th National Turbine Engine High Cycle Fatigue Conference,* Monterey, CA, April 16–19, 2003.
38. Ghiocel, D.M., Stochastic Modeling of Airfoil Thickness Spatial Variability due to Manufacturing Process, presented at the *6th National Turbine Engine High Cycle Fatigue Conference,* Jacksonville, FL, March, 5–8, 2001.
39. Ghiocel, D.M., Stochastic Field Models for Aircraft Jet Engines, *Journal of Aerospace Engineering,* ASCE, Special Issue on Reliability of Aerospace Structures, Vol. 14, No. 4, October 2001.

40. Griffiths, J.A. and Tschopp, J.A., Selected Topics in Probabilistic Gas Turbine Engine Turbomachinery Design, in *CRC Press Engineering Design Reliability Handbook,* Nikolaidis, E., Ghiocel, D.M., and Singhal, S., Eds., CRC Press, Boca Raton, FL, 2004.
41. Casenti, B., Probabilistic Blade Design, United Technologies — Pratt and Whitney Final Report, USAF Contract F33615-98-C-2928, June 2003.
42. Nair, P.B., Projection Schemes in Stochastic Finite Element Analysis, in *CRC Press Engineering Design Reliability Handbook*, Nikolaidis, E., Ghiocel, D.M., and Singhal, S., Eds., CRC Press, Boca Raton, FL, 2004.
43. Grigoriu, M., Formulation and Solution of Engineering Problems Involving Uncertainty, Invited presentation at the *Biannual SAE Probabilistic Method Committee Meeting*, Detroit, October 7–9, 2003.
44. McKay, D.J.C., *Bayesian Interpolation*, Neural Computation, 1992.
45. Robert, C.P. and Casella, G., *Monte Carlo Statistical Methods*, Springer-Verlag, Heidelberg, 1999.
46. Liu, J., Monte Carlo Strategies in Scientific Computing, *Spring Series in Statistics,* Springer-Verlag, Heidelberg, 2001.
47. Hanson, K.M., Markov Chain Monte Carlo Posterior Sampling with the Hamilton Method, *Proceedings SPIE 4322, Medical Image: Image Processing,* Sonka, M. and Hanson, M.K., Eds., San Diego, CA, 2001, 456–467.
48. Green, P.J., Monte Carlo Methods: An Overview, Invited Paper, *Proceedings of the IMA Conference on Complex Stochastic Systems and Engineering,* Leeds, U.K., September 1993.
49. Fismen, M., Exact Simulation Using Markov Chains, M.Sc. thesis, Norvegian Institute of Technology and Science, Trondheim, December 1997.
50. Yang, M.-T. and Griffin, J.H., A Reduced Order Model of Mistuning Using a Subset of Nominal Modes, *Journal of Engineering for Gas Turbines and Power,* 123(4), 893–900, 2001.
51. Keerti, A., Nikolaidis, E., Ghiocel, D.M., and Kirsch, U., Combined Approximations for Efficient Probabilistic Analysis of Structures, *AIAA Journal*, 2004.
52. Ghiocel, D.M., Combining Stochastic Networks, Bayesian Inference and Dynamic Monte Carlo for Estimating Variation Bounds of HCF Risk Prediction, presentation at the *7th National Turbine Engine High Cycle Fatigue Conference,* Monterey, CA, April 16–19, 2003.

21
Projection Schemes in Stochastic Finite Element Analysis

21.1	Introduction	21-1
21.2	Finite Element Formulations for Random Media	21-5
	Random Field Discretization • Spatial Discretization	
21.3	Polynomial Chaos Expansions	21-8
21.4	Polynomial Chaos Projection Schemes	21-10
	Nonlinear Dependence of $\mathbf{K}$ on $\boldsymbol{\xi}$ • Remarks	
21.5	The Stochastic Krylov Subspace	21-14
21.6	Stochastic Reduced Basis Projection Schemes	21-16
	Weak Galerkin Scheme • Strong Galerkin Scheme • Petrov-Galerkin Scheme • Nonlinear Dependence of $\mathbf{K}$ on $\boldsymbol{\xi}$ • Theoretical Analysis of Convergence	
21.7	Postprocessing Techniques	21-21
	Statistical Moments and Distributions • Reliability Analysis	
21.8	Numerical Examples	21-22
	Two-Dimensional Thin Plate • Analysis of Foundation on Randomly Heterogeneous Soil	
21.9	Concluding Remarks and Future Directions	21-29

Prasanth B. Nair
University of Southampton

21.1 Introduction

In traditional computational mechanics it is often assumed that the physical properties of the system under consideration are deterministic. This assumption of determinism forms the basis of most mathematical modeling procedures used to formulate partial differential equations (PDEs) governing the system response. In practice, however, some degree of uncertainty in characterizing virtually any engineering system is inevitable. In a structural system, deterministic characterization of the system properties and its environment may not be desirable due to several reasons, including uncertainty in the material properties due to statistically inhomogeneous microstructure, variations in nominal geometry due to manufacturing tolerances, and uncertainty in loading due to the nondeterministic nature of the operating environment. These uncertainties can be modeled within a probabilistic framework, which leads to PDEs with random coefficients and associated boundary and initial conditions governing the system dynamics. It is implicitly assumed here that uncertainty in the PDE coefficients can be described by random variables or random fields that are constructed using experimental data or stochastic micromechanical analysis.

The main focus of the area of computational stochastic mechanics is the development of numerical techniques for solving stochastic PDEs. Over the past decade, much progress has been made in combining well-known spatial discretization schemes such as finite elements, finite differences, spectral methods, and boundary elements with random field discretization techniques to solve this class of problems. In particular, there is a wide body of work that deals with the application of the finite element method (FEM) in conjunction with random field discretization techniques to solve stochastic PDEs; see, for example, the research monographs by Ghanem and Spanos [1], and Kleiber and Hien [2], and a comprehensive review of the state-of-the-art edited by Schuëller [3].

By combining conventional spatial discretization schemes with random field discretization techniques, it becomes possible to arrive at finite-dimensional approximations of stochastic PDEs as a system of ordinary differential equations (ODEs) with random coefficients. The system of ODEs can be converted into a system of random algebraic equations using a temporal discretization scheme or a frequency domain transform. In the case of steady-state stochastic PDEs, discretization in space and the random dimension directly lead to a system of random algebraic equations. Hence, efficient numerical schemes for solving random ODEs and algebraic equations are essential tools for tackling problems in computational stochastic mechanics.

Given a system of random ODEs or algebraic equations, the Monte Carlo simulation technique or its variants can be readily applied to approximate the response statistics to an arbitrary degree of accuracy [4, 5]. Simulation techniques are general-purpose in scope and hence are applicable to a wide range of complex problems [6]. In practice, however, this is the method of last resort because the attendant computational cost can be prohibitive for systems modeled with high fidelity. The perturbation method and the Neumann series offer computationally efficient alternatives and have been popularly applied to compute the first two statistical moments of the response quantities; see, for example, [2, 3, 7–10]. The major drawback of such local approximation techniques is that the results become highly inaccurate when the coefficients of variation of the input random variables are increased. The response surface method is another approximation technique that is usually applied to construct linear and quadratic models of the response quantities as a function of the basic random variables [11]. It is also possible to construct more general nonlinear models by leveraging techniques from the function approximation literature [12, 13]. However, such approximation techniques do not scale well to problems with a large number of variables due to the *curse of dimensionality*.[1] An overview of some alternatives to perturbation methods can be found in the recent monograph of Elishakoff and Ren [14].

In this chapter, we primarily focus on stochastic finite element analysis techniques that use polynomial chaos expansions and stochastic reduced basis representations coupled with projection schemes. We henceforth refer to this class of numerical methods as *stochastic subspace projection schemes*. Such projection schemes can be considered as a rational extension of existing numerical schemes for solving deterministic algebraic and differential equations. This connection allows for the possibility of leveraging and extending existing results in the literature for rigorous mathematical analysis. It is noted here that stochastic subspace projection theory is an evolving area of research. Numerical evidence accumulated thus far suggests that methods based on this idea offer significantly better accuracy than perturbation methods [1, 15]. Excellent overviews of more traditional approaches to stochastic finite element analysis based on local approximations and recent developments in this area can be found elsewhere in the literature; see, for example, [2, 3, 16, 17] and the references therein. Before delving into details of stochastic subspace projection schemes, we first outline the essential ideas used in these formulations.

The idea of using functional expansion techniques to represent stochastic processes was originally proposed by Wiener in 1938 [18]. Ghanem and Spanos [1] leveraged the notion of polynomial chaos (PC) expansions introduced by Wiener to develop a spectral approach for stochastic finite element analysis. Over the past decade, PC projection schemes have been applied to solve a wide variety of problems in computational stochastic mechanics. To illustrate the basic ideas used in functional expansion approaches (PC expansion can be viewed as a special case of this approach), consider the continuous stochastic operator problem

$$\mathcal{L}(\mathbf{x};\theta)u(\mathbf{x};\theta) = f(\mathbf{x};\theta), \tag{21.1}$$

---

[1]The curse of dimensionality arises from the fact that the number of hypercubes required to fill out a compact region of a M-dimensional space grows exponentially with M.

where $\mathcal{L}(\mathbf{x};\theta)$ is a stochastic differential operator (i.e., a randomly parameterized differential operator). For simplicity of presentation, consider the case when the operator $\mathcal{L}$ is parameterized in terms of a single random variable $\xi(\theta)$.[2] $f(\mathbf{x};\theta)$ is a random function, and $u(\mathbf{x};\theta)$ is the random solution process whose statistics are to be computed. Note here that we use the symbol θ to indicate the dependence of any quantity on a random dimension. For example, for each $\mathbf{x}\in\mathbb{R}^d$, $u(\mathbf{x};\theta):\Theta\to\mathbb{R}$ is a random variable on a suitable probability space $(\Theta,\mathcal{F},\Gamma)$, where Θ is the set of elementary events, $\mathcal{F}$ is the σ-algebra associated with Θ and Γ is a probability measure.

The main idea used in functional expansion techniques is to decompose the solution process $u(\mathbf{x};\theta)$ into separable deterministic and stochastic components by the *ansatz* (i.e., assumed form for a function)

$$u(\mathbf{x};\theta)\approx\hat{u}(\mathbf{x};\theta)=\sum_{i=1}^{m}u_i(\mathbf{x})\varphi_i(\xi(\theta)), \qquad (21.2)$$

where $u_i(\mathbf{x})\in\mathbb{R}$ is an undetermined deterministic function and $\varphi_i(\xi(\theta)):\Theta\to\mathbb{R}$ is a known stochastic basis function.[3] Henceforth, for notational convenience, we do not explicitly indicate the dependence of the random variable ξ on the random dimension θ.

It can be seen from Equation 21.2 that once the stochastic basis functions are chosen, the solution process boils down to computation of the undetermined functions $u_i(\mathbf{x})$, $i=1,2,\ldots,m$. Let us now substitute Equation 21.2 into the governing operator problem to arrive at the following stochastic residual error function

$$\varepsilon(\mathbf{x};\theta)=\mathcal{L}(\mathbf{x};\theta)\sum_{i=1}^{m}u_i(\mathbf{x})\varphi_i(\xi)-f(\mathbf{x};\theta). \qquad (21.3)$$

Equations governing the undetermined functions can now be derived by employing a projection scheme along the random dimension θ. Consider the case when the Galerkin projection scheme is employed, where the stochastic residual error is orthogonalized with respect to the approximating space $\varphi_i(\xi)$. That is, the inner product of $\varepsilon(\mathbf{x};\theta)$ with each basis function set to zero:

$$\sum_{i=1}^{m}\langle\varphi_j(\xi)\mathcal{L}(\mathbf{x};\theta)\varphi_i(\xi)\rangle u_i(\mathbf{x})-\langle\varphi_j(\xi)f(\mathbf{x};\theta)\rangle=0 \quad \forall j=1,2,\ldots,m, \qquad (21.4)$$

where $\langle\cdot,\cdot\rangle$ denotes the inner product in the Hilbert space of random variables [19]; that is,

$$\langle f(\theta)g(\theta)\rangle=\int f(\theta)g(\theta)d\Gamma(\theta) \qquad (21.5)$$

Because $\mathcal{L}(\mathbf{x};\theta)$ is a randomly parameterized differential operator, $\langle\varphi_j(\xi)\mathcal{L}(\mathbf{x};\theta)\varphi_i(\xi)\rangle$ is a deterministic differential operator.[4] Further, $\langle\varphi_j(\xi)f(\mathbf{x};\theta)\rangle$ is a deterministic function. Equation 21.4 therefore represents a set of m coupled deterministic operator problems which govern $u_i(\mathbf{x})$. Hence, by applying a functional expansion of the form given in Equation 21.2 in conjunction with the Galerkin projection scheme, we have arrived at a system of coupled deterministic operator problems, thereby increasing the dimensionality of the problem. Because Equation 21.4 is deterministic, it can be readily solved using conventional numerical techniques such as the FEM and $u_i(\mathbf{x})$ can be computed. Subsequently, it becomes possible to efficiently approximate the complete statistics of $u(\mathbf{x},\theta)$ in the postprocessing phase using

---

[2] For example, $\mathcal{L}(\mathbf{x};\theta)=\xi(\theta)\nabla^2$, where ∇^2 is the Laplacian operator.
[3] In the PC projection scheme of Ghanem and Spanos [1], Hermite polynomials are chosen as basis functions.
[4] For example, when $\mathcal{L}(\mathbf{x};\theta)=\xi\nabla^2$, $\langle\varphi_j(\xi)\mathcal{L}(\mathbf{x};\theta)\varphi_i(\xi)\rangle=\langle\varphi_j(\xi)\xi\varphi_i(\xi)\rangle\nabla^2$.

Equation 21.2. In summary, a functional expansion approach for solving stochastic operator problems involves four steps: (1) selection of a suitable set of stochastic basis functions $\varphi_i(\xi)$, (2) application of a projection scheme along the random dimension θ to arrive at a system of coupled equations, (3) numerical solution of the coupled system of deterministic equations to compute the undetermined functions $u_i(\mathbf{x})$, and (4) postprocessing to compute the statistics of interest after substituting the results obtained in the earlier step into the expansion in Equation 21.2.

An alternative approach can be formulated by first spatially discretizing the continuous governing Equation 21.1 using any conventional technique such as the FEM. This semidiscretization procedure essentially involves treating the discrete nodal values of the field variable $u(\mathbf{x}, \theta)$ as random variables.[5] This process ultimately leads to a system of random algebraic equations of the form

$$\mathbf{R}(\mathbf{u}(\xi); \xi) = 0, \tag{21.6}$$

where $\mathbf{u}(\xi) \in \mathbb{R}^n$ denotes a random vector composed of the values of the discretized field variable (i.e., the vector solution process) whose statistics are to be computed.

It is possible to apply the functional expansion approach outlined earlier to approximate the random vector $\mathbf{u}(\xi)$ in the following form

$$\hat{\mathbf{u}}(\xi) = \sum_{i=1}^{m} \mathbf{u}_i \varphi_i(\xi), \tag{21.7}$$

where $\mathbf{u}_i \in \mathbb{R}^n, i = 1, 2, \ldots, m$ are vectors of undetermined coefficients. These unknown vectors can be uniquely computed by substituting Equation 21.7 into Equation 21.6 and applying a Galerkin projection scheme along the random dimension θ. This results in the following system of deterministic algebraic equations with increased dimensionality (mn unknowns in comparison to Equation 21.6, which has only n unknowns)

$$\left\langle \varphi_j(\xi) \mathbf{R}\left(\sum_{i=1}^{m} u_i \varphi_i(\xi); \xi \right) \right\rangle = 0, \quad \forall j = 1, 2, \ldots, m. \tag{21.8}$$

It can be shown that the preceding system of algebraic equations is equivalent to a spatially discretized version of Equation 21.4. That is, the functional expansion approach can be applied either to the continuous form of the stochastic operator equation or its spatially discretized version.

However, because we are dealing here with a discretized problem where it is desired to approximate the random vector $\mathbf{u}(\xi)$ and not the random function $u(\mathbf{x}; \theta)$, it appears more natural to use a stochastic reduced basis approximation of the form

$$\hat{\mathbf{u}}(\xi) = \alpha_1 \psi_1(\xi) + \alpha_2 \psi_2(\xi) + \cdots + \alpha_m \psi_m(\xi) = \mathbf{\Psi}(\xi) \boldsymbol{\alpha}, \tag{21.9}$$

where $\mathbf{\Psi}(\xi) = [\psi_1(\xi), \psi_2(\xi), \ldots, \psi_m(\xi)] \in \mathbb{R}^{n \times m}$ denotes a matrix of known stochastic basis vectors and $\boldsymbol{\alpha} = \{\alpha_1, \alpha_2, \ldots, \alpha_m\} \in \mathbb{R}^n$ is a vector of undetermined coefficients. Note that the above representation has only m unknowns, whereas Equation 21.7 has a total of mn unknowns.

The main idea used here is to employ a rich set of (problem-dependent) stochastic basis vectors, which ensure that accurate approximations can be obtained for $m <<< n$. This idea of stochastic reduced basis representations was introduced recently in the literature [15, 20] in the context of solving large-scale linear random algebraic system of equations obtained from semidiscretization of stochastic PDEs. In contrast

---

[5]In the context of the FEM applied to static problems in structural mechanics, this means that the nodal displacements are treated as random variables. We cover this in more detail in the next section.

to the functional expansion scheme (where PC basis functions are typically used), the solution process is represented using basis vectors spanning the preconditioned stochastic Krylov subspace.

Substituting Equation 21.9 into Equation 21.6 and applying the Galerkin projection scheme, we arrive at the following reduced-order deterministic system of equations to be solved for the m unknown coefficients, $\alpha_1, \alpha_2, \ldots, \alpha_m$

$$\langle \boldsymbol{\Psi}^*(\xi) \mathbf{R}(\boldsymbol{\Psi}(\xi) \boldsymbol{\alpha}; \xi) \rangle = 0, \qquad (21.10)$$

where the superscript * denotes the complex conjugate transpose of a vector or matrix (if it is complex), or the transpose (if it is real).

Comparing Equation 21.8 with Equation 21.10, it can be observed that a key advantage of stochastic reduced basis representations is that the undetermined quantities can be efficiently computed compared to the functional expansion approach outlined earlier. This is because, in the stochastic reduced basis approach, application of the Galerkin projection scheme leads to a reduced-order system of equations. In contrast, the functional expansion scheme leads to system of equations with increased dimensionality. Clearly, the success of this reduced basis approach critically hinges on the choice of stochastic basis vectors. As we shall show later, using basis vectors spanning the preconditioned stochastic Krylov subspace, highly accurate results can be obtained using only a few basis vectors.

In this chapter we present the theoretical foundations of projection schemes that employ the PC expansion and the preconditioned stochastic Krylov subspace to approximate the solution of stochastic PDEs. The remainder of this chapter is organized as follows. In the next section we outline the steps involved in stochastic finite element analysis of random media. Section 21.3 presents the generalized PC expansion scheme that can be applied to represent the solution of stochastic PDEs. Section 21.4 outlines how the PC expansion scheme can be applied in conjunction with Galerkin projection to stochastic finite element analysis. In Section 21.5 we present the stochastic Krylov subspace as an alternative to PC expansions and outline some of its theoretical properties. Section 21.6 presents a number of stochastic reduced basis projection schemes that employ basis vectors spanning the preconditioned stochastic Krylov subspace for solving random algebraic equations. Section 21.7 outlines some procedures for postprocessing the solutions obtained using the projection schemes in order to compute the response statistical moments and conduct reliability assessment studies. Section 21.8 contains numerical studies on two-dimensional elasticity problems and the relative performance of various projection schemes are compared. Section 21.9 concludes the chapter and outlines some areas for further investigation.

21.2 Finite Element Formulations for Random Media

To illustrate the basic steps involved in stochastic finite element analysis of random media, consider a two-dimensional isotropic solid whose Youngs modulus is modeled as a random field, say $h(\mathbf{x}; \theta)$. That is, for each $\mathbf{x} \in \mathbb{R}^2$, $h: \Theta \to \mathbb{R}$ is a random variable. Because the Youngs modulus is represented by a random field, the elasticity matrix becomes a function of the spatial coordinates and a random dimension; that is,

$$\mathbf{D}(\mathbf{x}; \theta) = h(\mathbf{x}; \theta) \mathbf{D}_0, \qquad (21.11)$$

where $\mathbf{D}_0$ is the deterministic part of the elasticity matrix.

From a practical viewpoint, the elastic properties of a solid may be random due to the intrinsic stochastic inhomogeneity of the microstructure. Such situations may arise, for example, when studying the mechanical behavior of cellular solids such as metallic foams and bone [21], and granular media such as sand and soil [22]. In such cases, it is not sufficient to model the Youngs modulus alone as a random field because that may lead to underestimation of the response variability [23]. A more systematic approach would be to use stochastic homogenization techniques to derive random field models for the terms of the elasticity matrix; see, for example, [23–25]. Such models can be readily accommodated within the stochastic FEM. Further, it is also possible to include uncertainty in geometric variables, boundary conditions, and external loading in the formulation.

21.2.1 Random Field Discretization

To apply the FEM to problems wherein one or more of the physical quantities are modeled as random fields, we need to represent them first by a finite set of random variables. For a detailed exposition of random field modeling, the reader is referred to the text by Vanmarcke [26]. Various discretization techniques are available in the literature for approximating random fields, including the mid-point method, shape function methods, optimal linear estimation, weighted integral methods, orthogonal series expansion, and the Karhunen-Loève (KL) expansion scheme; see, for example, [27–29].

Let the correlation function of the random field $h(\mathbf{x}; \theta)$ be $R_h(\mathbf{x}, \mathbf{y})$. Then, a discretized version of the random field $h(\mathbf{x}; \theta)$ can be written in the general form

$$h(\mathbf{x}; \theta) = \langle h(\mathbf{x}; \theta) \rangle + \sum_{i=1}^{\infty} \xi_i(\theta) h_i(\mathbf{x}), \tag{21.12}$$

where $\xi_i(\theta)$, $i = 1, 2, \ldots, \infty$ are a set of uncorrelated random variables and $h_i(\mathbf{x})$, $i = 1, 2, \ldots, \infty$ are a set of basis functions used in the random field discretization procedure. $\langle h(\mathbf{x}; \theta) \rangle$ denotes the mean of the random field.

When the KL expansion scheme is employed, the i-th basis function can be written as $h_i(\mathbf{x}) = \sqrt{\lambda_i} \kappa_i(\mathbf{x})$, where λ_i and $\kappa_i(\mathbf{x})$ are the eigenvalues and the eigenfunctions, respectively, of a Fredholm integral equation of the second kind given below

$$\int_{\mathcal{D}} R_h(\mathbf{x}, \mathbf{y}) \kappa_i(\mathbf{x}) d\mathbf{x} = \lambda_i \kappa_i(\mathbf{y}). \tag{21.13}$$

Analytical solutions of the above integral eigenvalue problem can be obtained only for a special class of correlation functions (e.g., the exponential correlation function) defined on geometrically simple domains. For more general cases, numerical discretization schemes must be employed to compute the eigenvalues and eigenfunctions of $R_h(\mathbf{x}, \mathbf{y})$; see, for example, [1, 29, 30].

To implement the random field discretization scheme computationally, we truncate Equation 21.12 at the M-th term to arrive at the finite-dimensional approximation

$$h(\mathbf{x}; \theta) \approx \langle h(\mathbf{x}; \theta) \rangle + \sum_{i=1}^{M} \xi_i(\theta) h_i(\mathbf{x}). \tag{21.14}$$

If the eigenvalues of the covariance function $R_h(\mathbf{x}, \mathbf{y})$ decay rapidly, then only a few number of terms will be required to ensure an accurate representation of the random field. In the limiting case, when the correlation length of the random field tends to zero, the number of terms M will grow very rapidly toward infinity. In summary, the correlation length of the random field dictates the number of terms (M) required to ensure an accurate finite-dimensional representation.

21.2.2 Spatial Discretization

Substitution of the discretized random field representation into Equation 21.11 results in a representation of the elasticity matrix in terms of a finite number of random variables. This sets the stage for the application of the FEM to spatially discretize the governing equations. The starting point for the FEM is the weak form of the governing equations, which is obtained by multiplying the governing equation by a test function and integrating by parts; a detailed overview can be found in any standard text [31].

Subsequently, the domain is divided into a number of elements and the field variables are approximated within each element using a set of shape functions as

$$\tilde{u}(\mathbf{x}) = \sum_{i=1}^{n_e} u_i(\theta) N_i(\mathbf{x}), \qquad (21.15)$$

where N_i denotes the i-th shape function and $u_i(\theta)$ can be interpreted as a generalized field variable.

Substituting the above approximation into the weak form of the governing equations, we arrive at expressions for the element stiffness and mass matrices. For the case of linear structural systems, assembly of the element stiffness, mass and damping matrices leads to a system of coupled ODEs with random coefficients of the following form

$$\mathbf{M}(\boldsymbol{\xi}) \frac{d^2}{dt^2} \mathbf{u}(\boldsymbol{\xi}, t) + \mathbf{C}(\boldsymbol{\xi}) \frac{d}{dt} \mathbf{u}(\boldsymbol{\xi}, t) + \mathbf{K}(\boldsymbol{\xi}) \mathbf{u}(\boldsymbol{\xi}, t) = \mathbf{f}(\theta, t), \qquad (21.16)$$

where $\mathbf{M}(\boldsymbol{\xi}), \mathbf{C}(\boldsymbol{\xi})$, and $\mathbf{K}(\boldsymbol{\xi}) \in \mathbb{R}^{n \times n}$ denote the system mass, damping, and stiffness matrices, respectively. $\mathbf{f}(\theta, t) \in \mathbb{R}^n$ denotes the generalized force vector, which can be either deterministic or random, where $t \in \mathbb{R}^+$ refers to time. $\mathbf{u}(\boldsymbol{\xi}, t) \in \mathbb{R}^n$ is the random displacement vector whose statistics are to be computed. $\boldsymbol{\xi} = \{\xi_1, \xi_2, \ldots, \xi_M\} \in \mathbb{R}^M$ denotes the set of random variables arising from discretization of the random field representing uncertainty in the Youngs modulus and n is the total number of degrees of freedom (DOF).

To illustrate how the coefficient matrices in Equation 21.16 are computed, consider the case of static response analysis of a two-dimensional elastic solid with random Youngs modulus subject to deterministic loading.

Here, the element stiffness matrix can be written as

$$k^e = \int_{\mathcal{D}_e} \mathbf{B}^T \mathbf{D}(\mathbf{x}; \theta) \mathbf{B} \, d\mathbf{x}, \qquad (21.17)$$

where $\mathbf{B}$ is the strain-displacement matrix and $\mathcal{D}_e$ denotes the domain of the element.

Substituting the discretized version of the elasticity matrix $\mathbf{D}(\mathbf{x}; \theta)$ in the preceding equation, we arrive at the following expression for the stochastic element stiffness matrix

$$k^e(\theta) = k_0^e + \sum_{j=1}^{M} k_j^e \xi_j, \qquad (21.18)$$

where

$$k_0^e = \int_{\mathcal{D}_e} \langle h(\mathbf{x}, \theta) \rangle \mathbf{B}^T \mathbf{D}_0 \mathbf{B} \, d\mathbf{x}. \qquad (21.19)$$

and

$$k_j^e = \int_{\mathcal{D}_e} h_j(\mathbf{x}) \mathbf{B}^T \mathbf{D}_0 \mathbf{B} \, d\mathbf{x}. \qquad (21.20)$$

Standard numerical quadrature schemes can be used to evaluate the integrals in Equation 21.19 and Equation 21.20. A detailed discussion of these implementation issues can be found in the literature [1, 27].

Assembly of the element stiffness matrices and application of the specified boundary conditions result in the following system of linear random algebraic equations

$$\left(\mathbf{K}_0 + \sum_{i=1}^{M} \mathbf{K}_i \xi_i\right) \mathbf{u}(\xi) = \mathbf{f}, \tag{21.21}$$

where $\mathbf{K}_0 \in \mathbb{R}^{n \times n}$ and $\mathbf{K}_i \in \mathbb{R}^{n \times n}$ are deterministic matrices and $\mathbf{u}(\xi) \in \mathbb{R}^n$ is the random displacement vector. $\mathbf{f} \in \mathbb{R}^n$ denotes the force vector that we assume to be deterministic for simplicity of presentation. The preceding equation can be rewritten as

$$\left(\sum_{i=0}^{M} \mathbf{K}_i \xi_i\right) \mathbf{u}(\xi) = \mathbf{f}, \tag{21.22}$$

where $\xi_0 = 1$.

A similar set of equations can also be arrived at, for time-dependent problems, by applying a time-stepping scheme or a frequency domain transform to Equation 21.16. In the case of linear structural systems, the equations of motion in the frequency domain is a system of complex linear random algebraic equations of the form

$$[\mathbf{K}(\xi) - \omega^2 \mathbf{M}(\xi) + \Im \omega \mathbf{C}(\xi)] \mathbf{u}(\xi, \omega) = \mathbf{f}, \tag{21.23}$$

where ω is the frequency of excitation and $\Im = \sqrt{-1}$.

It is to be noted here that Equation 21.22 is strictly valid only when the stiffness matrix is a linear function of ξ. For the more general case when uncertainties exist in the material properties as well as the geometric parameters of the system, $\mathbf{K}(\xi)$ will not be a linear function of ξ. In the sections that follow, we will outline how the projection schemes developed for Equation 21.22 can be extended to tackle such general cases.

21.3 Polynomial Chaos Expansions

The idea of polynomial chaos (PC) representations of stochastic processes was introduced by Wiener [18, 32] as a generalization of Fourier series expansion. More specifically, Wiener used multidimensional Hermite polynomials as basis functions for representing stochastic processes. The basic idea is to project the process under consideration onto a stochastic subspace spanned by a set of complete orthogonal random polynomials. To illustrate the process of constructing PC expansions, let $\phi_i(\theta), i = 1, 2, \ldots, \infty$ denote a set of polynomials that form an orthogonal basis in $L_2(\Theta, \mathcal{F}, \Gamma)$. Then, a general second-order stochastic process (i.e., a process with finite variance) $h(\theta)$ can be represented as

$$\begin{aligned} h(\theta) = c_0 \Phi_0 &+ \sum_{i_1=1}^{\infty} c_{i_1} \Phi_1(\xi_{i_1}(\theta)) + \sum_{i_1=1}^{\infty} \sum_{i_1=2}^{i_1} c_{i_1 i_2} \Phi_2(\xi_{i_1}(\theta), \xi_{i_2}(\theta)) \\ &+ \sum_{i_1=1}^{\infty} \sum_{i_2=1}^{i_1} \sum_{i_3=1}^{i_2} c_{i_1 i_2 i_3} \Phi_3(\xi_{i_1}(\theta), \xi_{i_2}(\theta), \xi_{i_3}(\theta)) + \cdots, \end{aligned} \tag{21.24}$$

where $\Phi_p(\xi_{i_1}, \xi_{i_2}, \ldots, \xi_{i_p})$ denote the generalized polynomial chaos of order p, which is a tensor product of one-dimensional polynomial basis functions $\phi_i, i = 1, 2, \ldots, p$.

TABLE 21.1 Choice of Askey Polynomials for Different Random Inputs

Random Variables (ξ)	Weiner-Askey Chaos ($\varphi(\xi)$)	Support
Gaussian	Hermite-Chaos	$(-\infty, \infty)$
Gamma	Laguerare-Chaos	$[0, \infty]$
Beta	Jacobi-Chaos	$[a, b]$
Uniform	Legendre-Chaos	$[a, b]$

In the original work of Wiener [18], Φ_p is chosen to be a multidimensional Hermite polynomial in terms of a set of uncorrelated Gaussian random variables $\xi_1, \xi_2, \ldots, \xi_p$ that have zero mean and unit variance. The general expression for the Hermite chaos of order p can be written as

$$\Phi_p(\xi_{i_1}, \xi_{i_2}, \ldots, \xi_{i_p}) = (-1)^p e^{\frac{1}{2}\xi^*\xi} \frac{\partial^p}{\partial \xi_{i_1} \ldots \partial \xi_{i_p}} \left[e^{-\frac{1}{2}\xi^*\xi} \right]. \tag{21.25}$$

For example, if Hermite polynomials are used as basis functions, a second-order, two-dimensional PC expansion of $h(\theta)$ can be written as

$$h(\theta) = h_0 + h_1\xi_1 + h_2\xi_2 + h_3\left(\xi_1^2 - 1\right) + h_4\xi_1\xi_2 + h_5\left(\xi_2^2 - 1\right). \tag{21.26}$$

It can be seen from the above equation that the first term of the PC expansion represents the mean value of $h(\theta)$ because ξ_1 and ξ_2 are uncorrelated Gaussian random variables with zero-mean and unit variance. Another point worth noting here is that the number of terms in the expansion grows very quickly with the dimension of ξ and the order of the expansion.

More recently, Xiu and Karniadakis [33] proposed a generalized PC approach that employs basis functions from the Askey family of orthogonal polynomials, which form a complete basis in the Hilbert space. The Hermite chaos expansion appears as a special case in this generalized approach, which is referred to as Wiener-Askey chaos. The motivation for this generalization arises from the observation that the convergence of Hermite chaos expansions can be far from optimal for non-Gaussian inputs. In such cases, the convergence rate can be improved by replacing Hermite polynomials with other orthogonal polynomials that best represent the input. Table 21.1 shows alternative basis functions suitable for different distributions of ξ. Numerical studies that demonstrate the improvement in convergence due to the use of generalized PC expansions can be found in the literature [33, 34].

For notational convenience, Equation 21.24 can be rewritten as

$$h(\theta) = \sum_{i=0}^{\infty} h_i \varphi_i(\xi), \tag{21.27}$$

where there is a one-to-one correspondence between the functions $\Phi_p(\xi_{i_1}, \xi_{i_2}, \ldots, \xi_{i_p})$ and $\varphi_i(\xi)$. Also note here that $\varphi_0 = 1$ and $\langle \varphi_i \rangle = 0$ for $i > 0$. Because $\varphi_i(\xi), i = 0, 1, 2, \ldots, \infty$ form an orthogonal basis in $L_2(\Theta, \mathcal{F}, \Gamma)$

$$\langle \varphi_i(\xi)\varphi_j(\xi) \rangle = \langle \varphi_i^2(\xi) \rangle \delta_{ij}, \tag{21.28}$$

where δ_{ij} is the Kronecker delta operator and $\langle . \rangle$ is the ensemble average operator, that is,

$$\langle f(\xi)g(\xi) \rangle = \int f(\xi)g(\xi)W(\xi)d\xi, \tag{21.29}$$

where $W(\xi)$ is the weight function corresponding to the PC basis.

The weight function is chosen to correspond to the distribution of the elements of ξ; see Table 21.1. For example, when Hermite polynomials are used as basis functions, the weight function is given by the M-dimensional normal distribution

$$W(\xi) = \frac{1}{\sqrt{(2\pi)^M}} e^{-\frac{1}{2}\xi^*\xi}. \tag{21.30}$$

Cameron and Martin [35] proved the following result, which guarantees that the Hermite chaos expansion converges in a mean-square sense for any second-order stochastic process when the number of terms is increased.

Theorem 1: *The Hermite chaos expansion of any (real or complex) functional $h(\xi)$ of $L_2(\Theta)$ converges in the $L_2(\Theta)$ sense to $h(\xi)$. This means that if $h(\xi)$ is a second-order stochastic process, that is,*

$$\int |h(\xi)|^2 W(\xi) d\xi < \infty \tag{21.31}$$

then

$$\int \left| h(\xi) - \sum_{i=0}^{m} h_i \varphi_i(\xi) \right|^2 W(\xi) d\xi \to 0 \quad \text{as} \quad m \to \infty, \tag{21.32}$$

where h_i is the Fourier-Hermite coefficient

$$h_i = \int h(\xi) \varphi_i(\xi) W(\xi) d\xi \tag{21.33}$$

The Cameron-Martin theorem can be generalized to arrive at the result that expansion of any second-order stochastic process in terms of basis functions from the Weiner-Askey family converges in the L_2 sense [33]. Another standard fact noted earlier in the literature is that the convergence rate of PC expansions is faster than exponential. Further, it has also been shown that the error in the expansion decays as $\mathcal{O}(\frac{1}{(p+1)!})$, where p is the highest order of Hermite polynomials used in the basis. More specifically, Hou et al. [36] presented the following convergence estimate for a one-dimensional PC expansion

$$\left\| h(\xi) - \sum_{i=0}^{m} h_i \varphi_i(\xi) \right\| \leq \frac{C}{(m+1)!} \left\| \frac{\partial^{m+1} h}{\partial \xi^{m+1}} \right\|, \tag{21.34}$$

where C is a constant.

The above estimate can be extended to the case when multidimensional Hermite polynomials are used as basis functions.

21.4 Polynomial Chaos Projection Schemes

In this section we outline a weak Galerkin projection scheme that can be used in conjunction with a PC expansion of the response process to solve stochastic PDEs. We consider the case when the stochastic projection scheme is applied to the spatially discretized version of the governing equations. As outlined in the introduction, we can also apply a stochastic projection scheme directly to the continuous form of the governing equations. However, both these approaches ultimately lead to similar sets of equations.

Consider the two-dimensional elasticity problem described in Section 21.2, where semidiscretization of the governing equations leads to the system of linear random algebraic equations given in Equation 21.22. In the PC projection scheme of Ghanem and Spanos [1], the random nodal displacements are first expanded using a set of multidimensional Hermite polynomials. This results in the following expansion for the response process

$$\mathbf{u}(\boldsymbol{\xi}) = \sum_{i=0}^{P-1} \mathbf{u}_i \varphi_i(\boldsymbol{\xi}), \tag{21.35}$$

where $\mathbf{u}_i \in \mathbb{R}^n$, $i = 0, 1, 2, \ldots, P-1$ are sets of vectors formed from the undetermined coefficients in the PC expansions for each nodal displacement, and $\varphi_i(\boldsymbol{\xi})$ is a set of orthogonal Hermite polynomials. The number of terms in the expansion, P, is given by

$$P = \sum_{k=0}^{p} \frac{(M+k-1)!}{k!(M-1)!}, \tag{21.36}$$

where p is the order of the PC expansion (i.e., the highest order of the set of Hermite polynomials $\varphi_i(\boldsymbol{\xi})$).

Substitution of the PC expansion for $\mathbf{u}(\boldsymbol{\xi})$ into the governing random algebraic equations given in Equation 21.22 gives

$$\left(\sum_{i=0}^{M} \mathbf{K}_i \xi_i \right) \left(\sum_{j=0}^{P-1} \mathbf{u}_j \varphi_j(\boldsymbol{\xi}) \right) = \mathbf{f}. \tag{21.37}$$

As shown by Ghanem and Spanos [1], the undetermined terms in the PC expansion can be uniquely computed by imposing the Galerkin condition, which involves orthogonalizing the stochastic residual error to the approximating subspace as shown below

$$\langle \boldsymbol{\varepsilon}(\boldsymbol{\xi}), \varphi_k(\boldsymbol{\xi}) \rangle = 0, \quad k = 0, 1, 2, \ldots, P-1, \tag{21.38}$$

where the stochastic residual error vector $\boldsymbol{\varepsilon}(\boldsymbol{\xi}) \in \mathbb{R}^n$ is given by

$$\boldsymbol{\varepsilon}(\boldsymbol{\xi}) = \left(\sum_{i=0}^{M} \mathbf{K}_i \xi_i \right) \left(\sum_{j=0}^{P-1} \mathbf{u}_j \varphi_j(\boldsymbol{\xi}) \right) - \mathbf{f}. \tag{21.39}$$

Substituting Equation 21.39 into Equation 21.38, we arrive at the following system of deterministic equations

$$\sum_{i=0}^{M} \sum_{j=0}^{P-1} \mathbf{K}_i \mathbf{u}_j \langle \xi_i \varphi_j \varphi_k \rangle = \langle \varphi_k \mathbf{f} \rangle \quad k = 0, 1, 2, \ldots, P-1. \tag{21.40}$$

The above equation can be rewritten in a more compact fashion as

$$\sum_{j=0}^{P-1} \mathbf{K}_{jk} \mathbf{u}_j = \mathbf{f}_k \quad k = 0, \ldots, P-1, \tag{21.41}$$

TABLE 21.2 Values of P for Different Values of M and p.

M	\multicolumn{5}{c}{Order of PC(p)}				
	0	1	2	3	4
2	1	3	6	10	15
4	1	5	15	35	70
6	1	7	28	83	210

where

$$\mathbf{K}_{jk} = \sum_{i=0}^{M} \langle \xi_i \varphi_j \varphi_k \rangle \mathbf{K}_i \in \mathbb{R}^{n \times n} \tag{21.42}$$

and

$$\mathbf{f}_k = \langle \varphi_k \mathbf{f} \rangle \in \mathbb{R}^n. \tag{21.43}$$

The expectation operations in Equation 21.42 and Equation 21.43 can be readily carried out using the properties of Hermite chaos; see, for example, [1, 27]. Now, expanding the above equation about the subscripts j and k, we arrive at the following system of linear algebraic equations

$$\begin{bmatrix} \mathbf{K}_{0,0} & \mathbf{K}_{0,1} & \cdot & \mathbf{K}_{0,P-1} \\ \mathbf{K}_{1,0} & \mathbf{K}_{1,1} & \cdot & \mathbf{K}_{1,P-1} \\ \cdot & \cdot & \cdot & \cdot \\ \cdot & \cdot & \cdot & \cdot \\ \mathbf{K}_{P-1,0} & \mathbf{K}_{P-1,1} & \cdot & \mathbf{K}_{P-1,P-1} \end{bmatrix} \begin{bmatrix} \mathbf{u}_0 \\ \mathbf{u}_1 \\ \cdot \\ \cdot \\ \mathbf{u}_{P-1} \end{bmatrix} = \begin{bmatrix} \mathbf{f}_0 \\ \mathbf{f}_1 \\ \cdot \\ \cdot \\ \mathbf{f}_{P-1} \end{bmatrix}, \tag{21.44}$$

which is of the form $\tilde{\mathbf{K}} \tilde{\mathbf{u}} = \tilde{\mathbf{f}}$, where $\tilde{\mathbf{K}} \in \mathbb{R}^{nP \times nP}$ and $\tilde{\mathbf{u}}, \tilde{\mathbf{f}} \in \mathbb{R}^{nP}$.

Table 21.2 shows the values of P for different values of p (order of the polynomial chaos) and M (number of terms in the random field discretization). It can be seen that the computational complexity and memory requirements of the PC projection scheme grow rapidly when M and p are increased. The memory requirements can be reduced by precomputing and storing the ensemble average terms of the form $\langle \xi_i \varphi_j \varphi_k \rangle$ along with the matrices $\mathbf{K}_i, i = 0, 1, 2, \ldots, P-1$, instead of storing the matrices $\mathbf{K}_{jk}$ given in Equation 21.42. Further, the sparsity of the tensor products $\langle \xi_i \varphi_j \varphi_k \rangle$ can also be exploited to accelerate the computations. A detailed overview of numerical schemes that exploit the peculiar structure of Equation 21.44 can be found in the literature [37, 38]. Anders and Hori [39] proposed the idea of using block Jacobi iteration to solve Equation 21.44. Here, it is only required to factorize the diagonal blocks of the coefficient matrix $\tilde{\mathbf{K}}$. An attractive feature of this approach is that the solution procedure can be easily parallelized.

After solving Equation 21.44 and substituting the results into Equation 21.35, we arrive at an explicit expression for the response process. This enables the statistics of the displacements as well as other response quantities of interest to be efficiently computed in the postprocessing phase. A more detailed discussion of issues involved in postprocessing the final solution is presented later.

21.4.1 Nonlinear Dependence of K on ξ

In this section we consider the case when the stiffness matrix is a nonlinear function of the basic random variables. This can occur when the Youngs modulus is described by a nonGaussian distribution or when the geometrical variables are assumed to be uncertain. First, consider the case when uncertainty in the Youngs modulus is described using a lognormal random field; that is, the Youngs modulus is given by

$$E(\mathbf{x}; \theta) = \exp(h(\mathbf{x}; \theta)), \tag{21.45}$$

where $h(\mathbf{x}; \theta)$ is a Gaussian random field.

Discretization of the random field $h(\mathbf{x}; \theta)$ results in the following representation of $E(\mathbf{x}; \theta)$ in terms of a finite number of random variables

$$E(\mathbf{x}; \boldsymbol{\xi}) = \exp\left(\langle h(\mathbf{x}; \theta) \rangle + \sum_{i=1}^{M} \xi_i h_i(\mathbf{x})\right), \tag{21.46}$$

where $\xi_i, i = 1, 2, \ldots, M$ are a set of uncorrelated Gaussian random variables.

Because $E(\mathbf{x}; \boldsymbol{\xi})$ is a random function, it admits a PC decomposition of the form

$$E(\mathbf{x}; \boldsymbol{\xi}) = \sum_{j=0}^{N} E_i \varphi_i(\boldsymbol{\xi}), \tag{21.47}$$

where φ_i are multidimensional Hermite polynomials in $\xi_1, \xi_2, \ldots, \xi_M$, and E_i are expansion coefficients.

The expansion coefficients E_i can be computed analytically using the properties of Hermite polynomials [27, 40]. However, note here that there are two levels of approximations. First, the expansion of the random field $h(\mathbf{x}; \theta)$ is truncated at the M term. Second, only the first N terms in the PC expansion of $E(\mathbf{x}; \boldsymbol{\xi})$ are retained.

Using Equation 21.47 we can now derive an expression for the element stiffness matrix along the lines of the procedure outlined in Section 21.2.2. Hence, the governing linear random algebraic equations presented in Equation 21.22 can be rewritten in the more general form

$$\left(\sum_{i=0}^{N} \varphi_i(\boldsymbol{\xi}) \mathbf{K}_i\right) \mathbf{u}(\boldsymbol{\xi}) = \mathbf{f}. \tag{21.48}$$

Equation 21.48 can be readily solved using the PC projection scheme presented earlier. The only major difference is that ensemble averages of the form $\langle \xi_i \varphi_j \varphi_k \rangle$ must be replaced with $\langle \varphi_i \varphi_j \varphi_k \rangle$. For example, Equation 21.40 now becomes

$$\sum_{i=0}^{N} \sum_{j=0}^{P-1} \mathbf{K}_i \mathbf{u}_j \langle \varphi_i \varphi_j \varphi_k \rangle = \langle \varphi_k \mathbf{f} \rangle \quad k = 0, 1, 2, \ldots, P-1. \tag{21.49}$$

A similar approach can be employed when the stiffness matrix is a general nonlinear function of the random variable vector $\boldsymbol{\xi}$. Here, we need to first compute the PC decomposition of $\mathbf{K}(\boldsymbol{\xi})$ to arrive at a random algebraic system of equations of the form given in Equation 21.48.

21.4.2 Remarks

Over the past decade, the PC projection scheme has been successfully applied to solve a wide range of problems in stochastic mechanics, including elasticity problems [1], random vibration [41], soil mechanics [42], transport process in heterogeneous media [43], plasticity problems [39, 44], soil-structure interaction problems [45], fluid dynamics [46], mid-frequency structural dynamics [47], and wave propagation in random media [48]. The Weiner-Askey chaos proposed by Xiu and Karniadakis [33] has recently been applied to solve diffusion problems [49], fluid-structures interaction problems [50], the

Navier-Stokes equations [51], and heat transfer problems [52] in the presence of parameter uncertainty. For a more detailed overview of the theoretical foundations of PC projection schemes and related implementation issues, the reader is referred to [53–57].

More recently, Mathelin and Hussaini [58] presented a stochastic collocation approach that can be employed in conjunction with a PC expansion of the solution process. The main idea is to apply a collocation method to collapse the multidimensional summations that appear in the standard Galerkin scheme into a one-dimensional summation. Numerical studies were presented for fluid flow problems to show that the collocation approach is computationally more efficient than the Galerkin projection scheme.

21.5 The Stochastic Krylov Subspace

In this section we present an alternative approach where the solution of Equation 21.22 is approximated using a stochastic reduced basis representation. The idea of using a stochastic reduced basis representation to solve linear random algebraic equations was proposed recently by Nair [20] and Nair and Keane [15]. It was shown that highly accurate approximations for the response process can be computed using a set of basis vectors spanning the preconditioned stochastic Krylov subspace. This approach is essentially a stochastic generalization of Krylov subspace methods in the numerical linear algebra literature that have been popularly applied to solve large-scale deterministic linear algebraic equations; see, for example, the text by Saad [59] for a detailed exposition. It is also of interest to note that numerical methods based on the Krylov subspace have a history of nearly 50 years of existence, and they continue to be an area of extensive research; see Saad and Van der Vorst [60] for an historical overview.

In the context of linear random algebraic equations, the main idea used here is to approximate the response process $\mathbf{u}(\boldsymbol{\xi})$ using basis vectors spanning the stochastic Krylov subspace defined below

$$\mathcal{K}_m(\mathbf{K}(\boldsymbol{\xi}), \mathbf{f}) = \text{span}\{\mathbf{f}, \mathbf{K}(\boldsymbol{\xi})\mathbf{f}, \mathbf{K}(\boldsymbol{\xi})^2\mathbf{f}, \cdots, \mathbf{K}(\boldsymbol{\xi})^{m-1}\mathbf{f}\}. \tag{21.50}$$

This representation of the response process can be justified by the following theorem [15], which establishes the applicability of the stochastic Krylov subspace for solving Equation 21.22.

Theorem 2: *If the minimal random polynomial of a nonsingular random square matrix $\mathbf{K}(\boldsymbol{\xi})$ has degree m, then the solution to $\mathbf{K}(\boldsymbol{\xi})\mathbf{u}(\boldsymbol{\xi}) = \mathbf{f}$ lies in the stochastic Krylov subspace $\mathcal{K}_m(\mathbf{K}(\boldsymbol{\xi}), \mathbf{f})$.*

The degree of the minimal polynomial (m) of a random matrix depends on the distribution of its eigenvalues. More specifically, the number of basis vectors required to compute accurate approximations depends on the degree of overlap of the PDFs of the eigenvalues of the coefficient matrix $\mathbf{K}(\boldsymbol{\xi})$[15]. To ensure good approximations using a small number of basis vectors, it is preferable to use a preconditioner. That is, we premultiply both sides of Equation 21.22 with the preconditioner $\mathbf{P} \in \mathbb{R}^{n \times n}$ to arrive at the following system of equations

$$\left(\sum_{i=0}^{M} \mathbf{P}\mathbf{K}_i \xi_i \right) \mathbf{u}(\boldsymbol{\xi}) = \mathbf{P}\mathbf{f}, \tag{21.51}$$

The key idea here is to choose a matrix $\mathbf{P}$ such that the PDFs of the eigenvalues of the random matrix $\mathbf{P}\sum_{i=0}^{M}\mathbf{K}_i\xi_i$ numerically tend to have a high degree of overlap. In [15, 20], the deterministic matrix $\langle \mathbf{K}(\boldsymbol{\xi}) \rangle^{-1} = \mathbf{K}_0^{-1}$ is used as the preconditioner.[6] This choice is motivated by the observation that $\mathbf{K}_0^{-1}\mathbf{K}(\boldsymbol{\xi})$

---

[6]If the matrix $\mathbf{K}(\boldsymbol{\xi}_0)^{-1}$ is used as the preconditioner, then the error in the stochastic reduced basis representation will converge faster to zero near the point $\boldsymbol{\xi}_0$. This feature of SRBMs can be exploited in practice to accurately estimate the statistics of the extremes. For example, in reliability analysis problems, $\boldsymbol{\xi}_0$ can be chosen to be the most probable point of failure.

will numerically behave like a matrix with a small number of distinct eigenvalues, particularly when the coefficients of variation of $\xi_i, i = 1, 2, \ldots, M$ are small. Note that, in theory, convergence can be guaranteed as long as the preconditioner is invertible. However, by using the preconditioner suggested here, convergence can be significantly accelerated; in other words, it becomes possible to achieve high accuracy using around three to four basis vectors.

A stochastic reduced basis representation of the response process can be written as

$$\hat{\mathbf{u}}(\xi) = \alpha_1 \psi_1(\xi) + \alpha_2 \psi_2(\xi) + \cdots + \alpha_m \psi_m(\xi) = \Psi(\xi)\alpha, \tag{21.52}$$

where $\Psi(\xi) = \{\psi_1(\xi), \psi_2(\xi), \ldots, \psi_m(\xi)\} \in \mathbb{R}^{n \times m}$ is a matrix of basis vectors spanning the preconditioned stochastic Krylov subspace $\mathcal{K}_m(\mathbf{K}_0^{-1}\mathbf{K}(\xi), \mathbf{K}_0^{-1}\mathbf{f})$ and $\alpha = \{\alpha_1, \alpha_2, \ldots, \alpha_m\}^T \in \mathbb{R}^m$ is a vector of undetermined coefficients.

The numerical studies conducted by Nair and Keane [15] and Sachdeva et al. [61] suggest that using the first three basis vectors spanning the preconditioned stochastic Krylov subspace, highly accurate results can be obtained. Using Equation 21.50, the first three basis vectors spanning $\mathcal{K}_m(\mathbf{K}_0^{-1}\mathbf{K}(\xi), \mathbf{K}_0^{-1}\mathbf{f})$ can be written as

$$\psi_1(\xi) = \mathbf{K}_0^{-1}\mathbf{f} \tag{21.53}$$

$$\psi_2(\xi) = \mathbf{K}_0^{-1}\mathbf{K}(\xi)\psi_1(\xi) \tag{21.54}$$

$$\psi_3(\xi) = \mathbf{K}_0^{-1}\mathbf{K}(\xi)\psi_2(\xi) \tag{21.55}$$

Because $\mathbf{K}(\xi) = \mathbf{K}_0 + \sum_{i=1}^{M} \xi_i \mathbf{K}_i$, the basis vectors can be compactly rewritten as follows

$$\psi_1(\xi) = \mathbf{u}_0 \tag{21.56}$$

$$\psi_2(\xi) = \sum_{i=1}^{M} \mathbf{d}_i \xi_i \tag{21.57}$$

$$\psi_3(\xi) = \sum_{i=1}^{M} \sum_{j=1}^{M} \mathbf{e}_{ij} \xi_i \xi_j \tag{21.58}$$

where $\mathbf{u}_0 = \mathbf{K}_0^{-1}\mathbf{f}$, $\mathbf{d}_i = \mathbf{K}_0^{-1}\mathbf{K}_i\mathbf{u}_0$ and $\mathbf{e}_{ij} = \mathbf{K}_0^{-1}\mathbf{K}_i\mathbf{d}_j$.

It can be clearly seen from the above expressions that the basis vectors are random polynomials that can be written as explicit functions of ξ. Because of the recursive representation of the basis vectors, they can be efficiently computed given the factored form of the preconditioner $\mathbf{K}_0^{-1}$, which is readily available as a byproduct of deterministic analysis of the problem. Another point worth noting is that the basis vectors coincide with the Neumann series[7] when the matrix $\mathbf{K}_0^{-1}$ is chosen to be the preconditioner. However, when a general preconditioner $\mathbf{K}(\xi_0)^{-1}$ is chosen, this observation does not hold true. Further, when the stiffness matrix depends nonlinearly on ξ, the basis vectors become nonlinear functions of ξ. For such cases, using the general representation of the governing

---

[7]The Neumann series for the solution of Equation 21.22 can be written as $(\mathbf{I} - \mathbf{K}_0^{-1}\Delta\mathbf{K} + (\mathbf{K}_0^{-1}\Delta\mathbf{K})^2 + \cdots)\mathbf{K}_0^{-1}\mathbf{f}$, where $\Delta\mathbf{K} = \sum_{i=1}^{M} \mathbf{K}_i \xi_i$; see, for example, [1, 16].

linear random algebraic equations given in Equation 21.48, the basis vectors can be written in terms of PC basis functions as

$$\psi_2(\xi) = \sum_{i=1}^{N} \mathbf{d}_i \varphi_i(\xi) \tag{21.59}$$

$$\psi_3(\xi) = \sum_{i=1}^{N} \sum_{j=1}^{N} \mathbf{e}_{ij} \varphi_i(\xi) \varphi_j(\xi), \tag{21.60}$$

where N is the number of terms retained in the PC decomposition of $\mathbf{K}(\xi)$.

21.6 Stochastic Reduced Basis Projection Schemes

In this section, we present some projection schemes that can be employed in conjunction with the stochastic Krylov subspace representation of the solution process. Because the formulations presented in this section lead to reduced-order systems of equations, we refer to them as stochastic reduced basis methods (SRBMs).

21.6.1 Weak Galerkin Scheme

To compute the vector of undetermined coefficients $\boldsymbol{\alpha}$ using the Galerkin scheme, we first substitute Equation 21.52 into the governing random algebraic equations given in Equation 21.22 to arrive at the following stochastic residual error vector

$$\varepsilon(\xi) = \left(\sum_{i=0}^{M} \mathbf{K}_i \xi_i \right) \Psi(\xi) \boldsymbol{\alpha} - \mathbf{f} \in \mathbb{R}^n. \tag{21.61}$$

If we restrict our attention to self-adjoint stochastic PDEs, the matrices $\mathbf{K}_i$, $i = 0, 1, 2, \ldots, M$ are guaranteed to be symmetric positive definite (which is the case for the static problem described in Section 21.2). Hence, the undetermined coefficients in Equation 21.52 can be computed by enforcing the Galerkin condition

$$\sum_{i=0}^{M} \xi_i \mathbf{K}_i \Psi(\xi) \boldsymbol{\alpha} - \mathbf{f} \perp \psi_j(\xi), \quad \forall j = 1, 2, \ldots, m. \tag{21.62}$$

The condition in Equation 21.62 demands that the stochastic residual error vector $\varepsilon(\xi)$ be made orthogonal to the approximating subspace $\Psi(\xi)$. Hence, the Galerkin condition is also referred to as an orthogonal projection scheme. We first consider the case when the orthogonality condition is imposed using the definition of inner products in the Hilbert space of random variables; see Equation 21.5. Here, because only the ensemble average of the random functions $\psi_i^*(\xi) \varepsilon(\xi)$, $i = 1, 2, \ldots, m$ are set to zero, we refer to Equation 21.62 as a weak Galerkin condition. As we show later, SRBMs based on a stronger Galerkin condition can also be formulated.

Application of the weak Galerkin condition results in the following reduced-order $m \times m$ deterministic system of equations for $\boldsymbol{\alpha}$:

$$\left[\sum_{i=0}^{M} \langle \xi_i \Psi^*(\xi) \mathbf{K}_i \Psi(\xi) \rangle \right] \boldsymbol{\alpha} = \langle \Psi^*(\xi) \mathbf{f} \rangle. \tag{21.63}$$

Because explicit expressions for the stochastic basis vectors are available, the expectation operations required to compute the elements of the reduced-order terms in Equation 21.63 can be readily carried out. The deterministic reduced-order 3×3 system of equations for the second-order SRBM ($m = 3$) are given below for the case $\xi_i, i = 1, 2, \ldots, M$ are uncorrelated Gaussian random variables. Note that for the sake of compactness, we have used the Einstein repeated index notation; for example, a repeated index i indicates summation with respect to that index over the range $1, 2, \ldots, M$.

$$\begin{bmatrix} \mathbf{u}_0^* \mathbf{K}_0 \mathbf{u}_0 & \langle \xi_i^2 \rangle \mathbf{u}_0^* \mathbf{K}_i \mathbf{d}_i & \langle \xi_i^2 \rangle \mathbf{u}_0^* \mathbf{K}_0 \mathbf{e}_{ii} \\ & \langle \xi_i^2 \rangle \mathbf{d}_i^* \mathbf{K}_0 \mathbf{d}_i & \langle \xi_i \xi_j \xi_k \xi_l \rangle \mathbf{d}_i^* \mathbf{K}_j \mathbf{e}_{kl} \\ \text{sym} & & \langle \xi_i \xi_j \xi_k \xi_l \rangle \mathbf{e}_{ij}^* \mathbf{K}_0 \mathbf{e}_{kl} \end{bmatrix} \begin{bmatrix} \alpha_1 \\ \alpha_2 \\ \alpha_3 \end{bmatrix} = \begin{bmatrix} \mathbf{u}_0^* \mathbf{f} \\ 0 \\ \langle \xi_i^2 \rangle \mathbf{e}_{ii}^* \mathbf{f} \end{bmatrix}. \quad (21.64)$$

Note that the terms involving fourth-order products of the form $\langle \xi_i \xi_j \xi_k \xi_l \rangle$ can be readily computed using the identity

$$\langle \xi_i \xi_j \xi_k \xi_l \rangle = \delta_{ij} \delta_{kl} + \delta_{ik} \delta_{jl} + \delta_{il} \delta_{jk}, \quad (21.65)$$

where δ is the Kronecker delta function.

As we show later, three or four basis vectors are sufficient to ensure high accuracy. Solving the preceding reduced-order system of equations and substituting the computed value of $\boldsymbol{\alpha}$ in Equation 21.52, we arrive at an explicit expression for $\mathbf{u}(\boldsymbol{\xi})$. Hence, similar to the PC representation, this enables the application of efficient postprocessing techniques to compute the statistics and distribution functions of the response quantities of interest.

Note that when two basis vectors are used, the stochastic reduced basis approximation is of first-order because the second basis vector is a linear function of random variables. Similarly, when three basis vectors are used, the approximation is of second-order.

21.6.2 Strong Galerkin Scheme

In this section we show how the weak Galerkin condition in Equation 21.62 can be reinterpreted to derive a stronger condition. The main idea is to enforce the condition that, for *each realization* of the random variable vector $\boldsymbol{\xi}$, the residual error vector $\boldsymbol{\varepsilon}(\boldsymbol{\xi})$ is orthogonal to the m basis vectors $\psi_1(\boldsymbol{\xi}), \psi_2(\boldsymbol{\xi}), \ldots, \psi_m(\boldsymbol{\xi})$. That is, the set of random functions $\psi_i^*(\boldsymbol{\xi}) \boldsymbol{\varepsilon}(\boldsymbol{\xi}), i = 1, 2, \ldots, m$, is zero with probability one. This condition, which we refer to as a strong Galerkin condition, can be stated as follows:

$$P[\psi_i^*(\boldsymbol{\xi}) \boldsymbol{\varepsilon}(\boldsymbol{\xi}) = 0] = 1 \quad \forall i = 1, 2, \ldots, m. \quad (21.66)$$

It is straightforward to show that the strong Galerkin condition will be satisfied only when $\boldsymbol{\alpha}$ is computed by solving the following reduced-order $m \times m$ *random* algebraic system of equations

$$\left[\sum_{i=0}^{M} \xi_i \Psi^*(\boldsymbol{\xi}) \mathbf{K}_i \Psi(\boldsymbol{\xi}) \right] \boldsymbol{\alpha}(\boldsymbol{\xi}) = \Psi^*(\boldsymbol{\xi}) \mathbf{f}. \quad (21.67)$$

As can be seen from the preceding equation, to satisfy the strong Galerkin condition, we need to model the undetermined coefficients $\alpha_1, \alpha_2, \ldots, \alpha_m$ as functions of $\boldsymbol{\xi}$, that is, the stochastic reduced basis approximation in Equation 21.52 must be rewritten as

$$\hat{\mathbf{u}}(\boldsymbol{\xi}) = \alpha_1(\boldsymbol{\xi}) \psi_1(\boldsymbol{\xi}) + \alpha_2(\boldsymbol{\xi}) \psi_2(\boldsymbol{\xi}) + \cdots + \alpha_m(\boldsymbol{\xi}) \psi_m(\boldsymbol{\xi}) = \Psi(\boldsymbol{\xi}) \boldsymbol{\alpha}(\boldsymbol{\xi}). \quad (21.68)$$

Initial studies presented by Nair and Keane [15] suggest that by using Equation 21.67 to compute $\boldsymbol{\alpha}$, it is possible to derive higher-order approximations that are significantly more accurate than SRBMs and the PC projection scheme employing the weak Galerkin condition. Unfortunately, because $\boldsymbol{\alpha}$ now becomes a highly nonlinear function of $\boldsymbol{\xi}$, analytical expressions for the response statistics are difficult to obtain.

One way to relax the strong Galerkin condition in Equation 21.66 is to employ a PC decomposition of the undetermined coefficient vector $\boldsymbol{\alpha}(\boldsymbol{\xi})$, that is,

$$\boldsymbol{\alpha}(\boldsymbol{\xi}) = \sum_{i=0}^{P_1-1} \boldsymbol{\alpha}_i \varphi_i(\boldsymbol{\xi}), \tag{21.69}$$

where $\boldsymbol{\alpha}_i \in \mathbb{R}^m, i = 0, 1, 2, \ldots, P_1 - 1$ are undetermined vectors. To compute these undetermined coefficient vectors, we substitute Equation 21.69 into Equation 21.67 and apply the Galerkin projection scheme; that is, the stochastic residual error in satisfying Equation 21.67 is made orthogonal to the PC basis functions. This results in the following deterministic system of equations

$$\left\langle \varphi_k(\boldsymbol{\xi}) \left[\sum_{i=0}^{M} \xi_i \Psi^*(\boldsymbol{\xi}) \mathbf{K}_i \Psi(\boldsymbol{\xi}) \right] \sum_{j=0}^{P_1-1} \boldsymbol{\alpha}_j \varphi_j(\boldsymbol{\xi}) \right\rangle = \langle \varphi_k(\boldsymbol{\xi}) \Psi^*(\boldsymbol{\xi}) \mathbf{f} \rangle \quad \forall k = 0, 1, 2, \ldots, P_1 - 1. \tag{21.70}$$

The above equation can rewritten as

$$\sum_{i=0}^{M} \sum_{j=0}^{P_1-1} \langle \xi_i \varphi_k(\boldsymbol{\xi}) \varphi_j(\boldsymbol{\xi}) \Psi^*(\boldsymbol{\xi}) \mathbf{K}_i \Psi(\boldsymbol{\xi}) \rangle \boldsymbol{\alpha}_j = \langle \varphi_k(\boldsymbol{\xi}) \Psi^*(\boldsymbol{\xi}) \mathbf{f} \rangle \quad \forall k = 0, 1, 2, \ldots, P_1 - 1. \tag{21.71}$$

It can be seen that Equation 21.71 is essentially an $mP_1 \times mP_1$ system of deterministic linear algebraic equations. Hence, the strong Galerkin formulation based on PC expansion of $\boldsymbol{\alpha}$ is equivalent to employing the augmented set of basis vectors $[\Psi(\boldsymbol{\xi}), \varphi_1(\boldsymbol{\xi})\Psi(\boldsymbol{\xi}), \ldots, \varphi_{P_1-1}(\boldsymbol{\xi})\Psi(\boldsymbol{\xi})] \in \mathbb{R}^{n \times mP_1}$ in conjunction with the weak Galerkin condition. In comparison, the standard weak Galerkin SRBM presented in Section 21.6.1 results in an $m \times m$ system of equations. However, because the number of basis vectors $m <<< n$, the preceding system of equations is still significantly smaller than the system of equations arising from the standard PC projection scheme outlined earlier in Section 21.4.

The strong Galerkin scheme based on PC decomposition of $\boldsymbol{\alpha}$ can also be viewed as one possible way to hybridize the stochastic Krylov subspace representation with the PC projection scheme of Ghanem and Spanos [1]. Another important point worth noting is that the hybrid scheme presented here is a generalization of the weak Galerkin scheme. This is because if we set $P_1 = 1$ in Equation 21.69, we recover the weak Galerkin projection scheme presented in Section 21.6.1. We numerically demonstrate later that this hybrid approach gives better results than the weak Galerkin formulation of SRBMs as well as the PC projection scheme.

21.6.3 Petrov-Galerkin Scheme

An alternative formulation for computing the undetermined coefficients in SRBMs can be derived by employing the Petrov-Galerkin scheme. This is an oblique projection scheme that essentially involves enforcing the condition that the stochastic residual error vector is orthogonal to the subspace $\mathbf{K}(\boldsymbol{\xi})\Psi(\boldsymbol{\xi})$; that is,

$$\boldsymbol{\varepsilon}(\boldsymbol{\xi}) \perp \mathbf{K}(\boldsymbol{\xi})\psi_i(\boldsymbol{\xi}) \quad \forall i = 1, 2, \ldots, m. \tag{21.72}$$

Note that imposition of the above condition is equivalent to directly minimizing the L_2 norm of the residual error vector $\varepsilon(\xi)$. Enforcing the Petrov-Galerkin condition using the definition of orthogonality in the Hilbert space of random variables, we arrive at the following reduced-order $m \times m$ deterministic system of equations for $\boldsymbol{\alpha}$:

$$\langle \boldsymbol{\Psi}^*(\xi) \mathbf{K}^*(\xi) \mathbf{K}(\xi) \boldsymbol{\Psi}(\xi) \rangle \boldsymbol{\alpha} = \langle \boldsymbol{\Psi}^*(\xi) \mathbf{K}^*(\xi) \mathbf{f} \rangle. \tag{21.73}$$

Using Equation 21.22, we have

$$\left[\sum_{i=0}^{M} \sum_{j=0}^{M} \langle \xi_i \xi_j \boldsymbol{\Psi}^*(\xi) \mathbf{K}_i^* \mathbf{K}_j \boldsymbol{\Psi}(\xi) \rangle \right] \boldsymbol{\alpha} = \sum_{i=0}^{M} \langle \xi_i \boldsymbol{\Psi}^*(\xi) \mathbf{K}_i^* \mathbf{f} \rangle. \tag{21.74}$$

It can be seen from the preceding equation that in comparison to the weak Galerkin scheme, application of the Petrov-Galerkin condition leads to the requirement of computing higher-order ensemble averages. Note that it is also possible to formulate a strong Petrov-Galerkin condition along the lines of the approach used earlier in the strong Galerkin scheme. Another important point worth noting here is that the Petrov-Galerkin scheme ensures that the L_2 norm of the residual error, $\langle \boldsymbol{\varepsilon}^*(\xi) \boldsymbol{\varepsilon}(\xi) \rangle$, converges in a mean-square sense when the number of basis vectors is increased. As discussed in Section 21.6.5, the convergence of the Galerkin scheme can be proved only when $\mathbf{K}(\xi)$ is a Hermitian positive definite matrix. Hence, SRBMs based on the Petrov-Galerkin scheme are expected to work well for linear stochastic structural dynamic analysis in the frequency domain, because the dynamic stiffness matrix is nonHermitian; see [62] for a more detailed discussion of this point.

21.6.4 Nonlinear Dependence of K on ξ

Let us now apply SRBMs to the general case when $\mathbf{K}$ is a nonlinear function of ξ or a lognormal random field is used to describe uncertainty in the Youngs modulus. For such problems, a PC decomposition of $\mathbf{K}(\xi)$ can be used to arrive at a system of linear random algebraic equations of the form given earlier in Equation 21.48. As outlined in Section 21.5, for such a representation of the stiffness matrix, the basis vectors spanning the preconditioned stochastic Krylov subspace become functions of the PC basis. Using Equation 21.59 and Equation 21.60, the matrix of stochastic basis vectors can be written as

$$\boldsymbol{\Psi}(\xi) = \left[\mathbf{u}_0, \sum_{i=1}^{N} \mathbf{d}_i \varphi_i(\xi), \sum_{i=1}^{N} \sum_{j=1}^{N} \mathbf{e}_{ij} \varphi_i(\xi) \varphi_j(\xi), \cdots \right] \in \mathbb{R}^{n \times m}, \tag{21.75}$$

where the terms $\mathbf{d}_i$ and $\mathbf{e}_{ij}$ have been defined earlier in Section 21.5.

It is conceptually very straightforward to apply any of the projection schemes outlined earlier when the basis vectors are given in this form. However, from an implementation point of view, it may be preferable to first construct a PC decomposition of $\boldsymbol{\Psi}(\xi)$. That is, we compute a PC decomposition of the basis vectors spanning the preconditioned stochastic Krylov subspace. Fortunately, because the basis vectors are polynomials in ξ, the expansion coefficients can be calculated analytically. This leads to the following representation of the matrix of basis vectors used in SRBMs.

$$\boldsymbol{\Psi}(\xi) = \sum_{i=0}^{P_2-1} \boldsymbol{\Psi}_i \varphi_i(\xi), \tag{21.76}$$

where $\boldsymbol{\Psi}_i \in \mathbb{R}^{n \times m}, i = 0, 1, 2, \ldots, P_2 - 1$ are deterministic matrices.

Note that the PC expansion of $\boldsymbol{\Psi}(\xi)$ does not lead to an increase in the number of basis vectors. Now, applying the weak Galerkin scheme using the above set of basis vectors to Equation 21.48, we arrive at the following deterministic system of equations

$$\left(\sum_{i=0}^{P_2-1}\sum_{j=0}^{N}\sum_{k=0}^{P_2-1}\langle\varphi_i(\xi)\varphi_j(\xi)\varphi_k(\xi)\rangle\Psi_i^*\mathbf{K}_j\Psi_k\right)\boldsymbol{\alpha}=\sum_{i=0}^{P_2-1}\langle\varphi_i(\xi)\rangle\Psi_i^*\mathbf{f} \qquad (21.77)$$

The above equation is analogous to that presented earlier in Equation 21.40 in the context of the PC projection scheme. However, in contrast to the PC projection scheme, Equation 21.77 is a reduced-order $m \times m$ system of equations. Note that the derivation presented here can be extended to the case when the strong Galerkin condition or the Petrov-Galerkin scheme is employed to compute the undetermined coefficients in SRBMs.

21.6.5 Theoretical Analysis of Convergence

The Galerkin scheme has a number of interesting properties because it is an orthogonal projection scheme. Consider the case when the coefficient matrix $\mathbf{K}(\xi)$ is Hermitian positive definite. Further, let $\mathcal{K}_m$ denote the stochastic subspace spanned by the set of orthogonal[8] basis vectors $\psi_1(\xi), \psi_2(\xi), \ldots, \psi_m(\xi)$. Let us first rewrite the weak Galerkin condition given in Equation 21.62 as

$$\langle\{\mathbf{K}(\xi)\hat{\mathbf{u}}(\xi)-\mathbf{f}\}^*\psi_i(\xi)\rangle = 0, \quad \forall \; \psi_i(\xi) \in \mathcal{K}_m. \qquad (21.78)$$

Substituting $\mathbf{f} = \mathbf{K}(\xi)\mathbf{u}(\xi)$, the above condition becomes

$$\langle\{\hat{\mathbf{u}}(\xi)-\mathbf{u}(\xi)\}^*\mathbf{K}^*(\xi)\psi_i(\xi)\rangle = 0, \quad \forall \; \psi_i(\xi) \in \mathcal{K}_m. \qquad (21.79)$$

From the preceding equation, it can be seen that the weak Galerkin scheme ensures that the difference between the exact and the approximate solution is **K**-orthogonal to the approximating space $\mathcal{K}_m$. Now, using the elementary properties of orthogonal projectors [59], it can be shown that Equation 21.79 is the necessary and sufficient condition for $\hat{\mathbf{u}}(\xi)$ to be a minimizer of the error function $\langle\{\hat{\mathbf{u}}(\xi)-\mathbf{u}(\xi)\}^*\mathbf{K}\{\hat{\mathbf{u}}(\xi)-\mathbf{u}(\xi)\}\rangle$. This result, which establishes the optimality of the weak Galerkin scheme, can be stated as follows [15].

Theorem 3: Let $\hat{\mathbf{u}}(\xi) = \Psi(\xi)\boldsymbol{\alpha}$ be a stochastic reduced basis approximation to the solution of $\mathbf{K}(\xi)\mathbf{u}(\xi) = \mathbf{f}$, where $\mathbf{K}(\xi) \in \mathbb{R}^{n \times n}$ is a random Hermitian positive definite matrix, $\mathbf{u}(\xi), \mathbf{f} \in \mathbb{R}^n$ are random vectors, $\Psi(\xi) \in \mathbb{R}^{n \times m}$ is a matrix of stochastic basis vectors, and $\boldsymbol{\alpha} \in \mathbb{R}^n$ is a vector of undetermined coefficients. If the coefficient vector $\boldsymbol{\alpha}$ is computed by imposing the condition $\mathbf{K}(\xi)\Psi(\xi)\boldsymbol{\alpha} - \mathbf{f} \perp \Psi(\xi)$, then the following deterministic error function is minimized.

$$\Delta_m = \langle\{\mathbf{u}(\xi)-\hat{\mathbf{u}}(\xi)\}^*\mathbf{K}(\xi)\{\mathbf{u}(\xi)-\hat{\mathbf{u}}(\xi)\}\rangle. \qquad (21.80)$$

where Δ_m denotes the **K**-norm of the error.

An important corollary of Theorem 3 is that the weak Galerkin projection scheme used here ensures that $\{\mathbf{u}(\xi)-\hat{\mathbf{u}}(\xi)\}^*\mathbf{K}(\xi)\{\mathbf{u}(\xi)-\hat{\mathbf{u}}(\xi)\}$ converges in a mean-square sense when the number of basis vectors is increased (i.e., $\Delta_{m+1} \leq \Delta_m$) where Δ_{m+1} and Δ_m denote the **K**-norm of the error in the solution computed using $m+1$ and m basis vectors, respectively. Hence, it also follows that application of the strong Galerkin condition results in a lower value for the **K**-norm of the error compared to the weak Galerkin scheme. This is because, as shown in Section 21.6.2, application of the strong Galerkin condition is equivalent to using an augmented set of basis vectors. A similar set of results can be established for the L_2 norm of the residual for nonHermitian coefficient matrices when the Petrov-Galerkin scheme outlined in Section 21.6.3 is employed.

It is also worth noting that the **K**-norm of the error can be interpreted as an energy norm. This has an important practical ramification for stochastic structural systems. Because SRBMs employing the Galerkin scheme minimize the **K**-norm of the error, the results are bound to be more accurate for those

---

[8] Note that a orthogonal set of basis vectors spanning the preconditioned stochastic Krylov subspace can be computed using the stochastic version of Arnoldi's method presented in Nair [20].

DOF that contain the most strain energy. This is a very useful property because in many practical applications, such as reliability analysis, we are primarily interested in ensuring good approximations for the highly stressed regions of the structure.

21.7 Postprocessing Techniques

In this section we briefly outline how the final expressions for the solution process obtained using various projection schemes can be postprocessed to compute the statistics of interest.

21.7.1 Statistical Moments and Distributions

By virtue of the orthogonality of the basis functions in the PC representation, it becomes possible to derive explicit expressions for the mean and variance of the response once the undetermined coefficient vectors in the PC expansion are computed using a projection scheme. The mean of the response is simply given by $\mathbf{u}_0$ because, by definition, $\langle \varphi_0 \rangle = 1$ and $\langle \varphi_i(\xi) \rangle = 0$ for $i > 0$. The response covariance matrix can be computed as

$$\text{Cov}[\mathbf{u}(\xi), \mathbf{u}(\xi)] = \sum_{i=1}^{P-1} \langle \varphi_i^2 \rangle \mathbf{u}_i \mathbf{u}_i^* \tag{21.81}$$

In the case of SRBMs employing the weak Galerkin condition, we ultimately arrive at a random polynomial for the response process in the form

$$\hat{\mathbf{u}}(\xi) = \sum_{i=0}^{m} \alpha_i \psi_i(\xi). \tag{21.82}$$

Because the coefficients α_i are deterministic scalars, the mean and covariance of the response can be expressed in terms of the statistics of the basis vectors as follows:

$$\langle \hat{\mathbf{u}}(\xi) \rangle = \sum_{i=0}^{m} \alpha_i \langle \psi_i(\xi) \rangle, \tag{21.83}$$

and

$$\text{Cov}[\mathbf{u}(\xi), \mathbf{u}(\xi)] = \langle \Psi(\xi) \alpha \alpha^* \Psi^*(\xi) \rangle = \sum_{i=0}^{m} \sum_{j=0}^{m} \alpha_i \alpha_j \langle \psi_i(\xi) \psi_j^*(\xi) \rangle. \tag{21.84}$$

Similarly, the mean and covariance of the response can also be computed for SRBMs employing the strong Galerkin condition and the Petrov-Galerkin scheme. Note that the expectation operations in the preceding equations can be analytically carried out using the joint statistics of ξ; see, for example, McCullagh [63] and Chapter 4 of Ghanem and Spanos [1] for a detailed exposition on statistical analysis of random polynomials. It is also possible to derive expressions for the statistical moments of quantities such as stress and strain, because they are functions of the random displacements.

21.7.2 Reliability Analysis

Because all the projection schemes considered here lead to explicit expressions for the solution process, its complete statistical characterization becomes computationally feasible. For example, we can employ simulation schemes in conjunction with kernel density estimation techniques [64] to approximate the PDF of the response quantities given the final expression for the solution process in terms of a random polynomial. Alternatively, the wide body of numerical methods developed for reliability analysis such as the first-order reliability method (FORM), second-order reliability method (SORM) (see, for example, Chapter 14 on reliability analysis by Der Kiureghian in this handbook), and importance sampling (see, for example, the chapter 20 on Monte Carlo simulation in this handbook), can be applied to the explicit expressions for the solution process obtained using any of the projection schemes.

Sudret and Der Kiureghian [27, 65] have conducted a detailed investigation into the application of PC projection schemes to reliability analysis. They suggested the use of FORM to post-process the expression obtained using the PC projection scheme to compute the probability of failure. Further, it was shown that a sensitivity analysis approach can be used to efficiently compute the PDF of the response and derivatives of the reliability index. These procedures are also applicable to SRBMs.

21.8 Numerical Examples

In this section we apply SRBMs and PC projection schemes to compute the response statistics and PDFs for two example problems. The results obtained are benchmarked against those obtained using Monte Carlo simulation (MCS) with a sample size of 250,000. For all the problems considered, we use the following two-dimensional exponential correlation function to represent uncertainty in the random Youngs modulus:

$$R(\mathbf{x}, \mathbf{y}) = \exp\left(-\frac{|x_1 - x_2|}{b_1} - \frac{|y_1 - y_2|}{b_2}\right), \qquad (21.85)$$

where b_1 and b_2 are the correlation lengths of the random field.

Recollect that in the first-order, second-order, and third-order SRBMs, we use two, three, and four basis vectors, respectively. Henceforth, we refer to the first-order SRBM, second-order SRBM, and third-order SRBM as SRBMI, SRBMII, and SRBMIII, respectively. The first-order PC, second-order PC, and third-order PC schemes are referred to as PCI, PCII, and PCIII, respectively. We also present results for the strong Galerkin approach presented in Section 21.6.2, when the undetermined coefficients in the stochastic reduced basis representation are represented using a first-order PC expansion. These hybrid formulations are referred to as SRBMIPCI (where a first-order PC expansion is used in Equation 21.69 and two basis vectors are used) and SRBMIIPCI (where a first-order PC expansion is used in Equation 21.69 and three basis vectors are used). For comparison, we also show results obtained using the first- and second-order perturbation method, referred to as PERI and PERII, respectively.

The numerical studies presented here were conducted using the "SSFEM toolbox" developed by Sudret and Der Kiureghian [27, 65]. This toolbox contains routines for discretizing random fields, stochastic finite element formulation based on four-node quadrilateral elements and an implementation of the PC expansion scheme for solving random algebraic equations. Routines implementing SRBMs were integrated with this toolbox to enable a systematic comparison of the various projection schemes considered. Note that all the problems considered here were spatially discretized using four-node quadrilateral elements with two DOF per node. Further, the random field representing uncertainty in the Youngs modulus is discretized using the KL expansion scheme.

21.8.1 Two-Dimensional Thin Plate

The first problem considered is a thin square plate of unit length clamped at one edge and subjected to uniform inplane tension at the opposite edge (taken from [1]). The plate is discretized into 16 square elements as shown in Figure 21.1, which leads to a total of 50 DOF. The external loads are assumed to be deterministic and of unit magnitude. The Youngs modulus of the plate is modeled as a two-dimensional Gaussian random field with the exponential correlation model given in Equation 21.85 with $b_1 = b_2 = 1$. The random field is discretized using the KL expansion scheme and four terms are retained (i.e., $M = 4$).

Numerical studies were conducted to analyze the performance of the stochastic subspace projection schemes when the coefficient of variation of the random Youngs modulus is increased. Typical trends in the results are presented here for the case when the coefficient of variation is 0.2. The percentage errors in the mean and standard deviation of the displacement at point E (the top right-hand corner of the

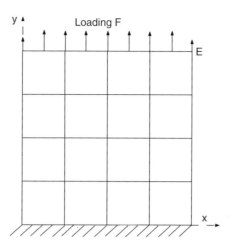

FIGURE 21.1 Static plate analysis problem.

plate shown in Figure 21.1) are shown in Figure 21.2 and Figure 21.3. It can be seen from the results that the errors in the projection schemes decrease rapidly when the number of terms in the expansion is increased. The performance of SRBMs is comparable to PC projection schemes of the same order. This is impressive, considering the fact that SRBMs solve a reduced-order system of equations, whereas the PC projection scheme involves the solution of a system of equations with increased dimensionality.

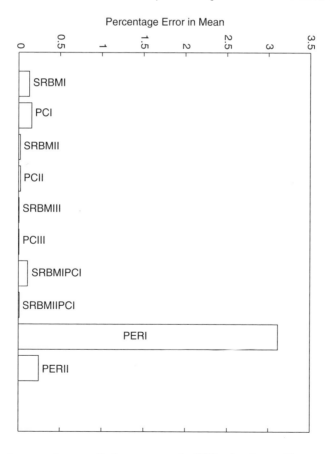

FIGURE 21.2 Percentage error in mean displacement at point "E" for the plate problem.

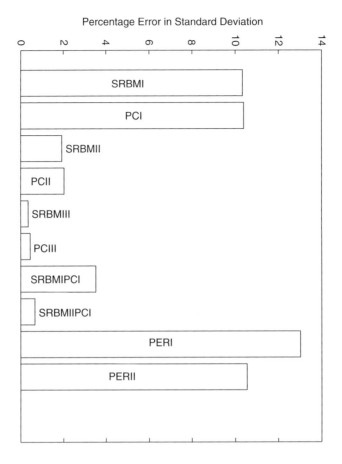

FIGURE 21.3 Percentage error in standard deviation of displacement at point "E" for the plate problem.

It can also be observed from Figure 21.2 and Figure 21.3 that the hybrid schemes motivated by the strong Galerkin condition give better results than the weak Galerkin scheme; that is, SRBMIPCI is better than SRBMI and SRBMIIPCI is better than SRBMII. In fact, the performance of SRBMIIPCI is comparable to SRBMIII and PCIII. The results obtained using PERII are comparable to SRBMI and PCI. However, as is well known, the rate of convergence of perturbation methods is very low compared to the projection schemes used here.

The PDFs of the displacement at point "E" obtained using second- and third-order projection schemes are shown in Figure 21.4. The displacement PDFs for the various projection schemes were calculated by applying a simulation scheme with a sample size of 250,000 to the obtained expressions for the response process. The accuracy of the approximations to the tail of the PDF is shown in Figure 21.5. These figures indeed confirm that the performance of SRBMs is comparable to the PC projection schemes. A more detailed comparison between the performance of SRBMs and PC projection schemes can be found in [61, 66].

21.8.2 Analysis of Foundation on Randomly Heterogeneous Soil

Next we consider a geotechnical problem that involves settlement analysis of a foundation on a randomly heterogeneous soil. This problem was studied by Sudret and Kiureghian [27, 65] in the context of reliability analysis. Consider an elastic soil layer of thickness t lying on a rigid stratum as shown in Figure 21.6. A uniform pressure P is applied over a length $2B$ of the free surface. The soil is modeled as an elastic linear isotropic material and plane strain analysis is carried out. Exploiting symmetry considerations, only one half of the structure is modeled by finite elements. The values of the soil parameters used in the computations are given in Figure 21.6.

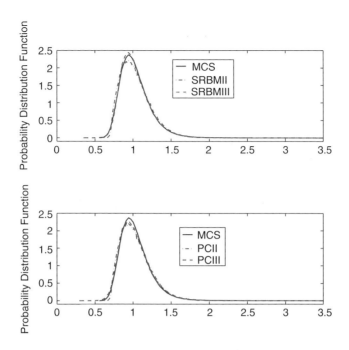

FIGURE 21.4 Comparison of displacement PDF at point "E" using second- and third-order SRBMs and PC projection schemes.

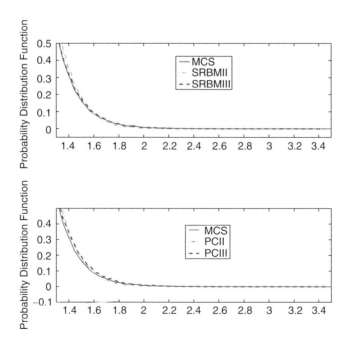

FIGURE 21.5 Comparison of tail of displacement PDF at point "E" using second- and third-order SRBMs and PC projection schemes.

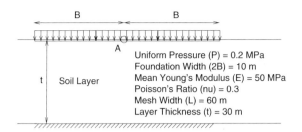

FIGURE 21.6 Soil foundation problem.

The Youngs modulus of the soil is assumed to be a homogeneous Gaussian random field. The Youngs modulus is considered to vary only in the vertical direction and hence a one-dimensional random field with exponential correlation function (see Equation 21.85) is employed. The correlation length b_2 is set at 30 m. The KL expansion scheme was used to discretize this random field and four terms are retained in the expansion (i.e., $M = 4$). An optimal mesh with 99 nodes and 80 elements developed by Sudret and Kiureghian [65] was used for the analysis.

Numerical studies were conducted to analyze the performance of the projection schemes when the coefficient of variation of the random Youngs modulus is set at 0.2. Results are presented here for the statistics of the vertical displacement at point "A", which is the center of the foundation as shown in Figure 21.6. The percentage errors in the mean and standard deviation of the displacement at point "A" are shown in Figure 21.7

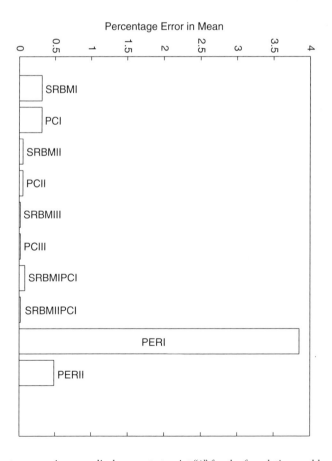

FIGURE 21.7 Percentage error in mean displacement at point "A" for the foundation problem.

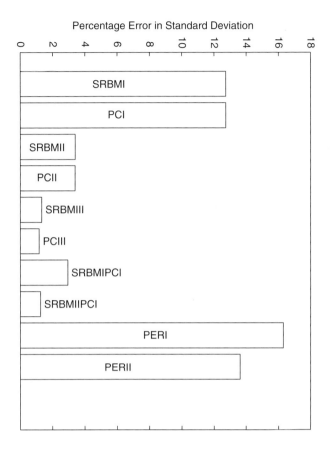

FIGURE 21.8 Percentage error in standard deviation of displacement at point "A" for the foundation problem.

and Figure 21.8. It can be seen that the trends in the results are similar to those obtained earlier for the plate problem.

The PDFs of the displacement at point "A" computed using second- and third-order projection schemes are shown in Figure 21.9. Figure 21.10 shows how well the stochastic subspace projection schemes approximate the tail of the PDF.

It can be observed from the results that the performance of SRBMs is comparable to PC projection schemes of equivalent order. In comparison, both the first- and second-order perturbation methods perform worse than the stochastic subspace projection schemes. The idea of hybrid schemes motivated by the strong Galerkin condition outlined earlier in Section 21.6.2 appear to work very well. For example, both SRBMII and SRBMIIPCI use the first three basis vectors spanning the preconditioned stochastic Krylov subspace. The only difference is that, in SRBMIIPCI, we use a first-order PC expansion of the vector of undetermined coefficients α; see Equation 21.69. Because four terms are used in the KL expansion ($M = 4$) and three basis vectors are employed ($m = 3$), it follows that $P_1 = 5$ in Equation 21.69. Hence, the SRBMIIPCI formulation leads to a 15×15 deterministic system of equations. In contrast, SRBMII solves a 3×3 system of equations. However, the results obtained using SRBMIIPCI are comparable to PCIII and better than those obtained using PCI and PCII. For reference, PCI, PCII, and PCIII projection schemes involve the solution of a 845×845, 2535×2535, and 5915×5915 system of equations, respectively. This suggests that SRBMIIPCI is capable of providing results of better accuracy than PCII at a significantly lower computational cost.

Finally, we applied the projection schemes to reliability analysis of the foundation, when the limit-state function is defined as

$$g(\xi) = u_{max} - u_A(\xi), \tag{21.86}$$

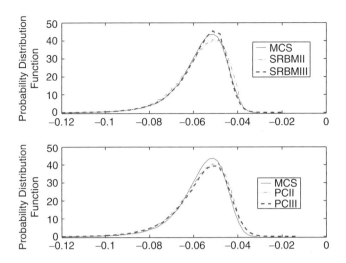

FIGURE 21.9 Comparison of displacement PDF at point "A" using second- and third-order SRBMs and PC projection schemes for the foundation problem.

where $u_A(\xi)$ is the displacement of the foundation at point "A" and u_{max} is an admissible threshold set to 10 cm.

The percentage errors in the probability of failure (P_f) computed using various methods are shown in Figure 21.11. P_f computed using MCS with sample size 250,000 is considered to be the benchmark result against which all methods are compared. The same sample size was also used for all the projection schemes. The percentage error in P_f using FORM has been taken from Sudret and Kiureghian [65] and is shown here for the sake of comparison. It can be seen that the performance of the third-order projection schemes is comparable to FORM. The hybrid scheme SRBMIIPCI gives the best result for this problem. This suggests that the hybrid scheme based on the strong Galerkin condition gives good approximations for the tail of the PDF. Note that none of the projection schemes make use of the most probable point (MPP) of failure. Further, it should be pointed out here that the performance of SRBMs in reliability

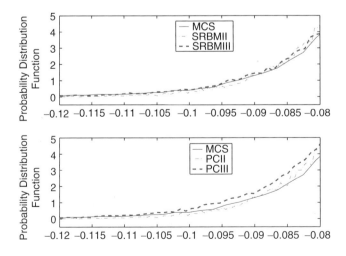

FIGURE 21.10 Comparison of tail of displacement PDF at point "A" using second- and third-order SRBMs and PC projections schemes for the foundation problem.

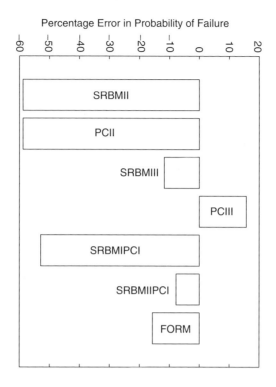

FIGURE 21.11 Comparison of various projection schemes and FORM for reliability assessment.

analysis can be potentially improved further using the inverse of the stiffness matrix computed at the MPP as a preconditioner.

21.9 Concluding Remarks and Future Directions

In this chapter we presented an overview of the theoretical foundations of stochastic subspace projection schemes in the context of stochastic finite element analysis. Attention focused on polynomial chaos (PC) expansion schemes, and stochastic reduced basis methods (SRBMs) that employ basis vectors spanning the preconditioned stochastic Krylov subspace. Numerical studies were presented for two-dimensional elasticity problems to compare the relative accuracies of the projection schemes. The results suggest that SRBMs based on the weak Galerkin condition give results that are comparable in accuracy to standard PC projection schemes, while incurring significantly lower computational cost. Further, it appears that the idea of hybridizing the preconditioned stochastic Krylov subspace representation with a PC decomposition of the undetermined coefficients can be used to develop highly accurate and efficient computational schemes for solving linear random algebraic equations. Hybrid projection schemes based on this idea are expected to be not only more accurate than PC projection schemes, but also orders of magnitude faster, particularly for large-scale problems with many random variables.

In principle, the projection schemes developed for linear random algebraic equations can serve as the essential backbone for solving nonlinear algebraic equations. For example, the projection scheme outlined in the introduction for the discrete case could be applied irrespective of whether the problem is linear or nonlinear. However, very few papers on the application of stochastic subspace projection schemes to complex nonlinear systems exist in the literature. In comparison, perturbation methods can be readily applied to static and dynamic analysis of nonlinear stochastic systems when the coefficients of variation of the random variables are small; see, for example, [67] and references therein. For a detailed exposition of the issues involved in applying PC projection schemes to nonlinear stochastic PDEs, the reader is referred to [39, 44, 46, 56, 68]. However, SRBMs are yet to be applied to nonlinear problems. It is hoped that the ideas developed in the context of the PC projection approach can be leveraged to design efficient stochastic reduced basis projection schemes for nonlinear problems.

It is worth noting here that all the projection schemes presented in this chapter were based on the implicit assumption that a complete probabilistic characterization of the system uncertainties is available. Hence, further research is required to investigate how projection schemes can be employed to compute spectral- and probability-distribution-free upper bounds on various probabilistic indicators of the response of stochastic systems [69]. Further work is also required to extend projection schemes to cases where a combination of probabilistic and possibilistic models are used for representing uncertainty. These theoretical developments can potentially enable the application of projection schemes to problems where insufficient experimental data is available to properly quantify the probability distributions of the system parameters.

Acknowledgments

This research was supported by a grant from the Faculty of Engineering, Science and Mathematics at the University of Southampton. The author thanks Sachin Sachdeva for his assistance in producing the results presented here.

References

1. R. Ghanem and P. Spanos, *Stochastic Finite Elements: A Spectral Approach*, Springer-Verlag, Heidelberg, 1991.
2. M. Kleiber and T.D. Hien, *The Stochastic Finite Element Method: Basic Perturbation Technique and Computer Implementation*, John Wiley & Son, Chichester, 1992.
3. G.I. Schueller (Editor), Special Issue—A state-of-the-art report on computational stochastic mechanics, *Prob. Eng. Mech.*, 12 (1997) 197–321.
4. M. Shinozuka and C.M. Jan, Digital simulation of random processes and its applications, *J. Sound Vibr.*, 25 (1972).
5. J.E. Hurtado and A.H. Barbat, Monte Carlo techniques in computational stochastic mechanics, *Arch. Comput. Meth. Eng.*, 5 (1998) 3–29.
6. R.Y. Rubinstein, *Simulation and the Monte Carlo Method*, Wiley, New York, 1981.
7. F. Yamazaki, M. Shinozuka, and G. Dasgupta, Neumann expansion for stochastic finite element analysis, *J. Eng. Mech.*, 114 (1988), 1335–1355.
8. T.D. Hien and M. Kleiber, Stochastic finite element modelling in linear transient heat transfer, *Comput. Meth. Appl. Mech. Eng.*, 144 (1997) 111–124.
9. M. Kaminski, Stochastic second-order perturbation approach to the stress-based finite element method, *Int. J. Solids Struct.*, 38 (2001) 3831–3852.
10. A. Haldar and S. Mahadevan, *Reliability Assessment Using Stochastic Finite Element Analysis*, John Wiley & Sons, New York, 2000.
11. M.R. Rajashekhar and B.R. Ellingwood, A new look at the response surface method for reliability analysis, *Structural Safety*, 12 (1993) 205–220.
12. P.B. Nair, A. Choudhury, and A.J. Keane, A Bayesian framework for uncertainty analysis using deterministic black-box simulation codes, AIAA Paper 2001–1676, April 2001.
13. P.B. Nair, A. Choudhury, and A.J. Keane, Some greedy algorithms for sparse regression and classification with Mercer kernels, *J. Mach. Learning Res.*, 3 (2002) 781–801.

14. I. Elishakoff and Y.J. Ren, *Finite Element Methods for Structures with Large Stochastic Variations,* Oxford University Press, Oxford, 2003.
15. P.B. Nair and A.J. Keane, Stochastic reduced basis methods, *AIAA J.,* 40 (2002) 1653–1664.
16. H.G. Matthies, C.E. Brenner, C.G. Bucher, and C.G. Soares, Uncertainties in probabilistic numerical analysis of structures and solids—stochastic finite elements, Structural Safety 19 (1997) 283–336.
17. G. Falsone and N. Impollonia, A new approach for the stochastic analysis of finite element modelled structures with uncertain parameters, *Comput. Methods Appl. Mech. Eng.,* 191 (2002) 5067–5085.
18. N. Wiener, The homogeneous chaos, *Am. J. Math.,* 60 (1938) 897–936.
19. M. Loeve, *Probability Theory, fourth edition,* Springer, New York, 1977.
20. P.B. Nair, On the theoretical foundations of stochastic reduced basis methods, AIAA Paper 2001–1677, April 2001.
21. L.J. Gibson and M.F. Ashby, *Cellular Solids: Structure and Properties, second edition,* Cambridge University Press, Cambridge, U.K., 1997.
22. K. Alzebdeh and M. Ostoja-Starzewski, On a spring-network model and effective elastic moduli of granular materials, *J. Appl. Mech.,* 66 (1999) 172–180.
23. L. Huyse and M.A. Maes, Random field modeling of elastic properties using homogenization, *J. Eng. Mech.,* 127 (2001) 27–36.
24. M. Ostoja-Starzewski, Micromechanics as a basis of continuum random fields, *Appl. Mech. Rev.,* 47 (1994) 221–230.
25. M. Ostoja-Starzewski and X. Wang, Stochastic finite elements as a bridge between random material microstructure and global response, *Comp. Meth. Appl. Mech. Eng.,* 168 (1999) 35–49.
26. E. Vanmarcke, Random Fields: Analysis and Synthesis, MIT Press, Cambridge MA, 1983.
27. B. Sudret and A. Der Kiureghian, Stochastic Finite Elements and Reliability: A State-of-the-Art Report, Technical Report No. UCB/SEMM-2000/08, University of California, Berkeley, 173 p.
28. C.-C. Li and A. Der Kiureghian, Optimal discretization of random fields, *J. Eng. Mech.,* ASCE, 119 (1993) 1136–1154.
29. S.P. Huang, S.T. Quek, and K.K. Phoon, Convergence study of the truncated Karhunen-Loeve expansion for simulation of stochastic processes, *Int. J. Numer. Meth. Eng.,* 52 (2001) 1029–1043.
30. K.E. Atkinson, *The Numerical Solution of Integral Equations of the Second Kind,* Cambridge University Press, Cambridge, MA, 1997.
31. O.C. Zienkiewicz and R.L. Taylor, *Finite Element Method: Volume 1,* Butterworth Heinemann, London, 2000.
32. N. Wiener, *Nonlinear Problems in Random Theory,* MIT Press, Cambridge, MA, 1958.
33. D. Xiu and G.E. Karniadakis, The Wiener-Askey polynomial chaos for stochastic differential equations, *SIAM J. Sci. Comput.,* 24 (2002) 619–644.
34. D. Lucor, D. Xiu, C.-H. Su, and G.E. Karniadakis, Predictability and uncertainty in CFD, *Int. J. Num. Meth. Fluids,* 43 (2003) 483–505.
35. R.H. Cameron and W.T. Martin, The orthogonal development of nonlinear functionals in series of Fourier-Hermite functions, *Ann. Math.,* 48 (1947) 385–392.
36. T.Y. Hou, H. Kim, B. Rozovskii, and H.-M. Zhou, Wiener chaos expansions and numerical solutions of randomly forced equations of fluid mechanics, in preparation.
37. R.G. Ghanem and R. Kruger, Numerical solution of spectral stochastic finite element systems, *Comput. Meth. Appl. Mech. Eng.,* 129 (1996) 289–303.
38. M. Pellissetti and R. Ghanem, Iterative solution of systems of linear equations arising in the context of stochastic finite elements, *Adv. Eng. Softw.,* 31 (2000) 607–616.
39. M. Anders and M. Hori, Three-dimensional stochastic finite element method for elasto-plastic bodies, *Int. J. Numer. Meth. Eng.,* 51 (2001) 449–478.
40. R. Ghanem, The nonlinear Gaussian spectrum of log-normal stochastic processes and variables, *J. Appl. Mech.,* 66 (1999) 964–973.

41. R. Ghanem and P. Spanos, A stochastic Galerkin expansion for nonlinear random vibration analysis, *Prob. Eng. Mech.*, 8 (1993) 255–264.
42. R. Ghanem and V. Brzkala, Stochastic finite element analysis of randomly layered media, *J. Eng. Mech.*, 122 (1996) 361–369.
43. R. Ghanem, Probabilistic characterization of transport in heterogeneous media. *Comput. Meth. Appl. Mech. Eng.*, 158 (1998) 199–220.
44. M. Anders and M. Hori, Stochastic finite element method for elasto-plastic body, *Int. J. Numer. Meth. Eng.*, 46 (1999) 1897–1916.
45. D. Ghiocel and R. Ghanem, Stochastic finite element analysis of seismic soil-structure interaction, *J. Eng. Mech.*, 128 (2002) 66–77.
46. O. Le Maitre, O. Knio, H. Najm, and R. Ghanem, A stochastic projection method for fluid flow: basic formulation, *J. Comput. Phy.*, 173 (2001) 481–511.
47. A. Sarkar and R. Ghanem, Mid-frequency structural dynamics with parameter uncertainty, *Comput. Meth. Appl. Mech. Eng.*, 191 (2002) 5499–5513.
48. G.D. Manolis and C.Z. Karakostas, A Green's function method to SH-wave motion in random continuum, *Eng. Anal. Boundary Elements*, 27 (2003) 93–100.
49. D. Xiu and G.E. Karniadakis, Modeling uncertainty in steady-state diffusion problems via generalized polynomial chaos, *Comput. Meth. Appl. Mech. Eng.*, 191 (2002) 4927–4948.
50. D. Xiu and G.E. Karniadakis, Stochastic modeling of flow-structure interactions using generalized polynomial chaos, *J. Fluids Eng.*, 124 (2002) 51–59.
51. D. Xiu and G.E. Karniadakis, Modeling uncertainty in flow simulations via generalized polynomial chaos, *J. Comput. Phy.*, 187 (2003) 137–167.
52. D. Xiu and G.E. Karniadakis, A new stochastic approach to transient heat conduction modeling with uncertainty, *Int. J. Heat Mass Transfer*, 46 (2003) 4681–4693.
53. R. Ghanem, Ingredients for a general purpose stochastic finite elements implementation, *Comput. Meth. Appl. Mech. Eng.*, 168 (1999) 19–34.
54. H.G. Matthies and C.G. Bucher, Finite element for stochastic media problems, *Comput. Meth. Appl. Mech. Eng.*, 168 (1999) 3–17.
55. M.K. Deb, I.M. Babuska, and J.T. Oden, Solution of stochastic partial differential equations using Galerkin finite element techniques, *Comput. Meth. Appl. Mech. Eng.*, 190 (2001) 6359–6372.
56. H.G. Matthies and A. Keese, Galerkin methods for linear and nonlinear elliptic stochastic partial differential equations, Informatikbericht Nr.: 2003–08, Insitute of Scientific Computing, Technical University Braunschweig, Brunswick, Germany, also submitted to *Comput. Meth. Appl. Mech. Engrg.* (2003).
57. A. Keese, A review of recent developments in the numerical solution of stochastic partial differential equations (stochastic finite elements), Informatikbericht Nr.: 2003–06, Insitute of Scientific Computing, Technical University Braunschweig, Brunswick, Germany, October 2003.
58. L. Mathelin and M.Y. Hussaini, A stochastic collocation algorithm for uncertainty analysis, NASA/CR-2003-212153, February 2003.
59. Y. Saad, *Iterative Methods for Sparse Linear Systems, second edition,* SIAM, Philadelphia, 2003.
60. Y. Saad and H.A. Van der Vorst, Iterative solution of linear systems in the 20th century, *J. Comp. and Appl. Math.*, 123 (2000) 1–33.
61. S.K. Sachdeva, P.B. Nair, and A.J. Keane, Comparative study of projection schemes for stochastic finite element analysis, *Comput. Meth. Appl. Mech. Eng.*, (2003) submitted for review.
62. M.T. Bah, P.B. Nair, A. Bhaskar, and A.J. Keane, Forced response analysis of mistuned bladed disks: a stochastic reduced basis approach, *J. Sound Vibr.*, 263 (2003) 377–397.
63. P. McCullagh, *Tensor Methods in Statistics,* Chapman & Hall, New York, 1987.
64. A.J. Izenman, Recent developments in nonparametric density estimation, *J. Am. Stat. Assoc.*, 86 (1991) 205–224.
65. B. Sudret and A. Der Kiureghian, Comparison of finite element reliability methods, *Prob. Eng. Mech.*, 17 (2003) 337–348.

66. S.K. Sachdeva, P.B. Nair, and A.J. Keane, "Hybridization of stochastic reduced basis methods with Polynomial chaos expansions" in preparation.
67. N. Impollonia and G. Muscolino, Static and dynamic analysis of nonlinear uncertain structures, *Meccanica*, 37 (2002) 179–192.
68. A. Keese and H.G. Matthies, Numerical methods and Smolyak quadrature for nonlinear stochastic partial differential equations, Informatikbericht Nr.: 2003–05, Insitute of Scientific Computing, Technical University Braunschweig, Brunswick, Germany, May 2003, also submitted to *SIAM J. Sci. Comput.*
69. G. Deodatis, L. Graham-Brady, and R. Micaletti, A hierarchy of upper bounds on the response of stochastic systems with large variation of their properties: random field case, *Prob. Eng. Mech.*, 18 (2003) 365–375.

22

Bayesian Modeling and Updating

Costas Papadimitriou
University of Thessaly

Lambros S. Katafygiotis
Hong Kong University of Science and Technology

22.1 Introduction 22-1
22.2 Statistical Modeling and Updating 22-3
22.3 Model Class Selection 22-5
22.4 Application to Damage Detection 22-6
22.5 Uncertainty Reduction Using Optimal
      Sensor Location 22-9
      Information Entropy Measure of Parameter
      Uncertainty • Asymptotic Approximation of Information
      Entropy for Large Number of Data • Design of Optimal
      Sensor Configuration • Computational Issues for Finding
      the Optimal Sensor Configuration • Optimal Sensor
      Configuration for Selecting a Model Class
22.6 Structural Reliability Predictions Based on Data........22-14
      Formulation as a Probability Integral • Computational
      Issues • Asymptotic Approximations for Large Number
      of Data
22.7 Conclusion 22-16

22.1 Introduction

The problem of structural model updating has gained much interest as finite element model capabilities and modal testing have become more mature areas of structural dynamics (e.g., [1–3]). The need for model updating arises because there are always errors associated with the process of constructing a theoretical model of a structure, and this leads to uncertain accuracy in predictive response. Moreover, a model updating methodology is useful in predicting the structural damage by continually updating the structural model using vibration data [4–12]. Such updated models obtained periodically throughout the lifetime of the structure can be further used to update the response predictions and lifetime structural reliability based on available data [13–16]. The updated information about the condition of the structure can be used to identify potentially unsafe structures, to schedule inspection intervals, repairs or maintenance, or to design retrofitting or control strategies for structures that are found to be vulnerable to possible future loads.

Structural model updating is an inverse problem according to which a model of a structure, usually a finite element model, is adjusted so that either the calculated time histories, frequency response functions, or modal parameters best match the corresponding quantities measured or identified from the test data. This inverse process aims at providing updated models and their corresponding uncertainties based on the data. These updated models are expected to give more accurate response predictions to future loadings, as well as allow for an estimation of the uncertainties associated with such response predictions. In practice, the inverse problem of model updating is usually ill-conditioned due to insensitivity of the response to changes in the model parameters, and nonunique [17–20] because of insufficient available data relative to the desired model complexity. Additional difficulties associated with the development of an effective model updating methodology include: (1) model error present due to the fact that the chosen class of structural models is unable to exactly model the actual behavior of the structure; (2) measurement noise in the dynamic data, especially for higher modes; (3) incomplete set of observed DOFs (degrees of freedom) due to the limited number of sensors available and the limited accessibility throughout the structure; (4) incomplete number of contributing modes due to limited bandwidth in the input and the dynamic response.

This chapter presents a Bayesian framework for statistical modeling and updating of structural models utilizing measured dynamic data, along with its use in making informed structural response predictions and reliability assessments. The Bayesian approach to statistical modeling uses probability as a way of quantifying the plausibilities associated with the various models and the parameters of these models given the observed data. The Bayes rule uses a prior distribution reflecting the beliefs about what values the model parameters might take, and then updates these beliefs based on measured data, resulting in a posterior distribution.

The Bayesian statistical system identification framework proposed by Beck [21] and Beck and Katafygiotis [11] is adopted, allowing for the explicit treatment of modeling and prediction uncertainties. The statistical methodology is used to provide more accurate representations of the uncertainties associated with the structural modeling, based on measured vibrational data and prior engineering information. It also allows for the explicit treatment of the ill-conditioning and nonuniqueness arising in the model updating problem. Probability distributions are used to quantify the various uncertainties and these distributions are then updated based on the information contained in the measured data. Two model updating cases are considered. The problem of comparing different models and selecting the best model from a specific class of models is first addressed. Next, the problem of selecting the optimal class of models from two or more alternative classes is addressed. Computational issues related to these problems are discussed. The application of the methodology in structural damage detection problems is illustrated.

The study also strongly focuses on the role of the sensor configuration in reducing the modeling uncertainties and, thus, improving the predictions of structural integrity, response and reliability. In the past, statistical-based approaches (e.g., [22–30]) have been developed to provide rational solutions to the optimal sensor location problem. In particular, the information entropy-based framework proposed recently by Papadimitriou et al. [27] for selecting the optimal sensor configuration for model updating using a single model class, is revisited in this study. Various theoretical and computational issues are addressed. Finally, the information entropy framework is extended to address the problem of selecting the optimal sensor configuration for model updating using multiple model classes.

The Bayesian framework is finally integrated with probabilistic structural analysis tools to obtain updated reliability predictions. Computational aspects for estimating the resulting probability integrals are addressed. Efficient asymptotic approximations valid for large number of data are proposed. Non-uniqueness and identifiability issues arising when computing the uncertainty in structural model parameters are briefly addressed, along with the computational tools that have been developed to estimate the uncertainty in the model parameters for both the identifiable [19] and the unidentifiable cases [12, 20].

The presentation in this chapter is organized as follows. Section 22.2 presents the Bayesian statistical system identification framework for uncertainty modeling and updating using measured data. The methodology is presented for the case where the data consist of input and output response time histories. However, the formulation is directly applicable to cases for which modal data is available. In Section 22.3 the Bayesian formulation is extended to address the problem of selecting the best class of models from a set of alternative model classes based on measured data. In Section 22.4 the model updating methodology is applied for damage detection, and a numerical example is given to illustrate the effectiveness of the Bayesian statistical framework in selecting the optimal class of models using modal measurements. The effect of various choices in sensor configuration on model updating and model selection is presented in Section 22.5. The problem of optimizing the sensor configuration (number and location of sensors) is addressed in detail. Section 22.6 addresses the problem of updating the robust reliability predictions based on measured data. Computational issues for evaluating the resulting multidimensional integrals are addressed. Efficient asymptotic approximations, valid for large number of data, are then given for estimating the structural reliability.

22.2 Statistical Modeling and Updating

Model updating is handled by embedding a class of deterministic structural models within a class of probability models so that the structural models give a predictable (systematic) part of the response and the probability models give the uncertain (random) part of the response due to modeling and measurement error. The class of structural models, designated by M, prescribes a functional relationship $q(m; \theta, Z_N, M)$ for the input-output behavior of the structure, where $\theta \in R^{N_\theta}$ are the free parameters that need to be assigned values in order to choose a particular model $M(\theta)$ from the class M. The quantity $q(m; \theta) \equiv q(m; \theta, Z_N, M)$ is the model output vector at N_d DOFs at time $t_m = m\Delta t$, $m = 1, \ldots, N$, where Δt is a prescribed sampling interval and $Z_N = \{z(m) \in R^{N_{in}}, m = 1, \ldots, N\}$ is the system input.

Let $Y_N = \{y(m) \in R^{N_0}, m = 1, \ldots, N\}$ be the measured response history at N_0 observed DOF of the structural model. The measured response and the model response predictions satisfy the equation

$$y(m) = L_0 q(m; \theta) + L_0 n(m; \theta) \tag{22.1}$$

where $n(m; \theta) \in R^{N_d}$ is the model prediction error due to modeling error and measurement noise. The matrix $L_0 \in R^{N_0 \times N_d}$ is an observation matrix comprised of zeros and ones that maps the model DOFs to the observed DOF. The locations of the N_0 observed model DOF are specified by the sensor configuration vector $\delta \in R^{N_d}$, with elements $\delta_j = 1$ if the j-th DOF is observed and $\delta_j = 0$ if the j-th DOF is not observed. It can be readily shown that $L_0^T L_0 = diag(\delta)$. The use of δ is needed in Section 22.5 dealing with the selection of the optimal sensor configuration.

The statistical system identification methodology presented in [11] is adopted to estimate the values of the parameter set θ and their associated uncertainties using the information provided by dynamic test data. According to the methodology, the prediction error $n(m; \theta)$ is modeled using a parameterized class of probability models P. The parameters of the probability model are selected to be the standard deviation of the underlying PDF (probability density function) of the prediction error $n(m; \theta)$. In the absence of information justifying the selection of a particular probability model over other probability models, the principle of maximum entropy is invoked according to which a Gaussian distribution, parameterized by its standard deviation σ, is chosen to model the prediction error $n(m; \theta)$. Among all probability models defined by a single parameter, the standard deviation of the distribution, the Gaussian model corresponds to the least prejudice model that maximizes the uncertainty in the predictions [11]. It must be noted that although herein we assume such a single-parameter simplified class of probability models, the extension to cases where the statistical assumptions regarding the prediction error are more involved, requiring more complex probability models, is straightforward.

The class of probability models M_P, which is defined by the selection of the classes M and P, is thus parameterized by the set $\alpha = [\theta, \sigma] \in S(\alpha) \subset R^{N_\alpha}$, where $N_\alpha = N_\theta + 1$. The utilization of the class M_P for response predictions involves the specification of uncertainties in the model parameters α. Herein, parametric uncertainties are quantified by modeling the parameters α by random variables with PDFs specifying the relative plausibilities of each of the corresponding models in the class M_P. Prior to the collection of any data, the uncertainty in the values of the parameters is specified, based on engineering judgment and experience, by choosing an initial (prior) PDF $p(\alpha) = \pi(\theta, \sigma)$ over the set α of possible parameter values. For simplicity and without loss of generality, it is assumed in this study that the parameters θ and σ are independent, that is, $\pi(\theta, \sigma) = \pi_\theta(\theta)\pi_\sigma(\sigma)$. The problem of model updating when posed within the adopted statistical framework amounts to calculating the updated (posterior) PDF of the parameters given the measured data.

For given dynamic data D_N consisting of sampled input and output time histories, Z_N and Y_N, respectively, Bayes theorem is used to convert the initial PDF $\pi(\theta, \sigma)$ for the model parameters to the updated (posterior) PDF, $p(\theta, \sigma | D_N, M_P)$, as:

$$p(\theta, \sigma | D_N, M_P) = \tilde{c}\, p(D_N | \theta, \sigma, M_P) \pi(\theta, \sigma) \qquad (22.2)$$

Note that the measured data D_N depends on the sensor configuration δ, that is, $D_N = D_N(\delta)$. Herein, this dependency on δ is omitted for brevity. However, in cases where the dependency on δ is important, such as when discussing the issue of optimal sensor configuration, this dependency will be made explicit in the notation. The updated PDF $p(\theta, \sigma | D_N, M_P)$ gives the relative plausibilities of each of the corresponding models in M_P based on the measured data D_N and on any available prior information. Measured data is accounted for in the updated estimates through the likelihood term $p(D_N | \theta, \sigma, M_P)$, while any available prior information is reflected in the term $\pi(\theta, \sigma)$. The relative likelihood of the various models in the class of models M_P depends on the probability model chosen for the prediction error $n(m; \theta)$. Using the Gaussian choice for the probability distribution of the prediction error, and assuming temporal and spatial independence of the prediction errors, this likelihood is readily obtained in the form

$$p(D_N | \theta, \sigma, M_P) = \frac{1}{(\sqrt{2\pi}\sigma)^{NN_0}} \exp\left[\frac{NN_0}{2\sigma^2} J(\theta; D_N, M)\right] \qquad (22.3)$$

where

$$J(\theta; D_N, M) = \frac{1}{NN_0} \sum_{m=1}^{N} \|y(m) - L_0 q(m; \theta)\|^2 \qquad (22.4)$$

represents the measure of fit between the measured response and the model response time histories, and $\|\cdot\|$ is the usual Euclidian norm.

Using the total probability theorem, the marginal probability distribution $p(\theta | D_N, M_P)$ for the structural model parameters θ is given by $p(\theta | D_N, M_P) = \int p(\theta, \sigma | D_N, M_P)\, d\sigma$. For a noninformative (uniform) prior distribution $\pi_\sigma(\sigma)$, the integration with respect to σ can be carried out analytically, yielding

$$p(\theta | D_N, M_P) = c^{-1} \left[J(\theta; D_N, M)\right]^{-N_J} \pi_\theta(\theta) \qquad (22.5)$$

where $N_J = (NN_0 - 1)/2$, and c is a normalizing constant chosen such that the PDF in Equation 22.5 integrates to 1, that is,

$$c = \int \left[J(\theta; D_N, M)\right]^{-N_J} \pi_\theta(\theta)\, d\theta \qquad (22.6)$$

For a general prior (initial) distribution $\pi_\sigma(\sigma)$, an asymptotic approximation [31] can be applied so that eventually the updated PDF is also obtained in the form of Equation 22.5 and Equation 22.6 with $\pi_\theta(\theta)$ replaced by $\pi_\theta(\theta)\, \pi_\sigma\left(\sqrt{J(\theta; D_N, M)}\right)$ [12].

It should be noted that the optimal model parameters, $\hat{\theta}$, correspond to the most probable model maximizing the updated PDF $p(\theta | D_N, M_P)$ or, equivalently, minimizing the function

$$g(\theta) \equiv g(\theta; D_N, M) = -\ln p(\theta | D_N, M_P) = N_J \ln J(\theta; D_N, M) - \ln \pi_\theta(\theta) + \ln c \qquad (22.7)$$

In particular, for a noninformative prior distribution $\pi_\theta(\theta)$, the optimal values of the model parameters θ correspond to the values that minimize the measure of fit $J(\theta; D_N, M)$ defined in Equation 22.4.

The Bayesian statistical system identification framework is applicable to both linear and nonlinear systems. Also, the formulation can handle both the cases of measured and unmeasured input encountered in applications for which measured dynamic data is to be taken from forced and ambient vibration tests, respectively. Applications to the identification of optimal models and corresponding uncertainties for linear structures using measured acceleration time histories can be found in [11, 12, 19–21]. Applications to the identification of nonlinear models for vehicle suspensions are presented in Metallidis et al. [32]. The case of ambient excitation, in which only output response time histories are available and the excitation is modeled as a stochastic process with known correlation properties, has also been addressed in [33–36]. Yuen and Beck [37] extended the formulation to the case of nonlinear models and unknown excitation modeled by stochastic process with known correlation structure.

The above Bayesian statistical framework can also be applied in the case where the measured data D_N consists of modal frequencies and modeshape values at N_0 measured DOFs. In this case, the functional relationship $q(m;\theta) \equiv q(m;\theta,M)$ in Equation 22.1 represents the predictions of the modal frequencies and modeshapes for the structural model with parameters θ; $y(m)$ represents the identified, through modal identification, modal frequencies and modeshape components; m is the number of measured modes (modal frequencies and modeshapes); and $n(m;\theta)$ represents the prediction error between the measured modal parameters and the modal parameters corresponding to a model $M(\theta)$ in the class M. Bayesian updating based on modal data has been developed for structural damage detection applications in [8, 9, 38–40].

Finally, the Bayesian statistical framework has recently been applied to other areas of engineering. Specifically, it has been used for leakage detection due to pipe rupture in water pipeline networks by updating hydraulic models of fluid flow using pressure and flow rate measurements [41, 42].

22.3 Model Class Selection

The above Bayesian framework is also attractive to address the problem of comparing two or more alternative classes of models and selecting the optimal class. Let M_i, $i=1,\ldots,\mu$ be μ classes of alternative models. Before the selection of data, each class has a probability $p(M_i)$ of being the appropriate class of models for modeling the structural behavior. Using Bayes theorem, the posterior probabilities of the various model classes given the data D_N is

$$p(M_i | D_N) = \frac{p(D_N | M_i) p(M_i)}{d} \quad (22.8)$$

where $p(D_N | M_i)$ is the probability of observing the data from the model class M_i, while d is given by

$$d = \sum_{i=1}^{m} p(D_N | M_i) p(M_i) \quad (22.9)$$

so that the sum of all model probabilities equals 1. The probability of selecting the model class M_i depends on the sensor configuration, as is shown in Equation 22.8 by the explicit dependence of $p(M_i | D_N)$ on δ.

It is assumed that each model class M_i is parameterized with a parameter set $\alpha_i = [\theta_i, \sigma_i]$. Also, given the model class M_i, the prior probability distribution of the model parameters α_i is assumed to be of the form $\pi(\theta_i, \sigma_i | M_i) = \pi_{\theta_i}(\theta_i) \pi_{\sigma_i}(\sigma_i)$. Using the Bayesian updating theory for each class of models, the quantity $p(D_N | M_i)$ is the inverse of the normalization constant $\tilde{c}$ defined in Equation 22.2 and, for the i-th model class, is given by

$$p(D_N | M_i) = \int p(D_N | \theta_i, \sigma_i, M_i) \, \pi(\theta_i, \sigma_i | M_i) \, d\sigma_i \, d\theta_i \quad (22.10)$$

Substituting $p(D_N | \theta_i, \sigma_i, M_i)$ from Equation 22.3 for the i-th model class M_i, using the simplification that $\pi(\theta_i, \sigma_i) = \pi_{\theta_i}(\theta_i) \pi_{\sigma_i}(\sigma_i)$ and carrying out the integration in Equation 22.10 over σ_i assuming a non-informative prior distribution $\pi_{\sigma_i}(\sigma_i) = c_\sigma$ for σ_i, one readily derives that

$$p(D_N | M_i) = c_0 \int \left[J(\theta_i; D_N, M_i) \right]^{-N_f} \pi(\theta_i | M_i) \, d\theta_i \quad (22.11)$$

where $J(\theta_i; D_N, M_i)$ is the measure of fit for the i-th model class M_i, and c_0 is a multiplicative constant that depends only on the number of data while it is independent of the model class.

An asymptotic approximation [31], based on Laplace's method of asymptotic approximation, can be used to give a useful and insightful estimate of the integral in the form

$$c_0^{-1} p(D_N | M_i) \sim I_i(\hat{\theta}_i) = (2\pi)^{N_\theta/2} \frac{\pi_\theta(\hat{\theta}_i | M_i) [J(\hat{\theta}_i; D_N, M_i)]^{-N_J}}{\sqrt{\det h(\hat{\theta}_i; D_N, M_i)}} \quad (22.12)$$

where $\hat{\theta}_i$ is the value that maximizes the integrand in Equation 22.11 or, equivalently, the most probable value that maximizes the updated PDF $p(\theta | D_N, M_p)$ given in Equation 22.5, and $h(\hat{\theta}_i; D_N, M_i)$ is the Hessian of the function $g(\theta_i; D_N, M_i)$ defined in Equation 22.7 for a fixed model class evaluated at $\hat{\theta}_i$. For simplicity, the asymptotic estimate in Equation 22.12 is based on the fact that only one optimal value of $g(\theta_i; D_N, M_i)$ exists. For multiple optimal values $\hat{\theta}_i^{(k)}$, $k = 1, \ldots, K$, the estimate in Equation 22.12 is replaced by $c_0^{-1} p(D_N | M_i) \sim \sum_{k=1}^{K} I_i(\hat{\theta}_i^{(k)})$, that is, a summation of expressions $I_i(\theta)$ evaluated at the optimal values for the class of models M_i. The estimate in Equation 22.12 assumes that the integrand in Equation 22.10 has a relatively fast decay in all directions as one moves away from the optimal points so that there is no significant overlapping of $I_i(\theta)$ between neighboring optimal points.

Consider the particular case for which $K = 1$ so that the estimate in Equation 22.12 is applicable. Also, assuming there is no prior preference as to what class of models we choose, we can set that $p(M_i) = 1/\mu$ in Equation 22.8. The result in Equation 22.8 with $p(D_N | M_i)$ given by Equation 22.12 suggests that among two model classes M_1 and M_2 that give the same fit to the data, that is, $J(\theta_1; D_N, M_1) = J(\theta_2; D_N, M_2)$, the most probable class is the one with the most uncertain parameter values as for this class the denominator in Equation 22.12 is smaller. As it will be seen later, the sensor location affects the uncertainty in the model parameters of each model class. A sensor configuration that minimizes the uncertainty in the parameter estimates may be critical in identifying the optimal class of models, especially for the case for which these models give approximately an equally good fit to the data.

Finally, it is worth mentioning that the ranking of the model classes also depends on the complexity of the model classes and the number of parameters involved in each model class. Model classes with increasing numbers of parameters are penalized in the selection of the best model class. A theoretical investigation of the dependence of the model class ranking on the number of parameters involved in each model class can be found in Yuen [43].

22.4 Application to Damage Detection

An application of the methodology to damage detection is next presented. It is assumed that damage in the structure causes stiffness reduction in the damaged part of the structure. Here we follow a substructure approach where we assume that the structure is comprised of a number of substructures. The objective is to identify the damaged substructure and the severity of the corresponding damages using measured dynamic data. For this, a family of μ model classes $M_1 \ldots M_\mu$ is introduced, where each model class allows for the modeling of damage in a given subset of substructures, while it assumes that the remaining substructures are undamaged. Thus, each model class M_i is assumed to be parameterized by a number of structural model parameters θ_i scaling the stiffness contributions of the possibly damaged substructures; all other substructures are assumed to have fixed stiffness contributions equal to those corresponding to the undamaged structure. Given dynamic measurements from the damaged structure, one is able to select the most probable class of models based on the theory presented in the previous section. The most probable model class will be indicative of substructures that are damaged, while the most probable values of the model parameters of the corresponding most probable model class will be indicative of the severity of damage in the identified damaged substructures.

For illustration purposes, it is assumed that the measured data $D_N = \{\hat{\omega}_m, \hat{\phi}_m \in R^{N_0}, m=1,\ldots,N\}$ consists of modal frequencies $\hat{\omega}_m$ and modeshape components $\hat{\phi}_m$ at N_0 measured DOFs, where N is the number of observed modes. Let $q(m;\theta) = \{\omega_m(\theta), \phi_m(\theta) \in R^{N_d}, m=1,\ldots,N\}$ be the predictions of the modal frequencies and modeshapes from a particular model $M(\theta)$ in the model class M. The prediction error $n(m;\theta) = [n_\omega(m;\theta)\ n_\phi(m;\theta)]$ between the measured modal data and the corresponding modal quantities predicted by the model $M(\theta)$ in the class M is given separately for the modal frequencies and the modeshapes by the prediction error equations:

$$\hat{\omega}_m = \omega_m(\theta) + \hat{\omega}_m n_\omega(m;\theta) \qquad (22.13)$$

$$\alpha_m \hat{\phi}_m = L_0 \phi_m(\theta) + L_0 \|\hat{\phi}_m\| n_\phi(m;\theta) \qquad (22.14)$$

where $n_\omega(m;\theta)$ and $n_\phi(m;\theta) \in R^{N_d}$ are, respectively, the prediction errors for the modal frequencies and modeshape components, and $\alpha_m(\theta) = \hat{\phi}_m^T L_0 \phi_m / \|\hat{\phi}_m\|^2$ is a normalization constant that accounts for the different scaling between the measured and the predicted modeshapes. Applying the Bayesian statistical framework, the updated PDF for the parameters of the model class M is given by Equation 22.5 with $J(\theta; D_N, M)$ given by

$$J(\theta; D_N, M) = \sum_{m=1}^{N} \left[\frac{\omega_m(\theta) - \hat{\omega}_m}{\hat{\omega}_m} \right]^2 + \sum_{m=1}^{N} \frac{\|L_0 \phi_m(\theta) - \alpha_m \hat{\phi}_m\|^2}{\|\hat{\phi}_m\|^2} \qquad (22.15)$$

In deriving Equation 22.15, it was assumed that the model prediction errors for the modeshape components and the modal frequencies are independent and that the ratio between the standard deviations of the prediction errors for the modeshape components and the modal frequencies is taken for simplicity to be equal to 1. In practice, the uncertainties associated with the frequency and modeshape prediction errors can be estimated directly through Bayesian modal updating [34, 35, 44]. The most probable model class given the data is the one that maximizes Equation 22.8 with $p(D_N | M_i)$ given by Equation 22.11 and approximated by Equation 22.12, where $J(\theta; D_N, M)$ is defined in Equation 22.15.

A simple illustrative numerical example is given next. Consider a structure represented by a 6-DOF chain-like spring-mass model, shown in Figure 22.1, with one end of the chain fixed at the base and the other end free. The nominal values of the stiffness and mass of each link in the chain are chosen to be $k_i = k_0$ and $m_i = m_0$, $i=1,\ldots,6$, respectively. The nominal model is considered to be the model of the structure at its undamaged condition.

A damage scenario corresponding to 40% stiffness reduction at the lower three links is considered. Simulated measured modal data is generated by the model with 40% reduced stiffness at the lower three links. To simulate the effects of measurement noise and modeling error, 2% and 5% noise, following a zero-mean Gaussian distribution, are respectively added to the modal frequencies and modeshapes simulated by the damaged model. This simulated noisy modal data $\hat{\omega}_m$ and $\hat{\phi}_m$ are then used in the methodology to compute the optimal class of models and to update the uncertainty in the parameters

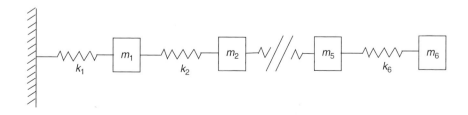

FIGURE 22.1 Six degree of freedom spring mass model.

TABLE 22.1 Values of the Ratio of Probabilities $p(M_1|D_N)/p_{max}$

Number of Modes N	Number of Sensors N_0					
	1	2	3	4	5	6
1	4.57e-12	1.98e-006	0.45*	15.48	1.89e+006	2.56e+010
2	0.63	1.44e-004*	0.2323*	1.76e+004	4.47e+010	6.28e+012
3	0.12*	0.36*	0.0130*	1.42e+009	1.80e+016	1.18e+016
4	3.05e-06*	5.03e+003	8.64e+005	5.34e+012	7.94e+013	1.46e+020
5	0.76e-02*	2.44e+006	6.35e+007	4.56e+016	5.21e+022	5.75e+025
6	29.53	2.69e+006	2.26e+007	3.30e+020	1.12e+028	9.36e+030

of the optimal class. For illustration purposes, the family of model classes $M_1 \ldots M_\mu$ is taken so that each model class M_i is parameterized by three parameters that account for the stiffness values of three out of the six links, while the stiffness and mass properties of the rest of the links equal the nominal values of the undamaged structure. The total number of distinct model classes involving three stiffness parameters is $\mu = 20$. It is expected that the application of the methodology should yield as the most probable model the one with the three parameters modeling the stiffness values of the lowest three links. For reference purposes, this optimal model class is denoted by M_1. The probability of each model class M_i is obtained by Equation 22.8. Herein, a noninformative prior probability distribution for the parameters of each model is considered.

The results for the ratio of the probability $p(M_1|D_N)$ of the model class M_1 to the highest probability $p_{max} = \max_{i \in [2,\mu]} p(M_i|D_N)$ among the rest of the models $M_2 \ldots M_{20}$ are given in Table 22.1 as a function of the number of measurement points and the number of contributing modes. The measurement locations for N_0 sensors are taken in all cases to be at the lowest N_0 masses. Values of the ratio greater than 1 indicate that the methodology correctly identifies the damage substructure and the locations of damage in the structure. The most probable values of the parameter set θ_1 of the model class M_1 are also given in Table 22.2 as a function of the number of measurement points and the number of contributing modes. It is seen in Table 22.1 that the effectiveness of the methodology in predicting the location of

TABLE 22.2 Most Probable Values of the Model Parameters θ_i, $i = 1, 2, 3$ Corresponding to Model Class M_1

Number of Modes N	θ	Number of Sensors N_0					
		1	2	3	4	5	6
1	θ_1	1.006	0.529	0.595	0.590	0.585	0.585
	θ_2	0.335	0.490	0.572	0.587	0.588	0.588
	θ_3	1.004	1.091	0.630	0.626	0.634	0.634
2	θ_1	0.524	0.609	0.613	0.644	0.583	0.585
	θ_2	0.507	0.588	0.579	0.583	0.584	0.597
	θ_3	1.017	0.586	0.601	0.567	0.641	0.626
3	θ_1	0.607	0.618	0.578	0.635	0.578	0.601
	θ_2	0.568	0.612	0.633	0.593	0.589	0.587
	θ_3	0.618	0.547	0.598	0.572	0.643	0.610
4	θ_1	0.604	0.619	0.579	0.622	0.589	0.601
	θ_2	0.565	0.612	0.633	0.593	0.591	0.588
	θ_3	0.626	0.547	0.599	0.592	0.622	0.609
5	θ_1	0.609	0.627	0.579	0.622	0.589	0.596
	θ_2	0.570	0.624	0.633	0.592	0.593	0.588
	θ_3	0.614	0.528	0.600	0.592	0.621	0.617
6	θ_1	0.610	0.613	0.591	0.621	0.588	0.596
	θ_2	0.571	0.598	0.613	0.592	0.593	0.588
	θ_3	0.611	0.568	0.600	0.593	0.623	0.617

damage depends on the number of modes and the number of measurement locations. Specifically, for sufficiently large number of observed modes and measurement points, the probability ratios in Table 22.1 are larger than 1, which indicates that the methodology correctly predicts the location of damage. In these cases, the results in Table 22.2 predict close to 40% stiffness reduction in all three lower links, which suggests that the severity of damage (stiffness reduction) is also correctly identified.

For very small number of modes and measurement locations (i.e., for the cases $(N, N_0) = [(1, 1), (1, 2), (2, 1)]$), the methodology does not reliably predict the location and severity of damage. This is due to the fact that in these cases, the amount of data is not sufficient to identify uniquely the three parameters of each model class. In fact, in these cases unidentifiability is encountered [12]; namely, there is an infinite number of parameter values that give equally good fit to the data. One of these infinite optimal parameter values for the model class M_1 is shown in Table 22.2 and do not correspond to the size of damage assumed. Methods for obtaining all optimal values of the model parameters are given in the work by Katafygiotis et al. [12]. However, this is not the main issue in this example and it is, therefore, not discussed further.

There is an intermediate range of the number of modes and measurement points, marked in Table 22.1 with the symbol $<*>$, for which although the model M_1 is not selected by the methodology as the most probable model, the most probable values of the parameter set θ_1 of the model class M_1 correctly predicts the severity of damage in the structure as shown in Table 22.2. The main reason for which model class M_1 is not chosen by the methodology as the most probable model class is that the uncertainty in identifying the model parameters of other model classes is orders of magnitude larger than the uncertainty in identifying the parameters of the model class M_1. This uncertainty considerably affects the factor $1/\sqrt{\det h(\hat{\theta}_i; D_N, M_i)}$ in Equation 22.12 and makes such model classes more probable than the model class M_1. The uncertainty in the model parameters depends on the number and location of sensors. Increasing the number of sensors, the uncertainty in the model parameters is reduced for all model classes and the model class M_1 is selected as the most probable model class as shown by the results in Table 22.1. For a fixed number of sensors, the uncertainty in the model parameter estimates can be reduced by appropriately selecting the location of the sensors in the structure. This optimal selection has not been performed in this example. Optimal sensor locations for identifying more than one class of models is therefore essential and a methodology for optimizing the sensor configuration is presented in Section 22.5.5. Another reason for which the model class M_1 is not selected as the optimal model class is that the number of data is not large enough to distinguish M_1 from other model classes. Specifically, the optimal models corresponding to some of the other model classes predict the data also very well. In all these cases, however, it has been observed that the parameter values of the model classes that have higher probability than the model class M_1, correspond to models having some of the substructures experiencing an increase of stiffness beyond the level corresponding to the undamaged structure. Assuming a prior distribution for the parameter set θ such that zero probability is assigned to parameter values corresponding to stiffness increase, the integration in Equation 22.11 is performed over a subdomain $0 \leq \theta_i \leq 1$ for all i. In this case, the model class M_1 is always obtained in this example as the most probable model.

22.5 Uncertainty Reduction Using Optimal Sensor Location

22.5.1 Information Entropy Measure of Parameter Uncertainty

The updated PDF $p(\theta|D_N, M_p)$ of the parameter set θ specifies the likelihood or the chance of the parameter set assuming each possible value. It provides a spread of the uncertainty in the parameter values based on the information contained in the measured data. A unique scalar measure of the uncertainty in the estimate of the structural parameters θ is provided by the information entropy, defined in [45]:

$$H(\delta, D_N) = E_\theta[-\ln p(\theta|D_N, M_p)] = -\int \ln p(\theta|D_N, M_p)\, p(\theta|D_N, M_p) d\theta \qquad (22.16)$$

where E_θ denotes mathematical expectation with respect to θ. Using the form of Equation 22.5 for the updated PDF $p(\theta|D_N, M_P)$, the information entropy takes the simplified form of

$$H(\delta, D_N) = \ln c(D_N) + N_I\, E_\theta[\ln J(\theta; D_N, M)] \qquad (22.17)$$

and depends only on the available data $D_N \equiv D_N(\delta)$ and the sensor configuration vector δ.

22.5.2 Asymptotic Approximation of Information Entropy for Large Number of Data

Next, an asymptotic approximation of the information entropy, valid for large numbers of data (i.e., as $N_I = (NN_0 - 1)/2 \to \infty$), is introduced which is useful in the experimental stage of designing an optimal sensor configuration. The asymptotic approximation is obtained by observing that the integrals defining the factors c and $E_\theta[\ln J(\theta; D_N, M)]$ involved in Equation 22.17 can be rewritten as Laplace-type integrals and then applying Laplace method of asymptotic expansion [46] to approximate these integrals. Specifically, it can be shown that for a large number of measured data (i.e., as $N_I \to \infty$), the following asymptotic results hold for the expressions appearing in Equation 22.17:

$$c \sim \pi_\theta(\hat{\theta}) \frac{(2\pi)^{N_\theta/2}\, \hat{\sigma}^{-2N_I}}{\sqrt{\det h(\hat{\theta}; \delta, M)}} \qquad (22.18)$$

and

$$E_\theta[\ln J(\theta; D_N, M)] = c \int \ln[J(\theta; D_N, M)][J(\theta; D_N, M)]^{-N_I} \pi_\theta(\theta) d\theta \sim \ln[\hat{\sigma}] \qquad (22.19)$$

where $\hat{\theta}$ is the optimal value of the parameter set θ that minimizes the measure of fit $J(\theta; D_N, M)$ given in Equation 22.4, $\hat{\sigma}^2$ is the optimal prediction error given by $\hat{\sigma}^2 = J(\hat{\theta}; D_N, M)$, and $h(\theta; \delta, M)$ is an $N_\theta \times N_\theta$ positive definite matrix defined by

$$h(\theta; \delta, M) = -\nabla_\theta \nabla_\theta^T \ln[J(\theta; D_N, M)]^{-N_I} \to \frac{1}{\hat{\sigma}^2} Q(\delta, \theta) \qquad \text{as } N_I \to \infty \qquad (22.20)$$

in which $\nabla_\theta = [\frac{\partial}{\partial \theta_1} \cdots \frac{\partial}{\partial \theta_{N_\theta}}]^T$ is the usual gradient vector. The matrix $Q(\delta, \theta)$ appearing in Equation 22.20 is a positive semidefinite matrix of the form

$$Q(\delta, \theta) = \sum_{j=1}^{N_d} \delta_j P^{(j)}(\theta) \qquad (22.21)$$

known as the Fisher information matrix [23] and containing the information from all measured positions. The matrix $P^{(j)}(\theta)$ is a positive semidefinite matrix given by

$$P^{(j)}(\theta) = \sum_{m=1}^{N} \nabla_\theta q_j(m; \theta)\, \nabla_\theta^T q_j(m; \theta) \qquad (22.22)$$

containing the information about the values of the parameters θ based on the data from one sensor placed at the j-th DOF. For given excitation characteristics, the matrix $P^{(j)}(\theta)$ depends only on the response of the optimal model at the particular DOF j, while it is independent of the sensor configuration vector δ. Substituting Equation 22.18 and Equation 22.19 into Equation 22.17 and simplifying, one finally derives that

$$H(\delta, D_N) \sim H(\delta, \hat{\theta}, \hat{\sigma}) = \frac{1}{2} N_\theta[\ln(2\pi) + \ln \hat{\sigma}^2] - \frac{1}{2} \ln[\det Q(\delta, \hat{\theta})] \qquad (22.23)$$

Bayesian Modeling and Updating

The importance of the asymptotic result in the experimental design of sensors will become evident in the discussion that follows in Section 22.5.3.

Using the structure of Equation 22.23 and the positive semidefiniteness of the matrices $P^{(j)}(\theta)$ and $Q(\delta, \theta)$, it can be readily shown that the value of the information entropy decreases as additional sensors are placed in a structure; that is,

$$H(\delta_1 + \delta_2, \hat{\theta}, \hat{\sigma}) \leq H(\delta_1, \hat{\theta}, \hat{\sigma}) \qquad (22.24)$$

Given the interpretation of the information entropy as a measure of the uncertainty in the parameter estimates, this should be intuitively expected, because adding one or more sensors in the structure will have the effect of providing more information about the system parameters. Moreover, it can be readily shown that the minimum (maximum) information entropy, $H_{\min}^{(L)}(H_{\max}^{(L)})$, for L sensors is an upper bound for the minimum (maximum) information entropy, $H_{\min}^{(L+R)}(H_{\max}^{(L+R)})$, for $L+R$ sensors with $R \geq 0$. That is,

$$H_{\min}^{(L+R)} \leq H_{\min}^{(L)} \quad \text{and} \quad H_{\max}^{(L+R)} \leq H_{\max}^{(L)} \qquad (22.25)$$

A direct consequence of the above statements is that the minimum (maximum) information entropy value corresponding to the optimal (worst) sensor configuration for L sensors is a decreasing function of the number of sensors L. This trend was also observed in the numerical results computed in the work by Papadimitriou et al. [27].

22.5.3 Design of Optimal Sensor Configuration

In experimental design, it is desirable to design the sensor configurations such that the resulting measured data are most informative about the condition of the structure. The information entropy, introduced in Equation 22.16 as a measure of the uncertainty in the system parameters, gives the amount of useful information contained in the measured data. The most informative test data gives the least uncertainty in the parameter estimates or, equivalently, minimizes the information entropy. Thus, the optimal sensor configuration is the one that minimizes the information entropy.

The problem of finding the optimal (or worst) sensor locations for a given number of sensors is formulated as a discrete optimization problem. The objective function to minimize is the information entropy given in Equation 22.16. The optimal (or worst) sensor locations minimize (or maximize) the information entropy. The discrete minimization variables are related to the location of sensors.

It should be emphasized that in the initial stage of designing the experiment, the test data is not available. Thus, the information entropy defined in Equation 22.16 is not specified completely because it depends explicitly on the details contained in the data D_N. To further process the information entropy, its explicit dependence on the data D_N must be removed. This can be accomplished by considering the limiting case of large number of data ($N_J \to \infty$), often arising in structural dynamics applications. The resulting asymptotic value of the information entropy, given in Equation 22.23, no longer depends explicitly on the measured response data D_N. The only dependence of the information entropy on the data comes implicitly through the optimal values

$$\hat{\theta} \equiv \hat{\theta}(\delta, D_N) = \arg\min_{\theta} J(\theta; D_N, M) \quad \text{and} \quad \hat{\sigma}^2 = J(\hat{\theta}; D_N, M) \qquad (22.26)$$

Consequently, the information entropy (Equation 22.23) is completely defined by the optimal value $\hat{\theta}$ of the model parameters and the optimal prediction error $\hat{\sigma}^2$ expected for a set of test data, while the time history details of the measured data do not enter explicitly the formulation.

Moreover, because the data are not available, an estimate of the optimal model parameters $\hat{\theta}$ and $\hat{\sigma}^2$ cannot be obtained from analysis. Thus, to proceed with the design of the optimal sensor configuration, this estimate must be assumed. In practice, useful designs can be obtained by taking the optimal model parameters $\hat{\theta}$ and $\hat{\sigma}^2$ to have some nominal values θ_0 and σ_0^2 chosen by the designer to be representative of the system. In this case, the entropy measure in Equation 22.23 takes the form (for large N)

$$H(\delta,\theta_0,\sigma_0^2) = \frac{1}{2}N_\theta\left[\ln 2\pi + \ln\sigma_0^2\right] - \frac{1}{2}\ln\left[\det Q(\delta,\theta_0)\right] \tag{22.27}$$

and depends on the sensor configuration vector δ and the chosen nominal values of the parameters θ_0 and σ_0^2 of the structural model. The aforementioned analysis provides a formal justification of the fact that the optimal sensor design is based only on a nominal structural model, ignoring the time history details of the measured data.

The optimal sensor configuration methodology has been extended to also optimize the actuator locations as well as the excitation characteristics, such as frequency content and amplitude of the excitation [29]. Applications of the methodology to the selection of the optimal excitation characteristics for nonlinear systems are reported in the work by Metallidis et al. [32].

22.5.4 Computational Issues for Finding the Optimal Sensor Configuration

For a structural model with N_d DOFs, the number of all distinct sensor configurations involving N_0 sensors is

$$N_s = \frac{N_d!}{N_0!(N_d - N_0)!} \tag{22.28}$$

which for most cases of practical interest can be an extremely large number. It is clear that an exhaustive search over all sensor configurations for the computation of the optimal sensor configuration is extremely time consuming and in most cases prohibitive even for models with a relatively small number of DOFs. Therefore, alternative approximate techniques must be used to solve the discrete optimization and obtain good estimates of sensor configurations that correspond to information entropy values close to the minimum information entropy.

Specifically, genetic algorithms (GA) [47, 48] are most suitable for solving the resulting discrete optimization problem and providing near-optimal solutions. The number of variables to be optimized equals the number of sensors, N_0. Details about various issues on the use of genetic algorithms for evaluating optimal sensor configuration, including convergence issues, are available in the work by Papadimitriou [30].

A more systematic and computationally very efficient approach for obtaining a good sensor configuration for a fixed number N_0 of sensors is to use a sequential sensor placement (SSP) algorithm as follows. The positions of N_0 sensors are computed sequentially by placing one sensor at a time in the structure at a position that results in the largest reduction in information entropy. The SSP algorithm can also be used in an inverse order, starting with N_d sensors placed at all DOFs of the structure and removing successively one sensor at a time from the position that results in the smallest increase in the information entropy.

The successive placement of each sensor in the structure requires the optimization of the information entropy with respect to only one sensor location. The solution is easily provided using an exhaustive search of the parameter space. Using the SSP algorithm, the total number of function evaluations for optimally placing the i-th sensor, given that $(i-1)$ sensors have already been placed in the structure, is equal to $(N_d - i + 1)$, where N_d is the total number of DOFs. Thus, the total number of function evaluations required for designing the optimal sensor configuration for N_0 sensors using the SSP algorithm is $\sum_{i=1}^{N_0}(N_d - i + 1) \leq N_0 N_d$. Moreover, the design of sensor configurations from 1 up to N_d sensors requires a total of no more than $N_d(N_d+1)/2$ function evaluations, which is an extremely small number compared to the number N_s, given in Equation 22.28.

The sequential sensor placement algorithm gives the optimal sensor configuration only in the case for which the optimal sensor positions for i sensors is a subset of the optimal sensor positions for $(i+1)$ sensors for all i from 1 to N_0. However, the last assumption does not hold in general and the sensor configuration computed by the SSP algorithm cannot be guaranteed to be the optimal one. The sensor configurations estimated from the SSP algorithm provide information entropy values that are upper bounds of the minimum information entropy. Numerical applications show that these bounds, in most cases examined, coincide with, or are very close to, the exact minimum information entropy. Consequently, the SSP algorithm is preferred over GAs because it maintains high levels of accuracy with minimal computational effort. Moreover, the solution provided by the SSP algorithm could be included in the initial population of a GA solver in order to accelerate convergence of the GAs and improve the optimal sensor location estimate.

Finally, it is worth pointing out that the maximum value of the information entropy corresponding to the worst sensor configuration is also useful because when it is compared with the minimum information entropy, it gives a measure of the reduction that can be achieved by optimizing the sensor configuration. Moreover, as shown in the subsequent section, the maximum value of the information entropy is needed in the estimation of the optimal sensor configuration for identifying the best model class from a number of alternative model classes. The worst sensor configuration can be obtained from the aforementioned algorithms by maximizing instead of minimizing the information entropy. In fact, the GA software used for finding the optimal sensor configuration can also be used with slight modifications to find the worst sensor configuration. Using the SSP algorithm, an approximation to the worst sensor configuration is obtained by placing successively one sensor at a time in the position that results in the smallest decrease in information entropy.

22.5.5 Optimal Sensor Configuration for Selecting a Model Class

The proposed optimal instrumentation depends on the chosen class of models, usually selected based on the type of states or damage scenarios to be monitored. For example, different types of structural and deterioration models may be required to monitor damage due to different environmental factors, such as corrosion, fatigue, severe earthquake and wind loading, etc. An optimal instrumentation should be capable of providing informative measurements for multiple classes of models $M_1, \ldots, M_\mu$ introduced in Section 22.3. Let $H_i(\delta) \equiv H_i(\delta, \theta_{0,i}, \sigma_{0,i}^2)$ be the information entropy measure for the i-th model class M_i. The optimal instrumentation problem reduces to a multi-objective optimization problem, with each objective accounting for the information entropy corresponding to the parameter uncertainty for each model class. A solution to the problem can be obtained using available multi-objective optimization algorithms (e.g., [49]) that provide the Pareto solutions. Alternatively, a solution to the problem can be obtained by defining an overall measure of the effectiveness of an instrumentation δ to monitor multiple classes of models. This measure is a weighted sum of the normalized information entropies for each model class [30]

$$G(\delta) = \sum_{i=1}^{\mu} w_i \, G_i(\delta) \tag{22.29}$$

where $G_i(\delta)$ is the i-th objective corresponding to the i-th model class, given by

$$G_i(\delta) = \frac{H_i(\delta) - H_i(\delta_{i,\text{opt}})}{H_i(\delta_{i,\text{worst}}) - H_i(\delta_{i,\text{opt}})} \tag{22.30}$$

where μ is the number of model classes; $\delta_{i,\text{opt}}$ and $\delta_{i,\text{worst}}$ are the optimal and worst sensor configurations, respectively, for the i-th model class; and w_i weights the importance with which the i-th model class is accounted for in the design of the optimal configuration. The value of $G_i(\delta)$ measures the effectiveness of the sensor configuration δ for the i-th model class. The most effective configuration for the i-th model class corresponds to a value of $G_i(\delta)$ equal to zero (0), while the least effective configuration for the i-th model class corresponds to a value of $G_i(\delta)$ equal to one (1). The aggregation rule given in Equation 22.29 is used to simultaneously consider in the design of the optimal sensor configuration the effect of multiple classes of models.

In the design stage, the weights could be chosen to be proportional to the prior probability of each model class. Model classes with low expected probability of being suitable classes for modeling well the structural behavior should be accounted less than model classes with high-expected probability. It is worth noting that the measure defined in Equation 22.29 and Equation 22.30 with $H_i(\delta)$ given by Equation 22.27 takes into account all possible model classes and parameter uncertainties within each class. For numerical examples illustrating the applicability of Equation 22.29 and Equation 22.30, the reader is referred to the work by Papadimitriou [30]. It should be emphasized that the SSP algorithm does not generally provide good estimates of the optimal sensor configuration for multiple classes of models. Genetic algorithms (GAs) are most suitable for solving the resulting discrete optimization problem. However, the optimal SSP solutions obtained by considering each class of models separately can be included in the initial GA population to speed up convergence.

22.6 Structural Reliability Predictions Based on Data

22.6.1 Formulation as a Probability Integral

Using the updated probability distribution of the model parameters, a methodology is next presented for updating the reliability predictions of structures, taking also into account the uncertainties in the excitation. A measure of structural safety is provided by the failure probability, the complement of structural reliability. Let F be the failure state of the structure and $\Pr(F|\theta, M_p)$ the conditional probability of failure given a particular model $M(\theta)$ in the class M. It is assumed herein that $\Pr(F|\theta, M_p)$ can be evaluated using probabilistic structural analysis tools or simulation methods. Using the total probability theorem, the failure probability of the system taking into account the uncertainties in the model parameters is a weighted average of the conditional failure probabilities $\Pr(F|\theta, M_p)$ for each model $M(\theta) \in M$, where the weights are given by the updated probability $p(\theta|D_N, M_p)$, quantifying the plausibility of each model $M(\theta)$.

Specifically, the updated failure probability of the system $\Pr(F|D_N, M_p)$, taking into account the information provided by the dynamic data D_N, is given by the weighted integral [14]

$$\Pr(F|D_N, M_p) = \int \Pr(F|\theta, M_p) p(\theta|D_N, M_p) d\theta \qquad (22.31)$$

over the whole parameter space. For the case where no dynamic data is available, the updated PDF $p(\theta|D_N, M_p)$ of the model parameters is replaced by the initial PDF $p(\theta|M) = \pi(\theta)$. The resulting initial probability of failure $\Pr(F|M_p)$ before the availability of measured data depends only on the class of models employed and the subjective judgment regarding the uncertainty in the model parameters. The updated failure probability (Equation 22.31), which takes into account the information about the structural state provided by the data, is expected to give an improved estimate of the lifetime failure probability.

22.6.2 Computational Issues

The estimation of the integral in Equation 22.31 requires the evaluation of the conditional probability of failure $\Pr(F|\theta, M_p)$. In certain cases, such as in the case of linear systems subjected to random excitations, the conditional probability $\Pr(F|\theta, M_p)$ can be evaluated analytically using available probabilistic structural analysis tools (see, e.g., [31, 50]). In more complicated cases where analytical results are not available, $\Pr(F|\theta, M_p)$ can be computed numerically using efficient simulation algorithms. For example, in the case of a linear MDOF system subjected to a single- or multiple-component random stationary or nonstationary Gaussian excitation, the probability of failure can be calculated in an extremely efficient manner using one of the following algorithms: importance sampling using elementary events [51, 52]; domain decomposition method [53]; or wedge simulation method [54]. Each of these methods requires only as few dynamic analyses of the structure as the number of independent random excitations driving the system, and therefore, the resulting computational effort is minimal. In the case of nonlinear structures or, more generally, in cases where the failure domain cannot be expressed as a union of linear failure domains, the above methods are not applicable. Monte Carlo simulations (MCS) offer a robust, although computationally expensive means of calculating $\Pr(F|\theta, M_P)$ in such cases. Variance reduction methods, such as importance sampling (IS), may lead to severely biased estimates when the number of random variables involved is very large, as is the case when modeling random excitations where several hundred (or even thousand) random variables are needed to characterize the random excitation process. Recently, promising simulation algorithms have been developed based on Controlled Monte Carlo simulations [55–57] and Markov chain simulations, such as Subset Simulation [58] and Spherical Subset Simulation [59]. The latter methods offer an increase of one to two orders of magnitude in computational efficiency over MCS. The smaller the value of $\Pr(F|\theta, M_P)$, the larger the computational savings.

As the dimension of the parameter space increases, the multidimensional numerical integration involved in computing the integral in Equation 22.31 becomes computationally prohibitive. This integration can be carried out efficiently using approximate numerical methods. One such method [31], providing a relatively rough estimate to the integral comparable to that of second-order reliability approximations [60, 61], involves finding the maxima of the integrand in Equation 22.31. Applications of the integral in Equation 22.31 to reliability-based optimal control design for passive tuned-mass dampers and active mass drivers can be found in [43, 62–64]. Alternatively, one can use a Markov chain approach to generate samples θ distributed according to the integrand function in Equation 22.28. Based on such samples, one can select an appropriate importance sampling density and proceed with importance sampling [65].

22.6.3 Asymptotic Approximations for Large Number of Data

For a large number N_I of available data, the updated PDF $p(\theta|D_N, M_P)$ of the structural model parameters θ given the model class M_P is concentrated in the neighborhood of a manifold S of dimension $m_S < N_\theta$ in the parameter space [12]. This is due to the fact that the exponent N_I in Equation 22.5 is a large number and, therefore, the relative posterior probabilities of the structural parameters θ are very sensitive to the corresponding values $J(\theta) \equiv J(\theta; D_N, M)$. Specifically, the updated PDF $p(\theta|D_N, M_P)$ becomes negligible everywhere, except in the immediate neighborhood of a region S of the parameter space where the corresponding values of $J(\theta)$ are very close to the global minimum $\hat{\sigma}^2 = \min\{J(\theta), \theta \in S\}$. Thus, the region of important probabilities extends immediately around the points that globally minimize $J(\theta)$. In a local identifiable case, S is comprised of a finite number of distinct optimal points $\hat{\theta}^{(k)}$, $k = 1, \ldots, K$ [11]. In a global identifiable case, $K = 1$. In such cases, the dimension m_S of the manifold S is equal to zero. On the other hand, in an unidentifiable (or almost unidentifiable) case [12, 66], there is a whole continuum of optimal (or almost optimal) solutions comprising the manifold S with dimension $0 < m_S \le N_\theta$. An algorithm for representing the manifold using a finite number of points θ_l, $l = 1, \ldots, L$ (in analogy to a local identifiable case) located on S is presented in [20].

It can be shown that the integral in Equation 22.28 over the entire parameter space can be approximated by a lower-dimensional integral over the manifold S, as follows:

$$\Pr(F|D_N, M_P) = \int_S w(\theta_S) \Pr(F|\theta_S, M_P) \, d\theta_S \qquad (22.32)$$

The weighting functions $w(\theta_S)$ express the probability volume of the space perpendicular to the manifold at point θ_S. Clearly, in a local identifiable case, the above integral reduces to a sum over all optimal solutions; while in an unidentifiable case, it can be further approximated using a representative finite set of points θ_l, $l = 1,\ldots,L$ on S:

$$\Pr(F \mid D_N, M_P) \sim \sum_{l=1}^{L} w_l \Pr(F \mid \theta_l, M_P) \tag{22.33}$$

Expressions for the weighting coefficients w_l can be found in the work by Katafygiotis et al. [12].

22.7 Conclusion

A Bayesian statistical framework was presented for structural modeling and updating using vibration measurements. The Bayesian formulation is attractive to use in selecting the best model from a model class as well as selecting the best model class from alternative model classes. The role of the sensor configuration in reducing the modeling uncertainties and, thus, improving the predictions of structural integrity, response, and reliability is emphasized. The methodology was applied to the identification of location and severity of damage of structures using measured modal data. It was demonstrated that model selection depends on the information contained in the data, the number of observed modes, and the number of measured locations. In particular, prediction of location and severity of damage can be improved by optimizing the sensor configuration in the structure. A methodology for optimally selecting the sensor configuration in the structure was proposed based on the information entropy measure. Useful asymptotic results were derived, valid for the case of a large number of data, which show that the time history details in the measured data can be ignored in the design of the optimal sensor configuration. This allows one to select the optimal sensor configuration in the initial design stage based only on the information provided by a nominal model.

The Bayesian statistical system identification framework was finally integrated with probabilistic structural dynamics tools for updating the reliability predictions of structures utilizing measured information. Computational aspects for estimating the resulting probability integrals were discussed. Efficient asymptotic approximations valid for large a number of data were also proposed for both identifiable and unidentifiable cases encountered in structural model updating.

The proposed Bayesian framework is applicable to a wide range of structural/mechanical systems and measurements encountered in engineering applications. In particular, it can handle linear and nonlinear models that are subjected to either measured or unmeasured excitations coming from forced and ambient vibration tests, respectively.

Acknowledgments

The work presented in this chapter was funded by the Greek General Secretariat of Research and Technology and the European Community Fund within the PENED 99 program framework under Grant 99ED580 and the Hong Kong Research Council under Grant HKUST6253/00E. Additional funding was provided by the Greek Earthquake Planning and Protection Organization. This support is gratefully acknowledged.

The authors also thank Maria Pavlidou, graduate student at the University of Thessaly, for her assistance in carrying out part of the numerical computations.

References

1. Masri, S.F., Miller, R.K., Saud, A.F., and Caught, T.K. (1987). Identification of Nonlinear Vibrating Structures. *ASME Journal of Applied Mechanics*, 54, 918–929.
2. Mottershead, J.E. and Friswell, M.I. (1993). Model Updating in Structural Dynamics: A Survey. *Journal of Sound and Vibration*, 167(2), 347–375.

3. Farhat, C. and Hemez, F.M. (1993). Updating Finite Element Dynamics Models Using and Element-by-Element Sensitivity Methodology. *AIAA Journal*, 31(9), 1702–1711.
4. Agbabian, M.S., Masri, S.F., Miller, R.K., and Caughey, T.K. (1991). System Identification Approach to Detection of Structural Changes. *Journal of Engineering Mechanics*, 117(2), 370–390.
5. Katafygiotis, L.S. (1991). *Treatment of Model Uncertainties in Structural Dynamics*. Technical Report EERL91-01, California Institute of Technology, Pasadena, CA.
6. Hemez, F.M. and Farhat, C. (1995). Structural Damage Detection via a Finite Element Model Updating Methodology. *The International Journal of Analytical and Experimental Modal Analysis*, 10(3), 152–166.
7. Sadeghi, M.H. and Fassois, S.D. (1997). A Geometric Approach to the Non-Destructive Identification of Faults in Stochastic Structural Systems. *AIAA Journal*, 35, 700–705.
8. Sohn, H. and Law, K.H. (1997). Bayesian Probabilistic Approach for Structural Damage Detection. *Earthquake Engineering and Structural Dynamics*, 26, 1259–1281.
9. Vanik, M.W. (1997). *A Bayesian Probabilistic Approach to Structural Health Monitoring*. Ph.D. thesis, EERL Report 97–07, Caltech, Pasadena.
10. Doebling, S., Farrar, C., Prime, M., and Shevitz, D. (1998). *Damage Identification and Health Monitoring of Structural and Mechanical Systems from Changes in Their Vibration Characteristics: A Literature Review*, Report LA-13070-MS, Los Alamos National Laboratory.
11. Beck, J.L. and Katafygiotis, L.S. (1998). Updating Models and Their Uncertainties — Bayesian Statistical Framework. *Journal of Engineering Mechanics (ASCE)*, 124(4), 455–461.
12. Katafygiotis, L.S., Papadimitriou, C., and Lam, H.F. (1998). A Probabilistic Approach to Structural Model Updating. *International Journal of Soil Dynamics and Earthquake Engineering*, 17(7–8), 495–507.
13. Yao, J.T.P. and Natke, H.G. (1994). Damage Detection and Reliability Evaluation of Existing Structures. *Structural Safety*, 15, 3–16.
14. Papadimitriou, C., Beck, J.L., and Katafygiotis, L.S. (2001). Updating Robust Reliability Using Structural Test Data. *Probabilistic Engineering Mechanics*, 16(2), 103–113.
15. Papadimitriou, C. and Katafygiotis, L.S. (2001). A Bayesian Methodology for Structural Integrity and Reliability Assessment. *International Journal of Advanced Manufacturing Systems*, 4(1), 93–100.
16. Igusa, T., Buonopane, S.G., and Ellingwood, B.R. (2002). Bayesian Analysis of Uncertainty for Structural Engineering Applications. *Structural Safety*, 24(2–4), 165–186.
17. Udwadia, F.E. and Sharma, D.K. (1978). Some Uniqueness Results Related to Building Structural Identification. *SIAM Journal of Applied Mathematics*, 34(1), 104–151.
18. Berman, A. (1989). Nonunique Structural System Identification. *Proc. 7th International Modal Analysis Conference*, pp. 355–359.
19. Katafygiotis, L.S. and Beck, J.L. (1998). Updating Models and Their Uncertainties—Model Identifiability. *Journal of Engineering Mechanics (ASCE)*, 124(4), 463–467.
20. Katafygiotis, L.S., Lam, H.F., and Papadimitriou, C. (2000). Treatment of Unidentifiability in Structural Model Updating. *Advances in Structural Engineering — An International Journal*, 3(1), 19–39.
21. Beck, J.L. (1989). Statistical System Identification of Structures. *Proceedings of the 5th International Conference on Structural Safety and Reliability*, ASCE, San Francisco, pp. 1395–1402.
22. Kammer, D.C. (1991). Sensor Placements for On-Orbit Modal Identification and Correlation of Large Space Structures. *Journal of Guidance, Control and Dynamics*, 14, 251–259.
23. Udwadia, F.E. (1994). Methodology for Optimal Sensor Locations for Parameter Identification in Dynamic Systems. *Journal of Engineering Mechanics (ASCE)*, 120(2), 368–390.
24. Kammer, D.C. and Yao, L. (1994). Enhancement of On-Orbit Modal Identification of Large Space Structures through Sensor Placement. *Journal of Sound and Vibration*, 171(1), 119–139.
25. Heredia-Zavoni, E. and Esteva, L. (1998). Optimal Instrumentation of Uncertain Structural Systems Subject to Earthquake Motions. *Earthquake Engineering and Structural Dynamics*, 27, 343–362.

26. Heredia-Zavoni, E., Montes-Iturrizaga, R., and Esteva, L. (1999). Optimal Instrumentation of Structures on Flexible Base for System Identification. *Earthquake Engineering and Structural Dynamics*, 28(12), 1471–1482.
27. Papadimitriou, C., Beck, J.L., and Au, S.K. (2000). Entropy-Based Optimal Sensor Location for Structural Model Updating. *Journal of Vibration and Control*, 6(5), 781–800.
28. Yuen, K.V., Katafygiotis, L.S., Papadimitriou, C., and Mickleborough, N.C. (2001). Optimal Sensor Placement Methodology for Identification with Unmeasured Excitation. *Journal of Dynamic Systems, Measurement and Control*, 123(4), 677–686.
29. Papadimitriou, C. (2002). Optimal Instrumentation Strategies for Structural Identification. *Proc. 4th Int. Conf. Computational Stochastic Mechanics*, Spanos, P.D. and Deodatis, G. (Eds), Millpress, Rotterdam, the Netherlands, pp. 463–472.
30. Papadimitriou, C. (2002). Applications of Genetic Algorithms in Structural Health Monitoring. *Proc. 5th World Congress on Computational Mechanics*, Vienna, Austria, http://wccm.tuwien.ac.at.
31. Papadimitriou, C., Beck, J.L., and Katafygiotis, L.S. (1997). Asymptotic Expansions for Reliability and Moments of Uncertain Dynamic Systems. *Journal of Engineering Mechanics (ASCE)*, 123(12), 1219–1229.
32. Metallidis, P., Verros, G., Natsiavas, S., and Papadimitriou, C. (2003). Fault Detection and Optimal Sensor Location in Vehicle Suspensions. *Journal of Vibration and Control*, 9(3–4), 337–359.
33. Katafygiotis, L.S., Yuen, K.V., and Chen, J.C. (2001). Bayesian Modal Updating by Use of Ambient Data. *AIAA Journal*, 39(2), 271–278.
34. Yuen, K.V. and Katafygiotis, L.S. (2001). Bayesian Time Domain Approach for Modal Updating Using Ambient Data. *Probabilistic Engineering Mechanics*, 16(3), 219–231.
35. Yuen, K.V. and Katafygiotis, L.S. (2001). Bayesian Spectral Density Approach for Modal Updating Using Ambient Data. *Earthquake Engineering and Structural Dynamics*, 30(8), 1103–1123.
36. Yuen, K.V. and Katafygiotis, L.S. (2002). Bayesian Model Updating Using Complete Input and Incomplete Response Noisy Measurements. *Journal of Engineering Mechanics (ASCE)*, 128(3), 340–350.
37. Yuen, K.V. and Beck, J.L. (2003). Updating Properties of Nonlinear Dynamical Systems with Uncertain Input. *Journal of Engineering Mechanics (ASCE)*, 129(1), 9–20.
38. Vanik, M.W., Beck, J.L., and Au, S.K. (2000). Bayesian Probabilistic Approach to Structural Health Monitoring. *Journal of Engineering Mechanics (ASCE)*, 126(7), 738–745.
39. Sohn, H. and Law, K.H. (2000). Bayesian Probabilistic Damage Detection of a Reinforced-Concrete Beam Column. *Earthquake Engineering and Structural Dynamics*, 29(8), 1131–1152.
40. Sohn, H. and Law, K.H. (2000). Application of Load Dependent Ritz Vectors to Bayesian Probabilistic Damage Detection. *Probabilistic Engineering Mechanics*, 15(2), 139–153.
41. Poulakis, Z., Valougeorgis, D., and Papadimitriou, C. (2001). Health Monitoring of Pipe Networks Using a Probabilistic Framework. *Proc. of 8th International Conference on Structural Safety and Reliability (ICOSSAR'01)*, Newport Beach, CA, June 17–22.
42. Poulakis, Z., Valougeorgis, D., and Papadimitriou, C. (2003). Leakage Detection in Water Pipe Networks Using a Bayesian Probabilistic Framework. *Probabilistic Engineering Mechanics*, 18(4), 315–327.
43. Yuen, K.V. (2002). *Model Selection, Identification and Robust Control for Dynamical Systems*. Ph.D. thesis, EERL Report 2002-03, Caltech, Pasadena, CA.
44. Katafygiotis, L.S. and Yuen, K.V. (2001). Bayesian Spectral Density Approach for Modal Updating Using Ambient Data. *Earthquake Engineering and Structural Dynamics*, 30(8), 1103–1123.
45. Jaynes, E.T. (1978). *Where Do We Stand on Maximum Entropy?*, MIT Press, Cambridge.
46. Bleistein, N. and Handelsman, R. (1986). *Asymptotic Expansions for Integrals*, Dover, New York.
47. Goldberg, D.E. (1989). *Genetic Algorithms in Search, Optimization and Machine Learning*. Reading, Addison-Wesley, MA.
48. Michalewich, Z. (1999). *Genetic Algorithms + Data Structures = Evolution Programs*, Springer, New York.

49. Zitzler, E. and Thiele, L. (1999). Multiobjective Evolutionary Algorithms: A Comparative Case Study and the Strength Pareto Approach. *IEEE Transactions on Evolutionary Computation*, 3(4), 257–271.
50. Lutes, L.D. and Sarkani, S. (1997). *Stochastic Analysis of Structural and Mechanical Vibrations*. Prentice Hall, Englewood Cliffs, NJ.
51. Au, S.K. (2001). *On the Solution of the First Excursion Problem by Simulation with Applications to Probabilistic Seismic Performance Assessment*. Ph.D. thesis, EERL Report 2001-02, Caltech, Pasadena, CA.
52. Au, S.K. and Beck, J.L. (2001). First Excursion Probabilities for Linear Systems by Very Efficient Importance Sampling. *Probabilistic Engineering Mechanics*, 16(3), 193–207.
53. Katafygiotis, L.S. and Cheung, S.H. (2003). On the Calculation of the Failure Probability Corresponding to a Union of Linear Failure Domains. *Fourth International Conference on Computational Stochastic Mechanics*, Spanos, P.D. and Deodatis, G. (Eds), Millpress, Rotterdam, the Netherlands, pp. 301–309.
54. Katafygiotis, L.S. and Cheung, S.H. (2003) Wedge Simulation Method for Calculation of Reliability of Linear Systems Subjected to Gaussian Random Excitations. *Fifth International Conference on Stochastic Structural Dynamics (SSD03)*, Hangzhou, China, May 26–28.
55. Pradlwalter, H.J. and Schueller, G.I. (1997). On Advanced Monte Carlo Simulation Procedures in Stochastic Structural Dynamics. *International Journal of Nonlinear Mechanics*, 32(4), 735–744.
56. Pradlwalter, H.J. and Schueller, G.I. (1999). Assessment of Low Probability Events of Dynamical Systems by Controlled Monte Carlo Simulation. *Probabilistic Engineering Mechanics*, 14(3), 213–227.
57. Proppe, C., Pradlwalter, H.J., and Schueller, G.I. (2003). Equivalent Linearization and Monte Carlo Simulation in Stochastic Dynamics. *Probabilistic Engineering Mechanics*, 18(1), 1–15.
58. Au, S.K. and Beck, J.L. (2001). Estimation of Small Failure Probabilities in High Dimensions by Subset Simulation. *Probabilistic Engineering Mechanics*, 16(4), 263–277.
59. Katafygiotis, L.S. and Cheung, S.H. (2002). MCMC Based Simulation Methodology for Reliability Calculations. *Fourth International Conference on Computational Stochastic Mechanics*, Spanos, P.D. and Deodatis, G. (Eds), Millpress, Rotterdam, the Netherlands, pp. 293–299.
60. Schueller, G.I. and Stix, R. (1987). A Critical Appraisal of Methods to Determine Failure Probabilities. *Structural Safety*, 4, 293–309.
61. Madsen, H.O., Krenk, S., and Lind, N.C. (1986). *Methods of Structural Safety*. Prentice Hall, Englewood Cliffs, NJ.
62. Papadimitriou, C., Katafygiotis, L.S., and Au, S.K. (1997). Effects of Structural Uncertainties on TMD Design: A Reliability-Based Approach. *Journal of Structural Control*, 4(1), 65–88.
63. May, B.S. and Beck, J.L. (1998). Probabilistic Control for the Active Mass Driver Benchmark Structural Model. *Earthquake Engineering and Structural Dynamics*, 27(11), 1331–1346.
64. Yuen, K.V. and Beck, J.L. (2003). Reliability-Based Robust Control for Uncertain Dynamical Systems Using Feedback of Incomplete Noisy Response Measurements. *Earthquake Engineering and Structural Dynamics*, 32, 751–770.
65. Beck, J.L. and Au, S.K. (2002). Bayesian Updating of Structural Models and Reliability using Markov Chain Monte Carlo Simulation. *Journal of Engineering Mechanics (ASCE)*, 128(4), 380–391.
66. Katafygiotis, L.S. and Lam, H.F. (2002). Tangential-Projection Algorithm for Manifold Representation in Unidentifiable Model Updating Problems. *Earthquake Engineering and Structural Dynamics*, 31, 791–812.

List of Symbols

D_N	Measured data
E_θ	Mathematical expectation with respect to θ
F	Failure state of a structure
$G_i(\delta)$	Normalized information entropy for the i-th model class
$G(\delta)$	Total normalized information entropy for μ model classes

$H(\delta, D_N)$	Information entropy given a sensor configuration δ and the data D_N
$H(\delta, \theta, \sigma)$	Asymptotic estimate of the information entropy given the sensor configuration δ and a model in the class M_P
$H_{\min}^{(L)}, H_{\max}^{(L)}$	Minimum and maximum information entropy for L sensors
$J(\theta; D_N, M)$	Measure of fit between measured and model predicted responses
L_0	Observation matrix
M_P, M_i	Classes of probability models
M	Class of structural model
$M(\theta)$	Particular model in model class M
N	Number of sampling points
N_d	Number of model degrees of freedom
N_0	Number of measured degrees of freedom
N_f	$(NN_0-1)/2$
N_α	Number of structural and prediction error model parameters
N_θ	Number of structural model parameters
P	Class of prediction error models
$\Pr(F \mid \theta, M_P)$	Conditional probability of failure given a model $M(\theta)$ in the class M_P
$\Pr(F \mid D_N, M_P)$	Updated structural failure probability given the data D_N
$P^{(j)}(\theta)$	Fisher information matrix based on one measurement at DOF j
$Q(\delta, \theta)$	Fisher information matrix based on measurements at DOFs specified by δ
S	Manifold in the space of parameters θ
$h(\theta_i; D_N, M_i)$	Hessian of the function $-\ln p(\theta_i \mid D_N, M_i)$
m	Time index
m_S	Dimension of manifold S
$n(m; \theta)$	Prediction error at time $t_m = m\Delta t$
$p(D_N \mid M_i)$	Probability of observing the data from the model class M_i
$p(M_i)$	Prior probability of model class M_i
$p(M_i \mid D_N)$	Posterior (updated) probability of model class M_i given the data D_N
$p(D_N \mid \theta, \sigma, M_P)$	Likelihood of observing the data by a model taken from class M_P
$p(\theta, \sigma \mid D_N, M_P)$	Posterior (updated) PDF of model parameters given the data D_N and the class of probability models M_P
$p(\theta \mid D_N, M_P)$	Marginal probability distribution for structural model parameters θ
$q(m; \theta)$	Predictions from model $M(\theta)$ at N_d degrees of freedom at time $t_m = m\Delta t$
$y(m)$	Vector of measurements at time $t_m = m\Delta t$
Δt	Sampling interval
α	$[\theta, \sigma]$
δ	Sensor configuration vector
θ	Parameter set of model class M
$\hat{\theta}$	Most probable value of the parameter set θ
μ	Number of model classes
$\pi_\theta(\theta)$	Prior PDF of structural model parameters
$\pi_\sigma(\sigma)$	Prior PDF of prediction error parameter
σ	Standard deviation of the prediction error
$\hat{\sigma}$	Most probable value of standard deviation of the prediction error
$\hat{\phi}_m, \phi_m(\theta)$	Measured and model predicted modeshape components
$\hat{\omega}_m, \omega_m(\theta)$	Measured and model predicted modal frequency

23
Utility Methods in Engineering Design

23.1	Introduction	23-1
23.2	An Engineering Design Example	23-2
	Engineering Applications	
23.3	Utility	23-3
	Mathematical Background • Lotteries • Multiattribute Utility	
23.4	Example	23-7
	Determining Utility Independence • Utilities Over Individual Attributes • The Multiattribute Utility Function • Expected Utility and Design under Uncertainty	
23.5	Conclusion	23-14

Michael J. Scott
University of Illinois at Chicago

23.1 Introduction

Utility analysis is a method, grounded in decision theory, to aid in the making of decisions. Utility analysis helps a decision maker to construct a real-valued function that encodes the decision maker's preferences for all possible outcomes. This function can then be used to determine which course of action, or decision, has the best expected result. Utility analysis is a general method, applicable in many different fields. Its application in engineering design is in the context of multiattribute optimization under uncertainty: the alternatives, or *candidate designs*, are given as vectors in $\mathbf{R}^n$, and the performance of each design can be described by a vector in $\mathbf{R}^m$. The application of utility theory in engineering design serves two important purposes:

1. Under reasonably general conditions, the utility function allows a designer to specify preferences independently with respect to a number of different attributes, and use that to construct a single-valued objective function that incorporates all the attributes and reconciles them on a single scale.
2. The utility function captures the decision maker's attitude toward risk. While it may be obvious that more of some attribute is always preferred to less (money, for instance), the utility function essentially tells how important it is to achieve a particular amount of that attribute. As a result, the utility function can be used to create a single-valued objective function based upon outcomes for each alternative that are given as probability distributions.

The two important uses of utility theory in engineering are thus (1) multiattribute optimization and (2) design under stochastic uncertainty.

23.2 An Engineering Design Example

To illustrate the concepts and their application, a utility analysis will be performed on a two-bar pin-jointed truss problem, previously presented by Azarm et al. [1]. The truss is shown in Figure 23.1. The design variables are the cross-sectional areas x_1 and x_2 of bars AC and BC, respectively, and the vertical distance y from the line connecting pins A and B to the point C where the load is applied. It can be considered as a two-objective optimization problem as follows:

$$\text{Minimize volume } V = x_1\sqrt{16+y^2} + x_2\sqrt{1+y^2}$$

$$\text{Minimize stress in AC } \sigma_{AC} = \frac{20\sqrt{16+y^2}}{x_1 y}$$

subject to

$$V \leq 0.1$$

$$\sigma_{AC} \leq 100000$$

$$\sigma_{BC} \leq 100000 \quad\quad\quad (23.1)$$

$$x_1, x_2 \geq 0 (m^2)$$

$$y \in [1, 3] (m)$$

where

$$\sigma_{BC} = \frac{80\sqrt{1+y^2}}{x_2 y}$$

The constraints are imposed in order to bound the optimization problem.

If the problem is considered as a multiobjective optimization problem, then the *Pareto frontier* may be of interest. The Pareto frontier is the set of all points that are not dominated by any other point; one point dominates another if the first point is superior with respect to both attributes. For the reduced problem presented here, setting $x_2 = 0.0008944$ and $y = 2$ pushes σ_{BC} to its constraint, and we can recover

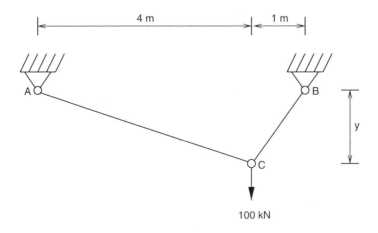

FIGURE 23.1 Two-bar truss example.

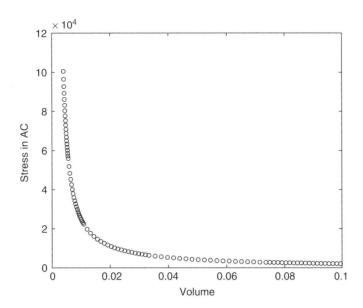

FIGURE 23.2 Pareto frontier for the truss example.

the Pareto frontier by varying x_1 (see Figure 23.2). There are various means of recovering a Pareto frontier for less tractable problems, but their discussion is not of relevance to this chapter. In multiobjective optimization, the Pareto frontier, or as much of it as possible, is calculated. The decision of which Pareto point to select, which clearly depends on the trade-off between the two objectives, is left to the designer.

As mentioned above, a utility theory approach will handle two common complications to this simple example. First, the trade-off between the two attributes will be formally specified through the assignment of a utility function, so that the optimization problem has a single objective. Second, with that utility function, it is possible to choose a best solution even if the design variables or model parameters are subject to random variations. The example will be worked in detail following an introduction to utility theory.

The truss example comes from the structural optimization literature, and it was presented here first as an optimization problem, in which the task is the generation of a Pareto frontier, not the decision of which point on that frontier to pursue. The application of utility theory to the problem does not provide any assistance in solving that optimization problem. Rather, it provides a means of defining a new, single-valued, objective function that more closely reflects the decision maker's true desires. It does not tell us how to numerically optimize that objective function. If the expected utility function is more easily optimized than the multiobjective problem of finding the entire Pareto frontier, then the utility approach can save computational resources.

23.2.1 Engineering Applications

There have been a number of published applications of utility theory to engineering design problems. Notable references include [2–5]. Thurston [6] also presents a nice overview of utility theory and decision making in engineering design.

23.3 Utility

The mathematical development of the concept of utility begins with von Neumann and Morgenstern [7], but the intuition that underlies the development can be traced back several centuries, when mathematicians studying gambling understood that people do not always act so as to maximize their expected profit. Bernoulli's Paradox is a good example: I will flip a fair coin. (There is no trickery, we both agree that the coin has equal chances of landing heads and tails.) The first head will appear on the ith flip, with probability

$$\left(\frac{1}{2^i}\right)$$

I will then pay you 2^i monetary units. Your expected payout is:

$$\sum_{i=1}^{\infty} \frac{2^i}{2^i} = \sum_{i=1}^{\infty} 1 = \infty$$

How much would you pay to take this gamble?

In practice, of course, no one will pay more than a small amount of money to enter Bernoulli's gamble. Indeed, if one pays $100, then there is a probability of less than 0.016 to make a profit; if one pays $1000, the probability falls below 0.002. This illustrates the basic idea that maximizing the expected value of profit may not be the best way to capture what a decision maker really wants when making decisions with uncertain outcomes.

23.3.1 Mathematical Background

While the shortcomings of expected profit as a decision-making tool were recognized early, the first formal development of utility is the work of von Neumann and Morgenstern [7], who set out to develop a new measure for preference. This measure they called *utility*, and its defining feature is that it is consistent with the mathematics of expectation: *expected* utility should correctly reflect the decision maker's desires. The mathematical development is informative, and is outlined in only the barest detail here. The interested reader is encouraged to refer to the original text.

First, von Neumann and Morgenstern define utility without resorting to a numerical representation, but simply as symbols u and v together with a natural relation "$\succ$" such that $u \succ v$ represents the case where an outcome with utility u is preferred to the outcome with utility v. This natural "$\succ$" is *not* assumed to be the usual "$>$" that applies to real numbers. They then seek a function ν which assigns real numbers to utilities. The crux of their argument is that they postulate the existence of a natural operation on utilities, which they call the "center of gravity," with respect to a scalar $\alpha \in (0, 1)$:

$$\alpha u \oplus (1-\alpha) v \qquad (23.2)$$

Note that this "$\oplus$" is not the usual "$+$" on real numbers either, yet, though $1 - \alpha$ is the usual real number. This operation simply means that if a decision maker has an opinion about two outcomes, he must also have an opinion about the "center of gravity" of the two outcomes with respect to any α on the unit interval. It is not too great an oversimplification to say that the requirement that

$$u \succ v \Rightarrow u \succ \alpha u \oplus (1-\alpha) v \succ v$$

is one of the axioms.

The major result of von Neumann and Morgenstern's work is that, given their axioms, utilities can be assigned real numbers in a unique way, up to a positive affine transformation. The symbols "$\succ$" and "$\oplus$" are then equivalent to the usual "$>$" and "$+$."

23.3.2 Lotteries

The center of gravity for utilities given in Equation 23.2 has a natural interpretation, and one that turns out to be quite useful in constructing utility functions. This is the concept of a *lottery* involving two outcomes. In a lottery, the decision maker will receive one outcome with probability α and the other with probability $(1-\alpha)$. If the utility for outcome a is $u(a)$, and the utility for outcome c is $u(c)$, then the utility for the lottery at α must be

$$\alpha u(a) + (1-\alpha)u(c) \qquad (23.3)$$

To construct a utility function, a decision maker is required to have preferences over outcomes, and over lotteries of outcomes, and over lotteries of lotteries, and so forth. This means that the decision maker can compare any two possibilities, including comparing an outcome to a lottery. Note that the decision maker never assigns a number, but only states, "I prefer outcome b to a lottery in which I receive outcome a with probability α and outcome c with probability $(1-\alpha)$." When a decision maker reports that a is preferable to b is preferable to c, that indicates only that

$$u(a) > u(b) > u(c)$$

To determine the utility function, the *lottery question* is asked: At what value of α are you indifferent between (1) receiving b with certainty, and (2) receiving a with probability α and c with probability $(1-\alpha)$?

In practice, it may be necessary to ask specific questions about several different values of α. Also, the need to specify probabilities has been criticized by some, and there are alternatives that do not require any knowledge of probabilities. See, for instance, Wakker and Deneffe [8]. In principle, however, the value α can always be found, and then it is known that

$$u(b) = \alpha u(a) + (1-\alpha)u(c)$$

As noted above, a utility function is only determined up to a positive affine transformation; if the lottery question is answered for all outcomes, then fixing the value of the utility function $u(\cdot)$ for any two outcomes determines the entire function.

23.3.3 Multiattribute Utility

If mathematical utility theory commenced with von Neumann and Morgenstern, Keeney and Raiffa [9] pioneered its practical application, especially the development of multiattribute utility theory.

We have, up to this point, been careful to define utility as a property or function of outcomes, not attributes. Technically speaking, one can only have a preference for an outcome. However, multiattribute utility theory clearly concerns itself with *attributes*, and an outcome is defined as a collection of values for all attributes. It is important in multiattribute utility theory to ask the right question. The wrong question is, how do we combine utilities for different attributes to get an overall utility function?

The right question is, given that there is an overall order of all the alternatives, when can we represent that order as a function of orders of the alternatives with respect to individual attributes, and if so, how?

A handful of results from utility theory address this question. Chapter 5 of [9] has a more detailed exposition. The results, in brief form, are:

- Utilities need to be somehow independent; while it is possible to consider attributes that are not utility independent, it no longer makes sense to speak of the utility of alternatives with respect to one of the attributes.
- Additive independence (a strong condition) leads to additive utility functions.
- Mutual utility independence (a somewhat weaker condition) leads to multilinear, or occasionally multiplicative, utility functions.

Let us turn now to the definitions and results.

23.3.3.1 Notation for Multiattribute Utility Theory

To this point, we have written u for a utility function, the domain of which is the set of possible outcomes $\{a, b, c, \ldots\}$, so that $u(a)$ is a real number. A slight change in notation will be convenient in order to assess outcomes using n different attributes $X_1, X_2, \ldots, X_n$. We write x_i for the level of attribute X_i (and we think of the x_i as belonging on the real line, even though that is not required). A particular outcome is then completely described by the vector $\vec{x} = (x_1, x_2, \ldots, x_n)$, and we are concerned with the utility function, which takes real values $u(\vec{x})$.

We now consider the case of two attributes, which we may call Y and Z rather than X_1 and X_2. The question to ask is, when can $u(y, z)$ be expressed as

$$u(y, z) = f(f_Y(y), f_Z(z))$$

where f_Y and f_Z are functions only of Y and Z, respectively?

In particular, when are f_Y and f_Z themselves utility functions u_Y and u_Z? And, of course, what form can f take? The question generalizes in a straightforward way to more than two attributes.

23.3.3.2 Independence

The first thing to note is that the idea of a utility function u_Y that depends only on the attribute Y only makes sense if our preferences for different values of Y are independent of the particular value of the other attribute(s) (in this case, Z). A conditional lottery is a question of the form, "If we fix $z = z_0$ and present the usual $y_1 \succ y_2 \succ y_3$, what value of α yields indifference between y_2 for certain and y_1 with probability α, y_3 with probability $(1 - \alpha)$?"

If the answers to conditional lotteries over Y given z do not depend on the fixed value z, then Y is *utility independent* of Z. If Y is utility independent of Z, then a utility function u_Y can be unambiguously defined. We need to be a little careful about what "unambiguously defined" means. If $u(\cdot, z_0)$ is a utility function on Y given z_0 and $u(\cdot, z_1)$ is a utility function on Y given z_1, there is no guarantee that $u(y, z_0) = u(y, z_1)$ in general. Rather, we know that the two are *strategically* equivalent; that is, they will always indicate the same choices by preserving value differences and expectations. Also, we know that each must be a linear transform of the other. In fact, for any fixed z_0, there are functions g and h such that $h(z) > 0$ and

$$u(y, z) = g(z) + h(z)u(y, z_0) \tag{23.4}$$

Equation 23.4 can be used as an alternative definition of utility independence.

Note that Y being utility independent of Z does not necessarily imply that Z is utility independent of Y. In fact, all combinations are possible.

The main result of multiattribute utility theory is the following theorem, proved in [9]: If Y and Z are mutually utility independent, then $u(y, z)$ is *multilinear*, specifically, it can be written as

$$u(y, z) = u(y, z_0) + u(y_0, z) + k u(y, z_0) u(y_0, z)$$

or as

$$u(y, z) = k_Y u_Y(y) + k_Z u_Z(z) + k_{YZ} u_Y(y) u_Z(z) \tag{23.5}$$

when suitably normalized.

There are two interesting special cases. The first is when $k = 0$ and $u(x, y)$ is a simple sum of the marginal utilities (the additive form). In this case, Y and Z are what we call *additive independent*. In two dimensions, this can be equated with the condition that a lottery with (y, z) and (y_0, z_0) is equivalent to one with (y, z_0) and (y_0, z) for all y and z and any arbitrary y_0 and z_0. Additive independence is always mutual, and additive independence implies mutual utility independence, but *not* vice versa.

Utility Methods in Engineering Design

The other special case is when $k \neq 0$. In that case, we can always apply a little algebra to get to a purely multiplicative form of the utility function

$$u'(y, z) = u'_Y(y) u'_Z(z)$$

where

$$u'(y, z) = ku(y, z) + 1$$
$$u'_Y(y) = ku(y, z_0) + 1$$
$$u'_Z(z) = ku(y_0, z) + 1$$

However, this multiplicative form obscures an interesting feature of the multilinear form, which is the value of the constant k. The parameter k characterizes the nature of the trade-off between the two attributes, as further discussed in [9]. However, it is fruitless to attempt to specify a value of k directly; it must be calculated.

23.4 Example

We will now exercise the example presented in Section 23.2 and shown in Figure 23.1. While the solution will vary depending on the decision maker, the procedure is consistent. The example was initially presented as a multi*objective* optimization problem, in which the task was to recover the entire Pareto frontier. We now think of it as a multi*attribute* problem, in which information about aspects of the problem, or attributes, will aid us in determining the one true objective.

For pedagogical purposes, the procedure may be thought of as a dialogue between a *utility analyst* and the *designer* or *decision maker*. It is the latter who has preferences, the former's sole role is to help specify the utility function that describes those preferences. The analyst asks questions of the designer. A designer who is familiar with utility theory may conduct such an analysis single-handed by playing both roles.

23.4.1 Determining Utility Independence

It was noted above that the usefulness of multiattribute utility theory depends on utility independence among the attributes. The two attributes in this problem are the volume V and the stress σ_{AC}. (As there is only one stress at this point, for simplicity let us write σ for σ_{AC}.) We have already stated that both attributes should be minimized, and the Pareto frontier shown in Figure 23.2 indicates that we can sensibly restrict discussion to the approximate ranges:

$$V \in [0.004, 0.1]$$

$$\sigma \in [2000, 100000]$$

In utility analysis, it is not necessary to restrict questioning to a known feasible range, but it is convenient.

To determine the utility functions over the individual attributes, the analyst will ask a number of lottery questions, of the form: at what value of α would you be indifferent between a stress of 50,000 kN for certain, and a lottery in which the stress in the design will be 2,000 kN with probability α and 100,000 kN with probability $(1 - \alpha)$? You may assume that the volume stays fixed.

The first task is to ensure that the designer understands the lottery question. Before answering any lottery questions, however, the designer should be able to state that the answer to any lottery question concerning stress will be the same for any fixed volume, and that the answer to any lottery question concerning volume will be the same regardless of the fixed level of stress. This checks mutual utility independence.

23.4.2 Utilities Over Individual Attributes

The next step is to specify the utility functions u_V and u_σ over the individual attributes. Consider the question asked above, comparing the three values 2,000, 50,000, and 100,000 for σ. Let us suppose that the designer replies that, even though 50,000 is approximately halfway between 2,000 and 100,000, in order to avoid the worst case, she is indifferent between getting 50,000 for certain and the lottery, which delivers 2,000 with $\alpha = 0.75$ and 100,000 with $(1 - \alpha) = 0.25$. As u_σ is determined only up to a positive affine transformation, we can arbitrarily set

$$u_\sigma(2000) = 1.0$$

$$u_\sigma(100000) = 0.0$$

Equation 23.3 tells us that the utility of the lottery above, and thus the utility of its certainty equivalent, is

$$u_\sigma(50000) = 0.75 u_\sigma(2000) + 0.25 u_\sigma(100000)$$

Several similar lottery questions must be asked, with different values of σ. For example, the questioning might proceed with the following triples:

Best	Middle	Worst	$\alpha = P(\text{Best})$
2,000	50,000	100,000	0.75
2,000	20,000	50,000	0.65
50,000	75,000	100,000	0.6
75,000	90,000	100,000	0.5
2,000	10,000	20,000	0.5
20,000	35,000	50,000	0.7
50,000	65,000	75,000	0.6

This gives the following values for u_σ, shown in Figure 23.3:

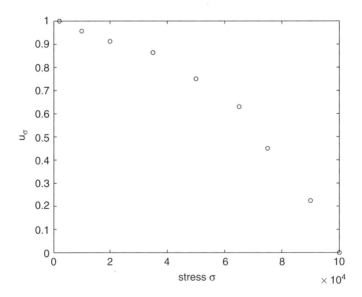

FIGURE 23.3 Preliminary values of u_σ.

Utility Methods in Engineering Design

σ	u_σ
2,000	1.0
10,000	0.95625
20,000	0.9125
35,000	0.86375
50,000	0.75
65,000	0.63
75,000	0.45
90,000	0.225
100,000	0

It is good practice to ask some redundant questions in order to check the consistency of the answers. For example, the following set of three stresses could be queried:

Best	Middle	Worst	$\alpha = P(\text{Best})$
35,000	50,000	65,000	0.5

This would give a value of $u_\sigma(50{,}000) = 0.7468$, which appears to confirm the original calculation of $u_\sigma(50{,}000) = 0.75$.

The plotted points must then be fit with a smooth curve. It can be unproductive to ask too many lottery questions. The points plotted in Figure 23.3 do not exhibit any great inconsistency, but neither do they appear to fall exactly on a smooth curve. The curve-fitting may be done by hand, or numerically, and the utility curve need not interpolate the points exactly. A rough second-order polynomial fit is given by

$$u_\sigma = -1.08 \cdot 10^{-10} \sigma^2 + 8 \cdot 10^{-7} \sigma + 1$$

and is shown in Figure 23.4.

A similar set of lottery questions is asked about preferences for different volumes. As volume of material is a direct surrogate for cost, this may well lead to a simple linear utility function such as

$$u_V = 1 - 10\,V$$

shown in Figure 23.5.

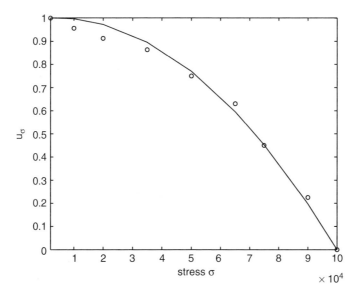

FIGURE 23.4 Rough second-order fit for u_σ.

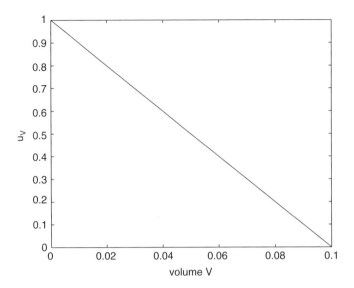

FIGURE 23.5 Linear utility function u_V.

The convex utility function u_σ is an example of a *risk-averse* attitude, while u_V exemplifies *risk-neutral* decision making. Some researchers have suggested attempting to determine risk attitude *before* fitting a utility function, with the idea that this will dictate a particular functional form and simplify the questioning.

23.4.3 The Multiattribute Utility Function

The two individual utility functions, u_σ and u_V, were calculated in such a way that all relevant utilities fall on the unit interval. We emphasize that the absolute value of the utilities provides no useful information. In particular, it makes no sense to compare u_σ and u_V; we cannot say, for instance, that the accidental equivalence

$$u_\sigma(71846) = u_V(0.05) = 0.5$$

means that the decision maker has the same preference for a stress of 71,846 kN as for a volume of 0.05 m³. Such comparison of two different attributes is not meaningful. The entire conditional utility function u_V could be multiplied by an arbitrary positive scalar, and that would change $u_V(0.05)$ without changing the decision maker's preferences for volume. One could compare three points, one with a stress of 100,000 kN and a volume of 0.1 m³, one with a stress of 71,846 kN and a volume of 0.1 m³, and one with a stress of 100,000 kN and a volume of 0.05 m³ to get information about relative preferences on stress and volume, but those preferences are independent of the incidental values of u_σ and u_V.

To uncover the multiattribute utility function, we again turn to lottery questions. Since the two attributes are mutually utility independent, the multilinear form of Equation 23.5 holds, and there are three constants to be determined. Note that the individual utility functions u_σ and u_V were specified such that

$$u_\sigma(100000) = u_V(0.1) = 0$$

$$u_\sigma(2000) = u_V(0) = 1$$

There is, of course, no physical design that can achieve $V = 0$, but that does not matter, and this will prove convenient in determining the multiattribute utility function $u(\sigma, V)$. We can start by setting

$$u(100000, 0.1) = 0 \tag{23.6}$$

$$u(2000, 0) = 1 \tag{23.7}$$

Then all feasible designs will have utilities somewhere on the unit interval. Referring to Equation 23.5,

$$u(\sigma, V) = k_\sigma u_\sigma(\sigma) + k_V u_V(V) + k_{\sigma V} u_\sigma(\sigma) u_V(V)$$

Equation 23.6 is trivially satisfied, while Equation 23.7 tells us that

$$k_\sigma + k_V + k_{\sigma V} = 1$$

We require two more equations to solve for the three constants, and thus require two more lottery questions. As before, it is useful to check consistency with some redundant questions.

Two convenient sets of three points include the two points above, where u takes the values 0 and 1, with a third point equal to either $(2000, 0.1)$, where $u_\sigma = 1$ and $u_V = 0$, or $(100{,}000, 0)$, where $u_\sigma = 0$ and $u_V = 1$. Note that the points to be compared need not be feasible. Let us suppose that our decision maker reports the following levels of α:

Best	Middle	Worst	$\alpha = P(\text{Best})$
(2,000, 0)	(2,000, 0.1)	(100,000, 0.1)	0.4
(2,000, 0)	(100,000, 0)	(100,000, 0.1)	0.7

From this, we can deduce that $k_\sigma = 0.4$, $k_V = 0.7$, and $k_{\sigma V} = -0.28$. The final multiattribute utility function is given by:

$$u(\sigma, V) = 0.4 u_\sigma(\sigma) + 0.7 u_V(V) - 0.28 u_\sigma(\sigma) u_V(V)$$

The parameters k_σ and k_V can be interpreted to mean that, within the region of interest, volume appears to be more important than stress to this decision maker. Also, $k_{\sigma V} < 0$ indicates that there is some *substitution* between the two attributes; an incremental improvement in both attributes at once is not worth double an incremental improvement in one or the other. Scaling the utilities of each attribute by a constant c results in a multiattribute utility that is less than that of the original design scaled by c.

A surface plot of the utility function as a function of σ and V is shown in Figure 23.6. Not all of the points displayed on the surface plot are feasible, but the utility function is readily optimized as a function of the design variables, with the best design point found at:

$$x_1 = 0.001402$$

$$x_2 = 0.0008944$$

$$y = 2$$

$$\sigma = 31898$$

$$V = 0.008270$$

$$u = 0.773$$

Here, utility theory has been used to construct a single-valued objective function for a deterministic design. This function combines and reconciles both performance attributes, indicating the desired trade-off between

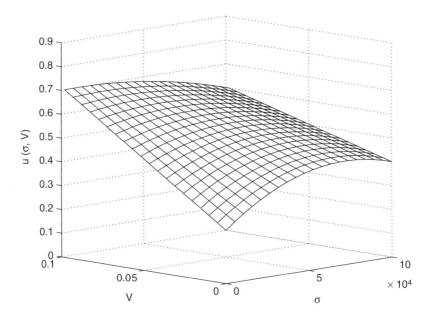

FIGURE 23.6 Overall utility function $u(\sigma, V)$.

the two. By applying numerical optimization to the utility function, we can converge immediately to the desired design, without the added computational expense of recovering the entire Pareto frontier.

The main use of utility theory in engineering, however, is for design in the presence of uncertainty.

23.4.4 Expected Utility and Design under Uncertainty

The original mathematical development of utility theory was undertaken with the express goal that the resulting utility functions should behave usefully when combined using the mathematics of expectation. Thus, the expected value of utility should provide a meaningful measure of rationality in design under uncertainty. In short, maximizing expected utility should allow a decision maker to choose the alternative that is most likely to give the most desired result (considered over all possible results). To accomplish this optimization under uncertainty, the same utility function is used as for the deterministic case. There is no need to provide any additional information about the utility function.

In design under uncertainty, variables and parameters that are treated as real numbers in the deterministic case are replaced with probability distributions. When sufficient data are available to support the assumption, it is standard practice to assume that these variables and parameters are normally distributed with known means and variances. In the example we have been discussing, the geometric variables x_1, x_2, and y would be assigned variances, or more commonly, standard deviations. Let us suppose that manufacturing variations can be assumed to result in standard deviations:

$$\sigma_{x_1} = 0.0001$$
$$\sigma_{x_2} = 0.0001$$
$$\sigma_{y} = 0.001$$

In utility theory, just as utilities or preferences must be expressed over *outcomes* in order to be meaningful, the calculation of expected utility relies upon knowing the probability distribution of outcomes for each alternative. The propagation of uncertainty from the variables and parameters to the performances can be a difficult or computationally costly task. In theory, one can integrate over all the probability distributions at each point. In practice, various techniques can be applied to reduce the computational load, including limiting the integration to $\pm 3\sigma$ for each distribution, Monte Carlo

Utility Methods in Engineering Design

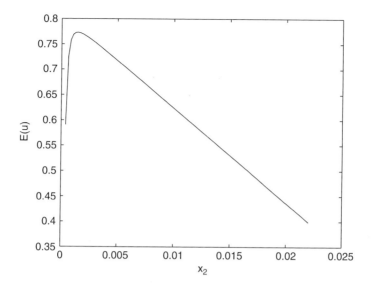

FIGURE 23.7 Expected utility $E(u)$ as a function of x_2, 100 Monte Carlo calculations per point.

sampling, and low-order Taylor approximations in cases when performance calculations are given as analytic expressions.

For the example discussed here, we saw above that the optimal deterministic designs are found by fixing $x_2 = 0.0008944$ and $y = 2$, and allowing x_1 to vary. Depending on the nature of the constraint in inequality (see Equation 23.1), it may be necessary to take a slightly smaller value of x_2, but still a fixed nominal value will be optimal. A quick Monte Carlo approximation of expected utility $E(u)$ as a function of x_2 is shown in Figure 23.7. Figure 23.8 zooms in on the portion of the curve where $E(u)$ is maximized. The variability in the Monte Carlo simulation is easily seen, but so is the trend of the plot. The optimal region is again right around 0.0014, perhaps shifted a bit to the right from the deterministic case.

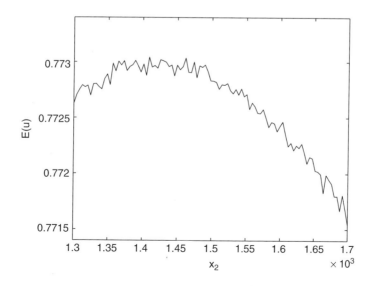

FIGURE 23.8 Expected utility $E(u)$ as a function of x_2, exploded view, 500 Monte Carlo calculations per point.

23.5 Conclusion

This chapter has presented an overview of utility theory and its application in engineering design. Utility theory provides a means of expressing the designer or decision maker's preferences as a scalar function that might be optimized. Classical utility theory is one of a number of preference-based methods that can be used to construct objective functions.

We saw in this chapter how the application of utility theory serves to assess attitudes toward risk, to combine and reconcile several competing and perhaps incommensurate performance measures, and to address the question of which designs are the best choices under uncertainty, when variables and parameters are expressed as normal probability distributions.

References

1. Azarm, S., Reynolds, B.J., and Narayanan, S., Comparison of two multiobjective optimization techniques with and within genetic algorithms, in *1999 ASME Design Engineering Technical Conference Proceedings*, DETC99/DAC-8584, ASME, 1999.
2. Bradley, S. and Agogino, A., An intelligent real time design methodology for catalog selection, in *Design Theory and Methods '91*, Vol. 31, Stauffer, L.A., Ed., ASME, 1991, pp. 201–208.
3. Thurston, D.L. and Carnahan, J., Fuzzy ratings and utility analysis in preliminary design evaluation of multiple attributes, *ASME J. Mechanical Design*, 114(4), 648–658, 1992.
4. Thurston, D.L., A formal method for subjective design evaluation with multiple attributes, *Res. Eng. Design*, 3(2), 105–122, 1991.
5. Wan, J. and Krishnamurty, S., Comparison-based decision making in engineering design, in *Proceedings of ASME 1999 Design Engineering Technical Conference*, DETC99/DTM-8746, ASME, 1999.
6. Thurston, D.L., Real and perceived limitations to decision based design, in *Proceedings of ASME 1999 Design Engineering Technical Conference*, DETC99/DTM-8750, ASME, 1999.
7. von Neumann, J. and Morgenstern, O., *Theory of Games and Economic Behavior*, 3rd ed., Princeton University Press, Princeton, NJ, 1953.
8. Wakker, P. and Deneffe, D., Eliciting von Neumann-Morgenstern utilities when probabilities are distorted or unknown, *Manage. Sci.*, 42(8), 1131–1150, 1996.
9. Keeney, R. and Raiffa, H., *Decisions with Multiple Objectives: Preferences and Value Tradeoffs*, John Wiley & Sons, New York, 1976.

24
Reliability-Based Optimization of Civil and Aerospace Structural Systems

24.1	Introduction	**24**-1
24.2	Problem Types	**24**-2
24.3	Basic Formulations	**24**-4
	Standard Optimization Problem • Reliability Index Approach • Performance Measure Approach	
24.4	Sensitivity Analysis in Reliability-Based Optimization	**24**-8
	General Remarks • Design Sensitivities for Reliability Index Approach • Design Sensitivities for Performance Measure Approach	
24.5	Optimization Methods and Algorithms	**24**-12
	Nested and "One-Shot" Optimization Methods • Algorithms	
24.6	Multicriteria and Life-Cycle Cost Reliability-Based Optimization	**24**-15
24.7	Examples in Civil Engineering	**24**-16
	Example 1: Material Cost-Based RBDO of Reinforced Concrete Girders [1] • Example 2: Life-Cycle Cost-Based RBDO of Deteriorating Reinforced Concrete Girders [2] • Example 3: Life-Cycle Cost-Based Repair Optimization of an Existing Deteriorating Bridge Using System Reliability [3] • Example 4: Life-Cycle Cost-Based Repair Optimization of an Existing Bridge Using System Survival Functions [4]	
24.8	Examples in Aerospace Engineering	**24**-21
	Composite Structures • Aerodynamics and Aeroelasticity • RBDO of Aircraft Systems	
24.9	Examples in MEMS	**24**-23
24.10	Conclusions	**24**-25

Dan M. Frangopol
University of Colorado

Kurt Maute
University of Colorado

24.1 Introduction

The main goal of an engineer is to design the best possible system. This is not an easy goal, since engineering systems are typically complex, involving multiple components and multiple failure modes, and their performance is time variant. Traditionally, structural designers used their intuition and experience to achieve their main goal. However, in 1960 the foundation of a new discipline called structural optimization was introduced by Schmit [5]. In general, the formulation of a structural optimization problem requires (1) an objective function, also called a cost function, that needs to be optimized (i.e., minimized or maximized), (2) a set of design variables whose values at the optimum need to be identified, and (3) a set of constraints to be satisfied. Design optimization is a formal, computer-oriented approach to solve the above formulation [6].

During the last four decades, theory and methods of structural optimization have developed significantly. The strong driving forces behind these developments include demands for lightweight structures (particularly in aerospace applications), efficient use of materials (particularly fiber-reinforced composites), enhanced performance of microelectromechanical systems, extending the service life of infrastructure systems, prioritizing maintenance interventions on deteriorating infrastructures, and energy conservation in various transportation systems [7, 8].

However, despite the demonstrated exponential growth in structural optimization research, most of it has been cast in deterministic format. In fact, a quick survey of about 2400 references [9] in structural optimization shows that more than 95% are deterministic-oriented. A major limitation of the deterministic optimization approach is that uncertainties in the loads, resistances, and structural responses are not included in the computation process. Therefore, irrespective of the level of sophistication in the approach used in the computational process, the optimum solutions are predicated on idealized deterministic assumptions. Consequently, these solutions are not computed under realistic conditions. These conditions have to include uncertainties associated with randomness, imperfect modeling, and estimation. In order to obtain optimal solutions recognizing all these uncertainties, the structural optimization process should be based on modern structural reliability theory by integrating optimization and reliability concepts and methods. Such integration is called reliability-based structural optimization (RBSO).

The methodologies used in the RBSO approaches for taking into account the stochastic nature of engineering systems can be classified into two groups: robust design optimization (RDO) and reliability-based design optimization (RBDO). As indicated in Frangopol and Maute [10], the RDO approach is based on purely deterministic analysis, and its purpose is dual: maximize the deterministic performance and minimize the sensitivity of this performance with respect to various uncertainties. Consequently, the impact of uncertainties on the RDO-based optimal solution is only captured in a qualitative sense. In contrast, the RBDO approach allows the design for a specific risk and target reliability level, accounting for the various sources of uncertainty in a quantitative sense.

The historical background (up to 1994) of reliability-based structural optimization based on the RBDO approach has been presented by Frangopol [8]. Advances in this field during the past decade have been summarized by Frangopol [11] and Frangopol and Maute [10], among others.

This chapter focuses on reliability-based optimization of civil and aerospace structural systems by using the RBDO approach. It presents a brief review of problem types (Section 24.2); a description of basic formulations, including the standard optimization problem, the reliability index approach, and the performance measure approach (Section 24.3); sensitivity analysis (Section 24.4); optimization methods and algorithms (Section 24.5); multicriteria and life-cycle-cost reliability-based optimization (Section 24.6); examples in civil engineering (Section 24.7); examples in aerospace engineering (Section 24.8); examples in microelectromechanical systems (Section 24.9); and conclusions (Section 24.10).

24.2 Problem Types

Frangopol and Moses [12] identified several types of RBDO problems that are presented below. In these problems, the designer is to find the design, s, that minimizes one of the following objective functions:

1. The total expected cost of a structural system, $C = C(s)$, expressed as

$$C = C_0 + C_F P \qquad (24.1)$$

where C_0 is the initial cost, which is a function of a vector of design variables s; C_F is the cost of failure; and P is the probability of failure, which is also a function of s.

2. The expected life-cycle cost of a structural system. According to Chang and Shinozuka [13], the life-cycle costs of systems under hazard risk, such as bridges under earthquake, can be divided into the following four major categories:

$$C_{LC} = C_{LC}^{(1)} + C_{LC}^{(2)} + C_{LC}^{(3)} + C_{LC}^{(4)} \tag{24.2}$$

where C_{LC} is the total life-cycle costs, $C_{LC}^{(1)}$ the planned and owner costs, $C_{LC}^{(2)}$ the user costs associated with $C_{LC}^{(1)}$, $C_{LC}^{(3)}$ the unplanned and owner costs, and $C_{LC}^{(4)}$ the user costs associated with $C_{LC}^{(3)}$. An alternative formulation will be presented in Section 24.6.

3. The total expected utility function, U, expressed as

$$U = B - C_0 - L \tag{24.3}$$

where B is the benefit derived from the existence of the structure, and L is the expected loss due to failure [14]. All monetary terms in Equation 24.1 and Equation 24.3 are expected present values, and all are functions of s. Conversion of future into present monetary values has to be done by using the discount rate of money v as follows:

$$C_p = \frac{C^*}{(1+v)^t} \tag{24.4}$$

where C_p = present value of cost, and C^* = cost at time t.

4. A cost function c of the system

$$\text{subject to} \quad P \leq P^0 \tag{24.5}$$

where P and P^0 are the probability of system failure and the allowable system failure probability, respectively.

5. The system failure probability P

$$\text{subject to} \quad c \leq c^0 \tag{24.6}$$

where c^0 is the allowable cost.

Other possible formulations of the RBDO problem are defined as follows: Find s that will minimize

6. A cost function c subject to multiple probability constraints such as

$$P_i \leq P_i^0 \quad i = 1, \ldots, m \tag{24.7}$$

$$P_j \leq P_j^0 \quad j = 1, \ldots, n \tag{24.8}$$

$$P_k \leq P_k^0 \quad k = 1, \ldots, p \tag{24.9}$$

where P_i is the system probability of unsatisfactory performance with respect to ultimate failure mode i (e.g., collapse, buckling, overturning), P_j is the component probability of unsatisfactory performance with respect to a specified limit state (i.e., ultimate or serviceability), P_k is the cross-section probability of unsatisfactory performance with respect to a specified limit state (e.g., yielding, crushing), and P_i^0, P_j^0, and P_k^0 are allowable failure probabilities.

7. One of the $m + n + p$ failure probabilities in Equation 24.7, Equation 24.8, or Equation 24.9, subject to specified performance and cost constraints.

The design variables in the vector s can be divided into several categories [15] as follows:

- Sizing design variables (e.g., cross-sectional dimensions)
- Geometrical or shape design variables (e.g., coordinates of joints in a truss or framed structure, fiber orientation in a composite material, height of a truss)
- Topological design variables (e.g., number of columns supporting a roof, number of layers in a fiber-reinforced composite bridge deck)
- Material design variables (e.g., mechanical properties of a composite material)
- Structural system design variables (e.g., frame or truss for a building, or cable stay or suspension for a bridge)

During the past decade, there has been tremendous progress in solving RBDO problems. Nevertheless, much work remains to be done until the designer will be able to obtain — for a specific complex and realistic structural system (e.g., high-rise building, long-span bridge, airplane) — a practical, global, optimum, reliability-based solution in which the type of structural system, material, system topology, geometry, and sizes of structural components are all treated simultaneously as design variables while minimizing the total expected cost — including inspection, maintenance, repair, rehabilitation, and user costs — during a specified time horizon.

24.3 Basic Formulations

24.3.1 Standard Optimization Problem

A standard optimization problem is formulated in terms of a cost function c; constraints $g_j, j = 1, \ldots, n_G$, and optimization variables $s \in \Re^{n_S}$, where n_S is the number of optimization variables and n_G denotes the number of constraints as follows:

$$\min_s \ c(s)$$
$$\text{subject to } g_j(s) \geq 0 \quad j = 1 \ldots n_G \qquad (24.10)$$
$$s_{lower} \leq s \leq s^{upper}$$

In structural optimization, for example, the objective and constraints are generally functions of the structural response v, which in turn is a function of the optimization variables s. Typical optimization criteria include weight, stiffness, displacements, stresses, eigenfrequencies, buckling loads, and cost. The structural response is in general governed by a set of nonlinear state equations, represented symbolically by:

$$R(s, v(s)) = 0 \qquad (24.11)$$

where R is the vector of residuals associated with a discretized structural system. Equation 24.11 can be added to the set of constraints of Equation 24.10. Here, the state equations are treated separately, as their mathematical properties and meaning differ from other design constraints.

In the presence of uncertainties, the design-optimization problem may include deterministic and probabilistic design criteria. Probabilistic criteria are often embedded as constraints restricting the failure probability, but they can also be used to formulate objectives, such as minimum life-cycle costs or maximum probability of reaching a target value [8, 12].

A generic reliability-based optimization problem can be formulated as follows, with r being the set of random parameters:

$$\min_{s} \quad P(c(s,r) > \bar{c}_t)$$
$$\text{subject to:} \quad \bar{P}_i - P(f_i(s,r) < 0) \geq 0 \qquad (24.12)$$
$$g_j^D(s) \geq 0$$
$$s_{lower} \leq s \leq s^{upper}$$

The objective is to minimize the probability of the cost function $c(s, r)$ being larger than a target value $\bar{c}_t$. One set of constraints limits the failure probability associated with the limit state functions $f_i(s, r)$, defining failure as $f_i < 0$. $\bar{P}_i$ is the maximum acceptable failure probability. A second set of constraints, g_j^D, contains all deterministic constraints. The state equations also depend on the random parameters:

$$R(s, r, v(s, r)) = \mathbf{0} \qquad (24.13)$$

RBDO methods can be distinguished, among other criteria, by the approach for determining the probability of an event. Formulations and algorithms in RBDO have been summarized in several review papers [8, 11, 12, 16–18]. In the following subsections, the most common approaches are discussed.

24.3.2 Reliability Index Approach

One of the most popular and well-explored approaches is based on the concept of characterizing the probability of survival by the reliability index and then performing computations based on first-order reliability methods (FORM). The reliability index β associated with the limit state function f_i is defined as the distance from the origin to the most probable point (MPP) of the failure surface in the standard normal space u (Figure 24.1). The probability of failure $P(f_i < 0)$, for example, is then found by integrating the first-order approximation of the limit state function $f_i = 0$ at the MPP:

$$P(f_i < 0) = \Phi(-\beta) \qquad (24.14)$$

where Φ is the standard normal cumulative distribution function (CDF). The feasibility of the mean-value point needs to be checked, to identify on which side of the failure curve the mean design point lies, i.e., whether $P(f_i < 0)$ in Equation 24.14 is the probability of failure or safety. The same procedure applies to the probability of the cost function $P(c > \bar{c}_t)$.

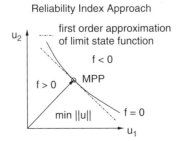 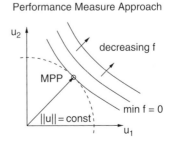

FIGURE 24.1 Representation of limit state function in the standard normal space with $f(u) = 0$ being the failure surface. Left: reliability index approach; right: performance measure approach.

TABLE 24.1 Transformations of Standard Normal Variables into Variables Following Common Probability Distribution Functions

Distribution Type	Transformation $T_u^{-1}(u_k)$
Normal (μ, σ)	$\mu + \sigma u_k$
Lognormal (μ, σ)	$e^{\mu + \sigma u_k}$
Uniform (a, b)	$a + (b-a)(0.5 + 0.5\, erf(u_k/\sqrt{2}))$
Gamma (a, b)	$ab\left(u_k \dfrac{1}{\sqrt{9a}} + 1 - \dfrac{1}{9a}\right)^3$

To evaluate the reliability index for the limit state function f_i by FORM, the random parameters r_k are mapped into standard normal space u_k:

$$u_k = T_u(r_k) \tag{24.15}$$

where T_u is generally a nonlinear mapping that depends on the type of random distribution of r. The inverse transformation, $r_k = T_u^{-1}(u_k)$ of some common distributions are listed in Table 24.1. In the standard normal space, the most probable point (MPP) is to be located by solving a nonlinear optimization problem subject to one equality constraint, with u being the vector of design variables:

$$\min_u \ (u^t u)^{\frac{1}{2}}$$
$$\text{subject to:} \quad f_i(u) = 0 \tag{24.16}$$

where the superscript t denotes the transpose of the vector or matrix. For convexification purposes, the above optimization problem can also be formulated augmenting the objective by a penalty term as follows:

$$\min_u \ (u^t u)^{\frac{1}{2}} + \text{pen}\, f_i(u)^2$$
$$\text{subject to:} \quad f_i(u) = 0 \tag{24.17}$$

where "pen" is a penalty factor. This formulation is particularly useful if a dual solution method is employed requiring a convex optimization problem [19].

FORM has inherent errors in approximating the failure curve as linear. Depending on the curvature of the limit state function, the failure probability will be underestimated or overestimated (see Figure 24.2). As long as the failure surface in the vicinity of the MPP is approximately linear, FORM leads to acceptable results.

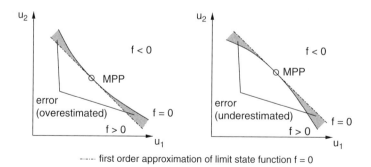

FIGURE 24.2 Approximation error of FORM: probability of failure is overestimated (left), probability of failure is underestimated (right).

Using FORM, the optimization problem of 24.12 can be rewritten in terms of reliability indices as follows:

$$\max_{s} \; \beta_c(c(s, r_c^*) > \bar{c}_t)$$
$$\text{subject to:} \; \beta_j(f_j(s, r_j^*) < 0) - \bar{\beta}_j \geq 0 \quad (24.18)$$
$$g_j^D(s) \geq 0$$
$$s_{lower} \leq s \leq s^{upper}$$

where β_c is the reliability index associated with the event defined between parentheses, β_j are the reliability indices associated with design constraints, $\bar{\beta}_j$ is the minimum required reliability index, and r_c^* and r_j^* are the values of the random parameters at the MPP of each reliability index. The above formulation is called the reliability index approach (RIA). Alternatively, the above problem can be written in terms of the failure probability $P(f_i(s, r) > 0)$, substituting the reliability index by the probability defined in Equation 24.14. However, as the failure probability may vary over several orders of magnitude during the optimization process, while the variations of the reliability index are usually small, from a numerical point of view, it is advantageous to use a formulation based on the reliability index.

Employing FORM within design optimization is attractive mainly for two reasons: (1) it requires only few evaluations of the system response, and (2) the gradients of the reliability index with respect to the design variables can be conveniently determined (see Section 24.4). On the other hand, it can not be guaranteed that FORM converges to the global optimum, leading to an incorrect approximation of the reliability. Checking that the global optimum has been found, for example by a hyperspace division method [20], is computationally costly and increases the algorithmic complexity. Furthermore, it can not be guaranteed that the RIA optimization problem has a solution for any given design during the design-optimization process. If the mean design is far away from the failure surface, there may be no solution u such that $f_i = 0$.

24.3.3 Performance Measure Approach

The performance measure approach (PMA) or target-performance approach [21, 22] offers an alternative to the RIA approach by also estimating the probability of failure by the reliability index. However, it avoids the problems of the RIA approach associated with the requirement to find a solution on the failure surface for any design during the optimization process. For a given target reliability index $\bar{\beta}_i$, the minimum value of the limit state function is computed by solving the following optimization problem in the standard normal space u:

$$\min \; f_i(u)$$
$$\text{subject to:} \; (u^t u)^{\frac{1}{2}} - \bar{\beta}_i = 0 \quad (24.19)$$

The value of the limit state function at the optimum of formulation 24.19 represents the worst possible performance for a required reliability index $\bar{\beta}_i$. In comparison with the RIA formulation 24.16, it can be guaranteed that the equality constraints in Equation 24.19 can be satisfied. The PMA formulation can be augmented by a penalty term in an analogous way as the RIA formulation 24.18.

Since the reliability requirement is already included in the reliability analysis, the PMA-based design-optimization problem takes on the following form:

$$\max_{s} \; f_c(s, r_c^*) = \bar{c}_t - c(s, r_c^*)$$
$$\text{subject to:} \; f_j(s, r_j^*) \geq 0 \quad (24.20)$$
$$g_j^D(s) \geq 0$$
$$s_{lower} \leq s \leq s^{upper}$$

where the performance f_c of the objective is measured with respect to a target reliability $\bar{\beta}_c$. The random parameters r_c^* and r_j^* are the solutions of the associated PMA problems (Formulation 24.19). It is also possible to apply RIA and PMA to different constraints of the same optimization problem.

24.4 Sensitivity Analysis in Reliability-Based Optimization

24.4.1 General Remarks

For realistic problems, the evaluation of the reliability by numerical simulation is often costly. If the design criteria and limit state functions are continuous and differentiable functions of the optimization variables and random parameters, respectively, gradient-based optimization algorithms are an appealing tool to solve the overall RBDO problem as well as the reliability analysis problem resulting from the FORM (Formulation 24.16) or the PMA formulation 24.19.

In comparison with gradient-free methods, such as genetic algorithms or simulated annealing methods, gradient-based algorithms require a smaller number of iterations. However, these algorithms call for the gradients of the design criteria with respect to the optimization variables. The gradients with respect to the optimization variables are called "design sensitivities," and the gradients with respect to the random parameters are called "imperfection sensitivities." These gradients are evaluated by a sensitivity analysis.

Sensitivity analysis can be subdivided into analytical methods and numerical approaches. The latter are easy to implement using existing software tools, as they require only additional function evaluations. A function q_j, $j = 1, \ldots, n_Q$, can be the objective or constraint of a design-optimization problem or the limit state function in a FORM reliability-analysis problem. As $q_j(v)$ depends in general on the state variables, each function evaluation calls for a structural analysis. If the system is random, a function evaluation requires a stochastic or reliability analysis.

For example, a central finite differencing scheme approximates the gradient of the design criterion q_j with respect to the parameter p_i, $i = 1, \ldots, n_P$, which represents either an optimization variable s_i or a random parameter u_i in the standard normal space, as follows:

$$\frac{dq_j}{dp_i} \approx \frac{q_j(s + \epsilon \Delta s) - q_j(s - \epsilon \Delta s)}{2\epsilon}, \quad \Delta s_k = \begin{cases} 1 & \text{for} \quad k = i \\ 0 & \text{for} \quad k \neq i \end{cases} \quad (24.21)$$

where ϵ is a perturbation factor. The above scheme calls for two additional function evaluations. This increases substantially the computational burden, in particular for RBDO problems of practical engineering systems, leading to often unacceptable turnaround times. In addition, the choice for the perturbation factor may affect the accuracy of the sensitivities, where ϵ should be as small as possible but sufficiently large to avoid round-off errors. The optimal value can only be found by trial and error and may change during the optimization process. A numerical approach that is less sensitive to the perturbation size is the complex-variable method [23]. This method, however, requires major software modifications and doubles the computational costs due to the complex arithmetic.

Analytical sensitivity-analysis methods overcome the computational and accuracy bottlenecks of numerical approaches and are the methods of choice for practical reliability analysis and RBDO problems. Two analytical approaches can be distinguished, namely the direct and the adjoint method. The derivation of these methods is outlined below. For a detailed review of parameter-sensitivity analysis, the reader is referred to Haug et al. [24] and Kleiber et al. [25].

The total derivative of q_j with respect to a parameter p_i can be written as:

$$\frac{dq_j}{dp_i} = \frac{\partial q_j}{\partial p_i} + \frac{\partial q_j^t}{\partial v}\frac{dv}{dp_i} \quad (24.22)$$

where the total derivative of the state variables v is obtained by differentiating the discretized governing equations (Equation 24.11):

$$\frac{dR}{dp} = \frac{\partial R}{\partial s_i} + \frac{\partial R^t}{\partial v} \frac{dv}{ds_i} = \mathbf{0} \qquad (24.23)$$

At equilibrium, the total derivative of the residual R vanishes. The partial derivative $\partial R/\partial v$ is the tangent stiffness matrix K_t at equilibrium. Solving Equation 24.23 for dv/dp and introducing this result in Equation 24.22 yields:

$$\frac{dq_j}{ds_i} = \frac{\partial q_j}{\partial s_i} - \frac{\partial q_j^t}{\partial v} K_t^{-1} \frac{\partial R}{\partial s_i} \qquad (24.24)$$

The above equation can either be solved by the direct or the adjoint method. Following the direct method, first the derivatives of the state variables are computed by solving

$$K_t \frac{dv}{ds_i} = -\frac{\partial R}{\partial s_i} \qquad (24.25)$$

and then, following Equation 24.22, postmultiplied with the derivative of the criterion with respect to the state variables and added to the partial derivatives of the criterion with respect to the parameter p_i. The direct approach requires the solution of the linear system (Equation 24.25), which corresponds to solving the linearized state equations with a fictitious load, as many times as there are parameters. Therefore, the direct approach is attractive if the number of parameters n_P is smaller than the number of functions n_Q.

In particular in the MPP search associated with FORM and PMA problems, there is only one performance criterion but often more than one random parameter. Therefore, it is computationally more efficient to evaluate the sensitivity equation (Equation 24.24) by the adjoint approach. First the so-called adjoint solution a_j is computed by solving

$$K_t^{\,t} a_j = \frac{\partial q_j}{\partial v} \qquad (24.26)$$

and then the total derivative of q_j is evaluated as follows:

$$\frac{dq_j}{dp_i} = \frac{\partial q_j}{\partial p_i} - a_j^t \frac{\partial R}{\partial p_i} \qquad (24.27)$$

This procedure has the distinct advantage that the linear system needs to be solved only n_Q times.

To limit the effort of implementing analytical sensitivity analysis, the partial derivatives in Equation 24.23 and Equation 24.25 (for the direct method) and Equation 24.26 and Equation 24.27 (for the adjoint method) can be evaluated by finite differencing. This combination of analytical and numerical approaches is called semianalytical sensitivity analysis. For details on the efficiency and accuracy of the semianalytical method, the reader is referred to Barthelemy [26], Mlejnek [27], and de Boer and van Keulen [28].

24.4.2 Design Sensitivities for Reliability Index Approach

One of the most distinct features of FORM is the simplicity of evaluating analytically the design sensitivities of reliability criteria [29, 30]. The derivative of a failure probability $P(f_i < 0)$, or any other probability calculated by FORM, with respect to a design variable s_i can be defined as:

$$\frac{dP(f_i < 0)}{ds_i} = \frac{d\Phi(-\beta)}{ds_i} = \frac{\partial \Phi(-\beta)}{\partial \beta} \frac{d\beta}{ds_i} \qquad (24.28)$$

Substitution of the expression for the CDF in Equation 24.28, Φ, yields:

$$\frac{\partial \Phi(-\beta)}{\partial \beta} = \frac{\partial \left(\frac{1}{2\pi} \int_{-\infty}^{-\beta} e^{-\frac{x^2}{2}} dx \right)}{\partial \beta} \qquad (24.29)$$

$$= -\frac{1}{2\pi} e^{-\frac{\beta^2}{2}} = -\varphi(-\beta)$$

where φ is the standard normal density function. Substituting Equation 24.29 into Equation 24.28, the sensitivity of the failure probability with respect to the design variable s_i is:

$$\frac{dP(f_i < 0)}{ds_i} = -\varphi(-\beta) \frac{d\beta}{ds_i} = -\frac{1}{2\pi} e^{-\frac{\beta^2}{2}} \frac{d\beta}{ds_i} \qquad (24.30)$$

To calculate the sensitivity of β with respect to a design variable s_i, $d\beta/ds_i$, one must first recall the definition of the reliability index:

$$\beta = (u^{*T} u^*)^{\frac{1}{2}}$$

The partial derivative of β with respect to u can then be written as:

$$\frac{\partial \beta}{\partial u} = \frac{1}{2} (u^{*T} u^*)^{-\frac{1}{2}} 2u^* = \frac{1}{\beta} u^* \qquad (24.31)$$

and the derivative of β with respect to design variable s_i follows as:

$$\frac{d\beta}{ds_i} = \frac{1}{\beta} u^* \frac{du^*}{ds_i} \qquad (24.32)$$

Based on the optimization problem within FORM, the most probable point u^* is derived from the Karush-Kuhn-Tucker (KKT) conditions as:

$$u^* = -\beta \frac{df/du}{\|df/du\|} \qquad (24.33)$$

Substituting Equation 24.33 into Equation 24.32, and rearranging the result, yields:

$$\frac{d\beta}{ds_i} = -\frac{df/du}{\|df/du\|} \frac{du^*}{ds_i} \qquad (24.34)$$

The derivative of the limit state function f with respect to u and the derivative of u with respect to the design variable s_i can be expanded in terms of the original random variables r as follows:

$$\frac{df}{du} = \frac{df}{dr} \frac{dT_u^{-1}(u)}{d_u} \qquad (24.35)$$

$$\frac{du^*}{ds_i} = \frac{dT(r)}{dr} \frac{dr^*}{ds_i} \qquad (24.36)$$

where dT_u^{-1} denotes the inverse mapping (Equation 24.15) from the standard normal space u into the space of the original random variables r.

The remainder of the derivation depends on whether the design variable s_i is stochastic or deterministic. In the deterministic case, the fact that the total variation of the limit state f with respect to the deterministic design variable s_i is zero at the most probable point u^* is used, together with Equation 24.34, to formulate the following expression:

$$\frac{d\beta}{ds_i} = \frac{1}{\|df/du\|}\frac{df}{ds_i} \qquad (24.37)$$

where the second term on the right-hand side is the gradient of the limit state function f with respect to the design parameter s_i at the MPP. Combining this result with Equation 24.30 gives the sought-after sensitivity value:

$$\frac{dP(f_i<0)}{ds} = -\frac{1}{2\pi}e^{-\frac{\beta^2}{2}}\frac{1}{\|df/du\|}\frac{df}{ds} \qquad (24.38)$$

A design parameter can also be nondeterministic. For example, the mean or the standard deviation of a random variable can be treated as a design variable. For nondeterministic design variables, the gradient of the reliability index can be derived from:

$$\frac{d\beta}{ds_i} = \frac{1}{\beta}u^*\frac{du^*}{ds_i} = \frac{1}{\beta}u^*\frac{dT(r^*,s)}{ds_i} \qquad (24.39)$$

where du^* is instead described by the derivative of the transformation of the associated distribution.

24.4.3 Design Sensitivities for Performance Measure Approach

In this subsection, the analytical formulation of the design sensitivities of the performance measure f in Formulation 24.19 is presented. The total derivative of the performance measure can be written as follows:

$$\frac{df}{ds_i} = \frac{\partial f}{\partial s_i} + \frac{\partial f}{\partial u}\frac{du}{ds_i} \qquad (24.40)$$

The partial derivative of f with respect to the design variables can be evaluated by differentiating analytically or numerically the discretized expressions. In order to compute the second term on the right-hand side, the necessary optimality conditions for the PMA optimization problem (Formulation 24.19) are evaluated:

$$\frac{\partial f}{\partial u} + \eta^t\frac{\partial \beta}{\partial u} = 0 \qquad (24.41)$$

where η is the vector of Lagrange multiplier associated with the equality constraint on the reliability index. Since at the MPP the reliability index $\beta = \bar{\beta}$ is constant, the performance measure does not depend on u, and the total derivative of the performance measure can be simply computed at the MPP as:

$$\frac{df}{ds_i} = \frac{\partial f}{\partial s_i} \qquad (24.42)$$

24.5 Optimization Methods and Algorithms

There is a wide variety of optimization methods and algorithms available for solving parameter-optimization problems resulting, for example, from design-optimization or FORM and PMA problems. Optimization methods differ in their underlying formulation of the RBDO and reliability-analysis problem. The algorithms are typically ordered by the type of optimization problems they are applicable to. The variety of problems is revealed by a coarse decision tree depicted in Figure 24.3. Any parameter optimization problem can be solved by algorithms for nondifferentiable, mixed-integer problems, such as genetic algorithms. However, when applying such algorithms, for example, to smooth, differentiable, unconstrained optimization problems, the computational costs are typically larger than when applying algorithms exploiting specific mathematical properties and features of the problem formulation. In this section, the basic types of optimization methods for RBDO problems are presented, and the most popular algorithms for different problem types are described.

24.5.1 Nested and "One-Shot" Optimization Methods

As described in Section 24.3, RBDO involves a hierarchy of models that is depicted in Figure 24.4. The hierarchy is based on the following functional dependency: the design-optimization problem calls for the evaluation of reliability-based and deterministic design criteria. In the reliability-analysis problem, reliability-based design criteria are evaluated, and in the structural model, the response is determined for a given set of design and random variables. The hierarchical structure is also revealed by the dependencies of the variables associated with each problem and model. The design variables s are independent, the random variables r may depend on s, and the structural-state variables $d(s, r)$ are functions of both the design and the random parameters.

The hierarchical structure of RBDO problems suggests a so-called nested-solution approach. Following the basic formulation (defined in 24.12), the design-optimization problem is formulated in terms of the objective, the design constraints, and the optimization variables. The equations and variables of the reliability- and structural-analysis problems are not explicitly considered in the formulation of the design-optimization problem; instead, they are satisfied for any given set of optimization variables. In other words, in each iteration of the optimization algorithm, the reliability- and structural-analysis problems are solved for the current optimization variables.

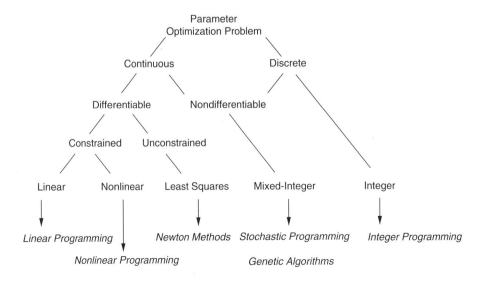

FIGURE 24.3 Classification of optimization algorithms.

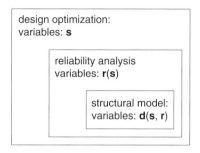

FIGURE 24.4 Models in reliability-based design optimization.

The nested approach allows the solution of the different subproblems — the design optimization, the reliability analysis, and the structural problem — by algorithms and software that are tailored to each subproblem. This flexibility is illustrated by Figure 24.5. The modularity of the approach allows the combination of different design-optimization techniques, reliability-analysis methods, and deterministic structural-analysis approaches, if needed.

The major shortcoming of the nested approach is the computational cost. If it is far away from the optimum design, the reliability-analysis problem needs to be solved with high accuracy. Otherwise, the governing reliability and structural equations are violated, thus affecting the convergence of the overall design-optimization procedure. For example, when using RIA or PMA, at each iteration of the design-optimization loop, the MPP search calls for the solution of another, inner optimization problem.

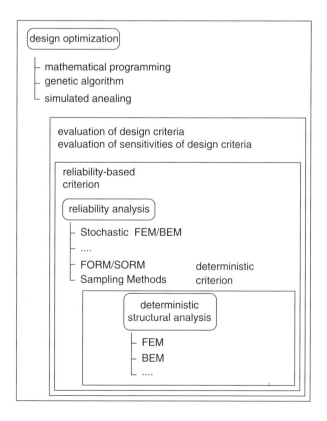

FIGURE 24.5 Nested solution in RBDO.

This bottleneck can be overcome for FORM-type formulations of the reliability-analysis problem by merging the outer design-optimization loop and the inner MPP search optimization loop. The design-optimization variables and the optimization variables of the MPP search are advanced simultaneously, and therefore this approach is called a one-shot or uni-level method.

The one-shot formulation based on the RIA approach (defined in 24.18) can be written by treating the uncertain variables projected into the standard normal space u and the optimization variables s as independent variables as follows:

$$\min_{s} \; z(s,r)$$
$$\text{subject to} \quad g_j^D(s) \geq 0 \quad \beta_j - \overline{\beta}_j \geq 0 \qquad (24.43)$$
$$\beta_j = \min_{u} (u_j^T u_j)^{\frac{1}{2}}$$
$$\text{subject to} \quad f_j = 0$$

Several methods have been proposed to merge the optimization problems in s and u, including the replacement of the inner MPP search by the associated KKT conditions. Alternatively, the objective function can be modified such that its minimum is the solution of both the design optimization and the MPP search problems. A multiplicative formulation is presented by Kharmanda et al. [31], leading to the following approach that requires only minor modification of existing reliability-based optimization software.

$$\min_{s,u} \left(z(s,r) \sum_j \left(u_j^T u_j \right) \right)$$
$$\text{subject to} \quad \beta_j = \left(u_j^T u_j \right)^{\frac{1}{2}} \qquad (24.44)$$
$$g_i^D(s) \geq 0 \quad \beta_j - \overline{\beta}_j \geq 0 \quad f_j \geq 0$$

In Kharmanda et al. [31], the overall computational cost is reported to be reduced by more than 80% for the one-shot method in comparison with a conventional nested approach.

24.5.2 Algorithms

A broad class of engineering design problems and FORM analysis problems can be formulated in terms of smooth, differentiable cost and constraint functions of the design and random parameters. For these types of problems, gradient-based optimization methods provide a computationally efficient tool. In this section, the most common gradient-based algorithms in use today are outlined. These algorithms are either applicable to generic design-optimization problems or tailored to RIA and PMA type of optimization problems.

Gradient-based optimization algorithms for constrained, nonlinear problems can be classified by the space in which they locate the optimum. Primal methods work on the optimization variables; dual methods work on the Lagrange multipliers associated with the constraints; and primal-dual methods, also called Lagrange methods, work on both the optimization variables and the Lagrange multipliers [32–35].

Among the primal approaches, the methods of feasible directions, such as gradient projection methods and generalized reduced gradient methods, are widely used for design-optimization problems [36]. The main advantage of these methods is that they generate feasible solutions in each iteration, but their computational cost increases with the number of optimization variables. For a large number of optimization variables but few constraints, dual methods in combination with local approximation methods have prevailed, such as the method of moving asymptotes (MMA) [37]. These algorithms are based on the following concept. In an outer loop, a sequence of convex explicit subproblems is constructed that can be solved efficiently in an inner loop by a dual algorithm. Today, Lagrange methods are considered to be the most universal algorithms, and they are used in such techniques as sequential linear programming (SLP) and sequential quadratic programming (SQP) [38]. Both methods advance simultaneously the optimization variables and the Lagrange multipliers to solve for the necessary optimality conditions. SLP methods generate a sequence of linear subproblems that can be solved either by classical SIMPLEX algorithms or by interior points' methods. SQP methods construct and solve a sequence of quadratic subproblems and belong to the most popular optimization methods today [39, 40]. While numerically efficient and robust for small and mid-size problems, Lagrange methods become inefficient for large problems, where large refers to the number of optimization variables and the number of constraints. In addition, SQP methods require accurate sensitivities, as the Hessian of the Lagrange function is approximated based on the first-order gradients of the objective and constraints.

The above methods are applicable to generic optimization problems. The optimization problems resulting from the RIA and PMA approaches based on FORM possess a specific structure that is exploited by tailored MPP search algorithms, such as the HL-RF iteration scheme [41, 42]. A study by Liu and Der Kiureghian [19], comparing generic optimization algorithms with the HL-RF scheme, demonstrates by numerical examples that both approaches are equally robust for FORM reliability analysis. The study recommends in particular the SQP and the HL-RF methods due to their efficiency for problems with large nonlinear numerical simulation models.

24.6 Multicriteria and Life-Cycle Cost Reliability-Based Optimization

In RBDO, there are often multiple objectives that must be considered simultaneously in the optimization process. For example, total expected cost and probability of system collapse are clearly in conflict. Therefore, to find the optimum solution (also referred to as the Pareto solution) multicriteria reliability-based optimization techniques (also referred to as vector, multiobjective, or Pareto optimization under uncertainty) have to be used.

The formulation of the multicriteria RBDO under uncertainty for structural systems was set forth by Frangopol and coworkers [43–46]. The problem is formulated as follows:

$$\min_{s} c(s) \quad \text{subject to} \quad s \in Y \quad (24.45)$$

where c and s are the objective and design variable vectors, respectively, and Y is the feasible space of s given by a set of constraints [47–49]. The main difference between this formulation and the basic formulation (24.10) is that c in formulation 24.45 is an objective vector and c in formulation 24.10 is a scalar objective.

Fu and Frangopol [46] and Frangopol and Iizuka [50–52] proposed multicriteria RBDO based on the minimization of a four-objective vector

$$c(s) = \left[V(s), P_{COL}(s), P_{YLD}(s), P_{DFM}(s)\right]^t \quad (24.46)$$

where V is the total volume of structural members, P_{COL} is the system probability of collapse, P_{YLD} is the system probability of first yielding, and P_{DFM} is the system probability of excessive deformation.

The multicriteria reliability-based optimization can also be used for damage-tolerant structures. As indicated in Frangopol and Fu [44], a structure has to have a substantial probability of performing its basic function even after it sustains a specified level of structural damage. Of course, this probability is decreasing with an increase in the level of structural damage. Frangopol and Fu [44] and Fu and

Frangopol [45, 46] formulated the damage-tolerant RBDO problem as the minimization of a three-objective vector as follows:

$$\min_{s} c(s) = \left[V(s), P_{INT}(s), P_{DMG}(s)\right]^{t} \tag{24.47}$$

in which P_{INT} is the probability of collapse of the intact structure, and P_{DMG} is the probability of collapse of the damaged structure.

The importance of life-cycle cost (LCC) for the analysis, design, maintenance, and management of infrastructure systems is now accepted by the civil and aerospace communities worldwide. The LCC optimization of these systems under uncertainties during their entire service life is a very complex computational process. This process involves minimizing the life-cycle cost of a deteriorating structure or a group of similar structures under uncertainties in loads, resistances, modeling of deterioration process, inspection, maintenance, and repair, among others. Frangopol et al. [2] formulated the reliability-based LCC optimization (RLCCO) problem as the minimization of the present value of the total expected life-cycle cost C_{ET} as follows:

$$C_{ET} = C + C_{PM} + C_{INS} + C_{REP} + C_{F} \tag{24.48}$$

where C, C_{PM}, C_{INS}, C_{REP}, and C_F are initial cost, routine maintenance cost, inspection cost, repair cost, and failure cost, respectively. All terms in Equation 24.48 are expected present values.

In recent years, the applicability of the RLCCO approach has been widened, and important progress has been made in civil and aerospace engineering [10]. There are clear indications that RLCCO is coming of age. However, both the databases and the computational efficiency must be improved for this approach to be effective.

24.7 Examples in Civil Engineering

Numerical examples presented up to 1994 in RBDO of structural components and systems were briefly reviewed by Frangopol and Moses [12] and Frangopol [8]. They concluded that the most common examples used in the RBDO literature up to 1994 were single-limit state sizing optimization of steel frames and trusses under static loading. However, during the period 1978–1994, there are also examples of more complex RBDO applications involving multiple-limit states [53, 54], time-varying loads [55], inspection and repair of structural systems [56], damage-tolerant structures [44, 57], multicriteria optimization [45, 46, 50, 52], shape optimization of structural systems [58, 59], and configuration of material systems [60, 61].

Since 1994, the range of applicability of reliability-based optimization in civil engineering has been widened, and much progress has been made in reliability-based life-cycle-cost optimization (RLCCO) of deteriorating structures. In this section, selected numerical examples describing reliability-based cost optimization of structural components and systems are briefly reviewed. Other examples of RLCCO can be found in various contributions such as Augusti et al. [62], Frangopol and coworkers [63–68], Estes and Frangopol [69], and Kong and Frangopol [70].

24.7.1 Example 1: Material Cost-Based RBDO of Reinforced Concrete Girders [1]

In this example, the total cost of steel and concrete of a reinforced concrete 60-ft (18.3 m) T-girder is adopted as the objective to be minimized. Structural deterioration is not included in the analysis. A total of 21 design variables and 12 random parameters are considered in the optimization process. Three optimum solutions are presented in Table 24.2 considering different minimum allowable reliability levels ($\beta^* = 3.0$ and $\beta^* = 3.5$), and these are compared with the optimum solution associated with bridge design

TABLE 24.2 Cost-Based Optimum Solutions of a Reinforced Concrete Girder

Optimum Solution	Design Variables									Optimum Total Cost
	s_1 (in.²)	s_2 (in.)	s_3 (in.)	s_4 (in.)	s_5 (in.)	s_6 (in.²)	s_7 (in.)	s_8 (in.)	s_9 (in.)	
Reliability-based optimum for $\beta^* = 3.0$	11.45	27.54	7.42	12.79	57.28	0.094	7.6	7.4	6.5	578.6
AASHTO-based optimum	13.27	31.51	7.55	14.78	47.33	0.131	8.7	8.5	6.3	609.3
Reliability-based optimum for $\beta^* = 3.5$	13.41	28.10	8.54	13.25	58.97	0.095	8.6	6.9	6.0	645.3

Note: s_1 = area of the tensile steel reinforcement; s_2 = width of flange; s_3 = depth of flange; s_4 = width of web; s_5 = depth of web; s_6 = area of the shear reinforcement; and s_7, s_8, and s_9 = distances between symmetrical shear reinforcement along the span of the girder.

Source: Adapted from Lin, K.-Y. and Frangopol, D.M., *Structural Safety*, 18, 239, 1996.

specifications in the U.S. (AASHTO) [71]. The solutions in Table 24.2 correspond to a steel-to-concrete unit cost ratio of 50. Because of the variations in the cost of materials, sensitivity analyses were presented in Lin and Frangopol [1] to quantify the effect of steel-to-concrete unit cost ratio on the optimum solution. The results show that when this ratio is varied from 10 to 90, the optimum cost associated with the AASHTO-based design [71] is always bounded by the optimum reliability-based design costs associated with $\beta^* = 3.0$ and $\beta^* = 3.5$.

24.7.2 Example 2: Life-Cycle Cost-Based RBDO of Deteriorating Reinforced Concrete Girders [2]

In order to implement the optimum maintenance strategy for a specific deteriorating structure, the minimum expected life-cycle-cost solution has to be found considering the reliability profile of the structure over a specified time horizon. This time-dependent approach has also to consider the costs of inspection, maintenance, and repair over the given time horizon. An approach including initial preventive maintenance, inspection, repair, and failure costs (see Equation 24.48) was proposed by Frangopol et al. [2]. This approach was illustrated by optimizing the reinforced concrete T-girders of a highway bridge. An optimal inspection/repair strategy was developed for these girders that are deteriorating due to corrosion in an aggressive environment.

As an example, Figure 24.6a and Figure 24.6b show the optimal inspection/repair strategies considering a rate of steel corrosion v for different damage-intensity factors η at which the inspection method has a 50% probability of detection ($\eta_{0.5} = 0.05$ and $\eta_{0.5} = 0.15$), a discount rate for money of 2%, a ratio of failure cost to unit cost of concrete of 50,000, a service life of 75 years, and a target lifetime-reliability index $\beta^* = 2$. In these figures, m and n are the number of inspections and repairs, respectively.

The effects of critical parameters such as rate of corrosion, quality of the inspection technique, and the expected cost of structural failure were all investigated along with the effects of both uniform and nonuniform inspection/repair time intervals. Regarding these effects for the specific example considered, the following conclusions were obtained:

- The optimum nonuniform time interval inspection/repair strategy is more economical and requires fewer lifetime inspections/repairs than that based on inspections at uniform time intervals.
- Numerical results indicate that the optimum number of inspections and the optimum expected total cost increase as the corrosion rate increases. Also, as the quality of the inspection method improves, the optimum number of inspections decreases.

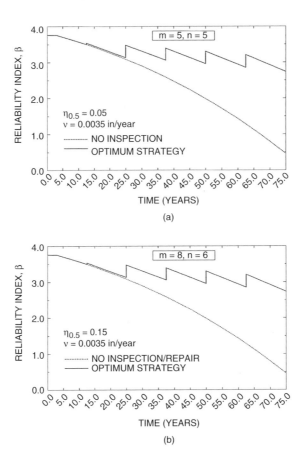

FIGURE 24.6 Optimum inspection/repair strategy for different inspection methods: (a) $\eta_{0.5} = 0.05$; (b) $\eta_{0.5} = 0.15$. (Adapted from Frangopol, D.M., Lin, K.-Y., and Estes, A.C., *J. Structural Eng.*, 123 (10), 1390–1401, 1997. With permission.)

- The cost of failure significantly affects the optimum inspection and repair strategy. A higher failure cost leads to an optimum solution requiring more inspections and repairs at a higher total cost.
- The minimum expected life-cycle cost is most sensitive to the corrosion rate and the cost of failure. This cost is relatively insensitive to the quality of inspection and the number of lifetime inspections above the optimum number.

24.7.3 Example 3: Life-Cycle Cost-Based Repair Optimization of an Existing Deteriorating Bridge Using System Reliability [3]

The general RLCCO methodology for optimizing the repair strategy of existing bridges was defined as follows [3]:

- Identify the relevant failure modes of the bridge. Decide which variables are random in nature and find the parameters (e.g., mean, standard deviation) associated with these random variables. Develop limit state equations in terms of these random variables for each failure mode. Compute the reliability with respect to the occurrence of each possible failure mode.
- Develop a system model of the overall bridge as a series–parallel combination of individual failure modes. Compute the system reliability of the bridge.

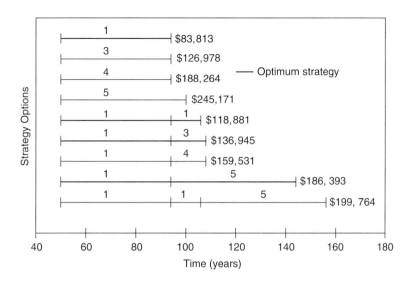

FIGURE 24.7 All feasible repair options for bridge E-17-AH using a series–parallel model requiring the failure of three adjacent girders. (From Estes, A.C. and Frangopol, D.M., *ASCE J. Structural Eng.*, 125 (7), 766–775, 1999. With permission.)

- Develop deterioration and live-load models that describe how the structure and its environment are expected to change over time. This will inevitably introduce new random variables. Compute the system reliability of the structure over time.
- Establish a repair or replacement criterion. Develop repair options and their associated costs.
- Using all feasible combinations of the repair options and the expected life of the structure, optimize the repair strategy by minimizing total lifetime repair cost while maintaining the prescribed level of system reliability.
- Develop a lifetime inspection program to provide the necessary information to update the optimum repair strategy over time.

This methodology was successfully applied to an existing three-span nine-girder Colorado bridge. For this bridge, the following repair actions and the associated costs were provided. The optimal lifetime repair strategy depends strongly on the structural system considered for the bridge under analysis. Figure 24.7 shows the optimal repair strategy for a system where system failure requires the failure of any three adjacent girders or failure of the deck or failure of the substructure. The limit state equations, random variables and their correlations, deterioration and system models, and the RLCCO methodology are all explained in Estes and Frangopol [3].

24.7.4 Example 4: Life-Cycle Cost-Based Repair Optimization of an Existing Bridge Using System Survival Functions [4]

The methodology used for optimizing maintenance strategies is adapted from that proposed by Estes and Frangopol [3]. It consists of the following nine steps:

1. Construct a system model of the overall structure as a series–parallel combination of individual components and establish a time horizon for the system.
2. Define the survivor function to be used for each component.
3. Compute the survivor function under no maintenance for the system model considered in the first step.
4. Establish a system reliability threshold, at which maintenance must be applied.
5. Determine all possible maintenance actions and their associated costs.

6. Determine all maintenance strategies (i.e., combination of several maintenance actions during the time horizon).
7. Compute the system survivor function for each maintenance strategy.
8. Compute the present values of lifetime cost for each maintenance strategy.
9. Determine the optimum solution based on minimum present value of lifetime cost.

This methodology was applied to the same bridge used in Example 3, but in this case, the time-dependent reliability was modeled using lifetime survivor functions [4, 72]. The service life of highway structures and their components is defined as the time taken for a significant defect to be recorded by an inspector.

Defects are classified [73] according to their severity as: severity 1, no significant defects; severity 2, minor defects of a nonurgent nature; severity 3, defects that shall be included for attention within the next annual maintenance program; and severity 4, severe defects for which urgent action is needed. Lifetime functions defining the deterioration process of the most significant components of common types of highways bridges are presented by Maunsell Ltd. [73].

In this example, failure of a component is defined as a defect of severity 4 being registered by a bridge inspector. In fact, only these defects are significant enough to justify the application of essential maintenance actions. The occurrence of such defects in the main components of the superstructure of the bridge is modeled using a Weibull function with shape and scale parameters κ and λ, respectively, that are assumed $\kappa = 2.37$ and $\lambda = 0.0077$/year for the deck, and $\kappa = 2.86$ and $\lambda = 0.0106$/year for each girder [73].

The maintenance actions and their associated costs considered in this study are defined in relation to the bridge used in Example 3 [3] as follows: (a) replace deck, $225,600; (b) replace exterior girders, $229,200; (c) replace deck and exterior girders, $341,800; and (d) replace superstructure, $487,100 (see Table 24.3).

The bridge under analysis is modeled considering that failure occurs if the deck slab fails, or if a combination of girders fails, or if both fail. Four system failure modes are considered for this bridge: (I) failure of any girder or deck failure cause the bridge failure; (II) failure of any external girder or any two adjacent internal girders or deck failure cause the bridge failure; (III) failure of any two adjacent girders or deck failure cause the bridge failure; (IV) failure of any three adjacent girders or deck failure cause the bridge failure. These system failure modes are denoted as I, II, III, and IV, respectively.

Using the system IV model, failure is defined as finding a defect in the deck or in any three adjacent girders. This system model is the most redundant system and, as a consequence, the first essential maintenance action is applied later than for any of the other systems. In this system, the deck is the most important component. As a result, the first maintenance action is replacement of the deck (option 1) at year 19 followed by replacement of superstructure (option 4) at year 36 (Figure 24.8). After year 36, there is a repetition of the lifetime function observed in the first 36 years. At year 72, a less expensive action (replace deck) is sufficient to extend the service life beyond the time horizon. The effect on the system probability of failure of replacing the deck at year 72 is smaller than it is at year 19, since at 72 years girders are more deteriorated and, as a result, are more important to the overall system probability of failure. The resulting total maintenance costs are $1,163,900, $526,453, $268,039, $149,320, and $88,949

TABLE 24.3 Repair Actions and Associated Costs

Option	Repair Option	Repair Cost (1996 US$)
0	Do nothing	0
1	Replace deck	225,600
2	Replace exterior girders	229,200
3	Replace exterior girders and deck	341,800
4	Replace superstructure	487,100
5	Replace bridge	659,900

Source: Adapted from Estes, A.C. and Frangopol, D.M., *ASCE J. Structural Eng.*, 125, 766, 1999.

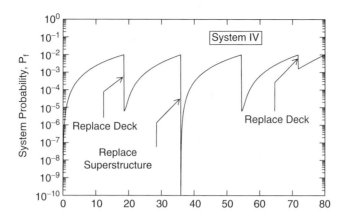

FIGURE 24.8 Time-dependent system failure probability associated with optimal maintenance strategy for bridge system IV. (From Yang, S.-I., Frangopol, D.M., and Neves, L.C., *Reliability Eng. System Safety*, 2004 [4]. With permission.)

for discount rates of money of 0, 2, 4, 6, and 8%, respectively. The resulting time-dependent system probability of failure is presented in Figure 24.8.

24.8 Examples in Aerospace Engineering

Aerospace systems inherently operate in extreme and variable environments, causing great uncertainty in their operating conditions. Reliability requirements for aerospace components and systems are severe due to the often costly and fatal consequences of failure. Simultaneously, aerospace systems are also subject to tough requirements on minimum weight and maximum performance. RBDO methods are an appealing approach for solving the resulting complex design problems in aerospace engineering.

The analysis of aerospace structures is in general complicated by complex loading and operating conditions, such as aerodynamic pressure and thermal loads. Often the interaction of one or multiple fields needs to be taken into account to predict the load acting on the structure. Furthermore, aerospace systems are often actively controlled, leading to another level of nonlinearity in the structural response. This complexity is amplified in RBDO, requiring a stochastic analysis of multifield and multidisciplinary systems.

24.8.1 Composite Structures

Composite materials allow for minimum weight design by tailoring the material composition to the loading conditions, and therefore they are frequently used in aerospace structures. However, it is crucial to account for uncertainties in material and geometric parameters and loading conditions, since composite structures typically fail abruptly.

Numerous approaches have been presented applying RBDO methods to the design of thin-walled composite structures. Probabilistic load conditions and material properties, including manufacturing uncertainties, have been considered, for example, in Yang and Ma [74], Miki et al. [75, 76], Thanedar and Chamis [77], Chao [78], and Mahadevan and Liu [79]. Composite structures with degradation models have been considered by Antonio et al. [80] and buckling instabilities by Su et al. [81]. The difference between deterministic optimization and RBDO results shows the importance of accounting for uncertainties, particularly in the design of composite structures [82].

24.8.2 Aerodynamics and Aeroelasticity

The performance of an aircraft depends to a large extent on its aerodynamic characteristics. Therefore, mostly robust design methods have been developed to improve the aerodynamics of an aircraft while considering uncertainties. These approaches are based on the first-order second-moment (FOSM) method and have been applied to airfoils only [83–85]. Robust design methods for optimizing the aerodynamics of entire aircraft configurations are lacking.

The performance of aircraft is further dominated by coupled multiphysics phenomena, such as aeroelasticity or thermo-aeroelasticity. While aeroelastic instabilities, such as divergence and flutter, lead to failure, fluid-structure interaction in general needs to be accounted for to achieve optimum performance.

Traditionally, only the deterministic case has been considered, first using linear flow theories and simplified structural models [86–89], and more recently employing nonlinear Euler and Navier-Stokes flows and detailed structural finite element models [90–93]. Only a few approaches have been presented that account for uncertainties [94, 95].

In addition to structural parameters, operating conditions are an important source of uncertainties. Stochastic variations in gust loads have been considered for RBDO of wings by Rao [96] and Yang et al. [97, 98] using elementary structural and gust models. Pettit and Grandhi [99] have presented a method for preliminary weight optimization of a wing structure for mean-gust response and aileron effectiveness reliability. Kuttenkeuler and Ringertz [100] and Stroud [101] have studied failure due to flutter considering uncertainties in material parameters. Flutter is predicted by a Doublet-Lattice method.

Employing high-fidelity simulation tools, an intermediate complexity wing is optimized under uncertainties by Luo and Grandhi [102], neglecting fluid-structure interaction. In order to account for uncertainties in predicting maneuver loads and the effect of control surfaces, an optimization method for minimizing the weight and maximizing the invariance to random load effects is presented by Zink et al. [103, 104]. The uncertainties of the aerodynamic loads are quantified by comparing the results of linear and nonlinear aerodynamics. Robust design methods using FOSM have been recently presented by Gumbert et al. [105, 106].

Considering uncertainties in structural parameters and flow conditions, a RIA method has been recently presented by Allen and Maute [107, 108] based on high-fidelity aeroelastic simulation and applied to the shape and material optimization of three-dimensional flexible wing structures. Therein, the authors show that it is not only important to consider uncertainties in a qualitative manner, but also the quantitative level of stochastic variations needs to be accounted for, as the optimum design changes with the level of stochastic variability. This is illustrated by the example in Figure 24.9. Following common practice, the structure of the wing is represented by a plate model made of a unidirectional composite material. The aerodynamic loads are predicted by finite volume discretization of the nonlinear Euler flow theory, accounting for the structural deformations. Following the generic RIA formulation (24.18), the probability of the lift/drag (L/D) ratio reaching a target value $(L/D)_{target}$ is maximized, subject to a probabilistic constraint on the maximum von Mises stress on the upper surface of the wing, by varying the degree of twist of the wing along the span and the orientation of the composite material of the wing. A positive twist angle increases the local effective pitch angle along the span. To account for the effects of random gust loading, the free stream angle of attack is modeled as Gaussian; random variations in

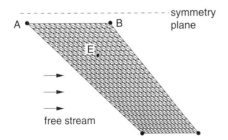
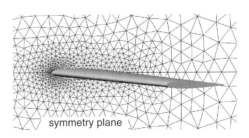

FIGURE 24.9 RBDO of aeroelastic wing.

TABLE 24.4 RBDO Results of Aeroelastic Wing for Different Coefficients of Variations of Structural and Operational Parameters

Case	Low Variability	Medium Variability	High Variability
	Coefficient of variation		
Angle of attack	0.075	0.1	0.125
Plate thickness	0.0375	0.05	0.625
	Design variables at optimum		
Wing twist (deg.)	0.004	0.612	0.180
Fiber orientation (deg.)	10.981	8.265	7.487

Source: Adapted from Allen, M. and Maute, K., Reliability based optimization of aeroelastic structures, in *9th AIAA/ISSMO Symposium on Multidisciplinary Analysis and Optimization*, AIAA 2002-5560, Atlanta, 2002.

geometrical and material properties are taken into account by describing the plate thickness by a Gaussian distribution.

This optimization problem has been studied for different levels of stochastic variation. The coefficients of variation of the angle of attack and the plate thickness are listed in Table 24.4. The results on the optimum configuration show clearly that the level of randomness in an aeroelastic system needs to be accounted for, in order to properly design the system for a target reliability level.

24.8.3 RBDO of Aircraft Systems

Today, only a few optimization methodologies have been presented for the design of aerospace systems under uncertainties. A generic framework for the stochastic multidisciplinary design has been proposed by Du and Chen [109] and Oakley [110]. For aircraft design, multidisciplinary approaches have been presented by Mavris and coworkers [111–113], accounting for various sources of uncertainties, such as modeling and economic variability, and aiming for system affordability. Uncertainty has been introduced primarily on the conceptual design level, where reliability analysis methods are combined with system-level deterministic analyses.

24.9 Examples in MEMS

Microsystems, or microelectromechanical systems (MEMS), are sensing and actuation devices integrated on a chip. One can experience the power of microsystems when using a tiny LCD projector with MEMS embedded optics. Other everyday products include accelerometers in airbags and nozzles in inkjet printers. Manufacturing technologies allow for the fabrication of single MEMS devices at very low costs. The broad spectrum of innovative applications and falling production costs make it likely that MEMS will soon become a major technological thrust.

However, MEMS devices are subject to a pervasive problem: reliability. Today, numerous MEMS systems are being designed in industry and at universities, but they are of limited use because they fail reliability requirements. Two factors largely control MEMS reliability. One factor is nanoscale progressive-failure mechanisms, such as sliding contact and adhesion of surfaces. The other factor is the stochastic nature of geometry, material properties, and operating conditions, all of which are exacerbated by the small length scale. The nondeterministic behavior may be further complicated by the interaction of one or more physical fields, such as electromechanical and fluid-structure interactions. More than 60% of today's MEMS devices are actuated by electrostatic forcing. Fluid-structure interaction is a factor in micropumps for on-chip cooling, chemical processing, medical applications, and microthrusters for miniaturized aircraft.

Conventional design optimization is frequently applied to MEMS to improve their deterministic performance [114–118]. Only a few approaches exist, however, that directly address MEMS reliability problems [119].

To overcome reliability problems due to surface contact in sliders and hinges, topology optimization is used to generate compliance mechanisms with elastic hinges. Topology-optimization methods are able to generate arbitrary complex geometries and do not require an initial design. Only the design domain, boundary conditions, design objective(s), and constraints have to be specified. The most general approach for formulating the topology-optimization problem relies on a material formulation describing the geometry of the structure by the material distribution in the design domain, which is usually discretized by a finite element mesh. For an introduction into topology optimization, the reader is referred to Rozvany et al. [120], Bendsøe [121, 122], Eschenauer and Olhoff [123], and Soto [124].

Optimizing for maximum compliance, topology optimization automatically forms mechanisms with elastic hinges. This is illustrated by the example in Figure 24.10. For a given forcing input F^{in} at point A, the horizontal displacement u^{out} to the left at point B is maximized, where a spring is attached. Considering a nonlinear kinematics, topology optimization leads to the mechanism depicted in Figure 24.10a.

Mature topology optimization procedures have been developed for the design of compliant mechanisms with linear and nonlinear structural response [125–132]. These approaches have been refined by taking into account piezoelectric effects [126, 133, 134] and electrothermal forcing, where the structure is deformed due to temperature loading that is produced by Joule heating [135, 136].

The above topology optimization approaches are based on a deterministic response of the MEMS device and do not account for stochastic variations of structural parameters and operating conditions, which are often significant for MEMS devices. Maute and Frangopol [137] have extended topology optimization to RBDO by combining (a) material topology optimization for compliant mechanisms undergoing large displacements and (b) design optimization under uncertainties using first-order reliability analysis methods and the PMA approach presented in Section 24.3.3. The effects of random support and load conditions are illustrated by the mechanism design in Figure 24.10b and Figure 24.10c. In comparison with the ideal case (Figure 24.10a), the structural redundancy is increased in the designs generated by the RBDO approach, leading to a different structural design with higher redundancy and increased member thicknesses.

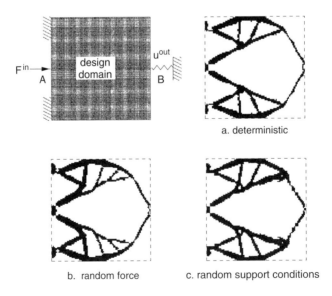

FIGURE 24.10 Topology optimization of compliant mechanism: a. deterministic, b. random loading, c. random support conditions (adapted from K. Maute and D.M. Frangopol. Reliability-based design of MEMS mechanisms by topology optimization. *Computers and Structures*, 81, 813–824, 2003.)

24.10 Conclusions

Reliability-based structural optimization methodology, based on the innovative melding of structural reliability and structural optimization, has reached the point where a solid foundation of theoretical understanding and computational experience is at hand. Twenty years ago, Schmit [7] arrived at the same conclusion, referring to the deterministic structural optimization methodology as the result of innovative melding of finite element analysis and mathematical programming algorithms.

It is important to account for the system reliability and its time variation in life-cycle optimization of new structures and in planning maintenance interventions on existing structures [10]. Indeed, studies on design of microelectromechanical systems (MEMS) and aerospace structures have demonstrated that optimum designs obtained from reliability-based optimization often outperform their deterministic counterparts.

While nobody can predict the existing and future uncertainties associated with civil and aerospace systems with perfect accuracy, engineers are often required to plan for the future and prepare economic assessments to support this plan. The reliability-based optimization methodology outlined in this chapter offers a rational and logical approach for optimizing the life-cycle cost of these systems.

Nomenclature

All bold symbols denote vectors; symbols with a bar or a star as superscript denote prescribed, given parameters. The following symbols are used in this chapter:

B	benefit
C	total expected cost
C_0	initial cost
C_F	cost of failure
C_P	present value of cost
C_{LC}	total life-cycle costs
C_{ET}	total expected life-cycle cost
C_{PM}	routine maintenance cost
C_{INS}	inspection cost
C_{REP}	repair cost
L	expected loss due to failure
P	probability
P_{COL}	system probability of collapse
P_{YLD}	system probability of first yielding
P_{DFM}	system probability of excessive deformation
P_{INT}	probability of collapse of the intact structure
P_{DMG}	probability of collapse of the damaged structure
Φ	standard normal cumulative distribution function
U	total expected utility function
V	total volume of structural members
c	vector cost functions of design-optimization problem
v	vector of state variables
s	vector of design-optimization variables
r	vector of random parameters
u	vector of random parameters in standard normal space
p	vector of parameters representing design and random variables
R	vector of residuals of governing equations
β	reliability index
ν	discount rate of money
c	cost function of design-optimization problem

q	function depending on state variables, such as design criterion and limit state function
g	design constraint
g^D	deterministic design constraint
f	limit state function
n_G	number of design constraints
n_Q	number of design criteria
n_S	number of design-optimization variables

Acknowledgments

The support of the National Science Foundation under grants CMS-9912525, CMS-0217290, and DMI-0300539 and by the Air Force Office of Scientific Research under grant F49620-01-1-052 is gratefully acknowledged. Several postdoctoral researchers and a number of former and current graduate students at the University of Colorado at Boulder contributed to the results presented in this chapter. Their contributions and assistance are greatly appreciated. The opinions and conclusions presented in this chapter are those of the writers and do not necessarily reflect the views of the sponsoring organizations.

References

1. Lin, K.-Y. and Frangopol, D.M., Reliability-based optimum design of reinforced concrete girders, *Structural Safety*, 18(2/3), 239–258, 1996.
2. Frangopol, D.M., Lin, K.-Y., and Estes, A.C., Life-cycle cost design of deteriorating structures, *J. Structural Eng.*, 123(10), 1390–1401, 1997.
3. Estes, A.C. and Frangopol, D.M., Repair optimization of highway bridges using system reliability approach, *ASCE J. Structural Eng.*, 125(7), 766–775, 1999.
4. Yang, S.-I., Frangopol, D.M., and Neves, L.C., Service life prediction of structural systems using lifetime functions with emphasis on bridges, *Reliability Eng. System Safety*, 86(1), 39–51, 2004.
5. Schmit, L.A., Structural design by systematic synthesis, in *Proceedings of Second Conference on Electronic Computation*, ASCE, Reston, VA, 1960, pp. 105–122.
6. Arora, J.S., Ed., *ASCE Manual on Engineering Practice No. 90: Guide to Structural Optimization*, ASCE, Reston, VA, 1997.
7. Schmit, L.A., Structural optimization — some key ideas and insights, in *New Directions in Optimum Structural Design*, Vol. 19, Atrek, E. et al., Eds., Wiley & Sons, Chichester, U.K., 1984, pp. 1–45.
8. Frangopol, D.M., Reliability-based structural design, in *Probabilistic Structural Mechanics Handbook*, Chapman & Hall, London, 1995, pp. 352–387.
9. Burns, S.A., Ed., *Recent Advances in Optimal Structural Design*, ASCE, Reston, VA, 2002.
10. Frangopol, D.M. and Maute, K., Life-cycle reliability-based optimization of civil and aerospace structures, *Comput. Struct.*, 81, 397–410, 2003.
11. Frangopol, D.M., Probabilistic structural optimization, *Progr. Structural Eng. Mater.*, 1(2), 223–230, 1998.
12. Frangopol, D.M. and Moses, F., Reliability-based structural optimization, in *Advances in Design Optimization*, Chapman & Hall, London, 1994, pp. 492–570.
13. Chang, S.E. and Shinozuka, M., Life-cycle cost analysis with natural hazard risk, *ASCE J. Infrastructure Syst.*, 1(3), 118–126, 1996.
14. Rosenblueth, E., Optimum reliabilities and optimum design, *Structural Safety*, 3(1), 69–83, 1986.
15. Kirsch, U., *Structural Optimization: Fundamentals and Applications*, Springer, Berlin, 1993.
16. Moses, F., Design for reliability-optimum concepts and applications, in *Optimum Structural Design*, Wiley & Sons, New York, 1973, pp. 241–265.

17. Frangopol, D.M., Structural optimization using reliability concepts, *ASCE J. Structural Eng.*, 111(11), 2288–2301, 1985.
18. Thoft-Christensen, P., On reliability-based structural optimization, *Lecture Notes Eng.*, 61, 387–402, 1991.
19. Liu, P.-L. and Der Kiureghian, A., Optimization algorithms for structural reliability, *Structural Safety*, 9, 161–177, 1991.
20. Katsuki, S. and Frangopol, D.M., Hyperspace division method for structural reliability, *ASCE J. Eng. Mech.*, 120(11), 2405–2427, 1994.
21. Tu, J., Choi, K.K., and Park, Y.H., A new study on reliability based design optimization, *ASME J. Mech. Design*, 121(4), 557–564, 1999.
22. Lee, J.-O., Yang, Y.-S., and Ruy, W.-S., A comparative study on reliability index and target performance-based probabilistic structural design, *Comput. Struct.*, 80, 257–269, 2002.
23. Newman, J.C., Anderson, W.K., and Whitfield, D.L., Multidisciplinary Sensitivity Derivatives Using Complex Variables, technical report, MSSU-COE-ERC-98-08, Mississippi State University, 1998.
24. Haug, E.J., Choi, K.K., and Komkov, V., *Design Sensitivity Analysis of Structural Systems*, Academic Press, Orlando, FL, 1986.
25. Kleiber, M., Antúnez, H., Hien, T.D., and Kowalczyk, P., *Parameter Sensitivity in Nonlinear Mechanics: Theory and Finite Element Computations*, Wiley & Sons, Chichester, U.K., 1997.
26. Barthelemy, B., Chon, C.T., and Haftka, R.T., Accuracy problems associated with semianalytical derivatives of static response, *Finite Elements Analysis Design*, 4, 249–265, 1988.
27. Mlejnek, H.-P., Accuracy of semianalytical sensitivities and its improvements by the "natural method," *Structural Optimization*, 4, 128–131, 1992.
28. de Boer, H. and van Keulen, F., Error analysis of refined semianalytical design sensitivities, *Structural Optimization*, 14, 242–247, 1997.
29. Hohenbichler, M. and Rackwitz, R., Sensitivities and importance measures in structural reliability, *Civil Eng. Syst.*, 3, 203–209, 1986.
30. Karamchandani, A. and Cornell, C., Sensitivity estimation within first and second order reliability methods, *Structural Safety*, 11(2), 95–107, 1992.
31. Kharmanda, G., Mohamed, A., and Lemaire, M., Efficient reliability-based design optimization using a hybrid space with applications to finite element analysis, *Struct. Multidisc. Optim.*, 24, 233–245, 2002.
32. Gill, P.E., Murray, W., and Wright, M.H., *Practical Optimization*, Academic Press, London, 1981.
33. Luenberger, D.G., *Linear and Nonlinear Programming*, Addison-Wesley, Reading, PA, 1984.
34. Arora, J.S., *Introduction to Optimum Design*, McGraw-Hill, New York, 1989.
35. Miller, R.E., *Optimization: Foundations and Applications*, Wiley, New York, 2000.
36. Zoutendijk, G., *Methods of Feasible Directions*, Elsevier, New York, 1960.
37. Svanberg, K., The method of moving asymptotes — a new method for structural optimization, *Int. J. Numerical Meth. Eng.*, 24, 359–373, 1987.
38. Schittkowski, K., Zillober, C., and Zotemantel, R., Numerical comparison on nonlinear programming algorithms for structural optimization, *Structural Optimization*, 7, 1–28, 1994.
39. Schittkowski, K., NLPQL: a FORTRAN subroutine for solving constrained nonlinear programming problems, *Ann. Operations Res.*, 5, 485–500, 1985.
40. Gill, P.E., Saunders, M.A., and Murray, W., SNOPT: an SQP Algorithm for Large-Scale Constrained Optimization, Report NA 97-2m, Department of Mathematics, University of California, San Diego, 1997.
41. Hasofer, A. and Lind, N., Exact and invariant second-moment code format, *J. Eng. Mech.*, 100, 111–121, 1974.
42. Rackwitz, R. and Fiessler, B., Structural reliability under combined load sequences, *Comput. Struct.*, 9, 484–494, 1978.
43. Frangopol, D.M., Multicriteria reliability-based structural optimization, *Structural Safety*, 3(1), 23–28, 1985.

44. Frangopol, D.M. and Fu, G., Optimization of structural systems under reserve and residual reliability requirements, in *Reliability and Optimization of Structural Systems*, Thoft-Christensen, P., Ed., Springer, Berlin, 1989, pp. 135–145.
45. Fu, G. and Frangopol, D.M., Reliability-based vector optimization of structural systems, *ASCE J. Struct. Eng.*, 116(8), 2143–2161, 1990.
46. Fu, G. and Frangopol, D.M., Balancing weight, system reliability and redundancy in a multiobjective optimization framework, *Structural Safety*, 7(2–4), 165–175, 1990.
47. Osyczka, A., *Multicriteria Optimization in Engineering*, Ellis Horwood, Chichester, U.K., 1984.
48. Duckstein, L., Multiobjective optimization in structural design: the model choice problem, in *New Directions in Optimum Structural Design*, Atrek, E. et al., Eds., Wiley & Sons, Chichester, U.K., 1984, pp. 459–481.
49. Koski, J., Multicriterion optimization in structural design, in *New Directions in Optimum Structural Design*, Atrek, A. et al., Eds., Wiley & Sons, Chichester, U.K., 1984, pp. 483–503.
50. Frangopol, D.M. and Iizuka, M., Pareto optimum solutions for nondeterministic systems, in *Proceedings of ICASP6, Sixth International Conference on Applications of Statistics and Probability in Civil Engineering*, Vol. 2, Esteva, L. and Ruiz, S.E., Eds., Mexico City, Mexico, 1991, Vol. 2. 1068–1075.
51. Frangopol, D.M. and Iizuka, M., Multiobjective decision support spaces for optimum design of nondeterministic systems, in *Probabilistic Safety Assessment and Management*, Vol. 2, Apostolakis, G., Ed., Elsevier, New York, 1991, pp. 977–982.
52. Frangopol, D.M. and Iizuka, M., Probability-based structural system design using multicriteria optimization, in *Proceedings of Fourth Symposium on Multidisciplinary Analysis and Optimization*, AIAA 92-4788-CP, AIAA/USAF/NASA/OAI, Part 2, Cleveland, Ohio, 1992, pp. 794–798.
53. Parimi, S.R. and Cohn, M.Z., Optimum solutions in probabilistic structural design, *J. Appl. Mech.*, 2(1), 47–92, 1978.
54. Hendawi, S. and Frangopol, D.M., Design of composite hybrid plate girder bridges based on reliability and optimization, *Structural Safety*, 15(1/2), 149–165, 1994.
55. Kim, S.H. and Wen, Y.K., Optimization of structures under stochastic loading, *Structural Safety*, 7(2–4), 177–190, 1990.
56. Thoft-Christensen, P. and Sørensen, J.D., Optimal strategy for inspection and repair of structural systems, *Civil Eng. Syst.*, 4, 94–100, 1987.
57. Liu, Y. and Moses, F., Bridge design with reserve and residual reliability constraints, *Structural Safety*, 11, 29–42, 1991.
58. Furuta, H., Fundamental Study on Geometrical Configuration and Reliability of Framed Structures Used for Bridges, Ph.D. thesis, Department of Civil Engineering, University of Kyoto, Kyoto, Japan, 1980.
59. Murotsu, Y. and Shao, S., Optimum shape design of truss structures based on reliability, *Structural Optimization*, 2(2), 65–76, 1990.
60. Shao, S., Reliability-Based Shape Optimization of Structural and Material Systems, Ph.D. thesis, Division of Engineering, University of Osaka Prefecture, Osaka, Japan, 1991.
61. Murotsu, Y., Miki, M., and Shao, S., Optimal configuration for fiber reinforced composites under uncertainties of material properties and loadings, in *Probabilistic Mechanics and Structural and Geotechnical Reliability*, Lin, Y.K., Ed., ASCE, Reston, VA, 1992, pp. 547–550.
62. Augusti, G., Ciampoli, M., and Frangopol, D.M., Optimal planning of retrofitting interventions on bridges in a highway network, *Eng. Struct.*, 20(11), 933–939, 1998.
63. Frangopol, D.M., Advances in life-cycle reliability-based technology for design and maintenance of structural systems, in *Computational Mechanics for the Twenty-First Century*, Topping, B.H.V., Ed., Saxe-Coburg Publ., Edinburgh, 2000, pp. 311–328.
64. Frangopol, D.M., Gharaibeh, E.S., Kong, J.S., and Miyake, M., Optimal network-level bridge maintenance planning based on minimum expected cost, *J. Transport. Res. Board, Transportation Res. Rec.*, 1696(2), 26–33, 2000.

65. Frangopol, D.M. and Guedes Soares, C., Eds., Reliability oriented optimal structural design, *Reliability Eng. Syst. Safety*, 73(3), 195–306, 2001.
66. Frangopol, D.M. and Furuta, H., Eds., *Life-Cycle Cost Analysis and Design of Civil Infrastructure Systems*, ASCE, Reston, VA, 2001.
67. Frangopol, D.M., Miyake, M., Kong, J.S., Gharaibeh, E.S., and Estes, A.C., Reliability- and cost-oriented optimal bridge maintenance planning, in *Recent Advances in Optimal Structural Design*, Burns, S., Ed., ASCE, Reston, VA, 2002, pp. 257–270.
68. Frangopol, D.M., Brühwiler, E., Faber, M.H., and Adey, B., Eds., *Life-Cycle Performance of Deteriorating Structures: Assessment, Design and Management*, ASCE, Reston, VA, 2004.
69. Estes, A.C. and Frangopol, D.M., Minimum expected cost-oriented optimal maintenance planning for deteriorating structures: application to concrete bridge decks, *Reliability Eng. System Safety*, 73(3), 281–291, 2001.
70. Kong, J.S. and Frangopol, D.M., Life-cycle reliability-based maintenance cost optimization of deteriorating structures with emphasis on bridges, *ASCE J. Structural Eng.*, 129(6), 818–828, 2003.
71. AASHTO, *ASSHTO Standard Specification for Highways Bridges*, 14th ed., American Association of State Highway and Transportation Officials, Washington, DC, 1992.
72. Yang, S.-I., Predicting Lifetime Reliability of Deteriorating Systems with and without Maintenance, Ph.D. thesis, Department of Civil, Environmental, and Architectural Engineering, University of Colorado, Boulder, 2002.
73. Maunsell Ltd., Serviceable Life of Highway Structures and Their Components — Final Report, Highways Agency, Birmingham, U.K., 1999.
74. Yang, L. and Ma, Z.K., Optimum design based on reliability for composite laminate layup, *Comput. Struct.*, 31(3), 377–383, 1989.
75. Miki, M., Murotsu, Y., and Shao, S., Optimum fiber orientation angle of multiaxially laminated composites based on reliability, *AIAA J.*, 31(5), 919–920, 1993.
76. Miki, M., Murotsu, Y., Tanaka, T., and Shao, S., Reliability-based optimization of fibrous laminated composites, *Reliability Eng. System Safety*, 56(5), 285–290, 1997.
77. Thanedar, P.B. and Chamis, C.C., Reliability considerations in composite laminate tailoring, *Comput. Struct.*, 54(1), 131–139, 1995.
78. Chao, L.-P., Multiobjective optimization design methodology for incorporating manufacturing uncertainties in advanced composite structures, *Eng. Optimization*, 25(4), 309–323, 1996.
79. Mahadevan, S. and Liu, X., Probabilistic optimum design of composite laminates, *J. Composite Mater.*, 32(1), 68–82, 1998.
80. Antonio, C.A.C., Marques, A.T., and Goncalves, J.F., Reliability based design with a degradation model of laminated composite structures, *Structural Optimization*, 12(1), 16–28, 1996.
81. Su, B., Rais-Rohani, M., and Singh, M., Reliability-based optimization of anisotropic cylindrical shells with response surface approximations of buckling instability, in *Proceedings of 43rd Structures, Structural Dynamics, and Materials Conference Proceedings*, AIAA 2002-1386, AIAA/ASME/ASCE/AHS/ASC, AIAA, Reston, VA, 2002.
82. Qu, X., Venkataraman, S., Haftka, R.T., and Johnson, T.F., Deterministic and reliability-based optimization of composite laminates for cryogenic environments, in *Proceedings of 8th Symposium on Multidisciplinary Analysis and Optimization*, AIAA 2000-4760, AIAA/USAF/NASA/ISSMO, AIAA, Reston, VA, 2000.
83. Huyse, L. and Lewis, M., Aerodynamic Shape Optimization of Two-Dimensional Airfoils under Uncertain Conditions, NASA/CR-2001-210648, ICASE Report No. 2001-1, ICASE NASA Langley Research Center, Hampton, VA, 2001.
84. Li, W., Huyse, L., and Padula, S., Robust Airfoil Optimization to Achieve Drag Reduction over a Range of Mach Numbers, technical report, Old Dominion University, Norfolk, VA, 2001.
85. Putko, M.M., Newman, P.A., Taylor, A.C., and Green, L.L., Approach for uncertainty propagation and robust design in CFD using sensitivity derivatives, in *Proceedings of 15th AIAA Computational Fluid Dynamics Conference*, AIAA 2001-2528, AIAA, Reston, VA, 2001.

86. Haftka, R.T., Structural optimization with aeroelastic constraints — a survey of U.S. applications, *Int. J. Vehicle Design*, 7, 381–392, 1986.
87. Bowman, K.B., Grandhi, R.V., and Eastep, F.E., Structural optimization of lifting surfaces with divergence and control reversal constraints, *Structural Optimization*, 1, 153–161, 1989.
88. Friedmann, P.P., Helicopter vibration reduction using structural optimization with aeroelastic/multidisciplinary constraints — a survey, *AIAA J. Aircraft*, 28, 8–21, 1991.
89. Barthelemy, J.-F., Wrenn, G.A., Dovi, A.R., and Hall, L.E., Supersonic transport wing minimum design integrating aerodynamics and structures, *AIAA J. Aircraft*, 31, 330–338, 1994.
90. Møller, H. and Lund, E., Shape sensitivity analysis of strongly coupled fluid–structure interaction problems, in *Proceedings of Eighth Symposium on Multidisciplinary Analysis and Optimization*, AIAA 2000-4823, AIAA/USAF/NASA/ISSMO, AIAA, Reston, VA, 2000.
91. Hou, G.J.-W. and Satyanarayana, A., Analytical sensitivity analysis of a statical aeroelastic wing, in *Proceedings of Eighth Symposium on Multidisciplinary Analysis and Optimization*, AIAA 2000–4824, AIAA/USAF/NASA/ISSMO, AIAA, Reston, VA, 2000.
92. Maute, K., Nikbay, M., and Farhat, C., Coupled analytical sensitivity analysis and optimization of three-dimensional nonlinear aeroelastic systems, *AIAA J.*, 39(11), 2051–2061, 2001.
93. Martins, J.R.R.A. and Alonso, J.J., High-fidelity aero-structural design optimization of a supersonic business jet, in *Proceedings of 43rd Structures, Structural Dynamics, and Materials Conference*, AIAA 2002-1483, AIAA/ASME/ASCE/AHS/ASC, AIAA, Reston, VA, 2002.
94. Pettit, C.L., Canfield, R.A., and Ghanem, R., Stochastic analysis of an aeroelastic system, in *15th ASCE Engineering Mechanics Conference Proceedings*, ASCE, Reston, VA, 2002.
95. Lindsley, N.J., Beran, P.S., and Pettit, C.L., Effects of uncertainties on nonlinear plate response in supersonic flow, in *Proceedings of 9th AIAA/ISSMO Symposium on Multidisciplinary Analysis and Optimization*, AIAA 2002-5600, AIAA, Reston, VA, 2002.
96. Rao, S.S., Automated optimum design of wing structures: a probabilistic approach, *Comput. Struct.*, 24, 799–808, 1986.
97. Yang, J.S., Nikolaidis, E., and Haftka, R.T., Design of aircraft wings subjected to gust loads: a system reliability approach, *Comput. Struct.*, 36(6), 1057–1066, 1990.
98. Yang, J.S. and Nikolaidis, E., Design of aircraft wings subjected to gust loads: a safety index based approach, *AIAA J.*, 29(5), 804–812, 1991.
99. Pettit, C. and Grandhi, R., Reliability optimization of aerospace structures for gust response and aileron effectiveness, in *Proceedings of Eighth International Conference on Structural Safety and Reliability*, 2001.
100. Kuttenkeuler, J. and Ringertz, U., Aeroelastic tailoring considering uncertainties in material properties, *Structural Optimization*, 15(3–4), 157–162, 1998.
101. Stroud, W., Krishnamurthy, T., Mason, B., Smith, S., and Naser, A., Probabilistic design of a plate-like wing to meet flutter and strength requirements, in *Proceedings of 43rd Structures, Structural Dynamics, and Materials Conference*, AIAA 2002-1464, AIAA/ASME/ASCE/AHS/ASC, AIAA, Reston, VA, 2002.
102. Luo, X. and Grandhi, R.V., ASTROS for reliability-based multidisciplinary structural analysis and optimization, *Comput. Struct.*, 62, 737–745, 1997.
103. Zink, P., Mavris, D., Love, M., and Karpel, M., Robust design for aeroelastically tailored/active aeroelastic wing, in *Proceedings of Seventh Symposium on Multidisciplinary Analysis and Optimization*, AIAA 98-4781, AIAA/USAF/NASA/ISSMO, AIAA, Reston, VA, 1998.
104. Zink, P., Raveh, D., and Mavris, D., Robust structural design for active aeroelastic wing with aerodynamic uncertainties, in *Proceedings of 41st Structures, Structural Dynamics, and Materials Conference and Exhibit*, AIAA 2000-1439, AIAA/ASME/ASCE/AHS/ASC, AIAA, Reston, VA, 2000.
105. Gumbert, C.R., Newman, P.A., and Hou, G.J.-W., Effect of random geometric uncertainty on the computational design of a 3-D flexible wing, in *AIAA 20th Applied Aerodynamics Conference Proceedings*, AIAA 2002-2806, AIAA, Reston, VA, 2002.

106. Gumbert, C.R., Hou, G.J.-W., and Newman, P.A., Reliability assessment of a robust design under uncertainty for a 3-D flexible wing, in *16th AIAA Computational Fluid Dynamics Conference Proceedings*, AIAA 2003-4094, AIAA, Reston, VA, 2003.
107. Allen, M. and Maute, K., Reliability based optimization of aeroelastic structures, in *Proceedings of 9th AIAA/ISSMO Symposium on Multidisciplinary Analysis and Optimization*, AIAA 2002-5560, AIAA, Reston, VA, 2002.
108. Allen, M. and Maute, K., Shape optimization of aeroelastic structures under uncertainties, in *Proceedings of 16th AIAA Computational Fluid Dynamics Conference*, AIAA 2003-3430, AIAA, Reston, VA, 2003.
109. Du, X. and Chen, W., Methodology of managing the effect of uncertainty in simulation-based design, *AIAA J.*, 38(8), 1471–1478, 2000.
110. Oakley, D.R., Sues, R.H., and Rhodes, G.S., Performance optimization of multidisciplinary mechanical systems subject to uncertainties, *Prob. Eng. Mech.*, 13(1), 15–26, 1998.
111. Mavris, D.N., Bandte, O., and Schrage, D.P., Application of probabilistic methods for the determination of an economically robust HSCT configuration, in *Proceedings of Sixth Int. Symposium on Multidisciplinary Analysis and Optimization*, AIAA/NASA/USAF/ISSMO, AIAA, Reston, VA, 1996.
112. Mavris, D.N., DeLaurentis, D.A., Bandte, O., and Hale, M.A., A stochastic approach to multidisciplinary aircraft analysis and design, in *36th Aerospace Science Meeting and Exhibit Proceedings*, AIAA 98-0912, AIAA, Reston, VA, 1998.
113. Mavris, D.N. and DeLaurentis, D., A stochastic design approach for aircraft affordability, in *Proceedings of 21st Congress of International Council on the Aeronautical Sciences*, ICAS, 1998.
114. Ye, W., Mukherjee, S., and MacDonald, N.C., Optimal shape design of an electrostatic comb drive in micro-electromechanical systems, *J. Micro Electromechanical Syst.*, 7, 16–27, 1998.
115. Ye, W. and Mukherjee, S., Optimal three-dimensional analysis and design of electrostatic comb drives using boundary element method, *Int. J. Numerical Meth. Eng.*, 45, 175–194, 1999.
116. Eggert, H., Guth, H., Jakob, W., Meinzer, S., Sieber, I., and Suess, W., Design optimization of microsystems, in *Technical Proceedings 1998 International Conference on Modeling and Simulation of Microsystems, MSM 98*, 1998.
117. Schneider, P., Huck, E., Reitz, S., Parodat, S., Schneider, A., and Schwarz, P., A modular approach for simulation-based optimization of MEMS, in *Proceedings of Conference on Design Modeling and Simulation in Microelectronics*, 2000, pp. 71–82.
118. Morris, C.J. and Forster, F.K., Optimization of a circular piezoelectric bimorph for a micro-pump driver, *J. Micromechanics Microengineering*, 10, 459–465, 2000.
119. Allen, M., Raulli, M., Maute, K., and Frangopol, D.M., Reliability-Based Analysis and Design Optimization of Electrostatically Actuated MEMS, *Comput. Struct.*, 2003. Vol. 82, 1007–1020, 2004.
120. Rozvany, G.I.N., Bendsøe, M.P., and Kirsch, U., Layout optimization of structures, *Appl. Mech. Rev.*, 48, 41–119, 1995.
121. Bendsøe, M.P., *Optimization of Structural Topology, Shape, and Material*, Springer, Berlin, 1995.
122. Bendsøe, M.P., Variable-Topology Optimization: Status and Challenges, technical report, Department of Mathematics, Tech. Univ. of Denmark, Lyngby, Denmark, 1999.
123. Eschenauer, H.A. and Olhoff, N., Topology optimization of continuum structures: a review, *Appl. Mech. Rev.*, 54(4), 331–389, 2001.
124. Soto, C.A., Applications of structural topology optimization in the automotive industry: past, present and future, in *Proceedings of Fifth World Congress on Computational Mechanics (WCCM V)*, Mang, H.A., Rammerstorfer, F.G., and Eberhardsteiner, J., Eds., Vienna University of Technology, Austria, 2002.
125. Ananthasuresh, G.K., Kota, S., and Kikuchi, N., Strategies for systematic synthesis of compliance MEMS, in *Proceedings of 1994 ASME Winter Annual Meeting*, ASME, New York, 1994, pp. 677–686.
126. Sigmund, O., On the design of compliant mechanisms, *Mech. Struct. Mach.*, 25, 493–524, 1997.

127. Frecker, M.I., Ananthasuresh, G.K., Nishiwaki, G.K., Kikuchi, N., and Kota, S., Topological synthesis of compliant mechanisms using multi-criteria optimization, *ASME J. Mech. Des.*, 119, 238–245, 1997.
128. Nishiwaki, S., Frecker, M., Min, S., and Kikuchi, N., Topology optimization of compliant mechanisms using homogenization method, *Int. J. Numerical Meth. Eng.*, 42, 535–559, 1998.
129. Bruns, T.E. and Tortorelli, D.A., Topology optimization of geometrically nonlinear structures and compliant mechanisms, in *Proceedings of 7th AIAA/USAF/NASA/ISSMO Symposium on Multidisciplinary Analysis and Optimization*, AIAA, Reston, VA, 1998, pp. 1874–1882.
130. Hetrick, J.A. and Kota, S., Topological and geometric synthesis of compliant mechanisms, in *Proceedings of 2000 ASME Design Engineering Technical Conferences*, DETC2000/MECH-14140, ASME, 2000.
131. Pedersen, C.B.W., Buhl, T., and Sigmund, O., Topology optimization of large displacement compliant mechanism, *Int. J. Numerical Meth. Eng.*, 50, 2683–2705, 2001.
132. Saxena, A. and Ananthasuresh, G.K., Topology synthesis of compliant mechanisms for nonlinear force-deflection and curved path specifications, *ASME J. Mech. Des.*, 123, 33–42, 2001.
133. Silva, E.C.N., Fonseca, J.S.O., and Kikuchi, N., Optimal design of piezoelectric microstructures, *Computational Mechanics*, 19, 397–410, 1997.
134. Li, Y., Xin, X., Kikuchi, N., and Saitou, K., Optimal shape and location of piezoelectric materials for topology optimization of flextensional actuators, in *Proceedings of 2001 Genetic and Evolutionary Computation Conference (GECCO-2000)*, 2001, pp. 1085–1089.
135. Sigmund, O., Design of multiphysics actuators using topology optimization — Part I: one-material structures, *Comput. Meth. Appl. Mech. Eng.*, 190(49–50), 6577–6604, 2001.
136. Sigmund, O., Design of multiphysics actuators using topology optimization — Part II: two–material structures, *Comput. Meth. Appl. Mech. Eng.*, 190(49–50), 6605–6627, 2001.
137. Maute, K. and Frangopol, D.M., Reliability-based design of MEMS mechanisms by topology optimization, *Comput. Struct.*, 81, 813–824, 2003.

25
Accelerated Life Testing for Reliability Validation

25.1	What It Is, and How It Is Applied	25-1
25.2	Accelerated Reliability Testing Models	25-2
25.3	Recommendations	25-2
25.4	Log-Log Stress-Life Model	25-2
	Example 1	
25.5	Overload-Stress Reliability Model	25-6
	Example 2	
25.6	Combined-Stress Percent-Life Model	25-7
	Example 3 • Applicability	
25.7	Deterioration-Monitoring Model	25-12
	What It Is • Test Procedure • Example 4 • Applicability	
25.8	The Step-Stress Accelerated Testing Model	25-14
	Parameters Determination for the Weibull Step-Stress Model Using the MLE Method • Example 5	

Dimitri B. Kececioglu
The University of Arizona

25.1 What It Is, and How It Is Applied

Accelerated life testing of products, components and materials is to get information *quickly* on specific lives, life distributions, failure rates, mean lives, and reliabilities. Accelerated testing is achieved by subjecting the test units to application and operation stress levels that are more severe than the stress levels applied during normal use in order to shorten their lives or their times to failure. If the results can be extrapolated to the stress levels encountered during normal use, they yield estimates of the lives and reliabilities under use stresses. Such tests provide savings in time and expense, since for many products, components, and materials, life under use conditions is so long that testing under those conditions is not timely or economically feasible.

Accelerated test conditions are typically produced by testing units at higher levels of temperature, voltage, wattage, pressure, vibration amplitude, frequency, cycling rate, loads, humidity, etc. (or some combination thereof) that are encountered under use conditions. Life data obtained from units tested at different elevated stresses are analyzed and extrapolated to obtain an estimate of the life and reliability at stress levels of typical use. The use of such accelerating variables for a specific product or material is dictated by experienced engineering practice. For example, when choosing the stress level at which a test will be performed, one must be cautious not to exceed the material limitations under the specific stress and to avoid introducing failure modes that would not be observed under use conditions.

25.2 Accelerated Reliability Testing Models

The following accelerated test models were covered by Kececioglu [1]: (1) Arrhenius; (2) Eyring; (3) inverse power law; (4) combination, when two or more accelerated stresses are applied, such as combinations of temperature and voltage; (5) generalized Eyring; (6) Bazovsky; and (7) Weibull stress life. This chapter covers the following additional accelerated testing models and their practical applications:

1. Log-log stress-life
2. Overload-stress
3. Combined-stress percent-life
4. Deterioration monitoring
5. Step-stress

The log-log stress-life model uses a log-log plot of stress-life data to develop curves for specific reliabilities at various functional stress levels. The overload-stress model presents a unique equation for reliability, which combines the Weibull and the log-log stress-life models' results to determine the reliability of units at their use-stress level, having determined their reliability at a higher stress level.

The combined-stress percent-life model analyzes the results when one unit is tested to failure at an elevated stress, and another unit is tested—first for a fraction of its life at the use-stress level, and subsequently at the high stress level used for the first unit—until this second unit fails. A first-approximation straight line is drawn to relate stress level and associated life, from which the use-stress life is determined. The deterioration-monitoring model involves monitoring, at both use and accelerated stress levels, some property that changes continuously during operation, and then determining that 1 h at accelerated stress is equivalent to some $m > 1$ h of test at use stress. Optimum step-stress tests enable the determination of life at constant use stress from step-stress life data. The MLE (maximum likelihood estimators) equations for the parameters of the step-stress model are derived for the Weibull case, and the methodology is explained.

Optimum accelerated life tests involve tests with and without censored units, with life test results obtained at only *two* different stress levels.

25.3 Recommendations

To apply these accelerated testing methods and models, the following facts should be adhered to:

1. Units identical to those to be used at the use-stress level should be tested at the accelerated-stress levels.
2. Only the stresses of acceleration should be applied; all other stresses should be kept constant.
3. The failure modes at accelerated stress should be the same as those observed under use-stress conditions.
4. Each sample tested at a specified stress level should be homogeneous.
5. The accelerated test results should not be extrapolated to stress levels beyond the range of applicability of the model used.
6. It should be ascertained that the accelerated test model used is applicable to the types of components and units being tested and the stresses applied.

25.4 Log-Log Stress-Life Model

Traditional engineering design employs the strength, S, vs. the cycles to failure, N, or the $S-N$ diagram, when designing components to withstand stresses that fluctuate over time or cause fatigue. The $S-N$ diagram can be improved by making it distributional, as in Figure 25.1 [2], by determining the cycles-to-failure distribution at each fixed-maximum alternating stress level. Figure 25.1 is constructed as follows:

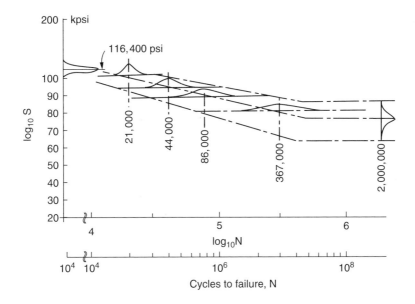

FIGURE 25.1 Statistical S–N surface for SAE 4340 steel wire, cold drawn and annealed, at fixed stress levels. (From Kececioglu, D.B. and Haugen, E., Interactions among the Various Phenomena Involved in the Design of Dynamic and Rotary Machinery and Their Effects on Reliability, 2nd technical report to ONR, Washington, DC, 1969. With permission.)

For each one of at least five stress levels, several identical components (a group) are tested. There should preferably be at least 35 components in each group. The mean of the distribution of the cycles to failure is found for each stress level and plotted on log-log scales, the assumption being that the loci of the means of the logarithms of the cycles to failure, particularly for steels in fatigue, fall on a straight line when using a log S ordinate and a log N abscissa scales. The equation of the straight line passing through these mean points at each stress level is

$$\overline{\log_{10} S} = m_1 \overline{\log_{10} N} + b_1 = m_1 \overline{N'} + b_1 \tag{25.1}$$

where

$\overline{\log_{10} S}$ = mean value of the logarithms of the maximum fluctuating, or alternating, stress levels
m_1 = slope of the mean S–N line on a log-log plot
$\overline{\log_{10} N}$ = mean of the component's life at that stress level calculated by averaging the logarithms of the cycles-to-failure data at that stress level
b_1 = stress required to cause failure when the stress does not fluctuate

The least-squares analysis can be used to find the best-fit straight line to the mean lives at the various stress levels.

Similarly the $\pm 3\sigma_{N'}$ limit-line equations about the mean life can be found from

$$\log_{10} S_+ = m_2(\overline{N'} + 3\sigma_{N'}) + b_2$$

and

$$\log_{10} S_- = m_3(\overline{N'} - 3\sigma_{N'}) + b_3$$

where m_i and b_i ($i = 2, 3$) are the empirically determined parameters of the $\overline{N'} \pm 3\sigma_{N'}$ limit lines, and $\sigma_{N'}$ is the standard deviation of the logarithms of the cycles-to-failure data at the respective stress levels.

It is assumed here that the cycles-to-failure distribution is well represented by the lognormal distribution, which is a good assumption; consequently, the logarithms of the cycles to failure are normally distributed.

From least-squares analysis,

$$b = \frac{\left(\sum \log_{10} S\right)\left[\sum (\log_{10} N)^2\right] - \left(\sum \log_{10} N\right)\left[\sum (\log_{10} S \cdot \log_{10} N)\right]}{n \sum (\log_{10} N)^2 - \left(\sum \log_{10} N\right)^2}$$

and

$$m = \frac{n \sum (\log_{10} S \cdot \log_{10} N) - \left(\sum \log_{10} N\right)\left(\sum \log_{10} S\right)}{n \sum (\log_{10} N)^2 - \left(\sum \log_{10} N\right)^2}$$

where n = number of stress levels, and N becomes $\overline{N'} \pm 3\sigma_{N'}$ for the $\pm 3\sigma_{N'}$-limit equations about the mean life, $\overline{N'}$.

25.4.1 Example 1

Using Figure 25.1, find the 95% reliable life, $N_{0.95}$, for SAE 4340 steel wire subjected to an 80-kpsi fluctuating stress.

Solution to Example 1

From Figure 25.1, at the 80-kpsi fluctuating stress level, using the experimental results, it is clear that

$$\overline{N'} = \log_{10}(8.5 \times 10^4) \text{ cycles}$$

Since it is assumed that the logarithms of the cycles-to-failure data are normally distributed, the standard deviation is given by Equation 25.2, where the $\overline{N'} \pm 3\sigma_{N'}$ are read off the corresponding limits indicated at the 80-kpsi stress level, or

$$\sigma_{N'} = \frac{(\overline{N'} + 3\sigma_{N'}) - (\overline{N'} - 3\sigma_{N'})}{6} \quad (25.2)$$

or

$$\sigma_{N'} = \frac{\log_{10}(2.5 \times 10^5) - \log_{10}(2.2 \times 10^4)}{6} = 0.1759 \quad (25.3)$$

The reliability is given by the probability, P, that the cycles to failure is equal to or greater than the $N_{0.95}$ life, defined as the life (the cycles of operation) by which only 5% of the components will fail and 95% will not fail (i.e., will survive), as illustrated in Figure 25.2. Then,

$$R(N_{0.95}) = P(N \geq N_{0.95}) = 0.95 = \int_{N_{0.95}}^{\infty} f(N_{S=80\,kpsi}) dN$$

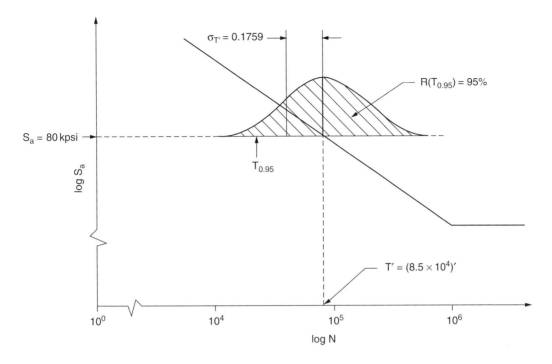

FIGURE 25.2 The cycles-to-failure distribution, the reliability sought, and the $T_{0.95}$ reliable life for Example 1.

or, utilizing the fact that $\log_e N = N'$ is normally distributed,

$$0.95 = \int_{\log_{10} N_{0.95}}^{\infty} f(N'_{S=80\,kpsi})\,dN'$$

and finally, utilizing the equivalent standard normal distribution,

$$0.95 = \int_{z(N'_{0.95})}^{\infty} \phi(z)\,dz$$

where

$f(N_{S=80\,kpsi})$ = cycles-to-failure distribution at the maximum fluctuating stress level of 80,000 psi
$f(N'_{S=80\,kpsi})$ = cycles-to-failure distribution of the logarithms of the cycles to failure at the maximum fluctuating stress level of 80,000 psi, with parameters $\overline{N'}$ and $\sigma_{N'}$

and

$$z(\overline{N'_{0.95}}) = \frac{\overline{N'}_{0.95} - \overline{N'}}{\sigma_{N'}} \tag{25.4}$$

From the standard normal distribution area tables for $R = 95\%$, the corresponding value of $z(\overline{N'_{0.95}})$ is 1.645. The 95% reliable life is then found from Equation 25.4, or

$$-1.645 = \frac{N'_{0.95} - \log_{10}(8.5 \times 10^4)}{0.1759}$$

Solving for $N'_{0.95}$ yields

$$N'_{0.95} = \log_{10} N_{0.95} = 4.64$$

and

$$N_{0.95} = 43{,}700 \text{ cycles}$$

25.5 Overload-Stress Reliability Model

A binomial test is applicable when there are only two possible test results, often taken to be success and failure. For this type of a test, a unique equation [3] relating the reliability at the accelerated stress level to the reliability at the use-stress level has been developed, as presented by Hornbeck [4]:

$$R_A = (R_U)^{(W)^{-\beta/m}}$$

or

$$R_U = (R_A)^{(W)^{\beta/m}} \qquad (25.5)$$

where

R_A = reliability at the accelerated stress level
W = overload stress factor = S_2/S_1
R_U = reliability at the use-stress level
S_i = stress level with $S_2 > S_1$
m = slope of the stress-life equation, such as that of Equation 25.1
β = Weibull slope of the times-to-failure distribution at a specific stress level

It is assumed that the Weibull slope remains invariant with change of stress level, as appears to be the case in accelerated testing, when the basic failure mode does not change from the use-stress level to the accelerated-stress level of testing.

The average, or the point estimate, of the reliability at a specific stress is calculated from

$$\bar{R} = \frac{N_S + 1}{N_T + 2} \qquad (25.6)$$

where N_S is the number of successful trials, and N_T is the total number of trials undertaken at that stress level.

25.5.1 Example 2

Units of a particular design are tested at 15% overload with the following results:

$$N_S = 14 \text{ successes}$$
$$N_T = 20 \text{ total units tested}$$

It is known that $\beta = 2.0$ and $m = -1/6$ from prior testing.
Find the reliability for this design under use-stress conditions, i.e., no overload.

Solution to Example 2

From Equation 25.6

$$\bar{R} = \frac{14+1}{20+2} = 0.682$$

Since the units are tested at 15% overload, the overload stress factor is

$$W = \frac{S_2}{S_1} = 1.15$$

Then, the average reliability at the use-stress level (no overload) from Equation 25.5 is

$$\overline{R}_U = (0.682)^{(1.15)^{2.0/(-1/6)}}$$

or

$$\overline{R}_U = 0.9309, \text{ or } 93.09\%$$

25.6 Combined-Stress Percent-Life Model

An interesting life behavior for a variety of products is that shown in Figure 25.3 [5]. These products include ball bearings, electric motors, electric drills, and light bulbs. Figure 25.3 is developed as follows. One unit is tested to failure at a low stress level S_1, and the time to failure is recorded. This is considered to be 100% of its life at stress level S_1. Another identical unit from the same production lot is tested at the chosen accelerated stress level S_2 until it fails, and the time to failure is recorded. This time, or life, is taken to be 100% of the unit's life at the accelerated stress level S_2. These points are then plotted on Cartesian paper, as shown in Figure 25.3, using the same scale for the ordinate and the abscissa. It has been found that if another unit from the same production lot is tested at stress level S_1, for a fraction of its life, or 100 α% of its life at S_1, and then tested at the higher stress level S_2 until it fails, the life at stress

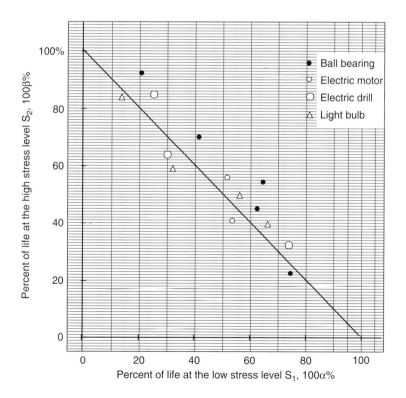

FIGURE 25.3 Experimental data verifying Equation 25.7.

level S_2, expressed as 100 β% of its life at S_2, falls approximately on a straight line drawn between the 100% life endpoints. A repetition of this process for additional units would then yield results similar to those shown in Figure 25.3. If the straight-line-fit assumption is acceptable, the equation of this line would then be

$$\alpha + \beta = 1 \tag{25.7}$$

The procedure for determining a component's life at the use-stress level S would be as follows. Take a unit and test it at stress level S_2 to failure. Designate this time to failure as T_2. Take a second unit, test it at use-stress level S_1 for a time T'_1 less than its full life at S_1. Then, increase the stress level to S_2 on this same unit and continue the test until this second unit fails. Designate this time to failure as T'_2. As illustrated in Figure 25.4, these results yield

$$\beta = \frac{T'_2}{T_2} \tag{25.8}$$

Now, α can be determined from Equation 25.7 as

$$\alpha = 1 - \beta = 1 - \frac{T'_2}{T_2} \tag{25.9}$$

But by definition

$$\alpha = \frac{T'_1}{T_1} \tag{25.10}$$

Consequently, T_1, the use-stress life, would be given from Equation 25.10 as

$$T_1 = \frac{T'_1}{\alpha} \tag{25.11}$$

which is the test's objective.

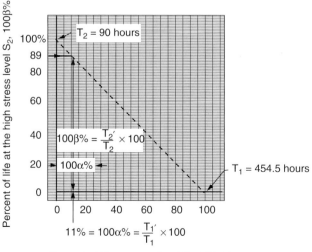

FIGURE 25.4 Illustration of Example 3.

If the original premise is applicable to the units tested, a substantial savings of test time is achieved. The total test time required for this type of test, T_A, is

$$T_A = T_2 + T_1' + T_2'$$

The test time to determine the life at the use stress, T_U, would be

$$T_U = T_1$$

Thus, the time saved using this accelerated test type is

$$T_{Saved} = T_U - T_A$$

This test method and the test time saved are illustrated in the next example (Example 3).

However, this method requires a rather large sample size, usually in the range of 20 to 100, if the scatter on the individual life times is large [6]. Rabinowicz [6] considered the life prediction error as follows. If N units are tested at stress 2 and the mean life is $\overline{T}_2'$; and if N more units are tested for a time $\alpha \overline{T}_1$ at stress 1 and then to failure at stress 2; and if the distributions of the lives of these units were normal with standard deviation at stress 2 of σ_{T_2}, then the standard error of the mean life for each case would be $\sigma_{T_2}/\sqrt{N-1}$. By extending Figure 25.5 for N units, it can be seen that the height of point A is $\overline{T}_2$, while the height of point C, $\overline{T}_2'$, is $(1-\alpha)T_2$ because

$$\frac{\alpha \overline{T}_1}{\overline{T}_1} + \frac{\overline{T}_2'}{\overline{T}_2} = 1$$

Then, the height difference between points A and C is

$$D = \overline{T}_2 - \overline{T}_2'$$

The mean value of D would then be

$$D = \alpha \overline{T}_2$$

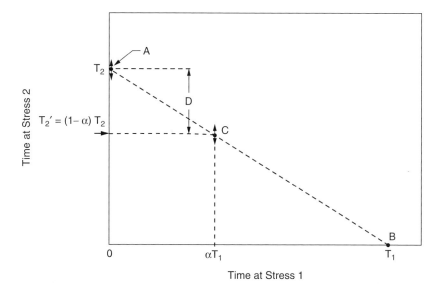

FIGURE 25.5 Effect of experimental scatter on the estimated life for two units.

and the standard error of D would be

$$\sqrt{2}\,\sigma_{T_2}/\sqrt{N-1}$$

The vertical difference in the heights of points A and C in Figure 25.5 can then be written as the mean ± the standard error, or

$$D \cong \alpha \overline{T}_2 \pm \sqrt{2}\,\sigma_{T_2}/\sqrt{N-1}$$

The slope of line AC is then given by

$$\frac{\Delta Y}{\Delta X} = -\frac{D}{\alpha \overline{T}_1} = -\frac{\alpha \overline{T}_2 \pm \sqrt{2}\,\sigma_{T_2}/\sqrt{N-1}}{\alpha \overline{T}_1}$$

or

$$\frac{\Delta Y}{\Delta X} = -\frac{\overline{T}_2}{\overline{T}_1} \mp \frac{\sqrt{2}\,\sigma_{T_2}/\sqrt{N-1}}{\alpha \overline{T}_1}$$

The proportional error of the slope, which, according to Rabinowicz, "essentially represents the proportional error of estimating the use-stress mean life, $\overline{T}_1$," can be written as

$$\text{Proportional error} = \frac{-\dfrac{\sqrt{2}\,\sigma_{T_2}/\sqrt{N-1}}{\alpha \overline{T}_1}}{-\overline{T}_2/\overline{T}_1}$$

or

$$\text{Proportional error} = \frac{\sqrt{2}}{\alpha\sqrt{N-1}}\frac{\sigma_{T_2}}{\overline{T}_2}$$

Rabinowicz [6] calculated the proportional error for different sample sizes of identical units he studied with

$$\alpha = 0.2$$

and

$$\frac{\sigma_{T_2}}{\overline{T}_2} = 0.4$$

The results are given in Table 25.1.

25.6.1 Example 3

A component is run at a high stress level S_2 only, and after $T_2 = 90$ h the component fails. Another component from the same lot is first run at a lower stress level S_1 for $T_1' = 50$ h. Subsequently, the stress

TABLE 25.1 Error in Estimating the Use-Stress Life Using the Combined-Stress Percent-Life Accelerated Test as a Function of Sample Size

Total Number of Bulbs Tested to Failure (2N)	Proportional Error (%)
10	141
20	95
50	57
100	40
200	28
500	14

level is increased to S_2 on this same component and it fails after $T_2' = 80$ h. What would the life, T_1, of these components be when run at the lower stress level S_1 only?

Solution to Example 3

This example is illustrated in Figure 25.4. From Equation 25.8,

$$\beta = \frac{T_2'}{T_2} = \frac{80}{90} = 0.89$$

and from Equation 25.9

$$\alpha = 1 - \beta = 1 - 0.89 = 0.11$$

Then, from Equation 25.11,

$$T_1 = \frac{T_1'}{\alpha} = \frac{50}{0.11} = 454.5 \text{ h}$$

or the life when operating at stress level S_1 only is $T_U = T_1 = 454.5$ h.

The total test time required for this test was

$$T_A = T_2 + T_1' + T_2' = 90 + 50 + 80 = 220 \text{ h}$$

whereas the total test time required in a nonaccelerated, or use-stress level, test would have been 454.5 h. This accelerated test thus saves

$$T_{Saved} = T_U - T_A = 454.5 - 220 = 234.5 \text{ h}$$

which is substantial.

25.6.2 Applicability

The combined-stress percent-life model can be applied with a minimum of two components, from the same lot, if quick life estimates are needed. Hence, it is the least costly and time-consuming accelerated test. However, the error in determining the use-stress life is the largest of the models presented here. To reduce this error to under 40%, two samples larger than 50 each need to be tested.

25.7 Deterioration-Monitoring Model

25.7.1 What It Is

This model is based essentially on the previous model, but it reduces the number of required test units while maintaining the same level of accuracy of prediction [6]. This method involves monitoring some property of the samples that varies continuously during operation and whose variation is directly related to deterioration. The property is measured both at use stress and at accelerated stress. Then, it can be determined that one hour at accelerated stress is equivalent to m hours at use stress.

25.7.2 Test Procedure

The life at accelerated stress is measured, and the life at use stress is estimated using the following test procedure:

1. Test at the use-stress level for t_1 time and measure a property that varies with time.
2. Calculate the rate of change of this property per unit time.
3. Change the stress on this same unit to a higher level and continue testing.
4. Calculate the rate of change of the same property, preferably near the beginning of the higher stress test.
5. Determine the use-stress life of this unit.

The actual determination of the mean life can be done by the graphical procedure shown in Figure 25.6 for Example 4 or by using Equation 25.12,

$$T_1 = t_1 + m t_2 \tag{25.12}$$

where

T_1 = predicted life at the use stress
t_1 = test time at the use stress
t_2 = test time at the high stress
$m = \gamma_2/\gamma_1$
γ_1 = property change rate at the use-stress level
γ_2 = property change rate at the high stress level

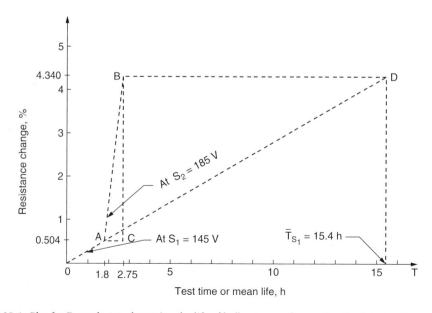

FIGURE 25.6 Plot for Example 4 to determine the life of bulbs at use voltage using the deterioration-monitoring accelerated test.

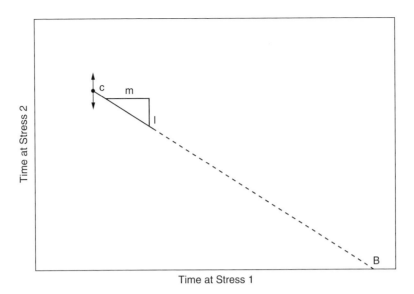

FIGURE 25.7 The error in estimating the position of B due to only the uncertainty in the position of C.

Rabinowicz [6] recommends using the γ_2 value determined near the beginning of the high stress life test.

In terms of the mean predicted life error, it can be seen from Figure 25.7 that the error in t_1 arises only from the standard error in determining the position of C, which is given in Figure 25.7, assuming that there is no error in determining m. This error of magnitude $\sigma_{T_2}/\sqrt{N-1}$ leads to a proportional error in estimating T_1 of $\sigma_{T_2}/T_2\sqrt{N-1}$.

Rabinowicz calculated the proportional error of this life-prediction method for the same units he studied for the combined-stress percent-life model for different sample sizes, and he obtained the results given in Table 25.2. Comparing the results given in Table 25.1 and Table 25.2, it can be seen that there is a significant improvement in reducing the proportional error of the mean predicted use-stress life using the deterioration-monitoring model.

25.7.3 Example 4

A batch of five bulbs was tested for $t_1 = 1.8$ h at $S_1 = 145$ V. During this period of time, their electrical resistance increased at the rate of $\Delta\Omega_1 = 0.28\%$ per h. They were then tested at $S_2 = 185$ V until failure. During the first few minutes, their resistance increased at a rate of $\Delta\Omega_2 = 4.0\%$ per h. Failure occurred

TABLE 25.2 Error in Estimating the Use-Stress Life Using the Deterioration-Monitoring Test as a Function of Sample Size

Number of Bulbs Tested to Failure (N)	Percent Error (%)
2	40
5	20
10	13
20	9
50	6

after an average time of $t_2 = 0.95$ h at 185 V. Thus, the ratio of the rate of increase of their electrical resistance at S_1 to that at S_2 is

$$m = \frac{\Delta\Omega_2}{\Delta\Omega_1} = \frac{4.0}{0.28} = 14.286$$

which means that a 1-h test at 185 V is equal to 14.286 h tested at 145 V. Therefore, the mean life at 145 V would be

$$T_1 = 1.8 + 14.826 \times 0.95 = 15.4 \text{ h}$$

The result can also be obtained by the graphical procedure shown in Figure 25.6, as follows. Draw line OA with a slope of 0.28% resistance change per 1 h of test time for a length ending at the coordinates of point A, or (1.8; 0.504). The value of 0.504 is obtained by multiplying the slope of 0.28% by the test duration for point A, which is 1.8 h, or $(0.28)(1.8) = 0.504$. At point A, draw another line with slope of 4.0% resistance change per 1 h of test time such that $CB\% = 4.0\%/h$, where AC is the additional test time at S_2 to failure of $t_2 = 0.95$ h. Then,

$$CB\% = 4.0\% (AC) = 4.0\% (0.95) = 3.80\%$$

Now the coordinates of point B become

$$(t_1 + t_2; FA + CB) \text{ or } (1.8 + 0.95; 1.8 \times [0.28\% + 3.80\%]) = (2.75 \text{ h}; 0.504\% + 3.80\%)$$
$$= (2.75 \text{ h}; 4.304\%)$$

Draw a horizontal line at point B, to intersect with the extension of line OA. Drop vertically down from this intersection and read off the corresponding abscissa value as $T_1 = 15.4$ h. The reason that the horizontal line drawn at point B, intersecting the resistance change line at use stress S_1, yields the use life is that point B represents the resistance change at failure. Consequently, point D — the intersection of the horizontal line drawn at point B (which point is the end of the extrapolated resistance-change line that leads to failure at point B) and of the resistance-change line at stress S_1 (extrapolated until it reaches this resistance change at failure) — represents the predicted life at stress S_1.

25.7.4 Applicability

The deterioration-monitoring model is an improvement over the combined-stress percent-life model, and it is applicable generally for nonthermal stresses. Table 25.2 shows that a proportional error in the life estimate of under approximately 10% would be obtained with a single sample of 20 or more units.

25.8 The Step-Stress Accelerated Testing Model

In step-stress accelerated testing, the test units are subjected to successively higher stress levels in predetermined stages. The units usually start at a lower level, and at a predetermined time or failure number, the stress is increased and the test continues. The test is terminated either: (a) when all units have failed (an uncensored test), (b) when a certain number of failures are observed, or (c) when a certain time has elapsed during which some units have not yet failed (a censored test). Step-stress testing shortens the reliability test's duration substantially. This is highly desirable in today's pressure on industry to shorten the introduction time for new products.

Depending on the units being tested, a step-stress test pattern will generally produce failures more quickly than a constant-stress test at the same stress level. However, as will be seen in this section, the analysis of the data using the step-stress model is much more complex than the analysis for the constant-stress methods.

Nelson [7] presents a statistical model for estimating a unit's life at the desired constant-stress level from step-stress test data. For many products, the life distribution used is the Weibull, with its characteristic life as an inverse power function of stress. These products include ball bearings, roller bearings, electrical insulation, dielectrics, and metal fatigue [7]. For this reason, the discussion will be centered on the Weibull distribution. However, the same methodology can also be applied to other distributions with modifications.

In this model, it is assumed that at a stress level V, the product's life follows the Weibull distribution and that the Weibull shape parameter, β, remains the same at all stress levels. This implies that at each successive stress level, the failure mode does not change, i.e., no new failure modes are introduced at the higher stress levels than those experienced at the use-stress level. It should be noted that for certain products, failures may occur at high stress levels that would not occur under normal operating conditions, i.e., the failure mode changes and β is not constant. It is therefore suggested that, before such a test is considered, data should be available to indicate whether (a) the units tested comply with this assumption and (b) the imposed stresses are within an *envelope* of stresses that comply with this assumption.

From the inverse power law, the scale parameter, η, of the Weibull distribution, can be expressed as an inverse power function of the stress, V, or

$$\eta(V) = \frac{1}{KV^n} \tag{25.13}$$

where K and n are parameters characteristic of the unit's design geometry, fabrication method, and test method employed.

The fraction of the units failing by time T under a constant stress V, is given by

$$F(T; V) = 1 - R(T; V) \tag{25.14}$$

where

$$R(T; V) = e^{-\left[\frac{T}{\eta(V)}\right]^\beta}$$

Combining Equation 25.13 and Equation 25.14 yields

$$F(T; V) = 1 - e^{-\left[\frac{T}{1/KV^n}\right]^\beta} = 1 - e^{-(KV^n T)^\beta} \tag{25.15}$$

To analyze the data from a step-stress test, a cumulative-exposure model is needed. Such a model relates the life distribution of the units, in this case the Weibull, at one stress level to the distribution at the next stress level. The same methodology can also be applied to the exponential distribution, with $\beta = 1$. This model assumes that the remaining life of the test units depends only on the *cumulative exposure* the units have seen. The units do not remember how such exposure was accumulated [7]. Moreover, since the units are held at a constant stress at each step, the surviving units will fail according to the distribution at the current step, but with a starting age corresponding to the total accumulated time up to the beginning of the current step.

Figure 25.8a presents a step-stress pattern with four steps at four different stress levels (V_1, V_2, V_3, and V_4), and the failure and censoring times of the test units. Figure 25.8b shows the *cdf* (cumulative distribution function) corresponding to each one of the four constant-stress levels. The combined *cdf* for the step-stress

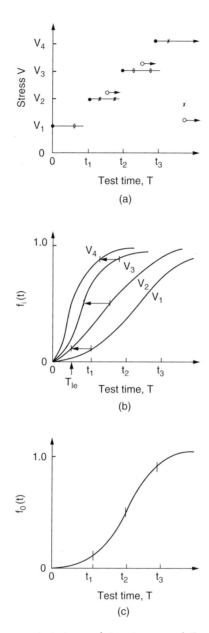

FIGURE 25.8 Relationship between constant-stress and step-stress cumulative distributions, as a function of test time. (Nelson, W.B., Statistical Methods for Accelerated Life Test Data: the Inverse Power Law, Technical Information Series Report 71-C-011, General Electric, 1970. With permission.)

pattern is a combination of the *cdf* at each stress. The first arrow in Figure 25.8b shows that the units first follow the *cdf* of the first stress level, V_1, until t_1, at which time the stress changes to V_2. When stress V_2 is applied, the surviving units continue along the life distribution of stress V_2, but starting at the fraction failed at stress V_1. The same is repeated for the higher stress levels, as can be seen for end times t_2 and t_3. Figure 25.8c shows the composite *cdf* for the step-stress pattern, which is made up of segments of the constant-stress *cdf*'s, $F_1(T)$, $F_2(T)$, etc., as shown in Figure 25.8b. This composite *cdf* represents the *cdf* for these units for the last step shown, in this case V_4. Such a composite *cdf* will be needed for each step in the derivations that follow, and these are denoted by $F_i(T; V_i)$, where i indicates the current step.

Accelerated Life Testing for Reliability Validation

To quantify this model, the following conventions are used:

1. N_i units are tested at each stress level V_i.
2. Stress levels are increased after a period of time t_i.
3. Recorded failures are denoted by $T_{i;j}$, where j indicates the failure number and i indicates the step number. Also, the times are recorded from the beginning of each step.
4. To reduce the variables that need to be carried through, and to simplify the model, the failures recorded in each step are recorded by the *total time* in that step and *not* the cumulative time from the beginning of the test. For example, consider a simple censored step-stress test with two stresses, V_1 and V_2. In general, the start of the test is at $t_0 = 0$ h. Now, assume that six units are in the test and that three failed at 50, 100, and 150 h. Immediately after the third failure is observed, the stress is increased to level V_2, and the test is continued. The remaining three units fail after the total cumulative test hours of 160, 170, and 185 h. In this model, the times to failure are recorded as follows:

Failures 1, 2, and 3 occur at $T_{1;1} = 50$, $T_{1;2} = 100$, and $T_{1;3} = 150$ h, respectively. Failures 4, 5, and 6 occur at stress level 2 after operating for 150 h at stress level 1, and these are recorded as $T_{2;1} = 10$, $T_{2;2} = 20$, and $T_{2;3} = 35$ h, respectively (noting that this step started at 150 h, or $T_{2;1} = 160 - 150 = 10$ h, and so forth). Also, for this case t_1, the time the units spend at level V_1, is 150 h, while t_2, the time the units spend at stress level V_2, is $t_2 = 185 - t_1 = 35$ h.

Now the fraction of the units failing under step 1 by time T is

$$F_1(T; V_1) = 1 - e^{-\left(KV_1^n T\right)^\beta}, \quad 0 \leq T \leq t_1$$

When Step 2 starts, the units have accumulated an equivalent age, T_{1e}, which would have produced the same fraction failed as seen at the end of Step 1, or

$$F_2(T_{1e}) = F_1(t_1)$$

or

$$1 - e^{-\left(KV_2^n T_{1e}\right)^\beta} = 1 - e^{-\left(KV_1^n t_1\right)^\beta} \qquad (25.16)$$

based on the conservation-of-reliability principle.

Solving Equation 25.16 for T_{1e} yields

$$e^{-\left(KV_2^n T_{1e}\right)^\beta} = e^{-\left(KV_1^n t_1\right)^\beta}$$

$$\left(KV_2^n T_{1e}\right)^\beta = \left(KV_1^n t_1\right)^\beta$$

$$V_2^n T_{1e} = V_1^n t_1$$

or

$$T_{1e} = \left(\frac{V_1^n}{V_2^n}\right) t_1 \qquad (25.17)$$

Then, the fraction of units failing at step 2 is given by

$$F_2(T; V_2) = 1 - e^{-\left\{KV_2^n\left[T + \left(\frac{V_1^n}{V_2^n}\right) t_1\right]\right\}^\beta}, \quad 0 \leq T \leq t_2$$

and, by the same reasoning as for Equation 25.16, using Equation 25.17 yields

$$F_3(T_{2e}) = F_2(t_2 + T_{1e}) = F_2\left[t_2 + \left(\frac{V_1^n}{V_2^n}\right)t_1\right]$$

or

$$1 - e^{-(KV_3^n T_{2e})^\beta} = 1 - e^{-\left\{KV_2^n\left[t_2 + \left(\frac{V_1^n}{V_2^n}\right)t_1\right]\right\}^\beta} \tag{25.18}$$

Solving Equation 25.18 for T_{2e} yields

$$e^{-(KV_3^n T_{2e})^\beta} = e^{-\left\{KV_2^n\left[t_2 + \left(\frac{V_1^n}{V_2^n}\right)t_1\right]\right\}^\beta}$$

$$\left[KV_3^n T_{2e}\right]^\beta = \left[KV_2^n\left(t_2 + \frac{V_1^n}{V_2^n}t_1\right)\right]^\beta$$

$$KV_3^n T_{2e} = KV_2^n\left(t_2 + \frac{V_1^n}{V_2^n}t_1\right)$$

$$V_3^n T_{2e} = V_2^n\left(t_2 + \frac{V_1^n}{V_2^n}t_1\right)$$

$$V_3^n T_{2e} = V_2^n t_2 + V_1^n t_1$$

or

$$T_{2e} = \frac{t_2 V_2^n + T_1 V_1^n}{V_3^n}$$

Finally,

$$F_3(T; V_3) = 1 - e^{-\left[KV_3^n\left(T + \frac{t_2 V_2^n + t_1 V_1^n}{V_3^n}\right)\right]^\beta}, \quad 0 \leq T \leq t_3$$

Extending this procedure to other steps, it can be seen that the general expression for $T_{1e}, T_{2e}, \ldots, T_{ie}$ is given by

$$T_{ie} = \frac{\sum_{k=1}^{i} V_k^n t_k}{V_{i+1}^n} \tag{25.19}$$

and for $F_i(T; V_i)$ is given by

$$F_i(T; V_i) = 1 - e^{\left[KV_i^n\left(T + T_{(i-1)e}\right)\right]^\beta} \tag{25.20}$$

Substituting Equation 25.19 into Equation 25.20 yields

$$F_i(T; V_i) = 1 - e^{\left[-KV_i^n \left(T + \frac{\sum_{j=1}^{i-1} V_j^n t_j}{V_i^n}\right)\right]^{\beta}} \tag{25.21}$$

Simplifying Equation 25.21 yields

$$F_i(T; V_i) = 1 - e^{-K^{\beta}\left[TV_i^n + \left(\sum_{j=1}^{i-1} V_j^n t_j\right)\right]^{\beta}} \tag{25.22}$$

Also, from Equation 25.22, the reliability at each step is

$$R_i(T; V_i) = 1 - F_i(T; V_i)$$

or

$$R_i(T; V_i) = e^{-K^{\beta}\left[TV_i^n + \left(\sum_{j=1}^{i-1} V_j^n t_j\right)\right]^{\beta}} \tag{25.23}$$

At this point the *cdf*, as well as the reliability, of each step in the step-stress model is quantified. Again, the fact that each T is timed from the time each step begins should be emphasized.

To effectively use this model, β, n, and K, which are taken to remain constant throughout the step-stress test, need to be determined. Once these parameters are determined, the fraction of units failing at any constant-stress level — and specifically at the use-stress level, V_u — can be determined by substituting the appropriate stress V along with the constants β, n, and K into

$$F(T; V) = 1 - e^{-(KV^n T)^{\beta}} \tag{25.24}$$

The reliability for any mission time, and at any stress level, can then be determined by

$$R(T; V) = e^{-(KV^n T)^{\beta}} \tag{25.25}$$

and the MTBF (mean time between failures) as a function of stress by

$$\text{MTBF}(V) = \left(\frac{1}{KV^n}\right)\Gamma\left(\frac{1}{\beta} + 1\right) \tag{25.26}$$

where

$$\left(\frac{1}{KV^n}\right) = \eta(V)$$

The same substitution for $\eta(V)$ can be performed in the Weibull failure-rate equation, as well as in the equations for the median, mode, etc. Note that all of the equations used thus far assume a two-parameter Weibull distribution, i.e., the Weibull location parameter, γ, is assumed to be zero and thus is omitted. The question that still remains is how to determine β, n, and K. The next section presents a method for determining these parameters.

25.8.1 Parameters Determination for the Weibull Step-Stress Model Using the MLE Method

The parameters β, n, and K can be determined using the maximum likelihood estimators (MLE) method. To proceed with the MLE method, the PDF (probability density function), $F_i(T; V_i)$, needs to be determined first. This is achieved by taking the derivative of Equation 25.22 with respect to T, or

$$f_i(T; V_i) = \beta K^\beta V_i^n \left[T V_i^n + \left(\sum_{j=1}^{i-1} V_j^n t_j \right) \right]^{\beta-1} \cdot e^{-K^\beta \left[T V_i^n + \left(\sum_{j=1}^{i-1} V_j^n t_j \right) \right]^\beta} \tag{25.27}$$

Equation 25.27 gives us the PDF at each stress level. To continue with the maximum likelihood estimation, it is now necessary to revert back to the original notation for each time to failure, T, with its identifying subscripts, i.e., $T_{i;j}$ where j indicates the failure number and i the step number. Substituting $T_{i;j}$ into Equation 25.27, readjusting the subscripts to reflect this new substitution, and simplifying, yields

$$f_i(T_{i;j}; V_i) = \beta K^\beta V_i^n \left[T_{i;j} V_i^n + \left(\sum_{q=1}^{i-1} V_q^n t_q \right) \right]^{\beta-1} \cdot e^{-K^\beta \left[T_{i;j} V_i^n + \left(\sum_{q=1}^{i-1} V_q^n t_q \right) \right]^\beta} \tag{25.28}$$

Then, the likelihood function, for an uncensored test, becomes

$$L(T; K, \beta, n) = \prod_{i=1}^{m} \prod_{j=1}^{r_i} f_i(t_{i;j}, V_i)$$

where m is the total number of steps, and r_i is the number of failures observed at each step i, or

$$L(T; K, \beta, n) = \prod_{i=1}^{m} \prod_{j=1}^{r_i} \beta K^\beta V_i^n \left[T_{i;j} V_i^n + \left(\sum_{q=1}^{i-1} V_q^n t_q \right) \right]^{\beta-1} \cdot e^{-K^\beta \left[T_{i;j} V_i^n + \left(\sum_{q=1}^{i-1} V_q^n t_q \right) \right]^\beta} \tag{25.29}$$

To avoid any confusion, it is necessary to point out that in Equation 25.29, the dummy variable i indicates the step number, and j indicates the jth failure in that step.

If the test is censored, i.e., terminated before all units failed, Equation 25.29 should be modified to reflect that. To do that, the following term should be included and multiplied with Equation 25.29:

$$[1 - F_l(T_l, V_l)]^s = [R_l(T_l, V_l)]^s \tag{25.30}$$

where the superscript s is the number of nonfailed units (surviving units), the subscript l indicates the last step, T_l indicates the test termination time, and V_l indicates the last stress level. From Equation 25.23,

$$[R_l(T_l, V_l)]^s = \left\{ e^{-K^\beta \left[V_l T_l + \left(\sum_{q=1}^{i-1} V_q^n t_q \right) \right]^\beta} \right\}^s$$

or

$$[R_l(T_l, V_l)]^s = e^{-sK^\beta \left[V_l T_l + \left(\sum_{q=1}^{l-1} V_q^n t_q\right)\right]^\beta} \tag{25.31}$$

Multiplying Equation 25.31 by Equation 25.29 yields

$$L_c(T; K, \beta, n) = e^{-sK^\beta \left[V_l T_l + \sum_{q=1}^{l-1} V_q^n t_q\right]^\beta} \cdot \prod_{i=1}^{m} \prod_{j=1}^{r_i} f_i(T_{i;j}, V_i)$$

or

$$L_c(T; K, \beta, n) = e^{-sK^\beta \left[V_l T_l + \sum_{q=1}^{l-1} V_q^n t_q\right]^\beta} \cdot \prod_{i=1}^{m} \prod_{j=1}^{r_i} \beta K^\beta V_i^n \left[T_{i;j} V_i^n + \left(\sum_{q=1}^{i-1} V_q^n t_q\right)\right]^{\beta-1} \cdot e^{-K^\beta \left[T_{i;j} V_i^n + \left(\sum_{q=1}^{i-1} V_q^n t_q\right)\right]^\beta} \tag{25.32}$$

where the subscript c is added to L to differentiate between the two likelihood functions, i.e., censored and uncensored. It should be noted that Equation 25.32 is the general expression for the likelihood function for both censored and uncensored tests, and it will be equal to Equation 25.29 when $s = 0$.

In the calculations that follow it will be assumed that the test is uncensored, or $s = 0$. Thus Equation 25.29 will be used. If needed, the reader can rederive the equations that follow, using the same procedure, but starting with Equation 25.32 if the test is censored.

Taking the natural logarithm of Equation 25.29 and simplifying yields

$$\log_e[L(T; K, \beta, n)] = \sum_{i=1}^{m} \sum_{j=1}^{r_i} \log_e f_i(T_{i;j}, V_i),$$

$$= \sum_{i=1}^{m} \sum_{j=1}^{r_i} \left\{ \begin{array}{l} \log_e(\beta K^\beta) + n \log_e V_i + (\beta-1) \log_e\left[T_{i;j} V_i^n + \left(\sum_{q=1}^{i-1} V_q^n t_q\right)\right] \\ -K^\beta \left[T_{i;j} V_i^n + \sum_{q=1}^{i-1} V_q^n t_q\right]^\beta \end{array} \right\}$$

or

$$\log_e[L(T; K, \beta, n)] = N \log_e \beta + N\beta \log_e K + n \sum_{i=1}^{m} r_i \log_e V_i$$

$$+ (\beta - 1) \sum_{i=1}^{m} \sum_{j=1}^{r_i} \log_e \left[T_{i;j} V_i^n + \sum_{q=1}^{i-1} V_q^n t_q\right]^\beta, \tag{25.33}$$

where

$$N = \sum_{i=1}^{m} r_i = \text{total number of observed failures}$$

Taking the partial derivative of Equation 25.33 with respect to each parameter, and setting each derivative equal to zero, yields the following:

1.
$$\frac{\partial \log_e[L(T; K, \beta, n)]}{\partial \beta} = \frac{N}{\beta} + N \log_e K + \sum_{i=1}^{m} \sum_{j=1}^{r_i} \log_e \left[T_{i;j} V_i^n + \sum_{q=1}^{i-1} V_q^n t_q \right]$$
$$- \sum_{i=1}^{m} \sum_{j=1}^{r_i} \left[\log_e \left(KT_{i;j} V_i^n + K \sum_{q=1}^{i-1} V_q^n t_q \right) \right] \cdot \left(KT_{i;j} V_i^n + K \sum_{q=1}^{i-1} V_q^n t_q \right) = 0,$$
(25.34)

2.
$$\frac{\partial \log e[L(T; K, \beta, n)]}{\partial K} = \frac{N\beta}{K} - \beta K^{\beta-1} \cdot \sum_{i=1}^{m} \sum_{j=1}^{r_i} \left[T_{i;j} V_i^n + \sum_{q=1}^{i-1} V_q^n t_q \right]^{\beta} = 0. \quad (25.35)$$

3.
$$\frac{\partial \log e[L(T; K, \beta, n)]}{\partial n} = \sum_{i=1}^{m} r_i \log_e V_i + (\beta - 1) \cdot \sum_{i=1}^{m} \sum_{j=1}^{r_i} \left[\frac{T_{i;j} V_i^n \log_e V_i + \sum_{q=1}^{i-1} V_q^n t_q \log_e V_q}{T_{i;j} V_i^n + \sum_{q=1}^{i-1} V_q^n t_q} \right]$$
$$- K^{\beta} \sum_{i=1}^{m} \sum_{j=1}^{r_i} \beta \left[T_{i;j} V_i^n + \sum_{q=1}^{i-1} V_q^n t_q \right]^{\beta-1} \cdot \left[T_{i;j} V_i^n \log_e V_i + \sum_{q=1}^{i-1} V_q^n t_q \log_e V_q \right] = 0.$$
(25.36)

Thus, we now have three nonlinear algebraic equations (Equation 25.34, Equation 25.35, and Equation 25.36) with three unknowns (β, K, and n). Any number of numerical methods can now be used to solve for these three unknowns. (See discussion in Example 5.)

It is very important to note that the summation in these equations is always carried out in the positive direction. For example, the following equation,

$$\sum_{q=1}^{i-1} V_q^n t_q$$

if typed in a computer software package, might, for the case where $(i-1) = 0 < q = 1$, cause the computer to sum in the negative direction, i.e., start at $q = 1$ and sum until $q = 0$, which is incorrect. During the first step, when $i = 1$, $i - 1 = 0$, and $q = 1$, the summation is zero, which correctly implies that no equivalent time has been accumulated before the first step. This fact should be kept in mind when using commercial software packages for the simultaneous solution of these equations.

Accelerated Life Testing for Reliability Validation

TABLE 25.3 Accelerated Test Data for Example 5

Step Level, i	Stress Level, V_i (kV)	Unit Number, j	Time to Failure, T_{ij} (h)	Step Time, t_i (h)
Step 1	10	1	150.0	312.5
	10	2	237.5	312.5
	10	3	312.5	312.5
Step 2	20	1	18.7	100.0
	20	2	32.5	100.0
	20	3	62.5	100.0
	20	4	100.0	100.0

25.8.2 Example 5

Seven units are subjected to a step-stress test. The use-stress level for these units is 5 kV. Table 25.3 gives the test results. The test method used, and the analysis of the results obtained, are as follows:

1. All seven units begin the test at a stress level of 10 kV. The stress level remains at 10 kV until three failures are observed.
2. Immediately after the third failure, the stress level on the remaining four units is increased to 20 kV until all units fail.

It should be noted that these units have already accumulated 312.5 h at the stress level of step 1, before continuing to operate at the stress level of step 2.

Assuming that the Weibull distribution governs, that the location parameter is zero, and that the failure mode remains the same (i.e., β is constant), determine the following:

1. Determine the parameters K, n, and β using the MLE method.
2. Determine the parameters of the Weibull distribution at the use-stress level.
3. Determine the reliability of these units for a mission of 1000 h at the use-stress level.
4. Determine the MTBF of these units at the use-stress level.
5. Determine the total clock-hours of test time consumed in this step-stress test.
6. Determine the approximate clock-hours of test time that would have been required if the complete test were done at the use-stress level.

Solutions to Example 5

1. Determine the parameters K, n, and β using the MLE method.
   There are several methods for obtaining the parameters. The simplest one is to have a computer program available that would automatically perform the calculations. This program can be written by the reader using the given equations and any valid method for solving nonlinear algebraic equations.
   If such a program is not available, numerical substitution into Equation 25.34, Equation 25.35, and Equation 25.36 can further reduce these equations to a more manageable level. However, one or more of the final equations, no matter how reduced, will end up in a form where the solution requires numerical methods, such as the Newton-Raphson method. Information about the Newton-Raphson method can be obtained at the following Web site: http://uranus.ee.auth.gr/Lessons/1/. To assist the reader in visualizing the values that should be substituted into Equation 25.34, Equation 25.35, and Equation 25.36 for this example, Table 25.3 has been prepared, which gives the values for all known variables where $N = 7$, $m = 3$, $r_1 = 3$, $r_2 = 4$.
   Substituting these values into Equation 25.34, Equation 25.35, and Equation 25.36 and solving numerically, using a commercial software package, yields $\beta = 2.07$, $n = 1.98$, and $K = 22.997 \times 10^{-6}$.
2. Determine the parameters of the Weibull distribution at the use-stress level.
   From the previous results, the Weibull pdf β parameter at the use-stress level of 5 kV (as well as all other stress levels, since β was assumed to be constant) is $\beta_u = 2.07$, and from the inverse

power law, η_u can now be calculated for the use stress of 5 kV, or

$$\eta_u = \frac{1}{KV^n} = \frac{1}{(22.997 \times 10^{-6})(5^{1.98})} = 1796.26 \text{ h}$$

3. Determine the reliability of these units for a mission of 1000 h at the use-stress level.

   The reliability for a mission of 1000 h, at the use-stress level, can be calculated either from the original Weibull reliability equation (since both parameters are now known) or by using Equation 25.25 in terms of K and n. Both methods will yield the same result. Then,

$$R(1000; 5\text{kV}) = e^{-(KV^n T)^\beta} = e^{-\left(\frac{T}{\eta_u}\right)^{\beta_u}} = e^{-\left(\frac{1000}{1796.26}\right)^{2.07}}$$

or

$$R(1000; 5 \text{ kV}) = 0.74269, \text{ or } 74.269\%$$

4. Determine the MTBF of these units at the use-stress level.

   The MTBF at the use-stress level can be calculated using Equation 25.26, or

$$\text{MTBF (5 kV)} = \left(\frac{1}{KV^n}\right)\Gamma\left(\frac{1}{\beta}+1\right)$$

$$= \eta_u \Gamma\left(\frac{1}{\beta}+1\right)$$

$$= 1796.26 \, \Gamma\left(\frac{1}{2.07}+1\right)$$

$$= 1796.26 \, \Gamma(1.48309)$$

$$= 1796.26 \, (0.8858)$$

or

$$\text{MTBF}(5 \text{ kV}) = 1591.13 \text{ h}$$

5. Determine the total clock-hours of test time consumed in this step-stress test.

   The test duration, T_d, for this step-stress test, in clock-hours, is

$$T_d = t_1 + t_2 = 312.5 + 100$$

or

$$T_d = 412.5 \text{ h}$$

6. Determine the approximate clock-hours of test time that would have been required if the complete test were done at the use-stress level.

   To determine the approximate test duration, if the test were conducted at the use-stress level, T_{d_u}, the acceleration factors, A_{F_i}, need to be determined for each step. For Step 1, the acceleration factor, A_{F_1} is

$$A_{F_1} = \left(\frac{V_1}{V_u}\right)^n,$$

$$= \left(\frac{10}{5}\right)^{1.98}$$

or

$$A_{F_1} = 3.945$$

and for Step 2, the acceleration factor, A_{F_2}, is

$$A_{F_2} = \left(\frac{V_2}{V_u}\right)^n$$
$$= \left(\frac{20}{5}\right)^{1.98}$$

or

$$A_{F_2} = 15.562$$

Then, the test duration, T_{d_u}, for this test, if performed at the use-stress level, is

$$T_{d_u} = t_1 A_{F_1} + t_2 A_{F_2}$$
$$= (312.5)(3.945) + (100)(15.562)$$

or

$$T_{d_u} = 2789 \text{ h}$$

It can be seen that the test time saved, using the step-stress method, is

$$T_{\text{Saved}} = T_{d_u} - T_d = 2789 - 412.5$$

or

$$T_{\text{Saved}} = 2376.5 \text{ h}$$

This is a large amount of time saved. Do use step-stress testing.

References

1. Kececioglu, D.B., *Reliability and Life Testing Handbook,* Vol. 2, DEStech Publications, Lancaster, PA, 2002.
2. Kececioglu, D.B. and Haugen, E., Interactions among the Various Phenomena Involved in the Design of Dynamic and Rotary Machinery and Their Effects on Reliability, 2nd technical report to ONR, Washington, DC, 1969.
3. Kececioglu, D.B., *Reliability and Life Testing Handbook,* Vol. 1, DEStech Publications, Lancaster, PA, 2002.
4. Hornbeck, R.W., *Numerical Methods,* Quantum Publishers, New York, 1975.

5. Rabinowicz, E.R., McEntire, R.H., and Shiralkar, B., A technique for accelerated life testing, *Trans. ASME*, 92B, 706–710, 1970.
6. Rabinowicz, E., Accelerated testing via deterioration monitoring, in *Proceedings of International Conference on Reliability, Stress Analysis and Failure Prevention*, ASME, pp. 247–250, 1980.
7. Nelson, W.B., Statistical Methods for Accelerated Life Test Data: the Inverse Power Law, Technical Information Series Report 71-C-011, General Electric, 1970.

26
The Role of Statistical Testing in NDA

Eric P. Fox
Veros Software

26.1	Introduction/Motivation	26-1
26.2	Statistical Distributions of Input Variables in Probabilistic Analysis	26-2
	Introduction/Importance of Correct Specification/Impact of Misspecification • Types of Statistical Distributions • Large Amount of Data • Small to Moderate Amount of Data • Little or No Data • Other Issues • Recommendations	
26.3	Output Variables in Probabilistic Analysis	26-18
	Statistically Validate Results • General Cases	
26.4	Statistical Testing for Reliability Certification	26-20
	Traditional Methods for Reliability Testing and Certification • Results of Probabilistic Analysis: Going from Output Variable to Reliability • Combination of Results: Bayesian Analysis	
26.5	Philosophical Issues	26-23
	Data Requirements for Probabilistic Analysis • Expertise Needed for Probabilistic Analysis	
26.6	Conclusions	26-24

26.1 Introduction/Motivation

Of course every chapter author would like to think that his or her chapter is most important — but this one really is the most important! All kidding aside, the role of the specification of the input variables in nondeterministic analysis (NDA) is an important one because it is precisely what makes a probabilistic analysis probabilistic (or, alternatively, what puts the "N" in NDA.). When the input distributions are specified properly, the very powerful results of NDA can be realized. Oftentimes, however, the attention to the selection of the appropriate input distributions is seriously lacking, which can lead to highly erroneous results.

The overall goal of this chapter is to give useful and meaningful recommendations on how to select appropriate distributions for input variables for an NDA. Recommendations in this chapter are based on a 90/10% rule — namely, spending 10% of one's time to get a "90% answer" is preferred to spending the remaining 90% of the time pontificating on why that "90% answer" is not 100% correct, and ultimately frustrating the readers that they can never get there. The author has also seen many cases of probabilistic practitioners working for that 100% answer only to be ignored by the end users, since meaningful results were needed very quickly.

To that end of the overall goal, this chapter is divided up into six sections. After the introduction in Section 26.1, Section 26.2 covers the importance of the correct specification of appropriate input distributions for cases where large, moderate, and little or no amounts of data exist. Section 26.3 discusses issues with the resulting output variables in generating statistically valid results. In Section 26.4, the use of NDA results in conjunction with test results for reliability certification is discussed. Section 26.5 addresses common philosophical issues in conducting a probabilistic analysis. Finally, Section 26.6 covers concluding remarks.

26.2 Statistical Distributions of Input Variables in Probabilistic Analysis

26.2.1 Introduction/Importance of Correct Specification/Impact of Misspecification

Perhaps the number one error that is made in conducting an NDA is in the correct specification of the statistical distributions of the input variables that go into the analysis. As in previous chapters, there are a wide variety of distributions that can be selected to describe the variability pattern for an input variable. One of the most common errors made in NDA is the lack of care in the proper selection of a reasonable statistical distribution for a given input variable. For example, some individuals always select a normal (Gaussian) distribution simply because it is convenient, widely understood, and easy to specify via its mean and standard deviation. In one example, a practitioner had specified a plate thickness to be normally distributed with a mean of 0.003 in. and a standard deviation of 0.001 in. When asked whether or not they knew that the normal distribution was giving a 0.001 probability of a negative thickness, they had no idea.

To illustrate the differences that can be obtained when care is not taken to specify correctly the statistical distribution of an input variable, consider a simple two-variable example that compares the tensile stress that a component sees during operation with its load-carrying capability or ultimate tensile strength. The interference region between these two variables is related to the failure probability. The distribution on the left (Figure 26.1) represents the tensile stress distribution, and the one on the right is the distribution of ultimate tensile strength. Thus, a failure will occur if the tensile stress is larger than the tensile strength. In this example, suppose that the tensile stress distribution is truly normal, with a mean $\mu = 80$ kpsi and a standard deviation $\sigma = 10$ kpsi, and the tensile strength distribution is truly beta, with minimum and maximum values of 100 kpsi and 140 kpsi, respectively, with parameters $\alpha = 2$ and $\beta = 2$. Values were simulated for both stress and strength from these two distributions. The resulting failure probability is then truly 0.0040 if the stress and strength distributions given above are correct.

But now suppose that the tensile strength distribution was incorrectly modeled as being normally distributed with a mean of $\mu = 120$ kpsi and a standard deviation $\sigma = 10$ kpsi. The resulting failure probability can be calculated in this case to be 0.0024. The first correct result yields a failure probability

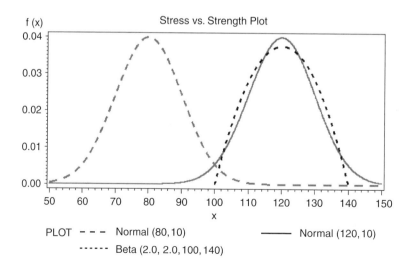

FIGURE 26.1 Stress vs. strength plot.

of 1 in 247 components, while for the latter a probability of 1 in 425 components. This is roughly a difference of nearly 2 times in magnitude, even though some might argue "a difference in the third decimal place." This difference may be surprising when both models appear to model roughly the same variation pattern in tensile strength. This underscores the importance of the distributional form. The bottom line illustrated by this simple example is that small differences in input distributions can lead to large differences in the resulting answer, in this case, the failure probability.

So what is a practitioner to do? Study a problem to death so as to select the most perfect distribution? Absolutely not! Rather, by looking at some basic properties of the variability of a given input distribution, it should be possible, with just a little bit of effort, to select an appropriate distribution for an NDA. In subsequent sections of this chapter, ways of choosing the appropriate form will be investigated so that the practitioner will be able to determine the best from among many potential distributional candidates for a specific input variable. Sections 26.2.3, 26.2.4, and 26.2.5 discuss how to do this for three distinct cases: when there is a large amount of data from which to fit a statistical distribution, when there is a small to moderate amount of data, and when there is little or no data.

26.2.2 Types of Statistical Distributions

Clearly, not all input variables in an NDA exhibit the same patterns in variability, and thus they should not all follow the same statistical distributions. Some input variables inherently exhibit a large amount of variability, and others have a very small scatter. Differences can also occur depending on whether or not the input variable distribution is symmetric about a nominal value or skewed in one direction. Some input variables are physically bounded at a minimum or maximum value, or both. Finally, some input variables have a large amount of data or experience from which to quantify their proper statistical distribution whereas others do not. Different probability models of these input variable distribution patterns are clearly required.

Before proceeding (going back to the 90/10% rule), although literally thousands of statistical distributions exist, only a handful will be considered here, since the majority of variation patterns that exist in almost any situation can be modeled properly with one of these distributions. The statistical distributions that will be considered here include: normal, lognormal, Weibull, uniform, beta, binomial, and Poisson.

The first class of distributions that can be used is the class of continuous distributions that are unbounded on at least one side. Examples of such distributions are the normal, lognormal, and Weibull. See Figure 26.2, Figure 26.3, and Figure 26.4 for sample plots of these distributions. The parameters for the normal are (μ, σ), for the lognormal are (ξ, σ), and for the Weibull are (β, η), and these are formally defined in Table 26.1.

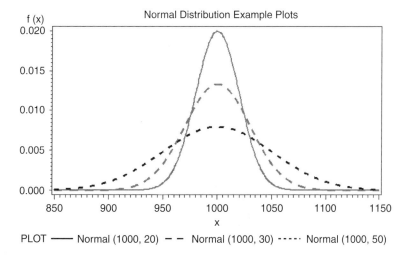

FIGURE 26.2 Normal distribution example plots.

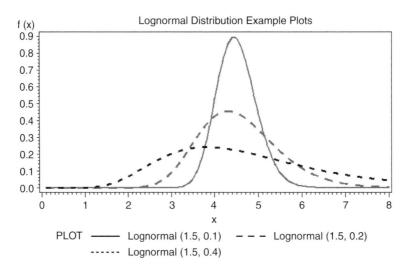

FIGURE 26.3 Lognormal distribution example plots.

Because these distributions are unbounded on at least one side, they often specify frequency information for a particular input variable in an extreme region where there is no observed data except with very large samples. As a result, these distributions should only be used in situations where there are large amounts of data or where a large experience base dictates that a particular distribution is most always correct. Certain situations fall into these categories. An example of the use of this class of distributions might be data collected within a statistical process control (SPC) program for a particular part dimension where hundreds of data points are available and there is a long history of collecting these data. Although this class of distributions is often widely used, especially the normal distribution, it is oftentimes, unfortunately, the least appropriate.

Another class of distributions is those continuous distributions that are bounded on both sides. The most common examples here are the uniform and beta distributions. See Figure 26.5 and Figure 26.6 for sample plots of these distributions. The parameters for the uniform are (a, b), and for the beta are (α, β, a, b), and these are formally defined in Table 26.1.

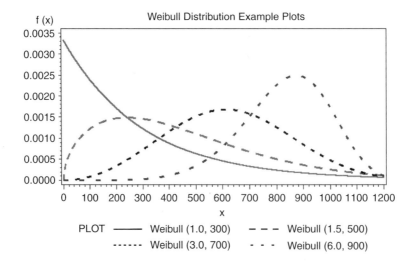

FIGURE 26.4 Weibull distribution example plots.

TABLE 26.1 Parameter Estimates for Selected Distributions

Statistical Distribution	Probability Distribution Function	Parameter Estimates
Normal	$f(x) = \dfrac{1}{\sqrt{2\pi\sigma^2}} \exp\left\{-\dfrac{(x-\mu)^2}{2\sigma^2}\right\}, \ -\infty < x < \infty$	$\hat{\mu} = \left(\dfrac{1}{n}\right)\sum_{i=1}^{n} x_i \equiv \bar{x}$ $\hat{\sigma}^2 = \left(\dfrac{1}{n}\right)\sum_{i=1}^{n}(x_i - \bar{x})^2 \equiv s^2$
Weibull	$f(x) = \dfrac{\beta x^{\beta-1}}{\eta^\beta} \exp\left\{-\left(\dfrac{x}{\eta}\right)^\beta\right\}, \ 0 < x < \infty$	$\hat{\beta} = n\left[\left(\dfrac{1}{\hat{\eta}}\right)^{\hat{\beta}} \sum_{i=1}^{n} x_i^{\hat{\beta}} \log(x_i) - \sum_{i=1}^{n} \log(x_i)\right]^{-1}$ $\hat{\eta} = \left[\dfrac{1}{n}\sum_{i=1}^{n} x_i^{\hat{\beta}}\right]^{\frac{1}{\hat{\beta}}}$
Lognormal	$f(x) = \dfrac{1}{x\sqrt{2\pi\sigma^2}} \exp\left\{-\dfrac{(\ln x - \xi)^2}{2\sigma^2}\right\}, \ 0 < x < \infty$	$\hat{\xi} = \left(\dfrac{1}{n}\right)\sum_{i=1}^{n} \ln(x_i) \equiv \bar{x}_{\ln}$ $\hat{\sigma}^2 = \left(\dfrac{1}{n}\right)\sum_{i=1}^{n}\left(\ln(x_i) - \bar{x}_{\ln}\right)^2$
Beta	$f(x) = \dfrac{\Gamma(\alpha+\beta)}{\Gamma(\alpha)\Gamma(\beta)(b-a)}\left(\dfrac{x-a}{b-a}\right)^{\alpha-1}\left(1-\dfrac{x-a}{b-a}\right)^{\beta-1},$ $a \leq x \leq b$	$\hat{\alpha} = \left(\dfrac{\bar{x}-a}{b-a}\right)\left[\dfrac{\left(\dfrac{\bar{x}-a}{b-a}\right)\left(1-\dfrac{\bar{x}-a}{b-a}\right)}{s^2} - 1\right]$ $\hat{\beta} = \left(1 - \dfrac{\bar{x}-a}{b-a}\right)\left[\dfrac{\left(\dfrac{\bar{x}-a}{b-a}\right)\left(1-\dfrac{\bar{x}-a}{b-a}\right)}{s^2} - 1\right]$ (a,b) usually specified
Uniform	$f(x) = \dfrac{1}{b-a}, \ a \leq x \leq b$	(a,b) usually specified
Binomial	$f(x) = \dbinom{n}{x} p^x (1-p)^{n-x}, \ x = 0, 1, \ldots, n$	$\hat{p} = \dfrac{x}{n}$
Poisson	$f(x) = \dfrac{\lambda^x \exp\{-\lambda\}}{x!}, \ x = 0, 1, 2, \ldots$	$\hat{\lambda} = \bar{x}$
Hyperdistribution	Problem specific	Problem specific

Since these distributions are bounded on both sides, their accuracy does not depend as much on having large amounts of data but, rather, in the accurate quantification of physical bounds. For many engineering applications, this information is readily available. Smaller amounts of data are required here because the use of the data is for specification of the most likely values between the physical bounds. In cases where it is impossible to declare one value more than another, the uniform distribution can be used. If some values have a greater chance of occurring than others, the beta distribution can be used. Many examples occur for the use of these types of models. An example for the use of this class of distributions might be for a temperature of a part where physical bounds are widely known and limited operational data exist. Although these distributions are not used as frequently as the former class, this class is oftentimes the most appropriate.

Another major category of statistical distributions includes those that are discrete. Common distributions of this type include the Poisson and binomial distributions. See Figure 26.7 and Figure 26.8 for sample plots of these distributions. The parameter for the Poisson is (λ), and parameters for the binomial

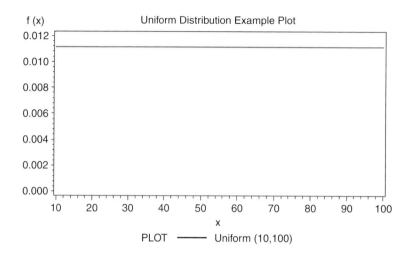

FIGURE 26.5 Uniform distribution example plot.

are (n, p), and these are formally defined in Table 26.1. Although these latter distributions are not often used in an NDA, they are included here for completeness. The Poisson distribution models the occurrence of some phenomenon such as the number of defects within a fixed region of space, such as a material specimen with particular dimensions. The binomial distribution models the number of defects or incidents in a fixed sample size with a fixed failure probability.

The final variation scheme that will be considered uses information from possibly all of the above-mentioned schemes and is referred to as hyperparameterization. Besides being an incredibly long word, this is the process whereby the parameters of the distribution are themselves allowed to vary. In other words, the parameters of the distribution are assigned distributions as well. See Figure 26.9 for an example plot of this distribution type.

Note that hyperparameterization is not necessary when there is a large amount of data or experience to quantify the distributional form. Rather, it is used when there is a small amount of data, poor quantification of physical bounds or lack thereof, or a large amount of uncertainty as to the most likely values. More will be discussed about this technique in Section 26.2.5.

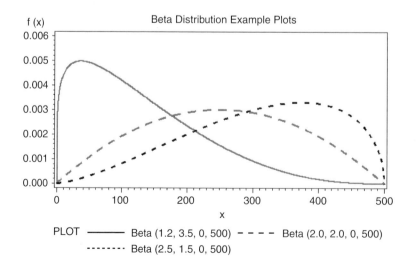

FIGURE 26.6 Beta distribution example plots.

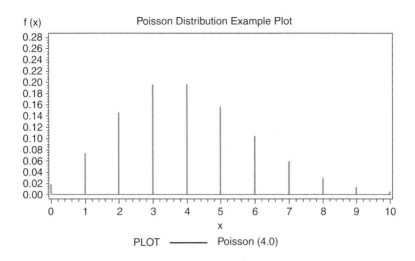

FIGURE 26.7 Poisson distribution example plot.

Table 26.2 summarizes the selection process for all classes of distributions.

As mentioned earlier, the statistical distributions mentioned above do not account for all possible variation schemes, but they do account for most. However, a quick mention is probably warranted for other types, such as those for modeling acoustics or amplitudes (Rayleigh), economic variables such as income (Pareto), very limited input information (triangular), and random processes in time such as meteorological (gamma). Other commonly known distributions such as chi-square distribution, F-distribution, and T-distribution are commonly used in the many derivations in statistics, but they are seldom used as variation patterns observed in typical input variables. The interested reader is referred to Johnson et al. [1–3] for an extensive discussion of many of these and other distributions of potential interest for less likely scenarios.

26.2.3 Large Amount of Data

Generally speaking, when there is a large amount of data available, it is most straightforward to select an appropriate statistical distribution to model the appropriate scatter in values for a given input variable.

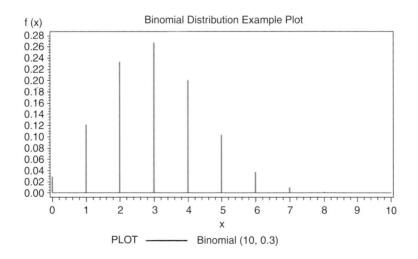

FIGURE 26.8 Binomial distribution example plot.

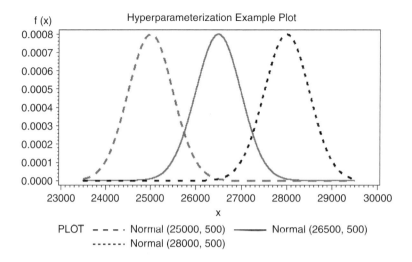

FIGURE 26.9 Hyperparameterization example plot.

For the purposes of this discussion, when it is stated that there is a "large amount" of data, it will be assumed that this will mean something on the order of many dozen to hundreds of values such as 25 to 100+. Here in this data scenario, since frequency is available in the tails of the distribution, any of the distributions could potentially be used, with the most likely being normal, Weibull, and lognormal.

An important comment to be made at this point is that the data must be relevant to the problem at hand. The importance of this assumption cannot be overstated. For example, suppose 100 data points are available for a hole diameter resulting from a drilling operation. These 100 data points could then be used to help define the appropriate distribution to use in an NDA for the hole diameter. It is important first to make sure that these 100 hole diameters were collected in conditions that are representative of the conditions assumed for the hole diameter distribution in the NDA. Were the drilling operations the

TABLE 26.2 Selection Process for All Classes of Distributions

Distribution	Typical Uses for Distribution
Normal	Large amounts of data Large amount of experience
Uniform	Small amount of data Good quantification of physical bounds Large amount of uncertainty of most likely values
Beta	Small amount of data Good quantification of physical bounds Good quantification of most likely values
Weibull	Moderate to large amounts of data Large amount of experience
Lognormal	Moderate to large amounts of data Large amount of experience
Poisson	Models total number of occurrences of some phenomenon during a fixed time period or within a fixed region of space
Binomial	Models number of defects or incidents in a fixed sample size with a fixed failure probability
Hyperdistribution	Small amount of data Poor quantification of physical bounds Large amount of uncertainty of most likely values

same for all of the data, and is this operation the one that will be used in the NDA? Are the materials the same? Does the experience of the operator have an effect? Does ambient temperature affect the operation? Is wear of the drill bit an issue? These are just a sampling of questions that would need to be addressed in this example prior to using the data (or culling out the nonrepresentative data points) to help define the appropriate distribution for the hole diameter.

Once the data are determined to be representative of the scatter of the input variable of interest, a formal analysis can be done to determine an appropriate statistical distribution to best model its variability. In many engineering applications, it should be noted that it is the *exception* rather than the rule that sufficiently large amounts of relevant data are available to discern between various distributional forms. Consider the example for the distribution of tensile strength given in Section 26.2.1 and the problem of trying to determine whether this should be modeled as a normal or a beta distribution. Again, a "large" amount of data will be defined as approximately 25 to 100+ or more data points. Such a large sample is required to *discriminate* between competing distributions. A goodness-of-fit test is the formal analysis that can be done. A goodness-of-fit test traditionally tests whether or not there is significant evidence in the data to reject the hypothesized distributional form for the data. The class of goodness-of-fit tests that will be discussed here are those that compare the empirical distribution function of the data with a given statistical distribution. The three most common of these are:

1. Kolmogorov-Smirnov goodness-of-fit test statistic (D)
2. Cramer-von Mises goodness-of-fit test statistic (W^2)
3. Anderson-Darling goodness-of-fit test statistic (A^2)

Prior to calculating them, some definitions are in order. First, define the data that are available as $x_1, x_2, \ldots, x_n$. Define the *ordered* values of these data as $x_{(1)} < x_{(2)} < \ldots < x_{(n)}$. Then the *empirical* cumulative distribution function of these data can be defined as

$$F_n(x) = \begin{cases} 0, & x < x_{(1)} \\ \dfrac{i}{n}, & x_{(i)} \leq x \leq x_{(i+1)} \\ 1, & x_{(n)} \leq x \end{cases} \quad (26.1)$$

All of the goodness-of-fit test statistics want to measure the difference between the empirical cumulative distribution function, $F_n(x)$, and the hypothesized cumulative distribution function, $F(x)$. The reader is reminded that the formal definition of the cumulative distribution function is

$$F(x) = \int_{-\infty}^{x} f(x)dx \quad (26.2)$$

where $f(x)$ is the standard probability distribution function. $F(x)$ can be thought of as giving the probability that a value of x or smaller would be observed with that distribution.

The above three goodness-of-fit statistics are defined formally as

$$D = \max_{x} |F_n(x) - F(x)|$$

$$W^2 = n \int_{-\infty}^{\infty} \{F_n(x) - F(x)\}^2 dF(x) \quad (26.3)$$

$$A^2 = n \int_{-\infty}^{\infty} \{F_n(x) - F(x)\}^2 \left(\frac{1}{F(x)[1 - F(x)]}\right) dF(x)$$

These statistics are all doing similar computations with slightly different emphasis.

TABLE 26.3 Anderson-Darling Goodness-of-Fit
Weighting vs. Percentile A^2

$F(x)$	$1/(F(x)[1-F(x)])$
0.01	101.0
0.10	11.1
0.25	5.3
0.50	4.0
0.75	5.3
0.90	11.1
0.99	101.0

The D statistic is simply looking for the maximum difference between a data point's actual percentile in its empirical distribution and its percentile in the hypothesized distribution. This statistic does not care whether this deviation occurs at a very large value of x, a very small one, or somewhere in the middle.

The W^2 statistic does something similar, except it penalizes a given deviation between a data point's actual percentile in its empirical distribution and its percentile in the hypothesized distribution by squaring the difference.

The A^2 statistic goes one step further from the W^2 statistic by penalizing this difference based on where it falls in the distribution. In particular, it scores a deviation more severely if it falls in the *tails* of the hypothesized distribution. To see this, examine Table 26.3, which computes the last term of A^2.

Note that with A^2, a given deviation that occurs at the 99th percentile of the hypothesized distribution is penalized 25 times more than the same deviation occurring at the median (101.0/4.0). A given deviation occurring at the 90th percentile of the hypothesized distribution is penalized 2.8 times more than the same deviation occurring at the median (11.1/4.0). As a result, this goodness-of-fit test statistic is particularly good for testing lack of fit in the tails of the hypothesized distribution.

As a recommendation, all goodness-of-fit tests should be used for the large data case, and the distribution that looks best according to all tests should be selected.

For computational purposes, if $z_{(i)} = F(x_{(i)})$, $i = 1, \ldots, n$, then

$$D = \max(D^+, D^-), \quad D^+ = \max_i \left\{ \frac{i}{n} - z_{(i)} \right\}, \quad D^- = \max_i \left\{ z_{(i)} - \frac{i-1}{n} \right\}$$

$$W^2 = \sum_{i=1}^{n} \left[z_{(i)} - \frac{2i-1}{2n} \right]^2 + \frac{1}{12n} \tag{26.4}$$

$$A^2 = -n - \left(\frac{1}{n}\right) \sum_{i=1}^{n} \left[(2i-1) \ln z_{(i)} + (2n+1-2i) \ln(1 - z_{(i)}) \right]^2$$

Once the three statistics are computed, what does one do with the values? The sample values are compared with tabled values (Table 26.4) of these statistics. Actually, the sample values are modified slightly to account for sample size, and if this modified value exceeds the value in the table at the indicated confidence level, the hypothesized distribution is rejected as being consistent with the data. If it does not exceed the value in the table, this gives evidence that the hypothesized distribution is consistent with the sample data. Note that this does *not* say that the data are distributed with the hypothesized distribution. It simply says that there is nothing about the data that would lead to the conclusion of rejecting the hypothesized distribution. Clearly, as the sample size becomes larger, when the hypothesized distribution is not rejected, accepting the hypothesized distribution as the actual "true" distribution is more reasonable. An excellent reference for goodness-of-fit methodologies is D'Agostino and Stephens [4].

TABLE 26.4 Critical Values for Goodness-of-Fit Test Statistics

Goodness-of-Fit Statistic	Modified Goodness-of-Fit Statistic	Confidence Level		
		90%	95%	99%
D	$D(\sqrt{n}+0.12+(0.11/\sqrt{n}))$	1.224	1.358	1.628
W^2	$(W^2-(0.4/n)+(0.6/n^2))(1.0+(1.0/n))$	0.347	0.461	0.743
A^2	No modification required	1.933	2.492	3.880

Calculation of $F_n(x)$ is straightforward in conducting goodness-of-fit tests. But how does one best go about determining the hypothesized distributions, $F(x)$? With a large amount of data, the first step in this process is to determine the distributions that should be tested. Then, in order to get specific numerical values for the parameters of these various distributions, the available data can be used to estimate these parameters. There are several methods of parameter estimation, with the most common being maximum likelihood estimation (MLE) and methods of moments (MOM). Table 26.1 shows the distributions that are being considered in this chapter along with their probability distribution functions and their parameter estimates by one of the two parameter estimation methods (all distributions in Table 26.1 are estimated with MLE with the exception of the beta distribution, which is MOM). Note that a common construct to show the estimate of a parameter, say θ, is to use the hat notation, $\hat{\theta}$. The variable θ is assumed to be an unknown population value that could be determined only with an enormous amount of data (in theory, an infinite amount), and $\hat{\theta}$ is the estimate of this unknown value with a finite sample of data. Alternatively, if there is a good physical reason for specifying a certain distribution, it can be selected as a candidate as well.

Once the parameters of the distributions have been estimated, the goodness-of-fit tests can be conducted. One issue with some of the parameter estimation techniques is that they are often very unstable or inaccurate when estimating the upper and lower bounds for some distributions. Examples include the upper and lower bounds for both the uniform and beta distributions. As a result, when distributions such as these are used, it is often necessary to specify the upper and lower bounds based on the physics of the problem and estimate the remaining parameters as indicated previously. It is often fairly straightforward to select a minimum value, below which a given input variable cannot fall, and a maximum value that the variable cannot exceed.

Example 1

Suppose that the following temperature data (in °F) exist for jet engine thermocouples:

651	680	651	655	655	628	609	629	643	680
613	663	632	657	693	653	655	676	661	675
630	671	648	664	612	668	629	624	670	660
664	639	644	661	689	649	660	626	650	612
664	623	648	621	627	677	639	671	644	698

A distribution of temperature is desired. Suppose three hypothesized distributions are selected. The first might be selected based on data physics that suggests that temperatures must fall between 600 and 700°F and should be symmetric about 650°F. That suggests a beta distribution with $\alpha = 2$, $\beta = 2$, $a = 600$, and $b = 700$. Second, a normal distribution (again symmetric) with a mean of $\mu = 650$ and a standard deviation of $\sigma = 15$ was chosen. Finally, a Weibull distribution with parameters of $\beta = 10$ and $\eta = 660$ was selected. As suggested earlier, one can also estimate the parameters of these three distributions using the maximum likelihood estimation or method of moments techniques. Using the computational equations above, the following *modified* values for D, W^2, and A^2 were obtained for each of the three distributions:

Goodness-of-Fit Test	Modified Test Statistic for Beta (2,2,600,700) Distribution	Modified Test Statistic for Normal (650,15) Distribution	Modified Test Statistic for Weibull (10,660) Distribution
Kolmogorov-Smirnov (D)	0.6487	1.2164	2.5997
Cramer-von Mises (W^2)	0.0441	0.3836	1.8391
Anderson-Darling (A^2)	0.3237	3.9888	9.3420

From the above results, it can be seen that the beta (2,2) distribution was not rejected as a distribution to model the temperature. In fact, the beta (2,2) was not even close to being rejected even at low confidence levels. This suggests that the beta (2,2) distribution is strongly consistent with the 50 data points. The normal (650,15) distribution is almost rejected at 90% confidence using D, is rejected at 90% confidence using W^2 (but not rejected at 95% confidence), and is rejected using A^2 at 95% confidence (and barely rejected at 99% confidence). This seems to suggest that although there is certainly evidence that the normal (650,15) distribution is not consistent with the data, it is just on the border of being rejected by these goodness-of-fit tests. Finally, the Weibull (10,660) distribution is strongly rejected by all three tests at the 99% confidence level. This suggests that the Weibull (10,660) is clearly inconsistent with these fifty data points and should not be used to model their variability. The bottom line from this analysis is that the beta (2,2,600,700) would clearly be the best of the three at modeling these data. As an aside, the temperature data were not actual temperature data at all. Rather, they were data that were randomly generated from a beta (2,2,600,700) distribution. It is reassuring to see that the goodness-of-fit tests picked up on the subtle differences between the distributions to select the "true" distribution of these data.

One quick mention of another tool is that of probability plots. Probability plots show both the empirical data as well as the statistical distribution on the same graph. The y-axis is the probability between 0 and 100%, and the x-axis is the input variable of interest. The y-axis is usually transformed to another grid (e.g., a log scale) so that when the distribution is plotted on the plot, it is linear. This makes it easier to see when there are deviations from the empirical distribution and statistical distribution. Figure 26.10 is an example of a lognormal probability plot of the 50 temperature data points in the previous example. The parameters of the lognormal distribution were estimated using maximum likelihood from the data and were $(\hat{\xi}, \hat{\sigma}) = (6.48, 0.0338)$. This lognormal distribution was then plotted next to the empirical distribution of the data. If the data fall close to the line, then the assumption is that the lognormal

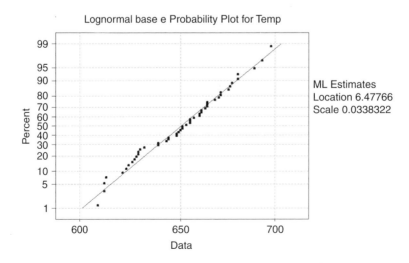

FIGURE 26.10 Lognormal probability plot.

distribution fits it well. Likewise, if the data do not fall close to the line, then the lognormal would not be a good choice to model the data. As can be seen in the probability plot, it appears that this distribution fits the data nicely, with no apparent deviation from the line. However, it is also evident that there is roughly a 0.01 chance that the lognormal distribution will produce a temperature less than 600°F and also roughly a 0.01 chance that it will produce a temperature greater than 700°F. This tool gives the user a visual perspective of the data that is useful in assessing whether or not the distribution fits well, especially in the tails. Probability plots can also be used in conjunction with goodness-of-fit tests.

26.2.4 Small to Moderate Amount of Data

Suppose now that instead of a large amount of data being available (25 to 100+ data points), that number is reduced to a much small number such as in the range of 5 to 25 data points. It is clear that there is no magic boundary between "large" and "small to moderate" amounts of data. Clearly 5 is poor and 100 would be very desirable. But what about 10? 15? 25? The answer is that it is case specific. Trying to hypothesize a very long-tailed distribution would require a much larger number of data points than trying to get acceptability with a truncated hypothesized distribution. The ranges here are meant simply as a guide. In this data scenario, any of the distributions could potentially be used, with the least likely being normal, to next likely being Weibull and lognormal, to most likely being beta and uniform.

The tools that one would use for large amounts of data are still applicable for small to moderate amounts of data. Namely, goodness-of-fit tests, parameter estimation by maximum likelihood methods or method of moments, and probability plots, all still can be used. A problem, however, arises in their use as the number of data points in the sample becomes reduced from the 25 to 100+ range down to the 5 to 25 range. To illustrate the problems that become evident with goodness-of-fit tests when the number of data points becomes small, consider the following example.

Example 2

Suppose that there are the following five values for a random variable: 2.55, 3.97, 4.74, 4.79, and 6.74. These five random values were generated from a uniform (0,10) distribution. The three hypothesized distributions are normal ($\mu = 5$, $\sigma = 1.5$), uniform ($a = 0$, $b = 10$), and Weibull ($\beta = 1.5$, $\eta = 5$). Which distribution is best? It is hoped that the uniform (0,10) distribution will be accepted and the other two will be rejected. As before, calculate the following *modified* values for D, W^2, and A^2 for each of the three distributions:

Goodness-of-Fit Test	Modified Test Statistic for Normal (5, 1.5) Distribution	Modified Test Statistic for Uniform (0, 10) Distribution	Modified Test Statistic for Weibull (1.5, 5) Distribution
Kolmogorov-Smirnov (D)	0.8555	0.7841	0.7387
Cramer-von Mises (W^2)	0.0438	0.1136	0.0918
Anderson-Darling (A^2)	0.4746	0.8106	0.7136

All of these test statistics are far below their critical values for all confidence levels in Table 26.4. This suggests that there is not significant evidence with the $n = 5$ data points for any of the goodness-of-fit tests to reject any of the hypothesized distributions. As the sample size increases, however, these tests would then begin to eliminate all but the correct distribution. Similar results would also be found for the probability plot with a small number of data points. Namely, all plots would look consistent with the observed data, and it would be difficult to eliminate any of the hypothesized distributions.

However, before assuming the worst when a small to moderate amount of data is available, try these approaches anyway. If there is a serious lack of fit with a small to moderate number of data points, these approaches will still flag them. They, however, will not be as robust as they would be in the presence of a large number of data points. As the number of data points reduces to a minimal level, they will become less and less effective until their use is of very little value.

If the goodness-of-fit test, maximum likelihood or method of moment estimates, and probability plots are not effective for a specific problem, what is the suggested approach to determine a statistical distribution for an input variable? First, the job of the practitioner becomes more difficult, since the role of expert opinion, historical information, and physics becomes more important to supplement the limited data. In some sense, this second case of "small to moderate amount of data" is really halfway between the "large amount of data" case and the "little or no data" case. The lines blur sometimes between the two cases. Tools that are clearly appropriate for the "large" case or for the "little or no" case *may* work for the "small to moderate" case.

In the case where a small to moderate amount of data is available, expert opinion should be used to help select the best distribution that would likely model the variation of the input variable if there were a large amount of data. If only the data are used, however, to estimate the parameters, large variability in the parameter estimates may result. Expert opinion can also be used to help quantify in what approximate range the parameters of the selected distribution should lie. Bayesian statistics is the mathematical method by which this expert opinion can be formally combined with the small to moderate amount of test data.

For simplicity, suppose that there is one parameter, θ, of a distribution, $f(x|\theta)$, for which there is expert opinion. The expert's opinion of the parameter θ is that it follows a distribution $\pi(\theta)$ or the *prior* distribution. Then Bayes' theorem says that the distribution of the parameter *after* having seen the small to moderate amount of data is given by

$$\pi(\theta|x) = \frac{f(x|\theta)\pi(\theta)}{\int_\Theta f(x|\theta)\pi(\theta)} \tag{26.5}$$

where Θ is the region over which θ is defined. This is also known as the *posterior* distribution.

This resulting distribution best describes the variation pattern of θ for the input variable of interest. A value for θ such as a mean or median of $\pi(\theta|x)$ can now be used in the original distribution. The interested reader is referred to Martz and Waller [5] for many examples of using Bayes' theorem, including when *all* of the parameters of the distribution have expert opinion about them. The interesting aspect about this methodology is that in the "large amount of data" case, the distribution *f* will receive most of the weight, and the effect of the prior will be minimal, as it should be. In the "little or no data" case, since there is minimal data, the effect of the prior will be very large. This case will balance the two.

Another methodology that is especially helpful in this case is a consistency check. This is important when expert opinion is involved. Namely, once a distribution of an input variable is obtained, how does one make sure that it is reasonable? One way is to ask the experts a couple of additional questions about it to ensure that it is reasonable. These additional questions involve asking what the expert would expect the probability to be for values exceeding large values of the input variable or falling below small values of the input variable. For example, if, with an expert's help, the distribution of temperature was determined to be normal with a mean of 600°F and a standard deviation of 30°F, one might ask the expert what he or she expected the probability of exceeding 700°F would be and what the probability of falling below 525°F would be. It turns out that the probability of exceeding 700°F is 0.00043, and the probability of falling below 525°F is 0.0062. If the expert's answers differed substantially, then the normal distribution that is being used is not consistent with the available information. If the expert is unable to determine probabilities that extreme, a few that are not as remote could be tried.

26.2.5 Little or No Data

Unfortunately, it is the exception rather than the rule that a large amount of "good" data exists to estimate the statistical distributions for the input variables. In this case, the number of data points is in the range of less than 5, possibly meaning zero! In these instances, the input variable distribution will have to be constructed consistent with the limited amount of information. The general rule is that the distribution must be consistent with the data, and where uncertainties exist, assumptions must be made in a conservative fashion. Here, the most likely distributions that will be used will be the beta and uniform distributions as well as hyperparameterization. The Weibull, lognormal, and normal distributions are quite difficult to justify using in this situation *unless* extremely large amounts of experience are available. A very good example of such a case exists with the Weibull distribution for failure modes in a jet engine. Extensive experience exists that the Weibull distribution accurately models these failure modes, even with a very small amount of data. The interested reader is referred to Abernethy et al. [6].

An extremely useful tool for this case is expert opinion. For example, suppose that a temperature spike in a mission profile is known to exist and varies from mission to mission. Three data points have been measured from engine tests. The measured values for these peaks are 1400, 1538, and 1727°F. Which distribution of temperature should be used? Here, nearly any hypothesized distribution that is not utterly absurd would not be rejected by the goodness-of-fit tests. Also, the maximum likelihood estimates of the parameter estimates of any distribution would also exhibit a large amount of uncertainty. So to solve this problem, a general distributional shape needs to be postulated. For example, experts familiar with the problem may agree that the most likely value (or peak of the distribution) will be the average of the observed values (i.e., 1555°F). Also, based on the physics of the problem, physical bounds exist at 1200 and 1800°F. A beta distribution will specify this type of variation pattern nicely. In Figure 26.11, a variety of beta distributions can be seen. Each of these beta distributions show the most likely value to be 1555°F, and all are bounded by 1200 and 1800°F. Only three of an infinite number of beta distributions with these properties are shown as examples. In order to select which distribution to use, the experts involved with the process feel that there is a nonremote chance that values near the extremes of 1200 and 1800°F can be obtained. Note that this eliminates distributions such as the one with $\alpha = 20.3$ and $\beta = 14.3$. Also, the experts feel that all values between 1200 and 1800°F are not equally likely, which eliminates distributions like the one with $\alpha = 1.3$ and $\beta = 1.2$. When questioned further about what was meant by "nonremote" as part of the consistency check mentioned earlier, the experts felt that there was on the order of a 1 in 100 chance that the temperature could be within 50°F of the lower physical bound of 1200°F or on the order of a 1 in 100 chance that the temperature could be within 25°F of the upper physical bound of 1800°F. After iterating with the parameters of the distribution, it was found that the beta distribution with parameters of $\alpha = 2.3$ and $\beta = 1.9$ was selected to represent this state of knowledge of temperature.

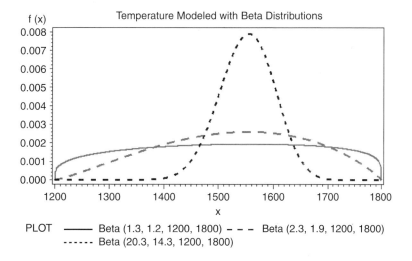

FIGURE 26.11 Temperature modeled with beta distributions.

A couple of points need to be made about the above example. First, the beta distribution with $\alpha = 2.3$, $\beta = 1.9$, $a = 1200°F$, and $b = 1800°F$ is probably not the ultimate *true* variation pattern of temperature. That is to say, if there existed 1000 temperature data points instead of three, it is likely that this particular beta distribution would not have been chosen. With 1000 data points, there would likely have been far less scatter and uncertainty in the tails of the distributions as was the case with this chosen distribution. This distribution, however, represents the best state of knowledge that exists at the time the NDA was conducted. Clearly, as more data or expertise becomes available, this distribution can be modified to reflect the new information.

The following is a second word of caution to those individuals who would like to model the above with a normal distribution. If the same data and expertise were available, a mean, μ, and standard deviation, σ, could be estimated from the above three data points. Those values would be $\hat{\mu} = 1555$ and $\hat{\sigma} = 134$. Now, when conducting a consistency check of this normal distribution, the probabilities that temperature is less than 1200°F can be calculated to be 0.0040, and the probability that temperature is greater than 1800°F is 0.0336. Unlike the beta distribution, which allows *no* probability to be beyond the physical limits, the normal distribution allows a total 0.0376 probability that the physical limits can be exceeded, a probability of 1 in 27! Note the concern is particularly troublesome in the upper tail. However, even the lower-tail probability should be of concern, as one temperature in 250 is allowed below the physical minimum. When computing overall failure probabilities in the 1/10,000 or 1/100,000 range, for example, this type of error can have a dramatic effect on the results. This is not a desirable outcome and again illustrates the problems of assuming normality in the absence of a large amount of data.

Another technique (or distributional type) that can be used in the "little or no data" situation is known as hyperparameterization. An example of this type of distribution has already been seen in Figure 26.9. Suppose that no data existed for this problem (data are rotor speeds in revolutions per minute (rpm)): but there was some expertise as to what the value would be. An expert might argue that, in an early design phase, it is not yet clear what the nominal rotor speed will be, and this nominal could go as low as 25,000 rpm and as high as 28,000 rpm, with no idea where in that range it would be. However, once a nominal is known, previous test programs always show that the speeds are normally distributed with a standard deviation (σ) of about 500 rpm. This then suggests that speed should be modeled in a hyperparametric fashion as normal (μ, 500) where the mean μ is uniform (25,000, 28,000). This captures the state of knowledge of the value of this particular input variable.

In the previous temperature example (Figure 26.11), if individuals were uncomfortable assuming fixed bounds of $a = 1200°F$ and $b = 1800°F$ of the beta distribution for temperature, bounds of simply a and b could be assumed where a and b followed other distributions. For example, a might be uniform (1175, 1225) and b might be uniform (1775, 1825). In any situation, if there is uncertainty in which assumption to make for distributional parameters in the absence of data or expertise, choose the distribution that is conservatively consistent with the available information.

The interested reader is referred to Fox and Safie [7] as an additional source of information regarding details of selecting input variables for statistical distributions.

26.2.6 Other Issues

This section discusses some other issues that sometimes come up and describes some of the traps that many practitioners fall into when selecting statistical distributions for input variables. First, not every variable is normally distributed! It has been previously noted that practitioners often select a mean and standard deviation for a normal distribution for every input variable in an NDA. Although the normal distribution is perhaps the most famous, the easiest to use, and the most widely understood of all distributions, it is also perhaps one of the least consistent with available information for many problems. The normal distribution is perhaps one of the least conservative distributions that can be used in many instances. This is because the tail of the normal distribution, compared with the tails of other common distributions, is much lighter. What this translates to is a much smaller likelihood of obtaining extreme values from the input variable distribution, which in turn usually translates to fewer "extreme" values of output variables, which implies a lower failure probability. Unless substantiated by a fairly large database (or experience base) that extends out into the tails of the distribution, the normality assumption is difficult to support. In most probabilistic analyses, the state of knowledge about many of the input variables would not support the normality assumption. Specifying every input variable as normally distributed, as some practitioners occasionally do, is usually a proxy for not thinking realistically about the problem.

Another issue worth mentioning is the common assumption that all statistical distributions for the input variables are independent of one another. This is not always the case. If knowing the value of one input variable will help in narrowing down the possible range of another input variable, the two input variables are not independent of one another. There are three common ways to account for the dependence among input variables. For purposes of discussion here, dependencies between two input variables will be discussed, although most of the techniques can be extended to multiple dependencies beyond two.

The first is to model the correlation coefficient between two input variables. The correlation coefficient measures the amount of *linear* association between two input variables and ranges from −1 to 1. A value of −1 would indicate a perfect negative relationship, while 1 would indicate a perfect positive relationship, and 0 would indicate no linear relationship. Of course, values like 0.5 would indicate that as one input variable increased, the other would increase on average as well, but it would be not a perfect one-to-one relationship. With knowledge of the correlation coefficient, ρ, one can model this structure between two input variable distributions. For example, if an input variable x_1 is normally distributed with mean μ_1 and standard deviation σ_1, and if input variable x_2 is normally distributed with mean μ_2 and standard deviation σ_2, correlated, but random, values of each can be generated as

$$\begin{aligned} x_1 &= \mu_1 + \sigma_1 z_1 \\ x_2 &= \mu_2 + \sigma_2 \left(\rho z_1 + z_2 \sqrt{1-\rho^2} \right) \end{aligned} \tag{26.6}$$

where z_1 and z_2 are random numbers from a standard normal distribution with a mean of 0 and a standard deviation of 1.

Another technique is to model the pair of input variables using a multivariate distribution function. The interested reader is referred to Kotz et al. [8] for an excellent reference on distributions such as multivariate normal, multivariate beta, etc. These distributions are general and extend beyond the two-input-variable case as well. Although conceptually simple, it is often difficult to determine complex relationships among multiple input distributions using multivariate distributions.

The third, and perhaps best, method to model the correlation structure between two input variables is to model it directly with the physics of the relationship. For example, suppose that it is known that two input variables, temperature (T) and the material property of ultimate tensile strength (S), are strongly related. In particular, ultimate tensile strength has a strong nonlinear relationship to temperature. Because the two are related in a nonlinear fashion, the correlation coefficient cannot be used to model their dependency. However, there is a data-based equation that relates these two input variables. In other words, there is a nonlinear equation that is $S(T)$. As a result, a random T can be generated, and with this value and the equation $S(T)$, a dependent S can be generated. Variability in the equation $S(T)$ for a given T could also be modeled if the relationship between S and T was not perfect.

26.2.7 Recommendations

The recommendations for the selection of which statistical distribution is best for a particular input variable are then simply a summary of the previous discussion. One can simply work from the "large amount of data" case to the "little or no data" case and implement techniques in this order in these sections. Namely, start first by estimating the parameters of the distribution by either maximum likelihood methods or method of moments and then performing goodness-of-fit tests. Probability plots can also give a visual perspective of the data to illustrate any apparent lack of fit. Expert opinion in conjunction with Bayesian analysis can also be tried for smaller amounts of data. Consistency checks can be performed in conjunction with the Bayesian analysis. Finally, hyperparameterization is a technique that can be employed when the amount of data is very small (or nonexistent), with large amounts of uncertainty.

26.3 Output Variables in Probabilistic Analysis

26.3.1 Statistically Validate Results

Actual data are often used to help determine the statistical distributions that will be used for some of the input distributions. These techniques were discussed in Section 26.2. Once an NDA is conducted, an output distribution on the output variables of interest is derived using one of the computational methods, as discussed in Chapters 14, 15, and 17–21 of this book. For example, if the output variable is stress, one can use the computed probability distribution of the stress to answer a question such as "what is the probability that stress exceeds a critical stress?" or "what is the probability that this stress exceeds the material strength?" One question that often arises is how one can "statistically validate" that the computed probability is correct. Without knowing anything else about the problem, all that can be generically stated is that this output distribution and risk statement is that which is warranted by the available information.

However, if data do exist on the output variable of interest, one can begin to address whether or not these data "statistically validate" that the resulting distribution is consistent with the data. Clearly, in a very preliminary design stage, it is unlikely that there would be any output test data available. In later states of design, the likelihood of data increases. In most cases, however, it is unlikely that there will be large amounts of data to "statistically validate" the results. There are three general cases that will be discussed in the following subsections.

26.3.2 General Cases

26.3.2.1 Data Do Not Reject NDA

There are two types of data that would typically be available in validating the results of the output distributions produced by an NDA. The first type of data is raw failure data. For example, if the output variable of interest was part lifetime and there existed actual failed times that were considered representative of the failure mode being modeled, these data could be considered. A second type of data is known as suspension data. Suspension data are lifetimes of these parts that had not yet failed. For example, if ten parts were being tested and eight of the ten had not yet failed, the current age of the eight nonfailed parts would be the suspension times, and the times on the other two parts would be two failure times.

The first case is that the suspension data do not reject the NDA. This case would occur when only suspension times occurred and were at times far less than the projected failure times from the NDA. As suspension times clearly are less than the failure times, all that can be said is that these times do not reject the NDA. This type of scenario is typical of many test programs of highly reliable products or products that take a long time to fail. See Figure 26.12 for an example of this type of data. In this example, it can be seen that the NDA predicts that the lifetimes where the failures will occur fall between roughly 700 and 1300 operating hours, whereas the test data suspension times occur in the 200 to 600-h operating range. These suspension times are certainly consistent with the failure distribution, but they do not confirm its accuracy. Confirmation would occur once these suspension times got into the 700 to 1300-h range and became failure times. The reason that they are only consistent with this failure distribution is that they are also consistent with a failure distribution where the lifetimes are double those in this example. As a result, suspension data are typically a weak source of information to validate the output distribution of the NDA.

One method to improve this situation includes pushing the suspension times further toward the failure region to see if, in fact, failures begin to occur. Many times, this is infeasible for a variety of reasons, including the cost of testing to that time duration, as the physical time that it takes for that component to accumulate that operating experience could be many months or years. In some situations, accelerated life testing is sometimes used, which tests the component in harsh (accelerated) use conditions (e.g., elevated temperature), gets an early failure in a timely fashion, and then extrapolates back to the intended use conditions via a physically based model. The interested reader is referred to Nelson [9] for a general introduction to the subject and to Meeker and LuValle [10] for an example of a very good physically based model.

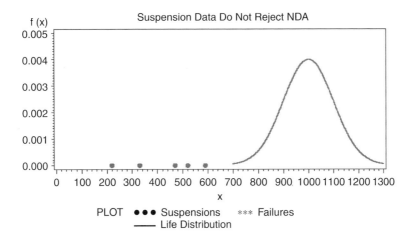

FIGURE 26.12 Suspension data do not reject NDA.

26.3.2.2 Data Reject NDA

The second case to be considered is that either the suspension data or the failure data reject the NDA. This case would occur either when suspension times occurred in the region where projected failure times were forecast from the NDA or when failure times occurred at times far less or far greater than the projected failure times from the NDA. See Figure 26.13 for an example of these types of data. In this example, it can be seen that the NDA predicts that the lifetimes where the failures will occur fall between roughly 700 and 1300 operating h, and the suspensions likewise occur in this range. In addition, the failures occur in a region between 200 and 600 h. These are both clearly at odds with the results of the NDA and invalidate the results of the NDA. Note that an assumption here is that the data are representative of the failure mode being modeled.

When the results of the NDA are invalidated by the relevant data, this suggests that errors in the NDA failure models must be corrected, additional information on input variables must be collected, and failure models must be improved. In this situation, both the failures and suspensions provide very useful information, but unfortunately, the conclusion in both cases is not a desirable one.

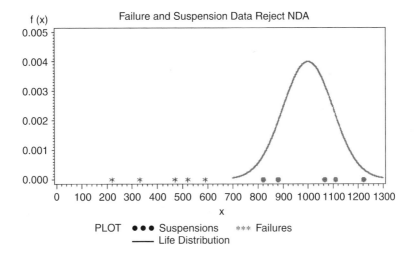

FIGURE 26.13 Failure and suspension data reject NDA.

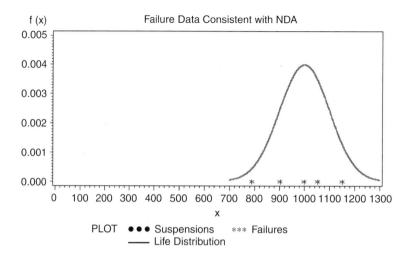

FIGURE 26.14 Failure data consistent with NDA.

26.3.2.3 Data Consistent with NDA

The final case is also the most desirable one. Namely, the failure data are consistent with the output-variable failure distribution from the NDA. This case would occur when the failure times occurred at times where the projected failure times were forecast from the NDA. See Figure 26.14 for an example of this type of data. In this example, it can be seen that the NDA predicts that the lifetimes where the failures will occur fall between roughly 700 and 1300 operating h, and the actual failures occur in a region between 700 and 1300 h also. This shows that the failure data are consistent with the NDA results. One question that then can be asked is, "How consistent is it?" or even, "Are the NDA results statistically validated?" To answer these questions, the previously discussed goodness-of-fit tests can be used to determine whether or not the distribution generated from the NDA would be rejected based on the available failure data. Clearly, if there were only a small number of failure times, it would be difficult to reject nearly *any* NDA output distribution. As the number of failure times increases, however, it would be easier for the data to reject those NDA output distributions that were not consistent. In this example, failures are a rich source of information in order to validate NDA results.

Failure data (and also suspension data, to a lesser degree) can also be combined with the NDA output distribution in a mathematically formal way using Bayesian statistical methods. More on this technique will be covered in Section 26.4.3.

26.4 Statistical Testing for Reliability Certification

26.4.1 Traditional Methods for Reliability Testing and Certification

Reliability certification is often done with testing using standard analysis techniques. Very simply, reliability certification is involved with putting some confidence on a reliability statement. For example, if 100 tests were run with one failure, no one would be comfortable saying that the reliability was certified at 99%. On the other hand, if more than 100 tests were run with one failure (say 200 or 300 or 400), then it seems that the 99% reliability would be certified at some confidence level.

The standard formula that is typically used is that if one runs n tests without failure, a reliability of R is demonstrated at $100(1-\gamma)\%$ confidence, where

$$R = \exp\left\{\frac{\ln(\gamma)}{n}\right\} \tag{26.7}$$

For example, if $n = 229$ tests were run without failure and a 90% demonstrated confidence interval on reliability was desired, this could be determined using the above equation as $R = 0.99$. This success/failure test plan is a binomial sampling scheme.

Similar formulas can be derived when one or more failures are obtained. See Lloyd and Lipow [11] for details. However, when one or more failures are observed, it becomes difficult to demonstrate *any* reasonably large reliability at a meaningful large confidence. Martz et al. [12] give a more elaborate implementation of this methodology by using testing at various subsystem and component levels and then aggregating it up to the system level for the final reliability certification number.

These results have not yet incorporated the results from the NDA. Section 26.4.3 will discuss how the results from the NDA can be combined with test data to produce a reliability certification number.

26.4.2 Results of Probabilistic Analysis: Going from Output Variable to Reliability

Prior to discussing the combination of NDA results and actual test data, a note is in order about how to calculate the actual reliability from the NDA. As was already seen in Section 26.2.1, something as simple as the tensile stress exceeding the ultimate tensile strength can allow a reliability to be calculated. The calculation of reliability does not have to assume that failure is a catastrophic event. In the examples in Section 26.3, reliability could be simply the probability that the life exceeds 700 h. As can be seen, once the final output distribution or distributions are obtained, one can then calculate the probability that a failure (or undesirable event) occurs. If it is the probability of success, it is referred to as the reliability. One minus reliability is, of course, the failure probability.

What if a *distribution* of reliability is desired? How does one compute this quantity? Clearly, if a single output distribution is computed, only one reliability or one failure probability can be generated from this scenario. In order to get a *distribution* of reliability, many output distributions of the same output variable must be generated. How are many output distributions generated? There are two standard methods.

The first method involves the use of hyperparameterization, as discussed earlier in Section 26.2.2 and Section 26.2.5. This is where some or all of the input distributions themselves have distributions on their parameters. By conducting the NDA in two stages, the desired results could be obtained. First, values of parameters would be selected for all distributions having distributions on those parameters. Once selected, this set of parameters would be used to generate all of the input variables of interest from which the distribution of the output variable or variables of interest would be determined. From this distribution, a reliability (or failure probability) could be calculated. Then another set of input variable parameters could be selected and the process repeated until numerous reliabilities are generated from which a distribution of reliability determined.

The second method is based on the consideration of all of the input distributions for which data were used in estimating the parameters of the distribution. If a large amount of data was used in estimating a particular input variable distribution, then the estimated parameters (formulas given in Table 26.1) of that distribution would have very little error in them. On the other hand, if a very small amount of data was used in estimating the parameters, they would have the potential for a large amount of error and uncertainty associated with them. For example, consider the normal distribution that is defined by the parameters (μ, σ). These are estimated by:

$$\hat{\mu} = \left(\frac{1}{n}\right)\sum_{i=1}^{n} x_i \equiv \bar{x}$$

$$\hat{\sigma}^2 = \left(\frac{1}{n}\right)\sum_{i=1}^{n} (x_i - \bar{x})^2 \equiv s^2$$

(26.8)

However, these estimates have uncertainty in them based on the finite amount of data (sample size of *n*) that was used in estimating them. It turns out that the distributions of these statistics are actually jointly distributed as multivariate normal; in particular, the joint distribution of $(\hat{\mu}, \hat{\sigma})$ turns out to be asymptotically multivariate normal. In fact, *any* distributional set of parameters that is estimated by the maximum likelihood approach will be asymptotically multivariate normal. The interested reader is referred to a text such as Lawless [13] for additional information on the specific formulas to implement this structure.

As a result, a random $(\hat{\mu}, \hat{\sigma})$ pair can be generated from the above distribution, and values of the input variable that is being modeled by this normal distribution in the NDA can be generated. A distribution of the output variable is then generated from this, and a reliability can be produced. Another random pair can then be generated, resulting in another reliability, and this process is continued until a distribution of reliability is complete.

A more complete reference on the distributions of parameters of distributions (both with asymptotic results and exact results) is Johnson et al. [1, 2]. As will be seen in Section 26.4.3, having a distribution of reliability is important in combining results from an NDA with actual test data.

26.4.3 Combination of Results: Bayesian Analysis

Now it is possible, using the previously developed information, to mathematically combine the results of the NDA with actual test data. The formal mathematical combination formula is known as Bayes' theorem. What will be covered in this section is just a brief introduction to a subject for which an entire book could be written. In fact, there are several fine books on Bayesian analysis, and one of the better ones for reliability work is by Martz and Waller [5].

Once an NDA is conducted, a distribution on an output variable is generated, and subsequently, a distribution on reliability (or failure probability) is generated. It is possible to combine this with actual test data to produce an updated distribution on reliability. Formally, let $\pi(p)$ be the *prior* distribution of the failure probability. The prior distribution of the failure probability is the same as the distribution that is generated from the NDA. It is called the prior distribution because it is the distribution of failure probability *prior to* seeing any actual failure data by which to assess whether or not the distribution is accurate. Define the distribution of the failure data as $f(x|p)$ as the likelihood of the failure data. The *posterior* distribution, denoted by $\pi(p|x)$, is the distribution of *p after* seeing all of the data. The posterior distribution can be calculated as

$$\pi(p|x) = \frac{f(x|p)\pi(p)}{\int_P f(x|p)\pi(p)} \tag{26.9}$$

where P is the region over which the prior distribution is defined.

With this posterior distribution, a demonstrated failure probability (or reliability) can be generated. For example, to generate a 90% confidence failure probability, the 90th percentile of the posterior distribution could be quoted. In general, the demonstrated failure probability *B* at 100 (1 − γ)% confidence would be the *B* that satisfies

$$\int_0^B \pi(p|x)dp = 1-\gamma \tag{26.10}$$

The demonstrated reliability at 100 (1 − γ)% confidence would then be 1 − *B*.

Example 3

Suppose that an NDA has been conducted to determine the reliability of a rocket launch. An NDA generated the result that the failure probability is uniform between 0 and 3%. In addition to the NDA results, there have been 69 successful rocket launches, with no observed failures. Ignoring the results of the NDA, what is the demonstrated reliability of this rocket at 90% confidence? Combining both sources of information, what is the updated demonstrated reliability at 90% confidence? What is the "value" of adding the NDA vs. just the 69 successful tests?

Using the equation given earlier in Section 26.4.1, and using only the 69 successful rocket launches with no observed failures, the demonstrated reliability at 90% confidence would be

$$R = \exp\left\{\frac{\ln(0.1)}{69}\right\} = 0.9672$$

Next, to account for the NDA results, the prior distribution of p can be defined as

$$\pi(p) = \frac{1}{0.03}, \quad 0 \leq p \leq 0.03$$

It should be noted that the likelihood of obtaining 69 tests with no failures is nothing more than a binomial distribution, which can be written as

$$f(x|p) = \binom{69}{0} p^0 (1-p)^{69-0} = (1-p)^{69}$$

The posterior distribution can then be written as

$$\pi(p|x) = \frac{(1-p)^{69} \frac{1}{0.03}}{\int_0^{0.03} (1-p)^{69} \frac{1}{0.03} dp} = 79.41(1-p)^{69}$$

To calculate the demonstrated reliability, the B is found that satisfies

$$\int_0^B 79.41(1-p)^{69} dp = 1 - 0.10$$

This is satisfied when $B = 0.0223$. Note that this is 90% *upper* confidence interval on the failure probability, so that in order to get a 90% *lower* bound on reliability, the result would $1 - B = 0.9777$. Note that although this integral could be evaluated in closed form, most will need to be computed numerically.

The "value" of the NDA vs. just the 69 successful tests is now clear. The demonstrated reliability with just the success testing was 0.9672, and that improved to 0.9777 with the addition of the results of the NDA. These differences could, of course, be traded off economically. By using the NDA, fewer than 69 successful tests without failure could be conducted in order to achieve the original 0.9672 demonstrated reliability. Or, as in this example, the 69 successful tests could still be used in conjunction with the NDA to obtain a higher demonstrated reliability. More details of this example can be found in Fox [14].

Besides the references mentioned earlier, the interested reader is also referred to work done at the Jet Propulsion Laboratory by Moore et al. [15] for additional information on the combination of the two sets of results.

26.5 Philosophical Issues

26.5.1 Data Requirements for Probabilistic Analysis

As its title suggests, this section will address those topics that are less technical in nature but nonetheless seem to be raised from time to time for the practitioner. One group of categories includes issues relating to data requirements for an NDA. One comment is that probabilistic analyses are meaningless without sufficient data to quantify the distributions of interest. This could not be further from the truth! That would be analogous to saying that standard deterministic analyses based on little or no data are meaningless. Clearly, a probabilistic analysis with thousands of data points to quantify all of the input variable distributions of importance will be more accurate and precise than one where large amounts of uncertainty exist because little or no data were available to conduct the same distributions. But because these models are quantifying the uncertainty in the resulting output variables of interest based on the current state of knowledge of the uncertainties going into the models, it is perfectly acceptable to conclude (and quantify) the resulting variability of the output variable as a result of the lack of data, for example, in some of the input variables.

Again, the distribution of a particular output variable such as stress or life is affected by numerous factors, including:

1. Natural or inherent variation in an input variable, such as due to manufacturing variability
2. Uncertainties in computer models
3. Uncertainties due to use of finite sample size of data
4. Measurement errors in collecting the data

In data-rich environments with mature and well understood manufacturing processes with stable environments and computer models, clearly the natural variation of the input variables (uncertainty 1) will drive the majority of the variability of the output-variable distribution. For data-poor environments or preliminary design stages with new processes in unstable environments that are not well understood, clearly uncertainties 2, 3, and 4 will dominate.

26.5.2 Expertise Needed for Probabilistic Analysis

Another common criticism of NDA is that the practitioner has to have a Ph.D. in order to actually implement the tools required for an NDA. If the question had been asked a decade or two ago, the state of affairs at that time would have been unfortunately true. Today, however, that is no longer the case. Several mainstream software packages have been developed that allow users to specify input-variable distributions, specify their models to compute the output variables of interest, link with computer models that calculate the output variables of interest, and produce the distributions of the output variables along with various probabilities and sensitivities. Examples of some of the better software on the market include Veros Software's VeroSOLVE (see Veros [16]), Applied Research Associates' ProFES (see Cesare and Sues [17]) and Southwest Research Institute's NESSUS (see Riha et al. [18]).

Can the user simply sit down and start using one of these probabilistic packages immediately with no training? Of course not. That would be analogous to asking if an engineer could sit down and immediately start working in a complex finite element package for a complex problem having never used such a package before. Depending on the level of knowledge desired, most probabilistic packages could be learned in a week or less, assuming that the class included an overview of the basic concepts of probability theory.

26.6 Conclusions

The bottom line in the selection of a statistical distribution for an input variable is not that it is a complex and impossible process. Indeed, quite the opposite is true. Very standard and reasonable checks should be followed to achieve good statistical distributions for all input variables, and these can then be used to obtain meaningful results for an NDA. One should always be a "conscience consumer" of any methodologies that are used rather than a "blind follower." Following all of the strategies discussed in this chapter, as well as researching the works cited in the References section, should allow for excellent results of an NDA. Such results will quantify the state of knowledge of the input variables at any point in time, ultimately allowing for meaningful and accurate statements about reliability.

References

1. Johnson, N.L., Kotz, S., and Balakrishnan, N., *Continuous Univariate Distributions*, Vol. 1, 2nd ed., John Wiley & Sons, New York, 1994.
2. Johnson, N.L., Kotz, S., and Balakrishnan, N., *Continuous Univariate Distributions*, Vol. 2, 2nd ed., John Wiley & Sons, New York, 1995.
3. Johnson, N.L., Kotz, S., and Kemp, A.W., *Univariate Discrete Distributions*, 2nd ed., John Wiley & Sons, New York, 1993.
4. D'Agostino, R.B. and Stephens, M.A., *Goodness-of-Fit Techniques*, Marcel Dekker, New York, 1986.
5. Martz, H.F. and Waller, R.A., *Bayesian Reliability Analysis*, Krieger Publishing, Malabar, FL, 1991.
6. Abernethy, R.B. et al., *Weibull Analysis Handbook*, AFWAL-TR-83-2079, Pratt & Whitney, West Palm Beach, FL, 1983.
7. Fox, E.P. and Safie, F., Statistical characterization of life drivers for a probabilistic design analysis, in *Proceedings of 28th Joint Propulsion Conference Proceedings*, AIAA-92-3414, AIAA/SAE/ASME/ASEE, 1992.
8. Kotz, S., Balakrishnan, N., and Johnson, N.L., *Continuous Multivariate Distributions*, Vol. 1, 2nd ed., John Wiley & Sons, New York, 2000.
9. Nelson, W., *Accelerated Testing — Statistical Models, Test Plans, and Data Analyses*, John Wiley & Sons, New York, 1990.
10. Meeker, W.Q. and LuValle, M.J., An accelerated life test model based on reliability kinetics, *Technometrics*, 37(2), 133–146, 1995.
11. Lloyd, D.K. and Lipow, M., *Reliability: Management, Methods, and Mathematics*, 2nd ed., American Society for Quality Control, Milwaukee, 1984.
12. Martz, H.F., Waller, R.A., and Fickas, E.T., Bayesian reliability analysis of series systems of binomial subsystems and components, *Technometrics*, 30, 143–154, 1988.
13. Lawless, J.F., *Statistical Models and Methods for Lifetime Data*, John Wiley & Sons, New York, 1982.
14. Fox, E.P., Confidence intervals for proportions and reliability that incorporate expert judgement, *SPES News*, 1(2), pp. 12–13, 1995.
15. Moore, N. et al., An Improved Approach for Flight Readiness Certification — Methodology for Failure Risk Assessment and Fatigue Failure Mode Case Study Examples, JPL D-9592, Jet Propulsion Laboratory, California Institute of Technology, Pasadena, CA, 1992.
16. Veros Software, Development of a multi-analysis external interface for VeroSOLVE, in *2nd Annual Probabilistic Methods Conference Proceedings*, PMC-2002-02-4, Veros Software, Irvine, CA, 2002.
17. Cesare, M.A. and Sues, R.H., ProFES probabilistic finite element system — Bring probabilistic mechanics to the desktop, in *Proceedings of 40th Structures, Structural Dynamics, and Materials Conference*, AIAA-99-1607, AIAA/ASME/ASCE/AHS/ASC, 1999.
18. Riha, D.S., Thacker, B.H., Enright, M.P., and Huyse, L., Recent advances of the NESSUS probabilistic analysis software for engineering applications, in *Proceedings of 43rd Structures, Structural Dynamics, and Materials Conference*, AIAA-2002-1268, AIAA/ASME/ASCE/AHS/ASC, 2002.

27
Reliability Testing and Estimation Using Variance-Reduction Techniques

David Mease
University of Pennsylvania

Vijayan N. Nair
University of Michigan

Agus Sudjianto
Ford Motor Company

27.1	Introduction	**27**-1
27.2	Virtual Testing Using Computer Models......................	**27**-2
27.3	Physical Testing..	**27**-4
27.4	A Review of Some Variance-Reduction Techniques.....	**27**-6
	Stratified Sampling • Importance Sampling • Other Schemes • Practical Issues	
27.5	Use of Variance-Reduction Techniques with Physical Testing...	**27**-10
	Background • Importance Sampling • Stratified Sampling	
27.6	Illustrative Application...	**27**-12
	Application of Importance Sampling • Application of Stratified Sampling	
27.7	Practical Issues on Implementation.............................	**27**-17
27.8	Concluding Remarks...	**27**-17

27.1 Introduction

Global competition as well as increasing customer demands are placing constant pressures on manufacturers to reduce product-development cycle times and costs. As a result, there is considerable interest in methods for developing new products quickly and cost-effectively while also ensuring high quality and reliability. Physical testing of prototypes is the traditional approach for assessing and improving product reliability during the design and development stage. Since this is costly and time consuming, there is now increasing reliance on computer-aided-engineering (CAE) techniques for analyzing, optimizing, and proving product reliability and robustness [1, 2].

The use of computer models is also crucial in situations where important design decisions must be made up front, even before physical prototypes can be built. For example, decisions on product architecture affect downstream design considerations about manufacturing facilities, design of interfacing systems, and so on. Reversing a decision at a later stage can result in major financial consequences as well as loss of product-development time. Because of these considerations, CAE models are used extensively in the engineering design environment to simulate various tasks and predict system behavior under a variety of design and noise conditions. These results are then used to make critical engineering and financial decisions.

Virtual testing using CAE techniques is clearly a very effective approach when the underlying models are valid. In practice, however, these models are only approximations of reality. Even though a lot of attention is paid to model validation, there will always be a considerable amount of uncertainty due to various sources of error. For example, in the application in Section 27.2, "failure" in virtual testing is defined as sealing force being below a certain threshold. This definition of failure is only a surrogate for leakage, the real failure mode that can be ascertained only by physical tests. Thus, there is still a need for physical testing before final decisions on product design and reliability can be made.

The goal of this chapter is to describe some variance-reduction techniques that use information from CAE models to increase the efficiency and reduce the cost of physical testing. We describe the use of two well-known techniques called importance sampling and stratified sampling for this purpose. It turns out that the use of optimal importance sampling and stratified sampling schemes requires knowledge of unknown quantities. However, knowledge of these quantities from CAE modeling and analysis can yield "prior information" that can be used to develop good sampling strategies. The actual estimators of the reliability parameter are unbiased, even if the prior information is not very good. When the prior information is reliable, however, the estimators turn out to have much smaller variance and hence can lead to a substantial reduction in the cost of physical testing.

27.2 Virtual Testing Using Computer Models

This section provides a mathematical description of virtual testing based on computer models. We will use a real application to motivate and discuss the ideas. Figure 27.1 is a visual representation of a finite-element analysis (FEA) model for deep thermal-shock simulation of a basic engine structure, including engine block, gasket, cylinder head, and bolts. It models, among other things, the rigidity and structure of the mating components so that one can analyze their impact on sealing performance and crack size. For internal combustion engines, the architecture of the components depends on the engine block and head designs. Choices of these designs are major determinants for downstream decisions on manufacturing facilities, which are expensive and time consuming to set up. Thus, simulation models for engine block and head-joint assembly are commonly employed to make structural decisions for optimizing reliability [3].

Physical testing to validate the design integrity of head and block sealing is typically conducted in a controlled laboratory environment by running the engines with (controlled) cyclical stresses obtained

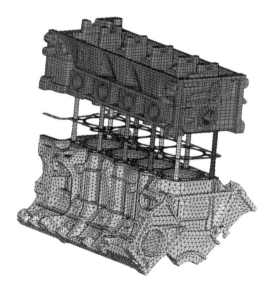

FIGURE 27.1 Finite-element analysis (FEA) model for deep thermal-shock simulation of a basic engine structure.

by varying cylinder pressure and coolant temperature. Running dynamometer tests, however, is costly and time consuming. Thus virtual testing using CAE models is an attractive alternative.

The mathematical formulation of virtual testing based on computer models can be described as follows. Let $x = (x_1, \ldots, x_p)$ denote the input variables in the computer model. This includes nominal values for the design parameters as well as noise variables that represent deviation from nominal design values, variation in manufacturing and customer-use conditions, and so on. The input space is usually very high-dimensional. In the engine thermal-shock simulation example in Figure 27.1, there were several hundred input variables corresponding to geometric features, material properties, thermal distributions, etc. Let the response of interest be denoted by $g(x)$. In our application, the responses are sealing force (unit line load of the joint sealing) and crack size.

The input–output relationship $g(x)$ is not known, but it can be represented (or approximated) through a CAE model. In most real applications, the computer model cannot be expressed analytically, but the value of $g(x)$ can be computed at chosen values of the input variables x through function evaluations. Even a single function evaluation can be very time consuming. For example, one run of the deep thermal-shock simulation with a basic engine structure (Figure 27.1) required 5 days of computing time on a high-end computing platform with 16 parallel processors.

Design engineers use CAE models and analysis for a variety of purposes. These include identifying the important input variables (factor screening), understanding how the response varies as a function of the input variables (analysis of the response surface), selecting the settings of the design variables to optimize performance, identifying regions of optimal or inadequate performance, and so on. In this chapter, we focus on another important goal, namely, that of estimating product reliability.

For the purposes of virtual testing using computer models, product reliability is defined as

$$R = P\{g(X) \geq T\}$$

for some fixed threshold value T. The probability distribution arises from the randomness in the input variables X, such as variation in materials, temperature, vibration, customer-use conditions, deviation from nominal design values, and so on. The distribution of X is usually assumed to be known. Letting $f(x)$ denote this distribution, we can write the reliability measure as an integral:

$$R = P\{g(X) \geq T\} = \int I[g(x) \geq T] f(x) dx \qquad (27.1)$$

We see from the definition of R that $\{x : g(x) \geq T\}$ defines the "success" (good) region of the input space, and its complement $\{x : g(x) < T\}$ is the "failure" (bad) region. Further, $\{x : g(x) = T\}$ is called the limit state [4]. From now on, without loss of generality, we will let $T = 0$.

Note that there is no concept of time associated with the reliability measure R. If the application is based on time, we can view R as the reliability associated with some fixed "time" point (say 3 years, 36,000 miles, or 1 million cycles of operations). Typically, however, one takes R to be the probability that the system or structure will function under selected extreme conditions (90th percentile customer-use or load conditions, accelerated environmental conditions, etc.).

There are a number of approaches in the literature for estimating the reliability R using numerical approximations and simulation techniques. If the failure region or the limit state can be specified easily, the reliability R can be computed using numerical integration. This is not the case in practice, so several approximate methods have been developed. These include first-order reliability method (FORM) [5] and second-order reliability method (SORM). These methods rely on computing the most probable (failure) point, or MPP (see Figure 27.2). The basic idea behind FORM is to approximate the function $g(x)$ by the best-fitting linear function at the MPP, use this to approximate the success/failure regions, and compute the reliability. SORM uses a second-order approximation. See Chapter 14 on first- and second-order reliability methods by Der Kiureghian in this book for more details. See also Du and Sudjianto [6] for a new technique based on saddlepoint approximation.

An alternative approach is to use Monte Carlo sampling [7–10]. The simplest or "naive" method selects n iid (independent and identically distributed) draws from the input space $X_1, \ldots, X_n$ according to the

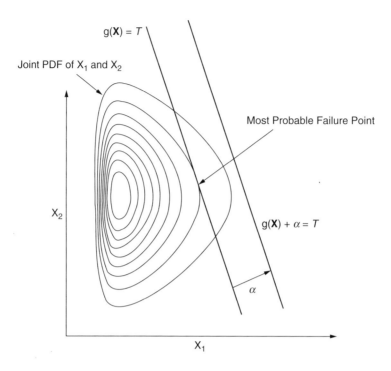

FIGURE 27.2 Computation of the most probable (failure) point.

distribution $f(x)$ and estimates R by

$$\hat{R} = \frac{1}{n}\sum_{i=1}^{n} I[g(X_i) \geq 0] \quad (27.2)$$

The variance of $\hat{R}$ is $[R(1-R)]/n$. In very-high-reliability applications where R is close to one (which, in practice, is the situation of most interest), this approach will require an excessively large number of function evaluations to get an estimate with reasonable precision. This has led to the use of "biased" sampling techniques, such as importance sampling and stratified sampling, in the literature [8, 9, 11, 12]. We provide a brief review of these in Section 27.4.

27.3 Physical Testing

For various reasons, there is a need to do physical testing even in the presence of computer models. For example, the model may not include all the important input variables; or even if it did, the functional form of $g(x)$ may not provide a good approximation. Furthermore, computer models are usually used to capture soft failures (sealing force below a threshold in the example in Figure 27.1), while we need to do physical tests to observe the actual failure mode (leakage in this case). The choice of the response $g(x)$ (sealing force) and the threshold are, in practice, surrogates for the true failure mode. Thus physical tests are needed as a way to validate the choice of the response and threshold in addition to assessing the usefulness of the computer model $g(x)$ itself.

While the output in CAE models is a continuous characteristic $g(x)$, the output in physical tests is often binary: pass/fail. This is the case with the leakage example. Even in situations where the failure can be characterized as a continuous response exceeding a threshold, one may not be able to observe the continuous response in physical tests due to measurement limitations or other practical considerations. Thus, we shall restrict attention in this chapter to physical tests with binary (pass/fail) outcomes.

Consider again the engine block and head example in Figure 27.1. Figure 27.3 shows the relationship between the CAE model and traditional physical tests in the block/head-joint sealing application.

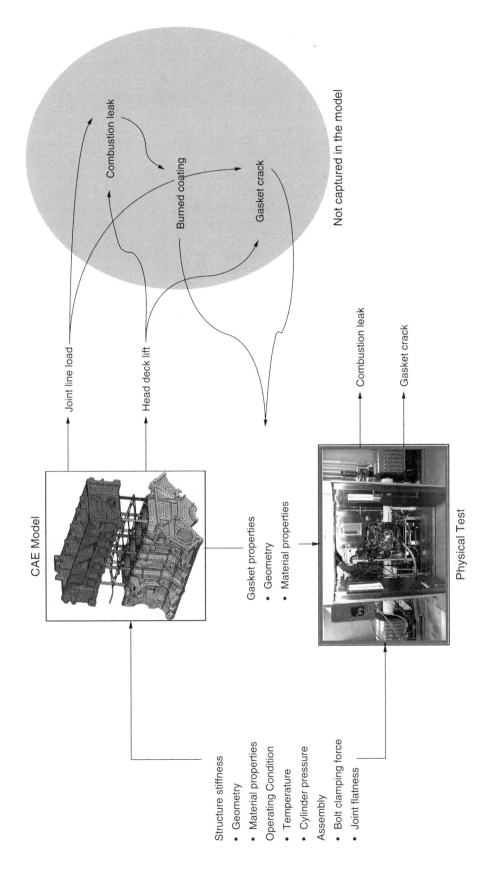

FIGURE 27.3 Relationship between the CAE model and traditional physical tests in the block/head-joint sealing application.

The input variables in the computer model include physical attributes of the product that can be measured prior to physical testing in a nondestructive manner (e.g., geometric dimension such as gasket thickness) as well as stress variables that can be controlled during the physical test (e.g., cylinder pressure and temperature). Some of the other input variables in the computer model, including material properties of the structure and gasket (such as load deflection curve), cannot be observed and controlled during physical tests. There are other variables whose effects are captured in physical tests but not necessarily in the computer model. For example, the nominal value of bolt clamp load is a design (input) variable, but its degradation over time due to thermal and cylinder pressure cycling was not captured in the computer model.

While the CAE model simulates engine operating cycles under various temperature and combustion cycles, the model does not predict leak or no leak. Instead, the computer model infers sealing failure from the joint line load and head deck lift. Reliability assessment using physical experiments, on the other hand, is based on the absence of leakage in the joint sealing for a given period of time (cycles).

Since physical testing is expensive and time consuming, we need to use prior information to the extent possible to reduce the amount of testing needed. A natural source of prior information here is the CAE models. We will describe how this can be combined effectively with variance-reduction techniques in Section 27.5.

27.4 A Review of Some Variance-Reduction Techniques

This section provides an overview of importance sampling and stratified sampling, two well-known strategies for variance reduction in the context of computer models and Monte-Carlo integration as well as survey sampling [8, 9, 13].

Let S be the p-dimensional input space, and let $f(x)$ be the distribution of the input variables. The goal is to estimate an integral of the form

$$\theta = \int k(x) f(x) dx \tag{27.3}$$

for some function $k(x)$. We assume that $k(x)$ is unknown but that we can compute $k(x)$ for any specified value of x using the computer model. A direct Monte Carlo estimate of θ can be obtained by drawing n iid samples $X_1, \ldots, X_n$ from the true (and assumed known) distribution $f(x)$, evaluating $k(X_i)$, and using

$$\hat{\theta} = \frac{1}{n} \sum_{i=1}^{n} k(X_i) \tag{27.4}$$

This is an unbiased estimator with variance

$$V = \frac{1}{n} \int [k(x) - \theta]^2 f(x) dx = \frac{1}{n} \left[\int k^2(x) f(x) dx - \theta^2 \right] \tag{27.5}$$

Our goal is to identify sampling strategies that improve upon this direct (or naive) Monte Carlo method.

For reliability estimation with virtual testing, $k(x)$ corresponds to the indicator function $I[g(x) \geq 0]$. If the outcome of the test is binary (pass/fail), the variance reduces to $[\theta(1 - \theta)]/n$.

27.4.1 Stratified Sampling

Stratified sampling is an old and extensively used technique in survey-sampling applications [13, 14]. It is well known that stratified sampling generally does much better than simple random sampling (which corresponds to the naive Monte Carlo above).

Stratified sampling can be used with computer models and Monte Carlo integration as follows:

1. Partition the input space S into L strata or cells $S_1, \ldots, S_L$. Let $p_\ell = P(X \in S_\ell)$ or relative size of the ℓth stratum.
2. Allocate the total simulation sample size n to the L strata. Let $n_1, \ldots, n_L$ be these sample sizes, with $\sum_{\ell=1}^{L} n_\ell = n$.
3. For $\ell = 1, \ldots, L$, take n_ℓ iid draws $X_1^\ell, \ldots, X_{n_\ell}^\ell$ from the ℓth stratum S_ℓ according to the density $f(x)/p_\ell$, the distribution of X truncated to be in the ℓth stratum.

Let $\hat{\theta}_\ell = \sum_{i=1}^{n_\ell} k(X_i^\ell)/n_\ell$ be the sample mean of the observations in the ℓth stratum. Then, we can estimate θ by

$$\hat{\theta}^{St} = \sum_{\ell=1}^{L} p_\ell \hat{\theta}_\ell \tag{27.6}$$

Note that

$$E(\hat{\theta}_\ell) = \mu_\ell = \frac{1}{p_\ell} \int I[x \in S_\ell] k(x) f(x) dx$$

is the conditional mean of the function $k(x)$ in the ℓth stratum. Thus, we see that $\hat{\theta}^{St}$ is an unbiased estimator of θ for any stratified-sampling scheme.

To assess the efficiency of stratified sampling, we have to consider the variance of $\hat{\theta}^{St}$. Let

$$\sigma_\ell^2 = \frac{1}{p_\ell} \int I[x \in S_\ell](k(x) - \mu_\ell)^2 f(x) dx$$

be the conditional variance within the ℓth stratum. Then, the variance of $\hat{\theta}^{St}$ is

$$V^{St} = \sum_{\ell=1}^{L} p_\ell^2 \frac{\sigma_\ell^2}{n_\ell} \tag{27.7}$$

This follows easily from the expression for $\hat{\theta}^{St}$ in Equation 27.6 and the fact that observations from the different strata are independent.

The *proportional allocation* scheme corresponds to the case with $n_\ell \propto p_\ell$. The variance of the proportional allocation estimator is

$$\frac{1}{n} \sum_{\ell=1}^{L} p_\ell \sigma_\ell^2$$

If all the within-stratum variances σ_ℓ^2 are the same and equal the overall population variance, then this is the same as the variance of the naive Monte Carlo estimator in Equation 27.5. In general, however, the within-stratum variances will be smaller, and hence the proportional allocation scheme will do better. It is well known in the statistical literature that stratified sampling with proportional allocation generally does at least as well as simple random sampling [13, 14].

For a given stratification scheme, the optimal allocation of the sample sizes can be determined easily. Let $\{p_\ell, \ell = 1, \ldots, L\}$ and $\{\sigma_\ell^2, \ell = 1, \ldots, L\}$ be the probabilities and variances associated with the strata.

Then, it is easily seen from Equation 27.7 that the allocation of sample sizes that minimizes the variance is given by $n_\ell \propto p_\ell \sigma_\ell$.

The only remaining (and most important) optimality issue is how to stratify or partition the input space S. We can see from Equation 27.7 that the variance is reduced by making the individual variances σ_ℓ^2 as small as possible. This can be accomplished by partitioning the input space so that the within-stratum variation in $k(x)$ is as small as possible. Of course, we do not know $k(x)$, but if we have some prior information or surrogates for the unknown response surface of $k(x)$, we can use it to select an appropriate stratification or partitioning scheme. Note that the estimator $\hat{\theta}^{St}$ remains unbiased regardless of the choice of the stratification scheme. If the stratification is based on good prior information, then the estimator will have much smaller variance than the naive Monte Carlo estimator.

Consider now the special case where $k(x)$ is the indicator function $I[g(x) \geq 0]$ and θ reduces to the reliability measure R in Equation 27.1. Suppose we stratify the input space into two strata and happen (by good fortune) to select $S_1 = \{x : g(x) \geq 0\}$ and $S_2 = \{x : g(x) < 0\}$. Since there is no variation in the value of $k(x) = I[g(x) \geq 0]$ within these two strata, $\sigma_\ell^2 = 0$ for $\ell = 1, 2$, and the variance of $\hat{\theta}^{St}$ is identically zero. Again, this is not practical, since the locations of these sets are unknown, but the goal is to use prior information to try and find strata that approximate these sets well.

27.4.2 Importance Sampling

The idea behind importance sampling is to sample from a "biased" distribution with density $h(x)$ instead of $f(x)$. Let X_i^h, $i = 1, \ldots, n$, be iid draws from $h(x)$, and let $k(X_i^h)$, $i = 1, \ldots, n$, be the corresponding function evaluations. Then, we can estimate θ by

$$\hat{\theta}^I = \frac{1}{n}\sum_{i=1}^{n} \frac{k(X_i^h)f(X_i^h)}{h(X_i^h)} \tag{27.8}$$

Under the assumption that $h(x) > 0$ on the support of $f(x)$, we see that $\hat{\theta}^I$ is an unbiased estimator, since

$$E(\hat{\theta}^I) = \int \frac{k(x)f(x)}{h(x)} h(x)dx = \theta$$

Further, the variance of $\hat{\theta}^I$ is given by

$$V^I = \frac{1}{n}\int \left[\frac{k(x)f(x)}{h(x)} - \theta\right]^2 h(x)dx = \frac{1}{n}\left[\int \frac{k^2(x)f^2(x)}{h(x)}dx - \theta^2\right] \tag{27.9}$$

The idea in importance sampling is to choose the biased sampling density $h(x)$ judiciously to reduce the variance V^I. The term *importance sampling* comes from the notion of giving more weight to the "important" regions (i.e., regions with more variability) of the input space of S.

To determine the optimal importance-sampling density $h(x)$, consider first the artificial situation where $k(x)$ is known. (Of course, if $k(x)$ is known, we can compute θ by direct or numerical integration, and there is no need for Monte Carlo methods.) Suppose we take $h(x) = [k(x)f(x)]/C$, where $C = \int k(x)f(x)dx$ is the proportionality constant, which happens to equal θ, the parameter of interest. (An implicit assumption here is that $k(x) > 0$, which is not a major constraint, since we can add an appropriate constant if necessary.) Then, the random variables

$$\frac{k(X_i^h)f(X_i^h)}{h(X_i^h)}$$

are all equal to the constant C. Hence, the variance of $\hat{\theta}^I$ will be identically zero. This can also be seen from the last expression in Equation 27.9 using the fact that $C = \theta$ and $\int \frac{k^2(x)f^2(x)}{h(x)} dx = \theta^2$.

This suggests that the best importance-sampling density for estimating θ should be proportional to $k(x)f(x)$, in contrast to the true density $f(x)$. This is not a practical solution in reality, since $k(x)$ is unknown. However, if we have some preliminary ideas about $k(x)$, we can use this information to construct reasonable approximations to $k(x)f(x)$ for carrying out importance sampling. This is the idea behind all practical applications of importance-sampling techniques.

27.4.3 Other Schemes

Although we have discussed the separate use of stratified-sampling and importance-sampling schemes, one can exploit the combined power of both schemes by first stratifying the input space into the L strata, allocating the total sample size to each stratum, and then using importance sampling within each stratum. The details are straightforward and omitted.

The variance-reduction schemes thus far deal with how to sample or select the units for testing. Another well-known class of schemes in the statistical literature deals with the use of variance-reduction techniques in the estimation stage. This includes ratio and regression estimators and the use of control variates. Readers are referred to Cochran [13] and Lohr [14] for details.

27.4.4 Practical Issues

As we have already noted, one needs some preliminary idea of the response surface of $k(x)$ in order to implement efficient stratified-sampling or importance-sampling schemes. It is possible to get such prior information from computer experiments of CAE models that are commonly used to explore the response surface. There is an extensive literature on computer experiments for approximating the response, including the use of Latin hypercube designs, space-filling designs, and scrambled nets [15–20]. See also Bingham and Mease [21] for a new method based on Latin hyper-rectangles. There are also methods for approximating the response surfaces using kriging or spline-based methods [15, 22]. Once we have a surrogate response surface $\hat{g}(x)$, we can use it to get approximately optimal stratified-sampling or importance-sampling schemes. Of course, the surrogate response $\hat{g}(x)$ can also be used to get initial estimates of the value of θ. The advantage of the (additional) random-sampling schemes is that they yield unbiased estimators.

There are also techniques in the literature for estimating the optimal importance-sampling density using adaptive or sequential methods. Specifically, we can update the choice of $h(x)$ sequentially as each new data point (function evaluation) becomes available. Various types of adaptive sampling strategies have been discussed in the literature. Evans and Swartz [9] use a "matching characteristics" algorithm that tries to match the moments of $h(x)$ at each stage. (See also Kloek and Van Dijk [23], Naylor and Smith [24], and Oh and Berger [25].) Oh and Berger [26] use a method that chooses $h(x)$ to minimize an estimate of the variance of the estimator, while Givens and Raftery [27] use kernel density estimates as importance-sampling functions at each iteration.

There are also a number of heuristic methods that have been discussed in the structural reliability literature for choosing $h(x)$ [28]. One idea is to take $h(x)$ to be the density $f(x)$ with its mean shifted to the most probable point (MPP). Another idea is to take $h(x)$ to be a multivariate normal density with mean and variance given by the corresponding values of $f(x)$ restricted to the failure region identified using FORM. The motivation is that sampling from these densities will give more failures, so fewer samples may be required, although there has been no formal justification for these methods.

When θ is an integral of an indicator function in Equation 27.1, we have $k(x) = I[g(x) \geq 0]$. In this case, one cannot take the optimal sampling scheme to be proportional to $k(x)f(x)$, as it does not satisfy the requirement of being strictly positive in the support of $f(x)$. For the estimator to be unbiased, any random sampling scheme has to have nonzero probability on the entire support of the input distribution $f(x)$. However, since one does not know the true $k(x)$ in any case, the optimal allocation question is only of theoretical interest. In practice, we can aim for a scheme that gives nonzero mass to the support of $f(x)$ while also trying to approximate $k(x)f(x)$ as well as possible.

27.5 Use of Variance-Reduction Techniques with Physical Testing

27.5.1 Background

As we have already noted, there are many important reasons for doing actual physical testing, even in the presence of computer models. Of course, physical testing is costly and time consuming, so there is a need for efficient and economical methods. In this section, we describe the use of prior information from CAE models to reduce variance and hence the cost of physical tests.

Many of the issues encountered in designing efficient physical testing plans are similar to those in virtual testing, in which the cost in time for evaluations of the computer model drive the need for efficiency. For instance, in virtual testing, a major difficulty in very high-reliability situations is that most of the samples drawn will not fall in the failure region. Similarly, with physical testing, high reliability implies that observing failures under standard test conditions is extremely unlikely.

The primary goal of this chapter is to describe methods for physical tests that take advantage of the ideas behind the methodology developed for variance reduction in virtual testing. Similar to the techniques described for virtual testing, efficient physical test plans will employ schemes that result in more failures than would be expected under standard test conditions. Further, these physical test plans will make use of the computer models used in virtual tests while at the same time being robust to the presence of inaccurate or incomplete models.

We consider first the technical issues. The practical details associated with implementation will be discussed in Section 27.7. As noted before, we will assume that the physical tests result in binary data: pass/fail outcomes.

In this section, we will let y denote the set of test variables that can be measured and manipulated during the physical test. These can be product attributes, test environment, and so on. They can include all or a subset of the input variables in the computer model as well as other variables. Define $p(y)$ as the probability of failure associated with the set of test variables y. Unlike virtual testing, where the probability of failure is either zero or one depending on whether the input values were in the success or failure region, we assume that the failure probability $p(y)$ in physical tests can take on any values in [0, 1]. For example, there may be some product attributes that cannot be measured without destructive testing or cannot be controlled or manipulated during tests. There may also be some environmental variables that can vary during physical tests but are not part of the computer model. Thus the failure probability $p(y)$ represents the average value obtained by integrating over the distribution of these other variables. This is illustrated in more concrete terms in Section 27.6. Finally, in the context of physical testing, $f_Y(y)$ denotes the true distribution of the values of y during manufacturing, under customer-use conditions, etc. Historical data on these variables must be used to get good estimates of the distribution.

The reliability measure R is now given by

$$R = 1 - \int p(y) f_Y(y) dy \qquad (27.10)$$

On the surface, this appears to be the same as the formulation in Section 27.4. However, in that section we were dealing with continuous responses $k(x)$, whereas now we are assuming that the outcome from physical tests is binary. As we will see below, this leads to a different optimal importance-sampling density and optimal strata than those derived in Section 27.4.

27.5.2 Importance Sampling

As before, the basic idea is to sample the values of y for physical tests from a biased distribution $h(y)$ instead of the true distribution $f_Y(y)$. Specifically, the test is based on n units or prototypes whose values y are selected according to *iid* draws from the density $h(y)$. (We will discuss practical issues associated with implementing such a scheme in Section 27.7.) Let Y_j^h, $j = 1, \ldots, n$, be these random variables. The binary responses will be denoted by Z_j, with $Z_j = 1$ if the unit fails and zero otherwise, for $j = 1, \ldots, n$. Note that the failure probability of Z_j depends on the value Y_j^h.

Then, as in Section 27.4, we can estimate the reliability by

$$\hat{R}^I = 1 - \frac{1}{n}\sum_{j=1}^{n} \frac{Z_j f_Y(Y_j^h)}{h(Y_j^h)} \quad (27.11)$$

As before, we can see that $\hat{R}^I$ is an unbiased estimator of the reliability R regardless of the choice of the importance-sampling density $h(y)$ as long as it is positive in the support of $f_Y(y)$. The variance of $\hat{R}^I$ is

$$\frac{1}{n}\left[\int \frac{p(y)f_Y^2(y)}{h(y)}dy - (1-R)^2\right]$$

Using the Schwartz inequality, we see that the optimal importance-sampling density that minimizes the variance is given by

$$h(y) \propto f_Y(y)\sqrt{p(y)}$$

The variance is reduced by sampling more heavily in regions where the failure probability is large. Note the difference between this optimal solution and that obtained with continuous observations $k(\cdot)$ in Section 27.4, where we saw that $h(\cdot) \propto f(\cdot)k(\cdot)$. This difference is due to the fact that the outcomes are now binary.

In order to find the optimal importance-sampling density $h(y)$, we can draw on information about the failure probability $p(y)$ from the computer model. Again, we emphasize that this will result in an unbiased estimator of the reliability R regardless of the accuracy of the computer model, and it will be effective at reducing the variance provided that the computer model leads to a reasonably accurate estimation of the conditional failure-probability function $p(y)$. An illustrative application is discussed in Section 27.6.

27.5.3 Stratified Sampling

We can use similar ideas for stratified sampling. Specifically, the input space of y is partitioned into L disjoint strata $S_1, \ldots, S_L$. Within the ℓth stratum S_ℓ, we draw n_ℓ independent units with values $Y_1^\ell, \ldots, Y_{n_\ell}^\ell$ from the conditional density $f_Y(y)/p_\ell$ within that stratum. From the physical tests, we observe Z_j^ℓ, which equals one if the product fails and zero if it works, for the jth observation from the ℓth stratum with value Y_j^ℓ. The estimator of the true reliability is the weighted count of successes within each stratum

$$\hat{R}_{St} = 1 - \sum_{\ell=1}^{L}\sum_{j=1}^{n_\ell} p_\ell Z_j^\ell/n_\ell$$

where $p_\ell = \int_{y\in S_\ell} f_Y(y)dy$ is the relative size (or probability) of the ℓth stratum. This estimator is unbiased regardless of the accuracy of the computer model.

The results in Section 27.4 can be used to get the optimal allocation and stratification. Since we are dealing with binary outcomes, the within-stratum variance σ_ℓ^2 is now given by $R_\ell(1-R_\ell)$, where R_ℓ is the reliability in the ℓth stratum, that is

$$R_\ell = \frac{1}{p_\ell}\int_{y\in S_\ell} [1-p(y)]f_Y(y)dy$$

Thus, the optimal allocation is given by

$$n_\ell \propto p_\ell \sqrt{R_\ell(1-R_\ell)}$$

and the optimal stratification problem reduces to finding strata $S_1, \ldots, S_L$ to minimize

$$\sum_{\ell=1}^{L} p_\ell \sqrt{R_\ell(1-R_\ell)}$$

It can be shown that the optimal strata that achieve this minimization have the form

$$S_1 = \{y : 0 \leq p(y) \leq c_1\}$$

and

$$S_\ell = \{y : c_{\ell-1} < p(y) \leq c_\ell\}$$

for $2 \leq \ell \leq L$ for some constants $c_1 < \ldots < c_L = 1$, where $p(y)$ is the failure probability. A proof for the case of $L = 2$ strata is given below. A similar argument can be used to establish the result for $L > 2$.

Fix R, p_1, and p_2. Writing R_2 as $(R - p_1 R_1)/p_2$, the problem reduces to finding the set S_1 to minimize

$$p_1 \sqrt{R_1 - R_1^2} + p_2 \sqrt{\frac{R - p_1 R_1}{p_2} - \left(\frac{R - p_1 R_1}{p_2}\right)^2} \quad (27.12)$$

where

$$R_1 = \frac{1}{p_1} \int_{y \in S_1} [1 - p(y)] f_Y(y) dy$$

Since Equation 27.12 is a concave function in R_1, it follows that it is minimized by choosing R_1 either as large or small as possible for the fixed value of p_1. This implies either

$$S_1 = \{y : 0 \leq p(y) \leq c_1\}$$

or

$$S_1 = \{y : c_1 < p(y) \leq 1\}$$

for some constant c_1. This constant is determined by the constraint

$$p_1 = \int_{y \in S_1} f_Y(y) dy$$

CAE modeling and analysis can be used to get good prior information on $p(y)$, which will lead to good stratification and sample size allocation schemes. This is demonstrated for an illustrative application below.

27.6 Illustrative Application

This section illustrates the methodology using a simple example. We will assume that the form of the function $g(x)$ is known and that failure occurs when $g(x) < 0$. However, in practice, the computer model $g(x)$ may be much more complex and have no analytical form. For such cases, one may view the $g(x)$

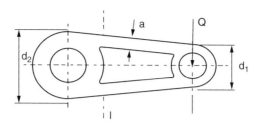

FIGURE 27.4 Automobile torque arm. (Adapted from Dimarrogons, A.D., *Machine Design: a CAD Approach*, Wiley, New York, 2001. With permission.)

here as a surrogate model for the true but unknown function obtained, for instance, by fitting a model to data from computer experiments.

The illustrative example involves an automobile torque arm pictured in Figure 27.4 and is adopted from Dimarrogons [29]. The failure mode of interest is yield failure, which occurs when $g(Q, l, d_2, a, S_y) < 0$, where

$$g(Q, l, d_2, a, S_y) = 1 - \frac{Q(2l - d_2)d_2}{4IS_y} \qquad (27.13)$$

and

$$I = \frac{a^2(d_2 - a)^2}{2} + \frac{a^4}{6}$$

Here l, d_2, and a are the dimensions pictured in Figure 27.4, Q is the force applied in the physical test, and S_y is the yield strength of the material.

The reliability of interest is that at Q equal to 6000 N, the largest force expected under standard use conditions. Unlike traditional accelerated stress testing, which would test at higher than normal levels of Q and extrapolate the results to use conditions using a regression model, we keep Q fixed at 6000 N throughout the testing. We will instead sample from a biased distribution on the product attributes d_2 and a. This can be viewed as inducing failures by testing "weaker" units.

We will initially let l be constant and equal to 120 mm. Further, we assume that S_y is normal with a mean of 170 N/mm² and a standard deviation of 4.0 N/mm², d_2 is normal with a mean of 55 mm and a standard deviation of 1.0 mm, and a is normal with a mean of 10 mm and a standard deviation of 0.25 mm. We will assume that all random variables here are independent. Using these distributions and the model above, we can compute the reliability as $R = 0.9917$. Consider now physical tests based on n test units. Traditional testing will yield an estimator with variance equal to $R(1 - R)/n$. Using the value of $R = 0.9917$ above, we get the variance as $0.0082/n$.

For designing efficient tests using importance sampling or stratified sampling, suppose that we can observe, measure, and manipulate the values of d_2 and a during physical testing but not the values of S_y. Hence, (d_2, a) correspond to the vector $\boldsymbol{y} = (y_1, y_2)$ in Section 27.5. Suppose failure/success is still given by $g(Q, l, d_2, a, S_y) < 0$, where $g(Q, l, d_2, a, S_y)$ is defined in Equation 27.13. However, since we can manipulate only d_2 and a in physical tests, the probability of failure $p(d_2, a)$ must be expressed in terms of (d_2, a) and is obtained by integrating out S_y from the failure region, i.e.,

$$p(d_2, a) = \int I[g(Q, l, d_2, a, s) < 0] f_{S_y}(s) ds.$$

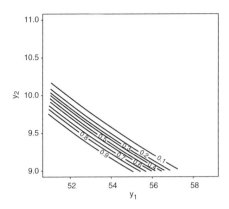

FIGURE 27.5 Contours of failure probability $p(d_2, a)$.

Recall that Q and l are assumed to be fixed constants. Using the form of $g(Q, l, d_2, a, S_y)$ in Equation 27.13 and the assumed normal distribution for S_y, we get

$$p(d_2, a) = F\left(\frac{6000(240 - d_2)d_2}{2a^2(d_2 - a)^2 + \frac{2}{3}a^4} \right) \quad (27.14)$$

where $F(x)$ is the normal cumulative distribution function (CDF) with mean 170 and standard deviation 4.0, i.e., $F(x) = \Phi[(x - 170)/4]$. The contours of $p(d_2, a)$ are shown in Figure 27.5.

We use our knowledge of $g(Q, l, d_2, a, S_y)$ to apply the variance-reduction ideas below. We will also consider sensitivity to model misspecifications. In situations where the CAE model is unknown, we can view the $g(Q, l, d_2, a, S_y)$ in Equation 27.13 as an approximation of the true model obtained from computer experiments.

27.6.1 Application of Importance Sampling

We see from Section 27.5.2 that the optimal importance-sampling density based on just (d_2, a) is given by

$$h(d_2, a) \propto f_Y(d_2, a)\sqrt{p(d_2, a)}$$

The contours of this density for this example are shown in the right panel of Figure 27.6. Comparing this with the density $f_Y(d_2, a)$ in the left panel as well as with the failure probability $p(d_2, a)$ in Figure 27.5 shows that the biased sampling density is shifted toward the (two-dimensional projection of) the failure region. In this respect, it is similar to importance sampling centered at the MPP.

Suppose we sample the attributes of (d_2, a) in *iid* fashion from this importance-sampling density. Straightforward calculations show that the variance of the importance-sampling estimator is $0.0005/n$, which is only 6% of the variance using traditional random sampling $(0.0082/n)$. In terms of costs of physical tests, this shows that we need only 6% of the test units compared with traditional physical tests.

Let us now consider some sensitivity analyses that examine how the results will change if our knowledge of the model is incorrect. We will examine the extent to which we can still achieve variance reduction for a couple of situations in which the model could be incorrect.

Suppose that our assumption that the variable l is constant is incorrect and that it in fact varies independently of the other variables according to a normal distribution with mean 120 mm and standard deviation 1.5 mm. This could be the case, for example, if $l = 120$ is the nominal design value, but the

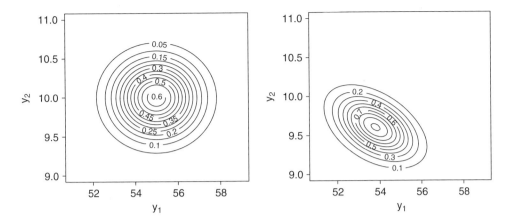

FIGURE 27.6 Contours of $f_Y(y)$ (left panel) and contours of $f_Y(y)\sqrt{p(y)}$ (right panel).

actual value of l varies during manufacturing. (This would then be an example of missing an important noise variable in the computer model.) The true reliability in this situation is $R = 0.9896$, which is less than the value of $R = 0.9917$ that would be obtained by virtual testing. Also, to get the true $p(d_2, a)$, we must integrate out the distribution of l in addition to that of S_y. Thus the specification of $p(d_2, a)$ in Equation 27.14 is incorrect. However, if we use this "wrong" guess for $p(d_2, a)$ as our prior information to get the importance-sampling density for physical tests, we still get an unbiased estimator with variance $0.0012/n$. This is still only about 10% of the variance of the estimator based on traditional physical testing, which is now $0.9896(1 - 0.9896)/n = 0.0103/n$.

A second case for sensitivity analysis is when the true function $g_T(\cdot)$ is misspecified. Specifically, suppose the true model is

$$g_T(Q, l, d_2, a, S_y) = \alpha + g(Q, l, d_2, a, S_y)$$

for some constant α where $g(Q, l, d_2, a, S_y)$ is given in Equation 27.13. There are several possible ways this can happen. For example, if the threshold constant for failure is wrong or known only approximately, this will lead to a shift α. Alternatively, this could also occur if the means of all the input variables are misspecified and the form of $g(\cdot)$ is approximately linear. Figure 27.2 provides a graphical representation of this type of model misspecification for the case where $g(x)$ is linear.

Table 27.1 provides a sensitivity analysis of the variances using traditional sampling and the (wrong) importance-sampling density based on Equation 27.14. For the values of α considered, the misspecified importance-sampling scheme still gives substantial variance reduction.

TABLE 27.1 Variances of the Importance-Sampling Estimator under Some Misspecified Models

α	Reliability (R)	Variance Using Traditional Sampling [$R(1 - R)$]	Variance Using Proposed Test Plan
0.04	0.9990	$0.0010/n$	$<0.0001/n$
0.02	0.9969	$0.0031/n$	$0.0001/n$
0.01	0.9949	$0.0051/n$	$0.0002/n$
−0.01	0.9870	$0.0128/n$	$0.0010/n$
−0.02	0.9795	$0.0201/n$	$0.0023/n$
−0.04	0.9540	$0.0438/n$	$0.0139/n$

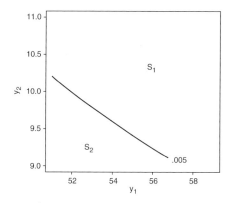

FIGURE 27.7 Optimal $L = 2$ strata.

27.6.2 Application of Stratified Sampling

We consider only $L = 2$ strata. Again, we assume that only the variables (d_2, a) can be manipulated, and so stratification can be based on only this two-dimensional space. Recall that the optimal strata have the form $S_1 = \{(d_2, a) : 0 \leq p(d_2, a) \leq c_1\}$ and $S_2 = \{(d_2, a) : c_1 < p(d_2, a) \leq 1\}$. Hence, we need to find only the value of c_1 that achieves smallest variance. A numerical search shows that this value is approximately $c_1 = 0.005$, which leads to $p_1 = 0.9248$ and $p_2 = 0.0752$.

The corresponding optimal stratification of the two-dimensional input space is shown in Figure 27.7. The conditional reliabilities within each stratum are $R_1 = 0.9998$ and $R_2 = 0.8922$. The optimal sample sizes in each stratum are $n_1 = 0.33\ n$ and $n_2 = 0.67\ n$. Based on this, we get the variance under this stratified-sampling scheme as $0.0012/n$. This is substantially smaller than the variance under traditional testing ($0.0082/n$), although the reduction is not as large as that with importance sampling.

We turn now to sensitivity analyses under misspecified models. Consider first the case where l is treated as a constant in the model but is in fact random (as described in the previous section). Stratified sampling based on the "wrong" $p(d_2, a)$ in Equation 27.14 results in a variance of $0.0022/n$. Again, this is a substantial reduction compared with traditional sampling ($0.0103/n$), although not quite as small as that using importance sampling ($0.0012/n$).

Table 27.2 provides the results of the sensitivity analyses when the true model $g_T(Q, l, d_2, a, S_y) = \alpha + g(Q, l, d_2, a, S_y)$ (as in the previous section). Again, for the for various values of α considered, the variances are substantially smaller than under traditional sampling, but not as small as with importance sampling.

As the example shows, the variance reduction achieved using stratified sampling with $L = 2$ strata is generally not as great as that obtained with importance sampling. However, it should be noted that in cases when $p(y)$ from the computer model is inaccurate enough to cause the variance of the importance-sampling

TABLE 27.2 Variances of the Stratified-Sampling Estimator under Some Misspecified Models

α	Reliability (R)	Variance Using Traditional Sampling $[R(1 - R)]$	Variance Using Proposed Test Plan
0.04	0.9990	$0.0010/n$	$0.0001/n$
0.02	0.9969	$0.0031/n$	$0.0003/n$
0.01	0.9949	$0.0051/n$	$0.0006/n$
−0.01	0.9870	$0.0128/n$	$0.0024/n$
−0.02	0.9795	$0.0201/n$	$0.0047/n$
−0.04	0.9540	$0.0438/n$	$0.0160/n$

estimator to be larger than $R(1-R)/n$, the stratified-sampling estimator still often gives substantial variance reduction. (This can be observed, for example, by increasing the standard deviation of l in the model where l is random.) In this sense, the optimal stratified-sampling plan is more robust to inaccurate CAE models, and consequently more conservative.

27.7 Practical Issues on Implementation

There are a number of practical issues in implementing the test plans described. For the variance-reduction ideas to be effective, there must exist variables y that can be measured, controlled, and manipulated during physical tests. In addition, these variables should have a significant impact on the failure probability $p(y)$. Otherwise, the variance-reduction techniques will not be very useful.

If the prior information about $p(y)$ is very poor, then one cannot expect the variance-reduction techniques to do well (although they are always unbiased). In the worst-case scenario, it is possible that the variance of the resulting estimators could be larger than that from traditional testing. However, as demonstrated in the last section, the variance-reduction techniques will yield considerable savings provided the model is not badly misspecified.

When the variables y to be manipulated during physical tests are environmental/stress variables (such as force Q in our illustrative application), it is straightforward to implement the importance-sampling and stratified-sampling schemes. However, when the variables to be manipulated are product attributes (such as product dimensions in our application in the last section), it is not as simple to implement the schemes. One must either make prototypes with these attributes or measure and select the products from the manufacturing line. This idea of inducing failure by testing such "weaker" units has not been exploited and has considerable potential.

Finally, we describe the details associated with implementing importance/stratified-sampling schemes. For both stratified sampling and importance sampling, we begin by sampling from $f_Y(y)$, the natural distribution for Y. For each product, the value of Y is measured. In stratified sampling, the products are simply sorted into strata based on their observed Y values until all strata have the required sample sizes. Once a stratum has the required number, any subsequent products with values of Y that correspond to that particular strata are not included for testing.

For importance sampling, the procedure is slightly more complex. We describe the use of a technique called *rejection sampling*. For each product sampled, a random variable U is generated uniformly from the interval $(0, 1)$. The product is then included for testing only if $U < \sqrt{p(Y)}$. It can be verified that this leads to sampling Y from the desired density $h(y) \propto f_Y(y)\sqrt{p(y)}$.

In some cases, it may be beneficial to modify these sampling strategies to be more efficient. When $p(y)$ tends to be extremely small except when $f_Y(y)$ is also extremely small, the techniques described above will result in sampling a large number of products that are not used for testing. A solution to this is to begin by sampling Y initially from a distribution that leads to larger failure probabilities $p(y)$ than under $f_Y(y)$. For instance, this could be done by manufacturing products at nominal values closer to the failure region determined by the CAE model. This manufacturing of "weaker" products would be feasible for products that are still in prototypes and are not yet in full-scale production. In order to achieve the precise sampling distributions prescribed in the previous section, rejection sampling can then be used.

27.8 Concluding Remarks

This chapter presents methods for combining information from CAE models with physical testing for efficient reliability estimation. We have demonstrated that the resulting methods yield estimators with substantially smaller variances than with traditional random testing. Further, the estimators are robust to inaccuracies in CAE models. The main reason the estimation schemes presented in this chapter are more effective than traditional testing is that they result in more failures, similar to popular methods of estimating reliability through virtual testing.

Increasing failure probability in physical testing is not new, and is common in *accelerated testing*. The key difference is that accelerated testing involves testing at more extreme stress conditions. However, physical tests are generally already designed to test at the upper limits of use conditions. Further extrapolation beyond these conditions is likely to illicit failure modes that are not of interest and would never occur in the field. The methodology presented here exploits the fact that the vector x includes strength variables in addition to stress variables, and that failures can be induced by testing "weaker" units or by placing more emphasis in the tails of the distribution of the strength variables. The notion of testing "weaker" units appears to be a new concept. By testing in the tails of the strength distribution rather than at extreme stress conditions, we can avoid extrapolating too far beyond standard use conditions and thus encountering irrelevant failure modes. Any failures that do occur in the test plan are quite relevant, as they represent failure due to problems inherent in the product itself rather than problems brought on by unrealistic stress conditions.

Acknowledgments

David Mease's work was supported in part by an NSF-DMS postdoctoral fellowship. Vijay Nair's work was supported in part by NSF-DMS grant 0204247.

References

1. Sudjianto, A. et al., Computer-aided reliability and robustness, *Int. J. Quality, Reliability, Safety*, 5, 181, 1998.
2. Welch, W. et al., Screening, predicting, and computer experiments, *Technometrics*, 34, 15, 1992.
3. Popielas, F., Chen, C., and Obermaier, M., CAE approach for multi-layer-steel cylinder head gaskets, in *Proceedings of 2000 SAE World Congress*, SAE technical paper series 2000-01-1348, SAE, Detroit, 2000.
4. Madsen, H.O., Krenk, S., and Lind, N.C., *Methods of Structural Safety*, Prentice Hall, Englewood Cliffs, NJ, 1986.
5. Hasofer, A. and Lind, N., Exact and invariant second-moment code format, *J. Eng. Mech.*, 100, 111, 1974.
6. Du, X. and Sudjianto, A., The First-Order Saddlepoint Approximation for Reliability Analysis, *AIAA J.*, 42, 1199, 2004.
7. Kalos, M.H. and Whitlock, P.A., *Monte Carlo Methods*, Wiley, New York, 1986.
8. Rubenstein, R.Y., *Simulation and the Monte Carlo Method*, Wiley, New York, 1981.
9. Evans, M. and Swartz, T., Methods for approximating integrals in statistics with special emphasis on Bayesian integration problems, *Statistical Sci.*, 10, 254, 1995.
10. Melchers, R.E., *Structural Reliability Analysis and Prediction*, Wiley, Chichester, U.K., 1999.
11. Bucher, C.G., Adaptive importance sampling—an iterative fast Monte Carlo procedure, *Structural Safety*, 5, 119, 1988.
12. Melchers, R.E., *Structural Reliability: Analysis and Prediction*, Wiley, New York, 1987.
13. Cochran, W.G., *Sampling Techniques*, Wiley, New York, 1977.
14. Lohr, S.L., *Sampling: Design and Analysis*, Duxbury Press, Pacific Grove, 1999.
15. Sacks, J. et al., Design and analysis of computer experiments, *Statistical Sci.*, 4, 409, 1989.
16. Sacks, J., Schiller, S., and Welch, W., Designs for computer experiments, *Technometrics*, 31, 41, 1989.
17. Tang, B., Orthogonal array-based Latin hypercubes, *J. Am. Statistical Assoc.*, 88, 1392, 1993.
18. Owen, A., Orthogonal arrays for computer experiments, integration and visualization, *Statistica Sinica*, 2, 439, 1992.
19. Owen, A., Scrambled net variance for integrals of smooth functions, *Annals Statistics*, 25, 1541, 1997.
20. Li, W., Sudjianto, A., and Ye, K., Algorithmic construction of optimal symmetric Latin hypercube designs, *J. Statistical Planning Inference*, 90, 145, 2000.
21. Bingham, D. and Mease, D., Latin Hyper-Rectangle Sampling for Computer Experiments, submitted for publication in *Technometrics*.

22. Friedman, J., Multivariate regression splines, *Annals Statistics*, 19, 1, 1991.
23. Kloek, T. and Van Dijk, H.K., Bayesian estimates of equation system parameters: an application of integration by Monte Carlo, *Econometrica*, 46, 1, 1978.
24. Naylor, J.C. and Smith, A.F.M., Econometric illustrations of novel integration strategies for Bayesian inference, *J. Econometrics*, 38, 103, 1988.
25. Oh, M.-S. and Berger, J., Adaptive importance sampling in Monte Carlo integration, *J. Statistical Computation Simulation*, 41, 143, 1992.
26. Oh, M.-S. and Berger, J., Integration of multimodal functions by Monte Carlo importance sampling, *J. Am. Statistical Assoc.*, 88, 450, 1993.
27. Givens, G.H. and Raftery, A.E., Local adaptive importance sampling for multivariate densities with strong nonlinear relationships, *J. Am. Statistical Assoc.*, 91, 132, 1996.
28. Melchers, R.E., Search-based importance sampling, *Structural Safety*, 9, 117, 1990.
29. Dimarrogons, A.D., *Machine Design: a CAD Approach*, Wiley, New York, 2001.

III

Applications

28 Reliability Assessment of Aircraft Structure Joints under Corrosion-Fatigue Damage *Dan M. Ghiocel and Eric J. Tuegel* 28-1
Introduction • Current Engineering Philosophy • Corrosion-Fatigue-Damage Modeling • Reliability of Aircraft Structure Joints Including Maintenance Activities • Illustrative Examples • Concluding Remarks

29 Uncertainty in Aeroelasticity Analysis, Design, and Testing *Chris L. Pettit* 29-1
Introduction • The Role of Uncertainty in Aeroelasticity • Probabilistic Aeroelasticity Analyses • Probabilistic Design for Aeroelasticity • Aeroelasticity Applications of Robust Control Concepts • Some Suggestions for Future Research • Final Remarks

30 Selected Topics in Probabilistic Gas Turbine Engine Turbomachinery Design *James A. Griffiths and Jonathan A. Tschopp* 30-1
Introduction • Traditional Reliability Engineering Approaches • Probabilistic Rotor Design/Fracture Mechanics • Fan, Compressor, Turbine Blade Probabilistic HCF Design • Overall Summary

31 Practical Reliability-Based Design Optimization Strategy for Structural Design *Tony Y. Torng* 31-1
Introduction • PADS/RELDOS Challenges • RELDOS Resolutions • Proposed RELDOS Analysis Procedure • Proposed RELDOS Theoretical Derivation • Tutorial Example • RELDOS Analysis for an 8-ft Cryogenic Tank • Summary and Conclusions

32 Applications of Reliability Assessment *Ben H. Thacker, Mike P. Enright, Daniel P. Nicolella, David S. Riha, Luc J. Huyse, Chris J. Waldhart, and Simeon H.K. Fitch* 32-1
Introduction • Overview of NESSUS • Overview of DARWIN • Application Examples • Conclusions

33 Efficient Time-Variant Reliability Methods in Load Space *Sviatoslav A. Timashev* 33-1
Reliability Analysis of Redundant Structures Subject to Combinations of Markov-Type Loads • Life and Reliability Assessment of Large Mechanical Systems Subjected to a Combination of Diffusion-Type Markov Loads by Generalized Pontryagin Equations • Examples • Conclusion

34 **Applications of Reliability-Based Design Optimization**   *Robert H. Sues,
Youngwon Shin, and (Justin) Y.-T. Wu* 34-1
Introduction • Overview of Reliability-Analysis Methods • Review of RBDO Methods
• RBDO Applications • Selected Aerospace Application Examples • Conclusions

35 **Probabilistic Progressive Buckling of Conventional and Adaptive Trusses**
Shantaram S. Pai and Christos C. Chamis 35-1
Introduction • Fundamental Approach and Considerations • Discussion of Results
• Adaptive/Smart/Intelligent Structures • Discussion of Results • Summary

36 **Integrated Computer-Aided Engineering Methodology for Various
Uncertainties and Multidisciplinary Applications**   *Kyung K. Choi,
Byeng D. Youn, Jun Tang, Jeffrey S. Freeman, Thomas J. Stadterman,
Alan L. Peltz, and William (Skip) Connon* 36-1
Introduction • Fatigue-Life Analysis and Experimental Validation • Reliability Analysis
and Reliability-Based Design Optimization • Conclusions

37 **A Method for Multiattribute Automotive Design under Uncertainty**
Zissimos P. Mourelatos, Artemis Kloess, and Raviraj Nayak 37-1
Introduction • A Quality Specification Structure • An Automotive Body Door Example
• An Automotive Bumper Standardization Example • Overview of RBDO Approaches
• Summary and Future Needs

38 **Probabilistic Analysis and Design in Automotive Industry**
Zissimos P. Mourelatos, Jian Tu, and Xuru Ding 38-1
Introduction • Common Reliability Methods • Nonparametric Metamodeling (Response
Surface) Methods • Variation Reduction in Robust Engineering: an Automotive Example
• Summary and Future Needs

39 **Reliability Assessment of Ships**   *Jeom Kee Paik and Anil Kumar Thayamballi* 39-1
Introduction • Historical Overview • Procedure for Reliability Assessment of Ships
• Target Reliability for Ships against Hull Girder Collapse • Ultimate Limit-State Equations
• Ultimate Hull Girder Strength Models • Extreme Hull Girder Load Models • Prediction
of Time-Dependent Structural Damage • Application to Time-Dependent Reliability
Assessment of Ships • Concluding Remarks

40 **Risk Assessment and Reliability-Based Maintenance for Large Pipelines**
Sviatoslav A. Timashev 40-1
Introduction • General Considerations • Selection of Adequate Analysis Schemes for
Pipeline Segments • Basic Theory of Pipeline Systems Quality, Reliability, Risk Control, and
Maintenance Optimization • Methods of Bringing Pipeline Safety Problems down to
Reliability Problems • Method of Assessing the Size of Damage due to Pipeline Failure • The
Central Maintenance Problem as Applied to a Pipeline Segment • Role of Diagnostics in
Reliability Monitoring • Optimal Cessation of Pipeline Segment Performance • Merger of
the CMP with the Optimal Cessation Procedure • The Basic Block-Module Model of
Pipeline Risk Control • Risk-Based Inspection/Maintenance • Risk Optimization
Flowchart (RAFT) • Application of MAST and RAFT Procedures • Prioritizing Pipeline
Segments for Repair: A Real Case Study • Conclusion

41 **Nondeterministic Hybrid Architectures for Vehicle Health Management**
Joshua Altmann and Dan M. Ghiocel 41-1
Introduction • Application 1: Vehicle Engine Prognostics Health Management (VEPHM)
• Application 2: DWPA Multiple-Band-Pass Demodulation and Automated Diagnostics
• Concluding Remarks

42 **Using Probabilistic Microstructural Methods to Predict the Fatigue Response
of a Simple Laboratory Specimen**   *Robert Tryon* 42-1
Introduction • Background • Damage-Accumulation Models • Estimate of Random
Variable Statistics • Monte Carlo Simulation of Each Stage of Damage
Accumulation • Fatigue-Life Predictions • Conclusions

43 Weakest-Link Probabilistic Failure *Brice N. Cassenti* 43-1
Introduction • Basic Theory for Static Structures • Failure-Location Predictions
• Extension to Time-Dependent Failure • Applications to Composite
Materials • Conclusions

44 Reliability Analysis of Composite Structures and Materials
Sankaran Mahadevan 44-1
Introduction • Strength Limit States (Laminate Theory) • Strength Limit States (Three-Dimensional Analysis) • Strength Limit State: Approximations • Fatigue Limit State: Material Modeling • Fatigue-Delamination Limit State • Creep Limit State • Conclusion

45 Risk Management of Composite Structures *Frank Abdi, Tina Castillo, and Edward Shroyer* 45-1
Introduction • Technical Approach • Conclusion

28
Reliability Assessment of Aircraft Structure Joints under Corrosion-Fatigue Damage

Dan M. Ghiocel
Ghiocel Predictive Technologies Inc.

Eric J. Tuegel
Air Force Research Laboratory

28.1 Introduction 28-1
28.2 Current Engineering Philosophy 28-2
    Deterministic Approach • Risk-Based Approach
28.3 Corrosion-Fatigue-Damage Modeling......................... 28-8
    Fatigue Damage • Corrosion Damage • Corrosion-Fatigue Damage
28.4 Reliability of Aircraft Structure Joints Including Maintenance Activities... 28-35
    Risk/Reliability-Based Condition Assessment
28.5 Illustrative Examples... 28-39
    Probabilistic Life Prediction Using Different Corrosion-Fatigue Models • Risk-Based Maintenance Analysis of a Lap Joint Subjected to Pitting Corrosion and Fatigue
28.6 Concluding Remarks.. 28-55

28.1 Introduction

A continuing challenge in the aviation industry is how to keep aircraft safely in service longer with limited maintenance budgets. Probabilistic methods provide tools to better assess the impact of uncertainties on component life and risk of failure. Application of probabilistic tools to risk-based condition assessment and life prediction helps managers to make better risk-informed decisions regarding aircraft fleet operation and airworthiness. In addition to assessing aircraft reliability, probabilistic methods also provide information for performing an analysis of the cost of continuing operation based on risks and their financial consequence.

Corrosion and fatigue, separately or in combination, are serious threats to the continued safe operation of aircraft. As a result, the U.S. Air Force, the U.S. Navy, the Federal Aviation Administration (FAA), and the European Joint Aviation Authorities (JAA) have guidelines on how aircraft should be designed and maintained to minimize the risk of failure from fatigue damage [1–5]. Although corrosion has a deleterious impact on structural integrity, the airworthiness regulations and requirements have limited instructions regarding corrosion, noting that each part of the aircraft has to be "suitably protected against deterioration or loss of strength in service due to any cause, including weathering, corrosion and abrasion" [6, 7]. The ability to assess the impact of future corrosion on structural integrity, alone or acting in concert with fatigue, is difficult. A framework to assess the effects of corrosion in combination with fatigue on structural integrity has been under development [8].

The parameters of primary interest to aircraft fleet managers are:

- Risk of failure for a single component on a single aircraft
- Failure risk of an individual aircraft (the sum of risks for all components)
- Hazard failure rates for individual aircraft and the aircraft fleet
- Cost-effectiveness of maintenance actions in reducing failure risk for individual aircraft and the fleet

The purpose of this chapter is to review key aspects of assessing the quantitative risk to airframe structures from concurrent corrosion and fatigue damage. Both the current engineering practice and new research developments are reviewed. The physics-based stochastic damage models necessary to make this risk assessment as well as the statistical data needed to construct these models are discussed. The emphasis is on physics-based stochastic modeling of corrosion-fatigue damage. Lack of data and engineering understanding of the physics of a damage process are highlighted. At the end of the chapter, various probabilistic results computed for different physics-based stochastic damage models and different corrosion severity conditions are illustrated for a typical aircraft lap joint.

28.2 Current Engineering Philosophy

Aircraft-structure joints are the most fatigue- and corrosion-susceptible areas on an aircraft. Loads are transferred from one structural detail to another through fasteners, with the attendant stress-concentrating holes making this a prime location for fatigue cracks to form. The tight fit of details and fasteners can trap moisture in the joint. Relative movement between the structural details and the fasteners, as well as the stress concentrations, can cause corrosion protection systems (anodize, primer, and topcoat) to crack and wear, allowing moisture to reach the aluminum parts and start the corrosion process. A typical example structure is a longitudinal skin joint on the pressurized fuselage of a transport aircraft (Figure 28.1). The loading of longitudinal skin joints, particularly those on or near the horizontal neutral axis of the fuselage, is simply the pressurization of the fuselage, which is approximately constant amplitude with a stress ratio (ratio between minimum over maximum stress) of zero. For illustration purposes, we assume that there is only a single pressurization stress cycle per flight.

28.2.1 Deterministic Approach

In the current USAF practice, when the aircraft is designed, a crack-growth analysis is performed for each critical location assuming a discrete 1.27-mm (0.05 in.) flaw or crack (Figure 28.2). This conservative assumption protects against the possibility of a rogue flaw at any one of the critical locations resulting in the loss of an aircraft or its crew. Different assumptions are allowed if the critical location is a cold-worked hole or interference-fit fastener. But for illustration, we will work with the 1.27-mm crack. The existence of a 1.27-mm (0.05 in.) flaw is a rare event that happens less than one in a million based upon back calculations from full-scale fatigue-test crack data [9].

Each critical location is to be inspected at half the component life, determined by the crack-growth analysis, after approximately 11,000 pressurization cycles for the example in Figure 28.2. In principle, half the life was chosen in order to cover scatter from the "mean" life given by the analysis. The condition of the structure in terms of amounts and severities of cracking, corrosion, fretting, etc., is determined with nondestructive inspections (NDI). The inspection should be accomplished with an NDI method capable of finding a crack less than or equal to the analytical crack length at half the component life from a 1.27-mm flaw.

The capability of NDI to find cracks, or other types of damage, is expressed in terms of the probability of detection (POD) curve. POD curves for fatigue cracks in standard geometries have been developed and compiled in handbooks [10–12] for a variety of NDI methods. An example of a POD curve for eddy-current inspection of a Boeing 737 lap joint is shown in Figure 28.3. Note that 1.27-mm fatigue cracks in the joint were found only about 5% of the time with this particular NDI setup. The USAF philosophy is to assume

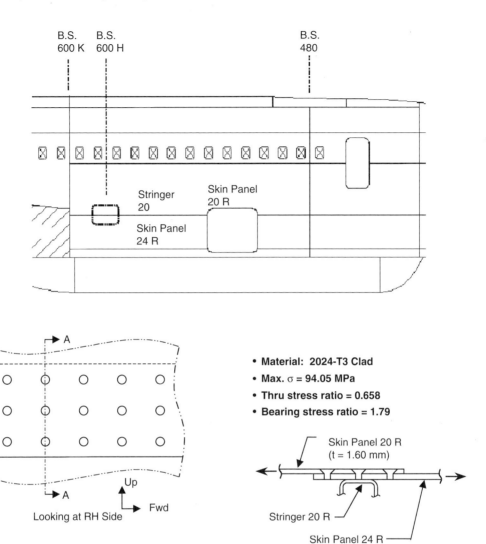

FIGURE 28.1 Details of joint selected as example.

after an inspection that there is a crack at the location just below the 90% detection with 95% confidence limit, denoted 90/95 value. For the aircraft splice joint in Figure 28.1, the 90/95 value of crack size is 2.39 mm, which would be adequate to find the almost 4-mm crack predicted for 1.27-mm starting crack at 11,000 hours (Figure 28.2). From the crack-growth analysis in Figure 28.2, it would take approximately 16,000 pressurization cycles for a 2.93-mm crack to grow to failure. Thus, if no cracks were found in the lap joint during the first inspection using the above eddy-current technique, the second inspection would need to be 8,000 cycles later, or after approximately 19,000 pressurization cycles. The times for subsequent inspections at this location are determined using this same procedure until a crack is found and repaired, or the aircraft is retired. After a repair, inspection intervals will be determined by the characteristics of the repair and its ability to prevent further damage and degradation to the structure. As an aircraft fleet becomes older, inspections can be required more frequently. These inspections can be a real burden to the maintainers and to the operators.

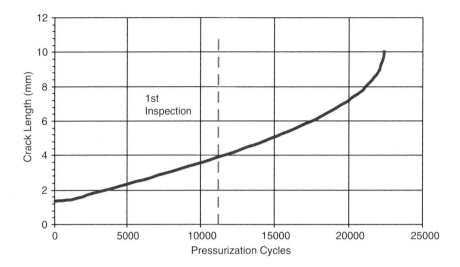

FIGURE 28.2 Example of crack-growth analysis and time to the first inspection.

Recent work has sought to quantify the capabilities of several NDI methods to find corrosion [13]. The major concern with NDI for corrosion is detecting corrosion that is buried between layers of built-up structure. Corrosion on a visible surface is best found visually in adequate lighting; however, this method does not reveal how deeply the corrosion penetrates. When looking for corrosion, the measured quantity is part thickness that is converted to thickness loss from the design specification. In general, eddy current and ultrasound are capable of determining the thickness of a part with reasonable accuracy when the accessible surface is uncorroded. A roughened surface due to corrosion creates difficulties for surface-contacting probes or probes that need an accurate standoff from the surface. X-rays can be used to measure part thickness with corrosion on either surface, provided that there is access to both sides of the part. The ability of any method to detect corrosion depends upon the size of the corroded area vs. the size of the area over which the NDI signal is averaged.

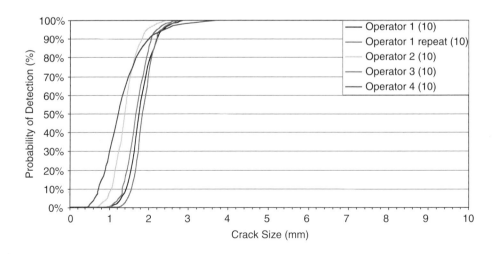

FIGURE 28.3 Results of probability-of-detection study for eddy-current inspection with 200-kHz probe of the fasteners in an unpainted 737 aluminum aircraft splice joint.

Multiple site damage (MSD) or widespread fatigue damage (WFD) should also be considered when evaluating failure risks of structural components. The above approach to aircraft maintenance was developed for discrete source damage, before the risks posed by MSD were fully recognized. The MSD scenario typically assumed when considering damage tolerance is a long, detectable crack emanating from a critical location with small, undetectable cracks at many of the adjacent fastener holes. These small cracks provide a low-energy path for the long crack to follow during fast fracture, much like perforations in paper make it easier to tear a sheet in a specific spot. This scenario cannot be identified with NDI. And the likelihood of MSD existing cannot be reliably estimated analytically because good models for estimating the distribution of small cracks in a structure do not exist. The issue of MSD will be left for another time, when it can be dealt with more thoroughly.

28.2.2 Risk-Based Approach

Typically, in order to determine the failure risk of an aircraft component, three pieces of information are needed:

1. The current "damage" condition of the component
2. The material capacity associated with the progressive "damage" mechanism, i.e., residual strength, or critical crack size, or fracture toughness
3. A predictive model of how the current "damage" condition will develop with continued usage

The maximum frequency of a structural failure leading to the loss of the aircraft acceptable to USAF is 10^{-7} event occurrences per flight [14, 15].

28.2.2.1 Risk-Based Condition Assessment

In a risk-based or risk-quantified approach to aircraft management, a distribution of crack sizes would be estimated, either analytically or based upon previous inspection experience, for a structure prior to an inspection. The crack size distribution would be modified after the inspection based upon the POD for the NDI method and the subsequent maintenance actions performed on the detected cracks.

Lincoln [15] discussed the utility of probabilistic approaches for assessing aircraft safety and for solving key reliability problems faced in practice, such as:

1. Potential-cracking problems are revealed, and the aircraft is beyond its deterministic damage-tolerance limits.
2. Aircraft cracking has occurred to the extent that the deterministic-damage-tolerance derived inspection intervals need to be shortened in order to preserve safety.
3. Aircraft have been designed to be fail safe, but (widespread) fatigue damage has degraded the aircraft structure such that the fail safety of the structure has been compromised.

One of the difficulties in managing aircraft fleets is tracking data from past aircraft inspections to refine the assessment of the current condition of each aircraft or the entire fleet. A good knowledge of the current state of a component or aircraft is important for accurately determining the risk of failure. Electronic databases make storing the data easier. The challenge is getting the data into the database.

Over the last decade, the USAF has developed the probability of fracture (PROF) software to compute the probability of a component fracturing during a single flight [16, 17]. Inputs to the program are based upon data that is readily available as a result of the USAF aircraft structural integrity program (ASIP). These inputs include: material fracture toughness, predicted crack size vs. flight hours for the usage spectrum, normalized stress intensity vs. crack length for the location of interest, distribution of crack sizes at that location throughout the fleet at some previous time, and the distribution of extreme loads the aircraft will experience.

PROF computes the single-flight probability of failure P_f by incorporating two independent failure events: (1) failure occurs when the effective crack size is larger than a prescribed maximum crack size (the residual strength of the component becomes unacceptably low), or (2) failure occurs when the effective crack size is smaller than critical size, a_c, but the maximum stress intensity factor is larger than material fracture toughness, K_c:

$$P_f = \int_0^a f(a)[1 - \text{Prob}(K(a) \leq K_c)]\,da + [1 - \text{Prob}(a \leq a_c)] \tag{28.1}$$

where $f(a)$ is the crack-size distribution function and a_c is the critical crack size.

USAF is continuing to improve the methods used to determine the probability of fracture and risks associated with operational aircraft fleets.

28.2.2.2 Local Failure Criteria

The effect of selecting different local failure criteria on the stress–strain curve is shown in Figure 28.4. Since the material toughness can be related to a critical crack size at failure for a given stress, the two failure criteria in PROF can be plotted together on the crack size–stress plane as in Figure 28.5. The residual strength of the component defines the limit of a component's ability to carry load [18] and can be simplistically thought of as a limit surface in the stress vs. crack size plane described by the minimum of the yield and fracture curves in Figure 28.5. When the structure is new and the sizes of any cracks are small, the net section stress must be less than the yield strength of the material. For a longitudinal lap splice subjected to only pressurization loading (Figure 28.1), the maximum net section stress is $\sigma/(1 - nd/W)$, where W is the width of the panel, d is the diameter of the fastener holes, and n is the number of fasteners in a row. As the component is fatigued, cracks form and grow. The residual strength of the component is the stress required to cause fracture. For a single crack, based on linear elastic fracture mechanics (LEFM), the local stress at the crack tip that defines the residual stress can be simply computed by the relationship

$$\sigma_c = \frac{K_c}{\beta\sqrt{\pi a}} \tag{28.2}$$

where K_c is the critical stress intensity factor or fracture toughness that causes material to fracture, β is the stress intensity geometry factor for the given crack, and a is the crack size.

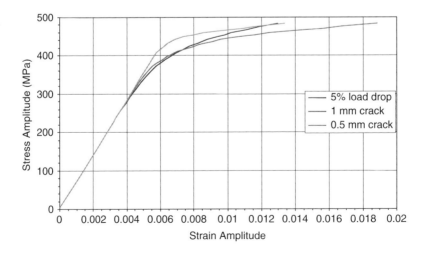

FIGURE 28.4 Cyclic stress–strain curve for 2024-T3 sheet.

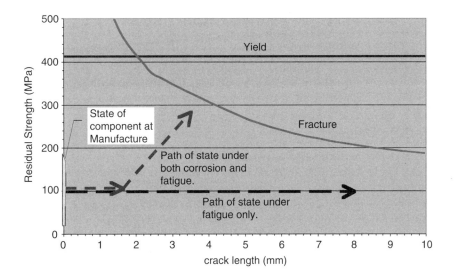

FIGURE 28.5 Pictorial description of residual strength space.

If the structure is designed such that load can be redistributed from the cracked component, or area of the component, to other components or areas, the determination of the critical stress for failure of the component is harder. As the component cracks, it becomes more compliant. Load redistributes to stiffer intact structure nearby, reducing the stress intensity at the crack and the likelihood that fast fracture will occur. Such structure is said to be fail safe.

Yield strength and K_c are variable from lot to lot and even locally within a given component; the limits in Figure 28.5 become zones of constant-probability contours. The values for yield strength used in aircraft design are based upon the A- or B-basis allowable in Mil-Hdbk-5 [19]. An A-basis allowable is the lower value of either a statistically calculated number or the specification minimum. The statistically calculated number indicates that at least 99% of the population is expected to equal or exceed the statistically calculated mechanical property value, with a confidence of 95%. A B-basis allowable indicates that at least 90% of the population of values is expected to equal or exceed the statistically calculated mechanical property value, with a confidence of 95%.

Plane-strain fracture toughness, K_{Ic}, is treated as being normally distributed, with mean values and standard deviations calculated on a rather small data set. However, the critical fracture toughness for a part, K_c, is dependent upon the thickness of the material. A number of K_c values are compiled by Skinn et al. [20] for 2024-T3 aluminum sheet. However, of over 140 tests reported, there were only 3 where the net section stress in the specimen did not exceed 80% of the yield strength and could be considered to have met the requirements of linear elastic fracture mechanics. The average of those three tests on 1.5-mm-thick sheets was 119.6 MPa$\sqrt{m}$, with a standard deviation of 9 MPa$\sqrt{m}$.

28.2.2.3 Uncertainty in Failure Criteria

In the above discussion of residual strength, failure is considered as a stepwise change of system state—from having structural integrity to having no structural integrity. In reality, the transition from a sound state to a "failed state" is smooth; the changes in system integrity occur gradually with small changes in time. It is difficult to define a distinct instant when "failure" occurred. Thus there is a lack of distinctness, or uncertainty, to the failure criteria. Several researchers have proposed using nondeterministic approaches, using either probabilistic or fuzzy approximation, to describe fatigue damage and subsequent failure [21, 22]. This approach has a certain appeal to it, but it still requires more development before being applied to practical situations.

28.3 Corrosion-Fatigue-Damage Modeling

At low homologous temperatures, fatigue damage accumulates with applied load cycles, regardless of how fast or slowly the cycles are applied. On the other hand, corrosion develops as a function of time, regardless of whether the structure is loaded or not. Putting these two mechanisms into the same model is challenging because of the different "time" scales at which "damage" develops.

28.3.1 Fatigue Damage

Fatigue changes crack size distribution as a result of applied loading only. A new structure starts out with few, if any, cracks. New cracks form at stress concentrations with applied loading as a result of local plasticity and microplasticity. Many microscopic cracks may form, but only a few become visible, macroscopic cracks. The portion of the fatigue life until the formation of a detectable crack, which is considered here to be about 2 mm, is denoted as crack nucleation. The portion of the fatigue life after a detectable crack is formed until the component fails is denoted as crack propagation or growth. Different mathematical models are used to analyze these two phases of the fatigue life, though it is likely that a single physical mechanism operates throughout the entire fatigue life [23].

28.3.1.1 Crack Initiation

A common model for estimating the load cycles until the development of a detectable crack is currently the semiempirical local strain-life approach [24–26]. The local strain-life method models the stress–strain history at the "root" of a stress concentration, or notch, from the cyclic stress–strain curve of the material and the notch (local plasticity) analysis. The number of constant-amplitude stress–strain (closed) cycles that is accumulated until the detectable crack size is reached is determined based on the strain-life curve of the material adjusted for the nonzero mean stress effects. For variable-amplitude cycle loading, the cumulative damage defined by crack size is then computed using the kinetic damage equation. To count stress–strain cycles, rainflow counting or other methods can be used. The growth and linkup of small cracks is included in crack initiation when the models are calibrated to the detection of a suitably long crack, so models for small crack growth are not needed.

The material for the aircraft structural joint shown in Figure 28.1 is 2024-T3 aluminum sheet. Examples of cyclic stress–strain curves for this material, found by putting a curve through tips of the stable hysteresis loops obtained during strain-controlled fatigue tests of smooth specimens, are shown in Figure 28.4. The differences between the curves are partially the result of using different failure criteria for the fatigue tests, which results in the hysteresis loops being defined as stable at different times [27]. Hysteresis loops are considered stable at half the cycles to failure. However, an alloy like 2024-T3, which is cold-worked prior to aging, can cyclically soften, i.e., the extreme stresses experienced at the extreme strain points decrease with increasing number of load cycles throughout the entire test, making the determination of the stable hysteresis loop somewhat imprecise.

The curve can be modeled using the Ramberg-Osgood equation

$$\frac{\Delta \varepsilon}{2} = \frac{\Delta \sigma}{2E} + \left(\frac{\Delta \sigma}{2K'}\right)^{\frac{1}{n'}} \quad (28.3)$$

where $\Delta \varepsilon / 2$ is the strain amplitude, $\Delta \sigma / 2$ is the stress amplitude, E is the elastic modulus, K' is the cyclic strain hardening coefficient, and n' is the cyclic strain hardening exponent. The values of K' and n' for the different curves are provided in Table 28.1 [27].

For the aircraft lap joint shown in Figure 28.1, a detailed analysis of the load transfer indicates that the most critical location is the first row of fasteners in the outer skin. It has the largest bypass, or through, stress, 61.9 MPa, and the largest bearing stress, 168.3 MPa. Applying the respective stress concentration factors for a hole in a plate [28] and adding the components together gives the maximum $K_t \sigma = 417.4$ MPa. The local stress–strain history can be determined by a finite-element analysis, which can be very time

TABLE 28.1 Cyclic Strain Hardening Coefficients and Exponents for Curves in Figure 28.4

Curve (Failure Criterion)	K'(MPa)	n'
5% load drop	843	0.109
1-mm crack	669	0.074
0.5-mm crack	590.6	0.040

consuming if the load history has a large number of load levels in it, or by using Neuber's equation, $K_t^2 = K_\sigma K_\varepsilon$, where K_σ is the stress concentration factor, $\sigma_{notch}/\sigma_{global}$, and K_ε is the strain concentration factor at the hole, $\varepsilon_{notch}/\varepsilon_{global}$. Solving for the maximum notch stress and strain yields the results in Table 28.2 for each of the cyclic stress–strain curves.

The strain amplitude and mean stress (or the maximum stress) are used to estimate the time to a detectable crack. First, the strain-life curve, determined with data from R = −1 (completely reversed) strain-control testing, is needed. The standard strain-life curve expresses alternating strain as a function of cycle life:

$$\varepsilon_a = \frac{\Delta \varepsilon}{2} = \frac{\sigma'_f}{E}(2N_f)^b + \varepsilon'_f(2N_f)^c \qquad (28.4)$$

Values of the coefficients and exponents for 2024-T3 using the three failure criteria above are listed in Table 28.3 [27]. The first term in Equation 28.4 characterizes high-cycle fatigue when macroscopic plastic deformation is not evident, while the second term characterizes low-cycle fatigue associated with macroscopic plastic deformation. The resulting strain-life curves are compared in Figure 28.6. The abscissa of the strain-life curves represent the number of applied load cycles at which 50% of the specimens tested at that strain amplitude would have failed.

The strain-life curves are for completely reversed loading with a zero mean stress. The loading for the fuselage lap joint has a stress ratio of zero, i.e., a mean stress of $\Delta\sigma/2$. So the strain-life curves need to be adjusted for a nonzero mean stress.

For evaluating the probabilistic crack initiation life, a local strain-life approach with randomized strain-life curve parameters can be used. Thus, the four parameters, σ'_f, b, $\varepsilon f'$, and c, are random material parameters. It is expected that the first pair of parameters that influences the short lives is statistically independent with respect to the second pair of parameters that influences the long lives. Within each of the two pairs of parameters, there is expected to be a certain level of statistical dependence.

The most popular numerical procedures used to correct the strain-life curve for the nonzero mean stress effects are [29]:

1. Morrow correction (MC): mean-stress effect in the elastic term

$$\varepsilon_a = \frac{\sigma'_f}{E}\left(1 - \frac{\sigma_m}{\sigma'_f}\right)(2N_f)^b + \varepsilon'_f(2N_f)^c \qquad (28.5)$$

TABLE 28.2 Notch Stress and Strain History for the First Row of Fasteners in the Joint of Figure 28.1

Curve (Failure Criterion)	σ_{max}(MPa)	$\Delta\varepsilon/2$	σ_{mean}(MPa)
5% load drop	388.7	0.00276	194.4
1-mm crack	392.3	0.00279	196.2
0.5-mm crack	412.8	0.00294	206.4

TABLE 28.3 Strain-Life Equation Coefficients and Exponents for 2024-T3 Sheet

Strain-Life Curve (Failure Criterion)	σ'_f (MPa)	B	ε'_f	c
5% load drop	835	−0.096	0.174	−0.644
1-mm crack	891	−0.103	4.206	−1.056
0.5-mm crack	1044	−0.114	1.765	−0.927

2. Modified Morrow correction (MMC): the mean-stress effect in the elastic and plastic strain terms

$$\varepsilon_a = \frac{\sigma'_f}{E}\left(1 - \frac{\sigma_m}{\sigma'_f}\right)(2N_f)^b + \varepsilon'_f \left(1 - \frac{\sigma_m}{\sigma'_f}\right)^{\frac{c}{b}} (2N_f) \qquad (28.6)$$

3. Smiths-Watson-Topper (SWT) approach changes the strain-life curve expression by

$$\sigma_{max}\varepsilon_a = \frac{(\sigma'_f)^2}{E}(2N_f)^{2b} + \sigma'_f \varepsilon'_f (2N_f)^{b+c} \qquad (28.7)$$

The mean and maximum stresses are denoted by σ_m and σ_{max} in the above equations. It should be noted that the selection of the mean-stress correction procedure has a large impact on the computed component lives.

For the 2024-T3 sheet material, the adjusted strain-life curves using MC procedure are compared in Figure 28.7. The range of strain amplitudes in Table 28.2 produces estimates of the time to 50% of the fastener holes in the first row having a detectable crack as 100,000 to 150,000 pressurization cycles.

If the loading stress history is variable amplitude instead of constant amplitude, then rainflow, range pair, or other cycle-counting procedures can be used to break the local stress–strain history into applied closed stress–strain cycles of different strain amplitudes, $\Delta\varepsilon/2$, and mean stresses, σ_m.

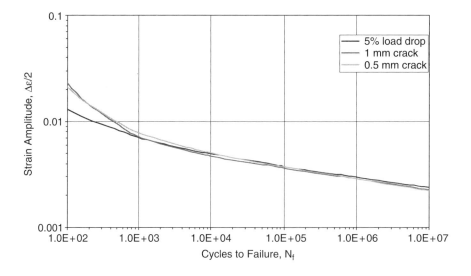

FIGURE 28.6 Strain-life curves for 2024-T3 sheet.

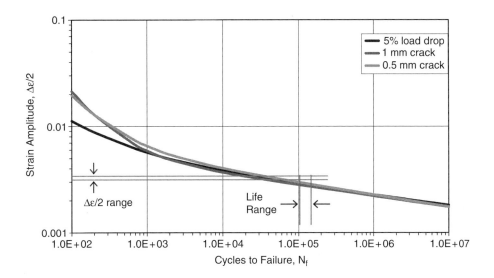

FIGURE 28.7 2024-T3 sheet strain-life curves adjusted for $R = 0$ loading with Morrow's equation.

For every cycle in the history, the number of cycles to failure of a smooth specimen under constant-amplitude loading is determined by solving Morrow's equation for N_f. Then, the cumulative damage is computed by solving the first-order differential kinetic damage equation of the form:

$$\frac{dD}{dN} = f(D, N, N_f(\varepsilon_a, \sigma_m), p) \tag{28.8}$$

where the letter p denotes the physical parameters of the cumulative damage model.

The total accumulated damage, D_T, due to cyclic loading can be directly computed by the convolution of the damage function, $D(X_{\min}, X_{\max})$, with cycle counting distribution $N_T(X_{\min}, X_{\max})$:

$$D_T = \int_T d(t)dt = \sum_{i \text{ for } v \leq u} D(v_i, u_i) = -\iint_{v \leq u} N_T(v, u) \frac{\partial^2 D(v, u)}{\partial v \partial u} dv du \tag{28.9}$$

The integral value is the summation of all elementary damages produced by the sequence of closed stress–strain hysteresis loops.

It was proven experimentally by Halford [30] that for a sequence of cycles with constant alternating stress and mean stress, the cumulative damage curve, the crack initiation life, $N_f(\varepsilon_m, \sigma_m)$, can be accurately constructed based only on two experiments for the extreme amplitude levels, i.e., maximum and minimum life levels. The greater the ratio between the (two) extreme life levels, the more severe damage interaction there is and the greater deviation from the linear-damage rule.

28.3.1.1.1 Linear-Damage Rule

The popular linear-damage rule (LDR) has the mathematical form:

$$D = \sum \frac{n_i}{N_i} = \sum r_i \tag{28.10}$$

where D is the damage, n_i is the number of cycles of ith load level, N_i is the fatigue life according the ith load level, and r_i is the cycle ratio of ith load level. In the linear-damage rule, the damage is measured by the cycle ratio. Failure occurs when the damage reaches unity.

The shortcoming of the popular linear-damage rule (LDR) or Miner's rule is its stress independence or load-sequence independence; it is incapable of taking into account the interaction of different load levels. There is substantial experimental evidence that shows that LDR is conservative under completely reversed loading condition for low-to-high loading sequences, $\sum r_i > 1.0$, and severely under conservative for high-to-low loading sequence, $\sum r_i < 1.0$. It should be noted that for intermittent low-high-low-high- ... -low-high cyclic loading, the LDR severely underestimated the predicted life, as indicated by Halford [30].

28.3.1.1.2 Damage Curve Approach

The damage curve approach (DCA) was developed by Manson and Halford [30]. The damage curve is expressed in the following form:

$$D = \left(\frac{n}{N}\right)^q = \left(\frac{n}{N}\right)^{\left(\frac{N}{N_{\text{ref}}}\right)^\beta} \tag{28.11}$$

where D is the accumulated damage, n is the number of cycles, and N is the fatigue life for the corresponding strain amplitude and mean stress. N_{ref} is the *reference* fatigue life. Parameter β is set equal to 0.40 for many alloys.

28.3.1.1.3 Double Damage Curve Approach

The double damage curve approach (DDCA) was developed by Manson and Halford by adding a linear term to the DCA equation [30]. The DDCA is defined by the relationship:

$$D = \left(\frac{n}{N}\right)\left[q_1^\gamma + \left(1 - q_1^\gamma\right)\left(\frac{n}{N}\right)^{\gamma(q_2 - 1)}\right]^{1/\gamma} \tag{28.12}$$

where

$$q_1 = \frac{0.35\left(\frac{N_{\text{ref}}}{N}\right)^\alpha}{1 - 0.65\left(\frac{N_{\text{ref}}}{N}\right)^\alpha}, \quad q_2 = \left(\frac{N}{N_{\text{ref}}}\right)^\beta$$

The parameters α and β are set to equal 0.25 and 0.40, respectively, for many alloys. Parameter study shows that the value of γ can be set at 5.00, which makes DCA a sufficiently close fit to DDCA. From the equations of DCA and DDCA, it can be seen that the exponent q in DCA and the parameters q_1 and q_2 are all stress-level dependent, and the interaction between different stress-levels can be adequately considered.

28.3.1.1.4 Stochastic Variability in Crack Initiation

Cracks nucleate in aluminum alloys at coarse slip bands inside large grains, with primary crystallographic slip planes oriented favorably to the applied loading so that there is microplastic deformation around large, hard constituent particles (or other phases), or at grain boundaries [31–38]. Variability in grain orientations and sizes and in particle sizes leads to variability in the time to nucleate a crack.

One source of data for crack nucleation in 2042-T3 is the results of the AGARD round cooperative test program for short-crack growth-rate data [32]. The number of cycles until a through crack developed was reported for many of the specimens. The specimens were 2.3 mm thick. The size is consistent with the definition of crack initiation stated earlier, so these data can provide an estimate of the scatter in the time to form a 2-mm crack. Single-edge notched tension (SENT) specimens shown in Figure 28.8 were used.

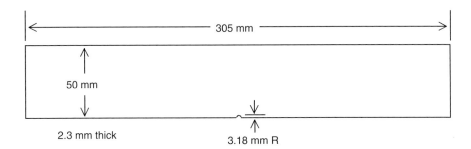

FIGURE 28.8 Single-edge notched tension specimen used in AGARD short-crack cooperative test program. (From Newman, J.C., Jr. and Edwards, P.R., Short-Crack Growth Behaviour in an Aluminum Alloy: an AGARD Cooperative Test Programme, AGARD R-732, 1988. With permission.)

Tests were conducted with constant-amplitude loading at three different maximum stress levels for each stress ratio, R, of −2, −1, 0, and 0.5. In addition, tests were carried out at three reference stress levels for two different load spectra: FALSTAFF (a standardized spectrum representative of the load-time history in the lower wing skin near the root of a fighter aircraft) and a Gaussian-type random load sequence. An example of the data along with cycles to initiation estimates using the three strain-life models is shown in Figure 28.9 for the constant-amplitude $R = 0$ tests.

Another useful data set for the variation in crack initiation time was done at Boeing in the 1970s [39]. The goal of the study was to develop sufficient fatigue data to identify the form of the life distributions, so that a probabilistic fatigue design method could be explored. Eight 2024-T3 panels, 914.4 mm wide by 3.18 mm thick, with 110 holes measuring 4.76 mm in diameter, were fatigue-tested under two different load spectra. The panels came from three different heats of material. A conductive-paint crack-detection circuit was used to detect cracks on the order of 0.5 mm from each hole. When a hole cracked, it was oversized to 9.53-mm diameter and cold-worked to inactivate that hole as a future crack site. Testing was continued until 10% to 20% of the holes had cracked, though in two instances testing continued until 50% of the holes cracked. The number of spectrum load points until crack detection during these tests is shown in Figure 28.10. The number for spectrum load points to cracking predicted with the

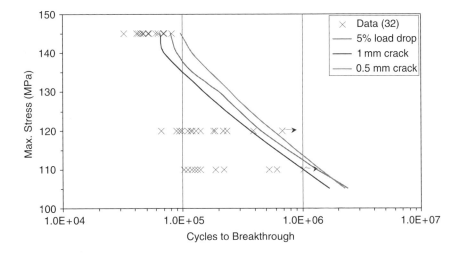

FIGURE 28.9 Cycles to development of a through-thickness crack in SENT specimens of 2024-T3.

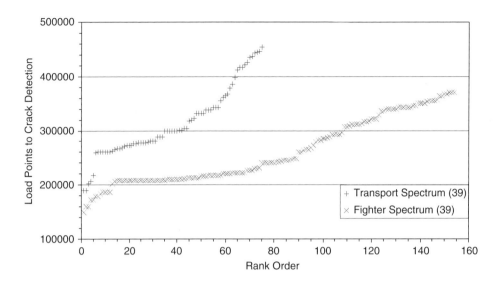

FIGURE 28.10 Cycles to detection of a crack at holes in 2024-T3 panels.

strain-life models presented earlier is approximately 1 million for the transport spectrum and about 440,000 for the fighter spectrum.

It should be noted that the most important source of uncertainty in probabilistic crack initiation life prediction comes from the strain-life curve uncertainty. The uncertainty in the shape of the damage curve model is of secondary significance in risk predictions.

28.3.1.2 Crack Propagation

Usually, in practice, the fatigue-crack-propagation models are based on linear elastic fracture mechanics (LEFM) theory. The limitation of the crack-propagation models based on LEFM theory is that they are applicable only to the propagation of long cracks. The small-crack growth below a given stress-intensity-range threshold is totally ignored. In fact, this is not true. Cracks nucleate at a microscale within the grains in plastic slip bands, and then, by accumulating strain energy, they penetrate the grain boundaries and start growing much faster. At each grain boundary there is potential for different crystallographic grain orientations in adjoining grains. If there is a significant difference in orientation, the small crack will stop until it can reform in the next grain. When many grains are penetrated and the crack is 1 to 2 mm, the crack becomes a long crack. Or in other words, a macrocrack was initiated. In the small-crack stage, the LEFM theory is not applicable, since the crack tip plastic zone occupies a large volume in comparison with the crack dimensions.

In this section, small-crack-growth modeling is covered by the cumulative damage models described earlier for crack initiation. No further discussion on the small-crack growth using micro- and meso-mechanics models is included here.

The rate of growth for long cracks, da/dN, is modeled as a function of the stress intensity range, ΔK, and some material behavior parameters. Crack size is denoted by its length, a, such that the current intensity of growth is uniquely defined by the increment per cycle or the crack growth rate, da/dN, expressed by a functional relationship of the form

$$da/dN = f(\Delta K, K_{max}, K_c, \Delta K_{th}, E, \nu, \sigma_Y, \sigma_U, \varepsilon_d, m) \quad (28.13)$$

where independent variables ΔK and K_{max} define the stress intensity range and maximum stress intensity, respectively, and E (elastic modulus), ν (Poisson's ratio), σ_y (yield strength), σ_u (ultimate

TABLE 28.4 Forman Equation Parameters for 2024-T3 Sheet

K_c (MPa√m)	C (mm/cycle)	m	n	p	Q
97.7	1.47×10^{-4}	0.39	1.66	0.93	0.54
R	0	0.5	0.7	−1	
ΔK_{th}	3.0	1.75	1.18	5.85	

strength), ε_d (ductility), m (hardening exponent), K_c (fracture toughness), and ΔK_{th} (threshold level) define the material properties. Several curve fits have been used to model empirical crack propagation.

28.3.1.2.1 Forman Model

One of the popular crack-propagation models is the generalized Forman fatigue-crack-growth model [40]:

$$\frac{da}{dN} = \frac{C(1-R)^m (\Delta K - \Delta K_{th})^p (\Delta K)^n}{((1-R)K_c - \Delta K)^q} \quad (28.14)$$

where R is the stress ratio, $\sigma_{min}/\sigma_{max}$; K_c is the critical stress intensity to cause fracture; ΔK_{th} is the threshold stress intensity as a function of the stress ratio; and C, m, n, p, and q are parameters used to fit the data. The values of the parameters for 2024-T3 sheet, based on data from 1.6- to 2.29-mm-thick sheet tested in lab air, dry air, or humid air, are given in Table 28.4, and the resulting crack growth-rate curves are shown in Figure 28.11.

The generalized Forman model describes the crack-growth behavior in all of the growth rate regimes. In the Region II, the above reduces to a linear equation in log–log space (Paris law).

The stochastic crack-growth model considers all the parameters as random quantities, but will include also two additional random factors for modeling uncertainties in the regions of low and high values of the rate da/dN in Regions I and III [40]:

$$\frac{da}{dN} = \frac{C(1-R)^m \Delta K^n (\Delta K - \lambda_{Kth} \Delta K_{th})^p}{[(1-R)\lambda_{Kc} K_C - \Delta K]^q} \quad (28.15)$$

The threshold random factor can be adjusted to simulate the uncertain small-crack growth.

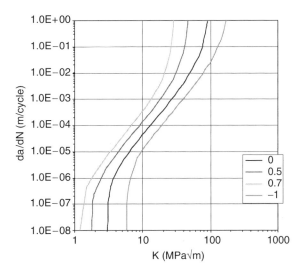

FIGURE 28.11 Corrosion rates for 2024-T3 sheet (1.63 mm thick) at four sites.

28.3.1.2.2 Hyperbolic Sine Model

The hyperbolic sine equation (SINH) model was developed by Pratt and Whitney Aircraft [41] to interpolate the crack-growth-rate data over a range of the four test variables T (temperature), R (stress ratio), F (frequency), and t_h (hold times). The SINH equation, which provides the basic sigmoidal shape and the constants to vary the shape of the curve and the inflection point, is given by the expression

$$\log(da/dN) = C_1 \sinh\{C_2[\log(\Delta K) + C_3]\} + C_4 \tag{28.16}$$

where da/dN is the crack growth rate per cycle, and ΔK is the stress-intensity-factor range. Parameters C_1 and C_2 are shape factors that "stretch" the curve vertically or horizontally, respectively, while C_3 and C_4 locate the inflection point horizontally and vertically, respectively. The slope of the curve at the inflection point is found to be $C_1 C_2$. The parameter C_1 is normally set to be 0.5 for many materials.

28.3.1.2.3 Modified Sigmoidal Model

The modified sigmoidal equation (MSE) model was developed by General Electric Company [41]. The basic MSE model is expressed as

$$\frac{da}{dN} = e^B \left(\frac{\Delta K}{\Delta K^*}\right)^P \left(\ln\left(\frac{\Delta K}{\Delta K^*}\right)\right)^Q \left(\ln\left(\frac{\Delta K_c}{\Delta K}\right)\right)^D \tag{28.17}$$

where da/dN is the crack growth rate per cycle, and ΔK is the stress intensity range. The equation has the general sigmoidal shape, with the lower asymptote ΔK^* representing the threshold value of ΔK. The equation involves six parameters, ΔK^*, ΔK_c, B, P, Q, and D. The parameter B controls the vertical motion of the entire curve. The parameter P provides the control of the slope at the inflection point of the sigmoidal curve. The vertical location of the inflection point is controlled by a combination of B, P, and ΔK^*.

28.3.1.2.4 Crack-Closure Model

A crack-propagation model based on crack-closure concepts was implemented in the FASTRAN code by Newman [42]. FASTRAN has been used to model small-crack propagation as well as long-crack propagation. However, the FASTRAN model does capture the effects of material microstructure on small cracks, e.g., the grain boundary effects on small-crack growth rates and orientation at the tip. FASTRAN has, however, been successfully used by different researchers to assess the fatigue life of aircraft components, starting from the initial size of a constituent particle to the final fatigue failure [43, 44].

The analytical crack-closure model is used to calculate crack-opening stresses (S_0) as a function of crack-length and load history. Based on the value of the crack-opening stress, the effective stress-intensity-factor range is computed, and consequently the crack growth rates are determined. The crack-propagation equation in FASTRAN [44] is

$$\frac{dc}{dN} = C_1 \Delta K_{\text{eff}}^{C_2} \frac{1 - \left(\frac{\Delta K_0}{\Delta K_{\text{eff}}}\right)^2}{1 - \left(\frac{K_{\text{max}}}{C_5}\right)^2} \tag{28.18}$$

where

$$\Delta K_0 = C_3 \left(1 - C_4 \frac{S_0}{S_{\text{max}}}\right) \quad K_{\text{max}} = S_{\text{max}} \sqrt{\pi c} \, F \quad \text{and} \quad \Delta K_{\text{eff}} = (S_{\text{max}} - S_0) \sqrt{\pi c} \, F$$

The crack-opening stress, S_0, is calculated from the analytical closure model. ΔK_{eff} is called effective stress intensity. ΔK_0 is the effective threshold stress-intensity-factor range. One of the advantages of using effective stress intensity is that the constants do not change at different stress ratios.

28.3.1.2.5 Stochastic Variability in Crack Propagation

The variability of fatigue-crack growth rate (FCGR) in aluminum alloys arises from changes in crystallographic texture along the crack path, the presence of microcracking at second-phase particles ahead of the crack, and the amount of transgranular vs. intergranular cracking. A significant data set for determining the variability in FCGR of 2024-T3 was produced by Virkler et al. [45]. Sixty-eight center-crack panels, 558.8 mm long by 152.4 mm wide, were cut from 2.54-mm-thick 2024-T3 sheet. Cracks were nucleated at a 2.54-mm-long electrodischarge-machined notch in the center of the panel and grown to 9.00 mm under controlled loading and environment. The number of cycles to reach specific crack lengths was then recorded for each panel under constant-amplitude ($R = 0.2$) loading with a maximum load of 23.4 kN at 20 Hz. Crack length vs. cycles data from a few select panels are presented in Figure 28.12 so that the individual curves can be identified more easily. The crack growth curves spread out as the cracks get longer, but they also cross each other in many places as a result of sudden increases and decreases in the growth rates. The corresponding FCGR data are plotted in Figure 28.13.

These data were generated at a single stress ratio; the scatter in crack growth rate data at other stress ratios may be different. In addition, the material was from a single lot. Lot-to-lot variations cannot be determined from these data. An estimate of the variation possible between different batches of materials can be made by comparing data collected from different test programs. Even then, the comparison is only over a limited range of stress intensities for a few stress ratios, and there are typically only a handful of specimens tested at each condition in any given program. An example of the data available from different test programs is compared with the curve given by the Forman equation in Figure 28.14 for $R = 0$ loading, which is of the most interest for the fuselage joint example.

It should be noted that, in Figure 28.12, the curves for crack length vs. load cycles have slightly different shapes, since the curves cross over. This indicates that an accurate probabilistic modeling would need to consider the random variation of the crack size evolution shapes. Typical stochastic crack-growth models [46]

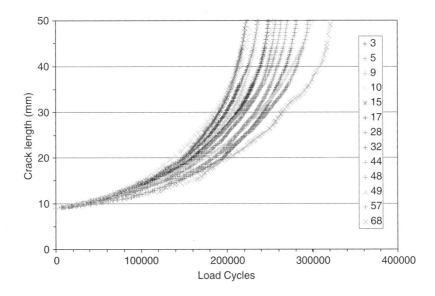

FIGURE 28.12 Crack length vs. load cycles for 2024-T3 sheet; $R = 0.2$, $P_{max} = 23.4$ kN. (Data from Virkler, D.A., Hillberry, B.M., and Goel, P.K., The Statistical Nature of Fatigue Crack Propagation, AFFDL-TR-78-43 [also DTIC ADA056912], 1978.)

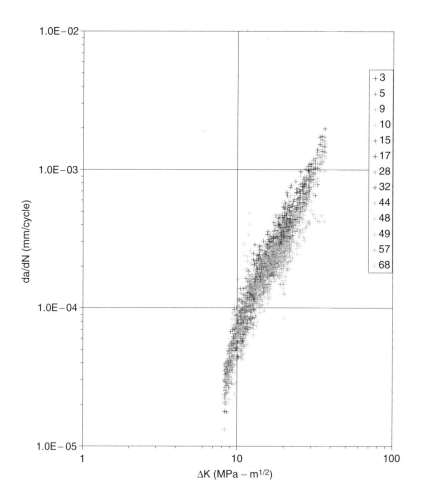

FIGURE 28.13 Fatigue-crack growth rate from curves in Figure 28.12.

assume that stochastic crack growth is composed of a median growth curve scaled by a positive random factor, neglecting the random fluctuating variations around the median shape. However, for a refined stochastic modeling, the crack growth process has to be idealized by a stochastic-process model rather that a random-variable model. The random shape variations indicate that the crack growth process has a finite correlation length. Correlation length is the distance for which the correlation between two points becomes lower than a threshold value. An infinite correlation length corresponds to a random-variable model.

Table 28.5 shows the statistics of number of cycles for four levels of crack sizes: 9.2 mm (close to initial flaw size of 9 mm), 14 mm, 29 mm, and 49.8 mm (considered critical crack size).

As shown in Table 28.5, between the random number of cycles measured for a 9.2-mm crack size and that measured for a 49.8-mm crack size, there is a correlation coefficient as low as 0.31. This indicates a relatively large departure from the usual perfect correlation or, equivalently, the infinite correlation length assumption.

Figure 28.15 shows the histograms from the Virkler data of the number of cycles at which the specified crack length was reached for two selected crack lengths. The numbers of cycles for two specimens are marked on the histogram plots with circles. It should be noted that the two specimens have crack evolutions that are quite different than other statistical crack evolutions. The position of the first specimen moves within the histogram of the crack population from a value in the far right tail at 9.2-mm crack size to a value close the mean at 49.8-mm crack size. The position of the second specimen moves from a value close to the mean at 9.2-mm crack size to a value in the far left tail at 49.8-mm crack size (it is

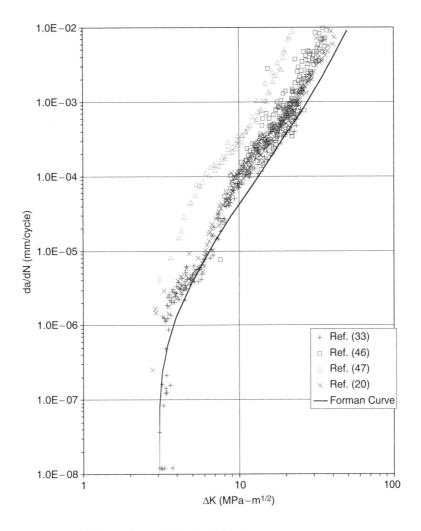

FIGURE 28.14 2024-T3 FCGR lot-to-lot variability, R = 0 loading.

the shortest-life crack path within the 68-specimen population). If there was a real perfect correlation between the random number of cycles measured at different crack lengths, then the two marked crack trajectories should maintain their position within the histogram of the crack population without migrating from one location to another.

The fact that the correlation for length of the crack growth process is not infinite adds more complexity to stochastic modeling of crack-propagation physics. For a constant-amplitude stress-cycle loading, the changes in the shape of the crack growth curve are a consequence of the local nonhomogeneities in the

TABLE 28.5 Statistics for the Number of Cycles for a Given Crack Size

Crack Size	Mean	Standard Deviation	C.O.V.	Correlation w/9.2 mm	Correlation w/14.0 mm	Correlation w/29.0 mm	Correlation w/49.8 mm
9.2 mm	7,304	1,560	0.210	1.00	0.52	0.38	0.31
14 mm	105,976	9,863	0.093	0.52	1.00	0.87	0.78
29 mm	212,411	14,252	0.067	0.38	0.87	1.00	0.96
49.8 mm	257,698	18,850	0.073	0.31	0.78	0.96	1.00

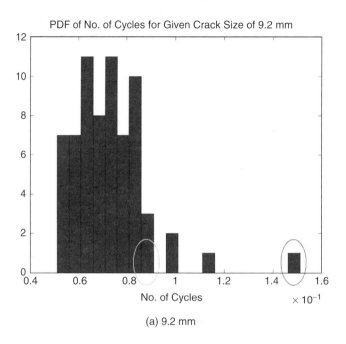

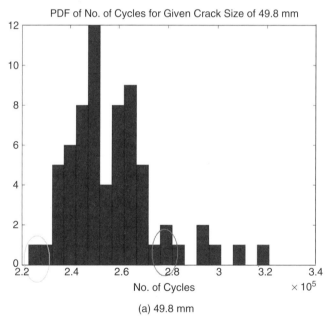

FIGURE 28.15 Histograms of number of cycles for given crack length: (a) 9.2 mm; (b) 49.8 mm.

material properties and resistance against crack growth. For random-amplitude loading, the changes in the shape of the crack growth curve are larger, since they also include the effects of the random fluctuation of the stress amplitude. The random effects due to material nonhomogeneity and variation in loading history are statistically independent.

Another key aspect for getting an adequate stochastic crack-growth model is to accurately consider the statistical correlation between the estimation of the fatigue-model parameter. For example, for a Paris-law model for Virkler data shown in Figure 28.12, the absolute value of statistical correlation coefficient between the estimate of the model coefficient and the estimate of the model exponent is as high as 0.90. Assuming statistical independence between the two random parameters of the Paris-law model produces significant modeling error that can affect the computed fatigue-failure risks by an order of magnitude.

28.3.2 Corrosion Damage

Corrosion in aluminum alloys can be broadly characterized into three types: pitting, general, and intergranular [47, 48]. The development of corrosion and its subsequent growth is less well described than is fatigue.

Pitting is a form of localized corrosion that takes the form of cavities on the surface of a metal. Pitting starts with the local breakdown of protective surface films. Pitting may cause the perforation of thin sections, as well as creating stress concentrations that may trigger the onset of fatigue cracking or other types of corrosion. Simplistic models for the progression of pitting corrosion are widely available.

Corrosion of aluminum alloys generally starts with pitting. Isolated pits are difficult to detect, but they have a significant effect on the fatigue life. Pits can occur on boldly exposed surfaces or on the faying surfaces of joints. Pits are stress concentrations where cracks can form; deep, narrow pits are essentially cracks. Even mild levels of pitting can significantly decrease the fatigue life of laboratory specimens. As pitting becomes widespread, a large area of material can become thinner, resulting in higher stress in that location. This increase in stress is generally less than the stress concentration at an isolated pit, but it is over a larger volume of material than with an isolated pit. These thinned regions will cause long cracks to grow faster, while pits will cause cracks to nucleate faster.

General corrosion is when pitting becomes so widespread that individual pits can no longer be identified. As a result, a significant area of the structure becomes thinner, resulting in higher stress at that location. This increase in stress is generally less than the stress concentration at an isolated pit, but it is over a larger volume of material than with an isolated pit. As a result, unless the structure is lightly loaded or used infrequently, cracking due to the interaction of fatigue and corrosion occurs well before general corrosion develops. If general corrosion occurs within a joint, the trapped corrosion products may cause bulging in the joint. Models for how general corrosion progresses are not readily available.

Intergranular corrosion develops out of pits as a result of preferential attack of the grain boundaries, as shown in Figure 28.16 [49]. Exfoliation and stress corrosion cracking (SCC) are special types of

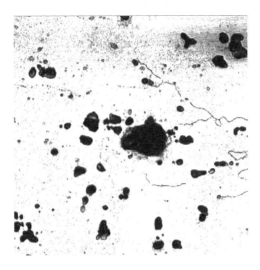

FIGURE 28.16 Photomicrograph of pits with intergranular corrosion on surface of 2024-T3 sheet (4 mm thick) after 4 h of exposure to 3.5% NaCl solution with no load applied (specimen PG-11). (From Bell, R.P., Huang, J.T., and Shelton, D., Corrosion Fatigue Structural Demonstration Program, Lockheed-Martin final report for AFRL/VASM, 2004, [49]. With permission.)

intergranular corrosion that occur in materials with directional grain structures (exfoliation) or under the influence of sustained tensile loads (SCC). The growth of intergranular corrosion is highly dependent upon chemical and metallurgical conditions and is not easily predicted.

28.3.2.1 Corrosion Pitting

In this subsection, two of the most accepted pitting models are described.

28.3.2.1.1 Power-Law Pit Model

For a boldly exposed surface, the depth of the deepest pit, a, as a function of exposure time, t, is typically described by a power law [50],

$$a = At^{1/n} \tag{28.19}$$

where A and n are empirically determined parameters, with n usually having a value between 2 and 4. This relationship does not mean that any one pit grows at this rate. Pits develop, sometimes rapidly, stagnate, and new pits begin. Rather, this equation represents how the maximum of the distribution of pit depth changes with time. An example of laboratory pitting data for 2024-T3 sheet in 3.5% NaCl solution is shown in Figure 28.17. The least-squares fit of the power-law equation to the data results in n equal to 2.52 and A equal to 20.07 μm. These pit depths were measured with either an optical microscope or a confocal microscope from the surface of the specimen. If the pit tunneled, as in Figure 28.18, this could not be determined until after the specimen was broken open.

28.3.2.1.2 Wei Pit Model

A spatial pit-growth model was proposed by Wei [51, 52]. This pit-growth model assumes that the pit shape is a hemispherical shape and that its size grows at a constant volumetric rate, dV/dt, given by

$$\frac{dV}{dt} = 2\pi a^2 \frac{da}{dt} = \frac{MI_{p0}}{nF\rho}\exp\left(-\frac{\Delta H}{RT}\right) \tag{28.20}$$

By integrating the above equation, the pit depth a at a given time t is given by

$$a = \left\{\left[\frac{3MI_{p0}}{2\pi nF\rho}\exp\left(-\frac{\Delta H}{RT}\right)\right]t + a_0^3\right\}^{1/3} \tag{28.21}$$

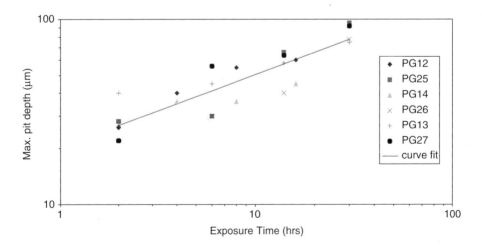

FIGURE 28.17 Maximum pit depth vs. time in 3.5% NaCl solution for 2024-T3 sheet (1.6 mm thick). (From Bell, R.P., Huang, J.T., and Shelton, D., Corrosion Fatigue Structural Demonstration Program, Lockheed-Martin final report for AFRL/VASM, 2004, [49]. With permission.)

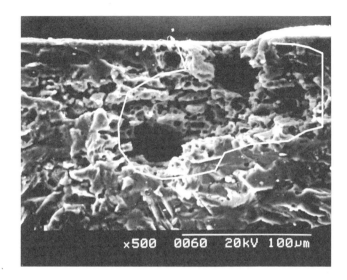

FIGURE 28.18 Example of pit that tunneled (specimen PG14). The pit is outlined by the white curve. (From Bell, R.P., Huang, J.T., and Shelton, D., Corrosion Fatigue Structural Demonstration Program, [51]. With permission.)

where a_0 is the initial pit radius, M is the molecular weight of the material, n is the valence, $F = 96,514$ C/mole is Faraday's constant, ρ is density of the material, ΔH is the activation energy, $R = 8.314$ J/mole-K is the universal gas constant, T is the absolute temperature, and I_{P0} is the pitting current coefficient.

In the Wei model, the assumption of a hemispherical shape of the pits introduces a significant modeling uncertainty. Pit shape is quite an important aspect. The pit depth alone does not adequately describe the stress concentration at a pit. Pit shapes are quite variable and can change with continuing corrosion. Pit shape is influenced by the microstructure of the material that developed as a result of prior thermomechanical processing, the environment (both mechanical and chemical), and the corrosion protection system. More-sophisticated models of pitting corrosion are clearly needed.

28.3.2.2 Stochastic Variability of Corrosion in Aluminum Alloys

28.3.2.2.1 Field Studies

Several long-term studies have been done to determine the statistical effects of environmental exposure on 2024-T3 aluminum materials [53, 54]. The first study [53] was conducted under the direction of the Atmospheric Exposure Test Subcommittee of ASTM Committee B-7 on Light Metals and Alloys. Several magnesium and aluminum alloys, including bare and clad 2024-T3 sheet (1.63 mm thick), were exposed at five test sites for periods of $1/2$, 1, 3, 5, and 10 years. The specimens included riveted joints as well as single-piece panels. The principal measurement in this test program was the change in tensile strength as a result of the exposure.

The second test program [54] involved four test sites for periods of 1, 2, and 7 years. The four tests sites represented rural marine (Kure Beach), industrial marine (Corpus Christi, TX), moderate industrial (Richmond, VA), and industrial (McCook, IL). In this test program, pit depths, mass loss, and changes in tensile strength were recorded. A plot of maximum pit depth vs. exposure time is presented in Figure 28.19. The results indicate that while corrosion in a seacoast environment may start more quickly, there is not much additional corrosion with continued exposure. After 7 years, all the panels have about the same maximum pit depth. The corrosion rate at each of the locations was determined from the total mass loss per unit area divided by the total days of exposure and reported as milligrams lost per square decimeter per day, mdd (Figure 28.20). It is interesting to note that while the marine environments initially had deeper pits than the industrial environments, the industrial environments had higher corrosion rates. The locations with higher

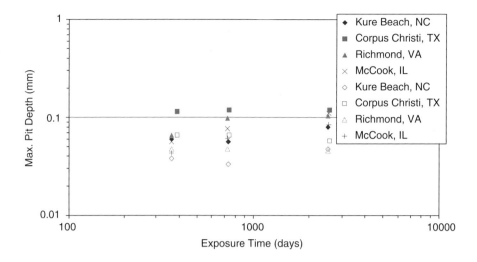

FIGURE 28.19 Maximum pit depths on exposed 2024-T3 sheet (1.63 mm thick). (Solid symbols are maximum depths; open symbols represent average of the deepest four pits.) (From Ailor, W.H., Jr., Performance of aluminum alloys at other test sites, in *Metal Corrosion in the Atmosphere*, ASTM STP 435, ASTM, 1968, pp. 285–307. With permission.)

corrosion rates likely had more pits per unit area, or there was more tunneling of the pits. These corrosion rates are for "boldly" exposed material.

In practical applications, the goal of a stochastic corrosion model is to obtain either the distribution of corrosion damage at any service time or the distribution of service times to reach any given level of corrosion. Different distributions may be required for corrosion on exposed surfaces and for corrosion in occluded areas such as joints. However, data on corrosion in occluded areas is just now becoming available.

28.3.2.2.2 Laboratory Studies

Numerous laboratory studies with accelerated protocols have looked at the distribution of corrosion pit sizes [55–57]. Pitting on exposed surfaces is primarily a function of the dispersion of constituent

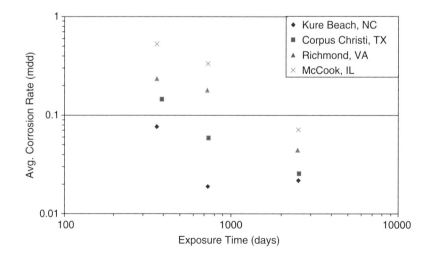

FIGURE 28.20 Corrosion rates for 2024-T3 sheet (1.63 mm thick) at four sites. (From Ailor, W.H., Jr., Performance of aluminum alloys at other test sites, in *Metal Corrosion in the Atmosphere*, ASTM STP 435, ASTM, 1968, pp. 285–307. With permission.)

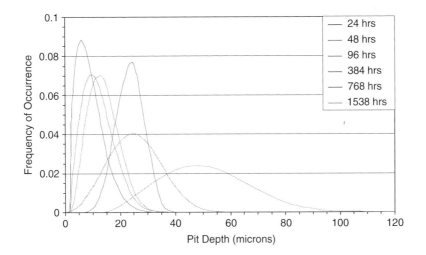

FIGURE 28.21 Three-parameter Weibull distributions of pit depth as a function of exposure time for 7075-T6.

particles in the material microstructure and not the environment. The stochastic descriptions of pitting developed during accelerated laboratory programs should be applicable to pitting on exposed surfaces in natural environments.

Sankaran et al. [55] estimated the distributions of pit dimensions on 7075-T6 as a function of time exposed per ASTM G85 Annex 2 from 200 randomly selected pits at each exposure time (Figure 28.21, Figure 28.22, Figure 28.23). The progression of these distributions with time exhibited a ratcheting behavior. This can be seen in the sequence of pit-depth distributions from 96-h exposure to 1538-h exposure shown in Figure 28.21.

In the studies of pitting on 2024-T3 [56, 57], statistics are reported on the projected area of the pits perpendicular to the loading direction. The Gumbel extreme-value distribution was used to describe the projected areas of the largest pits (Figure 28.24). Note that the area of these pits at 192 h of exposure is an order of magnitude greater than was the area of the pits in the 7075-T6 tests [55].

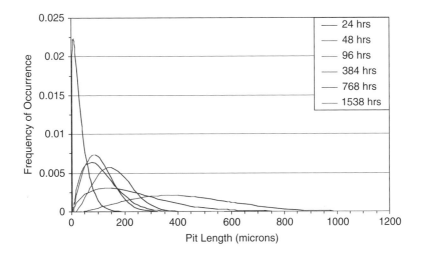

FIGURE 28.22 Three-parameter Weibull distributions of pit length (in rolling direction of sheet) as a function of exposure time for 7075-T6.

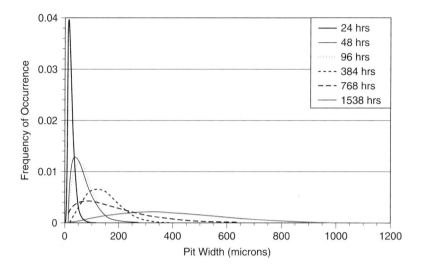

FIGURE 28.23 3-Parameter Weibull distributions of PitWidth (perpendicular to rolling direction of sheet) as function of exposure time for 7075-T6.

This is probably the result of the different severities of the environments and not any inherent material characteristics.

In subsequent fatigue analyses, the pits in the 2024-T3 materials were treated as semicircular surface cracks with a depth-to-width ratio of 0.5 and of equivalent area, which is the most severe case for these small "cracks." Data from the 7075-T6 tests demonstrate that a constant depth-to-width ratio is not realistic, as illustrated by the Weibull distributions shown in Figure 28.25. The nonanalytical estimated bivariate joint probability distribution of pit depth size and pit width is plotted in Figure 28.26. Engineering experience shows that the impact of ignoring the pit aspect ratio in fatigue-crack growth analyses is to potentially overestimate the stress intensity by about a factor of 2, which could lead to overestimating the crack growth rate by an order of magnitude or even more.

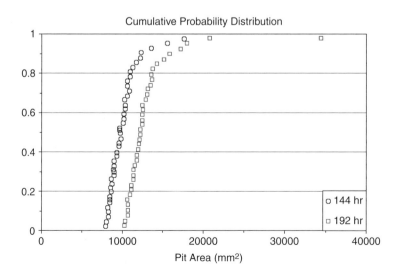

FIGURE 28.24 Extreme-value plots of pit area for largest 10% pits in 2024-T3 material. LT plane exposed to 3.5% salt water in alternate immersion for 144 h and 192 h. Pit area measured on ST plane.

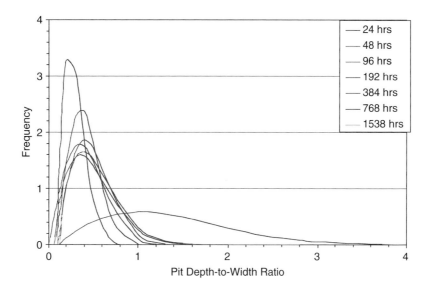

FIGURE 28.25 Three-parameter Weibull distribution of the pit depth-to-width ratio as a function of exposure time for 7075-T6.

28.3.2.2.3 Corroded Surface Topography

Corroded surface topography can have significant influence on corrosion progression and fatigue resistance due to its influence on the local stresses and stress intensity factors. Corroded surface topography incorporates all the key stochastic aspects of the random corrosion progression. At a global scale, in an average sense, the corrosion topography is defined by the general thickness loss, while at a local scale the corrosion topography is defined by the pitting geometry. The corrosion starts as pits on the surface at the boundaries between the aluminum matrix and constituent particles, and then grows with a rough spatial profile due to highly variable growth rates for individual pits. Finally the surface becomes slightly smoother as the pits broaden and link up to form a general corroded surface. Data on

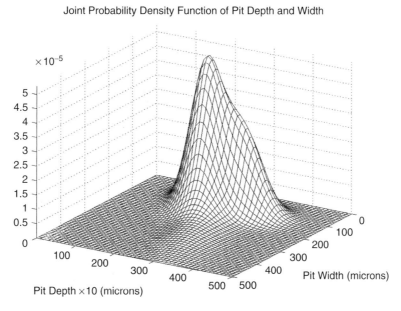

FIGURE 28.26 Joint PDF of pit depth and width after 768 h.

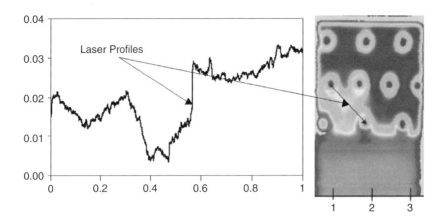

FIGURE 28.27 Thickness map of corroded surface and fine-detail line scan.

the time progression of corroded surfaces through these phases is lacking. Corrosion topography influences both the local stresses (through local pitting) and far-field stresses (through general thickness loss). A typical corroded surface and a cut-line (laser) profile through it is shown in Figure 28.27. Mathematically, stochastic corrosion surfaces can be handled using stochastic field-expansion models such as proper orthogonal decomposition or Karhunen-Loeve series expansion [58, 59].

28.3.3 Corrosion-Fatigue Damage

In this section the effects of corrosion of both crack initiation and propagation stages are discussed. Then, three corrosion-fatigue damage models are presented. These models are the Wei corrosion-fatigue (WCF) model [53, 54], the crack-closure corrosion-fatigue (CCCF) model [42], and the simultaneous corrosion-fatigue (SCF) model [8]. The WCF model replaces the crack-initiation model with a corrosion-pitting model, and after a crack is initiated, the corrosion has no effect on the fatigue cracking. The CCCF and SCF models incorporate the corrosion effects on fatigue cracking during both the crack-initiation and the crack-propagation stages.

Corrosion effects are of two kinds: (1) a local increase in stress near a corrosion pit and (2) a general increase of the far-field stress due to component thickness loss. Pitting can dramatically reduce component life, but it is only significant during the crack initiation phase. General thickness loss has a less dramatic effect on the time to form a crack, but the increased stress can speed up the growth of existing cracks.

Pitting corrosion usually shortens the time for cracks to form, in some cases eliminating the crack nucleation phase altogether. The reduction in the nucleation portion depends on the amount of corrosion, which in turn depends upon the length of exposure to the corrosive environment and the severity of the corrosive environment in relation to the rate at which load cycles are applied. It is difficult to simulate the effect of natural environments in the laboratory because of the time scale involved (20 to 30 years), and currently the relationship between the time scales in accelerated corrosion tests to the natural environment has not been established. So synchronizing the rate at which load cycles are applied with corrosion rate is impossible.

In the laboratory, cracks grow faster in aggressive environments. It is not clear how significant the environmental effects on crack growth are for aircraft structures. Many of the fatigue loads are applied when the aircraft is flying high, where conditions are cold and relatively dry. There can be condensation of moisture inside a transport aircraft. This creates at most a humid environment, so crack growth in humid air may be appropriate. Takeoff and landing loads can be applied in a warm and humid external environment, but this is a small fraction of each flight, unless the aircraft is used for short hops. Recent data gathered from coupons mounted in the wheel wells and vertical tails indicate that most corrosion in USAF transport aircraft occur while the aircraft is on the ground [60]. In general, crack growth rates in high-humidity air should be used for crack-propagation assessments on USAF aircraft.

Local corrosion, such as pitting, on the surface of a part does not greatly affect the growth of long cracks. Because the crack is "sampling" through the thickness of the material, small stress fluctuations at the surface affect only a small local portion of the crack. Not until corrosion becomes so widespread that there is a general loss of part thickness is crack growth affected. Then the crack growth rate increases because stress increases in the part.

28.3.3.1 Wei Corrosion-Fatigue (WCF) Model

The Wei model was developed over a period of several years [51, 52]. As described above, the Wei models assume two stages of corrosion-fatigue damage growth. The first stage is the corrosion stage due to initial pitting, which continues until a threshold level is reached (threshold corresponds to an equal growth rate of pit depth and crack depth, after which crack growth takes over). The second stage is the fatigue stage that ends with the material failure. The Wei models include two fatigue-crack stages, a surface-crack stage and then a through-crack stage. The surface-crack-fatigue part of the Wei models is based on simple Paris-law, with a ΔK_{th} of 3 MPa$\sqrt{m}$ for 2024-T3 aluminum. Thus, in the Wei model, corrosion is just the initiator of fatigue damage due to crack growth. After fatigue takes over (crack-depth growth rate is larger than pit-depth growth rate), corrosion has no further effect. This is a different concept than the SCF model that is presented in this section.

The time to failure, t_f is given by

$$t_f = t_{ci} + t_{tc} + t_{cg} \tag{28.22}$$

where, t_{ci} is the time required for a nucleated pit to grow and for a surface crack to initiate from it, t_{tc} is the time required for the surface crack to grow into a through crack, and t_{cg} is the time for a through crack to grow to a prescribed critical length, given as a part of a failure criterion.

Using the pit-growth equation given by Wei, the time for the pit to grow to a_{ci} is given by

$$t_{ci} = \frac{2\pi nF\rho}{3MI_{P0}}\left(a_{ci}^3 - a_0^3\right)\exp\left(\frac{\Delta H}{RT}\right) \tag{28.23}$$

The pit radius at which a crack is initiated, a_{ci}, can be expressed in terms of the threshold driving force ΔK_{th} via crack growth mechanism. For the sake of simplicity and computational expediency, the surface crack remains semicircular in shape, and the stress-intensity-factor range is given by

$$\Delta K_s = \frac{2.2}{\pi} K_t \Delta\sigma \sqrt{\pi a} \tag{28.24}$$

where $\Delta\sigma$ is the far-field stress range, K_t is the stress concentration factor resulting from the circular rivet hole, and the factor of $2.2/\pi$ is for a semicircular flaw in an infinite plate. Again, the surface crack is assumed to nucleate from a hemispherical corrosion pit when ΔK_s increases to ΔK_{th}. The corresponding crack length that satisfies this condition is easily found to be

$$a_{ci} = \pi\left(\frac{\Delta K_{th}}{2.2K_t\Delta\sigma}\right)^2 \tag{28.25}$$

The expression for t_{ci} can be found by substituting Equation 28.25 into Equation 28.26. The material parameters of the Wei corrosion model for 2024-T3 aluminum are shown in Table 28.6.

Within the Wei corrosion-fatigue model, a standard Paris law is assumed for crack propagation:

$$\left(\frac{da}{dN}\right)_C = C_C(\Delta K)^n \tag{28.26}$$

TABLE 28.6 Parameters Used in the Pit-Growth Model for 2024-T3

Parameters	2024-T3	Parameters	2024-T3
Density, ρ (gm/m³)	2.7×10^6	Initial pit radius, a_0 (m)	2×10^{-5}
Molecular weight, M	27	Pitting current constant, I_{p0} (C/sec)	0.5
Valence, n	3	Threshold, ΔK_{th} (MPa$\sqrt{m}$)	3.0
Activation energy, ΔH (J/mole)	50,000	Applied stress, $\Delta \sigma$ (MPa)	90
Temperature, T (K°)	293	Stress concentration factor, K_t	2.6

The driving force ΔK is considered to be of two different forms, according to whether the crack is a surface crack or a through crack. For a surface crack, ΔK equals ΔK_s given in Equation 28.24, and it remains so until the crack can be modeled as a through crack. When the crack becomes a through crack, ΔK is assumed to be equal to ΔK_{tc}, which has the following form:

$$\Delta K_{tc} = F_{tc}\left(\frac{a}{r_0}\right) \Delta \sigma \sqrt{\pi a} \tag{28.27}$$

where r_0 is the radius of the rivet hole. For ratios of a/r_0 in the interval from 0 to 10, inclusive, for an infinite plate under uniaxial tension containing a circular hole with a single through crack emanating from the hole perpendicular to the loading axis, the function $F_{tc}(a/r_0)$ can be numerically evaluated by

$$F_{tc}\left(\frac{a}{r_0}\right) = \frac{0.865}{(a/r_0) + 0.324} + 0.681 \tag{28.28}$$

The remaining question concerning the driving force ΔK is that of the transition from a surface crack to a through crack. It is assumed that the transition occurs at the crack length a_{tc}, which is defined by equating the geometry-dependent function from Equation 28.24 and Equation 28.28. Thus, the transition crack length a_{tc} is the solution of

$$F_{tc}\left(\frac{a_{tc}}{r_0}\right) = \frac{2.2}{\pi} K_t \tag{28.29}$$

which is easily found to be

$$a_{tc} = r_0 \left[\frac{0.865}{(2.2/\pi)K_t - 0.681} - 0.324 \right] \tag{28.30}$$

The final computation to be completed is for t_{tc} and t_{cg}. First consider the computation for the time between crack initiation and transition to a through crack, t_{tc}. Substituting Equation 28.24 into Equation 28.26 yields a simple differential equation in that the variables a and N can be separated, and an explicit solution can be found. Assuming that $N = \upsilon t$, where υ is the loading frequency, then

$$t_{tc} = \frac{2(\sqrt{\pi})^{n_C}[(\sqrt{a_{ci}})^{2-n_C} - (\sqrt{a_{tc}})^{2-n_C}]}{\upsilon(n_C - 2)C_C(2.2K_t \Delta \sigma)^{n_C}} \tag{28.31}$$

if $n_c \neq 2$. For aluminum alloys, typically n_c is not equal to 2.

The time between the through-crack initiation time and the final failure time is

$$t_{cg} = \int_{a_{tc}}^{a_f} \frac{1}{\upsilon C_c (\Delta \sigma \sqrt{\pi})^{n_C}} \left(\frac{0.324 r_0 + a}{1.086 r_0 \sqrt{a} + 0.681(\sqrt{a})^3} \right)^{n_C} da \tag{28.32}$$

TABLE 28.7 Parameters Used in the Crack Growth Model for 2024-T3

Parameters	2024-T3	Parameters	2024-T3
Fatigue coefficient, C_c (m/cycle)	3.3×10^{-10}	Final crack size, a_f (mm)	3.0
Crack growth exponent, n_c	3.0	Frequency, v (cycles/day)	2, 10
Radius of rivet hole, r_0, (mm)	3.0	Threshold, ΔK_{th} (MPa$\sqrt{m}$)	3.0

Source: Ailor, W.H., Jr., Performance of aluminum alloys at other test sites, in *Metal Corrosion in the Atmosphere*, ASTM STP 435, ASTM, 1968, pp. 285–307.

where a_f is the final crack size. In most cases, the t_{cg} can only be calculated by numerical integration. The parameters for crack growth that are shown in Table 28.7 are from the literature [52].

28.3.3.2 Crack-Closure Corrosion-Fatigue (CCCF) Model

The crack-closure fatigue model [42] was modified to include the effect of corrosion pitting on the local stress intensity. To include corrosion pit effects the effective-stress-intensity range, ΔK_{eff}, is amplified by a pitting factor as follows:

$$\Delta K'_{eff} = \psi(t) \Delta K_{eff} \tag{28.33}$$

The pitting factor $\psi(t)$ depends on pit size and crack size:

$$\psi(t) = \sqrt{1 + \frac{a_{pit}(t)}{a_{crack}(t)}} \tag{28.34}$$

Equation 28.33 and Equation 28.34 are equivalent to an additive corrosion-fatigue-damage model that superimposes linearly the corrosion damage, pit size, fatigue damage, and crack size in the stress-intensity analytical expression. Since, practically, there is no corrosion during flights, only the ground time is considered for evaluating the corrosion pit growth. The pit growth, Equation 28.34, is computed using Wei pit model Equation 28.21.

28.3.3.3 Simultaneous Corrosion-Fatigue (SCF) Model

The SCF model describes corrosion-fatigue damage occurring simultaneously [8]:

1. Cracks form during the crack nucleation phase. The time to form a crack can be decreased by corrosion that occurs during the crack nucleation phase.
2. Cyclic loading is interspersed with periods of pit growth, which increases the local stress concentration factor.
3. If load cycles are applied infrequently, pitting may transition to general corrosion and thickness loss, leading to an increase in the global stress.
4. Once a crack is formed, only the thickness loss due to general corrosion and the associated stress increase affect the growth rate of the crack.

The SCF model is implemented as follows: (1) for crack initiation, an additive incremental total damage model that linearly superimposes the pit-depth increment and the stress-concentration factor, and (2) for crack propagation, two time-variant corrosion-topography factors (pitting and thickness loss factors) that multiply the stress-intensity-factor range.

28.3.3.4 Comparative Results

In this subsection, computed results obtained using the deterministic corrosion-fatigue models described above are compared.

Figure 28.28 and Figure 28.29 compare the WCF model and the CCCF model for a plate with a hole [51]. The plate is 90 mm wide and 1.3 mm thick. The hole has a 3-mm radius and is located in the center of the plate. The initial pit size, corresponding to the size of the constituent particle from which

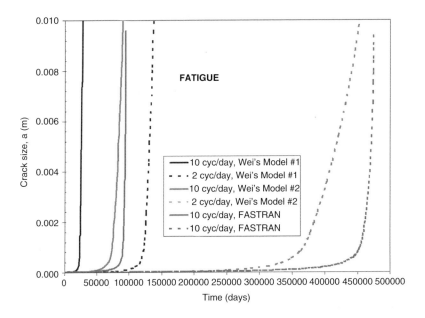

FIGURE 28.28 Results of the WCF and CCCF model (using FASTRAN) for pure fatigue.

the pit starts, is 20 μm. The constituent particle shape was assumed to be hemispherical. Two aircraft operating scenarios are considered here: (1) 10 load cycles/day, assuming 15-h flight and a 9-h stay on ground (for same location); and (2) 2 load cycles/day, assuming 3-h flight and 21-h stay on ground (for same location). The stress range was 90 MPa at temperature, and the notch factor was $K_t = 2.6$. The material was 2024-T3 aluminum. The fatigue coefficient, C_c, and the exponent, n_c, were assumed to be 3.95 E–11 and 3.55 for Wei model 1 [52] and 1.86 E–11 and 3.15 for Wei model 2, respectively.

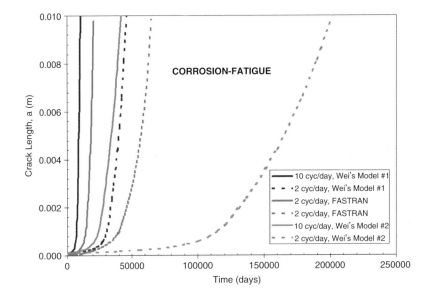

FIGURE 28.29 Results of WCF and CCCF model (using FASTRAN) for corrosion fatigue.

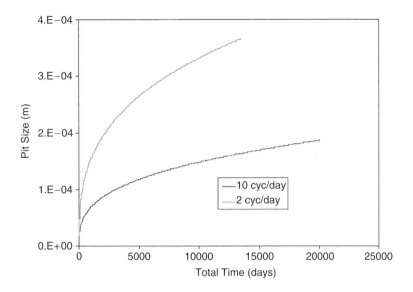

FIGURE 28.30 Pit-depth curves for the two operating scenarios for aircraft.

The pit-growth and the pitting-factor curves for the two corrosion scenarios are plotted in Figure 28.30 and Figure 28.31, respectively. The pit-growth curves are the same for the WCF and CCCF models. The pitting factors are applied only in conjunction with the CCCF model (Equation 28.33 and Equation 28.34).

Figure 28.28 shows the pure-fatigue lives computed for the WCF models and the CCCF model assuming 10 cycles/day and 2 cycles/day, respectively. Figure 28.29 shows the corrosion-fatigue lives using the same models. Only the time on ground was considered for corrosion growth. The computed lives are also included in Table 28.8.

It should be noted from Figure 28.28, Figure 28.29, and Table 28.8 that the range of results of the WCF model 1 and model 2 include the CCCF results. For pure fatigue, there is a poor matching between the lives computed using Wei model 1 and CCCF. The Wei model 2 matches quite well the FASTRAN results

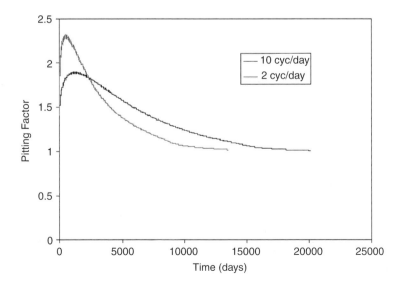

FIGURE 28.31 Pitting corrosion factor for the two operating scenarios for aircraft.

TABLE 28.8 Comparative Life Results Using WCF Models and CCCF Model

Investigated Case	WCF- Wei Model 1	WCF-Wei Model 2	CCCF
Fatigue, 10 cycles/day	27,560 days	90,647 days	94,905 days
Fatigue, 2 cycles/day	137,802 days	453,239 days	474,850 days
Corrosion fatigue, 10 cycles/day	10,185 days	41,320 days	20,018 days
Corrosion fatigue, 2 cycles/day	45,370 days	201,046 days	64,951 days

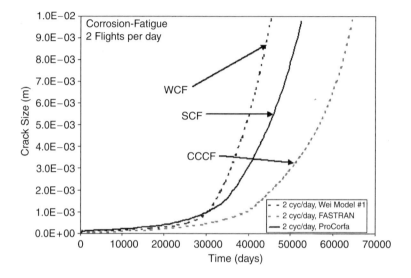

FIGURE 28.32 Comparative life predictions for 2 cycles/day using WCF model, SCF model, and CCCF model.

for pure fatigue, while the Wei model 1 agrees better than the Wei model 2 with CCCF results for the largest corrosion damage, namely, the 2-cycles/day case (21 h/day stay on ground).

Figure 28.32 and Figure 28.33 compare the WCF, CCCF, and SCF models for the two previous corrosion-fatigue scenarios. In this comparison, the SCF model uses the LDR for crack initiation combined with the

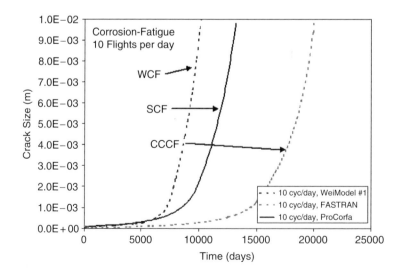

FIGURE 28.33 Comparative life predictions for 10 cycles/day using WCF model, SCF model, and CCCF model.

Forman model for crack propagation (Table 28.3 and Table 28.4). The main difference between the WCF model results and the SCF model results are due to the different fatigue-crack-propagation models used. WCF uses a truncated Paris-law model, and SCF uses the Forman model. The crack growth threshold, ΔK_{th}, was taken equal to 3 MPa√m. The Paris-law model was truncated at this threshold value.

28.4 Reliability of Aircraft Structure Joints Including Maintenance Activities

The reliability analysis concept is illustrated in Figure 28.34. It can be seen in the figure that the effect of corrosion on fatigue life is to increase the time-variant failure risk and to produce unscheduled maintenance events. In Figure 28.34, notation SME stands for scheduled maintenance events, and notation UME stands for unscheduled maintenance events. In the figure, the probability distributions of the crack size population before and after the inspections are also shown. The result of inspection is the replacement (or repair) of the components with larger cracks, which are most likely to be detected by inspections.

28.4.1 Risk/Reliability-Based Condition Assessment

Component risk/reliability-based condition assessment is usually based on three risk/reliability metrics: (1) instantaneous probabilistic failure risk that expresses the risk at any given time or damage level, (2) component remaining life (when no maintenance activity is included), and (3) future probabilistic failure risk that expresses the risk in the next time interval (this future interval is associated with a maintenance interval).

28.4.1.1 Physics-Based Reliability Engineering Approach

The physics-based reliability engineering approach integrates the structural reliability theory with the reliability engineering theory. The basic relationship that links the two theories is the relationship between

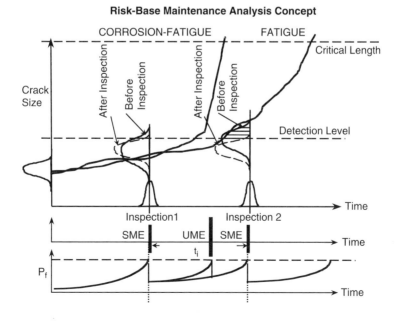

FIGURE 28.34 Risk/reliability-based maintenance analysis concept.

the computed instantaneous failure probability and the hazard failure rate at any given time:

$$\text{Prob}[T_f \leq t + \Delta t | T_f > t] = P_f(t) + [1 - P_f(t)] \exp\left[-\int_t^{t+\Delta t} h(x)dx\right] \quad (28.35)$$

The above equation expresses the failure probability within a time interval that defines the probability distribution of a component life, in terms of the instantaneous failure probability computed at the starting time of the interval, $P_f(t)$, and the variation of the hazard failure rate, $h(x)$, in the interval. Then, the reliability engineering metrics, such as MTBF (mean time between failures) that are required for maintenance cost analysis can be computed by integrating the reliability function (defined by unity minus the failure probability):

$$\text{MTBF} = \int_0^\infty [1 - P_f(t)]dt \quad (28.36)$$

Statistics and reliability metrics that are of interest to engineers and decision makers are:

1. Crack-length statistics evolution with no or multiple inspection intervals
2. Failure risk evolution with no or multiple inspection intervals
3. Reliability index evolution with no or multiple inspection intervals
4. Hazard failure rate evolution with no or multiple inspection intervals
5. Average hazard failure rates per inspection intervals
6. Number of failures (removals) per inspection intervals
7. PDF of the parent crack length population after each inspection
8. Equivalent Weibull failure (life) models
9. Posterior probability density function of life via Bayesian updating to incorporate failure data
10. Posterior PDF of crack size via Bayesian updating to include inspection data

An adequate risk/reliability-based condition assessment of an aircraft component with corrosion-fatigue damage needs to include the following analysis and modeling steps:

1. Stochastic modeling of operational loading condition and environmental conditions
2. Stochastic modeling of component loading, environmental surface conditions, and material and structural properties (This step may also include modeling of the component surface boundary conditions, such as contact-surface constraint effects, material property variations, manufacturing deviations from the baseline geometry, etc. These last aspects are not discussed here.)
3. Stochastic component stress/strain analysis to compute the stress/strain state in the component for given operating conditions that are time dependent
4. Stochastic modeling of component stress and strain histories at critical locations (This step includes the construction of principal-, component-, and equivalent-stress histories.)
5. Component reliability analysis or risk analysis for initial no-usage conditions (no deterioration due to progressive damage mechanisms) (This initial risk is due to stochastic variations in component design parameters, including manufacturing geometry deviations, material fabrication defects, assembly errors, etc. This time-invariant reliability problem is not discussed here.)
6. Reliability/risk-based condition assessment and life prediction (effect of maintenance is not included) based on stochastic damage models for both the crack-nucleation and crack-propagation stage
7. Reliability/risk-based maintenance analysis, including the effects of maintenance uncertainties on present failure risks, defined instantaneous failure risks, and future failure risks during the some time interval, typically selected to be the next inspection interval
8. Optimal-cost reliability/risk-based maintenance cost analysis, including the computation of overall maintenance costs vs. the component removal time based on reliability analysis results accounting for both scheduled maintenance events (SME) and unscheduled maintenance events (UME). (These postreliability analysis aspects are not discussed here.)

Two reliability analysis options are possible: (1) for a defined maintenance schedule and inspection techniques, the time-variant component risk/reliability (and unscheduled maintenance rates) can be computed, or (2) for a selected reliability level and selected inspection techniques, the required (scheduled) maintenance intervals can be determined.

28.4.1.2 Equivalent "Physics-Based" Weibull Failure Models

For practical purposes, equivalent Weibull component life models are determined based on the computational results of the physics-based reliability analysis. These equivalent "physics-based" Weibull life models have the advantage that they can be easily compared with the existing Weibull models developed from field failure data.

To compute the two parameters of the Weibull distribution, a least-squares error minimization technique is used to fit the random sample life data [61]. Before performing the least-squares fitting, a transformation of the coordinates is performed so that the Weibull distribution points are shown along a straight line.

For the equivalent Weibull model, the instantaneous failure probability is computed by

$$P_f(t) = 1 - e^{-\left(\frac{t}{\theta}\right)^\beta} \tag{28.37}$$

where β and θ are the shape and scale parameters of the Weibull distribution.

The Weibull hazard failure rate at time t is expressed by

$$h(t) = \frac{1}{\theta\beta\left(\frac{t}{\theta}\right)^{\beta-1}} \tag{28.38}$$

For a shape factor equal to unity, the Weibull distribution reduces to an exponential distribution that has a constant mean hazard failure rate.

28.4.1.3 Maintenance Inspection Uncertainties

To maintain an acceptable reliability level for a mechanical component, two strategies are available: (1) to design the component for a long life so that there is no need for any maintenance during service life, or (2) to allow maintenance through inspections during the component service life, with repairs as required. It is known that the second strategy corresponds to a more cost-effective approach and can help to extend the component service life. The key aspect for implementing such a strategy is to be able to accurately predict and control the evolution of the component's failure risk, including all maintenance activities and their associated uncertainties.

28.4.1.3.1 Nondestructive Inspection (NDI) Techniques

Inspection routines are adopted to detect and remove cracks with sizes larger than a rejection limit, resulting in the improvement of reliability toward an acceptable level. For a particular NDI technique, several factors randomly affect the inspection results. For aircraft components, the most important influencing factors are those related to the precision of the type of NDI used and the operator skill.

The detection probability is defined as the number of times a crack of size "a" has been detected, divided by the number of trials, with each trial being performed by a different inspector or inspection team using the same inspection technique. Crack sizing errors include a significant statistical uncertainty. The literature includes some outstanding references on the subject [62–64].

The rejectable crack size a_R is also an important parameter for component maintenance. This limit size is specified based on safety and economic aspects. The rejectable crack size a_R (corresponding to repair or replacement) represents the limit for maintenance action on a detected crack of either accepting (leave) or rejecting (fix) it.

The rejectable crack size can be used to evaluate the following probabilities, where independence between additive sizing error and detection is assumed [62]:

1. The probability $P_R(a)$ of rejecting a crack with size a, calculated as the product of the detection probability and the probability of sizing the detected crack larger than a_R:

$$P_R(a) = P_D(a)[1 - F_S(a_R - a)] \tag{28.39}$$

2. The probability $P_A(a)$ of accepting a crack with size a, calculated as the product of the detection probability and the probability of sizing the detected crack smaller than a_R, added to nondetection probability:

$$P_A(a) = P_D(a)F_S(a_R - a) + [1 - P_D(a)] = 1 - P_R \tag{28.40}$$

In Equation 28.39 and Equation 28.40, F_S is the cumulative probability distribution of the statistical crack sizing errors. For a given crack size a, the sum of these two probabilities equals unity, since a crack must always be either rejected or accepted. For a particular case where $a > a_R$, the function $P_R(a)$ is called the probability of correct rejection, while for $a < a_R$, the function $P_A(a)$ is called the probability of correct acceptance. It should be observed that both $P_R(a)$ and $P_A(a)$ depend on the reliability of the inspection technique and on the specified rejection limit a_R. These definitions can be used to evaluate four additional parameters quantifying the global effect of an inspection procedure [62]:

1. The total probability of correctly rejecting a crack

$$P_{CR} = \int_{a_R}^{-\infty} P_R(a) f_A(a) da \tag{28.41}$$

2. The total probability of incorrectly rejecting a crack

$$P_{IR} = \int_0^{a_R} P_R(a) f_A(a) da \tag{28.42}$$

3. The total probability of correctly accepting a crack

$$P_{CA} = \int_0^{a_R} P_A(a) f_A(a) da \tag{28.43}$$

4. The total probability of incorrectly accepting a crack

$$P_{IA} = \int_{a_R}^{-\infty} P_A(a) f_A(a) da \tag{28.44}$$

Obviously, the sum of the above probabilities is unity, $P_{CR} + P_{IR} + P_{CA} + P_{IA} = 1$. The function $f_A(a)$ is the probability density function of the crack-length population before inspection.

28.4.1.3.2 Brief Description of NDE Inspection Types

Nondestructive evaluation (NDE) tests are used in maintenance to avoid loss of aircraft due to aging effects. They are also used in manufacturing to assure the quality of the components. Nearly every form of energy is used in nondestructive tests, including all wavelengths of the electromagnetic spectrum as well as mechanical vibration. These tests are divided into the following basic methods: visual, liquid penetrant, radiographic, ultrasonic, eddy-current, microwave, and infrared.

28.4.1.3.3 Probability of Detection (POD) Curves

The capabilities of NDE techniques are typically quantified by plotting the probability of detection (POD) as a function of flaw size. Berens and Hovey [65] have shown that a lognormal formulation for the POD curve provides a reasonable model for the observed behavior of NDE data. The lognormal POD can be expressed as

$$\text{POD}(x) = \int_0^x \frac{1}{\sigma u \sqrt{2\pi}} \exp\left[\frac{-(\ln(u) - \mu)^2}{2\sigma^2}\right] du \qquad (28.45)$$

Studies have been conducted by various organizations to determine POD curves for various NDE techniques when applied to various selected aircraft components. Similar to the POD curves for crack detection, POD curves for thickness loss due to corrosion can be determined. Figure 28.3 shows the POD curve for detecting thickness loss using an eddy-current NDE inspection of an unpainted 737 aircraft splice joint. It can be noticed from that figure that the operator's skill can significantly affect the POD curve for a given eddy-current NDE technology.

28.4.1.4 Probabilistic Modeling for Crack Growth Process Including Multiple Inspections

Figure 28.35 shows the corrosion-fatigue-crack growth process with and without crack detection inspections. The plots show the time evolution of the PDF of crack length in an axonometric view and, using contour right-side plots, this corresponds to four NDE inspections at 4000 flight hours (FH) each. After each inspection, new cracks are born due to the repair or replacement of components with large cracks. The new crack populations are introduced by the removal of large cracks in the previous crack populations. An accurate stochastic modeling of the corrosion-fatigue-crack growth process, including inspections, has to include the presence of multiple statistical crack size populations. A nonnormal probabilistic mixture model is used for the crack size populations. For each crack size population, a nonnormal probability distribution is assumed.

28.5 Illustrative Examples

To keep the discussion simple, the illustrative examples presented in this section include only the effect of pitting corrosion on corrosion-fatigue life. The effects of other corrosion types, including intergranular corrosion in early stages or general thickness loss and pillowing in later stages, are not considered. No cladding was assumed. Also, the multiple site damage (MSD) or widespread fatigue damage (WFD) that usually produces the ultimate lap-joint system failures are not included. Only the local failure in critical locations is considered. However, both MSD and WFD are real threats to aircraft structural integrity and therefore they must be considered when evaluating the risk of failure for an actual aircraft structure.

Several examples of probabilistic life prediction and risk-based maintenance analysis against corrosion-fatigue damage are shown in this section:

1. Probabilistic life predictions using the WCF, CCCF, and SCF models
2. Risk-based optimal cost analysis for a typical component
3. Risk-based maintenance analysis for a typical aircraft lap joint (Figure 28.1), including the effect of randomly rotating the aircraft to different airfields

28.5.1 Probabilistic Life Prediction Using Different Corrosion-Fatigue Models

In the first example, the CCCF model (using a modified FASTRAN version) is used to compute the corrosion-fatigue life of thin 2024-T3 aluminum sheets [44]. The surface constituent particle size was statistically modeled using a lognormal probability distribution based on the results of Laz and Hillberry [44]. The simulated PDF of the particle size is plotted in Figure 28.36. Figure 28.37 shows the computed probabilistic fatigue life and corrosion-fatigue life assuming airport locations with different environmental severities. Figure 28.38 shows the fatigue life vs. the corrosion-fatigue life for all the airport locations

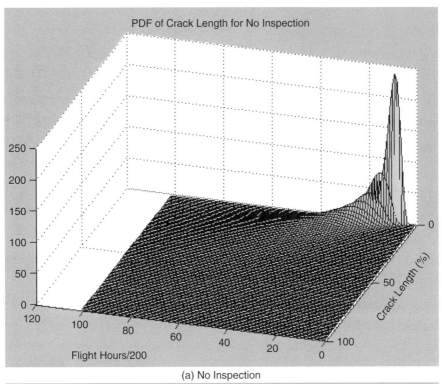
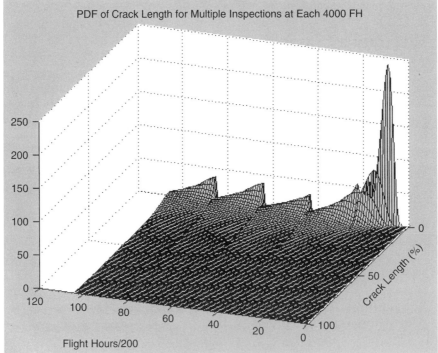

FIGURE 28.35 Evolution of crack-length development: (a) no inspection; (b) multiple inspections at each 4000 flight hours.

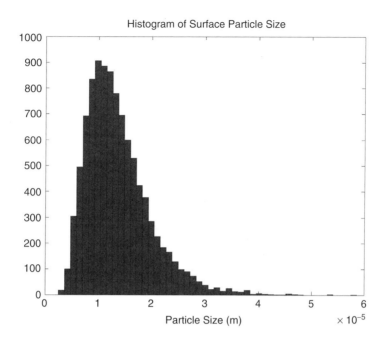

FIGURE 28.36 Simulated PDF of the surface constituent particle size.

considered. It should be noted that the corrosion effects reduce the component fatigue life up to 10 to 15 times. Figure 28.39 shows the plot of the corrosion-fatigue life vs. the particle size. As shown, there is a relatively weak negative statistical correlation between corrosion-fatigue life and particle size. This indicates that particle size may not be a governing parameter for the corrosion-fatigue-life prediction (the random corrosion effects influence the life more significantly). This negative correlation is much stronger between pure fatigue life and particle size, as shown in Figure 28.40. Thus, for pure fatigue damage, the role of the particle size on the predicted life is significantly greater than for corrosion-fatigue damage.

The second example is a comparison between the WCF model and the SCF model for assessing the probabilistic corrosion-fatigue life of an aircraft component. The material considered is 2024-T3 aluminum. The constituent surface particle sizes, the threshold stress intensity range ΔK_{th}, and the pit depth were assumed to be random variables for the probabilistic life prediction. The constituent particle size distribution is lognormal based on the data of Laz and Hillberry [43, 44], as shown in Figure 28.36. The stress-intensity-range threshold was modeled by a normal variable with mean of 3 MPa$\sqrt{m}$ and a coefficient of variation of 0.10. The pit depth at any arbitrary time was modeled by a random scale factor between 1 and 21 applied to a mean $I_{PO} = 0.5$ C/sec. The pit scale factor was introduced to simulate the different environmental severity conditions at various airport locations. The fatigue-crack-propagation models included in the WCF and SCF models are the truncated Paris-law crack-growth model and, respectively, the Forman crack-growth model. They are compared in Figure 28.41. From this figure it can be observed that for stress intensity ranges that only are slightly larger than the threshold of 3 MPa$\sqrt{m}$, the WCF model assumes much higher crack growth rates than the SCF model. This behavior is expected to reduce the fatigue life computed with the WCF model.

Figure 28.42 and Figure 28.43 show the probabilistic corrosion-fatigue life computed for the two aircraft operating scenarios of two load cycles/day and ten load cycles/day, respectively, assuming that the mean duration of one cycle (flight) is 1.5 hours. The probabilistic life estimations indicate that the WCF model overestimates the statistical variability of the corrosion-fatigue life due to two modeling

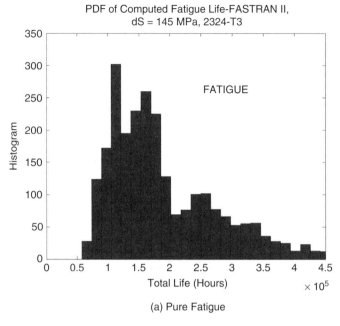

(a) Pure Fatigue

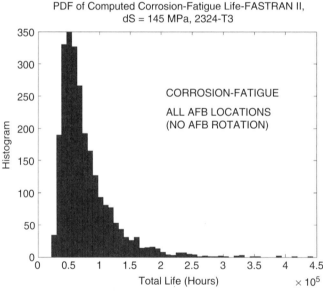

(b) Corrosion-Fatigue Including Different Airport locations

FIGURE 28.37 Predicted-life PDF computed using CCCF model: (a) pure fatigue; (b) corrosion fatigue including different airport locations.

effects: (1) it exaggerates the crack growth rates for ΔK slightly above ΔK_{th}, so that it produces a shorter life of some components, and (2) it does not include the effect of pitting on crack growth, so that it produces a longer life of some other components. The first effect, item 1, is stronger and more visible when the fatigue damage is greater, i.e., greater for ten cycles/day than for two cycles/day. The second effect, item two, is more visible when the *corrosion* damage is greater, i.e., greater for two cycles/day than for ten cycles/day.

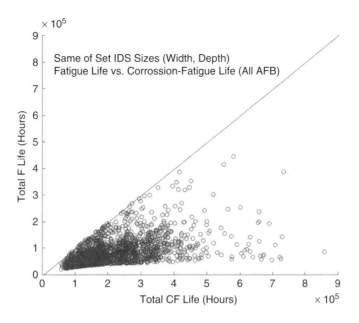

FIGURE 28.38 Fatigue life vs. corrosion-fatigue life.

28.5.2 Risk-Based Maintenance Analysis of a Lap Joint Subjected to Pitting Corrosion and Fatigue

The reliability analysis was performed for the aircraft lap joint shown in Figure 28.1. The major loading in the lap joint comes from the pressurization in the aircraft. Figure 28.44 shows the load transfer with the aircraft lap-joint components. The input random variables included in the reliability analysis are shown in Table 28.9.

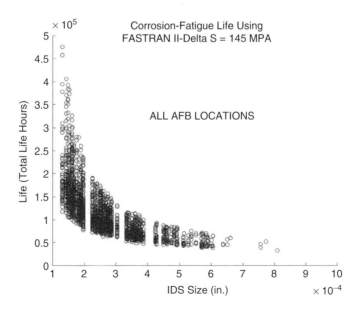

FIGURE 28.39 Corrosion-fatigue life vs. surface particle size.

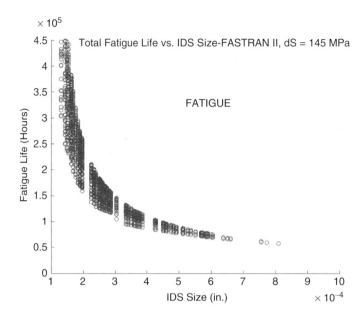

FIGURE 28.40 Pure fatigue life vs. surface particle size.

Figure 28.45 illustrates the stochastic history of pressure loading and environmental conditions of the aircraft. The elementary constituent of the stochastic history of the lap joint is the block that includes a single flight and a single stay on ground. It was assumed that the random pressure load is described by a single cycle for each flight. The environmental severity condition that drives corrosion was considered to randomly vary with the airport location. However, for the same location it was assumed that the environmental condition is a time-invariant quantity.

Figure 28.46 illustrates the simulated PDF of the pit growth volumetric rate based on the assumption shown in Table 28.9. The surface particles were assumed to be the initiators of the pits and microcracks.

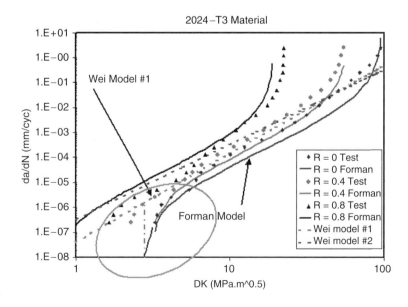

FIGURE 28.41 Comparison of fatigue-cracking models.

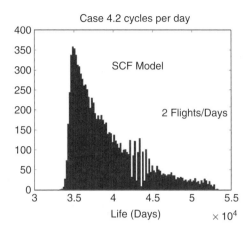 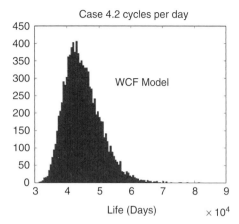

FIGURE 28.42 PDF of predicted life for the SCF and WCF models for 2 cycles/day.

From Figure 28.46, it should be noted that the environmental severity condition characterized by pit growth rate has a highly skewed probability distribution. Figure 28.46 indicates that the environmental severity conditions expressed by the pit growth rates are mild for most of the airport locations and severe for only a few locations. A truncated exponential distribution was used to fit the trend of the measured corrosion rate data at different airport locations [8, 60]. These large differences in values indicate that the crevice pits can grow up to ten times faster in some airport locations than in others.

Four flight scenarios were investigated for reliability analysis of the aircraft lap joint. The four scenarios were obtained by combining two aircraft operating scenarios with two flying scenarios. The two operating scenarios were (a) one flight/day and (b) three flights/day, and each of these was applied in two flying scenarios: (1) each aircraft flies from an airport location to the same airport location, without random rotation of the airport location, and (2) each aircraft flies randomly from an airport location to any other airport location, with random rotation of the airport location. In the last flying scenario, it was assumed that all airport locations are equally probable and that each individual aircraft can visit all airport locations. This is the ideal situation for reducing scatter of the corrosion effect, assuming a uniform distribution of the aircraft fleet across the airport location set.

To compute the probabilistic corrosion-fatigue life of the lap joint, both the crack-initiation and the crack-propagation stages were included. The stochastic strain-life curve and the stochastic Forman

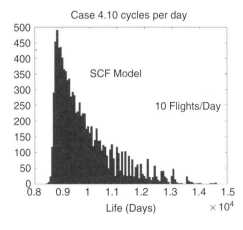 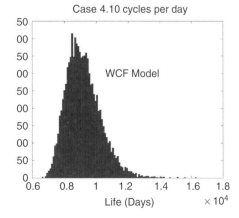

FIGURE 28.43 PDF of predicted life for the SCF and WCF models for 10 cycles/day.

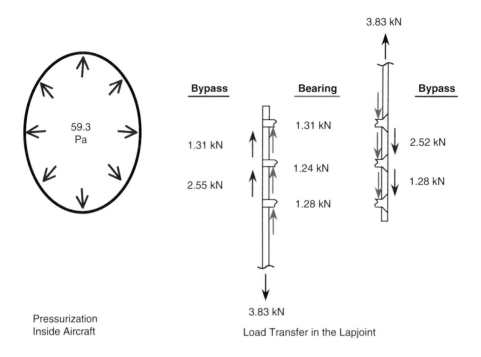

FIGURE 28.44 Pressure load transfer in the lap-joint components — fasteners and splices.

TABLE 28.9 Input Random Variables Included in the Reliability Analysis

Random Parameter	Mean	Standard Deviation	Probability Distribution
Uniform pressure inside aircraft, p(Pa)	59.3	2.97	normal
Single flight duration, d (h)	2.8	0.50	lognormal
Surface particle size, a_0 (μm)	13.66	6.02	Weibull (Figure 28.36)
Strain life curve exponents, b and c	−0.114, −0.927	0.00114, 0.00927	normal, normal
Strain life curve parameters, σ'_f (MPa) and ε'_f	1044 1.765	20.88 0.0353	normal normal
Stress-intensity-range threshold, ΔK_{th} (MPa√m)	3.00	0.15	normal
Toughness, K_c (MPa√m)	97.7	2.93	normal
Pit-growth parameter, I_{PO}, in Wei model variation due to different environmental conditions for different airport locations (C/sec)	14.08	22.26	truncated exponential 0.1–100 C/sec (Figure 28.46)

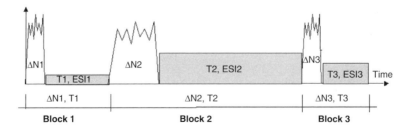

FIGURE 28.45 Stochastic history of loading and environmental conditions.

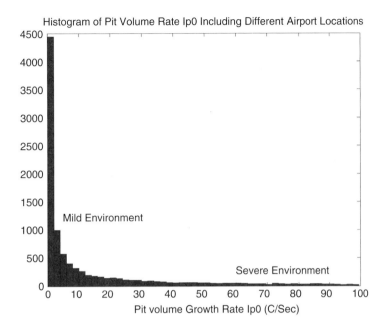

FIGURE 28.46 Simulated PDF of the pit-growth volumetric rate.

crack-propagation models were developed from the deterministic models based on the assumption that their parameters are random quantities, as shown in Table 28.9. To include the effect of pitting corrosion on the lap-joint fatigue life, a SCF model was employed.

Figure 28.47 and Figure 28.48 show the simulated pit depth growth curves for all airport locations assuming no rotation of airport locations. These pit curves were computed using Wei pitting model

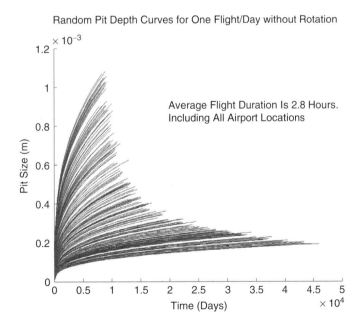

FIGURE 28.47 Simulated pit-growth curves for one flight/day without airport rotation.

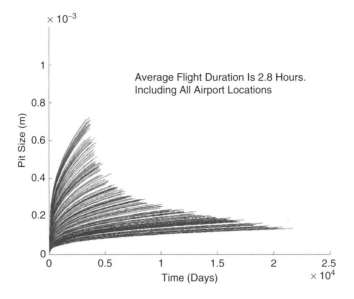

FIGURE 28.48 Simulated pit-growth curves for three flights/day without airport rotation.

(Equation 28.20). The pit growth curves shown in the figures stop at the failure times. Figure 28.47 is for the one-flight/day scenario and Figure 28.48 is for the three-flights/day scenario, respectively. Figure 28.49 and Figure 28.50 show the pit-growth curves for the same two scenarios with a random rotation of aircraft location. It was assumed that each aircraft has an equal probability to fly to any airport location. This means there is a high probability that each airport will be visited about the same number of times by each aircraft. Therefore, for the scenario with the airport rotation, the scatter of the pit growth

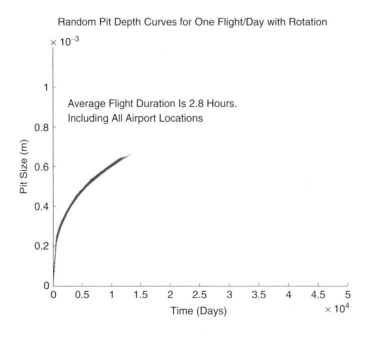

FIGURE 28.49 Simulated pit-growth curves for one flight/day with airport rotation.

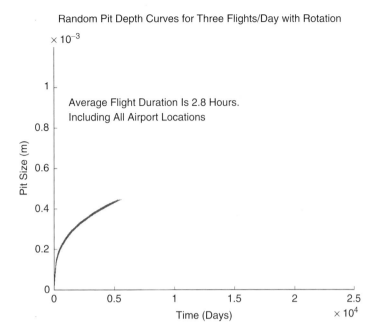

FIGURE 28.50 Simulated pit-growth curves for three flights/day with airport rotation.

drops significantly, converging in the limit to the (deterministic) mean pit growth for an infinite number of flights per aircraft.

The simulated crack-length curves are plotted in Figure 28.51 through Figure 28.54 for the four investigated scenarios. The computed histograms (with different incremental steps) of predicted corrosion-fatigue life of the four cases are shown in Figure 28.55. It should be noted that the mean

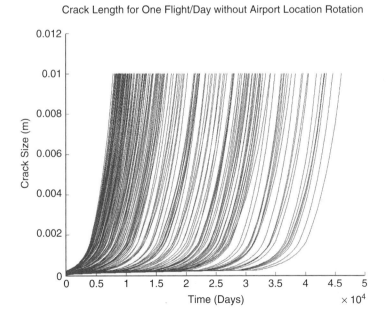

FIGURE 28.51 Simulated crack-size curves for one flight/day without airport rotation.

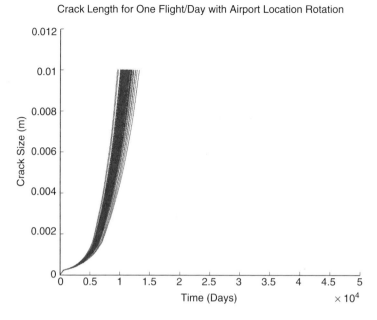

FIGURE 28.52 Simulated crack-size curves for one flight/day with airport rotation.

corrosion-fatigue life is about double for the one-flight/day scenario vs. three-flights/day scenario. Figure 28.56 and Figure 28.57 illustrate the probability density of the time until a 5.0-mm crack length is reached for the one-flight/day scenario, without airport rotation and with airport rotation, respectively. The computed probability densities (PDF) are compared with analytical densities, namely the lognormal and normal probability densities. It should be noted that for the case without rotation, the computed skewed density is far from the lognormal density, while for the case with rotation, the

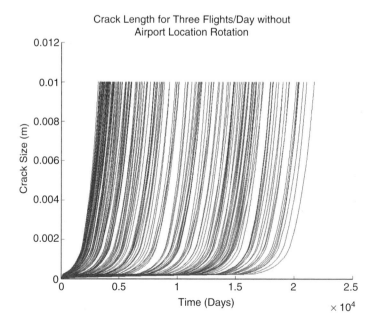

FIGURE 28.53 Simulated crack-size curves for three flights/day without airport rotation.

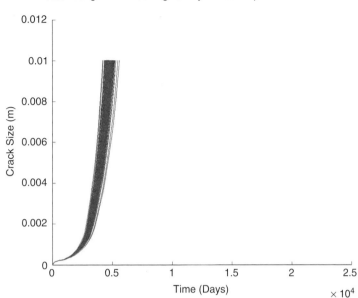

FIGURE 28.54 Simulated crack-size curves for three flights/day with airport rotation.

computed density is very close to normal density. For the former case, without rotation, the heavy right tail of the PDF shape is due to the fact that many airport locations have milder environmental-severity conditions, as indicated in Figure 28.46. For the latter case, the scatter of corrosion effects is reduced and the predicted-life probability density converges to the normal distribution in accordance with the central limit theorem.

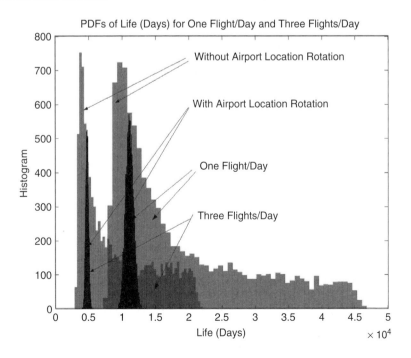

FIGURE 28.55 Corrosion-fatigue histograms (different steps) for the investigated scenarios.

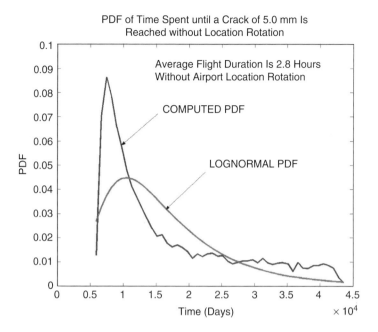

FIGURE 28.56 PDF of corrosion-fatigue life for one flight/day without airport rotation.

To consider the effect of maintenance, the uncertainties associated with the probability of crack detection for different standard NDE inspections were included using the appropriate POD curves. The eddy-current NDE technique with different operator skill classes was considered. The eddy-current POD curve was assumed to correspond to a lognormal distribution with a logarithmic mean and logarithmic standard deviation of (a) −4.73 and 0.98 for the best operator, (b) −3.75 and 0.70 for the average operator, and (c) −2.73 and 0.45 for the worst operator. No crack sizing error was included in addition to operator's

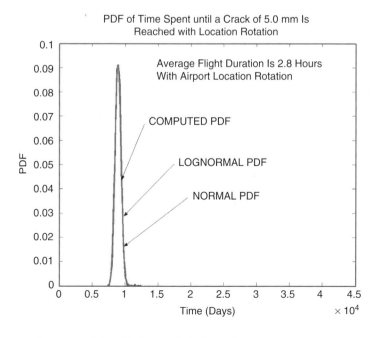

FIGURE 28.57 PDF of corrosion-fatigue life for one flight/day with airport rotation.

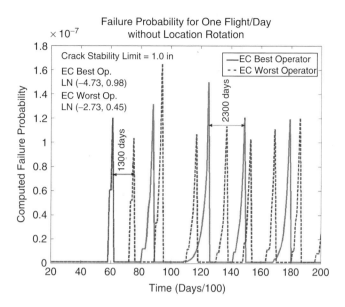

FIGURE 28.58 Risk-based inspection times for one flight/day, without rotation, for a given target risk of 2×10^{-7}: effect of the operator's skill.

skill variation. At each inspection time, the statistical crack population was filtered through the POD curve. Based on the computed probabilities of acceptance or rejection, each crack was randomly accepted or removed by replacing the cracked component. The repair effects were not considered for this illustrative example.

Figure 28.58 through Figure 28.61 indicate the inspection schedule required over 20,000 days (about 60 years) for maintaining the corrosion-fatigue damage risk under a reliability target defined by an upper-bound failure probability of 2×10^{-7}. Figure 28.58 and Figure 28.59 show the results computed for the

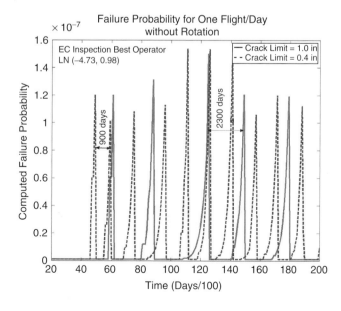

FIGURE 28.59 Risk-based inspection times for one flight/day, without rotation, for a given target risk of 2×10^{-7}: effect of crack-limit criterion.

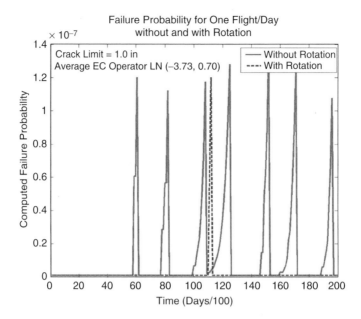

FIGURE 28.60 Risk-based inspection times for one flight/day without and with rotation.

one-flight/day scenario without airport rotation. Figure 28.58 compares results for different NDE operator's skills (best operator vs. worst operator), while Figure 28.59 compares results for different failure limit criteria (crack limit of 1.0 in. vs. crack limit of 0.40 in.). It should be noted that the minimum inspection interval drops from 2300 days (6450 flight hours [FH]) to 1300 days (3640 FH) due to the NDE operator's skill, and from 2300 days (6540 FH) to 900 days (2520 FH) due to the crack-limit criterion considered.

Figure 28.60 and Figure 28.61 compare the required inspection schedules for the two cases, without and with airport rotation, including both the one-flight/day scenario and three-flights/day scenario, assuming the same reliability target, an average operator's skill, and a 1.0-in. crack-limit failure criterion.

Without the airport rotation, the required inspection intervals in real time are about two or three times longer for the one-flight/day scenario than for the three-flights/day scenario. However, if the inspection intervals are measured in effective FH instead of days, this observation is not true. The minimum inspection intervals are 1600 days (4480 FH) for the one-flight/day scenario and 600 days (5040 FH) for the three-flights/day scenario. The increase of the inspection intervals expressed in flight hours from the one-flight/day scenario to three-flights/day scenario indicates that the effects of corrosion are more severe for one-flight/day when the time spent by an aircraft on ground is longer.

With the airport rotation, the minimum inspection intervals are much longer than those computed without airport location rotation. The minimum inspection intervals are 11,200 days (31,360 FH) for the one-flight/day scenario and 4,600 days (38,640 FH) for the three-flights/day scenario. This large benefit effect of the random rotation of airport locations is mainly a result of the large reduction in the statistical scatter of corrosion effects as a result of the central limit theorem.

The exclusive use of *instantaneous* failure probabilities to characterize aircraft reliability is insufficient for setting the risk-based maintenance strategy. This is because, from a risk-based-maintenance point of view, one is interested in the aircraft reliability over a period of time, not only at the critical instantaneous times. To illustrate the point, we can review the results in Figure 28.60. For the inspection schedule shown, the maximum risk is almost constant with a value of 1.2×10^{-7}. The maximum risk is bounded to 1.2×10^{-7} independent of the aircraft operating scenarios, without or with airport location rotation. However, the number of inspections is different, so that the number of times when the maximum failure

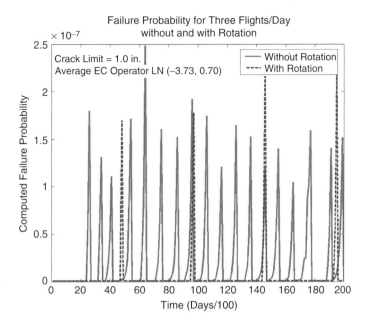

FIGURE 28.61 Risk-based inspection times for three flights/day without and with rotation.

risk is reached is different for the two operating scenarios. Thus, if the average hazard failure rates over a long period are computed, they are very different. For the results in Figure 28.60, if the average hazard failure rates are computed over the 20,000-day (about 60 years) period, these are 1.04×10^{-10} events/day and 7.97×10^{-12} for the cases without airport rotation and with rotation, respectively. This means that, in the long run, the average number of aircrafts having failures is about ten times higher for the case without airport rotation than the case with rotation, although the maximum instant risk is the same for the two cases. Thus, to maintain the same aircraft reliability for the two cases, it would be necessary to define different target instantaneous risks for the two cases, about an order of magnitude lower for the case without rotation, so that finally we end up with the same average hazard rates over the period of interest. Failure probabilities are good reliability metrics for time-invariant or instantaneous reliability problems, while average hazard failure rates are good reliability metrics for time-variant problems such as the risk-based maintenance problem.

The above discussion also indicates that the age distribution of the aircraft fleet plays an important role on the aircraft fleet reliability over a given period of time. Figure 28.60 shows that for different time periods, for example 7000 days (about 20 years), depending on the aircraft fleet age distribution and its variation in the selected periods, the individual aircraft risks can vary wildly. As a consequence of this, the average fleet hazard failure rates can vary about two orders of magnitude for different time periods and different fleet age distributions.

28.6 Concluding Remarks

This chapter presents an overview of the key engineering issues that are important for performing a reliability analysis of aircraft structure joints under corrosion-fatigue damage. The chapter focused on probabilistic modeling of stochastic cumulative damage due to corrosion fatigue. Different corrosion-fatigue models are reviewed in relative detail, and their results are compared. One model was then applied to the probabilistic life prediction and risk-based maintenance analysis of an aircraft fuselage lap joint. For this illustrative example, the loading stochasticity was limited to the small variability of aircraft pressurization. Variability-related structural modeling and analysis, as well as the stochasticity of the loading and load transfer through the structure to the individual fasteners, was ignored in order to keep the example simple.

Computational risk-based maintenance using physics-based stochastic damage models, carefully calibrated with the appropriate empirical data, provides a quantitative process for simultaneously maximizing aircraft availability and reducing maintenance costs while maintaining safety and airworthiness. The physics-based stochastic modeling tools and computational reliability methods are sufficiently mature to approach the difficult problem of aircraft fleet maintenance from a probabilistic risk-based perspective.

An important practical aspect, not discussed herein, is that probabilistic models need to be implemented so that they can incorporate new information and statistical data coming from lab tests, depot maintenance, and service history. Refinement of the probabilistic models in this way will make risk predictions sharper by reducing their statistical confidence intervals (reducing uncertainties due to modeling and lack of data).

References

1. U.S. Air Force, Aircraft Structural Integrity, AFPD 63-10, US Air Force Publications Distribution Center, Baltimore, 1997. (also www.e-publishing.af.mil)
2. U.S. Air Force, Aircraft Structural Integrity Program, AFI 63-1001, U.S. Air Force Publications Distribution Center, Baltimore, 2002. (also www.e-publishing.af.mil)
3. U.S. Air Force, *General Guidelines for Aircraft Structural Integrity Program*, MIL-STD-1530B, Aeronautical Systems Center, Wright-Patterson Air Force Base, OH, 2004. (also www.dodssp.daps.mil)
4. Damage Tolerance and Fatigue Evaluation of Structure, Federal Aviation Regulation, Part 25, Airworthiness Standards: Transport Category Airplanes, Sec. 25.571.
5. Damage Tolerance and Fatigue Evaluation of Structure, Joint Aviation Requirements, Part 25 Large Aeroplanes, Section 25.571. (available at http://www.jaa.nl)
6. Protection of Structure, Federal Aviation Regulation, Part 25, Airworthiness Standards: Transport Category Airplanes, Sec. 25.609.
7. Protection of Structure, Joint Aviation Requirement, Part 25, Large Aeroplanes, Section 25. 609. (available at http://www.jaa.nl)
8. Brooks, C.L., "ECLIPSE-Environmental & Cyclic Life Interaction Prediction Software," Aeromat 1999, Dayton, OH. (available at http://www.apesolutions.com/frm_link.htm)
9. Gallagher, J.P., Giessler, F.J., Berens, A.P., and Engle, R.M. Jr., *USAF Damage Tolerant Design Handbook: Guidelines for the Analysis and Design of Damage Tolerant Aircraft Structure*, AFWAL-TR-82-3073 (ADA153161), Defense Technical Information Center, Ft. Belvoir, VA, May 1984, p. 1.4.3.
10. Rummel, W.D., and Matzkanin, G.A., *Nondestructive Evaluation (NDE) Capabilities Databook*, ADA286978, 3rd Ed., Defense Technical Information Center, Ft. Belvoir, VA, 1997.
11. Yee, B.G.W., Chang, F.H., Couchman, J.C., Lemon, G.H., and Packman, P.F., *Assessment of NDE Reliability Data*, NASA-CR-134991 (ADA321309), Defense Technical Information Center, Ft. Belvoir, VA, 1974.
12. Matzkanin, G.A., and Yolken, H.T., *A Technology Assessment of Probability of Detection (POD) for Nondestructive Evaluation (NDE)*, NTIAC-TA-00-01 (ADA398282), 2001.
13. Hoppe, W., Pierce, J. and Scott, O., *Automated Corrosion Detection Program*, AFRL-ML-WP-TR-2001-4162 (ADA406600), Defense Technical Information Center, Ft. Belvoir, VA, 2001.
14. U.S. Dept. of Defense, Joint Service Specification Guide: Aircraft Structures, JSSG-2006, Appendix A.3.1.2, Requirement Guidance, Dept. of Defense, Washington, D.C., 1998. (also www.dodssp.daps.mil)
15. Lincoln, J.W., Risk assessments of aging aircraft, in *Proc. 1st Joint DoD/FAA/NASA Conf. on Aging Aircraft*, Vol. 1, Air Force Wright Aeronautics Laboratory, Wright-Patterson Air Force Base, OH, 1997, pp. 141–162.

16. Berens, A.P., Hovey, P.W., and Skinn, D.A., *Risk Analysis for Aging Aircraft Fleets*, Vol. 1, *Analysis*, WL-TR-91-3066 (ADA252000), Defense Technical Information Center, Ft. Belvoir, VA, 1991.
17. Hovey, P.W., Berens, A.P., and Loomis, J.S., *Update of the Probability of Fracture (PROF) Computer Program for Aging Aircraft Risk Analysis*, Vol. 1, *Modifications and User's Guide*, AFRL-VA-WP-TR-1999-3030 (ADA363010), Defense Technical Information Center, Ft. Belvoir, VA, 1998.
18. Broeck, David, Residual strength of metal structures, in *ASM Handbook*, Vol. 19, *Fatigue and Fracture*, Lampman, S.R., Ed., ASM International, Materials Park, OH, 1996, pp. 427–433.
19. U.S. Air Force, *Metallic Materials and Elements for Aerospace Vehicle Structures*, Mil-Hdbk-5H, Defense Area Printing Service, Philadelphia, 2001. (also www.dodssp.daps.mil)
20. Skinn, D.A., Gallagher, J.P., Berens, A.P., Huber, P.D., and Smith, J., *Damage Tolerant Design Handbook: a Compilation of Fracture and Crack Growth Data for High Strength Alloys*, Vol. 3, WL-TR-94-4054 (ADA311690), Defense Technical Information Center, Ft. Belvoir, VA, 1994.
21. Mishnaevsky, L.L., Jr., and Schmauder, S., Damage evolution and heterogeneity of materials: model based on fuzzy set theory, *Eng. Fracture Mechanics*, 57(6), 625–636, 1997.
22. Moller, B., Graf, W. and Beer, M., Safety assessment of structures in view of fuzzy randomness, *Comput. Struct.*, 81, 1567–1582, 2003.
23. Mitchell, M.R., Fundamentals of modern fatigue analysis for design, in *ASM Handbook*, Vol. 19, *Fatigue and Fracture*, Lampman, S.R., Ed., ASM International, Materials Park, OH, 1996, pp. 227–249.
24. Socie, D.F., and Morrow, JoDean, Review of contemporary approaches to fatigue damage analysis, *Risk and Failure Analysis for Improved Performance and Reliability*, Burke, J.J. and Weiss, V., Eds., Plenum Publishing Corp., New York, 1980, pp. 141–194.
25. Dowling, N.E., Estimating fatigue life, in *ASM Handbook*, Vol. 19, *Fatigue and Fracture*, Lampman, S.R., Ed., ASM International, Materials Park, OH, 1996, pp. 250–262.
26. Socie, D.F., Dowling, N.E., and Kurath, P., Fatigue life estimation of notched members, in *Fracture Mechanics: Fifteenth Symposium*, ASTM STP 833, Sanford, R.J., Ed., ASTM, Philadelphia, 1984, pp. 284–299.
27. Bucci, R.J., Nordmark, G., and Starke, E.A., Jr., Selecting aluminum alloys to resist failure by fracture mechanisms, in *ASM Handbook*, Vol. 19, *Fatigue and Fracture*, Lampman, S.R., Ed., ASM International, Materials Park, OH, 1996, pp. 771–812.
28. Peterson, R.E., *Stress Concentration Factors*, John Wiley & Sons, New York, 1974.
29. Bannantine, J.A., Comer, J.J. and Handrock, *Fundamentals of Metal Fatigue Analysis*, Prentice Hall, Englewood Cliffs, NJ, 1990.
30. Halford, G.A., Cumulative fatigue damage modeling — crack nucleation and early growth, *Int. J Fatigue*, 19(93), 253–260, 1997.
31. Lankford, J., The growth of small fatigue cracks in 7075-T6 aluminum, *Fatigue Eng. Mater. Struct.*, 5(3), 233–248, 1982.
32. Newman, J.C., Jr., and Edwards, P.R., *Short-Crack Growth Behaviour in an Aluminum Alloy — An AGARD Cooperative Test Programme*, AGARD R-732, National Technical Information Service (NTIS), Springfield, VA, 1988.
33. Edwards, P.R., and Newman, J.C., Jr., *Short-Crack Growth Behaviour in Various Aircraft Materials*, AGARD R-767, National Technical Information Service (NTIS), Springfield, VA, 1990.
34. Goto, M., Statistical investigation of the behaviour of small cracks and fatigue life in carbon steels with different ferrite grain sizes, *Fatigue Eng. Mater. Struct.*, 17(6), 635–649, 1994.
35. Pearson, S., Initiation of fatigue cracks in commercial aluminum alloys and the subsequent propagation of very short cracks, *Eng. Fracture Mech.*, 7, 235–247, 1975.
36. Kung, C.Y., and Fine, M.E., Fatigue crack initiation and microcrack growth in 2024-T4 and 2124-T4 aluminum alloys, *Metallurgical Trans. A*, 10A, 603–610, 1979.
37. Sigler, D., Montpettit, M.C., and Haworth, W.L., Metallography of fatigue crack initiation in overaged high-strength aluminum alloy, *Metallurgical Trans. A*, 14A, 931–938, 1983.

38. Li, P., Marchand, N.J., and Ilschner, B., Crack initiation mechanisms in low cycle fatigue of aluminum alloy 7075-T6, *Mater. Sci. Eng.*, A119, 41–50, 1989.
39. Butler, J.P., and Rees, D.A., Development of Statistical Fatigue Failure Characteristics of 0.125-inch 2024-T3 Aluminum Under Simulated Flight-by-Flight Loading, AFML-TR-74-124 (ADA002310), Defense Technical Information Center, Ft. Belvoir, VA, 1974.
40. Moore, N., Ebbeler, D. H., Newlin, L.E., Surtharshana, S., and Creager, M., "An Improved Approach for Flight Readiness Certification – Probabilistic Models for Flaw Propagation and Turbine Blade Fatigue Failure, Volume 1", NASA-CR-194496, Dec. 1992.
41. Vanstone, R.H., Gooder, O.C. and Krueger, D.D., Advanced Cumulative Damage Modeling, AFWAL-TR-88-4146 (ADA218555), Defense Technical Information Center, Ft. Belvoir, VA, 1988.
42. Newman, J.C., FASTRAN-II, A Fatigue Crack Growth Structural Analysis Program, NASA TM-104159, National Aeronautics and Space Administration, Langley, VA, 1992.
43. Laz, P.J. and Hilberry, B.M., The Role of Inclusions in Fatigue Crack Formation in Aluminum 2024-T3, presented at Fatigue 96: 6th International Fatigue Congress, Berlin, Germany, 1996.
44. Laz, P.J. and Hilberry, B.M., Fatigue life prediction from inclusion initiated cracks, *Int. J. Fatigue*, 20(4), 263–270, 1998.
45. Virkler, D. A., Hillberry, B.M., and Goel, P.K., The Statistical Nature of Fatigue Crack Propagation, AFFDL-TR-78-43 (ADA056912), Defense Technical Information Center, Ft. Belvoir, VA, 1978.
46. Liao, M. and Xiong, Y., A risk analysis of fuselage splices containing multi-site damage and corrosion, presented at *41st Structures, Structural Dynamics, and Materials Conference*, AIAA-2000-1444, American Institute for Aeronautics and Astronautics, Reston, VA, 2000.
47. Hollingsworth, E.H., and Hunsicker, H.Y., Corrosion of aluminum and aluminum alloys, in *ASM Handbook*, Vol. 13, *Corrosion*, ASM International, Materials Park, OH, 1987, pp. 583–608.
48. Cole, G.K., Clark, G., and Sharp, P.K., The Implications of Corrosion with respect to Aircraft Structural Integrity, DSTO-RR-0102, DSTO Aeronautical and Maritime Research Laboratory, Melbourne, Australia, 1997.
49. Bell, R.P., Huang, J.T. and Shelton, D., Corrosion Fatigue Structural Demonstration Program, final report, Defense Technical Information Center, Ft. Belvoir, VA, to be published.
50. Szklarska-Smialowska, Z., Pitting corrosion of aluminum, *Corr. Sci.*, 41, 1743–1767, 1999.
51. Harlow, D.G. and Wei, R.P. Probability approach for prediction of corrosion and corrosion fatigue life, *AIAA J.*, 32(10), 2073–2079, 1994.
52. Wei, R.P., Corrosion and Corrosion Fatigue of Airframe Materials, DOT/FAA/AR-00/22, National Technical Information Service, Springfield, VA, 2000.
53. Brandt, S.M. and Adam, L.H., Atmospheric exposure of light metals, in *Metal Corrosion in the Atmosphere*, ASTM STP 435, ASTM, Philadelphia, PA, 1968, pp. 95–128.
54. Ailor, W.H., Jr., Performance of aluminum alloys at other test sites, in *Metal Corrosion in the Atmosphere*, ASTM STP 435, ASTM, Philadelphia, PA, 1968, pp. 285–307.
55. Sankaran, K.K., Perez, R., and Jata, K.V., Effects of pitting corrosion on the fatigue behavior of aluminum alloy 7075-T6: modeling and experimental studies, *Mater. Sci. Eng.*, A297, 223–229, 2001.
56. Zamber, J.E. and Hillberry, B.M., Probabilistic approach to predicting fatigue lives of corroded 2024-T3, *AIAA J.*, 37(10), 1311–1317, 1999.
57. Burynski, R.M., Jr., Chen, G.-S., and Wei, R.P., Evolution of pitting corrosion in a 2024-T3 aluminum alloy, in *Structural Integrity in Aging Aircraft*, AD-Vol. 47, ASME, New York, 1995, pp. 175–183.
58. Ghiocel, D.M., Combining Stochastic Networks, Bayesian Inference and Dynamic Monte Carlo for Estimating Variation Bounds of HCF Risk Prediction, presented at the *7th National Turbine Engine High Cycle Fatigue Conference*, Monterey, CA, April 16–19, 2003.
59. Ghiocel, D.M., Refined Stochastic Field Models for Jet Engine Vibration and Fault Diagnostics, presented at *International Gas Turbine & Aeroengine Congress, 2000 TURBO EXPO*, Munich, Germany, May 8–11, 2000.

60. Kinzie, R. Jr., Corrosion Suppression: Managing Internal & External Aging Aircraft Exposures, presented at *6th Joint FAA/DoD/NASA Conf. on Aging Aircraft*, Aircraft Structures Session 1A, San Francisco, 2002.
61. Abernethy, R.B., *The New Weibull Handbook*, R.B. Abernethy, 1993.
62. Liao, M, Forsyth, D.S., Komorowski, J.P., Safizadeh, M., Liu, Z. and Bellinger, N., Risk Analysis of Corrosion Maintenance Actions in Aircraft Structures, presented at ICAF 2003, *Fatigue of Aeronautical Structures As An Engineering Challenge*, Lausanne, 2003.
63. Rocha, M.M., Time-Variant Structural Reliability Analysis with Particular Emphasis on the Effect of Fatigue Crack Propagation and Non-Destructive Inspection, report 32–94, University of Innsbruck, Institute of Engineering Mechanics, 1994.
64. Koul, A.K., Bellinger, N.C. and Gould, G., Damage-tolerance-based life prediction of aeroengine compressor discs: II — a probabilistic fracture mechanics approach, *Int. J. Fatigue*, 12(5), 388–396, 1990.
65. Berens, A.P. and P.W., Hovey, Evaluation of NDE Reliability Characterization, AFWAL-TR-81-4160 (ADA114467), Defense Technical Information Center, Ft. Belvoir, VA, 1981.

29
Uncertainty in Aeroelasticity Analysis, Design, and Testing

Chris L. Pettit
United States Naval Academy

29.1 Introduction 29-1
29.2 The Role of Uncertainty in Aeroelasticity 29-2
   Identification of Uncertainties • Sources of Aeroelasticity Uncertainty and Their Importance
29.3 Probabilistic Aeroelasticity Analyses 29-4
   Flutter of Aircraft Lifting Surfaces • Limit-Cycle Oscillation of Airfoils and Panels
29.4 Probabilistic Design for Aeroelasticity........................... 29-9
29.5 Aeroelasticity Applications of Robust Control Concepts 29-11
   Nonprobabilistic Methods for Analysis and Design • Uncertainty in Aeroelastic Tests
29.6 Some Suggestions for Future Research......................... 29-13
29.7 Final Remarks 29-15

29.1 Introduction

Computers have enjoyed the reputation of increasing aerospace design productivity at reduced cost for some time, and analytical techniques have proliferated in parallel with this increased computational capability. However, there is still little reliance on these analytical techniques when it comes to certifying aircraft structures. This is due primarily to lack of confidence in either the analytical method or the fidelity of the model. Hence, the aerospace industry accomplishes structural certification through a design process controlled by safety factors and a resource-intensive regimen of physical tests [1]. The current cost of performing these "building block" structural tests, from small specimens to full flight-testing, for certifying a military airframe is estimated to be around $250 million [2]. Additional costs are, of course, incurred to certify structural repairs and operational modifications during the life of every airframe design. These postproduction costs are particularly difficult to predict during design because most aircraft structures experience undesired structural dynamics behaviors — in operational test and evaluation or in the field — that were either underestimated or unanticipated during design.

Avoiding destructive aeroelastic phenomena is among the most important objectives in aircraft design; therefore, substantial flight-test resources are devoted to demonstrating aeroelastic stability throughout the flight envelope. Freedom from both static (e.g., divergence) and dynamic (e.g., flutter) aeroelastic instabilities must be guaranteed to ensure safe operation. Aeroelastic clearance of civil aircraft is primarily a concern during design and initial flight tests, but many military aircraft designs require flight tests throughout their useful life as their operational demands change (e.g., through the introduction of new external stores). As a consequence, the competing forces of increasingly constrained budgets and expanding requirements for operational flexibility have generated a compelling case for revolutionizing our approach to designing for and demonstrating aeroelastic stability.

This chapter describes recent efforts to employ uncertainty quantification (UQ) methods in this capacity. Each of these efforts was undertaken with at least one of the following goals in mind:

1. Improve understanding of aeroelastic performance sensitivity and variability in canonical aeroelastic systems.
2. Employ aeroelastic performance criteria in an uncertainty-based design framework.
3. Accelerate aeroelastic tests and enhance the insight they produce by improving the use of data generated during the tests.

An important goal that does not appear to have been pursued seriously in the available literature is the selection of risk-based design criteria for aeroelastic constraints. For example, U.S. military aircraft must satisfy a 15% flutter safety margin, a requirement that seems to be essentially empirical. This requirement has been in effect since at least the early 1960s [3], even though the interim has seen significant advances in computational aeroelasticity and testing. When viewed in the light of current technology, it seems reasonable to adjust aeroelasticity safety requirements, but no rational approach has been advanced for doing so. Section 29.6 provides additional comments on this important and timely topic.

Some explanations of terminology are appropriate before proceeding. In addition to the potentially severe consequences associated with primarily aeroelastic effects, difficulties often are encountered when the flight control system or thermal effects come into play; these are the domains of aeroservoelasticity and aerothermoelasticity, respectively. In the following sections, *aeroelasticity* will be used to encompass all of these disciplines unless a more specific term is needed. The presentation in this chapter generally assumes that the reader is familiar with fundamental aeroelasticity concepts; readers who lack this background should consult any of the commonly available textbooks on the subject (e.g., [4–6]). Also, all modern aeroelasticity analyses involve computer models, but *computational aeroelasticity* will signify the use of what commonly are referred to as high-fidelity or physics-based models. In the context of computational aerodynamics, this usually involves solving a discretized version of the Euler or Navier-Stokes equations to approximate the state of the fluid.

29.2 The Role of Uncertainty in Aeroelasticity

29.2.1 Identification of Uncertainties

Organized identification and quantification of uncertainty sources typically involves a specific taxonomy of uncertainty. Selecting an appropriate taxonomy in a multidisciplinary field like aeroelasticity is complicated by the conflicting perspectives of the various experts from the constituent disciplines as well as by the quality of information available in each discipline. In aeroelasticity, each of the primary disciplines — structures, aerodynamics, and controls — has a unique set of design variables, analytical methods, modeling pitfalls, and performance criteria. Furthermore, the information quality, availability, and needs of each discipline can differ at a given point in the design process. This diversity of technical priorities and vantage points has resulted in a jumble of methods and applications for mitigating uncertainty in aeroelastic design. This situation is common in the analysis of complex systems [7] and also reflects a continuation of the status quo in deterministic aeroelasticity.

The discussion in this chapter generally employs the commonly recognized uncertainty classes described by Melchers [8]: aleatory or irreducible uncertainty, the most common example being randomness in a system's parameter; epistemic uncertainty, which connotes limited knowledge and is often encountered as a lack of understanding about the physics that must be modeled; and uncertainty due to human error. This classification scheme is common in structural reliability analysis. More involved taxonomies are available (e.g., [9]), but the version described here is adequate for much of the following discussion. Aleatory uncertainty usually is amenable to probabilistic description, which therefore is the primary tool in most structural reliability methods. Consequently, probabilistic methods have also been the chosen approach in many of the published aeroelasticity studies that involve uncertainty. Epistemic uncertainty in aeroelasticity does not appear to have been considered directly in the available literature. However, one published application [10], which is described later in this chapter, considered uncertainty in nominally clamped panel boundaries by adding rotational springs with uncertain stiffness to the model; thus, epistemic uncertainty about the nature of traditionally ideal boundary conditions was converted to a source of aleatory uncertainty.

A useful nonprobabilistic representation of uncertainty has been developed in the control systems community under the rubric of robust control [11, 12] and has recently been employed to estimate aeroelastic stability [13]. The essence of this approach, which is often referred to as µ-analysis, is to partition the known parts of the system from the uncertain parts and pass information between them in a feedback-like connection. Bounds on the possible values of the uncertain elements are either assumed or based on existing data, and the system is judged to be robustly stable if it can be shown to be stable to variations within the assumed uncertainty bounds. This approach seems not to make a clear distinction between aleatory and epistemic uncertainty; instead, all uncertainties in the plant and the input are lumped under the heading of "model uncertainty." This usage differs from that employed in the structural reliability, where this phrase generally denotes a subclass of epistemic uncertainty that reflects uncertainty about the physical behavior the model must represent. Commonly encountered examples include specification of failure modes (i.e., what constitutes failure?) as well as any multiscale phenomenon for which several competing models can be shown to partially reproduce response quantities of interest. Space constraints do not permit a full description of µ-analysis here, but the publications of Lind and Brenner [13] and Balas et al. [12] are good starting points for further reading.

29.2.2 Sources of Aeroelasticity Uncertainty and Their Importance

It is impractical in the current context to develop a complete list of uncertainties in aircraft aeroelasticity, but the most important general sources include physics models, structural properties, and the operating environment. Typically, uncertainty in the physics or boundary-condition models is epistemic, whereas uncertainty in the operating environment or structural properties is aleatory, except perhaps for statistical uncertainty due to limited sample sizes. The author has commented elsewhere [14] that the relative importance in airframe design of these and other sources of uncertainty depends on several considerations, such as the

- Vehicle's layout, materials, load paths, and environment
- Vehicle's class and purpose
- Actual usage of each aircraft (e.g., some pilots stress their aircraft more than others)
- Risk aversion of the users
- Amount of design experience with component-level technologies and their integration into complex systems
- Level of experience with newer manufacturing processes

The interaction of system nonlinearities with the various sources of uncertainty must also be understood. For example, limit-cycle oscillations (LCO) induced by Hopf bifurcations have been observed in several aeroelastic systems (e.g., [15]). The response associated with a given Hopf bifurcation strongly depends on whether the bifurcation is supercritical or subcritical. The introduction of uncertainty into this problem results in an uncertain bifurcation point, and therefore a level of risk that depends critically on the nature of the bifurcation as well as the amount and type of uncertainty.

Hopf bifurcations can occur in aeroelastic systems that exhibit nonlinear behavior just in the structural stiffness operator (e.g., panel LCO). However, from the perspective of computational cost, model and mesh uncertainty, and physical complexity, nonlinear aerodynamics is in most cases the critical component of computational aeroelasticity that must be enhanced to enable dependable predictions of aeroelastic response and variability. Additional discussion of this topic can be found in Section 29.6.

UQ can also assume great prominence in aeroelastic tailoring, in which airframes are optimized specifically for aeroelastic performance. Two primary contributing factors can be cited here: (1) the relatively high variability in the properties of composite materials, and (2) the fact that model and parametric uncertainty can lead to unexpected and significant aeroelastic performance degradation in practice [16]. The latter is simply a specific realization of the common observation that optimum designs tend to be fragile, in that they often perform suboptimally in practice [17].

Any analysis, design, or testing process that makes explicit use of UQ should consider the points described in this subsection. Current program management and certification frameworks, such as those employed by the USAF, depend on iterative application of systems engineering methods to manage risk and promote safety while trying to meet performance goals. Multiple reviews conducted through system definition, design, and testing help to ensure that risks are identified and managed, but airframe design and certification as a whole includes little to no rigorous UQ, and therefore little quantitative risk assessment (QRA).

29.3 Probabilistic Aeroelasticity Analyses

29.3.1 Flutter of Aircraft Lifting Surfaces

Flutter is a dynamic aeroelastic instability that can destroy a lifting surface's structure if the resulting response is not sufficiently restricted, perhaps by nonlinear damping or stiffness in the system. Classical flutter prediction with linear structural and aerodynamics models can be pursued through various algorithms, the underlying goal being to solve a parametric or pseudo-eigenvalue problem to determine the flight speed and oscillation frequency at which flutter occurs. Specialized iterative algorithms are required because, except in the simplest case of quasi-steady aerodynamics, the aerodynamic forces depend nonlinearly on the oscillation frequency. Details of performing flutter analyses are available in the aeroelasticity references cited above.

Few published efforts focus on solving the classical flutter problem in an uncertain context. The basic problem was described by Poirion [18], who employed a first-order perturbation method to solve for the probability of flutter given uncertainty in the structural mass and stiffness operators, which were represented by a linear finite element model. Poirion observed that although this perturbation formulation of the stochastic eigenvalue problem produced an explicit estimate of the flutter probability, application to a realistic system model produced results that compared poorly with Monte Carlo simulation. The quality of Poirion's results was also difficult to evaluate because no details of the unsteady aerodynamics model were provided.

29.3.2 Limit-Cycle Oscillation of Airfoils and Panels

Beran and Pettit [19] studied a nonlinear incarnation of the standard aeroelastic system consisting of an airfoil with pitch (α) and plunge (h) degrees of freedom (DOF). Their deterministic structural and aerodynamic model was an extension of that employed by Lee et al. [20], in which the plunge DOF had linear stiffness but the pitch DOF included a third-order stiffness term in addition to the linear component. The third-order term, k_3, modeled aeroelastic limit-cycle oscillation (LCO) induced by a supercritical Hopf bifurcation. Beran and Pettit included a fifth-order pitch stiffness term, k_5, so that properly chosen stiffness coefficients would induce a subcritical Hopf bifurcation, which is more dangerous than the supercritical bifurcation, owing to the potential for an abrupt increase in LCO amplitude associated with subcritical bifurcations. This concept is illustrated qualitatively by the bifurcation diagram in Figure 29.1.

"Crude" Monte Carlo simulation and time integration were employed to quantify LCO amplitude variability induced by assumed parametric variability in the initial pitch angle (α_0), k_3, and k_5. The resulting

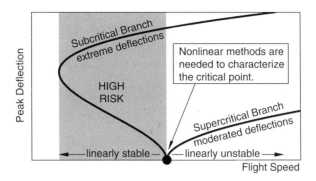

FIGURE 29.1 Subcritical and supercritical Hopf bifurcations.

"random" bifurcation diagram is shown in Figure 29.2, which shows the estimated LCO amplitude probability density function (PDF) for several reduced velocities. No attempt has been made to base the structural coefficients or their assumed variability on a physical system, so the primary purpose of this figure is to illustrate the potential of nondeterministic analysis of aeroelastic systems. Foremost is the ability to estimate the probability that LCO will exceed a specified threshold at a given reduced velocity. Such information could in the future be the basis for a risk-based flutter design criterion. For example, deterministic analysis of the baseline system predicts zero steady-state response (i.e., a stable focus) for $u = 5.65$, which is approximately 10% below the linear instability at $u = 6.26$ and also is below the lower limit point on the unstable LCO branch ($u = 5.95$). However, MCS predicted a small but finite probability (approximately 0.1%) of encountering LCO at this reduced velocity.

Several conference and journal papers have addressed the sensitivity of panel flutter and LCO to uncertain system parameters. This problem actually involves what is, in practice, a relatively benign nonlinear behavior in that it does not induce catastrophic failure, but for several reasons it has proven to be a productive vehicle for exploring the use of uncertainty quantification in aeroelasticity. Liaw and Yang [21, 22] examined flutter and LCO of laminated plates and shells with uncertainties in several structural and geometric parameters. In the study of aeroelastic stability, their work appears to represent

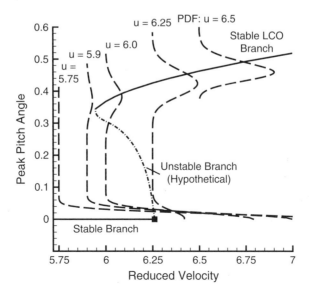

FIGURE 29.2 Response of pitch-and-plunge airfoil. The solid curve is LCO amplitude of baseline airfoil. Dashed curves show estimated probability density function from Monte Carlo simulation at each dynamic pressure.

the first published application in an archival journal of a stochastic finite element formulation. Their approach involved a second-moment, perturbation-based stochastic finite element model. They presented several figures that detailed the effects of variability on the likely range of responses, but because they employed a second-moment formulation, quantification of output variability was limited to the corresponding second moment also.

Lindsley et al. [23] also studied LCO of panels with spatial variability in the modulus of elasticity, but they employed Monte Carlo simulation (MCS) to better quantify the range of response variability in this nonlinear system. They limited their analysis to square panels and isotropic materials, but also included the influence of nonideal boundary conditions (BCs) so as to measure the relative importance of uncertainty in material properties and BCs. Figure 29.3 illustrates the output variability in the bifurcation diagram for LCO amplitude vs. dynamic pressure, with a separate nonparametric probability density function of response amplitude based on 200 panel realizations at each of several dynamic pressures. At dynamic pressures well above the baseline bifurcation point ($\lambda \approx 858$), Young's modulus variability had a relatively minor effect in that it simply added a random component to the steady-state LCO amplitude, but at $\lambda \approx 860$, modulus variability controlled the panel's long-term response, with the realizations approximately evenly split between LCO and zero deflection. Perhaps most important is the result that several realizations had a response amplitude much larger than the deterministic case. Figure 29.4 shows the proper orthogonal decomposition [24–26] of the elastic modulus fields that resulted in either LCO or no LCO. These plots strikingly demonstrate that LCO typically was associated with relatively low stiffness near the 3/4-chord region (measured from the panel's leading edge). Also, Figure 29.5 illustrates the parametric influence of nonideal BCs, where distributed torsion springs along each boundary were used to model flexibility in ideally clamped boundaries. At $\lambda \approx 850$, which is clearly below the deterministic bifurcation point, a moderate amount of boundary flexibility, parameterized by β, was seen to induce LCO.

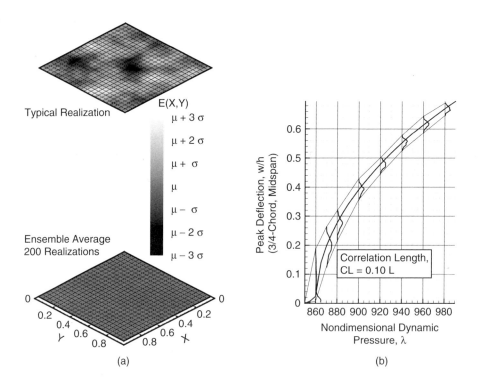

FIGURE 29.3 Monte Carlo simulation of stochastic panel. (a) Ensemble of panel Young's modulus realizations. (b) Bifurcation diagram. Dashed curve is LCO amplitude of baseline panel. Solid curves show estimated probability density functions from Monte Carlo simulation at each dynamic pressure, along with their associated envelope.

Uncertainty in Aeroelasticity Analysis, Design, and Testing

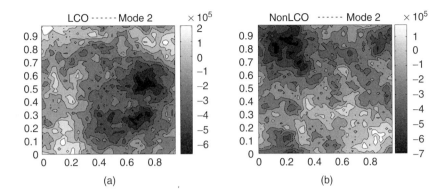

FIGURE 29.4 Mode 2 of proper orthogonal decomposition of Young's modulus field for LCO and nonLCO cases. Flow is from left to right.

More recent work by Lindsley et al. [10, 27] has extended the study of LCO of uncertain panels. In [10], they showed that the occurrence and amplitude range of LCO at a given dynamic pressure are relatively insensitive to the assumed ratio, l_c, of correlation length of the elastic modulus random field to panel side length, at least for $0.10 < l_c < 0.30$. They also examined the influence of assuming, in combination with a spatially variable Young's modulus, either a uniform or Weibull distribution for the rotational springs along each boundary.

Lindsley et al. [27] also studied aerothermoelastic effects through the addition of an uncertain thermal expansivity field, $\alpha(x, y)$, which was modeled by altering the baseline biaxial prestress induced by an assumed temperature change. Qualitative response classes are summarized in Figure 29.6, which shows the baseline response (i.e., $E(x, y) = E_0$ and $\alpha(x, y) = \alpha_0$) as a function of compressive prestress and dynamic pressure. The variety of responses exhibited by the baseline system are typified by the plots in

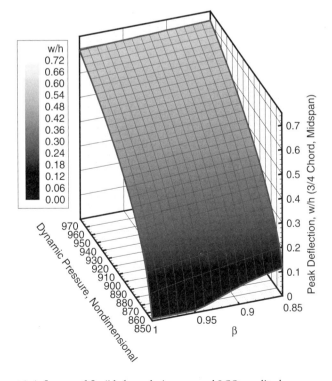

FIGURE 29.5 Parametric influence of flexible boundaries on panel LCO amplitude.

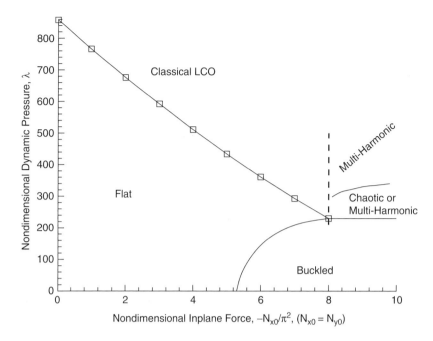

FIGURE 29.6 Baseline aerothermoelastic stability diagram.

Figure 29.7, which shows the response for several dynamic pressures and a dimensionless prestress of $-8\pi^2$. Realizations of $E(x, y)$ and $\alpha(x, y)$ were computed to measure their influence on the dynamical behavior of the panel. A wide range of behavior was observed, with several realizations falling into each of the categories depicted in Figure 29.6. In particular, stochastic results were generated for prestress and dynamic pressure values of $-8\pi^2$ and 230, a condition that results in a very weak attractor at zero deflection (see Figure 29.7). The vast majority of the realizations resulted in zero steady-state deflection, but a small number of realizations produced limit-cycle or chaotic behavior; three of these are depicted

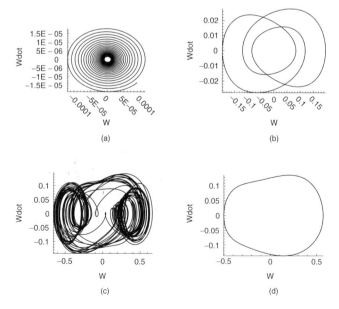

FIGURE 29.7 Phase-plane plots of the baseline panel for several dynamic pressures and a dimensionless prestress of $-8\pi^2$. Figures (a) through (d) correspond to $\lambda = 230, 235, 250, 290$, respectively.

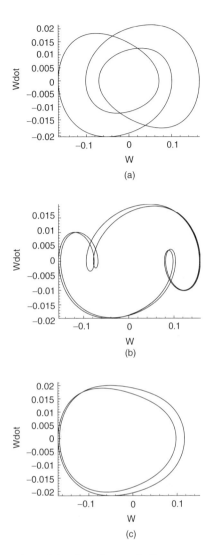

FIGURE 29.8 Representative phase-plane plots of aerothermoelastic response of the stochastic panel for prestress and dynamic pressure values of $-8\pi^2$ and 230.

in Figure 29.8. Comparing Figure 29.7 and Figure 29.8 shows that introduction of spatially variable thermoelastic properties in the panel can induce dynamic bifurcations of similar strength to those caused by significant changes in the dynamic pressure of the baseline system.

29.4 Probabilistic Design for Aeroelasticity

Basic concepts and methods of reliability-based design and optimization (RBDO) are described in the chapter on reliability-based design and optimization in this book (Chapters 24, 31, and 34). RBDO of preliminary wing designs for aeroelastic performance has been pursued recently, with published efforts beginning to appear in the mid-1980s. One of the earlier studies is by Yang and Nikolaidis [28], who employed a first-order reliability method (FORM) (see Chapter 21) in conjunction with system reliability optimization to design for gust loads a simplified wing structure composed of a segmented box beam. They compared their probabilistic design with a deterministic design based on Federal Aviation Administration (FAA) regulations and showed that optimization based on system reliability could be used to obtain either a lighter wing structure with the same implicit reliability as the design based on FAA requirements, or a

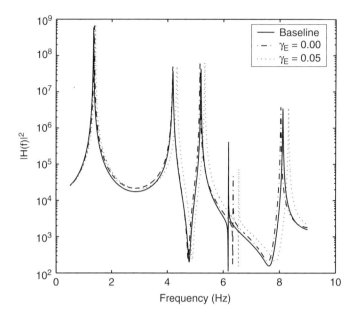

FIGURE 29.9 Squared magnitude of frequency response function (FRF) for root bending moment. The curve marked $\gamma_E = 0.00$ is for zero variability in Young's modulus, and the curve marked $\gamma_E = 0.05$ is for a Young's modulus coefficient of variation of 0.05.

higher reliability wing with the same weight as the FAA wing. However, their formulation could not be employed directly in practical design owing to the elementary nature of their structural model. Furthermore, a general concern that has received little attention in many aircraft RBDO studies is relevant here also: their approach did not address any empiricism or modeling limitations implicit in the assumptions upon which the FAA-specified design factors and requirements were based. Given that the existing FAA criteria for continuous gust design were proposed in the early 1970s [29] and that the supporting data, assumptions, and modeling methods are even older, it is reasonable to suggest that modeling advances and improvements in data acquisition and processing technology during the interim justify revising the design requirements to ensure that they reflect current knowledge and analysis capabilities.

Pettit and Grandhi [30, 31] developed a RBDO framework by combining ASTROS, a deterministic, linear finite-element-based structural optimization system that includes linear aeroelastic analysis, with DOT[TM1] to control optimization, and several MATLAB[TM2] functions that were written to implement a FORM that uses a two-point adaptive nonlinear approximation [32] of the limit-state function to accelerate convergence of the reliability index estimate. They performed RBDO of a relatively simple, fighterlike wing structure with uncertain element thickness values and elastic modulus to achieve specified reliability indices in several performance functions, including static deflection, aileron effectiveness, and gust response. Mean thickness values of the structural elements were taken as design variables, so that the mean weight served as the objective function. Redistribution of structural mass by the optimizer produced designs with improved aeroelastic performance reliability and relatively small weight penalties. Figure 29.9 shows the baseline and optimized frequency response functions (FRF) of the root bending moment. The first optimum design compensated for variability in the element thickness values, whereas the second optimum design also accounts for Young's modulus randomness. The changes in the FRF with increasing uncertainty agree with intuition in that they demonstrate that additional uncertainty in the structure's stiffness must be accommodated by increasing the mean stiffness (i.e., raising the mean modal frequencies) of the structure.

---

[1]Registered trademark of Vanderplaats Research and Development, Inc., Colorado Springs, CO.
[2]Registered trademark of The MathWorks, Inc., Natick, MA.

A primary impact of this study is that it helped to demonstrate the readiness of RBDO methods to be applied in preliminary design to ensure reliable aeroelastic performance. Moreover, comparison of the resulting reliability-based designs with deterministic designs produced according to traditional criteria could provide insight into how existing safety factors account implicitly for parametric variability. A study of this nature could perhaps spawn efforts to improve the rational basis of safety factors and required stability margins, but this path for implementing uncertainty-based design criteria has certain shortcomings (e.g., see Pettit [14] and the references cited therein) that could restrict its utility.

Allen and Maute [33] recently extended RBDO for aeroelastic performance to include computational fluid dynamics (CFD) in place of the doublet-lattice method for the computation of unsteady pressures. Their introduction of high-fidelity aerodynamics modeling adds an additional level of realism to the aeroelastic design problem and further supports the conclusion that RBDO is ready to begin being transitioned to industrial aeroelastic design applications. Allen and Maute also comment on the potential importance of accounting for parameter uncertainty in aeroelastic tailoring to achieve optimum cruise performance. They cite early research results that suggest that the solution of an aeroelastic tailoring problem is rather sensitive to uncertainty. At this time, they expect to conduct future studies to better understand this problem. Finally, the mere existence of their work further highlights the need described in Section 29.1 to establish probabilistic design criteria for airframe performance.

29.5 Aeroelasticity Applications of Robust Control Concepts

29.5.1 Nonprobabilistic Methods for Analysis and Design

Lind and Brenner have published a series of papers [34–36] and a monograph [13] that develop the application of nonprobabilistic robust control methods (e.g., μ-analysis) to aeroelastic stability. Although there is no reason why this approach could not be used in a purely analytical mode, in which it has been shown to compute "nominal" flutter margins that closely match those predicted by traditional flutter prediction methods [13], the primary applications have been in using wind tunnel, ground, and flight test data to update and validate aeroelastic models; for this reason, further discussion of their work will be delayed until Section 29.5.2.

Karpel et al. [37] developed an aeroservoelastic design process that includes robust control theory to accommodate structured uncertainty in the plant. They applied their process to design a flutter suppression system for a fighter wing structure that includes four control surfaces and a tip missile with uncertain inertial properties. The open-loop aeroelastic system was stable for the entire range of uncertainties in the missile's inertial properties, but the controller in the closed-loop system (i.e., including the actuated control surfaces) could not be designed to meet robust performance specifications. They showed that an additional structural design cycle could be performed to yield a new closed-loop system with the desired level of robustness. It should be noted that this application of UQ for the structure is fundamentally different from those described above because it captures uncertainty at a relatively high level (i.e., in the state-space representation of the system) instead of the physical level, which includes the material properties, dimensions, and boundary conditions. There presumably exists some unspecified mapping that could be used to connect the two perspectives.

Chavez and Schmidt [38] note that an important assumption in the use of μ-analysis for multidisciplinary systems is that an adequate uncertainty or variation model is available for each of the component analysis disciplines. They address this concern through employing a "systems approach" to characterize the anticipated uncertainty in finite-dimensional, linear, time-invariant models of flexible aircraft arising from unsteady aerodynamics, truncation of structural modes, and uncertain mass and stiffness properties. Chavez and Schmidt also address the possibility of interaction between the rigid-body dynamics and structural modes of highly flexible aircraft, which is an issue of potentially high importance in certain airframe structural design concepts.

29.5.2 Uncertainty in Aeroelastic Tests

Comprehensive wind tunnel and flight test programs are required to demonstrate that aeroelastic instabilities do not develop within the operational envelope. Of particular concern is the sometimes abrupt

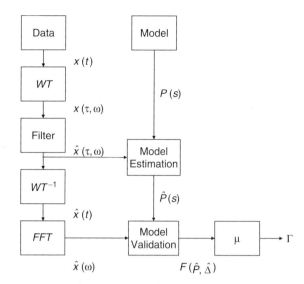

FIGURE 29.10 Flowchart of µ-analysis framework for estimating robust flutter boundaries. (Note: WT indicates a wavelet transform.) (Courtesy of Lind [55]. With permission)

occurrence of flutter when the damping in a particular aeroelastic mode decreases suddenly. Because the flight conditions that induce flutter are difficult to predict precisely, flight test programs generally are conducted to ensure that the flutter point is approached gradually, a practice that increases their duration and cost. These difficulties have motivated recent work to improve the integration between flight test data and model predictions so as to increase the overall efficiency and dependability of flutter margin estimation.

As noted above, Lind and Brenner have made significant progress in this area by applying robust control concepts (e.g., µ-analysis [11, 39]) to compute robust flutter speeds that represent worst-case flight conditions with respect to assumed forms of uncertainty in the plant and excitation. A key feature of their µ-analysis framework is its ability to accommodate both model and flight test information in an integrated manner. The flowchart in Figure 29.10 summarizes a reasonably general version of the algorithm, which includes the option of filtering the experimental data with a wavelet transform. Wavelets can be employed in this algorithm to reduce noise and unmodeled dynamics in the data and also to estimate model parameters [40]. In this approach, the uncertainty and modeling errors in system operators, inputs, and outputs are represented by a set of norm-bounded operators, Δ, and flutter speeds or dynamic pressures are computed that are robust in the sense that the system will be stable for modeling errors within Δ. As a result, the definition of an appropriate Δ is crucial to the success of this approach: a Δ that is too large will result in an overly conservative margin that could restrict the permitted flight envelope unnecessarily, whereas underestimating Δ will lead to an overestimate of the relative aeroelastic stability.

Although the µ-method provides a comprehensive and useful framework for aeroelastic stability estimation, it does have restrictions that could be important in certain applications. Because it is non-probabilistic by design, it cannot take full advantage of probabilistic information that might be available. Another restriction is imposed by the fact that the µ-framework is based entirely on linear operators. If the system in question cannot be represented adequately by linearized operators, the µ-method can accommodate nonlinearity in a coarse manner by associating sufficient uncertainty with the linear system model to cover errors that result from unmodeled nonlinear dynamics. Lind and Brenner [13] provide additional details to illustrate the implementation and utility of this formulation.

One shortcoming in the early implementation of the µ-method for estimating flutter margins is that it was not based on match-point flutter solutions. Recently, Lind [36] has described a refined application of this method that does compute robust match-point flutter margins on a model that includes a theoretical mass matrix, a stiffness matrix derived from ground vibration tests, and aerodynamic forces computed with a doublet-lattice program. He also suggests that the method can be extended readily to

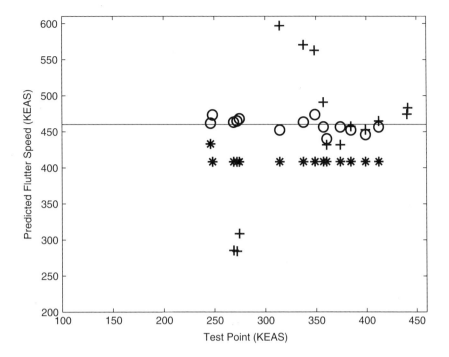

FIGURE 29.11 Flutter speeds predicted by damping extrapolation (+), original flutterometer (*), and updated flutterometer (o). (From Lind [55]. With permission.)

form a refined version of the flutterometer [35], which provides an on-line approach for robustly estimating and updating flutter speeds during flight tests. An important recent step is the improvement of estimates of model uncertainty through the use of Volterra kernels to isolate the linear component of the flight data [41, 42]. This approach has been particularly successful in reducing the potentially high conservatism of robust flutter margins. This is illustrated in Figure 29.11, which compares recent flutterometer results based on Volterra kernels, with flutter speed estimates computed using an earlier realization of the flutterometer as well as a traditional, damping-extrapolation method.

Borglund [43] has also recently employed µ-analysis in his study of the robust aeroelastic stability of a flexible slender wing model in a low-speed wind tunnel. The model includes a controllable trailing-edge flap, which can be used in conjunction with an existing optical measurement system to complete a closed-loop feedback control system for aeroelastic control studies. Borglund notes that the utility of a robust stability analysis for aeroservoelastic design depends on the availability of realistic uncertainty descriptions for the various components of this multidisciplinary problem. He focuses on uncertainty in the unsteady aerodynamic forces, which generally are the most difficult part of the problem to model accurately. Borglund used the "µ validation test" recommended by Lind and Brenner [13] to estimate the amount of uncertainty in the aerodynamic model, which was represented by uncertain coefficients in Fourier expansions of the spanwise loading for each aerodynamic "mode shape." The size of the aerodynamic uncertainty model was restricted by admitting uncertainty only in the lowest Fourier coefficients, which corresponded to aerodynamic mode shapes driven by the lowest structural modes, which dominated the structural motion.

29.6 Some Suggestions for Future Research

The ultimate goal of what can be called stochastic computational aeroelasticity (SCAE) is the efficient and accurate computation of performance estimates like those shown in Figure 29.2 for realistic structures and flight conditions. Several issues remain to be tackled before this goal can be approached with a reasonable hope of success.

The importance of system and model uncertainty in predicting and understanding aeroelastic response variability is amplified in many applications by the degree of nonlinearity exhibited by the system in the operational envelope. It was noted in Section 29.2.2 that aerodynamic model uncertainty often is substantially greater than uncertainty in the structure, and aerodynamics often is the dominant source of nonlinear behavior in aeroelastic systems. However, the majority of published applications continue to focus on structural uncertainties because they are easier to model and because structural solvers generally are orders-of-magnitude faster than fluid solvers. Key sources of uncertainty in the aerodynamics models employed in computational aeroelasticity remain to be explored in new or different ways. Perhaps the most commonly recognized are the standard sources of model uncertainty encountered in deterministic simulations, such as grid convergence, approximation of boundary conditions, and turbulence models. Uncertainty in the initial conditions of both the structure and the incoming flow also deserves substantial consideration because their selection, which is often based on the analyst's judgment, can govern the qualitative accuracy of computational aeroelasticity predictions.

Gust analysis (e.g., [44]) has for at least 50 years involved some probabilistic analysis in the form of traditional spectral analysis and random vibrations of linear systems, but the methods used to accommodate gust loads in airframe design still produce what might be called quasi-deterministic design loads that are the result of combining simple estimates of the gust-load spectral density with empirical safety factors. This patchwork approach predates the wide availability of computational linear aeroelastic tools for preliminary design. More recent work by Weishaupl and Laschka [45] has demonstrated the additional insight available just from deterministic computational aerodynamics in modeling unsteady loads caused by a nonuniform freestream. It is recommended that this work be extended to include aeroelastic interaction and the spatio-temporal statistical properties of atmospheric variability.

Recently propounded methods have begun to combine stochastic spectral projection with traditional computational fluid dynamics models. Published approaches generally employ polynomial chaos expansions, the basics of which are described by Ghanem and Spanos [46], to capture uncertainty in fluid properties, initial conditions, and boundary conditions. These methods are still in their infancy but show great promise for integrating uncertainty quantification with computational fluid dynamics and, eventually, computational aeroelasticity. Recent publications of interest here include Le Maitre et al. [47] and Xiu et al. [48, 49].

Reduced-order models (e.g., [26, 50]) also are currently under consideration for SCAE to improve the efficiency of time-accurate models; however, this work also is at a very early stage of development. Another potential source of improved computational efficiency in stochastic modeling that does not appear to have been examined rigorously in the literature is the diminishing marginal payoff associated with increasingly fine discretization of the structural and fluid domains, a trend that in some cases is yielding rapidly diminishing returns [51]. This conclusion is based on the observation that oftentimes meshes are refined to a level of numerical precision not justified by the uncertainties in the system parameters, initial conditions, or boundary conditions. In other words, squeezing additional "accuracy" out of a deterministic model is not justified when the numerical tolerances drop below the level of natural variability in the system. This wasted computational power that typically is used for deterministic point solutions could be more profitably invested to perform multiple realizations with less refined models. The author is not aware of any general methods that have been developed to identify a suitable "break-even" discretization level, but Marczyk [51] describes some elementary specific examples to illustrate the point.

As noted in Section 29.1, airframe reliability currently is enforced through empirically derived safety factors and required margins. The associated design and certification procedures tend to be cumbersome to implement and account rather indiscriminately for the various uncertainties and their concomitant risks, but can be made to suffice for conventional design concepts and structural technologies. However, their dependence on dated methods can be a serious impediment to the introduction of innovative designs, materials, and assembly technologies for which little history is available to justify the selection of safety factors and the prediction of likely failure modes [14]. Future designs based on unique structural concepts (e.g., joined wing [52, 53]) and incorporating innovative technologies (e.g., adaptive structures and distributed health-monitoring systems) will likely require improved methods for allocating system-level aeroelastic risks (e.g., flutter, divergence, and control-surface reversal) to the structural, control, and aerodynamic components of the system. Failure to develop appropriate decision-making methods and criteria to support this goal could render the design problem infeasible or impractical. Extensive research is needed to address this critical question in anticipation of the transition of cutting-edge structural technologies from research to technology demonstration and production. A potentially useful starting point suggested in Section 29.4 could be to study how the current flutter-margin requirement is allocated implicitly between parametric uncertainty and unmodeled structural and aerodynamic nonlinearities. This is a possible first step toward the rational selection of risk-based aeroelastic design criteria, which were suggested in Section 29.1.

The discussion above regarding justifiable precision in the presence of uncertainty suggests a more general need: modeling guidelines and best practices based on UQ demands in addition to traditional criteria. Effective implementation of uncertainty-based design criteria and decision processes for aeroelasticity or any other area of airframe design will require coordinated use of UQ modeling criteria to ensure that limited analysis and test resources will be invested appropriately to maximize understanding of the system's performance. The motivating goal should be to promote early recognition, characterization, and prioritization of uncertainty sources.

29.7 Final Remarks

This chapter has summarized key points in the quest to understand how various forms of uncertainty influence the prediction and understanding of aeroelasticity, which is one of many components of airframe design and certification that could benefit greatly from widespread use of uncertainty analysis. The author's recent conference paper [14] provides a broad review of the current status and some potential benefits of employing uncertainty quantification (UQ) throughout the airframe design process. Recent research, much of which is cited above, in elementary nonlinear aeroelastic systems has shown that relatively minor levels of variability in system parameters, loads, and boundary conditions can induce significant changes in the system stability.

Aeroelastic sensitivity to these sources of variability is likely to increase as the aircraft design community strives to achieve anything more than incremental improvements in the performance, range, endurance, and operational flexibility of future designs. Revolutionary advancements in these areas will require innovative aerodynamics, structural, and control systems, as well as new materials tailored to the demands placed on the structure. The active aeroelastic wing [54] is an early example of an aeroservoelastic system that breaks with traditional aeroelastic design methods and philosophy. Anticipated operational requirements call for design concepts that will demand even greater changes in design processes and perspectives. These advancements will not be possible within existing design and decision frameworks, which are historically based and therefore inherently biased against the efficient introduction of new technologies in all of the disciplines that influence aeroelastic performance. Because uncertainty quantification provides a consistent basis for assessing system performance and technological risks, it should be a cornerstone of aeroelasticians' efforts to stimulate fundamental advances in airframe design.

Acknowledgments

The author wishes to thank Dr. Rick Lind for providing Figure 29.10 and Figure 29.11 of this chapter and for providing some recent flutterometer results cited herein. The author is also grateful for the assistance of colleagues Drs. Frank Eastep, Tom Strganac, and Phil Beran in reviewing drafts of the manuscript and providing helpful comments on its technical content and scope.

Chris Pettit was at the United States Air Force Research Laboratory when he did the work presented in this chapter. He is now at the United States Naval Academy.

References

1. Harris, C.E., Starnes, J.H., and Shuart, M.J., Design and manufacturing of aerospace structures, state-of-the-art assessment, *J. Aircraft*, 29 (4), 545–560, 2002.
2. Veley, D.E., Pettit, C.L., and Holzwarth, R., Increased Reliance on Analysis for Certification of Aircraft Structures, presented at 8th International Conference on Structural Safety and Reliability, Newport Beach, CA, 2001.
3. Huttsell, L., private communication, Oct. 2002.
4. Bisplinghoff, R.L., Ashley, H., and Halfman, R.L., *Aeroelasticity*, Dover, New York, 1996.
5. Bismarck-Nasr, M.N., *Structural Dynamics in Aeronautical Engineering*, AIAA Education Series, AIAA, Reston, VA, 1999.
6. Hodges, D.H. and Pierce, G.A., *Introduction to Structural Dynamics and Aeroelasticity*, Cambridge University Press, Cambridge, U.K., 2002.
7. Ahl, V. and Allen, T.F.H., *Hierarchy Theory*, Columbia University Press, New York, 1996.
8. Melchers, R.E., *Structural Reliability Analysis and Prediction*, Wiley, New York, 1999.
9. Ayyub, B.M., *Elicitation of Expert Opinions for Uncertainty and Risks*, CRC Press, Boca Raton, FL, 2001.
10. Lindsley, N.J., Beran, P.S., and Pettit, C.L., Effects of Uncertainty on Nonlinear Plate Response in Supersonic Flow, AIAA-2002-5600, presented at 9th AIAA/ISSMO Multidisciplinary Analysis and Optimization Conference, Atlanta, 2002.
11. Dorato, P.E., Introduction to *Robust Control*, IEEE Press, New York, 1987.
12. Balas, G.J. et al., *μ-Analysis and Synthesis Toolbox*, The MathWorks, Natick, MA, 2001.
13. Lind, R. and Brenner, M., *Robust Aeroservoelastic Stability Analysis*, Springer-Verlag, London, 1999.
14. Pettit, C.L., Uncertainty Quantification for Airframes: Current Status, Needs, and Suggested Directions, 03M-123, presented at 2003 SAE World Congress, Detroit, 2003.
15. Dowell, E., Edwards, J., and Strganac, T., Nonlinear aeroelasticity, *J. Aircraft*, Vol. 40, No. 5, pp. 857–874, 2003.
16. Kuttenkeuler, J. and Ringertz, U., Aeroelastic design optimization with experimental verification, *J. Aircraft*, 35 (3), 505–507, 1998.
17. Marczyk, J., Beyond Optimization in Computer-Aided Engineering, International Center for Numerical Methods in Engineering (CIMNE), Barcelona, 2002.
18. Poirion, F., Impact of Random Uncertainties on Aircraft Aeroelastic Stability, presented at 3rd International Conference on Stochastic Structural Dynamics, San Juan, Puerto Rico, 1995.
19. Beran, P.S. and Pettit, C.L., unpublished report, July 2001.
20. Lee, B.H.K., Jiang, L.Y., and Wong, Y.S., Flutter of an airfoil with a cubic nonlinear restoring force, in *Proceedings of 39th Structures, Structural Dynamics, and Materials Conference*, AIAA 98-1725, AIAA/ASME/ASCE/AHS/ASC, Baltimore, MD, 1998.
21. Liaw, D.G. and Yang, H.T.Y., Reliability and nonlinear supersonic flutter of uncertain laminated plates, *AIAA J.*, 31 (12), 2304–2311, 1993.
22. Liaw, D.G. and Yang, H.T.Y., Reliability of uncertain laminated shells due to buckling and supersonic flutter, *AIAA J.*, 29 (10), 1698–1708, 1991.
23. Lindsley, N.J., Beran, P.S., and Pettit, C.L., Effects of uncertainty on nonlinear plate aeroelastic response, AIAA-2002-1271, in *Proceedings of 43rd Structures, Structural Dynamics, and Materials Conference*, AIAA/ASME/ASCE/AHS/ASC, Denver, CO.
24. Holmes, P., Lumley, J.L., and Berkooz, G., *Turbulence, Coherence Structures, Dynamical Systems and Symmetry*, Cambridge University Press, Cambridge, U.K., 1996.
25. Hall, K.C., Thomas, J.P., and Dowell, E.H., Proper orthogonal decomposition technique for transonic unsteady aerodynamic flows, *AIAA J.*, 38 (10), 1853–1862, 2000.
26. Pettit, C.L. and Beran, P.S., Application of proper orthogonal decomposition to the discrete Euler equations, *Int. J. Numerical Meth. Eng.*, 55 (4), 479–497, 2002.

27. Lindsley, N.J., Beran, P.S., and Pettit, C.L., Effects of Uncertainty on the Aerothermoelastic Flutter Boundary of a Nonlinear Plate, AIAA-2002-5136, presented at 11th AIAA/AAAF International Conference on Space Planes and Hypersonic Systems and Technologies, Orleans, France, 2002.
28. Yang, J.S. and Nikolaidis, E., Design of aircraft wings subjected to gust loads: a safety index based approach, *AIAA J.*, 29 (5), 804–812, 1991.
29. Federal Aviation Regulation Part 25, Appendix G; available online at www.faa.gov, 2002.
30. Pettit, C.L. and Grandhi, R.V., Reliability Optimization of Aerospace Structures for Gust Response and Aileron Effectiveness, presented at 8th International Conference on Structural Safety and Reliability, Newport Beach, CA, 2001.
31. Pettit, C.L. and Grandhi, R.V., Optimization of Aerospace Structures for Aeroelastic Response Reliability, presented at 4th ISSMO Congress of Structural and Multidisciplinary Optimization, Dalian, China, 2001.
32. Wang, L. and Grandhi, R.V., Improved two-point function approximations for design optimization, *AIAA J.*, 33 (9), 1720–1727, 1995.
33. Allen, M. and Maute, K., Reliability-Based Design Optimization of Aeroelastic Structures, AIAA-2002-5560, presented at 9th AIAA/ISSMO Multidisciplinary Analysis and Optimization Conference, Atlanta, 2002.
34. Lind, R. and Brenner, M., Analyzing aeroservoelastic stability margins using the μ-method, AIAA-98-1895, in *Proceedings of 39th Structures, Structural Dynamics, and Materials Conference*, AIAA/ASME/ASCE/AHS/ASC, Baltimore, MD, 1998.
35. Lind, R. and Brenner, M., Flutterometer: an on-line tool to predict robust flutter margins, *J. Aircraft*, 37 (6), 1105–1112, 2000.
36. Lind, R. and Brenner, M., Flight test evaluation of flutter prediction methods, in *Proceedings of 43rd Structures, Structural Dynamics, and Materials Conference*, AIAA-2002-1649, AIAA/ASME/ASCE/AHS/ASC, Denver, CO, 2002.
37. Karpel, M., Moulin, B., and Idan, M., Aeroservoelastic Design Process Using Structural Optimization and Robust Control Methods, AIAA 2000-4722, presented at 8th AIAA Symposium on Multidisciplinary Analysis and Optimization, Long Beach, CA, 2000.
38. Chavez, F.R. and Schmidt, D.K., Systems Approach to Characterizing Aircraft Aeroelastic Model Variation for Robust Control Applications, AIAA-2001-4020, presented at AIAA Guidance, Navigation, and Control Conference and Exhibit, Montreal, 2001.
39. Packard, A. and Doyle, J., The complex structured singular value, *Automatica*, 29 (1), 71–109, 1993.
40. Lind, R. and Brenner, M., Wavelet filtering to reduce conservatism in aeroservoelastic robust stability margins, AIAA-98-1896, in *Proceedings of 39th Structures, Structural Dynamics, and Materials Conference*, AIAA/ASME/ASCE/AHS/ASC, Baltimore, MD, 1998.
41. Prazenica, R.J., Lind, R., and Kurdila, A.J., Uncertainty estimation from Volterra kernels for robust flutter analysis, AIAA-2002-1650, in *Proceedings of 43rd Structures, Structural Dynamics, and Materials Conference*, AIAA/ASME/ASCE/AHS/ASC, Denver, CO, 2002.
42. Mortagua, J. and Lind, R., Accurate flutterometer predictions using Volterra modeling with modal parameter estimation, AIAA-2003-1405, in *Proceedings of 44th Structures, Structural Dynamics, and Materials Conference*, AIAA/ASME/ASCE/AHS/ASC, Norfolk, VA, 2003.
43. Borglund, D., Robust aeroelastic stability analysis considering frequency–domain aerodynamic uncertainty, in *Proceedings of 43rd Structures, Structural Dynamics, and Materials Conference*, AIAA-2002-1716, AIAA/ASME/ASCE/AHS/ASC, Denver, CO, 2002.
44. Hoblit, F.M., *Gust Loads on Aircraft: Concepts and Applications*, AIAA Education Series, AIAA, Reston, VA, 1988.
45. Weishaupl, C. and Laschka, B., Euler Solutions for Airfoils in Inhomogeneous Atmospheric Flows, AIAA 99-3587, presented at 30th AIAA Fluid Dynamics Conference, Norfolk, VA, 1999.
46. Ghanem, R. and Spanos, P.D., *Stochastic Finite Elements: a Spectral Approach*, Springer, Berlin, 1991.
47. Le Maitre, O.P., Knio, O.M., Najm, H.N., and Ghanem, R.G., A stochastic projection method for fluid flow: I. basic formulation, *J. Computational Phys.*, 173, 481–511, 2001.

48. Xiu, D. and Karniadakis, G.E., The Wiener-Askey polynomial chaos for stochastic differential equations, *SIAM J. Scientific Computing*, 24 (2), 619–644, 2002.
49. Xiu, D., Lucor, D., Su, C.H., and Karniadakis, G.E., Stochastic modeling of flow-structure interactions using generalized polynomial chaos, *ASME J. Fluids Eng.*, 124, 51–59, 2002.
50. Beran, P.S., Lucia, D.J., and Pettit, C.L., Reduced order modeling of limit-cycle oscillation for aeroelastic systems, in *Proceedings of 2002 ASME International Mechanical Engineering Congress and Exhibition*, IMECE2002-32954, New Orleans, LA, 2002.
51. Marczyk, J., *Principles of Simulation-Based Computer Aided Engineering*, FIM Publications, Madrid, 1999.
52. Livne, E., Aeroelasticity of joined-wing airplane configurations: past work and future challenges—a survey, AIAA-2001-1370, in *Proceedings of 43rd Structures, Structural Dynamics, and Materials Conference*, AIAA/ASME/ASCE/AHS/ASC, Seattle, WA, 2001.
53. Pettit, C.L., Canfield, R.A., and Grandhi, R.V., Stochastic Analysis of an Aeroelastic System, presented at 15th ASCE Engineering Mechanics Conference, New York, 2002.
54. Pendleton, E.W., Bessette, D., Field, P.B., Miller, G.D., and Griffin, K.E., Active aeroelastic wing flight research program: technical program and model analytical development, *J. Aircraft*, 37 (4), 554–561, 2000.
55. Lind, R., private communication, Jan. 2003.

30
Selected Topics in Probabilistic Gas Turbine Engine Turbomachinery Design

James A. Griffiths
GE Transportation

Jonathan A. Tschopp
GE Transportation

30.1	Introduction	30-1
30.2	Traditional Reliability Engineering Approaches	30-2
30.3	Probabilistic Rotor Design/Fracture Mechanics	30-4
	Introduction • Probabilistic Life Analysis for New Parts • Field-Management Risk Assessments • Future Probabilistic Efforts	
30.4	Fan, Compressor, Turbine Blade Probabilistic HCF Design	30-7
	Introduction • Background • Geometry Variation • Blade Frequency — Probabilistic Campbell Diagram • Mistuning • Forced-Response Prediction • Material Capability • Failure Probability Calculation • Updating of Predictions • Field Effects • Summary	
30.5	Overall Summary	30-19

30.1 Introduction

The use of probabilistic methods in gas turbine engine design is a rapidly growing field, with wide applications in various design disciplines such as performance, aero, heat transfer, and mechanical design. For example, probabilistics can be used to obtain a better understanding of engine performance and to define robust engine cycles that provide specified levels of performance in the presence of component/hardware variation. Probabilistic models can also be used to support aircraft/engine integration studies to optimize customer-based parameters such as the direct cost of fleet operations. This chapter focuses on the use of probabilistics in structural design, in particular, the design of rotating parts within the engine. Figure 30.1 provides a cross section of a modern commercial turbofan engine. The focus of this discussion is on the rotating parts — the fan, compressor, and turbine — with particular attention paid to design issues involving damage tolerance and fatigue. Given the complexities of the overall problem, this chapter can only provide an overview of the probabilistic methods available. The discussion begins with a summary of traditional reliability engineering methods.

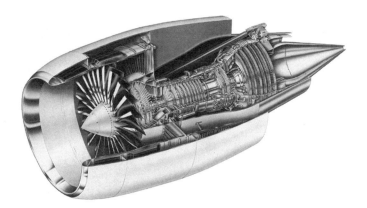

FIGURE 30.1 Trimetric of the GE90-94B turbofan engine. The chapter focus is on turbomachinery design, with particular emphasis on damage tolerance and fatigue behavior of fans and compressors.

30.2 Traditional Reliability Engineering Approaches

Reliability is defined as the probability that an item will perform its intended function for a specified interval under stated conditions. Failure is defined as the inability of the item to perform its function within previously specified limits. In general, such a definition transcends structural reliability, the focus of this chapter, and pertains to all aspects of engine reliability. Failures can be categorized based on their potential impact or hazard to the system. Category I failure hazards will lead to death or severe injury to personnel or total loss of the system. Category II hazards may cause personal injury, including death, or major system damage, or may require immediate corrective action for personnel or system survival. Category III hazards can be controlled without injury to personnel or major system damage, while Category IV hazards are highly unlikely to result in personal injury or system damage. Typical reliability metrics include shop visit rate and in-flight shutdown rate for both military and commercial engines, unscheduled engine removal rate and delay/cancellation rate for commercial engines, line replaceable unit (LRU) removal rate, in-flight abort rate, and mean time between failures for military engines.

System-level reliability is determined by a rollup of subsystem reliabilities. During the design phase, targets for subsystem reliability are established by allocating goal reliabilities for each subsystem based on the overall desired system reliability. Often, these subsystem allocations are based on historical experience with similar equipment. Figure 30.2 illustrates a general subsystem structure used for such analyses as adopted by the Air Transport Association (ATA). As data are developed for the actual subsystems, product reliability is updated by calculating from the bottom up, and it is measured at various subsystem and system levels vs. the allocations. Reliability block diagrams are used to roll up the reliability. Individual elements are considered to be in series or in parallel and their reliabilities combined appropriately. For two subsystems in series, the system operates only if both subsystems operate successfully. Thus system reliability is the product of the individual subsystem reliabilities.

$$R_{system} = R_A * R_B \tag{30.1}$$

where R_{system} is the combined reliability level, and R_A and R_B are the individual elements in series.

For parallel subsystems, where only one subsystem needs to remain operational for the system to operate, the system reliability can be expressed as

$$R_{system} = 1 - (1 - R_A) * (1 - R_B) \tag{30.2}$$

where R_A, R_B, and R_{system} are defined as above. Complex systems can be analyzed by using these series and parallel calculations in various combinations, leading to the system-level reliability.

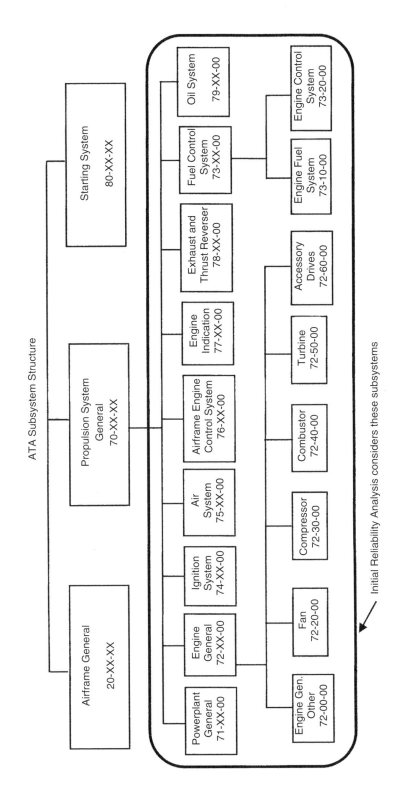

FIGURE 30.2 Portions of the Air Transport Association (ATA) consensus structure used as the framework to conduct reliability allocations and roll up reliability model results for aircraft engine subsystems.

A bottom-up reliability roll-up is based on failure modes effects and criticality analysis (FMECA). This is a detailed analysis of all potential failure modes, including description of the failure mode, its effects on the subsystem and engine, any compensating provisions, definition of failure detection method, hazard severity classification, and rate of occurrence. These analyses help identify high drivers of unreliability and serious hazards, and through the roll-up process, they help determine predicted reliability rates for the system.

System reliability typically improves over time as fixes to known failure modes are introduced to reduce their failure rate or as new failure modes are discovered and fixed. Often, reliability predictions are stated in terms of a "mature" rate, where maturity is defined at some amount of cumulative engine flight hours (EFH) for the fleet. Reliability growth can be characterized by plotting the cumulative number of failure events against cumulative EFH. Typically, a power function is fit to express the relationship between log(cumulative events) vs. log(cumulative EFH). The first derivative of this function then is used to determine the instantaneous failure rate.

Weibull analysis is typically used to characterize the statistical distribution of component life. The shape parameter for the Weibull function helps to identify age dependency of the failure rate; these include the classical bathtub curve segments of infant mortality (shape parameter < 1), random failures (shape parameter = 1), and wear out (shape parameter > 1). Commercial software is available for both Weibull and reliability growth analysis.

30.3 Probabilistic Rotor Design/Fracture Mechanics

30.3.1 Introduction

Two fundamental uses for probabilistic analyses on critical rotating components include design/life prediction of new parts and field-management risk assessments for fielded hardware. Field-management risk assessments are typically conducted to support development of field corrective-action plans in response to problems identified by either analytic predictions (a calculated life problem) or discovery of cracks in fielded components. Probabilistics methods used for field risk assessments can include Monte Carlo simulations for crack initiation and propagation lives, Weibull analysis of field experience, or probabilistic fracture mechanics assessments to address anomalies such as hard-alpha (nitrided titanium) in titanium components or alloy contaminants introduced during material processing.

Probabilistic analysis for the design of new rotating parts has evolved over time. Initially, probabilistic analyses were introduced to address specific issues that were determined to be best handled by a stochastic rather than a strict deterministic process. The primary introduction point for probabilistic analysis has been probabilistic fracture mechanics to address damage-tolerance criteria. Strategies have been developed limiting the use of random variables to only those variables thought to be key to the final results. Probabilistic fracture mechanics approaches typically focus on defect size and frequency as the primary random variables. The primary challenges have been understanding and modeling the fatigue behavior of the anomalies, development of the input anomaly size distribution, and validation and calibration to specimen data and field component experience. As these approaches mature, they are being applied earlier in the design process so that damage-tolerance considerations help shape the final design for the part rather than provide a postdesign assessment of damage risk.

Requirements for damage-tolerance analysis on commercial engines are described in the Federal Aviation Administration (FAA) Advisory Circular 33.14-1 [1]. The Rotor Integrity Subcommittee (RISC) of the Aerospace Industries Association (AIA) developed a strategy to address damage tolerance of critical titanium parts in response to the FAA Titanium Rotating Components Review Team Report in 1990 [2]. Key parts of this industry consensus strategy include (1) definitions of hard-alpha anomaly occurrence rates and size distributions in finished hardware, and (2) the definition of a design target risk (DTR) metric [3] to determine both the acceptability of new hardware designs as well as the appropriateness of proposed actions for fielded hardware. These methods are currently being extended to additional areas such as surface-anomaly damage tolerance and wrought nickel damage tolerance under the auspices of

RISC. The U.S. Department of Defense's description of damage-tolerance design best practices for military engines is set forth in the MIL-HDBK1783B handbook (Engine Structural Integrity Program or ENSIP [4]).

30.3.2 Probabilistic Life Analysis for New Parts

In performing a probabilistic life assessment, the distribution of potential anomalies (size and frequency distributions) and the behavior of such anomalies during field service must be defined. Typically, anomaly distributions are described by an overall frequency expressed as the number of anomalies of all sizes per unit volume or weight and a separate probability density function of anomaly size.

Anomaly fatigue behavior is described by crack-growth curves, assuming the anomaly acts like a crack under cyclic loading. Growth is a function of the initial size of the anomaly, the local stress and temperature fields, the number and cyclic content of the flight missions, and the material crack-growth resistance. In addition, surface and embedded anomalies can differ in crack-growth behavior, with surface anomalies typically growing faster under the same cyclic loading conditions.

The fundamental metric of interest is probability of fracture (POF), where POF is the predicted probability of component rupture per unit time/cycle (rate) or for a given interval in time or cycles (typically the component service life). This in turn is a function of two things: the chance of having an inclusion (inclusion probability) and the chance that the initial inclusion is of sufficient size to cause fracture within the specified life of the component (crack probability). POF is, in its simplest sense, the product of these two underlying probabilities.

Inclusion probabilities are dependent on the material, its manufacturing processes, and the sensitivity of production inspections to the presence of anomalies as a function of size and shape (probability of detection as a function of defect area). The AIA RISC has developed baseline hard-alpha anomaly distributions for titanium alloys, accounting for the several manufacturing and inspection steps in the process to go from ingot to finished parts [3], similar to that shown in Figure 30.3. Details of the process to determine the baseline anomaly distributions are provided in this reference. These distributions are based on both analytical models of the manufacturing process and correlation with historical data for commercial engine experience, such as dimensional data from detected anomalies and inspection capabilities. Based on these distributions, inclusion probability is estimated by taking the product of the volume of the material and the anomaly frequency.

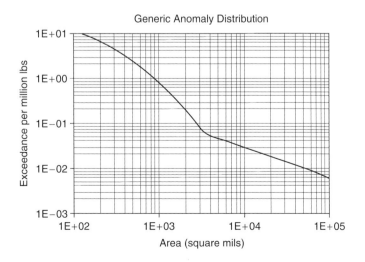

FIGURE 30.3 Generic anomaly size (area) distribution expressed as exceedance probability.

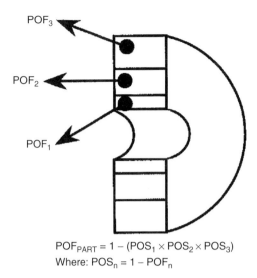

FIGURE 30.4 Schematic of summation of part subvolume probability based on binomial model of failure for combining failure probability.

Crack probabilities are determined in two steps. First, given a life goal — cycles to failure under a specified cyclic loading and environment — a maximum crack size that will survive to the life goal is determined using fracture mechanics crack-growth analysis. Given the critical crack size, one minus the cumulative probability (or exceedance probability) of the critical size from the anomaly size distribution determines the crack probability.

These analyses are conducted for a series of subvolumes within the component of interest to reflect the variation in stress, temperature, and geometry within the component and to reflect how crack-growth capability varies throughout the part. The part is divided into regions of near-constant life, and each region is analyzed. Probabilities of fracture are determined for each subvolume in the part, and these are subsequently combined statistically to get the total part probability of fracture schematically, as shown in Figure 30.4. These calculations are repeated for a range of service lives to create a part probability of fracture (POF) vs. life curve (predicted cumulative fracture probability distribution).

Figure 30.5 shows an example of a POF plot. The POF at the required service life from the curve is compared with the DTR, where DTR is an agreed upon standard (relative risk value) for accessing the acceptability of a calculated component POF. Typically, DTRs are validated against specimen data and component field experience. Parts with a calculated POF less than the DTR are considered acceptable. Parts that are predicted to exceed the DTR require action to reduce the POF below the DTR. Redesign or field inspection are the typical actions.

Much of this process has been automated in OEM specific analysis systems. In addition, the FAA has funded Southwest Research Institute (SWRI) for development of a commercially available computer code (DARWIN) to conduct these calculations [5].

A very important step in the use of probabilistic methods is the validation/calibration of such methods with observed field experience. Adamson [6] and an AIA RISC paper [3] provide examples of such validation/calibration efforts.

Several general conclusions can be drawn concerning the relative sensitivity of POF to key variables. First, POF is approximately directly proportional to volume, all other factors being equal. Second, factors affecting residual life, such as local stress, have a more significant effect on POF. For example, it is not uncommon for a 10% change in stress to yield a 30 to 40% increase in risk. Finally, the frequency of anomalies influences the nature of the regions driving risk. For low-frequency anomalies such as hard alpha in titanium, the POF is primarily driven by bulk stress regions in the part; large volumes of material

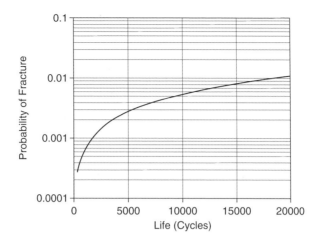

FIGURE 30.5 Example of a probability-of-fracture curve in which probability of failure is expressed as a function of cumulative cycles.

at "moderate" stress levels contribute significantly more risk than very highly local stress-concentration features. The contrary is true for alloys that have small inclusions at high frequencies. Here, stress-concentration features can primarily drive risk due to the comparatively high likelihood of anomalies in these small regions.

30.3.3 Field-Management Risk Assessments

Field-management risk assessments can be undertaken for such reasons as premature component cracking problems, hard-alpha inspection programs, or preplanned in-service inspection intervals on military engines. Techniques similar to those described above for life analysis are used, although in this case the focus is on the probability of fracture over a fixed interval of time in hours or cycles between in-service component inspections. Critical to this analysis is the probability of anomaly detection for in-service component inspection methods. This ultimately determines the distribution (size and frequency) of the larger anomalies that go undetected and can propagate during the next inspection interval.

30.3.4 Future Probabilistic Efforts

Probabilistics can be extended to creep, fatigue, and burst design tasks. Efforts to develop physics-based models and appropriate distributional data for such models may be substantial. As a benchmark, significant efforts by industry went into developing and validating the probabilistic fracture mechanics approach for damage tolerance. Extension of probabilistic methods to these alternative failure mechanisms will also require significant engineering effort, largely due to the need for validating failure predictions with field experience, as was the case with the probabilistic fracture efforts described above.

30.4 Fan, Compressor, Turbine Blade Probabilistic HCF Design

30.4.1 Introduction

The focus of the remaining discussion is on resonant-mode vibration of fans, compressors, and turbine blades and vanes. Resonant-mode vibration is a dominant cause of blade failure in high-cycle fatigue (HCF). The USAF is currently funding industrywide research into probabilistic methods for HCF failure probability assessment as part of its HCF initiative [7]. In conjunction with this, future engine development programs will likely face probabilistic HCF failure rate requirements.

Current deterministic-based methods are briefly described first. Following this, progress to date and future development needs and direction for probabilistic methods will be outlined. The primary focus of the discussion will be fan and compressor blades; the complex geometries, thermal stresses, and high-temperature effects associated with turbine blades and vanes provide additional complications for probabilistic HCF assessment.

30.4.2 Background

Integral-order blade stimuli are caused by other vanes, blades, and by airflow distortion. Rotating blades pass through pressure fluctuations due to these excitation sources. The frequency of stimulus is therefore related to the engine speed and the number of struts or vanes or the spatial content of the airflow distortion pattern. The fundamental harmonic of the excitation equals the product of the speed and order of the excitation (e.g., number of upstream vanes).

Traditionally, components are designed to either avoid resonances in the operating speed range of the engine associated with known engine-order excitation sources or to place resonances in lower-speed regions and away from mission points in the cycle where sustained operation is expected. The Campbell diagram, Figure 30.6, is used to display the excitation frequencies due to known sources and compare those frequencies with the variation in vibratory frequencies of the blades with engine speed. Blade vibratory frequencies can change with engine speed due to temperature changes and centrifugal stiffening of the blade. The intersection of engine-order excitation lines and blade-frequency lines indicate resonance at the crossing engine speed.

Historically, frequency placement has been based on frequencies associated with nominal blade geometries. For modes outside the operating range, a set percentage margin in frequency with allowance for maximum engine overspeed is used to account for blade frequency and engine differences. For modes with resonances within the operating range of the engine, maximum permissible blade vibratory stresses are set using factors of safety applied to minimum material HCF capability. Guidelines for maximum nominal steady-state stresses are used early in the design process to provide adequate allowable vibratory stresses. These measures are taken to provide confidence that measured stresses in the aeromechanical

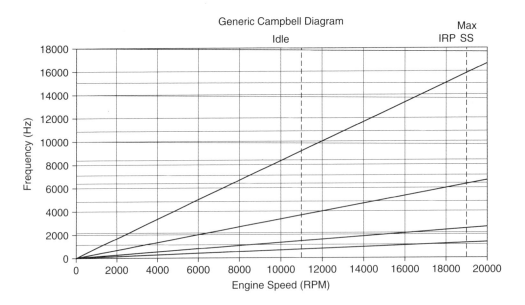

FIGURE 30.6 The Campbell diagram identifies vibratory modes with resonances in the engine operating range. "Horizontal" lines are individual blade vibratory modes. Driver lines are integral per-rev excitation sources: distortion, upstream/downstream blade or vane count, etc.

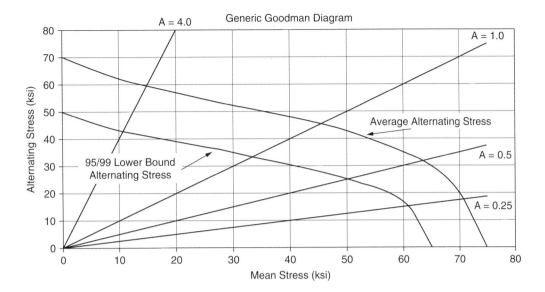

FIGURE 30.7 The Goodman diagram describes HCF alternating-stress capability as a function of mean or steady-state stress for a given number of HCF cycles. Average (mean) and lower-bound (95% confidence/99% of population exceedance) capabilities are indicated. The A-ratio lines are constant alternating stress/steady-state stress ratios.

qualification engine test would ultimately be acceptably low. The design is qualified based on the measured vibratory stress levels, not the predicted analytic values of stress.

Allowable vibratory stresses are expressed as a function of the local steady-state stress in the Goodman diagram, Figure 30.7. Allowable vibratory stress decreases with increasing steady-state stress.

Strain gauges monitoring blade vibratory stress are typically limited in quantity due to restrictions on the number of instrumentation leads from the rotating structure. Thus, there is uncertainty in the maximum blade vibratory stress, and one cannot be sure that the maximum responding blade was instrumented. Second, one or, at most, a very modest number of engines are instrumented for aeromechanical qualification tests. Therefore, factors of safety to account for sampling of blade stresses and engine-to-engine variation are used. In addition, both engine deterioration and damage to hardware occur in the fleet, potentially increasing vibratory stress, reducing material capability, or both. Historic factors of safety for vibratory stress have been used in the past to address all these issues based on experiences from successful designs.

With the advent of probabilistic methods, predictions of probability distributions for quantities like blade frequency and stress are replacing fixed design margins based on the nominal blade geometry and material properties. Variation in blade geometry, mode shape and frequency, and vibratory stress response can all be addressed explicitly. A number of deterministic tools developed in connection with the U.S. government's HCF initiative are being applied probabilistically to address critical aspects of the problem such as mistuning [8–10] (variation in blade-to-blade response due to subtle blade geometry differences on a single rotor) and forced-response vibratory stress prediction. As shown in Figure 30.8, geometry variation influences blade frequencies and mode shapes that, in turn, influence forced response.

Another important aspect of the probabilistic assessment methodology is the ability to update predictions as more data become available. Such data may include new analytic predictions, component or engine test results, and field experience. Probabilistic HCF analyses can be an ongoing process over the life of the engine design as new data become available. Quantifying uncertainty in predicted failure probability is desired, and such uncertainty should diminish throughout the product life cycle as additional data and field experience becomes available. A schematic for tracking such predictions is shown in Figure 30.9. This schematic is similar to that suggested by Los Alamos National Laboratory and their PREDICT [11] system.

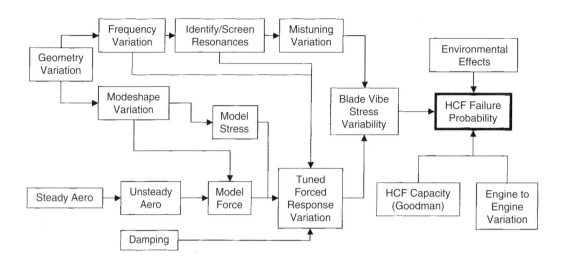

FIGURE 30.8 High-level analysis flowchart for prediction of HCF failure probability. Geometry variations drive mode shape and frequency, which, in turn, drive forced-response variation and HCF failure probability.

30.4.3 Geometry Variation

Probabilistic assessments are grounded on an understanding of the geometric variability of the blades. From variation in geometry and material properties, mode-shape and frequency data are derived. These, in turn, have a strong influence on mistuning effects, aero damping, and forced response. This discussion focuses on solid-blade geometries; the presence of internal cooling-air passages for turbine blades, for example, significantly complicates characterization and analysis of geometry variation, although in principle, the methods described below still apply.

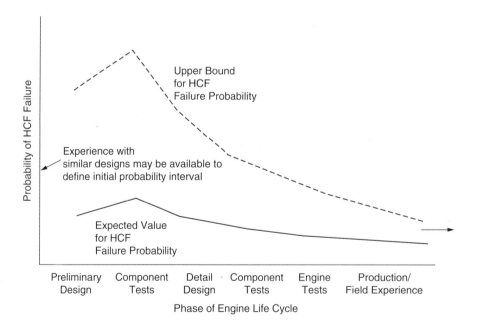

FIGURE 30.9 Predictions of HCF failure probability can be conducted throughout the engine life cycle. Uncertainty in failure probability should diminish with time as additional data become available. A one-sided confidence interval is shown in this example because the upper limit is of primary interest.

Two well-documented methods for statistical analysis of data are currently under study or in use in characterizing blade geometry variation. The first method is principal components analysis [12] (PCA), and the second is spatial statistics analysis (SSA) [13, 14]. The former method is available in a number of commercial software packages and is a technique used to simplify analysis when considering a large number of quantified variables, such as detailed geometry measurements. The latter is a more specialized technique arising from the mining industry; commercial software code is available for some applications.

In PCA, the data for various geometric features are assembled as a vector in which individual quantities are typically correlated, e.g., individual thicknesses on a grid over the blade. PCA constructs a series of independent linear combinations of the individual geometry variables (called components), which, when taken all together, explain all the variance in the original data. Each component is uncorrelated with all previous ones, facilitating subsequent generation of simulated random blades; the amount of variance explained with each additional component diminishes. The intent is to capture most of the original variance in a relatively small set of components, enabling rapid simulation of new random blades. The linear coefficients for each PCA component for all measured blades define distributions from which new sets of coefficients can be drawn independently. When working with physical geometry parameters, distributions of these coefficients are generally nonGaussian. One can either fit parametric models to individual-component coefficient distributions, or one can draw samples with replacement (bootstrap) if the database is sufficiently large. Due to the nonGaussian nature of the coefficient distributions, a mixture model of Gaussian distributions can be used, for example, to approximate a bimodal coefficient distribution. Significant amounts of actual hardware measurements are necessary to implement this method effectively. Figure 30.10 illustrates the process.

In spatial statistics, geometry variables at two locations are assumed correlated by a function of the distance between the points, closer points being more highly correlated than those farther apart. This correlation function with distance is characterized using the correlogram or a related function, the variogram, which tracks variance, not correlation, as a function of distance between points. For example,

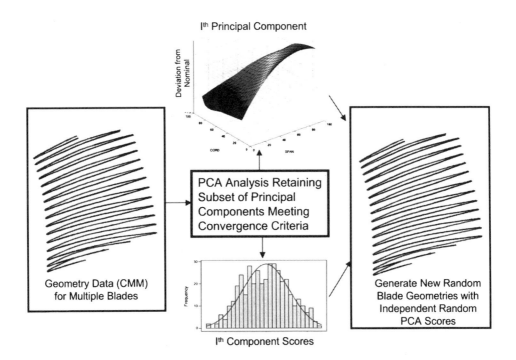

FIGURE 30.10 Principal components analysis (PCA) defines components ("manufacturing modes") containing most of the original variance in geometry and permits simple generation of new, random blade geometries.

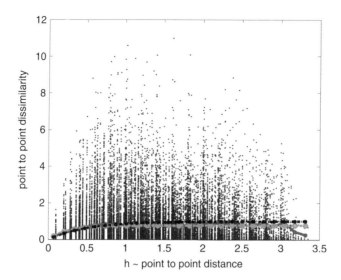

FIGURE 30.11 Example of experimental variogram cloud with fitted exponential theoretical variogram. Original data were generated from a given variogram. Raw average variogram and least-squares-fit exponential variograms are also shown as dashed lines. Data points are individual squared error values for grid point pairs.

blade thicknesses would be more highly correlated at adjacent locations than they are when well separated across the blade. Isotropic spatial statistics models are used for simplicity; correlation or variance is expressed as a single function of distance applied across the full blade.

Measurements can be used to construct an experimentally based variogram; the collection of raw variance vs. distance data is known as a variogram cloud. In this case, distances between measurement points are then divided into a number of bins of similar distance, and the average squared differences of the measurements within each bin is used to define the variogram. A number of traditional variogram functions are available, each described by a limited number of parameters, which can be fit to the average bin data using least-squares methods to provide a theoretical variogram. Figure 30.11 shows an example variogram cloud as well as an associated variogram model fit.

Spatial statistics models can be used early in the design stage to define generic correlation functions before hardware is manufactured and measured geometry data are available for PCA analysis. In this case, additional random blade geometries are generated using an appropriate theoretical variogram, likely based on similarity to a previous blade design and manufacturing method. A covariance matrix is constructed based on the variogram and a grid of measurement points; random blade geometries are then calculated using Cholesky decomposition of the covariance matrix. Spatial statistics and PCA analyses can therefore be used to complement each other in various stages of design evolution.

30.4.4 Blade Frequency—Probabilistic Campbell Diagram

PCA or SSA geometry models can be used to generate a series of random blades. Finite element models of these random blades would then be used to predict frequency and mode-shape variation. This can be done by either (1) building an ensemble of individual blade models and conducting modal analysis on individual random finite element models or by (2) modeling the nominal geometry and determining frequency and mode-shape sensitivities to geometry variation. Once sensitivities are known, random sets of deviations from nominal geometry can be processed to estimate blade frequency and mode-shape changes by linear superposition. Blair and Annis [15] provides examples of the application of SSA to predicting variation in blade frequency.

As an example of probabilistic analysis, consider a family of blades with a particular vibratory mode from a specified engine-order excitation predicted to have resonance above the engine operating range on the Campbell diagram. In the past, a factor of safety between redline engine speed (maximum overspeed) and nominal blade frequency would be used to avoid resonance for blades with lower-than-nominal frequency. Using the nominal blade frequency and frequency variations predicted by generating random blade geometries and frequencies about nominal by finite element or sensitivity analysis, one can now directly calculate the probability of lower blade frequencies resulting in a resonance at or below max speed. As opposed to a fixed-percentage frequency margin, this focuses attention on those modes for which blade frequency may be most sensitive to geometry variation and the variation in blade frequency is highest. By these techniques, one can produce a probabilistic Campbell diagram.

It is probable that some resonances will be within the engine operating range. Additional forced-response analyses can be used to assess the likelihood of potential HCF failure from these remaining modes. For these modes, the probabilistic Campbell diagram indicates the distribution of engine speed over which individual blades resonate.

30.4.5 Mistuning

Blades in a tuned system (all identical) share a common forced-response stress. However, blade geometric and material variations introduce structural dynamics effects known as mistuning, which result in variation in response from blade to blade. Responses of individual blades are coupled via the disk. In the extreme, energy can be localized in one or several blades, resulting in high response of those blade(s) substantially above the tuned response. Variation in blade response must be accounted for in predicting HCF failure probability. Mistuning analysis is used to predict the sensitivity of a stage's design to blade variation for a fixed excitation level.

It is not practical to build full rotor models repeatedly to determine mistuned responses. This has been done in isolated cases for single rotors to validate models, but it is not a practical design approach to characterize rotor-to-rotor variation. Reduced-order models [8–10] have been developed to provide a practical means of predicting blade-to-blade response variation on rotors due to blade differences. These models are based on cyclic-symmetry finite element models of the rotor. Cyclic-symmetry models reflect the fact that individual sectors (one blade plus associated share of the disk) on the rotor respond similarly, with a known phase relationship existing between the individual sectors based on the number of blades on the rotor and the excitation engine order. Mistuned modes of vibration may be described using linear combinations of tuned-system responses or blade-alone responses with constraint modes describing interactions with the disk at the blade–disk interface. The size of the reduced-order model is substantially smaller than the original number of degrees of freedom in the sector finite element model. Thus, reduced-order models, once built, can be used to rapidly characterize rotor-to-rotor variation.

Current advanced tools typically characterize blade differences with a frequency standard deviation parameter that is used to vary the stiffness of the individual mistuned sectors on the rotor. Stiffness varies with frequency squared. Individual rotor analysis provides blade-to-blade variability at a fixed excitation level. Monte Carlo simulations of a series of random rotors can also be conducted to determine the variability, engine to engine, due to mistuning structural dynamic effects at a constant modal force. Relative response of the maximum blade on each rotor, for example, is compared with a tuned rotor (all identical blades) and then typically displayed as a function of blade frequency standard deviation, as shown in Figure 30.12. However, the Monte Carlo simulations provide all blade responses for each rotor in the simulation, providing a wealth of information for subsequent probabilistic analysis.

Two methods can be used to store and use this full set of blade relative-response data in subsequent failure probability analysis. First, a parametric model can be selected for describing blade-to-blade vibratory stress variation on a single random rotor. Individual mistuned rotor stress variations can then be fit using the selected model, and distributions of the model parameters themselves can be captured for future use. There will be correlations between model parameters that must be accounted for in subsequent usage. Alternatively, one can retain a large sample of Monte Carlo-generated mistuned vibratory stress data and bootstrap random rotors from this ensemble of data. The former method requires less data storage but forces a specific upper-tail behavior on the data based on the chosen parametric model. The bootstrap

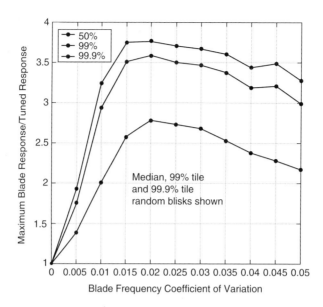

FIGURE 30.12 Amplification factors (ratio of maximum blade response on mistuned-blisk to tuned-blisk response) shown for higher responding random blisks. Amplification ratio depends on level of blade mistuning as characterized by blade frequency coefficient of variation.

method requires that substantially more data be stored, but it does not impose a specific upper-tail structure on the blade-to-blade variability coming from the mistuning analysis.

Typically, the worst mistuning responses (highest maximum-to-average ratio of blade response) occur in regions where there are closely spaced blade and "disk" modes that can interact (known as a veering region). A "disk" mode is a mode with significant disk activity. A tuned cyclic-symmetry sector finite element model analysis, containing a single blade and its share of the disk, can be used to identify those modes and engine-order stimuli that most likely result in higher mistuned response. One can then focus subsequent mistuning and forced-response analyses on those modes most likely to exhibit high variability in response.

30.4.6 Forced-Response Prediction

Blade resonant response can be expressed as

$$\sigma_v = \frac{MF * Q * \sigma_m}{(2\pi f)^2} \tag{30.3}$$

where σ_v is the blade vibratory stress, MF is the modal force normalized by modal mass, Q is a magnification factor (dependent on percent critical damping from aero and mechanical sources), σ_m is the modal stress at the location of interest, and f is frequency in Hz. Modal force is determined by the dot product of the unsteady pressure field with the mode shape of interest. Variations in frequency and mode shape have been determined by earlier geometry studies. Thus, variations in f and σ_m have been characterized. The additional complexities added for forced-response analysis are damping and unsteady pressure field variation. Both of these are challenging to characterize probabilistically.

Techniques are under development to predict the variation in unsteady pressure fields due to blade-geometry variation, with fixed boundary conditions on steady airflow. Adjoint methods [16, 17] can, in principle, be used to determine unsteady pressure variations due to blade-geometry variations characterized by methods described above. Although these methods would provide data concerning the variation in unsteady pressure due to geometry changes of the blades under study, they do not fully address all sources of unsteady pressure field variation.

The unsteady pressure field is a complex function of a large number of variables including, to name a few, engine operating line, variable geometry schedule, clearances, bleed rates, corrected speed, profiles and distortion, and stage boundary conditions determined by other stages. Aeromechanical testing is conducted over a range of operating conditions to survey vibratory stresses. However, characterizing engine-to-engine variability beyond available engine test data will be difficult. Where specific sources of excitation are linked to critical resonance, component tests and computational fluid dynamics (CFD) analyses might be used to understand sensitivity to critical variables of interest, although such analyses are typically difficult and expensive to conduct. Thus, in general, defining methods for predicting probabilistic forced response remains a critical research effort. Reliance on developing probability distributions for engine-to-engine variation in excitation levels will likely rely on distributions derived from expert opinion for the near future, augmented by test data and targeted analysis where feasible.

30.4.7 Material Capability

Allowable vibratory stress is determined using the Goodman diagram, where local mean stress is considered in setting allowable vibratory stress. Goodman diagrams can be linked to the traditional fatigue curve through the concept of equivalent stress. Equivalent uniaxial stresses [18, 19] can be defined for a given stress ratio, R, (minimum stress/maximum stress) or the related alternating stress ratio, A, (alternating stress/mean stress) for the test. Median and selected percentile uniaxial equivalent vibratory stress capability can be characterized by fitting a traditional fatigue curve. To be effective, such a fit generally requires substantial data covering a range of stress/life values and R (or A) ratio values. The fatigue curves can be translated then to a Goodman diagram using various R or A ratios based on mean stress to calculate the corresponding vibratory stresses. A method for transforming fatigue-curve data to Goodman curves is outlined as follows (in this example, for fully elastic behavior):

1. Select an A ratio of interest (multiple A ratios are eventually used to create the Goodman diagram).
2. Calculate the corresponding R ratio:

$$R = \frac{1-A}{1+A} \tag{30.4}$$

3. Extract the equivalent uniaxial stress at various percentile levels from the fatigue curve and convert to maximum stress using a preferred model. For the Walker model [18] for equivalent stress in the fully elastic regime, this is expressed using a material-dependent Walker exponent, m, as follows:

$$S_{eq} = \frac{S_{max}}{2} * (1-R)^m \tag{30.5}$$

4. Convert the maximum stress to alternating stress and plot on the Goodman diagram:

$$S_{alt} = \frac{A}{A+1} * S_{max} \tag{30.6}$$

Figure 30.13 shows a schematic of the process used to construct a Goodman diagram. Modifications to this basic procedure are necessary for higher steady-state stresses (A near 0), where cyclic elastic-plastic behavior may be involved.

Pascual and Meeker [20] introduced the random fatigue-limit model to characterize fatigue behavior. In this model, specimen lives have a lognormal probability distribution, with the mean parameter (log life)

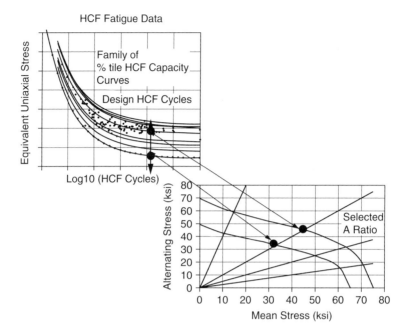

FIGURE 30.13 Example of a process for constructing a probabilistic Goodman diagram in a fully elastic regime, where more HCF fatigue failures typically occur.

linearly related to the log of the difference between the applied stress and a random fatigue limit for each specimen. A selection of probability-density models is available for modeling the fatigue-limit stress itself. Comparisons with data sets of equivalent stress derived from a series of varying R ratio tests indicate that this model can be useful in describing behavior of materials with fatigue-limit behavior. In particular, the model may represent well the large variation in fatigue life at lower stress levels approaching the fatigue limit.

Traditional LCF lifing methods have focused on life variability at a given level of stress. As discussed previously in probabilistic rotor design, LCF-driven design criteria focus on life (cycles) and, for probabilistic fracture mechanics in particular, the margin in life between inspection intervals. In the HCF regime, as typified by the random fatigue-limit model, there are order-of-magnitude larger variations in life at a given stress compared with the LCF regime. HCF design necessarily focuses on stress margin, not life.

It is assumed that the component will ultimately be subjected to sufficient cycles to cause failure. Typically, engine specifications stipulate a controlling number of cycles for which allowable stresses are defined, and these typically range from 10^7 to 10^9 cycles. The failure probability calculation then reduces to the calculation of the probability that the applied vibratory stress exceeds the allowable stress at the specified number of cycles as determined by the Goodman diagram. Of course, suitable factors can be applied either to vibratory or allowable stress to account for field degradation effects to either the hardware itself or to operating conditions within the engine.

30.4.8 Failure Probability Calculation

The probability of individual blade failure is determined at the limiting spot on the blade where the vibratory stress is the maximum percentage of local HCF capability based on the local steady-state stress; this location is known as the critical location. The probability that the critical location vibratory stress for a random blade exceeds material HCF capability can be expressed as the integral:

$$P_f = \int_0^\infty f_{\text{app}}(x) * F_{\text{allow}}(x) dx \qquad (30.7)$$

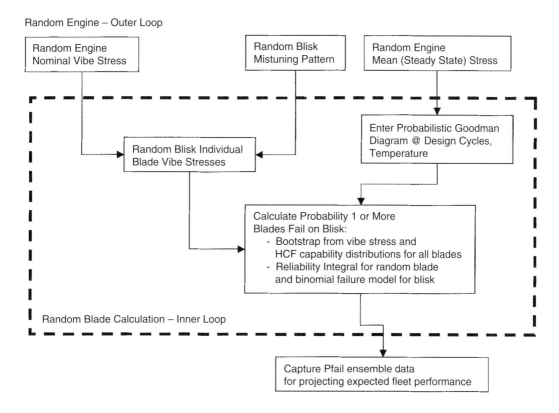

FIGURE 30.14 High-level Monte Carlo process schematic for calculating HCF failure probabilities based on typical forced-response distributional data. More-primitive variables such as geometry, frequency, and mode shape ultimately dictate the higher-level forced-response distributions used here.

where $f_{app}(x)$ is the probability density of critical location vibratory stress, and $F_{allow}(x)$ is the cumulative probability function for material HCF capability at the local steady-state stress on which the vibratory stress is superimposed. Simplified solutions for this integral exist for some pairs of distribution types, e.g., normal or lognormal for both distributions [21]. Otherwise, numerical integration is required, or Monte Carlo simulation is used to predict it.

To date, Monte Carlo techniques have been used primarily to assess HCF failure probability. A typical, but simplified, Monte Carlo simulation sequence (excluding field effects for now) would include the following steps (Figure 30.14 illustrates this process):

1. Select a nominal value of blade vibratory stress from a probability distribution representing anticipated engine-to-engine variability in average vibratory stress and uncertainty in that stress.
2. Generate a random blade-to-blade relative vibratory stress distribution from mistuning data (parametric or bootstrap models) and multiply by the selected nominal vibratory stress to obtain the distribution of individual blade vibratory stresses for a random engine.
3. Select mean stress level at the controlling critical location from a suitable probability density distribution reflecting the operating conditions at which the resonance occurs.
4. Using a probabilistic Goodman diagram, calculate the parameters for the allowable stress distribution (for example, a set of lognormal distribution parameters, μ and σ, can be calculated from the median and another specified lower-bound percentile curve on the Goodman diagram) based on mean stress from step 3.

5. Calculate the random blade failure probability in one of two ways:
   a. For parametric models of blade-to-blade stress variability and HCF capability, use the reliability integral to calculate failure probability of a random blade, P_{fRB}. Probability of failure of one or more blades on the rotor, P_{fRE}, can then be calculated using the binomial model by:

$$P_{fRE} = 1 - (1 - P_{fRB})^{N_{blade}} \tag{30.8}$$

   where N_{blade} is the number of blades on the rotor.
   b. For bootstrapped vibratory stress samples from the mistuning analysis, generate a random allowable stress for each blade on the rotor and compare it with the individual blade applied stress. One or more blades exceeding its allowable stress constitutes failure for the engine.
6. Collect ensemble data from the Monte Carlo simulation and interpret the results. For bootstrapped mistuning data, a count of the number of engines with failed blades provides an estimate of the failure probability of the fleet. With parametric models used for blade-to-blade mistuned stress, each random engine results in a failure probability for the random engine. The ensemble average of these engine-level failure probabilities is the expected value for fleet failure rate.

In the course of the Monte Carlo analysis, individual probability distributions may be scaled or additional variables defined as the product, sum, difference, or division of two random variables. Hahn and Shapiro [22] provide a ready reference for calculating such manipulations of independent random variables.

30.4.9 Updating of Predictions

Failure probability predictions can be updated with new information as it becomes available. Such data may include additional design analyses, component test results, engine test results, and field experience or failure data. At one extreme, distributions used previously can simply be ignored and new distributions defined based solely on the new data. An alternative and preferred approach is to reflect both old and new data in the latest predictions. Bayesian analysis [23, 24] is one well-known method of updating distributions where the final results reflect both the prior distributions (previous knowledge) and the new data.

To clarify, take a specific example. Assume, for convenience, that blade frequencies are normally distributed. Uncertainty in the actual population frequency distribution can then be expressed by probability distributions for the population mean and standard deviation themselves. Based on data to date, so-called "prior" distributions can be defined for the mean and standard deviation parameters themselves. Now assume that additional blades are tested. The likelihood of obtaining the actual test data can be expressed as a function of the normal distribution mean and standard deviation of the probability density function for the normal distribution. Posterior distributions for the parameters (combining previous knowledge and the new test data) are calculated using Bayes' theorem, a statement of conditional probabilities:

$$P(\theta, x) = \frac{P_{prior}(\theta) * L(x, \theta)}{\int_\theta P_{prior}(\theta) * L(x, \theta) d\theta} \tag{30.9}$$

where $P_{prior}(\theta)$ is the prior joint distribution for the parameters, and $L(x, \theta)$ is the likelihood of obtaining the actual test data, given the parameters, θ. One issue with Bayesian techniques is the balance between prior knowledge and new data influencing the posterior distributions. The literature abounds with references to "noninformative priors" to let the test data speak for itself. In our case, however, we want prior data to influence the final results. However, prior distributions that are too restrictive can result in their overpowering the associated test data. The balance on how best to combine prior data and new knowledge remains a significant area of research for HCF probabilistics.

The integral in the denominator of Equation 30.9 can be a difficulty. So-called conjugate pairs for prior distribution and likelihood have historically been used where the posterior distribution is in the same form as the prior distribution with altered hyperparameters [23, 24]. Use of these conjugate pairs restricts probability distributions used to model critical HCF data. More recently, Markov chain Monte Carlo methods [25, 26] have been developed, which eliminate the need to explicitly perform the integration in the denominator of Equation 30.9. The applicability of these methods for updating HCF-failure-related distributions is actively under investigation. Much effort remains to define best practices in updating distributions used for HCF analysis.

30.4.10 Field Effects

Alterations to both component hardware and engine operating conditions occur in the field. For example, engine temperatures may increase as the engine deteriorates. Physical damage can occur, such as foreign object damage (FOD), on leading edges of blades. The location, extent, and impact of such damage must be accounted for when making HCF reliability predictions. For example, in the case of FOD damage to the leading edge of blades on a fan, the rate at which FOD occurs, the distribution of FOD sites along the blade leading edge, and the geometry of the FOD site must all be described with suitable probability distributions. In addition, there must be analytic or empirical transfer functions relating FOD site location and geometry to a fatigue notch sensitivity factor, K_f, applied to the vibratory stress when calculating failure probability. Alternatively, a knockdown factor can be defined and applied to the HCF capability. FOD occurs on random blades, so the Monte Carlo simulation for FOD failure needs to address blade-to-blade stress variability as well. Griffiths and Kielb [27] provides an example of such a model considering FOD damage impact in calculating HCF failure probability for a generic fan blade design. Similar concerns exist for dovetail wear, LCF/HCF interaction, and other physical degradation mechanisms.

30.4.11 Summary

A logical progression of analyses culminating in the prediction of failure probability is presented. The intent is to identify those vibratory modes most likely to cause failure early on in the process using preliminary design tools whenever feasible, and then to concentrate the most intensive and expensive analyses on these modes of greatest concern.

Geometry variability is characterized, and variability in blade mode shape and frequency is derived. Potential resonance crossings for known engine excitation sources are then identified using the Campbell diagram, with the intent to place modes outside the operating range or at a relatively benign engine speed as much as possible. Historical experience and tuned-system modal analysis for a variety of engine orders of excitation can be used to screen for modes most likely to result in high stress due to structural dynamic effects known as mistuning.

Mistuning and forced-response analysis can then be used to predict nominal blade stresses for these most critical modes. Such analyses are used to characterize both blade-to-blade stress variability within an engine as well as from engine to engine due to structural dynamic considerations. Simulations of random rotors and engines are then conducted to generate random-engine applied and allowable vibratory stresses and assess failure probability.

The development of HCF probabilistics remains an exciting and ongoing area of research as an alternative design approach to fixed factors of safety applied to nominal designs. It is anticipated that, in the near term, both probabilistic methods and deterministic analysis with safety factors will be utilized together to assess designs, with eventual migration to probabilistic methods as the primary approach as confidence in and experience with such methods continue to grow.

30.5 Overall Summary

Probabilistic design methods for predicting component structural failures are proving, and will continue to prove, to be highly useful technologies for designing gas turbine engine components. Currently, probabilistic damage-tolerance methods, driven by industry and FAA development activities over the last decade, have a significant lead time over emerging probabilistic HCF failure prediction methods. An industry and USAF initiative for developing HCF failure prediction methodologies is underway to mature those technologies. Confidence in the use of these tools will grow as they are used to make design predictions and such predictions are then validated by component test results, engine test results, and ultimately by field experience. Much effort remains to mature and validate these tools and also to extend probabilistic methods to other failure mechanisms.

References

1. Damage Tolerance for High Energy Turbine Rotors, Advisory Circular (AC) 33.14-1, U.S. Dept. of Transportation, Federal Aviation Administration, Washington, DC, 2001.
2. Titanium Rotating Components Review Team Report, U.S. Federal Aviation Administration Aircraft Certification Service Engines and Propeller Directorate, Washington, DC, 1990.
3. Sub-team to the Aerospace Industries Association Rotor Integrity Subcommittee, The Development of Anomaly Distributions for Aircraft Engine Titanium Disk Alloys, American Institute of Aeronautics and Astronautics, Washington, DC, 1997.
4. Department of Defense, Engine Structural Integrity Plan (ENSIP), Department of Defense Handbook, MIL-HDBK-1783B, DOD, Washington, DC, 2002.
5. Wu, Y.-T., Millwater, H.R., and Enright, M.P., Probabilistic Methods for DARWIN™, paper presented at 4th Annual FAA/USAF Workshop, Application of Probabilistic Methods to Gas Turbine Engines, Jacksonville, FL, 1999.
6. Adamson, J.D., Validating the Pratt & Whitney Probabilistic Design System, paper presented at 6th Annual FAA/Air Force/NASA/Navy Workshop on the Application of Probabilistic Methods to Gas Turbine Engines, Solomon's Island, MD, 2003.
7. U.S. Air Force, High Cycle Fatigue (HCF) Science and Technology Program Annual Report, AFRL-PR-WP-TR-2001-2010, U.S. Air Force Research Laboratory, Dayton, OH, 2000.
8. Lim, S.-H., Bladh, R., Castanier, M.P., and Pierre, C., A compact, generalized component mode mistuning representation for modeling bladed disk vibration, AIAA 2003-1545, in *Proceedings of 44th Structures, Structural Dynamics, and Materials Conference and Exhibit*, AIAA/ASME/ASCE/AHS/ASC, Norfolk, VA, 2003.
9. Bladh, R., Pierre, C., Castanier, M.P., and Kruse, M.J., Dynamic response predictions for a mistuned industrial turbomachinery rotor using reduced-order modeling, *J. Eng. Gas Turbines Power*, 124 (2), 311–324, 2002.
10. Yang, M.-T. and Griffin, J.H., A Reduced Order Model of Mistuning Using a Subset of Nominal Modes, ASME 99-GT-288, presented at International Gas Turbine Institute Turbo Expo, Indianapolis, IN, 1999.
11. Meyer, M.A., Booker, J.M., Bement, T.R., and Kerscher, W.J., III, PREDICT: a new approach to product development, *R&D Mag.*, 41, 161, 1999.
12. Kshirsagar, A.M., *Multivariate Analysis*, Marcel Dekker, New York, 1972.
13. Cressie, N.A.C., *Statistics for Spatial Data*, rev. ed., John Wiley & Sons, New York, 1993.
14. Wackernagel, H., *Multivariate Geostatistics*, 2nd ed., Springer-Verlag, Berlin, 1998.
15. Blair, A. and Annis, C., Development of Probabilistic Methods for Campbell Diagram Frequency Placement, paper presented at 7th National Turbine Engine High Cycle Fatigue Conference, Palm Beach, FL, 2002.
16. Hall, K.C. and Thomas, J.P., Sensitivity Analysis of Coupled Aerodynamic/Structural Dynamic Behavior of Blade Rows, extended abstract for the 7th National Turbine Engine High Cycle Fatigue Conference, Palm Beach, FL, 2002.
17. Thomas, J.P., Hall, K.C., and Dowell, E.H., A Discrete Adjoint Approach for Modeling Unsteady Aerodynamic Design Sensitivities, IAAA 2003-0041, AIAA, Reston, VA, 2003.
18. Walker, K., The Effect of Stress Ratio during Crack Propagation and Fatigue for 2024-T3 and 7075-T6 Aluminum, ASTM STP 462, ASTM, Philadelphia, 1970.

19. Smith, R.N., Watson, P., and Topper, T.H., A stress-strain function for the fatigue of metals, *J. Mater. JMLSA*, 5, 767–778, 1970.
20. Pascual, F.G. and Meeker, W.Q., Estimating fatigue curves with the random fatigue-limit model, *Technometrics*, 41 (4), 277–289, 1999.
21. Dhillon, B.S., *Mechanical Reliability: Theory, Models, and Applications*, AIAA Education Series, American Institute of Aeronautics and Astronautics, Washington, DC, 1988.
22. Hahn, G.J. and Shapiro, S.S., *Statistical Models in Engineering*, John Wiley & Sons, New York, 1967.
23. Martz, H.F. and Waller, R.A, *Bayesian Reliability Analysis*, rev. ed., Krieger Publishing, Malabar, FL, 1991.
24. Gelman, A., Carlin, J.B., Stern, H.S., and Rubin, D.B., *Bayesian Data Analysis*, Chapman & Hall/CRC, Boca Raton, FL, 1997.
25. Robert, C.P. and Casella, G., *Monte Carlo Statistical Methods*, Springer-Verlag, New York, 1999.
26. Congdon, P., *Bayesian Statistical Modelling*, John Wiley & Sons, Chichester, U.K., 2001.
27. Griffiths, J. and Kielb, R., Monte Carlo Analysis Applied to HCF Failure Probability Prediction, paper presented at 7th National Turbine Engine High Cycle Fatigue Conference, Palm Beach, FL, 2002.

31
Practical Reliability-Based Design Optimization Strategy for Structural Design

Tony Y. Torng, Ph.D.
Boeing Phantom Works

31.1	Introduction 31-1
31.2	PADS/RELDOS Challenges 31-2
31.3	RELDOS Resolutions 31-3
	Data Problem • Certification Problem • Target Reliability Problem • RELDOS Efficiency Problem • RELDOS User-Friendliness Problem
31.4	Proposed RELDOS Analysis Procedure 31-5
	Data Preparation • Deterministic Optimization • Reliability Assessment • Reliability-Based Optimal Design
31.5	Proposed RELDOS Theoretical Derivation 31-7
	Background • Reliability-Based Design Optimization
31.6	Tutorial Example 31-10
	A Simple Analytical Problem • Deterministic Optimization • Reliability Assessment • Reliability-Based Design Optimization
31.7	RELDOS Analysis for an 8-ft Cryogenic Tank 31-12
	An 8-ft-Diameter Tank Introduction • Deterministic Optimization • Reliability Assessment • Reliability-Based Optimal Design • Observations
31.8	Summary and Conclusions 31-16

31.1 Introduction

Manufacturers are striving to develop high-performance, low-cost products that meet increasingly demanding reliability requirements and safety standards under stringent constraints for time and resources. In this environment, the traditional deterministic approach involving safety factors has become inadequate. What is needed are alternative methods that can provide greater accuracy and realism in the analysis and design process, thus eliminating unnecessarily high safety margins. This accuracy and realism can only be obtained by adequately accounting for uncertainty, usually by characterizing pertinent quantities as random variables, i.e., as quantities that are not deterministic and that actually exhibit statistical variations. Adequacy of performance is then measured directly in terms of reliability or its converse, probability of failure, rather than in terms of safety factors. The recognition of the inherent nondeterministic nature of basic physical phenomena led the Boeing Company to start a new project advancing the concept of probabilistic analysis and design system (PADS) in 1989.

FEBREL (finite element-based reliability) is a Boeing proprietary software package that contains a variety of advanced probabilistic analysis methods for performing PADS. Among these methods, some were originally developed by Boeing staff and can be referenced in the literature [1–7]. In particular, the X-based most-probable-point-search method has demonstrated its unique capability and has been patented.

To date, PADS has been applied to a variety of Boeing programs, such as the space shuttle, the international space station, the evolved expendable launch vehicle (EELV), cryogenic upper stage (CUS), X-33, Mir/shuttle docking [8–13], rotorcraft, C17, the recent space launch initiative program, etc. However, among these applications, most dealt with component failure probability calculation without design requirements.

In the year 2000, in response to NASA NRA 8-30 requirements and especially the loss-of-vehicle requirement of 1/1000, Boeing initiated an internal research and development program to address these requirements and to demonstrate our capabilities in solving these problems. In the past, structural-related risks were considered minimal compared with other risks, e.g., main engine, solid rocket booster, thermal protection system (TPS), etc. However, only recently are we beginning to understand how conservative we have been in the past. Our safety evaluation techniques do indeed minimize the risk of structural failure, but this comes at the cost of design requirements that lead to excessively overweight structures. To evaluate this problem, a practical design philosophy is required to ensure structural reliability while reducing the structural weight.

Based on this thought, we proposed to develop a practical RELDOS (reliability-based design optimization strategy) to demonstrate the ideas. Without a new design philosophy, the same safety factor approach will be applied and the structural weight will remain high. In the following sections, we will first discuss the challenges we faced throughout the development of RELDOS and then introduce the proposed RELDOS in details. At the end, we also discuss a detailed long-term RELDOS development plan. Several demonstration examples and an 8-ft cryogenic tank problem were solved using the current version of RELDOS Tool.

31.2 PADS/RELDOS Challenges

In recent years, PADS/RELDOS has been applied and promoted by our government agencies, e.g., NASA, air force, navy, army, and FAA. At present, PADS/RELDOS is still in its infancy stage within these government agencies because they meet the same kind of challenges we are facing today. By cooperating with these government agencies, it is possible that we will be able to standardize the PADS/RELDOS requirements. For example, for future manned space transportation systems, the vehicle risk number must be reduced from the current 1/1,000 to 1/10,000. How to establish an appropriate design strategy to meet these PADS/RELDOS requirements remains to be resolved.

If design engineers are to use the reliability-based design optimization strategy (RELDOS) for their product design, the RELDOS must be shown to be a practical tool that can solve a general reliability-based design optimization problem with a large number of random variables and reliability constraints within a reasonable and cost-effective time frame. To reach this stage, the following fundamental challenges must be overcome before design engineers can accept the proposed RELDOS:

1. Data problem: To achieve high-fidelity results from RELDOS, it is necessary to have high-fidelity random variables input. Therefore, it is important to know how to locate enough data and engineering judgments to support the development of high-fidelity data.
2. Target reliability problem: A standard procedure must be defined for determining the appropriate system reliability level for the design. In the traditional safety-factor approach, the target safety factor is based on experience. In the future, the system reliability requirements must be determined using a top-down approach in order to determine reliability requirements for subsystems and components.

3. Certification problem: There are three critical issues: (1) how to certify the calculated reliability using a limited number of tests; (2) how to apply past experience as the baseline data without rerunning the same tests; and (3) how to develop a high-fidelity deterministic model to simulate the actual behavior. Test by analysis is definitely the only way to certify system reliability.

In addition to these problems in identifying the optimal design, there are several technology challenges for the RELDOS development:

1. Efficiency problem: Designing a complicated structural system is a time-consuming problem because of the need for complicated finite element codes. Thus the need to reduce the computational effort or cycle time is a key issue for RELDOS.
2. Accuracy problem: Assuming that the given data are accurate, system reliability becomes an issue. The proposed probabilistic analysis methods must be robust enough to solve problems with a large number of random variables, extremely nonlinear failure models, and random variables with large coefficients of variation.
3. User-friendliness problem: If RELDOS is to be used to solve complicated structural design problem, the user-friendliness of the interface is another key issue. For most design engineers, the nondeterministic analysis (NDA) approach is not a routine procedure; therefore, it is important to develop proper guidelines and user-friendly tools to help engineers solve their design problem.

It is not possible to solve all of these problems without cooperating with other disciplines, e.g., multidisciplinary design optimization (MDO). In addition, it is not possible to baseline the proposed RELDOS approach without the support of government agencies. By working together with different disciplines, developers of RELDOS will gradually gain enough experience to establish a standardized design procedure that will become the mainstream design technology of the future.

31.3 RELDOS Resolutions

31.3.1 Data Problem

In the past, the safety factor was used to account for the uncertainties without truly understanding the impact of randomness and noise factors. Testing became the only viable way to check or verify a design's reliability. Based on this design philosophy, a potentially overweight and expensive product is anticipated.

If one is instead to calculate the reliability of the safety-factor approach, it is necessary to have high-fidelity random variable inputs. How to locate enough data and engineering judgments to support the development of high-fidelity data becomes critical for the success of the proposed RELDOS. In the past, even when we used the safety-factor approach, plenty of data were generated, e.g., the material properties, loads, tolerances, etc. However, only 3-σ or worst-on-worst data were extracted and applied in order to achieve a conservative design. If we save these data and use these data to develop the necessary random variable inputs, RELDOS can be used. Therefore, to resolve the data problem, we must develop a data bank to store all the valuable data and define how and why we select the statistical distributions.

For those cases without much information, RELDOS cannot be applied accurately; however, RELDOS still can be used to provide probabilistic sensitivity factors that can help the user to focus on the most important random variables. When only limited data are available for those important random variables, we need to find ways to obtain more data in order to reduce the noise level of this variable, or we may need to change our design to remove the importance of this variable.

31.3.2 Certification Problem

How to certify the calculated reliability, e.g., 0.999, by using a limited number of tests is an important issue for RELDOS. To resolve this problem, a building-block type of approach is required. At the beginning of RELDOS analysis, we have to rely on the success of the traditional safety-factor approach.

In other words, we want to calculate the corresponding reliability for the design achieved by the traditional safety-factor approach. Based on the reliability, we can then use RELDOS to calibrate the safety factor and certify the RELDOS results.

By comparing RELDOS with the traditional safety-factor approach, RELDOS results become more reliable. After we build up enough confidence, we can then introduce and develop an advanced computer-based testing program, i.e., a virtual testing capability. Why can RELDOS use virtual testing and not the safety factor approach? This is possible because RELDOS considers uncertainties and has random variables data available. With this capability, total testing cost can be reduced greatly when using RELDOS. However, to build up confidence, RELDOS needs to perform enough coupon- or component-level tests to verify the accuracy of component-level failure models. With accurate component models, we will be able to reduce system level testing by using the virtual testing capability.

31.3.3 Target Reliability Problem

As discussed in the previous section, RELDOS must rely on the success of the traditional safety-factor approach. By calculating the corresponding reliability for the design achieved by using the traditional safety-factor approach, we will have a rough idea of design reliability, which can be used to compare with the target reliability. But, how to determine the target system/component reliability level for the design from a top-down analysis is critical. The probabilistic risk assessment (PRA) strategy that was used in determining the risk for the space shuttle is a viable approach. Because there are thousands of components within a space vehicle, the PRA system provides a system of tools to properly define the component failure sequence and calculate the corresponding probability of failure distributions in order to define the overall system reliability distribution.

31.3.4 RELDOS Efficiency Problem

In general, designing a structure is a time-consuming problem, especially when a complicated structural analysis code is used, such as a finite element analysis code. RELDOS requires a double-loop analysis that includes reliability analysis within the first loop. In other words, many reliability analyses need to be performed before an optimal design can be found. Since reliability analysis itself requires a number of structural analyses already, the overall computational effort for RELDOS is very extensive. Thus, how to reduce the computational effort or design cycle time is a key issue for RELDOS.

To improve the efficiency, RELDOS must implement an efficient sensitivity calculation technique to develop an approximate reliability constraint for complicated problems that require the use of computationally intensive programs such as a finite element program. In addition, RELDOS may consider the following strategies:

- Develop screening technologies to reduce the total number of random variables as well as reliability constraints.
- Develop efficient search algorithms for optimal design search. For example, use a safety index function to perform a search within specified limits. Most probable point locus (MPPL) methods must be used to locate the most probable points.
- Develop a data bank to store data that have been calculated and use this to reduce the total computational effort.
- Develop a single-loop analysis strategy; in other words, combine reliability and optimization loops together.

31.3.5 RELDOS User-Friendliness Problem

A modularized RELDOS code, i.e., one that can easily be linked with other codes, must be developed. In other words, RELDOS must be able to use other specialized codes such as probabilistic analysis code, FEBREL, or other commercial package such as UNIPASS or NESSUS/FPI. Especially when considering the multidisciplinary design optimization issues, structural computational codes such as finite element codes and other specialized codes must be integrated with the RELDOS tool as well. It should be possible to integrate all the tools using a user-friendly wrapper code such as iSIGHT, Phoenix, RDCS, etc.

NASA's intelligent synthesis environment (ISE) program is actively promoting the development of an intelligent and user-friendly working environment for future product design to achieve faster, cheaper, better, and safer products.

31.4 Proposed RELDOS Analysis Procedure

The proposed RELDOS analysis procedure is shown in Figure 31.1. Most users are already familiar with deterministic design and deterministic design optimization. The key is the last step: reliability-based design optimization.

To perform a RELDOS analysis, as shown in Figure 31.1, users must have more information. Typical information required by the user can be obtained by answering the following questions:

- What is the objective? Weight or cost?
- What are the design variables?
- What are the nondesign variables, i.e., random variables or deterministic constants?
- How to collect statistical information for the selected random variables?
- What constraints should be considered?
  - What are the deterministic constraints?
  - What are the reliability constraints?
- What are the target component/system reliability goals?

31.4.1 Data Preparation

In addition to the above questions, the following steps are designed to help a design engineer to perform a RELDOS analysis.

31.4.1.1 Validation of Computer Aided Engineering (CAE) Models

This step validates the deterministic analysis models for the structure, which have been built for a deterministic design. The parametric modeling is required for the purpose of automation when performing probabilistic analysis and optimization. This step will be ignored if closed-form expressions are available for analyses.

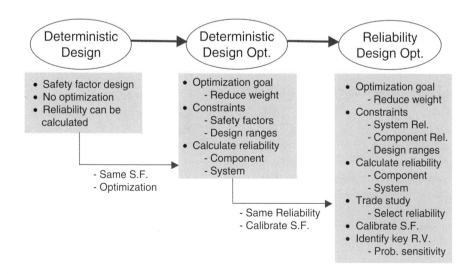

FIGURE 31.1 RELDOS analysis procedure.

31.4.1.2 Probabilistic Load Models

This step develops the probabilistic models for the loads involved. This is one of the critical tasks for the entire project. Since limited data are available in many cases, good engineering judgment is necessary to define the distribution types and parameters in the probabilistic load models.

31.4.1.3 Probabilistic Resistance Models

This step establishes the probabilistic models for the material properties such as Young's modulus and Poisson ratio. The probabilistic information on dimensions of the components should also be considered. The variability sources in material properties and sizes need to be identified, such as the materials themselves and manufacturing tolerances.

Another task for this step is to identify the critical failure modes for the structure. In many cases, the resistance capability of a failure mode is a function of basic random variables. Thus the probabilistic model for a critical failure mode can be obtained from such a functional relationship.

31.4.1.4 Generation of Response Surfaces

This step establishes response surfaces for the responses and load effects for reliability-based design problems that require the use of a computationally intensive program such as finite element code. The developed response surface will be used to replace the original complicated computation for reliability assessment and optimization design.

Usually, a software tool is needed to wrap up application analysis codes, design of experiments, and math tools to develop response surfaces. Advanced computer-aided optimization codes such as iSIGHT or the Boeing-developed robust design computational system (RDCS) should be used to serve this purpose. This step is ignored if closed-form expressions are available for analyses or if the level of computational effort can be tolerated.

31.4.2 Deterministic Optimization

This step uses the RELDOS tool to perform a deterministic optimal design, using the response surfaces and safety-factor constraints.

31.4.3 Reliability Assessment

RELDOS will be used to perform a reliability analysis for the optimal configuration obtained from deterministic optimization analysis (DOA). All the random variables will be used to calculate the reliability of each failure mode.

31.4.4 Reliability-Based Optimal Design

Based on the reliability results of all the selected failure modes, system reliability can be calculated as well. Depending on the dependency issue, the system reliability model must be established with care. When establishing the target system/component reliability levels, the most reasonable approach is based on the overall vehicle reliability requirements. From the top vehicle level to the component level, a minimum requirement can be established. For example, for a second-generation reusable launch vehicle program, the vehicle-level reliability requirement is 1/1000, so what is the requirement for a structural component?

A simple approach is to use the reliability level calculated based on the deterministic optimization. Because all the safety-factor constraints have been satisfied, the calculated reliability can be used as a reference number. When we use the same system reliability for reliability-based design optimization, there is a chance we can achieve weight/cost savings by obtaining a more uniformly distributed reliability level for all failure modes. More experience is required for this.

31.5 Proposed RELDOS Theoretical Derivation

31.5.1 Background

31.5.1.1 Fundamental Theory

The probabilistic optimization problem can be formulated as follows [18–23]:

$$\text{Minimize } E[f(x,\gamma)] \text{ or } \text{Var}[f(x,\gamma)]$$
$$\text{Subject to } P[g_i(x,\gamma) \leq 0] \leq 1 - \text{Re}_i \quad \text{for } i = 0, 1, \ldots, n \quad (31.1)$$

where x represents design variables, γ represents nondesign variables such as Young's modulus, $f(x, \gamma)$ is the objective function (such as weight or payload), E is the expectation operator, Var is the variance operator, P is the probability operator, $g_i(x, \gamma)$ is the limit state function, and Re_i is the desired reliability for the ith constraint.

In general, $f(x, \gamma)$ and $g_i(x, \gamma)$ have to be evaluated by a finite element analysis (FEA) code. Similar to the modern structural optimization technology, efficient reliability-based design algorithms require approximations of reliability constraints or their limit states. A number of FEA runs are required to obtain these approximations. Then the optimization operation is conducted subjected to these constraint approximations. The point for using constraint approximations is to avoid any FEA runs in the optimization process loop (or so-called redesign process loop). Once the optimum is found, the new approximations of reliability constraints or limit states will be generated through a different set of FEA runs. The total number of runs is determined by the structural reliability methods or the design-of-experiment methods used.

31.5.1.2 Optimization Analysis

Modern structural optimization technology has advanced to use about ten detailed finite element analyses for an optimal solution of a complex design optimization problem with thousands of design variables and millions of constraints [14, 15]. Keys to its success are primarily attributed to the use of (1) efficient techniques for design sensitivity calculations in FEA software packages and (2) approximation techniques for constraints in optimization algorithms. This cannot be accomplished using design-of-experiment and response-surface technologies.

31.5.1.3 Reliability-Based Design Optimization

The reliability-based design optimization strategy (RELDOS) is still in the R&D stage due to the nature of the intensive computation in the RELDOS problems. Much research work has been done to develop efficient algorithms to minimize the computational efforts.

At present, there are four major groups of RELDOS approach: gradient based, response-surface based, mixed gradient and response surface, and single-loop approaches. For the first group (e.g., Torng and Yang [17], Enevoldsen and Sorensen [18], Luo and Grandhi [19], Xiao et al. [20]), Equation 31.1 can be further simplified as follows:

$$\text{Minimize: } f(\bar{x}, \bar{\gamma})$$
$$\text{Subject to: } \beta_i \leq \beta_{0i} \quad (31.2)$$

where β_i and β_{0i} represent the safety index and target safety index for the ith constraint, respectively. Note that reliability index, β, is defined as an index to represent the shortest distance between the origin of the standard normal variable space to the failure surface.

In this group, the first-order reliability method (FORM) is used to estimate each reliability constraint. The advantage of using FORM is that it can provide the gradients of reliability index with respect to design variables at no cost. The second-order reliability method (SORM) can further improve the accuracy of the reliability constraint; however, it cannot offer those gradients as required by the optimizer without further computation.

The major FORM/SORM problem for calculation of reliability constraints is its large computational effort to search for the most probable failure points (MPP), which are defined as the shortest distance between the origin of the standard normal variable space to the failure surface. To identify an accurate MPP, several iterations may be required; and each iteration itself requires the calculation of the gradients of g-functions with respect to all random variables.

To resolve the problem, fast sensitivity analysis techniques, such as direct differentiation, need to be developed and implemented to reduce the overall computation time. Efficient search algorithms [1–4] and advanced numerical techniques need to be developed and implemented to minimize the intermediate computations during the MPP search process.

In the second group (e.g., Sues et al. [21], Xiao et al. [22]), reliability/optimization analyses are performed using the approximated response surfaces of the objective and limit-state functions. For this approach, the accuracy of the response-surface models controls the final outcomes, and these models need to be developed with care. As proposed in references [20–22], a refined response surface is generated to improve the optimal solution. Even with the response-surface models, efficient algorithms are still needed to reduce the number of refinements of the response surfaces.

In addition to the above two groups, some papers also proposed the mixed use of gradient-based and response-surface approaches to balance the accuracy and efficiency. In a recent publication [23], a traditional double-loop optimization problem had been converted to a single-loop optimization problem. Several problems were solved and compared.

31.5.2 Reliability-Based Design Optimization

31.5.2.1 RELDOS Analysis Framework

Five steps are implemented for reliability-based design optimization problems, as shown in Figure 31.2.

31.5.2.2 Deterministic Optimization

This step performs an optimization analysis with deterministic variables and constraints. The variability or randomness in manufacturing, material properties, and operational environment is not explicitly considered. The reliability remains unknown but is considered to be represented by the safety factor.

Automated design synthesis (ADS), developed by Vanderplaats, is used to perform an optimization analysis. It contains such algorithms as sequential linear and quadratic programming and the (modified) feasible direction method.

31.5.2.3 Reliability Assessment

The next step is to perform a reliability analysis for all of the failure modes (constraints) with consideration of variability or randomness information for all the selected random variables. The goal is to

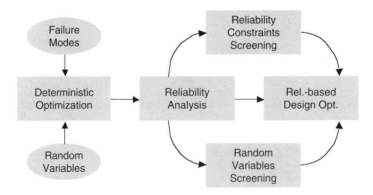

FIGURE 31.2 RELDOS analysis framework.

calculate the reliability levels for all of the failure modes as well as system reliability level for our design problem.

Currently, RELDOS implements a first-order reliability method (FORM) for reliability assessment of probabilistic constraints. Reliability is then calculated as follows,

$$R = \Phi(\beta) \tag{31.3}$$

where $\Phi(\beta)$ is a cumulative distribution function for a standard normal variable.

For system failure probability, various assumptions can be made. For this chapter, we assumed that the system is considered failed when one of the failure modes failed, i.e., the system failure probability is a union of a series of failure events. Based on the assumptions, both first-order and second-order bounds can be calculated. Furthermore, the Monte Carlo simulation techniques can be used to estimate the bounds if a more accurate result is desired.

At present, RELDOS uses the first order lower bound to obtain the system reliability.

$$\prod_{i=1}^{n} R_i \leq R_{sys} \leq \min_{i}(R_i) \tag{31.4}$$

where R_i is the reliability for the ith failure mode.

31.5.2.4 Random Variables Screening

To reduce the overall computational effort, both random variable screening and limit-state screening (discussed in the next section) are required. For random variable screening, the following procedure is used to rank the importance of all the random variables. For simplicity and efficiency, we use the unit normal vector at the mean values for such a purpose. The sensitivity factor at the mean values for the ith constraint is defined as:

$$\alpha_k = \frac{\dfrac{\partial g_i}{\partial x_k} \sigma_k}{\sqrt{\sum_j \left(\dfrac{\partial g_i}{\partial x_j} \sigma_j\right)^2}} \tag{31.5}$$

where the gradients are estimated at the mean values of the random variables.

A random variable is considered important when its calculated probabilistic sensitivity factor is greater than a user-defined small value, say 0.01. Otherwise, it is considered not to have much effect on the estimate of the probabilistic constraint. The design space is thus divided into two sets, deterministic and random variables.

For a complicated reliability-based design optimization problem, we are dealing with an optimization problem with multiple constraints. Therefore, a random variable is considered insignificant and converted to be a deterministic variable only if it is insignificant to *all of the constraints as well as the objective*. As a result, not many random variables are qualified to be insignificant ones. In general, the majority of the random variables may have to be treated as random variables. However, this step is still necessary and valuable toward reducing computational effort in the subsequent analyses.

31.5.2.5 Reliability Constraint Screening

A major benefit of performing reliability-constraint screening is to reduce the size of the problem and thus avoid unnecessary computation. To achieve this goal, an approach is implemented to reduce the number of the probabilistic constraints so that only active or nearly active constraints are considered in the RELDOS search process.

The proposed strategy is actually quite similar to that of the deterministic optimization process. The major difference is that we are dealing with the probabilistic constraints instead of deterministic ones. The goal is to use the least number of FEA runs to identify a working set of the probabilistic constraints, and the most efficient way is to use the reliability index of the first-order second moment (FOSM) method, β_i^{FOSM}, to screen out unnecessary reliability constraints. A probabilistic constraint is considered active or near active if ($\beta_{0i} - \beta_i^{FOSM}$) is less than a user-defined negative value, say -2.0. If the optimization is subject to the target system reliability only, β_{0i} will be taken as this target system-reliability index for a screening purpose.

In deterministic optimization problems, constraint screening can reduce the large number of the constraints down to a small number at around two or three times the number of design variables [16]. Similar effectiveness can be achieved for the probabilistic constraint screening. The working set of the active and nearly active constraints is updated during the search process at the end of each reliability/optimization analysis cycle. The savings in computational effort becomes more obvious when we use FORM/SORM for reliability calculation because the unnecessary computational effort on the MPP search can be avoided.

31.5.2.6 Reliability-Based Design Optimization

With random variables and limit-states screening strategies defined, we can now perform reliability-based design optimization strategy (RELDOS). The initial starting point for RELDOS must be from deterministic optimization (Section 31.5.2.2) and reliability assessment (Section 31.5.2.3).

However, one of the major challenging issues here is how to select the target component and/or system reliability levels as we discussed in Section 31.2. What target reliability levels should we use to design our products? As discussed in Section 31.3.3, we have discussed two resolutions that included a PRA approach and a safety-factor calibration approach.

For the chapter, we will use the safety-factor calibration approach. In other words, we will use the reliability level calculated for the design obtained by the deterministic design optimization. Based on the calculated reliabilities, a system-level reliability can be calculated and used as our target system reliability. The goal is to identify a design with more uniformly distributed component reliabilities.

A uniform reliability level can be achieved for different components corresponding to a failure mode, which results in a more reasonable design with weight savings. This cannot be achieved for a deterministic-based optimal design. The other benefit for using a safety-factor calibration approach is that the acceptance of safety-factor design can be directly applied to the reliability design without problems. The experience gained in relating the safety factor and reliability also provides valuable resources in determining future target reliabilities.

In addition to target system reliability, a target component reliability can be imposed if a user has a specific reliability requirement for one or more failure modes. RELDOS has the capability to do a probabilistic optimization analysis subject to one of three reliability constraint types: target component reliability levels, target component and system reliability levels, and target system reliability level.

31.6 Tutorial Example

31.6.1 A Simple Analytical Problem

Objective function:

$$\text{Minimize } (\pi X_1 * X_1 + X_2)$$

Subject to:

$$X_1^3 * X_2 \geq SF_1 * 95.5$$

$$\frac{X_1^2}{X_2}(4.5 * X_1 - 5.5) \leq \frac{1.0}{SF_2}$$

TABLE 31.1 Random Variables Input

Random Variable	Distribution Type	Mean	Standard Deviation
X_1	normal	1.72	0.15
X_2	normal	35.03	3.6

where safety factor 1, $SF_1 = 1.9$, and safety factor 2, $SF_2 = 1.2$. Design variables include X_1 and X_2, and their ranges are the following: $1.0 \leq X_1 \leq 4.0$ and $20.0 \leq X_2 \leq 50.0$.

31.6.2 Deterministic Optimization

By running RELDOS, the following results are obtained.

- Number of objective function evaluation = 5
- Number of g-function evaluation = 10
- Design variables for optimal solution: $X_1 = 1.72$ and $X_2 = 35.03$
- Objective = 44.324

31.6.3 Reliability Assessment

To calculate the reliability, it is necessary to determine the uncertainties of design variables or noise variables. Here, we assume that both design variables are random variables with fixed standard deviation. The random variables definition are shown in Table 31.1.

Reliability analysis results:

- For constraint one, the reliability = 0.9794
- For constraint two, the reliability = 0.99982

31.6.4 Reliability-Based Design Optimization

As shown in Section 31.3.2, the reliability level for the first constraint is relatively small. Assume that we need to have at least 0.99 reliability. In other words, the constraints will be redefined as follows,

$$P(g_1 \geq 0) \geq 0.99, \text{ where } \quad g_1 = X_1^3 * X_2 - 95.5$$
$$P(g_2 \geq 0) \geq 0.99, \text{ where } \quad g_2 = X_2 - \left(4.5 * X_1^3 - 5.5 X_1^2\right)$$

Based on the reliability constraints, RELDOS has the following results:

- Number of objective function evaluation = 35
- Number of g-function evaluation = 36
- Design variables for optimal solution: $X_1 = 1.9147$ and $X_2 = 26.934$
- Objective = 38.4524
- Reliability values: constraint one = 0.99 and constraint two = 0.9925

Note that deterministic design point is used as the initial starting point for further design search. As shown, both reliability constraints have met the minimum reliability requirement of 0.99, while the objective function has been further reduced to 38.45 from 44.324. In other words, a 13.25% reduction in the objective is achieved while improving the reliability. Note that reliability for constraint two has been reduced, and that is one of the major reasons for the reduction in objective function because we have equally distributed the resources.

In addition to the component reliability constraint, we can use system reliability constraint as well. Due to the use of common random variables used for both reliability constraints, dependency problems must be evaluated when considering system reliability constraint.

31.7 RELDOS Analysis for an 8-ft Cryogenic Tank

31.7.1 An 8-ft-Diameter Tank Introduction

This demonstration example applies RELDOS to search for a probabilistic optimal design configuration for a stiffener-base-skin section, which is a part of a SSTO LH2 8-ft-diameter test tank. It is an essentially all-composite structure. A simplified drawing of the LH2 8-ft-diameter test tank used for the design studies is shown in Figure 31.3. The FEM used for the design studies was an existing model developed by Boeing. The FEM model is a MSC/NASTRAN model with 2947 nodes and 3926 elements. Due to the symmetry of the cylindrical structure, a 30-degree section with the appropriate boundary conditions can represent the entire structure. The represented FEM model is shown in Figure 31.4.

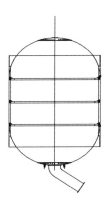

FIGURE 31.3 LH2 tank illustration.

The FEM developed by Boeing Tulsa had composite laminate properties that reflected some manufacturing anomalies; they represented the "as built" rather than the "as designed" laminate. For purposes of these design studies, the laminates were converted to the "as designed" properties described in the Tulsa report. This conversion changed the original PSHELL/MAT2 laminates to a ply-by-ply representation using PCOMP/MAT8. After this conversion, the laminate properties agreed exactly with the "as designed" properties listed in the referred report. This conversion to a ply-by-ply representation also allowed specifying ply thickness as a design variable. The material properties used for the design studies were measured at −423°F, as listed in the referred report.

The input for the probabilistic design approach used by Huntington Beach includes response surfaces for the design variables. We created a combination of FORTRAN computer codes and UNIX scripts to automatically update the NASTRAN FEM geometry for any specified values of the design variables, to perform a NASTRAN analysis, and to extract the response data. This enabled inserting the scripts into the iSIGHT program or the RDCS program to generate the response surfaces for the probabilistic design. The design variables used for the probabilistic studies included the height of the blade stiffener and the width of the base of the blade stiffener. The responses used for the design studies included the maximum strains in the individual plies. A lateral stability check for the blade stiffener was also a design response, as described in the referred report.

Based on the above data, the 8-ft tank, as shown in Figure 31.3, is redesigned using the proposed RELDOS analysis procedure.

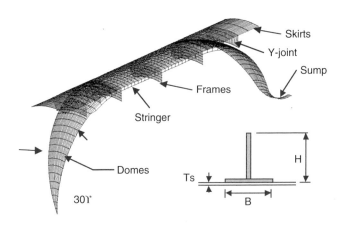

FIGURE 31.4 LH2 tank analytic model NASTRAN FEM.

TABLE 31.2 Deterministic Optimization Results

Design Variable	Height	Base	Thickness
Existing design	1.5	1.5	0.06
Deterministic optimal design	1.065	1.428	0.055

31.7.2 Deterministic Optimization

The first step of RELDOS is to perform a deterministic optimal design using the response surfaces and safety-factor constraints. The objective is to minimize the weight of the structure, and the constraints considered are shown as follows:

1. Skin strain limitation in fiber with safety factor (SF) of 1.4 at −423°F (Load case 1) at two critical locations (E7341/E7342)
2. Skin strain limitation in matrix with SF of 1.0 at −423°F (Load case 1) at two critical locations S
3. Lateral stability check of blade with SF of 1.4 at −423°F (Load cases 1 and 2)
4. Stiffener-skin crippling/buckling check with SF of 1.4 at 190°F (Load case 3)
5. Skin crippling check with SF of 1.4 at 190°F (Load case 3)

where
Load case 1 (LC1) is a 70-psi internal burst pressure
Load case 2 (LC2) is a 685-lb/in. axial compression skirt line load, combined with LC1
Load case 3 (LC3) is an 879-lb/in. axial landing compression skirt line load

For computational efficiency, second-order response-surface functions for the maximum strains at the two critical locations (E7341 and E7342) were developed using an orthogonal array. A total of 16 design points, various combinations of blade height, blade base, and skin layers were used to develop the response-surface model.

Based on the above deterministic design optimization problem setup, optimal design results were calculated and compared with existing deterministic design (without optimization), as shown in Table 31.2.

From the comparison, a total weight savings of 17.14% was demonstrated. For demonstration, the corresponding safety factors for the seven failure modes are shown in Table 31.3. From Table 31.3, we understand that blade lateral stability and blade crippling/buckling are the most critical constraints. Blade height is changed much more than base width and skin thickness, and this is because the safety factor (2.29 >> 1.4) for blade crippling/buckling is much larger than the required SF of 1.4.

31.7.3 Reliability Assessment

Next, RELDOS will perform a reliability analysis for the optimal configuration obtained from deterministic optimization analysis. All the random variables, listed in Table 31.4, will be used now to calculate

TABLE 31.3 Corresponding Safety Factors

Failure Mode	Designed SF	Existing Design SF	Deterministic Optimal Design SF
Strain 11 at E7341	1.4	N/A	3.1266
Strain 22 at E7341	1.0	N/A	2.8519
Strain 11 at E7342	1.4	2.89	2.8044
Strain 22 at E7342	1.0	1.37	1.1003
Blade lateral stability	1.4	1.43	1.4011
Blade crip./buck.	1.4	2.29	1.3981
Skin buckling	1.4	2.36	1.4508

TABLE 31.4 Random Variable Inputs for Material Properties, Loads, and Ultimate Strength (Strain) at −423°F

	Random Variable Name	Mean	COV
	Material Properties		
1	H_blade (inch)	1.065	0.05
2	B_base (inch)	1.428	0.05
3	E11_tape (psi)	2.04 E7	0.05
4	E22_tape (psi)	2.81 E5	0.05
5	G12_tape (psi)	1.16 E5	0.05
6	T_tape (inch)	0.005	0.05
7	E11_fab (psi)	1.16 E7	0.05
8	G12_fab (psi)	1.4 E6	0.05
9	T_fabric (inch)	7.3 E−3	0.05
10	theta_base (deg)	45	0.008
11	theta1_skin (deg)	65	0.008
12	theta2_skin (deg)	−65	0.008
	Loads		
13	Internal p (psi)	53.85	0.10
14	Compression (lb/in.)	428.13	0.20
15	Landing load (lb/in.)	549.38	0.15
	Ultimate Strength (Strain)		
16	Strain11_tape	1.2875 E−2	0.05
17	Strain22_tape	5.625 E−3	0.05
18	Strain11_fab	1.2375 E−2	0.05
19	Strain22_fab	1.175 E−2	0.05

the reliability of each failure mode. These random variables include loads, material properties, laminate (ply thickness and orientation), and geometry (manufacturing tolerance in sizing).

For these random variables, the following data and assumptions were used:

1. Mean values of material properties and ultimate strength are obtained from Torng et al. [1].
2. Engineering judgment is used to assign statistical parameters due to lack of data.
3. Internal pressure and landing load are assumed to follow lognormal distribution, while the rest follows normal distributions.

Based on these random variables, the reliability of the deterministic-based optimal design can be calculated and shown in Table 31.5.

The last four failure modes are identified to be critical for probabilistic optimization, since they have low reliability indices. The first-order reliability method implemented in RELDOS is used for reliability assessment. The first-order lower bound of system reliability is 0.999728, based on the above failure modes

TABLE 31.5 Reliability Results Summary

Failure Mode	Design SF	Existing SF	Deterministic Optimal SF	Beta	P_F
Strain 11 at E7341	1.4	N/A	3.1266	7.7991	3.1 E−15
Strain 22 at E7341	1.0	N/A	2.8519	7.6333	1.1 E−14
Strain 11 at E7342	1.4	2.89	2.8044	7.6305	1.2 E−14
Strain 22 at E7342	1.0	1.37	1.1003	3.7534	8.71 E−5
Blade lateral stability	1.4	1.43	1.4011	3.8056	7.07 E−5
Blade crippling/buckling	1.4	2.29	1.3981	3.6856	1.14 E−4
Skin buckling	1.4	2.36	1.4508	5.0761	1.93 E−7

TABLE 31.6 Probabilistic Optimization Results

Design Variable	Height	Base	Thickness	Weight Saving
Existing design	1.5	1.5	0.06	
Deterministic optimal design	1.065	1.428	0.055	17.14%
Reliability optimal design	1.069	1.3989	0.055	17.17%

31.7.4 Reliability-Based Optimal Design

Based on the calculated system reliability of 0.999728, a reliability-based optimal design can be performed. The objective is to minimize the weight of a stiffener-skin section with the reliability constraint: system reliability > 0.999728. System failure occurs when one of the failure modes fails. All the rest of constraints and random variables are considered the same, as discussed in Section 31.4.2. The first-order reliability method is used here to perform the reliability assessment. The results are shown in Table 31.6.

As shown, the weight saving for the probabilistic-based optimal design is 17.17%, which is only slightly better than the 17.14% obtained by the deterministic-based optimal design. The major reason for the close comparison can be explained by examining Table 31.7.

For both deterministic-based and probabilistic-based optimal designs, the corresponding safety factors for the most critical four are really close. At the same time, the corresponding probabilities of failure for these safety factors are also close and dominated by three out of four failure modes, as shown in Table 31.7 columns 5 and 8 rows 5 to 7. Therefore, there is not much room for further reduction in weight to achieve the same level of system reliability. In other words, the concept of equal level of reliability will not be applied here.

31.7.5 Observations

31.7.5.1 Probabilistic Sensitivity Factors

The probabilistic sensitivity factors for the four major modes and system reliability are shown in Figure 31.5. As shown, six random variables are significant to system reliability and have more impact on final optimal design than the rest of variables:

1. Blade height
2. Tape thickness
3. Fiber thickness
4. Tape strain strength
5. Internal pressure
6. Landing load

These six random variables can be used for fast search of the preliminary optimal design.

TABLE 31.7 Detailed Summary of Reliability-Based Design Optimization Results

Failure Mode	Design SF	Existin SF	Deterministic Optimal SF	Beta	P_F	Relative Optimal SF	Beta	P_F
Strain 11 at E7341	1.40	N/A	3.08	7.736	5.16 E−15	3.09	7.739	5.00 E−15
Strain 22 at E7341	1.00	N/A	2.84	7.550	2.18 E−14	2.83	7.539	2.38 E−14
Strain 11 at E7342	1.40	2.89	2.83	7.639	1.09 E−14	2.82	7.635	1.13 E−14
Strain 22 at E7342	1.00	1.37	1.11	3.769	8.21 E−05	1.11	3.758	8.58 E−05
Blade lateral stability	1.40	1.43	1.43	3.816	6.78 E−05	1.44	3.933	4.20 E−05
Blade crippling/buckling	1.40	2.29	1.40	3.690	1.12 E−04	1.34	3.481	2.50 E−04
Skin buckling	1.40	2.36	1.45	5.082	1.87 E−07	1.43	5.030	2.46 E−07

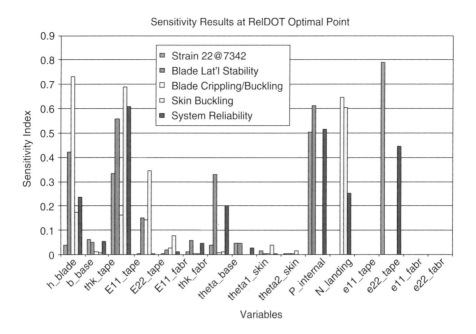

FIGURE 31.5 Probabilistic sensitivity factors.

TABLE 31.8 System Reliability Trade Study

System Reliability	Beta	Weight Saving (%)
0.99	2.3263	20.932
0.995	2.5758	20.405
0.999	3.0902	18.937
0.9995	3.2905	18.359
0.99973	3.4581	17.17

31.7.5.2 System Reliability Level Selection

As shown in Section 31.7.3, the same level of system reliability cannot produce much more benefits for the probabilistic-based optimal design. Therefore, it is interesting to see what will happen if we select different system reliability levels and compare the weight savings. In Table 31.8, a total of five system reliability levels is used.

As shown in Figure 31.6, greater weight savings can be achieved when a smaller system reliability is used. However, the change in weight here is not severe due to the case of equal reliability. For other examples, the potential benefits can be much larger.

31.8 Summary and Conclusions

Boeing has invested and accumulated valuable experience in various PADS tools (FEBREL and First Level PRObabilistic code) development as well as the PADS-based applications. The integration of PADS tools with deterministic analysis modules provides the capability for real-world applications of probabilistic technology such as RELDOS.

RELDOS is proposed as a means of changing the current design philosophy from a deterministic-based to a nondeterministic- or probabilistic-based approach. Many challenges have been identified, and their corresponding resolutions have been introduced from a top-level point of view. Much work remains

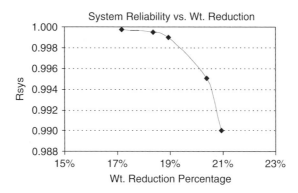

FIGURE 31.6 System reliability trade study.

to be done, and experience in real RELDOS application is required to enhance the proposed future use of RELDOS.

To solve design issues regarding an 8-ft-diameter cryogenic composite test tank, as shown by Torng [1], a practical RELDOS analysis procedure was proposed. It went from deterministic optimization, reliability assessment, to a reliability-based design optimization. The main goal for this approach is to help the traditional deterministic designer understand the differences between traditional deterministic-based design and reliability-based design and to highlight the benefits of this approach. The discussion examines the relationship between safety factor and reliability and shows how to interpret the reliability results to design a better product.

Additional research is needed to develop a more advanced, efficient, and user-friendly RELDOS tool by integrating PADS tools with advanced computer-aided optimization codes such as i-SIGHT or the Boeing-developed robust design computational system (RDCS), which has a user-friendly environment to integrate codes and analyze the results.

Deterministic and probabilistic optimization analyses were performed for a stiffener-skin section, a typical section in an 8-ft-diameter test tank. Optimal designs for both deterministic optimization and reliability-based optimization cases were achieved with a weight reduction of around 17%. Reliability-based optimal design shows a more uniformly distributed reliability level among failure modes than the deterministic design. Significant random variables can be identified using the reliability-based optimization approach. Still a primitive tool, RELDOT is being developed so that it can be integrated with other application codes.

Acknowledgments

Much of the work discussed in this chapter was performed under a Boeing internal research funding. The author would like to acknowledge program manager Dr. Diane Wiley for her support and guidance during the project execution and colleagues Dr. Qiang Xiao, Scott Zillmer, and Gene Rogers for their support.

References

1. Torng, T.Y., Xiao, Q., Zillmer, S., and Rogers, G., Practical reliability-based design optimization application, in *Proceedings of 42nd AIAA Structures, Structural Dynamics, and Materials Conference*, American Institute of Aeronautics and Astronautics, Washington, DC, 2001.
2. Wu, Y.-T., Torng, T.Y., and Khalessi, M.R., A New Iteration Procedure for Efficient Structural Reliability Analysis, paper presented at First International Symposium on Uncertainty Modeling and Analysis, College Park, MD, 1991.

3. Khalessi, M.R., Wu, Y.-T., and Torng, T.Y., Most-probable-point-locus reliability method in standard normal space, in *Proceedings of 9th Biennial ASME Conference on Reliability Stress Analysis and Failure Prevention*, ASME, 1991.
4. Lin, H.-Z. and Khalessi, M.R., Sensitivity calculation using X-space most-probable-point, in *Proceedings of 35th Structures, Structural Dynamics, and Materials Conference*, AIAA/ASME/ASCE/AHS/ASC, 1994.
5. Lin, H.-Z. and Khalessi, M.R., Identification of the most-probable-point in original space: applications to structural reliability, in *Proceedings of 34th Structures, Structural Dynamics, and Materials Conference*, AIAA/ASME/ASCE/AHS/ASC, 1993, pp. 2791–2800.
6. Khalessi, M.R. and Lin, H.-Z., Most-probable-point-locus structural reliability method, in *Proceedings of 34th Structures, Structural Dynamics, and Materials Conference*, AIAA/ASME/ASCE/AHS/ASC, 1993, pp. 1154–1162.
7. Khalessi, M.R., Lin, H.-Z., and Trent, D.J., Development of the FEBREL finite element-based reliability computer program, in *Proceedings of 34th Structures, Structural Dynamics, and Materials Conference*, AIAA/ASME/ASCE/AHS/ASC, 1993, pp. 753–761.
8. Torng, T.Y., B-1B aging aircraft risk assessment using a robust importance sampling method, in *Proceedings of 40th Structures, Structural Dynamics, and Materials Conference*, AIAA/ASME/ASCE/AHS/ASC, 1999.
9. Torng, T.Y., Lin, H.-Z., Khalessi, M.R., and Chandler F.O., Probabilistic thin shell structural design, in *Proceedings of 39th Structures, Structural Dynamics, and Materials Conference*, AIAA/ASME/ASCE/AHS/ASC, 1998.
10. Khalessi, M.R., and Lin, H.-Z., Structural reliability analysis using the MPPL Reliability method-case study of the HL-20 PLS Spacecraft, *Proceedings of the Predictive Technology Symposium*, Randolph, NJ, November 13–14, 1991.
11. Khalessi, M.R., Wu, Y.-T., and Torng, T.Y., A new most-probable-point search procedure for efficient structural reliability analysis, in *Proceedings of 32nd Structures, Structural Dynamics, and Materials Conference*, Part 2, AIAA/ASME/ASCE/AHS/ASC, 1991, pp. 1295–1304.
12. Khalessi, M.R., Lin, H.-Z., and Ghofranian, S., Probabilistic capture analysis of Mir/shuttle mission, in *Proceedings of 36th Structures, Structural Dynamics, and Materials Conference*, AIAA/ASME/ASCE/AHS/ASC, 1995.
13. Torng, T.Y., Funk, G.E., and Stephenson, R.M., Probabilistic on-orbit loads combination assessment, in *Proceedings of 40th Structures, Structural Dynamics, and Materials Conference*, AIAA/ASME/ASCE/AHS/ASC, 1999.
14. Vanderplaats, G.N., *Numerical Optimization Techniques for Engineering Design*, VMA Engineering, 1999.
15. Vanderplaats, G.N., Structural design optimization status and directions, *J. Aircraft*, 36 (1), 11–20, 1999.
16. VMA Engineering, *Genesis User Manual*, Vol. 1, ver. 5, VMA Engineering, 1998.
17. Torng, T.Y. and Yang, R.J., An advanced reliability based optimization method for robust structural system design, in *Proceedings of 34th Structures, Structural Dynamics, and Materials (SDM) Conference*, AIAA/ASME/ASCE/AHS/ASC, 1993.
18. Enevoldsen, J. and Sorensen, J.D., Reliability-based optimization in structural engineering, *Structural Safety*, 15, 169–196, 1994.
19. Luo, X. and Grandhi, R.V., ASTROS for reliability-based multidisciplinary structural analysis and optimization, in *Proceedings of 36th Structures, Structural Dynamics, and Materials Conference*, AIAA/ASME/ASCE/AHS/ASC, 1995.
20. Xiao, Q., Sues, R.H., Cesare, M., and Rhodes, G.S., Reliability-based multi-disciplinary optimization for aerospace structural design, in *Proceedings of 13th ASCE Engineering Mechanics Conference*, ASCE, Reston, VA, 1999.

21. Sues, R.H., Oakley, D.R., and Rhodes, G.S., Reliability-based optimization for design of aeropropulsion components, in *Proceedings of 37th Structures, Structural Dynamics, and Materials (SDM) Conference*, AIAA/ASME/ASCE/AHS/ASC, 1996.
22. Xiao, Q., Sues, R.H., and Rhodes, G.S., Multi-disciplinary wing shape optimization with uncertain parameters, in *Proceedings of 40th Structural Dynamics, and Materials (SDM) Conference*, AIAA/ASME/ASCE/AHS/ASC, 1999.
23. Yang, R.J. and Gu, L., Experience with approximate reliability-based optimization methods, in *Proceedings of 44th Structural Dynamics, and Materials (SDM) Conference*, AIAA/ASME/ASCE/AHS/ASC, 2003.

32
Applications of Reliability Assessment

Ben H. Thacker
Southwest Research Institute

Mike P. Enright
Southwest Research Institute

Daniel P. Nicolella
Southwest Research Institute

David S. Riha
Southwest Research Institute

Luc J. Huyse
Southwest Research Institute

Chris J. Waldhart
Southwest Research Institute

Simeon H.K. Fitch
Mustard Seed Software

32.1 Introduction 32-1
32.2 Overview of NESSUS 32-3
32.3 Overview of DARWIN 32-6
32.4 Application Examples 32-7
   Gas Turbine Engine Rotor Risk Assessment • Life Distribution of an Aircraft Lever • Crankshaft Reliability • Cervical Spine Impact Injury • Tunnel Vulnerability Assessment • Orthopedic Implant Cement Loosening • Explosive-Actuated Piston-Valve Assembly
32.5 Conclusions 32-28

32.1 Introduction

Recent programs of national significance are pushing numerical simulation to new levels. These include the Federal Aviation Administration (FAA) Turbine Rotor Material Design program aimed at reducing the risk of rotor fracture, the Nuclear Regulatory Commission (NRC) program to assess the long-term safety of the nation's first underground high-level radioactive waste repository, the Department of Energy (DOE) Stockpile Stewardship program to replace underground nuclear testing with computationally based full weapon system certification, and the Department of Defense (DOD) Efficient Certification Program to reduce the time and cost required to certify gas turbine engine designs. The common denominator of all these program areas is the need to compute — with high confidence — the reliability of complex, large-scale systems involving multiple physics, nonlinear behavior, and uncertain or variable input descriptions.

In addition to whose failures have high systems consequences, probabilistic analysis is also being employed to predict the reliability of engineered systems and identify important design and manufacturing variables for (1) one- or few-of-a-kind high-cost systems where there will be little or no full-system testing, (2) products that are manufactured in large numbers where warranty costs are prohibitive or unacceptable, and (3) products that are manufactured in large numbers where small changes in the manufacturing process can lead to large cost savings.

The basic problem being addressed by probabilistic analysis is illustrated in Figure 32.1. Model inputs are represented as random variables, leading to corresponding uncertainty in the model responses, which are usually related to some reliability measure, e.g., fatigue life. In addition, the probabilistic analysis identifies which input variables contribute the most (and the least) to the computed reliability.

Uncertainties enter a complex simulation from a variety of sources: inherent variability in input parameters, insufficient input data, human error, model simplification, and lack of understanding of the underlying physics. In general, all uncertainties can be categorized as being either inherent (irreducible) or epistemic (reducible). Epistemic uncertainty can, in principle, be reduced by gathering additional

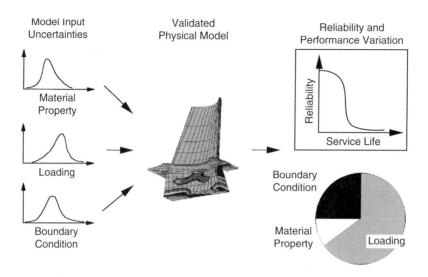

FIGURE 32.1 Basic process of input uncertainty propagation.

data, by implementing more rigorous quality control, or by using more sophisticated or higher-fidelity analysis. While it is well accepted that probabilistic methods are appropriate for characterizing inherent uncertainties, it is not as widely accepted to use a probabilistic approach to represent epistemic uncertainty. In many instances, however, a probabilistic approach is justified, especially when a variable is truly random but only limited data or expert opinion are available [1, 2].

To support the need for more-accurate simulations, analysts are developing higher-fidelity models that more closely represent the actual behavior of the system. Finite element models in excess of 1 million elements are not uncommon and often also involve multiple coupled physics, such as solid mechanics, structural dynamics, hydrodynamics, heat conduction, fluid flow, transport, chemistry, and acoustics. Even with the remarkable advances in computer speeds seen recently, simulations performed with these high-fidelity models can take hours or days to complete for a single deterministic analysis. Because probabilistic analysis methods, regardless of the particular method employed, require repeated deterministic solutions, efficient methods are needed now more than ever.

Beginning with the development of the NESSUS® (numerical evaluation of stochastic structures under stress) probabilistic analysis computer program [3], Southwest Research Institute (SwRI) has been addressing the need for efficient probabilistic analysis methods for nearly 20 years. Recently, SwRI has also focused on improving the NESSUS software to reduce the learning curve required for defining probabilistic problems [4], to improve support for large-scale numerical models [5], and to improve the robustness of the underlying probabilistic algorithms [6]. NESSUS can be used to simulate uncertainties in loads, geometry, material behavior, and other user-defined random variables to predict the probabilistic response, reliability, and probabilistic sensitivity measures of systems. Via automatic download from the NESSUS web site (www.nessus.swri.org), hundreds of copies of the software have been distributed to over 20 countries. Some of the applications currently being addressed by NESSUS users include aerospace structures, automotive structures, biomechanics, gas turbine engines, weapons, geomechanics, nuclear waste disposal and packaging, offshore structures, pipelines, and rotordynamics.

The framework of NESSUS allows the user to link probabilistic algorithms with analytical equations, external computer programs (including commercial finite element codes), and general combinations of the two to compute the probabilistic response of a system. NESSUS also provides a hierarchical model capability that allows the user to link different analysis packages and analytical functions. This capability provides a general relationship of physical processes to predict the uncertainty in the performance of the system. The powerful NESSUS Java-based graphical user interface (GUI) is highly configurable and allows tailoring to specific applications.

In 1995 SwRI initiated development of the DARWIN® probabilistic analysis software. As compared with NESSUS, which is a general-purpose probabilistic analysis code, DARWIN (design assessment of reliability with inspection) is a tailored code for performing probabilistically based damage-tolerance analysis of gas turbine engine rotor disks [7–9]. The software integrates finite element stress analysis, fatigue-crack-growth life assessment, material anomaly data, probability of detection by nondestructive evaluation (NDE), and inspection schedules to determine the probability of fracture of disks as a function of applied operating cycles. The program also identifies the regions of the disk most likely to fail and the risk reduction associated with inspections. DARWIN is currently being used by at least seven major aircraft engine companies worldwide, and several of these companies have already used DARWIN in support of FAA certification activities. In recognition of DARWIN's technology and acceptance by industry, the code received an R&D 100 award as "one of the 100 most technologically significant new products of the year" in 2000.

DARWIN includes tailored probabilistic methods adapted from NESSUS technology and an integral fracture mechanics module that includes NASGRO [10] technology. These sophisticated technology elements are integrated within a powerful custom graphical user interface (GUI) that makes the code extremely easy to use. DARWIN also permits ANSYS® models and ANSYS stress results to be directly input and displayed within the software. This, in turn, facilitates the rapid and highly efficient extraction of relevant geometry and stress information to support the probabilistic damage-tolerance calculations. DARWIN currently considers random variations in key input variables, including the initial defect size and frequency, NDE probability of detection as a function of damage size, time of inspection, crack-growth material properties, and applied stresses.

In the remainder of this chapter, an overview of the NESSUS and DARWIN codes is presented, followed by a series of application problems. Further information on the probabilistic methods employed in NESSUS and DARWIN can be found in the references.

32.2 Overview of NESSUS

NESSUS is a general-purpose tool for computing the probabilistic response or reliability of engineered systems. The software was initially developed by a team led by SwRI for the National Aeronautics and Space Administration (NASA) to assess uncertainties in critical space shuttle main engine components [11]. The framework of NESSUS allows the user to link traditional and advanced probabilistic algorithms with analytical equations, external computer programs including commercial finite element codes, and general combinations of the two. Eleven probabilistic algorithms are available, including traditional methods such as Monte Carlo simulation and the first-order reliability method (FORM) and advanced methods such as advanced mean value (AMV) and adaptive importance sampling (AIS). In addition, NESSUS provides a hierarchical modeling capability that allows linking of different analysis packages and analytical functions. This capability provides a general relationship of physical processes to predict the uncertainty in the performance of the system. A summary of the NESSUS capabilities is shown in Figure 32.2.

Most engineering structures can fail by multiple events, including multiple failure modes and components in which the nonperformance of one or a combination of events can lead to system failure. System reliability considers failure of multiple components of a system and multiple failure modes of a component. System reliability assessment is available in NESSUS via a probabilistic fault-tree analysis (PFTA) method [12]. System failure is defined through the fault tree by defining bottom (failure) events and their combination with "AND" and "OR" gates. Each bottom event considers a single failure (component reliability) and can be defined by a finite element model or analytical function. An example of a fault tree defined in the NESSUS GUI is shown in Figurē 32.3.

In the NESSUS GUI, an outline structure is used to define the required elements for the problem setup and execution. The user navigates through the nodes of the outline to set up the problem, define the analysis, and view the results. The outline structure for a typical problem is shown in Figure 32.4.

A powerful feature of NESSUS is the ability to link models together in a hierarchical fashion. In the problem statement window, each model is defined in terms of input/output variables and mathematical

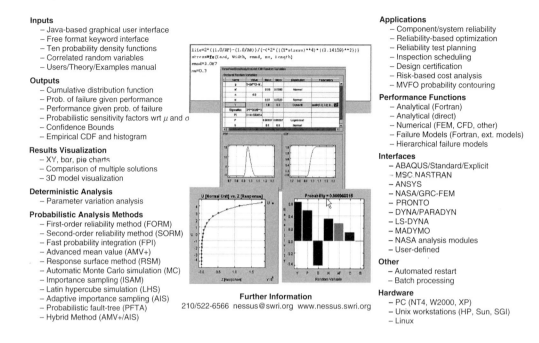

Inputs
- Java-based graphical user interface
- Free format keyword interface
- Ten probability density functions
- Correlated random variables
- Users/Theory/Examples manual

Outputs
- Cumulative distribution function
- Prob. of failure given performance
- Performance given prob. of failure
- Probabilistic sensitivity factors wrt μ and σ
- Confidence Bounds
- Empirical CDF and histogram

Results Visualization
- XY, bar, pie charts
- Comparison of multiple solutions
- 3D model visualization

Deterministic Analysis
- Parameter variation analysis

Probabilistic Analysis Methods
- First-order reliability method (FORM)
- Second-order reliability method (SORM)
- Fast probability integration (FPI)
- Advanced mean value (AMV+)
- Response surface method (RSM)
- Automatic Monte Carlo simulation (MC)
- Importance sampling (ISAM)
- Latin hypercube simulation (LHS)
- Adaptive importance sampling (AIS)
- Probabilistic fault-tree (PFTA)
- Hybrid Method (AMV+/AIS)

Applications
- Component/system reliability
- Reliability-based optimization
- Reliability test planning
- Inspection scheduling
- Design certification
- Risk-based cost analysis
- MVFO probability contouring

Performance Functions
- Analytical (Fortran)
- Analytical (direct)
- Numerical (FEM, CFD, other)
- Failure Models (Fortran, ext. models)
- Hierarchical failure models

Interfaces
- ABAQUS/Standard/Explicit
- MSC NASTRAN
- ANSYS
- NASA/GRC-FEM
- PRONTO
- DYNA/PARADYN
- LS-DYNA
- MADYMO
- NASA analysis modules
- User-defined

Other
- Automated restart
- Batch processing

Hardware
- PC (NT4, W2000, XP)
- Unix workstations (HP, Sun, SGI)
- Linux

Further Information
210/522-6566 nessus@swri.org www.nessus.swri.org

FIGURE 32.2 Summary of NESSUS 8.1 capabilities.

operators. This canonical description improves readability, conveys the essential flow of the analysis, and allows complex reliability assessments to be defined when more than one model is required to define the overall performance. A problem statement for a simple problem is shown in Figure 32.5. The performance referent is life (number of cycles to failure) given by an analytical Paris crack model, which requires input from other models. In this case, two stress quantities from an ABAQUS® finite element analysis are used in the life equation. Functions are also supported (ABAQUS in this example) and are defined in a subsequent screen. The function can be defined by an analytical equation, numerical model, preprogrammed subroutine, or regression model. This hierarchical model capability provides a general equation form to define the performance by linking results from numerical analysis programs and analytical equations.

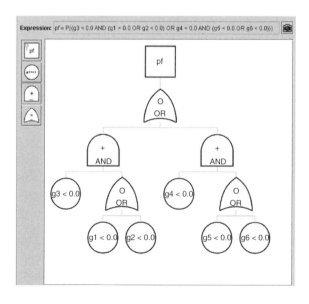

FIGURE 32.3 Example of a fault tree for system reliability analysis in NESSUS.

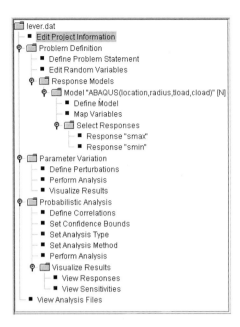

FIGURE 32.4 NESSUS outline structure for defining the analysis.

The random variable statistics are defined in the random variable definition window in NESSUS (shown in Figure 32.5). Tabular input is provided for distributions requiring parameters other than the mean and standard deviation, such as upper and lower bounds for truncated distributions. Plots of probability density and cumulative distribution functions provide a visual check of the defined random variable.

Any function defined in the problem statement is assigned in the response-model definition. The available definitions include analytical, regression, numerical, and predefined. The analytical model definition allows analytical models to be defined using standard mathematical operators. The numerical

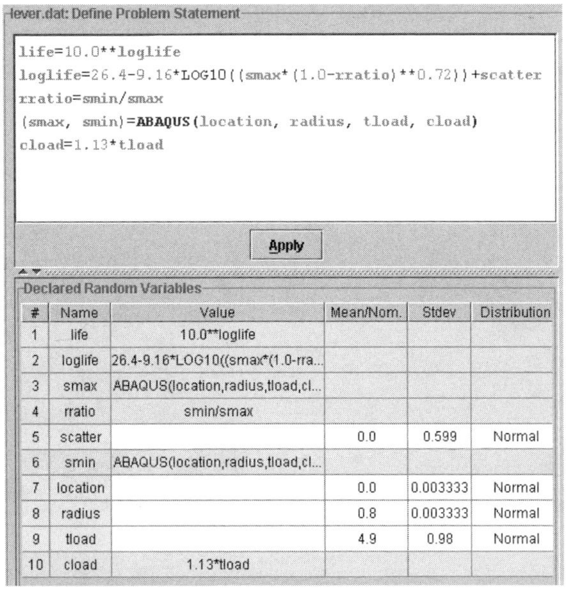

FIGURE 32.5 NESSUS problem statement window and random variable inputs.

model definition allows the use of finite element solvers such as ABAQUS, ANSYS, NASTRAN®, or a user-defined model. The regression model definition allows input of function coefficients or data sets that can be fit using linear regression to standard linear or quadratic functions. Finally, the predefined model definition allows the user to link user-programmed subroutines with NESSUS.

32.3 Overview of DARWIN

Traditional methodologies for life prediction of gas turbine rotors and disks in aircraft jet engines are based on nominal conditions that do not adequately account for material and manufacturing anomalies that can degrade the structural integrity of high-energy rotors. Occasional upsets can occur during processing of the premium grade titanium alloys used for fan and compressor rotors and disks, resulting in the formation of metallurgical anomalies referred to as hard alpha (HA). These anomalies are nitrogen-rich alpha titanium that are brittle and often have microcracks and microvoids associated with them. Although rare, these anomalies have led to uncontained engine failures (Figure 32.6) that resulted in fatal accidents, such as the incident at Sioux City, IA, in 1989. In a report issued by the FAA after the accident in Sioux City [13], it was recommended that a probabilistic damage-tolerance approach be implemented to explicitly address HA anomalies, with the objective of enhancing the conventional rotor life management methodology. This enhancement is intended to supplement, not replace, the current safe-life methodology.

In order to account for these anomalies, the Rotor Integrity Subcommittee (RISC) of the Aerospace Industries Association (AIA) recommended adoption of a probabilistic damage-tolerance approach to supplement the current safe-life methodology. The recommendation led to the development of a computer program called DARWIN, developed in collaboration with General Electric Aircraft Engines, Honeywell, Pratt & Whitney, and Rolls-Royce Allison [8, 9, 14]. DARWIN is a computer program that integrates finite element stress analysis, fracture mechanics analysis, nondestructive inspection simulation, and probabilistic analysis to assess the risk of rotor fracture. It computes the probability of fracture as a function of flight cycles, considering random defect occurrence and location, random inspection schedules, and several other random variables. Both Monte Carlo simulation and advanced fast integration methods are integral to the probabilistic driver. A fracture mechanics module, called Flight_Life [15],

FIGURE 32.6 Rare metallurgical anomalies can lead to uncontained engine failures. (From National Transportation Safety Board, Aircraft Accident Report: United Airlines Flight 232 McDonnell Douglas DC-10-10 Sioux Gateway Airport, Sioux City, Iowa, July 19, 1989, NTSB/AAR-90/06, NTSB, Washington, DC, 1990. With permission.)

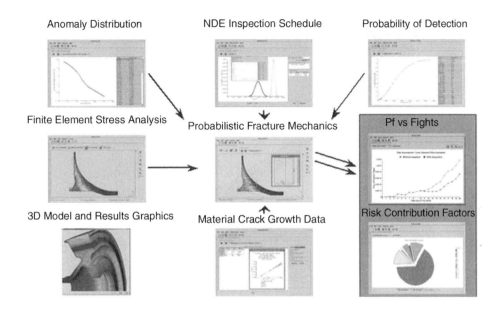

FIGURE 32.7 Overview of DARWIN probabilistic fracture mechanics computer program.

is also incorporated into the code. In addition, a user-friendly graphical user interface (GUI) is available to handle the otherwise difficult task of setting up the problem for analysis and viewing the results. An overview of the DARWIN computer program is shown in Figure 32.7.

DARWIN provides several sampling-based probabilistic analysis alternatives to predict the life of disks subjected to periodic inspection [16, 17]. Monte Carlo simulation provides accurate results, but it is inefficient because the failure limit state must be evaluated for each random sample using a fatigue-crack-growth algorithm. The life approximation function (LAF) in DARWIN creates deterministic life and crack-size arrays for a family of initial defects. During Monte Carlo simulation, the failure limit state is evaluated for each random sample using values interpolated from the deterministic arrays, thereby improving computational efficiency. The DARWIN importance sampling (IS) method focuses analysis on the initial conditions (defect size and other random variables) that would result in lives shorter than the specified design life. This approach reduces the size of the analysis region and may be significantly more efficient than Monte Carlo simulation. Recent enhancements [18, 19] provide additional computational efficiency.

The announcement of FAA Advisory Circular 33.14-1 (Damage Tolerance for High Energy Turbine Engine Rotors) [20] adds a new probabilistic damage-tolerance element to the existing design and life management process for aircraft turbine rotors. Use of DARWIN is an acceptable method for complying with AC33.14-1 and has the potential to reduce the uncontained rotor disk failure rate and identify optimal inspection schedules.

32.4 Application Examples

The NESSUS and DARWIN software have been used to predict the probabilistic response for a wide range of problems. Several problems are presented in this section to demonstrate the utility and flexibility of the codes and illustrate the current developments to support efficient probabilistic model development and support for large-scale problems.

32.4.1 Gas Turbine Engine Rotor Risk Assessment

Consider the aircraft rotor disk shown in Figure 32.8. The design life of the disk is 20,000 flight cycles. Internal stresses and temperatures are identified using finite element analysis based on operational loading

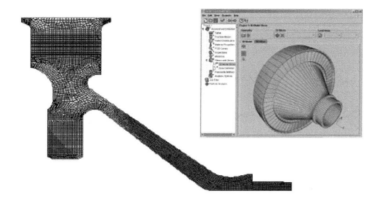

FIGURE 32.8 Aircraft rotor disk application — cross section.

TABLE 32.1 Titanium Aircraft Rotor Application Example — Random Variables

Random Variable	Median	COV (%)	Distribution
Stress scatter	1.0	20	lognormal
Life scatter	1.0	40	lognormal
Inspection time	10,000 cycles	20	normal

conditions. Five primary random variables are considered for probabilistic analysis. The main descriptors for three of these variables (stress scatter, life scatter, and inspection time) are indicated in Table 32.1. For the remaining two variables (defect area, probability of detection [POD]), empirical distributions (AIA POST95-3FBH-3FBH defect distribution, #3 FBH 1:1 Reject Calibration POD Curve) found in AC33.14-11 [20] were used for probabilistic fatigue life predictions. A total of 44 zones were used to model the disk. Additional details regarding the selection of random variables and associated distributions can be found in the literature [21, 22].

Failure probability results from DARWIN are shown in Figure 32.9 (100 samples per zone). It can be observed that the mean disk failure probability is below the target risk value specified by the FAA Advisory

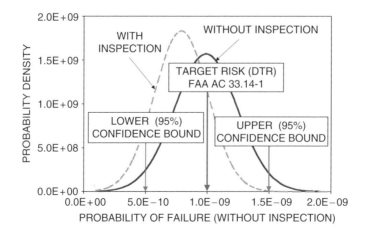

FIGURE 32.9 Upper confidence bounds on disk risk results for fixed number of zones initially do not satisfy FAA target risk [18].

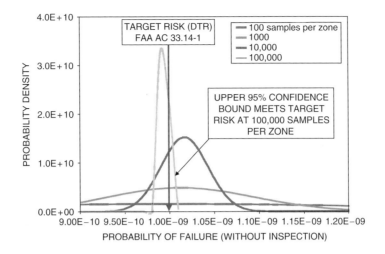

FIGURE 32.10 Increasing the number of zone samples reduces variance but is computationally expensive [18].

Circular 10 (1×10^{-9}). However, the upper confidence bound result (no inspection) is well above the target risk, so variance reduction is needed. As shown in Figure 32.10, the desired variance reduction can be achieved by increasing the number of samples in each zone to 100,000. However, this requires a total of over 4 million samples for the disk.

In Figure 32.11, a comparison of the confidence bounds vs. number of disk samples is shown for three sample allocation approaches: uniform (i.e., same number of samples in all zones), RCF (risk contribution factor), and optimal (see [18] for further details). For a specified number of samples, it can be observed that, compared with the uniform approach, the confidence bounds are narrower for the RCF and optimal approaches. The target risk can be achieved with the RCF and optimal approaches with approximately 40,000 samples (over 4 million samples are required for the uniform approach). It is interesting to note that the optimal method converges only slightly faster than the RCF approach.

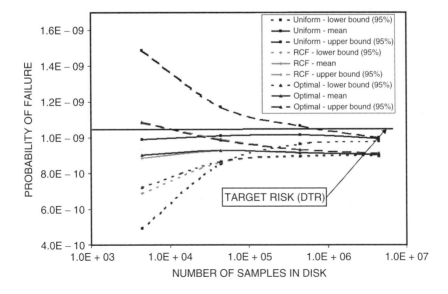

FIGURE 32.11 Comparison of three disk-variance-reduction techniques: uniform, RCF, optimal [18].

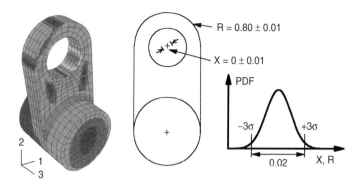

FIGURE 32.12 Finite element model, stress contours, and random variables for the aircraft lever example.

32.4.2 Life Distribution of an Aircraft Lever

The example problem is a lever that is representative of a critical aircraft structural component. A deterministic analysis of a similar lever was completed for a future military aircraft using standard methods of mechanical analysis. The lever transfers load between an actuator and a control surface and must survive extreme, limit, and normal operating loads without exceeding ultimate, yield, or fatigue strengths. A deterministic analysis may assume that all geometric variables are at the weakest extremes while loads are at their highest levels. The predicted stresses associated with the different load cases must be less than A-Basis material strengths, as defined in MIL-HDBK-5.

The finite element (FE) model of the lever showing deterministic stress contours is shown in Figure 32.12. The model consists of 44,844 degrees of freedom. Each analysis requires approximately 20 min on an HP 700 series workstation.

The random variables for the problem are listed in Table 32.2. The first two random variables affect the finite element model geometry. The hole location and radius random variables are shown in Figure 32.12. The tolerance for each dimension is assumed to be ±0.01 in., which represents three standard deviations from the mean.

The load is assumed random and models the worst-case tensile and compression loads through the range of motion of the lever. These loading extremes provide the stress range used for the fatigue-life computations.

The example considers the distribution of fatigue life based on S-N data for AerMet100, shown in Figure 32.13. The equation for fatigue life was developed based on the data

$$\log(N) = 26.4 - 9.16 \cdot \log(\sigma_{max}(1 - R)^{0.72}) + \psi \tag{32.1}$$

where N is the cycles to failure, σ_{max} is the maximum principal stress, R is the ratio of the maximum and minimum stress, and ψ is the scatter factor based on regression of the S-N data (Figure 32.13).

This problem requires linking a finite element model with several analytical relationships. First, the compressive load is defined as a function of the tensile load input. Both the tensile and compressive loads are mapped to the ABAQUS finite element model of the ABAQUS lever, which returns the minimum and maximum stresses at the critical location. These stresses are then used to compute the fatigue life of the lever, as shown in Figure 32.14.

TABLE 32.2 Random Variables for Aircraft Lever Example Problem

Name	Mean	Standard Deviation	Distribution Type
Hole location (in.)	0.0	0.003333	normal
Radius (in.)	0.8	0.003333	normal
Operating load (ksi)	4.9	0.98	normal
S-N scatter	0	0.6	normal

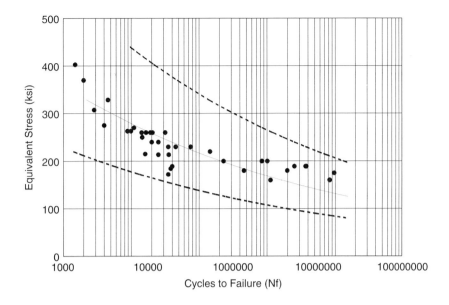

FIGURE 32.13 S-N curve for AerMet100 with 3σ bounds.

Thirteen points of the cumulative distribution function (CDF) were computed using the AMV+ method and required 101 ABAQUS FE analyses (approximately 35 CPU hours on a 700 series HP workstation). The AMV+ CDF is shown in Figure 32.15 and is compared with 100 Monte Carlo samples. The Monte Carlo solution provides points on the CDF between approximately 0.01 and 0.99. For about the same computational effort, the AMV+ solution provides a CDF range of 10^{-5} to 0.999 that covers the tail probabilities where the designer is most interested. The AMV+ CDF compares well with these limited Monte Carlo samples. Based on AMV+ results, the probability is 99.994% that the fatigue life of the lever will be at least 1 million cycles.

An important output of the probabilistic analysis is the probabilistic importance factors, which identify the variables that contribute the most to the reliability of the design (Figure 32.16). Closely related to the importance factors are the probabilistic sensitivity factors that describe the sensitivity of the computed probability of failure with respect to a change in the mean value or the standard deviation of each random variable. These sensitivity factors are shown in Figure 32.17 and allow the designer to evaluate the effect of the probability of failure due to a change in a design parameter. For this example, little importance is observed on the radius and hole location, and these may be reviewed for manufacturing process changes that may lead to loosening of tolerances and possible cost reduction. The importance of the scatter on the fatigue life suggests that it would be worthwhile to invest in improved characterization of this variable or to obtain a cleaner material.

Another failure mode of interest is exceeding the yield stress when the lever is subjected to the limit loading condition. The problem description is similar to the previous problem, except that the performance

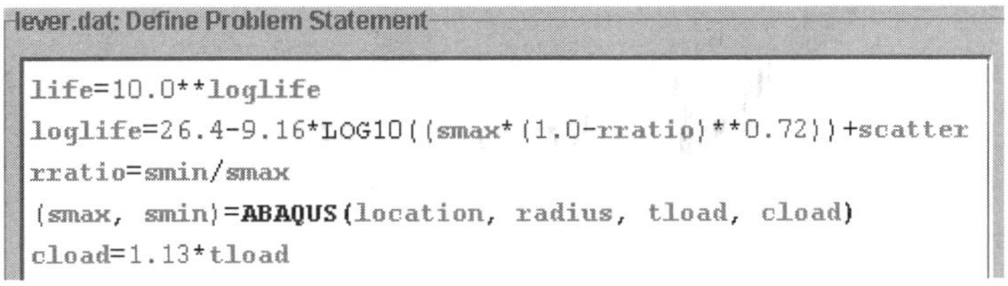

```
lever.dat: Define Problem Statement
life=10.0**loglife
loglife=26.4-9.16*LOG10((smax*(1.0-rratio)**0.72))+scatter
rratio=smin/smax
(smax, smin)=ABAQUS(location, radius, tload, cload)
cload=1.13*tload
```

FIGURE 32.14 NESSUS problem statement for aircraft lever life model.

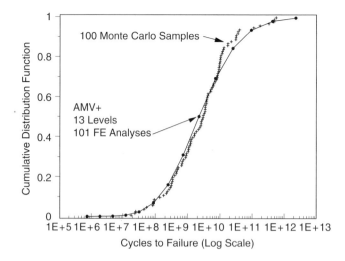

FIGURE 32.15 Cumulative distribution function for aircraft lever life model.

is now defined by the yield stress not being exceeded when the lever is subjected to the limit load condition. The problem statement and random variable definitions are shown in Figure 32.18.

The predicted reliability using the AMV+ method for the lever when subjected to the limit load is 99.999%. The sensitivity factors can be quickly viewed and exported from NESSUS, as shown in Figure 32.19.

32.4.3 Crankshaft Reliability

The crankshaft in an internal combustion engine is subjected to hundreds of millions of cycles of loading in its lifetime. High cycle fatigue under bending and torsional loading is a common failure mode in a crankshaft. The reliability of the crankshaft is predicted using NESSUS linked with the ANSYS finite element analysis program. This work is a continuation of a crankshaft reliability analysis performed by Shah et al. [23].

The model is a three-dimensional, parametric model of a single crank throw. Key parametric features in the ANSYS finite element model include the crank fillet radius and the pin fillet radius. The finite element model is shown in Figure 32.20.

Fatigue cracks usually initiate near the crank fillet region and propagate across the web under the crankshaft under bending and torsional loading. A stress life approach based on a Goodman diagram is

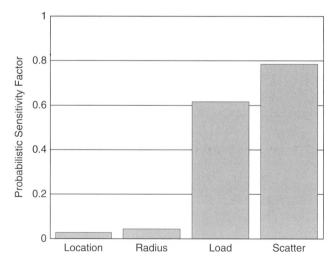

FIGURE 32.16 Probabilistic sensitivity factors for aircraft lever life model.

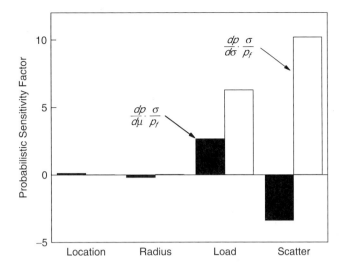

FIGURE 32.17 Sensitivity with respect to distribution parameters for aircraft lever life model.

used to develop the failure criteria and compute the fatigue margin. Failure is assumed when the fatigue margin is less than 1.0. The loading on the crankshaft during its entire cycle was evaluated, and the two crank angles that gave the worst fatigue loading were chosen for further analysis. The appropriate bending and torsional loads on the crankshaft are applied for these two load conditions and are considered as random variables (connecting rod force 1, force 2, and torque). The mean and alternating stresses are computed from these two load steps, and the Goodman diagram is used to evaluate the fatigue margin. The fatigue margin using the mean and alternating stress is computed directly in ANSYS. The random variables include the key geometric variables, loading, and fatigue properties as listed in Table 32.3.

The advanced mean value method is used to solve for the reliability and compute probabilistic sensitivity information. The probability of failure was computed to be 4% (the probability that the fatigue margin is less than 1.0). The probabilistic sensitivity factors are shown in Figure 32.21. From the figure, the torque and the endurance limit contribute most to the probability of failure. It is also clear that the crank and fillet radii have little influence on the probability of failure. These two parameters may be considered for manufacturing process changes by loosening tolerances to potentially cut costs.

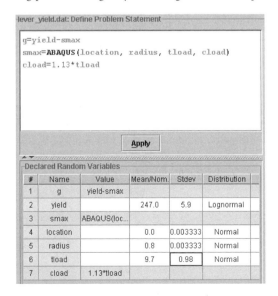

FIGURE 32.18 NESSUS problem statement for aircraft lever yield model.

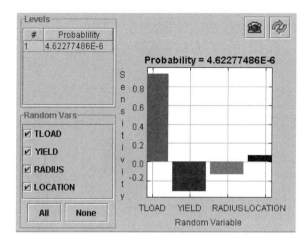

FIGURE 32.19 Probabilistic sensitivity factors for yield failure for aircraft lever life model.

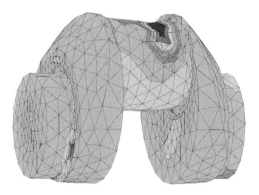

FIGURE 32.20 ANSYS finite element model of a crankshaft.

32.4.4 Cervical Spine Impact Injury

Cervical spine injuries occur as a result of impact or from large inertial forces such as those experienced by military pilots during ejections, carrier landings, and ditchings. Other examples include motor vehicle, diving, and athletic-related accidents. Reducing the likelihood of injury by identifying and understanding the primary injury mechanisms and the important factors leading to injury motivates research in this area.

Because of the severity associated with most cervical spine injuries, it is of great interest to design occupant safety systems to minimize probability of injury. To do this, the designer must have quantified knowledge of the probability of injury due to different impact scenarios, and also know which model parameters contribute the most to the injury probability. Stress analyses play a critical role in understanding the mechanics of injury and the effects of degeneration as a result of disease on the structural

TABLE 32.3 Random Variables for the Crankshaft Reliability Analysis

Name	Mean	COV(%)	Distribution
Crank fillet radius, Rc (mm)	13	0.5	normal
Pin fillet radius, Rp (mm)	11.5	0.5	normal
Connecting rod force 1, F1 (kN)	802	2	normal
Connecting rod force 2, F2 (kN)	11.8	2	normal
Torque, T (kN·m)	90	30	normal
Endurance strength, SIGe (MPa)	300	10	lognormal
Ultimate strength, SIGu (MPa)	1400	10	lognormal

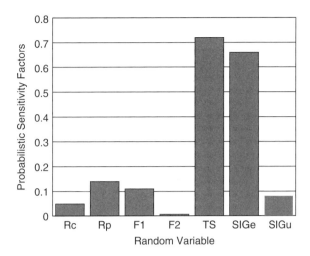

FIGURE 32.21 Probabilistic sensitivity factors for the crankshaft analysis.

performance of spinal segments, and finite element analysis (FEA) is the method of choice to conduct these analyses. However, in many structural systems, there is a great deal of uncertainty associated with the environment in which the spine is required to function. This uncertainty has a direct effect on the structural response of the system. Biological systems are an archetype example: uncertainties exist in the physical and mechanical properties and geometry of the bone, ligaments, cartilage, joint and muscle loads. Hence, the broad objective of this investigation is to explore how uncertainties influence the performance of an anatomically accurate, three-dimensional, nonlinear, experimentally validated finite element model of the cervical spine.

An experimentally validated three-dimensional finite element model of the C4-C5-C6 spinal segment developed at the Medical College of Wisconsin [24] was used to calculate the structural response of the lower cervical spine and to quantify the effect of uncertainties on the performance of the biological system [25]. The load-deflection response was validated against experimental results from eight cadaver specimens [26]. The moment–rotation response of the finite element model was validated against experimental results reported in the literature [27]. The model is shown in Figure 32.22.

Efficient probabilistic methods were used to calculate the probabilistic response of the cervical spine finite element model [28]. Biological variability was accounted for by modeling material properties and

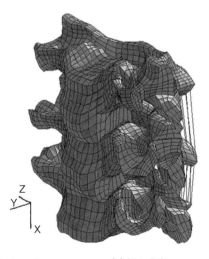

FIGURE 32.22 Probabilistic cervical motion segment model (C4–C6).

TABLE 32.4 Material Properties and Segment Loading Were Defined as Random Variables to Account for Biological Variability

No.	Name	Description	No.	Name	Description
1	MALLRV	Anterior longitudinal ligament deflection relationship	16	A56E	Annulus C56 Young's modulus
2	MPLLRV	Posterior longitudinal ligament deflection relationship	17	LM56E	Lusckha membrane C56 Young's modulus
3	MISRV	Interspinous ligament nonlinear spring force-deflection relationship	18	SM456E	Synovial membrane C45/C56 facet Young's modulus
4	MCLRV	Capsular ligament nonlinear spring force-deflection relationship	19	NP45FD	Nucleus pulposus C45 fluid density
5	MFLRV	Ligamentum flavum nonlinear spring force-deflection relationship	20	NP56FD	Nucleus pulposus C45 fluid density
6	C45AREA	45 disc cross-sectional area (rebar element) inner/middle/outer and in/out	21	FSF45FD	Facet synovial fluid C56 fluid density
7	C56AREA	56 disc cross-section area (rebar element) inner/middle/outer and in/out	22	FSF56FD	Facet synovial fluid C56 fluid density
8	C4PE	C4 posterior Young's modulus	23	LJ45FD	Luschka's joint C45 fluid density
9	C5PE	C5 posterior Young's modulus	24	LJ56FD	Luschka's joint C56 fluid density
10	C6PE	C6 posterior Young's modulus	25	CORTE	Cortical Young's modulus
11	C4CE	C4 cancellous Young's modulus	26	ENDPE	Endplate Young's modulus
12	C5CE	C5 cancellous Young's modulus	27	CARTE	Cartilage Young's modulus
13	C6CE	C6 cancellous Young's modulus	28	FIBRE	Fiber Young's modulus
14	A45E	Annulus C45 Young's modulus	29	FLEXLOAD	Flexion loading
15	LM45E	Lusckha membrane C45 Young's modulus			

spinal segment loading as random variables (Table 32.4). Where available, actual experimental data were used to generate the random variable definitions (e.g., the spinal ligaments' load-deflection behavior).

The probabilistic finite element model was exercised under flexion (chin down) loading by applying a pure bending moment of 2 N-m to the superior surface of the C4 vertebra. The inferior surface of the C6 vertebra was fixed in all directions. The resulting moment–rotation behavior was quantified by determining the rotation between the superior aspect of C4 and the fixed boundary of C6 and monitoring the reaction forces at the fixed boundary condition. Cumulative probability distribution functions, probability distribution functions, and probabilistic sensitivity factors were determined [27].

The probabilistic rotation response had an approximate mean of 3.82° and a standard deviation of 0.38°, resulting in a coefficient of variation of 10%. The cumulative distribution function (CDF) for rotation is shown in Figure 32.23. The CDF is used to determine probabilities directly, e.g., the cumulative probability at 4.2° is 82%.

Another useful output from the analysis is the probabilistic sensitivity factors. These sensitivities indicate which variables contribute the most (and the least) to the failure probability. The loading (FLEXLOAD) is the dominant variable (removed from the plot for clarity). Figure 32.24 shows the sensitivity information for the eight most significant random variables with FLEXLOAD removed so that the other variables can be seen more clearly. Not including FLEXLOAD, the most important variables are the: (1) annulus C45 and C56 Young's modulus, (2) interspinous ligament nonlinear spring force-deflection relationship, and (3) ligamentum flavum nonlinear spring force-deflection relationship. These results can be used to eliminate unimportant variables from the random variable vector and to focus further characterization efforts on those variables that are most significant.

The probabilistic methods used herein were initially developed for aerospace applications and are broadly applicable. Their use is warranted in situations where uncertainty is known or believed to have significant impact on the structural response. Furthermore, if an appropriate failure or injury metric is defined, the probability of failure or injury can be quantified. This emphasizes the additional information and advantage of performing a probabilistic analysis as compared with a deterministic analysis. For example, the cumulative probability at 4.2° rotation is 82%; thus, if failure is assumed when rotation

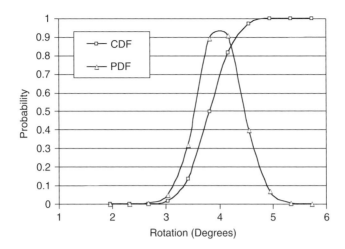

FIGURE 32.23 Cumulative distribution function and probability distribution function of the rotation response of the lower cervical spine segment subjected to pure flexion loading.

exceeds 4.2°, the probability of failure is 1 − 0.82 = 18%. With the mean value analysis, the computed rotation is less than the rotational failure limit (3.8° vs. 4.2°), thus indicating a noninjurious condition. However, as noted, by performing a probabilistic analysis, the likelihood or probability of exceeding the injury threshold is quantified (18%).

Progress in probabilistic biomechanics depends critically upon development of validated deterministic models, systematic data collection and synthesis to resolve probabilistic inputs, and identification and classification of clinically relevant injury modes. Future work in this area should include development of a high-fidelity continuum cervical spine model for assessing the combined likelihood of vertebral fracture, ligament sprain, and disc rupture. Additionally, research is needed to integrate random process loading and random field representations of geometrical variations and initial configuration into the probabilistic model. To be directly useful in design, the probabilistic methodology must be integrated into an occupant-safety design tool such that the sensitivity of design (controllable) parameters are related to probability of injury.

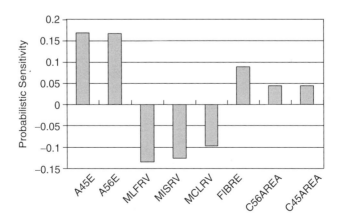

FIGURE 32.24 The eight most influential random variables for the rotation response of the lower cervical spine segment subjected to pure flexion loading (normalized scale on ordinate; variable FLEXLOAD removed for clarity).

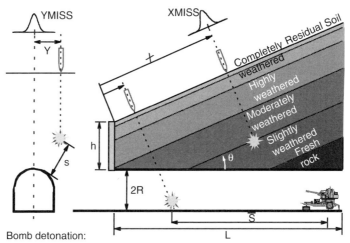

FIGURE 32.25 Conceptual model for the probabilistic tunnel vulnerability analysis.

32.4.5 Tunnel Vulnerability Assessment

Tunnels are ideally suited for storage and protection of high-value military assets, such as aircraft, material reserves, munitions, weapons of mass destruction, and control centers. The natural hardening provided by the surrounding geology renders tunnels in rock extremely difficult to attack and disable. Because construction costs are relatively low compared with aboveground facilities, the number of tunnels worldwide has grown significantly in recent years. Having the ability to functionally disrupt the activities in many of these tunnels has become one of the military's primary missions in the post Cold War era.

The probabilistic tunnel vulnerability assessment (PVTA) involves a tunnel built into a sloped, multilayered geology as shown in Figure 32.25. The tunnel vulnerability is measured as the probability that the tunnel will no longer be functional either by blocking ingress/egress or damaging equipment by airblast or fragmentation. The problem is solved by linking NESSUS with several analysis programs, including the PRONTO explicit dynamic finite element program. Additional details about this analysis can be found in the literature [29, 30].

Two scenarios are shown in Figure 32.25: one in which the weapon penetrates into the tunnel and another in which the weapon stops in the geology. Model inputs are listed in Table 32.5

The vulnerability of the tunnel is defined as the total probability of kill (P_k) considering all potential failure modes and their respective interactions. For the system analysis, a probabilistic fault tree (PFTA) is used to define the linkage between the bottom events through "OR" and "AND" gates. The PFTA method employs actual tunnel response models for each bottom event. Thus, correlations between failure modes due to common random variables are fully accounted for.

The probabilistic fault tree employed in the tunnel vulnerability assessment is shown in Figure 32.26. Note that each of the three failure modes is separated into distinct target response and weapons effects bottom events.

The NESSUS code integrated with the damage models was used to compute P_k for the three failure modes separately, as shown in Figure 32.27a. The comminution failure probability increases as the aimpoint is increased until hitting a maximum at $X = 8$ m, at which point P_k begins to decrease. The airblast and fragmentation failure probabilities decrease monotonically as the aimpoint is increased. The optimum aimpoint for these two failure mechanisms is at $X = 0$ m.

The system P_k computed using the NESSUS PFTA method is shown in Figure 32.27b. The curve labeled "assumed independent" is obtained by assuming that the component failure probabilities in Figure 32.27a are statistically independent, which does not account for common random variables between events.

TABLE 32.5 Probabilistic Tunnel Vulnerability Model Inputs

Random Variable	Identifier	Mean Value	COV (%)	Distribution
Layer 1 thickness	t1	0.5 m	25	lognormal
Layer 1 S number	s1	10	20	lognormal
Layer 2 thickness	t2	1.0 m	25	lognormal
Layer 2 S number	s2	6	20	lognormal
Layer 3 thickness	t3	2.0 m	25	lognormal
Layer 3 S number	s3	2	20	lognormal
Layer 4 thickness	t4	5.0 m	25	lognormal
Layer 4 S number	s4	1.5	20	lognormal
Layer 5 thickness	t5	7.0 m	25	lognormal
Layer 5 S number	s5	1	20	lognormal
Layer 6 S number	s6	0.8	20	lognormal
Slope angle	θ	33.81°	21	lognormal
Weapon initial velocity	V_p	335 m/sec	10	lognormal
Modeling error	B	1.0	10	normal
Miss in X-direction	XMISS	0	($\sigma = 2.37$ m)	normal
Miss in Y-direction	YMISS	0	($\sigma = 2.23$ m)	normal
Aimpoint	X	variable	0	
Retaining wall height	h	3 m	0	
Tunnel radius	R	2 m	0	
Weapon mass	M_p	8.9 kN	0	
Damage function curve-fit coeffs.	Ci	variable	0	
Rock fracture stress	TCUT	20 MPa	10	normal
Equiv. plastic volume failure strain	EPVFS	−0.05%	10	normal
Weapon yield	Vscale	1.0	10	normal

The other curve shows the results from the PFTA, in which the correlations between the individual failure modes are properly accounted for. The independent-event assumption predicts the optimum aimpoint to be at $X = 0$ m rather than at $X = 8$ m when correlations are accounted for. Thus, assuming independence can lead to significant error, depending on the degree of correlation between failure modes.

Another useful product of the NESSUS PFTA analysis is the identification of which failure modes and input parameters contribute most and least to the system P_k. The sensitivity of P_k with respect to changes

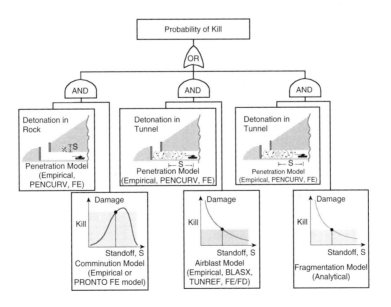

FIGURE 32.26 Probabilistic fault tree for the tunnel vulnerability analysis for three failure modes.

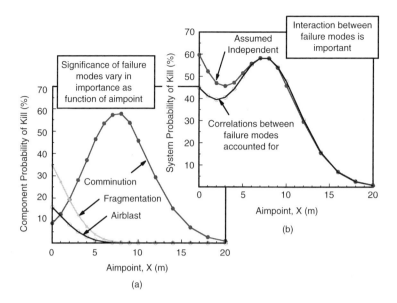

FIGURE 32.27 P_k vs. aimpoint for the three failure modes: (a) considered separately and (b) combined.

in the mean and standard deviation of the input parameters are shown in Figure 32.28a and b, respectively. From the figure, the mean value of the tensile strength (TCUT) along with several other geologic parameters contribute significantly to the P_k. Figure 32.28a shows that the standard deviation of the penetrator initial velocity as well as several geologic parameters are important.

The importance of the three failure modes at two different aimpoints is shown in Figure 32.29. At $X = 0$ m, the probability of the weapon detonating inside the tunnel is high, thereby resulting in a relatively high P_k due to fragmentation and airblast. At $X = 5$ m, the detonation is more likely to occur in the rock near the tunnel, which results in a high contribution by the comminution (and rock fall) failure mode.

Important findings revealed by the probabilistic systems analysis include: (1) the system model is efficient and easily adapted for component failure models of any sophistication; (2) the deterministic optimum aimpoint is, in general, different from the probabilistic optimum aimpoint; and (3) assuming

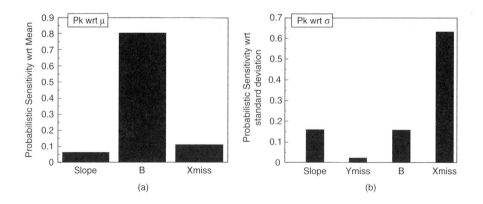

FIGURE 32.28 Probabilistic sensitivity factors for the tunnel vulnerability analysis with respect to mean and standard deviation. Variables with sensitivity less than 0.1 have been omitted for clarity.

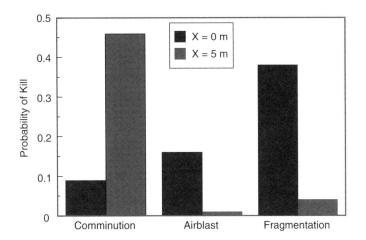

FIGURE 32.29 Failure mode probabilistic sensitivity for two different aimpoints.

the component failure modes to be independent leads to an incorrect probability of kill and optimum aimpoint location.

32.4.6 Orthopedic Implant Cement Loosening

Computational analysis methods such as finite element analysis are widely accepted in orthopedic biomechanics as an important tool used in the design and analysis of total joint replacements and other orthopedic devices. Much research in the use of computational methods in total hip arthroplasty (THA) has been used to improve the overall reliability (or some specific aspect of reliability) of orthopedic implants [31]. However, the actual probability of failure is not computed per se; rather, only quantities such as the stresses in the bone cement or at the implant-cement interface are computed and compared with a measure of the cement or interface strength in order to estimate the risk of failure. The objective of this investigation is to quantify the probability of failure of a cemented implant system in terms of specific failure modes. Model parameters such as the applied loads, material properties, and material strengths are modeled as random variables to account for the uncertainty and variability.

Advanced structural reliability methods were used to quantify the probability of failure of the cemented femoral component of a hip implant system [32]. Both failure of the bulk bone cement and failure of the cement–implant interface were investigated. For bulk cement failure, four performance functions are investigated:

1. Compressive failure:

$$g_1(X) = R_{UCS} - S_p(X) \qquad (32.2)$$

2. Shear failure:

$$g_2(X) = R_\tau - S_{Tr}(X) \qquad (32.3)$$

3. Fatigue failure:

$$g_3(X) = R_{FL} - S_{VM}(X) \qquad (32.4)$$

4. SED fracture criteria:

$$g_4(X) = SED_{cr} - SED(X) \qquad (32.5)$$

TABLE 32.6 Probabilistic Cemented Implant Failure Model Variables

Variable	Mean	Std. Dev.	COV(%)	Distribution
Cement shear strength MPa	30.0	2.7	9.0	normal
Cement compression strength MPa	81.4	2.14	2.6	normal
SED_{cr} (kJ/m^3)	75.7	7.57	10.0	normal
Cement fatigue limit (MPa) [a]	10.81	4.47	41.3	lognormal
Cement fatigue limit (MPa) [b]	10.17	3.9	38.4	lognormal
Cement fatigue limit (MPa) [c]	7.98	4.32	54.1	lognormal
ITS: early (MPa)	11.7	3.9	33.3	lognormal
ITS: dough (MPa)	7.9	1.6	20.3	lognormal
ISS (MPa)	33.3	17.6	52.9	lognormal

[a] Simplex P. [b] Zimmer LVC. [c] Zimmer Regular.

For the bone cement–implant interface failure, two performance functions are investigated:

1. Interface tensile failure:

$$g_1(X) = R_{IT} - S_\sigma(X) \tag{32.6}$$

2. Interface shear failure:

$$g_2(X) = R_{1\tau} - S_{VM}(X) \tag{32.7}$$

where R is a random variable describing the critical strength limit of the cement, and $S(X)$ is the computed response measure in the cement computed from a three-dimensional finite element model. Values of R are determined from the literature (Table 32.6). The cement responses were the minimum principal stress ($S_{\sigma1}$), maximum Tresca stress (S_{Tr}), the maximum von Mises (S_{VM}) stress, and the strain energy density (SED).

The three-dimensional finite element model [33] of a femur-implant system shown in Figure 32.30 was used to calculate the structural response ($S(X)$) of the implant system. The NESSUS probabilistic software was used to compute the probabilistic response. The variables listed in Table 32.7 were modeled as random variables.

For all failure criteria considered, the computed model response is less than the limit strength, indicating a deterministically safe design. However, the effect of variability is demonstrated in the probability of failure predictions. The most likely mode of bulk cement failure is local cement fatigue failure ($p_f = 0.448$ or 44.8%) in the region with the highest von Mises stress (see Figure 32.31). The differences in

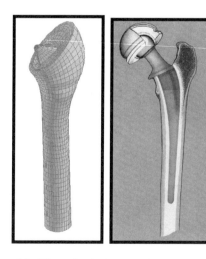

FIGURE 32.30 Finite element model of femur implant.

TABLE 32.7 Probabilistic Cemented Implant Model Variables

Variable	Mean	Std. Dev.	COV(%)	Distribution
Cortical bone E (Gpa)	20.3	2.3	11.3	lognormal
Cancellous bone E (GPa)	0.0	0.1	10.0	normal
Bone cement E (GPa)	1.95	0.15	7.7	lognormal
Joint load X (N)	1492.0	237.23	15.9	lognormal
Joint load Y (N)	915.0	408.09	44.6	lognormal
Joint load Z (N)	2925.0	731.25	25.0	lognormal
Muscle load X (N)	1342.0	335.5	25.0	lognormal
Muscle load Y (N)	832.0	208.0	25.0	lognormal
Muscle load Z (N)	2055.0	513.75	25.0	lognormal

reported fatigue limit stress for different cements significantly affects the computed probability of failure. An increase in the mean fatigue limit stress of 22% results in a 52% decrease in computed probability of failure (Zimmer Regular vs. Zimmer LVC). A 5.9% increase in the mean fatigue limit stress results in an 8.9% decrease in the probability of failure (Zimmer LVC vs. Simplex P). It should be noted, however, that the mean value, the standard deviation, and the probability distribution type of the random variable all contribute to the probability of failure. The computed probabilities of failure using the other failure criteria were negligibly small compared with the probability of cement fatigue failure. As with the bulk failure mode, the computed response for the interface failure criteria is less than the limit strength, indicating a deterministically safe design. However, the computed probability of failure is finite. Interface shear failure is more likely than interface tensile failure (Table 32.8). An increase in the tensile bonding strength from 7.9 MPa to 11.7 MPa (32% increase) results in a reduction of the probability of failure from 0.135 to 0.065 (52% reduction).

Probabilistic analysis methods offer a systematic technique to incorporate and account for variability and uncertainty in orthopedic systems and assess their impact on the biomechanical performance of these systems. System performance can be explicitly quantified in terms of risk of failure, and the important random variables can be identified so that limited resources can be better utilized by focusing on those variables that have the greatest impact on the implant's long-term reliability. In addition, probabilistic methods can be incorporated into a design framework in which the uncertainties are explicitly accounted for and the design goal is either to maximize implant reliability or to minimize the effect of variability.

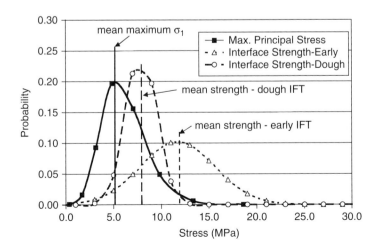

FIGURE 32.31 Probabilistic risk of failure results — cement–implant interface failure mode.

TABLE 32.8 Probabilistic Cemented Implant Probability of Failure Results

Failure Mode	Failure Criteria		Probability of Failure, p_f
Bulk cement	Compression failure		>1.00 E–6
	Shear failure		0.000015
	Fatigue failure	Simplex P	0.195
		Zimmer LVC	0.214
		Zimmer Reg.	0.448
	SED		0.0036
Interface	Tensile failure	Early IFT	0.065
	Tensile failure	Dough IFT	0.135
	Int. shear failure		0.408

32.4.7 Explosive-Actuated Piston-Valve Assembly

Enhanced evaluation capabilities are needed to determine the effect of possible anomalies that may arise in a weapon, e.g., due to aging mechanisms, and assess its performance, safety, and reliability. Experimental data and validated numerical models must be employed in determining the reliability of weapon components, including the weapon system. The validated numerical models must be based on accurate information regarding the component's geometry and material properties, e.g., in an aged condition. Once these variables are known, extrapolation of potential lifetime of the weapon can be computed with quantified accuracy and confidence.

The valve actuator uses a small amount of lead-styphnate explosive powder to propel a 0.5-in. (nominal) diameter piston down the valve barrel. An electrical signal is passed through a bridge-wire in the actuator, which in turn heats the lead-styphnate explosive, initiating a burn. The pressure buildup from the deflagration, within the actuator void space, exceeds the strength of a thin stainless steel diaphragm and subsequently allows the pressurized gas to expand and propel the piston down the valve barrel. An axisymmetric representation is shown in Figure 32.32.

Material characterization performed on both the valve body material 21-6-9 and Carpenter 455 stainless steel fall within the minimum specified properties. Structural failure of the piston constitutes a 360° fracture of the cutter ring (i.e., skirt) upon impact with the valve body ledge. Three possible failure conditions exist: (1) ductile failure due to the magnitude of net-section plastic strains at the failure strain of the material; (2) brittle fracture due to applied stress intensity factor at the crack tip and the material's

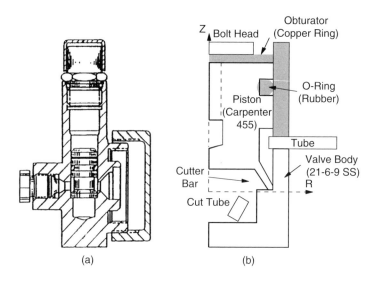

FIGURE 32.32 Axisymmetric representation of valve/piston and cutter bar assembly.

TABLE 32.9 Random variable input for the explosive actuated cutter bar assembly.

Part	Name	Mean	Standard Deviation	Distribution
Piston	E_p (psi)	29.0 E6	6.67E5	Lognormal
	S_{YP} (psi)	235775.0	23477.5	Lognormal
	E_{HP} (psi)	96375.0	9637.5	Lognormal
	M_p (g)	12.0	0.23	Normal
	Geometry	0.055	0.0022	Normal
Valve	E_V (psi)	29.0E6	6.67E5	Lognormal
	S_{YV} (psi)	45000.0	4500.0	Lognormal
	E_{HV} (psi)	209840.0	20984.0	Lognormal
Inertial	V_I (in/s)	−2750.0	131.0	Normal
	P_{RP} (psi)	12000.0	1200.0	Normal

fracture toughness; and (3) bimodal failure from ductile-to-brittle transition caused by void growth progressing to fracture.

Table 32.9 depicts the random variables used in this analysis along with the respective distributions. Impact loading parameters, such as initial velocity and reaction-product back-pressure, are normally distributed. Further analysis of lead-styphnate reaction is required to determine an accurate characterization of the burn-exponent and burn-constant parameters for better approximations of the loading distributions. Nevertheless, the distributions shown in Table 32.9 for the burn variables are considered acceptable.

Piston and valve material properties were assigned a lognormal distribution, which is consistent with material manufacturing observations. The piston cutter-ring thickness variation is shown in Figure 32.33, with the distribution as presented in Table 32.9.

The finite element analysis was performed using the ABAQUS/Explicit software. Three separate methods of probabilistic analyses were conducted to determine the efficiency and accuracy of the reliability algorithms within NESSUS. These were, MV (mean value), AMV (advanced mean value), and LHS (Latin hypercube sampling). The LHS method was included to verify the accuracy of AMV without performing a full Monte Carlo analysis to the ±3σ levels.

The information shown in Figure 32.34 represents the importance (i.e., sensitivity) of each random variable to the overall probability of exceeding the plastic shear strain, postulated to cause structural failure for the component. The most important variables leading to prediction of structural failure are (1) flow characteristics of piston, (2) flow characteristics of valve, (3) impact velocity, and (4) cutter-ring thickness. Of course, to a certain degree, the other random variables are important, although they may,

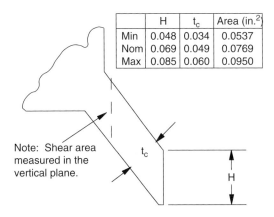

FIGURE 32.33 Piston ring thickness variability.

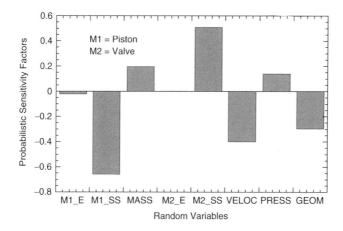

FIGURE 32.34 Importance factor for exceeding plastic shear strain.

in and of themselves, not lead to structural failure. Each component's Young's modulus, however, appears insignificant to the overall probabilistic response. As such, these two random variables could, in effect, have been maintained as deterministic variables without loss of generality in the results, thus reducing the number of finite element analyses conducted under MV, AMV.

Figure 32.35 gives the cumulative distribution functions (CDF) depicting the probability for the standard normal variate as a function of exceeding the plastic shear strain limit. The limit state (or g-function) is the failure strain of the material minus the computed strain from the finite element model. Thus, the g-function is the following:

$$g(X) = z_0 - Z(X) \tag{32.8}$$

where $Z(X)$ is the computed plastic shear strain and z_0 is the maximum allowable plastic shear strain. As shown, the AMV and LHS results follow closely throughout the CDF. Especially in regions away from the mean and nearing the $\pm 3\sigma$ levels, the AMV results are nearly identical to the LHS solution.

Figure 32.36 shows similar information as in Figure 32.35, but it is based on the standard normal variate probability function. The ordinate plots the standard deviations from the mean, while the abscissa plots the actual total strain from the FEA model. Therefore, at the mean (μ) or at a standard deviation of zero ($\sigma = 0$), the obtained result is that of exceeding 30% plastic shear strain.

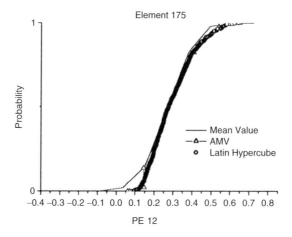

FIGURE 32.35 Cumulative distribution function for exceeding plastic shear strain.

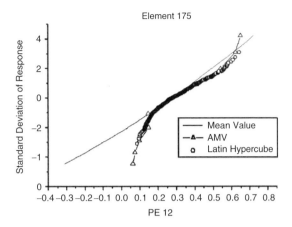

FIGURE 32.36 Standard normal cumulative distribution function for exceeding plastic shear strain.

TABLE 32.10 Number of Finite Element Analysis Solutions

Method	FEA Runs
MV	65
AMV	9
LHS	500

The number of finite element model solutions necessary for the MV, AMV, and LHS is shown in Table 32.10. It is evident from Figure 32.36 and Table 32.10 that the AMV method is quite efficient as compared with the LHS. The MV results appear to deviate into physically inadmissible regions, such that it predicts absolute-value shear strains that are negative. However, these are actually unconverged results that are corrected by the AMV method. The powerful and efficient methods employed in NESSUS make it extremely attractive in solving large models with the minimum number of numerical solutions.

Figure 32.37 shows a contour plot of the probability of exceeding 60% strain. This plot was generated by computing the probability of exceeding 60% strain at every node in the mesh and then plotting the

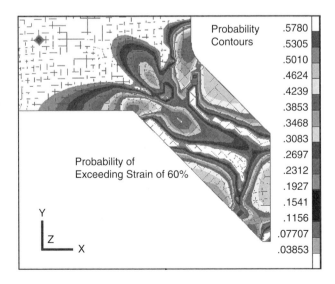

FIGURE 32.37 Probability of exceeding 60% strain.

results using contours. This information was computed efficiently by utilizing the MV results stored in the restart database.

32.5 Conclusions

The probabilistic methods available in NESSUS and DARWIN were initially developed for aerospace applications. However, the methods are broadly applicable, and their use is warranted in situations where uncertainty is known or believed to have a significant impact on the model response. The framework of NESSUS allows the user to link traditional and advanced probabilistic algorithms with analytical equations, commercial finite element programs, and "in-house" analysis packages to compute the probabilistic response or reliability of a system. The framework of DARWIN is specifically designed to perform risk assessment of gas turbine engine rotors. This capability is currently being extended to include additional materials and damage mechanisms.

For probabilistic analysis to be more widely accepted and utilized, probabilistic algorithms must be interfaced with commercial analysis packages such as ABAQUS, ANSYS, and NASTRAN, and they must also be easy to use via highly interactive graphical user interfaces. Integration with commercially available software allows the analyst to use the best tool for the problem, one in which the analyst is confident in using, or a tool that an organization has invested in for training or development. The development of the platform-independent graphical user interface capability in NESSUS and DARWIN makes the probabilistic analysis straightforward and efficient to set up and understand for simple, complex, and large-scale problems.

Several probabilistic analyses were presented that demonstrated the flexibility of the NESSUS and DARWIN software when linked with deterministic analysis packages such as ANSYS, ABAQUS, and NASTRAN. The advanced and efficient probabilistic analysis methods employed in the case studies allow the use of high-fidelity models to define the structure or system, even when each deterministic analysis may take several hours to run. In the application problems presented, the probabilistic results revealed additional information that would not have been available if traditional deterministic approaches were used.

Progress in probabilistic mechanics relies strongly on the development of validated deterministic models, systematic data collection and synthesis to resolve probabilistic inputs, identification and classification of failure modes, and the development of sophisticated analysis tools that are powerful yet easy to use.

Acknowledgments

The authors wish to acknowledge the support provided for the NESSUS and DARWIN probabilistic computer programs. NESSUS was supported by the National Aeronautic and Space Administration Lewis Research Center (now the Glenn Research Center at Lewis Field), The University of California Los Alamos National Laboratory, Southwest Research Institute Internal Research, and many others sources too numerous to mention. The Federal Aviation Administration supported DARWIN under cooperative agreement 95-G-041 and grant 99-G-016. The ongoing contributions of the Industry Steering Committee are also gratefully acknowledged.

References

1. Thacker, B.H. and Huyse, L.J., Probabilistic assessment on the basis of interval data, in *Proceedings of 43rd Structures, Structural Dynamics, and Materials (SDM) Conference*, AIAA/ASME/ASCE/AHS/ASC, Denver, CO, 2003, on CD.
2. Oberkrampf, W., Helton, J., Joslyn, C., Wojtkiewicz, S., and Ferson, S., Challenge Problems: Uncertainty in System Response Given Uncertain Parameters, technical report, Sandia National Laboratories, Albuquerque, NH, 2002.
3. Southwest Research Institute, NESSUS Reference Manual, ver. 7.5, 2002; available on-line at http://www.nessus.swri.org.

4. Riha, D.S., Thacker, B.H., Enright, M.P., Huyse, L., and Fitch, S.H.K., Recent advances of the NESSUS probabilistic analysis software for engineering applications, in *Proceedings of 42nd Structures, Structural Dynamics, and Materials (SDM) Conference,* AIAA-2002-1268, AIAA/ASME/ASCE/AHS/ASC, Seattle, WA, 2002, on CD.
5. Thacker, B.H., Rodriguez, E.A., Pepin, J.E., and Riha, D.S., Application of probabilistic methods to weapon reliability assessment, in *Proceedings of 42nd Structures, Structural Dynamics, and Materials (SDM) Conference,* AIAA 2001-1458, AIAA/ASME/ASCE/AHS/ASC, Seattle, WA, 2001, on CD.
6. Thacker, B.H., Riha, D.S., Millwater, H.R., and Enright, M.P., Errors and uncertainties in probabilistic engineering analysis, in *Proceedings of 42nd Structures, Structural Dynamics, and Materials (SDM) Conference,* AIAA 2001-1239, AIAA/ASME/ASCE/AHS/ASC, Seattle, WA, 2001, on CD.
7. Leverant, G.R. et al., A probabilistic approach to aircraft turbine rotor material design, in *Proceedings of ASME International Gas Turbine and Aeroengine Congress,* 97-GT-22, ASME, New York, NY, 1997.
8. Southwest Research Institute, Allied Signal, General Electric, Pratt and Whitney, Rolls-Royce Allison, and Scientific Forming Technologies, Turbine Rotor Material Design: Final Report, DOT/FAA/AR-00/64, Federal Aviation Administration, Washington, DC, 2000.
9. Leverant, G.R., McClung, R.C., Millwater, H.R., and Enright, M.P., A new tool for design and certification of aircraft turbine rotors, *ASME J. Eng. Gas Turbines Power,* 126(1), 155–159 (2003).
10. Southwest Research Institute, NASGRO User's Manual, ver. 4.0, 2003; available on-line at http://www.nasgro.swri.org.
11. Southwest Research Institute, Probabilistic Structural Analysis Methods (PSAM) for Select Space Propulsion System Components, final report, NASA contract NAS3-24389, NASA Lewis Research Center, Cleveland, OH, 1995.
12. Torng, T.Y., Wu, Y.-T., and Millwater, H.R., Structural system reliability calculation using a probabilistic fault tree analysis method, in *Proceedings of 33rd Structures, Structural Dynamics, and Materials (SDM) Conference,* Part 2, AIAA/ASME/ASCE/AHS/ASC, Dallas, TX, 1992, pp. 603–613.
13. National Transportation Safety Board, Aircraft Accident Report: United Airlines Flight 232 McDonnell Douglas DC-10-10 Sioux Gateway Airport, Sioux City, Iowa, July 19, 1989, NTSB/AAR-90/06, NTSB, Washington, DC, 1990.
14. Millwater, H.R., Fitch, S., Wu, Y-T., Riha, D.S., Enright, M.P., Leverant, G.R., McClung, R.C., Kuhlman, C.J., Chell, G.G., and Lee, Y.-D., A probabilistically based damage tolerance analysis computer program for hard alpha anomalies in titanium rotors, in *Proceedings of 45th ASME International Gas Turbine and Aeroengine Technical Congress,* ASME, Munich, Germany, 2000.
15. McClung, R.C., Leverant, G.R., Wu, Y-T., Millwater, H.R., Chell, G.G., Kuhlman, C.J., Lee, Y.-D., Riha, D.S., Johns, S.R., and McKeighan, P.C., Development of a probabilistic design system for gas turbine rotor integrity, in *Fatigue '99: The Seventh International Fatigue Conference,* Beijing, 1999, pp. 2655–2660.
16. Wu, Y.-T., Enright, M.P., and Millwater, H.R., Probabilistic methods for design assessment of reliability with inspection, *AIAA J.,* 40 (5), pp. 937–946, 2002.
17. Wu, Y-T., Millwater, H.R., and Enright, M.P., Efficient and accurate methods for probabilistic analysis of titanium rotors, in *Proceedings of 8th ASCE Specialty Conference on Probabilistic Mechanics and Structural Reliability,* PMC2000-221, ASCE, Reston, VA, 2000.
18. Enright, M.P. and Millwater, H.R., Optimal sampling techniques for zone-based probabilistic fatigue life prediction, in *Proceedings of 43rd Structures, Structural Dynamics, and Materials Conference, Non-Deterministic Approaches Forum,* AIAA/ASME/ASCE/AHS/ASC, Denver, CO, 2002.
19. Hulse, L. and Enright, M.P., 2003, Efficient statistical analysis of failure risk in engine rotor disks using importance sampling techniques, AIAA Paper 2003–1838, 44th AIAA/ASME/ASCE/AHS/ASC Structures, Structural Dynamics and Materials Conference, Norfolk, VA, 2003.
20. U.S. Department of Transportation, Damage Tolerance for High Energy Turbine Engine Rotors, advisory circular AC 33.14-1, Department of Transportation, Federal Aviation Administration, Washington, DC, 2001.
21. Southwest Research Institute, DARWIN™ User's Guide, ver. 3.5, 2002; available on-line at www.darwin.swri.org.

22. Aerospace Industries Association Rotor Integrity Subcommittee, The development of anomaly distributions for aircraft engine titanium disk alloys, in *Proceedings of 38th Structures, Structural Dynamics, and Materials Conference*, AIAA/ASME/ASCE/AHS/ASC, Kissimmee, FL, 1997, pp. 2543–2553.
23. Shah, C.R., Sui, P., Wang, W., and Wu, Y.-T, Probabilistic reliability analysis of an engine crankshaft, in *Proceedings of 8th International ANSYS Conference*, Vol. 11, ANSYS, Pittsburgh, PA, 1998, pp. 1212–1224.
24. Kumaresan, S., Yoganandan, N., Pintar, F.A., and Maiman D., Finite element modeling of the lower cervical spine; role of intervertebral disc under axial and eccentric loads, *Med. Eng. Phys.*, 21(10), 687–700, 1999.
25. Thacker, B.H., Nicolella, D.P., Kumaresan, S., Yoganandan, N., and Pintar, F.A., Probabilistic finite element analysis of the human lower cervical spine, in Math. Modeling Sci. Computing, 13(12), pp. 12–21, 2001.
26. Pintar, F.A., Yoganandan, N., Pesigan, M., Reinartz, J.M., Sances, A., and Cusik, J.F., Cervical vertebral strain measurements under axial and eccentric loading, *ASME J. Biomech. Eng.*, 117, pp. 474–478, 1995.
27. Shea, M., Edwards, W.T., White, A.A., and Hayes, W.C., Variations of stiffness and strength along the human cervical spine, *J. Biomech.*, 24(2), pp. 95–107, 1991.
28. Thacker, B.H., Wu, Y.-T., and Nicolella, D.P., *Frontiers in Head and Neck Trauma: Clinical and Biomechanical, Probabilistic Model of Neck Injury*, Yoganandan, N., Pintar, F.A., Larson, S.J., and Sances, A., Jr., Eds., IOS Press, Harvard, MA, 1998.
29. Thacker, B.H., Riha, D.S., and Wu, Y.-T., Probabilistic Structural Analysis of Deep Tunnels, Defense Nuclear Agency technical report TR-95-64, Southwest Research Institute, San Antonio, TX, 1996.
30. Thacker, B.H., Oswald, C.J., Wu, Y.-T., Patterson, B.C., Senseny, P.E., and Riha, D.S., A probabilistic multi-mode damage model for tunnel vulnerability assessment, in *Proceedings of 8th International Symposium on the Interaction of the Effects of Munitions with Structures*, Defense Special Weapons Agency, McLean, VA, 1997.
31. Huiskes, R. and Hollister, S.J., From structure to process, from organ to cell: recent developments of FE-analysis in orthopaedic biomechanics, *J. Biomech. Eng.*, 115, pp. 520–527, 1993.
32. Thacker, B.H., Nicolella, D.P., Kumaresan, S., Yoganandan, N., and Pintar, F.A., Probabilistic finite element analysis of the human lower cervical spine, *J. Math. Modeling Sci. Computing*, 13 (1–2), pp. 12–21, 2001.
33. Katoozian, H. and Davy, D.T., Effects of loading conditions and objective function on three-dimensional shape optimization of femoral components of hip endoprostheses, *Med. Eng. Phys.*, 22, pp. 243–251, 2000.

33
Efficient Time-Variant Reliability Methods in Load Space

Sviatoslav A. Timashev
Russian Academy of Sciences

33.1 Reliability Analysis of Redundant Structures Subject to Combinations of Markov-Type Loads......... 33-1
General Considerations • Introduction • Main Scheme of Reliability Analysis of Deformable Systems • General Theory • Reliability of Elastic-Plastic Frames Subject to a Combination of Random Loads • Reliability of a Frame Subject to Active Load and Unloading • On Design Values of Loads and Their Combinations

33.2 Life and Reliability Assessment of Large Mechanical Systems Subjected to a Combination of Diffusion-Type Markov Loads by Generalized Pontryagin Equations.. 33-15
Markov Loads • Generalized Pontryagin Equations • Methods of Downsizing the Dimension of the Reliability Problem • Transition from System Life to System Reliability

33.3 Examples 33-25
Case 1 • Case 2

33.4 Conclusion 33-26

33.1 Reliability Analysis of Redundant Structures Subject to Combinations of Markov-Type Loads

33.1.1 General Considerations

Methods for assessing structural reliability and life prediction have a long history. Structural reliability methods started as simple, modest applications of probability theory (random variables, RVs) to linear structural mechanics and dynamics. Authors that worked in this field in the early days of structural reliability were grafting the existing probability methods to the well-established methods of linear and nonlinear structural mechanics. Once this approach started, the internal logic of probability theory demanded the use of all its apparatus to the problems of linear and nonlinear structural mechanics, which are, by definition, multidegree-of-freedom problems. With time, this approach gained momentum and now has become the mainstream of research and development.

It was quickly observed that for redundant, multidegree-of-freedom structures/systems subjected to a combination of loads (and this is the central problem in structural design), reliability assessments require solving complex multidimensional integrals. Their dimensional size is actually the product of the number of structural elements (cross-sections) of the structure/system and the number of loads. Even for the simplest structures, these dimensions are forbiddingly high, starting from three to five and up to hundreds. This "curse of dimensionality" has ever since been present in all reliability studies. To the author's knowledge, the nature of the dimension of structural reliability problems was never questioned or addressed; it was considered as an intrinsic and therefore inevitable component of the problem.

In-depth analysis of the origin/source of the curse of dimensionality shows that it arises immediately when trying to solve the stochastic problem of structural mechanics directly. Once the input (usually, *independent* stochastic loads in the form of random variables [RVs]) is transformed by using the mathematical operator of the analyzed system (usually, algebraic, differential, and integral equations) into its internal properties (forces, stresses, strains, deflections, crack sizes, and the like), the latter immediately become statistically *interdependent*. Further, if the input is Gaussian, the output would be, in most practical cases, nonGaussian. Moreover, every next transformation in the chain that leads to the sought solution makes the interdependency more and more complicated and stochastically *nonlinear*. This means that the possibility of obtaining exact or closed solutions becomes more remote with each step.

Significant efforts were undertaken to overcome these difficulties. As a result, a plethora of methods were invented to mitigate/overcome (not eliminate) these difficulties, including the Rosenblatt transformation for nonGaussian RVs, the stochastic averaging methods, the moments' methods, etc. During this process, the original problem usually had to be simplified in order to obtain the needed numerical result. As a result, the level of sophistication of the statistical dynamics is substantially lower than the initial deterministic problem from which the stochastic problem originated. At the same time, the results are hard to interpret and visualize.

These considerations highlight the need to *decouple* the problem of structural mechanics from the stochastic problem in order to minimize the dimension and the nonlinearity of the stochastic problem. It was shown (Timashev [34]) that it is possible to reduce the dimension of the reliability-analysis problem to the dimension of the load vector. For instance, for a 1000-element system subjected to a two-component vector load, the dimension of the reliability integral in the conventional format would have to be at least 1000. This chapter presents and demonstrates a nonconventional approach that will make it possible to solve this kind of problem in a two-dimensional space.

33.1.2 Introduction

The problem of load combinations in structural analysis has always been a problem of great significance. Historically, loads were considered as being constant, and their maximum values (assessed through scarce statistics and common sense) were incorporated in the design codes to be on the conservative side. In the 1930s, however, it was realized that loads, in general, are stochastic. Since then, loads have been increasingly treated as random variables. The conventional methods of structural reliability assessment, namely the first- and second-order reliability methods (FORM and SORM), still use the RV concept and the probability transformation in order to reduce the reliability problem to that of calculating an integral over the failure region (formed by the limit-state surface) in the space of n standard normal variables. If the real limit-state surface is approximated by a hyperplane, one arrives at FORM; if it is approximated by a quadratic surface, one has the SORM. The mechanical transformation is usually an algorithm (for instance, the finite element method) that allows assessment of the stress and strain of a structure, given the loading and the boundary conditions.

From the above, it is clear that in order to use FORM and SORM it is necessary to:

- Deal with load, material properties and geometry of a structure as random variables that form the vector **V**
- Use direct mechanical transformations to get the load effects **U** (internal forces, stresses, strains, displacements)

- Use probabilistic transformations to derive from the arbitrary PDF (probability density function) of **V** a standard normal vector **U**
- Find the minimum distance point from the origin to the limit-state surface, which is known as the reliability index [1], as a solution of a constrained optimization problem

This scheme has been thoroughly studied, and many variations of each step have been proposed [2, 3]. Many models of different loads as RVs have been created, such as loads from snow [4, 5], wind [6], crane [7], loads on office buildings [8, 9], and even meteoroid loading on lunar structures [10].

The development of load requirements for structural codes extensively used the FORM and SORM [11–14]. Applications of the minimum-cost concept to structural load requirements using the SORM technique are given in the literature [9, 15]. In most practical situations, the design documents provide combinations of design loads multiplied by corresponding load factors [3, 11]. Currently, a partial load factor is applied to each load in the combinations [1, 12]. The load-combination problem using FORM and SORM, where loads were interpreted as RVs, was considered in [12–14, 16]. The process of development of probability-based design codes in the past 20 years in the U.S. is given in [17] and in Eastern Europe in [18].

In the 1970s, a better level of understanding was achieved. Loads started to be treated as random functions of time. Loads were described as homogeneous Gaussian differentiable functions, as different types of step-functions [19–21], as Poisson processes [22, 23], as renewal Markov functions [22, 24, 25], as birth-and-death (B.A.D.)-type Markov functions [26, 27], and as Markov nondifferentiable functions [28, 29]. Attempts were made to give a new insight into how to assess structural reliability using models of loads as random functions of time, and how to establish the design loads and load factors [30, 31]. FORM was used in cases when the structure experiences a combination of Borges-Castanheta types of loads [19, 32]. Analysis of the latest achievements is given by Low and Tang [33].

To evaluate the reliability of redundant systems when acted upon by a combination of loads, it is required to determine the probability of the multidimensional random process staying within the admissible region, which generally has a complex form. Solving this problem for general-form random processes presents great difficulties. However, by restricting the solution to processes possessing some special properties, comparatively simple methods of calculating this probability can be obtained that are of immediate practical interest. Such processes are Markov processes that completely describe the loads and are completely specified by a joint probability distribution of two variables.

The objective of this chapter is to solve the combination of load problem when loads are considered as random Markov functions of time. Two types of Markov functions are used: nonstationary homogeneous processes of the birth and death (B.A.D.) type (discrete states, continuous time) and diffusion-type processes (continuous states, continuous time).

33.1.3 Main Scheme of Reliability Analysis of Deformable Systems

For the central reliability problem of load combinations, the following reliability-analysis scheme is proposed [34]. This scheme consists of four steps. First, the system is modeled as a multi-input, multi-output system, and the input parameters' space **Q** and the output parameters' space **U** are selected. This permits the introduction of the system operator:

$$Lu = q; \quad u \in \mathbf{U}, \quad q \in \mathbf{Q}. \tag{33.1}$$

In the second step, single out in the operator L the elements χ, χ_0, χ_s, where χ and χ_0 are, respectively, the elements of the K and K_0 spaces of determinate properties of the system that are not and that are subject to optimization; χ_s stands for the elements of the space K_s of those properties of the system that are regarded as stochastic.

At the third step, by solving the *inverse* problem of mechanics, the permissible subspace **V** in the space **U** is determined and, through it, the admissible region $\Omega_0(\chi_s)$ in the space **Q**.

In the fourth step, the conditional reliability of the system is sought

$$R_c(t) = P[q(\tau) \in \Omega_s(\chi_s)], \quad 0 \leq \tau \leq t \tag{33.2}$$

and using Equation 33.2, the full reliability is defined as

$$R(t) = \int_{\chi_s} \cdots \int R_c(t) f(\chi_s) d\chi_s. \tag{33.3}$$

Thus, according to this scheme, the *reliability problem is always solved in the load space*. The admissible region is constructed according to the equation

$$\mathbf{v}_* = H(\mathbf{q}, \chi_s) \tag{33.4}$$

where $\mathbf{v}_*$ is the ultimate permissible value of the quality vector of the system, and operator H is the inverse to operator L. Operators L and H reflect the level of complexity of the deterministic structure problem (because the value of χ_s is fixed) that is solved with respect to the conventional state of structural mechanics and computer capabilities.

Such an approach allows constructing in the **Q** space partial admissible regions $\Omega_0^{(i)}(\chi_s)$ with respect to each *i*th quality criteria (Figure 33.1a). Their intersection yields the admissible region for all quality criteria simultaneously:

$$\Omega_0 = \bigcap_{i=1}^{N} \Omega_1^{(i)} \quad (N \text{ is the number of quality criteria}). \tag{33.5}$$

Using this method, it is easy to find regions $\overline{\Omega}$, where one, two, three, etc. types of failure take place simultaneously. For instance, according to Figure 33.1a:

$$\begin{aligned} \overline{\Omega}_i &= \Omega_0^{(i+1)} - \Omega_0^{(i)} \bigcap \Omega_0^{(i+1)}; \\ \overline{\Omega}_{(i+1,i+2)} &= \Omega_0^{(i)} - \Omega_0^{(i)} \bigcap \Omega_0^{(i+2)}. \end{aligned} \tag{33.6}$$

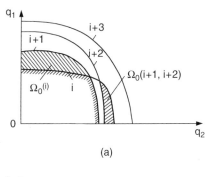

(a)

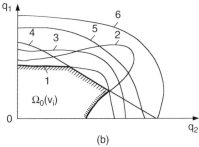

(b)

FIGURE 33.1 Admissible regions in load space.

In the case of a multielement system, the admissible regions can be constructed for each element according to some criterion (Figure 33.1b). Their intersection gives an admissible region for the system as a whole, according to the same criteria. Performing this procedure for all quality criteria, the problem of seeking the admissible region for a system according to all quality criteria is reduced to the scheme presented in Figure 33.1a.

The outlined approach is clear and simple to interpret. In fact, even *prior* to starting a reliability function computation, the engineer is aware of which of the quality criteria are the most stringent (in Figure 33.1a, these are the criteria i and $i+1$) and which elements do not participate in the formation of the admissible regions (for example, in Figure 33.1b these are elements 3, 5, 6). This allows singling out *excessively reliable* elements and mapping out constructive steps to bring down their reliability to a level that exerts no influence on the total reliability of the system.

Since the admissible regions are plotted for a fixed random vector value χ_s, the reliability function obtained using these regions is conditional. Integration of this function with weight $f(\chi_s)$ yields an unconditional reliability function. Thus, all randomness is concentrated in the admissible region.

It should be noted that solution of the above problem in the presented format requires random load models that enable one to perform probability calculations of load processes to outcross low levels. Not all types of random functions provide this ability. Best suited for this case are stepwise and diffusion (both nondifferentiable) Markov processes. Gaussian and nonGaussian differentiable processes do not provide such ability.

33.1.4 General Theory

When a structure is subject to a system of external loads $q_i(t)$, $i = 1, \ldots, n$ that can be described as nondifferentiable random processes, the above reliability problem is solved by using methods of Markov processes theory.

Consider some specifics of the reliability analysis of mechanical systems in the case when a structure is subjected to two vector-type loads. This is the so-called load-combination problem. Usually, both loads can be adequately represented as independent Markov nonstationary homogeneous processes of the "birth and death" (B.A.D.) type. In this case, the probabilities $P_{i,j}(t)$ that a process is in a fixed state (Figure 33.2) are found as the solution of the following system of differential equations:

$$P_{i,j} = P\{q_1(t) = i;\ q_2(t) = j\} \tag{33.7}$$

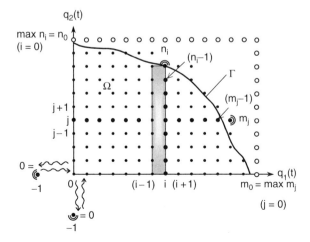

FIGURE 33.2 Rectangular mesh on which probabilities $P_{i,j}$ are calculated for bilateral assessments of the reliability function.

that is of the form [35]

$$\frac{d}{dt}P_{i,j}(t) = -(\lambda_i + \mu_i + \lambda'_j + \mu'_j)P_{i,j}(t) + \lambda_{i-1}P_{i-1,j}(t) + \mu_{i+1}P_{i+1,j}(t) + \lambda'_{j-1}P_{i,j-1}(t) + \mu'_{j+1}P_{i,j+1}(t) \quad (33.8)$$

where $i, j = 0, 1, \ldots, \lambda_{-1} = \mu_0 = 0; \lambda'_{-1} = \mu'_0 = 0$. Here the intensities of birth λ_i and death μ_i depend on the process $q_1(t)$ only, whereas the intensities λ'_j and μ'_j depend on the process $q_2(t)$ only.

Without loss of generality, the initial conditions for Equation 33.7 can be taken as

$$P_{0,0} = 1; \quad P_{i,j}(0) = 0; \quad i, j = 1, 2, \ldots, n \quad (33.9)$$

If a region Ω with a boundary Γ is singled out in $\{q_1, q_2\}$ space, and an auxiliary process $\bar{q}(t)$ is introduced such that for it $\bar{\lambda}_i = \lambda_i$, $\bar{\mu}_i = \mu_i$, $\bar{\lambda}'_j = \lambda'_j$, and $\bar{\mu}'_j = \mu'_j$ in points $(i, j) \in \Omega$ and $\bar{\lambda}_i = \bar{\mu}_i = \bar{\lambda}'_j = \bar{\mu}'_j = 0$ in points $(i, j) \in \Gamma$ (i.e., the boundary is an absorbing one; see Figure 33.3), then the probability that the process $\bar{q}(t)$ will not leave the region Ω will be equal to

$$R(t) = \sum_{i=0}^{m-1} \sum_{j=0}^{n_i} \bar{P}_{i,j}(t) \quad (33.10)$$

Here, $\bar{P}_{i,j}(t)$ satisfies the following system of differential equations:

$$\frac{d}{dt}\bar{P}_{i,j}(t) = -(\bar{\lambda}_i + \bar{\mu}_i + \bar{\lambda}'_{j-1} + \bar{\mu}'_j)\bar{P}_{i,j}(t) + \bar{\lambda}_{i-1}\bar{P}_{i-1,j}(t)$$
$$+ \bar{\mu}_{i+1}\bar{P}_{i+1,j}(t) + \bar{\lambda}'_{j-1}\bar{P}_{i,j-1}(t) + \bar{\mu}'_{j+1}\bar{P}_{i,j+1}(t), \quad (33.11)$$
$$\bar{\lambda}'_{-1} = \bar{\mu}'_0 = 0; \quad \bar{\lambda}_{-1} = \bar{\mu}_0 = 0; \quad (i, j) \in \Omega.$$

Equation 33.10 reflects the fact that the probability of the process $q(t)$ not leaving the region Ω is equal to probability of the process $\bar{q}(t)$ staying within the region Ω at a moment of time t.

Introduce into consideration the probabilities

$$P_i^{(m_j)}(t); \quad i = 0, 1, \ldots, m_j; \quad P_j^{(n_i)}(t); \quad j = 0, 1, \ldots, n_i \quad (33.12)$$

that express the probability that at a moment of time t the process $q_1(t)[q_2(t)]$ is in the $i(j)$ state, provided the state $m_j(n_i)$ is an absorbing one for the process $q_1(t)[q_2(t)]$. These probabilities can be found from

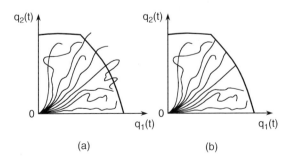

FIGURE 33.3 Graphical interpretation of a transparent boundary (a); absorbing boundary (b).

the following systems of differential equations:

$$\begin{aligned}
\frac{d}{dt}P_i(t) &= -(\bar{\lambda}_i + \bar{\mu}_i)P_i(t) + \bar{\lambda}_{i-1}P_{i-1}(t) + \bar{\mu}_{i+1}P_{i+1}(t), \\
i &= 0, 1, \ldots, m_j; \; \bar{\mu}_0 = 0, \mu_i = \mu_i; \; \bar{\lambda}_i = \lambda_i; \\
j &= 0, 1, \ldots, m_{j-1}; \; \bar{\lambda}_{m_j} = \bar{\mu}_{m_j} = 0; \\
P_0(0) &= 1, P_i(0) = 0, i = 1, 2, \ldots, m_j.
\end{aligned} \right\}$$ (33.13)

$$\begin{aligned}
\frac{d}{dt}P_j(t) &= -(\bar{\lambda}'_j + \bar{\mu}'_j)P_j(t) + \bar{\lambda}'_{j-1}P_{j-1}(t) + \bar{\mu}'_{j+1}P_{j+1}(t), \\
j &= 0, 1, \ldots, n_i; \; \bar{\mu}_0 = 0, \bar{\mu}_j = \mu_j; \; \bar{\lambda}_j = \lambda_j; \\
j &= 0, 1, \ldots, n_{i-1}; \; \bar{\lambda}_{n_i} = \bar{\mu}_{n_i} = 0; \\
P_0(0) &= 1, P_j(0) = 0, j = 1, 2, \ldots, n_i.
\end{aligned} \right\}$$ (33.14)

The problem of deriving a system of bilateral estimates for the reliability function $R(t)$ on the basis of solutions of simplified as compared with Equation 33.11 Equation 33.8, Equation 33.13, and Equation 33.14 was posed and solved by Timashev and Shterenzon [26]. From a practical point of view, this problem has been motivated by the circumstance that, even at comparatively small values of m_j and n_i, the solution of Equation 33.11 requires a considerable amount of computer time.

Introduce the following relation:

$$\widetilde{P}_{i,j} = P_i^{(m_j)}(t) \cdot P_j^{(n_i)}(t).$$ (33.15)

Lemma. Consider a region Ω such that $m_j \geq m_{j+1}$; $n_i \geq n_{i+1}$; $i = 0, 1, \ldots, m_0 - 1$; $j = 0, 1, \ldots, n_0 - 1$; and for any moment of time $t > 0$

$$H(t) = \sum_{i=0}^{m_j-1} \sum_{j=0}^{n_i-1} h_{i,j}(t) \leq 0$$ (33.16)

where $h_{i,j}(t)$ are errors for an arbitrary point $(i, j) \in \Omega$ that arise from the substitution of Equation 33.15 into Equation 33.16

$$h_{i,j}(t) = \bar{\lambda}_{i-1}P_{i-1}^{(m_j)}(t)\left[P_j^{(n_i)}(t) - P_j^{(n_{i-1})}(t)\right] + \bar{\mu}_{i+1}P_{i+1}^{(m_j)}(t) \\ \left[P_j^{(n_i)}(t) - P_j^{(n_{i+1})}(t)\right] + \bar{\lambda}'_{j-1}P_{j-1}^{(n_i)}(t)\left[P_i^{(m_j)}(t) - P_i^{(m_{j-1})}(t)\right] + \bar{\mu}'_{j+1}P_{j+1}^{(n_i)}(t)\left[P_i^{(m_j)}(t) - P_i^{(m_{j+1})}(t)\right].$$ (33.17)

In the conditions of the lemma, the following three theorems permitting construction of bilateral approximates for the reliability function hold true.

Theorem 33.1 (of lower-bound estimate). *For a region Ω such that $m_j \geq m_{j+1}$; $n_i \geq n_{i+1}$; $i = 0, 1, \ldots, m_0 - 1$; and $j = 0, 1, \ldots, n_0 - 1$, there exists a time T (finite or infinite) such that for $t < T$, the reliability function $R_1(t)$ estimate calculated according to Equation 33.18*

$$R_1(t) = \sum_{i=0}^{m_j-1} \sum_{j=0}^{n_i-1} P_i^{(m_j)}(t) P_j^{(n_i)}(t) = \sum_{i=0}^{m_j-1} \sum_{j=0}^{n_i-1} \widetilde{P}_{i,j}(t)$$ (33.18)

is not larger than the true reliability function R(t), determined by Equation 33.10 and Equation 33.11, i.e.,

$$R_1(t) \le R(t). \tag{33.19}$$

Corollary 1. If the region Ω is a rectangle, i.e., $m_j = m, j = 0, 1, \ldots, n_0 - 1; n_i = n, i = 0, 1, \ldots, m_0 - 1$, then

$$R_1(t) \equiv R(t). \tag{33.20}$$

Corollary 2. On the assumptions of theorem 1,

$$R_1(t) \xrightarrow[n_i \to \infty, \, m_j \to \infty]{} R(t). \tag{33.21}$$

Theorem 33.2 (of upper-bound estimate). In the conditions of theorem 1, for any $t > 0$, there holds the relation

$$R(t) \le R_2(t) \tag{33.22}$$

where $R(t)$ is the true reliability function (Equation 33.10), and

$$R_2(t) = \sum_{i=0}^{m_j-1} \sum_{j=0}^{n_j-1} P_i^{(m_0)}(t) P_j^{(n_0)}(t). \tag{33.23}$$

Theorem 33.3 (of lower-bound estimate). In the conditions of theorem 1,

$$R_3(t) \le R_1(t). \tag{33.23a}$$

Here, $R_1(t)$ is determined according to Equation 33.18, and

$$R_3(t) = \sum_{i=0}^{m_0-1} P^{(n_i)}(t) \left[P^{(n+1)}(t) - P^{(i)}(t) \right] \tag{33.24}$$

with

$$P^{(i+1)}(t) = \sum_{k=0}^{i} P_k^{(i+1)}(t) \qquad P^{(n_i)}(t) = \sum_{j=0}^{n_j-1} P_j^{(n_i)}(t). \tag{33.25}$$

It is useful to interpret the probability $R_3(t)$ in Equation 33.24 as the sum of the products of the probability that the process $q_1(t)$ will cross the level i at least once within the time t but never cross the level $i + 1$ by the probability that the process $q_2(t)$ will *not* cross the corresponding level n_i during the time t.

Henceforth, to obtain the estimate $R_3(t)$, it suffices to know only the probabilities that the $q_1(t)$ and $q_2(t)$ processes will not outcross the arbitrary levels. But this information is given by the distribution of the maxima within the time t for the processes $q_1(t)$ and $q_2(t)$ [16, 35]. Consequently, the approach to reliability analysis of mechanical systems that are subject to several loads (on the basis of the theory of random quantities), as practiced by some researchers, possesses an accuracy equal to the $R_3(t)$ estimation accuracy.

The results obtained in theorems 1 and 2 can by generalized for the case of more than two loads, as follows. For an r-dimensional load (quality) space, Equation 33.18, Equation 33.23, and Equation 33.24 will assume the form

$$R_1(t) = \sum_{i_1=0}^{n_1-1} \sum_{i_2=0}^{n_2-1} \cdots \sum_{i_r=0}^{n_r-1} P_{i_1}^{(n_1)}(t) P_{i_2}^{(n_2)}(t) \ldots P_{i_r}^{(n_r)}(t) \tag{33.26}$$

$$n_1 = n_1(i_2, i_3, \ldots, i_r); n_2 = n_2(i_3, i_3, \ldots, i_r); \ldots, n_r = n_r(i_2, i_3, \ldots, i_r); r = 2, 3, \ldots, q \tag{33.27}$$

where q is the number of loads in the considered case.

The probability $P_{i_k}^{(n_k)}$ characterizes the component $x_k(t)$ of the process $\bar{z}(t) = \{x_1(t), \ldots, x_r(t)\}$.

$$R_2(t) = \sum_{i_1=0}^{n_1-1} \sum_{i_2=0}^{n_2-1} \cdots \sum_{i_r=0}^{n_r-1} P_{i_1}^{(n_1^0)}(t) P_{i_2}^{(n_2^0)}(t) \ldots P_{i_r}^{(n_r^0)}(t) \tag{33.28}$$

$$\left.\begin{array}{l} n_1^0 = n_1(0, 0, \ldots, 0), \ n_2^0 = n_2(0, 0, \ldots, 0), \ldots \\ n_r^0 = n_r(0, 0, \ldots, 0); \ r = 2, 3, \ldots \end{array}\right\} \tag{33.29}$$

$$R_3(t) = \sum_{i_1=0}^{n_1-1} \{[P^{(i_1+1)}(t) - P^{(i_1)}(t)] \{ \sum_{i_2=0}^{n_2-1} [P^{(i_2+1)}(t) - P^{(i_2)}(t)] \times \ldots \times \sum_{i_{r-1}=0}^{n_{r-1}-1} \{ [P^{(i_{r-1}+1)}(t) - P^{(i_{r-1})}(t)] P^{(n_r)}(t) \underbrace{\}\} \ldots \}}_{r-1} \tag{33.30}$$

$$\left.\begin{array}{l} n_2 = n_2(i_1), \ n_3 = n_3(i_1, i_2), \ldots, \\ n_r = n_r(i_1, i_2, \ldots, i_{r-1}); \ r = 2, 3, \ldots \end{array}\right\} \tag{33.31}$$

Similar to the foregoing, $P^{(k)}(t)$ specifies $x^{(k)}(t)$.

Equation 33.26 can be further specialized for the case when the last rth coordinate of the r-dimensional process is a Poisson process for which the probability $P_{i_r}^{(n_r)}(t)$ is independent of the absorbing state. The modified Equation 33.26 is of the form

$$R_1(t) = \sum_{i_r=v}^{n_r-1} P_{i_r}(t) \sum_{i_1=0}^{n_1-1} \sum_{i_2=0}^{n_2-1} \cdots \sum_{i_{r-1}=0}^{n_{i-1}-1} P_{i_1}^{(n_1)}(t) P_{i_2}^{(n_2)}(t) \ldots P_{i_{r-1}}^{(n_{r-1})}(t). \tag{33.32}$$

In practice, only some of the components of multidimensional random processes are Markov birth and death (B.A.D.) processes. In this case, the reliability function $R(t)$ is estimated by synthesizing Equation 33.26 and Equation 33.30.

Let $\mathbf{z}(t) = \{x_1(t), \ldots, x_r(t)\}$ be an r-dimensional random process, and let K coordinates of the process $\mathbf{z}(t)$ be a Markov B.A.D.-type process. Without losing generality, it can be assumed that these components are $x_{r-k-1}(t), \ldots, x_r(t)$. The expression defining the reliability function now can be written as

$$R(t) = P\{\mathbf{z}(t) \in D^{(r)} | 0 \leq t \leq T\} \tag{33.33}$$

where the superscript r indicates the dimension of the D region. Using Equation 33.26 and Equation 33.30, we obtain

$$R(t) = \tilde{R}(t) = \sum_{i_1=0}^{n_1-1} \{[P^{(i_1+1)}(t) - P^{(i_1)}(t)]\}$$

$$\times \left\{ \sum_{i_2=0}^{n_2-1} \left[P^{(i_2+1)}(t) - P^{(i_2)}(t)\right] \ldots \left\{ \sum_{i_{r-k}=0}^{n_{r-k}-1} \left[P^{(i_{r-k}+1)}(t) - P^{(i_{r-k})}(t)\right] \underbrace{\cdots}_{r-k-1} \right\} \tag{33.34}$$

$$\times P\{\mathbf{z}_k(t) \in D^{(k)} | 0 \leq t \leq T\}.$$

Here, $\mathbf{z}_k(t) = \{x_{r-k-1}(t), \ldots, x_r(t)\}$ is a K-dimensional Markov process; $D^{(k)} = D^{(k)}{}_{i_1, \ldots, i_k}$ is a K-dimensional quality space region depending on the values of $i_1, \ldots, i_{r-k}$.

The expression

$$P\{\mathbf{z}_k(t) \in D^{(k)} | 0 \le t \le T\} \tag{33.35}$$

determines the probability that the K-dimensional Markov process will never extend beyond the region $D^{(k)}$ within the time $t \in [0, T]$. The value of Equation 33.35 can be calculated using Equation 33.26. The estimate using Equation (33.34) which is a synthesis of Equation 33.26 and Equation 33.30, is somewhat more cumbersome than Equation 33.30. However, the estimate $R_3(t)$ is less accurate than $R_1(t)$. Therefore, the estimate determined by Equation 33.34 is closer to the true value of the reliability function $R(t)$ than that obtained with the use of Equation 33.30 alone.

33.1.5 Reliability of Elastic-Plastic Frames Subject to a Combination of Random Loads

Allowing for the effect of a load combination on the reliability of an element or a strut system depends on how the loads are applied. If all loads act upon an element or structure in the same way (for example, all of them are compressive or bending loads), they produce in the element(s) the same kind of stress state. In this case, the reliability problem is solved in two steps. First, the loads are statistically summed, and then the generalized-load characteristics are used when estimating the system reliability.

The case when each load in a combination produces its own kind of stressed state in an element or system is a fundamental one. This case for statically determinate and indeterminate elastic frames was solved by Timashev and Shterenzon [26]. In that work (and in all the problems presented below), the calculation of frame system reliability is carried out in three steps. First, all the internal forces in the elements of the frame are determined (depending on the design scheme chosen), and then, in strict conformity with the norms in force (in our case, USSR Building Code II-6.3-72; Steel Constructions; Design Norms), reference sections of all elements are selected. Then, frame failure criteria are selected, and the admissible load-space regions (corresponding to the criteria) that were chosen for the reference sections at the first step are constructed. Finally, the reliability of a frame designed according to the norms is calculated.

Analysis shows that taking into account the distorted scheme of statically determinate frames leads to an insignificant *decrease* in design reliability [26]. For design schemes of statically determinate frames where each load acts upon its own element, an increase in the number of loads acting upon them leads, contrary to the conventional concept, to a *decrease* in their reliability. The use of usual failure criteria (fiber yield and instability of individual elements) for sufficiently rigid statically determinate frames allows disregarding of the initial deflections. Correct allowance for the randomness of the material yield stress *increases* abruptly the design reliability of a system. For the same kind of failure the reliability function for a statically indeterminate frame is much higher than that for a statically determinate one. Increasing the number of loads acting upon a statically indeterminate frame designed according to norms leads to a considerable *increase* in its design reliability.

Now consider the reliability of elastic–plastic frames. Start with a three-times statically indeterminate frame (Figure 33.4) designed according to norms for snow and wind loads. The design scheme accounts for the compression-bending effect of struts, shear strains, and the change of the strut length due to compression. For this kind of frame, the admissible load-space region (its boundary specifies the ultimate carrying capacity of a frame subject to active load) is represented in Figure 33.5 (curve 2). Curve 1 in the same figure represents the boundary of the admissible region plotted according to the criterion of elastic behavior of the frame. The values of the conditional reliability function $R_c(n)$ (with $n = 1$ year) are equal to 0.99752 and 0.99999961, respectively, (both values being obtained for the code value of the yield stress).

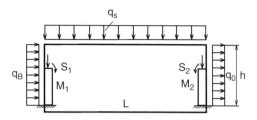

FIGURE 33.4 Computation scheme for three times statically indeterminate frame.

The unconditional reliability function is determined by integrating the conditional function over all the random strength parameters. If the admissible region is determined solely by the hazardous cross sections of one column, the unconditional reliability function is

$$R(nt) = \int_0^\infty R_c^{(n)}(t, \sigma_y) f(\sigma_y) d\sigma_y \qquad (33.36)$$

where $f(\sigma_y)$ is the PDF of the material yield stress (normal distribution with parameters $m = 270$ MPa and $s = 25$ MPa).

Figure 33.6 presents reliability function estimates constructed for the ultimate carrying capacity criterion, taking into account the randomness of the material yield stress according to Equation 33.12, Equation 33.17, and Equation 33.30. As ought to be expected, due to the modification of the failure criterion, the design reliability increases very appreciably.

Numerical calculations show that the ultimate carrying capacity of the frame is exhausted *earlier* than at least one plain hinge is formed. In this context, the criterion for the failure of frames — according to the type of their transformation into a mechanism due to the formation of a series of plastic hinges — is apparently less strict than that for the limiting carrying capacity.

33.1.6 Reliability of a Frame Subject to Active Load and Unloading

Consider the case of unloading. It may take place both when the frame is subject to active loads and when the external actions shrink the admissible region and its boundary Γ. In this format, the reliability function is no longer an unambiguous function of only the magnitude of loads, but also depends on the sequence of their application. For each loading sequence, there exists a unique admissible load-space

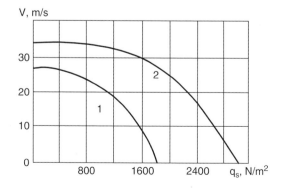

FIGURE 33.5 Admissible regions of a three times statically indeterminate frame, according to the criteria: 1-fibre yield; 2-ultimate carrying capacity.

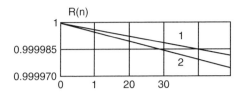

FIGURE 33.6 Reliability function for a three times statically indeterminate frame constructed according to the criterion of ultimate carrying capacity: 1-estimate from above; 2-estimate from below.

region. The only way to explore the reliability problem for similar systems with a "floating" admissible region is to use the Monte Carlo method.

We will examine the use of this method for a one-time statically indeterminate steel frame that is subject to snow and wind loads operating in elastic–plastic stage (Figure 33.7). The permissible region is constructed from the condition of the elastic work of the frame. The initial distribution of the probabilities of finding a two-dimensional Markov B.A.D.-type process outside the permissible region (whose coordinates are mutually independent one-dimensional Markov B.A.D.-type processes that simulate the snow and wind loads, see Figure 33.8) is plotted for time $t = 154$ days (the time of continuous existence of snow load in Ekaterinburg [26]). A total of 500 realizations were generated of the frame exposure to snow and wind loads *outside* the permissible region by the Monte-Carlo method. The intensities of the transition probabilities for the snow load and velocity of wind are respectively equal to [26]: $\lambda_i = \lambda = 0.160517$ 1/day; $\mu_i = i\mu$; $\mu = 0.023184$ 1/day; $\bar{\lambda}_0 = 3.184824$ 1/day; $\bar{\lambda}j = 1.772240$ 1/day; $j \geq 1$; $\bar{\mu}_j = \bar{\mu}$; $\bar{\mu} = 1.77408$ 1/day.

Studying the properties of the upcrossings of a two-dimensional B.A.D. load and the aftereffects produced by them permits estimating the reliability of the frame in consideration. The representative ensemble of loading trajectories and the corresponding results of the "response" of the frame as an elastic-plastic system are analyzed. As a result, the following histograms are constructed: the time τ for a two-dimensional Markov B.A.D. staying beyond the boundaries of the elastic behavior of the frame; the maximal deflection f of the collar beam; the maximal fiber strains ε; and the maximal residual strains ε_0.

Table 33.1 presents statistical estimates of the mathematical expectation, variance, and standard of the above four parameters. Only the time of the process staying outside the boundary of elastic behavior can be regarded as distributed exponentially (m/sec = 0.82). Typically, total fiber strains and, especially, total deflections feature a very small spread. For residual strains, conversely, the magnitude of the standard deviation is more than twice as large as the mathematical expectation.

33.1.7 On Design Values of Loads and Their Combinations

The overwhelming majority of real structures are subject to several loads during operation. It is for this reason that the central problem is to evaluate reliability under these conditions. The problem of assigning design loads and actions and their combinations arises when designing multielement structures.

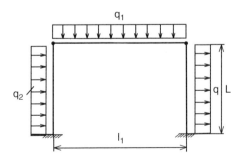

FIGURE 33.7 Design scheme for a one-time statically indeterminate frame.

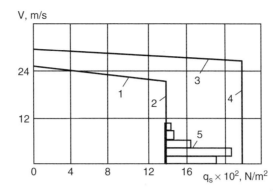

FIGURE 33.8 Permissible regions for the frame presented in Figure 33.7, according to criteria: fibre yield [Ω_1, lines (1), (2)]; ultimate carrying capacity [Ω_2, lines (3), (4)] and histogram (5) for the distribution of the process' {q_s, V} states beyond the permissible region.

The load-combination problem in terms of reliability theory in the CIS (Commonwealth of Independent States) was posed, using different approaches, by Pischikov, Rzanitsin, Baldin, and others; see for instance, Pischikov et al. [37]. The main concept is that design loads and their combinations should be assigned proceeding from the condition that the mechanical systems designed with reference to them possess optimal or normative reliability. Further, the design loads should have a clear and simple physical interpretation in order to help preserve the simple structure of existing codes.

A method of assigning design loads and their combinations is described below for mechanical systems. The method is based on in-depth analysis of the admissible region in the load space [34].

Consider an arbitrary multielement system subject to a combination of n random loads $\mathbf{q}(t) = \{q_1(t), q_2(t), \ldots, q_n(t)\}$ varying in time. The service life T and the concrete values of the intrinsic properties of the system, characterized by the vector χ_s^*, are given. The system failure criteria are also given. It is required to find the vector $\bar{\mathbf{q}}(t) = \{q_1, \ldots, q_n\}$ that ensures the optimal R_{opt} or design R^* reliability of the system designed to this load vector.

For the design reliability, one has the following equation:

$$R(t) = \varphi(\boldsymbol{\chi}, \boldsymbol{\chi}_0, \boldsymbol{\chi}_s, \mathbf{q}) = R^*. \tag{33.37}$$

The normative reliability R^* can be assigned from the following considerations. Previous investigations [26, 29] have established that the larger the number of loads acting upon structures designed according to norms, the higher is the reliability of these structures (with the design schemes and the materials used being the same). Thus, according to the norms, structures subject to one load possess the lowest reliability. Denote it by R^*.

On the other hand, the responsibility of a structure (the consequences of its failure) in general does not depend on the number of loads acting upon it. The responsibility of a structure depends only on its functional destination. Therefore, the quantity of R^* may, in its way, be regarded as a "spontaneously" inherent normative reliability, since no designed structure can have a reliability lower than this one.

TABLE 33.1 Statistical Estimates of Mean, Variance, and Standard for Four Parameters

Statistical Characteristics	Parameter Value			
	Time, τ (day)	Maximum Deflection, f (m)	Maximal Fiber Strain, ε ($\times 10^3$)	Maximal Residual Strain, ε_0 ($\times 10^4$)
Mean	5.94	0.21300	1.060	0.54
Variance	50.11	0.00017	0.013	1.28
Standard	7.08	0.01292	0.114	1.13

Since there are no logical grounds to require that systems of equal responsibility possess different reliability, one arrives at the natural conclusion that they should have the same reliability, it being possible to take R^* as this reliability. Thus, within the framework of the existing norms, it is possible to advance arguments for bringing the reliability level of equal-responsibility structures to R^*. It stands to reason that this will lead to appreciable savings in material and labor.

Having, from some considerations or other, the quantity R^*, one can find the optimal vector $\boldsymbol{\chi}_0$ value from the following optimization problem:

$$R(t) = R^*; \quad C = \psi[R(t)] = \psi[\psi(\boldsymbol{\chi}, \boldsymbol{\chi}_0, \boldsymbol{\chi}_s, \mathbf{q}, T)] \to \min_{\boldsymbol{\chi}_0}. \tag{33.38}$$

From the optimal vector $\boldsymbol{\chi}_0$ thus found, the admissible load-space region $\Omega_q(\overline{\boldsymbol{\chi}}_0, \boldsymbol{\chi}_s^*, \mathbf{u}_*)$ is plotted as the intersection of the admissible regions with respect to all the failure criteria

$$\mathbf{u}_* = H(\mathbf{q}_0, \boldsymbol{\chi}_0, \boldsymbol{\chi}_s^*) \tag{33.39}$$

where $\mathbf{u}^*$ is the maximum admissible value of the system quality vector, $\boldsymbol{\chi}_s^*$ is some fixed value of the vector of the random intrinsic properties of the system, and H is an operator inverse to the system operator. At $\boldsymbol{\chi}_s^*$, it is most natural to choose the design values of the intrinsic properties of the system (for example, the minimal yield stress, the maximum admissible initial deflections, etc.).

Taking the optimal reliability as a baseline, the latter can be found from the condition of minimizing the function of the system total cost with respect to the parameters $\boldsymbol{\chi}_0$

$$C(t) = \psi(\boldsymbol{\chi}, \boldsymbol{\chi}_0, \boldsymbol{\chi}_\sigma^*, T) = \min_{\boldsymbol{\chi}_0}. \tag{33.40}$$

Solution of Equation 33.40 gives the value of the optimal vector $\overline{\boldsymbol{\chi}}_0$. By substituting this into the dependence described in Equation 33.37, one obtains the optimal value of the reliability function

$$R_{opt} = R(t) = \varphi(\boldsymbol{\chi}, \overline{\boldsymbol{\chi}}_0, \boldsymbol{\chi}_s^*, \mathbf{q}, T). \tag{33.41}$$

Concrete values of the optimal reliability function (Equation 33.41) are obtained as the probability of a multidimensional random process reaching out beyond the admissible load-space region $\Omega_q(\boldsymbol{\chi}, \overline{\boldsymbol{\chi}}_0, \boldsymbol{\chi}_s^*)$. This region is also constructed similar to Equation 33.39.

Thus, in both cases considered, the admissible regions are plotted for optimal structures (design versions), since they possess reliability R_{opt} or R^*. The shape and dimensions of these regions depend substantially on the design scheme and the failure criterion chosen.

Denote the boundary of the admissible region by $\Gamma_q(\overline{\boldsymbol{\chi}}_0, \boldsymbol{\chi}_s^*, \mathbf{u}_*)$. In the two-dimensional case this boundary (for two types of failure criteria) is presented in Figure 33.9. Any point on the Γ_q boundary, by definition, satisfies Equation 33.39. At the same time, the coordinates of any point of the boundary are the values of the loads, i.e., of the vector $\mathbf{q}$ components that lead to the same unique optimal design version. Therefore, there exists, *in principle*, a *continuum* of load combinations that lead to the same structure possessing reliability R_{opt} (R^*).

From the above, it is clear that, generally speaking, it does not matter which point on the Γ boundary is selected to assign design loads and their combinations. However, from the point of view of understanding the essence of the problem and for correct interpretation of the results obtained, the representative-point selection procedure needs special substantiation. This can be done on the basis of the following agreement [38].

Choose in the space $\Omega_q(\overline{\boldsymbol{\chi}}_0, \boldsymbol{\chi}_s^*, \mathbf{u}_*)$ a point g whose coordinates are mean (normative) load values $g(<q_1>, \ldots, <q_n>)$ and draw through it a normal to the boundary $\Gamma_q(\overline{\boldsymbol{\chi}}_0, \boldsymbol{\chi}_s^*, \mathbf{u}_*)$. For the two-dimensional case, this situation (for each of the two failure criteria) is shown in Figure 33.9. Denote the intersection by the normal of the boundary Γ_q by r with the coordinates $[q_1^{(r)}, \ldots, q_n^{(r)}]$. Then the quantities $q_1^{(r)}, \ldots, q_n^{(r)}$

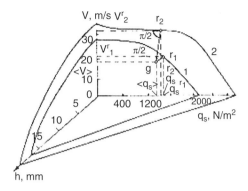

FIGURE 33.9 Admissible region for the frame presented in Figure 33.4 according to criteria: 1-fibre yield; 2-ultimate carrying capacity.

are implied to be the design values of loads $q_1, \ldots, q_n$ in a given combination (and for a given failure criterion). Now the design overload coefficients are

$$d_i = q_i^{(r)}/< q_i > \tag{33.42}$$

The combination coefficients are determined as the ratios of the design loads according to the number and character of the loads:

$$\beta_{ij} = q_{ij}^{r_j}/q_{i0}, \quad i = 1, 2, \ldots, n; \quad j = 1, 2, \ldots, m \tag{33.43}$$

where m is the number of the load combinations considered; q_{i0} is the design value of an ith load in a baseline (reference) combination; r_j is a representative point on the boundary $\Gamma_{q,j}(\overline{\chi}_0, \chi_s^*, \mathbf{u}_*)$ for a jth load combination. By a baseline combination, one understands a combination of the largest number of loads. In Equation 33.43, all the combinations are for the same failure criterion.

Equation 33.42 and Equation 33.43 can be readily generalized for the case when the vector χ_s is not fixed but is given by its distribution. Performing calculations for Equation 33.38 through Equation 33.43 for each set N (of similar structures) and averaging the results obtained

$$\alpha_* = \sum_{k=1}^{N} \alpha_i \gamma_k, \quad \beta_* = \sum_{k=1}^{N} \beta_i \gamma_k \tag{33.44}$$

(here γ_k is the "weight" of each version of a structure of a given type), one comes to the generalized values of overload and combination coefficients.

33.2 Life and Reliability Assessment of Large Mechanical Systems Subjected to a Combination of Diffusion-Type Markov Loads by Generalized Pontryagin Equations

33.2.1 Markov Loads

Examine a random process $X(t)$. If for any moments of time $t_1 < t_2 < \ldots < t_{n-1} < t_n$, then

$$f(x_n|x_1, \ldots, x_{n-1}) = f(x_n|x_{n-1}) \tag{33.45}$$

where $f(x_n|x_1, \ldots, x_{n-1})$ is the probability density function (PDF) of the random quantity x_n subject to the condition that the values of the ordinates of the random process at the moments of time $t_1, t_2, \ldots, t_{n-1}$ are known, then $X(t)$ is a Markov process.

It is known [39] that the multidimensional random Markov process $\{X_1(t), X_2(t), \ldots, X_n(t)\}$ satisfies the following system of ordinary differential equations into which random functions $\xi_j(t)$ possessing white-noise properties enter linearly:

$$dX_i/dt = \psi_i(t, X_1, \ldots, X_n) + \sum_{j=1}^{n} g_{ij}(t, X_1, \ldots, X_n)\xi_j(t), \quad i = 1, \ldots, n. \tag{33.46}$$

Here ψ_i and g_{ij} are known functions; $\xi_j(t)$ are mutually independent, and their correlation functions are of the form

$$K_{\xi,j}(\tau) = \langle \xi_j(t) \xi_j(t+\tau) \rangle = \delta(\tau).$$

Equation 33.46 permits formulation of the Focker-Planck-Kolmogorov (FPK) equation with respect to the probability density $f(t, x_1, x_2, \ldots, x_n)$ of a random event, which consists in that by the moment of time τ, the ordinates x_i will be within the intervals $x_i, x_i + dx_i$; $i = 1, 2, \ldots, n$, *without ever crossing*, during the time span (t, τ), the boundaries of some admissible region Ω, subject to the condition that at the moment of time $t < \tau$, the values of $x_i(t)$ are known (i.e., the vector $\{x_1^o, x_2^o, \ldots, x_n^o\}$ is assigned).

The FPK equation is of the form

$$\frac{\partial f}{\partial t} + \sum_{i=1}^{n} \frac{\partial}{\partial x_i}(a_i f) - \frac{1}{2}\sum_{i=1}^{n}\sum_{j=1}^{n} \frac{\partial^2}{\partial x_i \partial x_j}(b_{ij} f) = 0. \tag{33.47}$$

Here,

$$a_i(t, x_1, \ldots, x_n) = \psi_i(t, x_1, \ldots, x_n), \quad i = 1, 2, \ldots, n \tag{33.48}$$

$$b_{ij}(t, x_1, \ldots, x_n) = \sum_{k=1}^{n} g_{ik}(t, x_1, \ldots, x_n) g_{jk}(t, x_1, \ldots, x_n), \quad i, j = 1, 2, \ldots, n \tag{33.49}$$

are infinitesimal first- and second-increment moments of the vector $\{X_1(\tau), X_2(\tau), \ldots, X_n(\tau)\}$, respectively. Equation 33.47 is integrated with the following initial boundary conditions. When $\tau = t$

$$f(t, x_1^0, \ldots, x_n^0; \tau, x_1, \ldots, x_n) = \delta(x_1 - x_1^0) \ldots \delta(x_n - x_n^0). \tag{33.50}$$

Equation 33.50 reflects the fact that when $\tau = t$, ordinates x_i coincide with the values of x_i^0 $(i = 1, 2, \ldots, n)$. When $\tau > t$

$$f(t, x_1^0, \ldots, x_n^0; \tau, x_1, \ldots, x_n) = 0 \tag{33.51}$$

if the point with the coordinates $x_1, x_2, \ldots, x_n$ lies on the boundary of the region Ω.

The probability of the random vector $\{X_1(\tau), \ldots, X_n(\tau)\}$ not outcrossing the admissible region Ω during the time $T = \tau - t$, on the condition that at the initial moment of the time $\{x_1(t) = x_1^0, \ldots, x_n(t) = x_n^0\}$, is equal to

$$P(\tau; t, x_1^0, \ldots, x_n^0) = \int \ldots \int_\Omega f(t, x_1^0, \ldots, x_n^0; \tau, x_1, \ldots, x_n) dx_1, \ldots, dx_n. \tag{33.52}$$

Therefore, if a continuous random process is Markov, the problem of not outcrossing the permissible region (i.e., the reliability problem) is solved in a relatively simple manner.

In order to use the theory of Markov processes, it is necessary to (1) prove that real processes could be considered as Markov and (2) define the infinitesimal coefficients of the FPK equation, which is the cornerstone of this theory.

Diffusion Markov processes are nondifferentiable, while real load processes are always inertial and, therefore, differentiable. Because of this fact, real processes, as a rule, are not one-dimensional Markov processes. However, it is a well-established fact that the spectral density of most loads encountered in practice is sufficiently well approximated by the fraction-rational function. Therefore, by virtue of the second Doob's theorem [39], these loads can be regarded as components of a multidimensional Markov process.

Let $X(t)$ be a stationary random process whose spectral density $S(\omega)$ is a fraction-rational function

$$S(\omega) = \frac{|P_m(i\omega)|^2}{|Q_n(i\omega)|^2}, \quad m < n \tag{33.53}$$

with

$$Q_n(x) = x^n + \alpha_1 x^{n-1} + \ldots + \alpha_n \tag{33.54}$$

$$P_m(x) = \beta_0 x^m + \beta_1 x^{m-1} + \ldots + \beta_m \tag{33.55}$$

Now demonstrate that $X(t)$ can be regarded as a component of a multidimensional Markov process. It is known [40] that a process having a spectral density such as depicted in Equation 33.53 is a stationary solution to the differential equation

$$\frac{d^n X(t)}{dt^n} + \alpha_1 \frac{d^{n-1} X(t)}{dt^{n-1}} + \ldots + \alpha_n X(t) = \beta_0 \frac{d^m \xi(t)}{dt^m} + \beta_1 \frac{d^{m-1}\xi(t)}{dt^{m-1}} + \ldots + \beta_m \xi(t) \tag{33.56}$$

where $\xi(t)$ is a white-noise-type random function with the correlation function $K_\xi(\tau) = \delta(\tau)$.

Write Equation 33.56 as a system of first-order differential equations. Denoting $X(t) = X_1(t)$ and introducing $X_2(t), \ldots, X_n(t)$ according to the equations

$$\dot{X}_1(t) = X_2(t), \quad \dot{X}_{n-m}(t) = X_{n-m+1}(t) + C_{n-m}\xi(t)$$
$$\cdots \quad \cdots$$
$$\cdots \quad \cdots \tag{33.57}$$
$$\dot{X}_{n-m-1}(t) = X_{n-m}(t), \quad \dot{X}_{n-1}(t) = X_n(t) + C_{n-1}\xi(t)$$

Here $C_{n-m}, C_{n-m+1}, \ldots, C_{n-1}$ are arbitrary constants. They are selected in such a way that, upon exclusion from Equation 33.56 of the senior derivatives of functions $X(t)$ (by replacing them by the $X_i(t)$ functions according to Equation 33.57), a first-order equation is obtained, which contains no derivatives of $\xi(t)$.

Substituting Equation 33.57 into Equation 33.56 and equating the coefficients at the like derivatives of $\xi(t)$, we obtain a recurrent equation to determine the coefficients C_k:

$$C_k = \beta_{k+m-n} - \sum_{j=1}^{k+m-n} \alpha_j C_{k-j}, \quad k = n-m, n-m+1, \ldots, n \tag{33.58}$$

Equation 33.56 thus transforms into

$$\frac{dX_n}{dt} = -\sum_{j=1}^{n} \alpha_{n+1-j} X_j + C_n \xi(t) \tag{33.59}$$

and yields, jointly with Equation 33.57, a system of n first-order equations with n unknown functions $X_1, \ldots, X_n$ that contain the function $\xi(t)$ on the right-hand sides of the equations.

Equation 33.57 and Equation 33.59 represent a particular case of Equation 33.46, and using the dependences of Equation 33.48 and Equation 33.49, one obtains the infinitesimal coefficients of the FPK equation

$$a_{ij} = \begin{cases} X_{i+1} & \text{when } 1 \leq i \leq n-1 \\ -\sum_{j=1}^{n} \alpha_{n+1-j} X_j & \text{when } i = n \end{cases} \tag{33.60}$$

$$b_{ij} = \begin{cases} 0 & \text{when } 1 \leq i \leq n-m-1 \\ C_i C_j & \text{when } n-m \leq i \leq n, \ i \leq j. \end{cases} \tag{33.61}$$

Thus, it has been shown that a normal stationary process possessing a fraction-rational spectral density is a component of an n-dimensional diffusion Markov process, where $2n$ is the degree of the polynomial standing in the spectral density numerator. Proceed now to the task of constructing diffusion Markov models of specific loads that will be needed later.

33.2.1.1 Snow Load

Statistical processing of 46 years of meteorological data on snow loads performed by Timashev and Shterenzon [26, 41] shows that the correlation function of snow load can be approximated by the following expression:

$$K_s(\tau) \sigma_s^2 e^{-\alpha_s |\tau|}, \quad \alpha > 0 \tag{33.62}$$

where σ_s^2 is the process variance, and α_s is the exponent coefficient that describes behavior of the correlation function when $\tau \to \infty$.

The spectral density function (SDF) that corresponds to Equation 33.62 has the form of a fraction-rational function

$$S_s(\omega) = \frac{\sigma_s^2}{\pi} \cdot \frac{\alpha_s}{\omega^2 + \alpha_s^2}. \tag{33.63}$$

Comparing this equation with Equation 33.53 and Equation 33.54, one obtains $m = 0$, $n = 1$, $\beta_0 = \sigma \sqrt{2\alpha_s}$, and $\alpha_1 = \alpha_s$. In other words, load as a random function with the SDF of Equation 33.63 can be regarded as a diffusion Markov process. The coefficients of the FPK equation according to Equation 33.60 and Equation 33.61 will then be

$$a_1 = -\alpha_s x; \quad b_{11} = 2\sigma_s^2 \alpha_s. \tag{33.64}$$

Equation 33.47 for the snow load will take the form

$$\frac{\partial f}{\partial \tau} - \alpha_s \frac{\partial}{\partial x}(xf) - \alpha_s \sigma_s^2 \frac{\partial^2 f}{\partial x^2} = 0. \tag{33.65}$$

33.2.1.2 Wind Load

Analysis of a representative set of 37 year-long realizations of the average wind speed published in [42] shows that the wind load can be considered as a stationary Gaussian process with a correlation function that has the form

$$K_w(\tau) = \sigma_w^2 e^{-\alpha_B |\tau|} \cos \beta_w \tau, \quad \alpha > 0 \tag{33.66}$$

where β_w is the coefficient that characterizes the effective period of correlation, $T_{ef} = \pi/\beta_w$.

The SDF of the wind load is also fraction-rational and can be written in the form

$$S_w(\omega) = \frac{\alpha_w \sigma_w^2}{\pi} \frac{\omega^2 + \alpha_w^2 + \beta_w^2}{(\omega^2 - \beta_w^2 - \alpha_w^2) + 4\alpha_w^2 \omega^2} \tag{33.67}$$

$$S_w(\omega) = \frac{\alpha_w \sigma_w^2}{\pi} \frac{\left| i\omega + \sqrt{\alpha_w^2 + \beta_w^2} \right|^2}{\left| -\omega^2 + 2i\alpha_w \omega + \alpha_w^2 + \beta_w^2 \right|^2}.$$

In other words

$$n = 2, \; m = 1, \; a_1 = 2\alpha_w; \; \alpha_2 = \alpha_s^2 + \beta_s^2$$

$$\beta_0 = \sigma_w \sqrt{2\alpha_s}; \quad \beta_1 = \sigma_w \sqrt{2\alpha_w + \beta_w^2}.$$

Therefore, the wind load can be considered as a component of a two-dimensional Markov process. With this, the infinitesimal coefficients of the FPK equation, according to Equation 33.60 and Equation 33.61, are equal to

$$a_1 = x_2, \; a_2 = -(\alpha_w^2 + \beta_w^2)x_1 - 2\alpha_w x_2;$$

$$b_{11} = \sigma_w^2 2\alpha_w; \; b_{12} = \sigma_w^2 2\alpha_w \left(\sqrt{\alpha_w^2 + \beta_w^2} - 2\alpha_w \right);$$

$$b_{22} = \sigma_w^2 2\alpha_w \left(\sqrt{\alpha_w^2 + \beta_w^2} - 2\alpha_w \right)^2.$$

The FPK equation for the wind load obtains the following form:

$$\frac{\partial f}{\partial \tau} + \frac{\partial}{\partial x_1}(x_2 f) - \frac{\partial}{\partial x_2}\left\{ \left[(\alpha_w^2 + \beta_w^2)x_1 + 2\alpha_w x_2 \right] f \right\} - \alpha_w \sigma_w^2 \frac{\partial^2 f}{\partial x_1^2}$$

$$- 2\alpha_w \sigma_w^2 \left(\sqrt{\alpha_w^2 + \beta_w^2} - 2\alpha_w \right) \frac{\partial^2 f}{\partial x_1 \partial x_2} - \alpha_w \sigma_w^2 \left(\sqrt{\alpha_w^2 + \beta_w^2} - 2\alpha_w \right)^2 \frac{\partial^2 f}{\partial x_2^2} = 0. \tag{33.68}$$

33.2.1.3 Technological Loads

There are only a few statistical data on technological loads. Some simplest notions on the spectral content of technology loads can be found, for instance, in Prisedsky et al. [43].

Proceeding from the graphical form of an experimental correlation function for the output load on elements of a belt conveyer, it is approximated by the following analytical expression:

$$K_t(\tau) = \sigma_t^2 e^{-\alpha_T |\tau|} \left(\cos \beta_t \tau + \frac{\alpha_t}{\beta_t} \sin \beta_t |\tau| \right)$$

where α_t and β_t are, correspondingly, the exponent coefficient of the correlation strength and the coefficient that characterizes the periodicity of the load process. The corresponding SDF has the form

$$S_t(\omega) = \frac{2\alpha_t \sigma_t^2 (\alpha_t^2 + \beta_t^2)}{\pi\left[(\omega^2 - \alpha_t^2 - \beta_t^2)^2 + 4\alpha_t^2 \omega^2\right]}.$$

Following the general method described above, the following FPK equation can be brought into line with the above SDF:

$$\frac{\partial f}{\partial \tau} + \frac{\partial}{\partial x_1}(x_2 f) - \frac{\partial}{\partial x_2}\left\{\left[(\alpha_t^2 + \beta_t^2)x_1 + 2\alpha_t x_2\right]f\right\} - 2\alpha_t \sigma_t^2(\alpha_t^2 + \beta_t^2)\frac{\partial^2 f}{\partial x_2^2} = 0. \tag{33.69}$$

33.2.1.4 Combination of Random Loads

It was observed above that all real-life structures experience several loads simultaneously. For instance, an industrial building frame usually experiences snow, wind, and technology load, and it endures its own weight. Because of this, when solving reliability and longevity problems for real-life structures, one should use not only equations of the Equation 33.65 or Equation 33.68 type, but also FPK equations that are constructed for combinations of random loads.

In the case when acting loads are mutually independent, the FPK equations are derived simply. For instance, for the case of two snow-type loads

$$\frac{\partial f}{\partial \tau} - \alpha_1 \frac{\partial}{\partial x}(xf) - \alpha_2 \frac{\partial}{\partial y}(yf) - \alpha_1 \sigma_1^2 \frac{\partial^2 f}{\partial x^2} - \alpha_2 \sigma_2^2 \frac{\partial^2 f}{\partial y^2} = 0$$

where $f = f(\tau, x, y; t, x_0, y_0)$

When one load is of the snow type and the other is of the wind type, then

$$\frac{\partial f}{\partial \tau} - \alpha_1 \frac{\partial}{\partial x}(xf) + \frac{\partial}{\partial y}(yf) - \frac{\partial}{\partial x}\left\{\left[(\alpha_2^2 + \beta^2)y + 2\alpha_2 x\right]f\right\} - \alpha_1 \sigma_1^2 \frac{\partial^2 f}{\partial x^2} - \alpha_2 \sigma_2^2 \frac{\partial^2 f}{\partial y^2}$$

$$- 2\alpha_2 \sigma_2^2 \left(\sqrt{\alpha_2^2 + \beta^2} - 2\alpha_2\right)\frac{\partial^2 f}{\partial y \partial z} - \alpha_2 \sigma_2^2 \left(\sqrt{\alpha_2^2 + \beta^2} - 2\alpha_2\right)^2 \frac{\partial^2 f}{\partial z^2} = 0$$

where $f = f(\tau, x, y, z; t, x_0, y_0, z_0)$

If loads are dependent, the FPK equation will contain second mixed derivatives of f with infinitesimal coefficients that depend on mutual statistical characteristics of the random loads. In this chapter, only independent loads will be considered, as these are the cases most often encountered.

33.2.2 Generalized Pontryagin Equations

Now pass on to deriving a system of equations to determine the life distribution moments. The initial conditions are complicated, being in the form of δ-functions. Because of this, in the general case, fundamental difficulties arise in solving the FPK Equation 33.47 with respect to the transient probability $f(t, x_1^0, \ldots, x_n^0, \tau, x_1, \ldots, x_n) = f(t, x^0, \tau, x)$, with the initial and boundary conditions depicted in Equation 33.49 and Equation 33.51. To negotiate this complexity, the following approach will be used. Instead of Equation 33.47, use an equation that allows finding the conditional reliability $R(\tau, t, x^0)$. To this end, consider $f(t, x^0, \tau, x)$ as a function of variables $t, x_1^0, \ldots, x_n^0$. Then obtain an equation that is adjoint to Equation 33.47:

$$\frac{\partial f}{\partial t} + \sum_{i=1}^{n} a_i \frac{\partial f}{\partial x_i^0} + \frac{1}{2}\sum_{i=1}^{n}\sum_{j=1}^{n} b_{ij} \frac{\partial^2 f}{\partial x_i^0 \partial x_j^0} = 0 \tag{33.70}$$

where a_i and b_{ij} are also functions of variables $t, x_1^0, \ldots, x_n^0$.

Restricting attention to the case when the processes $X_i(t)$ ($i = 1, \ldots, n$) are stationary (a_i, b_{ij} do not depend implicitly on time), arrive at the result that the transient probability $f(t, x^0, \tau, x)$ depends only on the difference $(\tau - t)$. Therefore $\partial f/\partial t = -\partial f/\partial \tau$, and Equation 33.70 assumes the form

$$\frac{\partial f}{\partial \tau} = \sum_{i=1}^{n} a_i \frac{\partial f}{\partial x_i^o} + \frac{1}{2} \sum_{i=1}^{n} \sum_{j=1}^{n} b_{ij} \frac{\partial^2 f}{\partial x_i^o \partial x_j^o}. \tag{33.71}$$

The relation between conditional reliability and transient probability is given by Equation 33.52. Therefore, termwise integration of Equation 33.71 with respect to x yields the equation for conditional reliability

$$\frac{\partial R}{\partial \tau} = \sum_{i=1}^{n} a_i \frac{\partial R}{\partial x_i^0} + \frac{1}{2} \sum_{i=1}^{n} \sum_{j=1}^{n} b_{ij} \frac{\partial^2 R}{\partial x_i^0 \partial x_j^0} \tag{33.72}$$

with the following initial and boundary conditions:

$$R(t, x^0) = 1 \quad \text{when} \quad x^0 \in \Omega$$
$$R(\tau, x^0) = 0 \quad \text{when} \quad x^0 \in \Gamma \tag{33.73}$$

where Γ is the boundary of the region Ω.

Without loss of generality, it is possible to assume that the initial moment of time $t = 0$. In a number of cases, it is much more convenient to use Equation 33.72 instead of Equation 33.47. But even in this case, difficulties arise, especially when the dimension of a Markov process is large, when the permissible region is of a complex form, or when the coefficients of Equation 33.47 are variables of time.

The conditional reliability $R(\tau, x^0)$ is related to the conditional probability density $f(\tau, x^0)$ of the time a multidimensional random process stays (is located) in the admissible region by the equation

$$f(\tau, x^0) = -\partial R(\tau, x^0)/\partial \tau.$$

Here, the time of staying in a given region implies the conditional life of a system. Therefore, it is simpler to determine the boundary reaching time distribution moments, i.e., the system life PDF parameters

$$\langle T^k(\mathbf{x}^0) \rangle = \int_0^\infty \tau^k \hat{a}(\tau, \mathbf{x}^0) d\tau$$

or

$$\langle T^k(\mathbf{x}^0) \rangle = -\int_0^\infty \tau^k \frac{\partial R(\tau, \mathbf{x}^0)}{\partial \tau} d\tau. \tag{33.74}$$

Integration of the latter equation by parts yields

$$\langle T^k(\mathbf{x}^0) \rangle = k \int_0^\infty R(\tau, \mathbf{x}^0) \tau^{k-1} d\tau. \tag{33.75}$$

The total life distribution moments are determined as

$$\langle T^k \rangle = \int_0^\infty \langle T^k(\mathbf{x}^0) \rangle f(\mathbf{x}^0) d\mathbf{x}^0 \tag{33.76}$$

where $f(\mathbf{x}^0)$ is the probability density of the vector $\mathbf{x}$ at the initial moment of time.

Use the obtained equations. Multiplication of Equation 33.72 by $k\tau^{k-1}$ and termwise integration of this equation with respect to τ gives

$$k\int_0^\infty \tau^{k+1} \frac{\partial R}{\partial \tau} = \frac{1}{2}k \sum_{i=1}^n a_i \frac{\partial}{\partial x_i^0} \int_0^\infty \tau^{k-1} R(\tau, \mathbf{x}^0) d\tau + k \sum_{i=1}^n \sum_{j=1}^n b_{ij} \frac{\partial^2}{\partial x_i^0 \partial x_j^0} \int_0^\infty \tau^{k-1} R(\tau, \mathbf{x}^0) d\tau. \quad (33.77)$$

Allowing for Equation 33.75, obtain the following equations that provide solution for the moments of the conditional life PDF.

$$\sum_{i=1}^n \sum_{j=1}^n b_{ij} \frac{\partial^2 \langle T^k \rangle}{\partial x_i^0 \partial x_j^0} - \sum_{i=1}^n a_i \frac{\partial \langle T^k \rangle}{\partial x_i^0} = -k \langle T^{k-1} \rangle, \quad k = 1, 2, \ldots. \quad (33.78)$$

Equation 33.78 should satisfy the conditions of boundedness, continuity, and twofold differentiability within the region, and the boundary conditions

$$\langle T^k \rangle = 0 \text{ when } \mathbf{x} \in \Gamma \ (k = 1, 2, \ldots). \quad (33.79)$$

Equation 33.78 is a generalization of the well-known Pontryagin equation [44]. Indeed, when $k = 1$, Equation 33.78 convolutes into a single equation, related to average life, that was originally derived by Pontryagin [44]:

$$\frac{1}{2} \sum_{i=1}^n \sum_{j=1}^n b_{ij} \frac{\partial^2 \langle T \rangle}{\partial x_i^0 \partial x_j^0} + \sum_{i=1}^n a_i \frac{\partial^2 \langle T \rangle}{\partial x_i^0} = -1. \quad (33.80)$$

Having solved these equations using one method or the other, i.e., having determined $\langle T(x_1^0, \ldots, x_n^0) \rangle$, one can find in a similar way the second-order moment from Equation 33.78, with $k = 2$, by substituting the already known solution $\langle T(x_1^0, \ldots, x_n^0) \rangle$ into its right-hand side, and then find the third-order moment, etc.

33.2.3 Methods of Downsizing the Dimension of the Reliability Problem

A fundamental feature of the problem considered above is its multidimensionality. The dimension of the space in which the problem is solved is

$$N = \sum_{i=1}^n K_{ij} \quad (33.81)$$

where n is the number of Markov loads acting upon a system, and K_{ij} is the number of components in the ith Markov load.

As the number of loads increases, the problem becomes computationally vast. At the same time, the case of a large number of loads is of maximal practical interest. Therefore, it pays off to lower the dimension of the initial problem. This can be achieved by:

- Decreasing the number of acting loads by combining them into groups possessing the same statistical properties
- Simplifying the statistical nature of some loads (reducing them to purely Markov ones)
- Using computation methods that lower the dimension of the initial problem (the Kantorovitch method, the Vlasov method, or the straight-lines method)

It is noteworthy that, in contrast to the FPK equations, the use of generalized Pontryagin equations lowers at once the dimension of the problem by unity, due to the exclusion of time as a variable.

33.2.4 Transition from System Life to System Reliability

To find the conditional life distribution function from its known n moments, make use of the maximum-entropy principle. According to this principle, the *least predetermined* distribution function is the one that brings maximum to the following functional:

$$H = -\int_0^\infty f(\xi) \ln f(\xi) d\xi. \tag{33.82}$$

Proceeding from Equation 33.82, the problem is formulated as follows. It is necessary to find

$$\max_f \left\{ -\int_0^\infty f(\xi) \ln f(\xi) d\xi \right\} \tag{33.83}$$

for the conditions

$$\int_0^\infty \xi^k f(\xi) d\xi = \langle g^k \rangle, \quad k = 0, 1, 2, \ldots \tag{33.84}$$

where $\langle g^k \rangle$ stands for known probability density $f(\xi)$ moments.

In solving for Equation 33.83 and Equation 33.84 by the Lagrange indeterminate multipliers' method, one comes to the definition of the unconditional extreme value of the functional (at $k = 4$)

$$I = \int_0^\infty [-f(\xi) \ln f(\xi) + \lambda_0 f(\xi) + \lambda_1 \xi f(\xi) + \lambda_2 \xi^2 f(\xi) + \lambda_3 \xi^3 f(\xi) + \lambda_4 \xi^4 f(\xi)] d\xi. \tag{33.85}$$

The Euler equation for the functional depicted in Equation 33.85 is

$$dI/df = -\ell n f(\xi) - 1 + \lambda_0 + \lambda_1 \xi + \lambda_2 \xi^2 + \lambda_3 \xi^3 + \lambda_4 \xi^4 = 0$$

or

$$f(\xi) = \exp[-1 + \lambda_0 + \lambda_1 \xi + \lambda_2 \xi^2 + \lambda_3 \xi^3 + \lambda_4 \xi^4] = 0. \tag{33.86}$$

The entropy for the PDF in Equation 33.86 can be calculated using the following equation:

$$\max H = 1 - \lambda_0 - \lambda_1 \langle g_1 \rangle - \ldots - \lambda_m \langle g_m \rangle, \quad g_k = \langle g^k \rangle. \tag{33.87}$$

The functional conditional extremum problem has thus been reduced to the nonconditional extremum problem: to the function depicted in Equation 33.86 or, in other words, to solving the following system of integral equations:

$$F_i = e^{(\lambda_0 - 1)} \int_0^\infty \xi^k e^{(\lambda_1 \xi + \lambda_2 \xi^2 + \lambda_3 \xi^3 + \lambda_4 \xi^4)} d\xi - \langle g^k \rangle = 0; \quad k = 0, \ldots, 4. \tag{33.88}$$

After the roots of the system λ_i have been found, the conditional life function can be represented as the dependence expressed in Equation 33.86, where $\xi = \xi(x_0, y_0)$.

In order to obtain the unconditional life function, it is necessary to integrate the $f(\xi, \mathbf{x}_0)$ function with the weight $g(\mathbf{x}^0)$, where $g(\mathbf{x}^0)$ is the density of the probability of initial load distribution over the permissible region of the system Ω:

$$T(\xi) = \int_\Omega \cdots \int f(\xi, x_1^0, \ldots, x_N^o) dx_1 \cdots dx_N. \tag{33.89}$$

Practical calculations require knowledge of the density $g(\mathbf{x}^0)$. It can be found from data of actual measurements of "initial" loads (if they have a beginning in time), or it can be taken from physical considerations.

Having the unconditional life distribution function, it is simple to obtain the overall reliability function as

$$R(t) = 1 - \int_0^t T(\xi) d\xi. \tag{33.90}$$

The lines of argument adduced above hold for the cases when the number of loads does not vary during the entire period of operation. For most of the mechanical systems, however, the number of load combinations, as a rule, varies. To evaluate the reliability of structures in these cases, the following technique should be used. Introduce the following notation. Let n be the number of system operating cycles and r the number of combinations of loads acting within a cycle. Find separately the life functions for all of the r load combinations, and after that, seek the reliability at the end of the interval of action of the load combination that occurs first in the operation cycle

$$R(t_1) = 1 - \int_0^{t_1} T_1(s) ds \tag{33.91}$$

and then the reliability for the (chronologically) second combination, etc.:

$$R_2(t_2) = 1 - \int_{t_1}^{t_2} T_2(s) ds, \ldots, R_r(t_r) = 1 - \int_{t_{r-1}}^{t_r} T_r(s) ds. \tag{33.92}$$

The reliability of a structure within one cycle is

$$R(1) = R_1(t_1) R_2(t_2) \ldots R_r(t_r) \tag{33.93}$$

and the reliability function throughout the lifetime (n cycles) is

$$R_n = [R_1(t_1) \ldots R_r(t_r)]^n. \tag{33.94}$$

The distributions of material strength σ and geometric characteristics x of the cross sections of the system elements should be taken into account using the expression for the overall life moments

$$\langle T^k \rangle = \int_0^\infty \int_0^\infty \langle T^k(\sigma, x) \rangle f(\sigma, x) d\sigma dx \tag{33.95}$$

and, employing the procedure depicted in Equation 33.86 through Equation 33.90, one can calculate the overall life and reliability function.

33.3 Examples

33.3.1 Case 1

The method described above was used to calculate the reliability of a three-story single-span frame of height $H = 12$ m (each story is 4 m high) with a span of $\ell = 6$ m. The frame is subject only to snow load that is regarded as a one-component diffusion Markov process. The snow load has the mathematical expectation $\langle q_s \rangle = 6$ kN/m, the standard $\sigma_{q_s} = 1.2$ kN/m, and the attenuation factor $\alpha_s = 0.3 \cdot 10^{-6}$/sec. The frame is designed of grade M-400 concrete with fulcrum strength $R_{fs} = 0.17 \cdot 10^5$ kN/m², and of class A-III reinforced steel with a design resistance $R_a = R_{ac} = 0.34 \cdot 10^6$ kN/m². The geometric dimensions of the sections are standard ones: the protective layer $d = 0.03$ m; the width $b = 0.3$ m; $h = n \cdot 0.3$ m, where $n = 1$ for the upper struts, $n = 1.44$ for the lower columns, and $n = 1.26$ for all the other elements. The problem is solved using the Bubnov variational method and the finite differences method.

The results of calculus give the following dependence of frame reliability (with t being equal to 1 year) on the size of the admissible region (δ_1 is its upper boundary):

δ_1, kN/m	4.8	6.0	7.2	8.4	9.6	10.8	13.0
$R(1)$	0.171	0.594	0.790	0.939	0.987	0.998	0.9997

33.3.2 Case 2

Assess reliability for a hinge-supported rectangular in-plane cylindrical shell that has random initial median-surface form irregularities. The shell endures a uniformly distributed transverse load $q(x, y)$ (snow) and a tangential load $p(x)$ (wind) that is applied to two opposite sides of the contour. Both loads are diffusion Markov processes. The described shell can fail in two ways: it can buckle, or its deflection can exceed some ultimate permissible value.

For the case of buckling, the criterion of failure will be: $q(p) > q_k(p)$. For this kind of shell, buckling can occur only when $\xi_0 < \bar{\xi}_0$, where $\bar{\xi}_0$ is the limiting value of the initial-deflection amplitude at which the shell looses its ability to snap (buckle). This type of failure occurs when the load vector $\mathbf{Q}(p, q)$ outcrosses the admissible region Ω_1 constructed for the case of buckling.

The second failure criterion has the form: $\xi \geq \bar{\xi}$. In other words, failure of the system occurs if the deflection ξ exceeds some maximal admissible deflection $\bar{\xi}$. This kind of failure takes place when $\xi_0 \geq \bar{\xi}_0$. Failure occurs when the load vector $\mathbf{Q}(p, q)$ outcrosses the admissible region Ω_2 constructed for the case when $\xi_0 > \bar{\xi}_0$.

Assuming both kinds of failure to be equivalent, for the unconditional reliability function, one has

$$R(t) = \int_0^{\bar{\xi}_0} P_1(t, \xi_0) f(\xi_0) d\xi_0 + \int_{\bar{\xi}_0}^{\infty} P_2(t, \xi_0) f(\xi_0) d\xi_0 \tag{33.96}$$

where $f(\xi_0)$ is the PDF of the initial deflection ξ_0.

The numerical calculations are performed for a square-plane cylindrical panel ($k_x = 0$) with Poisson ratio $\mu = 0.3$; $k^* = k_y a^2/h = 24, 25, 26,$ and 27; and $h/a = 1/210$. Here, h is shell thickness, a is shell size, $k_y = 1/R_y$, and R_y is shell curvature radius. The PDF $f(\xi_0)$ is taken in the form of Rayleigh law. Results of calculus show that the reliability function is highly sensitive to an increase of the curvature parameter k^*: for $k^* = 24$ $R(t) = 0.74275$; for $k^* = 27$ $R(t) = 0.98767$ (for $t = 20$ years). For details, see Timashev [38].

33.4 Conclusion

This chapter described methods for solving the central problem of structural reliability in the load space. The proposed methods permit the following:

- Explicit accounting for time as a design parameter
- Dramatic decrease in the dimension of the problem for multielement structural systems
- Accounting for all nonlinear behavior of various structures and systems, as they are described in codes and direct comprehensive, sophisticated numerical analysis

The engineering agreement on how to define the unique design point in the space of loads *totally eliminates* the problem of and necessity for finding the minimum design point as a solution of a constrained optimization problem.

The methods of assessing structural reliability using Markov processes of time permits accounting for all basic random parameters that influence the overall reliability: scatter of the initial value of loads, material strength, overall geometry and cross section of structural elements, etc.

The method of constructing the PDF through its moments, based on the variational principle of maximum entropy, permits obtaining the least predetermined life distribution function and the reliability function with the greatest safety.

References

1. Kiureghian, A.D. and Lin, P.L., Optimization algorithms for structural reliability, *Structural Safety*, 9, 161–177, 1991.
2. Ditlevsen, O. and Bjeroger, P., Methods of structural system reliability, *Structural Safety*, 3, 195–230, 1986.
3. Madsen, H.O., Load Models and Load Combinations, Ph.D. thesis, University of Denmark, 1979.
4. Rzhanicyn, A.R., Snarskis, B.J., and Sukhov, Yu.D., Basis of probability-economic design method for engineering structures, *Stroitelnaya Mekhanika i Raschet Sooruzheniy*, 3, 67–71, 1979 (in Russian).
5. Timashev, S.A., *Recommendations on Reliability Assessment of Structures*, Sverdlovsk, Uralpromstroyniiproekt, 1974, p. 102 (in Russian).
6. Grigoriu, M., Estimates of extreme winds from short records, *ASCE J. Structural Eng.*, 117 (2), 375–390, 1991.
7. Pasternak, B., Rozmarynowsky, B., and Wen, Y.-K., Crane load modeling, *Structural Safety*, 17 (4), 205–224, 1996.
8. McGuire, R.K. and Cornell, C.A., Live Load Effects in Office Buildings, research report R73-28, Dept. of Civil Engineering, MIT, Cambridge, MA, 1973.
9. Stewart, M.G., Optimization of serviceability load combinations for structural steel beam design, *Structural Safety*, 18 (2/3), 78–89, 1997.
10. Steinberg, D. and Bulleit, W., Reliability analysis of meteoroid loading in lunar structures, *Structural Safety*, 15 (1-2), 51–66, 1994.
11. Ellingwood, B. et al., Probability-based load criteria: load factors and load combinations, *J. Struct. Div.*, 108 (5), 978–997, 1982.
12. Galambos, T.V. et al., Probability-based load criteria: assessment of current practice, *J. Struct. Div.*, 108 (5), 959–977, 1982.
13. ANSI A 58.1, Minimum Design Loads for Buildings and Other Structures, *Am. Nat. Inst. Stand.*, New York, 1982.
14. Rackwitz, R., Implementation of probabilistic concepts in design and organizational codes, in *Structural Safety and Reliability* (Proc. ICOSSAR), Elsevier, New York, 1981, pp. 593–614.
15. Kanda, J. and Ellingwood, B., Formulation of load factors based on optimum reliability, *Structural Safety*, 9 (3), 197–210, 1991.

16. Barucha-Rheid, A.G., *Elements of Theory of Markov Processes and Their Applications*, Nauka, Moscow, 1969.
17. Ellingwood, B., Probability-based codified design: past accomplishments and future challenges, *Structural Safety*, 13, 159–176, 1994.
18. Mrazik, I. and Krizma, M., Probability-based design standards of structures, *Structural Safety*, 19 (2), 219–234, 1997.
19. Manners, A., First-order reliability method for certain system and load combination calculation, *Structural Safety*, 6, 39–51, 1989.
20. Crespo-Minguillon, F. and Casas, J.R., A comprehensive traffic load model for bridge safety checking, *Structural Safety*, 19 (4), 339–359, 1997.
21. Ghosn, M. and Moses, F., Markov renewal model for maximum bridge loading, *ASCE J. Eng. Mech.*, 111 (9), 1093–1104, 1985.
22. Sniady, P., Sieniawska, R., and Zukowski, S., A train of pulses in load modelling, Wroclaw, Poland, *Structural Safety*, 13 (1-2), 29–44, 1993.
23. Wen, Y.K., Reliability-based design under multiple loads, Urbana, IL, U.S., *Structural Safety*, 13 (1-2), 3–20, 1993.
24. Nowak, A.S., Live load model for highway bridges, Ann Arbor, MI, U.S., *Structural Safety*, 13 (1-2), 53–66, 1993.
25. Rackwitz, R., On the combination of non stationary rectangular wave renewal processes, *Structural Safety*, 13 (1-2), 21–28, 1993.
26. Timashev, S.A. and Shterenzon, V.A., Engineering methods of reliability calculus of mechanical systems subjected to several random loads, in *Computer Research in Structural Design*, Lenpromstroyproekt, Leningrad, 1979, pp. 36–51 (in Russian).
27. Timashev, S.A., Reliability Analysis of Redundant Structures Subject to Combination of Markovian Type Loads, paper presented at 4th International Conference on Stochastic Dynamics, Notre Dame University, South Bend, IN, 1996.
28. Timashev, S.A., Estimation of Reliability of Large Mechanical Systems by Generalized Pontryagin Equations, paper presented at 4th International Conference on Structural Safety and Reliability (ICOSSAR), Kobe, Japan, 1985.
29. Timashev, S.A. and Kantor, S.L., Reliability assessment of engineering systems by generalized Pontryagin equations, in *Computer Research in Structural Design*, Lenpromstroyproekt, Leningrad, 1979, pp. 59–76 (in Russian).
30. Murzewsky, J.M., Combination of actions for codified design, Krakow, Poland, *Structural Safety*, 13 (1-2), 113–136, 1993.
31. Oslund, L., Load combination codes, Lund, Sweden, *Structural Safety*, 13 (1-2), 83–92, 1993.
32. Shiraki, W., Probabilistic load combinations for steel piers at ultimate limit states, Tottori, Japan, *Structural Safety*, 13 (1-2), 67–82, 1993.
33. Low, B.K. and Tang, W.H., Reliability analysis using object-oriented constrained optimisation, *Structural Safety*, 26 (1), 69–90, 2004.
34. Timashev, S.A., System approach to reliability assessment of mechanical systems, in *Computer Research in Structural Design*, Lenpromstroyproekt, Leningrad, 1979, pp. 5–24 (in Russian).
35. Gnedenko, B.V., Belyaev, Yu.K., and Solovyev, A.D., *Mathematical Methods in Reliability Theory*, Nauka, Moscow, 1965, p. 524, (in Russian).
36. Ellingwood, B. et al., Probability-based load criteria: load factors and load combinations, *J. Struct. Div.*, 108 (5), 978–997, 1982.
37. Pischikov, A.F. et al., Load combinations in *Transactions of TSNIISK*, Central Research Institute of Civil Structures, Moscow, 37–46, 1968 (in Russian).
38. Timashev, S.A., *Reliability of Large Mechanical Systems*, Nauka, Moscow, 1982 (in Russian); SEAG, PAVIA, 1984 (transl. to English).
39. Doob, G.L., *Random Processes*, IL-Inostrannaya Literatura, Moscow, 1956 (transl. to Russian).

40. Sveshnikov, A.A., *Applied Methods of Random Processes Theory*, Nauka, Moscow, 1968 (in Russian).
41. Timashev, S.A. and Shterenzon, V.A., Presentation of Snow Loads as a Birth-and-Death Type Process, presented at IV All-Union Conference on Reliability Problems in Structural Mechanics, Vilnius, Lithuania, 1975 (in Russian).
42. Timashev, S.A., and Shterenzon V.A., Presentation of wind speed as a homogeneous Markov process of the birth-and-death type, in *Reliability Problems of Reinforced Concrete Structures*, pp. 142–160, Kuybyshev Institute of Civil Engineering, Kuybyshev, 1977 (in Russian).
43. Prisedsky, G.V. et al., Formation of output loads in belt conveyer elements, in *Mine and Quarry Transport*, Nauka, Moscow, 1977 (in Russian).
44. Andronov, A.A., Pontryagin, L.S., and Vitt, A.O., On statistical consideration of dynamic system, in *Andronov A.A. Collected Papers*, Izd-vo AN SSSR, Moscow, 1956.

34
Applications of Reliability-Based Design Optimization

34.1	Introduction	34-1
34.2	Overview of Reliability-Analysis Methods	34-2
	Model Approximation Methods • Reliability Approximation Methods	
34.3	Review of RBDO Methods	34-5
34.4	RBDO Applications	34-5
34.5	Selected Aerospace Application Examples	34-6
	Shape Optimization of an Axial Compressor Blade • Shape Optimization of an Airplane Wing • Optimization of an Integral Airframe Structure Step Lap Joint • Transport Aircraft Wing Optimization	
34.6	Conclusions	34-21

Robert H. Sues
Applied Research Associates, Inc.

Youngwon Shin
Applied Research Associates, Inc.

(Justin) Y.-T. Wu
Applied Research Associates, Inc.

34.1 Introduction

Physics-based modeling combined with computer-aided design has increasingly become widely accepted by the design community to reduce product design and development time as well as testing requirements. To ensure high reliability and safety, uncertainties inherent to or encountered by the product during the entire life cycle must be considered and treated in the design process.

In general, there are three approaches to incorporate uncertainties into engineering design: reliability-based design, robust design, and safety-factor-based design. The RBDO (reliability-based design optimization) approach adopts a probability-based design optimization framework to ensure high reliability and safety. A typical formulation for reliability-based design is:

$$\text{Minimize: } F(\boldsymbol{d})$$
$$\text{Subject to: } P[g_j(X, \boldsymbol{d}) > 0] \geq R_j^*; \quad j = 1, J \qquad (34.1)$$
$$d_k^l \leq d_k \leq d_k^u; \quad k = 1, K$$

in which $F(\boldsymbol{d})$ is an objective function such as weight or expected life-cycle cost; $X(i=1, n)$ is a random variable vector; $\boldsymbol{d}$ is a design (or decision) variable vector with lower bound $\boldsymbol{d}^l$ and upper bound $\boldsymbol{d}^u$; $g_j(X, \boldsymbol{d}) = 0$ defines design-limit states, with each $g_j(X, \boldsymbol{d}) > 0$ corresponding to a successful event associated with multiple failure modes; and R_j^* depicts target reliabilities (i.e., 1 − allowable failure probabilities). The multiple failure events can be statistically correlated due to common random variables. A more general system-reliability definition requires the use of union and intersection events, e.g., $R = 1 - P\{[(g_1 \leq 0) \cup (g_2 \leq 0)] \cap [(g_3 \leq 0) \cup (g_4 \leq 0)]\}$.

Another useful formulation is to maximize reliability with resource constraints. A general RBDO formulation involves deterministic and probabilistic functions of multiple objective functions and multiple equality and inequality constraints. Design variables may be deterministic, e.g., tightly controlled geometry, or associated with random variables such as the mean value of a random variable.

The objective of robust design is to develop designs that are insensitive to input variations. An example formulation for robust design is:

$$\text{Minimize: } \frac{d(\sigma_{\text{performance}})}{d(\sigma_i)} \tag{34.2}$$
$$\text{Subject to: } d_k^l \leq d_k \leq d_k^u; \quad k = 1, K$$

While the concept recognizes and treats the uncertainties, it focuses on minimizing the performance variation, e.g., in the form of standard deviation, and does not explicitly address reliability requirements. The safety-factor approach treats uncertainties by adding safety/reliability margins in the design requirements and by attaching safety factors to the nominal values of the uncertainty variables. Even though the RBDO approach is more comprehensive and promises lower life-cycle cost, it has yet to become as widely used as the other two approaches, mainly because it requires more sophisticated analyses and skills with probabilistic analysis. RBDO may become more widely used when the methods are standardized and more commercial-grade, easy-to-use, RBDO design tools become available in the near future.

In addition to the need to generate probabilistic models, probabilistic analysis in RBDO generally requires a relatively larger number of deterministic analyses that may involve highly complicated and computationally time-intensive numerical models, such as finite element and CFD (Computational Fluid Dynamics) models. Fortunately, with recent advances in computational mechanics and the remarkable increase in computational power, including the use of parallel computing, high-fidelity numerical modeling thought to be infeasible a decade ago is now available on the engineer's personal computer. Moreover, a large amount of research work on the development of efficient reliability calculation algorithms has greatly increased the efficiency of the overall RBDO algorithms.

The remainder of this chapter is organized as follows. Section 34.2 provides an overview of the reliability-analysis methods, focusing on efficient approximation methods, followed by an overview of the RBDO formulation and solution strategies in Section 34.3. Section 34.4 gives an overview of RBDO applications. Section 34.5 presents several detailed RBDO application examples based on aerospace structural systems. Finally, the conclusions are given in Section 34.6.

34.2 Overview of Reliability-Analysis Methods

A straightforward approach to solve a RBDO problem, such as the formulation in Equation 34.1, is to conduct a double-loop optimization process in which the outer loop iteratively selects feasible designs that approach the minimum objective, while the inner loop evaluates reliability constraints for each selected design. However, for complicated g-functions and objective functions, the repeated inner-loop reliability analysis can cause the RBDO to be prohibitively time consuming.

The following subsections will provide a brief overview of reliability-analysis methods, focusing on efficient analysis strategies to reduce the computational burden of probabilistic analyses within the RBDO analysis framework. Section 34.3 will review the overall RBDO analysis strategy.

34.2.1 Model Approximation Methods

An increasingly popular approach for RBDO is to replace the computationally time-intensive models, using instead approximate, fast-running models that allow the use of Monte Carlo simulation for the entire RBDO analysis, thus reducing the user's need to learn advanced computational probabilistic analysis methods.

There are a wide variety of model approximation methods, including response surfaces, neural networks, and Taylor's series expansion approaches as well as various extensions or variations of these methods. For example, the versions of the response-surface approach include global, local (or moving), and hybrid global–local approximations, and Taylor's expansion method includes adding higher-order terms based on a few additional calculation points. Another useful and generic technique is to employ variable transformation methods to make the original function less nonlinear in the transformed variable space, before applying approximation methods.

Using the response-surface method, typically a design-of-experiment (DOE) approach is selected to define the layouts of the model-calculation points. There are many DOE methods to choose from, including Box-Benkehn, second-order central composite design, and factorial designs [1,2]. The selection of the method depends on several factors, including the number of variables, the nonlinear behavior of the model, and the acceptable modeling errors/residuals. The selection of the DOE points depends on the ranges of the input variables as well as the range of the responses that are of interest to designers.

Since the approximation model is built based on the selected points, the errors tend to be small only in a limited region and potentially large in the regions of extrapolation, especially if the model is highly nonlinear relative to the approximation model. It is therefore essential to conduct goodness-of-fit tests and check the model adequacy, including the use of residual plots to compare the exact and the approximate models at selected error-checking points. In addition, after the optimal design has been obtained using the approximate model, it would be useful to conduct a model update to confirm or improve the result.

As an example of using approximation models, Fu et al. [3] evaluated several response-surface models for car crashworthiness studies and applied a second-order polynomial regression model and moving least-squares regression for crash safety design optimization. Li [4] evaluated polynomial models of different orders and found that the second-order polynomial response surface was sufficient for many crash safety applications.

34.2.2 Reliability Approximation Methods

The objective of the reliability approximation methods is to efficiently produce accurate reliability results without using time-consuming Monte Carlo (MC)-type simulations. Typically, the efficiency is measured in terms of the number of times the g-function needs to be computed to generate the result. Many approximation methods, summarized below, are based on the most-probable-point (MPP) concept and focused on calculating the probability of failure (= 1 − reliability). For more details, see [5,6].

The probability of failure can be expressed as:

$$p_f = \int_\Omega \cdots \int f_X(x) dx \quad (34.3)$$

in which $f_X(x)$ is the joint probability density function (PDF), and the integration region is the failure domain, defined as $\Omega = \{x \mid g(x) \leq 0\}$.

The MC method computes p_f by drawing random samples of X and using the following formulation:

$$p_f = \int_\Omega \cdots \int I(x) f_X(x) dx = E[I(x)] = \frac{1}{K}\left[\sum_{i=1}^{K} I_i\right] \quad (34.4)$$

where $I(x)$ is the indicator function (i.e., $I = 1$ if $g(x) \leq 0$, otherwise $I = 0$) and K is the number of samples. This procedure is straightforward and highly robust. The drawback is that the number of samples must be large to reduce the variance of the p_f estimate. In safety-critical systems where the probability of failure must be very small, a large number of samples is typically required. Efficiency issues prompted the development of reliability approximation methods.

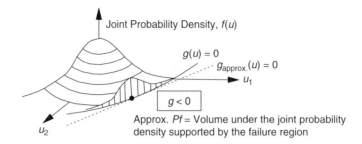

- In the standardized-normal space, find the most-probably-point (MPP) on the limit state that has the highest probability of occurrence
- Replace complicated g-function by simple g-function around the MPP
- Compute probability-of-failure using the simplified g-function
- Analytical local probabilistic sensitivities also available
- Probability can be updated by local sampling around MPP

FIGURE 34.1 The concept of MPP-based approximation.

The majority of the reliability approximation methods use the model approximation methods reviewed above combined with the idea of minimizing the model approximation errors in the high-risk region. These approximation methods are thus local, i.e., they focus around a point called the most probable point (MPP).

The standard MPP approach first transforms the original random variables to independent standard normal, or Gaussian, variables. As illustrated in Figure 34.1, in the transformed standard normal (reduced) u-space, MPP is the minimum distance point from the origin to the $g = 0$ surface. The minimum distance β is called the *safety index*, and the MPP is also called the β-point or the *design point*.

A linear or quadratic approximate model is typically developed using the tangent and curvatures at the MPP or using MPP-centered DOE and regression analysis. The approximation can be developed in the original X-space or the transformed u-space, depending on which one provides a better approximation. The widely used first-order reliability method (FORM) solution is based on the tangent surface in the u-space with the corresponding failure probability of:

$$p_f = \Phi(-\beta) \tag{34.5}$$

which provides a first-order estimate if the original limit-state surface is not significantly nonlinear. A second-order version, SORM, also in the u-space, is available by adding the curvature information [5–7]. The FORM and SORM approaches have been extended to time-variant reliability [8] and to system-reliability problems involving multiple limit states and multiple MPPs [5].

Searching for the MPP is a constrained optimization problem where the objective is to minimize the distance subject to $g = 0$. Several tailored methods have been developed to speed up the search [5, 6, 9]. It should be cautioned, however, that multiple MPPs may exist. In such cases, not only the global minimum distance point must be correctly identified [10, 11], but the accuracy of using a single MPP-based approximation should also be questioned. To help ensure that the correct MPP has been found and the p_f estimate is reasonably accurate, an error-checking procedure, such as applying the importance-sampling methods to update the p_f, should be considered [10, 12–15].

As a rule of thumb, assuming that the derivatives are computed by numerical differentiation, experience suggests that the required number of g-function calculations is on the order of five times $(n + 1)$ for FORM. For complicated models, the computation cost of the MPP approach depends on how many g-function calculations are needed to located the MPP and how many additional runs are needed to develop approximate models. For problems with a large number of variables, variable screening methods should be considered to reduce the problem dimension [3, 16].

Based on the above safety-index concept, a generalized safety index has been proposed using:

$$\beta_G = \Phi^{-1}(-p_f) \tag{34.6}$$

where p_f is the probability of failure calculated from any methods. In RBDO, sometimes the target reliabilities are defined using β or β_G.

34.3 Review of RBDO Methods

Even with the use of the above efficient methods, the double-loop procedure still demands a significant number of g-function evaluations, and the computation can be prohibitive when each function evaluation is computationally intensive. As a result, many approximate RBDO methods have been developed. Most of these methods are built on the MPP concept, the response-surface methodology, or a combination of both.

The most popular approach is perhaps to develop global response-surface models of the g-functions. With this approach, the reliability analysis can be conducted quickly using the methods in the previous section. While this approach is attractive, it is limited to well-behaved functions that can be well approximated by low-order polynomial regression models. More complicated response-surface models using multiple local surfaces may be more effective for highly nonlinear models, but the initial costs to develop the models could be much higher. It is a good practice to check the models using the methods mentioned in Section 34.2.1.

Yang and Gu [17] have reviewed several approaches that convert double loop to a single loop, including the single-loop–single-variable method by Chen et al. [18], the safety-factor approach by Wu et al. [19, 20], and the sequential optimization and reliability assessment (SORA) method by Du and Chen [21]. Another efficient single-loop approach based on adaptive response surfaces was developed by Sues et al. [22–24] and is described in Section 34.5. All of these methods simplify the reliability calculations.

34.4 RBDO Applications

Numerous examples of RBDO applications are available in the literature. An excellent reference that reviews such applications with emphasis on civil and aerospace structures is by Frangopol and Maute [25]. This section will provide additional examples of RBDO applications in the civil, aerospace, and mechanical engineering areas.

Davidson et al. [26] applied RBDO to the minimum weight designs of earthquake-resistant structures subjected to system-reliability constraints considering the uncertainties on the earthquake response spectra, the sizing design variables, and the material properties. Comparison was made between the optimal designs based on probabilistic response spectra and the design based on deterministic spectra that correspond to particular levels of the probability of system failure.

Feng and Moses [27] considered system-reliability constraints of both the damaged and original intact systems for statically indeterminate structures in which redundancy is added to increase the structural safety. Frames of several different configurations were optimized with sizing design variables.

Yang and Nikolaidis [28] applied RBDO to a preliminary design of the wing of a small commuter airplane subjected to gust loads that were modeled using probabilistic distributions. The system-reliability constraint considered the uncertainty in the material strengths, and the weight of the wing was optimized with the sizing design variables. The difference in the behavior of the reliability-based and the deterministic optimal designs were studied.

Pu et al. [29] applied RBDO to a typical frame of a small-waterplane-area twin-hull (SWATH) ship subjected to system-reliability constraints on structural failure criteria considering uncertainties on the loads and material strength. The influence of each sizing design variable to the system reliability was accessed by sensitivity information during RBDO, which recommended the most efficient way to improve the safety by identifying the critical design variable.

Stroud et al. [30] considered the probabilistic constraints of the material strength and the flutter speed in the design of a platelike wing. Finite element analysis and the doublet-lattice method were used to calculate the stresses and the aerodynamic loads for the flutter analysis under the presence of uncertainties in the thickness, flutter speed, load, and material strength. The study showed that improvement in the reliability was obtained in reliability-based optimal design with a small weight increase relative to the deterministic optimal design.

In the design of an intermediate complexity wing model, Pettit and Grandhi [31] included the aileron-effectiveness constraint in addition to the constraints on the wingtip displacement and the natural frequencies. With the consideration of uncertainties on the sizing design variables, reliability-based optimal design produced a final design that satisfied the target reliabilities with a small weight penalty added to the deterministic optimal design.

Grandhi and Wang [32] applied optimization to minimize the weight of a twisted gas turbine blade subjected to a probabilistic constraint on natural frequency with the consideration of uncertainties in the material properties and the thicknesses distributions.

Yang et al. [33] applied reliability-based optimal design to crashworthiness design of a full vehicle system in multicrash scenarios. Probabilistic constraints were imposed on the four impact modes: full frontal impact, roof crush, side impact, and offset impact. The optimization problem considered the uncertainties of the important design variables in local or global impact modes. They demonstrated that the weight could be reduced compared with a deterministic (baseline) design while satisfying the safety constraints.

34.5 Selected Aerospace Application Examples

In this section, selected application examples developed by the authors and their associates are presented along with details of RBDO formulations and solution strategies. The first three examples were solved using an in-house multidisciplinary stochastic optimization (MSO) shell code. The last example was solved using ProFES/MDO code, which was developed by Applied Research Associates (ARA).

An overview of the methodology implemented in the MSO shell is summarized in Figure 34.2 (for a complete description see Oakley et al. [24] or Sues et al. [34]). The methodology involves: (1) response-surface development for the objective and all the constraints; (2) stochastic optimization using the response

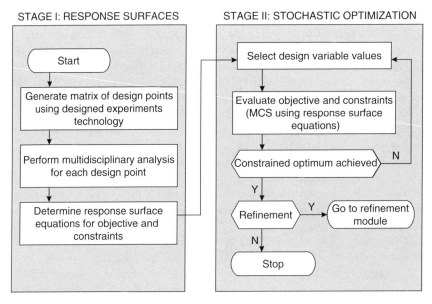

FIGURE 34.2 Flowchart of MSO shell.

surface, standard nonlinear programming, and Monte Carlo simulation; and (3) refinement of the response surfaces at the optimum for the objective and at the most probable failure points (MPPs) for each constraint. This process is repeated until convergence is achieved. The method is efficient because of the use of experimental design techniques for initial response-surface generation and the use of response surfaces for evaluation of system performance and constraints during the RBDO. The method is accurate because of the refinement procedure employed. Further, design and random variables are handled simultaneously so that the required number of g-function evaluations is a multiple of the number of design variables plus the number of random variables, as opposed to the number of design variables times the number of random variables. Unlike other typical reliability-based design optimization approaches wherein the goal is usually to minimize weight (or cost) for a structural configuration subject to a limiting weight, the MSO shell considers system performance based objective functions (e.g., cruise range, payload weight, revenue, etc.) and also considers aerodynamics and shape parameters. The MSO shell properly simulates the performance of a system that is a function of both single-occurrence and operational random variables via an efficient nested-loop algorithm. Single-occurrence random variables represent random conditions that occur only once during the lifetime of the system as well as random conditions that assume a fixed value once the system or component has been fabricated. Operational random variables represent the conditions that exhibit uncertain changes during operation of the system.

Figure 34.3 summarizes an overview of the methodology implemented in ProFES/MDO [35], which involves: (1) a multistage design-of-experiments variable-screening strategy to define the significant variable set; (2) probabilistic analysis using any of ProFES's probabilistic analysis methods and interaction with commercial CAE codes; (3) linearization of the constraints at the most probable failure points (MPPs) of each random variable, for each constraint; (4) development of a second-order response surface for the objective function at the mean values of all design variables; (5) nonlinear programming to find the optimum design using the response surfaces (via a public domain or commercial optimizer); and (6) updating of the MPPs for the current active constraint set using new values of the design variables. As shown, the process is repeated until convergence is achieved

34.5.1 Shape Optimization of an Axial Compressor Blade

The MSO shell has been used to perform a multidisciplinary stochastic shape optimization of an axial compressor blade [24, 34]. The overall goal is to determine a reliable design in which the blade twist and thickness distribution minimize cost as a function of the blade weight and of the expected efficiency during cruise conditions. Uncertainties in geometry, material properties, modeling error, operational rotor speed, and extreme off-design rotor speed conditions are considered, and reliability constraints against exceeding tip clearances, fatigue life, and yield stress at the blade root are imposed.

The entire geometry of the axial compressor is defined using Bézier curves. A Bézier curve is made of two endpoints and two control points. The hub and shroud geometry are simple surfaces of revolution from four-point Bézier curves, as illustrated in Figure 34.4, in which each curve's slopes at the two endpoints are tangent to the lines from the endpoints and the corresponding control points in the middle. In the figure, the four Bézier points are connected by straight lines.

The compressor blade is a skinned surface that is defined by a family of airfoil sections. At each radial station from hub to shroud, the airfoil section of the blade is defined by chord length, maximum thickness, twist angle, and leading-edge position. The airfoil section shape is based on a standard NACA (National Advisory Committee for Aeronautics) profile. The value of chord length, thickness, twist, and leading-edge position is defined by three-point Bézier curves spanning from hub to shroud. The baseline geometry for the analyses is based on a Bézier curve geometry that closely approximates the NAS R37 single-blade-row compressor geometry (see Figure 34.5).

34.5.1.1 RBDO Formulation

We want to determine the values of the design variables that define the shape of a blade. These variables are explained in Section 34.5.1.2.

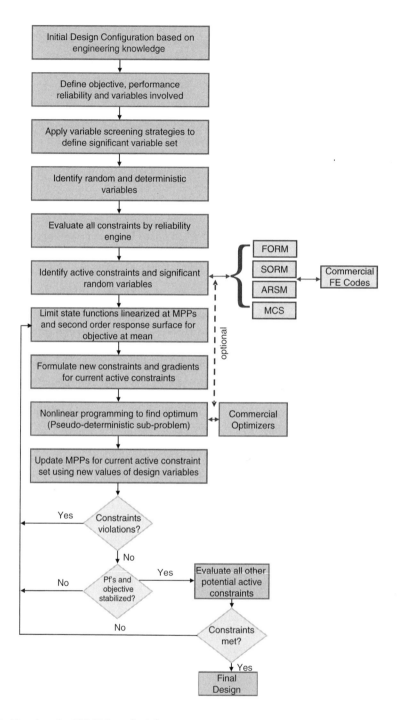

FIGURE 34.3 Flowchart for RBMDO methodology.

Objective: minimize cost as a function of the blade weight and of the expected efficiency during cruise conditions.

Constraints:

- $P[\sigma_{max} < \sigma_y] \geq 0.999$
- $P[\sigma_c < \sigma_{cr}] \geq 0.999$

FIGURE 34.4 Hub and shroud geometry for axial compressor example.

- P[blade extension in the radial direction < 0.050 mm] ≥ 0.999
- P[tip deflection in the axial direction < 0.055 mm] ≥ 0.99

The total cost is proportional to the inverse of the expected efficiency and a direct function of the weight. That is,

$$f = \frac{0.2}{E(\eta)} + C_W \qquad (34.10)$$

where f is the objective function, η is the efficiency, C_w is the coefficient of weight, and $E(\cdot)$ denotes the expected value of the enclosed quantity. The quantity 0.2 is a weighting factor that balances the relative contributions of the efficiency and weight terms to the overall cost.

The steady-state aerodynamic loads acting on the compressor blades as well as the blade row efficiency are determined using MTSB, a quasi-three-dimensional, inviscid turbomachinery analysis code [36]. The static response of the blade due to the aerodynamic and centrifugal loads is determined using NIKE3D, a nonlinear, implicit three-dimensional finite element code developed and distributed by Lawrence Livermore National Laboratory [37]. For this analysis, we couple the codes to account for the interaction between structural deflection and fluid load. Additionally, using a coupled analysis, the optimization provides the cold shape of the blade, which is required for manufacturing (i.e., an aerodynamics-only shape optimization produces the hot shape of the blade, and the cold shape would then be "backed out" by structural analyses).

34.5.1.2 Design Variables and Random Variables

Three random shape-design variables shown in Table 34.1 are considered in the optimization. Two Bézier parameters define the blade twist distribution using the following three-point Bézier curve:

$$\beta(r) = \beta_0(1-r)^2 + 2\beta_1 r(1-r) + \beta_2 r^2 \qquad (34.11)$$

where r denotes normalized radial station (i.e., $r = 0$ and $r = 1$ correspond to the root and tip of the blade, respectively). The first Bézier twist parameter β_0 is fixed at 50 degrees and controls the twist at the hub. The second and third Bézier twist parameters (i.e., β_1 and β_2) are the random design variables with design ranges shown in Table 34.1. β_1 and β_2 are taken to be random variables due to uncertainties in

FIGURE 34.5 Axial compressor geometry (NAS R37 single-blade-row compressor).

TABLE 34.1 Random Design Variables for Axial Compressor Blade

Variable Name	Minimum	Maximum	Standard Deviation
Bézier twist parameter (degrees), β_1	−55	−35	3.0
Bézier twist parameter (degrees), β_2	0.25	1.0	3.0
Thickness scaling factor, C_t	0.75	1.25	0.05

the manufacturing process. As a reasonable assumption, lognormal distributions with constant standard deviations of 3 degrees are used. The objective of RBDO is to optimize the mean values of β_1 and β_2.

A blade thickness scaling factor C_t is taken to be the third random design variable and is used to linearly scale the thickness distribution of the blade, which is defined using the following three-point Bézier curve:

$$T(r) = T_0(1-r)^2 + 2T_1 r(1-r) + T_2 r^2 \quad (34.12)$$

where thickness parameters T_0, T_1, and T_2 are fixed at values of 0.250, 0.185, and 0.176, respectively. All of the design variables have a lognormal distribution. The uncertainty in the design variables simulates manufacturing uncertainty.

In addition to the design variables, the problem includes seven additional random variables. These variables are given in Table 34.2, along with their mean and coefficient of variation (COV). A lognormal distribution is assumed for each of these variables. Inaccuracies in the mechanical modeling are simulated by applying the random error factors shown in Table 34.2 to the analysis results. These random variables are included to recognize the fact that the aerostructural modeling and solution procedures are idealizations of real-world phenomena. The MTSB error factor is applied to the predicted blade-row efficiency; the NIKE3D error factor is applied to the predicted stress and displacement results. These scaling factors and the three material-property random variables are treated as single-occurrence random variables. Variations in the operational rotor speed V_0 and uncertainties in the off-design or maximum rotor speed V_m represent two distinct random variables. Maximum off-design rotor speed uncertainty represents a single-occurrence random variable (since the maximum rotor speed occurs only once over the lifetime of the blade) that governs the constraints and therefore the reliability of the blade. Nominal rotor speed uncertainty represents an operational random variable that changes during a typical flight and affects the blade-row efficiency. Although it is possible to use a single experimental design to determine the blade response due to rotor speed, the design would need to cover a wide range to include the maximum rotor speed, and the accuracy of the resulting response-surface equations would likely be compromised. In order to properly simulate the affect of both single-occurrence and operational random variables, the nested-loop MCS algorithm described earlier is used.

TABLE 34.2 Random Variables for Axial Compressor Blade

Variable Name	Mean	COV
Young's modulus (GPa)	81.4	0.10
Yield stress (MPa)	550	0.100
Endurance limit[a] (MPa)	138	0.125
MTSB model error	1.00	0.030
NIKE3D model error	1.00	0.060
Operational rotor speed (rpm)	9,330	0.018
Maximum rotor speed (rpm)	10,500	0.016

[a] Endurance limit (10^7 cycles) taken from S-N diagram for Ti-6Al-4V at 482°C (900°F).

TABLE 34.3 RBDO Results for Axial Compressor Blade

Case	β_1 (degrees)	β_2 (degrees)	C_t	C_W	$E(\eta)$	f
Baseline	−40.0	−30.0	1.00	0.243	0.921	0.460
Deterministic	−54.7	−39.8	0.750	0.182	0.944	0.394
Stochastic	−55.0	−40.0	0.882	0.214	0.942	0.427
Safety factor	−35.0	−40.0	1.25	0.304	0.904	0.525

34.5.1.3 Results and Discussion

Results were obtained for four different cases:

1. Baseline case representing the initial geometry of the axial compressor
2. Deterministic optimization
3. Stochastic optimization
4. Deterministic optimization using safety factors

For deterministic optimization (Cases 2 and 4), the random variables take on their mean values. The second deterministic optimization case (Case 4) was performed using a safety factor of 2.00 on the stress and fatigue constraints and a safety factor of 1.67 on the deflection constraints. That is,

$\sigma_{max} < 0.5\, \sigma_y$
$\sigma_c < 0.5\, \sigma_{cr}$
$\delta_R < 0.030$ mm
$\delta_A < 0.033$ mm

These safety factors introduce additional conservatism, in much the same way that a deterministic design would be performed in practice, by reducing the maximum allowable stresses and deflections by factors of 2 and 1.67, respectively.

The final numerical results for each case are summarized in Table 34.3. The final optimum values for β_1, β_2, and C_t are tabulated along with the weight coefficient C_w, the expected efficiency $E(\eta)$, and the objective function value f. A plot of the twist distribution results is given in Figure 34.6, where the optimum twist angle is plotted as a function of the normalized radial position of the blade (i.e., 0 = blade root, 1 = blade tip). As shown in Table 34.3 and in Figure 34.6, the results for each case differ substantially from the baseline case and also from each other.

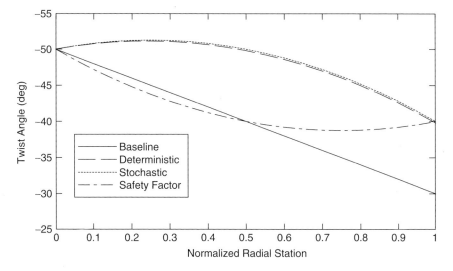

FIGURE 34.6 Optimum twist angle vs. normalized radius.

Considering the deterministic optimization results first, the twist increases to its upper limit to improve the efficiency. The increase in twist at the end of the blade also serves to reduce the radial deflection and helps satisfy the tip-clearance constraint. The blade thickness decreases to its lower limit to reduce the weight. The deterministic optimization leads to a design where the objective function (total cost) is 14% less than the baseline case. However, this design will not be reliable under variable rotor speed conditions and manufacturing uncertainties. The design does satisfy the stress and fatigue reliability goals; however, it only achieves reliabilities of 0.854 (vs. 0.990) and 0.998 (vs. 0.999) for the axial deflection and tip-clearance constraints, respectively.

For the stochastic optimization, all of the stress and deflection constraints are now reliability based. As in the deterministic case, the twist increases to its upper limit to improve the efficiency and reduce the radial deflection. As shown in Figure 34.6, the optimal twist distribution for the stochastic case is essentially the same as that for the deterministic case. As shown in Table 34.3, the blade thickness once again decreases to reduce the weight. However, this time it stops at a value of 0.882, rather than decreasing to its lower limit of 0.750, in order to satisfy the reliability-based constraints for tip clearance and axial deflection. Note that it would be possible to reduce the axial deflection by reducing β_2; however, this would significantly increase the radial deflection (which is more sensitive to β_2) and violate the tip-clearance constraint. As such, the optimizer chooses to increase the thickness, at the expense of a higher blade weight, in order to maintain maximum efficiency and meet the deflection constraints. Ultimately, the stochastic optimization results yield a design with an overall cost that is 8% higher than the deterministic case. However, this design satisfies the specified reliability goals and should therefore perform reliably under extreme "off-design" variable rotor speed conditions and manufacturing uncertainties.

For the safety-factor case, the blade thickness increases to its upper limit to satisfy the tighter constraints on stress and deflection. However, this increase in thickness is not sufficient to satisfy the tip-clearance constraint. Thus, to reduce the radial deflection, the optimizer drives twist parameters β_1 and β_2 to their lower and upper limits, respectively. Although this twist configuration reduces the radial deflection by nearly 30%, it increases the axial deflection by almost 9%. Ultimately, even with this reduced twist and the thicker blade, the tip-deflection constraints are not completely satisfied: the axial and radial tip deflections exceed the limits by 37 and 31%, respectively. The reduced twist and larger weight of the safety-factor design lead to an overall cost that is 14% higher than the baseline case. This is because the safety factors used in this example were chosen somewhat arbitrarily, and they actually represent a solution that is more conservative with respect to the deflection constraints than the stochastic optimum. That is, the reliability levels for the stochastic optimum are 0.9996 for tip clearance and 0.990 for axial deflection, which match the target reliabilities of these constraints, whereas those for the safety-factor solution (which, in fact, did not meet the safety-factor goals due to bounds on the design variables) are 0.99996 for tip clearance and essentially 1.00 for axial deflection. Thus, the safety-factor solution is overdesigned for deflection.

34.5.2 Shape Optimization of an Airplane Wing

This example considers a detailed airplane wing design problem [38]. The baseline airplane is a Mach 0.3, 20-seat transport with a payload capacity on the order of 5000 lb.

34.5.2.1 RBDO Formulation

Objective: Maximize expected cruise range
Subject to reliability constraints:
    P(upper surface root stress 1 $\leq$ yield) $\geq$ 99.0%
    P(upper surface root stress 2 $\leq$ yield) $\geq$ 99.0%
    P(lower surface root stress 1 $\leq$ yield) $\geq$ 99.0%
    P(lower surface root stress 2 $\leq$ yield) $\geq$ 99.0%
    P(takeoff distance $\leq$ 3000 ft) $\geq$ 99.0%
    P(wing area $\leq$ 600 ft^2) $\geq$ 50.0%

Applications of Reliability-Based Design Optimization

TABLE 34.4 Random Design Variables for Full-Featured Wing-Shape Optimization

Variable Name	Minimum	Maximum	COV
Aspect ratio, AR	6.0	12.0	0.0067
Taper ratio, λ	0.25	1.0	0.006
Semispan (ft), b	25.0	40.0	0.0002
Wingtip incidence (degrees), i_{tip}	−2	2	0.125
Structure skin thickness (in.)	0.07	0.12	0.01
Structure spar thickness (in.)	0.15	0.35	0.01
Wing sweep (degrees), Λ_{LE}	0.0	10.0	0.0002

34.5.2.2 Design and Other Random Variables

The wing design problem consists of random design variables, operational random variables, and single-occurrence random variables. In general, problems are formulated such that single-occurrence random variables affect constraint computations but not objective computations. The design variables themselves can be random, with the mean value being the designed value, and the achieved value being a function of a random distribution about the designed mean.

The wing optimization example includes seven random design variables, given in Table 34.4. The table includes the minimum and maximum values as well as the coefficient of variation. All of the design variables have a normal distribution. The uncertainty in the design variables simulates manufacturing uncertainty. The skin thickness is the average thickness of the upper and lower wing skin, and is used as the thickness of all finite elements used to model the skin. The spar thickness is the average thickness of the wing spar, and is used as the thickness of all finite elements used to model the spar. The remaining variables are basic wing geometric parameters and are defined graphically in Figure 34.7.

In addition to the seven design variables, the problem includes ten additional random variables. These variables are given in Table 34.5 and are marked as being single-occurrence or operational.

A driver computer program called Cpanel manages the complete end-to-end deterministic analysis. Cpanel includes the following features: (1) geometry and grid generation subroutines to parametrically generate aerodynamic and structural grids for numerical methods; (2) a coupled aerostructural analysis

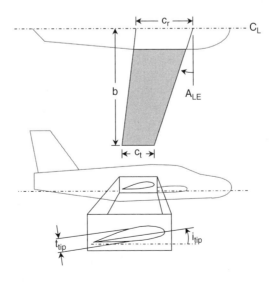

FIGURE 34.7 Wing geometric parameters.

TABLE 34.5 Random Variables for Full-Featured Wing-Shape Optimization

Variable Name	Mean	COV	Distribution
Single-occurrence			
Peak upward gust velocity (fps)	30.0	0.15	normal
Peak downward gust velocity (fps)	−30.0	0.15	normal
Skin Young's modulus (psf)	10.5 E+6	0.05	normal
Spar/rib Young's modulus (psf)	10.5 E+6	0.05	normal
Skin yield stress (psf)	5.76 E+6	0.10	lognormal
Material density (lb/in.^{3})	0.098	0.02	normal
Maximum takeoff altitude (ft)	4,000–6,000	0.20	truncated lognormal
Operational			
Cruise altitude (ft)	20,000–30,000	0.12	uniform
HP specific fuel consumption (lb/h/hp)	0.4–0.6	0.10	truncated lognormal
Payload (lb)	4,000–6,000	0.20	truncated lognormal

loop that uses aerodynamic and structural analysis programs to compute static aeroelastic deformation; and (3) analytic objective and constraint evaluation formulas.

The range objective is computed using the so-called constant-altitude, constant-airspeed flight program. The analytic function for this flight program requires the computation of maximum lift-to-drag ratio (E_{max}) as well as the lift-to-drag ratio (E_{cruise}), and lift coefficient ($C_{L,cruise}$) at the start of cruise. The latter is computed using the coupled aerostructural analysis code described later and is a direct function of the aircraft gross weight, the wing area, the Mach number, and altitude. All of these parameters are directly related to random design variables, operational random variables, and single-occurrence random variables, and so $C_{L,cruise}$ is also random. In computing the lift-to-drag ratios, we assume a parabolic drag polar and a "typical" zero-lift drag coefficient of 0.025. We approximate an induced drag coefficient using the method of approximate spanwise efficiency and the method of leading-edge suction.

The wing design problem places probabilistic constraints on wing area, takeoff ground-roll distance at a maximum lifetime takeoff altitude, and wing root stress in a maximum lifetime upward or downward gust. We compute the takeoff ground roll distance using a piston-prop formula [39]. We select an approximate value of 2.0 for $C_{L,max}$ in the piston-prop formula, based on carpet plot data presented by Raymer [40].

The driver program, Cpanel, performs a static aeroelastic analysis by calling an external aerodynamics code and an external structural finite element analysis code. The coupled analysis predicts (a) the slope of the flexible wing lift curve, $C_{L\alpha}$, required for induced drag calculations and (b) the stress at four distinct points on the wing skin under maximum lifetime upward and downward gust loads.

Prior to driving the coupled analysis, Cpanel first generates structural and aerodynamic grids or meshes for the analyses. Once the geometry is meshed, Cpanel executes a sequential set of steps for each static coupled analysis, using the PMARC code [41] to compute pressure loads on the wing and COMET-AR [42] to compute deformation of the wing. The aerodynamic analysis is performed on the deformed shape in order to compute a lift coefficient for the flexible, deformed wing.

Stress values at the wing root are read from the COMET-AR results for two extreme loading conditions. Two upper-surface stresses are read for a maximum lifetime upward gust velocity. Two lower-surface stresses are read for a maximum lifetime downward gust velocity. Cpanel performs the coupled analyses for these extreme loading conditions at an adjusted angle of attack equal to cruise angle of attack plus a perturbation due to a gust load perpendicular to the flight path. The analysis uses a gust alleviation factor described by Raymer [40] to account for the fact that the gust load is a transient effect that is not experienced instantaneously at full force.

TABLE 34.6 RBDO Results for Full-Featured Wing-Shape Optimization

Variable Name	Initial	Case A	Case B	Case C
Aspect ratio, AR	9.0	12.0	12.0	10.98
Taper ratio	0.625	0.62	0.26	0.58
Semispan, b (ft)	31.00	40.00	34.68	30.60
Wingtip incidence (degrees)	0.0	0.0	0.0	0.0
Structure skin thickness (in.)	0.095	0.070	0.0725	0.116
Structure spar thickness (in.)	0.25	0.35	0.35	0.26
Wing sweep (degrees)	5.0	5.0	5.0	5.0
Initial range (NM)	890.6	890.6	890.6	895.7
Optimal range (NM)	N/A	1024.7	984.7	974.9
Minimum constraint reliability (%)	99.9	38	96.0	99.0

34.5.2.3 Results and Discussion

Three cases for a wing-shape optimization are selected as described below:

- Case A: six deterministic constraints with no safety factor on yield stress
- Case B: six deterministic constraints with safety factor of 1.5 on yield stress
- Case C: six probabilistic constraints (there are seven random design variables and ten nondesign random variables, as described)

The optimization results for all three cases are summarized in Table 34.6. The results indicate that the RBDO strategy is able to successfully find an optimum solution that effectively balances performance and reliability. Comparing the three cases, we see that while Case A provides the best performance, there is no safety factor applied, so the design will not be reliable. In fact, the final row in the table shows that the reliability of one of the stress constraints is only 38% (i.e., probability of exceeding yield stress is 62%), which is obviously unacceptable. Case B, with a safety factor of 1.5 on yield stress, is also able to improve on the original design while meeting the more stringent constraints; however, the safety factor of 1.5 is inadequate to ensure an acceptable reliability. The final row in the table shows that the reliability of the stress constraint is only 96% (still less than the required 99%). This illustrates the pitfall of using deterministic safety factors. There is no information on the true reliability of the structure. Hence, even though a safety factor has been applied, the wing is actually underdesigned relative to our required reliability of 99%. Case C, designed using the RBDO approach, is able to improve on the original design and still meet the reliability-based constraint. Although, for this problem, the RBDO results in a design that is more costly than the deterministic optimization (because the deterministic MDO (Multidisciplinary Design Optimization) does not meet the reliability goal), this will not be true for all cases. It is just as likely (perhaps even more likely) that the RBMDO could result in a less costly design.

Table 34.7 summarizes the values of stresses, takeoff distances, and wing areas at the optimums for the three cases. For Case A, the allowable stress is the yield stress of 5.76×10^6 psf, and for Case B, the allowable stress is the yield stress divided by 1.5 (safety factor), or 3.84×10^6 psf. The results for Case C

TABLE 34.7 Information for Constraints at the Optimum Designs for Full-Featured Wing-Shape Optimization

Variable Name	Initial Design	Case A	Case B	Case C
Upper stress 1 (10^6 psf)	−3.09	−5.74	−3.76	−3.66
Upper stress 2 (10^6 psf)	−2.67	−4.99	−3.21	−3.18
Lower stress 1 (10^6 psf)	1.16	1.53	1.38	1.65
Lower stress 2 (10^6 psf)	1.06	1.40	1.25	1.47
Takeoff distance (ft)	1960.2	1404.6	2154.9	2751.2
Wing area (ft^2)	427.1	533.3	400.9	340.9

are obtained by using the average values for the random design variables and mean values for the nondesign random variables. In addition to advantages of the RBDO method discussed above, the RBDO approach also provides a better design than the one obtained with the deterministic approach (Case B). As can be seen from Table 34.7, the stress ranges at the upper and lower roots of the wing are narrower in Case C than in the first two cases. In other words, the load effects are more uniformly distributed to the components of the wing structures (skin and spar) compared with the deterministic design in Cases A and B. It can be seen from Table 34.6 that the skin and spar thicknesses are not as far apart as those in Cases A and B.

Thus, the RBDO results in a more balanced design. The objective is to maximize expected performance, accounting for all the uncertainties of the 17 random variables. This objective is subject to six probabilistic constraints, which again includes the uncertainties of the random variables related to these constraints. Therefore, the final wing-shape configuration from the RBDO-based design procedure considers all uncertainties and provides the best performance in a sense of expectation, and it meets all the safety and serviceability criteria in a sense of probabilities. These conditions generally force a more balanced design because the aircraft must perform over a range of operating and extreme conditions, as represented by the random variables. On the other hand, the deterministic-based design procedure for Cases A and B maximizes the cruise range using seven design variables, while the other ten nondesign variables are set as constants (either reduced by a safety factor, set at the mean, or set at an upper/lower bound for the worst combinations). Therefore, the deterministic-based design procedure can drive the optimal point to an extreme location in the global design space, thus resulting in an unbalanced product configuration like that in Case B.

34.5.3 Optimization of an Integral Airframe Structure Step Lap Joint

An integral airframe structure (IAS) panel step lap joint [43] is considered in this example. Developed and tested by NASA and Boeing, this lap-joint design (see Figure 34.8) adopted a new concept that takes advantage of the monolithic construction process of the panel.

There are two main failure modes to be considered for the joints: longitudinal fatigue cracks due to high hoop stresses caused by internal pressurization and circumferential fatigue cracks due to vertical bending of the fuselage. However, the possibility of circumferential crack propagation at the top of the crown is fairly low, since a stronger butt splice is normally used for the circumferential skin splice. Hence, the longitudinal splice is the critical component. Therefore, the focus here was to perform a reliability-based fatigue analysis of the step lap joint and then attempt to improve the current design (redesign) using reliability-based optimization techniques.

The NASA Langley COMET-AR structural analysis code [42] was chosen to model and analyze this component during RBDO. For the fatigue life of the step lap joint, a cyclic stress to failure (S/N) approach

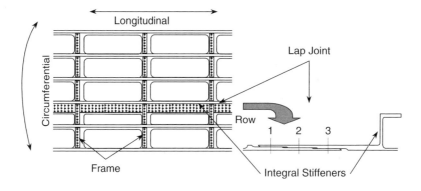

FIGURE 34.8 IAS panel with longitudinal lap joint.

Applications of Reliability-Based Design Optimization

TABLE 34.8 Random Design Variables for Lap-Joint Optimization

Variable Name	Minimum	Nominal	COV
lt_1 (in.)	0.06	0.06	0.01
lt_2 (in.)	0.06	0.085	0.01
lt_3 (in.)	0.06	0.11	0.01
lt_4 (in.)	0.06	0.17	0.01
rt_1 (in.)	0.06	0.06	0.01
rt_2 (in.)	0.06	0.085	0.01
rt_3 (in.)	0.06	0.11	0.01
rt_4 (in.)	0.148	0.17	0.01

is used that applies the maximum stresses (stress concentrations) near the rivets to the corresponding fitted equation [MIL-HDBK] [44]:

$$\log (N_f) = 10.0 - 3.96 \log (S_{eq}) + \varepsilon \quad (34.13)$$

where $S_{eq} = S_{max} (1.0 - R)^{0.64}$, N_f is the number of cycles to failure, and R is the stress ratio of the minimum to maximum cyclic stresses. S_{max} is the applied maximum stress (ksi). The error in the failure life estimation is given by ε.

34.5.3.1 RBDO Formulation

Objective: Minimize the weight
Subject to reliability constraint:
   P(Design life cycle $\geq$ 20,000) $\geq$ 99.9%

34.5.3.2 Design and Other Random Variables

This lap-joint problem considers uncertainties in the fatigue life of the base material and elastic material properties, the loading conditions, and the manufacturing tolerances.

The stepped lap-joint optimization example includes eight random design variables for the thicknesses, given in Table 34.8 and depicted in Figure 34.9. The table includes the minimum and nominal values as well as the coefficient of variation. All of the design variables have a lognormal distribution. The uncertainty in the design variables simulates manufacturing uncertainty.

In addition to the random design variables, the problem includes four additional random variables. The uncertainty in the fatigue law (cycles to failure), as quantified by the parameter ε (Equation 34.13), was obtained from the Military Handbook. The uncertainties for modulus (E) and Poisson's ratio (σ) were assumed to take on typical values of 5% coefficient of variation. The uncertainty of 5% on the load and of 1% on the thicknesses was obtained during discussion with the Boeing designers. All the random variables, with their mean values, coefficients of variation, and the distributions used for the analysis, are tabulated in Table 34.9.

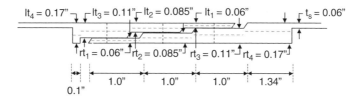

FIGURE 34.9 Current design dimensions of the stepped lap joint.

TABLE 34.9 Input Random Variables for Lap-Joint Optimization

Input Variable	Mean	COV	Distribution
ε	0.0	0.248^a	normal
P_{max}	18.0 ksi	0.05	lognormal
E	1.03×10^7 psi	0.05	lognormal
v	0.3205	0.05	lognormal

[a] Standard deviation.

34.5.3.3 Results and Discussion

The optimized values are shown in Figure 34.10. The objective function f_1 for the original design is 0.85, while the optimized objective function obtained was 0.631. The weight factor associated with the optimized design was 0.61104 (as opposed to the 0.7361 weight factor of the original design), which is a 17% reduction in the weight of the lap joint while maintaining the reliability requirement.

34.5.4 Transport Aircraft Wing Optimization

The reliability-based structural optimization of a full-scale transport aircraft wing [45] is considered, and the optimized design and performance characteristics obtained using RBDO are compared with those from deterministic optimization. The ACT S/RFI composite wing box of the advanced composite technology (ACT) wing from the Boeing Company was used for the RBDO analysis. The baseline aircraft selected for this demonstration is the D-3308-4 configuration of the proposed Boeing 190-passenger, two-class transport aircraft.

The design weights for this aircraft are maximum takeoff gross weight (MTOGW) = 180 kips and maximum landing weight (MLW) = 167.5 kips. The critical design conditions for this composite wing box were derived from the DC-10-10 and MD-90-30 aircraft loads for 2.5-g positive balance flight maneuver (upbending). The semispan composite wing external loads are applied to eight discrete actuator load points on the test article shown in Figure 34.11. The discrete loads are adjusted to best approximate the shears, moments, and torques of the flight and ground conditions. The semispan test article consists of upper and lower cover panels and front and rear spars, ribs, and bulkheads, as shown in Figure 34.12. The major components are the cover panels. Each contains skin, stringers, spar caps, and intercostal clips. These subcomponents are stitched together to form a single dry-fiber preform, which is then filled with resin and cured by the resin film infusion (RFI) process.

34.5.4.1 RBDO Formulation

Objective: Minimize expected weight of the wing
Subject to reliability constraints:
   P(maximum Von-Mises stress ≤ yield) ≥ target reliability
   P(maximum wingtip displacement ≤ tolerance) ≥ target reliability

34.5.4.2 Design and Other Random Variables

Probabilistic constraints were evaluated with MSC/NASTRAN using the CAD/CAE interface capability built in ProFES/MDO. The finite element model of the wing is shown in Figure 34.13.

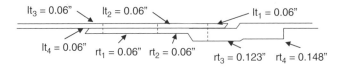

FIGURE 34.10 Optimized design configuration of the stepped lap joint.

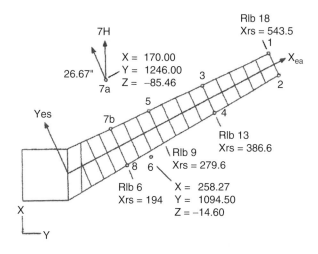

FIGURE 34.11 Actuator load points for semispan composite wing box.

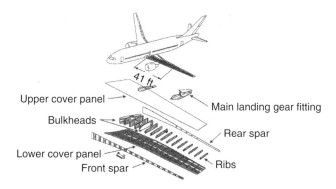

FIGURE 34.12 Baseline aircraft configuration and semispan structural arrangement.

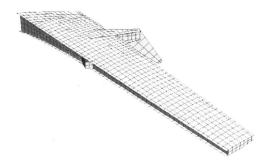

3804 FE nodes
3770 FE elements (2222 Shells + 1548 Beams)
22 material properties
47 shell element properties (PSHELL)

FIGURE 34.13 Finite element model of transport aircraft wing.

TABLE 34.10 Description of the Design Variables for Transport Aircraft Wing Optimization

Variable Name	Initial Design	Minimum	Maximum	Standard Deviation	Description
T1 (in.)	0.55	0.385	0.715	0.005	upper skin panel thickness
T2 (in.)	0.33	0.297	0.363	0.005	lower skin panel thickness
T3 (in.)	0.385	0.296	0.501	0.005	front spar panel thickness
T4 (in.)	0.385	0.296	0.501	0.005	rear spar panel thickness
T5 (in.)	0.149	0.104	0.194	0.005	rib panel thickness

The design random variables are chosen to be the thickness of the shell sections that contribute the most weight to the wing. The two design random variables T1 and T2 are the thickness of shell sections on the upper and lower skin panel, and they are the biggest weight contributors to the wing, with total weights of 251.38 and 152.44 lb, respectively. The other three random variables T3, T4, and T5 are the respective thicknesses of shell section on the front spar panel, the rear spar panel, and on the rib panels. They are the next weight contributors to the wing in their corresponding sections, with total weights of 125.03, 116.36, and 59.08 lb, respectively. Table 34.10 shows the minimum and maximum values of the five design random variables as well as their coefficients of variation. All of the design variables have a normal distribution. The uncertainty in the design variables simulates manufacturing uncertainty.

In addition to the five design variables, the uncertainty on material property (E_{11}) of the lower skin panels is considered, and it is shown in Table 34.11. For the material strength and displacement tolerances, the deterministic values are used. Figure 34.14 shows the finite elements in each section to which the design and random variables are assigned.

34.5.4.3 Results and Discussion

Three cases for reliability-based optimization are selected as described below:

- Case A: Minimize the weight while keeping the same reliability as the one in initial design
- Case B: Minimize the weight while increasing the reliability by a factor of 10
- Case C: Minimize the weight while increasing the reliability by a factor of 100

The optimization results for all three cases are summarized in Table 34.12.

The results in Table 34.12 show that the optimal design is achieved by transferring material to sections in the upper and lower skin panels from the rib and spar panels. It also shows that the optimal design reduces the weight by 25.02 lb from that of the initial design (i.e., 3.6% weight reduction in the five sections considered) while keeping the reliability of the new design same as that of the initial design. From the results for Cases B and C, the optimal design still reduced the weight from that of the initial design while improving the reliability of the design by a factor of 10 and 100, respectively.

Figure 34.15 shows comparison of the initial design with the three optimal designs described above. The numbers in the figure show the ratio of the weight for the optimal design as compared with the initial design for various factors of improvement in reliability. The ratios shown in the figure are for the weight of the five sections in the initial design (with the total weight of 704.29 lb) and the weight of the same five sections in the optimal designs. We should note that in this study, the weight of these five sections constitute only 8.5% of the total weight of the wing, and it would be a natural conclusion that more weight savings could be achieved if all shell sections as well as the beams in the model were treated as design random variables.

TABLE 34.11 Description of the Random Variables for Transport Aircraft Wing Optimization

Variable Name	Mean	COV	Distribution	Description (unit)
R1	1.21×10^7	1.21×10^5	normal	E_{11} of lower skin panels (psi)

Applications of Reliability-Based Design Optimization

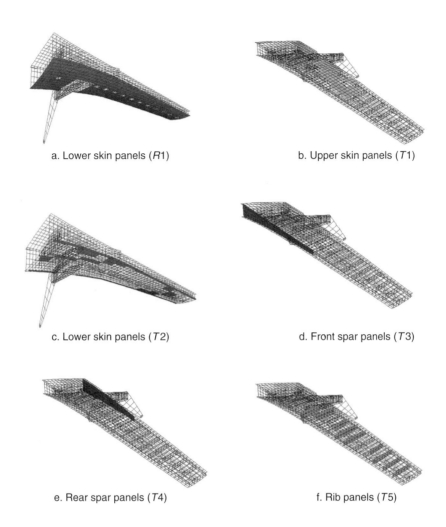

FIGURE 34.14 Design sections of transport aircraft wing.

34.6 Conclusions

Recent advances in reliability and RBDO computational methods have allowed designers to treat a wide range of application problems in many technical areas where there is a need to incorporate uncertainties into design optimization. This chapter summarized the methods, gave a brief overview of the RBDO

TABLE 34.12 Results of the ProFES/MDO for Transport Aircraft Wing Optimization

Variable Name	Initial Design	Optimal Design		
		Case A	Case B	Case C
T1 (in.)	0.55	0.5581	0.5512	0.5625
T2 (in.)	0.33	0.3328	0.3563	0.3630
T3 (in.)	0.385	0.3465	0.3465	0.3465
T4 (in.)	0.385	0.3465	0.3465	0.3466
T5 (in.)	0.149	0.1341	0.1341	0.1341
Objective (lb)	8275.22	8250.20	8257.90	8266.15
Weight of five sections (lb)	704.29	679.27	686.97	695.12

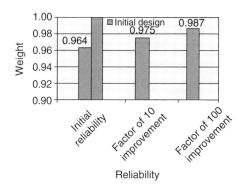

FIGURE 34.15 Comparison of initial design with optimal designs using the RBDO methodology on the semispan composite wing box subjected to 2.5-g upbending load.

applications, and presented several examples with some details. These examples demonstrated that RBDO produced more cost-effective and reliable designs than the deterministic-based designs.

The field of optimization has seen rapid advancement from single-discipline optimization to multidisciplinary optimization. However, even with efficient optimization/probabilistic-analysis algorithms and computational resources, the application of RBDO to large-scale multidisciplinary problems is still computationally demanding. Further research and development is needed to allow the practical incorporation of RBDO concepts to large-scale multidisciplinary optimization problems.

References

1. Montgomery, D.C., *Design and Analysis of Experiments*, John Wiley and Sons, New York, 1991.
2. NIST/SEMATECH, e-Handbook of Statistical Methods, http://www.itl.nist.gov/div898/handbook/, 2004.
3. Fu, Y., Chuang, C.H., Li, G., and Yang, R.J., Reliability-Based Design Optimization of A Vehicle Exhaust System, 2004-01-1128, presented at SAE World Congress and Exhibition, Detroit, MI, 2004.
4. Li, G. and Yang, R.J., Recent Applications on Reliability-Based Optimization of Automotive Structures, 2003-01-0152, presented at SAE World Congress and Exhibition, Detroit, MI, 2003.
5. Madsen, H.O., Krenk, S., and Lind, N.C., *Methods of Structural Safety*, Prentice Hall, Englewood Cliffs, NJ, 1986.
6. Ang, A.H.S. and Tang, W.H., *Probability Concepts in Engineering Planning and Design*, Vol. II, *Decision, Risk, and Reliability*, John Wiley & Sons, New York, 1984.
7. Der Kiureghian, A., Lin, H.-Z., and Hwang, S.-J., Second-order reliability approximations, *J. Eng. Mech.*, 113 (8), 1208–1225, 1987.
8. Wen, Y.K. and Chen, H.C., On fast integration for time variant structural reliability, *Probabilistic Eng. Mech.*, 2 (3), 156–162, 1987.
9. Liu, P.L. and Der Kiureghian, A., Optimization Algorithms for Structural Reliability Analysis, report UCB/SESM-84/01, Dept. of Civil Engineering, University of California at Berkeley, 1986.
10. Kuschel, N., Rackwitz, R.R., and Pieracci, A., Multiple Points in Structural Reliability, presented at 3rd International Conference on Computational Stochastic Mechanics, Thera-Santorini, Greece, 1998.
11. Der Kiureghian, A. and Dakessian, T., Multiple design points in first- and second-order reliability, *Structural Safety*, 20, 37–49, 1998.
12. Fujita, M. and Rackwitz, R., Updating first- and second-order reliability estimates by importance sampling, *Structural Eng./Earthquake Eng.*, 5 (1), 53–59, 1988.
13. Hohenbichler, M. and Rackwitz, R., Improvement of second-order reliability estimates by importance sampling, *ASCE J. Eng. Mech.*, 114 (12), 2195–2199, 1988.

14. Au, S.K. and Beck, J.L., First-excursion probabilities for linear systems by very efficient importance sampling, *Probabilistic Eng. Mech.*, 16 (3), 193–207, 2001.
15. Au, S.K. and Beck, J.L., Estimation of small failure probabilities in high dimensions by subset simulation, *Probabilistic Eng. Mech.*, 16 (4), 263–277, 2001.
16. Wu, Y.-T. and Mohanty, S., Variable screening and ranking using several sampling based sensitivity measures, in *Proceedings of 44th Structures, Structural Dynamics, and Materials Conference*, AIAA/ASME/ASCE/AHS, Norfolk, VA, 2003; submitted to *Reliability Eng. System Safety*, 2003.
17. Yang, R.J. and Gu, L., Experience with approximate reliability-based optimization methods, in *Proceedings of 44th Structures, Structural Dynamics, and Materials Conference*, AIAA/ASME/ASCE/AHS, AIAA 2003–1781, Norfolk, VA, 2003.
18. Chen, X., Hasselman, T.K., and Neill, D.J., Reliability based structural design optimization for practical applications, in *Proceedings of 38th Structures, Structural Dynamics, and Materials Conference*, AIAA/ASME/ASCE/AHS, AIAA 97–1403, Kissimmee, FL, 1997.
19. Wu, Y.-T. and Wang, W. Efficient probabilistic design by converting reliability constraints to approximately equivalent deterministic constraints, *J. Integrated Design Process Sci.* (JIDPS), 2 (4), 13–21, 1998.
20. Wu, Y.-T., Shin, Y., Sues, R.H., and Cesare, M.A., Safety-factor based approach for probability-based design optimization, in *Proceedings of 42nd Structures, Structural Dynamics, and Materials Conference*, AIAA 2001-1379, AIAA/ASCE/AHS/ASC, Seattle, WA, 2001.
21. Du, X. and Chen, W., Sequential Optimization and Reliability Assessment Method for Efficient Probabilistic Design, DETC2002/DAC-34127, presented at ASME Design Engineering Technical Conferences, Montreal, 2002.
22. Sues, R.H. and Rhodes, G.S., Portable Parallel Stochastic Optimization for the Design of Aeropropulsion Components, NASA-CR 195312, ARA Report No. 5786, 1993.
23. Sues, R.H., Oakley, D.R., and Rhodes, G.S., MDO of Aeropropulsion Components Considering Uncertainty, presented at the 6th AIAA/NASA/USAF Multidisciplinary Analysis and Optimization Symposium, Bellevue, WA, 1996.
24. Oakley, D.R., Sues, R.H., and Rhodes, G.S., Performance optimization of multidisciplinary mechanical systems subject to uncertainties, *Probabilistic Eng. Mech.*, 13 (1), 15–26, 1998.
25. Frangopol, D.M. and Maute, K., Life-cycle reliability-based optimization of civil and aerospace structure, *Comput. Struct.*, 81, 397–410, 2003.
26. Davidson, J.W., Felton, L.P., and Hart, G.C., On reliability-based structural optimization for earthquakes, *Comput. Struct.*, 12, 99–105, 1980.
27. Feng, Y.S. and Moses, F., Optimum design, redundancy and reliability of structural systems, *Comput. Struct.*, 24 (2), 239–251, 1986.
28. Yang, J.S. and Nikolaidis, E., Design of aircraft wings subjected to gust loads: a safety index based approach, *AIAA J.*, 29 (5), 804–812, 1991.
29. Pu, Y., Das, P.K., and Faulkner, D., A strategy for reliability-based optimization, *Eng. Struct.*, 19 (3), 276–282, 1997.
30. Stroud, W., Krishnamurthy, T., Mason, B., Smith, S., and Naser, A., Probabilistic design of a plate-like wing to meet flutter and strength requirements, in *Proceedings of 43rd Structures, Structural Dynamics, and Materials Conference*, AIAA 2002-1464, AIAA/ASCE/AHS/ASC, Denver, CO, 2002.
31. Pettit, C. and Grandhi R., Multidisciplinary Optimization of Aerospace Structures with High Reliability, in *Proceedings of 8th ASCE Joint Specialty Conference on Probabilistic Mechanics and Structural Reliability*, ASCE, Reston, VA, 2000.
32. Grandhi, R.V. and Wang, L., Reliability-based structural optimization using improved two-point adaptive nonlinear approximations, *Finite Elements Analysis Design*, 29, 35–48, 1998.
33. Yang, R.J., Gu, L., Tho, C.H., Choi, K.K., and Youn, B.D., Reliability-based multidisciplinary design optimization of a full vehicle system, in *Proceedings of 43rd Structures, Structural Dynamics, and Materials Conference*, AIAA 2002-1758, AIAA/ASCE/AHS/ASC, Denver, CO, 2002.

34. Sues, R.H., Oakley, D.R., and Rhodes, G.S., Portable Parallel Computing for Multidisciplinary Stochastic Optimization of Aeropropulsion Components, NASA contractor report 202307, NASA contract NAS3-27288, 1996.
35. Sues, R.H. and Cesare, M.A., An innovative framework for reliability-based MDO, in *Proceedings of 41st Structures, Structural Dynamics, and Materials Conference*, AIAA-2000-1509, AIAA/ASCE/AHS/ASC, Atlanta, GA, 2000.
36. Boyle, R.J., Haas, J.E., and Katsanis, T., Comparison between measured turbine stage performance and the predicted performance using quasi-3D flow and boundary layer analyses, *J. Propulsion Power*, 1, 242–251, 1985.
37. Hallquist, J., NIKE3D: an Implicit, Finite-Deformation, Finite Element Code for Analyzing the Static and Dynamic Response of Three-Dimensional Solids, report UCID-18822, Lawrence Livermore National Laboratory, University of California, Livermore, CA, 1984.
38. Xiao, Q., Sues, R.H., and Rhodes, G.S., Multi-disciplinary wing shape optimization with uncertain parameters, in *Proceedings of 40th Structures, Structural Dynamics, and Materials Conference*, AIAA 99-1601, AIAA/ASCE/AHS/ASC, St. Louis, MO, 1999.
39. Hale, F.J., *Aircraft Performance, Selection, and Design*, John Wiley and Sons, New York, 1984.
40. Raymer, D.P., *Aircraft Design: a Conceptual Approach*, AIAA Education Series, Reston, VA, 1992.
41. Ashby, D.L. et al., Potential Flow Theory and Operation Guide for the Panel Code PMARC, NASA TM-102851, 1991.
42. Stanley, G. et al., Computational Mechanics Testbed with Adaptive Refinement User's Manual, NASA contract report, 1998.
43. Fadale, T. and Sues, R.H., Reliability-based analysis and optimal design of an integral airframe structure lap joint, in *Proceedings of 9th International Space Planes and Hypersonic Systems and Technology Conference*, AIAA 99-1604, AIAA/ASCE/AHS/ASC, Norfolk, VA, 1999.
44. *Military Handbook 5G — Metallic Materials and Elements for Aerospace Vehicle Structures*, Vol. 1 and 2, 1994.
45. Aminpour, M.A., Shin, Y., Sues, R.H., and Wu, Y.-T., A framework for reliability-based MDO of aerospace systems, in *Proceedings of 43rd Structures, Structural Dynamics, and Materials Conference*, AIAA 2002-1476, AIAA/ASCE/AHS/ASC, Denver, CO, 2002.

35
Probabilistic Progressive Buckling of Conventional and Adaptive Trusses

	35.1	Introduction 35-1
	35.2	Fundamental Approach and Considerations 35-2
		Finite Element Model • Buckling of Columns • Probabilistic Model • Probabilistic Progressive Buckling of Conventional Trusses
	35.3	Discussion of Results 35-5
		Probabilistic Progressive Buckling: First Buckled Member • Probabilistic Progressive Buckling: Second/Third/Fourth Buckled Members • Probabilistic Truss End-Node Displacements • Probabilistic Buckling Including Initial Eccentricity
Shantaram S. Pai *National Aeronautics and Space Administration*	35.4	Adaptive/Smart/Intelligent Structures 35-15
Christos C. Chamis *National Aeronautics and Space Administration*	35.5	Discussion of Results 35-18
		Adaptive Structure • Smart Structure • Intelligent Structure
	35.6	Summary 35-26

35.1 Introduction

It is customary to evaluate the structural integrity of trusses by using deterministic analysis techniques and appropriate load/safety factors. Traditionally, these factors are an outcome of many years of analytical, as well as experimental, experience in the areas of structural mechanics/design. Load factors are used to take into account uncertainties in many different operating conditions, including the maximum loads. Safety factors are used to account for unknown effects in analysis assumptions, fabrication tolerances, and material properties.

An alternative to the deterministic approach is the probabilistic analysis method (PSAM) [1]. This method formally accounts for various uncertainties in primitive variables (fundamental parameters describing the structural problem) and uses different distributions such as the Weibull, normal, lognormal, etc. to define these uncertainties. Furthermore, PSAM assesses the effects of these uncertainties on the scatter of structural responses (displacements, frequencies, eigenvalues). Thus, PSAM provide a more realistic and systematic way to evaluate structural performance and durability. A part of PSAM is a computer code NESSUS (numerical evaluation of stochastic structures under stress), which provides a choice of solution for static, dynamic, buckling, and nonlinear analysis [2, 3].

In the recent past, NESSUS has been used for the analysis of space shuttle main engine (SSME) components. Representative examples include a probabilistic assessment of a mistuned bladed disk assembly [4] and an evaluation of the reliability and risk of a turbine blade under complex service environments [5]. Furthermore, NESSUS has also been used to computationally simulate and probabilistically evaluate a cantilever truss typical for outer-space-type structures [6] and quantify the uncertainties in the structural responses (displacements, member axial forces, and vibration frequencies). The objective of this chapter is to develop a methodology and to perform probabilistic progressive buckling assessment of space-type trusses using the NESSUS computer code. The space trusses evaluated include trusses with and without adaptive structural concepts: specifically, adaptive, smart, and intelligent structures. The implementation of each of these is described in detail in the Section 35.4, which discusses adaptive/smart/intelligent structures.

35.2 Fundamental Approach and Considerations

One of the major problems encountered in the analysis of space-type trusses is to come up with a stable and optimum configuration for given loading conditions and to be able to probabilistically analyze them to take into account the probable uncertainties in the primitive variables typical for environment conditions in outer space. Presently, it is a practice to design these trusses with cross bracings, thereby increasing the overall weight of the truss, the cost of fabrication, and the effort to deploy in space. Furthermore, the presently available methods/programs do not easily allow us to identify any local instability in any of the internal members of the truss during probabilistic buckling (eigenvalue) analysis and to calculate overall margins of safety of the truss.

A probabilistic methodology for analyzing progressive buckling has been developed using the NESSUS code. The method is described in this chapter.

35.2.1 Finite Element Model

A three-dimensional, three-bay cantilever truss is computationally simulated using a linear isoparametric beam element based on the Timoshenko beam equations. The element is idealized as a two-noded line segment in three-dimensional space. The cantilever truss is assumed to be made from hollow circular pipe members. The members are made up of wrought aluminum alloy (616-W) with modulus of elasticity (E) equal to 10 Mpsi. The outer and inner radii (r_o and r_i) of the tube are 0.5 and 0.4375 in., respectively. All six degrees of freedom are restrained at the fixed end (left side) nodes. Each bay of the truss is 5 ft wide, 8 ft long, and 6 ft high (Figure 35.1). The overall length of the truss is 24 ft. Six vertical and two longitudinal

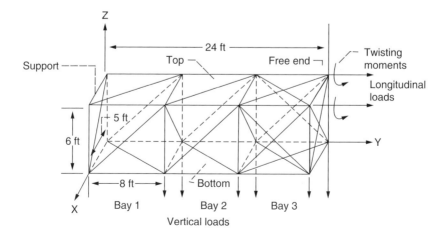

FIGURE 35.1 Solar-array panels mast — typical truss.

TABLE 35.1 Primitive Variables and Uncertainties for Probabilistic Structural Analysis of a Space Truss

	Primitive Variables	Distribution Type	Mean Value	Scatter, ± Percent
Geometry	Width	Normal	60 in.	0.5
	Length	Normal	96 in.	0.1
			192 in.	0.1
			288 in.	0.1
	Height	Normal	72 in.	0.2
Loads	Vertical	Lognormal	20 lb	6.3
	Longitudinal	Lognormal	20 lb	2.5
	Twisting moment	Lognormal	50 lb-in.	6.3
Material property	Modulus	Normal	10 Mpsi	7.5
Tube radii	Outer radius	Normal	0.5 in.	7.5
	Inner radius	Normal	0.44 in.	7.5

Note: Random input data.

loads are applied. In addition, twisting moments are applied at the truss-end nodes. The directions of the forces and moments are shown in Figure 35.1, and mean values are given in Table 35.1. The applied loads and moments are selected to represent anticipated loading conditions for a typical space-type truss.

In general, the finite element equation for motion is written as:

$$[M]\{\ddot{u}\} + [C]\{\dot{u}\} + [K]\{u\} = F(t) \tag{35.1}$$

where $[M]$, $[C]$, and $[K]$ denote the mass, damping, and stiffness matrices, respectively. It is important to note that these matrices are calculated probabilistically in the NESSUS code. Furthermore, $\{\ddot{u}\}$, $\{\dot{u}\}$, and $\{u\}$ are the acceleration, velocity, and vectors at each node, respectively. The forcing function vector, $\{F(t)\}$, is time independent at each node.

In this discussion, the static case is considered by setting the mass and damping matrices to zero and considering the forcing function to be independent of time in Equation 35.1, such that

$$[K]\{u\} = \{F\} \tag{35.2}$$

It is important to note that in the NESSUS code, a linear buckling analysis is carried out using a subspace iteration technique to evaluate the probabilistic buckling load. The matrix equation for the buckling (eigenvalue) analysis for a linear elastic structure is as follows:

$$\{[K] - \lambda[K_g]\}\{\Phi\} = 0 \tag{35.3}$$

In Equation 35.3, $[K]$ is the standard stiffness matrix, $[K_g]$ is the geometric stiffness matrix, λ is the eigenvalue, and Φ represents the eigenvectors.

Furthermore, the vibration frequency analysis is also carried by setting only the damping matrix to zero and using the following equation:

$$\{[K] - \lambda^2[M]\}\{\Phi\} = 0 \tag{35.4}$$

Finally, the NESSUS/FPI (fast probability integration) module extracts the response variables (buckling loads, vibration frequencies, and member axial forces) to calculate respective probabilistic distributions and respective sensitivities associated with the corresponding uncertainties in the primitive variables. The mean, distribution type, and percentage variation for each of the primitive variables are given in Table 35.1.

35.2.2 Buckling of Columns

In slender columns, a relatively small increase in the axial compressive forces will result only in axial shortening of the member. However, the member suddenly bows out sideways if the load level reaches a certain critical level. Large deformations caused by an increase in the induced bending moment levels may lead to the collapse of the member. On the other hand, tension members as well as short stocky columns fail when the stress in the member reaches a certain limiting strength of the material. According to Chajes [7], "Buckling, however, does not occur as a result of the applied stress reaching a certain predictable strength of the material. Instead, the stress at which buckling occurs depends on a variety of factors, including the dimensions of the member, the way in which the member is supported, and the properties of the material out of which the member is made." Chajes also describes the concept of neutral equilibrium that is being used to determine the critical load of a member such that, at this load level, the member can be in equilibrium both in the straight and in a slightly bent configuration. Furthermore, the Euler load (buckling load or critical load) is the smallest load at which a state of neutral equilibrium is possible or the member ceases to be in stable configuration. This above definition of buckling load is used to identify the probable truss members that contribute to the progressive buckling behavior of the cantilever truss.

35.2.3 Probabilistic Model

The following primitive variables are considered in the probabilistic analysis:

Nodal coordinates (X, Y, Z)
Modulus of elasticity (E)
Outer radius of the tube (r_o)
Inner radius of the tube (r_i)
Vertical loads (V)
Longitudinal loads (H)
Twisting moments (M)
Variables associated with adaptive/smart/intelligent structures, as described later

It is possible that the above primitive variables will vary continuously and simultaneously due to extreme changes in the environment when such trusses are used in upper Earth orbit for space-station-type structures. The normal distribution is used to represent the uncertainties in E, r_o, r_i, and X, Y, Z coordinates. The applied loads and moments are selected to represent an anticipated loading for a typical space-type truss. The scatter in these is represented by lognormal distributions. Initially, the NESSUS/FEM (finite element methods) module is used to deterministically analyze the truss for mean values of each of these primitive variables. In the subsequent probabilistic analysis, each primitive variable is perturbed independently and by a different amount. Usually, the perturbed value of the primitive variable is obtained by a certain factor of the standard deviation on either side of the mean value. It is important to note that, in the NESSUS code, a linear buckling analysis is carried out by making use of the subspace iteration technique to evaluate the probabilistic buckling load (see Equation 35.3). Finally, the NESSUS/FPI (fast probability integration) module extracts eigenvalues to calculate a probability distribution of the eigenvalues and to evaluate respective sensitivities associated with the corresponding uncertainties in the primitive variables. The mean, distribution type, and percentage variation for each of the primitive variables are given in Table 35.1.

35.2.4 Probabilistic Progressive Buckling of Conventional Trusses

Initially, the truss is deterministically analyzed for member forces and to identify the members in which the axial forces exceed the Euler load. These members are then discretized with several intermediate nodes, and a probabilistic buckling (eigenvalue) analysis is performed to obtain probabilistic buckling loads and respective buckled shapes. Furthermore, the sensitivity factors representing the impact of uncertainties in the primitive variables on the scatter of response variable (eigenvalue) are evaluated. Finally, any members that have buckled are identified, and the probabilistic buckled loads/moments at each probability level are obtained by multiplying the respective eigenvalues with the applied loads and moments.

In the subsequent analyses, the buckled members are removed from the original truss configuration, and the above-described analysis steps are repeated until the onset of collapse state. It is important to note that the mean values of the loads and moments are kept constant and are perturbed around their means during the probabilistic buckling analysis. The truss end-node displacements vs. the number of members removed are plotted to identify the onset of the truss collapse state. Finally, the minimum number of members needed to support the applied loads and moments are determined.

35.3 Discussion of Results

35.3.1 Probabilistic Progressive Buckling: First Buckled Member

Figure 35.2a through Figure 35.2f show the probabilistic progressive buckled mode shapes of the three-bay space truss as individual buckled members are sequentially removed from the original configuration until it reaches the onset of collapse. The probabilistic buckling analysis indicated that the first bay's front diagonal buckled first (Figure 35.2b). The corresponding probabilistic buckled loads and moments at 0.5 probability are shown in Figure 35.3. Probabilistic buckled loads and moments at different probability levels can also be obtained. Furthermore, Figure 35.4 shows a method of calculating the margin of safety (MOS) for specified probability by using known distributions for applied loads and moments and corresponding cumulative distribution function curves obtained from PSAM.

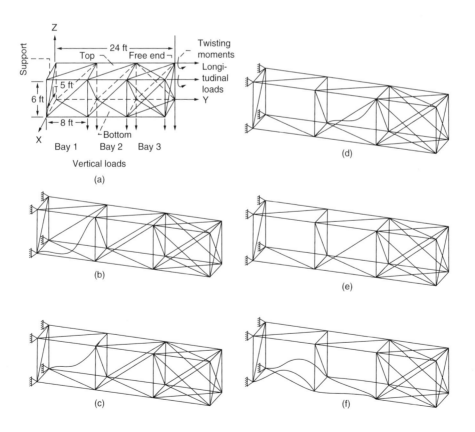

FIGURE 35.2 Probabilistic progressive buckling as buckled members are sequentially removed.

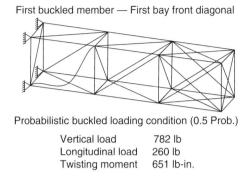

FIGURE 35.3 Probabilistic progressive buckling for first buckled member at 0.5 probability.

The sensitivity factors from Figure 35.5 suggest that the scatter in the bay length parameter (Y-coordinate) had the highest impact on the probabilistic distribution of the buckling load, followed by the bay height (Z-coordinate), bay width (X-coordinate), vertical and longitudinal loads, and finally twisting moments. Any slight variation in spatial (geometry) variables has a direct effect on the overall length of the members and thereby alters many terms in the stiffness matrix containing the length parameter. Finally, this has a definite effect on the probabilistic buckling loads that has been clearly observed in the results discussed above. However, it is important to note that even comparatively large variations in both member modulus (E) and area (r_o and r_i) (see Table 35.1) had very negligible impact. Similar conclusions can also be drawn for the probabilistic member force in the first buckled member (see Figure 35.6).

The variation in the resistance (mean area × mean yield strength) of the member was assumed to have a Weibull distribution and is shown in Figure 35.7. MOS calculations for stress exceeding strength using distribution curves for probabilistic member force and resistance as well as probabilistic buckling load and resistance indicate that the buckled member did satisfy the strength criteria condition. Therefore, it can be concluded from Figure 35.4 and Figure 35.7 that the member buckled when its axial force exceeded the Euler buckling load and when the stress due to tills load did not exceed the failure criteria.

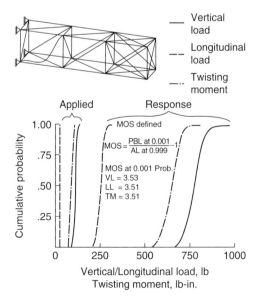

FIGURE 35.4 Probabilistic buckling load for first buckled member.

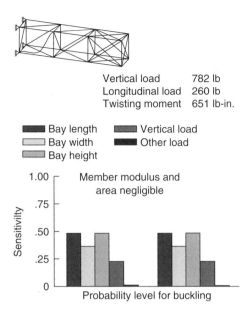

FIGURE 35.5 Sensitivity of probabilistic buckling load for first buckled member.

35.3.2 Probabilistic Progressive Buckling: Second/Third/Fourth Buckled Members

As described in the previous section, the deterministic analysis followed by the probabilistic analysis was performed with sequential removal of the first, second, third, and fourth buckled member from the truss. The probabilistic buckled loads and moments, sensitivity factors, and MOS values for stress were obtained for each stage.

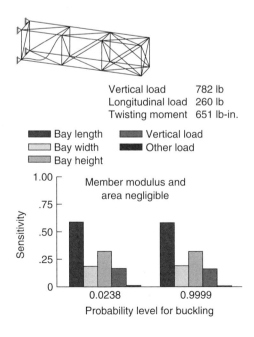

FIGURE 35.6 Sensitivity of probabilistic member force for first buckled member.

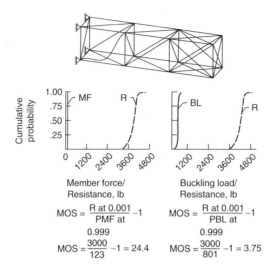

FIGURE 35.7 Probability of strength exceedence for first buckled member.

When the second member was buckled (see Figure 35.2c), the comparable results are shown in Figure 35.4 to Figure 35.6 and are described in Figure 35.8 to Figure 35.11. For these truss configurations, the MOS value decreased from 3.53 to 2.53. The similar details of the truss with the third buckled member (see Figure 35.2d) are given in Figure 35.12 to Figure 35.15. According to Figure 35.13, the MOS value further decreased to 1.62. Similarly, Figure 35.16 to Figure 35.19 present comparable results of the truss when the fourth member was buckled (see Figure 35.2e). It is important to note from Figure 35.18 and Figure 35.19 that the scatter in the bay height had much higher impact than scatter in either bay width or length on both the probabilistic buckling loads/moments and buckled member force.

Finally, the details of the onset of collapse state of the truss (Figure 35.2f) are shown in Figure 35.20 to Figure 35.22. When all the four buckled members were removed, the MOS value was equal to −3.75, which indicates that the onset of collapse was reached (Figure 35.21). Furthermore, the probabilistic buckling loads/moments at 0.001 probability value were equal to maximum applied loads/moments with assumed distributions (see Figure 35.20). In addition, at the collapse state, the uncertainties in both the bay length and bay height had sufficiently high impact on the distributions of the probabilistic buckling loads/moment (see Figure 35.22). In the various truss configurations discussed here, the uncertainties in the vertical loads had consistently the same impact on buckling loads/moment, whereas member modulus and area had negligible impact.

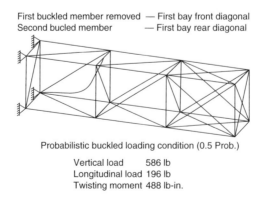

FIGURE 35.8 Probabilistic progressive buckling for second buckled member at 0.5 probability.

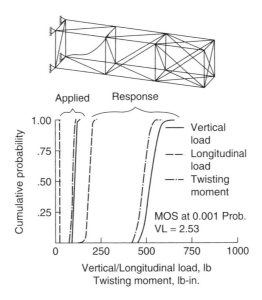

FIGURE 35.9 Probabilistic buckling load for second buckled member.

35.3.3 Probabilistic Truss End-Node Displacements

The truss end-node displacements (lateral and longitudinal) were also calculated during the deterministic analyses for each truss configuration, and these are shown in Figure 35.23. It is clear that there is no considerable change in either lateral or longitudinal displacement as each buckled member was sequentially removed. However, the truss end-node vertical displacement gradually increased up to the truss

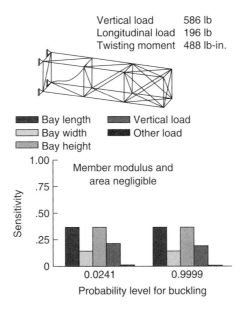

FIGURE 35.10 Sensitivity of probabilistic buckling load for second buckled member.

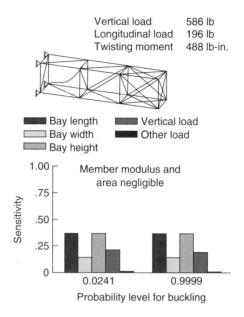

FIGURE 35.11 Sensitivity of probabilistic member force for second buckled member.

configuration with three buckled members removed, but it suddenly increased very rapidly when the fourth buckled member was removed, giving an indication of unbounded displacement growth, which suggests that the truss had reached the onset of its collapse state. This is due to the fact that the total vertical loads are six times higher than total longitudinal loads, and the perturbations in the vertical loads are higher than those of twisting moments.

Figure 35.24 and Figure 35.25 show, respectively, the relationships between the applied vertical loads and probabilistic buckling loads as well as probabilistic buckling loads and MOS values. The optimum truss configuration was reached with the fourth buckled member removed, whereby the probabilistic buckling load was equal to the applied vertical load at 0.001 probability level (see Figure 35.24). Similar conclusions can also be made for longitudinal loads and twisting moments. In addition, there is a gradual decrease in the MOS values as buckled members were sequentially removed, reaching a zero value when

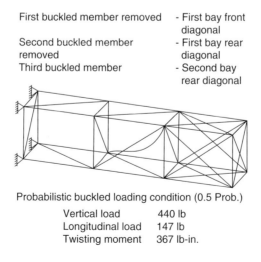

FIGURE 35.12 Probabilistic progressive buckling for third buckled member at 0.5 probability.

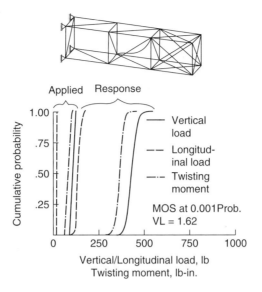

FIGURE 35.13 Probabilistic buckling load for third buckled member.

the optimum truss configuration was reached (see Figure 35.25). Again, similar conclusions can also be made for longitudinal loads and twisting moments.

35.3.4 Probabilistic Buckling Including Initial Eccentricity

In the probabilistic progressive buckling methodology discussed here, all the members were assumed to be initially perfectly straight. The buckled members were sequentially removed with the assumption that once the member buckled, it would yield and could not resist any additional loading, and thereby would

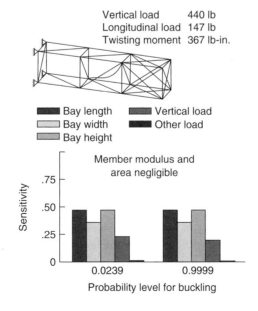

FIGURE 35.14 Sensitivity of probabilistic buckling load for third buckled member.

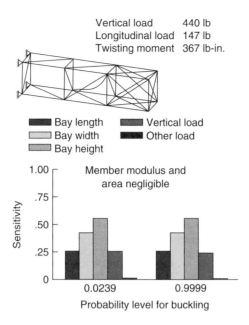

FIGURE 35.15 Sensitivity of probabilistic member force for third buckled member.

not contribute to the overall stiffness of the truss. In order to verify this assumption, we calculated the maximum eccentricity at which the yielding in the member (first-bay front diagonal) will take place due to the combined effects of axial and in-plane bending moments. Furthermore, this member was modeled to depict the buckled configuration of the member at which yielding will take place, using a parabolic distribution for the above-calculated eccentricity (see Figure 35.26).

The deterministic and subsequent probabilistic buckling analyses indicate, respectively, that the probabilistic buckling loads and moments did not change significantly from the original analysis (see

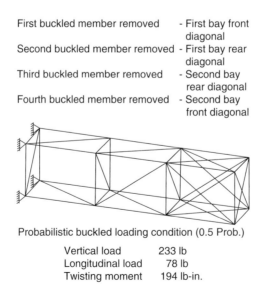

FIGURE 35.16 Probabilistic progressive buckling for fourth buckled member at 0.5 probability.

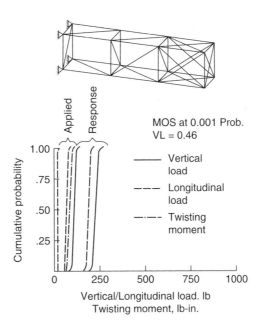

FIGURE 35.17 Probabilistic buckling load for fourth buckled member.

Figure 35.3), and that the first-bay rear diagonal has buckled (see Figure 35.27). However, as seen from Figure 35.5 and Figure 35.28 for probabilistic buckling loads and from Figure 35.6 and Figure 35.29 for probabilistic member forces, the sensitivity factors show some changes. Of particular note is the fact that the variation in bay width has the most dominant impact on both probabilistic buckling loads and moments (see Figure 35.28 and Figure 35.29). This is due to the fact that the member buckles in the

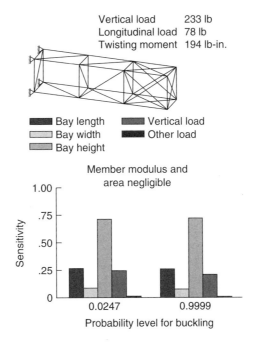

FIGURE 35.18 Sensitivity of probabilistic buckling load for fourth buckled member.

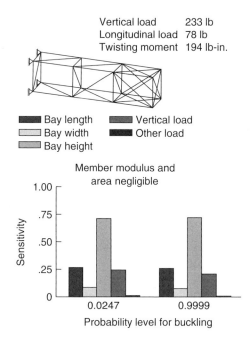

FIGURE 35.19 Sensitivity of probabilistic member force for fourth buckled member.

plane perpendicular to the direction of the loading. Nevertheless, it is important to note that the scatter in the spatial location accentuates the sensitivities of the bay length/width/height on the probabilistic load and diminishes that of vertical load. Once again, the variations in the member modulus and area have very negligible impact. These results justify the sequential removal of the buckled members during progressive buckling.

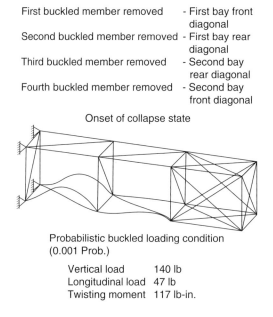

FIGURE 35.20 Probabilistic progressive buckling for onset of collapse state at 0.001 probability.

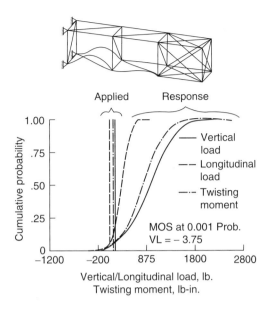

FIGURE 35.21 Probabilistic buckling load for onset of collapse state.

35.4 Adaptive/Smart/Intelligent Structures

NASA space missions have been advocating the use of adaptive/smart/intelligent structures for their spacecraft. These materials have great impact on the function of precision segmented reflectors, the control of large space truss structures, the manufacture of robotic assemblies/space cranes/manipulators, and the isolation of vibration frequencies. These materials also have a larger role in improving

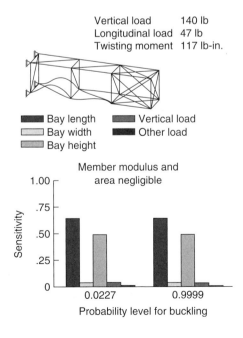

FIGURE 35.22 Sensitivity of probabilistic buckling load for onset of collapse state.

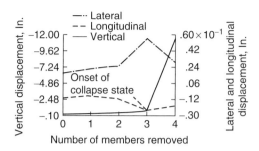

FIGURE 35.23 Progressive buckling leading to structural collapse, as indicated by unbounded displacement.

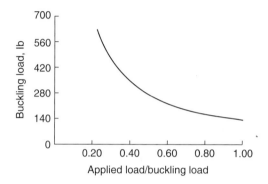

FIGURE 35.24 Buckling load vs. applied load/buckling load at 0.001 probability.

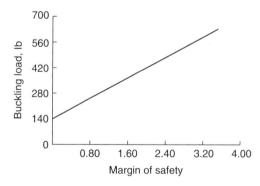

FIGURE 35.25 Buckling load vs. margin of safety at 0.001 probability.

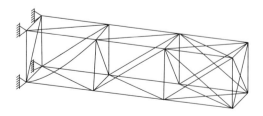

FIGURE 35.26 First-bay front diagonal with initial eccentricity.

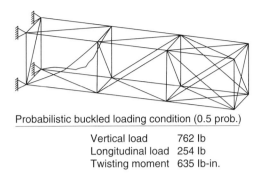

FIGURE 35.27 Probabilistic buckling: first-bay rear diagonal buckled.

the performance of aircraft and other commercial structures. However, the terminology, such as "adaptive," "smart," and "intelligent," is being used loosely and interchangeably in the research community [8].

Ahmad [8] defines the intelligent/smart materials and systems as those having "built in or intrinsic sensors, processors, control mechanisms, or actuators making it capable of sensing a stimulus, processing the information, and then responding in a predetermined manner and extent in a short/appropriate time and reverting to its original state as soon as stimulus is removed." Thus, the smart structures consist of sensors, controllers, and actuators. Furthermore, the intelligent materials usually respond quickly to environmental changes at the optimum conditions and modify their own functions according to the changes. Therefore, the intelligent truss structures are usually designed with active members early in the design process. In many instances, the trusses are designed with both active and passive members using an integrated design optimization procedure [9]. Finally, Wada [10] describes adaptive structures "as a structural system whose geometric and inherent structural characteristics can be beneficially changed to meet mission requirements either through remote commands and/or automatically in response to external stimulations." It is important to note that these structures have an in-built capability to geometrically

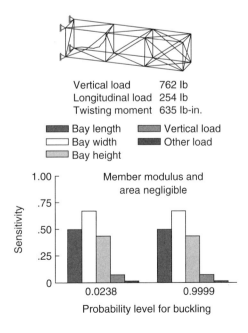

FIGURE 35.28 Sensitivity of probabilistic buckling load.

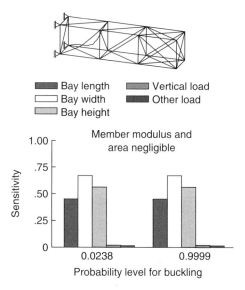

FIGURE 35.29 Sensitivity of probabilistic member force.

relocate critical points of the structure, when in space, to the desired positions through actuation of active members. Therefore, the configurations of such structures should have greater flexibility to move the critical locations.

For the evaluations performed herein, the following definitions are used:

1. When local instability is imminent, the length of that member is suitably controlled to prevent instability. This is referred to as an "adaptive structure."
2. When local instability is imminent, a redundant member engages in load sharing without weight penalty. This is referred to as a "smart structure."
3. When local instability is imminent as exhibited by bowing, the material induces a local restoring moment. This is referred to as an "intelligent structure" through the corresponding restoring action of the intelligent material.

The concepts discussed above are used for probabilistic structural analysis of adaptive/smart/intelligent structures typical for outer-space-type trusses using the NESSUS computer code. The individual analysis technique and respective results are discussed in the following section.

35.5 Discussion of Results

35.5.1 Adaptive Structure

The deterministic analysis indicated that the first sign of local buckling occurs in the first-bay front diagonal (Figure 35.30). Since buckling varies as the length squared, a decrease in length should prevent buckling at that load. Therefore, a suitable device or sensor can be attached to this truss member that will not only sense the local buckling in the member due to significantly high axial force, but also will automatically reduce the overall length of the member by predetermined increment. Thus, this member acts like an adaptive member, whereby its geometrical parameters were changed accordingly.

Figure 35.30 through Figure 35.35 show the cumulative distribution functions (CDF) and corresponding sensitivities of the probabilistic buckling loads and vibration frequencies of the truss and axial forces in the diagonal (member). By reducing the length of the member by 6 in., the probabilistic buckling loads increased by 6% (Figure 35.30). The sensitivity factors from Figure 35.31 show that the uncertainties in the bay height (Z-coordinate) had the highest impact on the probabilistic buckling loads. The probabilistic

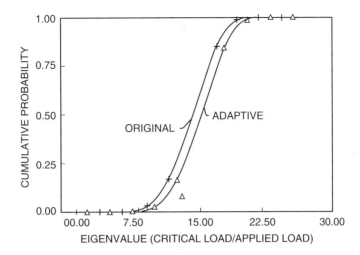

FIGURE 35.30 Probabilistic buckling load for adaptive structure.

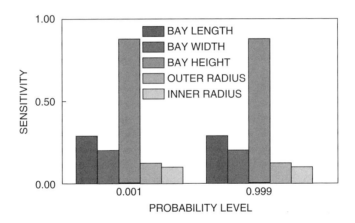

FIGURE 35.31 Sensitivity of probabilistic buckling load for adaptive structure.

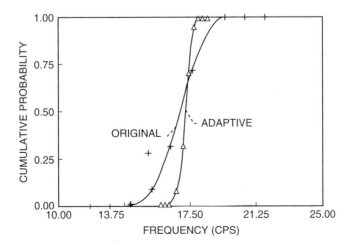

FIGURE 35.32 Probabilistic vibration frequency for adaptive structure.

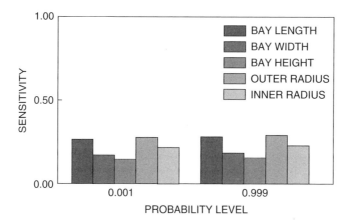

FIGURE 35.33 Probabilistic frequency sensitivities for adaptive structure.

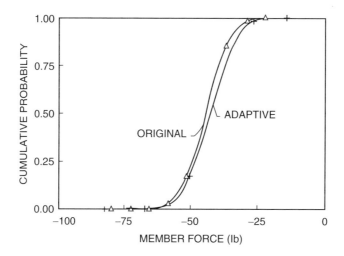

FIGURE 35.34 Probabilistic member force for adaptive structure.

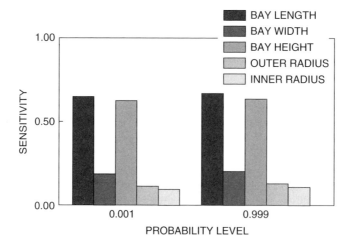

FIGURE 35.35 Sensitivity of probabilistic member force for adaptive structure.

vibration frequencies increased by 25 to 35% for lower probability levels (Figure 35.32). The scatter in the tube radii had equally significant impact on the probabilistic vibration frequencies (Figure 35.33). The magnitude of the probabilistic member axial forces decreased by 5% (Figure 35.34), and the scatter in bay length had the highest impact on the probabilistic member axial forces, followed by bay height (Figure 35.35). Thus, the adaptive structures are effective in increasing the buckling loads and vibration frequencies as well as controlling the member axial forces.

35.5.2 Smart Structure

As mentioned earlier, in the case of a smart structure, the original single hollow member (diagonal) was replaced with two hollow tubes. Once again, the outer tube was made up of wrought aluminum alloy (616-W), and the outer and inner radii of the tube were 0.5 and 0.468755 in., respectively. However, the inlet (inner) tube was modeled using a high-modulus-fiber–intermediate-modulus-matrix composite with 60% fiber–volume ratio. For this tube, the modulus of elasticity was equal to 36 Mpsi with 0.421875 and 0.384375 in. outer and inner radii, respectively. It was assumed that the inner composite tube can be inserted inside the outer tube without affecting the details of the member end connections. It is important to note that the composite tube not only reduces the overall weight of the truss, but also increases the stiffness, and it is assumed that this tube was made with tight tolerance. Therefore, the scatter in E and tube radii is not considered in the probabilistic analysis. In addition, the aluminum tube is also useful in protecting the composite tube from possible damage from orbital environmental debris.

Figure 35.36 shows that the probabilistic buckling loads increased by almost 30% at several probability levels. The sensitivity factors show that the uncertainties in the bay height had the highest impact on the probabilistic buckling loads (Figure 35.37). The probabilistic frequencies increased by 15% (Figure 35.38). The scatter in the inner-tube radii had equally significant impact on the probabilistic frequencies (Figure 35.39). Similarly, the magnitude of the probabilistic member forces increased (Figure 35.40), and the scatter in the bay length and height had equally significant impact on the probabilistic member axial forces (Figure 35.41). Once again, the smart structures can also be used to increase the probabilistic buckling loads and frequencies and to control the forces in the member.

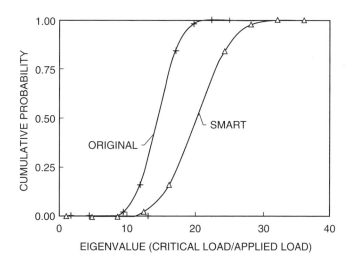

FIGURE 35.36 Probabilistic buckling load for smart structure.

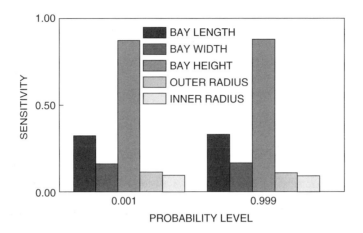

FIGURE 35.37 Sensitivity of probabilistic buckling load for smart structure.

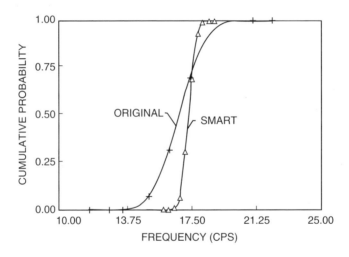

FIGURE 35.38 Probabilistic vibration frequency for smart structure.

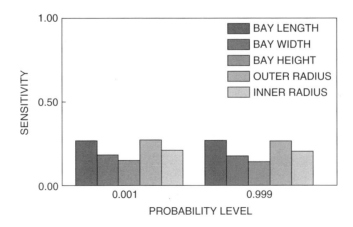

FIGURE 35.39 Probabilistic frequency sensitivities for smart structure.

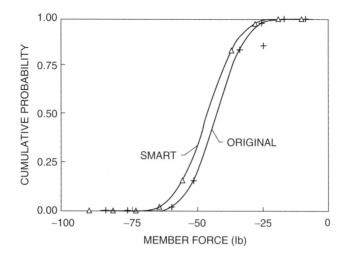

FIGURE 35.40 Probabilistic member force for smart structure.

35.5.3 Intelligent Structure

It is important to note that all of the truss members were assumed to be initially perfectly straight and that when any member buckled it would yield [5]. Therefore, the maximum eccentricity at which the yielding in the member (first-bay front diagonal) will take place due to the combined effects of axial and in-plane bending moments was calculated. Furthermore, this member (first-bay front diagonal) was modeled to represent the buckled configuration of the member at which yielding will take place by using a parabolic distribution with increased eccentricities. At the center of this diagonal (original shape), localized stress concentrators can be attached that will detect the local instability in the member and automatically apply a restoring moment of 15 lb-in. at the center of this diagonal. Thus, this diagonal acts like an intelligent structural member due to the action of intelligent material, and the loading parameters will be changed accordingly.

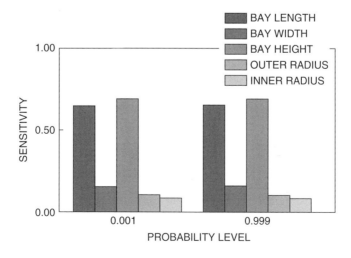

FIGURE 35.41 Sensitivity of probabilistic member force for smart structure.

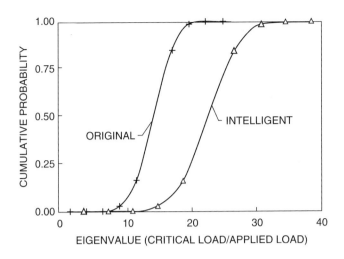

FIGURE 35.42 Probabilistic buckling load for intelligent structure.

It can be concluded that the probabilistic buckling loads increased by 50% (Figure 35.42) and that the scatter in the bay height had the highest impact on the probabilistic buckling loads (Figure 35.43). The probabilistic buckling frequencies increased by 20% only for the lower probability levels (Figure 35.44). Therefore, the level of scatter in the primitive variables did not increase the probabilistic frequencies at higher probability levels. The variations in the member radii had equally significant impact on the probabilistic vibration frequencies (Figure 35.45). The magnitude of the probabilistic member axial forces decreased by 40% (Figure 35.46), and the uncertainties in both bay length and bay height had a very significant impact on the probabilistic member forces (Figure 35.47). Finally, with the help of in-built intelligent materials, the intelligent structures are very useful in increasing the probabilistic buckling loads and decreasing the member axial forces.

Finally, the methodologies discussed above can be applied to more than one member at the same time, if the situation demands, and the probabilistic analysis can be carried to evaluate various CDFs and determine the structural performance and durability of the truss.

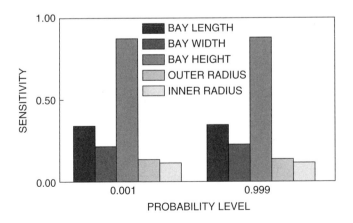

FIGURE 35.43 Sensitivity of probabilistic buckling load for intelligent structure.

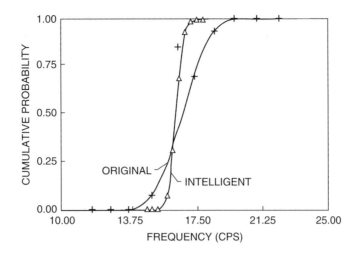

FIGURE 35.44 Probabilistic vibration frequency for intelligent structure.

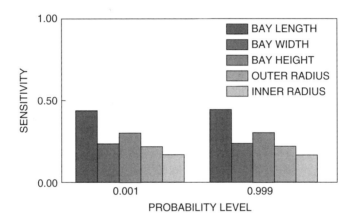

FIGURE 35.45 Probabilistic frequency sensitivities for intelligent structure.

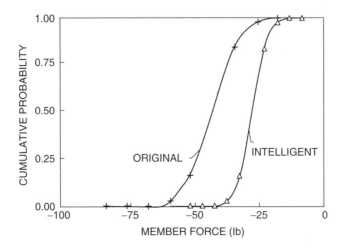

FIGURE 35.46 Probabilistic member force for intelligent structure.

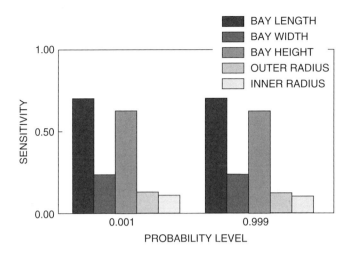

FIGURE 35.47 Sensitivity of probabilistic member force for intelligent structure.

35.6 Summary

The computational simulation of probabilistic evaluation for conventional and adaptive/smart/intelligent behavior of truss structures is demonstrated using the NESSUS computer code, and step-by-step procedures are outlined. Scatter of the probabilistic buckling loads, vibration frequencies, and member axial forces are evaluated, and the sensitivities associated with the uncertainties in the primitive variables are determined.

For the conventional truss, the results indicate that: (1) probabilistic buckling loads and margin-of-safety values decrease as buckled member(s) are sequentially removed; (2) the scatter in truss geometry (bay length/width/height) and vertical loads have considerable impact on the probability of the buckled load; (3) the member modulus and area parameters have negligible impact; and (4) initial eccentrics have negligible influence on the probabilistic buckling load but may influence the sensitivities.

For the adaptive/smart/intelligent truss, the results indicate that (1) the probabilistic buckling loads and vibration frequencies increase for each truss classification; however, they increase significantly for the case of intelligent structure; (2) the magnitude of the probabilistic member axial forces increases for the smart structure and decreases for both adaptive and intelligent structures, with a considerable decrease for the intelligent structure; (3) for each structure, the scatter in the bay height has the highest impact on the probabilistic buckling loads; (4) the scatter in member area parameters have equally significant impact on the probabilistic frequencies; (5) the uncertainties in the bay length/height have equally significant effects on the probabilistic member forces.

Collectively, the results indicate that all three structures can be used to increase the probabilistic buckling loads. However, the intelligent structure gives the highest increase. Furthermore, both adaptive and smart structures are recommended for controlling the frequencies, but this is not true for intelligent structures. Finally, the adaptive/smart/intelligent structures are recommended for controlling the member axial forces. Collectively, the results demonstrate that the probability of collapse of space-type trusses can be reliability assessed by the procedure described herein.

References

1. Chamis, C.C., Probabilistic structural analysis methods for space propulsion system components, in *Space System Technology Conference Proceedings*, AIAA, New York, 1986, pp. 133–144.
2. Dias, J.B., Nagtegaal, J.C., and Nakazawa, S., Iterative perturbation algorithms in probabilistic finite analysis, in *Computational Mechanics of Probabilistic and Reliability Analysis*, Liu, W.K. and Belytschko, Eds., ELME PRESS International, Lausanne, Switzerland, 1989, pp. 211–230.

3. Wu, Y.T., Demonstration of new, fast probability integration method for reliability analysis, in *Advances in Aerospace Structural Analysis*, Proceedings of ASME Winter Annual Meeting, Bumside, O.M., Ed., ASME, 1985, pp. 63–73.
4. Shah, A.R., Nagpal, V.K., and Chamis, C.C., Probabilistic analysis of bladed turbine disks and the effect of mistuning, in *Proceedings of 31st Structures, Structural Dynamics, and Materials Conference*, Part 2, AIAA, New York, 1990, pp. 1033–1038; also NASA TM-102564, NASA.
5. Shiao, M.C. and Chamis, C.C., A methodology for evaluating the reliability and risk of structures under complex service environments, in *Proceedings of 31st Structures, Structural Dynamics, and Materials Conference*, Part 2, AIAA, New York, 1990, pp. 1070–1080; also NASA TM-103244, NASA.
6. Pai, S.S., Probabilistic Structural Analysis of a Truss Typical for Space Station, NASA TM-103277, NASA, 1990.
7. Chajes, A., *Principles of Structural Stability Theory*, Prentice-Hall, Englewood Cliffs, NJ, 1974.
8. Ahmad, I., U.S.–Japan workshop on smart/intelligent materials and systems, *Onrasia Sci. Info. Bull.*, 15 (4), 67–75, 1990.
9. Manning, R.A., Optimum design of intelligent truss structures, in *Proceedings of 32nd Structures, Structural Dynamics, and Materials Conference*, Part 1, AIAA, New York, 1991, pp. 528–533.
10. Wada, B.K., Adaptive structures, in *Proceedings of 30th Structures, Structural Dynamics, and Materials Conference*, Part 1, AIAA, New York, 1989, pp. 1–11.

36
Integrated Computer-Aided Engineering Methodology for Various Uncertainties and Multidisciplinary Applications

Kyung K. Choi
College of Engineering, University of Iowa

Byeng D. Youn
College of Engineering, University of Iowa

Jun Tang
College of Engineering, University of Iowa

Jeffrey S. Freeman
University of Tennessee

Thomas J. Stadterman
U.S. Army Materiel Systems Analysis Activity (AMSAA)

Alan L. Peltz
U.S. Army Materiel Systems Analysis Activity (AMSAA)

William (Skip) Connon
U.S. Army Aberdeen Test Center

36.1 Introduction 36-1
36.2 Fatigue-Life Analysis and Experimental Validation 36-3
   Mechanical Fatigue Failure for Army Trailer Drawbar • Experimental Validation of Mechanical Fatigue • Design Optimization for Mechanical Fatigue
36.3 Reliability Analysis and Reliability-Based Design Optimization 36-12
   Reliability Analysis for Durability-Based Optimal Design • Reliability-Based Design Optimization for Durability • Results of Reliability-Based Shape Design Optimization for Durability of M1A1 Tank Roadarm
36.4 Conclusions 36-18

36.1 Introduction

Given the explosive growth in computational technology, computer-aided engineering (CAE) has long been used to analyze and evaluate product design. However, various uncertainties in an engineering system prevent CAE from being directly used for such purposes. Through the use of experimental validation and probabilistic methods, CAE will become an integral part of engineering product analysis and design. This chapter presents an advanced CAE methodology for qualitative, reliable, durable, and cost-effective product design under uncertainty that is composed of three key elements: CAE technology, experimental validation, and an uncertainty-based design [1–5].

CAE technology, such as simulation techniques, enables one to explore many different designs without building expensive prototype models. As shown in Figure 36.1, one must inevitably take account of physical input uncertainties, such as geometric dimensions, material properties, and loads. However, a simulation (or mathematical) model that departs from a prototype model introduces modeling uncertainty, uncertainty due to approximations in numerical algorithms, as well as any inherent physical

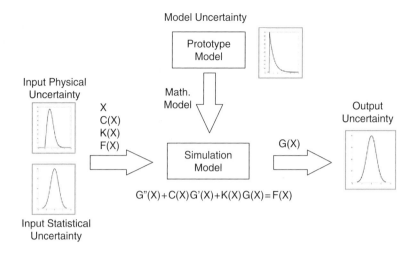

FIGURE 36.1 Uncertainty types existing in computer-aided design.

uncertainty in the structure. It is not possible to completely eliminate model uncertainty. Instead it could be more practical to minimize modeling uncertainty through experimental validation. Moreover, while modeling physical uncertainty, a lack of statistical information can lead to statistical uncertainty, such as uncertainty of the distribution type and its parameters, which could be modeled using Bayesian probability or possibility or evidence theory [6, 7]. In fact, any engineering uncertainty can be categorized within three general types: physical uncertainty, model uncertainty, and statistical uncertainty [8].

This chapter presents an integrated CAE methodology based on experimental validation and an uncertainty-based design, as shown in Figure 36.2. Assuming that there is enough statistical information so that statistical uncertainty is minimal, an advanced CAE methodology can be presented with experimental validation and reliability-based design. As a result, a high-fidelity model and analysis can be created that takes physical and model uncertainties into account.

Mechanical fatigue subject to external and inertial transient loads in the service life of a mechanical system often leads to structural failure due to accumulated damage [9]. A structural-durability analysis that predicts the fatigue life of a mechanical component subjected to dynamic stresses and strains is an intensive and complicated multidisciplinary simulation process, since it requires the integration of several CAE tools and a large amount of data communication and computation. Uncertainties in geometric dimensions and material properties due to manufacturing tolerances result in an indeterministic nature of fatigue life for the mechanical component. The main objective of this chapter is thus to demonstrate the possibilities of using advanced CAE methodology to predict structural durability with experimental

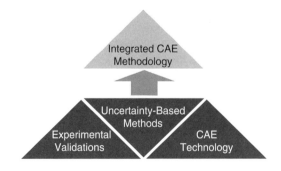

FIGURE 36.2 Integrated computer-aided engineering methodology.

validation and reliability-based design. In this way, it is possible to determine whether the modeling and simulation is feasible and ascertain whether an optimal design is reliable. One of the primary challenges in developing an integrated CAE methodology is to produce effective reliability-based design optimization (RBDO) methods [1, 4, 5].

36.2 Fatigue-Life Analysis and Experimental Validation

36.2.1 Mechanical Fatigue Failure for Army Trailer Drawbar

The U.S. Army trailer encountered a mechanical failure due to damage accumulation after driving 1,671 miles on the Perryman course no. 3 at a constant speed of 15 mph. The failure is illustrated in Figure 36.3. This failure initiated further research using a physics-of-failure model of the U.S. Army trailer that involved the validation of the simulation model, dynamic analysis, and durability analysis. The goal was to further improve the trailer's design to extend its overall fatigue life and minimize its weight.

36.2.2 Experimental Validation of Mechanical Fatigue

36.2.2.1 Validation of Simulation Model

The computer-aided design (CAD) model for the U.S. Army trailer is developed with Pro/Engineer to simulate a real trailer model (Figure 36.4). To achieve high fidelity, the CAD model goes through an experimental validation, which includes comparison of the mass/inertial properties and the natural frequencies and mode shapes of the model with physical measurements as seen in Figure 36.5(a).

Table 36.1 shows the results of the modal analysis. The results indicate that the simulation results are close to the experimental results obtained from modal analysis as shown in Figure 36.5(b). Meanwhile, other mechanical components — tire, axle, shock absorber, etc. — are appropriately modeled through experimental validation.

FIGURE 36.3 Army trailer and its structural failure.

 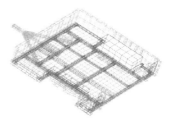

FIGURE 36.4 CAD and finite element models of trailer.

(a) Yaw Moment of Inertia (b) Experimental Modal Analysis

FIGURE 36.5 Experimental validations: (a) yaw moment of inertia; (b) experimental modal analysis.

TABLE 36.1 Results of Modal Analysis

Mode No.	Frequency (Simulation)	Frequency (Experiment)
1	18.92	17.73
2	21.55	22.14
3	26.30	29.07

36.2.2.2 Validation of Dynamic Analysis

A multibody dynamics model of HMMWV is created to drive the trailer on the Perryman course no. 3 at a constant speed of 15 mph (Figure 36.6). The trailer is modeled as a flexible dynamics model. A 30-sec dynamic simulation is performed with a maximum integration time step of 0.005 sec using the DADS dynamic analysis package [10]. To validate the dynamics model and analysis, dynamic strain (or stress) is measured by installing a (rosette) strain gauge at critical regions to collect strain time histories. As shown in Figure 36.7, a power spectral-density (PSD) curve of dynamic strain is used to compare testing data with simulation results, in addition to performing a statistical comparison of mean, root mean square, skewness, and kurtosis of the dynamic strain.

FIGURE 36.6 Dynamic model and analysis of trailer.

36.2.2.3 Validation of Durability Analysis

For a durability analysis, the fatigue life for crack initiation is calculated at those critical regions in the mechanical system that experience a short life span. The fatigue-life analysis consists of two primary computations: dynamic stress and fatigue life computations (Figure 36.8). Dynamic stress can be obtained from either a hardware prototype experiment in which sensors or transducers are placed on the physical component, or from numerical simulation. Using simulation, a stress influence coefficient (SIC) [11] obtained from quasistatic finite element analysis (FEA) using MSC NASTRAN is superposed with dynamic analysis results (including external forces, accelerations, and angular velocities) to compute the dynamic stress history. This history is then used to compute the crack-initiation fatigue life of the component.

Durability analysis is carried out using durability analysis and reliability workspace (DRAW) programs developed at the University of Iowa [11]. A preliminary durability analysis is executed to estimate the fatigue life of the U.S. Army trailer and to predict the critical regions that experience a low fatigue life.

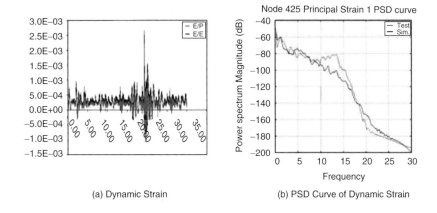

FIGURE 36.7 Validation of dynamic strain: (a) dynamic strain; (b) PSD curve of dynamic strain.

The critical regions on the drawbar assembly are clearly shown in Figure 36.9, excluding any fictitious critical regions that are the result of modeling imperfections due to applied boundary conditions. To compute the multiaxial crack initiation life of the drawbar, the equivalent von Mises strain approach is employed, which is described by the fatigue resistance and cyclic strength of the material as [12]

$$\frac{\Delta \varepsilon}{2} = \frac{\sigma'_f}{E}(2N_f)^b + \varepsilon'_f(2N_f)^c \quad \text{and} \quad \sigma = K'(\varepsilon_p)^{n'} \qquad (36.1)$$

with the empirical constants $(\sigma', \varepsilon', b, c, K', n')$. Details on durability analysis are discussed in Refs. [11, 12].

36.2.3 Design Optimization for Mechanical Fatigue

Because damage accumulation leads to structural fatigue failure in the drawbar assembly, durability design optimization for the U.S. Army trailer drawbar was carried out to improve its fatigue life and to minimize weight. The critical region where mechanical fatigue failure occurs is now taken into account in conducting the design optimization process. The integrated design optimization process involves: design parameterization [13], design sensitivity analysis (DSA) [13], and design optimization [14]. Design parameters of the drawbar assembly are carefully defined, taking geometric and manufacturing restrictions into account.

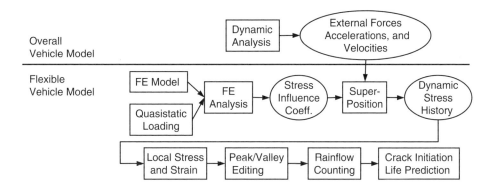

FIGURE 36.8 Computation process for fatigue life.

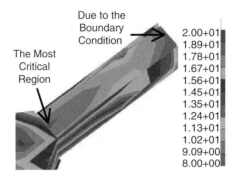

FIGURE 36.9 Fatigue-life contour of army trailer drawbar.

36.2.3.1 Design Parameterization

As shown in Figure 36.10, the drawbar assembly is composed of one central bar, two side bars, six side angles, two side attachments, and top and bottom plates. The optimum design of the drawbar assembly needs to be symmetric, and thus design parameterization is made to yield a symmetric design. Bars and attachments at the initial design have a uniform thickness. However, the thicknesses of those elements of the drawbar assembly that can be changed during the design optimization process are modeled as sizing design parameters. While maintaining the rectangular shape of the central and side bars, their height and width are considered as shape design parameters.

As shown in Table 36.2, seven design parameters are defined for the drawbar assembly. The first five are sizing design parameters, which include the thicknesses of the drawbar, side angles, and attachments. Two shape design parameters are defined as the width and height of the cross-sectional geometry of the drawbar.

36.2.3.2 Design Sensitivity Analysis for Fatigue Response

The sensitivity computational procedure for fatigue life is shown in Figure 36.11. First, quasistatic loadings need to be computed, which consist of inertial and reaction forces. For this problem, there are a total of 114 quasistatic loading cases. These cases are applied to the drawbar assembly to perform FEAs and to obtain the SICs, which are then used to compute a dynamic stress history of the current design. This dynamic stress history is used to predict fatigue life of the perturbed design. A continuum-based DSA of the SICs is also carried out [13, 15], which is then used to predict the dynamic stress history of the perturbed design. This perturbed dynamic stress history is then used to predict the fatigue life of the

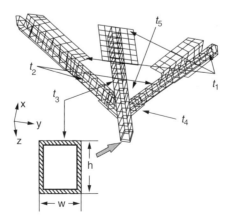

FIGURE 36.10 Design parameters of drawbar and attachments.

TABLE 36.2 Design Parameters

Design	Symbol	Description
d_1	t_1	thickness of six side angles
d_2	t_2	thickness of two side bars
d_3	t_3	thickness of center bar
d_4	t_4	thickness of two side attachments
d_5	t_5	thickness of top and bottom plates
d_6	w	cross-sectional width of three bars
d_7	h	cross-sectional height of three bars

perturbed design. Finally, the design sensitivity of fatigue life is computed by taking a finite difference of the original and perturbed fatigue life.

36.2.3.3 Durability Design Optimization

The design objective is to increase the fatigue life of the drawbar while minimizing the weight of the trailer's drawbar assembly. Due to restrictions placed on manufacturing and assembling processes, side constraints are generally imposed on the design parameters. Therefore, the design optimization problem

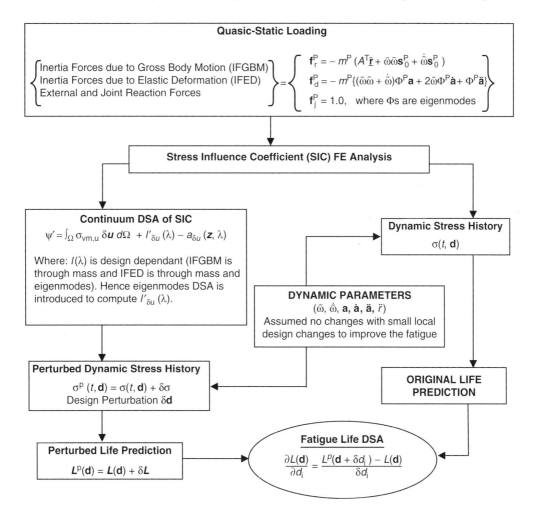

FIGURE 36.11 Computational procedure for design sensitivity analysis of flexible structural systems.

TABLE 36.3 Base Design and Its Bounds for Drawbar

Design Type	Design, d_j	Lower Bound, $\mathbf{d}^L$	Base Design	Upper Bound, $\mathbf{d}^U$
Sizing designs	d_1	0.100	0.250	0.500
	d_2	0.100	0.250	0.500
	d_3	0.100	0.250	0.500
	d_4	0.100	0.250	0.500
	d_5	0.100	0.250	0.500
Shape designs	d_6	1.000	2.000	5.000
	d_7	1.000	3.000	5.000

can be formulated as

$$\begin{aligned} \text{Minimize} \quad & W(\mathbf{d}) \\ \text{Subject to} \quad & g_i = 1 - L_i(\mathbf{d})/L_{\min} \leq 0, \quad i = 1, \ldots, nc \\ & \mathbf{d}^L \leq \mathbf{d} \leq \mathbf{d}^U, \qquad \mathbf{d} \in R^{ndv} \end{aligned} \qquad (36.2)$$

where $W(\mathbf{d})$ is the weight of the drawbar assembly, $L_i(\mathbf{d})$ is the fatigue life at the ith node, $L_{\min}$ is the required minimum fatigue life, g_i is the ith design constraint, and $\mathbf{d}^L$ and $\mathbf{d}^U$ are lower and upper bounds of the design parameters, respectively. In Equation 36.2, nc is the number of design constraints, and ndv is the number of design parameters.

For the seven design parameters ($ndv = 7$) defined in Table 36.2, the base design and design bounds are shown in Table 36.3. The side constraints need to be set by considering the restriction of manufacturing and assembling processes. For example, it is not possible for any upper bound of the sizing design parameter to be larger than half the size of a lower bound in the corresponding shape design parameter in the same cross section.

For optimization purposes, it will be extremely difficult to define fatigue-life constraints over the entire drawbar assembly in a continuum manner, since there could be an infinite number of design constraints. It is instead desirable to define a finite number of fatigue-life constraints that are limited to the critical regions. But if optimization is carried out only with respect to such critical regions, then it must be verified whether the fatigue life over the entire optimized drawbar assembly exceeds the required minimum fatigue life. As shown in Figure 36.12 and Table 36.4, the critical region is found on the central bar. Using symmetry, ten critical nodes ($nc = 10$) are selected along the center of the top and bottom of the central bar. The required minimum fatigue life is set at 3.0×10^8 cycles, which is more than 30 times the shortest life of the base design, which stands at 9.425×10^6 cycles.

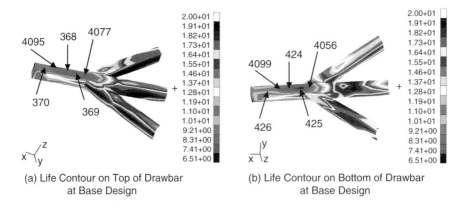

(a) Life Contour on Top of Drawbar at Base Design

(b) Life Contour on Bottom of Drawbar at Base Design

FIGURE 36.12 Fatigue-life contour on drawbar at base design: (a) life contour on top of drawbar at base design; (b) life contour on bottom of drawbar at base design.

TABLE 36.4 Critical Nodes at Base Design (Unit: Cycle)

Constraint ID	Node ID	Fatigue Life
1	425	9.425×10^6
2	424	7.148×10^7
3	426	1.115×10^{10}
4	368	4.927×10^9
5	369	3.595×10^9
6	370	3.056×10^{10}
7	4056	9.775×10^{11}
8	4077	2.161×10^{11}
9	4095	5.581×10^9
10	4099	6.137×10^9

As shown in Table 36.4, the fatigue life widely varies between 10^7 and 10^{12} cycles, resulting in a large difference (even in order of magnitude) in design constraints during the design optimization process. Therefore, design constraints are normalized by using the required minimum fatigue life, as shown in Equation 36.2 [14]. For design optimization, a modified feasible direction method is used [14]. It should be noted that another seemingly critical region appears at the tip of the drawbar on the base design. However, due to the boundary condition imposed at the tip of the drawbar, this region is fictitious.

36.2.3.4 Results of Design Optimization

As shown in Table 36.5, the optimum design is obtained in four iterations. The total mass is reduced by about 40% of its original size (from 58.401 to 35.198 lb), while all fatigue-life constraints are satisfied. As shown in Figure 36.12, the critical region at the base design is spread out over the front of the central bar. Among ten design constraints, only the first and second constraints (at nodes 425 and 424) are violated or active at the base design. On the other hand, at the optimum design, the first, third, and sixth design constraints (nodes 425, 426, and 370) appear to be active, as shown in Table 36.6.

TABLE 36.5 Design History in Optimization for HMT DRAW Durability Model

Iteration	Line Search	W	d_1	d_2	d_3	d_4	d_5	d_6	d_7
1	0	58.401	0.2500	0.2500	0.2500	0.2500	0.2500	2.0000	3.0000
	1	59.464	0.2494	0.2476	0.2762	0.2502	0.2503	1.9992	2.9966
	2	61.186	0.2485	0.2437	0.3186	0.2506	0.2507	1.9980	2.9910
	3	65.660	0.2462	0.2334	0.4297	0.2514	0.2519	1.9950	2.9764
	4	60.865	0.2487	0.2444	0.3107	0.2505	0.2506	1.9982	2.9920
2	0	65.656	0.2462	0.2334	0.4297	0.2514	0.2519	1.9950	2.9764
	1	35.712	0.1000	0.1011	0.3424	0.2283	0.2286	1.8110	2.5226
	2	24.990	0.1000	0.1000	0.2012	0.1910	0.1910	1.5134	1.7882
3	0	35.712	0.1000	0.1011	0.3424	0.2283	0.2286	1.8110	2.5226
	1	35.578	0.1000	0.1000	0.3417	0.2281	0.2284	1.8092	2.5183
	2	35.473	0.1000	0.1000	0.3405	0.2278	0.2281	1.8066	2.5115
	3	35.198	0.1000	0.1000	0.3375	0.2269	0.2272	1.7994	2.4937
	4	34.486	0.1000	0.1000	0.3297	0.2245	0.2248	1.7804	2.4471
	5	35.402	0.1000	0.1000	0.3405	0.2278	0.2281	1.8066	2.5115
4	0	35.198	0.1000	0.1000	0.3375	0.2269	0.2272	1.7994	2.4937
	1	19.704	0.1000	0.1000	0.1275	0.1622	0.1623	1.2888	1.2340
	2	24.044	0.1000	0.1000	0.1975	0.1837	0.1839	1.4590	1.6542
	3	29.212	0.1000	0.1000	0.2676	0.2053	0.2056	1.6292	2.0740
	4	32.190	0.1000	0.1000	0.3036	0.2164	0.2167	1.7168	2.2903
	5	34.649	0.1000	0.1000	0.3315	0.2250	0.2253	1.7848	2.4577
	6	34.713	0.1000	0.1000	0.3322	0.2252	0.2255	1.7864	2.4619
	7	34.900	0.1000	0.1000	0.3369	0.2267	0.2269	1.7976	2.4896
Optimum		35.198	0.1000	0.1000	0.3375	0.2269	0.2272	1.7994	2.4937

TABLE 36.6 Constraint History in Optimization for HMT DRAW Durability Model

Iteration		G_1	G_2	G_3	G_4	G_5	G_6	G_7	G_8	G_9	G_{10}
1	0	9.42×10^6	7.15×10^7	1.11×10^{10}	4.93×10^9	3.59×10^9	3.06×10^{10}	9.77×10^{11}	2.16×10^{11}	5.58×10^9	6.14×10^9
	1	2.09×10^7	9.47×10^8	3.26×10^{10}	2.16×10^{10}	1.88×10^{10}	1.28×10^{11}	2.49×10^{12}	4.51×10^{11}	8.73×10^8	1.16×10^{10}
	2	5.90×10^7	2.36×10^9	1.06×10^{11}	4.68×10^9	1.07×10^{11}	1.46×10^{10}	9.62×10^{12}	4.05×10^{12}	6.17×10^{10}	9.99×10^{19}
	3	1.07×10^9	1.28×10^{10}	2.09×10^{11}	2.10×10^{12}	1.13×10^{13}	2.00×10^{12}	2.08×10^{14}	2.57×10^{13}	8.48×10^{11}	9.99×10^{19}
	4	5.18×10^7	3.37×10^8	4.56×10^{10}	9.00×10^{10}	3.67×10^9	4.80×10^{11}	8.01×10^{12}	3.38×10^{12}	2.02×10^9	9.99×10^{19}
2	0	1.07×10^9	1.28×10^{10}	2.09×10^{11}	2.10×10^{12}	1.13×10^{13}	2.00×10^{12}	2.08×10^{14}	2.57×10^{13}	8.48×10^{11}	9.99×10^{19}
	1	7.47×10^9	1.25×10^{13}	2.36×10^9	2.00×10^{13}	1.28×10^{12}	9.76×10^9	1.62×10^{15}	1.00×10^{13}	9.73×10^{11}	2.50×10^{11}
	2	6.00×10^4	1.47×10^9	6.60×10^5	5.22×10^8	5.10×10^6	1.13×10^9	1.05×10^9	6.77×10^8	1.27×10^{10}	6.16×10^8
3	0	7.47×10^9	1.25×10^{13}	2.36×10^9	2.00×10^{13}	1.28×10^{12}	9.76×10^9	1.62×10^{15}	1.00×10^{13}	9.73×10^{11}	2.50×10^{11}
	1	5.08×10^8	6.75×10^{12}	4.18×10^8	4.44×10^{13}	3.86×10^{10}	9.07×10^{11}	5.79×10^{15}	2.91×10^{11}	1.60×10^{13}	1.44×10^{11}
	2	2.11×10^{10}	1.00×10^{13}	2.70×10^9	3.99×10^{13}	1.75×10^{11}	1.07×10^{10}	9.03×10^{14}	7.01×10^{12}	4.15×10^{10}	1.97×10^{11}
	3	8.99×10^8	3.82×10^{12}	1.19×10^9	1.12×10^{13}	3.05×10^{11}	1.97×10^9	1.78×10^{14}	3.68×10^{11}	7.43×10^{12}	1.17×10^{11}
	4	2.06×10^8	1.74×10^{12}	4.02×10^8	9.00×10^{12}	1.64×10^{11}	1.59×10^{11}	4.53×10^{13}	1.49×10^{12}	5.34×10^{12}	6.98×10^{10}
	5	4.59×10^8	3.67×10^{12}	3.66×10^8	2.69×10^{13}	6.23×10^{11}	4.42×10^{11}	1.87×10^{15}	2.81×10^{12}	1.69×10^{13}	1.13×10^{11}
4	0	8.99×10^8	3.82×10^{12}	1.19×10^9	1.12×10^{13}	3.05×10^{11}	1.97×10^9	1.78×10^{14}	3.68×10^{11}	7.43×10^{12}	1.17×10^{11}
	1	3.00×10^4	1.79×10^8	3.00×10^4	1.50×10^5	3.00×10^4	7.11×10^7	1.20×10^5	1.08×10^6	3.40×10^7	1.53×10^{10}
	2	3.00×10^4	1.29×10^9	4.20×10^5	4.79×10^8	2.55×10^6	8.49×10^8	4.08×10^8	3.05×10^8	1.69×10^{10}	6.81×10^8
	3	6.09×10^6	7.42×10^{10}	1.10×10^7	9.06×10^{10}	1.76×10^9	3.65×10^{10}	2.91×10^{11}	8.49×10^{10}	7.88×10^9	1.10×10^{10}
	4	6.58×10^7	4.04×10^{11}	5.56×10^7	3.65×10^{12}	2.11×10^{10}	7.27×10^{10}	7.89×10^{12}	6.80×10^{10}	3.55×10^{12}	2.02×10^{10}
	5	2.22×10^8	1.84×10^{12}	4.29×10^8	1.23×10^{13}	1.82×10^{11}	1.43×10^{11}	5.13×10^{13}	1.57×10^{12}	5.88×10^{12}	7.27×10^{10}
	6	2.39×10^8	1.94×10^{12}	4.48×10^8	7.19×10^{11}	1.88×10^{11}	1.54×10^{11}	5.67×10^{13}	1.68×10^{12}	5.52×10^{12}	7.60×10^{10}
	7	3.24×10^8	2.21×10^{12}	2.47×10^8	8.25×10^{11}	2.87×10^{11}	2.22×10^{11}	1.22×10^{15}	6.30×10^{12}	8.91×10^{12}	7.30×10^{10}
Optimum		8.99×10^8	3.82×10^{12}	1.19×10^9	1.12×10^{13}	3.05×10^{11}	1.97×10^9	1.78×10^{14}	3.68×10^{11}	7.46×10^{12}	1.17×10^{11}

TABLE 36.7 Design and Weight Changes between Base and Optimal Designs

Design	Base Design (in.)	Optimal Design (in.)	Change (%)
t_1	0.2500	0.1000	−60.0
t_2	0.2500	0.1000	−60.0
t_3	0.2500	0.3375	+35.0
t_4	0.2500	0.2269	−9.24
t_5	0.2500	0.2272	−9.12
h	2.0000	1.7994	−10.0
w	3.0000	2.4937	−16.9
Cost	Base Design (lb)	Optimal Design (lb)	Change (%)
Weight	58.401	35.198	−39.7

At the optimum design, all thicknesses decrease except for the central bar, and the width and height of all bars become smaller. Due to the decrease in some sizing design parameters and both shape parameters, about 40% of the mass is saved. The first two design parameters, b_1 and b_2, decrease slowly at the beginning of the optimization process, and then they rapidly decrease to the lower bound, since more-rigid side bars and angles penalize the central bar, resulting in a decrease in its fatigue life. Moreover, increasing the thickness of the central bar by 35% (d_3: from 0.25 to 0.3375 in.) further reinforces its strength and produces a longer fatigue life. At the optimum design, the fourth and fifth design parameters (triangular plates and side attachments) are reduced by about 9%, since the weight can be effectively reduced without reducing the fatigue lives at critical regions. With respect to shape design parameters, the width and height are reduced by about 10 and 17%, respectively. These design changes are summarized in Table 36.7.

Because optimization only considers the critical regions, the optimized design must be confirmed through reanalysis to determine whether the fatigue life over the entire drawbar assembly exceeds the required minimum specifications. As shown in the optimal design in Figure 36.13, the original critical region (nodes 425 and 424) at the base design seems to bifurcate into an original region at node 425 and another region around node 426. Except for the tip of the central bar shown in Figure 36.13a, all other areas satisfy the minimum requirements for fatigue life. Similar to the base design, a fictitious critical region is detected at the tip of the drawbar. As explained earlier, it is suspected that the boundary condition at the tip causes this fictitious condition.

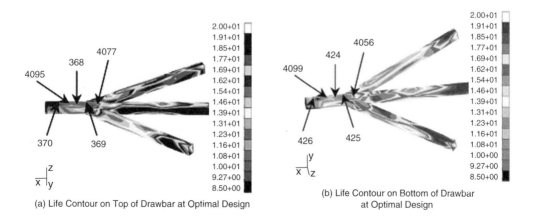

(a) Life Contour on Top of Drawbar at Optimal Design

(b) Life Contour on Bottom of Drawbar at Optimal Design

FIGURE 36.13 Fatigue-life contour on drawbar at optimal design: (a) life contour on top of drawbar at optimum design; (b) life contour on bottom of drawbar at optimal design.

TABLE 36.8 Fatigue Life with and without Considering Notch Effects

	Predicted Fatigue Life	Without Considering Notch Effects	Considering Notch Effects
Base	Driving cycle (block)	9.42×10^6	1.44×10^3
	Driving mile (mile)	1.18×10^6	180
	Driving time (hour)	78,500	12
Opt.	Driving cycle (block)	8.99×10^8	7.66×10^4
	Driving mile (mile)	1.12×10^8	9,580
	Driving time (hour)	7.49×10^6	638
	Life extension (times)	95.4	53.2

36.2.3.5 Results of Design Optimization Considering Notch Effects [16]

Having identified the region near node 425 on the trailer drawbar as the location of the shortest fatigue life, it is now necessary to apply the fatigue-strength reduction factor (K_f) to account for the effect of geometric discontinuities in the critical region, as shown in Table 36.8. A fatigue-strength reduction factor reduces the predicted fatigue life in a manner proportional to the severity of the geometric discontinuity. K_f is calculated from the stress intensity factor K_t and the notch-sensitivity factor q. Based on references in the literature [17], the values of K_t and q were estimated, and the associated K_f values were calculated to be 2.6. Using this K_f, the fatigue life on Perryman course no. 3 is estimated to be 9580 miles, which means that a fatigue crack will not initiate and grow to a 2-mm length until the trailer traverses 9580 miles of Perryman course no. 3 at 15 mph, or 638 h of continuous running. In contrast, the fatigue life of the base design was 180 miles or 12 h. This means that the fatigue life of the optimum design is 53.2 times greater than that of the base design. In this example we have tried to reduce the weight of the drawbar by imposing the constraint that the fatigue life be approximately equal to 30 times that of the baseline design (see Equation 36.2). A greater increase in fatigue life could be achieved if the fatigue life were constrained to be equal to that of the baseline.

36.3 Reliability Analysis and Reliability-Based Design Optimization

36.3.1 Reliability Analysis for Durability-Based Optimal Design

The reliability analysis of the durability-based optimum design is carried out to estimate the probability of failure as

$$P(L < 3 \times 10^8 \text{ cycles}) = \int \cdots \int_{L < 3 \times 10^8} f_X(\mathbf{x}) d\mathbf{x} \qquad (36.3)$$

For the reliability analysis, the uncertainty of each design parameter is modeled with a normal distribution and a 10% coefficient of variation. The optimal design turns out to be unreliable with a 49.7% ($\beta = 0.073$) probability of failure through the reliability analysis. Considering the variability of the design that lays on the design constraint boundaries, as shown in Figure 36.14, it is reasonable that the deterministic optimal design has only about 50% reliability. Because the deterministic optimal design is unreliable, it is necessary to perform a reliability-based design optimization (RBDO) [1–5] for reliable and durable designs. Among all random parameters, uncertainty in the third random parameter (central bar thickness) has the most significant effect on the probability of failure. Thus, without a new reliability-based optimum design, the thickness of the central bar must be manufactured more accurately to increase reliability, which will significantly increase the manufacturing cost.

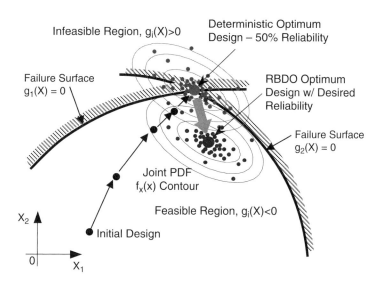

FIGURE 36.14 Overall procedure of PMA+ in RBDO.

36.3.2 Reliability-Based Design Optimization for Durability

As seen in the previous section, the various uncertainties in the mechanical system led to an unreliable deterministic design optimization. As shown in Figure 36.14, the design must move back to the feasible region to increase the reliability of design constraints while minimizing a design objective. The process is called reliability-based design optimization (RBDO), which will be explained in this section.

36.3.2.1 RBDO Model of Performance Measure Approach (PMA)

For any engineering application, the RBDO model [1–5, 18–20] can generally be formulated as

$$\begin{aligned} &\text{minimize} \quad \text{Cost}(\mathbf{d}) \\ &\text{subject to} \quad P(G_i(\mathbf{d}(\mathbf{X})) \le 0) - \Phi(\beta_t) \ge 0, \quad i = 1, 2, \ldots, nc \\ &\mathbf{d}^L \le \mathbf{d} \le \mathbf{d}^U \end{aligned} \qquad (36.4)$$

where $\mathbf{d} = [d_i^T] = \mathbf{\mu}(\mathbf{X}) \in R^{ndv}$ is the design vector, $\mathbf{X} = [X_i]^T \in R^{nrv}$ is the random vector, and ndv, nrv, and nc are the numbers of design parameters, random parameters, and probabilistic constraints, respectively. The probabilistic constraints are described by a probability constraint $P(\bullet) \ge \Phi(\beta_t)$ for a safe event $G_i(\mathbf{d}(\mathbf{X})) \le 0$.

The statistical description of the failure of the performance function $G_i(\mathbf{d}(\mathbf{X}))$ is characterized by the cumulative distribution function, $F_{G_i}(\bullet)$, as

$$P(G_i(\mathbf{X}) \le 0) = F_{G_i}(0) \ge \Phi(\beta_t) \qquad (36.5)$$

where the reliability of failure is described as

$$F_{G_i}(0) = \int \cdots \int_{G_i(\mathbf{X}) \le 0} f_\mathbf{X}(\mathbf{x}) dx_1 \ldots dx_n, \quad i = 1, 2, \ldots, nc \qquad (36.6)$$

In Equation 36.6, $f_\mathbf{X}(\mathbf{x})$ is the joint probability density function of all random parameters. Its evaluation requires a reliability analysis where multiple integrations are involved, as shown in Equation 36.6.

Some approximate probability integration methods have been developed to provide efficient solutions, such as the first-order reliability method (FORM) [1–5, 8] or the asymptotic second-order reliability method (SORM) [21, 22], with a rotationally invariant measure as the reliability. FORM often provides adequate accuracy and is widely used for design applications. In FORM, the reliability analysis requires a transformation **T** [23, 24] from the original random parameter **X** to the standard normal random parameter **U**. The performance function $G(\mathbf{X})$ in X-space can then be mapped onto $G(\mathbf{T}(\mathbf{X})) \equiv G(\mathbf{U})$ in U-space.

The probabilistic constraint in Equation 36.5 can be expressed as a performance measure through the inverse transformation of $F^G(\bullet)$ as [1, 3–5,18–20]:

$$G_{p_i}(\mathbf{d}(\mathbf{X})) = F_{G_i}^{-1}(\Phi(\beta_t)) \leq 0 \tag{36.7}$$

where G_{p_i} is the ith probabilistic constraint. In Equation 36.7, the probabilistic constraint in Equation 36.4 can be replaced with the performance measure. This is referred to as the performance measure approach (PMA) [1, 3–5, 18–20]. Thus, the RBDO model using PMA can be redefined as

$$\begin{aligned}
&\text{minimize} \quad \text{Cost}(\mathbf{d}) \\
&\text{subject to} \quad G_{p_i}(\mathbf{d}(\mathbf{X})) \leq 0, \quad i = 1, 2, \ldots, nc \\
&\mathbf{d}^L \leq \mathbf{d} \leq \mathbf{d}^U
\end{aligned} \tag{36.8}$$

36.3.2.2 Reliability Analysis Model of PMA

Reliability analysis in PMA can be formulated as the inverse of reliability analysis in the reliability index approach (RIA). The first-order probabilistic performance measure $G_{p,\text{FORM}}$ is obtained from a nonlinear optimization problem in U-space, defined as

$$\begin{aligned}
&\text{maximize} \quad G(\mathbf{U}) \\
&\text{subject to} \quad \|\mathbf{U}\| = \beta_t
\end{aligned} \tag{36.9}$$

where the optimum point on the target reliability surface is identified as the most probable point (MPP) $\mathbf{u}^*_{\beta=\beta_t}$ with a prescribed reliability $\beta_t = \left\|\mathbf{u}^*_{\beta=\beta_t}\right\|$. Unlike RIA, only the direction vector $\mathbf{u}^*_{\beta=\beta_t}/\left\|\mathbf{u}^*_{\beta=\beta_t}\right\|$ needs to be determined by exploring the spherical equality constraint $\|\mathbf{U}\| = \beta_t$.

General optimization algorithms can be employed to solve the optimization problem in Equation 36.9. However, a hybrid mean value (HMV) first-order method is well suited for PMA due to its stability and efficiency [1, 3–5,18–20].

36.3.2.3 Reliability Analysis Tools for PMA

Three numerical methods [1, 3–5, 18–20, 25] for PMA were used to solve Equation 36.9: the advanced mean value method [24] in Equation 36.10, the conjugate mean value method [1, 3–5, 18–20] in Equation 36.11, and the hybrid mean value (HMV) method [1, 3–5, 18–20] in Equation 36.12.

$$\mathbf{u}_{\text{AMV}}^{(1)} = \beta_t \mathbf{n}(0), \quad \mathbf{u}_{\text{AMV}}^{(k+1)} = \beta_t \mathbf{n}\left(\mathbf{u}_{\text{AMV}}^{(k)}\right) \quad \text{where} \quad \mathbf{n}\left(\mathbf{u}_{\text{AMV}}^{(k)}\right) = \frac{\nabla_U G\left(\mathbf{u}_{\text{AMV}}^{(k)}\right)}{\left\|\nabla_U G\left(\mathbf{u}_{\text{AMV}}^{(k)}\right)\right\|} \tag{36.10}$$

$$\mathbf{u}_{\text{CMV}}^{(0)} = \mathbf{0}, \quad \mathbf{u}_{\text{CMV}}^{(1)} = \mathbf{u}_{\text{AMV}}^{(1)}, \quad \mathbf{u}_{\text{CMV}}^{(2)} = \mathbf{u}_{\text{AMV}}^{(2)},$$

$$\mathbf{u}_{\text{CMV}}^{(k+1)} = \beta_t \frac{\mathbf{n}\left(\mathbf{u}_{\text{CMV}}^{(k)}\right) + \mathbf{n}\left(\mathbf{u}_{\text{CMV}}^{(k-1)}\right) + \mathbf{n}\left(\mathbf{u}_{\text{CMV}}^{(k-2)}\right)}{\left\|\mathbf{n}\left(\mathbf{u}_{\text{CMV}}^{(k)}\right) + \mathbf{n}\left(\mathbf{u}_{\text{CMV}}^{(k-1)}\right) + \mathbf{n}\left(\mathbf{u}_{\text{CMV}}^{(k-2)}\right)\right\|} \quad \text{for } k \geq 2 \quad \text{where} \quad \mathbf{n}\left(\mathbf{u}_{\text{CMV}}^{(k)}\right) = \frac{\nabla_U G\left(\mathbf{u}_{\text{CMV}}^{(k)}\right)}{\left\|\nabla_U G\left(\mathbf{u}_{\text{CMV}}^{(k)}\right)\right\|} \tag{36.11}$$

$$\mathbf{u}_{\text{HMV}}^{(k+1)} = \begin{cases} \mathbf{u}_{\text{AMV}}^{(k+1)} \text{ in Eq. (36.8)}, & \text{if } G\left(\mathbf{u}_{\text{HMV}}^{(k)}\right) \text{ is convex or sign } (\varsigma^{(k+1)}) > 0 \\ \mathbf{u}_{\text{CMV}}^{(k+1)} \text{ in Eq. (36.9)}, & \text{if } G\left(\mathbf{u}_{\text{HMV}}^{(k)}\right) \text{ is concave or sign } (\varsigma^{(k+1)}) \leq 0 \end{cases}$$

with $\varsigma^{(k+1)} = (\mathbf{n}^{(k+1)} - \mathbf{n}^{(k)}) \cdot (\mathbf{n}^{(k)} - \mathbf{n}^{(k-1)})$ (36.12)

$\text{sign}(\varsigma^{(k+1)}) > 0$: Convex type at $\mathbf{u}_{\text{HMV}}^{(k+1)}$ with respect to design $\mathbf{d}$

≤ 0 : Concave type at $\mathbf{u}_{\text{HMV}}^{(k+1)}$ with respect to design $\mathbf{d}$

Although the advanced mean value method performs well for the convex performance function in PMA, it was found to have some numerical shortcomings, such as slow convergence, or even divergence, when applied to the concave performance function. To overcome this difficulty, the conjugate mean value method was proposed [1, 3–5, 18–20]. The conjugate steepest descent direction significantly improves the rate of convergence as well as stability, as compared with the advanced mean value method for the concave performance function. However, the conjugate mean value method is not as efficient as the advanced mean value method for the convex function. Consequently, the hybrid mean value (HMV) method was proposed to attain both stability and efficiency in the MPP search for PMA [1, 3–5, 18–20]. The HMV method employs the criterion for the performance function type near the MPP. Once the performance function type is identified, either the advanced mean value or conjugate mean value method is adaptively selected for the MPP search. The numerical procedure of the HMV method is presented with some numerical examples by Youn et al. [1].

36.3.3 Results of Reliability-Based Shape Design Optimization for Durability of M1A1 Tank Roadarm [14]

A roadarm of the military tracked vehicle shown in Figure 36.15 is employed to demonstrate the effectiveness of PMA for a large-scale RBDO application. A 17-body dynamics model is created to simulate the tracked vehicle driven on the Aberdeen Proving Ground 4 at a constant speed of 20 mph. A 20-sec dynamic simulation is performed with a maximum integration time step of 0.05 sec using the dynamic analysis package DADS.

As shown in Figure 36.16, 310 20-node isoparametric finite elements (STIF95) and four beam elements (STIF4) of ANSYS are used to create the roadarm finite element model, which is made of S4340 steel. Finite element analysis is performed to obtain the stress influence coefficient of the roadarm using ANSYS by applying 18 quasistatic loads. The empirical constants (σ', ε', b, c, K', n') for fatigue material properties are defined in Table 36.9. It has been found [26] that geometric tolerances are normally distributed with about a 1% coefficient of variation (COV), while material properties are lognormally distributed with about 3% COV, except for negative quantities, such as the fatigue exponents b and c. The computation for fatigue-life prediction and for design sensitivity requires, respectively, 6950 and 6496 CPU sec (for eight design parameters) on an HP 9000/782 workstation.

FIGURE 36.15 Military tracked vehicle.

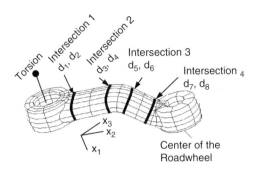

FIGURE 36.16 Random design parameters.

As shown in Table 36.9, eight design parameters are used to characterize the four cross-sectional shapes of the roadarm, while uncertainties in the dimensions and material properties of a structural component due to manufacturing tolerances are modeled using 14 random parameters. Vertical variations (in the x_1-direction) of cross-sectional shapes are defined as the random parameters d_1, d_3, d_5, and d_7 for intersections 1 to 4, respectively, and side variations (moving in the x_3-direction) of cross sectional shapes are defined using the remaining four random variables.

In design optimization for fatigue life, the number of design constraints could be very large if a fatigue-life constraint were defined at every point of the structural component. To make the problem computationally feasible for structural durability analysis, a preliminary fatigue-life analysis for a crack initiation is carried out (as shown in Figure 36.17) to detect those critical spots that have a short fatigue life and to define the design constraints. A refined durability analysis using the critical-plane method is then carried out at these critical spots to accurately predict fatigue life. The design constraints for durability in Equation 36.4 are defined as

$$G_i(\mathbf{d}(\mathbf{X})) = 1 - L_i(\mathbf{d}(\mathbf{X}))/L_t \tag{36.13}$$

where $L_i(\mathbf{d}(\mathbf{X}))$ is the crack-initiation fatigue life at the current design, and the target fatigue life L_t is set to 5 years. During this process, DRAW predicts the crack-initiation fatigue life, which is taken as the performance requirement.

TABLE 36.9 Definition of Random Parameters for Crack Initiation Fatigue-Life Prediction

	Parameters	Lower Bound	Mean	Upper Bound	COV (%)	Distribution Type
Design/ random (geometric tolerance)	d_1, X_2	1.3776	1.8776	2.0000	1	normal
	d_2, X_2	1.5580	3.0934	2.0000	1	normal
	d_3, X_3	2.5934	1.8581	3.2000	1	normal
	d_4, X_4	2.7091	3.0091	3.2000	1	normal
	d_5, X_5	2.2178	2.5178	2.7800	1	normal
	d_6, X_6	4.6500	2.9237	5.0000	1	normal
	d_7, X_7	2.6237	4.7926	3.0500	1	normal
	d_8, X_8	2.5000	2.8385	3.0000	1	normal
	Fatigue Material		Mean		COV (%)	Distribution Type
Random (material parameter)	X_9 cyclic strength coefficient, K'		1.358×10^9		3	lognormal
	X_{10} cyclic strength exponent, n'		0.12		3	lognormal
	X_{11} fatigue strength coefficient, σ'		1.220×10^9		3	lognormal
	X_{12} fatigue strength exponent, b		−0.073		3	normal
	X_{13} fatigue ductility coefficient, ε'_f		0.41		3	lognormal
	X_{14} fatigue ductility exponent, c		−0.60		3	normal

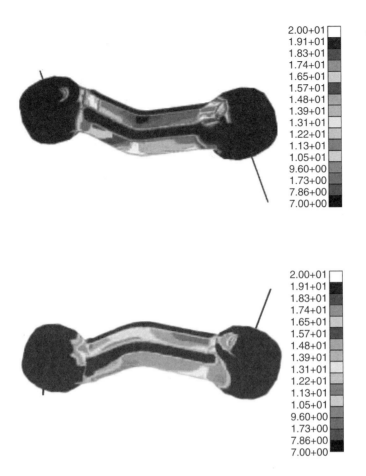

FIGURE 36.17 Preliminary fatigue-life analysis.

As illustrated in Figure 36.18, uncertainty of fatigue life is first determined only by considering uncertainty of geometric parameters, and then by considering uncertainty of both geometric and material parameters. It is shown that the RBDO process must consider uncertainty of material parameters, since it significantly affects uncertainty of the fatigue life [15]. Three-σ RBDO results of durability for the M1A1 tank roadarm are shown in Figure 36.19 and Figure 36.20. Note that small design changes are

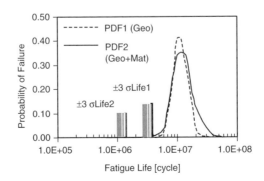

FIGURE 36.18 Uncertainty propagation to fatigue life.

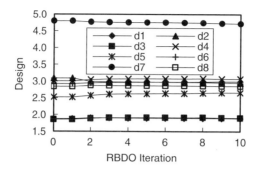

FIGURE 36.19 Design history in 3-σ.

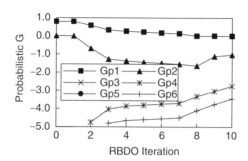

FIGURE 36.20 Probabilistic constraint history in 3-σ.

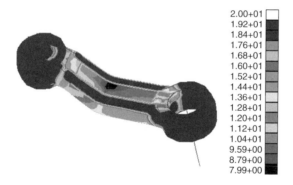

FIGURE 36.21 Fatigue-life contour at 3-σ optimum design.

made to satisfy the target fatigue life with target reliability levels, while increasing roadarm weight by 1%. The shortest fatigue life was 1.094 years at the initial design, but this increased to 5.017 years at the 3-σ optimum design. Figure 36.21 illustrates the fatigue-life contour of a 3-σ optimum design that has increased in overall fatigue life when compared with Figure 36.17.

36.4 Conclusions

This chapter presents the three key elements of an advanced CAE methodology: durability analysis, experimental validation, and an uncertainty-based design optimization. Since the CAE model is developed to simulate a prototype model, the former is exposed to a variety of natural uncertainties, which are categorized as model, physical, and statistical uncertainties. It was proposed that a high-fidelity model and analysis through experimental validation is necessary to accurately characterize the effect of input uncertainty. Furthermore, the RBDO method must be incorporated into CAE to take physical uncertainties into account. A lack of statistical information in modeling physical uncertainty creates greater uncertainty that can be dealt with by using a Bayesian possibility or evidence theory.

It has been found that experimental validation should be conducted to obtain high-fidelity models and analyses. The structural durability of a trailer drawbar was used to demonstrate CAE with experimental validation, and optimization was then carried out to improve the fatigue life of the trailer drawbar while minimizing its weight. To develop accurate CAE models for multibody dynamics, FEA, and durability analysis, CAD models were carefully compared with experimental results. Design optimization successfully increased the fatigue life of the drawbar 53.2 times and reduced the weight 40% when considering notch effects due to rivet holes in the critical region. To validate the deterministic optimum design, a reliability analysis evaluates the reliability of fatigue failure under manufacturing tolerances. The fact that its probability of failure is 49.7% underscores the need for reliability-based design optimization (RBDO) for a reliable and durable optimum design. In addition to experimental validation, the RBDO method must be incorporated into advanced CAE methodology. One of the challenges in developing an advanced CAE methodology is to create an effective RBDO method. PMA was used to carry out the RBDO process for large-scale multidisciplinary applications (e.g., M1A1 Tank roadarm durability), with emphasis on numerical efficiency and stability. Consequently, the advanced CAE methodology with experimental validation and uncertainty-based design optimization was successfully demonstrated by applying it to the computationally intensive and complicated multidisciplinary simulation process of predicting structural durability.

Acknowledgments

Research is supported by the Physics of Failure project sponsored by U.S. Army Material Systems Analysis Activity (AMSAA).

Nomenclature

L	crack initiation fatigue life
W	weight for design optimization
$\mathbf{d}$	design parameter; $\mathbf{d} = [d_1, d_2, \ldots, d_n]^T$
$g(\mathbf{d})$	design constraint of design parameters
$\mathbf{X}$	random vector; $\mathbf{X} = [X_1, X_2, \ldots, X_n]^T$
$\mathbf{x}$	realization of $\mathbf{X}$; $\mathbf{x} = [x_1, x_2, \ldots, x_n]^T$
$G(\mathbf{X})$	constraint of random parameters
$\mathbf{U}$	independent and standard normal random parameter
$\mathbf{u}$	realization of $\mathbf{U}$; $\mathbf{u} = [u_1, u_2, \ldots, u_n]^T$
P_f	probability of failure
f_X	probability density function of X
$\boldsymbol{\mu}$	mean of random vector $\mathbf{X}$; $\boldsymbol{\mu} = [\mu_1, \mu_2, \ldots, \mu_n]^T$
β	reliability index
K_f	fatigue-strength reduction factor
K_t	stress intensification factor
q	notch sensitivity factor

References

1. Youn, B.D., Choi, K.K., and Park, Y.H., Hybrid analysis method for reliability-based design optimization, *ASME J. Mechanical Design*, 125 (2), 221–232, 2003.
2. Enevoldsen, I. and Sorensen, J.D., Reliability-based optimization in structural engineering, *Structural Safety*, 15, 169–196, 1994.

3. Lee, T.W. and Kwak, B.M., A reliability-based optimal design using advanced first order second moment method, *Mech. Struct. Mach.*, 15 (4), 523–542, 1987/88.
4. Tu, J. and Choi, K.K., A new study on reliability-based design optimization, *ASME J. Mechanical Design*, 121 (4), 557–564, 1999.
5. Youn, B.D. and Choi, K.K., Selecting probabilistic approaches for reliability-based design optimization, *AIAA J.*, 42 (1), 124–131, 2004.
6. Shih, C.J., Chi, C.C., and Hsiao, J.H., Alternative α-level-cuts methods for optimum structural design with fuzzy resources, *Comput. Struct.*, 81, 2579–2587, 2003.
7. Bae, H.R., Grandhi, R.V., and Canfield, R.A., Structural Design Optimization Based on Reliability Analysis Using Evidence Theory, SAE 03M-125, presented at SAE World Congress, Detroit, 2003.
8. Madsen, H.O., Krenk, S., and Lind, N.C., *Methods of Structural Safety*, Prentice-Hall, Englewood Cliffs, NJ, 1986.
9. Fuchs, H.O. and Stephens, R.I., *Metal Fatigue in Engineering*, John Wiley & Sons, New York, 1980.
10. CADSI Inc., DADS User's Manual, rev. 7.5, CADSI, Oakdale, IA, 1994.
11. Center for Computer-Aided Design, DRAW: Durability and Reliability Analysis Workspace, College of Engineering, University of Iowa, Iowa City, 1994.
12. Yu, X., Choi, K.K., and Chang, K.H., Probabilistic structural durability prediction, *AIAA J.*, 36 (4), 628–637, 1998.
13. Haug, E.J., Choi, K.K., and Komkov, V., *Design Sensitivity Analysis of Structural Systems*, Academic Press, New York, 1986.
14. Arora, J.S., *Introduction to Optimum Design*, McGraw-Hill, New York, 1989.
15. Chang, K.H., Yu, X., and Choi, K.K., Shape design sensitivity analysis and optimization for structural durability, *Int. J. Numerical Meth. Eng.*, 40, 1719–1743, 1997.
16. Choi, K.K., Youn, B.D., and Tang, J., Structural Durability Design Optimization and Its Reliability Assessment, presented at 2002 ASME Design Engineering Technical Conferences: 29th Design Automation Conference, Chicago, 2003.
17. Sines, G. and Waisman, J.L., *Metal Fatigue*, McGraw-Hill, New York, 1959.
18. Youn, B.D. and Choi, K.K., A new response surface methodology for reliability-based design optimization, *Comput. Struct.*, 82, 241–256, 2004.
19. Choi, K.K. and Youn, B.D., An Investigation of Nonlinearity of Reliability-Based Design Optimization Approaches, presented at ASME Design Engineering Technical Conferences, Montreal, 2002.
20. Youn, B.D., Choi, K.K., and Yang, R.-J., Efficient Evaluation Approaches for Probabilistic Constraints in Reliability-Based Design Optimization, presented at 5th WCSMO, Lido di Jesolo-Venice, Italy, 2003.
21. Breitung, K., Asymptotic approximations for multi-normal integrals, *ASCE J. Eng. Mech.*, 110 (3), 357–366, 1984.
22. Tvedt, L., Distribution of quadratic forms in normal space-application to structural reliability, *ASCE J. Eng. Mech.*, 116 (6), 1183–1197, 1990.
23. Rackwitz, R. and Fiessler, B., Structural reliability under combined random load sequences, *Comput. Struct.*, 9, 489–494, 1978.
24. Hohenbichler, M. and Rackwitz, R., Nonnormal dependent vectors in structural reliability, *ASCE J. Eng. Mech.*, 107 (6), 1227–1238, 1981.
25. Wu, Y.T., Millwater, H.R., and Cruse, T.A., Advanced probabilistic structural analysis method for implicit performance functions, *AIAA J.*, 28 (9), 1663–1669, 1990.
26. Rusk, D.T. and Hoffman, P.C., Component Geometry and Material Property Uncertainty Model for Probabilistic Strain-Life Fatigue Predictions, presented at 6th Joint FAA/DoD/NASA Aging Aircraft Conference, San Francisco, 2002.

37
A Method for Multiattribute Automotive Design under Uncertainty

Zissimos P. Mourelatos
Oakland University

Artemis Kloess
General Motors Corporation

Raviraj Nayak
General Motors Corporation

37.1 Introduction **37**-1
37.2 A Quality Specification Structure **37**-2
  A Formal Preference Aggregation Theory
37.3 An Automotive Body-Door Example **37**-5
37.4 An Automotive Bumper Standardization Example **37**-8
  Bumper Optimization Using Preference Aggregation Theory
37.5 Overview of RBDO Approaches **37**-12
  Automotive Vehicle Side-Impact Example
37.6 Summary and Future Needs **37**-16

37.1 Introduction

Automotive design is usually performed in a multiattribute environment in the presence of inherent variability and uncertainty. The need to perform trade-off analysis among conflicting design criteria is common. Automotive quality is a major application area where multiattribute design under uncertainty is needed. For this reason, a multiattribute methodology for automotive design considering uncertainty is presented in this chapter with applications in automotive quality and component standardization.

Delivering reliable, high-quality products at low cost has become the key to survival in today's global economy. In general, quality can be defined as anything that enhances the product from the viewpoint of the customer. Products of high quality are distinguished by achieving this with high frequency. In an automotive sense, quality products means (1) vehicles manufactured to specifications, (2) vehicles whose characteristics are insensitive to variation, (3) vehicles whose specifications meet customer requirements, and (4) vehicles that people want to buy. The question is then, "How often and at what cost do vehicles meet these requirements for quality?" In a traditional costly approach to product design and quality assessment, specifications are set for a selected quality attribute, and components are designed to nominally meet the specifications. When problems are discovered through prototype testing, the design is modified and retested. In production, the design then encounters real-world variation, and the "in-spec" design may now fall "out-of-spec," resulting in even more effort and cost to make the necessary corrections. Since designing quality into the product is cheaper than trying to inspect and fix it after the vehicle hits production or, worse yet, after it gets to the customer, new approaches must be employed to design high-quality products at low cost. Furthermore, as vehicle development becomes increasingly math-based, design for quality in a math-based world means that the quality-conscious producer must utilize analytic methods to efficiently exploit customer information in setting vehicle specifications and explicitly account for uncertainty and process variation early in the vehicle development process.

A product design must satisfy a multitude of functional and engineering requirements. The functional requirements are deduced from the "voice of the customer." For an automotive door system, for example, they include among others, sealing against wind noise, water leaks, and low closing effort [1]. The "functional" requirements must be translated into measurable engineering requirements. The engineering solutions should be simple and should include manufacturing restrictions. Due to the inherent uncertainties associated with the "voice of the customer," manufacturing processes, material properties, etc., engineers must seek a robust design that is relatively insensitive to these uncertainties while meeting, of course, the engineering requirements. Robust design is a design approach for reducing variation in products and related processes [2].

Probabilistic design concepts are needed in robust design. In probabilistic design, the inherent uncertainties are accounted for so that the final design will meet specifications with a certain probability [3–5]. A probabilistic design requires accurate and efficient models (analytical or experimental) to measure compliance with each engineering requirement. In this increasingly "math-based" world, analytical tools are preferred because experimental setups are expensive to build as well as very time consuming to perform. Studies have found that probabilistic analysis and design methods can be used to address quality issues, since they quantify the influence of variation and uncertainty on the performance of automotive components [6, 7].

There are many ways to define automotive quality, but whether one is considering freedom from repair, fuel economy, safety, or even styling, they all essentially come down to "the **frequency** of meeting, or exceeding, customer desires." Not meeting customer desires can be a consequence of either unacceptable average performance or unacceptable performance of some vehicles in spite of an acceptable average performance of the fleet (too much variation). A high-quality vehicle requires, therefore, a good robust design if it is to frequently achieve desirable performance levels among various, usually competing, goals.

The design and production of a high-quality product requires knowledge of customer expectations, knowledge of component, assembly, and usage variation as well as knowledge of the impact of those variations in these on performance. A robust process is one that produces a product that is less sensitive to variations and therefore meets the specifications more often. In order to achieve desired quality levels, the design process must account for the variation present in manufacturing, assembly, and use. It must also create designs and utilize processes that together produce vehicles of specified quality. The use of a quality specification (QS) structure within a formal robust engineering process can be a driver in transforming the product development process.

Section 37.2 presents a multiattribute method based on a quality specification structure for addressing automotive quality during the vehicle development process. The method utilizes (1) a formal preference-aggregation theory for handling multiple, often conflicting, design criteria and (2) reliability-based design optimization (RBDO) for finding the optimal design under uncertainty. Two major applications demonstrate the method: an automotive body-door design presented in Section 37.3 and an automotive bumper standardization study presented in Section 37.4. The fundamentals of reliability-based design optimization are described in Section 37.5. The significance of RBDO in automotive design is highlighted using a vehicle side-impact example. Finally, the chapter closes with a summary of the presented work and some recommendations on future work.

37.2 A Quality Specification Structure

A major objective in product design is to drive the design process to consistently and predictably satisfy customer desires in an optimal fashion. For this reason, a proposed three-part quality specification (QS) structure is described in this section. It consists of an attribute target, a compliance-rate target, and a trade-off rate. The attribute target identifies the quality concern and the idealized goal (e.g., no customer reported wind-noise problems for automotive front doors). The compliance-rate target specifies the expected fraction of the vehicles produced that will meet the goal (e.g., 0.99). The trade-off rate is a measure of the value (or cost) of deviating from the specification. The combination of the first two components forces the design process to consider variation resulting from manufacturing and assembly, along with customer variation in both usage and expectation. The adoption of the quality specification structure will drive the design process to explicitly account for the variation in both manufacturing and use, while also encouraging rational trade-offs to achieve an optimal design.

In the quality specification structure, a third critical piece of information, the customer-driven trade-off rate, is added. This is necessary because the "value" of quality is critical in deciding what is feasible, or what to do if a difficulty arises in trying to achieve multiple compliance-rate targets. In normal practice, someone's qualitative judgment is the typical mechanism used for trade-off decisions. In design for six sigma (DFSS) [8], one numerically sums failures, regardless of their impact. By specifying trade-off rates up front, we push some of the specific design decisions down from the manager to the designer and inherently favor the multidiscipline design strategy rather than discipline-specific strategy. This enables creativity at the parts design level earlier in the product development process than we currently have. If cost is not an issue between competing designs, then relative-importance numbers are sufficient for trade-off work. Generally, however, because achievement of any given performance level will be cost dependent, it will be necessary to have some estimate of the dollar value of a change in either the specification or the compliance rate. This also provides a common basis in which to optimally balance competing goals. The methodology in this chapter contributes toward explicitly accounting for expected variation in design parameters and customer expectations as the design process becomes increasingly math-based.

The process of designing for quality based on the described QS structure is highlighted in Figure 37.1 using an automotive door wind-noise example. The same example is described in Section 37.3 in more detail. First, the quality specifications are defined at the customer level. Because the ultimate requirement of no door wind noise is statistically impossible, a compliance rate is specified instead. For example, 99% of the produced vehicles must be free of the door wind-noise problem. Considering the fact that improving on the wind-noise problem may deteriorate another performance attribute (e.g., door closing effort), a set of trade-off rates for all relevant performance measures is also provided. After the quality specifications are set at the customer level, an engineering design process is followed to determine the product design parameters such that the quality specifications are met. This is an iterative, math-based optimization process that accounts for inherent uncertainty, variation, and relative importance of the different

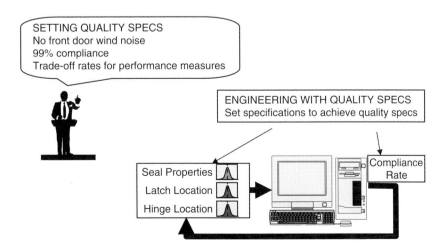

FIGURE 37.1 Design for quality process based on the quality specification structure.

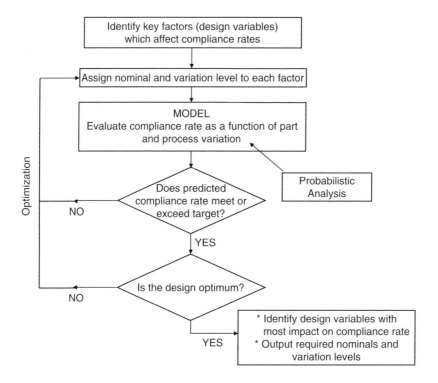

FIGURE 37.2 Reliability-based design optimization in quality specification structure.

performance measures using the trade-off rates. The quality specifications drive the engineering process to account for variation while also enabling optimal design trade-offs based on cost/benefit considerations.

Given a design and manufacturing/assembly process, the design variables are identified, including all key product and process characteristics. Physical models are used to provide the relationship between design variables, production tolerances, and performance measures. Uncertainty is accounted for by calculating compliance rates using a probabilistic analysis. In principle, a reliability-based design optimization (RBDO) problem is solved as shown in Figure 37.2. The details of existing RBDO formulations are given in Section 37.5. After the design variables that affect compliance rates are identified, the nominal value and the allowed variation for each design variable are determined by an optimization process so that the target compliance rates are met.

However, when we design for quality using the quality specification structure to probabilistically satisfy customer expectations, we must simultaneously address expectations and preferences among different performance measures. Furthermore, manufacturing process capabilities and different design alternatives must be also considered along with the performance measures. Not only do we have trade-offs between performance measures, but we also have trade-offs between performance, manufacturing capabilities, and different design alternatives. To handle these trade-offs, a formal aggregation procedure is needed to consider the customer preferences. The basics of a formal preference-aggregation theory are given in the following section.

37.2.1 A Formal Preference Aggregation Theory

The preference aggregation procedure used in this work is based on the method of imprecision [9, 10]. Otto and Antonsson [11] present a set of axioms or properties that an aggregation function must obey in order to be appropriate for rational design decision making. The set includes the annihilation, idempotency, monotonicity, commutativity, and continuity properties, which are mathematically described as

$$h[(0, w_1), (h_2, w_2)] = h[(h_1, w_1), (0, w_2)] = 0 \quad \text{(annihilation)} \tag{37.1}$$

$$h[(h_1, w_1), (h_1, w_2)] = h_1 \quad \text{(idempotency)} \tag{37.2}$$

$$h[(h_1, w_1), (h_2, w_2)] \leq h[(h_1, w_1), (h_2^*, w_2)] \quad \text{if} \quad h_2 \leq h_2^* \quad \text{(monotonicity)} \tag{37.3}$$

$$h[(h_1, w_1), (h_2, w_2)] = h[(h_2, w_2), (h_1, w_1)] \quad \text{(commutativity)} \tag{37.4}$$

$$h[(h_1, w_1), (h_2, w_2)] = \lim_{h_1^* \to h_1} h[(h_1^*, w_1), (h_2, w_2)] \quad \text{(continuity)} \tag{37.5}$$

where h_1, h_2 are the individual preference functions to be aggregated, w_1 and w_2 are importance weights that are assumed to be positive without loss of generality, and h is the aggregate preference function. All preference functions assume values between zero and one.

Scott and Antonsson [12] show that there is a family of aggregation operators h^s that satisfy these axioms. For a two-attribute design problem, h^s is given by

$$h^s[(h_1, w_1), (h_2, w_2)] = \left(\frac{w_1 h_1^s + w_2 h_2^s}{w_1 + w_2} \right)^{1/s} \tag{37.6}$$

where (h_1, w_1) and (h_2, w_2) are the individual preference functions to be aggregated and their corresponding importance weights. The parameter s can be interpreted as a measure of the level of compensation, or trade-off, and is sometimes referred to as the trade-off strategy. Higher values of s indicate a greater willingness to allow high preference for one criterion to compensate for lower values of another. Scott [10] has shown that if $s \to 0$, the aggregation of the two preferences provides maximum compensation. In this case, Equation 37.6 reduces to the following geometric product of the two preferences h_1, h_2

$$h^s = h^s_{prod} = \left[h_1^{w_1} h_2^{w_2} \right]^{1/w_1 + w_2}. \tag{37.7}$$

On the other hand, if $s \to -\infty$, the aggregation of the two preferences provides no compensation at all, and Equation 37.6 reduces to

$$h^s = \min(h_1, h_2). \tag{37.8}$$

When the parameters s and w are correctly chosen, the "best" design can be located by maximizing h^s. The described preference aggregation procedure has the advantage of a rigorous, provable procedure of "indifference points" [13] to determine those parameters. For more information, the interested reader can refer to the literature [10, 14].

37.3 An Automotive Body-Door Example

The described quality specification (QS) structure for satisfying customer expectations has been applied to the design of an automotive body-door subsystem. Three quality issues, classified as JD Power "hardy perennials" (wind noise, water leaks, and door closing effort), belong to the body-door subsystem. Trade-off scenarios can be conducted between door closing effort and door seal gap (which affects wind noise and water leaks), since a "critical to quality" specification for good sealing (e.g., thicker seal) can be detrimental to door closing effort. The conflicting door wind noise and door closing effort performance measures were traded off based on a customer-defined QS structure that provided a target compliance rate for both the door wind noise and door closing effort performance measures as well as a trade-off

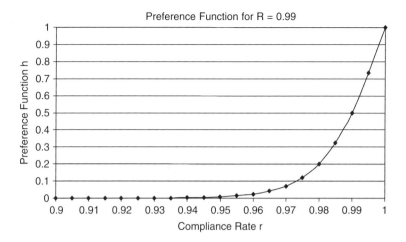

FIGURE 37.3 Preference function for a performance measure of automotive body-door example.

rate between them. The two performance measures were also traded off with manufacturing capability requirements, which were simply expressed with respect to tolerances of the system design variables. The objective was to optimally design the door subsystem considering (1) the trade-off between wind noise and closing effort and (2) the trade-off between good performance and manufacturing cost.

A preference function was constructed for each performance measure relating the expected cost of noncompliance to a number between zero and one. The preference function is constructed so that it is equal to 0.5 at the target compliance rate and increases almost linearly with increasing compliance rate above the target. However, it approaches zero exponentially for compliance rates below the target. In a similar way, a preference function was also constructed for each manufacturing tolerance and each design alternative. The performance preference functions were subsequently aggregated (based on the preference aggregation theory of Section 37.2.1) utilizing the QS trade-off rates in order to account for the relative importance of each performance measure. Similarly, the manufacturing and design alternative preference functions were also aggregated. Finally, the performance, manufacturing, and design alternative aggregate functions were further aggregated to an overall aggregate function, considering their relative importance and level of compensation. An optimization procedure maximized the overall aggregate function, resulting in optimum trade-offs between performance, manufacturing, and design alternatives while attempting to exceed the target compliance rates.

A representative preference function for a performance measure is shown in Figure 37.3 as a function of the compliance rate. A target compliance rate of 0.99 is used. The compliance rate for a particular performance measure k is related to an "expected" cost based on the following relationship

$$c_p(r) = A_k(1 - r_k) \tag{37.9}$$

where A_k is the cost per vehicle per customer complaint and $(1 - r_k)$ is the noncompliance rate for the k performance measure. The quantity A_k can be viewed as the impact cost for a particular performance measure.

The design of an automotive door must include variation in the locating scheme of the door, the properties of the seal, and the gap between the body and door. Complexities are naturally included in the physical problem, with both the nonlinear seal behavior and the dynamics of door closing. The physics-based models relating design-variable input to these two performance measures are detailed in the literature [15, 16]. Input to these deterministic models include hinge and latch location, seal-line location, spatial gap between the door and body, seal thickness, nominal seal compression, nominal load

TABLE 37.1 Random Variables Used in Models for Body-Door Seal Gap and Door-Closing Energy

Design Variable Name	Description
NOMGAP1, NOMGAP2, ... , NOMGAP8	Spatial variation in gap around the door-to-body seal line specified for eight locations
UHCC	Upper hinge location in cross-car direction
UHFA	Upper hinge location in fore-aft direction
UHUD	Upper hinge location in up-down direction
LHCC	Lower hinge location in cross-car direction
LATCC	Latch location in cross-car direction
LATUD	Latch location in up-down direction
STHICK	Seal thickness
DELBAR	Nominal seal compression
FLBAR	Nominal seal load at nominal compression

at nominal compression, stiffness of the body, and stiffness of the door. As an alternative to the computationally time-intensive physics-based models, kriging response-surface approximations for calculating body-door seal gap and door closing energy were also created [17–21]. These response-surface approximations are deterministic models relating design-variable input to performance-measure output, but with the advantage of extremely fast run times.

Variation was specified through random variables that include the mean value and the standard deviation of each design variable. From all the inputs to the deterministic models, 17 design variables were selected for the body-door seal gap, and 14 variables were selected for the door closing energy analysis. Their selection was based on the sensitivity of the response to the input and the ease in which variation could be included in the calculations. For the seal-gap model, an efficient finite element formulation was developed to solve the nonlinear contact problem between the door and body [16]. In addition to the selected design variables, the stiffness characteristics of the body and door were also considered. Table 37.1 describes the selected design variables. Figure 37.4 shows the location of each variable relative to the chosen finite element models.

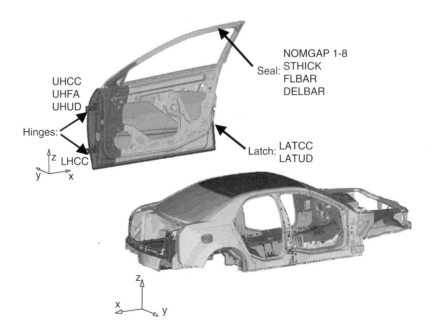

FIGURE 37.4 Finite element models and variable description of automotive body-door example.

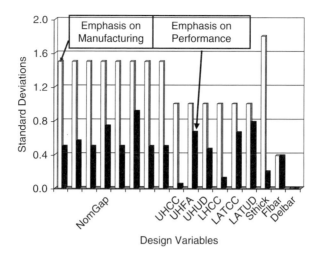

FIGURE 37.5 Design variable variation from optimum trade-off analysis of automotive body-door example. (The abbreviations for the design variables are described in Table 37.1.)

A limit-state value is also needed to calculate the probability of failure or the probability of noncompliance for each quality performance measure. For demonstration purposes, it was assumed that the presence of a gap anywhere along the door-to-body seal line would violate the quality specification of freedom from wind noise and water leaks. For the door closing effort, the door closing energy had to be below a threshold value.

Figure 37.5 shows the results of the optimum trade-off analysis in terms of the desired variation of each design variable. When emphasis is given to the manufacturing cost, the methodology pushes the tolerances (or standard deviations) of each design variable to their upper limit, since cost is inversely proportional to the manufacturing tolerance. However, when emphasis is given to the performance measures, the standard deviation of each design variable is much less than its upper limit. There is also a pronounced difference between the allowed variation of each design variable. The variation of the variables that are important to performance (such as UHCC, LHCC, and STHICK) is kept low.

37.4 An Automotive Bumper Standardization Example

In this section, another example is presented highlighting the application of the preference aggregation method to automotive multiattribute design. Automobile manufacturers can achieve substantial cost reduction through component reuse and standardization. If a component is common for a number of vehicle models, the cost associated with developing, manufacturing, and stocking it is greatly reduced. However, potential reuse and standardization must not compromise the vehicle's performance and quality. In general, different vehicle models may require different components to optimize their individual performance. Thus, the optimization of vehicle performance and the component standardization constitute potentially conflicting goals.

To standardize a component for a given application range determined by the product variety, we must find a small family of component designs covering the entire application range. In general, there are only "minor" differences among different family members. For example, the component width and thickness may be different but the cross-sectional topology may be the same. Therefore, the objective is to find a component family with a small number of members such that the performance variation among different vehicle models is minimal. The problem can be viewed in the context of robust design. In robust design, the impact of uncontrolled variations or noise is reduced. In the standardization problem, the needs of the different vehicle programs introduce variation. The goal is to develop a robust design that can accommodate this variation without significantly affecting performance. An optimization problem is formulated and solved for identifying a robust design that performs to the desired level on a variety of

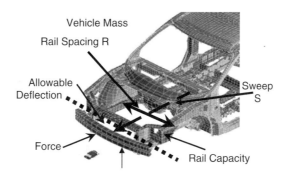

FIGURE 37.6 Design variables and nomenclature for vehicle–bumper interaction.

programs. The objective is to maximize the variety and simultaneously satisfy the performance requirements. A vehicle bumper design is used in this section to illustrate the standardization concept.

The primary function of a front bumper system is to protect the vehicle and its occupants during a crash. The main design variables and nomenclature are shown in Figure 37.6. The bumper must absorb some energy when deflected under a low-speed crash impact, thereby reducing the load on the rails. Furthermore, the deflection must be limited in order to prevent damage to the headlights and other damageable components in the engine compartment. For a vehicle configuration with a given mass, rail spacing and capacity, and sweep (determined by the styling of the vehicle), the bumper geometry (cross-sectional height, width, and thickness) is carefully chosen so that the load on the rails is less than the rail capacity and the bumper deflection is less than an allowable limit.

The vehicle parameters such as mass, sweep, rail spacing, rail capacity, and allowable deflections are generally different for different vehicle models. Even within the same vehicle model, different options for contents, such as the engine, will cause the mass of the vehicles to be different. If we optimize the bumper design for each individual vehicle, a higher cost is incurred. On the other hand, if a common bumper is simply reused, some of the performance requirements may not be met.

Before we describe the methodology to identify the reuse capacity of a bumper design and how to maximize it, the term *design bandwidth* must be defined. Design bandwidth refers to the flexibility of the design to accommodate changes without significant loss in performance or quality. The reuse potential of the design increases with increasing design bandwidth. The design bandwidth of a bumper system defines the maximum range of the vehicle parameters (mass, sweep, rail spacing) that a bumper beam cross section can accommodate without losing its effectiveness. If the bumper is used on any vehicle whose parameters are within the performance bandwidth of the bumper, then its performance under low-speed impact should meet the requirements. However, what is the performance bandwidth of a given bumper cross section? Once this question is answered, the right bumper cross section can be easily identified for the vehicle model under consideration.

To determine the performance band of a bumper cross section, a CAE model is needed to calculate the force on the rails and the bumper deflection during a low-speed crash. This can be done using an LS-DYNA crash analysis, a nonlinear finite element model. However, since each simulation is computationally very expensive and the system behavior is highly nonlinear, a kriging response-surface model (RSM) [17] was developed to approximate the results of the original CAE model. Using this RSM, the vehicle parameters can be easily varied in an optimization process to determine if the requirements are met.

37.4.1 Bumper Optimization Using Preference Aggregation Theory

Two different objectives must be simultaneously optimized: the variety measure that maximizes the component reuse and the performance measure. In this study, the two objectives are aggregated into one single measure using the preference aggregation method of Section 37.2.1. The variety measure,

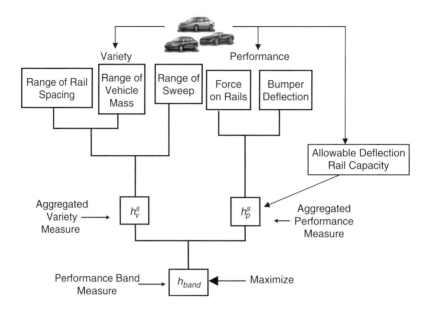

FIGURE 37.7 Aggregation of objectives for bumper example.

representing the range of vehicles on which the bumper can be used, must be balanced by the performance objective. The process to form the aggregate objective function is shown in Figure 37.7.

A separate preference function is assigned to each design parameter (range of rail spacing, range of vehicle mass, range of sweep, force on rails, and bumper deflection). Each preference function represents the level of preference or satisfaction throughout the application range of the parameter. For example, if the range of rail spacing that the bumper can accommodate is equal to the overall range of applications, the value of the preference function is 1 (100% satisfaction). If it is below the minimum acceptable range, the value of the preference function is 0. In this study, a linear preference function is used for intermediate values, as shown in Figure 37.8. Simplicity is the main advantage of the linear form. It has also been proved adequate for this application. For each design in the chosen variety range, the performance is calculated using the developed response-surface models. For example, the actual bumper deflection is predicted for a particular vehicle model within the variety range.

Assuming that all designs in the variety range are equally likely to occur, Monte Carlo simulations are conducted using the response-surface models to determine the probability distribution of the actual

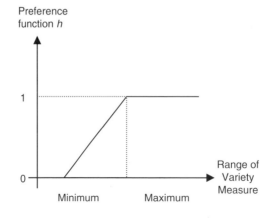

FIGURE 37.8 Preference function for bumper variety measures.

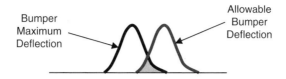

FIGURE 37.9 Probability of meeting bumper-deflection requirement.

bumper deflections over that range (Figure 37.9). The approximate distribution of the allowable bumper deflections for the considered vehicles is usually known. If less than 95% of the designs in the variety range meet the performance requirements, a zero value is assigned for the performance preference function. If 100% of designs meet the performance requirements, the preference function value is equal to one. A linear preference function is used in between, as shown in Figure 37.10. It should be noted that in Figure 37.9, the gray area is not equal to the probability of violating the requirement.

The different variety measures (range of rail spacing, range of vehicle mass, and range of sweep) are aggregated using assigned weighing factors based on priority. Two measures are aggregated at a time using Equation 37.6, where h is the preference function of a variety measure, w is its weighing factor, s is a chosen strategy to reflect the trade-off between the two measures (see Section 37.2.1), and h^s is the aggregated measure. The performance measures are aggregated in a similar fashion. The aggregation sequence is shown in Figure 37.7.

The aggregate variety measure h_v^s and the aggregate performance measure h_p^s are subsequently aggregated themselves to obtain a single measure h_{band} for the performance band of a bumper cross-sectional design (see Figure 37.7). The h_{band} measure will be 1 if the variety range is equal to the total range of application and the performance requirements are both met 100% of the time. Optimization is performed to obtain the best combination of variety and performance for a given bumper or, alternatively, to determine the best bumper design to maximize the variety without sacrificing the performance.

The aggregate h_{band} measure was maximized using the variety ranges and the bumper thickness as design variables. Since the bumper thickness can be easily changed without much cost implication, it was used as a design variable to tune the bumper design for a specific application. Therefore, the optimization process maximized the performance band of a bumper cross section, while using the bumper thickness as an adjustment factor. A genetic algorithm was used to carry out the optimization. As a result of this study, a relationship was obtained between on-the-shelf bumpers and their ranges of application. This method can be extended to develop a family of bumpers with a minimum number of cross sections that covers the entire application range. In general, preference aggregation methods can be applied to component standardization problems, achieving a balance between variety and performance objectives.

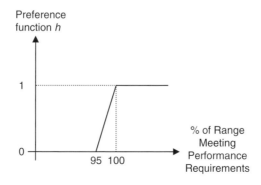

FIGURE 37.10 Preference function for bumper performance measures.

37.5 Overview of RBDO Approaches

In deterministic design we assume that there is no uncertainty in the design variables and modeling parameters. Therefore, there is no variability in the simulation outputs. However, there is inherent input and parameter variation that results in output variation. Deterministic optimization typically yields optimal designs that are pushed to the limits of design constraint boundaries, leaving little or no room for tolerances (uncertainty) in manufacturing imperfections or modeling and design variables. Therefore, deterministic optimal designs that are obtained without taking into account uncertainty are usually unreliable. In contrast, input variation is fully accounted for in reliability-based design optimization (RBDO). Probability distributions describe the stochastic nature of the design variables and model parameters. Variations are represented by standard deviations (typically assumed to be constant), and a mean performance measure is optimized subject to probabilistic constraints.

Reliability-based design optimization can be a powerful tool that can assist in decision making under uncertainty, since it provides optimum designs in the presence of uncertainty in design variables/parameters and simulation models. Reliability-based optimization has been extensively studied [22–32], but most of the published applications are limited to small problems. Accurate and efficient RBDO methods must be developed for large multidisciplinary systems with multiple objectives and multiple deterministic and probabilistic constraints at both the component and system levels. Robustness is also a very important issue in reliability-based design. Efficient and feasible methods to balance the reliability and robustness requirements of engineering systems are needed. Furthermore, the optimum design must be identified using an efficient reliability-based design methodology that integrates system-level and component-level requirements in the presence of multiple objectives and multiple reliability constraints.

Design under uncertainty occurs at several levels: (1) risk vs. cost trade-offs at component and system levels, (2) optimization of nominal values and tolerances when design variables are uncertain, and (3) design for system robustness (insensitivity to variations). The limit-state-based reliability methodology offers a pathway to integrating these requirements. The currently available RBDO methods involve nested optimization and reliability analysis, which may lead to prohibitive computational effort and convergence problems. A conventional RBDO problem is formulated as

$$\min_{d,\mu_X} f(d, \mu_X, p)$$
$$\text{subject to } P(G_i(d, X, p) \geq 0) \geq R_i, i = 1, \ldots, N \quad (37.10)$$

where d is the vector of deterministic design variables, X is the vector of random design variables, p is the vector of random and deterministic design parameters, and $f(\ )$ is the objective function. The desired reliability level for the ith constraint is denoted by $R_i = 1 - p_{f_i}$, where $p_{f_i} = P[G_i(d, X, p) < 0]$ is the target probability of violating the ith constraint, which is usually very small. Note that in the above RBDO formulation, the design variables include *only* the means of the random variables. The target probability of failure p_f is usually approximated by the following first-order relation

$$p_f \approx \Phi(-\beta_t) \quad (37.11)$$

where β_t is the target reliability index and Φ is the standard normal cumulative distribution function.

Based on Equation 37.10 and Equation 37.11, two different RBDO formulations are available; the reliability index approach (RIA) [26] and the performance measure approach (PMA) [22–24]. The RIA-based RBDO problem is stated as

$$\min_{d,\mu_X} f(d, \mu_X, p)$$
$$\text{subject to } \beta_{s_i} \geq \beta_{t_i}, i = 1, \ldots, N \quad (37.12a)$$

where $\beta_s = \min\|U\|$ represents the minimum distance of a point U on the limit state $G(U) = 0$ from the origin of the standard normal space. Therefore, it is calculated from the following reliability-minimization problem:

$$\min \|U\|$$
$$\text{subject to } G(U) = 0 \tag{37.12b}$$

The vector U denotes the random variables in the standard normal space. The entries of U are standard normal random variables.

Similarly, the PMA-based RBDO problem, which is practically the inverse of the RIA-based RBDO problem, is stated as

$$\min_{d,\mu_X} f(d, \mu_X, p)$$
$$\text{subject to } G_{p_i} \geq 0, i = 1, \ldots, N, \tag{37.13a}$$

where $G_p = \min G(U)$ is calculated from the following reliability-minimization problem

$$\min G(U)$$
$$\text{subject to } \|U\| = \beta_t. \tag{37.13b}$$

As shown from Equation 37.12 and Equation 37.13, both the RIA- and PMA-based RBDO formulations involve nested optimization loops, which may hinder their computational efficiency and convergence properties. To improve the computational efficiency, different optimization formulations have recently been proposed that either decouple the two nested optimization loops [27, 33–36] or, even better, merge them in a single optimization loop [28, 29], thus leading to significant efficiency and convergence benefits. Although these methods are very promising, they are also very new and will therefore require further development work before they are accepted in solving large practical RBDO applications.

37.5.1 Automotive Vehicle Side-Impact Example

In this example, a reliability-based design optimization (RBDO) methodology is applied to vehicle crashworthiness under a variety of side-impact constraints. RBDO provides an efficient analytical tool to account for uncertainty in the product development process. Uncertainties in structural design variables, material properties, and operating conditions, among others, are very important in automotive vehicle side-impact studies. For this reason, RBDO of vehicle crashworthiness has recently gained considerable attention [30].

The side impact is one of the safety requirements an automotive vehicle design must meet according to the National Highway Traffic Safety Administration (NHTSA). The side-impact procedure based on the European Enhanced Vehicle-Safety Committee (EEVC) is used in this example. The performance of the dummy in side impact, in terms of head injury criterion (HIC), chest V*C (viscous criterion), and rib deflections (upper, middle, and lower) must meet EEVC requirements.

The finite element model of the vehicle used in this study and the moving deformable barrier are shown in Figure 37.11. The total number of nodes and shell elements of the model are approximately 96,000 and 86,000, respectively. A finite element model of the dummy is also used. The velocity of the B-pillar at the middle point and the velocity of front door at the B-pillar are also considered in the side-impact design. The position of the moving deformable barrier is specified according to the EEVC side-impact procedure. All nodes of the moving barrier are assigned an initial velocity of 50 km/h.

In side-impact design, the increase of gauge design variables tends to improve the dummy performance. However, the vehicle weight is simultaneously increased, which is undesirable. For this reason, an optimization problem is formed by minimizing the vehicle weight subject to a number of safety constraints on the dummy according to the EEVC procedure. They include HIC, abdomen load, rib deflection or V*C, and pubic symphysis force.

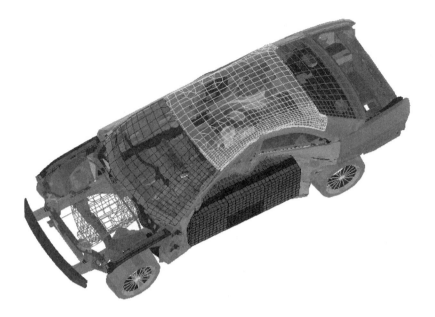

FIGURE 37.11 Finite element example for vehicle side-impact example.

In this example, the RBDO formulation for the vehicle side-impact problem is as follows:

$$\text{Minimize } \mu_w \text{ (mean of vehicle weight)}$$
$$\text{Subject to}$$
$$P[V_{B-Pillar} \leq 9.9(m/s)] \geq R_1$$
$$P[V_{front-door} \leq 15.69(m/s)] \geq R_2$$
$$P[F_{Abdomen-Load} \leq 1.0(kN)] \geq R_3$$
$$P[V*C_1 \leq 0.32(m/s)] \geq R_4$$
$$P[V*C_2 \leq 0.32(m/s)] \geq R_5 \quad\quad (37.14)$$
$$P[V*C_3 \leq 0.32(m/s)] \geq R_6$$
$$P[D_{Upper-Rib} \leq 32(m/s)] \geq R_7$$
$$P[D_{Middle-Rib} \leq 32(m/s)] \geq R_8$$
$$P[D_{Lower-Rib} \leq 32(mm)] \geq R_9$$
$$P[F_{Pubic-Symphysis} \leq 4.0(kN)] \geq R_{10}$$

and

$$\mu_{i_l} \leq \mu_i \leq \mu_{i_u}, i = 1, \ldots, 7$$
$$\mu_8, \mu_9 = 0.345 \text{ or } 0.192 \text{ (material properties)}$$
$$\mu_{10}, \mu_{11} = 0.0 \text{ (barrier height and position)}$$

TABLE 37.2 Description and Range of Design Variables for Vehicle Side-Impact Example

Random Design Variables and Parameters		Range		
Number	Description	Minimum	Nominal	Maximum
1	Thickness of B-pillar inner (mm)	0.5	1.0	1.5
2	Thickness of B-pillar reinforcement (mm)	0.45	0.9	1.35
3	Thickness of floor-side inner (mm)	0.5	1.0	1.5
4	Thickness of cross members 1 and 2 (mm)	0.5	1.0	1.5
5	Thickness of door beam (mm)	0.875	1.75	2.625
6	Thickness of door belt line reinforcement (mm)	0.4	0.8	1.2
7	Thickness of roof rail (mm)	0.4	0.8	1.2
8	Material of B-pillar inner	mild steel	mild steel	HS steel
9	Material of floor-side inner	mild steel	mild steel	HS steel
10	Barrier height (mm)	−30	0	+30
11	Barrier hitting position (mm)	−30	0	+30

where μ_i, $i = 1, \ldots, 7$ are the means of the seven random variables and μ_i, $i = 8, \ldots, 11$ are the means of four random parameters. The desired reliabilities for the ten probabilistic constraints are denoted by R_i, $i = 1, \ldots, 10$. In Equation 37.14, F represents force, V represents velocity, D represents rib deflection, and $V*C$ represents viscous criteria.

The random variables are sizing and material parameters, including the thickness of the B-pillar (inner and reinforcement), the thickness of floor side, the thickness of cross member, the thickness of door beam, the thickness of door belt-line reinforcement, and the thickness of roof rail. The random parameters include the material of the B-pillar (inner) and the floor side (inner) as well as the barrier height and barrier hitting position. A total of seven random variables and four random parameters are used. Table 37.2 shows sequentially the description of the random variables and parameters and their lower and upper bounds. The results from both a deterministic optimization and the RBDO formulation of Equation 37.14 are shown in Table 37.3 and Table 37.4. A 90% reliability is used for all ten constraints. All random variables and parameters are assumed normally distributed with standard deviations $\sigma_{1-4,6,7}$ = 0.03, $\sigma_5 = 0.05$, $\sigma_{8,9} = 0.006$, and $\sigma_{10,11} = 10.0$.

The deterministic optimization reduced the vehicle weight to 23.5. However, the reliability of the front-door velocity, lower rib deflection, and pubic force constraints is unacceptably low (see Table 37.4). The RBDO problem was solved with the double-loop approach of Equation 37.13 using the performance-measure approach for the inner loop. The reliability-based design optimization slightly increased the vehicle weight to 24.5 but satisfied all probabilistic constraints with at least a 90% reliability.

TABLE 37.3 Final Values of Design Variables for Vehicle Side-Impact Example

	Means of Design Variables	
Number	Deterministic Optimization	RBDO
1	0.5	0.5
2	1.226	1.31
3	0.5	0.5
4	1.187	1.294
5	0.875	0.875
6	0.914	1.2
7	0.4	0.4
8	0.345	0.345
9	0.192	0.192

TABLE 37.4 Final Values of All Constraints for Vehicle Side-Impact Example

Description	Deterministic Optimization		RBDO	
	Nominal Value	Reliability	Nominal Value	Reliability
B-pillar velocity	9.343	1.0	9.26	0.998
Front-door velocity	15.678	0.534	15.47	0.97
Abdomen load	0.573	1.0	0.484	1.0
VC1	0.23	1.0	0.233	1.0
VC2	0.203	1.0	0.212	1.0
VC3	0.293	1.0	0.29	1.0
Upper-rib deflection	29.372	0.999	29.56	0.998
Middle-rib deflection	27.66	0.996	27.14	0.998
Lower-rib deflection	32.0	0.5	31.17	0.9
Pubic force	4.0	0.5	3.95	0.9

37.6 Summary and Future Needs

Automotive design is usually performed in a multiattribute environment in the presence of inherent variability and uncertainty. The need to perform trade-off analysis among conflicting design criteria is common. The multiattribute method presented in this chapter is based on a quality-specification structure for addressing automotive quality during the vehicle development process. The method utilizes (1) a formal preference aggregation theory for handling multiple, often conflicting, design criteria and (2) reliability-based design optimization (RBDO) to find the optimal design under uncertainty. Two major applications were used to demonstrate the method: an automotive body-door design and an automotive bumper standardization study. The fundamentals of reliability-based design optimization were also described. The significance of RBDO in automotive design was highlighted using a vehicle side-impact example.

We believe that future research topics, related to the subject of this chapter, can be found in the areas of (1) preference aggregation under uncertainty for robust design, (2) computationally efficient reliability-based design optimization, (3) multidisciplinary optimization under uncertainty, and (4) reliability estimation of complex systems with highly nonlinear limit states.

Resolution of trade-offs in multicriteria optimization problems is very common in automotive applications. The weighted-sum approach [37] is commonly used for handling trade-offs, despite its serious drawbacks [38]. A multiobjective decision problem generally has a whole set of possible "best" solutions, known as the *Pareto* set. A complete decision model therefore requires specification of the degree of compensation between conflicting criteria in order to select the best of all Pareto points [12]. A preference aggregation method, based on a family of aggregation functions, is used in this chapter to formally model all possible trade-offs in engineering applications. An initial attempt has recently been made [39] to extend the preference aggregation methodology in the presence of uncertainty so that it can handle robust design problems. More research is needed in this area.

Reliability-based design optimization (RBDO) gradually becomes an important optimization tool in the presence of uncertainty [24, 27, 30] for automotive applications. However, the conventional double-loop approaches [24, 26] are almost impractical due to the required excessive computational effort. Efficient RBDO algorithms are therefore needed. Advances in this area are reported in the literature [27, 28]. Furthermore, the design and development of complex automotive systems typically requires the integration of multiple disciplines and the resolution of multiple conflicting objectives, also in the presence of uncertainty. Although recent advances have been made [31, 32], more research is needed in this area.

All reported RBDO formulations are based on the first-order relation of Equation 37.11 to calculate the probability of failure. Although Equation 37.11 is exact for linear limit states, it may introduce a large error for nonlinear limit states. For a variety of practical problems in the automotive industry, the failure domain is frequently defined by highly nonlinear limit states. Thus, efficient simulation-based methods are needed instead of the commonly used approximate analytical methods. Some recent work in this area is reported by Zou et al. [40].

Acknowledgments

The authors would like to thank Paul Meernik, John Cafeo, Jeff Robinson, R. Jean Ruth, and Mark Beltramo of the General Motors R&D Center for a variety of useful discussions on the material presented in this chapter. The continuous support and enthusiasm of Robert Lust and Mary Fortier of General Motors Corporation throughout the preparation of this chapter are also acknowledged.

References

1. Thomas, R.S. and Ehlert, G.J., An Alternative Approach to Robust Design: a Vehicle Door Sealing System Example, SAE 971924, Society of Automotive Engineers, Warrendale, PA, 1997.
2. Taguchi, G., *Introduction to Quality Engineering: Designing Quality into Products and Processes*, Asian Productivity Organization, Tokyo, Japan, 1986.
3. Madsen, H.O., Krenk, S., and Lind, N.C., *Methods of Structural Safety*, Prentice Hall, Englewood Cliffs, NJ, 1986.
4. Ang, A.H.-S. and Tang, W.H., *Probability and Concepts in Engineering Planning and Design*, Vol. 1, *Basic Principles*, John Wiley & Sons, New York, 1975.
5. Ang, A.H.-S. and Tang, W.H., *Probability and Concepts in Engineering Planning and Design*, Vol. 2, *Decision, Risk and Reliability*, John Wiley & Sons, New York, 1984.
6. Lust, R.V. and Wu, Y.-T., Probabilistic structural analysis: an introduction, *Experimental Tech.*, 22 (5), 1998.
7. Subramanian, R. and Lust, R.V., Application of probabilistic methods on automotive structures, in *Proceedings of 6th ISSAT International Conference on Reliability and Quality in Design*, 2000, pp. 154–159.
8. Chowdhury, S., *Design for Six Sigma: The Revolutionary Process for Achieving Extraordinary Profits*, Dearborn Trade, 2002.
9. Wood, K.L. and Antonsson, E.K., Computations with imprecise parameters in engineering design: background and theory, *ASME J. Mechanisms, Transmissions, Automation Design*, 111 (4), 616–625, 1989.
10. Scott, M.J., Formalizing Negotiation in Engineering Design, Ph.D. thesis, California Institute of Technology, Pasadena, CA, 1999.
11. Otto, K.N. and Antonsson, E.K., Trade-off strategies in the solution of imprecise design problems, in *Proceedings of International Fuzzy Engineering Symposium*, Vol. 1, Yokohama, Japan, 1991, pp. 422–433.
12. Scott, M.J. and Antonsson, E.K., Aggregation functions for engineering design trade-offs, *Fuzzy Sets Systems*, 99 (3), 253–264, 1998.
13. Scott, M.J. and Antonsson, E.K., Using indifference points in engineering decisions, in *Proceedings of ASME Design Engineering Technical Conferences (DETC)*, DETC2000/DTM-14559, Baltimore, MD, 2000.
14. Dai, Z., Scott, M.J., and Mourelatos, Z.P., Robust design using preference aggregation methods, in *Proceedings of ASME Design Engineering Technical Conferences (DETC)*, DETC2003/DAC-48715, Chicago, IL, 2003.
15. Zou, T., Mahadevan, S., Mourelatos, Z., and Meernik, P., Reliability analysis of automotive body–door subsystem, *Reliability Eng. System Safety*, 78, 315–324, 2002.
16. Kloess, A., Mourelatos, Z.P., and Meernik, P.R., Probabilistic analysis of an automotive body-door system, *Int. J. Vehicle Design*, 34(2), 101–125, 2004.

17. Sacks, J., Welch, W., Mitchell, T., and Wynn, H., Design and analysis of computer experiments, *Statistical Sci.*, 4, 409–435, 1989.
18. Jin, R., Chen, W., and Simpson, T.W., Comparative studies of metamodeling techniques under multiple modeling criteria, *Structural Multidisciplinary Optimization*, 23, 1–13, 2001.
19. Tu, J. and Jones, D.R., Variable screening in metamodel design by cross-validated moving least squares method, in *Proceedings of 44th Structure, Structural Dynamics, and Material Conferences*, AIAA-2003-1669, AIAA/ASME/ASCE/AHS/ASC, Norfolk, VA, 2003.
20. Eubank, R.L., *Spline Smoothing and Nonparametric Regression*, Marcel Dekker, New York, 1988.
21. Fan J. and Gijbels, I., *Local Polynomial Modeling and Its Applications*, Chapman & Hall, New York, 1996.
22. Tu, J., Choi, K.K., and Park, Y.H., A new study on reliability-based design optimization, *ASME J. Mechanical Design*, 121, 557–564, 1999.
23. Tu, J., Choi, K.K., and Park, Y.H., Design potential method for robust system parameter design, *AIAA J.*, 39 (4), 667–677, 2001.
24. Choi, K.K. and Youn, B.D., Hybrid analysis method for reliability-based design optimization, in *Proceedings of ASME Design Automation Conference*, Pittsburgh, PA, 2001.
25. Wu, Y.-T. and Wang, W., A new method for efficient reliability-based design optimization, in *Proceedings of 7th Special Conference on Probabilistic Mechanics and Structural Reliability*, 1996, pp. 274–277.
26. Reddy, M.V., Granhdi, R.V., and Hopkins, D.A., Reliability based structural optimization: a simplified safety index approach, *Comput. Struct.*, 53 (6), 1407–1418, 1994.
27. Du, X., Chen, W., Sequential optimization and reliability assessment method for efficient probabilistic design, in *Proceedings of ASME Design Engineering Technical Conferences (DETC)*, DETC2002/DAC-34127, Montreal, Canada, 2002.
28. Chen, X., Hasselman, T.K., and Neill, D.J., Reliability based structural design optimization for practical applications, in *Proceedings of 38th Structures, Structural Dynamics, and Materials Conference*, AIAA/ASME/ASCE/AHS/ASC, 1997.
29. Maglaras, G. and Nikolaidis, E., Integrated analysis and design in stochastic optimization, *Structural Optimization*, 2 (3), 163–172, 1990.
30. Yang, R.J., Gu, L., Tho, C.H., Choi, K.K., and Youn, B.D., Reliability-based multidisciplinary design optimization of a full vehicle system, in *Proceedings of 43rd Structures, Structural Dynamics, and Materials Conference*, AIAA/ASME/ASCE/AHS/ASC, Denver, CO, 2002.
31. Padmanabhan, D. and Batill, S.M., Decomposition strategies for reliability based optimization in multidisciplinary system design, in *Proceedings of 9th AIAA/SSMO Symposium on Multidisciplinary Analysis and Optimization*, Atlanta, GA, 2002.
32. Du, X. and Chen, W., Collaborative reliability analysis for multidisciplinary systems design, *Proceedings of 9th AIAA/SSMO Symposium on Multidisciplinary Analysis and Optimization*, Atlanta, GA, 2002.
33. Royset, J.O., Der Kiureghian, A., and Polak, E., Reliability-based optimal design of series structural systems, *J. Eng. Mechanics*, 127(6), 607–614, 2001.
34. Royset, J.O., Der Kiureghian, A., and Polak, E., Reliability-based optimal structural design by the decoupling approach, *Reliability Eng. System Safety*, 73, 213–221, 2001.
35. Wu, Y.-T., Shin, Y., Sues, R., and Cesare, M., Safety-factor based approach for probabilistic-based design optimization, in *Proceedings of 42nd Structures, Structural Dynamics, and Materials Conference*, AIAA/ASME/ASCE/AHS/ASC, Seattle, WA, 2001.
36. Wu, Y.-T. and Wang, W., Efficient probabilistic design by converting reliability constraints to approximately equivalent deterministic constraints, *J. Integrated Design Process Sci.*, 2 (4), 13–21, 1998.
37. Tang, X. and Krishnamurty, S., Performance estimation and robust design decisions, in *Proceedings of 41st Structures, Structural Dynamics, and Materials Conference*, AIAA/ASME/ASCE/AHS/ASC, Atlanta, GA, 2000.

38. Das, I. and Dennis, J., A closer look at drawbacks of minimizing weighted sums of objectives for Pareto set generation in multi-criteria optimization problems, *Structural Optimization*, 14 (1), 63–69, 1997.
39. Dai, Z., Scott, M.J., and Mourelatos, Z.P., Incorporating epistemic uncertainty in robust design, in *Proceedings of ASME Design Engineering Technical Conferences*, DETC2003/DAC-48713, Chicago, IL, 2003.
40. Zou, T., Mourelatos, Z.P., Mahadevan, S., and Tu, J., Component and system reliability analysis using an indicator response surface Monte Carlo approach, in *Proceedings of ASME Design Engineering Technical Conferences*, DETC2003/DAC-48708, Chicago, IL, 2003.

38
Probabilistic Analysis and Design in Automotive Industry

Zissimos P. Mourelatos
Oakland University

Jian Tu
General Motors R&D Center

Xuru Ding
General Motors Corporation

38.1	Introduction	**38**-1
38.2	Common Reliability Methods	**38**-2
	Component-Level Reliability Analysis • System-Level Reliability Analysis • Examples	
38.3	Nonparametric Metamodeling (Response Surface) Methods	**38**-10
	Parametric and Nonparametric Multiple-Regression Techniques • Uniform Sampling and Sample Partitioning in Computer Experiments • Cross Validation for Estimating Metamodel Prediction Error • Local Polynomial Fitting Using CVMLS Method • Variable Screening Based on CVMLS • Crash Safety Example for Illustrating CVMLS and Variable Screening	
38.4	Variation Reduction in Robust Engineering: an Automotive Example	**38**-19
38.5	Summary and Future Needs	**38**-22

38.1 Introduction

Delivering reliable, high-quality products at low cost has become the key to survival in today's global economy. The presence of uncertainty in the analysis and design of engineering systems has always been recognized. Traditional deterministic analysis accounts for these uncertainties through the use of empirical safety factors. These safety factors are derived from past experience and do not provide quantifiable measures of the frequency at which failure will occur.

Engineering design usually involves a trade-off between maximizing reliability at the component or system level while achieving cost targets. In contrast to the traditional deterministic design, probabilistic analysis provides the required information for optimum design and accomplishes both goals simultaneously. In the automotive industry, quality products are vehicles whose specifications, as manufactured, meet customer requirements. Given the uncertainties in loads, materials, and manufacturing, modern methods of reliability analysis should be used to ensure automotive quality in terms of reliability measures.

In large-scale systems, often encountered in the automotive and aerospace industries among others, reliability predictions based on expensive full-scale tests are not economically feasible. Efficient computational methods represent a far better alternative. The first requirement of a computational reliability analysis is to develop a quantitative model of the behavior of interest. Subsequently, the statistical behavior is defined for all random variables involved in the limit-state function that separates the failure and the safe regions. Finally, the reliability is estimated using a variety of methods. Both analytical and simulation-based methods are available for reliability analysis. The analytical methods are generally simple and efficient, but for complex problems, their accuracy cannot be guaranteed. In simulation-based methods, the accuracy can be controlled, but the efficiency is generally not satisfactory. An overview of the reliability methods is given in Section 38.2. The commonly used analytical and simulation-based methods are described at both the component and system level. Reliability examples involving an automotive liftgate and a glass-guidance design demonstrate how basic reliability principles are applied in automotive engineering.

To address the high computational costs associated with reliability and robust design assessment of complex automotive systems, the engineering community commonly relies on accurate metamodels, or response-surface models, as surrogates of the computationally demanding CAE models. Fast-running metamodels are typically developed to approximate the CAE model's input–output relationships from a set of pregenerated CAE simulations. Section 38.3 describes in detail the current state of the art of metamodeling in automotive reliability and robust design and illustrates the value of metamodeling with a practical automotive crash safety application.

The current competitive automotive market environment calls for high-quality vehicles in the presence of uncertainty in customer expectations, manufacturing tolerances, and operating conditions. Customers demand reliable and robust products. Robust design, originally proposed by Taguchi, is a method of improving the quality of a product by minimizing (without eliminating) the effect of the causes of variation. It is common to seek "robustness" of a design objective by simultaneously optimizing the mean performance and minimizing the performance variance. Although the importance of robust design has been recognized in the automotive industry, its high computational cost has hindered its wide-scale use by automotive designers. Section 38.4 highlights the importance and computational challenges of robust design by presenting a practical robust design procedure for an occupant-restraint system for frontal-impact performance.

The chapter closes with a summary of the presented work and some recommendations on reliability analysis needs from the automotive industry point of view.

38.2 Common Reliability Methods

Reliability analysis can be performed at either the component or the system level, depending on the number of limit states of the reliability problem. Analytical and simulation-based reliability methods are available for both component- and system-level reliability problems. The advantages and disadvantages of these methods are discussed in this section.

38.2.1 Component-Level Reliability Analysis

38.2.1.1 Analytical Reliability Methods

For a limit-state function g with multiple random variables represented by a vector X, whose joint probability density function is $f_X(\mathbf{x})$, the probability of failure is defined as:

$$P_f = \int \cdots \int_{g(\mathbf{x})<0} f_X(\mathbf{x}) d\mathbf{x} \tag{38.1}$$

In general, the exact calculation of the above multidimensional integral is difficult. Consequently, various analytical methods have been developed to estimate its value. They are mainly categorized as first-order reliability methods (FORM) or second-order reliability methods (SORM). FORM uses a linear approximation of the limit-state function at the most probable point (MPP), as shown in Figure 38.1. Consequently, for the nonlinear limit state shown, FORM will overestimate the probability of failure, since it considers the contribution of the region between the real limit state and the approximation in calculating the failure probability integral of Equation 38.1.

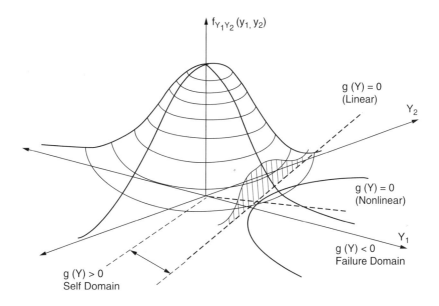

FIGURE 38.1 First-order reliability approximation.

In both FORM and SORM, the original random variables, which are generally nonnormal and correlated, are first transformed into an equivalent set of statistically independent normal variates. A general transformation for this purpose is the Rosenblatt transformation [1]. A two-parameter method suggested by Rackwitz and Fiessler [2] can also be used to find an equivalent set of normal variables. If the random variables are uncorrelated, the standard normal variables can be calculated as

$$Y = \frac{X - \mu_x}{\sigma_x} \tag{38.2}$$

where μ_x and σ_x are the equivalent normal mean and standard deviation, respectively, of random variable X.

The first-order reliability method (FORM) finds the minimum-distance point on the limit state to the origin of the standard normal space by solving an optimization problem, and it subsequently estimates the failure probability based on the minimum distance. The minimum-distance point on the limit state is also called the most probable point (MPP). The first-order reliability estimate is computed as

$$P_f = \Phi(-\beta) \tag{38.3}$$

where β is the distance from the origin to MPP, and Φ is the cumulative distribution function of a standard normal variable.

Various techniques can be used to calculate the most probable point (MPP). Rackwitz and Fiessler [2] used a Newton-type recursive formula to search for the MPP as

$$\mathbf{Y}_{k+1} = \frac{1}{|\nabla G(\mathbf{Y}_k)|^2} \{\nabla G(\mathbf{Y}_k)^T \mathbf{Y}_k - \nabla G(\mathbf{Y}_k)\} \nabla G(\mathbf{Y}_k) \tag{38.4}$$

where $\nabla G(\mathbf{Y}_k)$ is the gradient of the performance function at the kth iteration point $\mathbf{Y}_k$, defined in the standard normal space.

At each step of FORM, the sensitivity factors can be calculated as

$$\alpha_k = \left(\frac{\partial G(\mathbf{Y}_k)}{\partial y_k}\right) \Big/ |\nabla G(\mathbf{Y}_k)| \tag{38.5}$$

For problems with highly nonlinear limit states, the MPP search formula of Equation 38.4 may fail to converge. In such cases, other optimization methods can also be used to search for the MPP [3]. Among those, the sequential quadratic programming (SQP) is very common. In these methods, one needs to evaluate at least the first derivatives and, in some cases, the second derivatives, which brings additional computational cost for an implicit limit state containing a large number of random variables.

Even if the algorithm converges in searching for the MPP, the first-order reliability estimation may not be accurate for nonlinear limit states. Several more sophisticated reliability-analysis techniques have been developed to overcome the difficulty of FORM. The advanced mean value (AMV) method [4, 5] combines the information from the mean value first-order (MVFO) step and one additional deterministic analysis to obtain a substantially improved reliability estimate. Several second-order reliability methods (SORM) have also been developed [6–10]. In Breitung's SORM [6], the probability content in the failure domain is approximated using an asymptotic formula such as

$$P_f = \Phi(-\beta) \prod_{i=1}^{n-1} (1 + \beta \kappa_i)^{-1/2} \tag{38.6}$$

where β is the first-order reliability index, and κ_i represents the main curvatures of the failure surface at the design point.

Tvedt's SORM [8] improves over Breitung's method in two ways. First, it uses a full second-order Taylor series expansion at the design point to approximate the failure surface, and second, it numerically evaluates the probability content of the parabolic failure domain. Köylüoglu and Nielsen [7] proposed three closed-form approximations for SORM integrals, of which the one-term approximation is the simplest. Cai and Elishakoff [9] presented a refined second-order approximation, which is a series formula with three terms. Zhao and Ono [10] suggested a simple point-fitting SORM approximation and an empirical second-order reliability index for easy practical application of SORM.

38.2.1.2 Simulation-Based Reliability Methods

There is also a simulation-based category of reliability methods. The basic Monte Carlo method belongs in this category. It is simple to implement and can therefore be used for almost all problems, at any desired accuracy. However, the limit state needs to be evaluated a large number of times with randomly sampled input values of the design parameters, which can be time consuming and expensive for problems with implicit limit-state functions and high reliability. To address the high computational cost of the Monte Carlo method, several more-efficient simulation-based methods have been developed [11]. One of these is importance sampling.

The basic idea of importance sampling is to minimize the total number of sampling points by concentrating the sampling in the failure region. It can be reasonably efficient if a sampling density is used to generate sample points in the failure region, where the probability density is the greatest. However in many cases, it is difficult to know the shape of the failure region in advance. To overcome this difficulty, the concept of adaptive importance sampling (AIS) has been proposed [12, 13].

The AIS is based on the idea that the importance-sampling density function can be gradually refined to reflect the increasing state of knowledge of the failure region. The sampling space is adaptively adjusted based on the generated sampling points. Two versions of AIS have been developed. The first version uses an adaptive surface to approximate the limit state. Based on different adaptive surfaces, a radius-based method, a plane-based method, and a curvature-based method have been developed [14]. The curvature-based method is generally the most efficient among the three, although its efficiency and robustness rely on the accuracy of the initial MPP. The second version of AIS is called multimodal adaptive importance sampling [13, 15, 16]. It uses a multimodal sampling density to emphasize all important sample points in the failure domain, each in proportion to the true probability density at the particular sampling point [15, 16]. This method was applied to component and system reliability analysis of large structures [16–18] with very satisfactory results.

38.2.2 System-Level Reliability Analysis

The search for computationally efficient procedures to estimate system reliability has resulted in several approaches, most of which belong to the enumeration and analytical bounds category and the efficient MCS (Monte Carlo simulation) category.

In the first category, the branch-and-bound method provides a systematic procedure to identify the various failure sequences of a system. It mainly involves the branching and the bounding operations. In the second category, several importance-sampling schemes developed over the last decade are used to estimate the system reliability. These schemes can be divided into direct methods [13, 19], updating methods [20–22], spherical schemes [23, 24], and adaptive schemes [15, 25]. All of the above methods, except the adaptive schemes, require prior knowledge of the failure region. System-reliability techniques, which account for progressive damage over time, have also been developed [26–28].

38.2.2.1 Analytical Methods for System Reliability Analysis

The joint-failure probability of multiple failure events often needs to be computed during system reliability analysis. Due to the difficulty in determining the joint-failure probabilities of more than two failures except through MCS, approximations using bounds were developed first. The simplest bounds are Cornell's first-order bounds [29]

$$\max_{1 \leq i \leq n} P(E_i) \leq P\left(\bigcup_{i=1}^{n} E_i\right) \leq \sum_{i=1}^{n} P(E_i) \tag{38.7}$$

The general bimodal (second-order) bounds [30] can also be used to calculate the range of the system reliability as

$$P_1 + \max\left[\sum_{i=2}^{k}\left\{P_i - \sum_{j=1}^{i-1} P_{ij}\right\}; 0\right] \leq P_f \leq \sum_{i=1}^{k} P_i - \sum_{i=2}^{k} \max_{j<i} P_{ij} \tag{38.8}$$

where P_1 is generally selected as the maximum failure probability value among the k limit states, and P_{ij} is the joint probability of the ith and jth events. Ditlevsen [24] proposed a weakened version of the above bimodal bounds for Gaussian (normal) variables.

Hohenbichler and Rackwitz [20] and Gollwitzer and Rackwitz [31] developed numerical solutions for the multinormal integral that provide good approximations. Xiao and Mahadevan [27] proved that the upper bound for the intersection probability based on Ditlevsen's lower bound for the probability of union gives a rough upper bound, especially when the two least probable events are highly correlated. They also suggested that the numerical solutions of a multinormal integral be used if the reliability bounds do not have the desired accuracy.

38.2.2.2 Simulation-Based Methods for System Reliability Analysis

In addition to enumeration methods, many simulation-based methods have been developed for system-reliability analysis. In general, the multimodal adaptive importance sampling [15, 18] and the sequential conditional importance sampling (SCIS) [32] have been found efficient. In SCIS, the multinormal distribution function is expressed as a product of m conditional probability terms

$$\begin{aligned}\Phi_m(\mathbf{c}, \mathbf{R}) &= P\left[(X_m \leq c_m | \bigcap_{k=1}^{m-1}(X_k \leq c_k)\right] \\ &\times P\left[(X_{m-1} \leq c_{m-1} | \bigcap_{k=1}^{m-2}(X_k \leq c_k)\right], \ldots, \times \Phi(c_1) \\ &= \Phi(c_1)\prod_{k=2}^{m} E[I_k | I_1 = I_2 = \ldots = I_{k-1} = 1]\end{aligned} \tag{38.9}$$

where **R** is the covariance matrix, m is the number of random variables, and I_k is the indicator function ($I_k = 1$ for $X_k \leq c_k$, else $I_k = 0$). Equation 38.9 can be recast as a sequential simulation of random variables, $X_k \leq c_k$ ($k = 1, \ldots, m$), from the conditional normal density function, $\varphi_k(x_k|x_1, \ldots, x_{k-1})$, provided that $I_1 = I_2 = \ldots = I_{k-1} = 1$. This approach was found to lead to fairly accurate estimates of $\Phi_m(\mathbf{c}, \mathbf{R})$ with a relatively small number of simulations. Pandey and Sarkar [33] proposed the following approximation:

$$P\left[(X_{m+1-p} \leq c_{m+1-p}) \mid \bigcap_{k=1}^{m-p}(X_k \leq c_k)\right] \approx \Phi(c_{(m+1-p)|(m-p)}) \quad (38.10)$$

where $c_{(m+1-p)|(m-p)}$ is a conditional normal fractile, leading to the following approximation for the multinormal distribution function

$$\Phi_m(\mathbf{c}, \mathbf{R}) \approx \prod_{k=1}^{m} \Phi(c_{k|k-1}) \quad (38.11)$$

This is referred as product of conditional marginals (PCM). Both SCIS and PCM are based on the multinormal integral, which requires that the random variables have normal distributions and the limit-state functions be linear. Transformations can be used to extend these methods to nonlinear limit states with nonnormal random variables.

38.2.3 Examples

The application of these basic reliability principles in the automotive industry is illustrated with two examples. In the first example, the reliability of an all-glass liftgate design is assessed using the probability of the maximum deflection exceeding a certain target value. The second example provides a reliability study of a cable-drive glass guidance system that is widely used as the mechanism to operate the window glass of a vehicle.

38.2.3.1 Liftgate Reliability

The evenness and flushness of the gaps around the liftgate of a vehicle is important for the insulation of noise and water, as well as for customer-perceived quality. One source of variation in the gap flushness is the deflection of the liftgate panel. In this example, a method is presented for evaluating the reliability of the liftgate at the design stage, as it relates to the gap flushness.

Figure 38.2 shows the back view of a sport-utility type of vehicle, with a proposed all-glass liftgate design. The proposed liftgate is made of glass, with no metal frame around it. It is attached to the vehicle body with two hinges at the top. A pair of gas struts is glued to the liftgate to assist its opening. However, when the liftgate is closed, the gas struts exert a force on the liftgate, forcing it to deflect outward. This deflection increases the gap between the liftgate and the vehicle body, which may cause water leakage or an increased level of wind noise in the vehicle cabin.

The maximum deflection occurs where the gas struts are attached, and this can be evaluated using the ABACUS commercial finite element code. The maximum deflection of the liftgate varies due to the variation of glue properties such as modulus of elasticity E_{glue} and Poisson ratio μ_{glue}, the variation of the gas-strut forces due to temperature changes, and the variation of the glass modulus of elasticity E_{glass}. Our objective is to evaluate the reliability of the all-glass liftgate design, which is defined as the probability for the deflection not to exceed a certain maximum value. This reliability can be used as a performance measure to choose among alternative designs.

Figure 38.3 illustrates how the variation is transmitted. A linear static finite element model is created for predicting the deflections at the attachment locations of the gas struts. The parameters E_{glue}, E_{glass}, μ_{glue}, E_{steel} are assumed to be normally distributed, with means and standard deviations obtained from measurement data, best practices, or expert opinions.

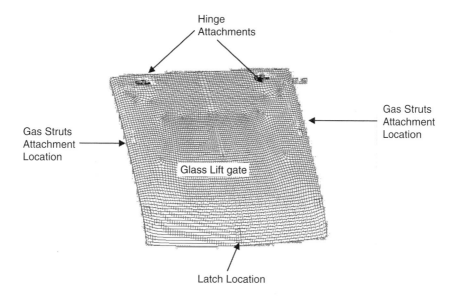

FIGURE 38.2 A CAD view of a glass liftgate at its closed position.

The limit-state function is defined by

$$d - D \leq 0, \tag{38.12}$$

where D is the target maximum displacement, and

$$d = f(E_{glue}, E_{glass}, \mu_{glue}, E_{steel}) \tag{38.13}$$

is the deflection at the gas-strut attachments when the liftgate is closed.

In this study, FORM was used to estimate the reliability of the design. Normal distributions were assumed, with large variation of the glue properties. For a target maximum displacement of $D = 1$ mm, it was found that the probability of d being greater than D is 0.15. This means that the reliability of the proposed design is 0.85. The relatively low reliability is due to the large deflection d resulting from the fact

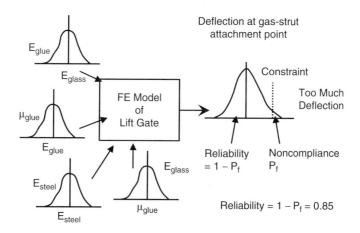

FIGURE 38.3 Schematic of liftgate reliability prediction.

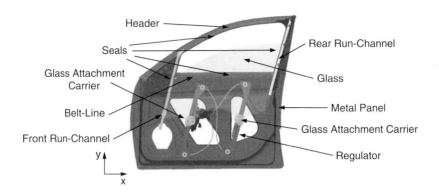

FIGURE 38.4 Schematic of a cable-drive glass guidance system.

that the nominal glass panel thickness does not provide adequate stiffness needed to counter the forces exerted by the struts. Another liftgate design, with a thin metal frame around the glass, was proposed as an alternative. Its reliability was found to be much higher than 0.85, and it was therefore chosen as the appropriate design.

In general, if an all-glass liftgate is preferred due to its aesthetic appeal, its reliability can be also improved by either increasing the thickness of the glass panel to provide more stiffness, or by reducing the gas strut forces. The strut forces are designed to hold the liftgate at the opening position at extremely low temperatures. A trade-off between the aesthetics and the hold-open requirement can be made to enable the use of an all-glass liftgate design.

38.2.3.2 Glass Guidance Design Reliability

A cable-drive glass guidance system [34–41], shown in Figure 38.4, is widely used as the mechanism to operate the window glass of a vehicle. This design has advantages in reducing weight and cost, compared with the commonly used cross-arm window-regulator design. However, the cable-drive design is more sensitive to system parameter variation. These variations affect the system reliability, which is traditionally assessed through hardware testing. Due to the complexity of the system integration, it is efficient and effective to use CAE simulations to assess the system reliability at the early stages of the product development process.

A glass guidance system consists of four major subsystems, the metal panel, seals, glass, and the regulator. The glass is driven by the regulator, which is powered by an electric motor. The motor torque is designed to overcome first the friction between the glass and the seal and second the glass weight. The required torque is therefore an important system performance metric. For a design to be valid, the available motor torque must be greater than the required torque to move the glass. Seal strips are attached to the metal panel in order to prevent airborne noise and water leakage. The window glass moves up and down through the space between the inner and outer belt-line seal lips, while its front and rear edges are embedded between the inner and outer seal lips of the run channels. The header seal provides insulation for the top edge of the glass at the fully up position. The glass movement is driven by a regulator subsystem through two attachment carriers, as shown in Figure 38.5. The carriers are driven by a cable along two guidance rails. An electric motor provides the torque needed to move the cable through a cable drum, as well as to overcome the friction between the cable and pulleys.

To analytically assess the reliability of the glass guidance system, a CAE model was created. A kinematic model was used to express the resultant forces at the two carriers, f_1 and f_2, as

$$f_1 = f_1(\delta_1, \delta_2, \mu_s, k) \tag{38.14a}$$
$$f_2 = f_2(\delta_1, \delta_2, \mu_s, k) \tag{38.14b}$$

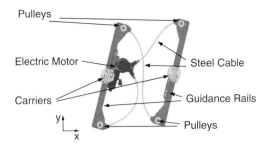

FIGURE 38.5 Schematic of a dual-cable regulator.

where δ_1 is the glass position in the cross-vehicle direction, δ_2 is the regulator position in the fore–aft direction, μ_s is the seal friction coefficient, and k is the seal stiffness. The required system torque T can be subsequently expressed as

$$T = F(f_1, f_2, p, \mu_r) \tag{38.15}$$

where p is the regulator cable pre-tension, and μ_r is the friction coefficient of the regulator components.

In order to handle the high computational cost associated with a reliability analysis, metamodels were developed for forces f_1 and f_2. Using a design of experiment (DOE) matrix, a number of simulations using the CAE model were first performed. Since there is a nonlinear relationship between the carrier forces and design parameters, a quadratic regression model that includes interaction terms was used. For efficiency reasons, 25 runs were performed, corresponding to a fractional factorial central composite DOE with four variables and three levels [42]. The developed metamodels for f_1 and f_2 are as follows:

$$\begin{aligned} f_1(\delta_1, \delta_2, \mu_s, k) &= 3.7 - 1.4\delta_1 - 0.1\delta_2 + 53.7\mu_s + 2.4k - 0.03\delta_1\delta_2 - 0.8\delta_1\mu_s + 8.9\delta_1 k \\ &\quad + 0.9\delta_2\mu_s + 0.1\delta_2 k - 51.4\mu_s k - 1.6\delta_1^2 - 3.2\text{e}{-}4\delta_2^2 - 51.4\mu_s^2 - 3.6k^2 \end{aligned} \tag{38.16a}$$

$$\begin{aligned} f_2(\delta_1, \delta_2, \mu_s, k) &= 40.6 - 2.5\delta_1 - 0.1\delta_2 - 45.3\mu_s + 3.8k + 0.02\delta_1\delta_2 + 8.8\delta_1\mu_s - 4.7\delta_1 k \\ &\quad - 1.1\delta_2\mu_s - 0.08\delta_2 k + 283.6\mu_s k + 1.9\delta_1^2 + 2.6\text{e}{-}5\delta_2^2 + 51.5\mu_s^2 + 0.7k^2 \end{aligned} \tag{38.16b}$$

Note that the metamodels include all linear and interaction terms as well as all squared terms.

For estimating the reliability of the glass guidance system, the six input variables in Equation 38.15 and the torque T_0 provided by the motor are considered normally distributed, independent random variables with means and standard deviations as given in Table 38.1. The probability distribution data have been obtained from the system manufacturers.

The limit-state function is given by $T_0 - T \leq 0$, where Equation 38.15 is used to calculate the motor torque T. For this particular example, the advanced mean value method (AMV) [4, 5] estimated the system reliability accurately and efficiently. Figure 38.6 shows the calculated cumulative distribution

TABLE 38.1 Description of Random Variables for the Glass Guidance System

Parameter	Assumed Distribution	Mean	Standard Deviation
Glass position, δ_1	normal	0.0	0.45
Regulator position, δ_2	normal	0.0	2.24
Seal friction coefficient, μ_s	normal	0.2	0.04
Seal stiffness, k	normal	1.0	0.32
Cable pretension, p	normal	33.0	1.82
Friction coefficient of regulator, μ_r	normal	0.1	0.03
Torque provided by motor, T_0	normal	7.0	0.5

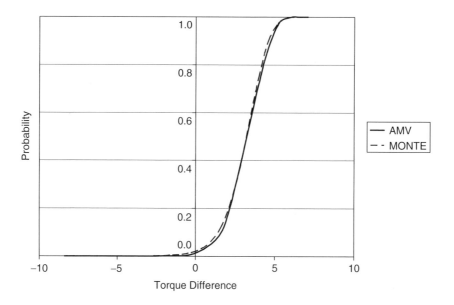

FIGURE 38.6 Reliability estimation for the glass guidance system.

function (CDF) with both the AMV and Monte Carlo methods, indicating the good accuracy of the AMV method.

If the estimated reliability is not satisfactory, a reliability-based design optimization (RBDO) [43–46] algorithm can be used to change the mean values of the six random variables so that the reliability requirement is met.

38.3 Nonparametric Metamodeling (Response Surface) Methods

Each probabilistic analysis of a CAE model often requires a large number of time-consuming CAE simulations. Moreover, each CAE simulation (such as a full-vehicle crash simulation using a large-scale finite element model) can be time consuming, even in the current high-performance computing environment. The broader applications of probabilistic analysis in engineering design [47, 48] only become feasible by using fast-running multivariate metamodels (also known as response-surface models or surrogate models) that can capture the generally nonlinear implicit input/output relationships in the time-consuming multidisciplinary CAE simulations [49, 50].

The traditional response-surface methodology using the design of experiments (DOE) techniques [51–54], such as fractional factorials, central composite, Plackett-Burman, and Box-Behnken designs, is usually computationally efficient. However, it often becomes inadequate in nonlinear multivariate metamodeling of complex CAE models. For this reason, some recent advances in nonparametric metamodeling are introduced in this section, along with the progressive space-filling sampling and variable screening techniques.

38.3.1 Parametric and Nonparametric Multiple-Regression Techniques

The multivariate metamodeling provides efficient input–output relationships. CAE simulations are used to calculate the output values at various points of the input variable space. Subsequently, multiple-regression techniques are used to construct a metamodel for best fitting the provided set of output values.

In traditional parametric regression, one assumes that the metamodel functional form is known (e.g., polynomials in the conventional least-squares regression). The goal of regression analysis is to determine the parameter values of the assumed functional form, so that the generated metamodel best fits the provided data set. A linear or nonlinear parametric metamodel is likely to produce good approximations only when the assumed functional form is close to the true underlying function. For this reason, nonparametric regression techniques have attracted a growing interest [55–62]. They only use a few general assumptions about the functional form of the metamodel, such as its smoothness properties. The functional form is not prespecified, but determined instead from the available data. The nonparametric approach is therefore, more flexible and is likely to produce accurate nonlinear approximations even if the true underlying function form is totally unknown.

Many nonparametric techniques have been proposed for univariate modeling, such as smoothing splines and local polynomial fitting [55–57]. They can be easily extended to multivariate cases. As a more general case of smoothing splines, nonparametric regression methods using Gaussian process models (similar to "kriging" in spatial statistics) and radial basis functions (RBF) have been used to fit data from computer experiments [58–60]. The local polynomial fitting and "kriging"/RBF techniques have demonstrated better predictive performance than the parametric regression techniques [61, 62]. Their main advantage is an automated metamodeling process, in which the hyperparameters in local polynomial fitting and "kriging" can be determined by minimizing the metamodel cross-validation error and maximizing the likelihood function, respectively [63–65]. The local polynomial fitting, using the cross-validated moving least squares (CVMLS) method, is discussed in detail in Section 38.3.5.

38.3.2 Uniform Sampling and Sample Partitioning in Computer Experiments

Fractional factorial designs are the most popular sampling schemes in design of experiments (DOE) using the traditional parametric multiple-regression techniques. However, they are very often limited to two or three levels, even for a small number of input variables, due to the rapidly increasing number of required computer experiments with increasing number of inputs. In creating a metamodel of a CAE model, we want to use a modest number of computer experiments with many levels to explore potentially nonlinear input/output relationships. Previous studies [66, 67] have found that uniformity is the most important criterion and, moreover, that the space-filling uniform design is preferable for robust design. The main advantage of the uniform design is its better representation of the input variable space with fewer samples for a large number of levels.

Perfectly uniform samples can be achieved by using a regular grid of sample points. However, the regular-grid sampling approach does not scale well to higher dimensions. In multivariate applications, random or quasi-random sampling techniques are therefore used to achieve optimum uniformity, which is commonly measured by a "minimum interpoint distance" criterion in terms of Euclidean distance normalized in $[0, 1]^n$ space. Instead of searching for the minimum interpoint distance among all sample pairs, a function φ_p is used to conveniently assess the minimum interpoint distance for a set of N samples $(\underline{x}^1 \ldots \underline{x}^N)$. The function φ_p is defined as $\varphi_p = \sum_{(i,j)} (1/(d(\underline{x}^i, \underline{x}^j))^p)$, where $p \geq 1$. Since the value of φ_p is mainly determined by the minimum interpoint distance, the criterion for optimum sample uniformity is equivalent to the minimization of φ_p (φ criterion).

The optimum uniform sample set is not necessarily unique and can be very difficult to find. It is, therefore, more practical to find the best sample set, in terms of the φ criterion, from a large number of either randomly or structurally generated sample sets. Although it is simple to find a better sample set by searching through a large number of randomly generated sample sets, the associated computing effort can be prohibitive. For this reason, some structured quasi-random sampling techniques, such as Latin hypercubes and randomized orthogonal arrays, have been used extensively in multivariate uniform sampling. The structured quasi-random samples are carefully constructed to give better uniform coverage of the sampled space while maintaining a reasonably random appearance. In Figure 38.7a, ten samples are randomly generated in a four-dimensional space, and their one-dimensional projections (normalized histogram plots on diagonal) and the two-dimensional projections (off-diagonal plots) are presented. As shown in Figure 38.7b, the quasi-random Latin hypercube (LH) sampling maintains the sample uniformity

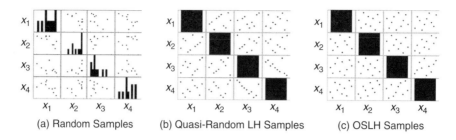

FIGURE 38.7 Comparison of uniform sampling techniques.

in their one-dimensional projection. For this reason, significantly less computing effort is required to find a much-improved sample set.

More recently, certain orthogonal properties were found in the so-called symmetric Latin hypercube (SLH) sampling, in which LH samples are constructed to be symmetric to the center of the sampled space. An efficient columnwise–pairwise exchange algorithm was proposed by Ye et al. [67] for finding optimal symmetric Latin hypercube (OSLH) samples in terms of certain uniformity criteria, such as the φ criterion. As shown in Figure 38.7c, the OSLH sample set provides much better uniform coverage compared with both a random sample set (Figure 38.7a) and an arbitrary quasi-random LH sample set (Figure 38.7b).

Because the uniform designs avoid close sample pairs, they are also preferred in estimating the metamodel prediction error, using a "leave-one-out" cross-validation procedure. The entire uniform sample set is divided into several groups so that the samples in each group are evenly scattered in the sampled space. The union of any groups maintains the uniformity or space-filling property. An intergroup exchange algorithm is used to minimize the sum of within-group φ measures. As shown in Figure 38.8, a 30-sample OSLH set in two dimensions is partitioned into two groups. Each group of samples uniformly covers the sampled space, even though it does not, in general, maintain the SLH properties by itself. The 30-sample OSLH set is the union of the two partitioned groups.

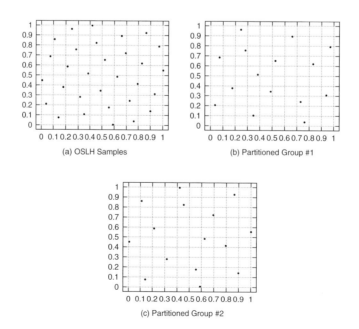

FIGURE 38.8 Illustration of uniform sample partitioning.

38.3.3 Cross Validation for Estimating Metamodel Prediction Error

For a given number of sample data, the metamodel prediction error can be reliably assessed using dedicated holdout samples. However, the holdout samples reduce the actual number of available samples for constructing the metamodel. For this reason, certain in-sample criteria are always used [52, 54].

To avoid overfitting in nonparametric regression techniques, cross validation and generalized cross validation [56, 68, 69] have been widely used for estimating metamodel prediction error. A K-fold cross-validation procedure is described in three steps. First, the known sample set is partitioned into K roughly equal-sized groups. Second, a metamodel is constructed using the union of $K-1$ groups, and the prediction error is calculated for the Kth group. Finally, the previous step is repeated for $k = 2, \ldots, K$, and all prediction errors are combined to form the cross-validation error. To minimize the cross-validation overhead, a "leave-one-out" cross-validation procedure can be used, in which every known sample is left out successively, and then its value is predicted back.

Although the cross-validation procedure generally provides a good indicator of the metamodel accuracy, it is often necessary to also check the metamodel's convergence with increasing sample size. We can generate a rather large uniform sample set and then uniformly partition it into many evenly scattered (space-filling) groups. All samples in each group are progressively evaluated using CAE simulations. Every time a new group of samples is added, the current metamodel's prediction error on the new samples is calculated. Convergence is achieved when the improvement on the metamodel prediction error with the newly added group is small. This progressive sampling strategy can also be used for a converged variable-screening procedure.

38.3.4 Local Polynomial Fitting Using CVMLS Method

The moving least squares (MLS) method originated in curve and surface fitting [70]. For a multivariate function $f(\underline{x})$, the notation $\underline{x}^i = [x_1^i \ x_2^i \ \ldots \ x_n^i]^T$ represents the ith point of N known sample points scattered in the n-dimensional space, where the sampling space is always normalized to the cube $[0, 1]^n$. The corresponding function value is denoted as $f^i = f(\underline{x}^i)$ for $i = 1, \ldots, N$. The function $f(\underline{x})$ can generally be approximated by $g(\underline{x})$, using a linear combination of polynomial basis functions [57], as

$$f(\underline{x}) \approx g(\underline{x}) = \sum_{j=1}^{m} a_j \cdot b_j(\underline{x}) = \underline{a}^T \cdot \underline{b}(\underline{x}) \tag{38.17}$$

where the terms in $\underline{b}(\underline{x})$ are the set of $m = n + 1$ linear polynomials

$$\underline{b}(\underline{x}) = [b_1(\underline{x}) \ b_2(\underline{x}) \ \ldots \ b_m(\underline{x})]^T = [1 \ x_1 \ \ldots \ x_n]^T. \tag{38.18}$$

It is assumed in MLS that the predicted function value at a point $\underline{x}$ should be most strongly influenced by the values of f^i at those points $\underline{x}^i$ that are closest to $\underline{x}$. This suggests that different weights should be assigned to each of the known sample points $\underline{x}^i$ according to their distance to the prediction point $\underline{x}$. We can then choose the coefficients $\underline{a}$ to minimize the weighted sum of residual-error squares

$$E_{\underline{x}}(g) = \sum_{i=1}^{N} w^i(\underline{x}) \cdot (g(\underline{x}^i) - f^i)^2 \tag{38.19}$$

where the positive weight function $w^i(\underline{x})$ has the property that its value decreases monotonically as the distance from $\underline{x}$ to $\underline{x}^i$ increases. A commonly used form for the weight function is

$$w(d(\underline{x}, \underline{x}^i)) = \exp(-\alpha(d(\underline{x}, \underline{x}^i))^2) = \exp\left(-\alpha \sum_{k=1}^{n} \left(x_k - x_k^i\right)^2\right) \tag{38.20}$$

where $d(\underline{x}, \underline{x}^i)$ is a measure of distance between $\underline{x}$ and $\underline{x}^i$. The ordinary least-squares formulation is a special case in which the weight function $w^i(\underline{x})$ is a constant. The nonnegative tuning parameter α controls the degree of localization in MLS local polynomial regression by scaling the slope of the weight function.

Since the weight function is defined around the prediction point $\underline{x}$ and its magnitude changes or "moves" with $\underline{x}$, the function that minimizes the error functional in Equation 38.19 is called the moving least squares (MLS) approximation $g_{mls}(\underline{x})$ of the original function $f(\underline{x})$. Because the weights $w^i(\underline{x})$ are functions of $\underline{x}$, the polynomial basis function coefficients $\underline{a}$ are also dependent on $\underline{x}$, i.e.,

$$f(\underline{x}) \approx g_{mls}(\underline{x}) = \sum_{j=1}^{m} a_j(\underline{x}) \cdot b_j(\underline{x}) = \underline{a}(\underline{x})^T \cdot \underline{b}(\underline{x}) \tag{38.21}$$

where the coefficients corresponding to any prediction point $\underline{a}$ can be obtained by solving m normal equations $\partial E_{\underline{x}}(g)/\partial a_j = 0$, for $j = 1, \ldots, m$. Details are given in the literature [57, 63–64].

The localized polynomial regression using the MLS procedure can be tuned by adjusting α in Equation 38.20. However, because the performance function can have a very different relationship with each input variable, the Euclidean distance in Equation 38.20 can be replaced by the following general parameterized distance formula:

$$d(\underline{x}, \underline{x}^i) = \sqrt{\sum_{k=1}^{n} \theta_k (x_k - x_k^i)^2} \tag{38.22}$$

where $\theta_1, \ldots, \theta_n$ are n positive weight function parameters. Thus, the weight function of Equation 38.20 can be expressed as

$$w(d(\underline{x}, \underline{x}^i))\big|_{\theta_1, \ldots, \theta_n} = \exp\left(-\sum_{k=1}^{n} \theta_k (x_k - x_k^i)^2\right). \tag{38.23}$$

The n nonnegative weight function parameters $\theta_1, \ldots, \theta_n$ can be tuned automatically in the metamodeling process by minimizing the metamodel cross-validation prediction error $E^{CV}(\theta_1, \ldots, \theta_n)$, where either the cross-validation root mean square error (CV-RMSE) or the average absolute error (CV-AAE) can be used. If we let $\hat{g}_{mls}^{-i}(\underline{x})$ denote the metamodel based on all data except the ith *left-out* sample, then

$$E_{RMSE}^{CV} = \sqrt{\frac{1}{N} \sum_{i=1}^{N} (\hat{g}^{-i}(\underline{x}^i) - f^i)^2} \tag{38.24a}$$

and

$$E_{AAE}^{CV} = \frac{1}{N} \sum_{i=1}^{N} |\hat{g}^{-i}(\underline{x}^i) - f^i| \tag{38.24b}$$

The resulting MLS metamodel will approximate the true performance function with minimum cross-validation prediction error based on available samples. This cross-validation-driven metamodeling technique is thus named the cross-validated moving least squares (CVMLS) method.

The leave-one-out cross-validation procedure effectively surveys the sampled space if the uniformity of the construction samples can be assured. However, the leave-one-out cross-validation procedure can be misleading if many samples appear in close pairs. When two samples are very close in input variable space, their performance output will also be very similar. When one sample from the pair is left out, the remaining one will dominate the predicted-back performance value, thus leading to an unrealistically low cross-validation error. A generalized cross-validation procedure [56, 68–69] has also been widely used for estimating nonparametric metamodel prediction error. It should be noted that although the generalized cross-validation procedure is closely related to the cross-validation procedure, it is not a special case of cross validation. Instead, the generalized cross-validation procedure can be viewed as a weighted version of cross validation [56]. To ensure the robustness of the cross-validation error estimate, and also to avoid redundant samples, a uniformly spaced sampling structure, such as the optimal symmetric Latin hypercube (OSLH) of Section 38.3.2, is recommended.

38.3.5 Variable Screening Based on CVMLS

The goal of the metamodel is to mimic the output of a complex, slow-running model in as much detail as possible while running much faster. In practical applications, on the order of tens to a few hundred input variables often need to be considered as likely contributors to the system performance. On the other hand, the number of metamodel samples obtained by expensive physics-based computer simulations is restricted by available computing resources and project time. While minimum sample requirements for constructing a numerically feasible metamodel with n inputs is $m = n + 1$ for ordinary least-squares regression using linear polynomials, N generally needs to be many times of m in order to capture potential nonlinear performance behaviors through localized regression using CVMLS.

For a metamodel to better characterize the relationship of important input variables, it is thus necessary to eliminate insignificant inputs that are small contributors to the targeted performance measures. A metamodel that uses a subset of all inputs essentially lumps the effect of eliminated inputs into the metamodel approximation error. The purpose of eliminating insignificant inputs is to reduce the metamodel prediction error by focusing on major contributors. Here a process is described where the screening of input variables is not a separate preprocessing step but an integral part of the metamodeling process. When N is less than or close to n in size, a main-effects estimate procedure using additive model with univariate CVMLS analysis is performed to eliminate insignificant inputs. Subsequently, if N is somewhat greater than the n and if the n-dimensional metamodel can be roughly constructed, a CVMLS backward-screening procedure is used for further input variable screening.

38.3.5.1 Main-Effects Estimate Using a CVMLS Additive Model

In practical applications, a few hundred input variables can often be considered as likely contributors to the system performance. Meanwhile, the number of metamodeling samples obtained by expensive CAE simulations is restricted by available computing resources and project cadence. All multiple-regression techniques face the so-called curse of dimensionality, according to which the required sample size must increase exponentially with the number of input variables. However, some dimensionality-reduction processes, such as additive modeling and the more general projection-pursuit modeling, can be used to circumvent the problem [56–57].

The additive model assumes an additive approximation of the form [56–57]

$$f(\underline{x}) \approx g_{add}(\underline{x}) = f^{mean} + \sum_{j=1}^{m} f_j(x_j) \qquad (38.25)$$

where $f_j(x_j)$ are unknown univariate functions of a single input variable x_j with zero mean. Therefore, in Equation 38.25, $f^{mean} = E[f^i]$, where the superscript i ($i = 1, \ldots, N$) indicates the ith sample point. Thus, when the additive model is accurate, the following relationship holds:

$$g_{add}^{mee}(x_j) = \iint \cdots \int_{\underline{x}_{-j}} g_{add}(\underline{x}) d\underline{x}_{-j} - f^{mean} = f_j(x_j) \qquad (38.26)$$

where "mee" stands for main-effects estimate. This suggests an iterative backfitting algorithm for computing all univariate functions, according to which a univariate CVMLS approximation can be used to fit $f_j(x_j)$ from the partial residuals in Equation 38.26, given $f_k(x_k), k \neq j$. The best additive model is obtained by cycling the backfitting procedure until it converges.

Since only univariate regression is used, the additive model overcomes the *curse of dimensionality*. Moreover, the univariate components provide easily interpretable main-effects estimates. Unlike the least-squares linear regression, the main-effects estimates using the CVMLS additive model can capture highly nonlinear input/output relationships. Cross-validation error can be used to measure the accuracy, thus avoiding potential overfitting associated with high-order least-squares polynomial regression.

The CVMLS additive model is recommended for input screening only in cases where N is less than or close to n in size or when an n-dimensional metamodel cannot be reasonably constructed without excessive cross-validation error. For other cases, a more robust backward-screening procedure should be used. This procedure is introduced in the next subsection.

38.3.5.2 Elimination of Insignificant Input Variables Using CVMLS Backward Screening

The main-effects estimate, using the CVMLS additive model of the previous subsection, cannot identify cross effects among input pairs. Therefore, the significance of some closely coupled inputs can be underestimated. In cases when an n-dimension metamodel can be roughly constructed even with large cross-validation error, the goal of eliminating insignificant inputs can be accomplished by reducing the metamodel approximation error measured through cross validation. This can be accomplished by a backward-screening procedure which is described in the following three steps:

1. Construct the n-dimension CVMLS metamodel with all inputs and obtain its cross-validation error E^{CV}.
2. Construct leave-one-input-out metamodels for each input x_j ($j = 1, \ldots, n$) and calculate its cross-validation error E^{CV}_{-j}.
3. Compare E^{CV}_{-j} ($j = 1, \ldots, n$) with E^{CV}, and then remove any input x_k from CVMLS basis functions in Equation 38.18 if its impact index is negative, i.e., $E^{CV}_{-k} < E^{CV}$.

The backward-screening process is more robust than "mee" because coupling effects, including cross effects among input pairs and other higher-order cross effects, are considered. Note that a leave-one-input-out metamodel ignores any potential effect from x_j. If x_j is an important contributor in $g_{cvmls}(\underline{x})$, the resulting approximation error in $g_{cvmls}(\underline{x}_{-j})$ will increase significantly. Conversely, the resulting metamodel approximation error will either not change or slightly decrease if x_j is an insignificant factor. The small decrease here reflects the benefit of the improved sample-to-input ratio.

38.3.6 Crash Safety Example for Illustrating CVMLS and Variable Screening

The performance of a driver restraint system in an NCAP (new car assessment program) crash test is considered. The test vehicle with a belted driver travels at 35 mph straight into a fixed rigid barrier that is perpendicular to the vehicle line of travel. The star rating of occupant-safety performance in NCAP is derived from a combined injury probability criterion P_{comb} based on two key injury numbers, HIC (head injury criteria) and Chest G (chest resultant acceleration in gs). The five-star rating is granted if P_{comb}, truncated to a full percentage point, is not more than 10%. Similarly, the four-star rating is granted if the P_{comb} number is larger than 10% but not more than 20%.

In this example, the vehicle restraint system and occupant motion are simulated in Madymo (a multibody dynamics and nonlinear finite element analysis code) to assess the crash dummy's injury numbers, as illustrated in Figure 38.9. There are 20 Madymo parameters that can potentially impact P_{comb} and therefore the star rating. Among them, six parameters can be adjusted within rather wide design ranges and are denoted as design variables d_1 to d_6. The other 14 variables are considered random noises and are denoted as random variables r_1 to r_{14}. The random noises are associated with the restraint system,

FIGURE 38.9 Model of occupant-restraint system.

test dummy setup, and vehicle deceleration pulse at any given design. The variation ranges for the 20 input variables are listed in Table 38.2. The outcomes of P_{comb} can be highly nonlinear in terms of input parameters.

Note that the Madymo simulation model is fundamentally deterministic. For a given set of inputs, the Madymo simulation produces a unique P_{comb} number, i.e.,

$$P_{comb} = P_{comb}(d_1, \ldots, d_6; r_1, \ldots, r_{14}) = P_{comb}(\underline{d}; \underline{r}) \tag{38.27}$$

We run 100 Madymo simulations in the 20-dimensional space using an optimal symmetric Latin hypercubes (OSLH) sampling design. The 100 Madymo samples will be used to demonstrate the main-effects estimate ("mee") using the CVMLS additive model and the backward-screening procedure. For convenience of illustration, all variable ranges will be normalized into a 20-dimensional space cube $[0, 1]^{20}$.

38.3.6.1 Case A: Main Effects Estimate Using CVMLS Additive Model

The main-effects estimate of each design variable is performed to provide an incomplete but intuitive view of the overall trend of P_{comb} with respect to each design variable in the variable space. Even in cases

TABLE 38.2 Madymo Model Input Variables

Input Variable	Lower Design Bound	Upper Design Bound
d_1	3	6
d_2	2	6
d_3	24	28
d_4	600	1000
d_5	127	635
d_6	0.03	0.07
Random Noise Variable	Lower Variation Bound	Upper Variation Bound
r_1	−0.22	−0.18
r_2	12	14
r_3	58	72
r_4	28	52
r_5	120	140
r_6	3033	3057
r_7	455	479
r_8	0.3385	0.4255
r_9	24.4	27.4
r_{10}	0.9	1.1
r_{11}	0.55	0.65
r_{12}	0.185	0.215
r_{13}	0.185	0.215
r_{14}	0.046	0.054

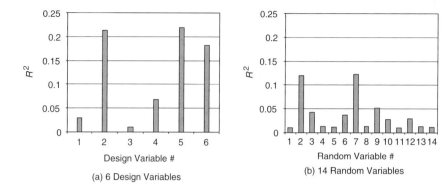

FIGURE 38.10 R^2 in CVMLS additive model of the NCAP example.

where the N-to-n ratio is close to or less than 1, "mee" provides a basis for eliminating relatively unimportant input variables. The R^2 results of six design variables and 14 random variables are compared in Figure 38.10a and Figure 38.10b, respectively. It shows clearly that design variables d_2, d_5, and d_6 are dominant controllable contributors to P_{comb}. Among the random variables, r_2, r_7, and r_9 have the most influence on P_{comb}.

When it is necessary to reduce inputs without regard to the controllability of the individual input, a threshold value, denoted as $\overline{R}^2$, can be specified. For instance, by setting $\overline{R}^2 = 0.05$, all inputs except a seven-dimensional subset $[d_2, d_4, d_5, d_6, r_2, r_7, r_9]$ will be eliminated.

38.3.6.2 Case B: Backward Screening (BS) for Elimination of Insignificant Input Variables

In this example, the N-to-n ratio is five prior to any input screening. This is generally adequate to construct a rough 20-dimensional metamodel. Because totally $n + 1$ metamodels need to be constructed in the backward screening, a fast variant of CVMLS procedure that uses only one weight function parameter in Equation 38.23 by setting $\alpha = \theta_1 = \ldots = \theta_n$ is recommended. In the so-called α-CVMLS procedure, the tuning of α therefore becomes a one-dimensional optimization problem. Using the fast α-CVMLS with linear polynomials, the baseline metamodel can be obtained with cross-validation average absolute error $E^{CV} = 0.0559$. The leave-one-input-out metamodels were constructed, and the impact index of the kth input variable was computed as

$$\text{Impact\_Index}(x_k) = \left(E^{CV}_{-k} - E^{CV}\right)/E^{CV} \tag{38.28}$$

As shown in Figure 38.11a and Figure 38.11b, the backward screening (BS) identifies that design variables d_2 and d_5 are the dominant factors related to the metamodel accuracy, while d_6 has much less

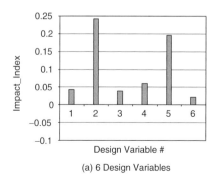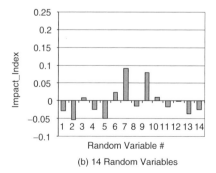

FIGURE 38.11 NCAP metamodeling using CVMLS.

TABLE 38.3 NCAP Metamodeling Using CVMLS

Metamodel	20-D (Full Set)		7-D (MEE Subset)		6-D (MEE/BS Subset)	
	CV-AAE	M-AAE	CV-AAE	M-AAE	CV-AAE	M-AAE
CVMLS Error	0.0509	0.0505	0.0445	0.0448	0.0435	0.0442

impact on the metamodel accuracy, even though it was suggested by "mee" as a significant variance contributor. This discrepancy implies that d_6 is an important but rather noisy factor. Among the random variables, only r_7 and r_9 are identified as moderate positive factors in metamodeling. Note that r_2 is closely associated with P_{comb}, but it has significant negative impact on metamodeling accuracy. Thus, r_2 is likely a big noise factor, and its analytical relationship with others cannot be adequately represented in the CVMLS metamodel.

Considering the above, r_2 can also be eliminated due to its negative impact on metamodeling accuracy. The resulting six-dimensional subset is $[d_2, d_4, d_5, d_6, r_7, r_9]$. In Table 38.3, the CVMLS metamodeling results using the full 20-dimensional set, the seven-dimensional mee subset, and the six-dimensional mee/BS subset inputs are obtained. For demonstration purpose, the true metamodeling errors in terms of M-AAE are obtained by comparing the metamodel results with new Madymo simulation results at 201 OSLH points in 20-dimensional input space that are different from the original 100 OSLH samples. Table 38.3 shows that metamodeling error decreases after elimination of insignificant or noisy input variables. Because of uniformly scattered and partitioned samples, CV-AAE is a close measure of M-AAE.

38.4 Variation Reduction in Robust Engineering: an Automotive Example

An automotive robust design example is presented in this section for the frontal-impact performance of an occupant-restraint system. It demonstrates how basic variation-reduction techniques can be used to achieve robust designs. An automotive occupant-restraint system consists of the air bag, seat belt, seat, steering wheel and column, knee bolster, as well as the driver or passenger. It is a very important part of the vehicle safety system, since it absorbs and dissipates energy in a crash situation to protect the occupant from severe injuries. The use of CAE models is common in the automotive industry for simulating a crash test. These models include dummies representing occupants of different sizes. The computer simulation helps in evaluating and improving the occupant-safety performance, which is measured by the "star rating," among other technical specifications [71]. Figure 38.9 shows a typical occupant-restraint system.

The occupant-restraint system is affected by a variety of parameters, as shown in Table 38.4. Some of them can be changed by design, such as the air-bag size and the air-bag inflator output. These variables are called design variables. Others may vary randomly in a test, such as the dummy's sitting position in the vehicle. These variables are called noise variables. Sometimes a design variable can also have small random variation around its nominal value. In that case, the nominal value is a design variable, and the

TABLE 38.4 Typical Variables in Frontal Occupant-Restraint System

Air-bag size and shape	Seat-belt pretension firing time
Air-bag tether length	Knee-bolster stiffness
Air-bag vent area	Vehicle crash pulses
Air-bag inflator output	Air-bag firing time
Steering-column stroke	Dummy position and orientation
Twist-shaft level	Frictions between dummy and seat belt and air bag
Seat-belt pretension spool	Friction between seat belt and routing rings/buckles

TABLE 38.5 Design Variables and Their Ranges

Variable	Range for Nominal
Air-bag tether-length scaling factor	0.86–1.48
Air-bag vent-area scaling factor	1.366–3.534
Twist-shaft-level scaling factor	0.62–1.49
Knee-bolster stiffness scaling factor	1.0–1.5
Air-bag inflation output scaling factor	1.0–1.4
Pretension spool scaling factor	2.0–4.4
Pretension firing-time scaling factor	0.5–1.0
Column strokes scaling factor	0.62–1.88

random variation around the nominal is a noise variable. The variation in the noise variables causes variation in safety performance, which can affect the "star rating" of the same vehicle model by one to two "stars." It is, therefore, a challenge for automotive manufacturers to produce vehicles with consistent safety performance.

In this example, the design and noise variables and their ranges are listed in Table 38.5 and Table 38.6, respectively. The range for each design variable is determined from packaging constraints, availability, and other performance requirements. The range for the noise variables is determined from available data and expert opinions.

A robust design for the occupant-restraint system must meet all requirements from the following three load cases:

1. The frontal NCAP (new car assessment program) test for occupant-safety "star rating." It is a test addressing 50th-percentile dummies, with the seat belt on, in both driver and passenger seats. The vehicle is crashed into a rigid barrier at 35 mph. Performance metrics include the head-injury criterion (HIC), chest acceleration (Chest G), and "star rating," which is a function of HIC and Chest G. The best and worst "star ratings" are five and one, respectively.
2. U.S. federal government requirements from FMVSS (Federal Motor Vehicle Safety Standards) 208 [71] for frontal crash, with unbelted 50th-percentile dummies, crashed at 25 mph.
3. U.S. federal government requirements from FMVSS 208 [71] for frontal crash, with unbelted fifth-percentile dummies, crashed at 25 mph.

All three load cases have common design and noise variables. For the second and third load cases, there are a number of requirements on head injury, chest acceleration, and neck injury, among others. A complete list of the specifications is given in the Federal Motor Vehicle Safety Standards and Regulations [71].

In this example, an evaluation of the safety-performance variation is first performed, and a robust optimization procedure is subsequently used to improve the probability of achieving high safety performance. Since each occupant-restraint-system simulation is computationally intensive, it becomes impractical to directly apply any probabilistic method. For this reason, a high-fidelity response-surface model (RSM) is built. Due to the highly nonlinear behavior of the system and the restriction on computation time, advanced sampling methods are also required. In this example, a sequential optimum symmetric Latin hypercube (OSLH) sampling method is used (see Section 38.3.2). It provides "space-filling" properties, which are desirable if a restricted number of samples are available due to the high computational

TABLE 38.6 Noise Variables and Their Ranges

Variable	Range
Air-bag firing time (sec)	±0.003
Friction coefficient with dummy	±0.35
Friction coefficient seat belt to buckle/ring	±0.2
Dummy position and orientation	Per FMVSS test procedure
Column build load (N)	±444

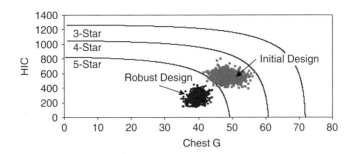

FIGURE 38.12 Comparison of the "star rating" ranges.

cost of the original simulation model. Nonlinear RSM techniques, such as stepwise regression [54] with nonlinear polynomials, kriging [72], and cross-validated moving least squares (see Section 38.3.4) can all generate response-surface models of decent fidelity. The required sample size is determined based on the needed accuracy, system nonlinear behavior, number of active variables in the range considered, sampling scheme, and the available computing resources. In this example, 150 simulations were first performed for each load-case using a Madymo model, and third-order polynomial stepwise regression RSM models were subsequently created for the key response metrics. All RSM models had a less than 5% average error.

The created RSM models are computationally efficient, since they are in polynomial form. Thus, Monte Carlo simulations were easily performed for evaluating the initial design performance. The used variation ranges for the design and noise variables, listed in Table 38.5 and Table 38.6, respectively, are based on test procedures, supplier tolerance specifications, test data, and engineering experience. All random variables were assumed to be uniformly distributed within their variation ranges. Monte Carlo simulations, with a 50,000 sample size, were used to assess the scatter in the safety performance. As shown in Figure 38.12, the "star rating" was mostly four star for the initial design.

The goal of this study was to produce a five-star design considering the variation in the design and noise variables. Therefore, the mostly four-star initial design "cloud," shown in Figure 38.12, had to move within the five-star domain. One way to achieve such a robust design is to move the mean and simultaneously reduce the variation (scatter) around the mean, so that the final design "cloud" is within the five-star domain. Therefore, both the mean design and its variation must change simultaneously. This can be achieved by maximizing the (mean -3σ) aggregate objective. If the (mean -3σ) of the "star rating" is five, the final design will be mostly within the five-star rating domain, considering the variation in the design and noise variables. The (mean -3σ) aggregate objective was mainly chosen for its simplicity. However, it was also chosen because the predictions of current CAE simulations of occupant-safety systems are not very accurate due to simplifying assumptions in the mathematical representation of the physical crash phenomena. Thus, a more rigorous reliability estimation using existing probabilistic techniques would not be meaningful [73].

The robust design is achieved by solving the following probability-based optimization problem

$$\max_{\substack{Design \\ Variables}} (\mu_{star} - 3\sigma_{star}) \tag{38.29}$$

subject to other constrains and requirements, where μ and σ denote the mean and standard deviation, respectively. Note that the "other constraints" can also be probability based. Due to the highly nonlinear behavior of the above objective, a global optimization search was utilized. A genetic algorithm was first used to identify the vicinity of a design point. A generalized reduced gradient algorithm was subsequently used to efficiently find the exact design point. For each function evaluation during the optimization process, 5000 Monte Carlo simulations were performed in order to evaluate the mean and standard deviation of the "star rating" and the constraints. The 5000 sample size for the Monte Carlo simulation

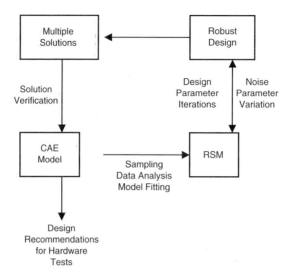

FIGURE 38.13 Process for robust design of occupant-safety system.

provided adequate accuracy in estimating both the mean and the standard deviation. Then the design parameters are changed based on the search algorithm. This process usually requires hundreds to thousands of iterations. In the end, a new design is chosen as a robust design. Figure 38.12 shows the "cloud" of the final robust "star rating" design. The performance improvement between the initial and final designs is due to an increase in the mean and a decrease in the standard deviation of the "star rating." The described optimization scheme produces multiple potential robust designs that are further evaluated using the original CAE simulation model.

The overall process to obtain a robust "star rating" design is summarized in Figure 38.13. It consists of the following steps.

1. Identify the design and noise variables and their ranges. Create sample points using space-filling sampling techniques, similar to those of Section 38.3.2. Use the CAE model of the occupant-restraint system to evaluate the response functions at all sample points.
2. Develop RSM models. Holdout samples are used to evaluate the accuracy of the developed RSM models.
3. Solve the robust optimization problem of Equation 38.29, using the RSM models.
4. Search for multiple alternative solutions.
5. Evaluate each alternative solution using the original CAE model.
6. Propose the best solution as the design recommendation.

In order to carry out the described robust design within the vehicle development process, it is essential to have numerically stable CAE models, high-fidelity RSM models, and an automated numerical process. The process automation is essential for fast execution and error avoidance for inexperienced engineers with minimal knowledge in sampling, RSM model generation, and optimization techniques.

38.5 Summary and Future Needs

This chapter presented an overview of the current state of the art in applying probabilistic methods in automotive engineering, and it described the commonly used reliability methods for addressing durability, performance, and quality issues. Some practical examples were used to demonstrate how the basic probabilistic design ideas are being used in the automotive industry to account for stochastic uncertainty. The automotive industry uses probabilistic methods mainly to improve vehicle quality and safety as well as to reduce inherent manufacturing variation in the context of robust engineering. The robust design of an occupant-restraint system's frontal-impact performance was studied using basic variation-reduction techniques.

Most of the applications in the automotive industry require large-scale, computationally intensive CAE models. It is therefore necessary to develop accurate and efficient metamodels of the actual CAE models in order to reduce the high computational cost of probabilistic analysis. For this reason, the latest advances in nonparametric metamodeling methods were presented along with efficient structured uniform sampling techniques, such as optimum symmetric Latin hypercube (OSLH). Furthermore, efficient variable screening procedures were described for effective elimination of insignificant input variables in practical problems with a large number of inputs. These techniques were demonstrated using an automotive crash-safety example.

We believe that future research is needed in the areas of:

- Component and system reliability of complex system with highly nonlinear limit states
- Computationally efficient reliability-based design optimization
- Multidisciplinary optimization under uncertainty
- Efficient sampling, metamodeling, and variable screening techniques
- Random fields
- Imprecision issues and uncertainty quantification and propagation with scarce data

For a variety of practical problems in the automotive industry, the failure domain is defined by highly nonlinear limit states. In such cases, the first-order reliability methods can overestimate or underestimate the true probability of failure. Thus, efficient simulation-based methods are needed instead of the commonly used approximate analytical methods. Some recent work in this area is reported by Zou [74, 75].

Reliability-based design optimization (RBDO) gradually becomes an important optimization tool in the presence of uncertainty [45, 46, 76] for automotive applications. However, the conventional double-loop approaches [43, 44, 46] are almost impractical due to the required excessive computational effort. Efficient RBDO algorithms are therefore needed. Advances in this area are reported in the literature [45, 77]. Furthermore, the design and development of complex automotive systems typically requires the integration of multiple disciplines and the resolution of multiple conflicting objectives in the presence of uncertainty. Although recent advances have been made [78–82], more research is needed in this area.

Almost exclusively, all probabilistic analysis and design methods in automotive industry are "random variable" based. However, a "random process" approach must be used to characterize uncertainty for a variety of automotive applications. Examples include, among others, the random vehicle road excitation and the combustion excitation in an internal combustion engine. Random field models can be typically used to characterize random variations in space and time. Advanced engineering applications of random field analysis have been reported in industries such as aerospace [83]. However, the authors are not aware of similar applications in the automotive industry.

Classical probabilistic analysis is used in automotive applications when sufficient data are available to quantify uncertainty using probability distributions. However, when sufficient data are not available or when there is a lack of information due to ignorance, the classical probability methodology may not be appropriate. During the early stages of vehicle development, quantification of the product's reliability or compliance with performance targets is practically very difficult due to insufficient data for modeling the uncertainties. Formal theories, such as evidence theory (or Dempster–Shafer theory) [84, 85] and possibility theory [86], can be used to handle uncertainty quantification and propagation with scarce data. Although such theories have been recently applied to engineering applications [87, 88], they are not well known in the automotive industry. Fundamental and applied research is needed in this area with emphasis on large-scale automotive applications.

Resolution of trade-offs in multicriteria optimization problems is very common in automotive applications. The weighted-sum approach [89, 90] is commonly used for handling trade-offs, despite its serious drawbacks [91]. A multiobjective decision problem generally has a whole set of possible "best" solutions, known as the Pareto set. A complete decision model, therefore, requires specification of the degree of compensation between conflicting criteria in order to select the best of all Pareto points [92]. A preference-aggregation method, based on a family of aggregation functions, is presented by Scott and Antonsson [92] to formally model all possible trade-offs in engineering design. An initial attempt has been recently made [93, 94] to extend the preference-aggregation methodology in the presence of uncertainty so that it can be used to handle robust design problems. More research is needed in this area.

Acknowledgments

The authors would like to thank Raviraj Nayak, Chun-Liang Lin, and Donald Jones of General Motors Corporation for providing information on some of the examples presented in this chapter. The continuous support and enthusiasm of Robert Lust and Mary Fortier of General Motors Corporation throughout the preparation of this chapter is also acknowledged.

References

1. Rosenblatt, M., Remarks on a multivariate transformation, *Ann. Mathematical Statistics*, 23 (2), 470-472, 1952.
2. Rackwitz, R. and Fiessler, B., Structural reliability under combined random load sequences, *Comput. Struct.*, 9 (5), 484–494, 1978.
3. Liu, P. and Der Kiureghian, A., Optimization algorithms for structural reliability, *Structural Safety*, 9, 161–177, 1991.
4. Cruse, T.A., Wu, Y.-T., Dias, J.B., and Rajagopal, K.R., Probabilistic structural analysis methods and applications, *Comput Struct*, 30, 163–170, 1988.
5. Wu, H., Millwater, R., and Cruse, T.A., Advanced probabilistic structural analysis method for implicit performance functions, *AIAA J.*, 28, 1663–1669, 1990.
6. Breitung, K. and Faravelli, L., Response surface methods and asymptotic approximations, in *Mathematical Models for Structural Reliability*, Casciati, F. and Roberts, J.B., Eds., CRC Press, Boca Raton, FL, 1996, chap. 5, pp. 237–298.
7. Köylüolu, H.U. and Nielsen, S.R.K., New approximations for SORM integrals, *Structural Safety*, 5, 119–126, 1988.
8. Tvedt, L., Distribution of quadratic forms in normal space-application to structural reliability, *ASCE J. Eng. Mech.*, 116 (6), 1183–1197, 1990.
9. Cai, G.Q. and Elishakoff, I., Refined second-order reliability analysis, *Structural Safety*, 14, 267–276, 1994.
10. Zhao, Y. and Ono, T., A general procedure for first/second-order reliability method (FORM/SORM), *Structural Safety*, 21 (2), 95–112, 1999.
11. Haldar, A. and Mahadevan, S., *Probability, Reliability and Statistical Methods in Engineering Design*, John Wiley & Sons, New York, 2000.
12. Bucher, C.G., Adaptive sampling—an iterative fast Monte Carlo procedure, *Structural Safety*, 5, 119–126, 1988.
13. Melchers, R.E., Improved importance sampling methods for structural system reliability calculation, in *Proceedings of International Conference on Structural Safety and Reliability (ICOSSAR)*, 1989, pp. 1185–1192.
14. Wu, Y.T., NESSUS/FPI User's and Theoretical Manuals, ver. 2.4, Southwest Research Institute, San Antonio, TX, 1998.
15. Karamchandani, A., Bjerager, P., and Cornell, C.A., Adaptive importance sampling, in *Proceedings of 5th International Conference on Structural Safety and Reliability*, Ang, A.H.-S., Shinozuka, M., and Schuëller, G.I., Eds., ASCE, New York, 1989, pp. 855–862.
16. Zou, T., Mahadevan, S., Mourelatos, Z., and Meernik, P., Reliability analysis of automotive body–door subsystem, *Reliability Eng. System Safety*, 78, 315–324, 2002.

17. Mahadevan, S. and Dey, A., Adaptive Monte Carlo simulation for time-variant reliability analysis of brittle structures, *AIAA J.*, 35 (2), 321–326, 1997.
18. Mahadevan, S. and Raghothamachar, P., Adaptive simulation for system reliability analysis of large structures, *Comput. Struct.* 77, 725–734, 2000.
19. Schuëller, G.I. and Stix, R., A critical appraisal of methods to determine failure probabilities, *Structural Safety*, 4 (4), 193–209, 1987.
20. Hohenbichler, M. and Rackwitz, R., A bound and an approximation to the multivariate normal distribution function, *Math. Japonica*, 30 (5), 821–828, 1985.
21. Fu, G. and Moses, F., Multimodal simulation method for system reliability analysis, *J. Eng. Mech.*, 119 (6), 1173–1179, 1993.
22. Fu, G., Reliability models for assessing highway bridge safety, in *Proceedings of International Conference on Structural Safety and Reliability (ICOSSAR)*, 1998, pp. 1883–1888.
23. Schuëller, G.I., Pradlwarter, H.J., and Bucher, C.G., Efficient computational procedures for reliability estimate of MDOF-systems, *Int. J. Nonlinear Mech.*, 26 (6), 961–974, 1991.
24. Ditlevsen, O., Narrow reliability bounds for structural systems, *J. Structural Mech.*, 7 (4), 453–472, 1979.
25. Wu, Y.-T., An adaptive importance sampling method for structural system reliability analysis and design, reliability technology 1992, in *Proceedings of ASME Winter Annual Meeting, AD-28*, Cruse, T.A., Ed., 1992, pp. 217–231.
26. Xiao, Q. and Mahadevan, S., Fast failure mode identification for ductile structural system reliability, *Structural Safety*, 13, 207–226, 1994.
27. Xiao, Q. and Mahadevan, S., Second-order upper bounds on probability of intersection of failure events, *J. Eng. Mech.*, 120 (3), 670–674, 1994.
28. Mahadevan, S. and Liu, X., Probabilistic optimum design of composite laminates, *J. Composite Mater.*, 32 (1), 68–82, 1998.
29. Cornell, C.A., Bounds on the reliability of structural systems, *J. Structural Div.*, 93 (ST1), 171–200, 1967.
30. Hunter, D., An upper bound for the probability of a union, *J. Appl. Probab.*, 3 (3), 597–603, 1976.
31. Gollwitzer, S. and Rackwitz, R., An efficient numerical solution to the multinormal integral, *Probabilistic Eng. Mech.*, 3 (2), 98–101, 1988.
32. Ambartzumian, R., Der Kiureghian, A., Ohanian, V., and Sukiasian, H., Multinormal probability by sequential conditioned importance sampling: theory and application, *Probabilistic Eng. Mech.*, 13 (4), 299–308, 1998.
33. Pandey, M.D. and Sarkar, A., Comparison of a simple approximation for multinormal integration with an importance sampling-based simulation method, *Probabilistic Eng. Mech.*, 17, 215–218, 2002.
34. Lin, C.-L., Virtual experimental design on glass guidance system with an integrated dual-cable regulator—Part I: CAE model validation, in *Proceedings of Seventh ISSAT International Conference on Reliability and Quality in Design*, ISSAT, 2001, pp. 237–241.
35. Lin, C.-L., Im, K.H., and Bhavsar, T.R., Virtual experimental design on glass guidance system with an integrated dual-cable regulator—Part II: design optimization, in *Proceedings of Eighth ISSAT International Conference on Reliability and Quality in Design*, ISSAT, 2002, pp. 287–291.
36. Lin, C.-L. and Im, K. H., System Design Process Integration of Cable-Drive Glass Guidance System Using Axiomatic Design, presented at Second International Conference on Axiomatic Design, Cambridge, MA, 2002.
37. Lin, C.-L., Virtual experimental design optimization on cable-drive glass guidance system, *Int. J. Reliability, Qual., Safety Eng.*, 9 (4), 317–328, 2002.
38. Chang, H.Y. and Song, J.O., Glass Drop Design for Automobile Windows—Design of Glass Contour, Shape, Drop Motion and Motion Guidance Systems, 951110, SAE, Warrendale, PA, 1995.
39. Singh, K., Zaas, C., and Newton, R., Engineering Moveable Glass Window Seals of Automotive Door Using Upfront CAE, 982382, SAE, Warrendale, PA, 1998.
40. Singh, K., Experimental Assessment of Door Window Glass Smooth Operation And Tracking, 1999-01-3161, SAE, Warrendale, PA, 1999.

41. Kanamori, M., Isomura, Y., and Suzuki, K., Dynamic Finite Element Analysis of Window Regulator Linkage System Using LS-DYNA, 980308, SAE, Warrendale, PA, 1998.
42. Lorenzen, T.J. and Anderson, V.L., *Design of Experiments*, Marcel Dekker, New York, 1993.
43. Reddy, M.V., Granhdi, R.V., and Hopkins, D.A., Reliability based structural optimization: a simplified safety index approach, *Comput. Struct.*, 53 (6), 1407–1418, 1994.
44. Wu, Y.-T. and Wang, W., A new method for efficient reliability-based design optimization, in *Proceedings of 7th Special Conference on Probabilistic Mechanics & Structural Reliability*, 1996, pp. 274–277.
45. Du, X. and Chen, W., Sequential optimization and reliability assessment method for efficient probabilistic design, in *Proceedings of ASME Design Automation Conference*, DETC-DAC34127, Montreal, Canada, 2002.
46. Choi, K.K. and Youn, B.D., Hybrid analysis method for reliability-based design optimization, in *Proceedings of ASME Design Automation Conference*, ASME, Pittsburgh, PA, 2001.
47. Tu, J., Choi, K.K., and Park, Y.H., A new study on reliability-based design optimization, *ASME J. Mechanical Design*, 121, 557–564, 1999.
48. Tu, J., Choi, K.K., and Park, Y.H., Design potential method for robust system parameter design, *AIAA J.*, 39 (4), 667–677, 2001.
49. Yang, R.J. et al., Metamodeling development for vehicle frontal impact simulation, in *Proceedings of 42nd Structures, Structural Dynamics, and Materials Conference*, AIAA/ASME/ASCE/AHS/ASC, 2001.
50. Gu, L., A comparison of polynomial based regression models in vehicle safety analysis, in *Proceedings of 42nd Structures, Structural Dynamics, and Materials Conference*, AIAA/ASME/ASCE/AHS/ASC, Seattle, WA, 2001.
51. Montgomery, D.C., *Design and Analysis of Experiments*, 5th ed., John Wiley and Sons, New York, 2001.
52. Myers, R.H., *Classical and Modern Regression with Application*, PWS-Kent, New York, 1990.
53. Myers, R.H. and Montgomery, D.C., *Response Surface Methodology: Process and Product Optimization Using Designed Experiments*, John Wiley and Sons, New York, 2002.
54. Draper, N.R. and Smith, H., *Applied Regression Analysis*, John Wiley and Sons, New York, 1981.
55. Stone, C.J., Consistent nonparametric regression, *Ann. Statistics*, 5, 595–645, 1977.
56. Eubank, R.L., *Spline Smoothing and Nonparametric Regression*, Marcel Dekker, New York, 1988.
57. Fan J. and Gijbels, I., *Local Polynomial Modeling and Its Applications*, Chapman & Hall, New York, 1996.
58. Sacks, J. et al., Design and analysis of computer experiments, *Statistical Sci.*, 4, 409–435, 1989.
59. Jones, D.R., A taxonomy of global optimization methods based on response surfaces, *J. Global Optimization*, 21, 345–383, 2001.
60. Dyn, N., Levin, D., and Rippa, S., Numerical procedures for surface fitting of scattered data by radial basis functions, *SIAM J. Scientific Statistical Computing*, 7, 639–659, 1986.
61. Jin, R., Chen, W., and Simpson, T.W., Comparative studies of metamodeling techniques under multiple modeling criteria, *Structural Multidisciplinary Optimization*, 23, 1–13, 2001.
62. Krishnamurthy, T. and Romero, V.J., Construction of response surface with higher order continuity and its application to reliability engineering, in *Proceedings of 43rd Structures, Structural Dynamics, and Materials Conference*, AIAA/ASME/ASCE/AHS/ASC, Denver, CO, 2002.
63. Tu, J., Cross-validated multivariate metamodeling methods for physics-based computer simulations, in *Proceedings of IMAC-XXI: A Conference and Exposition on Structural Dynamics*, 2003.
64. Tu, J. and Jones, D.R., Variable screening in metamodel design by cross-validated moving least squares method, in *Proceedings of the 44th Structure, Structural Dynamics, and Materials Conference*, AIAA-2003-1669, AIAA/ASME/ASCE/AHS/ASC, Norfolk, VA, 2003.
65. Tu, J. and Cheng, Y.-P., An Integrated Stochastic Design Framework Using Cross-Validated Multivariate Metamodeling Methods, 2003-01-0876, SAE, Warrendale, PA, 2003.
66. Bates, R.A., Buck, R.J., Riccomagno, E., and Wynn, H.P., Experimental design and observation for large systems, *J. Royal Statistical Soc. (Ser. B)*, 58, 77–94, 1996.
67. Ye, K.Q., Li, W., and Sudjianto, A., Algorithmic construction of optimal symmetric Latin hypercube designs, *J. Statistical Planning Inference*, 90, 145–159, 2000.

68. Stone, M., Cross-validatory choice and assessment of statistical predictions, *J. Royal Statistical Soc. (Ser. B)*, 36, 111–147, 1974.
69. Efron, B. and Tibshirani, R.J., *An Introduction to the Bootstrap*, Chapman & Hall, New York, 1993.
70. Lancaster, P. and Salkauskas, K., *Curve and Surface Fitting: an Introduction*, Academic Press, New York, 1986.
71. National Highway Traffic Safety Administration, Federal Motor Vehicle Safety Standards and Regulations, U.S. Department of Transportation, NHTSA, Washington, DC.
72. Jones, D.R., Schonlau, M., and Welch, W.J., Efficient global optimization of expensive black-box functions, *J. Global Optimization*, 13, 455–492, 1998.
73. Ding, X., Assessing Error in Reliability Estimates Obtained via CAE Simulations, 2003-01-0146, SAE, Warrendale, PA, 2003.
74. Zou, T., Mahadevan, S., Mourelatos, Z.P., and Meernik, P., Reliability analysis of automotive body–door subsystem, *Reliability Eng. System Safety*, 78 (3), 315–324, 2002.
75. Zou, T., Mourelatos, Z.P., Mahadevan, S., and Tu, J., Component and Monte Carlo approach, in *Proceedings of ASME Design Engineering Technical Conferences*, DETC2003/DAC-48708, ASME, 2003.
76. Yang, R.J., Gu, L., Tho, C.H., Choi, K.K., and Youn, B.D., Reliability-based multidisciplinary design optimization of a full vehicle system, in *Proceedings of 43rd Structures, Structural Dynamics, and Materials Conference*, AIAA/ASME/ASCE/AHS/ASC, Denver, CO, 2002.
77. Chen, X., Hasselman, T.K., and Neill, D.J., Reliability based structural design optimization for practical applications, in *Proceedings of 38th Structures, Structural Dynamics, and Materials Conference*, AIAA/ASME/ASCE/AHS/ASC, 1997.
78. Koch, P.N., Simpson, T.W., Allen, J.K., and Mistree, F., Statistical approximations for multidisciplinary design optimization: the problem of size, *J. Aircraft*, 36 (1), 275–286, 1999.
79. Oakley, D.R., Sues, R.H., and Rhodes, G.S., Performance optimization of multidisciplinary mechanical systems subject to uncertainties, *Probabilistic Eng. Mech.*, 13 (1), 15–26, 1998.
80. Padmanabhan, D. and Batill, S.M., Decomposition strategies for reliability based optimization in multidisciplinary system design, in *Proceedings of 9th AIAA/SSMO Symposium on Multidisciplinary Analysis and Optimization*, Atlanta, GA, 2002.
81. Du, X. and Chen, W., Collaborative reliability analysis for multidisciplinary systems design, in *Proceedings of 9th AIAA/SSMO Symposium on Multidisciplinary Analysis and Optimization*, Atlanta, GA, 2002.
82. Gu, X. and Renaud, J.E., Implementation study of implicit uncertainty propagation in decomposition-based optimization, in *Proceedings of 9th AIAA/SSMO Symposium on Multidisciplinary Analysis and Optimization*, Atlanta, GA, 2002.
83. Ghiocel, D.M., Stochastic field models for advanced engineering applications, in *Proceedings of 42nd Structures, Structural Dynamics, and Materials Conference*, AIAA/ASME/ASCE/AHS/ASC, Seattle, WA, 2001.
84. Klir, G.J. and Filger, T.A., *Fuzzy Sets, Uncertainty, and Information*, Prentice Hall, Englewood Cliffs, NJ, 1988.
85. Yager, R.R., Fedrizzi, M., and Kacprzyk, J., Eds., *Advances in the Dempster—Shafer Theory of Evidence*, John Wiley & Sons, New York, 1994.
86. Dubois, D. and Prade, H., *Possibility Theory*, Plenum Press, New York, 1988.
87. Wasfy, T.M. and Noor, A.K., Application of fuzzy sets to transient analysis of space structures, *AIAA J.*, 38, 1172–1182, 1998.
88. Akpan, U.O., Rushton, P.A., and Koko, T.S., Fuzzy probabilistic assessment of the impact of corrosion on fatigue of aircraft structures, in *Proceedings of 43rd Structures, Structural Dynamics, and Materials Conference*, AIAA/ASME/ASCE/AHS/ASC, Denver, CO, 2002.
89. Tang, X. and Krishnamurty, S., Performance estimation and robust design decisions, in *Proceedings of 41st Structures, Structural Dynamics, and Materials Conference*, AIAA/ASME/ASCE/AHS/ASC, Atlanta, GA, 2000.

90. Hacker, K. and Lewis, K., Robust design through the use of a hybrid genetic algorithm, in *Proceedings of 43rd Structures, Structural Dynamics, and Materials Conference*, AIAA/ASME/ASCE/AHS/ASC, Denver, CO, 2002.
91. Das, I. and Dennis, J., A closer look at drawbacks of minimizing weighted sums of objectives for Pareto set generation in multi-criteria optimization problems, *Structural Optimization*, 14 (1), 63–69, 1997.
92. Scott, M.J. and Antonsson, E.K., Aggregation functions for engineering design trade-offs, *Fuzzy Sets Systems*, 99 (3), 253–264, 1998.
93. Dai, Z., Scott, M.J., and Mourelatos, Z.P., Robust design using preference aggregation methods, in *Proceedings of ASME Design Engineering Technical Conferences*, DETC2003/DAC-48713, Chicago, IL, 2003.
94. Dai, Z., Scott, M.J., and Mourelatos, Z.P., Incorporating epistemic uncertainty in robust design, in *Proceedings of ASME Design Engineering Technical Conferences*, DETC2003/DAC-48715, Chicago, IL, 2003.

39
Reliability Assessment of Ships

39.1	Introduction	39-1
39.2	Historical Overview	39-2
39.3	Procedure for Reliability Assessment of Ships	39-3
39.4	Target Reliability for Ships against Hull Girder Collapse	39-4
39.5	Ultimate Limit-State Equations	39-5
39.6	Ultimate Hull Girder Strength Models	39-6
	Primary Failure Mode • Secondary Failure Mode • Tertiary Failure Mode • Quaternary Failure Mode • Effect of Corrosion • Effect of Fatigue Cracking • Effect of Local Denting • Effect of Combined Corrosion, Fatigue Cracking, and Local Denting	
39.7	Extreme Hull Girder Load Models	39-11
	Still-Water Bending Moment • Wave-Induced Bending Moment	
39.8	Prediction of Time-Dependent Structural Damage	39-12
	Time-Dependent Corrosion Wastage Model • Time-Dependent Fatigue Crack Model	
39.9	Application to Time-Dependent Reliability Assessment of Ships	39-20
	Scenarios for Operational Conditions and Sea States • Scenarios for Structural Damage • Reliability Assessment • Some Considerations Regarding Repair Strategies	
39.10	Concluding Remarks	39-36

Jeom Kee Paik
Pusan National University, Korea

Anil Kumar Thayamballi
Chevron Texaco Shipping Company LLC

39.1 Introduction

Failures in ship structures, including some total losses, continue to occur worldwide, in spite of continuous ongoing efforts to prevent them. Such failures can have enormous costs associated with them, including lost lives in some cases. One of the possible causes of such structural casualties is thought to be the inability of aging ships to withstand rough seas and weather, because the ship's structural safety decreases during later life, although it is quite adequate at the design stage and perhaps up to 15 years beyond. Corrosion and fatigue-related problems are considered to be the two most important factors potentially leading to such age-related structural degradation of ships and, of course, of many other types of steel structures. Local dent damage sometimes takes place in particular locations of certain merchant ships, e.g., inner bottom plates of bulk carriers.

In the design and operation of ship structures, there are a number of uncertainties that must be dealt with. Wherever there are uncertainties, a risk of failure exists. For a structure, the risk of failure for our purposes will be defined as the probability that the load-carrying capacity is smaller than the extreme or accidental load to which that the structure is subject. To minimize and prevent loss of life and financial exposure caused by ship structural casualties, it is of vital importance to keep the safety and reliability at an acceptable level.

It has been recognized that a reliability-based approach is potentially better than a deterministic approach for the design of ship structures as well as offshore platforms and land-based structures, since it can more rigorously deal with various types of uncertainties associated with the design variables [1]. The reliability-based approach is also equally useful in developing a damage-tolerant structure and for establishing strategies for repair and maintenance of age-degraded or otherwise damaged structural members.

The reliability of a structure is assessed based upon limit-state exceedence, which is defined as a condition in which the structure fails to perform an intended function. Four types of limit states can be considered, namely, serviceability limit state (SLS), ultimate limit state (ULS), fatigue limit state (FLS), and accidental limit state (ALS) [2]. SLS conventionally represents failure states for normal operations due to deterioration of routine functionality. ULS (also called ultimate strength) typically represents the collapse of the structure due to loss of structural stiffness and strength. FLS represents fatigue crack occurrence in structural details due to stress concentration and damage accumulation (crack growth) under the action of repeated loading. ALS represents excessive structural damage as a consequence of accidents, e.g., collisions, grounding, explosion, and fire, that affect the safety of the structure, environment, and personnel.

This chapter presents application examples for reliability assessment of merchant ships with the focus on hull girder ultimate limit state, taking into account the time-dependent effects of corrosion, fatigue cracking, and local denting. Some considerations for establishing a reliability-based repair and maintenance scheme are also made so as to keep ship hull girder strength reliability at an acceptable level, even in later life.

39.2 Historical Overview

The modern era of reliability-based structural design possibly started after World War II, when Freudenthal [3] suggested that statistical distributions of design factors should be taken into consideration when developing safety factors for engineering structures. Academic interest in structural reliability theory was further aroused in the 1960s, in part by the publication of another paper by Freudenthal [4].

The earliest works that used reliability-based methodologies in ship structural design were initiated by Mansour [5, 6] and Mansour and Faulkner [7]. Many such early efforts were primarily focused on developing reliability analysis methods [8–12], while the earliest applications to ship structures focused on the safety and reliability of ship hull girders subjected to wave-induced bending moments [10, 12–14].

The Ship Structure Committee (SSC; http://www.shipstructure.org) aims to promote safety, economy, education, and marine environment protection in the U.S. and Canadian maritime industries. Through its excellent research programs, which have been in place now for nearly 50 years, SSC has devoted considerable effort to the advancement of the safety and integrity of marine structures. Under sponsorship of the SSC, a series of research and development projects in areas related to ship structural safety and reliability, among others, has been carried out [15–23]. An historical review of the R&D activities of SSC in the area of reliability assessment of ship structures was made by Mansour et al. [23]. In addition, international effort has been made to develop ISO code 2394 [1], which deals with safety formats, partial load and resistance factors, load combinations, and properties of geometric and material parameters.

Ship structures are composed of individual structural components such as plating, stiffened panels, and support members. While some work continues in this area, there are today a number of useful methodologies for analyzing the safety and reliability of ship structural components. However, each structural component can fail in more than one way until the entire structure reaches the state where it fails to perform its intended function, and failure of a single structural component may or may not lead to overall system collapse. One cannot achieve target reliabilities for the global system on the basis of individual component reliability alone because of the different ways in which such individual components may connect and interact prior to and during the failure process.

In order to appreciate whether ship failures achieve levels consistent with acceptable targets, it is of crucial importance to assess the safety and reliability of ship structures at the level of the global system. Application examples presented in this chapter focus on the ultimate hull girder strength reliability of ships at the global level.

39.3   Procedure for Reliability Assessment of Ships

The procedure for the reliability assessment of ship structures is similar to that adopted for other types of steel structures as presented elsewhere in this handbook. For our purposes, the associated risk can be written as follows:

$$\text{Risk} = P_f = \text{Prob}(C \leq D) \quad (39.1a)$$

where P_f = probability of failure, C = load-carrying capacity, and D = demand. To be safe, the following criterion should then be satisfied:

$$P_f \leq P_{f0} \quad (39.1b)$$

where P_{f0} = target value of risk or probability of failure.

The safety and reliability of a structure is the converse of the risk, i.e., the probability that it will not fail, namely

$$\text{Reliability} = \text{Prob}(C > D) = 1 - P_f \quad (39.2)$$

The result of a standard reliability calculation, normally carried out after transformation of design variables into a standardized normal space, is a reliability index β that is related to the probability of failure P_f by

$$\beta = -\phi^{-1}(P_f) \quad (39.3a)$$

where ϕ is the standard normal distribution function. To be safe, the following criterion should then be satisfied:

$$\beta \geq \beta_0 \quad (39.3b)$$

where β_0 = target value of reliability index = $-\phi^{-1}(P_{f0})$. It should be emphasized that in order for Equations 39.3a and 39.3b to be valid, all nonnormal variates should be transformed to normal variates when calculating the reliability indices.

It should be noted here that, broadly speaking, a formal risk assessment involves consideration of not only the probability of failure, but also the consequences of failure, ideally in quantitative terms (see for example [24, 25]). In this chapter, however, reliability assessment is taken as the converse of risk analysis. This practice has been common in the field of structural reliability for some time. A major aim of structural risk assessment for merchant cargo vessels is normally to determine (a) the level of risk in terms of probability as related to total loss from structural causes and, increasingly, (b) the possibility of environmental pollution. The main tasks needed for reliability assessment in a "design by analysis" approach are generally as follows:

1. Specify the required target value of the reliability index (or a target level of risk).
2. Identify all likely failure modes of the structure or fatal events.
3. Formulate a limit-state function for each failure mode identified in item two.

4. Identify the probabilistic characteristics (mean, variance, distribution) of the random variables in each limit-state function.
5. Calculate the reliability for each limit state with respect to each failure mode of the structure.
6. Assess if the determined reliability is greater than or equal to the target reliability.

Repeat the above steps as required after changes to relevant design parameters.

39.4 Target Reliability for Ships against Hull Girder Collapse

The required level of structural safety and reliability may vary from one industry to another, depending on various factors such as the type of failure, the seriousness of its consequence, or perhaps even the cost of adverse publicity and other intangible losses. Appropriate values of target safety and reliability are not readily available and are usually determined by surveys or by examinations of the statistics of failures, although the fundamental difference between a risk assessment and a reliability analysis needs to be acknowledged when interpreting such results.

The methods to select the target safeties and reliabilities can be categorized into the following three groups:

1. Guesstimation: A reasonable value as recommended by a regulatory body or professionals on the basis of prior experience. This method can be employed for new types of structures for which a statistical database on past failures does not exist.
2. Analysis of existing of design rules: The level of risk one has traditionally lived with is estimated by calculating the reliability that is implicit in existing design rules that have been successful. This method is often used for the revision of existing design rules; particularly from a traditional experience-based format to a reliability-based format.
3. Economic value analysis: The target value of safety and reliability is selected to minimize the total expected costs during the service life of the structure. This is perhaps the most attractive approach, although it is often difficult to undertake in practice.

Figure 39.1 shows the reliability indices of some types of ships based on data that have been obtained by different investigators using different calculation methods, as a function of the year of publication [26].

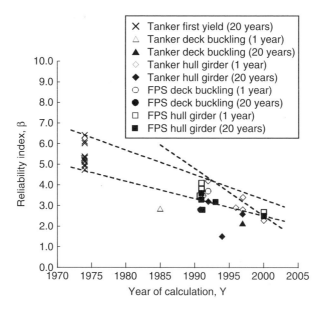

FIGURE 39.1 Variation of the calculated notional reliability indices over the passage of the years for ships from 1974 to 2000 (FPS = floating, production, and storage unit).

The categories of vessels, governing failure modes, and service life are identified. For ships with 20 years of service life, the effect of corrosion damage was accounted for in calculating the β values. It is seen from Figure 39.1 that the calculated reliability index has decreased for the assessments conducted in more recent years. The trend shown in Figure 39.1 does not necessarily mean that vessels considered in more recent years are becoming less reliable, at least not to the extent implied by the trends shown. Of course, it is to be expected that ship structures have become more efficient over time because of the increasing availability of technology. Also, some of the calculated results are perhaps more notional than others, and the considered failure modes may be more sophisticated as well. It is true that most calculations tend to use a design wave environment based on nominal values and thus ignore actual experience that may be more benign. It is, of course, challenging to pin down the true wave environment for trading vessels.

An example is the calculations made by Mansour in 1974 [13]. His early results demonstrate higher notional β values. This appears to be mainly a consequence of obtaining the probability of failure under a wave load criterion that is different from, and less than, the lifetime extreme load. The same investigator has, over time, vastly refined and further developed his early calculation methodology, and has even successfully developed unified calculation procedures that are applicable to wave load criteria [23] ranging from a mild storm to the most severe during a vessel's life. Also, many early pioneering calculations typically used 'first yield' as the failure criterion.

The first-yield criterion ignores loss of plate effectiveness due to any propensity for buckling, and so the location of the neutral axis of a ship's hull girder during the actual ultimate failure process will not be correctly simulated. This can result in somewhat lower levels of stress being determined for the compression region of the hull girder in addition to the basic panel strength being too high in some cases, implying a higher predicted hull girder bending strength and similarly higher reliability when compared with a reliability based on a more refined prediction of ultimate hull girder bending strength.

Many of the prior studies ignore age-related degradation effects, which will decrease the β values further in comparative terms. Illustrations shown later in this chapter include such effects. That improvement not withstanding, there is no escaping the fact that even today's calculations result in reliability indices that are not anything other than notional and comparative, mostly because of the uncertainties in the loads involved, and this situation is expected to continue into the foreseeable future. We can, however, improve the value of comparative and notional reliability measures further by appropriately taking advantage of continuing advances in load prediction and ultimate strength assessment procedures to higher levels of refinement, while also considering age-dependent strength degradation and other types of structural damage considered.

Regarding the more recent results of Figure 39.1, it is of some interest to note that whereas the β values determined in 1991 average around 3.5, those calculated in 2000 average 2.5. Based on the above varied results, and for purposes of use with evolving and recent (advanced) methodologies for ultimate hull girder strength calculations, it is considered that $\beta = 2.5$ may be a speculative but good target reliability index to aim for in respect to ultimate hull girder strength. For a more elaborate description of how to determine a target value of reliability index for ship structures, the reader can refer to the SSC report by Mansour et al. [23].

39.5 Ultimate Limit-State Equations

For hull girder ultimate strength reliability assessment, we will define four levels of failure mode that need to be considered, which we will term the primary, secondary, tertiary, and quaternary failure modes.

The primary failure mode will be taken to represent a condition where hull girder collapse takes place, involving buckling collapse of the compression flange (i.e., deck panel in sagging or bottom panel in hogging) and yielding of the tension flange (i.e., bottom panel in sagging or deck panel in hogging). The secondary failure mode will be a condition wherein the flange (stiffened panel) in compression reaches its ultimate strength. The tertiary failure mode is one in which support members in the compression flange fail, e.g., lateral-torsional buckling or tripping of stiffeners. The quaternary failure mode is a condition

TABLE 39.1 Samples of Mean and COV of Variables Related to Modeling Uncertainties and Load Combination Factors

Variable	Distribution	Mean	COV
x_u	normal	1.0	0.1
x_{sw}	normal	1.0	0.05
x_w	normal	1.0	0.15
k_{sw}	fixed	1.0	—
k_w	fixed	0.9	—

when plating between support members in the compression flange reaches its ultimate strength. The last three failure modes do not necessarily result in the total loss of a ship, except in very special cases. However, a ship's function will be presumed to be totally lost if the primary failure mode occurs.

The ultimate limit-state equation for each failure mode noted above can be written as follows:

Primary failure mode:

$$F_I = x_{1u} M_{1u} - (x_{sw} k_{sw} M_{sw} + x_w k_w M_w) \leq 0 \quad (39.4a)$$

Secondary failure mode:

$$F_{II} = x_{2u} M_{2u} - (x_{sw} k_{sw} M_{sw} + x_w k_w M_w) \leq 0 \quad (39.4b)$$

Tertiary failure mode:

$$F_{III} = x_{3u} M_{3u} - (x_{sw} k_{sw} M_{sw} + x_w k_w M_w) \leq 0 \quad (39.4c)$$

Quaternary failure mode:

$$F_{IV} = x_{4u} M_{4u} - (x_{sw} k_{sw} M_{sw} + x_w k_w M_w) \leq 0 \quad (39.4d)$$

where M_{iu} = hull girder capacity for the ith failure mode; M_{sw} = still-water bending moment; M_w = wave-induced bending moment; k_{sw}, k_w = load combination factors related to still-water bending moment and wave-induced bending moment, respectively, assuming that the corresponding variables they are applied to are the extreme values that do not necessarily occur simultaneously in time; and x_{iu}, x_{sw}, x_w = service-experience-dependent modeling parameters reflecting uncertainties associated with hull girder capacity, still-water bending moment, and wave-induced bending moment, respectively.

Table 39.1 proposes selected mean values and COVs (coefficients of variation) of variables related to modeling uncertainties and load combinations, broadly based on prior cited studies related to the reliability assessment of merchant cargo ships. The tabulated values will be used later in the illustrative examples of reliability analyses. It is to be noted that the mean values in particular have been mostly set to unity for convenience, although in particular studies they have been varied by investigators, and correctly so, depending on the particular circumstances involved. For purely illustrative purposes, all reliability calculations have been conducted with $k_{sw} = 1.0$ and $k_w = 0.9$ in this chapter.

39.6 Ultimate Hull Girder Strength Models

The largely simplified yet somewhat sophisticated formulations for predicting hull girder capacity are now presented for the four failure modes noted in Section 39.5. In predicting ultimate hull girder strength, two types of stiffened panel structural idealizations are relevant, namely the plate-stiffener combination model and the plate-stiffener separation model, as shown in Figure 39.2. The examples are selected from past work by the authors and also other investigators [2, 27, 28].

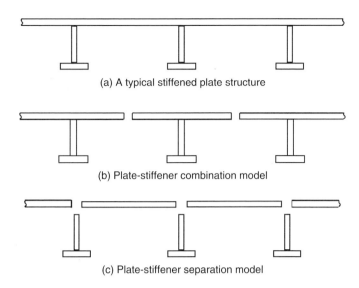

FIGURE 39.2 Three types of structural idealization for a stiffened panel. (From Paik, J.K. and Thayamballi, A.K., *Ultimate Limit State Design of Steel-Plated Structures*, John Wiley & Sons, Chichester, U.K., 2003. With permission.)

39.6.1 Primary Failure Mode

The ultimate bending moment capacity of a ship's hull (with positive sign for hogging and negative sign for sagging) will be calculated [2] by

$$M_{1u} = \sum_C \sigma_i A_{ei}(z_i - g_u) + \sum_T \sigma_j A_j(z_j - g_u) \tag{39.5}$$

where

$g_u = [\sum_C \sigma_i A_{ei} z_i + \sum_T \sigma_j A_j z_j] / [\sum_C \sigma_i A_{ei} + \sum_T \sigma_j A_j]$

σ_i = longitudinal bending stress in the ith structural member, negative for compression and positive for tension, which is given by

$\sigma_i = (z_i - g)\sigma_{Yeqd}/(D - g)$ for hogging

$\sigma_i = (g - z_i)\sigma_{Yeqb}/g$ for sagging

z_i = coordinate of the ith element measured from the baseline ($z_i = 0$ at the baseline)

g = neutral axis, which is given as

$g = [\sum_C A_{ei} z_i + \sum_T A_j z_j]/[\sum_C A_{ei} + \sum_T A_j]$

$\sum_C ()$, $\sum_T ()$ = summation for the part in compression or tension, respectively

A_{ei} = effective cross-sectional area of the ith element in compression

A_j = cross-sectional area of the jth element in tension

σ_{Yeqd}, σ_{Yeqb} = average equivalent yield stresses at upper deck or outer bottom panels, respectively

D = depth of the ship

In calculating the longitudinal bending stress value defined in Equation 39.5, the following criteria should be satisfied:

$$\sigma \leq \sigma_{Yeq} \text{ for tension elements} \tag{39.6a}$$

$$|\sigma| \leq |\sigma_{1u}| \text{ for compression elements} \tag{39.6b}$$

where σ_{Yeq} = equivalent yield stress for both plating and attached stiffeners and σ_{1u} = ultimate compressive stress of the structural element.

When a continuous, longitudinally stiffened plate structure is idealized as an assembly of plate-stiffener combination units, as shown in Figure 39.2b, the ultimate compressive strength of each element can be approximated by the so-called Paik-Thayamballi formula, as follows [2]:

$$\sigma_u = -\frac{\sigma_{Yeq}}{\sqrt{0.995 + 0.936\lambda^2 + 0.170\beta_x^2 + 0.188\lambda^2\beta_x^2 - 0.067\lambda^4}} \quad \text{and} \quad |\sigma_{1u}| \leq \frac{\sigma_{Yeq}}{\lambda^2} \quad (39.7)$$

where

$$\beta_x = \frac{b}{t}\sqrt{\frac{\sigma_Y}{E}}$$
$$\lambda = \frac{L}{\pi r}\sqrt{\frac{\sigma_{Yeq}}{E}}$$

with L = length of the unit, E = Young's modulus, r = radius of gyration of the plate-stiffener combination, b = breadth of plating between stiffeners, t = plate thickness, and σ_Y = yield stress of plating.

39.6.2 Secondary Failure Mode

The hull girder strength formula based on the secondary failure mode will be taken as

$$M_{2u} = \sigma_{2u} Z_e \quad (39.8)$$

where σ_{2u}, Z_e = ultimate compressive strength and elastic section modulus, respectively, at the compression flange (stiffened panel) of the ship.

The ultimate compressive strength of the compression flange may again be predicted by Equation 39.7, assuming that the deck or bottom panel concerned is idealized as an assembly of plate-stiffened combination units, as shown in Figure 39.2b.

39.6.3 Tertiary Failure Mode

The hull girder strength formula based on the tertiary failure mode will be taken as

$$M_{3u} = \sigma_{3u} Z_e \quad (39.9)$$

where σ_{3u} = ultimate compressive strength of the stiffeners (support members) in the compression flange, and Z_e is as defined in Equation 39.8.

The ultimate compressive strength σ_{3u} in the tertiary failure mode is approximately represented by the elastic-plastic tripping strength of stiffeners in the compression flange. Useful closed-form formulae for predicting the tripping strength of typical stiffener profiles (e.g., flat-bar, angle and T-bar) can be found in the literature [2].

39.6.4 Quaternary Failure Mode

The hull girder strength formula based on the quaternary failure mode will be approximated by

$$M_{4u} = \sigma_{4u} Z_e \quad (39.10)$$

where σ_{4u} = ultimate compressive strength of plating between stiffeners in the compression flange, and Z_e is as defined in Equation 39.8.

The ultimate compressive strength of plating between stiffeners can be predicted by [29].
For $\frac{a}{b} \geq 1$:

$$\frac{\sigma_{4u}}{\sigma_Y} = \begin{cases} 0.032\beta_x^4 - 0.002\beta_x^2 - 1.0 & \text{for } \beta_x \leq 1.5 \\ -1.274/\beta_x & \text{for } 1.5 < \beta_x \leq 3.0 \\ -1.248/\beta_x^2 - 0.283 & \text{for } \beta_x > 3.0 \end{cases} \quad (39.11a)$$

For $\frac{a}{b} < 1$:

$$\frac{\sigma_{4u}}{\sigma_Y} = -\frac{a}{b}\frac{\sigma_{xu}}{\sigma_Y} - \frac{0.475}{\beta_y^2}\left(1 - \frac{a}{b}\right) \quad (39.11b)$$

where

σ_Y = yield stress of the plating
$\beta_x = \frac{b}{t}\sqrt{\frac{\sigma_Y}{E}}$
$\beta_y = \frac{a}{t}\sqrt{\frac{\sigma_Y}{E}}$

with a = plate length (in the ship's longitudinal direction), and b = plate breadth (in the ship's transverse direction). The term σ_{xu} is taken as $\sigma_{xu} = \sigma_{4u}$, but using Equation 39.11a together with replacement of β_x by β_y.

39.6.5 Effect of Corrosion

Corrosion wastage in ship plates can reduce their ultimate strength. Two types of corrosion damage are usually considered, namely, general (or uniform) corrosion and localized corrosion. General corrosion reduces the plate thickness uniformly, while localized corrosion such as pitting appears nonuniformly in selected regions such as the vessel bottom in cargo tanks of crude oil carriers.

The ultimate strength of a steel member with general corrosion can be easily predicted by excluding the plate thickness loss due to corrosion. On the other hand, it is proposed that the ultimate strength prediction of a structural member with pitting corrosion can be made using a strength-knockdown-factor approach. The following ultimate strength-knockdown factor may be relevant [30]

$$R_r = \frac{\sigma_u}{\sigma_{uo}} = \left(\frac{A_0 - A_r}{A_0}\right)^{0.73} \quad (39.12)$$

where R_r = ultimate strength reduction factor due to pitting corrosion; σ_u, σ_{uo} = ultimate compressive strengths for plating with and without (intact) pitting corrosion; A_0 = cross-sectional area of the intact member; A_r = largest cross-sectional area of plating lost to pitting corrosion (see Figure 39.3).

39.6.6 Effect of Fatigue Cracking

Under the action of repeated loading, fatigue cracks can be initiated at stress concentrations in the structure. Initial defects or cracks can also be formed in the structure by usual fabrication procedures and may conceivably remain undetected for some time. In addition to propagation under repeated cyclic

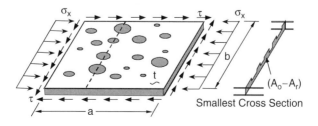

FIGURE 39.3 A schematic for localized pitting corrosion and definition of the smallest cross-sectional area (A_o = cross-sectional area of the intact plate).

loading, cracks can also grow in an unstable way under monotonically increasing extreme loads, a circumstance that eventually can lead to catastrophic failure of the structure. This possibility is, of course, usually tempered by the ductility of the material involved and also by the presence of reduced-stress-intensity regions in a complex structure that can serve as crack arresters in an otherwise monolithic structure.

For residual strength assessment of aging steel structures under extreme loads as well as under fluctuating loads, it is thus often necessary to take into account a known or estimated crack as a parameter of influence. The ultimate strength of a structural member with fatigue cracking can be predicted, somewhat pessimistically, as follows [2]:

$$R_c = \frac{\sigma_u}{\sigma_{uo}} = \frac{A_o - A_c}{A_o} \tag{39.13}$$

where R_c = ultimate strength reduction factor due to fatigue cracking; σ_u, σ_{uo} = ultimate tensile or compressive strengths for a cracked member or an intact (uncracked) member; A_o = cross-sectional area of the intact member; and A_c = cross-sectional area affected by fatigue cracking.

39.6.7 Effect of Local Denting

Plate panels in ships and offshore structures can suffer mechanical damage in many ways, depending upon where such plates are situated. At inner bottom plates of cargo holds of bulk carriers, mechanical damage can take place by indelicate loading or unloading of cargoes. Inner bottom plates can suffer mechanical damage during loading of iron ore when the iron ore strikes the plates. In unloading of bulk cargoes such as iron ore or coal, the excavators used can result in impacts to the inner bottom plates. Deck plates of offshore platforms may be subjected to impacts due to dropped objects from cranes. Such mechanical damage normally exhibits various features such as denting, cracking, residual stresses or strains due to plastic deformation, and coating damage.

In calculating ultimate hull girder strength, therefore, the effect of local denting of plating may need to be taken into account, where significant. An ultimate strength reduction factor for local denting is useful for this purpose, such as [31]

$$R_d = \frac{\sigma_u}{\sigma_{uo}} = C_3 \left[C_1 \ln\left(\frac{D_d}{t}\right) + C_2 \right] \tag{39.14a}$$

where R_d = ultimate strength reduction factor due to local denting; σ_u, σ_{uo} = ultimate compressive strengths of the dented member or the intact (undented) member, D_d = depth of the dent; and t = plate thickness. The coefficients C_1 through C_3 are empirically determined by regression analysis of computed results [31] as follows:

$$C_1 = -0.042 \left(\frac{d_d}{b}\right)^2 - 0.105 \left(\frac{d_d}{b}\right) + 0.015$$

$$C_2 = -0.138 \left(\frac{d_d}{b}\right)^2 - 0.302 \left(\frac{d_d}{b}\right) + 1.042 \tag{39.14b}$$

$$C_3 = -1.44 \left(\frac{H}{b}\right)^2 + 1.74 \left(\frac{H}{b}\right) + 0.49$$

where d_d = diameter of dent, b = plate breadth, and $H = h$ for $h \leq b/2$, $H = b - h$ for $h > b/2$, and h = y-coordinate of the center of denting (Figure 39.4).

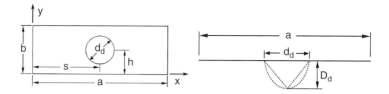

FIGURE 39.4 Geometric parameters for local denting on a plate surrounded by support members.

39.6.8 Effect of Combined Corrosion, Fatigue Cracking, and Local Denting

When pitting corrosion, fatigue cracking, and local denting exist simultaneously, the ultimate strength of a plate member is postulated to be given, again perhaps pessimistically (by virtue of a multiplicative model), as follows:

$$\sigma_u = R_r R_c R_d \sigma_{uo} \quad (39.15)$$

where R_r, R_c, R_d are as defined in Equation 39.12, Equation 39.13, and Equation 39.14, respectively.

39.7 Extreme Hull Girder Load Models

39.7.1 Still-Water Bending Moment

M_{sw} in Equation 39.4 is taken as the maximum value of the still-water bending moment resulting from the worst load condition for a ship, considering both hogging and sagging. The related detailed distribution of the still-water moment along the ship's length can be calculated by double integration of the difference between weight and buoyancy, using simple beam theory.

For convenience, the mean value of M_{sw} will be taken from an empirical formula that has been suggested as a first-cut estimate of the maximum allowable still-water bending moment by some classification societies in the past. That approximate formula is given by

$$M_{sw} = \begin{cases} +0.015 C L^2 B (8.167 - C_b) \text{ (kNm) for hogging} \\ -0.065 C L^2 B (C_b + 0.7) \text{ (kNm) for sagging} \end{cases} \quad (39.16)$$

where

$$C = \begin{cases} 0.0792 L & \text{for } L \leq 90 \\ 10.75 - \left(\dfrac{300-L}{100}\right)^{1.5} & \text{for } 90 < L \leq 300 \\ 10.75 & \text{for } 300 < L \leq 350 \\ 10.75 - \left(\dfrac{L-350}{150}\right)^{1.5} & \text{for } 350 < L \leq 500 \end{cases}$$

with L = ship length (m), B = ship breadth (m), and C_b = block coefficient at summer load waterline.

The COV associated with still-water bending moment of a merchant cargo vessel is normally large, perhaps as high as 0.4 [32]. The variation in still-water bending moment is usually assumed to follow the normal distribution.

39.7.2 Wave-Induced Bending Moment

For reliability assessment of newly built ships, M_w in Equation 39.4 is normally taken as the mean value of the extreme wave-induced bending moment that the ship is likely to encounter during its lifetime.

This is given for unrestricted worldwide service by IACS (International Association of Classification Societies), as follows:

$$M_w = \begin{cases} +0.19CL^2BC_b \text{ (kNm) for hogging} \\ -0.11CL^2B(C_b + 0.7) \text{ (kNm) for sagging} \end{cases} \quad (39.17)$$

where C, L, B, C_b are as defined in Equation 39.16.

For the safety and reliability assessment of damaged ship structures in particular cases, short-term-based response analysis can be used to determine M_w when the ship encounters a storm of specific duration (e.g., 3 h) and with certain small encounter probability. The MIT sea-keeping tables developed by Loukakis and Chryssostomidis [33] are useful for predicting the short-term-based wave-induced bending moment of merchant cargo vessels, and these are used in our study. The MIT sea-keeping tables are designed to efficiently determine the root-mean-square value of the wave-induced bending moment given the values of significant wave height (H_s), B/T ratio (B = ship breadth, T = ship draft), L/B ratio (L = ship length), ship operating speed (V), the block coefficient (C_b), and sea-state persistence time.

The most probable extreme value of the wave-induced loads, M_w, i.e., mode, which we may refer to as a mean for convenience, and its standard deviation, σ_w, can then be computed based on up-crossing analysis as follows [34]:

$$M_w = \left[\sqrt{2\lambda_o \ln N} + \frac{0.5772}{\sqrt{2\lambda_o \ln N}}\right]\rho g L^4 \times 10^{-16} \text{ (GNm)} \quad (39.18a)$$

$$\sigma_w = \frac{\pi}{\sqrt{6}}\sqrt{\frac{\lambda_o}{2\ln N}}\rho g L^4 \times 10^{-16} \text{ (GNm)} \quad (39.18b)$$

$$\text{COV} = \frac{\sigma_w}{M_w} \quad (39.18c)$$

where $\sqrt{\lambda_o}$ is the nondimensional root-mean-square value of the short-term wave-induced bending moment process, which can be estimated using the MIT sea-keeping tables. N is the expected number of wave bending peaks, which can be estimated as follows [34]:

$$N = \frac{S}{\sqrt{13H_s}} \times 3600 \quad (39.18d)$$

where H_s = significant wave height (m), and S = storm persistence time (h).

The USAS-L computer program, which automates the vertical bending moment calculation procedure using the MIT sea-keeping tables as well as by Equation 39.16 and Equation 39.17, can be downloaded from an Internet Web site [2]. The COV associated with the wave-induced bending moment can be defined by Equation 39.18c, on which the short-term response is based, although it is often assumed to be 0.1 when M_w is predicted from Equation 39.17.

39.8 Prediction of Time-Dependent Structural Damage

As a ship gets older, corrosion and fatigue cracking are expected to be the two most common types of structural degradation that will affect safety and reliability. Such strength degradation is time-dependent in nature. To make time-dependent reliability assessment possible under these circumstances, therefore, it is essential to establish time-dependent models for predicting corrosion and fatigue cracking.

This section presents some useful models for predicting the amount of corrosion and fatigue cracking, each as a function of time (ship age). While based largely on the authors' own related past work, the models presented do make use of several pertinent studies by other investigators as well. It should be noted that when time-dependent strength effects are considered, the loads used for reliability analysis in the illustrative examples presented herein are assumed to be time-independent. Also, it is assumed that local dent damage is not a function of time once it has occurred.

39.8.1 Time-Dependent Corrosion Wastage Models

The corrosion characteristics in a ship structure are influenced by many factors such as type of corrosion protection, type of cargo, temperature, humidity, and so on. This means that it should be possible to estimate corrosion depths for different structural members grouped by type and location for various types of ships or cargoes and other pertinent factors.

Figure 39.5 represents a plausible schematic of the proposed corrosion process model for a coated area in a marine steel structure. The corrosion behavior in this model is categorized into three phases: (a) corrosion on account of durability of coating, (b) transition to visibly obvious corrosion, and (c) progress of such corrosion [2]. The curve showing corrosion progression as indicated by the solid line in Figure 39.5 is convex, but it may in some cases be concave (dotted line). The convex curve indicates that the corrosion rate (i.e., the curve gradient) is initially relatively steep but decreases as the corrosion progresses. This type of corrosion process may be typical of a static immersion environment in seawater, because the relatively static corrosion scale at the steel surface tends to inhibit the corrosion progression. On the other hand, the concave curve represents a case where the corrosion rate is accelerating as the corrosion process proceeds. This type of corrosion progression may happen in changing immersion conditions at sea, particularly in dynamically loaded structures, where flexing continually exposes additional fresh surface to corrosion effects.

The life (or durability) of a coating will, in a specific case, correspond to the time when a predefined and measurable extent of corrosion starts after (a) the time when a newly built ship enters service, (b) the application of a coating to a previously bare surface, or (c) repair of a failed coating area in an existing structure to a good intact standard. The coating life typically depends on the type of coating system used, details of its application (e.g., surface preparation, film thickness, humidity and salt control during application, etc.), and relevant maintenance, among other factors. While the coating life to a predefined state of breakdown must essentially be a random variable, it is often treated as a constant parameter.

After the effectiveness of a coating is lost, some transition time, i.e., duration between the time of coating effectiveness loss and the time of corrosion initiation, can be considered to exist before the corrosion initiates over a large enough and measurable area. The transition time is often considered to be an exponentially distributed random variable.

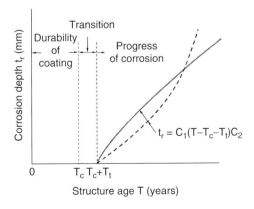

FIGURE 39.5 Schematic of a proposed corrosion process model for marine structures.

Three types of corrosion models, of perhaps increasing level of sophistication and sometimes increasing difficulty in use, have been suggested as follows:

Paik and Thayamballi [2, 35, 38, 39]:

$$t_r = C_1(T - T_c - T_t)^{C_2} \tag{39.19a}$$

$$r_r = C_1 C_2 (T - T_c - T_t)^{C_2 - 1} \tag{39.19b}$$

where t_r = depth of corrosion (mm); r_r = annualized corrosion rate (mm/year); T_c = coating life (year); T_t = transition time (year) between coating durability and corrosion initiation; T = structure age (year); and C_1, C_2 = coefficients taking account of the characteristics of corrosion progress.

Guedes Soares and Garbatov [36]:

$$t_r = t_{r\infty}\left[1 - \exp\left(-\frac{T - T_c}{T_t}\right)\right] \tag{39.20a}$$

$$r_r = \frac{t_{r\infty}}{T_t} \exp\left(-\frac{T - T_c}{T_t}\right) \tag{39.20b}$$

where $t_{r\infty}$ = depth (mm) of corrosion when the corrosion progress stops, with other terms as previously defined.

Qin and Cui [37]:

$$t_r = t_{r\infty}\left\{1 - \exp\left[-\left(\frac{T - T_{st}}{\eta}\right)^\alpha\right]\right\} \tag{39.21a}$$

$$r_r = t_{r\infty} \frac{\alpha}{\eta}\left(\frac{T - T_{st}}{\eta}\right)^{\alpha - 1} \exp\left[-\left(\frac{T - T_{st}}{\eta}\right)^\alpha\right] \tag{39.21b}$$

where T_{st} = time (year) when accelerating of the corrosion process stops, and α, η = coefficients to handle the corrosion decelerating, with the other terms as previously defined.

Figure 39.6 shows means and COVs of the coefficients C_1 in Equation 39.19 for tanker corrosion groups (defined by location, category, and corrosion environment of member), when it is assumed that $C_2 = 1$, $T_t = 0$, and $T_c = 7.5$ years. Although proposed for double-hull tank vessels, the results were obtained by statistical analysis of corrosion measurement data of 230 aging single-hull tankers that carried crude oil or petroleum product oil. A total of 33,820 measurements for 34 different member groups, which comprised 14 categories of plate parts, 11 categories of stiffener webs, and 9 categories of stiffener flanges, were considered [38]. Figure 39.6a represents the most probable (average) value of the annualized corrosion rates, while Figure 39.6b gives the severe (95 percentile) values.

A similar model has also been proposed as being relevant for bulk carriers, as shown in Figure 39.7 [39]. A total of 12,446 measurements for 23 different member groups (defined by location and category of member), which comprised 9 categories of plate parts, 7 categories of stiffener webs, and 7 categories of stiffener flanges, were obtained and made available for the statistical analysis. Figure 39.7a and Figure 39.7b show the means and COVs of the average and severe levels of the coefficient C_1 for the 23 member location/category groups of a bulk carrier structure, respectively, when it is assumed that $C_2 = 1$, $T_t = 0$, and $T_c = 7.5$ years, except for IBP (inner bottom plating) and LSP (lower sloping plating), which may take a shorter coating life, i.e., $T_c = 5$ years.

With the coefficient C_1 known, the corrosion depth of any structural member location in a tanker, a ship-type FPSO (floating, production, storage, and off-loading unit), or a bulk carrier can be predicted

Reliability Assessment of Ships

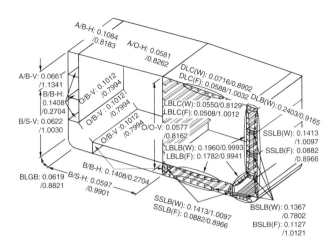

FIGURE 39.6A Mean and COV of the average (most probable) value of the coefficient C_1 for the 34-member location/category groups, proposed for double-hull tanker or ship-shaped FPSO (floating, production, storage and off-loading unit) structure. (From Paik, J.K. et al., *Marine Technol.*, 40 (3), 201–217, 2003. With permission.)

from Equation 39.19 as a function of ship age, assuming that $C_2 = 1$, $T_t = 0$, and $T_c = 7.5$ or 5 years, as specified.

39.8.2 Time-Dependent Fatigue Crack Model

Fatigue cracking is a primary source of costly repair work for aging ships. Crack damage is typically found in welded joints and in local areas of stress concentrations, e.g., in hull sides at the weld intersections of longitudinals, frames, and girders. Much of this is, of course, attributable to initial defects that can be formed in the structure by fabrication procedures and that conceivably remain undetected over time. Under cyclic loading or even large monotonic loading, cracking can originate and propagate from such defects and become larger with time.

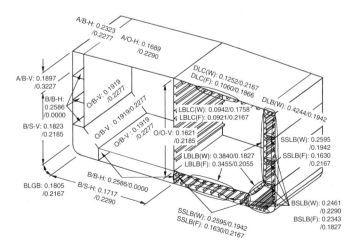

FIGURE 39.6B Mean and COV of the upper bound (severe) value of the coefficient C_1 for the 34-member location/category groups, proposed for double-hull tanker or ship-shaped FPSO (floating, production, storage and off-loading unit) structure. (From Paik, J.K. et al., *Marine Technol.*, 40 (3), 201–217, 2003. With permission.)

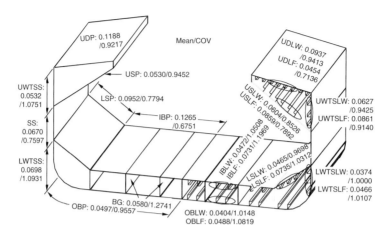

FIGURE 39.7A Mean and COV of the average level of the coefficient C_1 for the 23-member location/category groups of a bulk carrier structure. (From Paik, J.K. et al., Int. J. Maritime Eng., 145 (A2) 61–87, 2003. With permission.)

Since cracks can conceivably lead eventually to catastrophic failure of the structure, it is essential to properly consider and establish relevant crack-tolerant design procedures for ship structures, in addition to implementation of appropriate close-up survey strategies and maintenance philosophies. For design-stage reliability assessment of an eventually aging ship structure under extreme loads, it is often necessary to include a known (existing or postulated or anticipated) crack as a parameter of influence in the ultimate limit-state analysis. To make this possible, it is necessary to develop a relatively simple time-variant fatigue-crack model that can predict crack damage in location and size as the ship gets older.

Figure 39.8 shows a schematic of fatigue cracking progress as a function of time (age) in steel structures. The fatigue cracking progress can be separated into three stages, namely the crack initiation stage (stage I), the crack propagation stage (stage II), and the failure (fracture) stage (stage III).

It is assumed in the model shown that no initial defects exist so that there is no cracking damage until the time T_I. While the fatigue damage is affected by many factors, for a given structural location, service profile and other characteristics, the local stress concentration effects and the number of stress range cycles leading to detectable initiation of cracking can be evaluated by fatigue analysis. Fatigue analyses typically use small-specimen laboratory data in which the crack size at what we here call initiation is

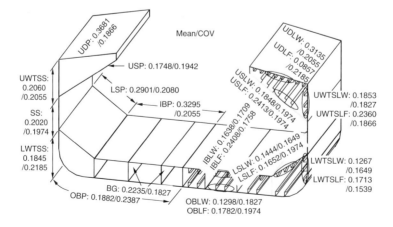

FIGURE 39.7B Mean and COV of the severe level of the coefficient C_1 for the 23-member location/category groups of a bulk carrier structure. (From Paik, J.K. et al., Int. J. Maritime Eng., 145 (A2) 61–87, 2003. With permission.)

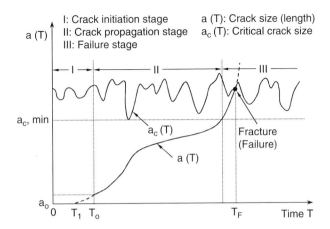

FIGURE 39.8 Schematic for crack initiation and growth in a steel structure with time.

typically not controlled or measured. On the other hand, when any crack is detected in an existing structure at time T_o during an inspection or survey process, it is normally of a size (length), denoted by a_o and called the initial crack size, that must be detectable.

For various reasons, there is the potential for mismatch between what the analysis predicts as "life to initiation" and the cracks that may be detected at the earliest survey in a real structure. For the assessment of time-dependent risks of a structure in the present illustrative examples, however, it is assumed that the initial crack size is small, say 1.0 mm. It should be noted that cracks of such a small size may not normally be detectable on structures in service using the predominantly visual inspection methods that are currently deployed.

Fatigue cracking will propagate with time. Crack propagation is mostly progressive in ductile materials but will be unstable in brittle materials. Crack propagation is in fact affected by many parameters, such as initial crack size, history of local nominal stresses, load sequence, crack retardation, crack closure, crack growth threshold, and stress intensity range in addition to the stress intensity factor at the crack tip, which depends on material properties and geometry. Further, in seawater exposure situations, even the loading frequency and the form of loading may affect crack growth. The fracture-mechanics approach is often used to analyze the behavior of crack propagation, but usually in much simpler terms than reality.

As illustrated in Figure 39.8, the crack growth process with time can be categorized into three stages, namely crack initiation, propagation, and failure. In this regard, for purposes of predicting the effects of such a process on the structural capacity, the time-dependent cracking damage model can also be composed of three separate stage models as follows:

1. A model for crack initiation assessment and detected cracks
2. A model for crack growth assessment
3. A model for failure assessment

When a structure is designed, the crack initiation life at critical structural details is usually theoretically assessed using the S-N curve approach (S = fluctuating stress range, N = associated number of stress range cycles). In this approach, the so-called Palmgren-Miner cumulative damage rule can be applied together with the relevant S-N curve. This normally follows three steps, namely, (a) define the histogram of cyclic stress ranges, (b) select the relevant S-N curve, and (c) calculate the cumulative fatigue damage and judge whether the time to initiation of crack meets the required target fatigue-life value. In a reliability analysis involving this same phenomenon, the relevant design variables including the S-N parameters must be characterized by their probability distributions, including mean, COV and form (type of distribution).

For an existing structure, cracking at critical joints and details is detected, and their size (length) denoted by a_o can be measured. Typically, the crack size that can be detected visually needs to be larger than a certain amount, say 15 to 30 mm, or sometimes more. When the integrity of existing aged-ship structures is considered, it is assumed that the crack of length a_o at a critical joint or detail has initiated at a known ship age of T_o years.

The crack growth can in general be assessed by the fracture-mechanics approach. This considers one or more postulated cracks in the structure, and predicts the fatigue damage during the process of crack propagation, including any coalescence and through-thickness cracking propagation of a through-thickness crack, and subsequent failure. In this approach, a major task is to preestablish the relevant crack growth equations or 'laws' as a function of time (year).

Two types of fracture-mechanics approaches can be considered, namely, physical and empirical models. In the physical model, which will be used in the present illustrative examples, the crack growth rate is expressed as a function of the stress intensity factor at the crack tip, under the assumption that the yielded area around the crack tip is relatively small. The so-called Paris-Erdogan law is often used for this purpose and is expressed as follows:

$$\frac{da}{dN} = C(\Delta K)^m \tag{39.22}$$

where the left-hand side of the equation represents the incremental growth per cycle of a crack of length a; ΔK = stress intensity factor range at the crack tip; and C, m = material constants to be determined from tests.

For steel structures with typical types of cracks, stress-intensity-factor formulae can be found in the literature [2, 40]. In ship stiffened panels, cracks are observed along the weld intersections between plating and stiffeners. For a plate with a crack, ΔK can be represented over a small enough period of time by

$$\Delta K = F\Delta\sigma\sqrt{\pi a} \tag{39.23}$$

where $\Delta\sigma$ = stress range (or double amplitude of applied fatigue stress), and F is a geometric parameter depending on the loading type and configuration of the cracked body. For plates with typical types of cracks and under axial tension, as shown in Figure 39.9, F is approximately given as follows [2]:

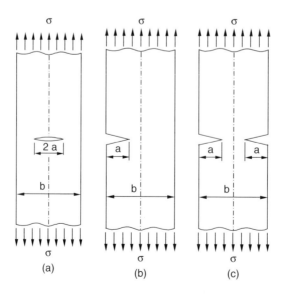

FIGURE 39.9 Typical crack locations in a plate under tensile stress: (a) center crack, (b) crack to one side, (c) crack to both sides.

For a center crack as shown in Figure 39.9a:

$$F = \left(\sec\frac{\pi a}{b}\right)^{1/2} \tag{39.24a}$$

For a crack on one side as shown in Figure 39.9b:

$$F = 30.38\left(\frac{a}{b}\right)^4 - 21.71\left(\frac{a}{b}\right)^3 + 10.55\left(\frac{a}{b}\right)^2 - 0.23\left(\frac{a}{b}\right) + 1.12 \tag{39.24b}$$

For cracks on both sides as shown in Figure 39.9c:

$$F = 15.44\left(\frac{a}{b}\right)^3 - 4.78\left(\frac{a}{b}\right)^2 + 0.43\left(\frac{a}{b}\right) + 1.12 \tag{39.24c}$$

It should be noted that the above simplified situations use idealized boundary conditions. The effect of stiffening is also neglected. There are other, more-refined stress-intensity-factor solutions and methods of calculating improved stress intensity factors in particular situations, e.g., by FEA or the Green's function technique [40].

The crack length, $a(T)$, as a function of time, T, can then be calculated by integrating Equation 39.22 with respect to stress range cycles, N. In the integration of Equation 39.22, it is sometimes assumed that the geometric parameter, F, is constant, i.e., assuming that the geometric parameter, F, is unchanged as the crack propagates. Note, however, that this assumption is only reasonable as long as the initial crack size, a_0, is small. In such a case, the integration of Equation 39.22 after substitution of Equation 39.23 results in

$$a(T) = \begin{cases} \left[a_o^{1-m/2} + \left(1 - \frac{m}{2}\right) C(\Delta\sigma F\sqrt{\pi})^m (T - T_o)\omega\right]^{\frac{1}{1-m/2}} & \text{for } m \neq 2 \\ a_o \exp\left[C\Delta\sigma^2 F^2 \pi (T - T_o)\omega\right] & \text{for } m = 2 \end{cases} \tag{39.25}$$

where ω = number of cycles per year. It is often considered in ships that a wave load cycle occurs (very approximately) once in every 6 to 10 sec, and hence $N \approx (T - T_o) \times 365 \times 24 \times 60 \times 60/10 = \omega \times (T - T_o)$, so $\omega \approx 365 \times 24 \times 60 \times 60/10$, where T = ship age in years, already defined above.

Figure 39.10 shows a sample application of Equation 39.25 and compares its results with results from a direct integration of Equation 39.22, which accounts correctly for the effect of crack growth on the geometric parameter, F. It is seen from Figure 39.10 that Equation 39.25 slightly overestimates the fracture life as the crack propagates, as would be expected because Equation 39.25 was derived under the assumption that the geometric parameter, F, remains constant over time. Notwithstanding, it is seen in this case that the difference for a small initial crack size is not significant, so Equation 39.25 apparently provides a reasonable tool for crack growth assessment.

In reliability analysis involving this same phenomenon, the relevant design variables, including the crack growth rate equation parameters, must be characterized by their probability distributions, including mean, COV, and form (type of distribution).

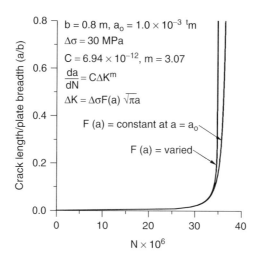

FIGURE 39.10 A comparison of Equation 39.25 with a direct (numerical) integration of Equation 39.23 for a small initial crack size.

39.9 Application to Time-Dependent Reliability Assessment of Ships

The methodology of reliability assessment of ships described above was automated within the GUI-based computer program TRAAS (time-dependent reliability assessment of aging ships). The following subsections describe application examples of the time-dependent reliability assessment of three types of hull girders: a 105,000-dwt double-hull tanker, a 170,000-dwt bulk carrier, and a 113,000-dwt ship-shaped FPSO (floating, production, storage, and off-loading unit). The effects of age-related structural degradation (corrosion and fatigue cracking) and local denting are included. The purpose is to illustrate some of the previously discussed concepts. In the present application examples, only the primary failure mode among the four hull-girder failure modes previously described is considered. Figure 39.11 shows schematic representations of the mid-ship sections of the three ships considered. Table 39.2 indicates the principal dimensions of the ships.

39.9.1 Scenarios for Operational Conditions and Sea States

It is important to realize that hull girder loads depend on the operational conditions, vessel speed, vessel heading, and the sea states. Hence seamanship is a strong factor affecting hull girder loads in particular

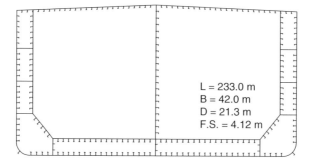

FIGURE 39.11A Mid-ship section of a hypothetical 105,000-dwt double-hull tanker with one center-longitudinal bulkhead (L = ship length, B = ship beam, D = ship depth, F.S. = frame spacing).

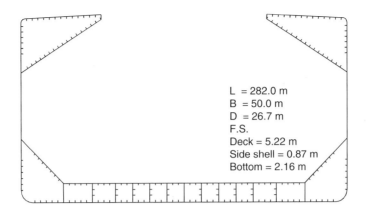

FIGURE 39.11B Mid-ship section of a 170,000-dwt single-sided bulk carrier (L = ship length, B = ship beam, D = ship depth, F.S. = frame spacing).

situations. In this regard, it is necessary to correctly establish relevant scenarios for operational conditions and sea states in predicting hull girder bending moments.

The still-water bending moments, M_{sw}, of the ships in this example are estimated from the empirical formula given as Equation 39.16. This, in a short-term sense, assumes that the vessels are loaded in the most onerous permitted still-water condition. Where long-term reliability analysis is planned, such as in a design context, the entire range of still-water bending moments possible and the effects of different types of hull behavior (hogging, sagging) must be correctly accounted for.

For the present illustrative purposes, on the other hand, the wave-induced bending moment, M_w, is predicted from Equation 39.18a using the short-term response analysis, which involves the operational conditions and sea states defined in Table 39.3. It is noted that for wave-load prediction purposes, the ship-shaped FPSO is assumed to have an equivalent operational speed of 10 knots in waves, while it usually remains at a specific location once installed.

The present scenarios associated with the operational conditions and sea states are adopted for illustrative purposes. The results of such reliability analyses are then indicative, i.e., only notional probabilities of failure, conditional on the specific storm condition noted above, and conditional on the vessel being loaded in that storm in the very onerous way, as also noted.

Figure 39.12 shows the variations of wave-induced bending moments as a function of significant wave height. Table 39.4 compares the wave-induced bending moments for the three ships computed by the

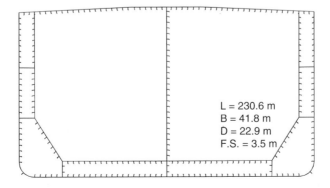

FIGURE 39.11C Mid-ship section of a 113,000-dwt ship-shaped FPSO (floating, production, storage, and offloading unit) (L = ship length, B = ship beam, D = ship depth, F.S. = frame spacing).

TABLE 39.2 Hull Sectional Properties for the Three Example Vessels

	Item	Double-Hull Tanker	Bulk Carrier	FPSO
LBP, L (m)		233.0	282.0	230.6
Breadth, B (m)		42.0	50.0	41.8
Depth, D (m)		21.3	26.7	22.9
Draft, d (m)		12.2	19.3	14.15
Block coefficient, C_b		0.833	0.826	0.831
Design speed (knots)		16.25	15.15	15.4
DWT or TEU (dwt)		105,000	170,000	113,000
Cross-sectional area (m²)		5.318	5.652	4.884
Height to neutral axis from baseline (m)		9.188	11.188	10.219
I (m⁴)	vertical	359.480	694.307	393.625
	horizontal	1,152.515	1,787.590	1,038.705
Z (m³)	deck	29.679	44.354	31.040
	bottom	39.126	62.058	38.520
σ_Y	deck	HT32	HT40	HT32
	bottom	HT32	HT32	HT32
M_p (GNm)	vertical moment	11.930	20.650	12.451
	horizontal moment	19.138	31.867	19.030

Note: FPSO = floating, production, storage, and offloading system; I = moment of inertia; Z = section modulus; σ_Y = yield stress; M_p = fully plastic bending moment; HT32 = high-tensile steel with σ_Y = 32 kgf/mm²; HT40 = high-tensile steel with σ_Y = 40 kgf/mm²; LBP = length between perpendiculars; TEU = twenty-foot equivalent unit.

TABLE 39.3 Scenarios for Operating Condition and Sea States of the Three Example Vessels

Parameter	Double-Hull Tanker	Bulk Carrier	FPSO
Operating speed	0.6 × design speed = 9.75 knots	0.6 × design speed = 9.09 knots	10 knots [a]
Significant wave height, H_s	$1.1\sqrt{L}$, L in ft = 9.27 m	$1.1\sqrt{L}$, L in ft = 10.198 m	$1.1\sqrt{L}$, L in ft = 9.222 m
Storm duration [b]	3 h ($N \approx 1000$)	3 h ($N \approx 1000$)	3 h ($N \approx 1000$)

[a] This is meant as an equivalent design speed, because a FPSO does not have an operating speed as such.
[b] N = number of wave peaks.

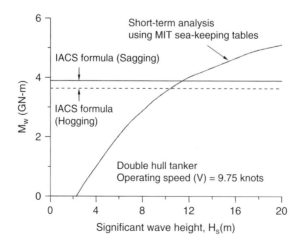

FIGURE 39.12A Variation of the wave-induced bending moment as a function of significant wave height for the 105,000-dwt double-hull tanker.

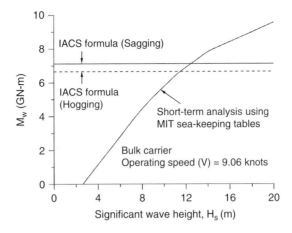

FIGURE 39.12B Variation of the wave-induced bending moment as a function of significant wave height for the 170,000-dwt bulk carrier.

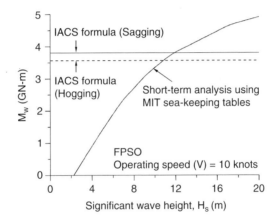

FIGURE 39.12C Variation of the wave-induced bending moment as a function of significant wave height for the 113,000-dwt ship-shaped FPSO.

IACS formula (Equation 39.17) with those computed by Equation 39.18a using the MIT sea-keeping tables. These tables are based on the short-term response during a storm persistence time of 3 h for a particular sea state, as noted above. While the IACS formula is independent of the operational conditions and sea states, it is seen from Table 39.4 that the IACS formula values are larger than the short-term-based calculations by 8 to 20% for the present specific scenarios. With different scenarios of operational conditions and sea states, the opposing trend can appear as well.

TABLE 39.4 Wave-Induced Bending Moments, M_w, for the Three Example Vessels (GN·m)

M_w (GN-m)	Double-Hull Tanker		Bulk Carrier		FPSO	
	Sagging	Hogging	Sagging	Hogging	Sagging	Hogging
IACS formula [a]	−3.895	3.634	−7.121	6.655	−3.805	3.564
MIT sea-keeping tables [b]	−3.352	3.352	−5.927	5.927	−3.172	3.172

[a] Based on a long-term analysis.
[b] Based on a short-term analysis.

39.9.2 Scenarios for Structural Damage

Age-related structural degradation and its effects need to be dealt with as a function of a ship's age, while mechanical damage can be considered to be time invariant. The results obtained from the present approach can be sensitive to the underlying assumptions. Specifically, the vessel is considered to be under the most onerous still-water condition and subject to the conditions of a given, reasonably severe, short-term storm; the vessel may be of varying age and subject to certain generic patterns of corrosion and certain idealized crack scenarios, as will be soon described below.

39.9.2.1 Corrosion Damage

In Section 39.8.1, corrosion wastage models for different structural member groups by type and location, considering plating, and stiffener webs and flanges, were presented. These models can be used to predict the corrosion depth in primary members as the ship ages. As previously noted, the corrosion-progress characteristics of a tanker structure are considered to be similar to those of a ship-shaped FPSO structure as long as the corrosion environment is similar.

In the present reliability assessment, a most probable (average) level of corrosion wastage is considered. While it is assumed that corrosion starts immediately after the breakdown of coating, the coating life of all structural members in the three ships considered is assumed to be 7.5 years, except for the inner bottom plating and the lower sloping plating of the bulk carrier, with coating lives of 5 years. Figure 39.13 shows the progress of corrosion depth for selected members as the vessels age. The figure neglects any effect of steel renewal after inspections and surveys; some consideration of repair of heavily corroded members is made in later illustrations.

As noted earlier, several types of corrosion are possible for mild- and low-alloy steels used in marine applications. While the so-called general (or uniform) corrosion, which reduces the member thickness over large areas, is normally regarded as an idealized type of corrosion in today's ships, localized corrosion such as pitting is more likely to be observed in ship structures. As discussed in Section 39.6.5, the ultimate strength behavior of ship structures with pitting corrosion is different from that of general corrosion. For more realistic assessment of the reliability, therefore, it is of importance to take into account pitting as well as general corrosion.

In the present reliability assessment, it is assumed that the most heavily pitted cross section of any structural member extends over the plate breadth. This may provide a somewhat pessimistic evaluation of residual strength, but it is a practical approach for the reliability assessment. The mean and COV

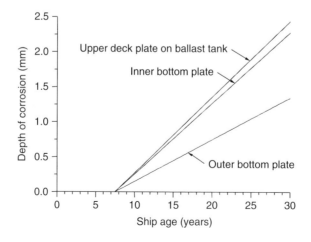

FIGURE 39.13A Progress of corrosion depth for selected members in the 105,000-dwt double-hull tanker and the 113,000-dwt FPSO.

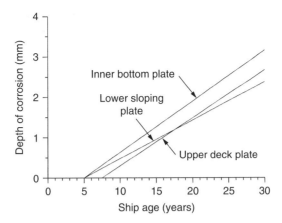

FIGURE 39.13B Progress of corrosion depth for selected members in the 170,000-dwt bulk carrier.

values of annualized corrosion rates for individual structural members are defined in Figure 39.6 and Figure 39.7. It should further be noted that, in reality, pits will be repaired once they reach certain depths and extents, regardless of the related strength criteria, and this is not accounted for in the present illustrative calculations.

39.9.2.2 Fatigue Cracking

In Section 39.8.2, a time-dependent fatigue cracking model was established. The crack length of any critical area is predicted by a closed-form formula as a function of ship age. In the present application examples, it is assumed that cracking initiates in all stiffeners and plating when the ship is 5 years of age. The initial crack size is considered to be 1.0 mm. These are simply assumptions for illustrative purposes, as previously discussed. Cracks normally start as surface cracks and then progress through the thickness of the plating. They certainly do not simultaneously occur at all stiffeners and plating. Also, it has been assumed that the structure had been designed based on the fatigue limit state using crack initiation technology to start with, and that they start at a specific time (namely 5 years in this case). While the constants of the Paris-Erdogan equation can usually be considered the same at all joints, the crack growth characteristics can be different because the stress ranges affecting the stress intensity factors at individual joints vary for reasons of geometry, location of crack, and any differences in load effects that may apply.

Fatigue loading characteristics are random in nature, and their sequence is normally unknown, while the long-term distribution of the fatigue loading is at most known. Therefore, some refined methodologies are used to generate a random loading sequence. With fatigue loading sequence and amplitude known, the dynamic stress range, $\Delta\sigma_i$, at the ith joint can then be given by

$$\Delta\sigma_i = 2 \times \sigma_{xi} \times SCF_i \times k_f = 2 \times \sigma_{xi}^* \times SCF_i \tag{39.26}$$

where σ_{xi} = cyclic "peak" stress amplitude acting on the ith structural element, which can be given by $M_w z/I$; M_w = wave-induced bending moment; I = time-dependent moment of inertia; z = distance from the time-dependent neutral axis to the point of stress calculation; and $\sigma_x^* = k_f \times \sigma_x$. SCF_i in Equation 39.26 is the stress concentration factor at the ith critical joint. In the present illustrative examples, it is assumed that the SCF at all joints between plating and stiffeners (or support members) is 2.1, except for hold-frame connections to upper or lower wing tanks and the side shell of bulk carrier, where the SCF is 3.75, and hopper knuckle joints, where the SCF is much larger at 10.5. Note that more-refined calculations can be conducted to determine the SCF values for different joints. The SCF values used here are simply illustrative assumptions based on related proposals by some classification societies, e.g., Det Norske Veritas [41]. The variable k_f is a

TABLE 39.5 Samples of the Probabilistic Characteristics for Random Variables at a Given Age of the Vessels

Parameter	Definition	Distribution Function	Mean	COV
E	elastic modulus	normal	205.8 GPa	0.03
σ_Y	yield stress	lognormal	as for each member	0.10
t_p	thickness of plating	fixed	as for each member	—
t_w	thickness of stiffener web	fixed	as for each member	—
t_f	thickness of stiffener flange	fixed	as for each member	—
T	ship age	fixed	as for each age	—
T_c	coating life	normal	5.0 years	0.40
			7.5 years	0.40
C_1	corrosion rate	Weibull	as for each member	as for each member
a_o	initial crack size	normal	1.0 mm	0.20
C	$\dfrac{da}{dN}=C(\Delta K)^m$	lognormal	6.94E–12	0.20
m	$\dfrac{da}{dN}=C(\Delta K)^m$	fixed	3.07	—
d_d	diameter of local dent	normal	$0.3b$	0.10
			$0.5b$	0.10
			$0.8b$	0.10
D_d	depth of local dent	normal	$1.0t_p$	0.10
			$2.5t_p$	0.10
			$5.0t_p$	0.10

Note: a = crack size, b = plate breadth, N = number of stress cycles, ΔK = stress intensity factor.

knockdown factor accounting for the dynamic stress cycles and is assumed to be 0.25 for the present illustrative purposes.

At a given age of the vessels, the ultimate strength of structural members with known (or assumed) fatigue cracking damage can be predicted by the strength-knockdown-factor approach, as previously noted, while it is considered that fracture takes place if the crack size (length) of the member reaches the critical crack size, which can be assumed to be the smaller of the plate breadth and the stiffener web height.

Table 39.5 indicates example probabilistic characteristics (mean, COV, distribution) of the random variables used for the present illustrative purposes. It is important to realize that the probabilistic characteristics of random variables will normally be different for different types of ship structures, operating scenarios, and applications.

39.9.2.3 Local Denting

Inner bottom plates of bulk carriers usually have local dent damage caused by mishandled loading and unloading of dense cargo such as iron ore, while other types of merchant ships such as tankers or FPSOs may not suffer such mechanical damage at the same or similar locations or of the same severity.

In this regard, it is assumed that inner bottom plating of only the bulk carrier has local dent damage after 5 years. The size of a local dent is assumed to be the same for all inner bottom plating, namely, $D_d/t = 2.5$, $d_d/b = 0.5$, and $h/b = 0.5$, with the parameters as defined in Equation 39.14. The ultimate strength of plating with local dent damage is then predicted by the simplified formulae previously presented. The COVs associated with the local dent-related parameters are assumed as defined in Table 39.5.

39.9.3 Reliability Assessment

The ship hull ultimate strength formula is eventually expressed as a function of design parameters related to all relevant geometric and material properties. When time-variant structural degradation (e.g., corrosion, fatigue cracking) and local denting areas are considered, the value of member thickness at any particular time is a function of such damage. Thus we have

$$M_{us} = M_{us}(E_i, \sigma_{Yi}, t_{pi}, t_{wi}, t_{fi}, T, T_{ci}, C_{1i}, a_{oi}, C_i, m_i, D_{di}, d_{di}) \quad (39.27a)$$

$$M_{uh} = M_{uh}(E_i, \sigma_{Yi}, t_{pi}, t_{wi}, t_{fi}, T, T_{ci}, C_{1i}, a_{oi}, C_i, m_i, D_{di}, d_{di}) \quad (39.27b)$$

where t_p = thickness of plating, t_w = thickness of stiffener web, t_f = thickness of stiffener flange, T = ship age, C_1 = corrosion rate, and the other variables are as previously defined. The subscript i represents the ith member, and M_{us}, M_{uh} = ultimate hull girder moments in sagging or hogging.

In a reliability assessment, all the parameters noted in Equation 39.27 are treated as random variables, with the probabilistic characteristics (i.e., mean, COV, and distribution function) as defined in Table 39.5. The structural-damage scenarios are divided into the following five groups:

1. Intact (undamaged)
2. Localized corrosion damage alone
3. Localized corrosion and local dent damage
4. Localized corrosion and fatigue cracking damage
5. Localized corrosion, fatigue cracking, and local dent damage

Local dent damage is considered only for the bulk carrier in hogging. Figure 39.14 to Figure 39.16 show the effects of the above damage scenarios on the time-dependent characteristics of ultimate hull girder strength and reliability of the three object vessels when no repair or renewal is made. As the vessels age, the corrosion depth and cracking size (length) increase, and thus the ultimate hull girder strength and reliability index decrease (or failure probabilities increase).

The reliability indices for the three vessels against hull girder collapse in the intact condition are about 2.5, which may be considered adequate in light of the target value previously noted for merchant cargo vessels. At the age of around 15 years, the safety and reliability of the three vessels reduces to less than 90% of the original (as-built) states. If repair and maintenance are not properly carried out, the levels of reliability can decrease rapidly.

39.9.4 Some Considerations Regarding Repair Strategies

To maintain the ship's safety and reliability at a certain target level or higher, a proper, cost-effective scheme for repair and maintenance must be established. In this regard, some considerations for repair strategies of structural members postulated to be heavily damaged by corrosion, fatigue cracking, and

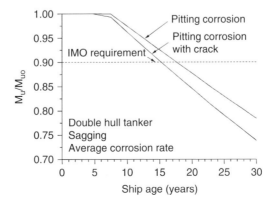

FIGURE 39.14A Time-dependent ultimate hull girder strength of the 105,000-dwt double-hull tanker in sagging.

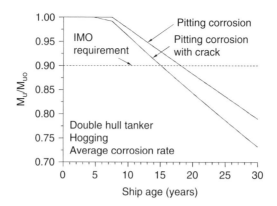

FIGURE 39.14B Time-dependent ultimate hull girder strength of the 105,000-dwt double-hull tanker in hogging.

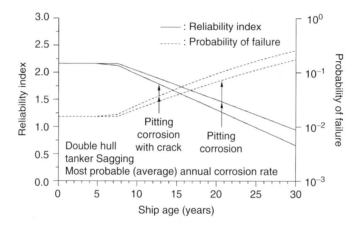

FIGURE 39.14C Time-dependent reliability of the 105,000-dwt double-hull tanker associated with hull girder collapse in sagging.

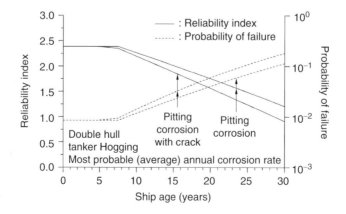

FIGURE 39.14D Time-dependent reliability of the 105,000-dwt double-hull tanker associated with hull girder collapse in hogging.

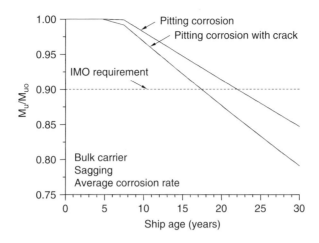

FIGURE 39.15A Time-dependent ultimate hull girder strength of the 170,000-dwt bulk carrier in sagging.

local denting are now illustrated. The International Maritime Organization (IMO) [42] requires that one should keep the longitudinal strength of an aging ship at the level of at least 90% of the initial state. While the IMO requirement is in fact based on the ship's section modulus, in the present illustrative examples it is extended as a device for establishing a more sophisticated maintenance and repair scheme based on hull girder ultimate strength. The aim of the illustrated scheme is that the ultimate hull girder strength of an aging ship must always be at least 90% of the initial, as-built vessel value.

Figure 39.17 to Figure 39.19 show the time-dependent hull girder ultimate strength and reliability values for the object vessels after repair of postulated heavily damaged structural members so that the ultimate hull girder strength is always at least 90% of its original value. In these illustrations, the renewal criterion for any damaged member is based on the member's ultimate strength rather than member thickness, as is traditionally done. This is advantageous because the latter can not reveal the effects of pitting corrosion, fatigue cracking, or local dent damage adequately, even though it may handle the thickness-reduction effects of uniform corrosion reasonably well. On the other hand, the former

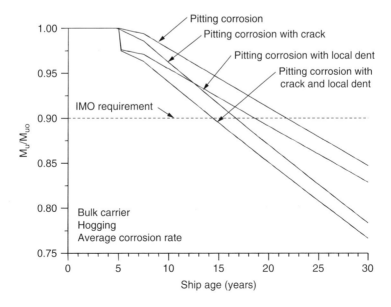

FIGURE 39.15B Time-dependent ultimate hull girder strength of the 170,000-dwt bulk carrier in hogging.

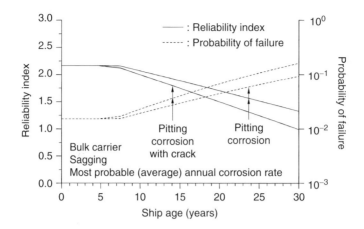

FIGURE 39.15C Time-dependent reliability of the 170,000-dwt bulk carrier associated with hull girder collapse in sagging.

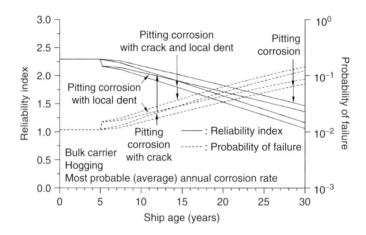

FIGURE 39.15D Time-dependent reliability of the 170,000-dwt bulk carrier associated with hull girder collapse in hogging.

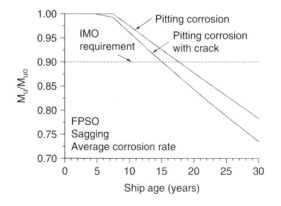

FIGURE 39.16A Time-dependent ultimate hull girder strength of the 113,000-dwt FPSO in sagging.

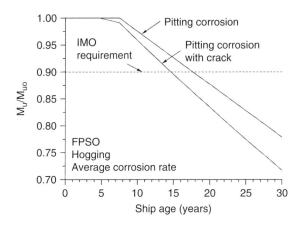

FIGURE 39.16B Time-dependent ultimate hull girder strength of the 113,000-dwt FPSO in hogging.

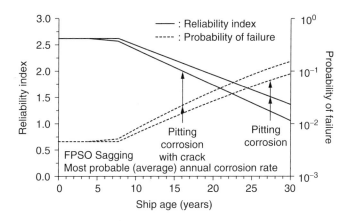

FIGURE 39.16C Time-dependent reliability of the 113,000-dwt FPSO associated with hull girder collapse in sagging.

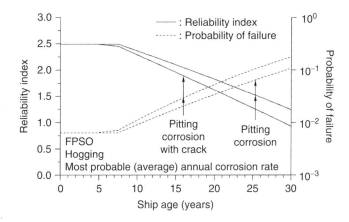

FIGURE 39.16D Time-dependent reliability of the 113,000-dwt FPSO associated with hull girder collapse in hogging.

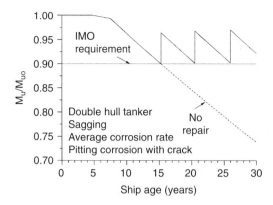

FIGURE 39.17A Repairs and the resulting time-dependent ultimate hull girder strength of the 105,000-dwt double-hull tanker in sagging.

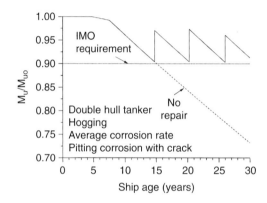

FIGURE 39.17B Repairs and the resulting time-dependent ultimate hull girder strength of the 105,000-dwt double-hull tanker in hogging.

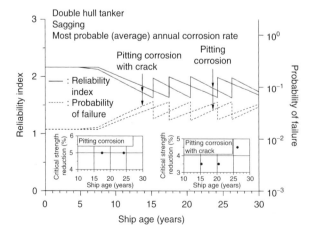

FIGURE 39.17C Repairs and the resulting time-dependent reliability of the 105,000-dwt double-hull tanker associated with hull girder collapse in sagging.

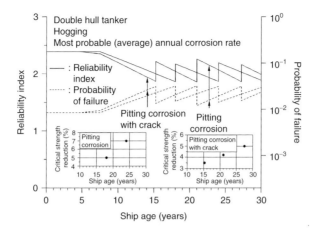

FIGURE 39.17D Repairs and the resulting time-dependent reliability of the 105,000-dwt double-hull tanker associated with hull girder collapse in hogging.

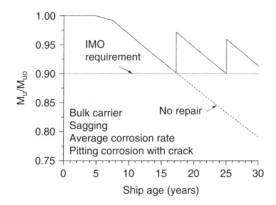

FIGURE 39.18A Repairs and the resulting time-dependent ultimate hull girder strength of the 170,000-dwt bulk carrier in sagging.

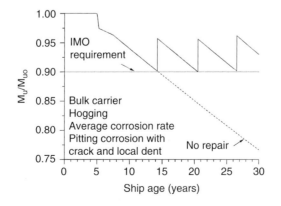

FIGURE 39.18B Repairs and the resulting time-dependent ultimate hull girder strength of the 170,000-dwt bulk carrier in hogging.

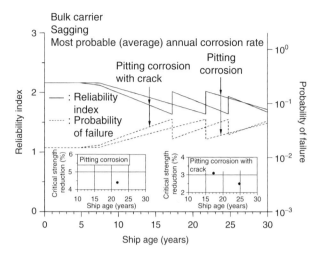

FIGURE 39.18C Repairs and the resulting time-dependent reliability of the 170,000-dwt bulk carrier associated with hull girder collapse in sagging.

(i.e., member's ultimate strength as defined) is adequate and better equipped to deal with all types of structural damage.

As the illustrations imply, the more heavily damaged members need to be renewed (or repaired) to their as-built state immediately, before the ultimate longitudinal strength of an aging ship reduces to a value less than 90% of the original ship.

It is evident from Figure 39.17 through Figure 39.19 that the structural safety and reliability of aging vessels can be controlled by proper repair and maintenance strategies. It is also seen that the repair criterion based on member ultimate strength can provide a potential improvement to better control the age-dependent degradation of a ship's longitudinal strength. It is seen in the illustrations that the percentage reduction in critical ultimate strength of structural members that need to be repaired is not constant, as might be expected, and is in the range of 2 to 7% of the as-built state.

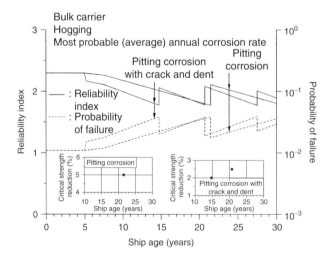

FIGURE 39.18D Repairs and the resulting time-dependent reliability of the 170,000-dwt bulk carrier associated with hull girder collapse in hogging.

Reliability Assessment of Ships

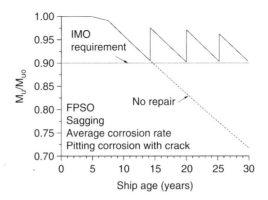

FIGURE 39.19A Repairs and the resulting time-dependent ultimate hull girder strength of the 113,000-dwt FPSO in sagging.

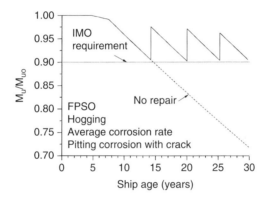

FIGURE 39.19B Repairs and the resulting time-dependent ultimate hull girder strength of the 113,000-dwt FPSO in hogging.

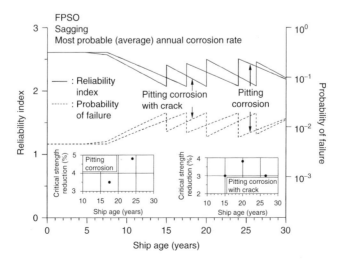

FIGURE 39.19C Repairs and the resulting time-dependent reliability of the 113,000-dwt FPSO associated with hull girder collapse in sagging.

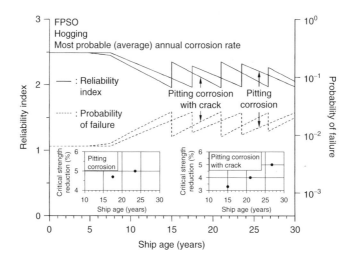

FIGURE 39.19D Repairs and the resulting time-dependent reliability of the 113,000-dwt FPSO associated with hull girder collapse in hogging.

39.10 Concluding Remarks

It is recognized that reliability technology is a powerful tool for a more rigorous assessment of the integrity of ship structures as well as other types of steel structures such as offshore platforms and land-based structures. The use of reliability methods is required for many applications in the maritime industry [23]: (a) to develop probability-based design-code requirements, (b) to estimate reliability in existing ship structures, (c) to perform failure analysis that investigates the cause of structural failure, (d) to compare alternative designs that compete with existing or conventional design concepts, (e) to support economic value analysis (cost-benefit analysis) that identifies the trade-off between cost and risk so as to minimize total expected life-cycle cost, and (f) to develop optimal maintenance strategies of aging structures, leading to minimum cost without reducing the reliability below a specified level.

The present chapter describes a methodology for performing the reliability assessment of aging ships with the focus on hull girder ultimate limit state, accounting for corrosion, fatigue cracking, and local dent damage. Application examples of the methodology to two merchant cargo ships and one ship-shaped FPSO are presented. Time-dependent reliability indices with respect to ultimate strength limit state of three hull girders under a set of presumed damage scenarios are evaluated as they age, indicating that reliability methods can be very useful for identifying a realistic level of failure probability of hull girder collapse. It is also apparent that reliability methods are useful for establishing a cost-effective scheme for repair and maintenance of aging or damaged structures.

During the last two decades, significant developments in the reliability assessment of ships have been achieved. However, a number of problem areas still remain in applying reliability methods to ship structures, although it seems that the mathematical algorithms needed for such analysis have almost been established. These remaining problems are mainly due to the difficulties of quantifying probabilistic characteristics of random variables or the properties involved in structural reliability assessment, which include structural failure types, failure consequences, fabrication-induced initial imperfections, age-related damage (general or localized corrosion, fatigue cracking), mechanical damage (local denting), accident-induced damage (collision/grounding, fire, explosion), extreme sea states, and operational conditions. It is also not an easy task to determine relevant levels of target reliability index of ship structures for various types of limit states [2].

More effort to resolve such problem areas should be undertaken in this regard so that the reliability-based approach can be more widely employed in shipbuilding industry practice, in place of the deterministic approach currently used for ship structural design.

References

1. ISO 2394, General Principles on Reliability for Structures, 2nd ed., International Organization for Standardization, Geneva, 1998.
2. Paik, J.K. and Thayamballi, A.K., *Ultimate Limit State Design of Steel-Plated Structures*, John Wiley & Sons, Chichester, U.K., 2003.
3. Freudenthal, A.M., The safety of structures, *Trans. ASCE*, 112, 125–180, 1947.
4. Freudenthal, A.M., Garrelts, J.M., and Shinozuka, M., The analysis of structural safety, *ASCE J. Structural Div.*, 92 (ST1), 267–325, 1966.
5. Mansour, A.E., Methods of computing the probability of failure under extreme values of bending moment, *J. Ship Res.*, 16 (2), 113–123, 1972.
6. Mansour, A.E., Probabilistic design concepts in ship structural safety and reliability, *SNAME Trans.*, 80, 64–97, 1972.
7. Mansour, A.E. and Faulkner, D., On applying the statistical approach to extreme sea loads and ship hull strength, *RINA Trans.*, 114, 273–314, 1972.
8. Ang, A.H.S., Structural risk analysis and reliability based design, *ASCE J. Structural Div.*, 100 (ST9), 1775–1789, 1973.
9. Ang, A.H.S. and Cornell, C.A., Reliability bases of structural safety and design, *ASCE J. Structural Eng.*, 100 (9), 1755–1769, 1974.
10. Stiansen, S.G., Mansour, A.E., Jan, H.Y., and Thayamballi, A.K., Reliability methods in ship structures, *RINA Trans.*, 121, 381–406, 1979.
11. Ayyub, B.M. and Haldar, A., Practical structural reliability techniques, *ASCE J. Structural Eng.*, 110 (8), 1707–1724, 1984.
12. White, G.J. and Ayyub, B.M., Reliability methods for ship structures, *ASNE Naval Engineers J.*, 97 (4), 86–96, 1985.
13. Mansour, A.E., Approximate probabilistic method of calculating ship longitudinal strength, *J. Ship Res.*, 18 (3), 201–213, 1974.
14. Mansour, A.E., Jan, H.Y., Zigelman, C.I., Chen, Y.N., and Harding, S.J., Implementation of reliability methods to marine structures, *SNAME Trans.*, 92, 353–382, 1984.
15. Kaplan, P., Benatar, M., Benton, J., and Achtarides, T.A., Analysis and Assessment of Major Uncertainties Associated with Ship Hull Ultimate Failure, Ship Structure Committee report SSC-322, Washington, DC, 1984.
16. Mansour, A.E., An Introduction to Structural Reliability Theory, Ship Structure Committee report SSC-351, Washington, DC, 1990.
17. Nikolaidis, E. and Kaplan, P., Uncertainties in Stress Analysis on Marine Structures, Ship Structure Committee report SSC-363, Washington, DC, 1991.
18. Mansour, A.E., Lin, M., Hovem, L., and Thayamballi, A.K., Probability-Based Ship Design Procedures: a Demonstration, Ship Structure Committee report SSC-368, Washington, DC, 1993.
19. Pussegoda, L.N., Dinovitzer, A.S., and Malik, L., Establishment of a Uniform Format for Data Reporting of Structural Material Properties for Reliability Analysis, Ship Structure Committee report SSC-371, Washington, DC, 1993.
20. Mansour, A.E. and Thayamballi, A.K., Probability Based Ship Design; Loads and Load Combination, Ship Structures Committee report SSC-373, Washington, DC, 1994.
21. Hughes, O.F., Nikolaidis, E., Ayyub, B., White, G.J., and Hess, P.E., Uncertainty in Strength Models for Marine Structures, Ship Structure Committee report SSC-375, Washington, DC, 1994.
22. Basu, R.I., Kirkhope, K.J., and Srinivasan, J., Guideline for Evaluation of Finite Elements and Results, Ship Structure Committee report SSC-387, Washington, DC, 1996.

23. Mansour, A.E., Wirsching, P., Luckett, M., and Plumpton, A., Assessment of Reliability of Ship Structures, Ship Structure Committee report SSC-398, Washington, DC, 1997.
24. Ayyub, B.M., *Elicitation of Expert Opinions for Uncertainty and Risks*, CRC Press, Boca Raton, FL, 2001.
25. Ditlevsen, O. and Madsen, H.O., *Structural Reliability Methods*, John Wiley & Sons, New York, 1996.
26. Paik, J.K. and Frieze, P.A., Ship structural safety and reliability, *Progr. Structural Eng. Mater.*, 3 (2), 198–210, 2001.
27. Paik, J.K., Wang, G., Kim, B.J., and Thayamballi, A.K., Ultimate limit state design of ship hulls, *SNAME Trans.*, 110, 21.1–21.31, 2002.
28. Paik, J.K., A guide for the ultimate longitudinal strength assessment of ships, *Mar. Technol.*, 41 (3), 122–139, 2004.
29. Paik, J.K., Thayamballi, A.K., and Lee, J.M., Effect of initial deflection shape on the ultimate strength behavior of welded steel plates under biaxial compressive loads, *J. Ship Res.*, 48 (1), 45–60, 2004.
30. Paik, J.K., Lee, J.M., and Ko, M.J., Ultimate compressive strength of plate elements with pit corrosion wastage, *J. Eng. for the Maritime Environment*, 217 (M4), 185–200, 2003.
31. Paik, J.K., Lee, J.M., and Lee, D.H., Ultimate strength of dented steel plates under axial compressive loads, *Int. J. Mech. Sci.*, 45, 433–448, 2003.
32. Mansour, A.E. and Hovem, L., Probability based ship structural safety assessment, *J. Ship Res.*, 38 (4), 329–339, 1994.
33. Loukakis, T.A. and Chryssostomidis, C., Seakeeping series for cruiser stern ships, *SNAME Trans.*, 83, 67–127, 1975.
34. Paik, J.K. and Faulkner, D., Reassessment of the M.V. *Derbyshire* sinking with the focus on hull girder collapse, *Marine Technol.*, 40(4), 258–269, 2003.
35. Paik, J.K. and Thayamballi, A.K., Ultimate strength of ageing ships, *J. Eng. Maritime Environ.*, 216 (M1), 57–77, 2002.
36. Guedes Soares, C. and Garbatov, Y., Reliability of maintained, corrosion protected plates subjected to nonlinear corrosion and compressive loads, *Mar. Struct.*, 12, 425–445, 1999.
37. Qin, S.P. and Cui, W.C., Effect of corrosion models on the time-dependent reliability of steel plated elements, *Mar. Struct.*, 16, 15–34, 2003.
38. Paik, J.K., Lee, J.M., Hwang, J.S., and Park, Y.I., A time-dependent corrosion wastage model for the structures of single-and double-hull tankers and FSOs and FPSOs, *Mar. Technol.*, 40 (3), 201–217, 2003.
39. Paik, J.K., Lee, J.M., Park, Y.I., and Hwang, J.S., A time-dependent corrosion wastage model for bulk carrier structures, *Int. J. Maritime Eng.*, 145 (A2), 61–87, 2003.
40. Broek, D., *Elementary Engineering Fracture Mechanics*, Martinus Nijhoff, Dordrecht/Boston/Lancaster, 1986.
41. DNV, Fatigue Assessment of Ship Structures, classification notes 30.7, Det Norske Veritas, Høvik, Norway, January 2001.
42. IMO, SOLAS XI/2, Recommended Longitudinal Strength, MSC.108(73), Maritime Safety Committee, International Maritime Organization, London, 2000.

40
Risk Assessment and Reliability-Based Maintenance for Large Pipelines

40.1	Introduction	**40**-1
40.2	General Considerations	**40**-2
40.3	Selection of Adequate Analysis Schemes for Pipeline Segments	**40**-3
40.4	Basic Theory of Pipeline Systems Quality, Reliability, Risk Control, and Maintenance Optimization	**40**-4
40.5	Methods of Bringing Pipeline Safety Problems down to Reliability Problems	**40**-8
40.6	Method of Assessing the Size of Damage Due to Pipeline Failure	**40**-10
40.7	The Central Maintenance Problem as Applied to a Pipeline Segment	**40**-11
40.8	Role of Diagnostics in Reliability Monitoring	**40**-13
	Identification of the Pipeline State • Identification of the Defect-Growth Processes • Inspection of Defect Growth	
40.9	Optimal Cessation of Pipeline Segment Performance	**40**-17
40.10	Merger of the CMP with the Optimal Cessation Procedure	**40**-21
40.11	The Basic Block-Module Model of Pipeline Risk Control	**40**-22
40.12	Risk-Based Inspection/Maintenance	**40**-24
40.13	Risk Optimization Flowchart (RAFT)	**40**-25
40.14	Application of MAST and RAFT Procedures	**40**-28
40.15	Prioritizing Pipeline Segments for Repair: A Real Case Study	**40**-31
40.16	Conclusion	**40**-35

Sviatoslav A. Timashev
Russian Academy of Sciences

40.1 Introduction

The pipeline industry is critically important in all countries that produce and transport oil/gas and liquid chemical products. It is also a potentially dangerous industry because it comprises high-energy containment systems, failures of which can lead to casualties, pollution of the environment, and great economic losses. Pipelines may also be prime targets for a terrorist attack. Pipeline companies must therefore have a comprehensive, sophisticated maintenance system in place, utilizing the best of current science and technology. Such maintenance systems provide for pipeline integrity while also providing the added benefit of driving down the overall expenses.

Existing maintenance management systems are of three main types: (1) qualitative, based on subjective pipeline index [1]; (2) based on relative-risk numbers for prioritizing pipeline segments for inspection/maintenance [2]; and (3) based on previous statistics [3, 4] without accounting for intrinsic individual properties of each pipeline in consideration. These systems do not take into account safety levels that have to be assured, or if they account for them at all, they consider them as constraints. In the latter case, their values are not given.

The problem of ensuring the safety of large, potentially dangerous systems (LPDS) is increasingly drawing the attention of scientists, industry people, and governmental bodies, especially after each large-scale disaster with heavy consequences involving casualties, loss of ecological balance, and property losses. As a rule, all LPDS are vitally important or even strategic objects. A classical example of a civilian LPDS is a main oil-gas-product pipeline.

The ratio of maintenance and management costs to the initial cost for typical large machine aggregates and systems can be 5 to 12 and up to 100 for more complex PD systems. By accounting for such accompanying factors as fines, ecological losses, economic (technological, energy) losses, and decrease of state's political power, this ratio may exceed 100. Given these potential consequences, the development of scientifically proven methods of operating LPDS safely is becoming strategically important.

To solve the problem of safety of complex systems, an interdisciplinary, cross-functional approach is needed. It has to include fundamental and applied research along with a worldwide analysis of safe operation of large pipeline systems.

The use of risk management in the pipeline industry is in its early stages. There are five leading causes of pipeline accidents: excavation (or other outside source of pipe damage), corrosion, pipe defects, and pipeline operator errors. Now we have to add terrorist attacks to our concerns. This chapter addresses the particular problem of assessing corrosion and cracklike pipe defects and their overall influence on the residual lifetime and maintenance optimization of a pipeline. Assessing these problems correctly provides:

- A comprehensive pipeline- and company-specific risk management program based on implementing the newest concept of integrity maintenance and risk management
- Disclosure of the role and current state of the art of quantitative assessment of pipeline defects (metal loss, cracks, and cracklike defects)

The risk-based pipeline integrity management program described in this chapter permits:

- Ranking pipeline segments in terms of risk
- Prioritizing and optimizing monitoring/inspection/maintenance/repair actions
- Defining an optimal frequency and timing of conducting the above activities

It also helps in:

- Quantifying the process of in-line inspection (ILI)
- Getting the most out of the inspection (having a "yardstick" for choosing the most cost-effective ILI instrument and provider)
- Getting the most precise assessments of the residual lifetime of pipeline segments

40.2 General Considerations

The ultimate goal of pipeline maintenance, including integrity and risk control, is to minimize overall maintenance/monitoring/inspection costs and breakdown consequences. In order to achieve this goal, some maintenance principles have to be used. There are three main principles of systems maintenance

[5, 6]. One is based on the notion of inherent resource, which is a random variable or function. The other two use reliability level or the current condition of the system.

Historically, pipelines were first used with respect to their normative resource (the least economical method). This uneconomical approach was chosen because pipelines are highly critical, and at that time they were not monitored. The breakdown of pipelines immediately leads to great economic loss (including pollution of the environment) and, sometimes, noneconomic losses, such as injuries and loss of human lives. The total admissible (normative) time of operation of each type of pipeline was chosen based on experimental results and theoretical analysis. When the admissible time of performance was used up, the pipeline segment was to be decommissioned and replaced by a brand-new piece.

Pipelines should not be utilized up to their reliability level because they are highly critical systems. In other words, they should not be allowed to fail in a manner that causes spill or pollution of the environment, injuries, or casualties. Therefore, reactive maintenance is the worst choice of maintenance for pipelines, although this approach is still used by a number of pipeline companies.

The overwhelming percentage of pipelines should be used with respect to their current condition and purpose of usage, which is the most progressive way of using not only machinery, but pipelines as well. Pipelines are maintained only when their condition becomes a hazard or when there is a warning issue. This is proactive maintenance. Among the different kinds of proactive maintenance—preventive, reliability based, condition based, predictive, etc.—the Integrated Optimal Maintenance IOM® based on reliability and residual life monitoring [6, 7] is the most suited (with some adjustments that take into account the main specifics of pipelines).

In this chapter, risk is considered as the product of the conditional probability of failure and the total cost of consequences of such failure. The scientific basis of solving the risk problem embraces a range of theories, including the theory of controlled random processes, reliability theory of large renewable systems, theory of monitoring physically unmeasurable quality parameters (such as conditional probabilities of failure, remaining lifetime) of complex systems, general theory of technical risk analysis, and human factor analysis.

It is important to start the process with a quantitative assessment of pipeline defects and maintenance optimization in the context of system approach to pipeline integrity. The remaining lifetime (RL) is a conditional random variable (RV), since it depends on many parameters random by nature and on a set of restrictions, rules, and decisions imposed by the pipeline performance and ILI technology. For a full description of a RV, it is necessary and sufficient to construct its probability density function (PDF).

By definition, RL is the random time of transition of a pipeline segment from its present state to a limit state. For pipelines, all limit states can be subdivided into deformation-type limit states (elastic-plastic deformation) and integrity-loss limit states (leak, rupture). For risk analysis (which is conducted throughout the chapter), the main limit state is the case of leak/rupture. In order to construct the RL PDF, it is necessary to have the following: degradation models, algorithms for calculating the conditional PDFs, a full group of future performance scenarios (in the form of event/decision/fault trees), and the relevant database. Maximum economic value is obtained by assessing the remaining lifetime (RL) for each individual defective pipe run.

40.3 Selection of Adequate Analysis Schemes for Pipeline Segments

Analysis of causes that may lead to leak and rupture of a pipeline shows that considerable quantitative input into the overall stress and strain state of a specific pipe run can be brought in by:

- Stresses and strains induced in a pipeline run during its manufacture, transportation, and construction
- Nonhomogeneity of its foundation
- Seasonal upheaval and sedimentation (settlement) of the underlying soil
- Macro- and microseismic forces
- Slope movement and slides
- Temperature-induced loss of longitudinal stability

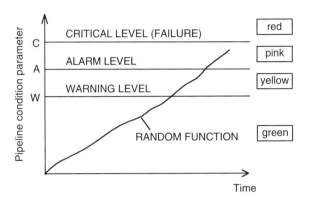

FIGURE 40.1 Levels of control parameters in maintenance (C and W represent a two-level maintenance strategy; C, A, and W represent a three-level maintenance strategy; colors in rectangles allow an operator to visually assess the current quality of pipeline performance).

Depending on the environmental specifics, each of the above components can be the main source of pipeline failure.

Pipeline reaction to external forces differs dramatically, depending on its actual design (overall geometrical form, materials, technology of manufacture and installation) and applied external forces. It is impossible to get an adequate estimate of the RL of the segment without using a precise-as-possible structural design scheme of the pipeline run that correctly accounts for the main specifics of the surrounding environment and product pumping technology. Pipe runs also have different kinds of defects that lead, by themselves, to different corrosion, fatigue, and fracture-mechanics analysis schemes. Given all of these considerations, it is clear that identifying the one design and analysis scheme that is the best fit for a specific pipeline segment among the plethora of alternatives is a difficult problem that can be correctly solved only by taking into account the results of overall stress and strain analysis, monitoring, inspection, testing, and diagnostics.

When predictive maintenance was first introduced, it did not stand up to the expectations [8] because there were no means created to define the levels of quality that should trigger the maintenance action (Figure 40.1). The comprehensive theory of how to assign the levels that trigger inspection/monitoring and maintenance was developed by Timashev and Copnov [5, 9, 10] and is briefly outlined below. Their work showed that maintenance can be treated as a control of random functions of deterioration and of structure renewal, with no restriction on the level of its complexity, which amounts to the control of its reliability.

40.4 Basic Theory of Pipeline Systems Quality, Reliability, Risk Control, and Maintenance Optimization

There is a strong correlation between structural reliability control and maintenance, and control of renewable pipelines can be achieved by maintenance. Optimal control leads to optimal maintenance and vice versa. Maintenance theory of pipelines and mechanical systems started developing in the late 1960s [5, 11, 12].

The overwhelming percentage of structures should be used with respect to their current condition and purpose of usage. Structures that are maintained only when their condition becomes a warning or hazard issue are examples of proactive maintenance. Among the different kinds of proactive maintenance — preventive, reliability based, condition based, etc. — predictive maintenance is the leading edge in contemporary maintenance theory and practice because it allows quantitative planning of future maintenance.

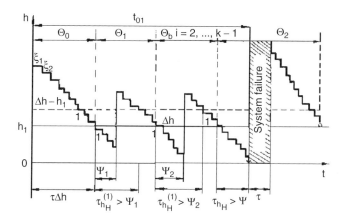

FIGURE 40.2 Diagram of system reliability control in terms of corrosion.

Consider the case when pipe wear (pipe wall and insulation erosion, corrosion damage, etc.) is controlled. The control is achieved by close interval surveys (CIS), intelligent pigging, ultrasound techniques (UT), etc. Assume that the corrosion of the insulation has the form of a random stepwise process and occurs step by step after random time intervals ξ. These intervals have the distribution $K(\xi)$ and the mean $\langle \xi \rangle$. An order for repairs (painting, reconditioning the protective layer) of scope Δh is initiated when the protective layer (or the pipe wall), as it thins, becomes equal to $h_1 > 0$. It is carried out during the random time ψ, which has the distribution function $L(\psi)$ and the mean $\langle \psi \rangle$. It stands to reason that $\Delta h \geq h_1$.

If, upon initiation of an order for repairs of scope Δh, corrosion destroys the h_1 layer before the order is implemented, the element fails. After failure, the pipe segment is restored during a random time τ, and corrosion again commences, reducing the wall and the protective layer thickness. The time τ obeys the distribution function $V(\tau)$ with the mean $\langle \tau \rangle$.

This corrosion process, illustrated in Figure 40.2, is a regeneration process, with the random quantities $\{\xi_i\}$, $\{\psi_i\}$, $\{\tau_i\}$, $i = 1, \ldots, \infty$ being independent of one another. Regeneration points of this corrosion process are the moments of initiation of lost-quality restoration orders.

The mean value of one generation cycle (the time span between adjacent regeneration points Θ) is given by Timashev [5] and is

$$\langle \Theta \rangle \leq \Delta h \langle \xi \rangle + q(h_1 \langle \xi \rangle + \langle \tau \rangle) \tag{40.1}$$

where

$q = 1 - p$
$p = P(\tau_{1,0} > \psi)$
$p = \int_0^\infty [1 - K_1(t)] dL(t)$

$K_1(t)$ is an h_1-fold convolution of $K(t)$.

The mean time prior to first failure $\langle t_{f_1} \rangle$ estimate is

$$\langle t_{f_1} \rangle = \Delta h \langle \xi \rangle / q - (\Delta h - h_1) \langle \xi \rangle \tag{40.2}$$

It follows from Equation 40.2 that when $q \to 0$, $t_{f_1} \to \infty$.

Considering the behavior of $\langle t_{f_1} \rangle$ with q tending to zero, one can ascertain that

$$P\{q \cdot t_{f_1} / \Delta h \langle \xi \rangle \geq t\} \to \exp(-t) \quad \text{when } q \to 0 \tag{40.3}$$

At small q, the reliability function is

$$R(t) \cong \exp[-qt/\Delta h \langle \xi \rangle] \qquad (40.4)$$

Equation 40.2 and Equation 40.4 explicitly relate the statistical characteristics of the corrosion process $[K(\xi), \langle \xi \rangle]$ and of the restoration method $[L(\psi), \langle \psi \rangle]$ to the control parameters $h_1, \Delta h$.

The steady-state value of the quality level (in the present case, the wall and protective layer thickness $h(t)$) is

$$\langle h(t) \rangle = \frac{\Delta h \langle \xi \rangle (\Delta h + 1)/2 + q h_1 \langle \xi \rangle (h_1 + 1)/2 - \Delta h \langle \psi \rangle + \tau_{\Delta h}}{\Delta h \langle \xi \rangle + q[h_1 \langle \xi \rangle + \langle \tau \rangle]} \qquad (40.5)$$

$$\tau_{\Delta h} = \int_0^\infty [1 - L(v)] K_{h_1}(v) dv$$

The cost of pipeline-segment + control function can be constructed in the following way. Introduce the cost of damage C_d due to failure per unit time; then assume that the rate of expenditure to maintain quality at level i is proportional to that level:

$$C_m = k_m \cdot i \qquad (40.6)$$

The cost of order to restore lost quality is $C(\Delta h)$. Then the rate of losses C, when controlling the reliability of a pipeline segment, is

$$C = \frac{C_d \langle \tau \rangle + k_m \langle X(\Theta) \rangle + C(\Delta h)}{\Delta h \langle \xi \rangle + q[h_1 \langle \xi \rangle + \langle \tau \rangle]} \qquad (40.7)$$

where $\langle X(\Theta) \rangle$ is determined by the following formula:

$$\langle X(\Theta) \rangle = \frac{\Delta h \langle \xi \rangle}{2}[\Delta h + 1] + q \frac{h_1 \langle \xi \rangle}{2}(2\Delta h + h_1 + 1) - \Delta h(\langle \psi \rangle + \tau_{\Delta h}) \qquad (40.8)$$

Equation 40.4 and Equation 40.7 permit computation and optimization of the reliability function of the pipeline segment, where its maintenance is considered as its operator.

The single most important case of pipe deterioration is crack (defect) propagation. Consider a crack whose depth $d(t)$ increases in time (passing from one state i to another $i + 1$) in random time intervals ξ. The crack also has a distribution function $K(\xi)$ and a finite mean $\langle \xi \rangle$. The moment that the increasing crack depth $d(t)$ reaches some level d_1, an order is initiated to repair the pipe segment (by eliminating the crack of d_1 depth). This repair occurs after a random time ψ, whose distribution function (DF) is $L(\psi)$ and whose finite average is $\langle \psi \rangle$. A case is possible when, upon initiating the repair, the crack depth grows and becomes critical (d_{cr}) before the repair occurs, thus resulting in a failure of the pipe segment (leak before break [LBB] or rupture). After the onset of pipe failure, it is restored during some random time τ, and the process of crack forming and propagation starts again, increasing from zero (there can be a latent time, when there is no visible crack while it develops at a submicrometer level). The time τ has a distribution function (DF) $V(\tau)$ and a finite average $\langle \tau \rangle$. A graphic interpretation of the situation outlined is presented in Figure 40.3.

In the conditions of the premises formulated above, the FCG (fatigue-crack growth) process $d(t)$ is a regenerative one, and the random quantities $\{\xi_i\}_{i=1}^{i=\infty}, \{\psi_i\}_{i=1}^{i=\infty}, \{\tau_i\}_{i=1}^{i=\infty}$ are independent of one another. Moments of initiating orders for pipe repair serve as regeneration points.

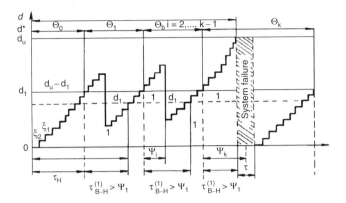

FIGURE 40.3 Diagram of system reliability control in terms of crack depth.

The mathematical expectation of a regeneration cycle Θ (time between adjacent regeneration points) is

$$\langle \Theta \rangle = p[(d_u - d_1)\langle \xi \rangle - \langle \zeta \rangle] + d_u \langle \xi \rangle + \langle \tau \rangle \tag{40.9}$$

where

$$p = \int_0^\infty [1 - K_{u-1}(t)]dL(t)$$

and K_{u-1} is a $(d_u - d_1)$-fold convolution of $K(t)$.

The evaluation of the mean time prior to first failure $\langle t_{f_1} \rangle$ is as follows:

$$\langle t_{f_1} \rangle = d_1 \langle \xi \rangle / q + p(2d_1 - d_u)\langle \xi \rangle \tag{40.10}$$

The reliability function when q tends to zero, which is typical of highly reliable systems with rapid order implementation and low dispersion, is

$$R(t) = P\{t_{f_1} > t\} \approx \exp[-qt/d_1 \langle \xi \rangle] \tag{40.11}$$

where the time prior to first failure is

$$T_1 = d_1 \langle \xi \rangle / q \tag{40.12}$$

It can be seen that the reliability function (Equation 40.11) explicitly relates the statistical characteristics of the crack depth $[\xi, L(\xi)]$ to the control parameters d_1, d_u. The steady-state value of the quality level, in this case that of the crack depth, is

$$\lim_{t \to \infty} \langle d(t) \rangle = \langle \xi \rangle d_1 (4d_u - 3d_1 - 1)/2 + \langle \xi \rangle q[(d_u - d_1)^2 - (d_u - d_1)]/2$$
$$- \langle \xi \rangle (\langle \psi \rangle - r_{u-1} p[(d_1 - d_u)\langle \xi \rangle - \langle \tau \rangle] + d_u \langle \xi \rangle + \langle \tau \rangle) \tag{40.13}$$

$$r_{u-1} = \int_0^\infty [1 - L(v)]K_{u-1}(v)dv$$

Now introduce the cost of damage C_d due to failure per unit time, assuming that the rate of expenditures to maintain quality at level i is proportional to the level $C_m = -k_m(i + b)$, that $k_m = b/d_{cr}$, and that the cost to restore lost quality is $C(d_1)$. Then the steady-state rate of losses when controlling the reliability of a pipeline segment with a crack will be

$$C = \frac{C_d \langle \tau \rangle - k_m \langle X(\Theta) \rangle + b + C(d_i)}{p[(d_i - d_u)\langle \xi \rangle - \langle \tau \rangle] + d_u \langle \xi \rangle + \langle \tau \rangle} \tag{40.14}$$

Equation 40.11, Equation 40.12, and Equation 40.14 permit computation and optimization of the reliability function for a controlled (monitored) pipeline segment. More-sophisticated models of corrosion and crack propagation can be found in the literature [9, 10].

If the degradation (corrosion) is a Markov death process, then the mean cycle value at a steady-state regime of pipeline performance can be expressed as

$$MX_c = M\Theta^{-1} \sum_{i=1}^{n} iM\omega_i \tag{40.15}$$

where ω_i represents intervals of a cycle Θ (phases of the process) in which $X_c(t)$ is equal to i.

The steady-state rate of expenditures, s, is

$$s(m, n) = M\Theta^{-1}[(C_1 + C_0 M\sigma)p + k_c MX_c + d(n - m)] \tag{40.16}$$

where C_0 is the cost of damage due to failure per unit time, C_1 is the cost of pipe segment restoration, k_c is a proportional coefficient, $d(n - m)$ is the cost of order to restore lost quality, and σ is the time needed for system restoration after its failure. The optimal parameters m and n are determined at the condition of minimum of Equation 40.16.

The reliability function corresponding to the optimal control by two-level policy (at small values of p) is

$$R_{opt}(t) = \exp[-pt(MW)^{-1}] \tag{40.17}$$

where MW is the mean value of a cycle without failure that corresponds to determined value of m and n by minimizing the result obtained from Equation 40.16.

The theory outlined in this section shows that maintenance can be treated as a control of random functions of deterioration and renewal of a system, with no restrictions on the level of its complexity. The theory allows one to find the system regeneration cycle, the mean time before first failure, the reliability function, the steady-state quality level and steady-state rate of generalized losses, and the optimal triggering levels, depending on the parameters of the two/three-level control (maintenance) policy. Here one sees for the first time that in order to correctly solve a maintenance problem, we have to simultaneously consider the probabilistic and economic issues of system usage. These ideas will be used below to define the central maintenance problem as applied to pipelines. But before doing this, we will have to learn how to translate a pipeline safety problem to a reliability problem.

40.5 Methods of Bringing Pipeline Safety Problems down to Reliability Problems

Risk analysis and reliability analysis of large pipeline systems are closely related problems. They have a similar scientific base and use the same mathematical tools (probability theory and theory of random functions). Given this similarity, there is a possibility of expanding the reliability problems to safety problems or vice versa.

To bring safety problems down to reliability problems, it is necessary to estimate the number of casualties/fatalities when a disaster or catastrophe develops, and then to assign an economical equivalent for human traumas, casualties, and fatalities (the "cost of life"). There are at least two general ways to assess the "cost of life" [5]. One of them is connected with statistical assessments of insurance sums or compensations for a person's death, paid by insurance companies, by court rulings, or by the state. The second method is by assessing the economic equivalent of noneconomical losses. In other words, assessing the statistical average of societal losses connected with medical treatment of a traumatized (injured) person and the discounted gross national product (GNP) that would have been produced by the deceased person before retirement.

Having estimated the average cost of an injury and life, it is possible to optimize the reliability (safety) function of a pipeline system. The problem comes down to optimizing the system reliability function, but an additional component — the economical equivalent of noneconomic losses — is introduced into the cost function C:

$$C_e(T) = \alpha \int_0^T \langle N \rangle S_m(t, T_0) \dot{R}(t) dT \qquad (40.18)$$

where α is the coefficient accounting for the time that people are staying in the zone of destruction, $\langle N \rangle$ is the mathematical expectation of the number of casualties, $S_m(t, T_0)$ is the economic societal loss from one casualty, $\dot{R}(t)$ is the derivative of the corresponding reliability function with respect to time; and T_0 is the baseline year.

The optimization problem is posed as follows [5, 13]. The reliability function $R(t) = \varphi_0(\chi, \chi_0, \chi_c, q)$ of the pipeline system and its cost function C are given. It is needed to find such a vector χ_0 such that

$$C = \varphi_0(\chi, \chi_0, \chi_c, q) = \min_{\chi_0} \qquad (40.19)$$

where χ is the vector of deterministic parameters of the system, χ_0 is the vector of the optimized parameters of the system, χ_c is the vector of the random parameters of the system, and q is the vector of forces and actions on the pipeline system.

By inserting the optimal value of χ_0 (the solution of the problem of Equation 40.19) into the expression for $R(t)$, one finds the optimal value of the reliability function, i.e., the one that delivers minimal cost and the corresponding optimal level of risk. In order to expand the reliability problem to the safety problem, it is necessary to consider the studied system together with its monitoring/inspection/diagnostics/maintenance/repair/rehabilitation subsystems. The global minimum of the cost of the pipeline system C_Σ together with its protection subsystem is sought as

$$C_\Sigma = C_k + C_d + S'T - \int_0^T \beta^{-t} s_d(t, T_0) \dot{R}(\langle \psi \rangle, \langle \eta \rangle, \langle \tau \rangle, t) dt \to \min_{\substack{p_w, p_a, p_u \\ \langle \psi \rangle, \langle \eta \rangle, \langle \tau \rangle}} \qquad (40.20)$$

Here C_k and C_d are the capital expenditures for creating the LPDS itself and the subsystem that is responsible for its safety; $s_d(t, T_0)$ is the cost of damage due to failure of the system; β^{-t} is the discount coefficient; S' represents the expenditures that are needed to keep the system's quality on a certain level; $\langle \psi \rangle$ is the average time needed to increase the system quality; $\langle \eta \rangle$ is the average time it takes for the system to degrade from the level $w(a)$ to failure level; $\langle \tau \rangle$ is the average time of liquidation of failure; T is the design lifetime of operation of the system; and p_w, p_a, p_u are correspondingly the warning, alarm, and ultimately permissible conditional probability of failure of the leak/rupture/explosion/fire type. The values $R(t)$ and C_k depend on $p_w, p_a, p_u, \langle \psi \rangle$; and C_d is a function of $\langle \psi \rangle$. Equation 40.20, apparently for the first time, is connecting directly the problem of system safety with the optimization of its maintenance infrastructure (including repair/inspection/diagnostics/monitoring facilities). Now it is possible to pose a large number of various risk optimization problems, depending on initial conditions. Most important, it is possible to correctly pose the problem of optimizing the reliability function for systems with noneconomical consequences (i.e., which failure can lead to casualties):

$$C_\Sigma(\chi_0, p_w, p_a, p_u, \langle\psi\rangle, \langle\eta\rangle, \langle\tau\rangle) = \text{const}$$
$$R_C(\chi_0, p_w, p_a, p_u, \langle\psi\rangle, \langle\eta\rangle, \langle\tau\rangle, R(t), t) = \max_{\chi_0, p_w, p_a, p_u, \langle\psi\rangle, \langle\eta\rangle, \langle\tau\rangle} \quad (40.21)$$

Using this kind of optimization problem does not require assessing and codifying risk, because solution of Equation 40.21 allows spreading optimally available resources while automatically achieving the maximum possible safety (for a given amount of resources). This problem can be also posed as follows:

$$C[R(t)] = \text{Const}, \qquad P[\tau_{wc} - \tau_{tr} > 0] \to \max \quad (40.22)$$

where the first equation (full cost of operation and maintenance of the pipeline as a function of its reliability and of parameters of control policy) is a restriction. The pipeline owner/operator usually knows how much he or she can afford to put into safety measures. The second equation maximizes the probability that the renewal process will require less time than for the degradation process to bring the system to a failure. This equation permits minimization of the risk of occurrence and the number of failures that involve casualties while avoiding direct assessment of the noneconomic damage altogether. In Equation 40.22

$$\tau_{tr} = \tau_d + \tau_p + \tau_r \quad (40.23)$$

where τ_{tr} is the total renewal time for a pipeline after the warning failure. It depends on the ability of the monitoring/diagnostics/renewal subsystems to reveal damage and repair it within the necessary time span; τ_d is the time needed to detect the alarm (warning) failure; τ_p is the time needed for preparation of the repair crew and the equipment for repair of the failure damage; τ_r is the time actually spent on performing the specific repair task (without stopping performance of the pipeline); τ_{wc} is the time needed for the risk function $r(t)$ to reach the failure level from the moment it reaches the warning level on the condition that the pipeline system during this time is not repaired (maintained). All components in Equation 40.23 are considered to be stochastic.

40.6 Method of Assessing the Size of Damage Due to Pipeline Failure

The cost of damage—the first component of risk—can be quantitatively assessed by computer simulation of a full group of scenarios of a structural failure developing into a full-scale pipeline disaster/catastrophe. The average cost of damage is obtained by averaging the computer results over a set of realizations.

The main ideas behind the method of risk analysis described below are presented in the literature [14–16]. The method is based on sophisticated and comprehensive simulation of scenarios of development of such failures as leak/rupture/explosion/fire into a disaster/catastrophe, including all the technical, economic, and ecological consequences and the number of injuries and casualties. The level of detail of the breakdown depends on the level of desired precision of the damage assessment.

Failure is considered as a physical process event that takes place through the accumulation of a certain number of defects (residual plastic deformation, corrosion pitting and metal loss, cracks, changes in the metal structure, etc.) without creating a critical failure situation, such as a partial or full fracture of the system leading to a critical failure (breakdown).

The second part of the problem includes investigation of not only the precritical failures, but also of the whole process of transition of the pipeline into a breakdown state. Here all the previous experience of operating such systems is utilized.

The failure simulation is conducted in several steps. In the first step, the mechanism of the pipeline failure developing into a full-blown catastrophe is studied. Generally it consists of a chain of events: fracture of a critical pipeline component, development of a breakdown situation, damage of the environment, and fatalities/casualties. This analysis involves solution of a chain of interrelated problems of fracture, diffusion, combustion, spread, filtration, etc. As a result, zones of destruction (ZOD) (location, form, size) are found, level of damage of live and nonlive objects in the ZOD is characterized, and the probability of such destruction is evaluated. In step two, the economic losses due to pipeline breakdown are calculated as well as the number of injuries and casualties. In step three, the considered pipeline breakdown is repeatedly simulated with different realizations of input random variables; for each realization, the level and size of damage of the objects located in the ZOD is assessed. In the fourth step, the magnitude of economic and noneconomic damages for each realization is evaluated. In step five, by averaging over the set of realizations, the statistical parameters of damage size are calculated. In step six, generalized figures of damage size are derived (for coding purposes) by averaging over an ensemble of similar pipelines.

This method increases the precision of each of its components without loss of generalization. With further development of fracture mechanics and penetration theory; with stricter solutions of the problems of combustion, spread, filtration, etc.; and with accumulation of statistical information on repair/rehabilitation costs, it is possible to obtain more precise assessments of damage.

40.7 The Central Maintenance Problem as Applied to a Pipeline Segment

Throughout the whole life cycle of a pipeline, it is subjected to two mutually antagonistic random processes: the degradation process and the renewal process. The degradation process is a combination of a continuous random process and a discrete-step-like random process and depends on the natural and technological environment in which the structure (in our case, a pipeline) operates. The renewal process is a discrete random process and depends on maintenance policy (strategy), quality of inspection/diagnostics, and quality of repair/maintenance. Thus the renewal process largely depends on the human factor: repair crew skill, diagnostician qualification, and decision-making person (DMP) leadership skills/abilities. The interaction of the degradation and renewal processes creates a regeneration process. Its analysis, performed by using the renewal theory and the theory of regeneration processes, allows one to solve the so-called central maintenance problem.

The typical central maintenance problem as applied to a pipeline segment is shown one-dimensionally for the sake of simplicity in Figure 40.4, where the segment's condition or reliability parameter is deteriorating in time. In order to implement this method of maintenance, it is necessary to know how the pipeline parameters change with time, i.e., to monitor them. Maintenance can intervene when the parameter reaches the warning (w), alarm (a), or failure (f) level. This provides for three levels of safety. The intervention boils down to

- Run-to-failure (the parameter reaches level f with no maintenance, and no w or a levels are needed, Figure 40.4a). In this case, the segment's maintenance consists merely of replacing the failed unit, which requires time m_1^f, a random variable that depends on many factors, including maintenance crew skills.
- Action when parameter crosses level w or a. This is reactive preventive segment maintenance, after which the parameter is brought again into the safe space below level w (Figure 40.4b). Here maintenance is more complicated. Note that $m_1^a < m_1^w$ due to the fact that corrective actions after crossing level a are or, should be, faster than when crossing level w.

Predictive maintenance involves (a) having an algorithm that can predict, with reasonable accuracy, the future behavior of the random function of the parameter, (b) knowledge of the times of crossing level w or a with a high degree of accuracy, and (c) the ability to initiate actions when the parameter

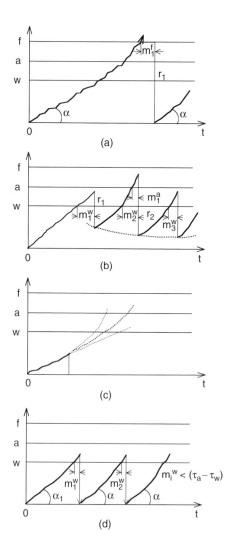

FIGURE 40.4 Graphical interpretation of the central maintenance problem.

reaches level w or a. From the above, one can conclude that the CMP embraces all of the three main types of maintenance (condition-based, reliability-based, and allowed-time-based). Figure 40.4c demonstrates usage of proactive (predictive) maintenance by means of close monitoring of the pipeline condition parameter.

Continuous monitoring of a pipeline creates a swiftly accumulating database that is invaluable for:

- Actuarial analysis (evaluation of past condition)
- Diagnostics (current condition)
- Prognosis (future condition)

It also provides information for assessment of reliability and residual life. Having accumulated a significant amount of data, it is possible to predict, with reasonable accuracy, the future behavior of the pipeline segment.

The model outlined above is able to take into consideration both predictable (i.e., corrosion) and unpredictable damage to pipelines (sudden one-time third-party interactions). In the latter case, the pipeline must be equipped with watchdog systems that monitor and diagnose on-line such kinds of impacts. Such systems are now being developed (see, for instance, Timashev [28], [29]).

40.8 Role of Diagnostics in Reliability Monitoring

The goal of diagnostics in reliability and residual-life monitoring is to evaluate as precisely as possible the current condition of the pipeline segment. In other words, diagnostics minimizes the entropy of the information about the current state of the pipeline (PL). Contemporary diagnostics usually give qualitative judgments, i.e., identify which part of the pipeline segment is damaged. These methods are still not comprehensive enough to give a qualitative assessment of the amount of damage.

Early diagnosis, as compared with conventional diagnosis, expands the time needed for analysis of the situation and for action. The warning and alarm levels that are assigned based on experimental results and theoretical analysis serve the same purpose. Early diagnostic techniques allow more time for continuous monitoring of the evolving condition of the pipeline. The ultimate goal is individual prognosis of the pipeline as a whole, of a pipeline segment, or of a pipeline cross section in which a defect or a set of dangerous defects is located.

40.8.1 Identification of the Pipeline State

The goal of identifying the pipeline state can be reached only if the pipeline damage (defects) is correctly and quantitatively diagnosed (i.e., assessed and described). The state of performing pipelines is usually identified by direct measurements using NDI (nondestructive inspection) and NDT (nondestructive testing) methods.

Identification of a performing pipeline is not like machinery identification, mainly because a pipeline is a distributed large system and all the processes are quasi-dynamic, with no inertia terms to be accounted for. Most often it is not economically feasible to make special identification tests or to dismount the system in order to obtain numerical values of some important parameters of the system mathematical operator.

Pipeline identification can be classified as *local* identification and *general* identification. General (or total) identification deals with identification of the general design scheme of the performing pipeline section (possession of geometrical, physical, or structural nonlinearity; beam or cylindrical shell span; type of supports; constraints on boundaries; types of loading and their parameters; material properties; soil reaction; residual stresses; etc.). Local (or partial) identification deals most often with only specific cross sections of the pipeline (macro- and microgeometric characteristics, loading parameters, material properties). To achieve a precise lifetime assessment, usually both types of identification should be performed.

The real geometry of the pipeline can be extracted from design blueprints or found by direct measurements and special GPS (global positioning system) field survey. The crack sizes and their distribution are found using intelligent pigging. The cycling loading parameters are obtained through direct measurements.

Pipeline material properties are identified by using material certificates provided by the contractor, by NDT, or by destructive testing of material specimens taken directly from the performing pipe. Although the latter method is the most precise, it is also the most difficult and expensive to perform. It should be noted that all measurements bring inevitable instrument errors, which are stochastic by nature. Due to the fact that cracks are of random nature, one should work with their probability density functions (PDF). The latter are constructed using statistical data gathered by relevant measurements. Other factors that give their input into the PDF of crack sizes are the instrumental bias and the interpretation errors, which could be considerable.

Thus, local identification is done by direct measurements using NDT and NDI and material destructive testing. The output of the local identification should be in the form of probability density functions PDF or, at least, their first (four) moments. The general identification procedure is much more complicated (see, e.g., Malyukova and Timashev [17]).

40.8.2 Identification of the Defect-Growth Processes

The procedures outlined above dealt with models that describe the stress and strain state of the pipeline. There is, however, the degradation process (here, the FCG process) that should also be identified. Because of the uniqueness of the solution of a Cauchey problem, which usually describes a defect growth in time, it seems that there is no need to have any kind of such identification procedure. But the evolution of the crack shape influences the FCG curve in a stochastic manner.

The algorithms are usually built on the assumption that the initial crack originated on the pipe wall instantaneously without any prehistory. Actually, during intelligent pigging or direct assessment of a performing pipeline, one discovers (diagnoses) a crack that is a result of some history of loading, usually unknown to the inspector (diagnostic operator). Therefore, it is not the initial crack at $N = 0$. It is a crack with parameters somewhere on one of the curves after $N(t)$ load cycles. Therefore, in order to estimate the residual longevity, an identification procedure first should be worked out and used.

It is obvious that this identification procedure has to compare the real-life curve with the theoretical ones. A section of the real curve can be constructed only if a crack-growth-monitoring procedure is established, so that a part of the "$b/a - N$" curve could be constructed. Having this part of the curve, it could be extrapolated backward to the zero moment of time and forward to the time N that corresponds with leak/rupture. Using this technique, the type of deterioration can be identified. Other methods of stochastic FCG identification are given by Bogdanoff and Kozin [18].

In conventional problems, the future loading and environmental impact is predetermined (given), because it greatly simplifies the solution. In real-life problems, one should construct a full set of possible scenarios of future performance of the pipelines in consideration. This is done using a set of event and decision trees (EDT) for each type of possible failure and for every pipeline segment. These in their turn, can be constructed if the company has a deep enough vision of the worldwide competition, of product (oil, gas) consumption and future prices, and of pipeline current quality of performance, and if the company has worked out its inspection and maintenance principles and policies (Figure 40.5 through Figure 40.7).

According to the outlined graphic schemes of pipeline performance under monitoring/inspection/repair/replacement, it is most convenient to use probabilistic measures of system performance quality, namely conditional probability of failure $p_c(f)$ with respect to a certain set of criteria (crack propagation, buckling, plastic deformation, leak before break [LBB], rupture, etc.). This parameter is also chosen because it is one of the main components in risk analysis.

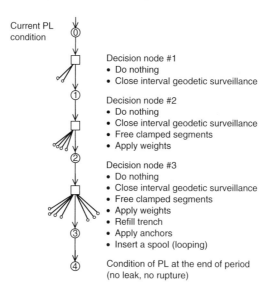

FIGURE 40.5 Event decision-tree analysis: case of general buckling.

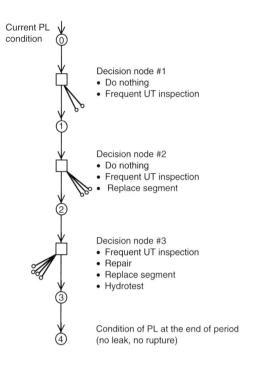

FIGURE 40.6 Event decision-tree analysis: case of FCG.

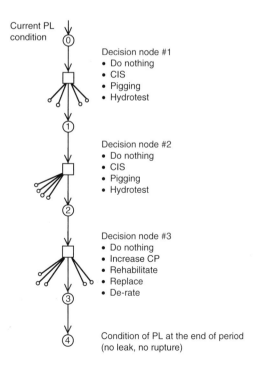

FIGURE 40.7 Event decision-tree analysis: case of external corrosion.

Following Figure 40.4, one can observe the following:

- Parameter α is complex because it connects pipeline output, pumping regime, and environmental impact (e.g., rate of corrosion, fatigue-crack growth [FCG], stress corrosion cracking [SCC]). After each regeneration cycle, the value of the parameter α_i may change in a random way, depending on actual deterioration of the pipe and on the scenario of its future usage. Therefore α is a RV and has a PDF, which will be here denoted as $f_\alpha(\alpha)$.
- The times m_i reflect the ability of the maintenance system to perform after the need for maintenance action (repair/replacement) is triggered by levels w, a, or f. The times m_i are also complex because they depend on the financial and technical ability of the maintenance crew to provide needed maintenance actions. Times m_i are obviously also RVs, and their PDF will be denoted as $f_m(m)$.
- Parameter r describes the quality of repair/correction and is also a RV. Nondestructive evaluation (NDE) should be performed after each repair in order to assess the variance of r and to avoid cases where the pipeline is worse off after rehabilitation.

The monitoring system and maintenance crew should have as their goal the monitoring of the levels w, a, and f that trigger relevant maintenance actions. Their ability to do this depends largely on the level of precision of the instruments used, on the level of sophistication of the algorithms involved, and on the cost of monitoring the physically unmeasurable parameter, $p_c(f)$.

One of the bedrock assumptions in the described theory is that the conditional probability of failure is calculated for an individual pipeline. Consider one regeneration cycle of pipeline performance (Figure 40.8).

Due to the fact that the processes of degradation in pipelines are generally relatively slow, the cycle can embrace several years (2 to 5 years and more). During these years, some routine monitoring, surveillance, and inspection/diagnostics are done. All of these monitoring/inspection procedures serve only one purpose: to enhance the precision of measurements and thus the assessment of current technical condition of the pipeline and its conditional probability of failure. It is obvious that the regeneration cycle can be formed in a number of ways that make sense. They will differ in cost and in the amount of accumulated information. Therefore, the best cycle is the one that gets the necessary and sufficient amount of information for making a right decision at the minimum cost.

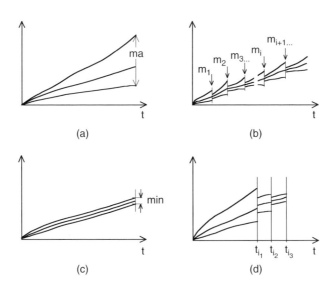

FIGURE 40.8 Effect of monitoring/inspection/testing on precision of assessing pipeline current condition: (a) no monitoring/inspection, (b) periodic monitoring/inspection/testing, (c) continuous monitoring/inspection, (d) sequence of different technique inspections that produce more-precise assessment of current pipeline condition.

40.8.3 Inspection of Defect Growth

The combination of inspection for defects (intelligent pigging) and prediction of crack growth using linear or nonlinear fracture mechanics (NLFM) allows one to predict the conditional probability of failure (CPF) (leak/rupture) in pipelines leading to spill, fire, explosion, environmental pollution, casualties, and other safety implications. Magnetic flux leakage (MFL) and ultrasound inspection identify internal and external cracks, their location, and their size. NLFM will then predict how the clusters of identified defects of different types will grow in time. This will allow one to assess the conditional probability of failure (leak/rupture) as a function of time. Usually computer programs are used for this purpose. They allow for the present crack size, far-field stress and prevailing stress field at the crack, material properties and temperature, environmental impact (corrosion: external, internal, uniform, and local), and type of cyclic loading [19–22]. The procedure of updating the probabilistic parameters of the FCG process separately for each type of defect by using NLFM and sequential multistep Bayesian estimate of a defect-type PDF is given in Figure 40.9.

For pipelines, failure mechanisms other than crack propagation must be considered in predicting CPF: coating degradation, corrosion (internal, external, SCC), corrosion fatigue, local buckling (wrinkling), overall buckling, local plasticity, mechanical damage by third party, etc. The technology of inspecting pipeline for condition relative to some of these failure mechanisms (buckling, plastic hinge, coating discontinuity, etc.) is less precise than in the case of crack propagation using the high-resolution (HR) MFL pigging. Use of algorithms to predict residual resource, based on pipeline operating conditions, has seen some success (see for instance, Malyukova and Timashev [19–21]).

40.9 Optimal Cessation of Pipeline Segment Performance

Currently, the two-level policy model of risk control is the one most often used for pipelines. This model relates to two main levels of safety. For many pipelines, two layers of safety are not enough, and in these cases it is more appropriate to use the three-level policy (warning, alarm, failure). To further maximize the probability of preserving pipeline integrity, it is expedient to additionally use the optimal cessation of performance that, in its turn, leads to four main layers of safety. The problem of optimal cessation of nonobservable processes is posed as follows [23].

The prerequisite information is: history of loading; information about initial properties of the material and design of the pipeline's critical components and how they change in time; relevant theory of damage accumulation; method of prognosis of remaining life; specific operation costs function; and an optimal cessation rule for a random sequence of realizations of the mathematical expectation of specific expenditures on each cycle of pipeline performance.

Denote the cost of preventive cessation of operation as $(c + m)$ and the cost of cessation due to pipeline failure as d (where m is the cost of monitoring the technical state of pipeline, c is the cost of basic maintenance), and $d > (c + m)$. Restricting ourselves to the case when the pipeline operation can be shut down only in discrete moments of time t_i ($i = 1, 2, \ldots$), it is natural to define the moment of optimal cessation by minimizing the average intensity of losses (expenditures) that are related to the pipeline operation. These expenditures at the moment of cessation can be written as

$$\gamma_i = \sum_{j=1}^{i} \frac{c+m+d}{t_j} I\{t_{j-1} < T \le t_j\} + \frac{c+m}{t_j} I\{T > t_i\} \tag{40.24}$$

where T is current time, and $I\{\bullet\}$ is the indicator function of the event. Thus, the problem comes down to defining the moment of cessation of a random sequence (Equation 40.24).

In order to define the optimal cessation rule, introduce the probability space (Ω, F, P). All points of this space $\omega \in \Omega$. Furthermore, it is assumed that the probability pairs (γ_i, F_i) $i = 1, 2, \ldots$ form an integrable stochastic sequence of σ-algebras, such that $F_i \subset F_{i+1} \subset F$ and γ_i are measurable with respect

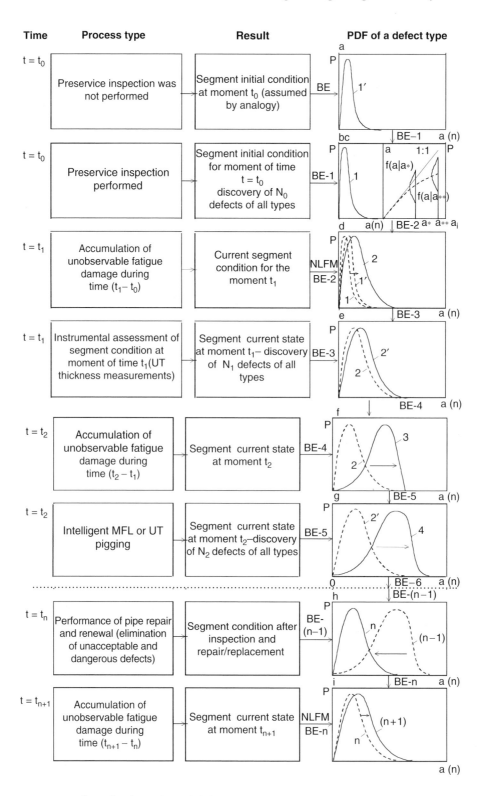

FIGURE 40.9 Procedure of updating the probabilistic parameters of the FCG process on a pipe segment by using regular inspections, nonlinear fracture mechanics (NLFM), and sequential multistep Bayesian estimate (BE) of defect PDF.

to F_i, and the mathematical expectation $E(\gamma_i)$ exists and is finite. F_i includes all the sets of all possible events that can occur up to the time t_i. For instance, events that are described in Equation 40.24 by indicator functions belong to F_i. In this case, measurability of γ_i means that up to the moment t_i, the monitoring system is giving unbiased information about a failure being present/absent.

Using the martingale approach, define the Markov moment of cessation as a random variable τ such that for each i

$$\{\tau \leq i\} \in F_i, \quad P\{\tau < \infty\} = 1 \tag{40.25}$$

e.g., τ is measurable on F_i and finite.

Therefore it is necessary to find such a rule at which

$$E(\gamma^*) = \inf_\tau E(\gamma_\tau) \tag{40.26}$$

According to Chow et al. [24], for the case when degradation is described by a random nondecreasing process up to the moment of making a decision, there is a stochastic sequence (γ_i, F_i) that satisfies the following condition for each i:

$$E(\gamma_{i+1}|F_i) \geq \gamma_i \rightarrow E(\gamma_{i+2}|F_{i+1}) \geq \gamma_{i+1} \tag{40.27}$$

where $E(\gamma_{i+1}|F_i)$ is the mathematical expectation of γ_{i+1} relative to F_i.

Now the rule of optimal cessation can be written as

$$\tau = \{\text{the least } i, \text{ such that } E(\gamma_{i+1}|F_i) \geq \gamma_i\} \tag{40.28}$$

Apply Equation 40.28 to the considered case. Write γ_i in the following form:

$$\gamma_{i+1} = \gamma_i + \frac{c+m+d}{t_{i+1}} I\{t_i < T \leq t_{i+1}\} + \frac{c+m}{t_{i+1}} I\{T > t_{i+1}\} - \frac{c+m}{t_i} I\{T > t_i\} \tag{40.29}$$

According to the rules of calculating the conditional mathematical expectations,

$$E(\gamma_{i+1}|F_i) = \gamma_i + \frac{c+m+d}{t_{i+1}} P\{t_i < T \leq t_{i+1}\} + \frac{c+m}{t_{i+1}} P\{T > t_{i+1}\} - \frac{c+m}{t_i} P\{T > t_i\} \frac{I\{T > t_i\}}{P\{T > t_i\}} \tag{40.30}$$

From this follows

$$E(\gamma_{i+1}|F_i) = \gamma_i + \frac{c+m+d}{t_{i+1}} P\{t_i < T \leq t_{i+1}\} + \frac{c+m}{t_{i+1}} P\{T > t_{i+1}\} - \frac{c+m}{t_i} P\{T > t_i\} \frac{I\{T > t_i\}}{P\{T > t_i\}} \tag{40.31}$$

$$E(\gamma_{i+1}|F_i) \geq \gamma_i \leftrightarrow R_i = \frac{P\{t_i < T \leq t_{i+1}\}}{P\{T > t_{i+1}\}} \geq \frac{c+m}{di} \tag{40.32}$$

Now the optimal cessation rule takes the form

$$\tau^* = \left\{ \text{the least integer } i, \text{ such that } R_i \geq \frac{c+m}{di} \right\} \tag{40.33}$$

where R_i is the conditional probability of failure.

The obtained solution is easily generalized for the case when pipeline degradation is described by a sequence of degradation processes that differ by the velocity of deterioration due to changes in operation technology and environment. These processes can be directly or indirectly measured or are the product of solution of corresponding problems of fracture mechanics, wear, etc.

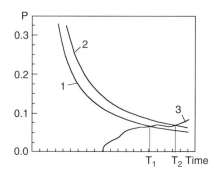

FIGURE 40.10 Graphical presentation of the optimal cessation procedure: (a) ultimate permissible probability of failure; (b) ultimate permissible probability of failure taking into account the cost of monitoring; (c) individual cumulative conditional probability of failure due to cycle loading (fatigue).

All of the above is related to operational risk control of one of pipeline critical component. Existing methods of engineering analysis of large systems allow reducing all PDS to a relatively small number of critical components. By monitoring their behavior in parallel in time (a monitoring system is needed for this), it is possible to apply to each critical pipeline component the optimal cessation procedure. As a result, one gets the following set of cessation moments:

$$\tau_k^* = \left\{ \text{the least integer } i_k \text{ such that } R_{i,k} \geq \frac{c_k + m_k}{d_k i} \right\} \tag{40.34}$$

where $k = 1, 2, \ldots, N$, and N is the total number of critical components in the considered pipeline. Note that in the general case, the parameters c_k, m_k, and d_k will be different for each kth pipeline critical component.

The above-bedrock result of the solution as applied to a pipe-run unit can be put as follows:

A pipe run must be shut down at the first moment of time, τ, when the conditional probability of its failure is equal to the ultimate permissible probability of failure (see Figure 40.10). In mathematical terms,

$$\tau = \{\text{first } t_n, \text{ such that } (1 - R_n) \geq (m + c)/d \cdot n\} \tag{40.35}$$

where c is the intensity of normal maintenance expenses (visual inspection, etc.); m is the intensity of monitoring expenses (various NDE inspection, etc.); d is the breakdown expense (pipeline damage, spill, leak, rupture, casualties, environmental pollution, etc.) in the current year; n is the number of time cycles of pipeline performance (usually, in years); $(m + c)/d \cdot n$ is the ultimate permissible failure rate $p_{c,u}$ as a function of time; and $(1 - R_n)$ is the conditional probability of pipeline failure on the condition that at cycle t_{n-1} of performance it did not fail. Actually, pipelines are not shut down. The optimal cessation time indicates the safe period of the pipeline segment operation. All the maintenance/repair should be performed before that time interval expires.

Analyze the ultimate permissible risk of failure $p_{c,u} = (c + m)/d \cdot n$. Note that as more is spent on monitoring, $p_{c,u}$ rises. Therefore, a higher risk of performance is justified, $T_2 > T_1$ (Figure 40.10, curve 2).

Using precursors of failure (like levels w and a) or monitoring minor failures (say, loss of coating integrity), one can start manipulating with larger $p_{c,u}$. For these types of failure, $c + m \geq d$ (maintenance and monitoring costs are greater than or equal to precursor failure expenses). When $c + m = d$, the ultimate permissible conditional probability of failure (CPF) curve is $1/n$. If $c + m > d$, no cessation of performance or inspection is needed while $p_{c,u} \geq 1.0$. For such pipeline failures as rupture, large plastic deformations are usually $(c + m)/d \approx 10^{-1} \ldots 10^{-5}$ and less.

40.10 Merger of the CMP with the Optimal Cessation Procedure

The theory outlined above illustrates the general concept of how to find the "triggering" values of parameters, which provides the possibility of controlling the system reliability (and risk of performance) at minimal expense. As was noted above, the same logic can be applied to the case when the probability of "failure" is used as the triggering parameter (Figure 40.11b). It can be shown that there is a unique relation between the relevant physical parameters and the conditional probability of failure, but the latter, being complex, allows one to convolute multiparameter systems to one-parameter systems, thus avoiding the "curse of dimension" (Figure 40.11a, b).

In Figure 40.11, the $p_{c,u}$ is shown as a constant. In reality, as it was shown in the previous section, $p_{c,u}$ is a decreasing hyperbolic function of time and depends on the ratio $(c + m)/d$.

It is expedient to merge the central maintenance problem (as applied to the probability parameter) with the optimal cessation procedure, as shown in Figure 40.12. In order to merge these two problems, the CMP should also be posed in the conditional probability of failure space.

This merger gives all the missing links in all predictive maintenance systems: (1) the triggering levels, (2) the ultimate permissible level of conditional probability of failure, and (3) optimal time between inspections. Having these components, one can organize the inspection/testing/maintenance procedure by calculating the relevant conditional probabilities $p_{c,w}$ and $p_{c,a}$ having the $p_{c,u}$ (hyperbolically diminishing function of time) as an upper moving boundary.

It is important to note that the growth of the p_c is controlled by the inspection procedures and different maintenance procedures (Figure 40.12, curve 1). For instance, such inspection procedures (IP) as high-resolution magnetic flux leakage (HR MFL), ultrasound techniques (UT), etc. allow one to diagnose the current condition of the pipeline (Figure 40.12, curve 1).

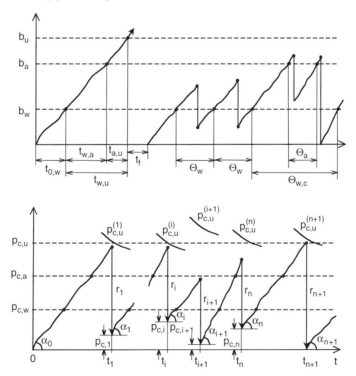

FIGURE 40.11 Unique relation between (a) physical failure parameters b_w and b_a and (b) probabilistic parameters $p_{c,u}$ (ultimate permissible conditional probability [failure probability]), $p_{c,a}$ (alarm level conditional probability), and $p_{c,w}$ (warning level conditional probability).

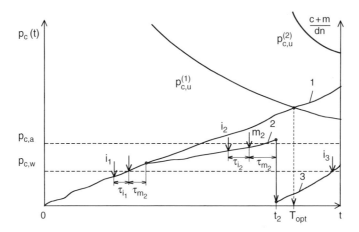

FIGURE 40.12 Central maintenance problem embedded in optimal cessation procedure (τ_{i1}, τ_{i2} = time of inspections; τ_{m1}, τ_{m2} = time of maintenance actions; vertical arrows indicate the starting time of inspection or maintenance).

Maintenance procedures (MP) can be divided into two categories. Such MPs as reduction of operational pressure, temperature, etc. reduce, after their implementation, the rate of deterioration and, hence, the accumulation of p_c of failure (Figure 40.12, curve 2). MPs like reconditioning, repair, replacement of a pipe run, etc. abruptly drop the p_c down to a level close to zero, depending on the quality of repair/replacement. After that, on the second cycle, p_c again starts growing (curve 3), but it is now ruled by the $p_{c,u}^{(2)}$ which, in its turn, starts from the moment t_2 that the maintenance action m_2 is fulfilled. Such multistep optimal cessation (MAST) procedures must "cover" the whole life cycle of the pipeline.

Calculation of the conditional probability of failure is done continuously in order to assess the time when the physically unmeasurable parameter p_c reaches levels $p_{c,w}$, $p_{c,a}$, and $p_{c,u}$. When no action is taken, p_c is calculated, using relevant algorithms with input taken from previous statistical data. When inspection and maintenance procedures are taking place, the p_c is calculated more precisely, taking into account the statistical results of IP and MP of the individual pipeline (run).

40.11 The Basic Block-Module Model of Pipeline Risk Control

Analysis of the three key problems (determination of the trigger levels, optimal cessation, and optimal time between inspections) of pipeline condition and maintenance control makes it possible to understand the place and the level of importance of different isolated problems related to stress and strain analysis, non-destructive evaluation and diagnostics, prognosis, reliability and remaining lifetime assessment, failure scenario analysis, monitoring technique in pipeline integrity, and maintenance control (Figure 40.13).

As a result of the merger described in Section 40.10, the basic block-module model of risk control is formed. The model makes it possible to:

- Solve all main problems of pipeline maintenance optimization using the risk criteria
- Unite and blend intrinsically different problems into a single algorithm of monitoring and risk control
- Assess the level of expediency of using each and every block and module relevantly toward the ultimate goal of maintenance control: provision of pipeline integrity and minimization of overall maintenance costs
- Create hardware and software systems that realize optimal maintenance (control) according to the developed theory

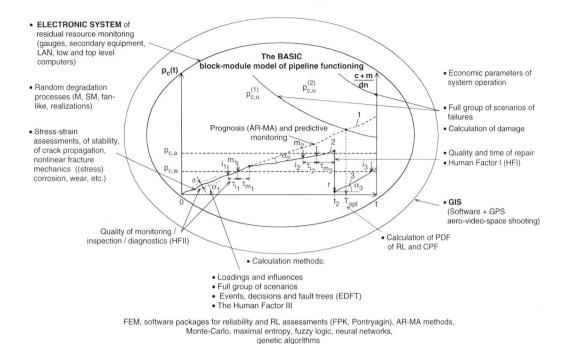

FIGURE 40.13 Monitoring system of residual resource and safety of pipelines.

The model consists of the following three main blocks: degradation assessment block (DAB), ultimate permissible conditional probability of failure block (UPFB), and renewal assessment block (RAB). Each block has its own peculiarities and models.

The degradation-assessment (DA) block comprises:

- Adequate algorithms for degradation assessment on relevant event and decision trees (EDT)
- Algorithms of accounting for the probabilistic influence of inspection/diagnostics results on the size and number of defects and, consequently, on the PDF of residual lifetime and the conditional probability of failure
- Prognostication algorithms

Note that the conditional CPF, the main parameter of risk, is at the same time that the convolution parameter allows convoluting the multiparameter problem into a single degree of freedom problem (together with the method of solving the reliability problem in the space of loads [5]).

The ultimate permissible conditional probability of failure (UPF) block, in its turn, convolutes all of the economic parameters and the economic equivalent of noneconomic parameters (cost of life) into one function.

The renewal assessment (RA) block contains all of the renewal components:

- Levels p_w and p_a, which trigger action of the protection system
- Parameters m^a, m^w, m^f, r, d, and α, which characterize the human factor (HF): HF1 = work force (repair team) (m^a, m^w, m^f, r); HF2 = diagnostician qualification (d); and HF3 = qualification of decision-making person (α)

This quantification of human skills is a new discipline that deals with the psychological and physiological aspects of human behavior and its influence on precision, accuracy, and speed of different task performance.

Concluding this section, one should say that implementation of the model described above requires practical fusion of heterogeneous data, which is not a trivial problem. This is generally achieved using methods that boil down all solutions of different problems and parameters to a single dimension.

40.12 Risk-Based Inspection/Maintenance

Risk can be expressed as a product of the conditional probability of failure and the consequences of the same failure:

Risk = Conditional probability of a given type of failure (CPF) × Failure (FC) Consequences

$$R = CPF \times FC \tag{40.36}$$

Therefore, risk = expected failure cost. The failure consequences include downtime cost, loss in efficiency, cost of repairing damage created by failure (including damage induced by environmental pollution), and casualties.

The main factors that raise the importance of risk-based methods in pipeline operation and maintenance are aging of pipes and worldwide competition. As pipelines age (most of them were constructed in the 1950s, 1960s, and 1970s), the conditional probability of their failure constantly increases (Figure 40.14). Strong competition limits gross revenue and profit, thereby changing the financial part of the equation.

Competition and aging create two competing financial pressures: a reduction in the cost of product transportation and an increase in the consequential cost of pipeline failure. Postponing maintenance contributes to profitability but increases the risk of failure when the cost of resulting failure starts to exceed the savings in reduced maintenance expenditures. Performing major maintenance too is a misuse of resources.

Conditional probability of failure (CPF) represents the traditional maintenance point of view. If a pipeline failure occurs, maintenance staff can expect the blame. As a result, maintenance staff are motivated to reduce CPF and seek to control or minimize the frequency of pipeline failures. This responsibility or motivation addresses one element of risk.

Failure consequence (FC) represents the perspective of operations staff. They are less concerned with CPF than with what effect failures have on operation and their ability to make the pipeline perform. The consequence of forced downtime for a particular pipeline segment will generally depend on time as a result of expected changes in market share, market volume, prices, etc. Unplanned forced downtime for a pipeline has the following associated costs: cost of repair, loss of profitable revenue, cost of providing alternative sources to meet a contract or avoid penalty, and intangible costs (perceived lack of reliability by regional authorities and clients).

Consequences vary from pipeline to pipeline with season and other measures of time relative to a business cycle. The consequences are also influenced by current demand for pipeline output and by revenue from pipeline output. When there is substantial overcapacity for current demand, the consequences may be small, even for critical pipelines, (Figure 40.14, curves 1, 2). When loss of certain pipelines directly translates into a reduction of capacity for which there is demand, it directly impacts the financial bottom line (Figure 40.14, curves 3, 4).

The third viewpoint (of the corporate CFO) may be sufficient for instances that combine the CPF and the consequences of that failure. Traditional engineers use a worst-case consequential cost for decision making. Advanced engineers use a probability-weighted cost (expected cost of forced downtime) and

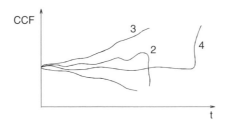

FIGURE 40.14 Conditional consequence of failure vs. time.

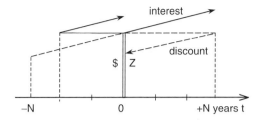

FIGURE 40.15 Interest and discount functions.

manipulate with both elements of maintenance risk. Combining risk with time yields a basis for communicating the value of maintenance decisions and their timing to those who control the corporate resources and decide on the best target for and timing of expenditure of money. Rate of growth used in analyzing the effects of time is typically called the discount rate and accounts for the expected or needed revenue from competing investments (Figure 40.15). By using the time value of money, one puts the engineering decision analysis on an equal basis with alternative corporate financial analyses that use money as the measure of value.

40.13 Risk Optimization Flowchart (RAFT)

CPF and risk in the absence of maintenance action (and the associated cost of that action) tends to grow with time (Figure 40.16, curve 1). At the point of maintenance action, there is a step drop in CPF. Maintenance action shifts curve 3 to the right and transforms it into a curve similar or equal to curve 2 (in dot ellipses). This was discussed earlier in other terms (Figure 40.2 to Figure 40.4, and Figure 40.12).

Curves 1, 2, 3 in Figure 40.16 introduce decision year as an optimization parameter. In analyzing the timing of maintenance action (MA), a series of alternative curves are to be compared on the basis of net present value NPV or some other acceptable corporate evaluation criterion. A decision analysis using this maintenance optimization addresses two issues: whether to take the maintenance action, and when to take the maintenance action.

A practical decision analysis and synthesis tool for pipeline maintenance optimization, which stems out of the theory described above, is the risk optimization flowchart (RAFT). The risk optimization flowchart is an array of connected nodes (Figure 40.17). RAFT organizes the flow of information from a decision or other inputs (multiple decisions), which form a full group of events (FGE) on the top of the chart to a single output on the bottom of the chart, namely net present value (NPV). Its construction

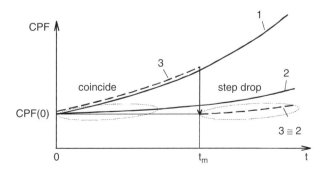

FIGURE 40.16 Influence of maintenance action on the growth of CPF and risk in time as a function of time of action: 1 = no maintenance action (basic function); 2 = maintenance action at time zero; 3 = maintenance action scheduled at time t_m.

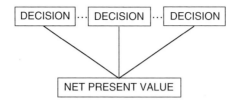

FIGURE 40.17 Risk optimization flowchart (RAFT).

is best accomplished by moving from the bottom (NPV) upward (to the decisions) in the "risk gravity" space. NPV is one of the most robust measures of value to the corporation. NPV is effective in optimization procedures and across multiple projects. However, its determination is more complicated than other measures, such as: benefit/cost ratio, payback period, internal rate of return (IRR) (which are effective in single projects in the yes/no decision making). In addressing timing and multiple competing projects with multiyear financial and safety constraints, NPV is a more accurate and effective measure. This type of economic measures is widely used in machinery maintenance [25]. Figure 40.18 shows the simplest case for RAFT, with

$$\text{NPV} = B - C \tag{40.37}$$

where B = benefit = expected cost of failure with no MA, and C = cost of failure with MA.

If one denotes C_{fna} as the cost of no action (expected failure cost without maintenance), $C_{f,a}$ as the cost of failure with action (expected failure cost with maintenance), and C_a as the cost of action (cost of maintenance), then

$$\text{NPV} = C_{fna} - (C_{fa} + C_a) \tag{40.38}$$

Equation 40.38 compares two scenarios with different outcomes (losses). Therefore, NPV > 0 means that MA yields a smaller loss and should be chosen as an alternative. The sole purpose of maintenance is to achieve a reduction in overall costs

$$C_{f,a} + C_a < C_{fna} \tag{40.39}$$

The year of maintenance action influences the failure cost and the cost of maintenance itself (time value of money, inflation, taxes, etc.).

RAFT can serve as a framework for organizing the problem and as a tool for problem definition and communication (see Figure 40.19). Assume that the following are given: period (depth) of analysis, N, years; interest rate r_g; inflation rate r_i; tax rate r_t; composite tax rate r_{ct}; conditional probability p_c curve

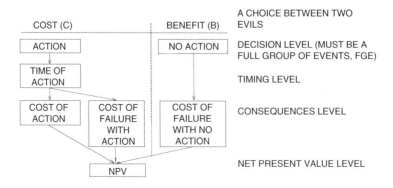

FIGURE 40.18 RAFT for the simplest case.

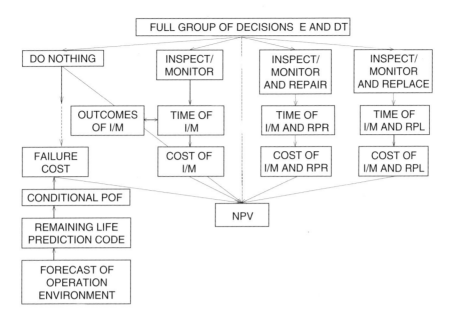

FIGURE 40.19 Dates of inspection/monitoring (I/M), repair (RPR), and replacement (RPL) prioritized for maximum net present value (NPV).

for a given type of pipeline failure without maintenance action and after maintenance action p'_c; average cost of different types of maintenance/repairs/replacement $C_m/C_{rp}/C_{rl}$ combinations (see Figure 40.5 through Figure 40.8); the discount factor $d_f = (1 + r_i)/(1 + r_g)$; and year M of maintenance action (repair/replacement), $M = 0, \ldots, N$. Then the expected year-by-year net increment/decrement of NPV for that year will be

$$\text{NPV}(M) = \sum_{i=0}^{N} C_{fna}^{(i)} d_i - \left(\sum_{i=0}^{M-1} C_{fna}^{(i)} d_i + C_m^{(M)} d_i + \sum_{i=M}^{N} C_{fna}'^{(i)} d_i \right) \quad (40.40)$$

where $C_m^{(M)}$ is the cost of maintenance that occurs in that year; $C_{fna}^{(i)}$ is the expected cost of pipeline failure in the year i on the condition that no maintenance is taken up to the year $M - 1$; and $C_{fna}'^{(i)}$ is the same, but after the year M of maintenance action.

$$C_{fna}^{(i)} = p_c^{(i)} \cdot C_f \quad (40.41)$$

$$d_f = (1 + r_i)/(1 + r_g) \quad (40.42)$$

The cash flow without maintenance action is given by the term $p_c \cdot C_f$, and the increment for an individual year is given by $p_c^{(i)} \cdot C_f$. The maintenance cost C_m is a relatively well-known quantity (through past experience, recent quotations for pipeline replacement, cost of unavailability of the analyzed pipeline)

$$C_m = M_p + C_u \quad (40.43)$$

where M_p is maintenance cost proper, and C_u is the cost of pipeline unavailability during maintenance action (it could be uncertain, for instance, in cases when the damage is discovered only after uncovering of the pipeline). The cash flow when maintenance action is taken is given by $p'_c \cdot C_f$, where $p'_c \ll p_c$, because the conditional probability of failure will be greatly reduced to the no-maintenance-action conditional probability curve when the pipeline was new (depending on the quality of repair, see Figure 40.4).

The actual cash-flow curve with maintenance action equals the cash-flow curve without maintenance action until the maintenance action year. It then drops down to the calculated curve with maintenance action cash flow after the maintenance-action year.

Equation 40.38 through Equation 40.43 allow one to make a decision about the timing of performing maintenance. This decision influences the year at which the maintenance expenditures will be incurred and when the jump between the two curves occurs for expected cost. Delaying this action benefits the maintenance cost and NPV. But at the same time, it keeps NPV connected to the higher conditional probability of failure without maintenance action. This situation brings up the conditional optimization problem. A maintenance budget limit (MBL) and a permissible-reliability limit R_u must be given in most maintenance optimization studies.

The constraints are the given amount of resources for inspection/maintenance actions

$$\text{MBL} = \text{Const} \tag{40.44}$$

and safety factor

$$R(t) \geq R_u(t) < 1, 0 \tag{40.45}$$

Absolute safety ($R = 1.0$) needs infinite resources.

By calculating the NPV(M) curve, one can find the NPV_{max} or NPV_{sup}. The year that corresponds to these values, t_{max} or t_{sup}, is the one that should be taken as the best time for maintenance action, with but one most important additional consideration: at time M of maintenance action t_{max} or t_{sup}, the pipeline must still have full integrity (no leak, no rupture), or in other words, the pipeline reliability has to be

$$R(M) \geq R_u(M) \tag{40.46}$$

where R_u is the ultimate permissible safety level with respect to leak and rupture type of failures. The $R_u(M)$ value is obtained, using the optimal cessation of performance procedure. Therefore, in order to get the conditional optimal maintenance year, one must compare M with T_{opt} and shift it to T_{opt}. This has to be done every time a decision is made on the real-life event/decision tree (Figure 40.5 through Figure 40.8) using the MAST procedure. From the above, it is clear that predicting the CPF and how it is influenced by inspection/maintenance action is the most important single issue in implementation of risk-based maintenance methods.

Existing methods of generating pipeline conditional probability of failure fall into following categories:

- Statistical analysis and inference from actuarial data
- Expert assessments (knowledge and experience)
- Inspection of defects
- Simulation of flaw/crack growth
- Life prediction based on future operating conditions (scenarios)
- Prognosis of meteorological, hydrodynamic, and geodynamic loads on the deteriorating pipeline
- Operational monitoring and failure-mode analysis
- Analytical calculations

Note that in this chapter, only inspection of FCG is discussed.

40.14 Application of MAST and RAFT Procedures

The risk-based maintenance optimization using MAST and RAFT procedures was applied to a main oil pipeline. A branch of inspection/maintenance of an event tree is shown in Figure 40.20. The year-by-year increments of conditional PF are given in Figure 40.21. The calculations were conducted using a software tool developed by Timashev [26]. The values of the basic parameters involved in calculation are given in Figure 40.22, which also shows the input/output display of the MAST and RAFT procedure as it is seen

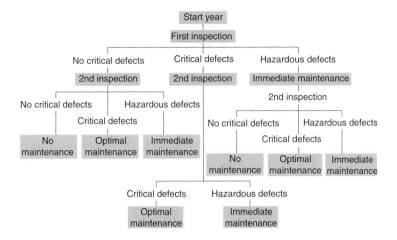

FIGURE 40.20 RAFT application for pipeline inspection/maintenance.

on the PC monitor. All calculations were made without accounting for the constraints imposed by Equation 40.32 and Equation 40.33. According to Figure 40.23, with increase of N, the year of optimal maintenance (YOM) shifts away from the origin. With increase of the discount factor, YOM shifts to the origin (Figure 40.24). Inflation has the opposite effect (Figure 40.25). With increase of the ratio C_m/C_f, the YOM shifts away from the origin (Figure 40.26). The influence of maintenance quality on YOM is given in Figure 40.27. The higher the quality of maintenance, the sooner the maintenance should be performed. In Figure 40.27, the CPF curve before maintenance in all cases is assumed to be linear ($k = b = 25 \cdot 10^{-6}$). The CPFs after maintenance are: for low-quality maintenance $k = b = 25 \cdot 10^{-4}$; for average-quality maintenance $k = b = 25 \cdot 10^{-6}$; for high-quality maintenance $k = b = 25 \cdot 10^{-8}$.

Table 40.1 to Table 40.3 allow one to perform event-tree cost analysis (Table 40.1), to compare the CPF for the YOM with R_u (Table 40.2), and to quantify the discounted inspection/maintenance costs for the optimal scenario of inspection/maintenance (Table 40.3). Analysis of Table 40.2 shows that CPF for all branches (except for branch 10) at the end of each maintenance (CPF M1, CPF M2) and at the end of each branch (CPF end) is lower than $p_{c,u}$ calculated by the optimal cessation procedure (CPF$_u$ = 0.05). Therefore, in order to approach the optimal solution, it is necessary to reshape (shorten) branch 10 to make CPF $p_{c,u} \leq 0.05$ and reshape (expand) other branches to make CPF at their end closer to $p_{c,u} = 0.05$.

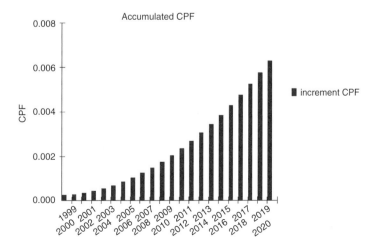

FIGURE 40.21 CPF function.

40-30   *Engineering Design Reliability Handbook*

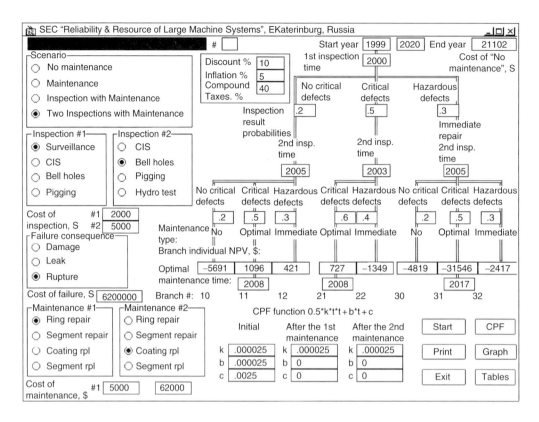

FIGURE 40.22 Input/output display of the MAST/RAFT procedure.

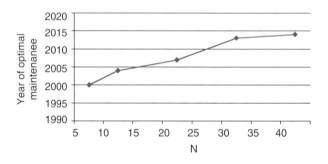

FIGURE 40.23 Dependence of YOM vs. period of analysis.

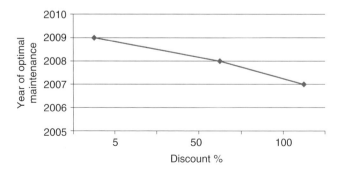

FIGURE 40.24 YOM vs. discount curve.

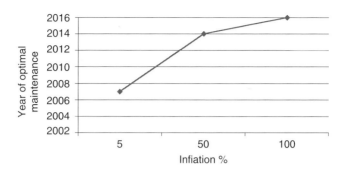

FIGURE 40.25 YOM vs. inflation curve.

40.15 Prioritizing Pipeline Segments for Repair: A Real Case Study

The method and algorithms outlined above were used to design a software package for prioritizing pipeline segments (PPS) with defects discovered by intelligent pigging. The software calculates the order in which defects should be repaired between inspections.

The software package for PPS was first implemented when considering the Ta-Mu main oil pipeline located in far northwestern Siberia. The pipeline was investigated by ultrasound high resolution pigging tool. Built in 1972, the pipeline is 720 mm in diameter and is made out of 17G2 grade steel with a working pressure of 5.5 MPa. The pigging revealed that there were dangerous defects that had to be (and were) repaired immediately, as well as specific defects that were potentially dangerous. "Potentially dangerous" means that the defects must be repaired in the near future, before the next inspection. The pipeline company's annual maintenance budget is always restricted. Therefore, there is always is a strong necessity to prioritize pipeline segments for maintenance and repair, given the next time of inspection. After the time span between the just completed and the next closest inspection is established (given by codes prescribed by some high authority or calculated using one of the methods outlined above), it is necessary to create a knowledge-based plan of maintenance/secondary inspection/repair during this time interval, t_{b_i}. Such a plan is created in several steps.

1. Collect the full Geo Information System information (pipeline plan, profile, soil properties stratigraphy, seasonal temperature fields, pumping hydraulics, material and weld properties of pipeline segment, distribution of stress tensor components along the pipeline longitudinal axis, etc.). The hydraulics diagram is calculated from the output of one oil pumping station (OPS) to the input

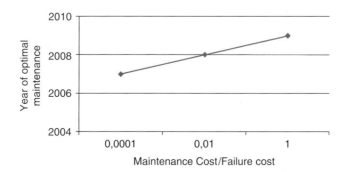

FIGURE 40.26 YOM vs. C_m/C_f.

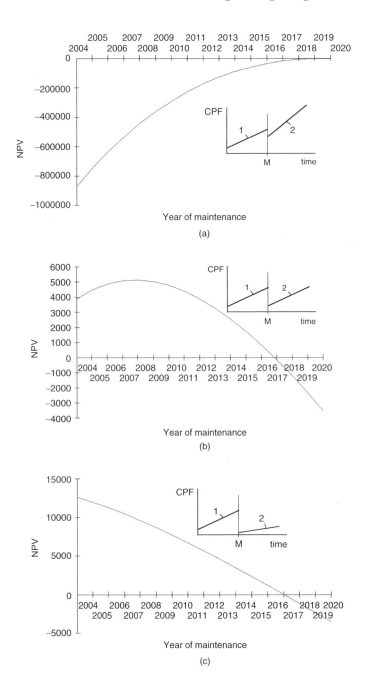

FIGURE 40.27 Influence of maintenance quality on the optimal maintenance year: (a) low maintenance quality, (b) average maintenance quality, (c) high maintenance quality (1 = before maintenance, 2 = after maintenance, M = moment of maintenance).

of the next OPS, between which the specific defects are located. A full group of possible pumping regimes for the time t_{b_i} is formed, which gives the maximum possible average pressure that can be expected during time t_{b_i} and the distribution of pressure.

2. The mapping of diagnosed specific defects (Figure 40.28) is adjusted with the full pipeline GIS. This allows one to complete the initial data set needed to pose a full set of mechanical problems for determining the residual lifetime of each cross section with one or several specific defects.

TABLE 40.1 Event-Tree Cost Analysis

Branch No.	Action No.	Year	Action	Content	Action Cost (U.S.$)
10	1	2000	inspection 1	surveillance	2,000
10	2	2005	inspection 2	bell holes	5,000
11	1	2000	inspection 1	surveillance	2,000
11	2	2005	inspection 2	bell holes	5,000
11	3	2007	maintenance 1	ring repair	5,000
12	1	2000	inspection 1	surveillance	2,000
12	2	2005	inspection 2	bell holes	5,000
12	3	2005	maintenance 1	ring repair	5,000
21	1	2000	inspection 1	surveillance	2,000
21	2	2003	inspection 2	bell holes	5,000
21	3	2003	maintenance 1	ring repair	5,000
22	1	2000	inspection 1	surveillance	2,000
22	2	2000	inspection 2	bell holes	5,000
22	3	2005	maintenance 1	ring repair	5,000
30	1	2000	inspection 1	surveillance	2,000
30	2	2000	maintenance 1	ring repair	5,000
30	3	2005	inspection 2	bell holes	5,000
31	1	2000	inspection 1	surveillance	2,000
31	2	2000	maintenance 1	ring repair	5,000
31	3	2005	inspection 2	bell holes	5,000
31	4	2010	maintenance 2	coating replacement	62,000
32	1	2000	inspection 1	surveillance	2,000
32	2	2000	maintenance 1	ring repair	5,000
32	3	2005	inspection 2	bell holes	5,000
32	4	2005	maintenance 2	coating replacement	62,000

TABLE 40.2 Accumulated CPF

Branch No.	CPF M1	CPF M2	CPF End	CPF_u
10	—	—	0.06287	0.05
11	0.01250	—	0.02112	0.05
12	0.00850	—	0.02812	0.05
21	0.01250	—	0.02112	0.05
22	0.00550	—	0.03612	0.05
30	0.00287	—	0.05000	0.05
31	0.00287	0.01250	0.01250	0.05
32	0.00287	0.00312	0.02812	0.05

TABLE 40.3 Branch Discounted Cost

Branch No.	Branch Cost (U.S.$)	Branch Probability
10	5,691	0.04
11	9,137	0.1
12	9,473	0.06
21	9,506	0.3
22	10,211	0.2
30	10,464	0.06
31	47,630	0.15
32	57,363	0.09

Note: Total discounted cost = $19,538, including inspections = $5,875 + maintenance = $13,663.

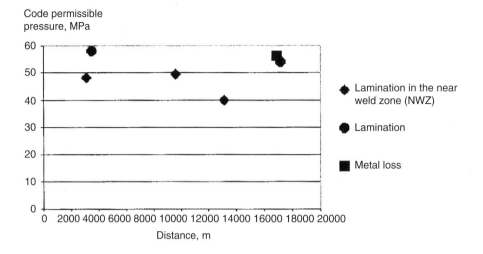

FIGURE 40.28 Specific defects along the 0 to 20-km segment of Ta-Mu pipeline.

3. Residual lifetime (RL) is a random variable. Therefore, in order to correctly solve the problem, it is necessary to build the probability density function (PDF) of the RL of each cross section of the pipeline with specific defects, accounting for randomness of loads, material properties, and the size of the specific defects as measured by the intelligent pigging tool [26, 27]. Depending on the type of defect and type of pipeline failure (leak before break [LBB], rupture, excessive plastic deformation), one should choose the right algorithm. It could be: low-cycle fatigue, combination of low- and high-cycle fatigue, crack propagation, combination of fatigue-crack propagation and corrosion, exhaustive consumption of pipeline material plasticity before rupture, etc. The algorithm of PDF calculation is given by Malyukova and Timashev [19, 20]. Having the PDFs of the defects' lifetimes, one can calculate the conditional probabilities of failure (CPF) of each defect for every year in the time internal t_{b_1}: 1, 2, 3, ..., t_{b_i}, on the condition that the defect did not fail during the previous time span (Figure 40.29).

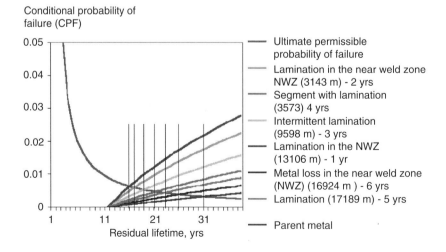

FIGURE 40.29 The ultimate permissible conditional probability of pipeline failure and optimal times of cessation of performance for different specific defects.

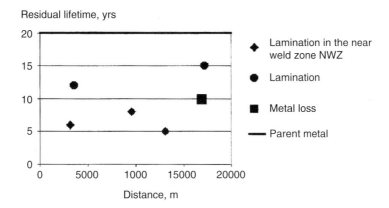

FIGURE 40.30 Lifetime of specific defects on the 0 to 20-km segment of Ta-Mu main pipeline.

4. The curve for the ultimate permissible conditional probability of failure should be plotted (Figure 40.29) using real-life costs for all types of pipeline survey, conventional maintenance, repair rehabilitation, all types of damage (economical, of environment), and casualties.
5. The optimal cessation-of-performance time for each specific defect is calculated using results of steps 3 and 4, as shown above in the sections (40.9, 40.10) dedicated to maintenance optimization. The results are plotted on a chart as a curve with "defect coordinate" vs. "optimal residual performance time" (Figure 40.30). This chart is the key instrument for prioritizing pipe runs with respect to maintenance and repair. Indeed, by drawing horizontal lines, say, with a time increment of 0.5 to 1.0 year, the chart space is divided into strips. The defect content of each strip is the number of defects that must be eliminated or repaired in this concrete time span. Graphical visualizations such as this are very easy to grasp and understand. Having this chart, the decision-making person can draw a sound and safe maintenance plan having, as a constraint, the company's yearly maintenance budget plan.

The mode of repair boils down either to permanent composite repair, to the clock-spring technology (polyester resin wrap reinforced by glass filament, tightly wound and bonded to pipe), or to defective-spool replacement (by cutting out the defective part of pipe and replacing it with a brand new section) (see, for instance, Timashev [26]). In general, the severity of the pipe-run condition can be assessed by the defect with the least residual lifetime. In order to account for a lack of full knowledge about the loads, environment, future pumping regimes, pipeline material degradation, etc., a safety coefficient $k_s > 1.0$ can be used. Its implementation into the chart (dividing the relevant residual lifetimes by k_s, as seen in Figure 40.29) brings a conservative estimate of the residual operational lifetime of a pipe run in consideration.

40.16 Conclusion

The risk-based maintenance optimization method described in this chapter provides the necessary tools for a pipeline company to develop a comprehensive proactive risk management program (RMP). The RMP allows the creation of a computerized in-house segment-ranking tool that supports planning and scheduling of rehabilitation projects. It also provides all necessary support for an alternative regulatory approach that allows pipeline companies greater flexibility to define company- and pipeline-specific risks and to develop company-specific solutions. The outlined maintenance system is an attempt to institutionalize an integrated, cross-functional approach to identifying and managing pipeline operating risk. The outlined method is a part of an expert system for monitoring the structural safety and reliability of main pipelines. The proposed system can be considered as a risk management technology benchmark because it provides a level of protection that is above and beyond the current regulations.

1. The theory of optimal control of pipeline operating risk is in its formative phase. Its further development is possible only on the basis of an interdisciplinary approach, with full utilization of the achievements of fundamental and applied sciences.
2. Probabilistic risk analysis and structural reliability theory can be merged into one discipline. To achieve this, the reliability problem should be expanded to account for the influence of monitoring/diagnostic/repair systems on the level of pipeline safety and to use the economic equivalent of noneconomic losses in reliability optimization problems.
3. Considering maintenance as control of random functions of deterioration and renewal of a pipeline system allows one to solve the important but elusive problem of predictive maintenance — defining the system quality levels that trigger maintenance action. The central maintenance problem is defined as a three-level policy control problem. The solution explicitly and simultaneously takes into consideration the probabilistic and economic issues of pipeline usage.
4. Application of the optimal cessation-of-performance problem allows solving the task of prioritizing pipe runs with respect to their maintenance and repair, based simultaneously on economic and safety considerations.
5. The multistep optimal cessation (MAST) procedure merges the central maintenance problem with the optimal-cessation procedure. This merger allows one to link the trigger levels with the ultimate permissible level of conditional probability of failure to activate risk-based optimal maintenance planning. At the same time, it ensures the integrity of the pipeline and minimizes the risk of its performance.
6. The developed basic-block module of pipeline risk control allows one to find and/or optimize the following: the trigger levels w (warning) and a (alarm); time interval between inspections; renewal times; reliability function; quality of maintenance. At the same time, it uses adequate diagnostics means and methods for defect assessment in the form of a software package that provides a system approach to statistical analysis and interpretation of the ILI measurement data.
7. Application of MAST/RAFT methods allows one to start on the wide application of the process of pipeline maintenance optimization. It also shows the real value of refining the information-gathering process. It graphically links the engineering and financial functions of a pipeline company, showing all decision elements from input to output, and it can be reused with new inputs to screen a full group of possible scenarios fed by monitoring/inspection/repair/replacement data as inputs to make the best future decision, thus casting predictive maintenance as an investment rather than a cost.
8. Risk-based approaches to maintenance planning coupled with optimization algorithms provide the most complete corporate view of the benefit of inspection/maintenance expenditures and can bring the pipeline company multimillion dollar savings while following a financially sound process.
9. Predicting the CPF and how it is influenced by inspection/maintenance action is the most important single issue in implementation of risk-based maintenance methods. The assessments of the PDF of remaining lifetime and of the conditional probability of failure and how they change due to inspection/diagnostics/renewal/maintenance are the first component of risk and, at the same time, the most important component of any contemporary pipeline maintenance system based on optimization of pipeline operating risk.

The methods described above were used to create a sophisticated and comprehensive software package for optimizing pipeline inspection and maintenance. The Russian oil and gas industry is using a combination of hardware and software to monitor pipeline residual life and safety systems. The proposed theoretical approach reduces maintenance cost for pipelines in arctic and subarctic regions by up to 12%, with the average reduction being around 5 to 7%.

References

1. Davis, M.J., Tenneco's Efforts for Verifying Pipeline Integrity, presented at ASA Distribution/Transmission Conference, Toronto, Canada, 1988.

2. Kiefner, J.F. et al., Methods for Prioritizing Pipeline Maintenance and Rehabilitation, final report on Project PR3-919 to American Gas Association, AGA, Battelle, Columbus, OH, 1990.
3. Kulkarni, R.B. and Conroy, J.E., Pipeline Inspection and Maintenance Optimization System (PIMOS), presented at Pipeline Risk Assessment, Rehabilitation and Repair Conference, Houston, 1994, pp. 1–13.
4. Kulkarni, R.B. and Conroy, J.E., Development of a Pipeline Inspection and Maintenance Optimization System, GRI, Chicago, IL, 1998.
5. Timashev, S.A., *Reliability of Large Machine Systems*, Nauka, Moskva, 1982.
6. Timashev, S.A., Machinery diagnostic and residual life monitoring system, in *Proceedings of Machinery Reliability Conference*, Industrial Communications, Knoxville, TN, 1998, pp. 176–188.
7. Timashev, S.A., Elements of optimal machinery maintenance theory, in *COMADEM'98 Proceedings*, Monash University, Melbourne, AU, 1998, pp. 833–845.
8. Pagliari, L., Vibration analysis, in *Predictive Maintenance and Machine Vibration*, Coastal Video Communications Corp., Vera Cruz, CA, 1997.
9. Timashev, S.A, On a problem of reliability control, in *Third International Vilnius Conference on Probability Theory and Mathematical Statistics, Abstracts of Communications*, Lithuanian Academy of Science, Vilnius, Lithuania, 1981, pp. 186–187.
10. Timashev, S.A. and Copnov, V.A., Optimal death process control in two-level policies, in *Fourth International Vilnius Conference on Probability Theory and Mathematical Statistics, Abstracts of Communications*, Lithuanian Academy of Science, Vilnius, Lithuania, 1985, pp. 308–309.
11. Andronov, A.M. et al., *Reliability and Efficiency of Machinery*, Vol. 8 of 10, *Maintenance and Repair*, Mashinostroeniye, Moscow, 1990 (in Russian).
12. Barzilovich, E.Yu., *Maintenance Models for Complex Systems*, Visshaya Shkola, Moscow, 1982 (in Russian).
13. Timashev, S.A., Optimization of thin-walled designs by criteria of reliability, in *Vilnius Conference on Problems of Optimization in the Mechanics of a Firm Deformable Body, Abstracts of Communications*, Vol. 2, Vilnius Institute of Civil Engineering, Vilnius, Lithuania, 1974, pp. 55–56.
14. Timashev, S.A., Reliability control of mechanical systems, in *Reliability Problems: Materials of All-Union Conference on Optimization and Reliability Problems in Structural Mechanics*, MEI, Moscow, 1979, pp. 161–163.
15. Timashev, S.A., The system approach to estimation of mechanical systems reliability, in *Research in the Field of Reliability of Engineering Structures*, Lenpromstroyproekt, St. Petersburg, Russia, 1979, pp. 5–24.
16. Timashev, S.A. and Vlasov, V.V., Method of assessment of damages size due to various kind of structural failures, in *Research in the Field of Reliability of Engineering Structures*, Lenpromstroyproekt, St. Petersburg, Russia, 1979, pp. 25–35.
17. Malyukova, M.G. and Timashev, S.A., PDF of residual life of pipelines with longitudinal initial cracks, *Int. Electr. J.*, 6 (6), 37–50, 2000.
18. Bogdanoff, J. and Kozin, F., *Probabilistic Models of Cumulative Damage*, Wiley & Sons, New York, 1984.
19. Malyukova, M.G. and Timashev, S.A., Residual life of pipelines with longitudinal initial cracks, in *ASME/JSME PVP'98*, Vol. 373, ASME, Boston, MA, 1998, pp. 99–104.
20. Malyukova, M.G. and Timashev, S.A., Probabilistic longevity of an oil pipeline with crack subjected to internal corrosion, in *ASME PVP'99*, ASME, New York, NY, 1999.
21. Malyukova, M.G. and Timashev, S.A., Computation of Remaining Life PDF of Pipeline with Crack Subjected to a Combination of Cyclic Loading and Corrosion, presented at 5th U.S. National Congress on Computational Mechanics, Computational and Probabilistic Fracture Mechanics Symposium, Boulder, CO, 1999.
22. Lambert, Y. et al., Application of the J-concept to fatigue crack growth in large-scale welding, in *Proceedings of 19th Symposium ASTM STP 969*, American Society for Testing and Materials, Philadelphia, 1988, pp. 318–329.

23. Timashev, S.A. and Kopnov, V.A., Optimal resource assessment with respect to fatigue crack formation criteria, *J. Machine Building Machinery Reliability Problems*, Vol. 12, No. 1, 65–70, 1990 (in Russian).
24. Chow, Y.S., Robbins, H., and Siegmung, D., *Great Expectation: The Theory of Optimal Stopping*, Houghton Mifflin, Boston, 1971.
25. Smalley, A.J. and Mauney, D.A., Risk-Based Maintenance of Turbo Machinery, in *Proceedings of the 26th Turbomachinery Symposium*, ASME, Texas, 1997, pp. 177–187.
26. Timashev, S.A., Diagnostics and Maintenance of Pipelines, in *Intensive Short Course Material*, Monash University, VIC, Australia, 1998.
27. Vieth, P.H., Rust, S.W., and Johnson, E.R., Statistical analysis methods for ILI metal loss data, *Corrosion Prevention Control*, Vol. 42, No. 2, February, 20–30, 1998.
28. Timashev, S.A., Khopersky, G.G., and Chepursky, V.N., Reliability and Residual Resource Monitoring System of Oil Pumping Stations Equipment, in Truboprovodny Transport Nefti, #8, 5–8 and #9, 8–13, 1997 (in Russian).
29. Timashev, S.A., Machinery Diagnostic and Residual Life Monitoring System, in *Machinery Reliability Conference*, Charlotte, NC, Industrial Communications, Inc., Knoxville, TN, 176–188, 1998.

41
Nondeterministic Hybrid Architectures for Vehicle Health Management

41.1	Introduction..	**41**-1
41.2	Application 1: Vehicle Engine Prognostics Health Management (VEPHM)...	**41**-2
	Stochastic VEPHM Tools Development • Nondeterministic Adaptive Network-Based Fuzzy Inference Models • Feature Vector of Engine Behavior • Sparse Sensor Array Studies • Comparison of ANFIS Model Fault-Detection Sensitivity • Multivariate Reliability Analysis Using Stochastic Fault-Basin Concept • Decision-Level Fusion	
41.3	Application 2: DWPA Multiple-Band-Pass Demodulation and Automated Diagnostics...............	**41**-19
	Implementation of ANFIS for Automatic Feature Extraction • Performance of ANFIS Feature Extraction • Automated Fault Classification • Multiple-Band-Pass Fault-Severity Index	
41.4	Concluding Remarks..	**41**-27

Joshua Altmann
Vipac Engineers & Scientists Ltd.

Dan M. Ghiocel
Ghiocel Predictive Technologies Inc.

41.1 Introduction

This chapter presents two nondeterministic applications where hybrid architectures have been employed to provide enhanced health-management reasoning for the detection, diagnosis, and prognosis of faults (that are typically defined by some loss of system functionality).

The first application illustrates a prognostic health management (PHM) system capable of predicting faults of air or ground vehicle engines under highly transient in-operation conditions. The system's predictions also include the associated confidence or risk levels. To adequately address the complex problem of probabilistic in-operation diagnostics and prognostics, a hybrid stochastic-neuro-fuzzy inference system was developed that is a combination of stochastic parametric and nonparametric modeling techniques. This hybrid nondeterministic inference system, named StoFIS, is an integration of multivariate stochastic space-time process models with adaptive network-based fuzzy inference system models using clustering techniques [1]. StoFIS provides a hierarchical data-fusion modeling to maximize the extracted information used for diagnostic–prognostic reasoning. StoFIS is used to quantify the fault risks of an engine system at any given time and project their risk evolution in the future for risk-based prognostics.

The second nondeterministic application illustrates the application of discrete wavelet analysis in conjunction with an adaptive network-based fuzzy inference system to provide automated fault detection and diagnosis of rolling-element bearings. The proposed method involves the automatic extraction of wavelet packets containing bearing-fault-related features from the discrete wavelet packet analysis representation of machine vibrations. The resultant signal extracted by this technique is essentially an optimal multiple-band-pass filter of the high-frequency bearing impact transients. The discrete wavelet packet analysis multiple-band-pass filtering of the signal results in improved signal-to-noise ratios, with an exceptional capacity to exclude contaminating sources of vibration.

41.2 Application 1: Vehicle Engine Prognostics Health Management (VEPHM)

Many VEPHM technologies over the last decade have focused on the ability to classify engine performance faults as predicted by either gas-path-analysis models or maintenance personnel experience. Different technologies implemented in the past have included various advanced techniques, such as neural-network architectures, expert systems, fuzzy inference systems, empirical-based lifing algorithms, and more recently, probabilistic or stochastic modeling techniques. Each of the implementations of these VEPHM technologies brings benefits in terms of their capabilities for detection, diagnostics, and prognostics of engine faults. None of these technologies provides a complete solution to the challenge of developing a VEPHM system that is robust, reliable, and yet sensitive. The focus of more recent work in the field of VEPHM has shifted to utilizing a combination of the aforementioned technologies, and providing diagnostic reasoning based on data fusion. As greater knowledge of system behavior becomes available with the increased data collection and dissemination, improved system integration will become feasible, including the data and feature fusion of nondeterministic performance-based and vibration-based diagnostic reasoning.

At this time, there is a significant need to move to in-operation-capable VEPHM systems that can be used on a regular ongoing basis, thus providing a proactive approach rather than a reactive approach to vehicle engine maintenance. In-operation engine diagnostics and prognostics offer critical information on the engine state and functionality that is of key importance for quick, cost-effective decisions of the vehicle pilot. Also, in-operation diagnostics offer extremely useful information to maintenance engineers for preventive actions and, in some potentially catastrophic situations, to vehicle pilots for avoiding the accidental loss of the vehicle.

The application presented in this section addresses the air VEPHM problem. The previous generation of air VEPHM systems suffered from several shortcomings, which limited their use to ground test environments. In some cases, air VEPHM systems assumed that the relationship between engine parameters and rotor operating speed (or corrected speed) could be represented by one-dimensional high-order polynomial functions [2]. This polynomial fitting is suited for simple, well-controlled engine tests with constant or slowly varying operational conditions, but not for a realistic engine environment. In a realistic engine environment characterized by large and rapid variations in speed and engine inlet conditions, the polynomial fitting provides a poor approximation for the in-operation engine problem. Another significant shortcoming of the previous generation of air VEPHM systems that needs to be addressed is the limited prognostic capabilities available. This was partially due to the diagnostic reasoning of these systems concentrating on the type of fault detected, with limited assessment of the fault severity. Prognostic capabilities are thus usually limited to trending of nondimensional trend parameters taken under specific operating conditions.

These issues are addressed by the development of stochastic models that treat the progression of engine performance faults as movement into and along fault basins of attraction. The fault basins are based on multivariate stochastic modeling of performance parameter deviations from normal operating condition for given engine faults or combinations of faults.

41.2.1 Stochastic VEPHM Tools Development

The development of a probabilistic framework for engine diagnostics and prognostics, based on parameter deviations from transient gas-path-analysis (GPA) engine models, provides the basis for in-operation risk-based assessment of engine condition at any given time. By incorporating transient engine models, appropriate feature extraction, data filtering, and probabilistic reasoning, an assessment of engine condition in terms of risk can be ascertained given prior risk-association data (risk associated with given fault type and severity).

To enable robust and sensitive system performance, regardless of whether the system is operating in a ground test environment or a highly transient flight profile, nondeterministic GPA models were developed using adaptive network-based fuzzy inference system (ANFIS) models. These ANFIS GPA models were incorporated into the probabilistic reasoning model. This was in recognition of the fact that functional variations are significantly greater than the random variations in performance parameters from their normal operating conditions. The input parameters selected were based on a typical analytical GPA model, with additional parameters (power lever angle) used as substitutes for parameters that are not currently measured (fuel flow rate). The remaining error patterns described by uncertain parameter fluctuations were substantially a result of random quantities, although some degree of functional variability remains due to input factors not modeled and a limited size of the statistical database. A robust probabilistic diagnostic-prognostic VEPHM system has to consider both the functional and random aspects of the fault patterns. Three different classes of engine performance models were developed, namely the overall quasi-stationary model (OQS), the overall transient model (OT), and the partial transient model (PT).

Three levels of data fusion are embedded into StoFIS. The data- and feature-level fusions are nondeterministic data-fusion processes, whereas the decision-level fusion method is deterministic. This procedure enables in-operation variability and data uncertainty to be taken into account, and provides a robust diagnostic-prognostic output of engine health. The three major components that form the basis for the StoFIS development are outlined below.

41.2.1.1 Feature Vector Extraction

The first component of StoFIS is a data-level fusion procedure, where data from multiple sensors are fused into ANFIS engine models prior to feature extraction. This fusion process utilizes data from a sparse sensor array to track performance parameter deviations from normal operating conditions under typical real-life environments. Basically, in StoFIS, engine behavior is nondeterministically modeled using an ANFIS model, with the system design derived from a physics-based quasi-stationary GPA model. It involves a two-step computational process; the initial membership functions (likelihood functions) and fuzzy-logic rules are formed through subclustering, followed by a fine-tuning of the system with a loosely coupled procedure using least squares and a backward-propagation neural network. The procedure enables system adaptation to include *engine–engine variability*, as well as the provision for *quasi-stationary*, highly *transient*, and *partial-engine transient* implementations of the models. Synergies derived from the multiple modeling procedures enable more-sensitive fault detection under adverse conditions, increased confidence in diagnostics, and improved multiple-fault discrimination.

41.2.1.2 Multivariate Reliability Analysis

The second component of StoFIS involves the reliability analysis using multivariate stochastic feature vectors extracted from the ANFIS engine models. Multisensor stochastic feature vectors are used to characterize the system and arrive at a diagnostic-prognostic output for each fault type. This is the most critical step in the fusion process, with diagnostic and prognostic reasoning based on stochastic parameter deviations due to off-specification conditions. This stochastic modeling has been termed as the stochastic fault basin, as it models the stochastic deviations in multidimensional parameter space, with the development of faults treated as progression into and along a fault basin (not a specific center point for a given severity level) of attraction. Prognostics are based on the location with respect to and rate of progression toward a fault basin of attraction. The outputs from the stochastic feature-level fusion are a reliability index (RI), used for

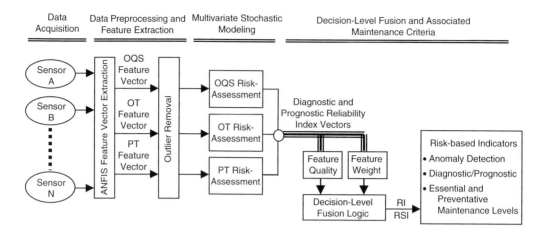

FIGURE 41.1 Illustration of StoFIS architecture.

diagnostics, and a reliability sensitivity index (RSI), used for prognostics. These two reliability indices can be associated with the essential maintenance and preventive maintenance events, respectively. Separate outputs are available from the three basic nondeterministic ANFIS GPA engine models.

41.2.1.3 Decision-Level Fusion

The third component of StoFIS is the decision-level fusion, where diagnostic outputs from separate engine models, OQS, OT, and PT, are fused after diagnostic reasoning to enhance the confidence of the final output. The purpose of this procedure is to take advantage of the synergies provided by using the multiple engine models. The method employed is a heuristic fusion procedure based on a scoring method that incorporates knowledge-based rules. The reliability indices computed using the three separate engine models are weighted based on the fault-detection index and the fault-discrimination sensitivity index of the engine models for given fault conditions. The combined effects of the data-level and decision-level fusion technologies developed during this project have indicated improvements in the fault-diagnostic resolution for in-operation conditions of between 100 and 1000%, depending on the location and type of fault in the engine compared with corrected rotor-speed-based VEPHM systems. Figure 41.1 shows a schematic illustration of the StoFIS architecture.

41.2.2 Nondeterministic Adaptive Network-Based Fuzzy Inference Models

The adaptive network-based fuzzy inference system, which is utilized in both of the studies illustrated in this chapter, is a transformational model of integration where the final fuzzy inference system is optimized via artificial neural-network training. ANFIS has the ability to either incorporate expert knowledge or use subtractive clustering to form its initial rule base. In both cases, ANFIS maintains system transparency while allowing tuning of the fuzzy inference system via neural training to ensure satisfactory performance. The validity of the expert knowledge and the suitability of the input data chosen can then be verified by examining the structure and the performance of the final fuzzy inference system. This section describes the design and operation of an ANFIS [3].

The initial membership functions and rules for the fuzzy inference system can be designed by employing human expertise about the target system to be modeled. ANFIS can then refine the fuzzy if–then rules and membership functions to describe the input/output behavior of a complex system.

If human expertise is not available, it is possible to intuitively set up reasonable membership functions and then employ the neural training process to generate a set of fuzzy if–then rules that approximate a desired data set. Sugeno-type fuzzy inference systems have been used in most adaptive techniques for constructing fuzzy models because they are more compact and provide a more computationally efficient

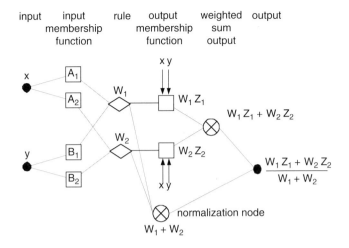

FIGURE 41.2 ANFIS Sugeno fuzzy model.

representation of data than the Mamdani or Tsukamato fuzzy systems. A typical fuzzy rule in a zero-order Sugeno fuzzy system has the form:

$$\text{If } x \text{ is } A \text{ and } y \text{ is } B, \text{ then } z = c \tag{41.1}$$

where A and B are fuzzy sets in the antecedent, and z is a crisply defined function in the consequent. It is frequently the case that the singleton spike of the crisply defined consequent is completely sufficient to cater to a given problem's needs. If required, the more general first-order Sugeno can be employed by setting the consequent to $z = px + qy + c$. Higher-order Sugeno systems add an unwarranted level of complexity, with minimal remuneration. A zero-order Sugeno fuzzy inference system is used in this investigation. The equivalent ANFIS architecture for a Sugeno fuzzy inference system is illustrated in Figure 41.2.

The nodes in the *input membership function layer* are adaptive. Any appropriate membership functions can be used to describe the input parameters. The outputs of this layer, $\mu_{A_i}(x)$ and $\mu_{B_i}(x)$, are the membership values of the premise where x and y are the node inputs, and A_i and B_i are the associated fuzzy sets. The output of the fixed nodes in the *rule layer* represents the fuzzy strengths of each rule. Either the product or minimum rules can be used to calculate the weighting function for the fuzzy operator "AND" of a Sugeno fuzzy inference system.

$$\text{Product: } W_i = \mu_{A_i}(x) \times \mu_{B_i}(x) \tag{41.2}$$

$$\text{Minimum: } W_i = \min\{\mu_{A_i}(x), \mu_{B_i}(x)\} \tag{41.3}$$

The adaptive nodes in the *output membership function layer* calculate the weighted output of the consequent parameters, as given by

$$W_i Z_i = W_i C_i \tag{41.4}$$

The *weighted-sum output layer* consists of a single fixed node. The weighted-sum output is the summation of the weighted output of the consequent parameters,

$$\sum_i W_i Z_i \tag{41.5}$$

The final *output layer* is the normalized weighted output given by

$$\frac{\sum_i W_i Z_i}{\sum_i W_i} \tag{41.6}$$

The *normalization node* connects the rule layer to the output layer in order to normalize the final output. The normalization factor is calculated as the sum of all weight functions.

$$\sum_i W_i \tag{41.7}$$

Although any feed-forward network can be used in an ANFIS, Jang and Sun [4] implemented a hybrid learning algorithm that converges much faster than training that relies solely on a gradient-descent method. During the forward pass, the node outputs advance until the output membership function layer, where the consequent parameters are identified by the least-squares method. The backward pass uses a back-propagation gradient-descent method to upgrade the premise parameters, based on the error signals that propagate backward. Under the condition that the premise parameters are fixed, the consequent parameters determined are optimal. This reduces the dimension of the search space for the gradient-descent algorithm, thus ensuring faster convergence. This hybrid learning system is used in the training of the fuzzy inference systems used for both applications presented. A more-detailed explanation of ANFIS can be found in the literature [3, 4].

41.2.3 Feature Vector of Engine Behavior

Figure 41.3 shows a sketch of the investigated turbofan jet engine including typically installed sensors. Figure 41.4 and Figure 41.5 show pressure variations as a function of the high-pressure shaft speed for testing data and in-operation conditions, respectively. It is obvious from these figures that the pressure closely follows a nonlinear relationship with shaft speed for slowly varying testing conditions. However, for in-operation conditions, the pressure deviates from this nonlinear path due to highly transient conditions and significant changes in the inlet conditions, namely inlet pressure, temperature, and mass flow. This means that using deviations from a fitted polynomial regression line for diagnostics, as is commonly used in engine health-monitoring applications based on test data, is not suited to in-operation conditions. In fact, a large stochastic

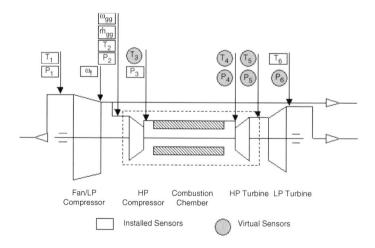

FIGURE 41.3 Schematic of a typical sparse sensor array. (T-temperature, P-pressure, ω_f-fan rotor speed, ω_{gg}-gas generator rotor speed, $\dot{m}_{gg}$-gas generator mass flow rate.)

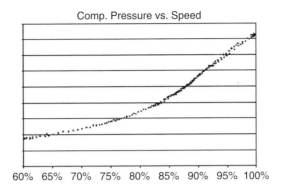

FIGURE 41.4 P_3 vs. ω_{gg} for ground tests.

variability projected on the pressure–speed plane in Figure 41.5 is apparent. This large variability is mostly due to the transient variations induced by the vehicle's pilot maneuvers. A key aspect for getting realistic predictions for in-operation conditions is to separate the *true statistical variabilities* (random part) from the *functional variabilities* introduced by engine transient behavior. For real, in-operation transient conditions, the functional dependence between engine performance parameters becomes complex and highly nonlinear. If these transient complex functional dependencies between multiple parameters are ignored, then the statistical variability is overestimated and the computed fault risks are unreliable.

Physics-based, analytical GPA models only cater to quasi-stationary engine operation, which provides limited ability to track engine parameters during the transitory operating conditions encountered during typical operation of a vehicle. An alternative scheme capable of including the highly transient in-flight conditions has been developed based on deviations from physics-based empirical GPA models of the performance parameters. This leads to the formation of a stochastic engine GPA model, developed by building and calibrating a generic analytical GPA model for a given engine and then tuning its stochastic input–output using the nondeterministic ANFIS GPA model based on a statistical training data set from typical operational data.

The ANFIS model was developed based on subtractive clustering using the modified mountain technique [5]. Subtractive clustering is a fast one-pass algorithm for estimating the number of clusters and the cluster centers in a set of data and provides a powerful tool to form the initial fuzzy inference system (FIS). Then, using a neural network (NN), the FIS was fine-tuned using least-squares training on the forward-pass and back-propagation on the backward pass, resulting in a final FIS in the ANFIS model that includes the NN-based correction. The final ANFIS GPA engine model provides an engine-specific

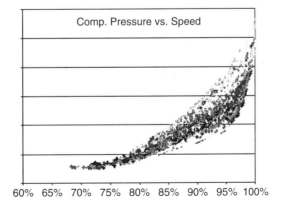

FIGURE 41.5 P_3 vs. ω_{gg} for in-operation (flight) conditions.

physics-based empirical GPA model that is capable of accurately predicting normal-condition pressures and temperatures at each section of the engine under transient operating conditions. The ANFIS GPA models were developed and tested with nine different operating-condition spectra. Three operating spectra were used for the NN training; three were used to check the fuzzy inference system for overtraining; and three were used to test the final system. The developed ANFIS models performed robustly for all of the data sets considered, with only minor flight-profile dependencies. The ANFIS models provide two functions: (1) an indicator of engine or sensor malfunction based on statistical deviations from measured performance parameters; and (2) virtual sensors in the event that physical sensors are not mounted at certain compartments in production models of an engine. Virtual sensors are normally used to provide lifing estimates for the critical engine components.

Updating of ANFIS to cater to variations due to engine-to-engine variability for engines within a given engine class can be accommodated by fine-tuning based on tests prior to in-operation implementation. Alternatively, calibration can be performed based on in-operation data collected from a specific engine. In this case, the vehicle's pilot-to-pilot variability in addition to operation-to-operation variability can be included. The deviations from the original ANFIS GPA models can be used to rate an engine's performance in relation to the standard engine characteristics for a given engine class. This provides an important quality-control function for the engine-manufacturing process. The deviation from the calibrated ANFIS GPA models can be used to detect, diagnose, and provide prognosis of deterioration in the performance of the specific engine.

41.2.4 Sparse Sensor Array Studies

Three key issues need to be resolved in the modeling of the engine performance: (1) maximization of the useful information expressed in terms of system outputs available for diagnostic–prognostic reasoning; (2) incorporation of multiple engine faults for detection; and (3) isolation of their respective contributions for diagnostic–prognostic purposes. Due to production engines currently having a very limited number of onboard sensors, it was of interest to develop a sparse-sensor-array implementation that could satisfy these three key issues. For example, we assume that a typical production engine can be confined to an output parameter space of P_2, T_2, P_3, T_6, ω_f, and $\dot{m}_{gg}$. In order to maximize the available information, the three nondeterministic ANFIS GPA models, OQS, OT, and PT, were used.

The two overall engine models (OT and OQS) have a qualitatively different behavior. The OT model eliminates feedback of downstream faults on the predicted performance parameters, while the OQS model includes this feedback effect, thus providing additional parameters and more-complex interactions for diagnostic reasoning. Detection of multiple cascaded engine faults is not easily achieved when limited to overall engine models, especially when a downstream engine fault is shadowed by an upstream engine fault. Thus a partial-engine model (PT) was also introduced. The functional basis for the three GPA models is described in the following three subsections.

41.2.4.1 Overall Quasi-Stationary (OQS) Engine Model

The OQS engine model is similar to measuring performance parameters as a function of corrected high-pressure rotor speed. It has the advantage of additional outputs compared with the transient models above, in the form of mass-flow-rate and low-pressure shaft speed, however it does not perform as well under highly transient operating conditions. The following relationship was assumed in the OQS model:

$$P_n, T_n, \dot{m}_{gg}, \omega_f = f(P_1, T_1, \omega_{gg}, \text{PLA}) \tag{41.8}$$

Parameters P_n and T_n represent the compartment static pressure and temperature parameters indicated in Figure 41.3. Significant functional errors were present due to the inability of the quasi-stationary model to cater to the highly transient conditions encountered in operation. To reduce the function errors in the quasi-stationary model, the available parameters were assessed for their ability to decrease the residual errors. The power lever angle (PLA) was the only parameter that significantly reduced the model errors.

41.2.4.2 Overall Transient (OT) Engine Model

The OT engine model takes transients into account by including the low-pressure shaft speed and flow rates. The air mass flow is denoted by $\dot{m}_{gg}$, and ω_f and ω_{gg} denote the fan and gas-generator rotor speeds, respectively. The following relationship was assumed in the OT model:

$$P_n, T_n = f(P_1, T_1, \dot{m}_{gg}, \omega_f, \omega_{gg}) \tag{41.9}$$

Significant functional errors were present in the postcombustion chamber the OT GPA model-predicted temperatures. This error is due to the lack of a fuel flow-rate measurement, which currently cannot be directly measured. Unlike the quasi-stationary model, PLA only provided a marginal reduction in the residual errors, with the errors in some parameters increasing.

41.2.4.3 Partial-Engine Transient (PT) Model

The PT model uses the previous temperature and pressure available in the sensor array as inputs to the ANFIS GPA model. This isolates contributions to deviations in the measured parameters from model predictions due to faults at earlier stages of the engine. The benefits of the PT model are highly dependent on the number of sensors installed on the engine. As the number of sensors is increased, the capability of the PT model to distinguish between faults and to identify multiple faults in the engine is augmented significantly. The PT models can be written in their functional form as:

$$P_n, T_n = f(P_{n-p}, T_{n-q}, \dot{m}_{gg}, \omega_f, \omega_{gg}) \tag{41.10}$$

where $n\text{-}p$ and $n\text{-}q$ represent the location of the previous sensor in the array. Although the PT models do not provide a true compartmentalization of the turbofan engine, they still bring important complementary information that adds to the overall engine models and can be used to enhance engine diagnostics.

Figure 41.6 and Figure 41.7 illustrate the mean deviations in the measured parameters for each of the models for two different fault conditions of a turbofan engine. Two important benefits of the multiple-model approach are illustrated in Figure 41.6. First, although the OQS model indicates a probable fault, it would not provide as accurate an indicator of severity as the OT model due to the high scatter in

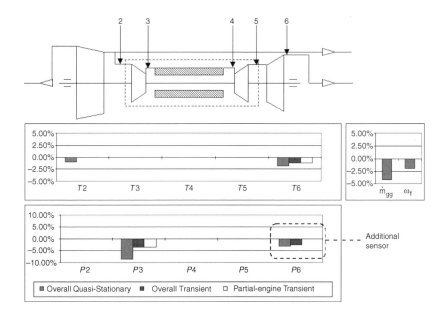

FIGURE 41.6 Sparse sensor array mean parameter deviations for 2% drop in HPC capacity.

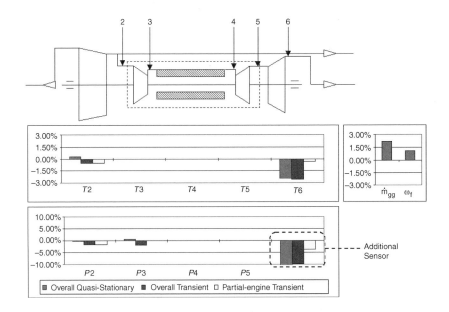

FIGURE 41.7 Sparse sensor array mean parameter deviations due to a combined fault (2% drop in fan capacity and a 1% drop in HPT efficiency).

parameter deviations. In addition, the OQS model would not sufficiently discriminate between the following faults: High Pressure Compressor (HPC) capacity drop, Low Pressure Compressor (LPC) capacity/efficiency drop, and High Pressure Turbine (HPT)/Low Pressure Turbine (LPT) capacity increase. This is due to the relatively small, normalized sensitivity indices of the OQS model for these faults. In contrast, the OT and PT models clearly isolate the fault as being located in the HPC compartment, although they are not as sensitive in distinguishing between an efficiency- and capacity-related fault. A logical combination of these results allows the fault to be isolated as a HPC capacity drop.

The benefits of the multiple-model approach at detecting and discriminating multiple faults are illustrated in Figure 41.7. In this example, the OT and PT models strongly support the supposition that there is a drop in fan capacity, while the OQS model is less definitive due to the feedback from the HPT efficiency drop confusing the issue. In this situation, the characteristic deviations for the OQS model would lie somewhere between the drop in fan-capacity and HPT-efficiency basins of attraction in multidimensional parameter space. This could result in a reduction in confidence of the diagnostics based on the OQS model due to potential confusion with other faults or combinations of faults.

The benefit of this is that the OQS model indicates that there is a high probability of faults existing in more than one compartment of the engine. As the OT model is relatively insensitive to the HPT fault, the OT model provides a more robust indicator of the drop in fan capacity, but it is unable to detect the existence of a second fault. The PT model, on the other hand, is able to isolate the drop in fan capacity and indicates either a drop in LPT/HPT efficiency or an increase in LPT/HPT capacity. Linking this knowledge with the deviations in the OT and OQS model, it is possible to eliminate LP capacity increase as a potential fault. Thus diagnostic reasoning after preliminary fusion of the three models could isolate the primary fault as a drop in fan capacity, with a secondary fault located in the HPT compartment (either efficiency drop or capacity increase).

41.2.5 Comparison of ANFIS Model Fault-Detection Sensitivity

Figure 41.8 illustrates the normalized sensitivity index of the three ANFIS models for fault detection under typical transient in-operation conditions, based on the sparse sensor array illustrated in Figure 41.3. The normalized sensitivity index is an indicator of each model's sensitivity in detecting the presence of an engine fault in a given compartment, relative to corrected rotor-speed-based measurements. The fault-sensitivity

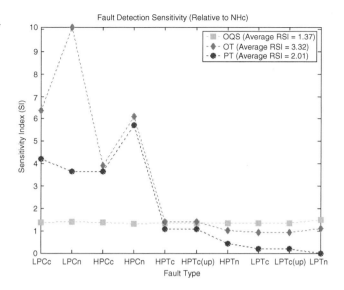

FIGURE 41.8 Relative model sensitivity: OQS, OT, and PT.

index is illustrated for changes in capacity and efficiency of each engine compartment, with the average model sensitivity (no weighting given for the importance of fault location) displayed in the legend.

The sensitivity index (*SI*) is based on the Euclidean distance between the mean deviation vectors in transformed space. The definition of SI for a given fault (f) condition is as follows:

$$SI(f) = \sqrt{\sum_{i=1}^{n} (\mu(f)_i^* - \mu(nc)_i^*)^2} \tag{41.11}$$

where i represents the measurable parameters for fault detection (i.e., pressures, temperatures, rotor speed, and flow rate), $\mu(f)^*$ is the mean parameter deviation in transformed space for a given fault condition, and $\mu(nc)^*$ is the mean parameter deviation in transformed space for an engine in normal condition (≈ 0).

The transformed space is based on the relative standard deviation of the normal-condition parameter errors for the three ANFIS models. The transformation enables the mean parameter variations to be considered with an equal weighting of importance, with the models using a common reference σ_{ref} to allow the performance of the models to be compared directly. The definitions of μ_i^* and σ_{ref} are given as:

$$\mu_i^* = \frac{\sigma_{ref}}{\sigma(nc)_i} \tag{41.12}$$

$$\sigma_{ref} = \min\{\sigma(nc)_{i,\text{Model}}/; i \in 1{:}n, \text{Model} \in [\text{OQS}, \text{OT}]\} \tag{41.13}$$

The fault-detection capabilities of the models are dependent on both the type of fault and the number/location of the onboard sensors. The OT and PT models are more sensitive that the OQS model for detecting compressor faults, whereas the OQS model is the most sensitive for turbine faults. This indicates the critical need for an in-operation VEPHM system to use intelligent fusion of the three models to take advantage of the provided synergies.

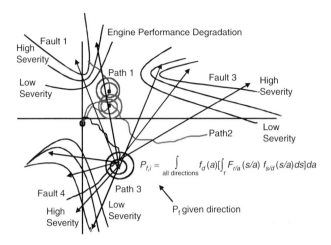

FIGURE 41.9 Engine performance degradation.

41.2.6 Multivariate Reliability Analysis Using Stochastic Fault-Basin Concept

As discussed in the previous section, engine parameters measured on-line include pressures, temperatures, and fuel flows in different compartments of an engine. The proposed stochastic fault-diagnostic–prognostic procedure is illustrated in Figure 41.9 using a two-dimensional stochastic parameter space representation [1, 2]. As shown in Figure 41.9 for the usage path 3, the engine condition at a given time can be diagnosed by evaluating all the risks of potential engine faults. Figure 41.10 shows the engine performance degradation from usage point P1 to usage point P2. This degradation is shown in the original parameter space, X-space, and in transformed standard Gaussian space, U-space that is used typically for reliability estimate calculations. Herein, the engine reliability is measured by the fault-reliability index, which is similar in concept to the traditional reliability index used in structural reliability theory computed in a transformed standard Gaussian parameter space.

To compute the fault-reliability index, the performance safety margin or the performance function first needs to be defined. The performance safety margin in engine-parameter space was simply defined by the stochastic distance between the measurement-variability ellipsoid (cluster) and the fault-variability ellipsoid (cluster), as shown in Figure 41.11. Figure 41.11 shows that this distance can be defined in two ways: (1) safety margin of Type A, a linear distance between the two multidimensional ellipsoids, and (2) safety margin of Type B, an arc length defined by the curvilinear usage trajectory. The curvilinear

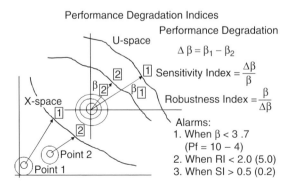

FIGURE 41.10 Standard space reliability model.

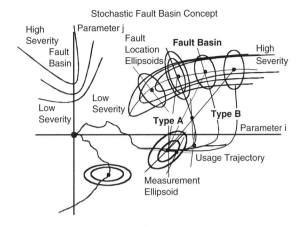

FIGURE 41.11 Illustration of stochastic fault-basin concept.

safety margin B, in comparison with the approach using linear safety-margin A, ensures a more complete compatibility between the measurement and the fault-complex patterns.

There are two important aspects related to stochastic fault diagnostics:

1. *Single fault pattern*: The single-fault-pattern modeling needs a continuous representation in the parameter space. This stochastic representation is called herein the stochastic fault-basin (SFB) concept. The fault-point location model is a truncated representation that can produce erroneous diagnostics and prognostics. More appropriately, faults should be represented in the parameter space as basins of attraction rather than point locations, as illustrated in Figure 41.11. Thus, the reliability index has to be computed with respect to a continuous fault path from low-severity to high-severity levels, and not with respect to a particular fault location.
2. *Multiple-fault pattern*: If multiple faults with different severities are simultaneously present, then it is necessary to decompose the multivariate statistical measurement in the fault patterns before defining the fault safety margins. The stochastic representation for multiple faults is called herein the stochastic multiple-fault map (SMFM) concept. Figure 41.12 illustrates that directly using the statistical measurement reference point M for reliability computations may hide the existence of simultaneous faults. The safety margins computed for location M are quite large, and there is no imminent fault detected. If the measurements in the two fault patterns are decomposed, the two reference measurement points M1 and M2 are determined, with one reference point for each fault

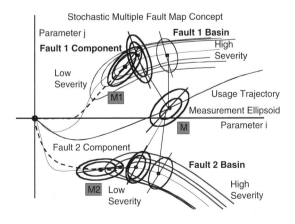

FIGURE 41.12 Illustration of multiple-fault-basin map concept.

(measurement fault component). If the reliability is computed using the two reference points M1 and M2 instead of the single measurement point M, the results are very different. In this last situation, the simultaneous faults 1 and 2 are detected with different severities.

To diagnose the engine fault and provide prognostics for in-operation conditions, reliability indices were computed for any point on the predicted trajectory within the fault basin of attraction. The reliability index computed for current measurement location is used for fault diagnostics. Reliability indices with computed locations on the future projected usage trajectory, from a predicted location to the fault location, are used for prognostics. The associated fault diagnostic and prognostic probabilities, P_f, in the multidimensional parameter space are approximated based on the computed reliability index, β

$$P_f \approx \Phi(-\beta) \tag{41.14}$$

where $\Phi(\cdot)$ is the standard Gaussian cumulative distribution function.

To determine the usage rate in probabilistic terms (measurement location speed on the trajectory), the reliability index gradients are required. Specifically, two reliability-index sensitivity measures are introduced: a cumulative sensitivity index and an evolutionary sensitivity index. The cumulative reliability sensitivity index (CRSI) is defined by the "global" nondimensional variation of the reliability index, β (the relation between "failure probability," here read as fault-diagnostic probability, and reliability index is discussed on the next page) from initial state, at 0, to the final state, at t (over the interval $[0, t]$):

$$C_{0,t} = -\frac{\beta_t - \beta_0}{\beta_0} = -\frac{\Delta\beta_{ot}}{\beta_0} \tag{41.15}$$

The evolutionary reliability sensitivity index (ERSI) is defined by the "local" nondimensional variation of the reliability index from an intermediary state, at time ti, to another intermediary state, at time $ti+1$ (over the interval $[ti, ti+1]$)

$$E_{ti,ti+1} = -\frac{\beta_{ti+1} - \beta_{ti}}{\beta_{ti}} = -\frac{\Delta\beta_{ti,ti+1}}{\beta_{ti}} \tag{41.16}$$

These two reliability-sensitivity indices indicate (in percentage) the changes in engine reliability. A zero value indicates no safety (performance) degradation, while a positive value indicates a safety (performance) degradation, and a negative value indicates a safety improvement. Robustness indices (RI) can be defined as the inverse of sensitivity indices (SI). After model calibration, "red" alarms can be set to a lower bound of reliability index of 3.70 (equivalent to fault probability of 0.0001). A CRSI of 0.5 or, equivalently, a CRRI of 2.0, and an ERSI of 0.2 or, equivalently, an ERRI of 5.0, can be set as "yellow" alarms.

Figure 41.13 through Figure 41.16 show the computed reliability index and cumulative sensitivity reliability index (CRSI) for a fan fault that produces a 3% efficiency drop. The reliability computations are done using the dimensional variations of engine parameters. It should be noticed from these figures that the reliability index value is zero and the CRSI value is unity for this fault type. The computed values indicate that the fan fault defined by 3% efficiency has certainly occurred if the measurement and fault location for a given overlap (probability of occurrence is 50%).

It is interesting to note — by comparing Figure 41.13 and Figure 41.14 with Figure 41.15 and Figure 41.16 — the different qualitative behavior of the OQS and the OT models. The OQS GPA model incorporates a significant "feedback" effect that is visible in the reliability estimate profile computed along the engine profile (Fault 7 is at inlet, Fault 1 is at outlet). This is due to the fact that transients are not captured well by this model, so that they add larger variability in the statistical deviations of engine parameters.

In contrast, the OT GPA model shows as a "forward" model for which the reliability estimate profile indicates a monotonic growth that shows a gradual increase of statistical deviations. Higher reliability indices indicate that the statistical deviations are much larger than those that correspond to fault-severity levels between 1 and 3% in the compartments other than the fan compartment. This can be a critical

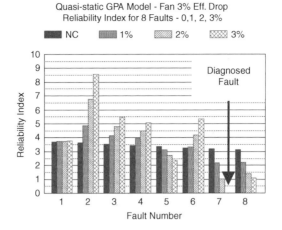

FIGURE 41.13 Computed reliability index using the OQS GPA model.

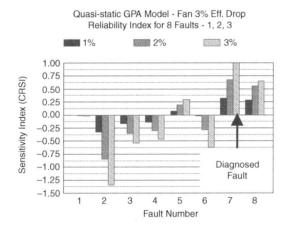

FIGURE 41.14 Reliability sensitivity index using the OQS GPA model.

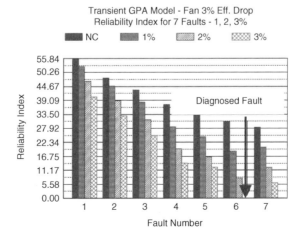

FIGURE 41.15 Computed reliability index using the OT model.

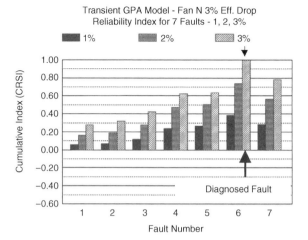

FIGURE 41.16 Reliability sensitivity index using the OT model.

issue when an incipient fan fault exists, since this can project unreal, "phantom" faults in the other compartments. This aspect can be fixed by using the OT model in conjunction with the PT model.

One important aspect that can reflect on the accuracy of a fault diagnostic–prognostic procedure is the fact that statistics of different flight profiles are quite different, and they are also different than the statistics of the overall ensemble of all flight-profile types (Table 41.1). In statistical terms, this indicates that the ensemble of flight profiles (including all profile types) is not ergodic. Thus, the fault diagnostics–prognostics have to be approached by operational spectrum-type statistical modeling.

41.2.7 Decision-Level Fusion

The decision-level fusion presented here is based on diagnostic–prognostic reasoning from the individual models. The benefit of this approach is that it maintains separation of the models, thus providing a plug-and-play scenario, where one, two, or all three modeling approaches could be used to assess the engine's health, depending on the engine's requirements. The decision-level fusion uses the diagnostic outputs from the separate engine models, (OQS, OT, PT) and combines them using a heuristic scoring approach to enhance the confidence of the final output. The purpose of the fusion procedure is to take advantage of the synergies provided by using the multiple engine models. For example, the PT model is good for detecting multiple faults, although it is limited in its ability to discriminate between certain turbine faults; the OT model is good for detecting and discriminating between compressor faults; and the OQS model is good for detecting turbine faults.

The reliability-based indices from three separate engine models are weighted, based on the fault-detection and fault-discrimination sensitivity indices of the engine models for given fault conditions, to provide a joint reliability index (RI). The scoring method used employs the sensitivity and normalized sensitivity indices for the three models. This applies weighting to the model outputs based on their fault-detection and fault-discrimination capabilities. The scoring model implemented computes the sum

$$S(RI_i) = \frac{1}{C_i} \sum_{j=1}^{3} RI_{ij}\left(\frac{3}{4} SI_{ij} + \frac{1}{4} NSI_{ij}\right) \quad (41.17)$$

where C_i is the normalization constant.

An example based on the single-fault implementation of the fault-basin approach is presented in Figure 41.17 through Figure 41.20 to illustrate the benefits of decision-level fusion in the diagnostic process. In this case, the OQS model performs relatively consistently across the faults considered. However, the reliability indices are relatively low with the exception of HPT efficiency and LPT capacity faults.

TABLE 41.1 Mean and Standard Deviation of Aircraft Engine Parameters

				Table of Mean Values						
Profile Type	P2	P3	P4	P5	P6	T2	T3	T4	T5	T6
All Types	0.000	0.000	−0.032	0.024	0.024	−0.048	0.720	−0.162	0.000	−0.141
Type A	−0.003	0.046	−0.009	0.012	0.012	−0.038	0.299	−1.217	−0.955	−0.672
Type B	0.002	−0.017	−0.015	0.010	0.010	−0.054	0.604	−1.873	−1.494	−1.484
Type C	−0.001	−0.054	−0.050	0.015	0.015	−0.072	1.100	−0.283	−0.126	−0.159
Type D	0.005	0.091	0.008	0.046	0.046	0.035	0.660	−0.131	0.053	0.700
Type E	0.005	−0.522	−0.057	0.074	0.074	−0.082	0.607	0.783	0.940	0.792
Type F	0.005	0.095	0.044	0.042	0.042	−0.043	0.975	−0.189	0.025	−0.605
Type G	0.005	−0.181	−0.227	0.007	0.007	−0.145	0.847	2.561	2.271	1.218
Type H	0.000	0.019	−0.016	0.031	0.031	−0.026	0.795	0.189	0.299	−0.113
Type I	−0.009	−0.024	−0.017	0.005	0.005	−0.049	0.566	−0.369	−0.153	−0.299
Maximum	0.005	0.095	0.044	0.074	0.074	0.035	1.100	2.561	2.271	1.218
Minimum	−0.009	−0.522	−0.227	0.005	0.005	−0.145	0.299	−1.873	−1.494	−1.484
Mean	0.001	−0.061	−0.038	0.027	0.027	−0.053	0.717	−0.059	0.096	−0.069
STD	0.005	0.192	0.077	0.023	0.023	0.048	0.240	1.247	1.074	0.844
				Table of Standard Deviations						
Profile Type	P2	P3	P4	P5	P6	T2	T3	T4	T5	T6
All Types	0.054	0.406	0.369	0.129	0.129	0.468	1.481	6.367	5.290	4.854
Type A	0.060	0.237	0.259	0.099	0.099	0.473	0.931	7.198	5.874	4.885
Type B	0.041	0.260	0.242	0.083	0.083	0.354	1.114	6.181	5.032	4.995
Type C	0.047	0.312	0.289	0.104	0.104	0.425	1.640	6.072	5.033	4.474
Type D	0.066	0.383	0.336	0.132	0.132	0.507	1.531	5.879	4.768	4.837
Type E	0.047	0.298	0.326	0.090	0.090	0.776	1.659	7.775	6.747	5.251
Type F	0.046	0.329	0.292	0.103	0.103	0.353	1.591	5.075	4.501	4.362
Type G	0.063	0.703	0.627	0.194	0.194	0.624	1.642	6.855	5.649	5.515
Type H	0.055	0.408	0.395	0.140	0.140	0.418	1.525	5.798	4.864	4.411
Type I	0.054	0.450	0.388	0.155	0.155	0.379	1.496	6.307	5.232	4.869
Maximum	0.066	0.703	0.627	0.194	0.194	0.776	1.659	7.775	6.747	5.515
Minimum	0.041	0.237	0.242	0.083	0.083	0.353	0.931	5.075	4.501	4.362
Mean	0.053	0.376	0.351	0.122	0.122	0.479	1.459	6.349	5.300	4.844
STD	0.009	0.141	0.116	0.036	0.036	0.141	0.258	0.811	0.690	0.387

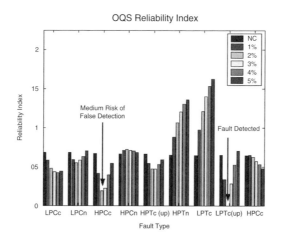

FIGURE 41.17 OQS reliability indices LPT 2% capacity increase.

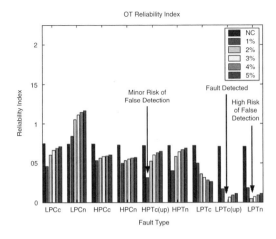

FIGURE 41.18 OT reliability indices LPT 2% capacity increase.

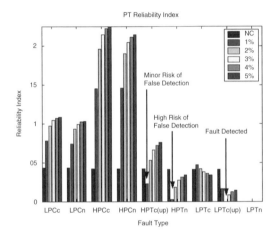

FIGURE 41.19 PT reliability indices LPT 2% capacity increase.

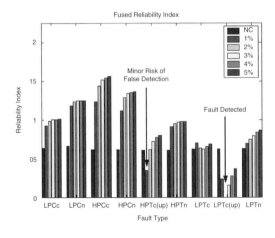

FIGURE 41.20 Fused reliability indices LPT 2% capacity increase.

TABLE 41.2 Benefits of Feature Vector Extraction (ANFIS)

Category of Benefit	General Benefit	Operational Advantages
Feature extraction	Converts data into useable features that describe the state of the engine	Provides deviations from normal operating conditions
Robust operational performance	In the event of a reduced sensor array (one or more nonoperational sensors), the engine models still extract features. Loss of output sensor results in reduced dimensionality of feature space. Loss of input sensor results in a reduction of model performance	Allows continued operation despite reduced sensor array (input or output)
Engine–engine variability	Models can be readily tuned to fit a specific engine, based on initial generic ANFIS models for a given engine class	More accurate model of a specific engine enables assessment of engine performance relative to fleet
In-flight capabilities	Transient engine models extend the capabilities of air VEHM systems to be a true flight-capable system. Quasi-stationary models are improved about 40% through the inclusion of PLA	More up-to-date information regarding engine condition means less reliance on ground-based testing, leading to a reduction in maintenance costs
Improved detection	Increased sensitivity index of engine models results in earlier and more confident fault detection	Earlier detection of fault onset
Improved diagnosis	Increased fault discrimination of the engine models results in improved diagnostic output	Accuracy of fault diagnostics enables maintenance and parts to be planned prior to servicing

The OT model provides additional evidence against a Fan/LPC fault; however, it is unable to clearly isolate the severity of the fault, and there is potential for false detection of an HPT capacity or an LPT efficiency fault. The PT model clearly eliminates the fan/compressor sections as a source of the fault. However, it has difficulties discriminating between the source of the turbine fault and its severity. In this case, the fused reliability indices provide clear advantages, with the source of the fault identified as resulting from a 2% increase in the LPT capacity. There is still a small probability of false detection of a 1% HPT capacity increase. Nevertheless, the fused reliability indices perform significantly better than the individual models, both at fault diagnosis and determination of the fault severity.

Embedded into StoFIS are three component technologies. The first is the feature vector extraction (data-level fusion), where data from multiple sensors are fused into the nondeterministic ANFIS GPA engine models prior to feature extraction. The second is multivariate stochastic analysis of the extracted feature vectors (feature-level fusion), where parameter deviations from the ANFIS models are used to characterize the system and arrive at a diagnostic–prognostic output. The third is decision-level fusion, where diagnostic outputs from separate nondeterministic ANFIS GPA engine models are fused after diagnostic reasoning to enhance the confidence of the final output. The benefits of the three component technologies of StoFIS are described in Table 41.2 through Table 41.4.

41.3 Application 2: DWPA Multiple-Band-Pass Demodulation and Automated Diagnostics

Rolling-element bearings are the most common cause of small machinery failure, and overall vibration-level changes are virtually undetectable in the early stages of deterioration. However, due to the characteristics of rolling-element bearing faults, vibration analysis techniques have proven to be an effective tool for the detection and diagnosis of incipient faults. The demodulated spectrum is the most common technique used for the detection of localized bearing faults. However, for low-speed rolling-element bearings, demodulation can be unreliable for the detection and diagnosis of faults [6]. Difficulties include spectral smearing due to speed fluctuations and skidding of rolling elements, poor performance under

TABLE 41.3 Benefits of Multivariate Stochastic Analysis (Fault Basin)

Category of Benefit	General Benefit	Operational Advantages
Robust operational performance	Multidimensional mapping of parameter deviations is robust in the event of the loss of one or more parameters	Allows continued operation despite reduced sensor array (input or output)
More-refined feature mapping	Extraction of stochastic features provides more-refined diagnostics and prognostics (e.g., two faults may have similar mean deviations, but their ellipsoid clusters may have significantly different correlation structure and thus orientation in space)	Earlier detection of anomalies and improved fault discrimination
Incorporates data uncertainty	Transformation into standard Gaussian space enables engine reliability to be assessed in terms of risk	Enables calculation of the risks of potential engine faults
Fault basins of attraction	The reliability index is computed with respect to a continuous fault path from low-severity to high-severity levels and not with respect to a particular fault location	Improved robustness and more accurate indicator of fault severity; more-refined and accurate diagnostics/prognostics
Reliability-based reasoning	Diagnostic–prognostic output is provided in terms of reliability-based indices	Operational risks can be assessed, thus assisting in operational status and maintenance decisions for aircraft
Multiple fault detection	Scanning of multiple fault deviations in multidimensional parameter space provides the ability to assess the risk of multiple faults	Improved robustness for multiple faults
Prognostic output	Projected usage trajectories in multidimensional space are used to provide reliability-based prognostic output of fault degradation	Provides knowledge of the speed of progression for a given fault and assists with maintenance planning

high levels of noise, and difficulties in identifying and extracting the regions of bearing resonance. This section presents a nondeterministic method to surmount these problems by combining several techniques, including time-frequency decomposition, autoregressive (AR) stochastic process spectral analysis, and nondeterministic ANFIS.

TABLE 41.4 Benefits of Decision-Level Fusion

Category of Benefit	General Benefit	Operational Advantages
Robust operational performance	Model redundancy allows for the situation where one model is unable to adequately detect and diagnose a fault In the event of a reduced sensor array (one or more nonoperational sensors), the most robust model can provide information while others are not performing adequately. Adaptive weighting can be used to reflect the relative performance of each of the models under adverse conditions or reduced sensor arrays	Facilitates the best possible diagnostic accuracy for each fault class Enables optimum operation for a given array of sensors
Increased fault-detection sensitivity	Sensitivity of fault detection is limited by the most sensitive model for a given fault condition. Depending on the fault location and type, this may be the OQS, OT, or PT model	Earlier detection of performance degradation
Improved system reliability	Two or three models can confirm the same engine fault condition	Reduced probability of false detection
Reduced ambiguity	Reduced probability of uncertainty in fault diagnosis, as each of the models provides different levels of fault discrimination from other fault sources	Increased plausibility in the fault type diagnosed and corresponding severity level

The extraction of attenuated resonant vibrations due to impacts from localized faults in rolling-element bearings is normally achieved by high- or band-pass filtering of the vibration signal. The main problem with this approach is the difficulty in choosing an appropriate filter range of interest. This section presents an alternative to traditional approaches, which enables the automation of the selection and the inclusion of multiple frequency bands of interest.

The method for the extraction of high-frequency transients due to bearing impact resonance is achieved at an optimal time-frequency resolution via best-basis discrete wavelet packet analysis (DWPA) representation, using the Daubechies-20 wavelet [7]. Selection of the frequency band or bands of interest is achieved by analyzing the characteristics of each of the wavelet packets. The selection process is automated through the use of an ANFIS model, thus removing the need for the analyst to manually identify the bands of interest.

The best-basis DWPA provides an optimal time-frequency decomposition of the signal and facilitates the extraction and reconstruction of wavelet packets containing bearing fault-related information. For a signal component composed of wide-band transients, high time resolution and low frequency resolution would be required. On the other hand, a slowly varying narrow-band component of a signal requires better frequency resolution, with its time resolution being less important. By obtaining the best-basis DWPA of the signal prior to extraction of the wavelet packets of interest, two important objectives are accomplished. The most important of these is the improved time resolution of the bearing transients while maintaining isolation from other signal components. This improves the ability to resolve low-amplitude transient features over the noise floor level. Second, subsequent processing required to extract the relevant wavelet packets is substantially reduced. A detailed explanation of optimal time-frequency decompositions as well as the choice mother wavelet can be found in the literature [7–9].

DWPA multiple-band-pass filtering surmounts the problem of extracting regions of bearing resonance that are intertwined with continuous signals. Figure 41.21 illustrates how this method facilitates the extraction of bearing-fault-related components from a signal while rejecting the unwanted harmonics. The wavelet packets identified by the ANFIS model as containing bearing-fault-related features are indicated. To visualize the rejection of wavelet packets containing unwanted continuous signal components, the power spectral density is plotted along the vertical axis of the DWPA representation. Wavelet packets that contained the harmonic peaks present in the power spectral density plot were rejected by the adaptive network-based fuzzy inference system as containing excessive levels of signal contamination.

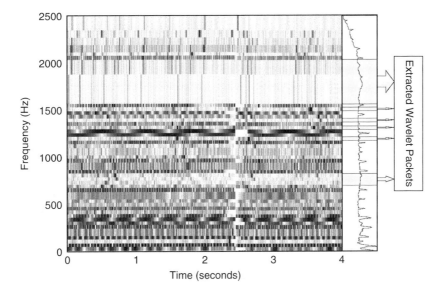

FIGURE 41.21 Selection and extraction of wavelet packets containing bearing-failure-related features. The extracted wavelet packets are (3, 7), (4, 10), (6, 17), (6, 26), (6, 27), (6, 30), (6, 49), (6, 50), and (6, 53).

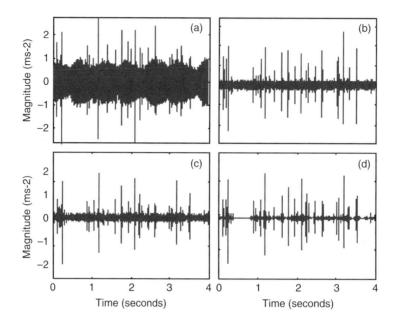

FIGURE 41.22 (a) High-pass filtered signal (order 40, [F > 500 Hz]); (b) FIR band-pass filter (order 40, [1500 Hz < $F_{band\text{-}pass}$ < 2500 Hz]); (c) reconstruction of extracted wavelet packets; (d) hard-threshold denoised reconstruction of extracted wavelet packets.

This clearly demonstrates the ability of DWPA multiple-band-pass filtering to extract only the wavelet packets composed predominantly of bearing-fault-related vibrations.

The ability of the multiple-band-pass technique to select more than a single band of interest enables sections of the signal that contain predominantly noise or contaminating sources of vibration to be excluded. This results in the extraction of a cleaner bearing-fault signal. Figure 41.22 shows a comparison between high-pass, manually optimized band-pass filters, and DWPA multiple-band-pass filtering (with and without noise reduction). Visual examination of the filtered signals indicates a significant decrease in the contaminating effects of noise and other sources of vibration when the DWPA multiple-band-pass filtering is applied. The DWPA reconstructed signal has a marginally lower level of sinusoidal contamination than the best possible band-pass filter, and the bearing-fault-related transients are also stronger. Hard threshold denoising almost eliminated the remaining polluting sources of vibration. This further enhances the ability of DWPA multiple-band-pass enveloped spectra to accurately diagnose the location and magnitude of bearing defects.

41.3.1 Implementation of ANFIS for Automatic Feature Extraction

In order to design and develop a robust and reliable nondeterministic identification of wavelet packets of interest, an ANFIS was used to provide a statistical best-estimate based on the input parameters of the model. The parameters chosen must enable the neuro-fuzzy network to make intelligent decisions regarding the extraction of wavelet packets containing bearing-fault-related information. The input parameters that were chosen for this process were kurtosis and the spectrum peak ratio (SPR).

Kurtosis is an effective measure of the spikiness of a signal. A high kurtosis level indicates that the wavelet packet is impulsive in nature, as would be expected from a wavelet packet that contains bearing-fault-related features. Kurtosis is defined as:

$$\text{Kurtosis} = \frac{1}{NS_y^4} \sum_{i=1}^{N} (y(i) - \bar{y})^4 \qquad (41.18)$$

S_y is the standard deviation, and $\bar{y}$ is the mean of data sample y. Kurtosis was chosen over other measures of spikiness (crest factor, impulse factor, and shape factor) due to its statistically robust nature.

The spectrum peak ratio is defined as the sum of the peak values of the defect frequency and its harmonics, divided by the average of the spectrum [10]. Shiroishi [10] used the spectrum peak ratio as a trending parameter to indicate the presence of localized bearing defects, which was found to be more robust than considering just the defect frequency.

$$\text{SPR} = \frac{N \times \sum_{h=1}^{n} P_h}{\sum_{k=1}^{N} A_i} \tag{41.19}$$

P_h is the amplitude of the peak located at the defect frequency harmonic; A_i is the amplitude at any frequency; and N is the number of points in the spectrum. In order to differentiate between wavelet packets belonging to different classes of bearing faults, three autoregressive-based peak ratios are employed: spectrum peak ratio inner (SPRI), spectrum peak ratio outer (SPRO), and spectrum peak ratio rolling-element (SPRR). Calculation of the spectrum peak ratios was based on Yule-Walker autoregressive spectral estimates of the reconstructed wavelet packets using a model order of 125, equivalent to one shaft revolution. Autoregressive spectral analysis was used in preference to the FFT (fast Fourier transform), as this method has been shown to reduce the effect of spectral smearing and skidding for low-speed rolling-element bearings [11].

Seeded faults in a low-speed test rig and mathematical models of bearings containing localized faults [12] were used to construct a database of 2810 wavelet packets. These wavelet packets were individually assessed as to whether they contained bearing-fault-related features by visual examination of their time series and the envelope AR spectrum. They were then categorized for each fault class as containing fault-related features (1), probably containing fault-related features (0.66), probably not containing fault-related features (0.33), or not containing fault-related features (0). The wavelet packet data set included 444 containing inner-race fault-defect information, 221 containing rolling-element fault information, and 162 containing outer-race fault information. The wavelet packets were split into three data sets: a training data set of 1000 wavelet packets, a checking data set of 1000 wavelet packets, and a testing data set of 810 wavelet packets.

Given the training and checking input/output data sets, the membership function parameters were adjusted using a back-propagation algorithm in combination with a least-squares method. The checking data were used to cross-validate and test the generalization capability of the fuzzy inference system. This was achieved by testing how well the checking data fits the fuzzy inference system at each epoch of training, and the final membership functions were associated with the training epoch that has a minimum checking error. This was an important task, as it ensured that the tendency for the fuzzy inference system to overfit the training data, especially for a large number of epochs, was avoided.

Two different membership function structures were compared in this study. The first consisted of the kurtosis and spectrum peak ratio input amplitudes being transformed by two membership functions (small and large), and the second was split into three membership functions (small, medium, and large). The parameters for the initial input membership functions were determined by expert knowledge of the system being modeled. Figure 41.23 illustrates a three-membership function structure.

41.3.2 Performance of ANFIS Feature Extraction

A substantial reduction of the sum of squared errors for the testing data was apparent after training of the neuro-fuzzy systems. This reduction is quantified in Table 41.5, with an average decrease in the sum of squared errors of 34.9% for the neuro-fuzzy systems. The errors in the testing data are defined as the sum of the squared differences between the output of the fuzzy inference system and the fault-categorization

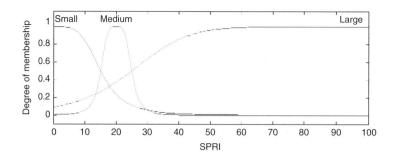

FIGURE 41.23 SPRI membership function (small, medium, and large).

scheme based on visual inspection. In order to classify the individual wavelet packets as either containing or not containing bearing-fault-related features, it is necessary to define a minimum crisp output value that would indicate the presence of high-frequency bearing transients. The minimum crisp output value was set as 0.5 for the correct classification rates presented in Table 41.6. The 3/3 neuro-fuzzy system had a marginally better correct classification rate than the 2/2 neuro-fuzzy system.

41.3.3 Automated Fault Classification

A multiple-band fusion technique was implemented to provide an automated diagnostic tool based on the DWPA feature-extraction process. This is similar in concept to the multisensor fusion used by Loskiewicz and Uhrig [13], where multiple sensor data were fused in order to increase the confidence factor for the final diagnosis compared with single-sensor diagnosis. The difference in this case is that the data-fusion concerns multiple bands of interest for a single sensor, with the intention of enhancing the confidence of correct diagnosis based on single-sensor vibration analysis.

The wavelet packet selection process using the nondeterministic ANFIS identified the existence and fault class of bearing transients for individual wavelet packets. The output for each wavelet packet can be viewed as the confidence factor relating to the existence of a particular bearing-fault class for the wavelet packet being examined. Through fusion of the confidence factors for each wavelet packet that indicated the presence of bearing-fault-related transients, a more confident assessment of the fault classification can be ascertained, with the signals requiring further manual analysis for definitive classification easily identified. Manual classification can be achieved by visual examination of the demodulated spectrum of the multiple-band-pass-filtered signal [14]. The use of multiple-band fusion of confidence factors for fault classification reduces both the probabilities of missing or of incorrectly classifying a stochastic faulty bearing.

Two parameters were used in the proposed fault-classification scheme: the average confidence level of fault classification and the number of wavelet packets that contained bearing-fault-related features. The average confidence level (ACL) was based on the average output from the nondeterministic ANFIS models

TABLE 41.5 Testing Sum of the Squared Errors for the Initial and Final Fuzzy Systems

	2/2 Fuzzy		3/3 Fuzzy	
	Initial	Trained	Initial	Trained
IRF	0.1239	0.0806	0.1383	0.0829
REF	0.1380	0.0926	0.1639	0.0958
ORF	0.0501	0.0390	0.0591	0.0380

IRF-inner race fault; REF-rolling-element fault; ORF-outer race fault.

TABLE 41.6 Correct Classification Rate (%) for the Neuro-Fuzzy Systems

	2/2 Fuzzy	3/3 Fuzzy
Fault	97.98%	98.85%
Probably fault	83.53%	82.18%
Probably no fault	90.45%	87.52%
No fault	99.96%	99.96%

for the wavelet packets that were identified as containing bearing-fault-related transients, and it was calculated for each fault class (inner-race fault, rolling-element fault, and outer-race fault).

$$\text{ACL} = \frac{\sum_{i=1}^{n} \text{ANFIS Output}}{n} \qquad (41.20)$$

where n is the total number of wavelet packets that contain bearing-fault-related features. Table 41.7 shows typical outputs obtained from this process.

Box plots of the outputs from the fusion of the multiple-band-pass fuzzy confidence factors are presented in Figure 41.24 for each of the fault classes. Box plots give you an idea of the distribution of data, especially in terms of symmetry and scale. The limits of the box correspond to the first and the third quartiles (Q1 and Q3), and the fences, respectively, to Q1 − 1.5(Q3-Q1) and Q3 + 1.5(Q3-Q1). The minimum and maximum data points are printed with a black circle. The distinction between each of the fault classes is evident. In the case of the combined inner-race/rolling-element fault class, the average confidence level was lower for both fault classes than for an individual fault. There were two factors at play that result in the reduced confidence levels. First, there was the interference due to the presence of bearing transients from both fault classes in the extracted wavelet packets. This reduced the spectral peak ratio for each fault class, and may have in turn reduced the fuzzy confidence factor indicating the presence of a fault for the individual wavelet packets. The second factor was a consequence of variation in the frequency and relative amplitude of the bearing impact transients, which was found to have a degree of dependence on the impact location. Due to this, certain bands contained bearing transients of significantly reduced amplitude from one of the fault classes, and in some cases contained transients from only one fault class. Although this resulted in a reduction of the average confidence level after fusion of the outputs from the ANFIS model, the existence of multiple faults was clearly indicated in each case.

Fault classification through fusion of multiple-band-pass fuzzy confidence factors resulted in a successful classification rate of 100% for the signals examined, with 95% requiring no additional analysis for verification of the fault classification. The vibration signals where the average confidence level was considered inadequate for a positive diagnosis to be made where limited to the condition of combined inner-race/rolling-element faults.

TABLE 41.7 Typical Outputs Obtained from Fusion of Multiple-Band-Pass Fuzzy Confidence Factors, and the Number of Wavelet Packets on Which the Calculations Were Based

Fault Type	Number of Wavelet Packets	ACL (IRF)	ACL (REF)	ACL (ORF)
IRF	6	98%	14%	0%
ORF	7	0%	0%	99%
REF	7	0%	93%	0%
IRF/REF	8	65%	85%	0%

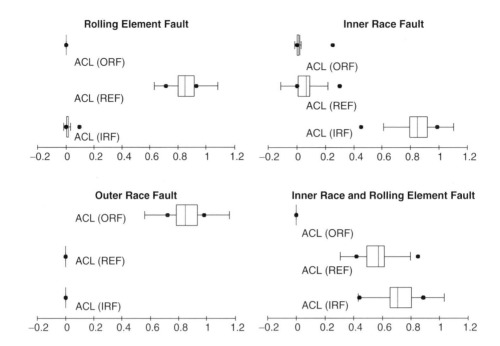

FIGURE 41.24 Box plots of the outputs obtained from the fusion of multiple-band-pass fuzzy confidence factors for each fault class considered.

41.3.4 Multiple-Band-Pass Fault-Severity Index

Obtaining an accurate determination and trending of fault severity is an integral part of a successful predictive maintenance program. Trending parameters can be used to indicate the general health of a machine, or that of a specific machine element. In order for trending parameters to be effectual, they must be sensitive enough to pick up changes in the condition of machine elements, yet not so sensitive that small variations in operating conditions trigger an alarm. A bearing-specific trending parameter, the wavelet peak index (WPI), based on the multiple-band-pass DWPA feature-extraction technique, is presented for this purpose.

The wavelet peak index is defined as the peak level of the combined reconstructed wavelet packets that contain bearing-fault-related features. The time-domain-based peak level was favored over trending of spectral peaks, as this allowed the modulating effects of load to be ignored, thus allowing direct comparison between the severity of different types of faults (inner-race, outer-race, and rolling-element faults). To test the performance of this trending parameter, a series of tests involving the introduction of an artificial crack on the inner race of a low-speed cylindrical rolling-element bearing were performed, simulating wear-out by deepening and widening the groove in several stages. The tests were conducted at 20, 60, and 120 rpm. The proposed trending parameter was then compared with a number of commonly used trending parameters for bearing-health determination, peak level, RMS (root mean square), crest factor, and kurtosis.

Figure 41.25 displays the results of the trending parameters tested for each of the series of low-speed artificial wear-out tests. For each series of tests, the wavelet peak index provided a clear and sensitive trend of the deteriorating condition of the bearing as the fault width was increased. This was true at all operating speeds, even for the smallest of the fault widths considered (0.38 mm). At operating speeds of 20 and 60 rpm, no discernible increase in either peak level or crest factor was noted for fault widths below 0.67 mm, and RMS provided no indication of the deteriorating bearing condition for any of the fault widths considered. The wavelet peak index is indisputably more sensitive to bearing transients of low amplitudes than either the peak level or crest factor, with the wavelet peak index more closely coupled

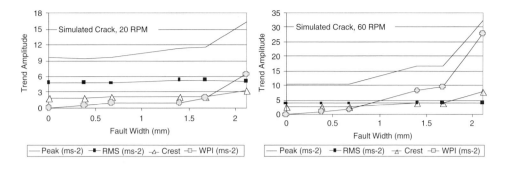

FIGURE 41.25 Trend parameters vs. crack size.

with the width of the bearing fault. Unlike the peak level, which is an overall indicator including vibration components unrelated to the bearing fault, the index WPI is specific to bearing-related faults, as it is the peak of the multiple-band-pass-filtered signal.

41.4 Concluding Remarks

Two nondeterministic hybrid architectures have been demonstrated to provide enhanced vehicle health-management reasoning for the detection, diagnosis, and prognosis of faults. The nondeterministic hybrid approaches have enabled more robust and comprehensive solutions to the challenges of developing vehicle health-management systems capable of predicting faults with associated confidence or risk levels. The utilization of advanced signal processing and multiple levels of data fusion to maximize the extracted information used for diagnostic-prognostic reasoning have also been illustrated.

References

1. Ghiocel, D.M. and Altmann, J., A Hybrid Stochastic-Neuro-Fuzzy Model-Based System for In-Flight Gas Turbine Engine Diagnostics, presented at 55th Machine Failure Prevention Technology Meeting, Society for Machinery Prevention Technology, Virginia Beach, VA, 2001.
2. Ghiocel, D.M., A new perspective on health management using stochastic fault diagnostic and prognostic models, *Int. J. Adv. Manuf. Syst.*, 4 (1), PAGE, 2001.
3. Jang, J.-S.R., ANFIS: adaptive network-based fuzzy inference systems, *IEEE Trans. Syst., Man, Cybernetics*, 23 (3), 665–685, 1993.
4. Jang J.-S.R. and Sun C.-T., Neuro-fuzzy modelling and control, *Proc. IEEE*, 83 (3), 378–406, 1995.
5. Ghiocel, D.M. and Altmann, J. Conceptual and Tool Advances in Machinery Preventive Diagnostics, presented at 56th Machine Failure Prevention Technology Meeting, Society for Machinery Failure Prevention Technology, Virginia Beach, VA, 2002.
6. Mathew, J. et al., Incipient damage detection in low speed bearings using the demodulated resonance technique, in *Proceedings of International Tribology Conference*, Monash University, Melbourne, Australia, 1987, pp. 366–369.
7. Daubechies, I., The wavelet transform, time-frequency localisation and signal analysis, *IEEE Trans. Inf. Theory*, 36 (5), 961–1005, 1990.
8. Misiti, M. et al., *Wavelet Toolbox MATLAB*, The MathWorks, Natick, MA, 1996.
9. Altmann, J. and Mathew J., Optimal configuration of the time-frequency representation of vibration signals, *Machine Condition Monitoring Res. Bull.*, 8, 13–24, 1996.
10. Shiroishi, J. et al., Bearing condition monitoring via vibration and acoustic emission measurements, *Mech. Syst. Signal Process.*, 11 (5), 693–705, 1997.
11. Mechefske, C., Fault detection and diagnosis in low-speed rolling-element bearings; 1: the use of parametric spectra, *Mech. Syst. Signal Process.*, 6, 297–307, 1992.

12. Altmann J. and Mathew J., Analytical modelling of vibrations due to localised defects in rolling-element bearings, in *Proceedings of COMADEM '98*, Monash University, Launceston, Australia, 1998, pp. 31–40.
13. Loskiewicz, A. and Uhrig, E., Decision fusion by fuzzy-set operation, in *Proceedings of IEEE World Congress on Computational Intelligence, Fuzzy Systems Conference*, IEEE, Orlando, FL, 1994, pp. 1412–1417.
14. Altmann J. and Mathew J., Multiple band-pass autoregressive demodulation for rolling-element bearing fault diagnosis, *Mech. Syst. Signal Process.*, 15 (5), 963–977, 2001.

42

Using Probabilistic Microstructural Methods to Predict the Fatigue Response of a Simple Laboratory Specimen

42.1	Introduction...	**42**-1
42.2	Background...	**42**-2
	Various Stages of Fatigue • Scatter in Fatigue Life • Probabilistic Micromechanics	
42.3	Damage-Accumulation Models.....................................	**42**-5
	Overview of a Probabilistic Microstructural Fatigue Model • Crack Nucleation • Microstructurally Small-Crack Growth • Long-Crack Growth	
42.4	Estimate of Random Variable Statistics.......................	**42**-19
	Grain Size • Grain Orientation • Frictional Shear Stress	
42.5	Monte Carlo Simulation of Each Stage of Damage Accumulation..	**42**-25
	Crack Nucleation • Small-Crack Growth • Long-Crack Growth	
42.6	Fatigue-Life Predictions...	**42**-29
	Crack Nucleation • Small-Crack Growth • Predicted Total Fatigue Life of a Test Specimen	
42.7	Conclusions..	**42**-33

Robert Tryon
Vextec Corporation

42.1 Introduction

This chapter gives an example of how probabilistic methods can be used to predict the variability expected in the fatigue test of simple laboratory specimens such as smooth round bars. The variability observed in laboratory experiments is discussed. The physical source of the variability is identified. Fracture-mechanics-based models that account for the physical mechanism occurring during fatigue are introduced. Using these models, the governing material parameters likely to have significant uncertainty are identified, and the uncertainty is quantified. A Monte Carlo simulation routine is developed using the mechanistic model and the governing random variables to predict the scatter in the smooth bar response.

For the purpose of this chapter, fatigue is defined as the entire range of damage-accumulation sequences, from crack nucleation of the initially unflawed bar to final fast fracture.

42.2 Background

42.2.1 Various Stages of Fatigue

Current fatigue-life prediction methods in metallic components consider three stages: crack initiation, long-crack propagation, and final fracture. Long-crack propagation and final fracture are the stages of damage accumulation that are well characterized using linear elastic or elastic-plastic fracture mechanics. Crack initiation is the early stage of damage accumulation where small cracks (cracks with depths less than several grain diameters) have been observed to deviate significantly from predicted long-crack fracture mechanics behavior [1]. The deviation is attributed to the heterogeneous media in which small cracks evolve [2].

The crack-initiation stage can be broken down into two phases: crack nucleation and small-crack growth. Crack nucleation is the locally complex process of crack formation on the microstructural scale. Crack nucleation is characterized by smooth fracture surfaces at angles inclined to the loading direction. This type of failure is indicative of shear stress Mode II (sliding mode) fracture. Although loading has been shown to affect the nucleation size [3, 4], experimental evidence suggest that the nucleation size is on the order of the grain size [5–7].

Small-crack growth is characterized by fracture-surface striations perpendicular to the loading direction. This type of failure is indicative of tensile stress Mode I (opening mode) fracture. The behavior of small cracks tends to transition to linear or elastic-plastic fracture mechanics behavior when the crack depth reaches about ten mean grain diameters [8]. Crack nucleation and small-crack growth must be modeled separately because different mechanisms control each phase. The relative importance of the crack-nucleation stage on overall fatigue life depends on several factors. Materials that exhibit a strong preference for planar slip show a strong correlation between the crack causing final fracture and the earliest nucleated cracks [9]. Materials that prefer cross slip showed almost no correlation between the crack causing final fracture and the earliest nucleated cracks [9]. The relative importance of the crack nucleation may also depend on the loading condition. If the loading is relatively low (high-cycle fatigue), the majority of life will be spent in the nucleation of a crack. If the loading is high (low-cycle fatigue), cracks may nucleate early and spend the remainder of the fatigue life in the crack-growth stages. However, high-strength materials have been shown to spend the majority of fatigue life in the crack-nucleation stage, even during low-cycle fatigue [10].

42.2.2 Scatter in Fatigue Life

The coefficient of variation (COV) of fatigue-life tests range widely, depending on the material alloy and load level. Even for well-controlled laboratory tests of annealed smooth specimens at room temperature, the COV varies from less than 10% [9] to over 500% [11] for different steel alloys. This indicates that the fatigue reliability experienced by components in the field may be substantially attributed to the material behavior.

Sasaki et al. [9] compare the variation in crack-nucleation life of mild steel, pure copper, and stainless steel. They found that the COV clearly depends on the stacking fault energy denoted by γ. Relatively low COV was found for mild steel, which has high γ (wavy slip), and high COV was found for stainless steel, which has low γ (planar slip). Copper, which has an intermediate γ, was found to have a value of the COV between mild and stainless steel.

Figure 42.1 shows the relative scatter in the different stages of crack growth for 16 mild steel specimens exposed to high-cycle fatigue [12]. The specimens were shallow-notched and tested in rotating bending. The cracks nucleated at a size roughly equal to the mean grain size (0.07 mm). The small-crack regime for these data extends from the initiation event until the crack reaches about eight times the mean grain

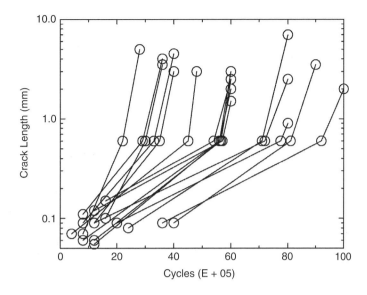

FIGURE 42.1 Crack growth curves for mild steel emphasizing the large scatter in the early stages.

size (0.6 mm). The long-crack growth regime extends from 0.6 mm to failure. Figure 42.1 illustrates the larger amount of scatter in the early stages of crack growth, with relatively little scatter (similar slopes) in the large-crack growth stage. The variation is attributed to the heterogeneous media in which small cracks evolve.

Table 42.1 shows the scatter factors for a NiCrMoV steel turbine rotor shaft material data [13]. The specimens were shallow-notched round bars tested in rotating bending low-cycle fatigue. The scatter factors are for a 99.87% (-3σ) reliability at 90% confidence level using the life-reduction model found in Fu-Ze [14], which assumes a lognormal life distribution. The behavior in Figure 42.1 and Table 42.1 is observed for different stress amplitudes and materials [9, 12, 15, 16].

The values in Table 42.1 show the importance of material response scatter. The scatter factors are life-reduction factors for material-response scatter only. They do not account for variations in loading, geometry, environment, temperature, or size effect. The potential for improved design through the reduction of material scatter is great. When you consider that fatigue is estimated to account for at least 90% of all service failures due to mechanical causes [17], understanding the statistical aspects of fatigue becomes paramount.

Limited experimental work has been performed with regard to the effect of microstructural variation on other aspects of material response. Gokhale and Rhimes [18] prepared pure aluminum to give several grain size distributions with the same mean grain size but different variances. They found that the scatter in grain size played a primary role in controlling yield stress, ultimate tensile stress, reduction in area, and the area under the stress–strain curve. The parameters were much more sensitive to changes in the grain size variance than to changes in the mean grain size. Their research emphasizes the importance of microstructural variation with regard to any type of plastic behavior that is likely to be important to crack initiation and growth kinetics.

TABLE 42.1 Scatter Factors for Stainless Steel in Low-Cycle Fatigue

Failure Definition	Number of Samples	Mean Life (cycles)	Scatter Factor	-3σ Life (cycles)
Nucleation	36	26,700	44.2	604
Small crack	36	52,400	15.5	3,380
Final fracture	14	82,400	2.78	29,600

42.2.3 Probabilistic Micromechanics

Most crack-nucleation models are empirically based macrostructural models [19]. They reduce crack nucleation to simple parametric functions of macrostress and macrostrain variables. As such, the macrostructural models assume the material to be homogeneous and isotropic. The models are necessarily approximate because they cannot represent the heterogeneous media in which the damage processes occur. In contrast to macrostructural models, micromechanical models establish material behavior based on the explicit response of the microelements, such as dislocations and slip planes. Micromechanics have successfully explained the qualitative behavior of crack initiation. However, a theoretical crack-initiation model that explicitly relates the microstructure to the macroresponse has not been developed because too many complex micromechanical processes are operating simultaneously [20].

Statistical concepts have been used to develop empirical fatigue-life models in which the independent variable (applied stress or strain) is considered deterministic, and the dependent variable (life) is considered random [21]. The models do not account for the mechanisms that regulate fatigue damage, and thus the source of the scatter is unknown and must be attributed to incomplete data and missing parameters. The models cannot be used to accurately describe materials and loading conditions that are not explicitly part of the data-based test program.

This study addresses the statistical aspects of fatigue using a fundamentally different approach. The fatigue mechanisms are considered, and the independent variables, which include material variables that govern response, are recognized as random variables. The approach identifies the sources of uncertainty and quantitatively links the variation in the material microstructure to the scatter in the fatigue response.

The research is based on the concepts of probabilistic mesomechanics [22], which provides the relationships between the microstructural material properties and noncontinuum mechanics [23]. In this research, the mesoelements are defined as the individual grains of a polycrystalline aggregate. Each grain is considered a single crystal with homogeneous (although not isotropic) properties. The properties are considered to vary from grain to grain. The macrostructure is modeled as an ensemble of grains. The material properties of the ensemble of grains are defined using the appropriate statistical distributions. Mesomechanical modeling is an approximation of the actual material because certain properties will vary within a grain. However, it is believed that mesomechanics is a better approximation of the true material characteristics than macromechanics.

Mesomechanics also recognizes the multiple stages of fatigue-damage accumulation such as crack nucleation, small-crack growth, and long-crack growth. Each stage is driven by different mechanisms and must be distinctly modeled. The stages must be quantitatively linked because the crack grows successively from one stage to the next. In this research, a theoretical micromechanical model is used to determine the number of cycles necessary to nucleate a crack in the individual grains. A combination of models based on empirical observations and theoretical micromechanics are used to determine the number of cycles necessary to grow the cracks from nucleation to the long-crack regime. An empirically based (Paris law) model is used to determine the number of cycles necessary for the crack to grow through the long-crack regime to the critical crack size. Failure of the macrostructure is defined by the first crack to nucleate and grow beyond the critical crack size. The statistical distribution of fatigue life for the macrostructure is determined using Monte Carlo simulation methods. The probabilistic mesomechanical model provides a direct quantitative link between the variations in the material microstructure to the scatter in the fatigue behavior.

Many material and structural design factors influence component reliability in terms of the defined durability problems. From a material performance standpoint, many of these factors are at work in the durability "size effect." The size effect was first reported by Peterson [24] when he noticed that the mean fatigue life and variation in fatigue life were a function of the stressed area. The size effect has a fundamental role in controlling reliability because damage accumulation starts on a small scale and grows through various characteristic sizes, each with its own geometric complexities, constitutive laws, and heterogeneities. Fatigue behavior cannot be fully understood and predicted without obtaining information about each of the characteristic sizes, or what can be called mesodomains [23]. Nested models can link each of the mesodomains to determine the response of the macrodomain [25].

TABLE 42.2 Mesodomains of a Mechanical Component

Mesodomain	Sources of Variation	Damage-Accumulation Mechanisms
Dislocation level	vacancies, interstitials	dislocation pileup
Subgrain level	slip bands, microvoids, second-phase particles	slip-band decohesion
Grain level	crystallographic orientation, twins, inclusions	crack nucleation
Specimen level	surface finish, cracks, notches	small-crack growth
Component level	cracks, notches, processing, geometry, machining	large crack growth
Fleet level	heat treatment, service duty, applications	NDE inspection screening, life distributions

The concepts of mesomechanics can be used to explicitly examine each of the characteristic sizes or mesodomains. For example, a fleet of simple polycrystalline metallic components can be divided into six mesodomains, as shown in Table 42.2. At each level, heterogeneities can be introduced from various sources, and fatigue damage can accumulate via various mechanisms.

The true primary mesodomain is below the dislocation level. The most primitive variables controlling fatigue may be at the atomic level. Modeling at such levels is not yet possible and is not required for the purposes of the present study. The slip process, which takes place on slip steps typically 0.1 μm wide, is of continuum scale with respect to the atomic size of 0.5 nm [26]. The smallest practical mesodomain depends on the material, the loading, and the available information-gathering techniques. The research has focused on the use of slip-band models from the subgrain mesodomain, together with probabilistic variables being defined at the grain-size mesodomain. These models are used to predict behavior for the specimen mesodomain.

The specimen level is an artificial mesodomain because there are no specimens in the fleet. Specimens are generally prepared to limit the introduction of heterogeneities. However, the bulk of the information used in design is usually gathered from specimen testing, so it is important to understand the characteristics of this level. Specimen testing can identify scale effects, defect origins, and processing influences on crack initiation.

A large component such as an aeroengine fan blade will have several mesodomains between the grain size and the component level. The airfoil and the dovetail would be two component-scale mesodomains. Properties such as the grain size, material properties, and surface finish are different in these two mesodomains. The delineation of the mesodomains is specific to the material, geometry, loading, and failure mode.

The overall fatigue response at the fleet level is predicted by nesting the individual mesoscale models. The lowest level model uses the appropriate mesoscale parameters to determine the initial state of the next level. This level uses the results from the previous level along with the appropriate parameters to determine the initial state of the next level and so on. Through the use of nested models, fleet reliability can be linked to the heterogeneities at each mesodomain. Additionally, by modeling each level of the fatigue process individually, and rigorously linking the levels, various size effects are included.

42.3 Damage-Accumulation Models

42.3.1 Overview of a Probabilistic Microstructural Fatigue Model

The probabilistic fatigue modeling algorithm discussed in this chapter concentrates on the regimes in which the effects of the local variation in the microstructure are assumed dominant. The purpose of the algorithm is to assess only the microstructural effects. Variations in the applied loading and global geometry are not considered. The fatigue process is divided into three phases. The first phase is the crack-nucleation phase, in which damage accumulates to form a crack in an initially uncracked structure. The second phase is the small-crack-growth phase, in which the crack size is on the order of the microstructure. The third phase is the long-crack-growth phase, in which the crack size is large compared with the microstructure.

The algorithm uses microstructural models that predict crack-nucleation life and crack nucleation size. These models have been proposed in the literature as disclosed in this section. These models predict damage accumulation through irreversible dislocation pileup at microstructural obstacles. Cracks nucleate when the critical strain energy is exceeded.

Although the material fatigue response algorithm is able to model each of the crack-nucleation mechanisms described above, of particular interest to initially undamaged components is the slip-band cracking within the grain. Many of the high-strength materials currently used in safety critical structures accumulate damage through the creation of persistent slip bands. The theory of continuously distributed dislocations is used to model the persistent slip band within a grain. Dislocations pile up at the grain boundaries with each load cycle. When the energy associated with the dislocation pileup exceeds a critical value, a crack forms along the slip band the size of the metallic microstructural grain.

The second phase is the small-crack-growth phase in which the crack size is on the order of the microstructure. The crack growth rate is modeled as a function of the crack-tip-opening displacement (CTOD). The theory of continuously distributed dislocations is used to model the CTOD. The plastic zone is modeled as dislocations distributed ahead of the crack tip. The tip of the plastic zone is either propagating freely within a grain or blocked at the grain boundary. The CTOD depended on the relative location of the crack tip and the plastic-zone tip. The crack grows in the small-crack-growth phase until the plastic zone has spanned many grains, so that local microstructural variations have little effect.

The third phase is the long-crack-growth phase. The crack growth rate is modeled using Paris law. The microstructural variations are not explicitly considered. All variation in long-crack-growth is modeled by allowing the Paris law coefficient to be a random variable.

Fatigue test of an ensemble of smooth round bar (SRB) specimens is used to show the capabilities of the method. The overall fatigue response of the SRB is predicted by nesting the individual mesoscale models. The crack-nucleation model uses the appropriate mesoscale parameters to determine the initial state of the small-crack-growth model. This model uses the results from the previous crack-nucleation model along with the appropriate parameters to determine the initial state of the long-crack-growth model. Using nested models, specimen reliability can be linked to the heterogeneities at each mesodomain. Additionally, by modeling each level of the fatigue process individually, and rigorously linking the levels, various size effects are included.

The local microstructural variables considered random in this chapter are: grain size, grain orientation, microstress, and the frictional stress needed to move dislocations. The grain-size statistical distribution is determined from data in the literature. The grain orientation statistical distribution is determined for both surface and interior grains using theoretical considerations. The microstress statistical distribution is determined from a Voronoi cell finite element model [27]. The variables are common to both the crack-nucleation and small-crack-growth models. The grain shape is assumed equiaxial, and the grain orientation is untextured and described using the face-centered cubic slip system. The loading and material properties within a grain are homogeneous although not isotropic. The material properties vary from grain to grain.

42.3.2 Crack Nucleation

42.3.2.1 Micromechanics of Crack Nucleation

Fatigue-crack nucleation is a complex and obscure process. The mechanisms for crack nucleation change with material, loading, temperature, and environment. One overriding observation is that cracks tend to nucleate near the free surface. For many loading conditions, the highest loads are at the surface. However, even when the nominal stress is constant throughout, cracks tend to nucleate at the surface because deformation of each grain is allowed to concentrate on a preferred crystallographic plane. In the interior, deformation on a single crystallographic plane is hampered by the constraints of the surrounding grains.

Experimental evidence clearly shows that defects in the material can cause fatigue-crack nucleation by acting as stress concentrations, and the cracks tend to nucleate along the preferred slip plane [28].

Examples of defects include surface pores, ceramics inclusions, second-phase particles, and microcracks. The fatigue resistance of many alloys has been improved by decreasing the size and number of defects. However, slip-band decohesion also causes crack nucleation even when no apparent defect is present. The surface grains must be favorably oriented for slip-band decohesion to occur, but not all favorably oriented grains have cracks. Slip along preferred planes plays an important role in crack nucleation. When annealed metals are exposed to cyclic loading, they strain harden. Strain hardening is one of the earliest mechanical responses to fatigue. Initial hardening is rapid and controlled by multiplication of dislocations in the atomic lattice. When the material is first cycled, dislocations glide freely to accommodate large plastic strains. Eventually, the dislocations interact and start to create a substructure of pinned dislocations [29].

The substructure consists of veins for low-strain amplitude and cells for higher strain amplitudes [30]. The veins and cell walls consist of high dislocation density, while the volume between the veins and cell walls has a much lower dislocation density. The dislocations can only glide freely in the volume of low density. As the substructure develops, hardening will result because the increased interaction of dislocations constrain their movement. If the cyclic strain amplitude is increased, the cell size decreases, which reduces the volume between cells, and hardening continues. Fine slip lines appear on the surface as the dislocation density increases [31].

The rate of hardening gradually decreases until the flow stress becomes constant. The dislocation substructure is saturated and can no longer accommodate strain. Saturation is accompanied by the formation of coarse slip bands that roughen the surface of the grain with extrusions and intrusions. If the surface is polished, small vacancy pits are found in the slip bands. If the specimen is again cycled, the same slip bands roughen the surface. These bands are referred to as persistent slip bands [31].

The persistent slip bands have a distinctive substructure of walls of high dislocation density [32]. The walls are perpendicular to the primary slip direction and stretch across the thickness of the band. The distance between the walls is fairly constant. This substructure is often referred to as a ladder structure [33].

Slip-band behavior is not well understood, and many different theories exist to explain how the bands accommodate strain [32–35]. However, it is recognized that the strain is localized in the persistent slip bands, and very little strain is accommodated by the volume of material between slip bands. Upon further cycling, cracks form in the persistent slip bands. The cracks are believed to be the combined result of vacancy creation, repulsive dislocation stresses, and surface-roughening stress concentrations. Experimental evidence shows that if the strain amplitude is lower than the saturation point, the strain is accommodated by fine slip associated with presaturation dislocation substructure, and no cracking takes place [36]. Thus, persistent slip bands are essential to fatigue damage and must be addressed by crack-nucleation models.

There are two fundamentally different types of slip-band-induced crack nucleation. One is Forsyth's well-known Stage I crack nucleation, in which a very small crack (much smaller than the grains size) nucleates along the slip plane very early in life. A crack is evident from crack-opening displacement when a static load is applied [1]. The size of the plastic zone is relatively small, being equal to or less than the crack size [37]. The crack propagates in Mode II until it reaches an obstacle, often the grain boundary. This type of crack nucleation has been observed in age-hardened aluminum [1] and alloy single crystals [31]. Elastic-plastic crack growth models have been successful in modeling the mean behavior of such alloys down to a very small crack size [38].

The more prevalent though less recognized slip-band-induced crack nucleation is sudden crack nucleation. In sudden crack nucleation, a slip band that stretches across the grain forms very early in life, but no crack is formed. The lack of a crack is evident from no crack-opening displacement when a static load is applied [1]. The slip band is not associated with crack beneath the surface [35, 39]. Upon continued cycling, the slip band is blocked by the grain boundary and does not grow in length. However, the depth and the width of the slip band increase slightly until suddenly a crack is formed. This slip-band crack-nucleation behavior is observed in many alloys, including steel, aluminum, and brass [1, 12, 40]. The crack-nucleation model developed in this chapter addresses sudden crack nucleation.

42.3.2.2 Micromechanical Crack Nucleation Models

Models used to predict scatter in crack-nucleation response must have two attributes. They must be quantitative with regard to the number of cycles needed to produce a crack to a specific size if they are to be used for lifetime predictions. The models must also be able to address the microstructural parameters in order to provide a physical link between the microstructure and the fatigue behavior. Although the literature contains numerous expressions for modeling the propagation rate of fatigue cracks as discussed later, only a limited number of analytical crack-nucleation models exist. Most of these models use dislocation theory [41] to model fatigue-damage accumulation as the buildup of a continuous array of dislocations [42].

Microstructural models that predict crack-nucleation life and crack-nucleation size have been proposed independently by Tanaka and Mura [34] and Chang et al. [43]. Both of these models predict damage accumulation through irreversible dislocation pileup at microstructural obstacles. Cracks nucleate when a critical strain energy is exceeded.

Models are available to address a wide variety of crack-nucleation mechanisms, including:

- Slip-band cracking within a grain [34]
- Grain boundary cracking [34]
- Matrix/inclusion interface cracking [34]
- Cracks emanating from inclusions [34, 43]
- Cracks emanating from notches [44]

These models have been modified to account for partial reversibility and random load amplitudes [45]. The crack-nucleation models had to be modified by the authors to include the effects of additional random variables. The models are consistent with the Coffin-Manson relationship for fatigue-crack initiation [34, 46], the Petch equation for the grain-size dependency of the fatigue strength [34], and the Palmgren-Miner law of damage accumulation for variable-amplitude loads.

The crack-nucleation model used in this study is based on one proposed by Tanaka and Mura [34], in which the forward and reverse plastic flow within the persistent slip band of a surface grain is related to the creation of dislocations of opposite signs on closely spaced planes. This model is applicable to metallic components for which crack nucleation takes place by transgranular shear stress fracture and is outlined below.

As a load greater than the local yield stress is applied to grain with diameter d, dislocations are generated and move along the slip plane as shown in Figure 42.2. The dislocations pile up at the grain boundary, which acts as an obstacle to dislocation movement. The dislocation movement is assumed to be irreversible such that when the reverse load is applied, dislocations of the opposite sign pile up on a closely spaced plane. Since the residual load from the back stress of the positive dislocations act in the same direction as the reverse applied load, unloading will cause negative dislocation movement. During each of the subsequent load cycles, the number of dislocations monotonically increases.

On the first loading, the equilibrium condition can be expressed as

$$\tau_1^D + (\tau_1 - k) = 0 \tag{42.1}$$

where k is the frictional stress that must be overcome to move dislocations, τ_1 is the applied shear stress (τ_1 must be greater than k for damage to accumulation to occur), and τ_1^D is the back stress caused by the dislocations. If the dislocation density $D_1(x)$ along the slip plane is assumed to be continuous [42]

$$\tau_1^D = A \int_{-r}^{r} \frac{D_1(x')}{x - x'} dx' \tag{42.2}$$

$A = G/2\pi(1-\upsilon)$ for edge dislocations

$A = G/2\pi$ for screw dislocations

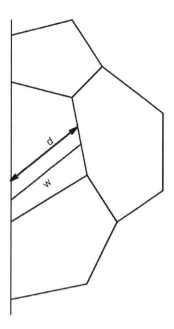

FIGURE 42.2 Slip band of width w interacting developing on grain of size d.

where r is the grain radius, G is the shear modulus, and v is Poisson's ratio. Substituting Equation 42.2 into Equation 42.1 creates a singular integral equation that is solved for $D_1(x)$ using the inversion formula of Muskhelishvili [47] for unbounded dislocation density at the grain boundary

$$D_1(x) = \frac{(\tau_1 - k)x}{\pi A \sqrt{r^2 - x^2}} \tag{42.3}$$

The incremental increase of dislocation density $\Delta D(x)$ with each load cycle is

$$D_1(x) = \frac{(\Delta \tau_1 - 2k)x}{\pi A \sqrt{r^2 - x^2}} \tag{42.4}$$

The slip displacement $\phi(x)$ due to the increment $\Delta D(x)$ is

$$\phi(x) = \int_x^r \Delta D(x) dx \tag{42.5}$$

The plastic strain increment $\Delta \gamma$ is

$$\Delta \gamma = \int_{-r}^{r} \phi(x) dx = \frac{(\Delta \tau - 2k)r^2}{2A} \tag{42.6}$$

such that the constitutive equation is

$$\gamma = \frac{(\tau - 2k)r^2}{2A} \tag{42.7}$$

which describes the stress–strain hysteresis loop in Figure 42.3.

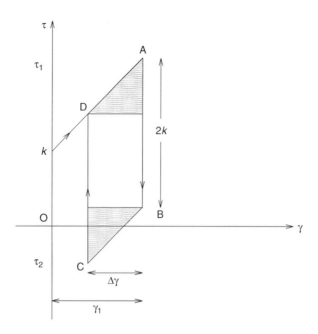

FIGURE 42.3 Stress–strain hysteresis loop.

During the first forward loading of stress τ_1, the material hardens for any stress above k. On reverse loading to τ_2, the path ABC is followed. On subsequent forward loading, the path CDA is followed. The amount of plastic strain increment is a linear function of $(\tau - k)$. The dislocation strain energy is the same for forward and reverse loading except for the first loading. The incremental stored dislocation strain energy ΔU corresponds to the shaded area of Figure 42.3.

$$\Delta U = \Delta\gamma(\Delta\tau - 2k) \qquad (42.8)$$

The energy associated with the unshaded area of Figure 42.3 is the dissipated work against the frictional stress k. Crack nucleation takes place when the total stored energy after N_n cycles is equal to the fracture energy of the grain.

$$N_n \Delta U = 2rW_s \qquad (42.9)$$

$$N_n = \frac{4GW_s}{(\Delta\tau - 2k)^2 \pi(1-\nu)d} \qquad (42.10)$$

where d is the grain diameter and W_s is the specific fracture energy per unit area.

The probabilistic mesomechanical fatigue model calculates the crack-nucleation life using Equation 42.10 in a slightly modified form. The modification is that $\Delta\tau$ is replaced with $\Delta\sigma/M$, where $\Delta\sigma$ is the applied axial stress, and M is the grain orientation factor.

The model has several assumptions and limitations.

1. *The grain is homogeneous.* The dislocations are free to move to the grain boundaries, i.e., no subgrain structure exists to pin or disrupt dislocation movement. Although the grain is homogeneous it is not isotropic.
2. *Damage accumulates on a single planar slip system.* In general, grains within a polycrystalline aggregate are not free to deform but are constrained by neighboring grains. Crack nucleation takes place on the surface grains, which are not as constrained as grains embedded in the interior. Surface grains are able to accommodate more strain on the primary slip system [48]. The model can only be directly used on alloys that show planar slip.

3. *The crack-nucleation size is equal to the grain size.* Although loading has been shown to affect the nucleation size (the crack size at Mode II to Mode I transition) [3, 4], experimental evidence suggest that the nucleation size is on the order of the grain size [5–7].
4. *The dislocation movement is irreversible and dipoles pile up monotonically at the grain boundaries.* It is reasonable to expect some of the dislocations to move back into the interior of the grain upon reverse loading or be annihilated by back stresses. Theoretical investigations on how to account for the partial reversibility of slip-band formation have been inconclusive [45, 49, 50]. However, there is experimental evidence that this reverse movement is small [34].
5. *The number of cycles to saturation are negligible.* This assumption is reasonable for many materials under certain loading conditions [7, 31].

Equation 42.10 is necessarily a simplification of the complex phenomenon of slip-band cracking. It does not directly address the effect of vacancy creation or the stress concentration of the surface roughening. However, the model is attractive because the fatigue life is inversely proportional to the square of the plastic strain amplitude, which is in agreement with the Coffin-Manson empirical equation for fatigue. Equation 42.10 can be rewritten as

$$\Delta \tau = 2k + \left(\frac{4GW_s}{\pi N_n (1-v)} \right)^{\frac{1}{2}} d^{-\frac{1}{2}} \quad (42.11)$$

which is in the form of the Hall-Petch equation for the dependence of fatigue strength on grain size.

42.3.3 Microstructurally Small-Crack Growth

42.3.3.1 Micromechanics of Small-Crack Growth

The experimentally observable parameter that has been correlated to the varying rate of small-crack growth is the crack-opening displacement (COD) [37]

$$\frac{da}{dN} = C'(\Delta COD)^{n'} \quad (42.12)$$

where a is the crack length, N is the number of cycles, and C' and n' are empirical constants based on material testing. The COD is a measure of the amount of damage associated with the crack tip. The larger the COD, the higher is the crack growth rate. This phenomenon was first observed by Laird and Smith [51] and has been well established in long-crack growth [52, 53]. The exponent n' has been found to have a value near unity when the COD is replaced by crack-tip-opening displacement (CTOD)

$$\frac{da}{dN} = C'\Delta \phi_t \quad (42.13)$$

where the CTOD, denoted by ϕ_t, is measured at the location of crack extension for the previous cycle. The direct proportionality of Equation 42.13 has been observed in small-crack growth of aluminum, nickel, and titanium alloys [54]. Equation 42.13 has also been shown to correlate the behavior between small- and long-crack growth [55]. Nisitani and Takao [1] showed that small-crack arrest could be associated with no CTOD. Tanaka et al. [56] showed that regressing data to Equation 42.13 showed much less scatter than exponential models based on ΔK or ΔJ. Also, developing models for three different materials — copper, mild steel, and stainless steel — produced very similar values for C' and n'. (The exponent n' was not unity because the COD measurements were made on the specimen surface at the center of the crack.) It appears as though the relationship between da/dN and $\Delta\phi_t$ is more of an intrinsic material behavior than models based on ΔK or ΔJ.

Determining C' for small-crack growth has been performed through direct microscopic observations [1]. However, there has been limited success in using ΔK or ΔJ data to determine C'.

CTOD can be shown to be related to the J integral through

$$\phi_t = \alpha \frac{J}{\sigma_0} \qquad (42.14)$$

where σ_0 is the bulk yield strength and α is nearly unity [57]. Assuming elastic perfectly plastic behavior, CTOD can be related to K through

$$\phi_t = \frac{2K^2}{\pi\sigma_0 v} \qquad (42.15)$$

where v is Poisson's ratio. Combining Equation 42.13 and Equation 42.15

$$\frac{da}{dN} = \frac{C'\Delta K^2}{2\pi\sigma_0 v} \qquad (42.16)$$

which is the form of a second-order Paris equation. Determining C' using da/dN vs. ΔK data is straightforward for alloys that are governed by a second-order Paris relationship. Donahue et al. [58] have compiled an extensive list of data that fit a second-order Paris equation. They find that a C' value of about 0.1 fits most of the data. However, in general the Paris exponent is not expected to be 2.

It is interesting to note that McEvily [59] presents data from several sources that show that a well-defined region of constant slope is seldom found in the Paris fit. The slope was found to vary with ΔK and to have a value of 2 at low ΔK (the region of interest in the present study) that increases at higher ΔK. Also, there has been some success [60] with correlating data to

$$\frac{da}{dN} = C''\Delta K_{eff}^2 \qquad (42.17)$$

where ΔK_{eff} is the effective ΔK, which is the applied ΔK minus the ΔK when the crack first opens. However, the measurement of ΔK_{eff} still requires direct observations at the crack tip. The relationship in Equation 42.13 is assumed to be valid and will be used in the present study.

42.3.3.2 Small-Crack-Growth Models

Two basic approaches have been used to model small-crack growth behavior: modify a continuum mechanics-based stress intensity factor, K, to account for the microstructural heterogeneity, or explicitly model the damage ahead of the crack tip using dislocation theory. Hobson [61] presented a simple continuum-based model in which the crack growth rate is related to the distance between the crack tip and the nearest grain boundary. All of the model parameters are determined by fitting the model to experimental data. Chan and Lankford [37] modified K to account for grain size and orientation using a simple analytical approach. Chan and Lankford [62] later used a more rigorous analytical approach to modify K for microstructural effects and large-scale yielding. They introduce the concept of microstructural dissimilitude, which accounts for the fact that small cracks actually lie in relatively few grains.

Similitude can be assumed when the crack front interrogates enough grains such that the material properties averaged along the crack front have the same value as the bulk material properties. When the crack front interrogates relatively few grains, the average material properties at the crack front can vary significantly from the bulk properties, hence, microstructural dissimilitude. The number of grains interrogated by the crack front necessary to assume similitude depends on the amount of scatter in the local material properties. Also, by using an equivalent-properties model that effectively averages the microstructural environment interrogated by the two-dimensional crack front, Chan and Lankford [37] were able to reduce small-crack growth to a one-dimensional problem.

Gerberich and Moody [63] used a modified continuum approach to address the semicohesive zone associated with selective cleavage in the microstructure at the crack tip for Ti-6Al-4V. Using this model, they were able to predict the mean grain size effect on threshold for titanium alloys.

Bilby and Eshelby [42] described the damage ahead of the crack tip using the theory of continuous dislocations. The models are equivalent to the Dugdale [64] model found by a different method. Weertman [65] used the model of Bilby et al. [66] to develop a fatigue-crack-growth law and later used dislocation theory to develop a K for short cracks [67].

Several researches have extended Bilby's model to account for microstructural effects. Taira et al. [68] obtained a model for a crack-tip slip band blocked by a grain boundary. Tanaka et al. [69] extended Taira's model to slip-band propagation through grain boundaries. The CTOD predicted by the model was found to be equivalent to that predicted by both Morris et al. [70] and de los Rios et al. [71]. Navarro et al. [72] used a model equivalent to Tanaka's to describe small and short crack growth. The models predicted the bounds on the variation in small-crack growth.

Many statistical and probabilistic crack-growth models can be found in the literature. The Markov-based models [73] describe the crack-growth-rate scatter as a process in which the amount of crack extension for each cycle is a random function. The Paris-based models describe the crack-growth-rate scatter by allowing the material-property parameters to be random [74]. A common feature of these models is that the random nature of the crack growth is not related to microstructural variables. Thus, these models are not useful in understanding small-crack growth behavior.

Limited work has been reported on models that directly address the statistical aspects of small-crack growth. Morris et al. [75] used Monte Carlo simulation to model the crack-initiation behavior of aluminum smooth round bars. They used the crack-nucleation model of Chang et al. [43] and a modified, continuum-based-K, small-crack-growth model. The random variables included crystallographic orientation, grain diameter, inclusion diameter, and an experimentally determined material parameter associated with the fracture strength of the inclusion. The statistical distributions of the random variables were not discussed. The predicted results compared favorably with the experimental observations. Tanaka et al. [56] used Monte Carlo simulation to predict the general behavior of small-crack growth. They used the small-crack-growth model of Tanaka et al. [69]. The random variables included grain size, grain frictional stress, and an independent grain boundary strength. A two-parameter Weibull distribution was assumed for all of the random variables. They extended the model to include two-phase materials. Trends predicted by the model compared favorably with general trends observed in small-crack growth behavior.

42.3.3.3 A Model Based on Continuously Distributed Dislocations

The model developed in the present study follows the approach used by Tanaka et al. [69]. The approach is outlined below for Mode II (sliding) crack growth. The solution for Mode I (tensile) crack growth is obtained through the simple transformations discussed at the end of this section.

Assume a crack has length a and the crack tip lies within a grain, as shown in Figure 42.4. As load is applied, dislocations are emitted from the crack tip, creating a slip band with dislocation density ($D(x) > 0$) as represented by the length w in Figure 42.4. For low stress (where the applied stress is less than the stress needed to move dislocations) and the slip-band tip far from the gain boundary ($c < d$), a condition called the *equilibrium slip band* exists.

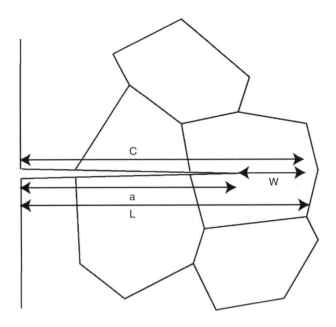

FIGURE 42.4 Crack tip in a grain.

The solution for the equilibrium slip band was obtained by Bilby et al. [66]. The equilibrium condition can be expressed as

$$\tau^D + \tau^0 = 0$$

$$\tau^D = A \int_{-c}^{c} \frac{D(x')}{x-x'} dx'$$

$$A = \begin{cases} G/2\pi(1-v) & \text{for edge dislocations} \\ G/2\pi & \text{for screw dislocations} \end{cases} \quad (42.18)$$

$$\tau^0 = \begin{cases} \tau & x < a \\ \tau - k_1 & a < x < c \end{cases}$$

where τ is the resolved applied shear stress, τ^D is the back stress caused by the dislocations, G is the shear modulus, and k_1 is the friction stress (which must be overcome to move dislocations) of the grain in which the crack tip and slip band lie. The size of the slip band w is determined from the condition of vanishing dislocation density at the slip-band tip.

$$w = c - a \quad (42.19)$$

$$\frac{a}{c} = \cos\left(\frac{\pi \tau}{2k_1}\right) \quad (42.20)$$

The dislocation density $D(x)$ is obtained by solving the singular integral Equation 42.18 using the inversion formula of Muskhelishvili [47] for unbounded dislocation density at the crack tip.

$$D(x) = \frac{k_1}{\pi^2 A} f(x; c, a)$$

$$f(x; c, a) = \ln\left|\frac{x\sqrt{c^2 - a^2} + a\sqrt{c^2 - x^2}}{x\sqrt{c^2 - a^2} - a\sqrt{c^2 - x^2}}\right| \quad (42.21)$$

The crack-tip sliding displacement (CTSD) is

$$\phi_t = \frac{2k_1 a}{\pi^2 A} \ln\frac{c}{a} = \frac{2k_1 a}{\pi^2 A} \ln\left[\sec\left(\frac{\pi\tau}{2k_1}\right)\right] \qquad (42.22)$$

As the crack grows, the tip of the slip band will eventually be blocked at the grain boundary. This condition is called the *blocked slip band*. The solution for the blocked slip band was obtained by Taira et al. [68]. The size of the slip band is simply

$$w = c - a$$
$$c = L \qquad (42.23)$$

The dislocation density $D(x)$ is obtained by solving the singular integral Equation 42.18 using the inversion formula of Muskhelishvili [47] for unbounded dislocation density at the crack tip and the slip-band tip.

$$D(x) = \frac{\beta\tau}{\pi A}\frac{x}{\sqrt{c^2-x^2}} + \frac{k_1}{\pi^2 A} f(x; c, a)$$
$$\beta = 1 - \frac{2k_1}{\pi\tau}\arccos\frac{a}{c} \qquad (42.24)$$

The microscopic stress intensity factor K_m at the tip of the slip band is similar to the crack-tip stress intensity factor and is defined as

$$K_m = \pi A\sqrt{2\pi}\lim_{x\to c}\left[\sqrt{c-x}D(x)\right] = \beta\tau\sqrt{\pi c} \qquad (42.25)$$

The CTSD is

$$\phi_t = \frac{\beta\tau}{\pi A}\sqrt{c^2-x^2} + \frac{2k_1 a}{\pi^2 A}\ln\frac{c}{a} \qquad (42.26)$$

As the crack grows, K_m increases. For the crack to overcome the grain-boundary obstacle and propagate into the subsequent grain, K_m must exceed the critical microscopic stress intensity factor K_{mc} provided by the grain boundary. If as $a \to c$, K_m does not exceed K_{mc}, then the CTSD $\to 0$ and the crack growth arrests. If K_m exceeds K_{mc}, the slip-band tip propagates into the next grain, and a condition called the *propagating slip band* exists.

The solution for the propagated slip band was obtained by Tanaka et al. [69]. The equilibrium condition is the same as Equation 42.18 except

$$\tau^0 = \begin{cases} \tau & x < a \\ \tau - k_1 & a < x < L \\ \tau - k_2 & L < x < c \end{cases}$$

where k_2 is the frictional stress in the second grain. The size of the slip-band zone is determined from the condition of vanishing dislocation density at the slip-band tip.

$$\arccos\frac{a}{c} + \left(\frac{k_2}{k_1} - 1\right)\arccos\frac{L}{c} = \frac{\pi\tau}{2k_1} \quad (42.27)$$

The dislocation density and CTSD are determined in a similar manner as before.

$$D(x) = \frac{k_1}{\pi^2 A} f(x; c, a) + \frac{(k_2 - k_1)}{\pi^2 A} f(x; c, L)$$

$$\phi_t = \frac{2k_1 a}{\pi^2 A} \ln\frac{c}{a} + \frac{(k_2 - k_1)}{\pi^2 A} g(a; c, L)$$

$$g(a; c, L) = L \ln\left|\frac{\sqrt{c^2 - L^2} + \sqrt{c^2 - a^2}}{\sqrt{c^2 - L^2} - \sqrt{c^2 - a^2}}\right| \quad (42.28)$$

$$- a \ln\left|\frac{a\sqrt{c^2 - L^2} + L\sqrt{c^2 - a^2}}{a\sqrt{c^2 - L^2} - L\sqrt{c^2 - a^2}}\right|$$

Tanaka et al. [56] solve for the case in which the slip band extends over several grains, as shown in Figure 42.5. If the crack tip is in the jth grain and the slip band is in the nth grain, the equilibrium condition is the same as Equation 42.18, except

$$\tau^0 = \begin{cases} \tau & x < a \\ \tau - k_1 & a < x < L_j \\ \vdots & \vdots \\ \tau - k_2 & L_{n-1} < x < c \end{cases}$$

The size of the slip-band zone can be found from

$$0 = \frac{\pi\tau}{2} - k_j \arccos\frac{a}{c}$$

$$- \sum_{i=j+1}^{n} (k_i - k_{i-1}) \arccos\left(\frac{L_{i-1}}{c}\right) \quad (42.29)$$

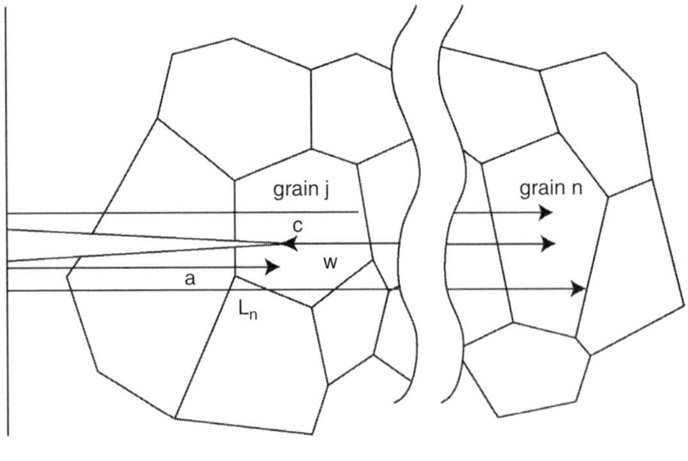

FIGURE 42.5 Crack-tip slip band in multiple grains.

The CTSD is given by

$$\phi_t = \frac{2k_j a}{\pi^2 A}\ln\frac{c}{a} + \sum_{i=j+1}^{n}\frac{(k_i - k_{i-1})}{\pi^2 A}g(a;c,L_{i-1}) \qquad (42.30)$$

For the crack tip in the jth grain and the slip band blocked in the nth grain, the size of the slip-band zone is

$$w = L_n - a \qquad (42.31)$$

The CTSD is given by

$$\begin{aligned}\phi_t &= \frac{\beta\tau}{\pi A}\sqrt{c^2 - a^2} + \frac{2k_j a}{\pi^2 A}\ln\frac{c}{a} \\ &\quad + \sum_{i=j+1}^{n}\frac{(k_i - k_{i-1})}{\pi^2 A}g(a;c,L_{i-1}) \\ \beta &= 1 - \frac{2k_1}{\pi\tau}\arccos\frac{a}{c} \\ &\quad - \sum_{i=j+1}^{n}\frac{2(k_i - k_{i-1})}{\pi\tau}\arccos\left(\frac{L_{i-1}}{c}\right)\end{aligned} \qquad (42.32)$$

The microscopic stress intensity factor is

$$K_m = \beta\tau\sqrt{\pi c} \qquad (42.33)$$

The above model allows for grain-to-grain variation in grain size and frictional stress.

In the present study, the model by Tanaka et al. [69] is extended to include the variation in the microstress and the grain orientation by allowing grain-to-grain variation in the applied resolved shear stress τ. Consider a crack with the crack tip in the jth grain and the slip-band tip in the nth grain. The equilibrium condition is the same as Equation 42.18, except

$$\tau^0 = \begin{cases} \tau_j & x < a \\ \tau_j - k_j & a < x < L_j \\ \vdots & \vdots \\ \tau_n - k_n & L_{n-1} < x < c \end{cases}$$

For the propagated slip band, the size of the slip-band zone can be found from

$$\begin{aligned}0 &= \frac{\pi\tau_j}{2} - k_j\arccos\frac{a}{c} \\ &\quad - \sum_{i=j+1}^{n}((\tau_{i-1} - k_{i-1}) - (\tau_i - k_i))\arccos\left(\frac{L_{i-1}}{c}\right)\end{aligned} \qquad (42.34)$$

The CTSD is given by

$$\phi_t = +\frac{2k_j a}{\pi^2 A}\ln\frac{c}{a}$$
$$+\sum_{i=j+1}^{n}\frac{(\tau_{i-1}-k_{i-1})-(\tau_i-k_i)}{\pi^2 A}g(a;c,L_{i-1}) \quad (42.35)$$

For the blocked slip band, the size of the slip-band zone is

$$w = L_n - a \quad (42.36)$$

The CTSD is given by

$$\phi_t = \frac{\beta\tau}{\pi A}\sqrt{c^2-a^2} + \frac{2k_j a}{\pi^2 A}\ln\frac{c}{a}$$
$$+\sum_{i=j+1}^{n}\frac{(\tau_{i-1}-k_{i-1})-(\tau_i-k_i)}{\pi^2 A}g(a;c,L_{i-1})$$
$$\beta = 1 - \frac{2k_1}{\pi\tau_j}\arccos\frac{a}{c}$$
$$-\sum_{i=j+1}^{n}\frac{2((\tau_{i-1}-k_{i-1})-(\tau_i-k_i))}{\pi\tau_j}\arccos\left(\frac{L_{i-1}}{c}\right) \quad (42.37)$$

The microscopic stress intensity factor is

$$K_m = \beta\tau\sqrt{\pi c} \quad (42.38)$$

The solution for mode I loading is easily obtained through the following substitutions:

$$\tau \to c$$
$$\text{CTSD} \to \text{CTOD}$$

42.3.4 Long-Crack Growth

The linear elastic crack growth is modeled using the Paris law representation of a surface crack in a semi-infinite body subjected to a constant stress cycle.

$$\frac{da}{dN} = C\Delta K^n$$
$$\Delta K = \beta\Delta s_{xx}\sqrt{a} \quad (42.39)$$

where a is the crack length, N is cycles, ΔK is the stress intensity factor, Δs_{xx} is the stress range, β is the geometry constant ($1.12\sqrt{\pi}$), and C and n are material properties.

Expanding ΔK and integrating both sides,

$$C(\beta \Delta s_{xx})^n \int_0^{N_g} dN = \int_{a_i}^{a_f} a^{-\frac{n}{2}} da \qquad (42.40)$$

$$N_g = \frac{a_i^{1-\frac{n}{2}} - a_f^{1-\frac{n}{2}}}{C \Delta s_{xx}^n \beta^n \left(\frac{n}{2} - 1\right)}; \quad n \neq 2 \qquad (42.41)$$

where, N_g is the number of cycles needed for the crack to grow to failure, a_i is the initial crack size, and a_f is the failure crack size.

If $n > 2$ and $a_i \ll a_f$, then $a_i^{1-\frac{n}{2}} \gg a_f^{1-\frac{n}{2}}$, and Equation 42.41 can be written as

$$N_g = \frac{a_i^{1-\frac{n}{2}}}{C \Delta s_{xx}^n \beta^n \left(\frac{n}{2} - 1\right)}; \quad n \neq 2 \qquad (42.42)$$

42.4 Estimate of Random Variable Statistics

The variables in the damage-accumulation models that are considered random in the probabilistic microstructural fatigue algorithm are:

- Grain size
- Surface grain orientation
- Interior grain orientation
- Frictional shear stress
- Paris law coefficient

The mean values of the random variables such as grain size and frictional strength can vary significantly with heat treatment and other processes. The mean values can often be easily determined from the information traditionally reported in the literature. However, information needed to determine the statistical variation and the distributions types of the random variables is usually not reported. Nor is this information gathered during routine material characterizations performed by the aerospace industries. Luckily, specific studies have been performed to gather the needed information. These studies are used to estimate the random variables for the Phase I effort and are discussed below.

42.4.1 Grain Size

Empirical observations have indicated that the scatter in grain size for *natural grain growth*, i.e., cast polycrystalline structures, is insensitive to material. This has been observed in pure metals, complex alloys, and inorganic ceramics [76]. This phenomenon has been attributed to the well-behaved kinetics that determines natural grain growth.

Kurtz and Carpay [77] performed extensive grain-volume measurements on Ni-Zn ferrites by way of planar sectioning, and they describe the grain size by the equivalent spherical volume diameter. They determined the grain size distribution for seven different mean grain size microstructures measuring several thousand grains for each microstructure. They found the COV to vary between 0.3 and 0.4. The COV was independent of mean grain size. The distribution was accurately described by a lognormal distribution with the maximum-to-mean-grain-size ratio of about 2.7.

Kumar et al. [78] investigated the grain size distribution using Monte Carlo techniques and a Voronoi tessellation technique, which closely model the grain topography of polycrystals. They found a COV of

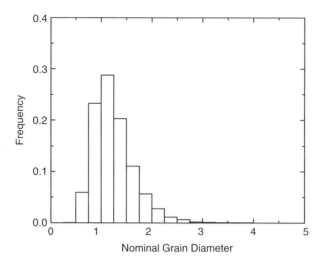

FIGURE 42.6 Grain diameter distribution.

about 0.4. The lognormal distribution was a good fit up to about 5000 grains. Simulating greater than 100,000 grains, they found the gamma distribution to be a best fit. However, the gamma distribution may well be an artifact of the modeling technique and not intrinsic to the grain size distribution. Voronoi tessellation uses the Poisson process to generate the grain geometry, and the gamma distribution is directly related to the Poisson process [79]. In the present study, a lognormal distribution with a COV of 0.4, as shown in Figure 42.6, will be assumed.

The bulk of the grains measured in the above research were interior grains. A distinction must be made between the size of the surface grains and those in the interior. Although the surface grains may account for only a small fraction of the total grains within a component, understanding the properties of the surface grains is paramount because they play an important role in crack nucleation and small-crack growth.

The surface effectively slices each grain in a random manner such that

$$l = d_s \cos\theta \qquad (42.43)$$

where l is the surface length, d_s is the grain diameter, and θ is the random angle of incidence, as shown in Figure 42.7. For an arbitrary cut through the grain, θ is uniformly distributed between 0 and $\pi/2$. The distribution of d_s can be determined from

$$l = d_s \cos\left(\frac{\pi}{2}u\right) \qquad (42.44)$$

where d_s is lognormally distributed, as shown in Figure 42.6, and u is the standard uniform random variable.

Monte Carlo simulation was used to evaluate the distribution of d_s, as shown in Figure 42.8. Comparing Figure 42.8 with Figure 42.6 shows that the average diameter of the surface grains is smaller than the interior grains. The increased width of the surface-grain distribution indicates that the scatter in the surface-grain diameters is larger than that of the interior grains.

42.4.2 Grain Orientation

In metallic structures, slip can occur on many planes in several directions. All orientations for cubic structures can be defined within the standard stereographic triangle [80].

The orientation dependence of the reciprocal Schmid factor is shown in Figure 42.9. Figure 42.9 represents axially loaded grains that are free to deform such that slip occurs on a single plane. However, grains within a polycrystalline aggregate are not free to deform but are constrained by neighboring grains.

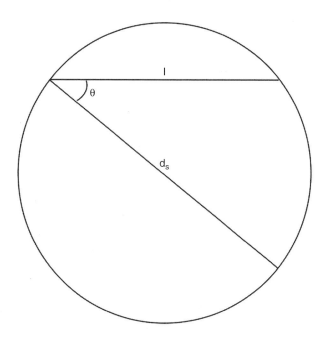

FIGURE 42.7 Line cutting across the grain.

Taylor [81] determined the equivalent to the reciprocal Schmid factor, M, for axisymmetric flow in a face-centered cubic polycrystal. He assumed the frictional stress was the same for each slip system. He also assumed that each grain is acted upon by the same applied strain as the macroscopic strain. Taylor assumed that the active slip systems are those for which the sum of the shear strains is a minimum. Bishop and Hill [82] later showed that the Taylor analysis is equivalent to a maximum-work principle. This quantity is referred to as the Taylor factor. Chin and Mammel [83] developed a computer model based on the Taylor analysis and found the Taylor factor for slip on other orientations in cubic polycrystals.

Crack nucleation takes place on the surface grains. It is difficult to determine M for the surface grains, which are not as constrained as grains embedded in the interior. Surface grains are able to accommodate more strain on the primary slip system. Although slip-band formation and crack nucleation on secondary

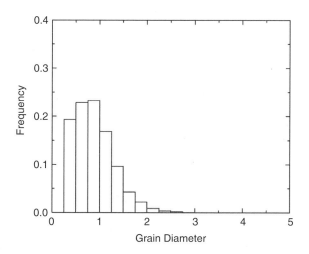

FIGURE 42.8 Surface grain size distribution.

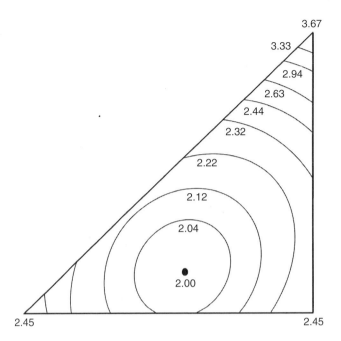

FIGURE 42.9 Orientation dependence of reciprocal Schmid factor.

planes is not uncommon, experimental evidence shows that cracks tend to nucleate on the primary slip plane [84]. In reality, the deformation of surface grains is somewhere between free deformation and fully constrained. The reciprocal Schmid factor is used to describe the surface grains, and the Taylor factor is used to describe the interior grains in the present study.

The orientations of the grains of an untextured polycrystalline material can be expressed as a uniform distribution of points within the stereographic triangle. To determine the statistical distribution of the Schmid factor for the randomly oriented grains, response surface was generated to approximate the Schmid factor within the stereographic triangles of Figure 42.10. A Monte Carlo simulation was used to

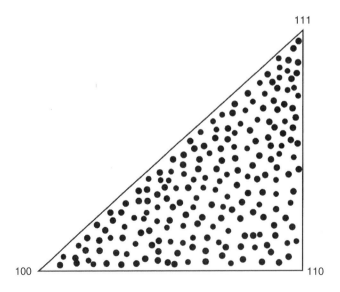

FIGURE 42.10 Uniform distribution of points throughout the stereographic triangle.

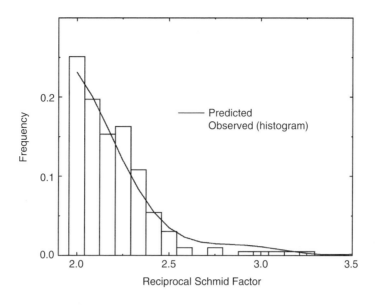

FIGURE 42.11 Probability density function of the reciprocal Schmid factor.

generate points uniformly throughout the stereographic triangle, as shown in Figure 42.10. (A similar technique has since been published by Ono et al. [85].) The response-surface use that was used to determine the Schmid factor for each of these points and the scatter among the Schmid factor is determined. The probability density function (PDF) of M is shown in Figure 42.11. The mean value of the distribution is 2.21, which is in agreement with other analytical findings [86]. The predicted probability mass function (PMF) of M compares favorably with the experimentally determined value for 203 surface grains of an untextured pure iron [87], as shown in Figure 42.11.

Using the same Monte Carlo technique, the PDF of the Taylor factor for an untextured polycrystal is shown in Figure 42.12. The mean value of the distribution is 3.07, which is in agreement with analytical and experimental findings [88]. The predicted PDF of M determined in the present study compares favorably

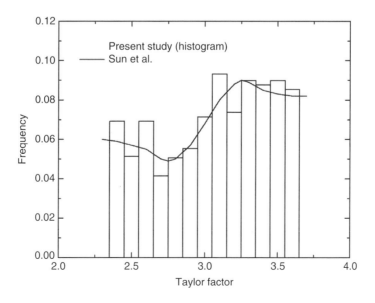

FIGURE 42.12 Probability density function of the Taylor factor.

with the analytical results of Sun et al. [89]. They determined the PDF of the Taylor factor using the computer solution of Chin and Mammel [83] and considering all possible crystallographic orientations.

42.4.3 Frictional Shear Stress

The frictional stress is the stress that must be overcome for dislocations to move within a grain. The frictional stress can be thought of as the local yield stress. Because of the crystallographic orientation of the grain, yielding takes place on well-defined planes (in the low Γ planar slip alloys). Experimental observations have shown that the frictional stress is nearly uniform across the grain [90].

There is little direct data available in the literature on the statistical distribution of the frictional stress. A rigorous numerical determination of the grain-to-grain scatter in frictional stress has not been made. However, empirical observations provide some insight into the behavior of the scatter.

Taira et al. [3] experimentally observed the minimum cyclic stress for which slip bands formed in three different mean-grain-size microstructures of low-carbon steel. The applied stress was below the fatigue limit, and slip bands formed in very few grains. They found that the minimum frictional stress is independent of mean grain size. The minimum frictional stress was nearly equal to the frictional stress predicted by the Petch relationship for the fatigue limit, expressed as

$$\sigma_{fl} = k_{fl} + \frac{K_m}{\sqrt{d}} \tag{42.45}$$

where σ_{fl} is the fatigue limit, k_{fl} is the frictional stress of the grains participating in fatigue, and K_m is the microscopic stress intensity factor.

Taira et al. [3] used the Petch relationship for flow stress to determine the frictional stress for applied loads up to 5% plastic strain. As the load increased, more and more grains produced slip bands. An indication of the scatter can be made by comparing the frictional stress determined at high applied load with the frictional stress at low applied load. At high applied load, many grains produce slip bands, and this frictional stress may be thought of as the frictional stress of the nominal grain. At low applied load, only a few grains produce slip bands, and this frictional stress may be thought of as the frictional stress of the weakest grain.

This method is not rigorous because the variation in the microstress is not taken into account. In addition, it is difficult to determine the shape of the distribution. In the present study, a two-parameter Weibull distribution is assumed and fitted to the data of Taira et al. [3]. The parameters of the Weibull distribution are determined by taking the frictional stress determined from the 5% plastic-strain test, $k_{0.5}$ = 340 MPa, to be the frictional stress of the 50th-percentile grain, and the frictional stress from the fatigue-limit test, k_{fl} = 110 MPa, is taken to be the frictional stress of the first-percentile grain. This gives a normalized Weibull distribution with a shape factor β_k = 3.7 and a characteristic value η_k = 1.12 that yields a mean value of 1 and a COV of 0.3. Tanaka et al. [56] indicate that a two-parameter Weibull distribution with COV between 0.3 and 0.7 can be used to describe the frictional stress.

42.4.3.1 Applied Microstress

Because each grain acts as an anisotropic single crystal, the actual loading on an individual grain is caused by the deformation of the surrounding grains, which are in turn loaded by the deformations of each of their surrounding grains. Therefore, the microstress distribution is a function of the anisotropic deformation of all of the grains that compose the structure.

Barenblatt [91] proposed a theoretical model to describe the microstress field. Many simplifying assumptions were necessary to make the model tractable. Zhao and Tryon [92] investigated the microstress field for a single-phase nickel alloy using finite elements and Voronoi tessellation to produce a model that closely approximates the microstructure. They modeled a statistical volume element with 200 grains. Each grain was modeled as an anisotropic single crystal with several hundred finite elements per grain. An elastic analysis was performed in which a uniaxial macroscopic load was applied. The von

Mises stress at the grain interiors could be described by a normal distribution with a mean equal to the applied macroscopic stress and a COV of 0.25. The stress distribution of the surface grains was found to be the same as the interior grains. The COV was a function of the elastic anisotropy of the material.

In this chapter, the microstress will be assumed to have a normal distribution with a mean value equal to the applied load s_{xx} and a COV of 0.25.

42.5 Monte Carlo Simulation of Each Stage of Damage Accumulation

The Monte Carlo simulation of the fatigue-damage accumulation is used to determine the probability that a component will fail at a given number of cycles. The component to be modeled is a smooth round bar subjected to constant cyclic load amplitude. The global loading is assumed to be uniform throughout the gauge section. A single representative volume element (RVE) is also assumed to be valid throughout the gauge section. The material is assumed to be a single-phase polycrystal such as a nickel superalloy or high-strength steel. This analysis simulates an empirical fatigue characterization test in which specimens are cycled to failure and the number of cycles at failure is recorded. The Monte Carlo simulation produced the same data. A bar is chosen (simulated) and cycled to failure. The number of cycles at failure are recorded.

The basic flow of the simulation is as follows

1. A bar is chosen and the grains of the bar are generated using the statistical distributions described above. Each grain has a distinct size, orientation, frictional strength, and microstress.
2. A crack is nucleated in a surface grain, and the cycles to failure are found using the model described earlier.
3. Each crack then grows through the microstructure via the small- and then long-crack-growth models described earlier.
4. Another grain is chosen, and steps 2 and 3 are repeated.
5. Step 4 is repeated multiple times until all surface grains are considered. However, step 3 is not repeated for all of the surface grains. A method of choosing only the grains that initiate a crack that is likely to lead to failure is employed.
6. The life of the component is set equal to the minimum life of all of the cracks.

The details of the simulation are described in the following subsections.

42.5.1 Crack Nucleation

A surface grain is simulated by generating grain diameter, d, the applied microstress, σ, the frictional stress needed to move dislocations, k, and the Schmid grain orientation factor, M_s, from the appropriate distributions shown in Table 42.3, which are typical values for a high-strength single-phase alloy. The microscopic stress intensity factor, K_m, is determined according to Equation 42.25. If K_m is less than the critical stress intensity factor needed to overcome the grain boundary, K_{mc}, then the crack arrests at the grain boundary. The next surface grain is generated by repeating the process. If K_m is greater than K_{mc}, the number of cycles needed for crack nucleation is determined from one of the models presented in Section 42.3.2.2, and the crack will continue into the small-crack phase.

42.5.2 Small-Crack Growth

In the small-crack phase, the microstructure surrounding the crack-nucleating grain is simulated by considering a crack nucleating at X_0 in Figure 42.13. Zone 1, directly in front of the crack, is simulated first. The properties d, σ, k, and M_s of grain g_{11}^s are generated using the appropriate distributions. The grain is a

TABLE 42.3 Input to the Monte Carlo Simulation

Variable	Description	Distribution Type	Distribution Parameters		Average	COV
C	Paris law coefficient	lognormal	$\lambda = -0.034$	$\zeta = 0.30$	4.4×10^{-9} MPa$\sqrt{m}$	0.30
C'	CTOD law coefficient	deterministic	N/A		0.10	N/A
d	Grain diameter	lognormal	$\lambda = -0.076$	$\zeta = 0.39$	55.8 μm	0.40
da	Small-crack growth interval	deterministic	N/A		0.5	N/A
G	Bulk shear modulus	deterministic	N/A		76×10^{-3} MPa	N/A
k	Frictional strength	Weibull	$\eta = 1.12$	$\zeta = 3.7$	69 MPa	0.30
K_{crit}^M	Critical microstructural stress intensity factor	deterministic	N/A		769 MPa$\sqrt{m}$	N/A
M_S	Schmid orientation factor		curve fit		2.21	0.08
M_T	Taylor orientation factor		curve fit		3.07	0.13
n	Paris law exponent	deterministic	N/A		3	N/A
W_S	Specific fracture energy	deterministic	N/A		440 kN/m	N/A
v	Poisson ratio	deterministic	N/A		0.3	N/A
σ	Applied microstress	normal	$\mu = 1$	$\sigma = 0.3$	variable[a]	0.30

[a] The stress level of interest is a user input.

surface grain, therefore the Schmid factor orientation is assumed. The arc length l_{arc1} is determined by

$$l_{arc1} = \frac{\pi}{2}\left(\frac{L_0}{2} + \frac{d_{11}^s}{2}\right) = \frac{\pi}{4}\left(L_0 + d_{11}^s\right) \quad (42.46)$$

If $d_{11}^s > l_{arc1}$, then the grain fills the zone, and d_{11}^s is set equal to l_{arc1}. The properties of zone 1 are those generated for g_{11}^s. If $d_{11}^s < l_{arc1}$ then the grain does not fill the space. A gap remains of size

$$gap = l_{arc1} - d_{11}^s \quad (42.47)$$

The other surface grain g_{12}^s in zone 1 (see Figure 42.13b) is generated with the appropriate properties. The size of the grain d_{12}^s is compared with the gap. If $d_{12}^s > gap$, then the grain fills the gap, and d_{12}^s is set

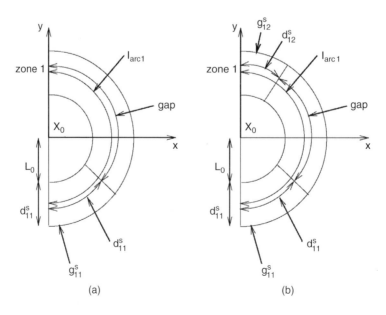

FIGURE 42.13 Representation of the simulated microstructure.

equal to the gap. The effective properties of zone 1 are calculated as a weighted average with respect to the grain volume.

$$P_n = \frac{\sum_i^j P_i d_i^2}{\sum_i^j d_i^2} \qquad (42.48)$$

where P_n is the property of interest (grain orientation, microstructural stress, or fictional strength). If $d_{12}^s <$ gap, then l_{arc1} is recalculated based on the average diameter of the two grains such that

$$l_{arc1} = \frac{\pi}{4}\left(L_0 + \frac{d_{11}^s + d_{12}^s}{2}\right) = \frac{\pi}{4}\left(L_0 + d_{1avg}^s\right) \qquad (42.49)$$

A gap remains of size

$$gap = l_{arc1} - d_{11}^s - d_{12}^2 \qquad (42.50)$$

An interior grain is now generated with the appropriate properties. The Taylor analysis is used for the orientation factor. If the grain fills the gap, then the grain size is set equal to the gap, and the effective zone properties are calculated. If the grain does not fill the gap, then the arc length and the gap are recalculated, and grains are generated until the gap is filled. The effective properties of the subsequent zones are generated using the same technique. The simulated microstructure can be represented by Figure 42.14.

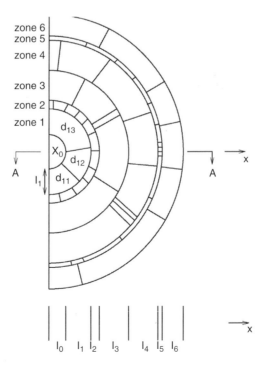

FIGURE 42.14 Microstructure created by the Monte Carlo simulation.

Zones are generated until the effective material properties are within ±10% of the average material properties for three successive zones. Thus, the microstructure is generated until microstructural similitude is achieved. The number of zones and the total area of the zones are random and depend on the variation of the microstructural properties. Microstructures with small variations will have a smaller region of dissimilitude than microstructures with large variations.

Once the microstructure surrounding the nucleating grain is simulated, the small-crack growth is simulated. The equations governing small-crack growth depend on the condition at the tip of the slip band. The theory and mathematical models used in the small-crack-growth analysis were outlined in the last bimonthly report. The following discussion describes the flow of the Monte Carlo simulation.

42.5.2.1 Propagated Slip Band

First, the propagated slip-band phase is considered. The crack tip is at l_0 in Figure 42.14. The slip-band tip is in zone 1. If the effective microstructural stress, σ_{eff}, is greater than the effective frictional stress, k_{eff}, then the zone has yielded, and the slip-band tip has traversed to the next zone boundary. In this case, the propagated slip-band phase requires zero cycles, and the next phase of blocked slip band is considered.

If $\sigma_{eff} < k_{eff}$, then the slip band has not yet reached the next zone boundary, and the position of the slip-band tip is determined. An iterative technique is needed to solve for the location of the slip-band tip because the equation describing its location is not in closed form. Newton's method is used with the convergence criterion that successive values be within 0.1 da, where da is the crack growth increment, which is an input variable. (The variable da must be some small fraction of the average grain diameter.) Once the location of the slip-band tip has been determined, the crack-tip-opening displacement, ϕ_t, is evaluated. The number of cycles needed for the crack tip to traverse a distance da is calculated by

$$dN = \frac{C'\phi_t}{da} \qquad (42.51)$$

where ϕ_t is the crack-tip-opening displacement. The new position of the crack tip is $a + da$. The process is repeated until the slip-band tip reaches the next zone boundary. At this point, the blocked slip-band phase begins.

42.5.2.2 Blocked Slip Band

The crack tip is located at a, determined from the slip-band routine propagated above. The slip-band tip is blocked at the zone boundary. The microscopic stress intensity factor K_m is calculated. If $K_m > K_{mc}$, then the slip-band tip successfully penetrates the zone boundary. In this case, the blocked slip-band phase requires zero cycles, and the propagated slip-band phase is considered for the next zone. If $K_m < K_{mc}$, then ϕ_t is calculated. The number of cycles needed for the crack tip to traverse a distance da is calculated using Equation 42.51. The new position of the crack tip is $a + da$. The process is repeated until the slip-band tip penetrates the zone boundary or the crack growth arrests.

The slip-band tip will penetrate the zone boundary when $K_m > K_{mc}$. With each successive iteration, the crack tip grows by da, and K_m increases. However, as the crack tip approached the grain boundary, ϕ_t approaches zero. This causes dN to approach infinity in Equation 42.51. Therefore, if $K_m < K_{mc}$ when the crack tip reaches the grain boundary, then the crack growth is arrested. In other words, if the crack tip reaches the zone boundary and the stress intensity factor is still less than critical, then the crack stops growing.

If the crack growth is arrested, then the next surface grain is generated. If the crack growth does not arrest, then the slip-band tip successfully penetrates the zone boundary, and the propagated slip-band phase is considered for the next zone.

The crack continues to grow through propagated and blocked phases of successive zones until the crack is arrested or the slip-band tip has reached the end of the microstructural dissimilitude region, at which point the crack enters the long-crack growth phase.

42.5.3 Long-Crack Growth

Once the crack has reached the long-crack growth stage, there is no mechanism for blockage. The number of cycles spent in the long-crack growth stage is calculated using Paris law.

42.6 Fatigue-Life Predictions

42.6.1 Crack Nucleation

The crack-nucleation model has previously been investigated by Tryon and Cruse [93, 94]. Tryon et al. [95] developed a first-order reliability method (FORM) to investigate scatter in fatigue-crack nucleation. The results of the investigation are briefly presented here. The predicted shape of the crack-nucleation life distribution was similar to the experimentally observed shapes found in the literature. The COV of crack-nucleation life was predicted to increase as the applied load decreased, and the COV was independent of the mean grain size. These predictions are in agreement with the experimental findings. Tryon and Cruse [94] use FORM to investigate the sensitivity of fatigue to the random variables and how the sensitivity changes with reliability level.

42.6.2 Small-Crack Growth

Small-crack growth behavior is modeled using the small-crack-growth phase of the Monte Carlo FORTRAN software. The results are determined using the parameters in Table 42.3. These values are characteristic of a stainless steel. The results are compared with trends in the experimental data from the literature. The comparisons show that the small-crack-growth model is able to predict the significant aspects of small-crack growth behavior.

One aspect of small-crack growth behavior is that the average crack growth rate is much higher than what would be predicted based on long-crack-growth data and applied ΔK. Phillips and Newman [96] showed that not only do small cracks grow much faster than ΔK-equivalent long cracks, but that da/dN vs. ΔK increases with applied stress. Thus, ΔK-based similitude is not valid for small-crack growth. Figure 42.15 shows the predicted average crack growth rate as a function of applied ΔK. The results compare favorably with experimental results of Phillips and Newman [97].

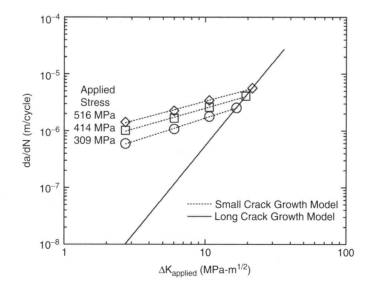

FIGURE 42.15 Predicted applied stress effect on crack growth rate.

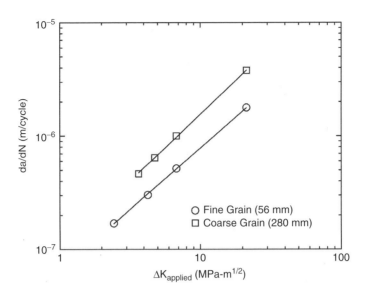

FIGURE 42.16 Predicted grain size effect on small-crack growth rate.

Another aspect of small-crack growth behavior is that the mean crack growth rate for coarse-grain microstructures is higher than the crack growth rate for fine-grain microstructures of the same alloy [98]. This is in contrast to long-crack growth-rate behavior that shows that, in general, the crack growth rate for coarse-grain microstructures is lower than that of fine-grain microstructures. These observations are significant because the assumption has been made, based on long-crack growth data, that the coarse-grain materials produce better fatigue resistance. However, for cracking in service application, the small-crack-growth regime must be considered. The overall fatigue resistance may be driven by the small-crack growth behavior [99].

Figure 42.16 shows the predicted average crack growth rate for two microstructures in which the average grain size has been changed. The figure shows that the crack growth rate is lower for the fine-grain microstructure. In small cracks, the low growth rate for fine-grain microstructures is attributed to the fact that there are more grain boundaries available to retard crack growth than what would be available over the same distance in a coarse microstructure. The difference in long-crack growth rate for different-grain-size microstructures is attributed to two factors: closure and intergranular (grain boundary) cracking. The fine-grain material has a smoother fracture surface, allowing for less opposing crack-face roughness-induced closure. An element of intergranular cracking is observed as the plastic zone becomes large compared with the grain size [100]. Intergranular crack growth rates are generally higher than transgranular [101]. The introduction of the intergranular cracking will take place at a lower ΔK for the fine-grain material. A combination of less closure and more intergranular cracking causes a reduction in the overall fatigue resistance for the fine-grain material.

Figure 42.17 shows the predicted crack growth rate as a function of crack length for five cracks that have successfully penetrated the first grain boundary. The crack growth rate seems to vary haphazardly as the crack interacts with the microstructure. Similar behavior has been observed experimentally by many investigators [102, 103]. The only obvious correlation that has been made between the crack growth rate and the microstructure is that the large jumps in crack growth rate can be associated with the crack tip nearing the grain boundary.

The small-crack-growth equations indicate that several factors govern the interaction between the crack growth rate and the microstructure:

1. Crack length
2. Local effective resolved shear stress at the crack tip
3. Local effective frictional stress at the crack tip

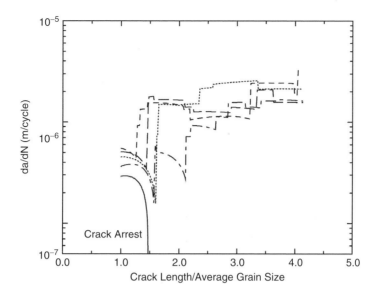

FIGURE 42.17 Predicted small-crack growth behavior of five cracks.

4. Slip-band length
5. Effective resolved shear stress along the slip band
6. Effective frictional stress along the slip band
7. Whether the slip-band tip is propagating or blocked

To illustrate these interactions, consider the predicted small-crack growth rate vs. crack-length curve for a single crack chosen at random shown in Figure 42.18. Grain boundaries are located at O, D, and F. The average frictional stress is 70 MPa. The average resolved shear stress is 41.4 MPa. The local effective properties are shown in the figure. The following list describes the conditions governing crack growth for the various regimes in Figure 42.18.

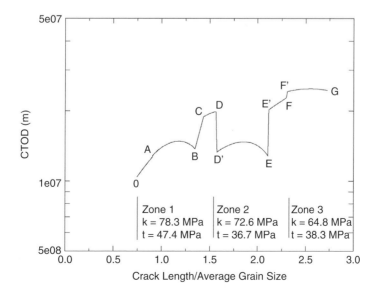

FIGURE 42.18 Predicted small-crack growth behavior.

1. *Crack tip between O and A.* The crack tip is in zone 1. The slip-band tip is propagating in zone 1.
2. *Crack tip between A and B.* The crack tip is in zone 1. The slip-band tip is blocked at the boundary between zones 1 and 2.
3. *Crack tip between B and C.* The crack tip is in zone 1. The slip-band tip is propagating in zone 2.
4. *Crack tip between C and D.* The crack tip is in zone 1. The slip-band tip is blocked at the boundary between zones 2 and 3.
5. *Crack tip between D′ and E.* The crack tip is in zone 2. The slip-band tip is blocked at the boundary between zones 2 and 3.
6. *Crack tip between E′ and F.* The crack tip is in zone 2. The slip-band tip is blocked at the boundary between zones 3 and 4.
7. *Crack tip between F′ and G.* The crack tip is in zone 3. The slip-band tip is blocked at the boundary between zones 3 and 4.

As the crack grows, the slip band becomes large and spans several grains. The crack continues to grow in such a manner until the effective properties of the material between the crack tip and the slip-band tip approach the bulk properties.

42.6.3 Predicted Total Fatigue Life of a Test Specimen

The most thorough investigation of the scatter in fatigue life available in the open literature is that of Bastenaire [11], who performed an investigation of the scatter in fatigue life for five different grades of low-alloy steel. Steels may nucleate cracks by mechanisms other than slip-band cracking, depending on the alloy composition and the impurities. However, the trend in the scatter in steel data has been observed in other metallic alloys [9]. Bastenaire performed rotating bending fatigue experiments for many stress levels for each grade of steel with several hundred specimens for each stress level.

Figure 42.19 shows the trends in the scatter exhibited in Bastenaire's data. (The curves are replotted from the data in Figure 7 of Bastenaire.) The general trend is that the COV (indicated by the slope of

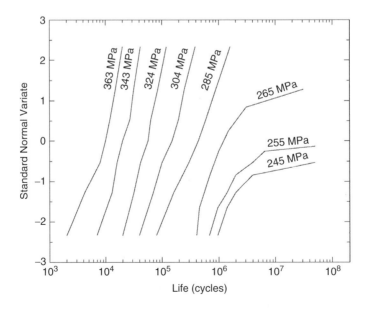

FIGURE 42.19 Fatigue-life test data plotted on lognormal paper. (Data from Bastenaire, F.A., in *Probabilistic Aspects of Fatigue*, Heller, R.A., Ed., ASTM STP 511, ASTM, Philadelphia, PA, 1972, pp. 3–28.)

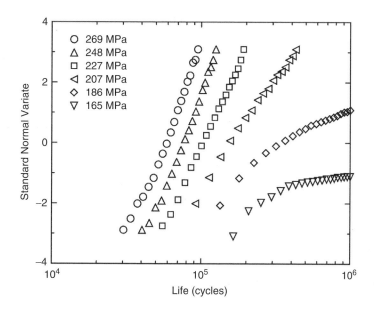

FIGURE 42.20 Predicted fatigue-life distribution plotted on lognormal paper.

the curves) is fairly constant for applied stresses well above the fatigue limit (363 to 324 MPa). As the applied stress decreases, the COV starts to increase (304 to 285 MPa). As the applied stress approaches the fatigue limit, run-outs start to occur. The right tail of the distribution becomes heavy, which causes a line through the data to bend to the right (265 to 245 MPa). If data plots as a nonstraight line in Figure 42.19, then the lognormal distribution is no longer valid. The 363-MPa data curves slightly to the left, indicating that the distribution has a short right tail and the data can also be fitted to the normal distribution. As the applied stress decreases, the curvature shifts to the right.

Comparison of Figure 42.19 with the results in Figure 42.20 shows that the model predicts all of the above trends observed in the experimental data.

42.7 Conclusions

This paper showed the feasibility of probabilistic microstructural fatigue modeling. Information was found in the open literature on the mechanism that drives crack nucleation and small-crack growth for many alloys. Damage-accumulation models were developed that address the most important factors driving microstructural crack growth. The random variables that govern fatigue were estimated from empirical data and theoretical considerations. Statistical volume element FEM models based on Voronoi cells were able to accurately represent the local stress response. Simple closed-form equations were developed to approximations of statistical volume element models based on the crystallographic anisotropy. Software for a Monte Carlo simulation was written that combined all of the above with the damage-accumulation models to correctly predict:

- Shape of the crack-nucleation life distribution
- Multiple cracks
- Applied global stress effects and the mean grain size effect on the COV of crack-nucleation life
- Applied global stress effects and the mean grain size effect on small-crack growth rate
- Variation in small-crack growth rate
- Shape of the total fatigue-life distribution
- Applied global stress effects on the shape of the total fatigue-life distribution

Acknowledgments

The author would like to thank C. Chamis of NASA Glenn Research Center and J. Jira, A. Rosenberger, and S. Russ of the Air Force Research Laboratory for their financial support. The author would also like to thank T. Cruse, professor emeritus of Vanderbilt University, for his technical support.

References

1. Nisitani, H. and Takao, K.-I., Significance of initiation, propagation and closure of microcracks in high cycle fatigue of ductile metals, *Eng. Fracture Mech.*, 15 (3), 445–456, 1981.
2. McDowell, D.L., An engineering model for propagation of small cracks in fatigue, *Eng. Fracture Mech.*, 56, 357–377, 1997.
3. Taira S., Tanaka, K., and Hoshina, M., Grain size effect on crack nucleation and growth in long-life fatigue of low-carbon steel, in *Fatigue Mechanisms*, Fong, J.T., Ed., ASTM STP 675, ASTM, Philadelphia, PA, 1979, pp. 135–173.
4. Floreen, S., High temperature crack growth structure property relationships in nickel base super-alloys, in *Creep-Fatigue-Environment Interactions*, Pelloux, R. and Slotoff, N., Eds., Metallurgical Society, Warrendale, PA, 1980, pp. 121–128.
5. Tokaji, K., Ogawa, T., and Ohya, K., The effect of grain size on small fatigue crack growth in pure titanium, *Fatigue*, 16, 571–578, 1994.
6. Gayda, J. and Miner, R., Fatigue crack initiation and propagation in several nickel-based superalloys at 650°C, *Int. J. Fatigue*, 5 (3), 135–143, 1983.
7. Lerch, B., Microstructural Effects on the Room and Elevated Temperature Low Cycle Fatigue Behavior of Waspaloy, NASA CR 165 497, NASA, Cleveland, OH, 1982.
8. Lankford, J. and Davidson, D.L., The role of metallurgical factors in controlling the growth of small fatigue cracks, in *Small Fatigue Cracks*, Ritchie, R.O. and Lankford, J., Eds., Metallurgical Society, Warrendale, PA, 1986, pp. 51–71.
9. Sasaki, S., Ochi, Y., Ishii, A., and Hirofumi, A., Effects of material structures on statistical scatter in initiation and growth lives of surface cracks and failure life in fatigue, *JSME Inter. J., Ser. I*, 32 (1), 155–161, 1989.
10. Bataille, A. and Magnin, T., Surface damage accumulation in low cycle fatigue: physical analysis and numerical modelling, *Acta Metall. Mater.*, 42 (11), 3817–3825, 1994.
11. Bastenaire, F.A., New method for the statistical evaluation of constant stress amplitude fatigue-test results, in *Probabilistic Aspects of Fatigue*, Heller, R.A., Ed., ASTM STP 511, ASTM, Philadelphia, PA, 1972, pp. 3–28.
12. Goto, M., Statistical investigation of the behaviour of small cracks and fatigue life in carbon steels with different ferrite grain sizes, *Fatigue Fracture Eng. Mater. Struct.*, 17 (6), 635–649, 1994.
13. Ishii, A., Ochi, Y., Sasaki, S.K., and Nakamura, H., Effect of microstructure on statistical scatter of crack initiation and growth lives in NiCrMoV cast steel, *J. Soc. Mater. Sci., Japan/Zairyo*, 40 (452), 568–574, 1991.
14. Fu-Ze, Z., The fatigue scatter factors and reduction factors in the design of aircraft and helicopter's structural lives, in *47th Annual Forum Proceedings of American Helicopter Society*, Vol. 911984, SAE, Warrendale, PA, 1991, pp. 173–178.
15. Goto, M., Statistical investigation of the behaviour of microcracks in carbon steel, *Fatigue Fracture Eng. Mater. Struct.*, 14 (8), 833–845, 1991.
16. Sasaki, S.K., Ochi, Y., and Ishii, A., Statistical investigation of surface fatigue cracks in large-sized turbine rotor shaft steel, *Eng. Fracture Mech.*, 28 (5/6), 761–772, 1987.
17. Dieter, G.E., *Mechanical Metallurgy*, 3rd ed., McGraw-Hill, New York, 1986.
18. Gokhale, A.B. and Rhimes, F.N., Effect of grain volume distribution on the plastic properties of high purity aluminum, in *Microstructural Science*, Vol. 11, DeHoff, R., Braum, J., and McCall, J., Eds., Elsevier, New York, 1983, pp. 3–11.

19. Halford, G.R., Evolution of creep-fatigue life prediction models, creep-fatigue interactions at high temperatures, AD Vol. 21, Haritos, G.K. and Ochoa, O.O., Eds., ASME, 1991, pp. 43–57.
20. Schijve, J., Fatigue predictions and scatter, *Fatigue Fracture Eng. Mater. Struct.*, 17 (4), 381–396, 1994.
21. Little, R.E. and Jebe, E.H., *Statistical Design of Fatigue Experiments*, Applied Science Publ., Essex, U.K., 1975.
22. Axelrad, D.R., The mechanics of discrete media, in *Continuum Models of Discrete Systems (CMDS3)*, Kroner, E. and Anthony, K.-H., Eds., Univ. Waterloo, Waterloo, Canada, 1980, pp. 3–34.
23. Haritos, G.K., Hager, A.K., Salkind, M.J., and Wang, A.S.D., Mesomechanics: the microstructure-mechanics connection, *Int. J. Sol. Struct.*, 24 (11), 1081–1096, 1988.
24. Peterson, R.E., Methods of correlating data from fatigue test of stress concentration specimens, in *Contributions to the Mechanics of Solids*, Macmillan, New York, 1939, pp. 179–183.
25. Fong, J.T., Statistical aspects of fatigue at microscopic, specimen, and component levels, in *Fatigue Mechanisms*, ASTM STP 675, ASTM, Philadelphia, PA, 1979, pp. 729–758.
26. Smith, R.A., Short fatigue cracks, in *Fatigue Mechanism: Quantitative Measurements of Physical Damage*, Lankford, J., Davidson, D.L., Morris, W.L., and Wei, R.P., Eds., STP 811, ASTM, Philadelphia, PA, 1983, pp. 264–297.
27. Ghosh, S. and Moorthy, S., Elastic-plastic analysis of arbitrary heterogeneous materials with Voronoi Cell finite element method, *Comp. Meth. Appl. Mech. Eng.*, 121, 373–409, 1995.
28. Fine, M.E. and Kwon, I.B., Fatigue crack initiation along slip bands, in *Small Fatigue Cracks*, Ritchie, R.O. and Lankford, J., Eds., Metallurgical Society, Warrendale, PA, 1986, pp. 29–40.
29. Laird, C., Mechanisms and theories of fatigue, in *Fatigue and Microstructure*, ASM, Metals Park, OH, 1979, pp. 149–203.
30. Kuhlmann-Wilsdorf, D., Dislocation behavior in fatigue, *Mater. Sci. Eng.*, 27, 137–156, 1977.
31. Forsyth, P., *The Physical Basis of Metal Fatigue*, American Elsevier Publ., New York, 1969.
32. Brown, L.M., Dislocation substructure and the initiation of cracks by fatigue, *Metal Sci.*, Vol. 11, Aug./Sept., 315–320, 1977.
33. Grobstein, L.L., Sivashankaran, S., Welsch, G., Panigrahi, N., McGervery, J.D., and Blue, W., Fatigue damage accumulation in nickel prior to crack initiation, *Mater. Sci. Eng.*, A138, 191–203, 1977.
34. Tanaka, K. and Mura, T., A dislocation model for fatigue crack initiation, *ASME J. Appl. Mech.*, 48, 97–103, 1981.
35. Laird, C. and Duquette, D.J., Mechanisms of fatigue nucleation, in *Proceedings of International Corrosion Fatigue Conference*, Devereux, O., McEvily, A.J., and Steable, R.W., Eds., 1972, pp. 88–117.
36. Laird, C., Recent advances in understanding the cyclic deformation of metals and solid solutions, in *Work Hardening in Tension and Fatigue*, Metallurgical Society, Warrendale, PA, 1977, pp. 150–176.
37. Chan, K.S. and Lankford, J., A crack tip strain model for the growth of small fatigue cracks, *Scripta Metall.*, 17, 529–532, 1983.
38. Newman, J.C., Swain, M.H., and Phillips, E.P., An assessment of the small-crack effect for 2024-T3 aluminum alloy, in *Small Fatigue Cracks*, Ritchie, R.O. and Lankford, J., Eds., Metallurgical Society, Warrendale, PA, 1986, pp. 427–452.
39. Grosskreutz, J.C., The mechanism of metal fatigue, *Physica Status Solidi*, 47b, 359–396, 1971.
40. Sasaki, S. and Ochi, Y., Some experimental studies of fatigue slip bands and persistent slip bands during fatigue process of low-carbon steel, *Eng. Fracture Mech.*, 12, 531–540, 1979.
41. Head, A.K. and Louat, N., The distribution of dislocations in linear arrays, *Aus. J. Phys.*, 8, 1–7, 1955.
42. Bilby, B.A. and Eshelby, J.D., Dislocations and the theory of fracture, in *Fracture*, Vol. 1, Liebowitz, H., Ed., 1968, pp. 99–182.
43. Chang, R., Morris, W.L., and Buck, O., Fatigue crack nucleation at intermetallic particles in alloys: a dislocation pile-up model, *Scripta Metall.*, 13, 191–194, 1979.
44. Mura, T. and Tanaka, K., A dislocation dipole model for fatigue crack initiation, in *Mechanics of Fatigue*, AMD Vol. 47, ASME, Philadelphia, PA, 1981, pp. 111–132.
45. Kato, M. and Mori, T., Statistical consideration of fatigue damage accumulation, *Mech. Mater.*, 13, 155–163, 1992.

46. Cooper, C.V. and Fine, M.E., Coffin-Manson relation for fatigue crack initiation, *Scripta Metall.*, 18, 593–595, 1984.
47. Muskhelishvili, N.I., *Singular Integral Equations*, Noordhoff Inter., Boston, MA, 1977.
48. Davidson, D.L., Small and large fatigue cracks in aluminum alloys, *Acta Metall.*, 38 (8), 2275–2282, 1988.
49. Mura, T. and Nakasone, Y., A theory of fatigue crack initiation in solids, *J. Appl. Mech.*, 57, 1–6, 1990.
50. Venkatraman, G., Chung, Y.-W., Nakasone, Y., and Mura, T., Free energy formulation of fatigue crack initiation along persistent slip bands: calculation of S-N curves and crack depths, *Acta Metall. Mater.*, 38, 31–40, 1990.
51. Laird, D. and Smith, G.C., Crack propagation in high stress fatigue, *Philos. Mag.*, 7, 847–857, 1962.
52. Weertman, J., Fatigue crack propagation theories, in *Fatigue and Microstructure*, ASM, Metals Park, OH, 1979, pp. 279–206.
53. Kikukawa, M., Jono, M., and Adachi, M., Direct observation and mechanics of fatigue crack propagation, in *Proceedings of an ASTM-NBS-NSF Symposium*, Fong, J.T., Ed., ASTM STP 675, ASTM, Philadelphia, PA, 1979, pp. 234–253.
54. Hicks, M.A. and Brown, C.W., A comparison of short crack growth behaviour in engineering alloys, in *Fatigue 84*, Engineering Materials Advisory Services Ltd., London, U.K., 1984, pp. 1337–1347.
55. Hudak, S.J. and Chan, K.S., In search of a driving force to characterize the kinetics of small crack growth, in *Small Fatigue Cracks*, Ritchie, R.O. and Lankford, J., Eds., Metallurgical Society, Warrendale, PA, 1986, pp. 379–406.
56. Tanaka, K., Kinefuchi, M., and Yokomaku, T., Modelling of statistical characteristics of the propagation of small fatigue cracks, in *Short Fatigue Cracks*, Miller, K.J. and de los Rios, E.R., Eds., ESIS 13, Mechanical Engineering Publications, London, 1992, pp. 351–368.
57. Ewalds, H.L. and Wanhill, R.J.H., *Fracture Mechanics*, Edward Arnold Publ., New York, 1991.
58. Donahue, R.J., Clark, H.M., Atanmo, P., Kumble, R., and McEvily, A.J., *Int. J. Fracture Mech.*, 8, 209, 1972.
59. McEvily, A.J., *Microstructure and Design of Alloys*, Vol. 2, Inst. of Metals, London, U.K., 1974.
60. Kikukawa, M., Jono, M., and Tanaka, K., Fatigue crack closure behavior at low stress intensity lever, in *Proceedings of ASM Mech. Behavior of Mater.*, ASM, Boston, 1976, pp. 716–720.
61. Hobson, P.D., The formulation of a crack growth equation for short cracks, *Fatigue Eng. Mater. Struct.*, 5, 323–327, 1982.
62. Chan, K.S. and Lankford, J., The role of microstructural dissimilitude in fatigue and fracture of small cracks, *Acta Metall.*, 36 (1), 193–206, 1988.
63. Gerberich, W.W. and Moody, N.R., A review of fatigue fracture topology effects on threshold and growth mechanisms, in *Fatigue Mechanisms*, ASTM STP 675, ASTM, Philadelphia, PA, 1979, pp. 292–341.
64. Dugdale, D.S., Yielding of steel sheets containing slits, *J. Mech. Phys. Sol.*, 8, 100–104, 1960.
65. Weertman, J., Rate of growth of fatigue cracks calculations from the theory of infinitesimal dislocations distributed on a plane, *J. Fracture Mech.*, 2, 460–467, 1966.
66. Bilby, B.A., Cottrell, A.H., and Swinden, K.H., The spread of plastic yield from a notch, *Proc. Roy. Soc.*, A272, 304–314, 1963.
67. Weertman, J., Crack tip stress intensity factor of the double slip plane crack model: short cracks and short short-cracks, *Int. J. Fracture*, 26, 31–42, 1984.
68. Taira, S., Tanaka, K., and Nakai, Y., A model of crack tip slip band blocked by grain boundary, *Mech. Res. Comm.*, 5 (6), 375–381, 1978.
69. Tanaka, K., Akiniwa, Y., Nakia, Y., and Wei, R.P., Modeling of small fatigue crack growth interacting with grain boundary, *Eng. Fracture Mech.*, 24 (6), 803–819, 1986.
70. Morris, W.L., James, M.R., and Buck, O., A simple model of stress intensity range threshold and crack closure stress, *Eng. Fracture Mech.*, 18, 871–877, 1983.
71. de los Rios, E.R., Tang, Z., and Miller, K.J., Short crack fatigue behavior in a medium carbon steel, *Fatigue Fracture Eng. Mater. Struct.*, 7, 97–108, 1984.

72. Navarro, A. and de los Rios, E.R., A microstructurally short fatigue crack growth equation, *Fatigue Fracture Eng. Mater. Struct.*, 11 (5), 383–396, 1988.
73. Newby, M.J., Markov models for fatigue crack growth, *Eng. Fracture Mech.*, 27 (4), 477–482, 1987.
74. Yang, J.N., Salivar, G.C., and Annis, C.G., *Constant Amplitude Fatigue Crack Growth at Elevated Temperatures*, Vol. 1 of *Statistics of Crack Growth in Engine Materials*, AFWAL-TR-82-4040, WPAFB, WPAFB, OH, 1982.
75. Morris, W.L., James, M.R., and Buck, O., Computer simulation of fatigue crack initiation, *Eng. Fracture Mechs.*, 13, 213–221, 1980.
76. Smith, C.S., *A Search for Structure*, MIT Press, Cambridge, MA, 1981.
77. Kurtz, S.K. and Carpay, F.M.A., Microstructure and normal grain growth in metals and ceramics, Part I, theory, *J. Appl. Phys.*, 51 (11), 5725–5744, 1980.
78. Kumar, S., Kurtz, S.K., Banavar, J.R., and Sharma, M.G., Properties of a three-dimensional Poisson-Voronoi tessellation: a Monte Carlo study, *J. Statistical Phys.*, 67 (3/4), 523–551, 1992.
79. Ang, A.H.-S. and Tang, W.H., *Probability Concepts in Engineering Planning and Design*, Vol. 1, John Wiley and Sons, New York, 1975.
80. Barrett, C.S., *Structure of Metals*, 2nd ed., McGraw-Hill, New York, 1952.
81. Taylor, G.I., Plastic deformation in metals, *J. Inst. Metals*, 62, 307–24, 1938.
82. Bishop, J.F.W. and Hill, R., *Philos. Mag.*, 42, 1298–1307, 1951.
83. Chin, G.Y. and Mammel, W.L., Computer solution of the Taylor analysis for axisymmetric flow, *Trans. Metall. Soc.*, 239, 1400–1405, 1967.
84. Davidson, D.L. and Chan, K.S., The crystallography of fatigue crack initiation in coarse grained astroloy at 20°C, *Acta Metall.*, 37 (4), 1089–1097, 1989.
85. Ono, N., Kimura, K., and Watanabe, T., Monte Carlo simulation of grain growth with the fill spectra of grain orientation and grain boundary energy, *Acta Mater.*, 47, 1007–1017, 1999.
86. Backofen, W.A., *Deformation Processing*, Addison-Wesley, Reading, MA, 1972, pp. 72–82.
87. Tanaka, T. and Kosugi, M., Crystallographic study of the fatigue crack nucleation mechanism in pure iron, in *Basic Questions in Fatigue*, Vol. 1, ASTM STP 924, ASTM, Philadelphia, PA, 1988, pp. 98–119.
88. Backofen, W.A., *Deformation Processing*, Addison-Wesley, Reading, MA, 1972, pp. 72–82.
89. Sun, Z., de los Rios, E.R., and Miller, K.J., Modelling small fatigue cracks interacting with grain boundaries, *Fatigue Fracture Eng. Mater. Struct.*, 14 (2/3), 277–291, 1991.
90. James, M.R. and Morris, W.L., The effect of microplastic surface deformation on the growth of small cracks, in *Small Fatigue Cracks*, Ritchie, R.O. and Lankford, J., eds., Metallurgical Society, Warrendale, PA, 1991, pp. 145–156.
91. Barenblatt, G.I., On a model of small fatigue cracks, *Eng. Fracture Mech.*, 28 (5/6), 623–626, 1987.
92. Zhao, Y. and Tryon, R.G., Automatic Simulation of Polycrystalline Metallic Materials, presented at ISSAT International Conference on Reliability and Quality in Design, Anaheim, CA, 2002.
93. Tryon, R.G. and Cruse, T.A., Probabilistic mesomechanical fatigue crack nucleation model, *ASME J. Eng. Mater. Technol.*, 119, 257–267, 1997.
94. Tryon, R.G. and Cruse, T.A., A reliability-based model to predict scatter in fatigue crack nucleation life, *Fatigue Fracture Eng. Mater. Strength*, 21, 257–267, 1998.
95. Tryon, R.G., Cruse, T.A., and Mahadevan, S., Development of a reliability-based fatigue life model for gas turbine engine structures, *Eng. Fraction Mech.*, 53 (3), 807–828, 1996.
96. Phillips, E.P. and Newman, J.C., Impact of small-crack effects on design-life calculations, *Exp. Mech.*, 29 (2), 221–225, 1989.
97. Phillips, E.P. and Newman, J.C., Impact of small-crack effects on design-life calculations, *Exp. Mech.*, 29 (2), 221–225, 1989.
98. Gerdes, C., Gyser, A., and Lutjering, G., Propagation of small surface cracks in Ti-alloys, in *Fatigue Crack Growth Threshold Concepts*, Davidson, D.L. and Suresh, S., Eds., AIME, Warrendale, PA, 1984, pp. 465–478.

99. Brown, C.W. and King, J.E., The relevance of microstructural influence in the short crack regime to overall fatigue resistance, in *Small Fatigue Cracks*, Ritchie, R.O. and Lankford, J., Eds., Metallurgical Society, Warrendale, PA, 1986, pp. 73–95.
100. Gayda, J. and Miner, R.V., The effect of microstructure on 650°C fatigue crack growth in P/M astroloy, *Metall. Trans. A*, 14A, 2301–2308, 1983.
101. Lerch B.A., Jayaraman, K., and Antolovich, S.D., A study of fatigue damage mechanisms in Waspaloy for 25 to 800°C, *Mater. Sci. Eng.*, 66, 151–166, 1984.
102. Tokaji, K. and Ogawa, T., The growth behaviour of microstructurally small fatigue cracks in metals, in *Short Fatigue Cracks*, Miller, K.J. and de los Rios, E.R., Eds., ESIS 13, Mechanical Engineering Publications, London, 1992, pp. 85–99.
103. Reed, P.A. and King, J.E., Comparison of long and short crack growth in polycrystalline and single crystal forms of Udimet 720, in *Short Fatigue Cracks*, Miller, K.J. and de los Rios, E.R., Eds., ESIS 13, Mechanical Engineering Publications, London, 1992, pp. 153–168.

43
Weakest-Link Probabilistic Failure

43.1	Introduction	43-1
43.2	Basic Theory for Static Structures	43-1
43.3	Failure-Location Predictions	43-3
43.4	Extension to Time-Dependent Failure	43-5
43.5	Applications to Composite Materials	43-8
43.6	Conclusions	43-9

Brice N. Cassenti
Pratt & Whitney

43.1 Introduction

In 1939, Weibull [1] proposed a model for the failure of materials that could be used to predict the increased strength observed in bending specimens when compared with tensile specimens. The model was based only on the statistical scatter in experimental results and the assumption that the weakest point in the structure governs the failure of the entire structure. The weakest-link assumption implies that there is a greater probability for failure in larger structures than in equally stressed smaller structures. This follows from the fact that there is a greater probability for a critical flaw in the volume of the larger structure. Of course, weakest-link theory will result in conservative predictions for the failure probabilities, since stress can be redistributed when the weakest point fails. Hence, the conservatism may significantly overpredict the failure probability in metallic structures, but it should be accurate in ceramic structures. Composite material structures should lie intermediate between metallic and ceramic structures.

Many aspects of the effects of size on strength have been examined. Harter [2] has produced a review of the early literature on size effects in material structures. Weakest link has been extended to the prediction of failure locations [3], time dependent failure [4], multi-axial stress states [5], and composite material structures [6].

This chapter will first present the basic theory, and then this will be extended to the prediction of failure locations. Time-dependent failure will then be discussed, and the results will be extended to composite materials.

43.2 Basic Theory for Static Structures

The development of weakest-link theory for static structures presented here will follow the discussion provided in the literature [6]. Consider an infinitesimal volume element, δV, at a point x in a total volume V, and let the stress be σ_{ij} at point x. Assume that the infinitesimal probability for failure, δf, in volume element δV is proportional to the volume element and is a function of the stress state. Then

$$\delta f = \psi[\sigma_{ij}(x)]\delta V. \qquad (43.1)$$

The variable ψ will be referred to as the failure parameter and is a function of the stress, which is in turn a function of the location. The probability for the volume element δV to survive the stress σ_{ij} is

$$\delta S = 1 - \delta f = 1 - \psi[\sigma_{ij}(x)]\delta V. \tag{43.2}$$

We now apply the basic assumption of weakest-link theory: if any volume δV fails in the total volume V, then the entire structure occupying volume V fails. Now consider the structural volume V to be divided into N volume elements, δV_k, located at point x_k. Then the total volume is given by

$$V = \lim_{N \to \infty} \sum_{k=1}^{N} \delta V_k = \int_V dV. \tag{43.3}$$

This is just a Reimann integral. Note that we will assume that, within the limit, all of the volume elements, δV_k, approach zero as the number of volume elements approach infinity.

Applying the basic assumption of weakest-link theory, the probability for volume V to survive is

$$S = \lim_{N \to \infty} \prod_{k=1}^{N} \delta S_k = \lim_{N \to \infty} \prod_{k=1}^{N} \{1 - \psi[\sigma_{ij}(x)]\delta V_k\}, \tag{43.4}$$

where

$$\prod_{k=1}^{N} a_k = a_1 a_2 a_3, \ldots, a_N. \tag{43.5}$$

The product can be converted to a sum by taking the natural logarithm of each side of Equation 43.4. Then

$$\ln(S) = \lim_{N \to \infty} \sum_{k=1}^{N} \ln\{1 - \psi[\sigma_{ij}(x)]\delta V_k\}. \tag{43.6}$$

Since the volume element is infinitesimal, we can use the Taylor series expansion

$$\ln(1+x) \approx x,$$

for $x \ll 1$. Equation 43.4 now becomes

$$\ln S = \lim_{N \to \infty} \sum_{k=1}^{N} \{-\psi[\sigma_{ij}(x)]\delta V_k\} = -\int_V \psi[\sigma_{ij}(x)]dV_k, \tag{43.7}$$

or

$$S = \exp\left\{-\int_V \psi[\sigma_{ij}(x)]dV_k\right\}, \tag{43.8}$$

is the probability for volume V to survive. Hence, the probability for the volume to fail, f, is

$$f = 1 - S = 1 - \exp\left\{-\int_V \psi[\sigma_{ij}(x)]dV_k\right\}. \tag{43.9}$$

As an example, consider a tensile member of volume V subject to the uniaxial stress σ. Take the failure parameter to be

$$\psi = A\sigma^m = \frac{1}{V_0}\left(\frac{\sigma}{\sigma_0}\right)^m. \qquad (43.10)$$

Note that this is a two-parameter Weibull distribution [7]. The parameter V_0 can be fixed to the size of a standard specimen size. Equation 43.9 gives, for the failure probability,

$$f = 1 - \exp\left\{-\frac{V}{V_0}\left(\frac{\sigma}{\sigma_0}\right)^m\right\}. \qquad (43.11)$$

Note that in Equation 43.11, the probability of failure depends on the volume of the structural component.

Consider now a rectangular cross section of a structural member of equal volume, V, but now subject to a pure bending moment. The stress is now a function of the through-thickness direction, z, and is given by

$$\sigma = \sigma_m\left(\frac{2z}{h}\right),$$

where h is the thickness of the rectangular cross section and $-\frac{h}{2} \leq z \leq \frac{h}{2}$. Assume the material is a ceramic and fails in tension according to Equation 43.10, but fails at essentially infinite loads in compression. See Cassenti [6] for a discussion of materials with finite unequal failure loads in tension and compression. From Equation 43.9, the failure probability is

$$f = 1 - \exp\left\{-\frac{V}{2(m+1)V_0}\left(\frac{\sigma_m}{\sigma_0}\right)^m\right\}. \qquad (43.12)$$

Note that Equation 43.12 indicates that the failure probability is less for the same maximum stress in pure bending than in uniform tension. This is due to the fact that there is considerably less volume at, or near, the maximum stress than the uniformly stressed volume represented by Equation 43.11. This is a direct consequence of the size effect in material failure.

43.3 Failure-Location Predictions

Weakest-link theory not only provides predictions for the probability for failure, but can also be used to predict probability densities for the failure location [6]. Consider an infinitesimal increase in the stress state, $d\sigma_{ij}$, in Equation 43.1. Then the change in the failure probability is

$$d(\delta f) = \psi,_{\sigma_{ij}}(\sigma_{ij})d\sigma_{ij}\delta V, \qquad (43.13)$$

where $\psi,_{\sigma_{ij}}(\sigma_{ij}) = \partial\psi/\partial\sigma_{ij}$ at the point x, where the stress state is σ_{ij} and repeated indices are summed. The quantity $\psi,_{\sigma_{ij}}(\sigma_{ij})$ will be written as $\psi,_{ij}$. The following equation is shown in the literature [6]:

$$d(\delta f) = \delta(df).$$

The basic assumption of weakest-link theory implies that only one infinitesimal volume element can fail in the total structural volume. Hence, the probability to observe a failure, f^o, as the stress increases from σ_{ij} to $\sigma_{ij} + d\sigma_{ij}$ in volume δV is

$$d(\delta f^o) = \delta(df^o) = S\psi,_{\sigma_{ij}}(\sigma_{ij})\delta Vd\sigma = \exp\left[-\int_V \psi dV\right]\psi,_{\sigma_{ij}}(\sigma_{ij})\delta Vd\sigma_{ij}. \qquad (43.14)$$

Equation 43.14 can be integrated over the change in stress to yield the probability density to observe a failure

$$\frac{\delta f^o}{\delta V} = \int_0^{\sigma_{ij}} \exp\left[-\int_V \psi dV\right] \psi,_{\sigma_{ij}}(\sigma_{ij}) d\sigma_{ij}. \tag{43.15}$$

Note that the integrand in Equation 43.15 is not an exact differential, and therefore the probability density to observe a failure depends on the loading history.

As an example, consider a uniaxial tensile specimen. Equation 43.15 yields for this case

$$\delta f^o = \frac{\delta V}{V}[e^{-\psi(0)V} - e^{-\psi(\sigma)V}]. \tag{43.16}$$

If $\psi(0) = 0$, then Equation 43.16 yields

$$\delta f^o = \frac{\delta V}{V}[1 - e^{-\psi(\sigma)V}]. \tag{43.17}$$

In addition, if $\psi \to \infty$ as $\sigma \to \infty$, then

$$\delta f^o = \frac{\delta V}{V}. \tag{43.18}$$

Hence, if all the specimens are loaded to failure, then the distribution in failure locations should be uniform.

Since the probability to observe a failure at any location is independent of any other location, we can integrate Equation 43.17 over the volume to yield

$$f^o = \int_V \delta f^o = 1 - e^{-\psi(\sigma)V}. \tag{43.19}$$

The survival probability is

$$S = 1 - f^o = e^{-\psi(\sigma)}. \tag{43.20}$$

Equation 43.20 agrees exactly with Equation 43.8. This indicates that δf^o can be integrated over the volume to yield survival probabilities. In reference [4], it is shown that

$$S + \int_V \frac{\delta f^o}{\delta V} dV = 1. \tag{43.21}$$

Hence, the observed probabilistic failure density, $\delta f^o/\delta V$, can be integrated over the volume to find the total probability of failure for the volume V.

Three-point-bend specimens also provide an instructive example. In reference [6], it is shown that if all the specimens are taken to failure, then the probability to observe a failure between x and $x + \delta x$ is given by

$$\frac{\delta f}{\delta(x/L)} = 2^m(m+1)F^m(x), \tag{43.22}$$

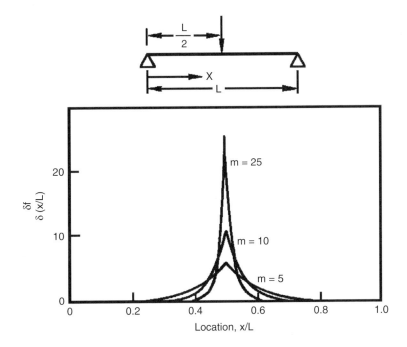

FIGURE 43.1 Three-point-bending failure locations for $\psi = (\sigma/\sigma_0)^m$. (From Cassenti, B.N., *AIAA J.*, 22, 103–110, 1984. With permission.)

where L is the length of the beam, m is the Weibull slope in Equation 43.10, and

$$F(x) = \begin{cases} x/L & 0 \leq x \leq L/2 \\ 1 - x/L & L/2 \leq x \leq L \end{cases}. \tag{43.23}$$

In Figure 43.1, the failure location can be seen to have a sharp peak under the applied load. Note that increased scatter in the failure load (small values for m) results in a wider distribution in the failure location.

In reference [6], loading-history effects are illustrated in a four-point-bend specimen. Vasko and Cassenti [3] describe the design of a dog-bone four-point-bend ceramic specimen. Six original specimens were used to find the Weibull parameter m in Equation 43.10. Although only one specimen failed in the test section, it was still possible to find the Weibull parameter m. A finite element code was modified to predict the probabilistic failure location distribution, and this was used to verify the statistics. The model agreed with the six measurements. A new specimen was designed using the modified code, and it was predicted that 93% of the specimens would fail in the test section. The new specimen became the standard and resulted in 92% failures in the test section.

43.4 Extension to Time-Dependent Failure

We can now assume that the failure can occur at any time, and once failure occurs at one time, the volume of interest has failed for all times. Although this seems to be reasonable, it should be noted that if plastic deformation is considered as a failure, the failed volume may continue to support loads and may actually increase in strength at later times. Following the work described in reference [4], this weakest-link assumption over time yields

$$\delta f = \psi \delta V \delta t, \tag{43.24}$$

where $\dot{\psi} = \dot{\psi}[\sigma_{ij}(\vec{x}, t)]$. Note that the quantity $\dot{\psi}$ is not in general a derivative with respect to time, but it is a function of location and time. In a manner similar to the development in the section on static failure, we can show that

$$S = \exp\left\{-\int_V \int_0^t \dot{\psi}[\sigma_{ij}(x), \tau] d\tau dV_k\right\}. \tag{43.25}$$

Again, it can be shown that the probability to observe a failure in the interval from t to $t + \delta t$ and in volume δV is

$$\delta f^\circ = S \delta f = S \dot{\psi} \delta V \delta t, \tag{43.26}$$

and, integrating over time,

$$\frac{\delta f^\circ}{\delta V} = \int_0^t S\dot{\psi} dt. \tag{43.27}$$

Three approaches can be taken to find the functional forms for $\dot{\psi}$. They include: (1) taking the parameters in the distribution to be dependent on time or number of cycles, (2) using damage models, and (3) using fatigue or creep rupture data directly. Some damage models are summarized by Wilson and Walker [8]. In reference [9], damage models and probabilistic failure are combined to describe the failure of ceramic matrix composite material structures. We can also take the parameter σ_0 to be time dependent in Equation 43.10. The parameter m cannot be time dependent, since we cannot ensure that the probability for failure will always increase over time.

An extremely important example is fatigue [4]. If the number of cycles to failure, N, is governed by a power law in the maximum applied stress, σ, then

$$CN^n \sigma^m = k(S_N). \tag{43.28}$$

The constant can be taken so that

$$\left(\frac{V}{V_0}\right)\left(\frac{N}{N_0}\right)^n \left(\frac{\sigma}{\sigma_0}\right)^m = k(S_N), \tag{43.29}$$

where, for example, V_0 and N_0 can be a chosen fixed value.

Solving Equation 43.29 for S_N

$$S_N = \exp(-\psi_N V) = k^{-1}\left[\left(\frac{V}{V_0}\right)\left(\frac{N}{N_0}\right)^n \left(\frac{\sigma}{\sigma_0}\right)^m\right]. \tag{43.30}$$

Equation 43.30 can be used to find ψ_N. The change in the failure parameter over one cycle is

$$\psi_{N+1} - \psi_N = \Delta \psi_N = \frac{1}{V_0}\left[\left(\frac{N+1}{N_0}\right)^n - \left(\frac{N}{N_0}\right)^n\right]\left(\frac{\sigma}{\sigma_0}\right)^m. \tag{43.31}$$

For $N \gg 1$, Equation 43.31 is

$$\Delta\psi_N = \frac{n}{V_0 N_0}\left(\frac{N}{N_0}\right)^{n-1}\left(\frac{\sigma}{\sigma_0}\right)^m. \qquad (43.32)$$

Equation 43.33 is equivalent to integrating over one cycle (while the stress is positive). Therefore

$$\Delta\psi_N = \int_0^\sigma d\psi = \frac{nm}{V_0 N_0 \sigma_0}\left(\frac{N}{N_0}\right)^{n-1}\int_0^\sigma \left(\frac{\sigma_1}{\sigma_0}\right)^m d\sigma_1. \qquad (43.33)$$

Equating the differentials in Equation 43.33,

$$\dot\psi = \frac{nm}{V_0 N_0 \sigma_0}\left(\frac{N}{N_0}\right)^{n-1}\left\langle\frac{\sigma}{\sigma_0}\right\rangle^{m-1}\left\langle\frac{\dot\sigma}{\sigma_0}\right\rangle, \qquad (43.34)$$

where $\langle x \rangle = \begin{cases} 0 & x<0 \\ x & x\geq 0 \end{cases}$ is the unit ramp function.

We can remove the cycle dependence in Equation 43.34 by solving for the number of cycles in Equation 43.30, and we can change σ from the maximum stress to the current stress by introducing a multiplicative constant K. We can now express $\dot\psi$ as a function of ψ [4] through

$$\dot\psi = \frac{nmK}{V_0 N_0}(V_0\psi)^{\frac{n-1}{n}}\left\langle\frac{\sigma}{\sigma_0}\right\rangle^{m-1}\left\langle\frac{\dot\sigma}{\sigma_0}\right\rangle. \qquad (43.35)$$

Integrating Equation 43.35 over one cycle, it can be shown that [4]

$$K = \frac{1}{n}. \qquad (43.36)$$

We now have an equation describing the evolution of ψ as

$$\dot\psi = \frac{m}{V_0 N_0}(V_0\psi)^{\frac{n-1}{n}}\left\langle\frac{\sigma}{\sigma_0}\right\rangle^{m-1}\left\langle\frac{\dot\sigma}{\sigma_0}\right\rangle. \qquad (43.37)$$

In Figure 43.2, the exact representation for failure in N cycles is compared with the approximation developed from the difference in Equation 43.37. It can be seen that except for a very small number of cycles to failure, the approximation is accurate. Figure 43.2 also displays the life for various statistical parameters, m and n, and maximum stress σ/σ_0.

Figure 43.3 compares the probability of survival for tension, pure bending, and three-point-bend specimens and clearly illustrates the effect of size. The pure-tension specimen subjects all of the specimen to the maximum stress and so has the smallest probability of survival for a given maximum stress, while three-point bending has the smallest volume subject to the highest stresses and so has the largest probability of survival.

In reference [4], a creep rupture example is also discussed. Creep provides a good example of how to convert a deterministic law into a probabilistic formulation.

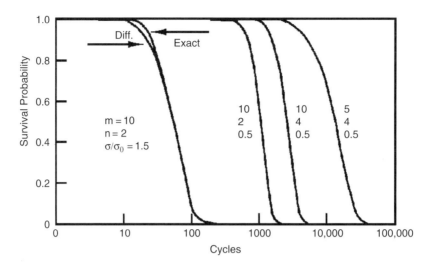

FIGURE 43.2 Fatigue survival probability for exact (30) and difference (37) forms. (From Cassenti, B.N., *AIAA J.*, 29, 127–134, 1991. With permission.)

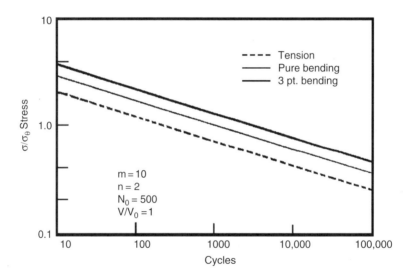

FIGURE 43.3 Size effect in fatigue life for 50% survival probability. (From Cassenti, B.N., *AIAA J.*, 29, 127–134, 1991. With permission.)

43.5 Applications to Composite Materials

The reliability of structures is clearly dependent on the materials that are used. Metallic structures can redistribute the load from regions where the material yields to regions that have not yet yielded. Hence, weakest-link theory is not a valid representation of plastic yielding. This is clearly an advantage, but metals have a relatively high density and so are not efficient for aerospace applications. Ceramic material structures have lower densities, high strengths, and can survive to extremely high temperatures, but they are extremely sensitive to flaws, making the weakest-link assumption accurate. Composite material structures are strong, lightweight, and intermediate in their sensitivity to flaws. The weakest-link assumption is not as accurate, complicating the probabilistic strength evaluations. Two additional factors add to the complications: (1) composite materials are not isotropic, and (2) the tension and compression failure responses differ, as in ceramic materials. This section summarizes probabilistic models for failure differences in tension and compression, and then continues with a summary of the extensions to anisotropic failure. The details can be found in the literature [6].

The tension response is usually considered to be linear elastic to failure, and the tensile failure stress is well represented by probabilistic models, but compression failure is more complicated. In reference [6], three compression failure models are considered:

1. The material does not fail in compression but continues to support the maximum compressive stress (e.g., fiber buckling).
2. The compression failure load is infinite (i.e., very large compared with the tensile failure load). This is equivalent to the case of a ceramic material.
3. The compression failure load satisfies weakest-link theory and is independent of the tension failure load.

Reference [6] shows that the survival probability has less scatter in bending with a deterministic compressive failure load than it would if the deterministic compressive failure load were infinite. Of course, the maximum bending moment that can be supported is lower for a finite maximum compressive load. Reference [6] also demonstrates that independent probabilistic failure in tension and compression is highly dependent on the characteristic failure loads, σ_0, in Equation 43.10. For typical composite materials, the compressive failure loads are relatively low compared with the tension failure loads and may actually control the probabilistic failure in bending [6].

Multiaxial failure presents additional complications. Not only are properties required for tensile and compressive failure, but the properties also change with material direction. We must also include the fact that the material can fail due to shear, and each of these failure modes can be connected. In reference [6], five separate failure loads are considered. They are:

1. Failure in tension along the fiber direction
2. Failure in compression along the fiber direction
3. Failure in tension perpendicular to the fiber direction
4. Failure in compression perpendicular to the fiber direction
5. Failure in shear, regardless of the direction

These failures are each assumed to represent a different failure surface. The surface intersections are continuous, but the slopes at the intersections are not. The surface in stress space is assumed to be a discontinuous ellipsoid (actually the surface can be a hyper-ellipse) with a unit ramp function used to represent differences in tension and compression. The characteristic failure stresses are taken to be probabilistic. A second approach can also be used, where each of the five different failure modes is taken to be independent. We can now use survival probabilities as in Equation 43.8 for each of the five modes and take their product to be the overall survival probability [6].

43.6 Conclusions

Weakest-link failure theory has direct application to the prediction of structural reliability. It provides, through its basic assumption, a conservative estimate of the structural reliability. This makes it ideal for predicting the reliability of ceramic structures, but it will usually result in overly conservative estimates for the reliability of metallic structures. The basic assumption of the theory naturally includes the effect of size.

The predictions not only include static failure, but have also been adapted to time-dependent failure modes such as fatigue or creep rupture. Other extensions to the original theory can be used to predict failure location probability densities. This prediction of failure locations is not only useful in the design of test specimens, but is an important aid to the engineer in improving the design of structures in general.

Finally, weakest-link failure can be readily applied to the design of composite material structures. Although the accuracy for composite structures is not as good as for ceramic structures, it still provides considerable accuracy and can be readily adapted to represent the large number of competing failure mechanisms present. An excellent guide for developing these models is the use of existing damage models.

References

1. Weibull, W., A Statistical Theory of the Strength of Materials, report 151, Royal Swedish Academy of the Engineering Society, Stockholm, Sweden, 1939.
2. Harter, H.L., A Survey of the Literature on the Size Effect on Material Strength, AFFDL-TR-77-11, Air Force Flight Dynamics Laboratory, Wright-Patterson AFB, OH, 1977.
3. Vasko, T.J. and Cassenti, B.N., The statistical prediction of failure location in brittle test specimens, in *Proceedings of 27th Structures, Structural Dynamics, and Materials Conference*, AIAA/ASME/ASCE/AHS, San Antonio, TX, 1986, pp. 757–763.
4. Cassenti, B.N., Time dependent probabilistic failure of coated components, *AIAA J.*, 29, 127–134, 1991.
5. Batdorf, S.B., Fracture statistics of polyaxial stress states, in *Proceedings of the International Symposium on Fracture Mechanics*, University of Virginia Press, Charlottesville, 1978, pp. 579–691.
6. Cassenti, B.N., Probabilistic static failure of composite material, *AIAA J.*, 22, 103–110, 1984.
7. Weibull, W., A statistical distribution function of wide applicability, *J. Appl. Mech.*, 18, 63–75, 1951.
8. Wilson, D.A. and Walker, K.P., Constitutive Modeling of Engine Materials, AFWAL-TR-84-7073, Materials Laboratory, Wright Patterson AFB, OH, 1983.
9. Cassenti, B.N. and Koenig, H., Life prediction for composite material structures, in *Proceedings of ASME Winter Annual Meeting*, ASME, Atlanta, GA, 1996.

44
Reliability Analysis of Composite Structures and Materials

Sankaran Mahadevan
Vanderbilt University

44.1	Introduction	44-1
44.2	Strength Limit States (Laminate Theory)	44-2
	Ply-Level Limit States and Reliability • System Failure Probability • Numerical Example	
44.3	Strength Limit States (Three-Dimensional Analysis)	44-8
	Shear Deformation Theory and Analysis • Ply-Level Limit States • Stiffness Modification for Progressive Failure Analysis • Numerical Example (Composite Plate)	
44.4	Strength Limit State: Approximations	44-13
	Fast Branch-and-Bound Method • Deterministic Initial Screening • Weakest-Link Model • Critical-Component Failure • Numerical Example: Aircraft Wing	
44.5	Fatigue Limit State: Material Modeling	44-18
	Fatigue-Damage Modeling • Experimental Results and Model Performance	
44.6	Fatigue-Delamination Limit State	44-22
	Failure Mechanism and Model • Random Variables • Response-Surface Modeling of the Limit State • Verification of Methodology	
44.7	Creep Limit State	44-26
	Creep of Undamaged Composites • Creep Analysis with Stochastic Fiber Fracture • Failure Criterion • Numerical Example	
44.8	Conclusion	44-32

44.1 Introduction

Composite materials are being widely used in modern structures, such as aircraft and space vehicles, due to their high performance, high temperature resistance, tailoring facility, and light weight. Considerable research on the design and failure analysis of composite structures is being conducted. The results of experiment and research into composite materials show large statistical variations in their mechanical properties. Therefore, probabilistic analysis has to play an important role in structural assessment. This chapter presents reliability analysis methods for composite structures, considering several types of failure criteria: ultimate strength, fatigue, delamination, and creep.

From the perspective of strength limit states, composite laminate failure may be considered in two major stages, first-ply failure (FPF) and last-ply failure (LPF). The first-ply failure usually corresponds to the commencement of matrix cracking failure; structural ultimate failure or last-ply failure consists of a series of the ply-level component failures, such as matrix cracking, delamination, and fiber breakage, from the first one to the last one. Well-known methods such as the first-order reliability method (FORM), second-order reliability method (SORM), or Monte Carlo simulation can be combined with finite element analysis to compute the component failure probability. The branch-and-bound method can then be employed to search for the significant system failure sequences. When each component failure occurs, the structural stiffness is modified to account for this damage, and the damaged structure is reanalyzed. This proceeds until system failure occurs. Based on the identified significant failure sequences, the system failure probability is determined by means of bounding techniques. In the composite structure, multiple sequences are found to be highly correlated, leading to efficient approximations in the failure probability computation. Section 44.2 to Section 44.4 present these methods and illustrate them with simple examples, with a practical application for a composite aircraft wing presented in Section 44.4.5.

The characteristics of fatigue-damage growth in composite materials are different from those of damage growth in homogeneous materials. Continuum damage mechanics concepts have been used to evaluate the degradation of composite materials under cyclic loading. Damage-accumulation models that capture the unique characteristics of composite materials are presented in Section 44.5. The predictions from the models are compared with experimental data.

Fatigue leads to delamination in several laminated composite applications, affected by anisotropic material properties of various plies, ply thicknesses and orientations, loads, and boundary conditions. The limit state can be formulated in terms of the strain energy release rate computed based on the virtual crack-closure technique. The critical value of the strain energy release rate, obtained from test data, is seen to be a random function of life of the structure. Once the delamination initiation probability is estimated, the propagation life until system failure is estimated through the exploration of multiple paths through the branch-and-bound enumeration technique. The analysis is then repeated for multiple initiation sites, and the overall probability of failure is computed through the union of the multiple initiation/growth events. These techniques are presented and illustrated in Section 44.6 using a practical application—fatigue-delamination analysis of a helicopter rotor component.

Composite materials are particularly attractive for high-temperature applications, as in engine components. Therefore, creep reliability models for high-temperature composites are presented in Section 44.7. Time-dependent reliability analysis, including the effect of broken fibers, is discussed. The final section discusses new research needs in assessing the reliability of composite materials and structures.

44.2 Strength Limit States (Laminate Theory)

Several studies have developed static strength reliability assessment methods for fibrous composites [1, 2] using the first-ply failure (FPF) assumption; that is, if any of plies in a laminate fails, the entire laminate is considered to be in failure. Other studies [3, 4] estimated ultimate strength reliability using the last-ply-failure (LPF) assumption; that is, the laminate fails only if all plies fail. This latter analysis includes the search for the dominant failure sequences, and the laminate reliability is approximated through the union of the dominant failure sequences. Two options have been pursued in modifying the system definition during the progressive damage analysis. In the first option, a ply is completely removed from the system when it fails, since the material used in such a system is usually brittle [3]. The second option is to realistically consider the consequence of different types of failure, such as matrix cracking and fiber breakage [4]. For instance, after the matrix failure of a ply occurs, the ply's stiffness along the matrix direction is reduced to zero, but the fiber is not yet broken and thus could have load-bearing capacity. If the fiber fails first, the ply's stiffness along the fiber direction is reduced to zero, and the matrix is still able to carry the loading. Thus if any one basic failure event occurs, then the corresponding stiffness matrix terms are modified; the ply is not removed until both matrix cracking and fiber failure occur.

A probabilistic progressive failure model can be developed as follows: the first-order reliability method (FORM) is used to compute the component reliability. The structural stiffnesses are modified to reflect damage during the simulation of the progressive failure process. The significant failure sequences are identified based on the branch-and-bound method, and the system failure probability is determined as the failure probability of the union of the significant failure sequences.

44.2.1 Ply-Level Limit States and Reliability

In-plane failure of the ply can generally be classified into two major failures: matrix failure and fiber breakage. In this section, the following limit-state functions are used:

Fiber failure [5]:

$$g_f = 1 - \left\{ f_{11} \sigma_1^2 + f_1 \sigma_1 \right\} \tag{44.1}$$

Matrix failure, based on Tsai-Wu criterion [6]:

$$g_m = 1 - \left\{ f_{11} \sigma_1^2 + f_{22} \sigma_2^2 + f_{66} \sigma_{12}^2 + f_{12} \sigma_1 \sigma_2 + f_1 \sigma_1 + f_2 \sigma_2 \right\} \tag{44.2}$$

The coefficients, f_{11}, f_{22}, etc. in Equation 44.1 and Equation 44.2 are related to material strengths, such as

$$f_{11} = \frac{1}{(X_T * X_C)} \quad f_{22} = \frac{1}{(Y_T * Y_C)} \quad f_{66} = \frac{1}{S^2}$$

$$f_1 = \frac{1}{X_T} - \frac{1}{X_C} \quad f_2 = \frac{1}{Y_T} - \frac{1}{Y_C} \quad f_{12} = -\sqrt{f_{11} - f_{22}}$$

where X_T is the tension strength in 1-1 direction; X_C is the compression strength in 1-1 direction; Y_T is the tension strength in 2-2 direction; Y_C is the compression strength in 2-2 direction; S is the shear strength; and σ_1, σ_2, and σ_{12} are stresses in 1-1, 2-2, and 1-2 direction, respectively, as shown in Figure 44.1. The limit-state functions in Equation 44.1 and Equation 44.2 imply failure when $g_f \leq 0$ or $g_m \leq 0$.

In the reliability evaluation, the stress resultants $\{N\}$, the stress couples $\{M\}$, and the strength parameters X_T, X_C, Y_T, Y_C, S are considered as the basic random variables. These basic random variables are expressed as a vector $X = \{X_1, X_2, \ldots, X_m\}^T$. In FORM, the component reliability index is obtained as

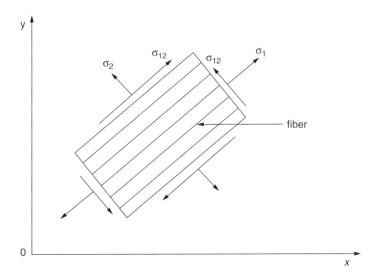

FIGURE 44.1 Laminate stresses.

$\beta = (y^{*T} y^*)^{1/2}$ where y^* is the point of minimum distance from the origin to the limit state $G(Y) = 0$, where Y is the vector of equivalent uncorrelated standard normal variables. The first-order approximation of the failure probability is computed as $P_f = \Phi(-\beta)$, where Φ is the standard normal cumulative distribution function (CDF). The transformation of the random variables from the X-space to Y-space is achieved by any of the well-known methods [7].

The most probable failure point y^* is found using the following iterative formula proposed by Rackwitz and Fiessler [8]:

$$y_{i+1} = \left[y_i^T \sigma_i + \frac{G(y_i)}{|\nabla G(y_i)|} \right] \alpha_i^T \quad (44.3)$$

where $\nabla G(y_i)$ is the gradient vector of the limit-state function at y_i, and α_i is the unit vector normal to the limit-state surface away from the origin. It has the following form:

$$\alpha_i = -\nabla G(y_i)/|\nabla G(y_i)| \quad (44.4)$$

The computation of $\nabla G(y_i)$ is achieved by using the chain rule of differentiation on Equation 44.1 or Equation 44.2.

44.2.2 System Failure Probability

The laminate system is assumed to collapse only when all the plies have failed. Consider a laminate system that has n plies under the external loads $\{N\}$ and $\{M\}$. Since there are two basic failure events for each ply (matrix cracking and fiber breakage), $2n$ component failure events exist in the system. Based on the branch-and-bound concept, a procedure to identify the significant failure sequences for composite laminates can be formulated as follows:

1. Suppose the laminate is originally in its intact state. For each basic event, compute its reliability index β using the component-level first-order reliability method (FORM) and then obtain its corresponding failure probability. Order the $2n$ values of failure probabilities, and the maximum value of failure probability is recorded as $P_{f\,max}^1$. The superscript 1 means the first stage. Select the failure event that has the largest failure probability as the first one to occur. At the same time, the components that have failure probability greater than a specified fraction of $P_{f\,max}^1$ are saved for exploration (branching), and the other components that have failure probability less than the specified fraction of $P_{f\,max}^1$ are discarded.
2. Modify the stiffness corresponding to the first failure event. If the event is corresponding to matrix failure, the modulus E_2 and G_{12} of the ply are reduced to zero. If the event is corresponding to fiber failure, the modulus E_1 of the ply is reduced to zero. The loads are globally redistributed in the damaged laminate in accordance with the modified stiffness. The same calculations as in step 1 are repeated. The event that has the largest path probability $P_{f\,max}^2$ at the second stage is taken as the next event to fail. This proceeds until system failure occurs.
3. After the first failure sequence is identified, step 2 is repeated to consider the other branched events until all the possible significant failure sequences are found. The failure sequences that have path probability lower than a prespecified value are considered insignificant and are not explored.

44.2.2.1 Single Failure Sequence

For one failure sequence with m individual failure events, the failure probability can be computed as

$$P_f^k = P\left(\bigcap_{j=1}^m E_j^k\right) \quad (44.5)$$

where E_j^k is the jth basic failure event along the kth failure sequence under the condition that the first $(j-1)$ basic failure events have occurred.

In general, it is difficult to evaluate Equation 44.5 numerically when $m \geq 3$. In practical problems, approximate solutions may be obtained. The second-order upper bound suggested by Murotsu [9] is a good approximation for the estimation of the joint failure probability:

$$P\left(\cap_{j=1}^{m} E_j\right) \leq \min_{i \neq j}[P(E_i \cap E_j)] \quad (44.6)$$

In the case of composite structures, usually there are one or two components that have much smaller failure probabilities than others, as will be shown later in the numerical examples. In that case, a simpler formula [10] can be used to approximate the joint probability:

$$P\left(\cap_{j=1}^{m} E_j\right) \leq P(E_1 \cap E_2) \quad (44.7)$$

where E_1 and E_2 are the two least probable events in a failure sequence.

In general, some of the failure events are correlated; hence, the calculations of the joint probabilities in Equation 44.6 and Equation 44.7 remain difficult, especially when the basic random variables are non-normal and the limit-state functions are nonlinear. Mahadevan et al. [11] proposed an approximate method for this problem, as shown in Figure 44.2. The joint failure region is approximately bounded by the hyperplanes at the point y_{12}^*, which is the closest point (from the origin) on the intersection of the limit-state surfaces $G_1(y) = 0$ and $G_2(y) = 0$ in the Y-space. The value for y_{12}^* is found by solving the following optimization problem:

$$\text{minimize} \quad \sqrt{Y^T Y} \quad (44.8)$$

such that

$$G_1(Y) = 0$$
$$G_2(Y) = 0$$

In the following discussion, sequential quadratic programming is used to solve this problem. After y_{12}^* is found, the limit states $G_1(y) = 0$ and $G_2(y) = 0$ are linearized at y_{12}^* so that the correlation coefficient between the two limit states ρ_{12} can be obtained as the product of the two unit gradient vectors, α_1 and α_2, of the linear approximations at y_{12}^*. That is,

$$\rho_{12} = \sum_{r=1}^{n} \alpha_{1r} \alpha_{2r} \quad (44.9)$$

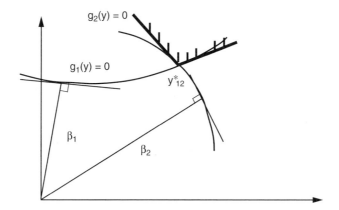

FIGURE 44.2 Joint failure probability with nonlinear limit states.

where α_{1r} and α_{2r} are the gradients similar to that defined in Equation (44.4), and n is the number of random variables. The joint-failure probability is then calculated using the two-dimensional standard normal cumulative distribution as

$$P(E_1 \cap E_2) = \Phi(-\beta_1, -\beta_2, \rho_{12}) \tag{44.10}$$

44.2.2.2 Multiple Failure Sequences

As mentioned before, the overall failure probability requires the computation of probability of union of the multiple failure sequences. Since the performance functions of the failure sequences are not available, the Cornell first-order bound [12] is a suitable candidate for an approximate estimate. If l significant failure sequences are identified, the system failure probability, P_f, can be approximately computed as:

$$\max_{k=1,l} P_f^k \leq P_f \leq \sum_{k=1}^{l} P_f^k \tag{44.11}$$

In the case where all failure sequences are fully dependent, it follows directly that the weakest failure sequence will govern the failure probability. Hence, the system failure probability is equal to the lower bound in Equation 44.11. For independent failure sequences, the system failure probability corresponds to the upper bound of Equation 44.11.

44.2.3 Numerical Example

A typical graphite/epoxy (T300/5208) laminate with configuration [90°/45°/–45°/0°] and equal ply thickness (0.25 mm) is studied. This laminate system can be subjected to stress resultants $\{N\}$ and/or the stress couples $\{M\}$, as shown in Figure 44.3. Two in-plane loading cases are considered for illustration: (1) uniaxial loading, and (2) biaxial loading.

The material strength parameters X_T, X_C, Y_T, Y_C, S and loading parameters $\{N\}$ and $\{M\}$ are considered as the basic random variables. In the current example, the strength parameters are assumed to have Weibull distributions, and the loads $\{N\}$ and $\{M\}$ are assumed to have Type I extreme value distributions.

A cutoff value of 0.3 is used in the branch-and-bound search for the significant failure sequences. That is, components with failure probabilities greater than 0.3 of the maximum component failure probability are saved for exploration, and others are discarded. For the laminate [90°/45°/–45°/0°], there are four

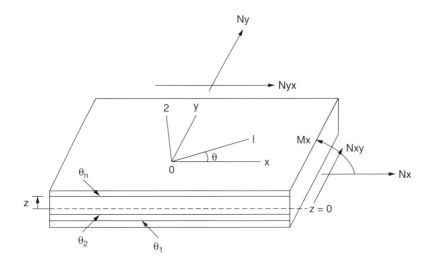

FIGURE 44.3 Composite laminate.

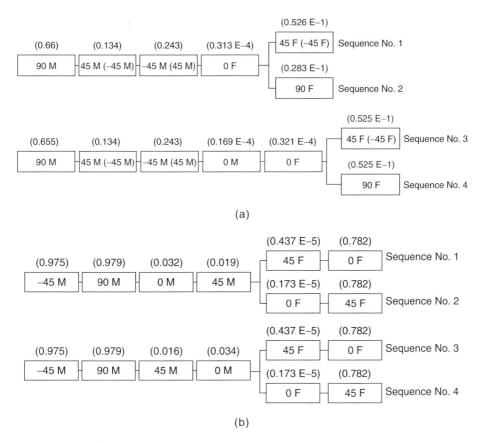

FIGURE 44.4 Dominant failure sequences: (a) axial loading only, (b) general in-plane loading.

plies with different orientations, and two basic events (matrix failure event and fiber breakage event) for each ply. Thus, there are eight basic events. Figure 44.4 shows the dominant failure sequences for the two loading cases. The failure events are denoted using ply orientation and the type of failure event. For example, $90M$ stands for the matrix failure event of the $90°$ ply, and $90F$ stands for its fiber breakage event.

Refer to Mahadevan et al. [4] for details of this numerical example. Three results are important for future discussion:

1. Figure 44.4 shows that, in general, for the same laminate configuration, different loading combinations have different significant failure sequences. That is because component failure probabilities depend on stress responses of components, which change with loading combinations. Also, it is seen from Figure 44.4 that matrix cracking failures usually occur first, and then fiber failures occur. These kinds of failure sequences are consistent with experimental results.
2. The overall failure probability of a single sequence is found to greatly depend on the least probable event, owing to the strong correlation among the components in the sequence. The least probable event is usually the first failed fiber. This indicates that the failure of the first fiber can approximately be considered as system collapse.
3. The significant failure sequences under a given load are found to be quite similar. That implies strong correlations among the significant failure sequences. Therefore, considering multiple sequences, the system failure probability may be very close to the Cornell first-order lower bound.

These observations are useful in devising computationally efficient schemes for larger systems considered in Section 44.3 and Section 44.4.

44.3 Strength Limit States (Three-Dimensional Analysis)

The analysis in Section 44.2 was based on classical laminate theory, and the composite laminate was considered as a system consisting of individual plies as components. A general composite structure may consist of different laminate configurations in different parts of the structure, and it may need to be analyzed using a finite element method and with a more advanced behavior model than classical laminate theory. Therefore, this section extends the probabilistic progressive failure analysis methodology to general three-dimensional composite structures, and presents several practical techniques to address issues with respect to system failure definition, progressive damage analysis, and probabilistic computation. The proposed methodology is applied to a numerical example of a composite plate.

44.3.1 Shear Deformation Theory and Analysis

The first-order shear deformation theory for laminated anisotropic plates is used in this section [13–15]. That is, normals to the centerplane are assumed to remain straight after deformation, but not necessarily normal to the centerplane.

The displacement field is of the form

$$u_x(x, y, z) = u_0(x, y) + z\psi_x(x, y)$$
$$u_y(x, y, z) = v_0(x, y) + z\psi_y(x, y) \qquad (44.12)$$
$$u_z(x, y, z) = w(x, y)$$

where u_x, u_y, u_z are the displacements in the x, y, z directions, respectively; u_0, v_0, w are the associated midplane displacements; and ψ_x and ψ_y are the rotations of normals to midplane about the y and x axes, respectively. In Equation 44.12, considering that the transverse normal stress is on the order of $(h/a)^2$ times the in-plane normal stresses, the assumption that w is not a function of the thickness coordinate is justified.

For any elastic body, the strain-displacement equations describing the functional relations between the elastic strains in the body and its displacements are given by

$$\varepsilon_{ij} = \frac{1}{2}(u_{i,j} + u_{j,i}) \qquad (44.13)$$

where $i, j = x, y, z$ in a Cartesian coordinate system, and the comma denotes partial differentiation with respect to the coordinate denoted by the symbol after the comma. Explicitly, the relations are:

$$\varepsilon_x = \frac{\delta u_x}{\delta x}$$
$$\varepsilon_y = \frac{\delta u_y}{\delta y}$$
$$\varepsilon_z = \frac{\delta w}{\delta z}$$
$$\varepsilon_{xz} = \frac{1}{2}\left(\frac{\delta u_x}{\delta z} + \frac{\delta w}{\delta x}\right) \qquad (44.14)$$
$$\varepsilon_{yz} = \frac{1}{2}\left(\frac{\delta u_y}{\delta z} + \frac{\delta w}{\delta y}\right)$$
$$\varepsilon_{xy} = \frac{1}{2}\left(\frac{\delta u_x}{\delta y} + \frac{\delta u_y}{\delta x}\right)$$

Substituting Equation 44.12 into Equation 44.14 results in

$$\varepsilon_x = \frac{\delta u_0}{\delta x} + z\frac{\delta \psi_x}{\delta x} = \varepsilon_x^0 + zk_x$$

$$\varepsilon_y = \frac{\delta v_0}{\delta y} + \frac{\delta \psi_x}{\delta y} = \varepsilon_x^0 + xk_y$$

$$\varepsilon_x = 0$$

$$\varepsilon_{xy} = \frac{1}{2}\left(\frac{\delta u_0}{\delta y} + \frac{\delta v_0}{\delta x}\right) + \frac{z}{2}\left(\frac{\delta \psi_x}{\delta y} + \frac{\delta \psi_y}{\delta x}\right) = \varepsilon_{xy}^0 + zk_{xy} \qquad (44.15)$$

$$\varepsilon_{xz} = \frac{1}{2}\left(\psi_x + \frac{\delta w}{\delta x}\right)$$

$$\varepsilon_{yz} = \frac{1}{2}\left(\psi_y + \frac{\delta w}{\delta y}\right)$$

For a plate of constant thickness h and composed of thin layers of orthotropic material, the constitutive equations can be derived [16] under the assumption that each layer possesses a plane of elastic symmetry parallel to the x–y plane, as

$$\begin{Bmatrix}\sigma_1\\\sigma_2\\\sigma_6\end{Bmatrix} = \begin{bmatrix}Q_{11} & Q_{12} & 0\\Q_{12} & Q_{22} & 0\\0 & 0 & 2Q_{66}\end{bmatrix}\begin{Bmatrix}\varepsilon_1\\\varepsilon_2\\\varepsilon_6\end{Bmatrix} \qquad (44.16)$$

$$\begin{Bmatrix}\sigma_4\\\sigma_5\end{Bmatrix} = \begin{bmatrix}2Q_{44} & 0\\0 & 2Q\end{bmatrix}\begin{Bmatrix}\varepsilon_{23}\\\varepsilon_{31}\end{Bmatrix} \qquad (44.17)$$

where all above quantities are in the principal material directions (1, 2, 3) of the layer. To relate these relationships to the x-y-z coordinate system, a transformation is performed. The result is

$$\begin{Bmatrix}\sigma_x\\\sigma_y\\\sigma_{xy}\end{Bmatrix} = \begin{bmatrix}\overline{Q}_{11} & \overline{Q}_{12} & 2\overline{Q}_{16}\\\overline{Q}_{12} & \overline{Q}_{22} & 2\overline{Q}_{26}\\\overline{Q}_{16} & \overline{Q}_{26} & 2\overline{Q}_{66}\end{bmatrix}\begin{Bmatrix}\varepsilon_x\\\varepsilon_y\\\varepsilon_{xy}\end{Bmatrix} \qquad (44.18)$$

$$\begin{Bmatrix}\sigma_{yz}\\\sigma_{xz}\end{Bmatrix} = \begin{bmatrix}2\overline{Q}_{44} & 2\overline{Q}_{45}\\2\overline{Q}_{45} & 2\overline{Q}_{55}\end{bmatrix}\begin{Bmatrix}\varepsilon_{yz}\\\varepsilon_{xz}\end{Bmatrix} \qquad (44.19)$$

The ply stresses along the material axes can be obtained using finite element analysis. Then, they are substituted into a suitable strength failure criterion for reliability analysis of ply-level component failure modes.

44.3.2 Ply-Level Limit States

Failure criteria are based on four types of theories: (1) limit theory, (2) polynomial theory, (3) strain energy theory, and (4) direct-mode determining theory. The limit theory compares the value of each stress or strain component to a corresponding ultimate value. The polynomial theory uses a polynomial of stress terms to describe the failure surface. The strain energy theory uses a nonlinear energy-based criterion to define failure. Finally, the direct-mode determining theory uses polynomials of stress terms and uses separate equations to describe each mode of failure. In progressive failure analysis, the direct-mode determining failure criteria are most widely used because they automatically determine the mode of failure (for instant, matrix cracking failure mode) so that the stiffness can be reduced in the correct manner. However, it is to be noted that the direct failure criteria are based mainly on empirical reasoning. In this chapter, Lee's [13] simple empirical criteria are used to derive component performance functions for the sake of illustration, as follows:

For fiber breakage:

$$g_f = 1 - \sigma_1/X_T, \quad \sigma_1 > 0$$
$$g_f = 1 + \sigma_1/X_T, \quad \sigma_1 < 0 \quad (44.20)$$

For matrix cracking:

$$g_m = 1 - \left[\frac{\sigma_2^2}{Y_T Y_C} + \frac{\sigma_{12}^2}{S^2}\right] \quad (44.21)$$

For delamination:

$$g_d = 1 - \left(\frac{1}{S_z}\right)^2 \left(\sigma_{23}^2 + \sigma_{31}^2\right) \quad (44.22)$$

In the above equations, X_T and X_C are the tensile and compressive strengths in the fiber direction; Y_T and Y_C are the tensile and compressive strengths in the direction transverse to fibers; and S and S_z are the shear strengths in the x–y plane and in the z direction, respectively. As in Section 44.2, $g < 0$ indicates the failure of the component.

44.3.3 Stiffness Modification for Progressive Failure Analysis

Once the ply-level failure probabilities are determined using the limit states in Equation 44.20 to Equation 44.22, the significant sequences to laminate (system) failure are identified using the branch-and-bound technique. This requires progressive failure analysis similar to that in Section 44.2.2. The stiffness matrix terms corresponding to each type of failure are modified as follows.

When fiber failure occurs, the ply stiffness terms contributed by the fiber (related to material direction "1"), i.e., E_{11}, G_{12}, and G_{31}, are reduced to zero, and $[Q]$ becomes

$$\begin{bmatrix} 0 & 0 & 0 & 0 & 0 \\ 0 & E_{22} & 0 & 0 & 0 \\ 0 & 0 & 0 & 0 & 0 \\ 0 & 0 & 0 & 2G_{23} & 0 \\ 0 & 0 & 0 & 0 & 0 \end{bmatrix} \quad (44.23)$$

When matrix cracking occurs, the ply stiffness terms related to material direction "2," i.e., E_{22}, G_{12}, and G_{23}, are reduced to zero, and $[Q]$ becomes

$$\begin{bmatrix} E_{11} & 0 & 0 & 0 & 0 \\ 0 & 0 & 0 & 0 & 0 \\ 0 & 0 & 0 & 0 & 0 \\ 0 & 0 & 0 & 0 & 0 \\ 0 & 0 & 0 & 0 & 2G_{31} \end{bmatrix} \quad (44.24)$$

When delamination failure occurs, the ply stiffness terms related to material direction "3," i.e., G_{23} and G_{31}, are reduced to zero, and $[Q]$ becomes

$$\begin{bmatrix} E_{11} & E_{12} & 0 & 0 & 0 \\ E_{12} & E_{22} & 0 & 0 & 0 \\ 0 & 0 & 2G_{12} & 0 & 0 \\ 0 & 0 & 0 & 0 & 0 \\ 0 & 0 & 0 & 0 & 0 \end{bmatrix} \quad (44.25)$$

44.3.4 Numerical Example (Composite Plate)

This example deals with a simply supported, square, cross-ply laminated plate subjected to a uniform pressure p_0. The structure and configuration are shown in Figure 44.5. The composite material is a 0°/90°/90°/0° graphite/epoxy (T300/5208). There are totally 18 basic random variables. They are strength properties X_T, X_C, Y_T, Y_C, S, S_Z and material properties E_X, E_Y, E_Z, G_{XY}, G_{YZ}, G_{ZX}, and v_{XY}, the loading p_0, and the configuration properties (ply orientations and thicknesses). See Xiao and Mahadevan [17] for details of this numerical example. All the variables are assumed to be normally distributed, for the sake of illustration. This is not a limitation of the method. Other types of distribution can also be used.

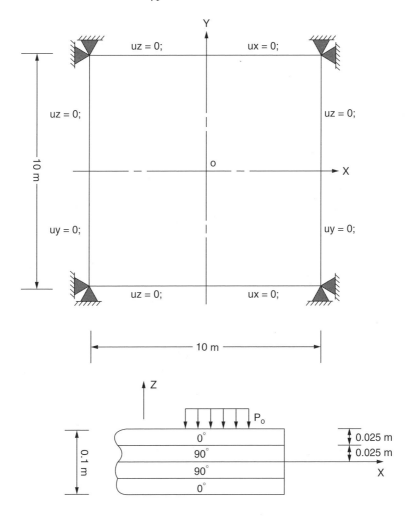

FIGURE 44.5 Composite plate with transverse loading.

The stresses are computed using finite element stress analysis, using the software ANSYS. The eight-node shell element SHELL99 is used, which is capable of modeling the multiple plies in the laminate. The stresses, σ_X, σ_Y, σ_Z, σ_{XY}, σ_{YZ}, σ_{ZX}, used in the failure criteria are average values of stresses at the element nodes in the middle of each layer. The component performance functions in Equation 44.20 to Equation 44.22 are used corresponding to the three ply-failure modes of fiber breakage, matrix cracking, and delamination.

The first-order reliability method using the Rackwitz-Fissler algorithm is used to estimate the failure probability of each component. In searching for the significant failure sequences, the branch-and-bound method is employed, and a cutoff value of $\lambda = 0.5$ is selected. System failure (structure collapse) is assumed if fiber breakages take place across the entire width of the plate.

Figure 44.6 and Figure 44.7 show the structural damage in the most dominant failure sequence and the corresponding failure probabilities, respectively. It is seen that the damage of the composite plate

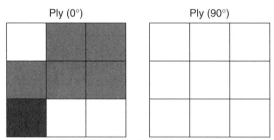
Damaged state when the first fiber breakage occurs

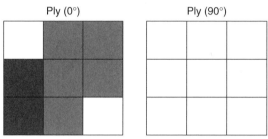
Damaged state when the second fiber breakage occurs

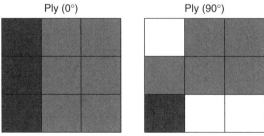
Damaged state when the third fiber breakage occurs (system failure)

FIGURE 44.6 Progressive damage in composite plate.

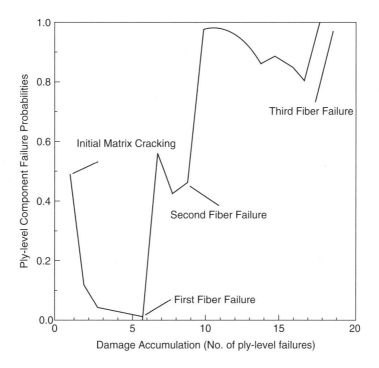

FIGURE 44.7 Ply-level failure-probability change with progressive damage.

begins with matrix cracking in some plies and elements. After some damage accumulation, the first fiber breakage failure occurs. Further structural damage becomes faster due to higher failure probabilities of the failed components after the first fiber breaks. When the second fiber breakage takes place, the structural damage becomes much faster due to much higher failure probabilities. When the third fiber breaks, the whole structure is regarded to be failed because fiber failures have crossed the entire width of the plate at this state. Figure 44.7 shows that component failure probability changes with progressive structural damage. It can be seen that after each fiber failure, the failure probability of the next component shows a remarkable increase. This is consistent with the fact that fiber failure is a major failure event, leading rapidly to the overall structural system failure.

Several other significant sequences are also obtained, but it is seen that they are very similar to the first identified failure sequence. The computed results indicate that these significant sequences consist of common components with slight changes in the order of failed components. Thus, it is reasonable to assume that there are high correlations among these failure sequences and that the overall system failure probability can be approximated by the probability of the first identified failure sequence.

44.4 Strength Limit State: Approximations

The reliability method followed in Section 44.2 and Section 44.3 is applicable in general to the probabilistic ultimate strength analysis of any structure. However, this procedure is computationally time consuming for large structures with many components, where the structural analysis is carried out through a finite element code with numerous finite elements. In the case of composite structures, the brittleness of the failures and the correlation between different failures and sequences can be used advantageously to introduce several techniques and approximations that make the method efficient and practical for ultimate strength reliability analysis. These are discussed in this section.

44.4.1 Fast Branch-and-Bound Method

When there are a large number of components in a structural system, which is the case for composite structures, the basic branch-and-bound method described in Section 44.2 becomes time consuming and tedious. In the original method, only one component failure is imposed at each damaged stage, so that a large number of steps are required to complete a failure sequence. This makes the basic branch-and-bound method difficult to apply in practice.

The following strategy can be used to speed up the enumeration procedure. The different component limit states share many common random variables related to loading and material properties. Therefore, there is correlation among the component limit-state functions in any structure. This implies that if a component fails, then other components that are highly correlated with this component may also fail subsequently with high probability. Therefore, in the failure sequence enumeration, several component failures (instead of only one failure) can be imposed together as a group at each damage stage.

For a group of strongly correlated components, the component having the highest failure probability is selected to be representative of the group at the current damage state. Other components in the group are identified using a conditional probability criterion:

$$P(E_i/E_k) \geq \lambda_0 \quad i = k+1, \ldots, n \quad (44.26)$$

where λ_0 is a chosen cutoff value, E_k stands for the event of failure of the component k, E_i is the event of failure in the ith component among the remaining $n-k$ components, and $P(E_i/E_k)$ is the probability of event E_i given that the event E_k has occurred. The failure probability for the group of components is approximated by the failure probability of the representative component. All the components in the group are removed at the same time, and the structure is reanalyzed with the remaining components.

The other steps in the failure sequence search are the same as that in the original branch-and-bound method. This concept of grouping can drastically reduce the number of damage states and hence the number of structural reanalysis. Therefore, this method can be referred to as a fast branch-and-bound method. Xiao and Mahadevan [18], who proposed this method, found that for even a small problem with 18 possible component failures, the fast branch-and-bound method requires only 0.6% of the computational time taken by the original branch-and-bound method. The savings in computational time grows with the size of the problem.

44.4.2 Deterministic Initial Screening

Any component failure could be the starting point of a failure sequence. However, in the branch-and-bound enumeration, only component failures with high probability of occurrence are generally the starting points of the dominant failure sequences. The computation of the probability $P(g < 0)$ corresponding to each component failure involves several iterations of structural analysis to find the minimum distance point, if FORM or SORM are used. (Monte Carlo simulation requires many more structural analyses.) For large structures with numerous components, this first step in failure sequence enumeration is quite time consuming. Therefore, an efficient idea is to observe that the components with higher failure probabilities are, in general, likely to have the g values closer to zero when the structure is analyzed at the mean values of the random variables. Therefore, the starting points of the dominant failure sequences may be selected by using deterministic structural analysis at the mean values, and then choosing only those component failures with g values below a cutoff value. Note that this strategy is simply to select the starting points of the sequences, not for the final probability computation. Therefore, it will provide significant reduction in computational effort, with minimal impact on the accuracy of the probability result.

Note that the probability-based criterion for all component failures is much more rigorous and accounts for the variation in random variable sensitivities. That is, some component failures that may appear insignificant with a deterministic criterion may become significant with a probabilistic criterion. Therefore, the deterministic screening should be done carefully so that probabilistically significant events do not get discarded. Mahadevan and Liu [19] adopted the following strategy. The limit-state value, calculated at the mean values of random variables, is used for selecting the 20 most likely component failures at each stage. Then, the 20 component failures are subjected to probabilistic analysis, and only the component failure with the highest probability among these 20 is used to start the fast branch-and-bound enumeration of the most probable failure sequence. Thus, the failure sequence enumeration is, in fact probabilistic, except for the initial screening to ignore component failures that have very low probability.

44.4.3 Weakest-Link Model

As seen in Section 44.4.2, the various significant failure sequences are very similar in the list of failed components, which implies that there are strong correlations among these failure sequences. When the sequences are very similar, the system failure probability can be approximated using the most probable sequence, i.e., by Cornell's first-order lower bound. In that case, only one significant failure sequence might provide an adequate estimate of system failure probability. Therefore, instead of enumerating many failure sequences, it may be adequate to stop after the first significant failure sequence is identified. This provides tremendous savings in the computational effort of structural reanalysis corresponding to numerous steps of progressive failure in multiple sequences.

44.4.4 Critical-Component Failure

Strictly speaking, a structural system failure is defined to be the collapse of the entire structure. Section 44.4.3 has shown that a failure sequence from initial failure to system final failure involves a large number of ply-level components. Thus the search for even a single complete sequence is quite time consuming. Therefore, Mahadevan and Liu [19] proposed another approximation to avoid the tedious and expensive computation.

It is assumed that there exists a critical component for the structure. This critical component should have a failure probability much lower than the other components that fail after the critical component. The components with higher failure probabilities will easily fail once the critical one fails. In other words, the critical-component failure is not far from the entire system failure. As a result, the structural system failure can be approximately defined to be the failure of the critical component.

The assumption of a critical failure to approximate system failure is quite reasonable for laminated composites. The overall probability of the sequence is computed as the probability of intersection of the component events in that sequence. The probability of intersection of several events is dominated by the low-probability events [17], which are usually present before the first critical failure. After this critical failure, the probabilities of subsequent failures are much higher, and thus do not have a significant contribution to the probability computation. As an example, consider the plate structure made of composite laminates as considered in Section 44.4.3. It is seen that after the first fiber failure (which has the lowest probability), the probabilities of the subsequent events are significantly increased. The overall probability of the sequence is dominated by the low-probability events up to the first fiber failure, and it is closely approximated by the probability of intersection of the first fiber failure and a few matrix-cracking failures immediately preceding it. Therefore, considering probability computation, it appears adequate to terminate the exploration of the failure sequence at the first critical failure.

With the above assumption, a large amount of computation after the failure of the critical component can be avoided, and the prediction of ultimate failure probability will have adequate accuracy. The grouping concept, deterministic screening, and the weakest-link approximations of the previous subsections provide additional computational efficiency. Thus, the approximations can be summarized as follows:

1. Start from the intact structure. Calculate the values of performance functions (g functions) for all the components and sort the values in descending order. Select the first several components with small g values to be explored, using a suitable cutoff value for g (deterministic initial screening).
2. Estimate the component failure probability of the selected components and arrange them in the descending order of probability values. Select the component with the highest failure probability and other components with high correlation to this first component (fast branch and bound).
3. Simulate the damage in the selected components by modifying the corresponding structural stiffness terms, and reanalyze the damaged structure to compute the failure probabilities of the remaining components.
4. Repeat steps 2 and 3 until the critical component fails.
5. Approximate system failure probability using the failure probability of the sequence identified in steps 2, 3, and 4.

44.4.5 Numerical Example: Aircraft Wing

An aircraft composite wing composed of skin and stringer components and consisting of center, leading edge, and trailing edge is shown in Figure 44.8. The stringer is constructed in both longitudinal and transverse directions. The data for this example was taken from Shiao and Chamis [1], who computed first-ply-failure probability. Mahadevan and Liu [19] computed the system failure probability using the methods in this section.

In Figure 44.8, the left end (section A-A) is fixed, and the right end (section B-B) is free. The air pressure is assumed to be triangularly distributed along transverse direction of the wing and linearly distributed along the longitudinal direction. The ply-level stiffness properties, material strengths, ply thicknesses and orientations, and the pressure loads are all assumed to be random variables. A laminate configuration of $[0°/-45°/90°/45°/0°]$ is used for the skin and $(0°)_5$ for the stringer. The statistical distribution of a variable is assumed to be lognormal for the strength parameters and normal for the rest, with a coefficient of variation of 0.10. A standard deviation of 2° is used for the ply angles. See [19] for detailed data, finite element modeling, and results of this numerical example.

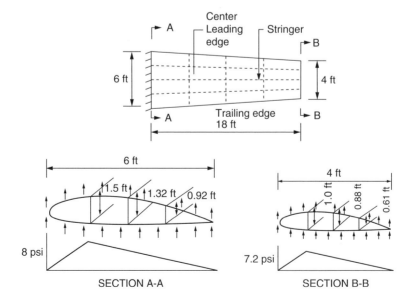

FIGURE 44.8 Composite aircraft wing: geometry and loading.

TABLE 44.1 Structural Progressive Damage Stages

Damage Stage	Failed Components	Representative Failure Probability
1st	$85_{2M}, 85_{4M}, 86_{4M}, 86_{2M}$	0.2820
2nd	$37_{2M}, 37_{4M}, 87_{4M}, 38_{2M}, 73_{2M}, 97_{4M}$	0.0903
3rd	$87_{2M}, 49_{4M}, 97_{2M}, 25_{4M}, 98_{2M}$	0.0451
4th	$98_{4M}, 25_{2M}, 50_{4M}, 49_{2M}, 26_{4M}, 73_{4M}, 26_{2M}, 38_{4M}$	0.0338
5th	$74_{2M}, 85_{3M}, 39_{2M}, 74_{4M}, 88_{4M}, 89_{2M}, 88_{2M}, 86_{3M}$	0.0156
6th	$99_{4M}, 39_{4M}, 50_{2M}$	0.0036
7th	$51_{4M}, 13_{2M}, 13_{4M}$	0.0029
8th	$37_{3M}, 14_{4M}, 75_{2M}, 14_{2M}, 27_{4M}, 85_{1F}, 89_{4M}, 85_{5F}$	0.0010

For the fast branch-and-bound method, $\lambda = 0.4$ is used in this example, only for the sake of demonstration. The choice of λ has to be based on a trade-off between accuracy and efficiency. A larger value of λ reduces the number of failures in the grouping operation, and therefore makes the computation more expensive but more accurate. For specific applications, the variation of failure probability estimate and computational effort with λ can be investigated, and an optimum value can be chosen.

The first significant failure sequence identified using the fast branch-and-bound method is used to estimate the probability of ultimate strength failure using the weakest-link model. The following notation is used to identify the ply-level failures: for example, 85_{2M} stands for finite element number 85, ply number 2, and matrix cracking failure mode; 37_{1F} stands for finite element number 37, ply number 1, and fiber breakage failure mode.

At each stage of damage, a few highly correlated components are chosen to fail together. Structural failure proceeds through progressive damage accumulation, as shown in Table 44.1. Damage accumulates in the zones near the fixed end and at the center of the skin. The structure experiences eight stages of damage before experiencing the first fiber failure. In the damage process, the failure probability of each damage stage decreases with damage accumulation up to the first fiber breakage occurrence. Probabilistic analysis was also carried out for the ninth damage stage. Its representative failure probability was increased to 0.028, which is much higher than the one prior to it. This indicates that the first fiber failure is likely to be a severe or critical failure for the entire structure. After this critical event, the structure will rapidly proceed to final failure. This implies that the exploration of the failure sequence may be terminated after this first severe failure.

Three damage levels with the lowest failure probabilities, namely the sixth, seventh, and eighth damage levels, dominate the computation of the system failure probability. The computed correlation coefficient matrix between the limit states representing these three events is:

$$[\rho_{ij}] = \begin{bmatrix} 1 & \cdots & Sym \\ 0.99 & 1 & \cdots \\ 0.71 & 0.70 & 1 \end{bmatrix} \quad (44.27)$$

The corresponding individual-event failure probabilities and the two-event joint failure probabilities are obtained as:

$$[P_{ij}] = \begin{bmatrix} 0.0036 & \cdots & Sym \\ 0.0026 & 0.0029 & \cdots \\ 0.0003 & 0.0003 & 0.001 \end{bmatrix} \quad (44.28)$$

where P_{ij} refers to the joint probability of the ith and jth failures (the off-diagonal terms in the matrix). The diagonal terms in the matrix ($i = j$) refer to the individual failure probabilities.

Using only the two-event joint probabilities, the second-order upper-bound formula suggested by Murotsu [9] is used to estimate the system failure probability as

$$P_f = 2.977 \times 10^{-4} \tag{44.29}$$

Using the three-event joint probability, the third-order estimate for the system failure probability is obtained as

$$P_f = 2.957 \times 10^{-4} \tag{44.30}$$

These two results are quite close to each other, as expected. For this reason, most system reliability studies report only second-order estimates.

44.5 Fatigue Limit State: Material Modeling

Many applications of composite materials involve components that are subjected to cyclic loading. Cyclic loading causes damage and material property degradation in a cumulative manner. Considerable research on fatigue behavior has been carried out for monolithic materials such as metals, and progress has been made in devising fatigue-resistant materials as well as in developing methodologies for life prediction. For composite materials, fatigue analysis and consequent life prediction become difficult because the material properties of the constituents of the composite are quite different. The fatigue behavior of one constituent may be significantly affected by the presence of other constituents and by the interfacial regions between the fibers and matrix. Fatigue properties of composites can vary significantly due to the large difference in the properties between the fibers and matrix of the composite and the composition of constituents.

Many experimental studies have been reported for obtaining the fatigue properties of different types of composite materials since the 1960s. Based on these experimental results, empirical S-N curves have been derived between stress (S) and fatigue life (N). These relationships have been suggested for use in design [20]. Both linear and nonlinear S-N curves have been proposed based on the experimental results [21–23]. A nonlinear curve between strain and fatigue life is also used to predict the fatigue life of composite materials [24–26]. The following relationship is widely used:

$$S = m \log N + b \tag{44.31}$$

where m and b are parameters related to material properties.

With the predicted fatigue life under constant cyclic loading, fatigue damage can be evaluated after a given number of cycles with a fatigue-damage-accumulation model. The composite is assumed to fail when the accumulated damage exceeds the critical level of damage.

44.5.1 Fatigue-Damage Modeling

For homogeneous, or monolithic materials with isotropic material properties, damage is accumulated at a low growth rate in the beginning, and a single crack propagates in a direction perpendicular to the cyclic loading axis. On the other hand, in composite materials, especially for structures with multiple plies and laminates, the fracture behavior is characterized by multiple damage modes, such as crazing and cracking of the matrix, fiber/matrix decohesion, fiber fracture, ply cracking, delamination, void growth, and multidirectional cracking. These modes appear rather early in the fatigue life of composites.

The mechanisms of crack initiation and crack growth are quite complex for composite materials. Even for unidirectional reinforced composites under the simple loading case such as tension along the direction of fibers, cracks can initiate at different locations and in different directions. Cracks can initiate in the matrix, perpendicular to the direction of loading. Cracks can also initiate in the interface along the directions of fibers between the fibers and matrix due to debonding. Many experimental fatigue tests have been carried out to study the crack growth in composites when there is only one dominant crack that is propagating. The crack propagates in the same plane and direction as the initial crack. The Paris law has been used to describe this fatigue-crack-propagation behavior. But this is limited to unidirectionally aligned fiber-reinforced composites. For more general laminates, a similar mode of crack propagation cannot be obtained, even under simple loading. Thus, traditional fracture mechanics cannot be used for the fatigue analysis of composite materials.

The concept of damage accumulation can be used as a more suitable approach to predict the fatigue life of structures of composite materials. However, fatigue damage cannot be measured directly. Therefore, for quantitative evaluation of fatigue damage, change in Young's modulus or stiffness is often used to evaluate the fatigue damage due to cyclic loading. For example, fatigue damage can be defined in terms of Young's modulus as [27]:

$$D = \frac{E_0 - E}{E_0 - E_f} \quad (44.32)$$

where D_1 is the accumulated fatigue damage, E_0 is initial Young's modulus of the undamaged material, E_f is the Young's modulus when fracture occurs, and E is the Young's modulus at any stage. According to Equation 44.32, the accumulated damage will be in the range between 0 and 1.

As previously mentioned, the complexity of composites leads to the presence of many modes of damage. These modes appear at the early stages of the fatigue life. The damage accumulates rapidly during the first few cycles. During this stage, microcracks initiate in multiple locations in the matrix. Debonding occurs at the weak interfaces between fibers and matrix. Also, some fibers with low strength may break during this stage. The next stage shows a slow and steady damage growth rate. Finally, the damage again grows rapidly during the last stage before the fracture occurs. Figure 44.9 shows schematically the comparison of damage accumulation in composite materials and homogeneous materials as a function of fatigue cycle ratio. The Young's modulus measured from fatigue tests also shows the same characteristics of damage accumulation [28–30]. Figure 44.9 is plotted in terms of damage index vs. cycle ratio, where the damage index is defined as in Equation 44.32. The cycle ratio is the number of cycles at a given instant divided by the fatigue life.

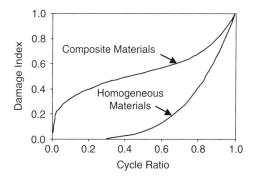

FIGURE 44.9 Fatigue-damage accumulation: metals vs. composites.

A linear damage summation model was first used to evaluate the fatigue behavior of composite materials by Nicholas and Russ [31]. Halverson et al. [28] used a power function in terms of the cycle ratio to evaluate the remaining strength of the material and to calculate the fatigue life.

$$F_r = 1 - (1 - F_a)\left(\frac{n}{N}\right)^j \tag{44.33}$$

where F_r is the normalized remaining strength (normalized by the undamaged static strength), F_a is the normalized applied load (also normalized by the undamaged static strength), j is a material constant, and n is the number of cycles of applied load. N is the fatigue life of a constant load. Then, according to the definition of damage in Equation 44.32, the mathematical function for damage accumulation will also be a power function of cycle ratio.

$$D = (1 - F_a)\left(\frac{n}{N}\right)^j \tag{44.34}$$

Once the residual strength is computed, the degradation of the material can be described with Equation 44.33 or Equation 44.34. Other nonlinear damage-accumulation functions have also been used. These nonlinear damage-accumulation functions are able to capture the characteristics of rapid damage growth either at the early stages of life or near the end of life, but not both. For example, the damage model of Subramanian et al. [25] explains the fast damage growth during early loading cycles, but it does not accurately describe the rapid damage growth close to the material fracture. Halverson et al. model the characteristics of rapid damage growth at the end of fatigue life of the material. But the model is not accurate during the early loading cycles.

Mao and Mahadevan [27] proposed a versatile new damage-accumulation model, for accuracy in both the early and final stages of life, as

$$D = q\left(\frac{n}{N}\right)^{m_1} + (1 - q)\left(\frac{n}{N}\right)^{m_2} \tag{44.35}$$

where D is the normalized accumulated damage; q, m_1, and m_2 are material-dependent parameters; n is the number of applied loading cycles; and N is the fatigue life at the corresponding applied load level. The characteristics of rapid damage accumulation during the first few cycles can be captured with the first term, with $m_1 < 1.0$. The second term shows the fast damage growth at the end of fatigue life, with $m_2 > 1.0$.

The parameters in Equation 44.35 are defined in terms of fatigue life of interest as

$$q = \frac{A\left(\frac{N_0}{N}\right)^\alpha}{1 - (1 - A)\left(\frac{N_0}{N}\right)^\alpha} \tag{44.36}$$

$$m_1 = \left(\frac{N_0}{N}\right)^\beta \tag{44.37}$$

$$m_2 = \left(\frac{N}{N_0}\right)^\gamma \tag{44.38}$$

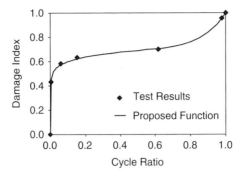

FIGURE 44.10 Experiment vs. model: damage index for 810 O laminates (75% loading level).

where N_0 is the reference fatigue life. The parameters α, β, and γ are material-dependent constants. These parameters can be obtained with fatigue experimental data. Once the damage indices are obtained during the fatigue tests, regression analysis can be carried out to obtain the parameters q, m_1, and m_2. Then, parameters α, β, and γ can be calculated using Equation 44.36 to Equation 44.38.

44.5.2 Experimental Results and Model Performance

Two sets of experimental results are used here to demonstrate the performance of the damage-accumulation model in Equation 44.35. The first set is with 810 O laminates and the second one is with a woven composite.

Subramanian et al. [25] obtained two sets of data for 810 O laminates, under two different loading levels (75 and 80% of ultimate strength of the laminate). These were symmetric $(0, 90_3)$ cross-ply laminates. Apollo fibers and HC 9106-3 toughened epoxy matrix were used. The fibers received 100% surface treatment and were sized with a thermoplastic material. The fatigue tests were performed at R ratio = 0.1 and 10-Hz frequency. The Young's moduli of the laminated composite were measured after different numbers of cycles of tensile fatigue loading, and the corresponding damage indices were computed according to Equation 44.32.

The experimental results and the proposed model are plotted in Figure 44.10 and show excellent agreement. Next, parameters for the damage model at the 80% loading level are obtained with the values of α, β, and γ obtained at the 75% loading level. Figure 44.11 shows the comparison between the damage measured at 80% loading level and the predicted damage with the proposed model with the parameters obtained at 75% loading level. It is seen that the predicted results agree well with the experimental results in this example for the tested composite material.

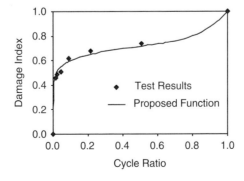

FIGURE 44.11 Experiment vs. model: damage index for 810 O laminates (80% loading level).

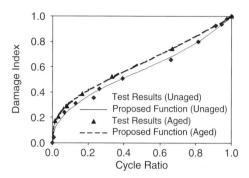

FIGURE 44.12 Experiment vs. model: damage index for AS4/PR500 woven composite.

Kumar and Talreja [30] conducted tension–tension fatigue experiments on the AS4/PR500 5 harness satin weave composite laminates at a frequency of 10 Hz and R ratio of 0.1 to study the fatigue behavior of the composite material. Tension fatigue tests were conducted on the symmetric $(0/90_{2w})$ laminates. Two types of specimen, unaged and 6000-h aged, were tested. Young's modulus of the material was measured after different numbers of fatigue cycles.

The damage-accumulation model of Equation 44.35 and experimental results are compared in Figure 44.12, and it is seen that the proposed function does an excellent job of capturing the characteristics of the damage evolution in composite materials. This is important in order to develop confidence in model-based prediction of fatigue life and reliability.

44.6 Fatigue-Delamination Limit State

Fatigue loads can initiate progressive failure in a composite laminate by way of successive delamination, matrix cracking, fiber waviness, etc. A probabilistic analysis framework to predict the fatigue-delamination reliability of composite structures can be developed along the following steps. Finite element analysis is used to compute the global and ply-level response of the structure at the damage sites. A suitable failure model is used to evaluate the fatigue-delamination limit state. A response-surface approximation is constructed for the limit state, in terms of basic random variables related to input loads, material properties, geometry, etc. This response surface is then used with FORM, SORM, or Monte Carlo simulation to estimate the failure probability.

44.6.1 Failure Mechanism and Model

Consider a helicopter rotor hub test specimen shown in Figure 44.13. Hingeless, bearingless helicopter rotor hubs are being designed using laminated composite materials to reduce weight, drag, and the number of parts in the hub. During flight, the rotor hub arm experiences centrifugal loads as well as bending in the flapping flexure region. An effective elastic hinge is designed integrally to the composite rotor yoke by incorporating a tapered region between thick and thin regions. The varying thickness of the tapered region is accomplished by dropping internal plies, as shown in Figure 44.13. The thick-taper-thin geometry is tailored to give the proper flapping flexure.

Generalized test specimens have been developed to understand the basic response of the composite rotor hub yoke [32]. The specimens are geometrically simple with thick, thin, and tapered regions, as shown in Figure 44.13, and approximately one-quarter symmetry of the geometry of section A-A is considered for analysis. The specimens consist of an outside fabric layer denoted as F, continuous 0° belt plies denoted as B1 through B5, and discontinuous ±45° plies denoted by D1 through D4. The specimens are subjected to a constant axial tensile load (P) to simulate the centrifugal load, and a cyclic bending

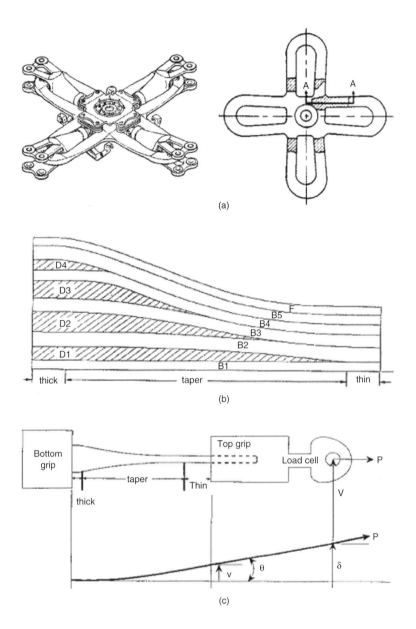

FIGURE 44.13 Composite helicopter rotor: (a) rotor hub assembly, (b) half of the symmetric section of the test specimen, (c) typical test setup.

load (V) to simulate the interaction of the rotor passage with the fuselage. The cyclic load (V) induces an angular displacement (θ) that simulates the flexural bending in the yoke.

The ply drop in the laminate creates geometrical and material discontinuities that cause large interlaminar stresses and initiate delamination. The failure mechanism observed from this type of loading is an initial tension crack between the internal delamination at the thick-to-taper transition, where internal ply drop-offs occur.

Numerous studies have investigated computational models for delamination failure. The discussion below is limited to tapered laminates. Some studies have used stress-based criteria for modeling delamination failure [33–36]. Others have used a strain-energy release-rate approach [37–42]. Most of these studies have only considered delamination under pure tension, bending, or torsion loads. Very few studies

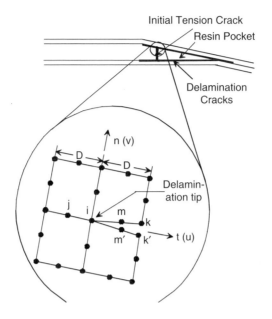

FIGURE 44.14 Computation of strain energy release rate.

have considered the combined effect of bending and tension on delamination of tapered laminated composites [32, 43].

The strain-energy release rate, G, computed in these studies is associated with edge delamination. Finite element analysis has been used to show that once the delamination progresses beyond a distance equal to the thickness of a few plies from the edge, G reaches a constant plateau [44]. However, delamination in composite laminates may interact with other damage mechanisms such as matrix cracking and fiber bridging, resulting in stable growth behavior [45]. Because a delamination is constrained to grow between individual plies, both interlaminar tension and shear stresses are commonly present at the delamination front. Therefore, delamination is often a mixed-mode fracture process. The strain-energy release rate, G, generally equals to the sum of G_I, G_{II}, and G_{III}, which are interlaminar tension, sliding shear, and scissoring shear, respectively. The effect of G_{III} is assumed to be small and can be ignored, following Murri et al. [32].

A virtual crack-closure technique (VCCT) is used to calculate G at the delamination tip, as shown in Figure 44.14 [46, 47], such that

$$G = G_I + G_{II} \tag{44.39}$$

where, G_I and G_{II} are the mode I and mode II strain-energy release rates, computed as

$$G_I = -\frac{1}{2\Delta}[F_{ni}(v_k - v_{k'}) + F_{nj}(v_m - v_{m'})]$$

$$G_{II} = -\frac{1}{2\Delta}[F_{ti}(u_k - u_{k'}) + F_{nj}(u_m - u_{m'})]$$

where u and v are tangential and perpendicular nodal displacements, respectively, and F_t and F_n are the tangential and perpendicular nodal forces, respectively, computed from finite element analysis. Here, t refers to the direction tangential to the crack, and n refers to the direction perpendicular to the crack. In Figure 44.14, node i is the delamination tip, and node j is the next node to which the delamination will advance.

Delamination onset is assumed to occur when the calculated G exceeds a critical value G_{crit} derived from material coupon delamination tests [32, 37, 38]. See [48] for details of finite element stress analysis and delamination probability estimation. A few pertinent observations are summarized below.

44.6.2 Random Variables

Based on the material data, seven random variables were identified for probabilistic analysis with the finite element model of the helicopter rotor hub. These are material property variables E_{11}, E_{22} (elastic moduli along directions 1 and 2); ν_{13} (Poisson ratio); G_{13} (shear modulus); the oscillatory bending angle θ caused by the cyclic loading; the magnitude of the axial load P; and the limiting value of the strain-energy release rate G_{crit}. All the random variables are represented by Gaussian (normal) distribution. No statistical correlation is assumed among the material property variables. (This is not realistic; however, as shown by the analysis results, only E_{11} is significant. Therefore, this assumption is not critical in this problem.) The seventh variable, G_{crit}, is discussed below.

The objective is to estimate the probability of initiation of delamination at the required life. This is assumed to occur when the strain-energy release rate exceeds the limiting value G_{crit}, which is a function of load cycles, N. Figure 44.15 shows the plot of G_{crit} vs. N obtained from material test data. Figure 44.15 also shows the best-fit line through the data, assuming G_{crit} is a linear function of $\log_{10}(N)$. This is shown by the solid line marked "mean." The lines corresponding to mean plus and minus one standard deviation are shown above and below this line, marked "+ sigma" and "− sigma" respectively. For the given data, the relationship between G_{crit} and N is obtained using regression analysis as

$$G_{crit} = 448.56 - 58.571 \log_{10}(N) \tag{44.40}$$

where G_{crit} has units of J²/m. Based on statistical analysis of the G_{crit} vs. N data, it was determined that G_{crit} is a Gaussian (normal) random variable, with its mean value described by Equation 44.40 and a constant standard deviation of 36.6 J²/m. The corresponding performance function is

$$g = G_{crit} - G \tag{44.41}$$

In this case, both G_{crit} and G are random variables that are dependent on basic random variables such as material properties, geometric properties, loading conditions, etc., which have inherent scatter in their definition.

44.6.3 Response-Surface Modeling of the Limit State

The limit-state function in Equation 44.41 is not available as a closed-form function of the basic random variables. It can only be computed through a nonlinear finite element stress analysis, combining several analysis codes. Thus, it is an implicit function of the random variables. In such a case,

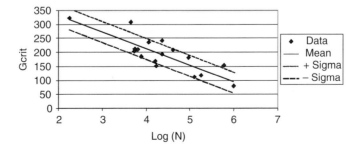

FIGURE 44.15 Relationship between G_{crit} and cycles of life.

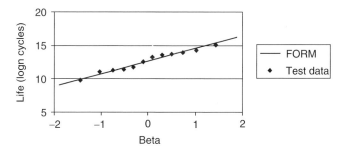

FIGURE 44.16 Predicted fatigue-delamination life vs. test data.

the response-surface approach can be used to develop an approximate closed-form expression of the limit-state function, and then the first-order reliability method (FORM) can be used to estimate the delamination probability.

The literature on design of experiments is quite vast, and many methods are available [49, 50]. Using a central composite design of experiments, the following second-order approximation of the limit-state function is constructed for illustration:

$$g = G_{crit} - G = G_{crit} - 175.344(0.569 - 0.0861E_{11} + 0.0231P - 0.117\theta \\ - 0.000546P^2 + 0.00376\theta^2 + 0.0046P\theta) \tag{44.42}$$

where the units of G and G_{crit} are J/m^2.

44.6.4 Verification of Methodology

Having developed and implemented the probabilistic analysis framework, the next step is to compare the predicted results with test data and to validate the proposed method. Twelve test specimens were tested by Bell/Textron Helicopter, which gave a mean life of $\mu = 1{,}559{,}694$ cycles and standard deviation of $\sigma = 7{,}725{,}401$ cycles, which gives a large coefficient of variation of 4.95. The life data from the 12 specimens was plotted on lognormal probability paper and compared with the FORM reliability analysis with the response surface in Equation 44.42.

Figure 44.16 shows that the statistical distribution of life predicted by the reliability analysis showed excellent agreement with the test data. The reliability analysis predicts a nominal life similar to the experimental observations. However, currently available deterministic analysis may be capable of predicting nominal behavior. What is more important is that the reliability analysis also predicts a lognormal distribution with a coefficient of variation similar to the experimental observations (5.66 vs. 4.95). Zhang and Mahadevan [51] developed a Bayesian validation technique for model-based reliability prediction and found that the data gave strong support for the model. Using the same data, Mahadevan and Zhang [52] also developed a test planning methodology that integrates both model-based and test-based information. Such a methodology was found to offer significant savings in testing cost.

44.7 Creep Limit State

In certain elevated-temperature applications, composites undergo creep deformation, which is a function of loading, operating environments, and the properties of the constituents. Creep is one of the principal damage mechanisms for materials operating at elevated temperatures. If the temperature is less than about 30% of the melting temperature of the fiber material, creep of fibers can be neglected. The composite longitudinal creep behavior is largely controlled by the fracture behavior of the fibers. It has been shown [53] that, at low applied load in the fiber direction, very few fibers fracture during creep.

The composite creep strain is limited by the elastic strain of the fibers. However, when a relatively large load is applied, a substantial number of fibers can fracture, and damage initiates during the initial loading and creep of the matrix. In addition, some defects introduced during the manufacture of fibers can also cause fracture. When fibers break, the load on the cross section is redistributed. The load shedding from the broken fibers is shared by the remaining intact fibers and the matrix, which increases the creep strain and causes more fibers to break.

A number of experimental and theoretical studies have been conducted on the creep of composites. Cell models have been widely used to study the creep behavior of composites [54–60]. Bullock et al. [61] proposed an analytical model to predict the creep rate by assuming that both the fibers and matrix creep according to a power law. McLean [62] proposed a model assuming that the composite consists of creeping matrix and elastic fibers. In this model, the effect of the broken fibers is ignored. A simple model that accounts for fiber breakage in a uniaxial tension test has been presented by McLean [63]. Goto and McLean [55] also modified the above model by introducing a strain-hardening third phase at the interface, deforming according to a power law. The McLean model gives very accurate predictions [64] for undamaged composites compared with finite element analysis results. Theoretical studies on the fiber failure within the framework of global load sharing, whereby the load shed from broken fibers is shared nearly equally among all intact fibers, have been carried out by Curtin [65] for composites with weak interface. Recently, Du and McMeeking [58] made use of Curtin's results and studied the effects of fiber breakage and the consequential stress relaxation in the broken fibers. The simulation method is also used to predict the lifetime of composites with broken fibers [66, 67]. The above models are limited to longitudinal creep behavior.

There have also been some studies carried out theoretically and experimentally on the transverse creep behavior of composites [68–71]. Song et al. [64] analyzed the effect of transverse loading and plasticity of matrix on the longitudinal creep behavior. It was observed that the applied transverse tension can reduce the composite creep strain and the normal stress in the fibers; at the same time, the applied transverse compression increases the creep strain and the normal stress in the fibers.

All of these models are developed to carry out the deterministic stress and strain analysis of the composites, or to predict the creep life of the composites. However, it is well known that there is large scatter in the material strength and other mechanical properties. The strength of the fibers is usually described with the weakest-link model, and the Weibull distribution is commonly used.

The purpose of this section is to develop a method for the probabilistic analysis of longitudinally reinforced composites under creep, with and without fiber breaks. Sensitivity analysis is also carried out to investigate the importance of different stochastic parameters on the failure probability of composites. First the time-dependent creep behavior of longitudinal composites without broken fibers is discussed, followed by creep behavior of composites with fiber fractures. Next, different limit-state functions (failure criteria) are established for the reliability analysis of composites.

44.7.1 Creep of Undamaged Composites

Fiber-reinforced composites exhibit creep deformation at high temperatures [63, 72]. In the range of operating temperatures, the fibers may not creep, but the matrix can. In this section, it is assumed that the composite consists of elastic fibers and a creep matrix. It is assumed that the fibers neither creep nor fracture, but deform elastically at a rate governed by the surrounding creep matrix. The total strain includes elastic, plastic, and creep strains. Previous experience [64] shows that matrix plasticity has a very limited effect on the creep behavior of composites. Therefore, the plasticity effect is not included here.

The governing equations for the longitudinal stress and strain in the composite can be expressed as

$$\sigma_f = E_f \varepsilon \tag{44.43}$$

$$\dot{\varepsilon} = \frac{\dot{\sigma}_m}{E_m} + B\sigma_m^n \tag{44.44}$$

$$\sigma = f\sigma_f + (1-f)\sigma_m \tag{44.45}$$

where σ_f and σ_m are the fiber and matrix stresses, respectively; E_f and E_m are the Young's modulus of the fibers and the matrix, respectively; f is the fiber volume fraction; ε and $\dot{\varepsilon}$ are the strain and strain rate of the composite, respectively; and n and B are the creep exponent and creep constant of the matrix, respectively.

The performance of the composite can be evaluated based on this model. Assume that the composite is subjected to a fixed stress loading. Solving Equation 44.43 to Equation 44.45, the stress in the matrix is given by

$$\sigma_m(t) = \left[\left(\frac{E}{E_m\sigma}\right)^{n-1} + \frac{f(n-1)E_f E_m B t}{E}\right]^{-\frac{1}{n-1}} \tag{44.46}$$

The stress in the fibers is given by

$$\sigma_f = \frac{\sigma}{f} - \frac{1-f}{f}\left[\left(\frac{E}{E_m\sigma}\right)^{n-1} + \frac{f(n-1)E_f E_m B t}{E}\right]^{-\frac{1}{n-1}} \tag{44.47}$$

The strain is given by

$$\varepsilon = \frac{\sigma}{fE_f} - \frac{1-f}{fE_f}\left[\left(\frac{E}{E_m\sigma}\right)^{n-1} + \frac{f(n-1)E_f E_m B t}{E}\right]^{-\frac{1}{n-1}} \tag{44.48}$$

And the strain rate is

$$\dot{\varepsilon} = \frac{(1-f)E_m B}{E}\left[\left(\frac{E}{E_m\sigma}\right)^{n-1} + \frac{f(n-1)E_f E_m B t}{E}\right]^{-\frac{n}{n-1}} \tag{44.49}$$

In this model, as time proceeds, the fibers increasingly sustain the loads formerly carried by the creeping matrix. Eventually, when the matrix stress is completely relaxed, all of the loads are carried by the intact fibers, and the strain approaches a steady state. The creep strain rate decreases as the stress in the matrix relaxes due to the matrix creep. For the situation where the composite is subjected to a low level of stress, the above model predicts creep strain with reasonable success [64]. However, when a relatively large load is applied to the composite, damage initiates in the form of isolated or localized fiber breaks [53, 72]. In addition, the defects introduced in the fibers during manufacturing can also lead fibers to break. McLean's model ignores these effects and therefore underestimates the creep strain and fails to predict the creep ruptures resulting from the failures of fibers. Therefore, creep analysis with fiber break is considered next.

44.7.2 Creep Analysis with Stochastic Fiber Fracture

The randomness in the strength of the fibers can be modeled using the Weibull distribution, which is an extreme value distribution of the smallest values, following the weakest-link approach. For a composite with the stress in fibers, σ_f, some of the fibers will break. The failure probability of the fibers can be estimated by

$$p = 1 - \exp\left(-\left(\frac{\sigma_f}{\sigma_c}\right)^m\right) \tag{44.50}$$

where m is referred to as the Weibull modulus that characterizes the shape of the distribution, and σ_c is the characteristic strength of the fibers.

Here, p can be considered as the fraction of the broken fibers. Equation 44.50 can be approximated and simplified as

$$p \cong \left(\frac{\sigma_f}{\sigma_c}\right)^m \tag{44.51}$$

For a composite with broken fibers, the load carried by the fiber near the end of fiber break is released and transferred to the matrix and to the intact fibers through the interface. From the end of a broken fiber to a certain length known as the stress-recovery length, the stress in the fiber increases from zero to nominal stress. The shear stress is set equal to the yield stress of the interface τ_y at the end of fiber break. The stress-recovery length can be obtained by the equilibrium condition. The average fiber stress over the stress-recovery length is given by

$$\overline{\sigma}_f = \left(1 - \frac{1}{2}p\right)\sigma_f \tag{44.52}$$

From Equation 44.43, Equation 44.51, and Equation 44.52, the average stress in the fibers is obtained as

$$\overline{\sigma}_f = \left(1 - \frac{1}{2}\left(\frac{E_f \varepsilon}{\sigma_c}\right)^m\right) E_f \varepsilon \tag{44.53}$$

From Equation 44.45, the matrix stress is obtained as

$$\sigma_m = \frac{1}{1-f}\left[\sigma - f\left(1 - \frac{1}{2}\left(\frac{E_f \varepsilon}{\sigma_c}\right)^m\right) E_f \varepsilon\right] \tag{44.54}$$

and the matrix stress rate as

$$\dot{\sigma}_m = \frac{-f}{1-f}\left(1 - \frac{1+m}{2}\left(\frac{E_f \varepsilon}{\sigma_c}\right)^m\right) E_f \dot{\varepsilon} \tag{44.55}$$

By substitution of the matrix stress and matrix stress rate into Equation 44.44, the creep strain rate is obtained as

$$\dot{\varepsilon} = \frac{B\left(\sigma - f\left(1 - \frac{1}{2}\left(\frac{E_f \varepsilon}{\sigma_c}\right)^m\right) E_f \varepsilon\right)^n}{\left(1 + \frac{f}{1-f}\left(1 - \frac{1+m}{2}\left(\frac{E_f \varepsilon}{\sigma_c}\right)^m\right)\frac{E_f}{E_m}\right)(1-f)^n} \tag{44.56}$$

The creep behavior of the composite is governed by Equation 44.53, Equation 44.54, and Equation 44.56. The strain is a function of time. It can be obtained from the integration of Equation 44.56. However, analytical integration is difficult; instead, a numerical solution can be obtained. The applied external stress produces an initial instantaneous elastic response. The instantaneous elastic strain of the composite is obtained as

$$\varepsilon_0 = \frac{\sigma}{fE_f + (1-f)E_m} \tag{44.57}$$

The corresponding instantaneous stresses in the fibers and the matrix are

$$\sigma_{f0} = \frac{E_f \sigma}{fE_f + (1-f)E_m} \tag{44.58}$$

$$\sigma_{m0} = \frac{E_m \sigma}{fE_f + (1-f)E_m} \tag{44.59}$$

The creep strain rate at time t_0 is obtained from Equation 44.56.

With broken fibers taken into account, it can be seen that the average fiber stress is not only a function of the strain of composites and the Young's modulus of fibers, but also a function of the Weibull distribution parameters of the fiber strength. Due to the broken fibers, the stress in the remaining fibers in the cross section is larger than the average stress. If we assume that the load previously carried by the broken fibers is shared by the intact fibers in the cross section, then the stress in the intact fibers can be estimated as

$$(\sigma_f)_{\max} = \frac{\sigma_f \sigma_c^m}{\sigma_c^m - \sigma_f^m} \tag{44.60}$$

44.7.3 Failure Criterion

In order to evaluate the failure probability of composites at different service times, it is necessary to establish limit-state functions based on the failure criteria for the composite. Failure criteria can be established simply based on strength or deformation. Here, two models can be postulated based on the fiber stress or the strain. In one model, the composite is assumed to fail when the fiber stress exceeds the allowable stress. This can be termed as the stress-based failure model. The limit-state function can be given as

$$g(\sigma_f, \sigma_{f\_\lim}) = \sigma_{f\_\lim} - \sigma_f \tag{44.61}$$

where $\sigma_{f\_\lim}$ is the limiting strength of the fibers.

In this formula, when the fiber stress is larger than the limiting value, $g(\sigma_f, \sigma_{f\_\lim}) < 0$ denotes the failure of the composite. When the fiber stress is less than the limiting value, $g(\sigma_f, \sigma_{f\_\lim}) > 0$ denotes that the composite is safe. And $g(\sigma_f, \sigma_{f\_\lim}) = 0$ is known as the limit state that separates the safe and failure states. The probability of failure of the composite is

$$p_f = P(\sigma_{f\_\lim} < \sigma_f) = P(g(\sigma_f, \sigma_{f\_\lim}) < 0) \tag{44.62}$$

For the composite without fiber fracture, the function $g(\sigma_f, \sigma_{f\_\lim})$ can be written as

$$g(\sigma_f, \sigma_{f\_\lim}) = \sigma_{f\_\lim} - \left\{ \frac{\sigma}{f} - \frac{1-f}{f} \left[\left(\frac{E}{E_m \sigma} \right)^{n-1} + \frac{f(n-1)E_f E_m Bt}{E} \right]^{-\frac{1}{n-1}} \right\} \tag{44.63}$$

For the composite with broken fibers taken into account, the function $g(\sigma_f, \sigma_{f\_\lim})$ can be written as

$$g(\sigma_f, \sigma_{f\_\lim}) = \sigma_{f\_\lim} - \frac{\sigma_f \sigma_c^m}{\sigma_c^m - \sigma_f^m} \tag{44.64}$$

The second model assumes that the composite fails when the strain exceeds the limiting value. This is termed as a strain-based failure model. The corresponding limit-state function is written as

$$g(\varepsilon, \varepsilon_{\lim}) = \varepsilon_{\lim} - \varepsilon \tag{44.65}$$

where $\varepsilon_{\lim}$ is the limiting value of the composite strain.

The probability of failure of the composite with this limit-state function is

$$p_f = P(\varepsilon_{\lim} < \varepsilon) = P(g(\varepsilon, \varepsilon_{\lim}) < 0) \tag{44.66}$$

For the composite without fiber fractures, the function $g(\varepsilon, \varepsilon_{\lim})$ can be written as

$$g(\varepsilon, \varepsilon_{\lim}) = \varepsilon_{\lim} - \left(\frac{\sigma}{fE_f} - \frac{1-f}{fE_f} \left[\left(\frac{E}{E_m \sigma} \right)^{n-1} + \frac{f(n-1)E_f E_m Bt}{E} \right]^{-\frac{1}{n-1}} \right) \tag{44.67}$$

For the composite with broken fibers taken into account, the function $g(\varepsilon, \varepsilon_{lim})$ can be written as

$$g(\varepsilon, \varepsilon_{\lim}) = \varepsilon_{\lim} - \varepsilon \tag{44.68}$$

where ε is obtained by integrating Equation 44.56.

44.7.4 Numerical Example

In this example, the probabilistic creep analysis of SiC fiber-reinforced Ti-6Al-4V matrix metal-matrix composite is demonstrated with the stress-based failure criterion. The constituent properties of the composite are obtained from experimental data on specimens [73]. The limiting fiber strength and the Young's moduli of the fibers and the matrix are treated as random variables. Other constituent properties of the composite are taken as deterministic (e.g., the volume fraction of the composite, creep exponent [$n = 3$], and creep constant of the matrix $B = 4.12 \times 10^{-31}$ Pa^{-3}sec^{-1}). The reliability indices at different times for composites with different fiber volume fraction and different shape parameter of the Weibull distribution of the fiber strength are calculated. See Mao and Mahadevan [73] for details of the numerical example.

First, the reliability indices at different time instants are calculated with FORM. The effect of broken fibers on the reliability is investigated. Figure 44.17 presents the reliability index β vs. time with and without considering the effect of broken fibers. From the results, it can be seen that the broken fibers have a large contribution to the reduction of reliability index. Also, the degradation in reliability over time is larger when the broken fibers are considered.

In Figure 44.17, the reliability index decreases slowly at first, followed by a rapid decrease in the second stage and a slow decrease in the third stage. At the initial time, all of the loads are carried by both fibers and matrix, and there are no broken fibers. Therefore the composite creeps slowly. After some fibers break, the load previously carried by broken fibers is shared by intact fibers and matrix, the composite creeps faster, and the stress in the fiber consequently increases. As more loads are sustained by fibers, the creep rate decreases. The reliability index still decreases, but more slowly. After a certain time, the

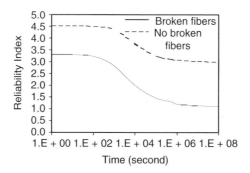

FIGURE 44.17 Creep reliability vs. time.

reliability index almost ceases to decrease. At this stage, most of the stress is carried by fibers. The stress in the matrix is so small that the creep strain rate is very low, and the stress increase in the fibers is also very slow and can be neglected.

Mao and Mahadevan [73] also observed that the applied load, fiber volume fraction, and the Weibull strength parameters m and σ_c in Equation 44.50 all had a strong effect on the reliability of the material. The larger the load is, the lower is the reliability of the material, and the earlier is the start of the rapid degradation stage. When the volume fraction is small ($f = 0.1$), the reliability index starts low and begins to decrease rapidly (which means the failure probability increases rapidly), and fracture will happen in a short time. When the fiber volume fraction is large, fracture may not happen for the same applied load.

It can be seen that the fiber characteristic strength has the most influence on the reliability of the material. When the reliability is estimated at different service times, the sensitivity indices show that the fiber characteristic strength has even more effect on the reliability with increasing service time, while the sensitivity indices for Young's moduli of fiber and matrix become much smaller. Therefore, for reliability estimation after a long lifetime, the Young's moduli of fiber and matrix can be treated as deterministic parameters.

The characteristic strength of the fiber follows the Weibull distribution that is described by two parameters: characteristic value and shape parameter. A lower value of the shape parameter m gives a higher failure probability under the same load. Also, the reliability index was observed to decrease rapidly slightly earlier for a lower value of m [73].

44.8 Conclusion

This chapter discussed reliability evaluation methods for composite structures under strength (matrix cracking and fiber breakage), fatigue, delamination, and creep limit states. In all cases, the essential steps of individual limit-state reliability evaluation are: structural analysis, failure definition, random variable data collection, uncertainty propagation through the structural and limit-state models, and reliability analysis. For multiple limit states, the overall failure probability can be computed through the union of individual failures. In the case of progressive damage, dominant failure sequences are identified, and the system failure is computed through the union of the failure sequences. Several approximations to reduce the computational effort, based on practical considerations, were also identified and illustrated.

Most of the discussion in this chapter was focused on laminated composites, partly due to the widespread use and availability of data for such materials. As composite material technology expands with increased use of other materials such as woven composites, particulate composites, etc., reliability evaluation methods will also follow. However, the essential steps are similar to those outlined in this chapter.

One important need that has not been addressed yet is the effect of interaction between different failure mechanisms on reliability. This first requires development of deterministic models. For example, creep-fatigue interaction modeling has been developed for metallic materials [74]. Similar models are yet to be explored for composite materials. The progressive failure analysis of Section 44.2 to Section 44.4 considers the effect of multiple limit states on system reliability, but all of them are related to strength, and it is therefore easy to compute the effect of one failure on another through structural reanalysis. The interaction of qualitatively different mechanisms such as creep, fatigue, etc. requires experimental and analytical study before reliability methods can be applied to such situations.

References

1. Shiao, M.C. and Chamis, C.C., Reliability of composite structures with multi-design criteria, in *Proceedings of 35th Structures, Structural Dynamics, and Materials Conference*, AIAA/ASME/ASCE/AHS/ASC, Hilton Head, SC, 1994, p. 606.
2. Murotsu, Y. and Miki, M., Reliability design of fiber reinforced composites, *Structural Safety*, 15 (1–2), 35, 1994.
3. Zhao, H. and Gao, Z., Reliability analysis of composite laminate by enumerating significant failure modes, *J. Reinforced Plast. Composites*, 14, 427, 1995.
4. Mahadevan, S., Liu, X., and Xiao, Q., A probabilistic progressive failure model of composite laminates, *J. Reinforced Plast. Composites,* 16 (11), 1020, 1997.
5. Tan, S.C., A progressive failure model for composite laminates containing openings, *J. Composite Mater.*, 25, 557, 1991.
6. Tsai, S.W. and Wu, E.M., A general theory of strength for anisotropic materials, *J. Composite Mater.*, Jan., 58, 1971.
7. Haldar, A. and Mahadevan, S., *Probability, Reliability, and Statistics in Engineering Design,* John Wiley & Sons, New York, 2000.
8. Rackwitz, R. and Fiessler, B., Structural reliability under combined random load sequences, *Comput. Struct.*, Vol. 9, No. 5, 484, 1978.
9. Murotsu, Y. et al., Reliability assessment of redundant structures, in *Proceedings of the 3rd International Conference on Structural Safety and Reliability, ICOSSAR* Norway, Trondheim, 1981, p. 315.
10. Liu, X. and Mahadevan, S., Ultimate strength failure probability estimation of composite structures, *J. Reinforced Plast. Composites*, 19 (5), 403, 2000.
11. Mahadevan, S. et al., Structural reanalysis for system reliability computation, in *Reliability Technology 1992, ASME Winter Annual Meeting*, AD–Vol. 28, ASME, Anaheim, CA, 1992, p. 169.
12. Cornell, C.A., Bounds on the reliability of structural systems, *ASCE J. Structural Division*, 93, 171, 1967.
13. Lee, J.D., Three dimensional finite element analysis of damage accumulation in composite laminate, *Comput. Struct.*, 15 (3), 335, 1982.
14. Reddy, Y.S. and Reddy, J.N., Three-dimensional finite element progressive failure analysis of composite laminates under axial extension, *J. Composites Technol. Res.*, 15 (2), 73, 1993.
15. Cruse, T.A., Review of Recent Probabilistic Design in Composite Structures, CAME, Consultants in Applied Mechanics and Engineering, Nashville, TN, 1994.
16. Vinson, J.R. and Sierakowski, R.L., *The Behavior of Structures Composed of Composite Materials*, Martinus Nijhoff Publishers, Leiden, The Netherlands, 1986.
17. Xiao, Q. and Mahadevan, S., Second-order upper bounds on probability of intersection of failure events, *J. Eng. Mech.*, 120 (3), 670, 1994.
18. Xiao, Q. and Mahadevan, S., Fast failure mode identification for ductile structural system reliability, *Structural Safety*, 13 (4), 207, 1994.
19. Mahadevan, S. and Liu, X., Probabilistic analysis of composite structure ultimate strength, *AIAA J.*, 40 (7), 1408, 2002.
20. Harris, B. et al., Constant-stress fatigue response and life-prediction for carbon-fiber composites, in *Progress in Durability Analysis of Composite Systems,* Cardon, A.H., Fukuda, H., and Reifsnider, K., Eds., A.A. Balkema Publisher, Rotterdam, 1995, p. 63.

21. Yang, J., Fatigue and residual strength degradation for graphite/epoxy composite under tension-compression cyclic loading, *J. Composite Mater.*, 12 (1), 19, 1978.
22. Nicholas, S., An approach to fatigue life modeling in titanium-matrix composites, *Mater. Sci. Eng. A*, 200, 29, 1995.
23. Cardon, A.H., Fukuda, H., and Reifsnider, K., Eds., *Progress in Durability Analysis of Composite Systems*, A.A. Balkema Publisher, Rotterdam, 1995.
24. Halford, G.R. et al., Proposed framework for thermomechanical fatigue life prediction of metal-matrix composites (MMCs), in *Thermomechanical Fatigue Behavior of Materials*, Sehitoglu, H., Ed., ASTM STP 1186, American Society for Testing and Materials, Philadelphia, 1993, p. 176.
25. Subramanian, S., Reifsnider, K.L., and Stinchcomb, W.W., A cumulative damage model to predict the fatigue life of composite laminates including the effect of a fiber-matrix interphase, *Int. J. Fatigue*, 17 (5), 343, 1995.
26. Reifsnider, K., Case, S., and Duthoit, J., The mechanics of composite strength evolution, *Composites Sci. Technol.*, 60, 2539, 2000.
27. Mao, H. and Mahadevan, S., Fatigue damage modeling of composite materials, *Composite Struct.*, 58, 405, 2002.
28. Halverson, H.G., Curtin, W.A., and Reifsnider, K.L., Fatigue life of individual composite specimens based on intrinsic fatigue behavior, *Int. J. Fatigue*, 19 (5), 269, 1997.
29. Chawla, K.K., *Composite Materials — Science and Engineering*, 2nd ed., Springer-Verlag, New York, 1998.
30. Kumar, R. and Talreja, R., Fatigue damage evolution in woven fabric composites, in *Proceedings of 41st Structures, Structural Dynamics, and Materials Conference and Exhibit*, AIAA/ASME/ASCE/AHS/ASC, Atlanta, GA, 2000.
31. Nicholas, S. and Russ, R., Elevated temperature fatigue behavior of SCS-6/Ti-24Al-11Nb, *Mater. Sci. Eng. A*, 153, 514, 1992.
32. Murri, G.B., O'Brien, T.K., and Rousseau, C.Q., Fatigue Life Prediction of Tapered Composite Laminates, presented at 53rd Annual Forum, American Helicopter Society, Virginia Beach, 1997.
33. Fish, J.C., and Lee, S.W., Tensile strength of tapered composite structures, in *Proceedings of 30th Structures, Structural Dynamics, and Materials Conference*, AIAA 88-2252, AIAA/ASME/ASCE/AHS, Williamsburg, VA, 1988, p. 324.
34. Hoa, S.V., Daoust, J., and Du, B.L., Interlaminar stresses in tapered laminates, *Polymer Composites*, 9 (5), 337, 1988.
35. Llanos, A.S., Lee, S.W., and Vizzini, A.J., Delamination prevention in tapered composite structures under uniaxial tensile loads, in *Proceedings of 31st Structures, Structural Dynamics, and Materials Conference*, AIAA 90-1063, AIAA/ASME/ASCE/AHS, Long Beach, CA, 1990, p. 1242.
36. Hoa, S.V. and Daoust, J., Parameters affecting interlaminar stresses in tapered laminates under static loading conditions, *Polymer Composites*, 10 (5), 374, 1989.
37. Murri, G.B., Salpekar, S.A., and O'Brien, T.K., Fatigue delamination onset prediction in unidirectional tapered laminates, in *Composite Materials: Fatigue and Fracture*, 3rd vol., ASTM STP 1110, O'Brien, T.K., Ed., ASTM, Philadelphia, 1991, p. 312.
38. Murri, G.B., O'Brien, T.K., and Salpekar, S.A., Tension fatigue of glass/epoxy and graphite/epoxy tapered laminates, *J. Am. Helicopter Soc.*, 38 (1), 29, 1993.
39. Winsom, M.R., Delamination in tapered unidirectional glass fibre-epoxy under static tension loading, in *Proceedings of 32nd Structures, Structural Dynamics, and Materials Conference*, AIAA 91-1142, AIAA/ASME/ASCE/AHS, Baltimore, MD, 1991, p. 1162.
40. Trethewey, B.R., Jr., Gillespie, J.W., Jr., and Wilkins, D.J., Interlaminar performance of tapered composite laminates, in *Proceedings of American Society for Composites*, 5th technical conference, American Society for Composites, East Lansing, MI, 1990, p. 361.
41. Armanios, E.A. and Parnas, L., Delamination analysis of tapered laminated composites under tensile loading, in *Composite Materials: Fatigue and Fracture*, ASTM STP 1110, Vol. 3, O'Brien, T.K., Ed., ASTM, Philadelphia, 1991, p. 340.

42. Salpekar, S.A., Raju, I.S., and O'Brien, T.K., Strain energy release rate analysis of delamination in a tapered laminate subjected to tension load, in *Proceedings of American Society for Composites*, 3rd technical conference, American Society for Composites, Seattle, WA, 1988, p. 642.
43. O'Brien, T.K. et al., Combined tension and bending testing of tapered composite laminates, *Appl. Composite Mater.*, 1 (6), 401, 1995.
44. Rybicki, E.F., Schmueser, D.W., and Fox, T., An energy release rate approach for stable crack growth in the free-edge delamination problem, *J. Composite Mater.*, 11, 470, 1977.
45. Russell, A.J. and Street, K.N., Delamination and Debonding of Materials, ASTM STP 876, ASTM, Philadelphia, 1985, p. 349.
46. Rybicki, E.F. and Kanninen, M.F., A finite element calculation of stress-intensity factors by a modified crack-closure integral, *Eng. Fracture Mech.*, 9, 931, 1977.
47. Raju, I.S., Calculation of strain energy release rates with higher order and singular finite elements, *Eng. Fracture Mech.*, 28, 251, 1987.
48. Mahadevan, S. et al., Reliability analysis of rotorcraft composite structures, *ASCE J. Aerospace Eng.*, 14 (4), 140, 2001.
49. Myers, R.H. and Montgomery, D.C., *Response Surface Methodology: Process and Product Optimization Using Designed Experiments*, John Wiley & Sons, New York, 1995.
50. Khuri, A. and Cornell, J.A., *Response Surfaces: Design and Analyses*, Marcel Dekker, New York, 1997.
51. Zhang, R. and Mahadevan, S., Bayesian methodology for reliability model acceptance, *Reliability Eng. System Safety*, 80, 95, 2003.
52. Mahadevan, S. and Zhang, R., Fatigue test planning using reliability and confidence simulation, *Int. J. Mater. Prod. Technol.*, 16 (4–5), 317, 2001.
53. Ohno, N. et al., Creep behavior of a unidirectional SCS-6/Ti-15-3 metal matrix composite at 450°C, *ASME J. Eng. Mater. Technol.*, 116, 208, 1994.
54. Dragone, T.L. and Nix, W.D., Geometric factors affecting the internal stress distribution and high temperature creep rate of discontinuous fiber reinforced metals, *Acta Metallurgica Materialia*, 38 (10), 1941, 1990.
55. Goto, S. and McLean, M., Role of interfaces in creep of fiber reinforced metal matrix composites, II: short fibers, *Acta Metallurgica Materialia*, 39 (2), 165, 1991.
56. Bao, G., Hutchinson, J.W., and McMeeking, R.M., Particle reinforcement of ductile matrices against plastic flow and creep, *Acta Metallurgica Materialia*, 39 (8), 1871, 1991.
57. McMeeking, R.M., Power law creep of a composite material containing discontinuous rigid aligned fibers, *Int. J. Solids Struct.*, 30, 1807, 1993.
58. Du, Z.Z. and McMeeking, R.M., Creep models for metal matrix composites with long brittle fibers, *J. Mech. Phys. Solids*, 43 (5), 701, 1995.
59. Aravas, N., Cheng, C., and Ponte Castaneda, P., Steady-state creep of fiber-reinforced composites: constitutive equations and computational issues, *Int. J. Solids Struct.*, 32 (15), 2219, 1995.
60. Cheng, C. and Aravas, N., Creep of metal-matrix composites with elastic fibers, Part I: continuous aligned fibers, *Int. J. Solids Struct.*, 34 (31–32), 4147, 1997.
61. Bullock, J., McLean, M., and Miles, D.E., Creep behavior of a Ni-Ni$_3$Al-Cr$_3$C$_2$ eutectic composite, *Acta Metallurgica*, 25, 333, 1977.
62. McLean, M., Creep deformation of metal matrix composites, *Composite Sci. Technol.*, 23, 37, 1985.
63. McLean, M., *Directionally Solidified Materials for High Temperature*, Materials Society, London, 1989.
64. Song, Y., Bao, G., and Hui, C.Y., On creep of unidirectional fiber composites with fiber damage, *Acta Metallurgica Materialia*, 43 (7), 2615, 1995.
65. Curtin, W.A., Theory of mechanical properties of ceramic matrix composites, *J. Am. Ceramic Soc.*, 74 (11), 2837, 1991.
66. Ibnabdeljalil, M. and Phoenix, S., Creep rupture of brittle matrix composites reinforced with time dependent fibers: scalings and Monte Carlo simulations, *J. Mech. Phys. Solids*, 43 (6), 897, 1995.
67. Iyengar, N. and Curtin, W.A., Time-dependent failure in fiber-reinforced composites by matrix and interface shear, *Acta Materialia*, 45 (8), 3419, 1997.

68. Binienda, W.K. and Robinson, D.N., Creep model for metallic composites based on matrix testing, *J. Eng. Mech.*, 117, 624, 1991.
69. Lee, S.C. et al., Modeling of transverse mechanical behavior of continuous fiber reinforced metal-matrix composites, *J. Composite Mater.*, 25, 536, 1991.
70. Chun, H.J. and Daniel, I.M., Creep characterization of unidirectional SiC/Al under transverse loading, in *Proceedings of SEM*, Society of Experimental Mechanics, Grand Rapids, MI, 1995, p. 279.
71. Chun, H.J. and Daniel, I.M., Transverse creep behavior of a unidirectional metal matrix composite, *Mech. Mater.*, 25, 37, 1997.
72. Weber, C.H., Du, Z.Z., and Zok, F.W., High temperature deformation and fracture of a fiber reinforced titanium matrix composite, *Acta Materialia*, 44 (2), 683, 1996.
73. Mao, H. and Mahadevan, S., Probabilistic analysis of creep of metal-matrix composites, *J. Reinforced Plast. Composites*, 21 (7), 587, 2001.
74. Mahadevan, S., Mao, H., and Ghiocel, D., Probabilistic simulation of engine blade creep fatigue life, in *Proceedings of 43rd Structures, Structural Dynamics, and Materials Conference*, AIAA 2002-1384, AIAA/ASME/ASCE/AHS/ASC, Denver, CO, 2002.

45
Risk Management of Composite Structures

Frank Abdi
Alpha STAR Corporation

Tina Castillo
Alpha STAR Corporation

Edward Shroyer
Boeing Integrated Defense Systems

45.1 Introduction 45-1
  Background • Problem • Methodology • New-Product Development • Manufacturing: Process Control and Inspection, Disposition of Anomalies • Mission: Manufacturing and Mission Failures • Benefits
45.2 Technical Approach 45-13
  Building-Block Verification Strategy • Life-Prediction Methodology
45.3 Conclusion 45-39

45.1 Introduction

45.1.1 Background

The two aspects of risk assessment for an event are likelihood of occurrence of the event and the consequence of the event given its occurrence. A typical risk matrix where several events would usually be plotted is shown in Figure 45.1.

Composite structures are often used in situations where they are subjected to multiple environments and combined loads. They can fail in a variety of modes, some of which have dire consequences. Knowledge of the risk-likelihood/consequence space is required for proper risk management. For example, space vehicles face combined environment loading that includes dynamic/acoustic/static thermal-mechanical loads applied in a wide range of random atmospheric (e.g., hygral) conditions that gradate to the vacuum of space during a mission. These complex conditions and the consequences of possible system failure make risk management to achieve mission-readiness quality both costly and time consuming. This complexity is present to varying degrees in most military and commercial products employing advanced composite designs. Much of the cost and time of assessing/managing risk is due to physical testing. Physical testing is usually done in the design stage to obtain an adequate design and again during production to ensure that the design criteria have been met. Tests that give useful failure information (destructive tests) are usually limited to small portions of a system and a few failure modes. Acceptance-type tests that do not involve failure give minimal information about either the likelihood or consequence of failure for multiple complex loadings and failure mechanisms. The penalty is paid in cost, weight, and quality trade-offs—limited risk management.

45.1.2 Problem

The present method of structural design is based on tailoring structural components and joints to provide the required strength and stiffness response based on finite element stress analyses. Safety factors are

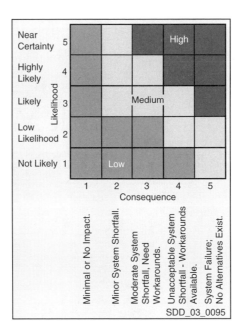

FIGURE 45.1 Typical risk matrix.

used locally to provide sufficient strength for each component to remain safe under occasional overloading. The factors of safety for each failure mode are usually rules of thumb based on limited experience and test data. The focus is often on local rather than on system considerations. For example, during the initial design phase, the effects of possible local failures due to overloads and discrete source damage are not considered with regard to global structural response and damage tolerance.

Addressing combined environments, such as those described for spacecraft, usually involves modifying (increasing) the factors of safety. There is generally little data to define how much to increase the factor of safety. The risk strategy/culture of the enterprise or regulatory agency may drive the decision, or "similar" experience with metals may be used. This drives design risks such as cost and weight that can result in a product that cannot satisfy the customer or market requirements (market failure). Another aspect of the risk is designing products that perform satisfactorily for a required period of time. This involves not only the design, but the fabrication and maintenance of the product. Managing these risks involves trading weight, cost, envelope, and performance and life criteria to achieve a safe and marketable product. Faced with a complex spectrum of loading conditions over the life of the product and a wide variety of design options for which little information is available, the task of assessing/managing risk becomes extremely difficult. This is true for metal structures having nontraditional designs and subjected to severe combined environments with large uncertainties thrown in for good measure. For composite structures, the problem is even worse, since there are often no similar experiences to draw on. Ideal solutions would use analysis methods that gave the exact likelihood and consequence of the risk item or physical testing of the complete system in the service environment for the expected life a sufficient number of times to determine the uncertainty. Even this would not eliminate the risk of the occurrence of an event with an extremely small probability. Even if such ideal systems were available, the cost and time required for analysis and testing would be prohibitive. Thus, the best we can do is a structured approach using a combination of analysis and physical tests to determine the likelihood and consequences of the various risks and then take steps to manage them by various risk-mitigation strategies during design, production, and service life.

Since aerospace missions provide a wide range of dynamic/acoustic/static hygral loads in diverse environments, uncertainties in these loads combined with uncertainties in geometry, material, responses, and failure mechanisms have a profound effect on risk. These uncertainties must be quantified and

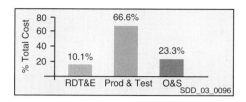

FIGURE 45.2 Typical space system: percent production and test cost of spacecraft. (From Systems based on Aerospace Corp. report April 1997, Costs of Space Launch and Ground Systems. With permission. [1])

reduced to achieve accurate safety design margins. Costly and time-consuming experimental testing of structure and engine systems provides insufficient data for meaningful analyses of useful remaining life.

Experimental testing to evaluate the mission-readiness quality of aerospace systems and their components increasingly involves prohibitive costs and times. As shown in Figure 45.2, for space applications the typical cost driver is production and testing, which represent 66.6% of the total cost (75% of that is production testing cost).

Experimental testing is currently the preferred risk-mitigation tool. However, risk assessment by probabilistic and progressive-failure analysis has the potential to significantly reduce the cost and uncertainties in development of next-generation aerospace systems. To achieve this, virtual-testing software that will enable the detection of prematurely failing components and lead to the extension of safe life limits for flight-critical hardware is required. Models of useful remaining life are needed to provide useful information to the crew and maintenance support staff. Such models are also required to significantly reduce the time of component testing and increase the reliability of mission-readiness evaluations. A typical risk management flow diagram is shown in Figure 45.3.

The life of a typical product is shown in Figure 45.4. The design of the product is such that the useful life, i.e., the constant failure rate due to random causes usually associated with uncertainty in the geometry, the material, and the environment, is below a specified level. The onset of fatigue and the number of fatigue failures at various points in the expected life (i.e., 20% failed after two lifetimes) is dealt with in the design for fatigue life. The initial portion of the curve, manufacturing defects, is dealt with partially by design for manufacturing and assembly (DFMA) and partially through an inspection/test program during assembly and prior to mission deployment.

Many aircraft and spacecraft have advanced designs based on a well-matured and systematic testing technology. However, the manufacture and service-life risks still remain. In production, testing to reduce risk due to manufacturing defects is performed at many steps up to and including the entire assembly/integration system. Even with all this testing, costly failures still occur, and these are increasingly prone to public scrutiny. Table 45.1 shows the possible mission outcomes given a test prior to service. These

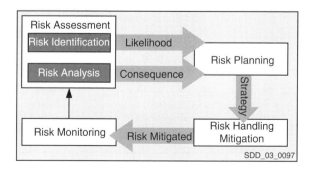

FIGURE 45.3 Vehicle application cost driver will be reduced by use of virtual testing for risk management. Current spacecraft testing is expensive, and bad outcomes can occur.

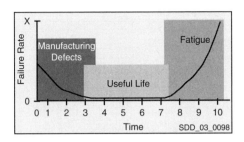

FIGURE 45.4 Typical failure rate over product life.

tests could be at the part, subcomponent, component, subassembly, assembly, and product levels. As the system moves through production, the cost of executing the mitigation strategy increases due to the increased time and labor required for disassembly and reassembly. Composites are particularly vulnerable to increased risk from the execution of this mitigation strategy from both the additional testing that the adjacent structure will be subjected to and the risk of damage due to increased handling. Sometimes, the additional damage done by repeated ground tests is not considered in the overall remaining-life assessment of the system.

Possible causes of bad outcomes are:

- Wrong test performed: leads to outcomes 3 and 4
- Tested to the wrong level: leads to outcomes 3 and 4
- Tested the wrong item: leads to outcomes 3 and 4
- Wrong amount of testing done: leads to outcomes 3 and 4; adds cost to outcomes 1 and 2

In the early days of powered flight and rocketry, failures were common and successes rare. Even into the 1950s, launch successes were celebrated and failures accepted by the public as part of the space program. Today, success is the expected norm, and the occasional failure comes under public scrutiny due to both the cost and expectation. Costly failures still occur, even though extensive testing is done at each stage of manufacture, assembly, and integration. From 2000 through 2002, there were approximately 210 launches. Nine of these ended in mission failure, with a substantial number of delays due to component failures on the ground that had to be repaired prior to launch [2]. A list of the mission failures is shown in Table 45.2.

As is typical for launch failures, structural evidence is usually destroyed, while the software/systems failures are documented due to onboard instrumentation/feedback. Similarly, many of the anomalies discovered during ground testing are due to software and systems problems (some manifested in structural malfunction). The implication for composite structures is again the retests of surrounding structure (often without additional subsurface inspection) that consumes additional portions of useful life and introduces the risk of damage initiated by additional handling during removal and reinstallation.

TABLE 45.1 Possible Outcomes of the Item Tests

Outcome	Test Results	Mission Outcome	Consequence/Mitigation
1	passes	will pass	success
2	fails	would have failed	replace item and retest
3	fails	would have passed	replace item and retest
4	passes	will fail	mission failure

Note: Outcome 1 is good (but fairly costly); Outcome 2 is good but costly; Outcome 3 is bad and is very costly; Outcomes 2 or 3 still leave any of the four outcomes possible, with the new item and other items to which the replaced item is attached being subjected to multiple tests that may increase the subsequent likelihood of Outcome 3; Outcome 4 is the most costly outcome.

TABLE 45.2 Summary of Launch Failures, 2000–2002

Mission	Date	Problem (Failure)
Astro-E	2/10/2000	M-5 rocket veered off course shortly after launch ($105 million payload loss)
ICO Teledesic 1	3/11/2000	Software glitch led to a valve failure on Zenit 3SL; mission abort at 467 sec ($108 million payload loss)
Strela 3S	12/28/2000	Tsyklon 3 third-stage engine failure; payload failed to separate
X-34A	6/2/2001	Pegasus air-launched booster veered off course; a fin broke off, causing loss of control and destruction of booster
Artemis BSAT-2b	7/12/2001	Ariane 5 had fuel-mix malfunction, resulting in premature engine shutdown and low orbit insertion
OrbView 4, QuickTOMS	9/21/2001	Drive shaft on the Orion 50S solid rocket malfunctioned, causing Tarus rocket to veer off course; payloads did not achieve stable or proper orbit
Photon-M1 capsule	10/15/2002	Component failure caused engine shutdown; Soyuz U fell back to launch pad and exploded
Astra 1K	11/25/2002	Payload did not achieve proper orbit when the Proton K upper stage separated prematurely
Hot Bird 7, STENTOR	12/11/2002	Ariane 5 destroyed at 456 sec due to erratic flight; pressure drop in Vulcain 2 main-engine faring caused jettison at incorrect altitude

Before incorporation into a vehicle system, a component design must have been individually and globally validated for durability and reliability for the intended mission. This is very critical for the success of a mission, since the malfunctioning of even a simple component can cause total mission failure. This can occur, for example, because of a stuck fuel valve, or a leak due to a defective gasket seal, or a fatigue crack in a fuel line. As the number of critical components increases, the durability and reliability requirements of these components increases significantly. However, validation of component and global design iterations by experimental testing to determine durability and reliability is so costly and time consuming that design optimization suffers.

45.1.3 Methodology

The general methodology for risk management was shown in Figure 45.1 and Figure 45.3. The analysis/test aspects of risk management were stated, and the problem of complexity and high cost of analysis and testing was stated. The reality is that the physical testing required to introduce a new material system into an existing product is so costly that sometimes the benefits of the system cannot be realized due to the difficulty in making a business case for the test expenditure. The high-cost penalty incurred in production testing was also discussed. The cost of mission failure was shown for spacecraft launch failures. As stated in the discussion of Table 45.1, the most costly outcome was the deployment of a product that failed. The loss of a satellite or an aircraft is expensive not only from a direct cost point of view, but also from the cost in terms of human life, insurance, and business loss. If there are several identical or similar products in service, the observation of one failure brings about the need for a risk-mitigation strategy to prevent fleet failure. The risk strategy requires a definition of whether the observed failure is an isolated event or is representative of a systemic failure. The determination of likelihood and consequence as well as an assessment of the effectiveness of any proposed risk-mitigation strategy requires analysis and testing.

A major advance in efficient risk assessment/management is the use of virtual testing methodology that performs progressive-failure and time-dependent uncertainty analyses to identify potential mission-critical failures, failure modes, and their influences on mission success. An example of the use of virtual-testing/progressive-failure analysis software in the risk-mitigation process for design, manufacture, and mission aspects of new material system/design implementations is shown in Figure 45.5. The virtual-testing/GENOA (general optimizer analysis) output portions can be replaced by physical testing and traditional analyses, but only at a high cost in both money and insertion time and with a substantial loss of information. (GENOA is developed, marketed, and supported by AlphaStar Corporation, 5199 E. Pacific Coast Highway, Suite 410, Long Beach, CA 90804; http://WWW.ALPHASTARCORP-GENOA.COM/.) Using the

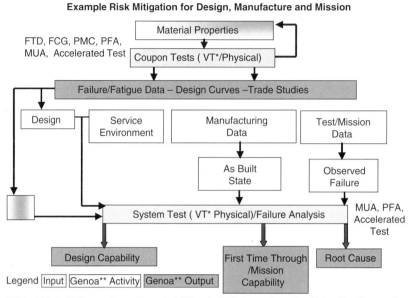

FIGURE 45.5 Example of virtual testing/analysis software for risk mitigation.

methodology described in Figure 45.5, critical failures can be defined, tracked, and targeted for elimination (risk mitigation).

The virtual-testing (VT) approach is enabled by a computer code that allows the interrogation of each local design option with regard to the overall product performance, structural integrity, durability, and reliability. Most companies that are involved with product design and manufacture using composites either have an in-house code or use one of the commercially available codes to aid in the design activity. When properly configured, most of these codes or some combination of codes can be used to increase the efficiency of risk reduction or management for the general cases previously described. Examples given in this section are based on the use of the general optimizer analysis (GENOA) computer code that is referenced in Figure 45.5. The virtual-test/analysis software must be able to evaluate the load levels and combinations that would initiate damage. Additionally, it should have the ability to simulate (1) the microscopic damage initiation, (2) the stable crack growth per applicable requirements (Air Force 0.05-in. crack and Navy 0.01-in. crack initiation), and (3) progression of damage until the loss of air-vehicle structural integrity occurs. The virtual-test/analysis code should be able to simulate structural damage processes under combined loads that are applicable to the product. For the aerospace industry, this would include static, dynamic, impact, creep, thermal, low- and high-cycle fatigue, and random power spectral-density loads. The effects of hygral and thermal environments might also be considered in the simulation of damage evolution. The virtual-test process generally uses material test data, and material uncertainty at the local level, to simulate structure-level damage initiation, growth, and propagation processes and to determine the true global safety factors associated with each component design. Changes in structural load paths with damage progression must be accurately identified based on physics and material properties. Additionally, manufacturing-process variability and anomalies (tolerance voids, defects, etc.) can be accounted for as part of life-prediction simulation. Virtual structural-health monitoring and nondestructive evaluation (i.e., inspection) can be monitored by simulation of exhausted-energy release rate.

The example code integrates (1) finite element structural analysis with nonlinear stress–strain definitions and plasticity, (2) composite-micromechanics or fracture-mechanics options, (3) damage-progression tracking, and (4) probabilistic risk assessment and material characterization to scale up the effects of local damage mechanisms to the structure level to evaluate overall performance and integrity. Most codes have finite element structural-analysis modules that perform automated remeshing to refine critical points and track the initiation and crack growth of any size. There are significant advantages to the use of virtual testing/analysis in the design process. One is that the number of actual physical tests at the component and substructure levels can be substantially reduced and made more efficient and effective. Another is that tests can be run to failure in the computer to give significantly more information than nondestructive tests. Physical testing of major components to failure is usually prohibitive. Another advantage to using virtual-testing software, even when physical testing is being done, is the ability to obtain information that allows the proper placement of instrumentation and the proper monitoring of data so that the desired information is obtained from the physical test. Often an expensive physical test is conducted, and a critical observation is missed because of a faulty or misplaced transducer.

The use of virtual-testing/analysis software in design, manufacturing, or mission-anomaly disposition requires risk management on the virtual-test methodology. In general, some combination of traditional analysis/physical testing will be used with the virtual-test methodology. The mix of this combination will depend on the risk acceptance/aversion of the stakeholders. To take maximum advantage of the cost, time, and information benefits of virtual testing, perceived risk must be lowered. The risk-mitigation strategy for using a virtual-test software is to validate the results of the virtual test by physical testing to the degree necessary to satisfy the stakeholders that the position of virtual testing on the risk-likelihood/consequence matrix is sufficiently low. It is also possible to design a specific experimental program for risk reduction using a specific combination of physical and virtual testing to reduce cost and time by an amount commensurate with a specific risk-acceptance/aversion attitude. This concept will be illustrated later in the example of a square experimental design.

As stated earlier, risk management is required at all stages of the product life cycle from concept to disposal. Composite structures must be addressed not only on a component basis but also as part of the overall system. For example, the disposal of a product when it has fulfilled its intended purpose may be governed by regulations. The use of some composites could result in the product being treated as hazardous waste (increasing disposal cost), or the substitution of a an environmentally friendly composite for a metal component that requires a hazardous protective coating could reduce disposal cost. Listed below are some areas of the product life cycle that require special attention and that can benefit from a judicious combination of actual and virtual testing for risk management.

45.1.4 New-Product Development

New-product design and development is based on a combination of analysis and tests. The analysis and testing activities range from concept feasibility determination to qualification of the final design. Usually the designers do not start with a blank sheet of paper. Substantial reliance on past experience, similar products or components, and rules of thumb are used to mitigate the need for a complete set of analyses and tests from scratch. Even with the past experience, either developing a new product or doing a major modification of a current design is time consuming and expensive. Often the market belongs to those who can come up with the best designs, at the lowest cost, in the fastest time. Designs (either new or modified) must meet a variety of requirements. These include regulatory agency qualification requirements (prove to the agency that you meet the requirements by test/analysis), customer requirements (prove to the customer that you meet or exceed the customer requirements such as acquisition cost, life-cycle cost, performance, having a specific feature), and market requirements (prove to yourself that you meet cost, performance, reliability, and durability levels). During the design process, the environment (loads) that the product will see must be specified for the various missions that it will be required to perform. The number and duration of these missions should also be known. Under these conditions, the designer then comes up with a product that meets the weight, cost, performance, durability, reliability, etc., requirements. As stated before, the environments are usually complex combinations with a fair amount of uncertainty in magnitude, location, and duration. The uncertainty in the load/environment is a major factor in risk management. The design also has uncertainties in the geometry and material

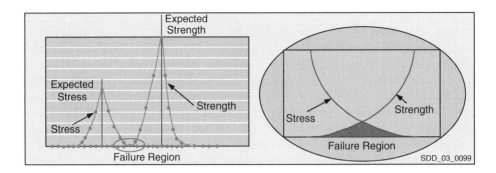

FIGURE 45.6 Failure region for a single-load case design.

properties. Uncertainty in the load/environment also impacts the uncertainty in geometry and material properties for various mission segments. The design/load space then involves the expected values and variance of the load (stress) and the design (strength) and the probability of failure when the stress exceeds the strength. This is illustrated in Figure 45.6.

Figure 45.6 illustrates a single-load case, where the strength is calculated for a single failure mode. The shaded area is the area of risk that must be addressed. For this case, the risk can be reduced by increasing the expected value of the strength (usually add weight) and by reducing the uncertainty. When several load environments are involved and, as is common with composites, competing failure modes are involved, the complexity of the problem increases dramatically. This illustrates the need for a probabilistic analysis code and testing throughout the range of possible values. The methodology that is proposed, using a combination of virtual testing and physical testing, allows a more thorough evaluation of the design space. This results in a design that is tailored to the desired risk rather than to a worst-case scenario. In general, the prudent method is to use the virtual testing to interpolate rather than to extrapolate physical test data. This conservatism, however, is very expensive, since the maximum points usually imply a test either to failure or very near to failure. If testing to failure is not feasible, it is recommended that virtual testing be used to define the level and mode at which failure occurs, and then a physical test can be designed to verify this prediction either by testing close to failure or by calibration with a "dummy" test specimen.

The above discussion deals with multiple loads and their levels. The reliability/durability aspects of risk require testing at mission loads for mission time periods, or else a test that simulates this. These tests are not only expensive, but the time involved can run into years. The virtual-test software can be used to help design an accelerated test. Key elements of such a test are:

- Development of a model and analysis that predicts observed events
- Ability to take the part to failure using the mission profile's failure modes
- Ability to determine sensitivity of damage and total damage-energy release rate of the various test forcing functions
- Development of equivalent test forcing functions, i.e., those that produce the same failure mode and damage
- Development of a cost-effective accelerated test

A key issue in this test design, especially for composites, is the understanding of the failure modes under different combined loading conditions. The virtual-test code must be able to track the various failure modes over time, determine the contribution of each, and define which one produces the final failure. For example, the matrix may craze during the test, which can allow the stress in the fibers to redistribute. The matrix crazing and fiber failure are competing, but not independent, modes. An example of an accelerated test that was designed to produce the same failure mode that the product would exhibit in the full-length test is shown in Figure 45.7.

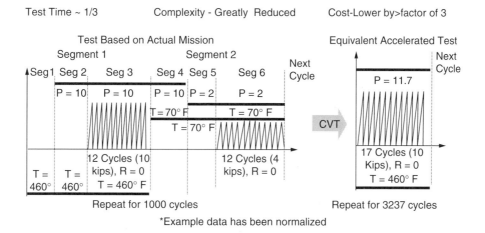

FIGURE 45.7 Example of accelerated-system test development.

As part of the design, a new material system may be indicated. Similar to the overall product, new material systems need to be qualified and their properties documented prior to use in a design. This is often cost prohibitive due to the high cost of the physical tests that are involved. The tests usually involve an experimental design that allows the evaluation of some material parameter with the variation of one or more parameters. Virtual testing can be used to effectively reduce the cost of such qualifications. For composites, some cells in the design can be filled with data that are already available. However, the failure-mode issues described above need to be considered in the design. In general, a given layout would involve only one failure mode. As mentioned before, the combination of physical and virtual testing can be tailored to the risk-acceptance/aversion attitude of the user. The cost savings and increasing risk is shown in Figure 45.8 for a typical square experimental design.

The new-product design/qualification risk-management methodologies described above address the useful life and the fatigue region of the product-life curve (shown in Figure 45.4) by providing adequate strength and adequate tolerance specification. This addresses the risks that are illustrated for a simple case in Figure 45.6. The failure rate is held to a low enough level; the onset of fatigue is at an adequately long time; and the rate of fatigue failure is such that a sufficient number of products survive for a specified time period. The design does not address the "infant mortality or manufacturing defect" region. The product is designed such that every unit will work (except for the observation of a failure at the useful life rate) at the start of service. The risk that must now be addressed is the risk introduced by manufacturing the product.

45.1.5 Manufacturing: Process Control and Inspection, Disposition of Anomalies

The manufacturing process introduces risk, given a good design for cost and performance, due to anomalies that occur due to poor process capability, out-of-control processes, or mistakes. The life curve of Figure 45.4 is repeated in Figure 45.9, along with the tests that are used to mitigate risk in each area. Manufacturing factors can influence all three regions.

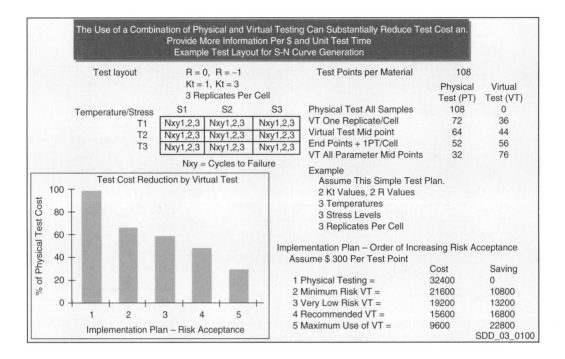

FIGURE 45.8 Reduction of qualification test cost vs. risk acceptance.

The importance of design for manufacturability will be apparent with the application of the virtual-test software using possible manufacturing variances. For purposes of this discussion, it is assumed that the design allows the use of processes that are capable of producing the product to the desired tolerance and that its assembly can be mistake-proofed, in the sense of lean manufacturing. The problem now becomes one of manufacturing process control and inspection. Such process-based manufacturing requires monitoring of the process parameters rather than 100% test for manufacturing defects.

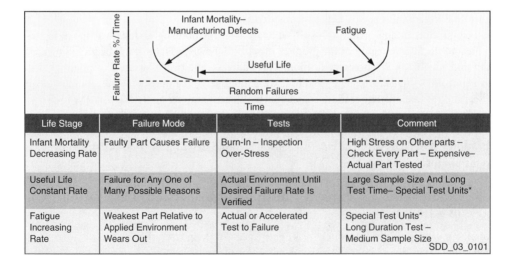

FIGURE 45.9 Product failure minimization and testing.

Determining the parameters to monitor and the acceptable test limits requires a long and expensive test program for most processes. Virtual testing reduces both the cost and the time required for the test program while actually allowing more information to be collected and incorporated into process control than with physical tests. For a new or existing process to establish process control parameters that will identify risk likelihood and consequence for various levels of compliance, the following items should be included:

Determine variations in geometry, material properties, and anomalies (voids, thin spots, solder splash, etc.)
Apply physical test loads and accumulate damage for the number of cycles or length of time required
Record failure/nonfailure, with the type and extent of damage known
Take a specimen to failure to determine when or at what level it would have failed and what the failure mode would have been
Provide Monte Carlo simulations to give useful life (random failure rate)
Determine onset of fatigue life and failure rates due to fatigue

If the stakeholders believe that there is an existing inspection program that gives adequate risk mitigation, then the process parameters need only be related to the inspection test level via a combination of physical and virtual testing, as shown in Figure 45.8. The link to system risk is implied by acceptance of the inspection criteria.

When the process parameters are not within the control limits or when inspection of parts results in one or more parts that do not meet the inspection criteria, a disposition of the part with the anomaly must be made. A similar disposition would be required if the part were damaged during manufacture.

When an anomaly is encountered, engineering and management usually need to provide a disposition. Often the disposition is to repair the anomaly or replace the defective unit. This is followed by testing of the new or repaired unit, usually with the surrounding structure. This results in double testing (or more) of the structure/units adjacent to the replaced or repaired unit. Since the inspection criteria are often based on limited experience, rules of thumb, or the level of anomaly that can be detected at a given point in time, the product that did not pass inspection may perform adequately for the desired product life. This situation corresponds to item 3 in Table 45.1. Physically testing the unit to failure is not an option unless one is faced with a large number of rejected parts and the test is on a subset of these parts. Even in this situation, virtual testing can provide answers to the pertinent questions for risk management, such as:

- Is the unit with the anomaly likely to fail in service?
- Is the repair adequate or more than adequate for full life?
- How much damage has been done to adjacent structure or units?
- Will additional testing reduce the useful life of the adjacent structure or units below acceptable levels?

The above discussion has focused on manufacturing anomalies that, although they impact useful life and fatigue life, usually show up as manufacturing defects. The mitigation strategy is to eliminate the manufacturing defect portion of the curve by removing or repairing "defective" products before they are placed in service. This is a major issue with products using composites. The question of hidden damage incurred in fabrication and unseen handling during product manufacture and assembly has long been an issue with many customers. As shown above, virtual-test software can address the impact of actual anomalies throughout the product life. Use of the risk-mitigation methodology shown in Figure 45.5 with a validated virtual-testing tool not only allows rational decision making on risk issues based on the physics of the problem, but may also decrease some of the risk aversion of the user community.

45.1.6 Mission: Manufacturing and Mission Failures

Figure 45.5 and Figure 45.9 show that even with all the testing during design and manufacturing, failures do occur in service. When products are in service and a failure occurs either in service or during checkout prior to implementation in service, the risk that must be resolved concerns the nature of the observation. The likelihood of faulure, the consequence, and the mitigation strategy are different for a systemic failure as opposed to a random occurrence. For the case of the checkout failure, the questions are: "What is the likelihood of failure for the units in service and for other similar units that are in production?" and "What is the expected time to failure?" Similarly, for in-service failures, the question is, "What is the likelihood of other units currently in service failing and when?" If the failure is systemic, then the root cause of the failure must be determined and corrected. If identical products that have not been in service are available, they are usually tested to the level experienced by the observed unit (if known). If there are identical products in service, they may be inspected to determine the root cause of the failure. In any case, the perceived likelihood of occurrence is usually increased by the observation of a failure until it can be shown otherwise.

Risk management requires the determination of what portion of the life curve (Figure 45.9) the failure came from. The test implications are shown in the figure. Again, virtual testing can be a great aid. Even though virtual testing will not give a definitive answer to the root cause question unless all of the pertinent parameters are known, it allows for the cost-effective examination of product life under a wide variety of conditions. Through probabilistic analysis, one can determine whether each of the conditions that produces the virtual failure is likely to occur with a given probability. This allows the risk analyst to focus on failures that might occur at the three- or four-sigma level rather than those that occur at the six-sigma level (one sided). Given the complexity of the design variables involved with composite design (fiber, matrix, orientation, number of plys, material compatibility, etc.) and the number of possible failure mechanisms, it is clear that managing risk for an observed failure for composites using only physical testing and traditional analysis will require a good deal of experience and luck. The combined virtual-test methodology described in Figure 45.5 will not eliminate the problem, but it will make it more manageable.

Another important aspect of risk management that is particularly well suited to composites is health monitoring of a product. A transducer is either mounted in the structure or the structure is used as a transducer (fibers used to measure strain, etc.) to determine the response to load or the environment of the structure. The information is then transmitted to the user, who determines the "health" of the product. Some health-monitoring devices simply transmit the occurrence of a failure or other related event. The information provided by the transducers defines the position of the structure in the risk matrix along the likelihood axes. The consequences of the event given the occurrence have already been determined. Again, virtual test software can aid in several ways. Virtual testing can help define the information that would be most meaningful and the best locations for the transducers to obtain that information. Once the information is obtained, virtual testing can help refine the location of the risk on the likelihood axis given the actual service history to date. If the risk is the likelihood/consequence portion of the curve that requires a risk-mitigation strategy, virtual testing can help in the development of a strategy and in determining the effectiveness of the particular strategy.

45.1.7 Benefits

The methodology described above and illustrated in Figure 45.5 has included a qualitative description of the benefit of using a properly selected combination of physical and virtual testing to mange risk in composite applications. Much of the discussion is also applicable to metals, electronic systems, etc. Virtual testing has been validated by comparing failure predictions with known failures in critical aircraft, missile, satellite, and launch-vehicle components. The example virtual-test software validation was validated by simulating structural damage-evolution details prior to the testing for over 30 aerospace composite structures. The virtual-testing software is estimated to significantly contribute toward risk-reduction goals and will reduce product-development and process costs (30 to 50%), time to market (10 to 25%), and product quality testing (30%). Examples showing reduction in test time and cost by accelerated testing and qualification by reduction of physical test costs were shown in Figure 45.7 and Figure 45.8.

While the monetary benefits of the combined virtual and physical testing for composites is very attractive, perhaps the major benefit is the understanding gained by this type of risk management. The relationship between the physics of the problem and risk are more easily understood. "What if" questions can be answered with minimal cost and time. New-material systems can be investigated and their capability examined in a timely and cost-effective manner. The practice of designing with well-known traditional materials and ply-layup arrangements to minimize risk can be replaced by cost-effective designs with well-defined risk analyses and mitigation plans. Lastly, but perhaps most important, is the ability of virtual-test software with visualization capability to provide a well-communicated, shared understanding of the design capability and the risk analysis/mitigation strategy to stakeholders who are not skilled in current composite technology theory. The benefit of having informed decision makers rather than those that react to old or incorrect information can reduce the risk aversion to composite designs and so allow better designs to be developed and validated.

The next section will detail the underlying technical approaches that make the combined physical–virtual test risk-management methodology viable for composite structures.

45.2 Technical Approach

The objective of the design process is to demonstrate the composite material and structure risk assessment by life-prediction simulation, which evaluates progressive fracture of composite structures subjected to static loading and fatigue cyclic loading (including high-cycle fatigue under dynamic loading). It is best to perform the material risk evaluation up front in the design process before committing resources of time and money. This synergistic approach of materials, design, and analysis leads to quantification of reliability and can be performed at various steps of the design process. A sample of a qualification process for a typical material candidate is shown in Figure 45.10, with the highlighted areas indicating where a life-prediction simulation can be used.

45.2.1 Building-Block Verification Strategy

The building-block strategy encompasses a step-by-step verification approach to assure that the micro- and macromechanical behavior is accurately simulated. The approach calibrates the fiber matrix properties using an integrated composite analysis at the material-constituent level modeled on the American Society of Testing Materials (ASTM) standard tests. Once calibrated, the material constituents are then used in subelemental, component, and full-scale or assembly models for further verification. Figure 45.11 shows an example of the building-block strategy envisioned by the FAA for a typical wing assembly. The incorporation of this philosophy in GENOA is the cornerstone for achieving accuracy in simulations (Figure 45.12). Figure 45.13 shows aspects that GENOA code can address at each simulation level.

45.2.2 Life-Prediction Methodology

Commercial finite-element-based models such as GENOA (generalized optimizer and analyzer), ABAQUS (Figure 45.13), NASTRAN, and ANSYS have been enhanced to perform composite-structure life assessments. These codes are able to accurately simulate structural-damage process under static, dynamic, impact, creep, stability, thermal, low- and high-cycle fatigue, and random power spectral-density loads [3]. The effects of hygral and thermal environments are also considered in the simulation of damage evolution. The GENOA software virtual testing (VT) process capability uses material test data and uncertainty at the micro level to simulate structure-level damage initiation, growth, and propagation processes, and the software determines the true global safety factors associated with each component design. Changes in structural load paths with damage progression are accurately identified based on physics and material properties. Additionally, manufacturing process variabilities, such as anomalies (voids, defects, etc.) can be accounted for in the life-prediction simulation. Virtual structural-health monitoring and nondestructive evaluation (i.e., inspection) are monitored via exhausted-energy release rate.

The GENOA software answers critical structural design questions:

- Where is failure occurring?
- When will failure occur?

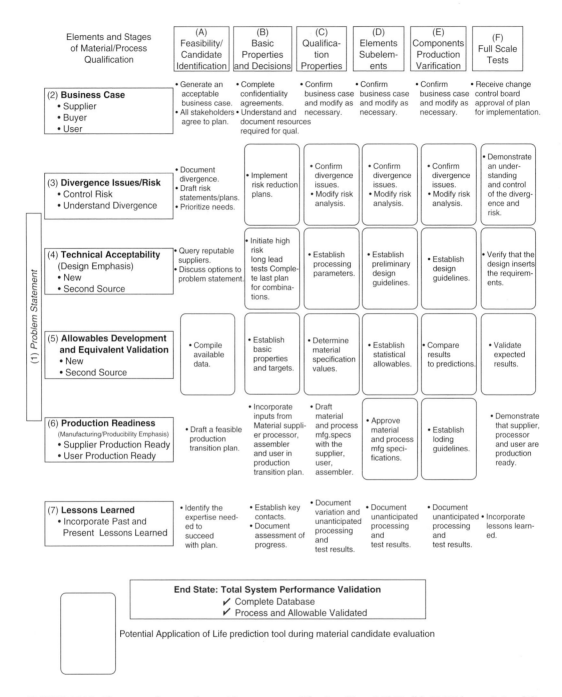

FIGURE 45.10 Elements and stages of material or process qualification. (From Mil-Hndbk-17. With permission. [4])

- Why is failure occurring?
- What can be done to fix the failure?
- Where and how can money be saved?

The GENOA code (Figure 45.13) integrates: (1) finite element structural analysis with nonlinear stress–strain and plasticity considerations, (2) composite micromechanics and fracture mechanics options, (3) damage-progression tracking, (4) probabilistic risk assessment, (5) minimum-damage-design

Risk Management of Composite Structures

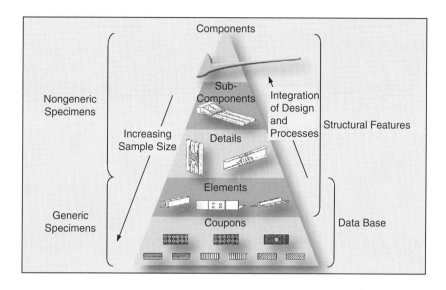

FIGURE 45.11 Schematic diagram of building-block tests. (Courtesy of FAA. With permission. [5–7])

optimization, and (6) material characterization codes to scale up the effects of local damage mechanisms to the structure level to evaluate overall performance and integrity. The GENOA finite element structural-analysis module performs automated remeshing to zoom in on critical points and track the initiation and growth of cracks of any size. Cracks as small as 0.01 in. long in composite aircraft structures have been successfully detected and tracked by GENOA in quantifying structural-damage tolerance. A significant advantage of using VT in the design process is that the number of experimental tests at the

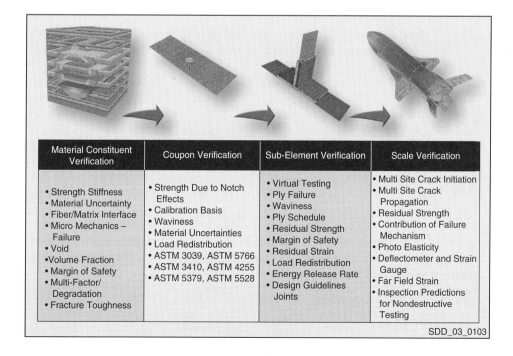

FIGURE 45.12 Building-block verification strategy. (From FAR-025. With permission. [8, 9])

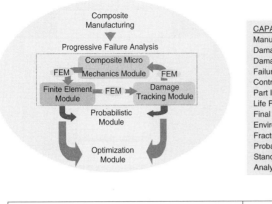

FIGURE 45.13 Functionality of GENOA software.

component and substructure levels can be substantially reduced, and the experimental testing that is done is made more efficient and effective.

The GENOA software is an integrated composite design tool that automatically iterates and performs: (1) material-constituent analysis (MCA) [10], (2) material-uncertainty analysis (MUA), (3) progressive-failure analysis (PFA), (4) probabilistic progressive-failure analysis (PPFA), and (5) design optimization to achieve damage minimization under hygral conditions.

The benefits of the GENOA software modules are as follows:

Material-constituent analysis (MCA) predicts the composite lamina and laminate properties under manufacturing and environmental conditions (e.g., humid environment) [10].

Material-uncertainty analysis (MUA) predicts the percent contribution of composite mechanical properties at the constituent, lamina, and laminate levels to potential failure mechanisms such as delamination.

Progressive-failure analysis (PFA) has the ability to simulate (1) microscopic damage initiation, (2) U.S. Navy requirement of 0.01-in. crack initiation, (3) U.S. Air Force requirement of stable crack growth of 0.05-in. crack, and (4) progression of damage until the loss of structural integrity occurs. PFA computes the inspection intervals, the multisite location of damage, contribution of damage mechanisms to failure, margin of safety, and design for minimum joint damage in service by progressive-failure optimization technique. It can predict, under all operational conditions (temperature/thermal gradient, humidity, oxidative/corrosive agents): (1) constituent/ply mechanical properties and (2) property degradation by implementing the multifactor interaction model.

Time-dependent reliability (TDR) utilizes the probabilistic progressive-failure analysis to predict the composite structure probability of failure and the sensitivity effects of bond-line strength/thickness/voids on joint failure. Table 45.3 summarizes the benefits and functionality of the life-prediction software process.

The use of computer-based computational modeling to evaluate the service life (i.e., durability and reliability) of Polymer Matrix Composite (PMC) structures has been under development by NASA Lewis/Langley [11–16] for several years, with the result that computer codes have become available for evaluation of PMC structures under service conditions. These codes have been modified and integrated by Alpha STAR into a unified software system [17], namely GENOA (Figure 45.13), with a verified capability to evaluate PMC structural durability and reliability [18, 19] via simulation of crack initiation and progression

TABLE 45.3 Benefits of Life-Prediction Software for Use in the Design, Building, and Testing of Composite Components

Module	Benefit
Material-constituent analysis (MCA)	Useful during the early phases of concept/product development in evaluating the impact of changes in volume fraction involved in deciding on an appropriate fabrication approval or assessing environmental effect, degradation of material properties to environmental effects (moisture, thermal), manufacturing (defects, residual strains), etc.
Material-uncertainty analysis (MUA)	Useful during the early phases of concept/product development in evaluating the impact of changes in fiber architecture, volume fraction, and material mechanical properties on potential fracture damage involved.
Progressive-failure analysis (PFA)	Ability to predict the fractional contribution of failure mechanisms during the critical damage events (i.e., initiation, final failure), recommend inspection intervals under certain type of loading environmental conditions, and predict the effect of residual stress and defects introduced during manufacturing cooldown process on structural integrity.
Progressive-failure optimization (PFO)	Ability to perform parametric trade study of design parameters (volume fraction, plyangle, thickness) by minimizing the damage under certain types of loading environmental conditions.
Time-dependent reliability (TDR) by probabilistic progressive-failure analysis (PPFA)	Identification and sensitivity of progressive-damage parameters; uncertainty evaluation of material strength to material parameters; determination of sensitivities of failure modes to design parameters to facilitate targeting design parameter changes that will be most effective in reducing probability of a given failure mode from occurring; and prediction of probability of failure.

to structural failure under static, cyclic fatigue, thermomechanical and creep, impact, and random loadings. The concept and functionality of GENOA is shown in Table 45.4.

Fourteen detailed failure mechanisms are evaluated to identify the damage mechanisms and their contribution to the structure's behavior. Included are mechanisms for determination of delamination (Table 45.5).

TABLE 45.4 Concept/Methodology and Functionality of GENOA Software

Concept/Methodology	Functionality
Nodal-based Mindlin–Reisner layered shell or layered solid element	Five through-thickness stress at each node Reduced number of elements for transition, fillets, ply drop-offs (accepts four material properties per element)
Transverse shear generation by shell/solid element	Interlaminar shear, interlaminar tension
Out-of-plane stress (S33) by shell element	Surface traction, equilibrium state Lekhnitskii closed-form solution, Kedward closed-form solution
Peel stress input for material	Switch the material coordinate system where zz is xx, xx is zz, and yy remains yy
Updated/total Lagrangian	Geometrical nonlinearity, including the crack volume
Property degradation	Material nonlinearity at fiber/matrix or lamina level
Adaptive meshing if lamina/laminate damaged	Singularity conditioning, required crack size detection, dynamic ability to refine the mesh at key locations according to damage, allows the use of coarse model
Mixed iterative Finite Element Model	Minimize residual error conditioning, acts similar to p element formulation
Fourteen failure mechanisms	Flexibility for damage growth (three-dimensional) space
Percent contribution of failure modes to fracture	Identify fracture for each mode
Strain energy rate: local and global	Damage and fracture monitoring
Progressive stochastic evaluation	Random damage initiation/propagation, sensitivity of parameters, failure probability
Interface to FEM and CAD	Translation of FEM to many commercialized FEM and CAD
Graphical user interface	Animation, post-process of damage from macro to micro levels, two-dimensional/three-dimensional graphics
Availability	PC-based computer, NT, Windows 2000, Linux, and Unix

TABLE 45.5 Damage Mechanisms Considered in GENOA

Mode of Failure	Description
Longitudinal Tensile	Fiber tensile strength, fiber volume ratio
Longitudinal Compressive	1) Rule of mixtures based delaminations, 2) Fiber microbuckling, and 3) fiber Crushing
Transverse Tensile	Matrix modulus, matrix tensile strength,
Transverse Compressive	Matrix compressive strength, matrix modulus, and fiber volume ratio.
Normal Tensile	Plies are separating due to normal tension
Normal Compressive	Due to very high surface pressure i.e. crushing of laminate
In-Plane Shear (+)	Failure due to Positive in plane shear with reference to laminate coordinates
In-Plane Shear (−)	Failure due to negative in plane shear with reference to laminate coordinates
Transverse Normal Shear (+)	Shear Failure due shear stress acting on transverse cross section that is taken on a transverse cross section oriented in a normal direction of the ply
Transverse Normal Shear (−)	Shear Failure due shear stress acting on transverse cross section that is taken on a negative transverse cross section oriented in a normal direction of the ply
Longitudinal Normal Shear (+)	Shear Failure due shear stress acting on longitudinal cross section that is taken on a positive longitudinal cross section oriented in a normal direction of the ply
Longitudinal Normal Shear (−)	Shear Failure due shear stress acting on longitudinal cross section that is taken on a negative longitudinal cross section oriented in a normal direction of the ply
Modified Distortion Energy	Combined stress failure criteria used for isotropic materials
Relative Rotation Criterion	Considers failure if the adjacent plies rotate excessively with one another

Table 45.6 presents the GENOA software prediction capabilities. Table 45.7 shows the advantages and disadvantages of other D&DT (durability and damage tolerance) prediction methods.

A demonstration of the GENOA life-prediction analytical procedure was previously used on a NASA White Sands composite overwrapped pressure vessel (COPV) design application. It predicted: (1) composite overwrapped pressure vessel (COPV) ply schedule and thickness distribution in agreement with NASA test data, (2) critical-damage events, and (3) damage mechanisms contributing to failure. A GENOA-based filament-winding manufacturing module has been used to simulate the COPV manufacture and pressurization of a filament-wound cylinder (6.045-in. inner diameter, 13-in. length) consisting of an aluminum 6061-T6 liner over-wrapped with a combination of hoop and helical windings of graphite-epoxy composite (T-300 fibers). The predicted COPV results for ply angles, ply schedule, and thickness distribution were then incorporated in a PFA simulation (Figure 45.14).

TABLE 45.6 GENOA Progressive-Failure Analysis Capabilities [20, 21]

Material Properties	Virtual Testing of Structure	Durability of Structure	Reliability of Structure
Mechanical properties prediction considering uncertainties	Rosettes/strain gauges, deflectometer	Residual strength	What is the sensitivity of crack initiation to design and service parameters?
Degradation of material in service	Photoelasticity contours	Damage initiation, matrix microcrack, interfacial debonding	What can be done to stop/turn cracks?
Laminated/woven/braided/stitched fiber architecture	Crack detect or initiation, propagation, residual strength	Damage location, damage accumulation	What parameters are critical to failure?
Effects of manufacturing defects	Fracture toughness, stress intensity factor	Damage progression	What parameters are most sensitive to design changes?
Coupling of damage effects to life prediction	Margins of safety	Service life prediction	What design changes produce the greatest results?
Environmental effects	Damage energy release rate	Contributions to failure mechanisms	What is the probabilistic distribution?

Risk Management of Composite Structures

TABLE 45.7 Advantages and Disadvantages of D&DT Prediction Methods [22]

Method	Advantages	Disadvantage
GENOA progressive fracture—a new approach	Reduce experimental testing Reduce design time Reduce design cost Computer code available Verified accuracy Most powerful of methods	Requires significant computer resources
Simplified equations	Rapid analysis Promotes design optimization	Accuracy is limited
R-curve	Well-established method	Little predictive capability for fracture propagation in PMC Requires extensive testing R-curves are part specific
Linear elastic fracture mechanics (LEFM)	Accurate prediction of tensile strength if matrix cracking and delamination are minimal Good for long damage cuts	Not good if matrix cracking and/or delamination are significant Needs case-specific fracture toughness parameters
Damage-energy release rate (DERR) [23]	Computational simulation method Indicates structural resistance to damage propagation	
Nonlinear response [24]	Analytical study suggests accurate prediction of stiffened shell response to damage	Insufficient experimental verification Limited effort in this area

Source: Moon, D., Abdi, F., and Davis, A.B., Discrete Source Damage Tolerance Evaluation of s/RFI Stiffened Panels, presented at Sampe 99, Long Beach, CA, 1999. With permission.

Material Characterization	Virtual Manufacturing Filament Winding	Progressive Failure Analysis	Time Dependent Reliability, Damage Minimization
Predict Lamina Properties	Mandrel Type, Pressure	Global Multi-site Crack Initiation	Manufacturing Sensitivity
Predicts Laminate Properties	Tape Pretension	Global Multi-site Crack Propagation	Residual Strength Sensitivity
Material Uncertainty	Wet/Sticky Tap Wind	Matrix/Fiber/Ply Failure	Probability of Failure
Micro Properties Effect on Failure	Polar/Helical/Hoop Winding	Inspection Predictions for NDE	Ply Thickness Sensitivity to Leakage
Failure Mechanisms	Slippage in Dome	Contribution of Failure Mechanism	Ply Angles Sensitivity to Leakage
Void Inclusion Effect	Tape Failure	Leakage/Permeation	Sensitivity of Tank Load to Failure
Manufacturing Cool-down Effect	Residual Stress Distribution	Deflectometer and Strain Tracking	Design Guidelines
Fiber Volume Fraction Effect	Thickness Distribution	Stress Redistribution	Optimized Design
Moisture Diffusivity/Expansion	Ply Schedule Distribution	Energy Release Rate	
	Ply Angle Deviation	Margin of Safety	
		Residual Strength	SDD_03_0104.1

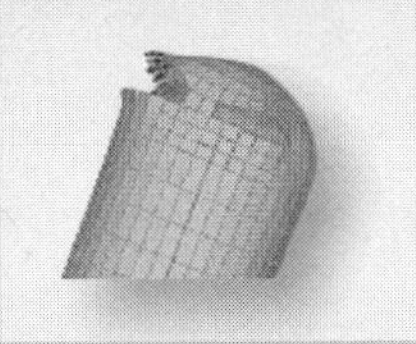

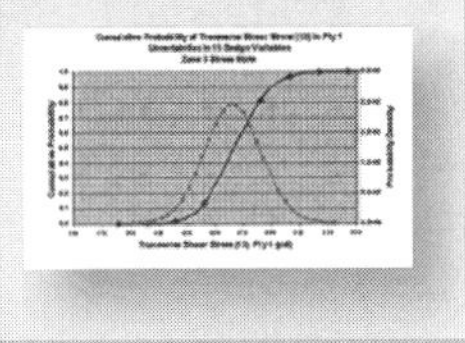

FIGURE 45.14 Design approach for composite pressure vessel is based on building-block strategy, a step-by-step verification.

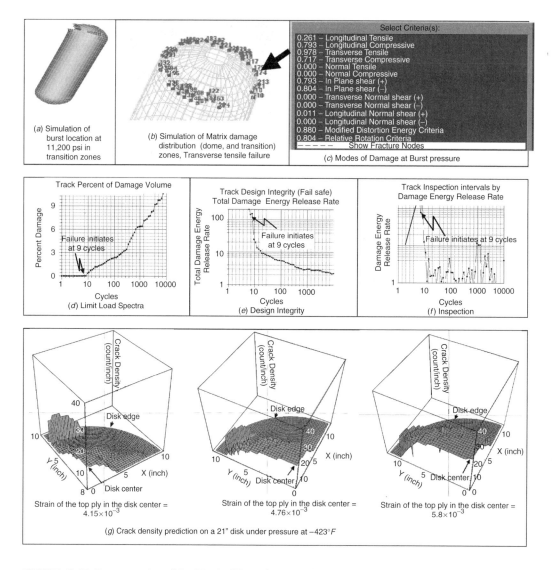

FIGURE 45.15 Demonstration of the GENOA life prediction analytical procedure: (a) Simulation of burst location, (b) Simulation of Matrix damage distribution (dome, and transition) zones, Transverse tensile failure, (c) Modes of Damage at Burst pressure, (d) track percent of damage volume, (e) track design integrity (fail safe design) by total damage energy release rate, (f) track inspection intervals by damage-energy release rate, (g) Exhaustion of Energy Demonstration as Life Prediction Analytical Procedure

The life prediction by GENOA-PFA simulated the composite prepreg tapes over a thin metal shell. The assessment of vessel durability and damage tolerance (D&DT) is shown in (Figure 45.15). Figure 45.15a shows the final failure location under pressure. Figure 45.15b shows the extent of matrix damage in transition zone. Figure 45.15c shows the tank fractional contribution of damage modes to final failure. Probabilistic material sensitivity analyses can also be conducted by the GENOA-MUA module to show the contribution of each material constituent property to the lamina's strength. The exhausted energy release rate plots is generated by the software, the analyst can use these plots to perform the predictive capability. An example of the percent change in damage volume is shown in Figure 45.15d, Another example of tracking the design integrity is shown in Figure 45.15e, as Total Damage Energy Release Rate (T-DERR). Tracking the inspection interval by damage energy release rate (DERR) is shown in Figure 45.15f, the

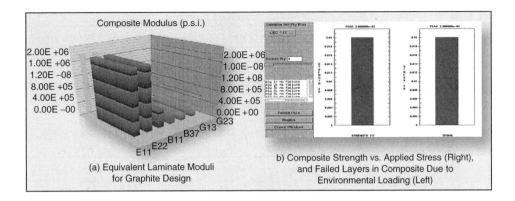

FIGURE 45.16 Functionality of material-constituent analysis: example of MCA analysis.

rise and fall in the curves shows the points on the pressurization where there are major damage events. Figure 45.15g shows exhaustion of Energy as crack density vs. strain. The proposed efforts will initiate guidelines and methodologies that contribute to accelerating and assisting the design activity and inspire detailed numerical analyses that rapidly leads to configurations with high prospects of success in the form of structural integrity and durability. Several of key Alpha Star team members have published and presented widely on the general subject of design, life prediction, building and testing of composite over-wrapped Pressure Vessel (COPV).

45.2.2.1 Material-Constituent Analysis (MCA) and Prediction of Lamina Properties under Manufacturing and Environmental Condition

Material-constituent analysis (MCA) predicts the composite lamina and laminate properties under manufacturing and environmental conditions (e.g., humid environment). MCA is useful during the early phases of concept/product development in (a) evaluating the impact of changes in volume fraction on material properties before seeking fabrication approval or (b) assessing environmental effect on degradation of material properties (moisture, thermal), manufacturing (defects, residual strains), etc.

Material-constituent analysis (MCA) predicts the equivalent constituent properties, including (1) composite modulus (Figure 45.16a), (2) thermal expansion coefficient, (3) moisture expansion coefficients, (4) heat conductivity, (5) laminate strength, (6) Poisson's ratio, (7) moisture diffusivity, (8) ply stress, and (9) ply strength. MCA provides important design information (e.g., margin of safety) needed to optimize composite material systems, and it facilitates designers in understanding fiber and matrix stress distributions in a composite system under load. Figure 45.16b shows an example of the ultimate ply strength prediction vs. the applied stress to determine the margin of safety and provide information relative to failed layers in a composite.

45.2.2.1.1 Example of MCA

45.2.2.1.1.1 Material System—The PMC material used in this study, IM7/PETI-5, consisted of a high-strength, intermediate-modulus, continuous carbon fiber in a thermoplastic polyimide matrix. All test materials were laminated composites fabricated at the NASA Langley Research Center [25].

The typical reusable launch vehicle (RLV) laminates are 13-ply quasi-orthotropic laminate [45/90$_3$/-45/0$_3$/-45/90$_3$/45] (RLV(0)) and [-45/0$_3$/45/90$_3$/45/0$_3$/-45] (RLV(90)) of IM7/PETI-5. The fiber volume fraction in each ply was assumed to be 60%. The coupon laminate analyses include unidirectional unnotched-tension and notched-tension specimens to simulate the effect of flaws on the laminate strength. The simulation was compared (Table 45.8 through Table 45.10) with tests that were conducted by NASA-Langley [26–28].

TABLE 45.8 Tensile Properties of UN-RLV Laminates: Simulation vs. Tests

UN-tension	Simulation-Strength (ksi)		Test-Strength (ksi)		Simulation-Stiffness (msi)		Test-Stiffness (msi)	
	RLV(0)[a]	RLV(90)[b]	RLV(0)	RLV(90)	RLV(0)	RLV(90)	RLV(0)	RLV(90)
450°F	95.1	143.3	107.7	157.9	8.8	14.4	9.7	14.4
400°F	88.6	147.1	111.4		8.6	14.1	9.4	
350°F	82.9	136.5	109.3	179.8	8.6	14.1	10.2	17.2
75°F	97.1	152.9	108.4	180.8	8.5	13.6	8.2	13.0
−320°F	97.1	151.9	115.6	164.6	8.3	13.2	8.6	14.1
−423°F	95.2	151.9	111.9	171.3	9.0	14.3	8.5	14.4

[a] RLV(0), [45/90$_3$/-45/0$_3$/-45/90$_3$/45].
[b] RLV(90), [-45/0$_3$/45/90$_3$/45/0$_3$/-45].

TABLE 45.9 Tensile Strengths of Notched RLV Laminates: Simulation vs. Tests

N-Tension Strength	Simulation (ksi)		Test (ksi)	
	RLV(0)	RLV(90)	RLV(0)	RLV(90)
450°F	44.5	62.5	53.1	87.4
400°F	42.3	68.2	54.1	85.4
350°F	40.2	63.9	55.7	82.8
75°F	48.5	72.5	49.4	78.2
−320°F	49.2	72.2	50.7	85.6
−423°F	47.4	71.4	53.2	83.6

Calibration of lamina properties of IM7/PETI-5 at −423, −320, 75, 350, 400, and 450°F was conducted using the GENOA-PMC module, which is a back-calculation process of lamina properties, by selecting the correct constituent (fiber/matrix) properties to match the test lamina or ply properties. The objective of calibration is to obtain a reliable constituent (fiber/matrix) databank as the base for laminate and structural analysis. The calibration process included matching the test-observed lamina mechanical properties: (1) longitudinal modulus, (2) transverse modulus, (3) shear modulus, (4) longitudinal tensile strength, (5) longitudinal compressive strength, (6) transverse tensile strength, (7) transverse compressive strength, (8) shear strength, and (9) coefficients of thermal expansion (Figure 45.17).

45.2.2.1.1.2 Simulation Results—The simulation results show:

1. Stiffness of RLV(0) and RLV(90) of simulation is consistent with test data.
2. The simulated tensile strengths of both RLV(0) and RLV(90) at high temperature (from 350 to 450°F) are lower than those at low temperature. Compared with the test results, the deviation of the simulation is 10 to 24%.

TABLE 45.10 Ratio of the Tensile Strengths of Notched RLV to the Tensile Strengths of Unnotched RLV at Various Temperatures

Temperature		450°F	400°F	350°F	75°F	−320°F	−423°F
RLV(0)	Simulation (%)	47	48	48	50	51	50
	Test (%)	49	49	51	46	44	48
RLV(90)	Simulation (%)	44	46	47	47	48	47
	Test (%)	55		46	43	52	49

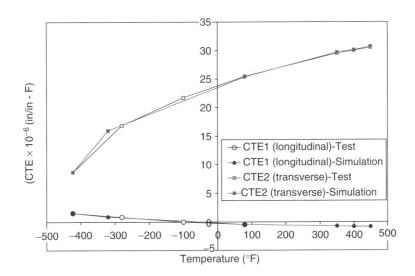

FIGURE 45.17 Calibrated coefficients of thermal expansion of IM7/PETI-5 vs. test data at six temperatures.

3. Simulated tensile strengths of notched RLV(0) and RLV(90) match the test well at room and cryogenic temperatures, but show 25% deviation from tests at high temperatures.
4. The notch reduces the tensile strength of RLV(0) and RLV(90) by 50%.
5. Usually, the more damage energy that is dissipated, the greater is the damage that is caused in the laminate, and the greater is the leakage. For RLV(90), as the service temperature decreases, the laminate tends to have more leakage. For RLV(0), the laminate behavior is not clear.

45.2.2.2 Material-Uncertainty Analysis (MUA) and Prediction of Composite Mechanical Properties to Potential Failures

Material-uncertainty analysis (MUA) predicts the percent contribution of composite mechanical properties at the constituent, lamina, and laminate levels to potential failure mechanisms such as delamination. MUA is useful during the early phases of concept/product development in evaluating the impact of changes in fiber architecture, volume fraction, and material mechanical properties on potential fracture damage involved.

An application of material-uncertainty analysis (MUA) might be to perform a probabilistic failure analysis under thermomechanical loading at room and high temperatures to show the sensitivity of variation that each constituent property contributes to the failure mechanisms (e.g., delaminations, debonds) [29, 30]. MUA performs identification and sensitivity of progressive-damage parameters, uncertainty evaluation of material strength to material parameters, and determination of sensitivities of failure modes to design parameters. These functions are helpful in targeting the design parameter changes that will be most effective in reducing the probability of a given failure mode from occurring and thus the probability of failure. It predicts:

- Uncertainty evaluation of material strength vs. material parameters
- Sensitivities of design requirements to design parameters
- Degree to which design parameters contributed to failure
- Cumulative distribution functions (CDF) for failure-strength evaluation
- Probability of failure
- Margin of safety

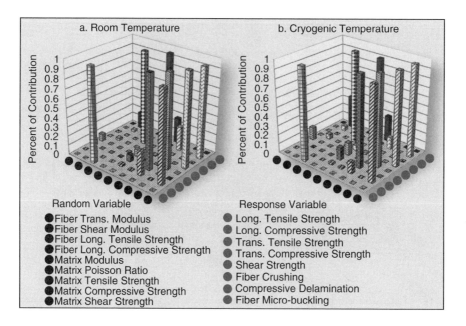

FIGURE 45.18 Probabilistic sensitivity analysis of lamina strengths (5% variation of each random variable) at room temperature and cryogenic temperature.

Probabilistic sensitivity analysis has been conducted with the GENOA-MUA module to show the contribution of each material-constituent property to the lamina's strength at room and cryogenic temperatures (Figure 45.18). The uncertainty distribution of the material was chosen as 5%.

45.2.2.2.1 Example of MUA Analysis
Sensitivity analysis of the lamina properties of IM7/PETI-5 to constituent properties was performed. The contribution of constituent properties to lamina properties of polymer matrix composites can be usually described as follows:

Lamina longitudinal properties are mainly determined by the fiber properties
Lamina transverse properties depend on the matrix properties

If the fiber properties are not sensitive to the temperature change, the above dependency of the lamina properties on constituent properties still holds with varying temperatures. Therefore, the sensitivity of IM7/PETI-5 is similar at different temperatures. As a representative for those at various temperatures, the sensitivity of IM7/PETI-5 at room temperature is shown in Figure 45.19. The sensitivity was obtained by GENOA-MUA with a 5% variation in each constituent property used as the standard deviation for a normal distribution.

In Figure 45.19, the definition of each parameter is: Ef22, fiber transverse modulus; SfT, fiber tensile strength; SfC, fiber compressive strength; Em, matrix modulus; SmT, matrix tensile strength; SmC, matrix compressive strength; SmS, matrix shear strength; Vf, fiber volume fraction; S11T, lamina longitudinal tensile strength; S11C, lamina longitudinal compressive strength; S22T, lamina transverse tensile strength; S22C, lamina transverse compressive strength; S12, lamina shear strength.

45.2.2.3 Progressive-Failure Analysis (PFA) and Evaluation of Failure in Structure

Progressive-failure analysis calculates the damage initiation, propagation, and final failure of engineering materials and structures. Thus, it is a powerful tool for damage-tolerance evaluation of engineering materials and structures. PFA has the ability to simulate the (a) microscopic damage

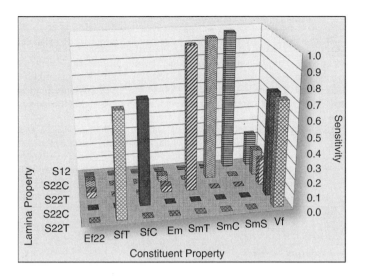

FIGURE 45.19 Sensitivity of IM7/PETI-5 lamina properties to the constituent properties at room temperature.

initiation, (b) U.S. Navy requirement of 0.01-in. crack initiation, (c) U.S. Air Force requirement of stable crack growth of 0.05-in. crack, and (d) progression of damage until the loss of structural integrity occurs. PFA computes the inspection intervals, the multisite location of damage, the contribution of damage mechanisms to failure, the margin of safety, and the design for minimum composite damage in service by progressive-failure optimization technique. By implementing the multifactor interaction model, it can predict, under all operational conditions (temperature/thermal gradient, humidity, and oxidative/corrosive agents), both the constituent/ply mechanical properties and the property degradation.

PFA predicts the fractional contribution of failure mechanisms during the critical damage events (i.e., initiation, final failure) under certain types of loading environmental conditions, and it suggests appropriate inspection intervals. The effects of residual stress and defects on structural integrity are introduced during the manufacturing cooldown process.

The PFA simulation concept and functionality involves: (1) ply-layering methods using FEM model with through-thickness representation, transverse shear, adaptive meshing, updated Lagrangians (geometrical nonlinearities), (2) nonlinear material degradation due to service-loading spectrum (e.g., moisture, corrosion), (3) manufacturing anomalies relative to voids, fiber waviness, and residual stress, (4) simulation of the initiation and growth of cracks to failure under static, cyclic fatigue, random thermal acoustic fatigue, creep, and impact loads, (5) identification of various material-failure modes involved in critical damage events, and (6) determining sensitivities of failure modes to design parameters as fiber volume fraction, ply thickness, fiber orientation, and thickness of adhesive bonds.

The methodology builds on the existing GENOA virtual-evaluation capability, under operational environments (Table 45.11), to determine: (1) durability and damage tolerance based on degradation of material properties due to crack initiation, location, and growth of damage, (2) the contributions of various damage mechanisms to failure, (3) predictions of optimum inspection intervals, (4) incipient damage locations, (5) margin of safety, (6) probabilistic failure based on sensitivities to identified progressive damage parameters, (7) evaluation of uncertainty of material parameters, (8) determination of sensitivities of failure modes to design parameters to facilitate targeting design-parameter changes that will be most effective in reducing the probability of a given failure mode from occurring, (9) the probability of failure, and (10) evaluation of material fracture toughness without the need for conducting costly and time-consuming ASTM standard testing.

TABLE 45.11 Existing Virtual-Evaluation Capabilities

Functions	Simulation	Simulation Benefit over Test
Deflectometer	Plot of far-field applied load vs. deflection	Multiple deflectometers
Far-field stress/strain	Plot of far-field applied stress vs. strain	Reduce need for MIL-HDBK-5, 17
Strain gauge	Plot of applied load vs. strain	Multiple strain-gauge/rosette
Photoelastic fringe	Isochromatic, isoclinic, thermograph	Simulation at no cost
Nondestructive evaluation	Local/global energy release rates vs. applied loads	Simulation at no cost
Margin of safety	Plot of applied load vs. % damage volume	Not available in test
Fracture toughness	Plot of applied stress vs. material thickness	Helps in ASTM standard for composite
Residual strength	Plot of fracture stress vs. crack length	Helps in ASTM Standard for composite
S-N curve	Plot of stress vs. cycles	Helps in ASTM E-466 procedures
Multisite damage	Location of crack initiation and crack growth	Eliminates the need for sectional cut
Fatigue-crack growth	Plot of da/dN vs. stress intensity factor range	Reduces the need for ASTM fatigue test

45.2.2.3.1 Damage-Evolution Metrics

In addition to the failure criteria discussed in Table 45.5, eight damage-related functions are calculated (Figure 45.20) [30–40]:

1. Strain energy.
2. Percent damage volume (DAMAG) vs. load, representing margin of safety, is computed from the ratio of the total damage volume of the defect to the total volume of the composite structure, multiplied by 100. This is an overall indicator of structural degradation.
3. Damage-energy release rate (DERR) is used for acoustic emission, inspection, nondestructive evaluation. It is the ratio of the incremental work done by external forces to the incremental volume of damage created during a load increment that causes damage.
4. Total damage-energy release rate (TDERR) is used for global failure evaluation. It displays the ratio of total energy expended to the total damage produced.
5. Damage energy or cumulative damage energy is computed based on the exhausted strain energies of damage mechanisms. It is an overall index of degradation similar to the percent damage.

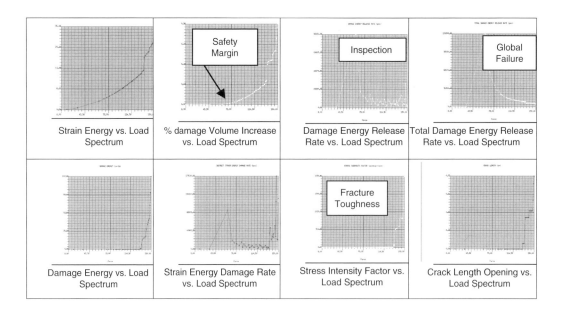

FIGURE 45.20 Damage-evolution metrics as derivatives of energy release rate.

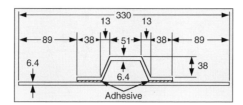

FIGURE 45.21 Cross section through stiffened graphite/epoxy composite of 48 plies $[0/+45/90]_{a6}$ (all dimensions are in millimeters).

6. Strain energy damage rate (SEDR) is the ratio of incremental strain energy to incremental damage volume and is computed as the incremental work done by external forces minus the damage energy exhausted during the same load increment. Very low levels of SEDR indicate critical loading levels where most of the external energy is exhausted.
7. Stress intensity factor (Kcr) corresponds to mode I from fracture mechanics, where Kcr is the average material stress multiplied by the square root of π times the simulated crack length.
8. Crack growth length for both crack opening and through-thickness accumulation.

45.2.2.3.2 Examples of PFA

45.2.2.3.2.1 Effects of Adhesive Thickness—The adhesive thickness plays an important role on the durability of bonded joints. For adhesively bonded joints, it is not realistic to expect perfectly uniform adhesive thickness to be achieved in the fabrication process. Therefore, it is necessary to quantify the effects of uncertainties in the adhesive thickness on structural durability and damage tolerance. To determine the sensitivity of composite damage progression to adhesive thickness, it is proposed to represent the adhesive thickness probabilistically.

45.2.2.3.2.2 Example of Current Adhesive Bond-Line Simulation—A PMC (graphite/epoxy) panel (Figure 45.21) stiffened by an adhesive-bond hat-type stringer was simulated in GENOA progressive-failure analysis (PFA) to determine damage initiation/growth and the fracture path in a panel with an adhesive-bond joint thickness of Type 1 (0.00521 in.). A finite element model was prepared to consider a dimensioned cross section through the adhesively bonded stiffened panel.

Axial tension and compression were applied by imposing a gradually increasing uniform axial load at the clamped left edge of the panel. Damage initiation and progression were monitored as the panel was gradually loaded. Damage was initiated by ply failure in the top surface of the skin near the end of the stiffener. Figure 45.22 indicates that the stiffened panel experiences damage initiation sooner for a thicker

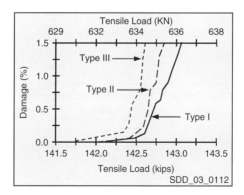

FIGURE 45.22 Damage progression under tension of 48-ply graphite/epoxy: Type I, Type II, and Type III correspond to 0.00521-, 0.01042-, and 0.02084-in. thicknesses adhesive bond, respectively.

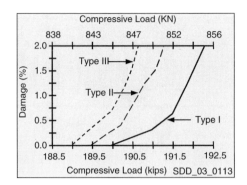

FIGURE 45.23 Damage progression under compression of 48-ply graphite/epoxy: Type I, Type II, and Type III correspond to 0.00521-, 0.01042-, and 0.02084-in. thicknesses adhesive bond, respectively.

adhesive bond. However, the effect of adhesive thickness on the damage initiation load is very small under tension. DERR levels for the three adhesive thicknesses investigated indicates that, although a thicker adhesive bond has a lower damage initiation load, structural resistance to damage progression under tension is considerably higher for the thicker bond.

Figure 45.23 shows damage initiation and progression under compressive loading. Similar to tensile loading, a thicker adhesive bond experiences damage initiation sooner. Unlike tensile loading, DERR levels do not monotonically increase with adhesive thickness. Type II (0.01042 in.) bond with an adhesive thickness of two plies has the best DERR performance. However, if the thickness is increased to Type III (0.02084 in.), with an adhesive thickness of four plies, the DERR level is significantly degraded.

45.2.2.3.2.3 Example of Progressive Fracture Simulation in PMCs in a Hygrothermal Environment—The influence of hygrothermal environmental conditions on the durability of composite structure was investigated via computational simulation of a composite plate made of T-300/epoxy with a laminate configuration of +15/−15/−15/+15 degrees. Uniaxial loading of the plate was simulated at service temperatures of 70, 200, and 300°F with moisture contents of 0 and 1%. Damage initiation, damage growth, fracture progression, and global structural fracture results are shown in Table 45.12 at four stages of loading under two hygrothermal conditions. The initial fracture load is defined as the load to initiate a through-thickness laceration of the composite at one node. The secondary fracture load is the load that causes either propagation of the initial fracture or initiation of fracture at another location. The critical load occurs at the minimum value of the strain energy release rate (SERR) during fracture propagation. Figure 45.24 shows the relationship between the applied loading and the resulting initial damage. It indicates that the overall strength of the composite structure is reduced with increasing temperature and moisture. The simulations also showed marked reductions in the first natural frequencies as load-induced damage occurred.

TABLE 45.12 T-300/epoxy (+15/−15)s Hygrothermal Environment and Loads [26]

Temperature (°C)	Moisture (%)	Initial Fracture Load (kN)	Secondary Fracture Load (kN)	Critical Load (kN)	Global Fracture Load (kN)
21.11	0	23.13	26.29	27.58	27.58
21.11	1	23.13	26.29	27.58	27.97
93.33	0	23.13	25.42	26.94	27.10
93.33	1	23.13	24.71	26.20	26.42
148.89	0	19.18	22.34	24.31	24.83
148.89	1	18.38	19.44	22.69	22.69

Source: Minnetyan, L., Murthy, P.L.N., and Chamis, C.C., *Int. J. Damage Mech.*, 1 (1), 60–79, 1992. With permission.

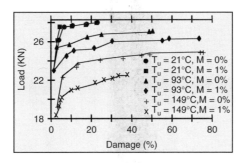

FIGURE 45.24 T-300/epoxy (+15/−15)s created damage with loading.

45.2.2.3.4 *Example of PFA*

Tensile strength predictions of un-notched RLV(0) and RLV(90) at −423, −320, 75, 350, 400, and 450°F are compared with NASA observed tests (Figures 45.25 through Figure 45.28). Simulation [40] reveals that:

1. Stiffness of RLV(0) and RLV(90) in the simulation is consistent with test data, except at 350°F. At 350°F, neither the lamina longitudinal modulus nor lamina transverse modulus is higher than those at other temperatures. Therefore, simulated stiffness of RLV(0) and RLV(90) at 350°F does not show higher magnitudes than those at other temperatures, in contrast to the test data.
2. The laminate strengths of both RLV(0) and RLV(90) at high temperature (from 350 to 450°F) are lower than those at low temperature in the simulation results. Compared with the test results, the deviation of the simulation results is 10 to 15%, except at 400 and 350°F, at which the deviation is 20 and 24%, respectively.
3. Sensitivity analysis shows that the upper limits of the tensile strengths of RLV(0) and RLV(90) reach the test results, with a 5% variation of constituent properties (moduli, strength, and coefficient of thermal expansion (CTE)) as standard deviation for normal distribution (see Figure 45.29).

Tensile strength simulations of notched RLV(0) and RLV(90) at −423, −320, 75, 350, 400, and 450°F were conducted. The open-hole model was used to analyze the effect of defects on the tensile strength of RLV(0) and RLV(90) at six temperatures by PFA-GENOA. The FEM model of the open-hole specimen

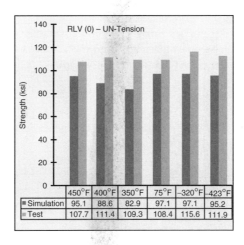

FIGURE 45.25 Comparison of simulation and testing tensile strengths of unnotched RLV(0).

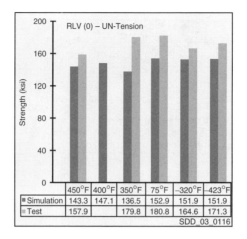

FIGURE 45.26 Comparison of simulation and testing tensile strengths of unnotched RLV(90).

applied temperature, thermal load, as well as mechanical load. The specimen was assumed to undergo uniform temperature change from glass-transition temperature (512°F) to service temperature. It was also assumed that there is no temperature gradient through the specimen thickness. Figure 45.30 and Figure 45.31 are, respectively, the tensile strengths of notched RLV(0) and RLV(90) at −423, −320, 75, 350, 400, and 450°F.

45.2.2.4 Time-Dependent Reliability by Probabilistic Failure Analysis and Evaluation of Composite Damage Propagation

Probabilistic failure analysis and evaluation of composite damage propagation account for the uncertainties in material properties, the composite fabrication process, and the global structural parameters. All the relevant design variables are quantified to determine their effects on fracture initiation and progression. The composite mechanics, finite element structural simulation, and fast probability integrator (FPI) have been integrated into the probabilistic analysis code. FPI, contrary to the traditional

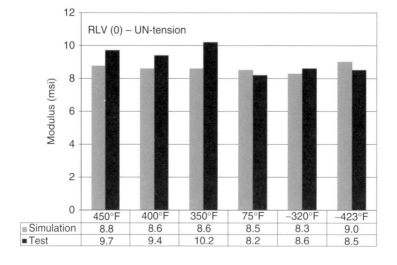

FIGURE 45.27 Comparison of simulation and testing tensile stiffness of unnotched RLV(0).

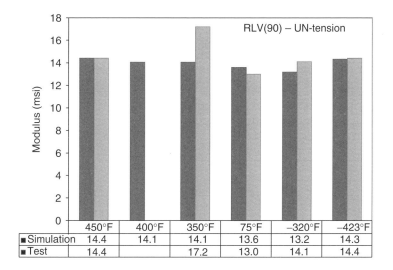

FIGURE 45.28 Comparison of simulation and testing tensile stiffness of unnotched RLV(90).

Monte Carlo simulation, makes it possible to achieve orders-of-magnitude computational efficiencies that make probabilistic analysis acceptable for practical applications. A probabilistic analysis cycle starts with defining uncertainties in material properties at the most fundamental composite scale, i.e., fiber/matrix constituents (see Table 45.13).

The uncertainties are progressively propagated to those at higher composite scales, such as subply, ply, laminate, and structural. The damaged/fractured structure and ranges of uncertainties in design variables, such as material behavior, structural geometry, supports, and loading, are input to the probabilistic analysis module. Consequently, probability density functions (PDF) and cumulative distribution functions (CDF) can be obtained at the various composite scales for the structure response and sensitivities. Input data for probabilistic analysis are generated from the degraded composite model that is developed as progressive damage and fracture stages are monitored.

45.2.2.4.1 Time-Dependent Reliability (TDR)

Time-dependent reliability (TDR) utilizes the probabilistic progressive-failure analysis predictions, the composite system's probability of failure, and the sensitivity effects of bond-line strength/thickness/voids on failure. Identification and sensitivity of progressive-damage parameters, uncertainty evaluation of material strength to material parameters, and sensitivities of failure modes to design parameters will identify the changes in targeting design parameters that will be most effective in reducing the probability of a given failure mode from occurring and thus the probability of failure.

45.2.2.4.2 Examples of Time-Dependent Reliability

Probabilistic analysis of tensile properties of un-notched RLV(0) and RLV(90) specimens were conducted [40]. The random design variables are normally distributed with a 5% variation. They are (1) fiber volume fraction, (2) fiber tensile and compressive strength, (3) fiber thermal expansion coefficient, (4) fiber modulus, (5) matrix modulus, (6) matrix thermal expansion coefficient, and (7) matrix tensile, compressive, and shear strength. The response variables are final tensile strength of RLV(0) and RLV(90). The simulations were conducted at six service temperatures: −423, −320, 75, 350, 400, and 450°F.

Sensitivity of the tensile strengths of RLV(0) and RLV(90) to their constituent properties are shown in Figure 45.32 and Figure 45.33, respectively, while Figure 45.34 and Figure 45.35 are, respectively,

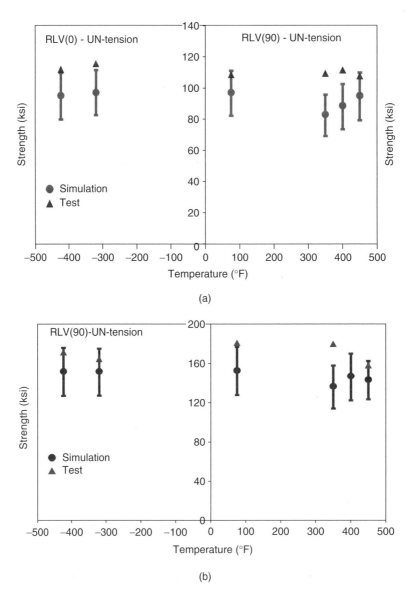

FIGURE 45.29 Variation of the tensile strength of RLV(0), and RLV(90) with 5% of variation of constituent properties (moduli, strength, and CTE) — sensitivity analysis.

cumulative distribution functions (CDF) and probability density functions (PDF) of the tensile strengths of RLV(0) and RLV(90).

TDR simulation revealed that:

1. Sensitivity of the tensile strengths of RLV(0) and RLV(90) to their constituent properties are very similar at different temperatures. Among all the constituent properties, fiber longitudinal tensile strength and fiber volume fraction are most effective on the tensile strengths of RLV(0) and RLV(90).
2. Fiber and matrix moduli and CTE also affect the tensile strengths of RLV(0) and RLV(90), though not significantly, due to the constraints of the deformation between the plies in the laminates.

CDF and PDF curves of the tensile strengths of RLV(0) and RLV(90) at different temperatures are similar.

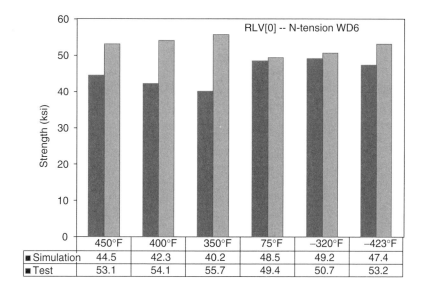

FIGURE 45.30 Comparison of simulation and testing tensile strengths of notched RLV(0) at various temperatures.

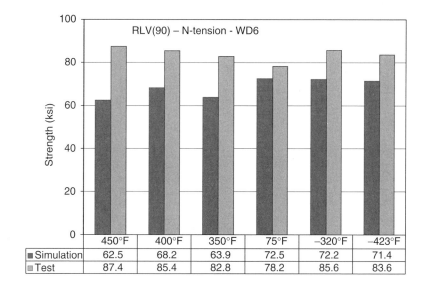

FIGURE 45.31 Comparison of simulation and testing tensile strengths of notched RLV(90) at various temperatures.

TABLE 45.13 Geometrical and Material Uncertainties Considered

Longitudinal strain	Transverse strain
In-plane shear strain	Longitudinal stress
Transverse stress	In-plane shear stress
Longitudinal tensile strength	Longitudinal compressive strength
Transverse tensile strength	Transverse compressive strength
In-plane shear strength	MDE failure criterion
Hoffman's failure criterion	Inter-ply delamination failure criterion
Fiber-crushing criterion (compressive strength)	Delamination criterion (compressive strength)
Fiber-microbuckling criterion (compressive strength)	Longitudinal normal shear stress
Transverse normal shear stress	Transverse normal shear strength

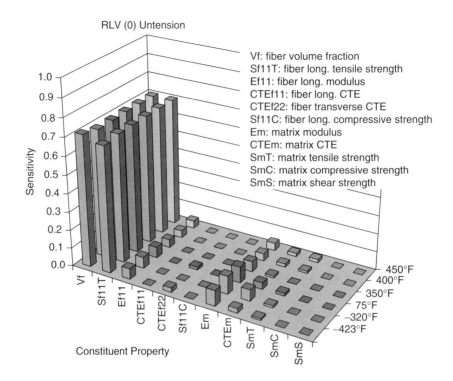

FIGURE 45.32 Sensitivity of the tensile strength of RLV(0) to its constituent properties at six temperatures.

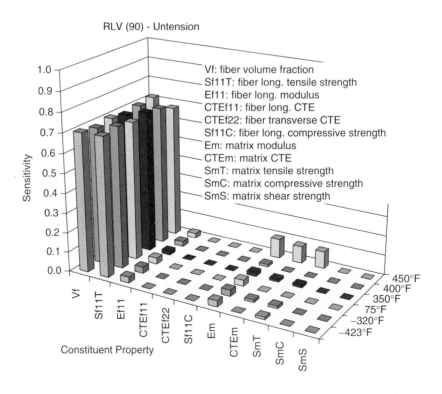

FIGURE 45.33 Sensitivity of the tensile strengths of RLV(90) to its constituent properties at six temperatures.

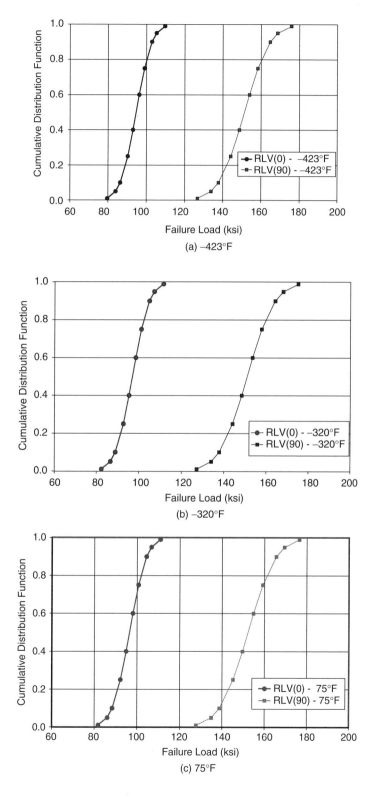

FIGURE 45.34 Cumulative distribution functions (CDF) of the tensile strengths of RLV(0) and RLV(90) at various temperatures.

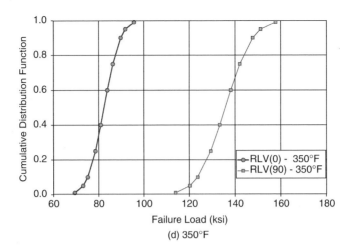
(d) 350°F

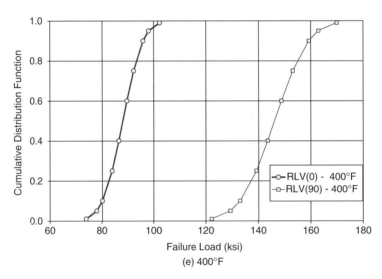
(e) 400°F

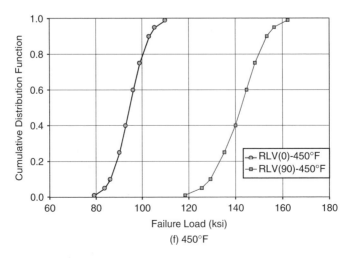
(f) 450°F

FIGURE 45.34 (*Continued*)

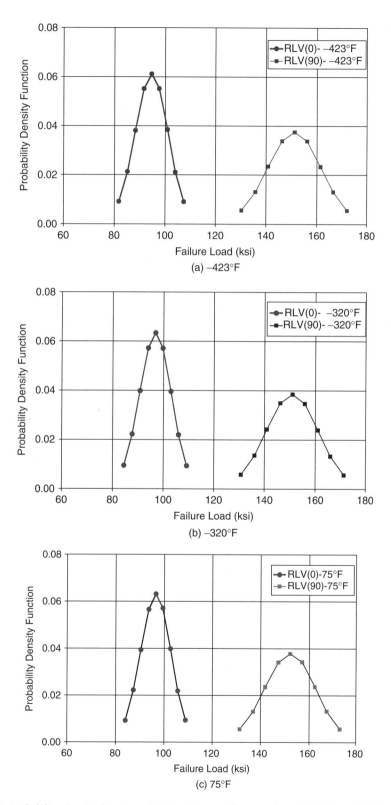

FIGURE 45.35 Probability density functions (PDF) of the tensile strengths of RLV(0) and RLV(90) at various temperatures.

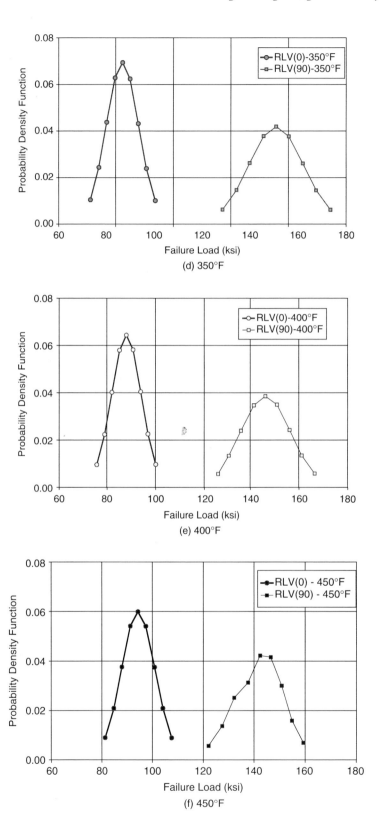

FIGURE 45.35 (*Continued*)

45.3 Conclusion

The present method of aircraft structural design is based on tailoring structural components and joints to provide the required strength and stiffness response based on finite element stress analysis. Safety factors are used locally to provide sufficient strength for each component to remain safe under occasional overloading. However, during initial design, the effects of possible local failures due to overloads and/or discrete source damage are not considered with regard to global structural response and damage tolerance. Significant costs are incurred by both aerospace manufacturing companies and the government due to the formal processes for the certification of air vehicles. The certification process is preceded by multiple coupon tests, subcomponent tests, and full-scale structural tests to verify design failure loads. Aerospace companies spend an estimated $350 million per year on experimental tests alone. The progressive-failure analysis (PFA) and virtual testing (VT) capabilities will cut the cost of these tests by an estimated 50%.

It is envisioned that all aerospace structures will be designed and maintained with the help of VT in the future. The composite air-vehicle structure will be more robust, congruent, and economical if VT becomes part of its design process.

Acknowledgment

The authors are grateful to Dr. Thomas Gates of NASA Langley for his technical guidance and for providing the composite material properties.

References

1. Aerospace Corp. Report April 1997, "Costs of Space Launch and Ground Systems."
2. Aviation Week and Space Technology, January 13, 2003.
3. Chamis, C.C. and Minnetyan, L., Progressive Fracture of Polymer Matrix Composite Structure: a New Approach, presented at 14th Annual Energy-Source Technology Conference and Exhibition, Houston, TX, 1992.
4. Composite Material Handbooks (MIL-HDBK-17) Chapter 2.1 from Volume 1 of the Polymer Matrix Composite (PMC).
5. Federal Aviation Regulations, Part 25, Airworthiness Standards: Transport Category Airplanes, Federal Aviation Administration, Washington, DC, 1993.
6. Damage-Tolerance and Fatigue Evaluation of Structure, Advisory Circular 25.571-1A, Federal Aviation Administration, Washington, DC, 1986.
7. Composite Aircraft Structure, Advisory Circular 20-107A, Federal Aviation Administration, Washington, DC, 1984.
8. Abdi, F., Shroyer, E., "Virtual Testing Validation Requirements by Building Block Approach," Collaborative Virtual Testing (CVT) for Flight Qualified Spacecraft Application, Contract No. NAS-01067, August 2004.
9. Design Considerations for Minimizing Hazards Caused by Uncontained Turbine Engine and Auxiliary Power Unit Rotor and Fan Blade Failures, Advisory Circular 20-128, Federal Aviation Administration, Washington, DC, 1988.
10. Cox, B.N. and Flanegan, G., Handbook of Analytical Methods for Textile Composites: Binary Model of Textile Composites, ver. 1, Rockwell Science Center, Thousand Oaks, CA, 1996.
11. Murthy, P.L.N. and Chamis, C.C., Integrated Composite Analyzer (ICAN): Users and Programmers Manual, NASA tech. paper 2515, National Aeronautics and Space Administration, Washington, DC, 1986.
12. Minnetyan, L., Murthy, P.L.N., and Chamis, C.C., Composite structure global fracture toughness via computational simulation, *Comput. Struct.*, 37 (2), 175–180, 1990.
13. Irvine, T.B. and Ginty, C.A., Progressive fracture of fiber composites, *J. Composite Mater.*, 20, 166–184, 1986.

14. Minnetyan, L., Chamis, C.C., and Murthy, P.L.N., Structural behavior of composites with progressive fracture, *J. Reinforced Plast. Composites*, 11 (4), 413–442, 1992.
15. Minnetyan, L., Murthy, P.L.N., and Chamis, C.C., Progressive fracture in composites subjected to hygrothermal environment, *Int. J. Damage Mech.*, 1 (1), 60–79, 1992.
16. Minnetyan, L., Chamis, C.C., and Murthy, P.L.N., Structural durability of a composite pressure vessel, *J. Reinforced Plast. Composites*, 11 (11), 1251–1269, 1992.
17. Minnetyan, L., Murthy, P.L.N., and Chamis, C.C., Progressive fracture in composites subjected to hygrothermal environment, *Int. J. Damage Mech.*, 1 (1), 69–70, 1992.
18. Huang, D., Computational Simulation of Damage Propagation in Three-Dimensional Woven Composites, doctoral thesis, Clarkson College, Potsdam, NY, 1997.
19. Abdi, F., Blade Manufacturing Process Technology Development of Preform Composite Net Shapes, Goldsworthy and Associates, White Paper Report, 1998.
20. Shah, A.R., Shiao, M.C., Nagpal, V.K., and Chamis, C., Probabilistic evaluation of uncertainties and risks in aerospace components, AIAA, Reston VA, 1992, chap. 10.
21. Abdi, F., Lorenz, R., and Hadian, J., Progressive Fracture of Braided Composite Turbomachinery Structures, NASA SBIR Phase I report NAS3-27334, NASA, Washington, DC, 1996.
22. Moon, D., Abdi, F., and Davis, A.B., Discrete Source Damage Tolerance Evaluation of s/RFI Stiffened Panels, presented at Sampe 99, Long Beach, CA, 1999.
23. Abdi, F. and Minnetyan, L., Progressive Fracture of Braided Composite Turbomachinery Structures, NASA SBIR Phase II report NAS3-97041, NASA, Washington, DC, 1999.
24. Starnes, J.H., Britt, V.O., and Rankin, C.C., Nonlinear Response of Damaged Stiffened Shells Subjected to Combined Internal Pressure and Mechanical Loads, AIAA 951462-CP, AIAA, Reston, VA.
25. Whitley, K.S. and Gates, T.S., Thermal/mechanical response and damage growth in polymeric composites at cryogenic temperatures, in *Proceedings of 43rd Structures, Structural Dynamics, and Materials Conference*, AIAA-2002-1416, AIAA/ASME/ASCE/AHS/ASC, 2002.
26. Johnson, T.F. and Gates, T.S., Temperature polyimide materials in extreme temperature environments, in *Proceedings of 42nd Structures, Structural Dynamics, and Materials Conference*, AIAA/ASME/ASCE/AHS/ASC, Reston, VA, 2001.
27. Whitley, K.S. and Gates, T.S., Thermal/mechanical response and damage growth in polymeric composites at cryogenic temperatures, in *Proceedings of 43rd Structures, Structural Dynamics, and Materials Conference*, AIAA-2002-1416, AIAA/ASME/ASCE/AHS/ASC, Reston, VA, 2002.
28. Johnson, T.F. and Gates, T.S., Temperature polyimide materials in extreme temperature environments, in *Proceedings of 42nd Structures, Structural Dynamics, and Materials Conference*, AIAA/ASME/ASCE/AHS/ASC, Reston, VA, 2001.
29. Murthy, P.L.N. and Chamis, C.C., Integrated Composite Analyzer (ICAN): Users and Programmer's Manual, NASA technical paper 2515, NASA, Glenn Research Center, 1986.
30. Cox, B.N., Failure Models For Textile Composites, NASA contractor report 4686, contract no. NAS1-19243, NASA, Cleveland, OH, 1995.
31. Chamis, C.C., Murthy, P.L.N., and Minnetyan, L., Progressive fracture of polymer matrix composite structures, *Theor. Appl. Fracture Mech.*, 25 (1), 1–15, 1996.
32. Abdi, F. and Lorenz, R., Concurrent Probabilistic Simulation of High Temperature Composite Structural Response, NASA Lewis SBIR Phase II final report, NASA, 1995.
33. Shah, A.R., Shiao, M.C., Nagpal, V.K., and Chamis, C., Probabilistic evaluation of uncertainties and risks in aerospace components, 1992.
34. Minnetyan, L., Murthy, P.L.N., and Chamis, C.C., Progressive fracture in composites subjected to hygrothermal environment, *Int. J. Damage Mech.*, 1 (1), 60–79, 1992.
35. Chamis, C.C. and Minnetyan, L., Progressive Fracture of Polymer Matrix Composite Structure, a New Approach, presented at 14th Annual Energy-Source Technology Conference and Exhibition, Houston, TX, 1992.
36. Li, Q., Minnetyan, L., and Chamis, C.C., Structural Durability and Fatigue of Composites in Acoustic Environment, presented at Sampe 2001, Long Beach, CA, 1999.

37. Li, Q., Computational Simulation of Damage Propagation of Composite Structure under PSD Fatigue Loading Condition, M.S. thesis, Clarkson University, Potsdam, NY, 2000.
38. Khatiblou, M., Huang, L.D., and Abdi, F., Impact and Tension after Impact of Composite Launch Space Structure, presented at Sampe 2001, Long Beach, CA, 1999.
39. Abdi, F. et al., Composite Cryogenic Tank Permeability, (Material Characterization, Virtual Testing, Accelerated testing), report no. TA2CT-DD-02-005, Feb. 2002.
40. Abdi, F. and Sue, X., Progressive failure analysis of RLV laminates of IM7/PETI-5 at high, room, and cryogenic temperatures, in *Proceedings of 44th Structures, Structural Dynamics, and Materials Conference*, AIAA-2003-7553, AIAA/ASME/ASCE/AHS/ASC, Reston, VA, 2003.

Index

A

Accelerated life testing, **25**-1–26
   combined-stress percent-life model, **25**-7–11
   deterioration-monitoring model, **25**-12–14
   log-log stress-life model, **25**-2–6
   models, **25**-2
   overload-stress reliability model, **25**-6–7
   step-stress accelerated testing model, **25**-14–25
   test procedure, **25**-12–13
   variation reduction, robust engineering, **25**-1–26
   Weibull step-stress model, **25**-20–22
Adaptive trusses, buckling, **35**-1–27
   adaptive structure, **35**-18–21
   buckling of columns, **35**-4
   conventional trusses, probabilistic progressive bucking of, **35**-4
   finite element model, **35**-2–3
   first buckled member, **35**-5–6
   initial eccentricity, probabilistic buckling, **35**-11–14
   intelligent structure, **35**-23–25
   mathematical models, validation of, **35**-4–5
   probabilistic model, **35**-4
   probabilistic truss end-node displacements, **35**-9–11
   second/third/fourth buckled members, **35**-7–8
   smart structure, **35**-21–22
Advanced simulation
   advanced human-computer interfaces, **2**-15
   components of, **2**-13–15
   intelligent tools, facilities, **2**-14
   nontraditional methods, **2**-15
Adviser-expert
   in expert knowledge reliability characterization, **13**-4
   identifying, in expert knowledge reliability characterization, **13**-6–7
Aeroelasticity
   design, testing, **29**-1–18
   robust control concepts, **29**-11–13
   structural system optimization, **24**-21–23

Aerospace system reliability-based optimization, **24**-1–32, **34**-6–21
   aerodynamics, **24**-21–23
   aeroelasticity, **24**-21–23
   aircraft systems, **24**-23
   algorithms, **24**-12–15
   composite structures, **24**-21
   design sensitivities, **24**-11
   life-cycle cost-based, **24**-15–24
   material cost-based, **24**-16–17
   multicriteria optimization, **24**-15–16
   nested optimization, **24**-12–14
   one-shot optimization method, **24**-12–14
   optimization methods, **24**-12–15
   performance measure approach, **24**-7–8
   reliability index approach, **24**-5–7, **24**-9–11
   sensitivity analysis, **24**-8–11
   standard optimization problem, **24**-4–5
Aggregate uncertainty, generalized information theory, **9**-26
Aggregation of expert judgement, in expert knowledge reliability characterization, **13**-26–27
Aircraft engine field support, Weibull models, **3**-8-9
Aircraft lever, reliability assessment, life distribution, **32**-10–12
Aircraft structure joints
   corrosion-fatigue damage, **28**-1–58
   engineering philosophy, **28**-2–7
   equivalent "physics-based" Weibull failure models, **28**-37
   inspection types, **28**-38
   inspection uncertainties, **28**-37–39
   maintenance activities, **28**-35–39
   modeling, **28**-8–35
   nondestructive inspection (NDI) techniques, **28**-37–38
   physics-based reliability engineering approach, **28**-35–37
   probabilistic modeling, **28**-39
   probability of detection curves, **28**-38–39
   reliability assessment, **28**-1–58
   risk/reliability-based condition assessment, **28**-35–39

I-1

Airframe structure step lap joint, **34**-16–18
Airplane wing, shape optimization, reliability-based design optimization, **34**-12–16
Aleatory uncertainty, **8**-6, **8**-16–17
Aluminum alloys, stochastic variability of corrosion in, **28**-23–28
Ambiguity reduction, nondeterministic hybrid architectures, vehicles, **41**-20
Analysis of expert judgement, expert knowledge reliability characterization, **13**-5
Analyst, in expert knowledge reliability characterization, **13**-4
Analytical reliability methods, **38**-2–4
Applications of reliability assessment, **32**-1–30. *See also under* specific application
    aircraft lever, life distribution, **32**-10–12
    cervical spine impact injury, **32**-14–18
    crankshaft reliability, **32**-12–13
    design assessment of reliability with inspection (DARWIN), **32**-3
    examples, **32**-7–28
    gas turbine rotor risk assessment, **32**-7–9
    numerical evaluation of stochastic structures under stress (NESSUS), **32**-2
    orthopedic implant cement loosening, **32**-21–23
    piston-valve assembly, explosive-actuated, **32**-24–28
    tunnel vulnerability assessment, **32**-18–21
Approximation errors, interval methods, **12**-17–18
Automated fault classification, nondeterministic hybrid architecture, **41**-24–25
Automatic feature extraction, nondeterministic hybrid architecture, **41**-22–23
Automotive design, **5**-1–18
    analytical reliability methods, **38**-2–4
    body-door, **37**-5–8
    bumper-deflection requirement, probability of meeting, **37**-11
    bumper performance measures, preference function for, **37**-11
    bumper standardization, **37**-8–11
    competitive uncertainties, **5**-12
    component-level reliability analysis, **38**-2–4
    cost uncertainties, **5**-12
    crash safety, **38**-16–19
    decision analysis cycle, **5**-4–16
    description, and range of design variables, **37**-15
    deterministic phase, **5**-7–9
    door seal selection, **5**-5–9, **5**-11–12, **5**-15–16
    door sealing system, **5**-6–7
    economic uncertainties, **5**-12
    emergence, decline of suppliers, **5**-12
    exchange rates, fluctuations in, **5**-12
    final values, design variables, **37**-15
    finite element, vehicle side-impact example, **37**-14
    formal preference aggregation theory, **37**-4–5
    glass guidance design reliability, **38**-8–10
    information phase, **5**-13–16
    insignificant input variable elimination, backward screening, **38**-16
    liftgate reliability, **38**-6–8
    local polynomial fitting, **38**-13–15
    main-effects estimate, **38**-15–16
    mathematical model validation, **37**-16
    metamodel prediction error, cross validation, **38**-13
    model validation, **5**-13–15
    nonparametric metamodeling (response-surface) methods, **38**-10–19
    nonparametric multiple-regression techniques, **38**-10–11
    objectives, aggregation of, bumper, **37**-10
    parametric multiple-regression techniques, **38**-10–11
    preference aggregation theory, bumper optimization using, **37**-9–11
    preference function for performance measure, **37**-6
    probabilistic analysis, **38**-1–28
    purchase price, parts, **5**-12
    quality specification structure, **37**-2–5
    range of design variables, **37**-15
    raw material prices, fluctuations in, **5**-12
    reliability-based design optimization, **37**-4
    robust engineering, variation reduction in, **38**-19–22
    sample partitioning, computer experiments, **38**-11–12
    simulation-based methods, system reliability analysis, **38**-5–6
    system-level reliability analysis, **38**-5–6
    translation uncertainties, **5**-12
    uncertainty, multiattribute, **37**-1–19
    uncertainty characterization methods, **5**-10–11
    unemployment levels, **5**-12
    uniform sampling, computer experiments, **38**-11–12
    validation, mathematical models, **5**-13
    variable screening, **38**-15–16
    vehicle-bumper interaction, design variables, **37**-9
    vehicle development process, **5**-2, **5**-3
    vehicle side-impact, **37**-13–16
Axial compressor blade, shape organization, reliability-based design optimization, **34**-7–12

B

Basic probability assignments, evidence theory, **10**-17–20
Bayesian modeling **22**-1–16
    asymptotic approximations, **22**-10–11, **22**-15–16
    computational issues, **22**-12–15
    damage detection, **22**-6–9
    design, **22**-11–12
    distribution function formulation of, **9**-10–11
    empirical Bayes, **7**-6
    family of, **7**-6–7
    formulation, as probability integral, **22**-14
    generalized information theory, **9**-9–10
    hierarchial Bayes, **7**-6–7
    information entropy measure, parameter uncertainty, **22**-9–10
    model class selection, **22**-5–6
    sensor configuration for selecting model class, **22**-13–14
    sensor location, **22**-9–14
    statistical modeling, updating, **22**-3–5
    in statistical testing for reliability certification, **26**-22–23
    structural reliability predictions, **22**-14–16
    symbols, **22**-19–20
    updating, **22**-1–20

Belief
    construction of, for system response, 10-20–24
    evidence theory, 10-4–6
    generalized information theory, 9-22
Beta distribution, 26-5
    example plots, 26-6
Bias, in expert knowledge reliability characterization, 13-17–20
Binomial/beta reliability example, generalized information theory, 9-11–12
Binomial distribution, 26-5
    example plot, 26-7
Blade, turbine, probabilistic design, 30-7–19
Block-module model, pipeline risk control, 40-22–23
Block slip band, fatigue response prediction, 42-28
Body of evidence, generalized information theory, 9-23
Bounding approximation, interval methods, rounding errors, 12-17–19
Buckling, progressive, conventional, adaptive trusses, 35-1–27
    adaptive structure, 35-18–21
    buckling of columns, 35-4
    conventional trusses, probabilistic progressive bucking of, 35-4
    finite element model, 35-2–3
    first buckled member, 35-5–6
    initial eccentricity, probabilistic buckling, 35-11–14
    intelligent structure, 35-23–25
    mathematical models, validation of, 35-4–5
    probabilistic model, 35-4
    probabilistic truss end-node displacements, 35-9–11
    second/third/fourth buckled members, 35-7–8
    smart structure, 35-21–22

C

Calculus of probability, 9-8
Cantilever with uncertain load, design of, 11-2–11
CCCF model. *See* Crack-closure corrosion-fatigue
Certain distribution, generalized information theory, 9-31
Certification, reliability, 1-3, 25-1–25, 26-20–21, 27-1-18
Cervical spine impact injury, reliability assessment, 32-14–18
Choquet capacities, generalized information theory, 9-3
Civil structural system optimization, 24-1–32
    aeroelasticity, 24-21–23
    algorithms, 24-12–15
    composite structures, 24-21
    cost-based, 24-17–24
    design sensitivities, 24-11
    life-cycle cost reliability-based optimization, 24-15–16
    material cost-based, 24-16–17
    multicriteria optimization, 24-15–16
    nested optimization, 24-12–14
    one-shot optimization methods, 24-12–14
    optimization methods, 24-12–15
    performance measure approach, 24-7–8
    reliability index approach, 24-5–11
    sensitivity analysis, 24-8–11
    standard optimization problem, 24-4–5

Cognitive bias, in expert knowledge reliability characterization, 13-18–19
Collaborative strategy, defining, in expert knowledge reliability characterization, 13-7–8
Column buckling, 35-4
Combined-stress percent-life model, accelerated life testing, 25-7–11
Commercial software systems, 2-13
Community of practice, domains, in expert knowledge reliability characterization, 13-4
Competitive uncertainties, in automotive design, 5-12
Complementary cumulative function, evidence theory, 10-6–7
Component level, mesodomain, mechanical component, 42-5
Component-level reliability analysis, probabilistic analysis, design, 38-2–4
Composite materials, weakest-link probabilistic failure, 43-8–9
Composite structures
    composite plate example, 44-11–13
    creep limit state, 44-26–31
    critical-component failure, 44-15–16
    deterministic initial screening, 44-14–15
    experimental results, 44-21–22
    failure mechanism, 44-22–25
    fast branch-and-bound method, 44-14
    fatigue-damage modeling, 44-18–21
    fatigue-delamination limit state, 44-22–26
    fatigue limit state: material modeling, 44-18–22
    laminate theory, 44-2–7
    materials, 44-1–36
    multiple failure sequences, 44-6
    numerical example, 44-6–7
    numerical example: aircraft wing, 44-16–18
    ply-level limit states, 44-3–4, 44-9–10
    random variables, 44-25
    response-surface modeling, limit state, 44-25–26
    shear deformation theory, 44-8–9
    single failure sequence, 44-4–6
    stiffness modification, progressive failure analysis, 44-10–11
    stochastic fiber fracture, creep analysis with, 44-28–30
    strength limit state approximations, 44-13–18
    structural system optimization, 24-21
    system failure probability, 44-4–6
    three-dimensional analysis, 44-8–13
    undamaged composite, creep of, 44-27–28
    verification, methodology, 44-26
    weakest-link model, 44-15
Compressor, turbine blade, probabilistic design, 30-7–19
    blade frequency-probabilistic Campbell diagram, 30-12–13
    failure probability calculation, 30-16–18
    field effects, 30-19
    forced-response prediction, 30-14–15
    geometry variation, 30-10–12
    material capability, 30-15–16
    mistuning, 30-13–14
    predictions, updating of, 30-18–19

Computer-aided engineering methodology, **36**-1–20
   decision maker, **36**-9–10
   design optimization, **36**-5–10
   durability analysis, **36**-4–5
   durability design optimization, **36**-6–7
   dynamic analysis, **36**-4
   experimental validation, mechanical fatigue, **36**-3–5
   fatigue-life analysis, experimental validation, **36**-3–10
   mechanical fatigue failure, army trailer drawbar, **36**-3
   model, performance measure approach, reliability analysis, reliability-based design optimization, **36**-13–14
   notch effects, **36**-9
   optimal design, durability-based, reliability analysis, reliability-based design optimization, **36**-10–12
   parameterization, design, **36**-5
   reliability analysis, **36**-10–18
   results, design optimization, **36**-7–9
   sensitivity analysis, design, **36**-5–6
   simulation model, **36**-3
   tank roadarm durability, reliability analysis, reliability-based design optimization, **36**-15–18
Computer models, virtual testing using, **27**-2–4
Confidence interval modeling, **3**-17–18
Conflict, nonspecificity, **8**-11
Continuously distributed dislocations, model based on, fatigue response prediction, **42**-13–18
Convexity, generalized information theory, **9**-8
Corroded surface topography, corrosion-fatigue-damage modeling, **28**-27–28
Corrosion damage
   laboratory studies, **28**-24–26
   Wei pit model, **28**-22–23
Corrosion-fatigue damage, **28**-8–35
   aircraft structure joints, **28**-1–58
   aluminum alloys, stochastic variability, **28**-23–28
   comparative results, **28**-31–35
   corroded surface topography, **28**-27–28
   corrosion damage, **28**-21–28
   corrosion pitting, **28**-22–23
   crack-closure corrosion-fatigue model, **28**-31
   crack-closure model, **28**-16–17
   crack initiation, **28**-8–14
   crack propagation, **28**-14–21
   damage curve approach, **28**-12
   double damage curve approach, **28**-12
   field studies, **28**-23–24
   forman model, **28**-15
   hyperbolic sine model, **28**-16
   laboratory studies, **28**-24–26
   linear-damage rule, **28**-11–12
   maintenance activities, **28**-35–39
   modified sigmoidal model, **28**-16
   power-law pit model, **28**-22
   simultaneous corrosion-fatigue (SCF) model, **28**-31
   stochastic variabilities, **28**-12–14, **28**-17–21
   Wei corrosion-fatigue (WCF) model, **28**-29–31
   Wei pit model, **28**-22–23
Corrosion wastage models, **39**-13–15
Cost uncertainties, in automotive design, **5**-12
Crack-closure corrosion-fatigue model, **28**-31
Crack-closure model, corrosion-fatigue-damage modeling, **28**-16–17
Crack initiation, corrosion-fatigue-damage modeling, **28**-8–14
Crack nucleation
   fatigue response prediction, **42**-6–11, **42**-25
   micromechanics of, fatigue response prediction, **42**-6–7
Crack propagation, corrosion-fatigue-damage modeling, **28**-14–21
Crankshaft reliability, reliability assessment, **32**-12–13
Crash safety, probabilistic analysis, **38**-16–19
Creep limit state, composite structures, materials, **44**-26–31
   creep analysis with stochastic fiber fracture, **44**-28–30
   numerical example, **44**-31–32
   undamaged composite, creep of, **44**-27–28
Crisp possibility distribution, **9**-31
Crispness histogram, generalized information theory, **9**-31–34
Cumulative distribution, generalized information theory, **9**-7
Cumulative function, evidence theory, **10**-6–7

D

Damage-accumulation models, fatigue response prediction, **42**-5–19
Damage-accumulation stages, Monte Carlo simulation, fatigue response prediction, **42**-25–29
Damage curve approach, corrosion-fatigue-damage modeling, **28**-12
Damage due to pipeline failure, assessing, **40**-10–11
DARWIN. *See* Design assessment of reliability with inspection
Data congeries, **7**-8–9
Data uncertainty, multivariate stochastic analysis, **41**-20
Decision analysis cycle, in automotive design, **5**-4–16
Decision-level fusion
   benefits of, **41**-20
   fault-detection sensitivity, **41**-20
   nondeterministic hybrid architectures, vehicles, **41**-4
   system reliability, **41**-20
Decision maker, in expert knowledge reliability characterization, **13**-4
Decline of suppliers, in automotive design, **5**-12
Decomposable measure, **9**-21
Defect growth, pipeline risk assessment, **40**-17
Defect-growth processes, identification of, pipeline risk assessment, **40**-14–16
Definitions of uncertainty, **2**-3
Delphi approach, reliability analysis, **13**-22–23
Dempster combination, **9**-23
Dempster-Shafer evidence theory, **9**-2, **9**-22–26
Dependency, interval methods, **12**-5
Dependent nonnormal random variables, **14**-5–6
Dependent normal random variables, **14**-4
Design assessment of reliability with inspection (DARWIN), **32**-3
Design optimization strategy, reliability-based, **31**-1–19
   analysis procedure, **31**-5–6
   certification problem, **31**-3–4
   cryogenic tank, **31**-12–16

data preparation, **31**-5–6
data problems, **31**-3
deterministic optimization, **31**-6, **31**-11, **31**-13
example, **31**-10–11
generation of response surfaces, **31**-6
probabilistic load models, **31**-6
probabilistic resistance models, **31**-6
probabilistic sensitivity factors, **31**-15
reliability assessment, **31**-6, **31**-11, **31**-13–14
reliability-based design optimization, **31**-11
reliability-based optimal design, **31**-6, **31**-15
simple analytical problem, **31**-10–11
system reliability level selection, **31**-16
target reliability problem, **31**-4
validation of CAE (CAE) models, **31**-5
Design space exploration, product development, **4**-5–8
Design under uncertainty, **23**-12–13
Deterioration-monitoring model, accelerated life testing, **25**-12–14
test procedure, **25**-12–13
Deterministic phase, in automotive design, **5**-7–9
Deterministic systems
current methods, **6**-2, **6**-4–5
input, **6**-2–4
research needs, trends, **6**-2–7
stochastic input, **6**-4–7
Dislocation level, mesodomain, mechanical component, **42**-5
Ditlevsen bounds, **15**-14–15
DMC. *See* Dynamic Monte Carlo
Documentation, in expert knowledge reliability characterization, **13**-22
Domains, in expert knowledge reliability characterization, **13**-3, **13**-4
Door seal selection, in automotive design, **5**-5–7
deterministic phase, **5**-8–9
information phase, **5**-15–16
probabilistic phase, **5**-11–12
Door sealing system, in automotive design, **5**-6–7
Double damage curve approach, corrosion-fatigue-damage modeling, **28**-12
Drawbar, trailer, computer-aided engineering methodology, **36**-3
Durability design optimization, computer-aided engineering methodology, **36**-6–7
Dynamic Monte Carlo simulation in high-dimensional stochastic spaces, **20**-22–25

E

Early developments, history of nondeterministic analysis, **3**-8
Economic uncertainties, in automotive design, **5**-12
Efficient time-variant reliability methods in load space, **33**-1–28
active load, frame subject to, **33**-11–12
deformable system, **33**-3–5
design values of loads, combination, **33**-12–15
downsizing dimension, reliability problem, **33**-22–23
elastic-plastic frames, subject to combination of random loads, **33**-10–11
examples, **33**-25
large mechanical systems, subjected to combination of diffusion-type Markov loads, by generalized Pontryagin equations, **33**-15–25
Markov loads, **33**-15–20
Pontryagin equations, **33**-20–22
random loads, combination of, **33**-20
redundant structures, subject to combinations of Markov-type loads, **33**-1–15
snow load, **33**-18
system life, system reliability, transition, **33**-23–25
technological loads, **33**-19–20
unloading, frame subject to, **33**-11–12
wind load, **33**-19
Empirical Bayes method, **7**-6
Engine-engine variability, feature vector extraction, **41**-19
Engineering decision under uncertainty, **1**-3
Enhancing modeling, **2**-7–9
Ensembled upcrossing rate approach, time-variant reliability, time-variant reliability estimation techniques, **18**-9–10
Epistemic uncertainty, **8**-6, **8**-17–18
Equally correlated elements, series systems with, **15**-15–16
Error checking, response surfaces for reliability assessment, **19**-14–15
Error estimation, **3**-16–17
EUR. *See* Ensembled upcrossing rate
Evidence theory, **10**-1–30
analysis using, **10**-17–24
application of, **10**-26
basic probability assignments, **10**-17–20
belief, **10**-4–6
belief construction, for system response, **10**-20–24
combination of evidence, **10**-14–15, **10**-18–19
complementary cumulative function, **10**-6–7
cumulative function, **10**-6–7
example problem, **10**-12–26
fundamentals, **10**-4–12
input/output uncertainty mapping, **10**-8–10
interpretation of results, **10**-25–26
plausibility construction, for system response, **10**-20–24
probabilistic response, construction of, **10**-15–16
problem description, **10**-12–13
results, comparison, **10**-25–26
simple conceptual examples, **10**-10–12
traditional analysis, **10**-14–16
Exchange rates, in automotive design, fluctuations in, **5**-12
Expert elicitation approach, criteria for, **13**-5–6
Expert judgement, defined, **13**-14–15
Expert judgement elicitation, in expert knowledge reliability characterization, **13**-5
Expert knowledge reliability characterization, **3**-7, **3**-10, **13**-1–29
analyzing expert judgement, **13**-23–29
definitions, **13**-3–14
Delphi approach, reliability analysis, **13**-22–23
expert judgement elicitation, **13**-14–22
fundamental relationship representation, **13**-13
grouped representation, **13**-12

information integration, for reliability, **13**-28–29
model development, eliciting expertise for, **13**-8–13
model population, **13**-14–23
ontology for missile design, **13**-13
philosophy, **13**-3–5
problem definition, **13**-6–8
problem identification, **13**-5–6
reliability block diagram, **13**-11
representativeness, **13**-18
star graph, **13**-12
stoplight chart, **13**-14
uncertainties, characterizing, **13**-23–27
Experts, in expert knowledge reliability characterization, **13**-4, **13**-17
Extension principle, generalized information theory, **9**-26
Extreme value distribution, **8**-15

F

Failure-location predictions, weakest-link probabilistic failure, **43**-3–5
Failure mechanisms, identification of, **15**-26–38, **44**-30–31
   assessment, system reliability, **15**-27–32
   brittle elements, two-story braced frame with, **15**-34–36
   ductile elements, two-story braced frame with, **15**-32–34
   elastic plastic framed structure, **15**-36–38
   examples of, **15**-32–38
Failure of series systems, probability of, **15**-12–13
Family of Bayesian methods, **7**-6–7. *See also* Bayesian modeling
Fan, probabilistic design, **30**-7–19
   blade frequency-probabilistic campbell diagram, **30**-12–13
   failure probability calculation, **30**-16–18
   field effects, **30**-19
   forced-response prediction, **30**-14–15
   geometry variation, **30**-10–12
   material capability, **30**-15–16
   mistuning, **30**-13–14
   predictions, updating of, **30**-18–19
Fatigue crack model, **39**-15–19
Fatigue-delamination limit state, composite structures, materials, **44**-22–26
   failure mechanism, **44**-22–25
   model, **44**-22–25
   random variables, **44**-25
   response-surface modeling, limit state, **44**-25–26
   verification, methodology, **44**-26
Fatigue-life analysis, experimental validation, computer-aided engineering methodology, **36**-3–10
   decision maker, **36**-9–10
   design optimization, **36**-5–10
   durability analysis, **36**-4–5
   durability design optimization, **36**-6–7
   dynamic analysis, **36**-4
   experimental validation, mechanical fatigue, **36**-3–5
   mechanical fatigue failure, army trailer drawbar, **36**-3
   notch effects, **36**-9
   parameterization, design, **36**-5
   results, design optimization, **36**-7–9

sensitivity analysis, design, **36**-5–6
simulation model, **36**-3
Fatigue limit state, **44**-18–22
   experimental results, **44**-21–22
   fatigue-damage modeling, **44**-18–21
   model performance, **44**-21–22
Fatigue response prediction, **42**-1–38
   applied microstress, **42**-24–25
   block slip band, **42**-28
   continuously distributed dislocations, model based on, **42**-13–18
   crack nucleation, **42**-6–11, **42**-25
   damage-accumulation models, **42**-5–19
   frictional shear stress, **42**-24–25
   grain orientation, **42**-20–24
   grain size, **42**-19–20
   long-crack growth, **42**-18–19
   mathematical model validation, **42**-5–6
   mesodomains, mechanical component, **42**-5
   micromechanical crack nucleation models, **42**-8–11
   micromechanics, **42**-11–12
   microstructural probabilistic methods, **42**-1–38
   microstructurally small-crack growth, **42**-11–18
   Monte Carlo simulation, damage accumulation stages, **42**-25–29
   nucleation, crack, micromechanics, **42**-6–7
   overview, **42**-5–6
   probabilistic micromechanics, **42**-4–5
   propagated slip band, **42**-28
   random variable statistics, estimate of, **42**-19–25
   scatter, in fatigue life, **42**-2–3
   small-crack growth, **42**-12–13, **42**-25–28
   stages of fatigue, **42**-2
Fault basins of attraction
   multivariate stochastic analysis, **41**-20
   nondeterministic hybrid architectures, vehicles, **41**-20
Feature mapping, multivariate stochastic analysis, **41**-20
Feature vector extraction
   benefits of, **41**-19
   nondeterministic hybrid architectures, vehicles, **41**-3
Finite element formulation
   probabilistic analysis, **3**-10–11
   random media, **21**-5–8
   spatial discretization, **21**-6–8
Finite element reliability analysis, **14**-22
First-order polynomial, response surfaces for reliability assessment, **19**-8–9
First-order reliability method, **14**-6–15
   component reliability by FORM, **14**-6–8
   example series system reliability analysis of frame by FORM, **14**-12–13
   FORM importance, sensitivity measures, **14**-13–15
   frame after proof test, reliability updating, example, **14**-13
   importance, sensitivity measures for column, example, **14**-15
   reliability analysis of column by FORM, example, **14**-9
   system reliability by FORM, **14**-9–13
First passage probability, time-variant reliability, **18**-3–4
Fleet level, mesodomain, mechanical component, **42**-5
Focal set, **9**-23

Forman model, corrosion-fatigue-damage modeling, **28**-15
Framing decision, uncertainty and, **8**-13
Frictional shear stress, fatigue response prediction, **42**-24–25
Fuzzified Dempster-Shafer theory, **9**-3
Fuzzy arithmetic, generalized information theory, **9**-17–19
Fuzzy interval, **9**-18
Fuzzy logic, generalized information theory, **9**-16
Fuzzy measures, **9**-21–22
Fuzzy number, **9**-18
Fuzzy quantity, **9**-17
Fuzzy sets
   generalized information theory, **9**-14–16
   type II, level-II, **9**-3
Fuzzy subsets, generalized information theory, **9**-15
Fuzzy systems
   generalized information theory, **9**-14–20
   probability, comparing, **9**-17
   theory of, **9**-2
Fuzzy theory, in transition assessment, **3**-12

G

Galerkin scheme, stochastic reduced basis projection schemes, **21**-16–18
Gas turbine engine, probabilistic, turbomachinery design, **30**-1–21
   blade frequency-probabilistic Campbell diagram, **30**-12–13
   failure probability calculation, **30**-16–18
   fan, turbine blade probabilistic design, **30**-7–19
   field effects, **30**-19
   field-management risk assessments, **30**-7
   forced-response prediction, **30**-14–15
   future probabilistic efforts, **30**-7
   geometry variation, **30**-10–12
   material capability, **30**-15–16
   mistuning, **30**-13–14
   new parts, rotor design/fracture mechanics, probabilistic, **30**-5–7
   predictions, updating of, **30**-18–19
   rotor design/fracture mechanics, probabilistic, **30**-4–7
   traditional reliability engineering, **30**-2–4
Gaussian processes, time-variant reliability, **18**-7–8
Gaussian vectors
   stochastic finite element analysis, **20**-506
   stochastic simulation, **20**-506
General dynamic system, probabilistic analysis, **17**-2–10
   crossings of level during period, **17**-9
   dynamic systems, classified under following categories:, **17**-7
   failure analysis, **17**-8–10
   fields, random, **17**-2–3
   first passage time, **17**-10
   impulse response, or Green's function, **17**-7
   modeling random processes, **17**-2–8
   parameters, random, time invariant, **17**-7
   parameters deterministic, **17**-7
   probabilistic model of random processes, **17**-3–8
   probability density function, peaks, **17**-9–10
   random processes, **17**-2–3
   response, calculation of, **17**-5–8
   time variant system, parameters, random, **17**-7
Random sets, general, **9**-25
Generalized information theory, **9**-1–40, **13**-26
   aggregate uncertainty, **9**-26
   Bayes theorem, **9**-9–10
   belief, **9**-22
   binomial/beta reliability example, **9**-11–12
   body of evidence, **9**-23
   calculus of probability consists of:, **9**-8
   call such monotone measure distributional or decomposable, **9**-21
   certain distribution, **9**-31
   Choquet capacities imprecise probabilities, **9**-3
   classical approaches to, **9**-3–12
   consistency, **9**-31–34
   convex combinations of probability measures, **9**-38
   convexity, **9**-8
   crisp possibility distribution, **9**-31
   crispness histogram, **9**-31–34
   cumulative distribution, **9**-7
   Dempster combination, **9**-23
   Dempster-Shafer evidence theory, **9**-2, **9**-22–26
   distributions, **9**-10–11, **9**-18, **9**-29–31
   extension principle, **9**-26
   focal element, **9**-23
   focal set, **9**-23
   fuzzified Dempster-Shafer theory, **9**-3
   fuzzy arithmetic, **9**-17–19
   fuzzy interval, **9**-18
   fuzzy logic, **9**-16
   fuzzy measures, **9**-21–22
   fuzzy number, **9**-18
   fuzzy quantity, **9**-17
   fuzzy sets, **9**-3, **9**-14–16
   fuzzy subset, **9**-15
   fuzzy systems theory, **9**-2, **9**-14–20
   general random set, **9**-25
   generalized information theory (GIT), **9**-2
   historical development, **9**-12–13
   imprecise probabilities, **9**-38
   infinite order Choquet capacities, **9**-38
   information content of random set, **9**-25–26
   interpretations of probability, **9**-8–9
   interval analysis, **9**-2, **9**-4–6
   likelihood approaches for probability, **9**-9–10
   linguistic variable, **9**-20
   logical theoretical approach, **9**-3–4
   membership function, **9**-14
   Mobius inversion, **9**-23
   monotone measures, **9**-21–22
   monotone or fuzzy measure, **9**-2
   multiplication, **9**-8
   necessity measure, **9**-30
   normalization condition, **9**-6
   operators, **9**-13–14
   P-boxes, **9**-27–29
   plausibility, **9**-22
   possibilistic histogram, **9**-31–34
   possibility distribution, **9**-29
   possibility measure II, **9**-29

possibility measures, **9**-29–31
possibility theory, **9**-2, **9**-29–31
probabilistic, possibilistic concepts, relations, **9**-33–34
probability density, **9**-18
probabilistic representations, **9**-6–12
probability density function, **9**-7
probability space, **9**-7
probability theory, **9**-2
proper possibility distribution, **9**-31
random intervals, **9**-27–29
random sets, **9**-22–26
random variable, **9**-24
rough sets, **9**-2
set theoretical approach, **9**-2–4
specific random sets, **9**-26
triangular conorm, **9**-14
triangular norm, **9**-14
uncertainty-based information theory, **9**-1–3
uncertainty quantification, **9**-2
GIT. *See* Generalized information theory
Glass guidance design reliability, **38**-8–10
Grain level, mesodomain, mechanical component, **42**-5
Grain orientation, fatigue response prediction, **42**-20–24
Grain size, fatigue response prediction, **42**-19–20

H

Health management system, **3**-6
Hierarchial Bayes method, **7**-6–7
History of nondeterministic analysis, **3**-7–10
    civil engineering-based reliability developments, **3**-9
    early developments, **3**-8
    probability integration algorithms, **3**-9–10
    Weibull models, aircraft engine field support, **3**-8–9
Hull girder collapse, target reliability for ships against, **39**-4–5
Hull girder load models, **39**-11–12
    still-water bending moment, **39**-11
    wave-induced bending moment, **39**-11–12
Hull girder strength models, **39**-6–11
    corrosion, effect of, **39**-9, **39**-11
    denting, local, effect of, **39**-10, **39**-11
    fatigue cracking, effect of, **39**-9–11
    primary failure mode, **39**-7–8
    quaternary failure mode, **39**-8–9
    secondary failure mode, **39**-8
    tertiary failure mode, **39**-8
Hull sectional properties, example vessels, **39**-22
Human-computer interfaces, **2**-15
Human errors
    information for, **8**-18
    uncertainty due to, **8**-10
Hybrid architectures, vehicles, nondeterministic
    ambiguity reduction, **41**-20
    automated fault classification, **41**-24–25
    automatic feature extraction, **41**-22–23
    data uncertainty, incorporation of, **41**-20
    decision-level fusion, **41**-4, **41**-16–20
    fault basins of attraction, **41**-2, **41**-20
    fault-detection sensitivity, **41**-10–11
    feature mapping, **41**-20
    feature vector, engine behavior, **41**-6–8
    feature vector extraction, **41**-3, **41**-19
    in-operation conditions, predictions for, **41**-7
    mathematical model validation, **41**-4–6
    multiple-band-pass demodulation, **41**-19–27
    multiple-band-pass fault-severity index, **41**-26–27
    multiple fault detection, **41**-20
    multiple-fault pattern, stochastic fault diagnostics, **41**-13
    multivariate reliability analysis, **41**-3–4, **41**-12–16
    multivariate stochastic analysis, benefits of, **41**-20
    nondeterministic adaptive network-based fuzzy inference models, **41**-4–6
    partial-engine transient model, **41**-9–10
    prognostic output, **41**-20
    quasi-stationary engine model, **41**-8
    reliability-based reasoning, **41**-20
    robust operational performance, **41**-20
    single fault pattern, stochastic fault diagnostics, **41**-13
    sparse sensor array studies, **41**-8–10
    stochastic fault diagnostics, aspects related to, **41**-13
    stochastic tools development, **41**-3–4
    transient engine model, **41**-9
Hybrid uncertainty, **11**-22–25
    info-gap robustness as decision monitor, **11**-22–23
    nonlinear spring, **11**-23–25
Hyperbolic sine model, corrosion-fatigue-damage modeling, **28**-16
Hyperdistribution statistical distribution, **26**-5
Hyperparameterization example plot, **26**-8

I

Identifying experts, in expert knowledge reliability characterization, **13**-16–17
Importance sampling, **27**-10–11
    variance-reduction, **27**-8–9, **27**-14–15
Imprecise probabilities, **9**-38
Impression management, in expert knowledge reliability characterization, **13**-20
In-flight capabilities, feature vector extraction, **41**-19
In-operation conditions, predictions for, nondeterministic hybrid architectures, vehicles, **41**-7
Inductive research approach, **7**-5–7
Infinite order Choquet capacities, **9**-38
Info-gap decision theory, **11**-1–30
    cantilever with uncertain load, design of, **11**-2–11
    clash with performance-optimal design, **11**-7–8
    hybrid uncertainty, **11**-22–25
    info-gap robustness as decision monitor, **11**-22–23
    information-gap models of uncertainty, **11**-28–29
    maneuvering, **11**-11–16
    model uncertainty, **11**-11–12
    nonlinear spring, **11**-23–25
    optimal identification, **11**-17
    performance optimization, **11**-3–5
    performance optimization with best model, **11**-12–13
    robustness, **11**-17–19
    robustness function, **11**-13–15
    system identification, **11**-17–21
    uncertain load, **11**-5–6, **11**-9–11

uncertainty, **11**-17–19
vibrating system with uncertain dynamics
Information
  expert knowledge combining with, in expert knowledge reliability characterization, **13**-28–29
  types of, **8**-14–15
Information aggregation, **7**-7–9
  confidence intervals, **7**-8
  data congeries, **7**-8–9
Information content of random set, generalized information theory, **9**-25–26
Information fusion, **3**-2
Information integration, analysis, in expert knowledge reliability characterization, **13**-5
Information phase, in automotive design, **5**-13–16
Input/output uncertainty mapping, evidence theory, **10**-8–10
Input variables, statistical distribution of, **26**-2–17
Insignificant input variable elimination, backward screening, **38**-16
Integrated computer-aided engineering methodology, **36**-1–20
  decision maker, **36**-9–10
  design optimization, **36**-5–10
  design parameterization, **36**-5
  design sensitivity analysis, **36**-5–6
  durability, **36**-13–15
  durability analysis, **36**-4–5
  durability-based optimal design, **36**-10–12
  durability design optimization, **36**-6–7
  dynamic analysis, **36**-4
  experimental validation, mechanical fatigue, **36**-3–5
  fatigue-life analysis, experimental validation, **36**-3–10
  mechanical fatigue failure, army trailer drawbar, **36**-3
  notch effects, **36**-9
  performance measure approach, **36**-13–14
  reliability analysis, reliability-based design optimization, **36**-10–18
  results, design optimization, **36**-7–9
  simulation model, **36**-3
  tank roadarm durability, **36**-15–18
Intelligent structure, probabilistic progressive buckling, conventional, adaptive trusses, **35**-23–25
Intelligent tools, facilities, **2**-14
Interpretations, **9**-19–20, **9**-33
Interval analysis, generalized information theory, **9**-2, **9**-4–6
Interval arithmetic
  operations of, **12**-23–24
  in transition assessment, **3**-12–13
Interval finite element methods, **12**-9–17
Interval methods, computing, **12**-1–24
  applications, **12**-2–5
  approximation errors, **12**-17–18
  bounding approximation, rounding errors, **12**-17–19
  definitions, **12**-2–5
  dependency, **12**-5
  future developments, **12**-19–21
  interval arithmetic operations, **12**-23–24
  interval finite element methods, **12**-9–17
  linear interval equations, **12**-6–8

major areas of mathematics-computations, **12**-2
mathematics-computations field, **12**-2
matrices, **12**-6
predicting system response due to uncertain parameters, **12**-8–17
rounding-off errors, **12**-18–19
scientific, engineering modeling, **12**-3
scope, **12**-2–5
sensitivity analysis, **12**-9
uncertainty modeling, **12**-2–8
vectors, **12**-6
Intuitive heuristics, in expert knowledge reliability characterization, **13**-18

K

Knowledge concepts, in expert knowledge reliability characterization, **13**-3–4
Knowledge model, in expert knowledge reliability characterization, defined, **13**-8–9
Knowledge representation techniques, for modeling expertise, **13**-10–13
Krylov subspace, stochastic finite element analysis, **21**-14–16
  theorems, **21**-14–15

L

Laminate theory, composite structures, materials, **44**-2–7
  multiple failure sequences, **44**-6
  numerical example, **44**-6–7
  ply-level limit states, **44**-3–4
  reliability, **44**-3–4
  single failure sequence, **44**-4–6
  system failure probability, **44**-4–6
Life-cycle cost-based structural system optimization, **24**-15–24
Life-cycle cost reliability-based optimization, **24**-15–16
Liftgate reliability, **38**-6–8
Linear-damage rule, corrosion-fatigue-damage modeling, **28**-11–12
Linear interval equations, **12**-6–8
Linear model, response surfaces for reliability assessment, **19**-4–5
Linguistic information, **8**-14
Linguistic variables, **9**-20
Load combinations, time-variant reliability, **18**-11–13
Load space, time-variant reliability methods in, **33**-1–28
  active load, frame subject to, **33**-11–12
  deformable system, **33**-3–5
  design values of loads, combination, **33**-12–15
  downsizing dimension, reliability problem, **33**-22–23
  elastic-plastic frames, subject to combination of random loads, **33**-10–11
  examples, **33**-25
  large mechanical systems, subjected to combination of diffusion-type Markov loads, by generalized Pontryagin equations, **33**-15–25
  Markov loads, **33**-15–20
  Pontryagin equations, **33**-20–22

random loads, combination of, **33**-20
redundant structures, subject to combinations of Markov-type loads, **33**-1–15
snow load, **33**-18
system life, system reliability, transition, **33**-23–25
technological loads, **33**-19–20
unloading, frame subject to, **33**-11–12
wind load, **33**-19
Loads, as processes, time-variant reliability, **18**-2–3
Local polynomial fitting, probabilistic analysis, **38**-13–15
Log-log stress-life model, accelerated life testing, **25**-2–6
Logical approach, generalized information theory, **9**-2
Logical theoretical approach, generalized information theory, **9**-3–4
Lognormal distribution, **26**-5
example plots, **26**-4
Long-crack growth, fatigue response prediction, **42**-18–19
Lotteries, utility method, **23**-4–5

M

Main-effects estimate, **38**-15–16
Maintenance strategies, reliability-based, **15**-41–43
Managing uncertainties, **2**-5
Material cost-based structural system optimization, **24**-16–17
Mathematical model validations, **8**-5, **8**-9, **42**-5–6
Mathematics-computations, interval methods, **12**-2
Matrices, interval methods, **12**-6
Mechanical design
business case, **4**-2–3
low-cost development, **4**-4
multidisciplinary analysis, **4**-11–12
nondeterministic technologies, role, **4**-1–14
probabilistic analysis approaches, **4**-9–11
sensitivity analysis, **4**-8–9
software implementation, nondeterministic technologies, **4**-12–13
space exploration, product development, **4**-5–8
technology advances, **4**-13–14
technology transition, **4**-12–13
Mechanical fatigue, fatigue-life analysis, experimental validation, **36**-3–5
Membership function, **9**-14
Mesodomains, mechanical component, **42**-5
fatigue response prediction, **42**-5
Metamodel prediction error, cross validation, **38**-13
Micromechanical crack nucleation models, fatigue response prediction, **42**-8–11
Microstress in fatigue response prediction, **42**-24–25
Microstructural probabilistic methods, fatigue response prediction, **42**-1–38
applied microstress, **42**-24–25
block slip band, **42**-28
continuously distributed dislocations, model based on, **42**-13–18
crack nucleation, **42**-6–11, **42**-25
damage-accumulation models, **42**-5–19
frictional shear stress, **42**-24–25
grain orientation, **42**-20–24
grain size, **42**-19–20

long-crack growth, **42**-18–19
mathematical model validation, **42**-5–6
mesodomains, mechanical component, **42**-5
micromechanical crack nucleation models, **42**-8–11
microstructurally small-crack growth, **42**-11–18
models, small-crack growth, **42**-12–13
Monte Carlo simulation, damage accumulation stages, **42**-25–29
overview, **42**-5–6
probabilistic micromechanics, **42**-4–5
propagated slip band, **42**-28
random variable statistics, estimate of, **42**-19–25
scatter, in fatigue life, **42**-2–3
small-crack growth, **42**-11–12, **42**-25–28
Misspecification, impact of, **26**-2–3
Mobius inversion, **9**-23
Model approximation methods, reliability-based design optimization, **34**-2–3
Model development, in expert knowledge reliability characterization, **13**-5
Model updating, **3**-3
Model validation, in automotive design, **5**-13–15
Modeling environments, components of, **2**-13–15
Modeling uncertainty, **8**-15
Models of uncertainty, **8**-18
Modified sigmoidal model, corrosion-fatigue-damage modeling, **28**-16
Monotone measures
fuzzy, **9**-2
generalized information theory, **9**-21–22
Monte Carlo simulation, damage accumulation stages, fatigue response prediction, **42**-25–29
Motivational bias, in expert knowledge reliability characterization, **13**-19–20
Motivational biases, in expert knowledge reliability characterization, **13**-19
Multiattributable utility function, utility method, **23**-10–12
Multiattribute automotive design under uncertainty, **37**-1–19
body-door, **37**-5–8
bumper, preference function, **37**-10, **37**-11
bumper standardization, **37**-8–11
deflection requirement, **37**-11
finite element, vehicle side-impact example, **37**-14
mathematical model validation, **37**-16
objectives, aggregation of, bumper, **37**-10
preference aggregation theory, bumper optimization using, **37**-9–11
quality specification structure, **37**-2–5
vehicle-bumper interaction, design variables, **37**-9
vehicle side-impact, **37**-13–16
Multiattribute utility, **23**-5–7
independence, **23**-6–7
notation, **23**-6
Multicriteria optimization, structural system, **24**-15–16
Multiple-band-pass demodulation, nondeterministic hybrid architectures, vehicles, **41**-19–27
Multiple-band-pass fault-severity index, nondeterministic hybrid architectures, vehicles, **41**-26–27
Multiple-fault pattern, stochastic diagnostics, **41**-13

Multivariate reliability analysis, nondeterministic hybrid architectures, vehicles, **41**-3–4, **41**-12–16
Multivariate stochastic analysis
   benefits of, **41**-20
   nondeterministic hybrid architectures, vehicles, benefits of, **41**-20

N

Nataf distribution, random variables with, **14**-4-5
Nature of uncertainty, taxonomies according to, **8**-10–11
NDA. *See* Nondeterministic analysis
Necessity measures, **9**-30
NESSUS. *See* Numerical evaluation of stochastic structures under stress
Nested structural system optimization, **24**-12–14
Network-based fuzzy inference models, nondeterministic adaptive, **41**-4–6
Neural networks, in transition assessment, **3**-11
Nondeterministic analysis
   advances in, **1**-1–3
   applications of, **1**-4
   bounding uncertainties in, simulation models, **2**-6–7
   categories, **2**-5–7
   civil engineering-based reliability developments, **3**-9
   complex systems, **2**-7
   confidence interval modeling, **3**-17–18
   design environment, **3**-2–6
   design experiments, **3**-13
   design technology, **3**-6–7
   early developments, **3**-8
   error estimation, **3**-16–17
   expert systems, **3**-13
   finite element based probabilistic analysis, **3**-10–11
   fuzzy theory, **3**-12
   health management system, **3**-6
   history of, **3**-7–10
   information fusion, **3**-2
   interval arithmetic, **3**-12–13, **12**-1–21
   method validation, **3**-14–16
   model updating, **3**-3
   neural networks, **3**-11
   optimization, **3**-13
   perspectives on, **2**-1–19
   probabilistic data expert, **3**-4
   probabilistic error bounds, **3**-5
   probabilistic mesomechanics, **3**-5
   probabilistic methods, **3**-6–11
   probabilistic model helper, **3**-5
   probabilistic updates, **3**-4
   probability integration algorithms, **3**-9–10, **3**-16
   quality control, **2**-7
   reliability assessment, **3**-18–19
   research, learning network, **2**-15–17
   research engineering design transition, **3**-1–24
   response surface generator, **3**-5
   response surface methods, **3**-13
   safety assessment methods, **3**-18–19
   status of, **1**-1
   system reliability interface, **3**-6
   tool sets in design environment, **3**-2
   transition assessment, **3**-11–13
   uncertainty management, **2**-7
   unknown-unknown errors, **3**-17
   validation errors, **3**-17
   verification, **3**-14–16
   Weibull models, aircraft engine field support, **3**-8–9
Nondeterministic hybrid architectures, **41**-1–28
   ambiguity reduction, **41**-20
   automated diagnostics, **41**-19–27
   automated fault classification, **41**-24–25
   automatic feature extraction, **41**-22–23
   data uncertainty, incorporation of, **41**-20
   decision-level fusion, **41**-4, **41**-16–20
   fault basins of attraction, **41**-20
   fault-detection sensitivity, **41**-10–11
   feature mapping, **41**-20
   feature vector, engine behavior, **41**-6–8
   feature vector extraction, **41**-3, **41**-19
   functional variabilities, **41**-7
   fuzzy inference models, **41**-4–6
   in-operation conditions, predictions for, **41**-7
   mathematical model validation, **41**-4–6
   multiple-band-pass demodulation, **41**-19–27
   multiple-band-pass fault-severity index, **41**-26–27
   multiple fault detection, **41**-20
   multiple-fault pattern, stochastic fault diagnostics, **41**-13
   multivariate reliability analysis, **41**-3–4, **41**-12–16
   multivariate stochastic analysis, benefits of, **41**-20
   partial-engine transient model, **41**-9–10
   prognostic output, **41**-20
   quasi-stationary engine model, **41**-8
   reliability-based reasoning, **41**-20
   robust operational performance, **41**-20
   single fault pattern, stochastic fault diagnostics, **41**-13
   sparse sensor array studies, **41**-8–10
   statistical variabilities, (random part) from functional variabilities, **41**-7
   stochastic fault diagnostics, aspects related to, **41**-13
   stochastic tools development, **41**-3–4
   transient engine model, **41**-9
NonGaussian vectors, stochastic simulation, **20**-6–8
Nonparametric metamodeling (response-surface) methods, probabilistic analysis, **38**-10–19
Nonparametric multiple-regression techniques, **38**-10–11
Nonspecificity
   conflict and, **8**-11
   vagueness and, **8**-11
Normal distribution example plots, **26**-3
Normal statistical distribution, **26**-5
Normalization condition, **9**-6
Notch effects, computer-aided engineering methodology, **36**-9
Number line, in expert knowledge reliability characterization, **13**-24
Numerical evaluation of stochastic structures under stress (NESSUS), **32**-2
Numerical simulations, verification of, **2**-9
Numerical solution of predictive models, **8**-18

O

One-dimensional random variables, stochastic simulation, **20**-3–5
   bivariate transformation, **20**-4
   density decomposition, **20**-4–5
   inverse probability transformation, **20**-3–4
   sampling acceptance-rejection, **20**-4
   uniform random numbers, **20**-3
One-shot optimization, in structural system optimization, **24**-12–14
Orthopedic implant cement loosening, reliability assessment, **32**-21–23
Outcrossings, time-variant reliability, **18**-3
Overload-stress reliability model, accelerated life testing, **25**-6–7

P

P-boxes, generalized information theory, **9**-27–29
Parallel systems, reliability assessment of, **15**-18-26
   assessment of probability of failure of parallel systems, **15**-19–20
   with equally correlated elements, **15**-24–26
   equivalent linear safety margins for parallel systems, **15**-22–24
   reliability bounds for parallel systems, **15**-20–22
   with unequally correlated elements, **15**-26
Parameter estimates, **26**-5
Parametric multiple-regression techniques, probabilistic analysis, **38**-10–11
Partial-engine transient model, nondeterministic hybrid architectures, vehicles, **41**-9–10
Partitioning, sample, computer experiments, **38**-11–12
PDE. *See* Probabilistic data expert
PEB. *See* Probabilistic error bounds
Performance measure approach, in structural system optimization, **24**-7–8, **36**-13–15
Performance-optimal design, **11**-7–8
Petrov-Galerkin scheme, stochastic reduced basis projection schemes, **21**-18–19
Philosophy of local knowledge, **13**-3
Pipeline state, identification of, **40**-13
Pipeline systems quality, **40**-4–8
Pipelines, risk assessment, reliability-based maintenance, **40**-1–38
   block-module model, pipeline risk control, **40**-22–23
   central maintenance problem, pipeline segment, **40**-11–13
   damage due to pipeline failure, assessing, **40**-10–11
   defect growth, **40**-17
   defect-growth processes, identification of, **40**-14–16
   diagnostics in, **40**-13–17
   maintenance, risk-based, **40**-24–25
   maintenance optimization, **40**-4–8
   optimal cessation, pipeline segment performance, **40**-17–20
   pipeline state, identification of, **40**-13
   pipeline systems quality, **40**-4–8
   repair, prioritizing pipeline segments for, **40**-31–35
   risk control, **40**-4–8
   risk optimization flowchart, **40**-25–28
Piston-valve assembly, explosive-actuated, reliability assessment, **32**-24–28
Plausibility
   evidence theory, **10**-4–6, **10**-20–24
   generalized information theory, **9**-22
PMA. *See* Performance measure approach
Poisson distribution, **26**-5
   example plot, **26**-7
Polyhedral models, response surfaces for reliability assessment, **19**-10
Polynomial chaos expansions, stochastic finite element analysis, **21**-8–10
   theorems, **21**-10
Polynomial chaos projection schemes, stochastic finite element analysis, **21**-10–14
   nonlinear dependence, **21**-12–13
Polynomial chaos series-based simulation models, **20**-15–16
Polynomial fitting, probabilistic analysis, **38**-13–15
Possibilistic histogram, **9**-31–34
Possibility distributions, **9**-29
Possibility measures, **9**-29
   generalized information theory, **9**-29–31
Possibility theory, **9**-2
   generalized information theory, **9**-29–31
Power-law pit model, corrosion-fatigue-damage modeling, **28**-22
Predicting outcome of action, **8**-13–14
Predicting system response due to uncertain parameters, interval methods, **12**-8–17
Probabilistic, possibilistic concepts, relations, **9**-33–34
Probabilistic analyses, **4**-9–11, **17**-1–19, **29**-4–9
   analytical methods, **38**-2–5
   classification, dynamic system, **17**-7
   component-level reliability analysis, **38**-2–4
   crash safety, **38**-16–19
   crossings of level during period, **17**-9
   damage, due to cyclic loading, probability of fatigue failure, **17**-16–17
   data requirements, **26**-23–24
   density function, peaks, **17**-9–10
   design, automotive industry, **38**-1–28
   deterministic parameters, **17**-7
   dynamic system, **17**-1–19
   expertise, **26**-24
   failure analysis, **17**-8–10, **17**-12–17
   fatigue failure, **17**-15–16
   first excursion failure, **17**-12–15
   first passage time, **17**-10
   flutter, aircraft lifting surfaces, **29**-4
   general dynamic system, **17**-2–10
   glass guidance design reliability, **38**-8–10
   impulse response, or Green's function, **17**-7
   insignificant input variable elimination, backward screening, **38**-16
   liftgate reliability, **38**-6–8
   limit-cycle oscillation, airfoils, panels, **29**-4–9
   linear system, stochastic response failure analysis, **17**-10–17

local polynomial fitting, **38**-13–15
main-effects estimate, **38**-15–16
metamodel prediction error, cross validation, **38**-13
model of random processes, **17**-3–8
modeling random processes, **17**-2–8
nonparametric metamodeling (response-surface) methods, **38**-10–19
nonparametric multiple-regression techniques, **38**-10–11
nonstationary inputs, response to, **17**-11–12
parametric multiple-regression techniques, **38**-10–11
probabilistic design, **29**-9–11
random parameters, time invariant, **17**-7
random processes, **17**-2–3
reliability methods, **38**-2–16
response, calculation of, **17**-5–8
robust engineering, variation reduction in, **38**-19–22
sample partitioning, computer experiments, **38**-11–12
simulation-based methods, **38**-4–6
stationary inputs, response to, **17**-11
stochastic response, **17**-11–12, **17**-17
system-level reliability analysis, **38**-5–6
time variant system, parameters, random, **17**-7
uncertain strength, reliability assessment of systems with, **17**-18
uniform sampling, computer experiments, **38**-11–12
variable screening, **38**-15–16
Probabilistic data expert, **3**-4
Probabilistic density, **9**-18
Probabilistic error bounds, **3**-5
Probabilistic failure, weakest-link, **43**-1–10
composite materials, applications to, **43**-8–9
failure-location predictions, **43**-3–5
static structures, theory, **43**-1–3
time-dependent failure, extension to, **43**-5–8
Probabilistic gas turbine engine turbomachinery design, **30**-1–21
blade frequency-probabilistic Campbell diagram, **30**-12–13
compressor, probabilistic design, **30**-7–19
failure probability calculation, **30**-16–18
fan, turbine blade probabilistic design, **30**-7–19
field effects, **30**-19
field-management risk assessments, **30**-7
forced-response prediction, **30**-14–15
future probabilistic efforts, **30**-7
geometry variation, **30**-10–12
material capability, **30**-15–16
mistuning, **30**-13–14
new parts, rotor design/fracture mechanics, probabilistic, **30**-5–7
predictions, updating of, **30**-18–19
rotor design/fracture mechanics, probabilistic, **30**-4–7
traditional reliability engineering, **30**-2–4
turbine blade, probabilistic design, **30**-7–19
Probabilistic mesomechanics, **3**-5
Probabilistic micromechanics, **42**-4–5
Probabilistic microstructural methods, **42**-1–38
applied microstress, **42**-24–25
block slip band, **42**-28

continuously distributed dislocations, model based on, **42**-13–18
crack nucleation, **42**-6–11, **42**-25, **42**-29
damage-accumulation models, **42**-5–19
fatigue-life predictions, **42**-29–33
frictional shear stress, **42**-24–25
grain orientation, **42**-20–24
grain size, **42**-19–20
long-crack growth, **42**-18–19, **42**-29
mathematical model validation, **42**-5–6
mesodomains, mechanical component, **42**-5
micromechanical crack nucleation models, **42**-8–11
micromechanics, **42**-4–5
microstructurally small-crack growth, **42**-11–18
Monte Carlo simulation, damage accumulation stages, **42**-25–29
overview, **42**-5–6
propagated slip band, **42**-28
random variable statistics, estimate of, **42**-19–25
scatter, in fatigue life, **42**-2–3
small-crack growth, **42**-11–13, **42**-25–32
stages of fatigue, **42**-2
total fatigue life, predicted, **42**-32–33
Probabilistic model helper, **3**-5
Probabilistic phase, in automotive design, **5**-10–12
Probabilistic progressive buckling, adaptive trusses, **35**-1–27
adaptive structure, **35**-18–21
buckling of columns, **35**-4
conventional trusses, probabilistic progressive bucking of, **35**-4
finite element model, **35**-2–3
first buckled member, **35**-5–6
initial eccentricity, probabilistic buckling, **35**-11–14
intelligent structure, **35**-23–25
mathematical models, validation of, **35**-4–5
probabilistic model, **35**-4
probabilistic truss end-node displacements, **35**-9–11
second/third/fourth buckled members, **35**-7–8
smart structure, **35**-21–22
Probabilistic representations, generalized information theory, **9**-6–12
Probabilistic response, construction of, evidence theory, **10**-15–16
Probabilistic updates, **3**-4
Probability
density function, **9**-7
of detection (POD) curves, **28**-38–39
distribution function, **26**-5
integration algorithms, **3**-9–10, **3**-16
interpretations of, generalized information theory, **9**-8–9
likelihood approaches for, **9**-9–10
Probability space, **9**-7
Probability theory, **9**-2
in expert knowledge reliability characterization, **13**-26
Problem definition, in expert knowledge reliability characterization, **13**-5
Problem description, evidence theory, **10**-12–13
Problem identification, in expert knowledge reliability characterization, **13**-5

Process for eliciting expertise, in expert knowledge reliability characterization, 13-9–10
Projection schemes in stochastic finite element analysis, 21-1–33
   analysis of foundation on randomly heterogeneous soil, 21-24–29
   finite element formulations for random media, 21-5–8
   Krylov subspace, 21-14–15
   nonlinear dependence, 21-12–13, 21-19–20
   numerical examples, 21-22–29
   Petrov-Galerkin scheme, 21-18–19
   polynomial chaos expansions, 21-8–10
   polynomial chaos projection schemes, 21-10–14
   postprocessing techniques, 21-21–22
   random field discretization, 21-6
   spatial discretization, 21-6–8
   statistical moments, distribution, 21-21
   stochastic Krylov subspace, 21-14–16
   stochastic reduced basis projection schemes, 21-16–21
   strong Galerkin scheme, 21-17–18
   theoretical analysis of convergence, 21-20–21
   two-dimensional thin plate, 21-22–24
   weak Galerkin scheme, 21-16–17
Propagated slip band, fatigue response prediction, 42-28
Propagation of uncertainty, 1-2–3
Purchase price of parts, in automotive design, 5-12

Q

Quantum mass ratio
   logarithmic variance, 16-4–9
   probability distribution of, 16-2–4
Quantum-physics-based probability models, 16-1–13
   applications, 16-11–12
   geometric mean, 16-2
   logarithmic variance, 16-4–9
   probability distribution of, 16-2–4
   quantum size ratio, probability distribution, 16-9–11
Quantum size ratio, probability distribution, 16-9–11
Quasi-stationary engine model, nondeterministic hybrid architectures, vehicles, 41-8
Question phrasing, in expert knowledge reliability characterization, 13-20–22

R

RAFT. *See* Risk optimization flowchart
Raleigh distribution, 8-15
Random intervals, 9-27
   in expert knowledge reliability characterization, 13-24
   in generalized information theory, 9-27–29
Random sets, generalized information theory, 9-22–26
Random variable statistics, estimate of, fatigue response prediction, 42-19–25
Randomly heterogeneous soil, analysis of foundation on, 21-24–29
Raw material prices, in automotive design, fluctuations in, 5-12
RBDO. *See* Reliability-based design optimization
Reduced basis projection schemes, stochastic finite element analysis, 21-16–21
   nonlinear dependence, 21-19–20
   Petrov-Galerkin scheme, 21-18–19
   strong Galerkin scheme, 21-17–18
   theoretical analysis of convergence, 21-20–21
   weak Galerkin scheme, 21-16–17
Rejection sampling, variance-reduction, 27-17
RELDOS. *See* Reliability-based design optimization strategy
Reliability assessment. *See also under* specific assessment strategy
   brittle elements, two-story braced frame with, 15-34–36
   Ditlevsen bounds, 15-14–15
   ductile elements, two-story braced frame with, 15-32–34
   elastic plastic framed structure, 15-36–38
   equivalent linear safety margins for parallel systems, 15-22–24
   examples, 15-32–43
   failure mechanisms, identification of, 15-26–38
   formal representation of systems, 15-6–10
   fundamental systems, 15-2–4
   gas turbine rotor risk assessment, 32-7–9
   levels, 15-27–32
   modeling of systems, 15-4–6
   parallel systems, 15-18–26
   reliability-based optimal design of steel-jacket offshore structure, 15-40
   reliability-based optimal maintenance strategies, 15-41–43
   simple bounds, 15-13–14
   structural system modeling, 15-2–10
   tubular joint, reliability of, 15-38–40
Reliability-based design optimization, 24-1-26, 31-1–19, 34-1–24. *See also under* specific design
   aerospace application examples, 34-6–21
   airplane wing, shape optimization, 34-12–16
   analysis procedure, 31-5–6
   axial compressor blade, shape organization, 34-7–12
   certification problem, 31-3–4
   cryogenic tank, 31-12–16
   data preparation, 31-5–6
   data problem, 31-3
   design system, 31-2
   deterministic optimization, 31-6, 31-8, 31-11, 31-13
   efficiency problem, 31-4
   example, 31-10–11
   fundamental theory, 31-7
   generation of response surfaces, 31-6
   model approximation methods, 34-2–3
   optimization analysis, 31-7
   overview, reliability-analysis methods, 34-2–5
   probabilistic analysis, 31-2
   probabilistic load models, 31-6
   probabilistic resistance models, 31-6
   probabilistic sensitivity factors, 31-15
   random variables screening, 31-9
   reliability approximation methods, 34-3–5
   reliability assessment, 31-6, 31-8–9, 31-11, 31-13–14
   reliability constraint screening, 31-9–10
   simple analytical problem, 31-10–11

system reliability level selection, **31**-16
target reliability problem, **31**-4
theoretical derivation, **31**-7–10
user-friendliness problem, **31**-4–5
validation of CAE (CAE) models, **31**-5
Reliability-based reasoning, multivariate stochastic analysis, **41**-20
Reliability bounds, series systems, **15**-13–15
Reliability certification, **1**-3, **26**-4, **27**-1–18
Reliability index approach, in structural system optimization, **24**-5–7
Reliability testing **27**-1–18
Research, **7**-1–15
   aggregation, information, **7**-7–9
   confidence intervals, **7**-8
   data congeries, **7**-8–9
   empirical Bayes, **7**-6
   engineering design, transition, **3**-1–24
   hierarchial Bayes, **7**-6–7
   inductive approach, **7**-5–7
   information, **7**-11–13
   learning network, nondeterministic approache, **2**-15–17
   new approach, need for, **7**-3–5
   sensitivity analysis, **7**-13
   time dependent processes, **7**-9–11
Response, predicting, **8**-4
Response-surface methods, probabilistic analysis, **38**-10–19
Response surfaces, **19**-1–24
   adaptation of response surface, **19**-14–15
   analysis of variance, **19**-5–8
   basic formulation, **19**-3–4
   choice of method, **19**-14
   design of experiments, **19**-10–13
   error checking, **19**-14–15
   expotential relationships, **19**-9–10
   first-order polynomial, **19**-8–9
   generator, **3**-5
   linear model, **19**-4–5
   linear response surface, **19**-15–17
   methodology, **2**-11
   models, response surface, **19**-3–10
   nonlinear finite element structure, **19**-19–21
   nonlinear response surface, **19**-17–19
   polyhedral models, **19**-10
   redundant designs, **19**-12–13
   regression, **19**-4–5
   reliability computation, **19**-14–15
   saturated designs, **19**-11–12
   second-order polynomial, **19**-8–9
   transformations, **19**-10–11
Risk-based approach, **28**-5–7
   local failure criteria, **28**-6–7
   risk-based condition assessment, **28**-5–6
   uncertainty in failure criteria, **28**-7
Risk management process, **2**-12
Risk optimization flowchart, pipeline, **40**-25–28
Roadarm durability, tank, computer-aided engineering methodology, **36**-15–18
Robust control
   overview, **29**-11–13
   uncertain load, **11**-5–6
Robust engineering
   variation reduction, **25**-1–26
   variation reduction in, **38**-19–22
Robust operational performance
   feature vector extraction, **41**-19
   multivariate stochastic analysis, **41**-20
   nondeterministic hybrid architectures, vehicles, **41**-20
Robustness, **11**-18
Roles in expert knowledge reliability characterization, **13**-4
Rotor design/fracture mechanics, probabilistic, **30**-4–7
   field-management risk assessments, **30**-7
   future probabilistic efforts, **30**-7
   new parts, probabilistic life analysis for, **30**-5–7
Rough sets, **9**-2
Rounding-off errors, interval methods, **12**-18–19
RSG. *See* Response surface, generator

S

Safety assessment methods, **3**-18–19. *See also* Reliability
Sample partitioning, computer experiments, **38**-11–12
Satisficing, **11**-2
Saturated designs, response surfaces for reliability assessment, **19**-11–12
Scatter, in fatigue life, **42**-2–3
Scientific, engineering modeling, interval methods, **12**-3
Second-order polynomial, response surfaces for reliability assessment, **19**-8–9
Second-order reliability method, **14**-15–19
   reliability analysis of column by SORM, example, **14**-19
Sensitivity analysis, **4**-8–9, **7**-13
   interval methods, **12**-9
   in structural system optimization, **24**-8–11
Sensor array studies, nondeterministic hybrid architectures, vehicles, **41**-8–10
Set-theoretical approach, **9**-2–4
Ships, reliability assessment, **39**-1–38
   corrosion damage, **39**-24–25
   corrosion wastage models, **39**-13–15
   denting, local, **39**-26
   effect of corrosion, **39**-9, **39**-11
   effect of denting, local, **39**-10, **39**-11
   effect of fatigue cracking, **39**-9–11
   extreme hull girder load models, **39**-11–12
   fatigue crack model, **39**-15–19
   fatigue cracking, **39**-25–26
   historical overview, **39**-2–3
   hull girder collapse, target reliability for ships against, **39**-4–5
   hull sectional properties, example vessels, **39**-22
   operational conditions, scenarios for, **39**-20–23
   primary failure mode, **39**-7–8
   quaternary failure mode, **39**-8–9
   repair strategies, **39**-27–35
   secondary failure mode, **39**-8
   still-water bending moment, **39**-11
   structural damage, scenarios for, **39**-24–26
   tertiary failure mode, **39**-8
   time-dependent reliability assessment, **39**-20–35
   time-dependent structural damage, prediction of, **39**-12–19

ultimate hull girder strength models, **39**-6–11
ultimate limit-state equations, **39**-5–6
wave-induced bending moment, **39**-11–12
Simple conceptual examples, evidence theory, **10**-10–12
Simulation, components of, **2**-13–15
Simulation-based methods, system reliability analysis, **38**-4–6
Simulation technologies, enhancing modeling, **2**-7–9
Simultaneous corrosion-fatigue (SCF) model, **28**-31
Single fault pattern
    nondeterministic hybrid architectures, vehicles, stochastic fault diagnostics, **41**-13
    stochastic diagnostics, **41**-13
Small-crack growth
    fatigue response prediction, **42**-25–28
    micromechanics of, fatigue response prediction, **42**-11–12
Small-crack-growth, fatigue response prediction, **42**-12–13
Smart structure, probabilistic progressive buckling, conventional, adaptive trusses, **35**-21–22
Specific random sets, generalized information theory, **9**-26
Specimen level, mesodomain, mechanical component, **42**-5
Stages in expert knowledge reliability characterization, **13**-5
Stages of fatigue, **42**-2
Star graph, expert knowledge reliability characterization, **13**-12
Static structures, weakest-link probabilistic failure, **43**-1–3
Statistical modeling, updating, Bayesian modeling, updating, **22**-3–5
Statistical testing, nondeterministic analysis, **26**-1–25
    Bayesian analysis, **26**-22–23
    beta distribution example plots, **26**-6
    beta statistical distribution, **26**-5
    binomial distribution, **26**-5
    binomial distribution example plot, **26**-7
    data requirements, probabilistic analysis, **26**-23–24
    expertise, for probabilistic analysis, **26**-24
    hyperdistribution statistical distribution, **26**-5
    hyperparameterization example plot, **26**-8
    input variables, statistical distribution of, **26**-2–17
    lack of data, **26**-14–16
    large amount of data, **26**-7–13
    little data, **26**-14–16
    lognormal distribution, **26**-5
    lognormal distribution example plots, **26**-4
    misspecification, impact of, **26**-2–3
    normal distribution example plots, **26**-3
    normal statistical distribution, **26**-5
    output variables, **26**-18–20
    parameter estimates, **26**-5
    philosophical issues, **26**-23–24
    Poisson distribution, **26**-5
    Poisson distribution example plot, **26**-7
    probability distribution function, **26**-5
    selection process, classes of distribution, **26**-8
    small to moderate amount of data, **26**-13–14
    statistical testing for reliability certification, **26**-20–23
    stress *vs.* strength plot, **26**-2
    traditional methods, **26**-20–21
    types of statistical distribution, **26**-3–7
    uniform distribution, **26**-5

uniform distribution example plot, **26**-6
Weibull distribution example plots, **26**-4
Weibull statistical distribution, **26**-5
Statistically independent random variables, **14**-3–4
Steel-jacket offshore structure
    reliability-based optimal design of, **15**-40
    system reliability, **15**-40
Step-stress accelerated testing model, accelerated life testing, **25**-14–25
    Weibull step-stress model, **25**-20–22
Stochastic fault diagnostics, **41**-13
    multiple-fault pattern, **41**-13
    nondeterministic hybrid architectures, vehicles, multiple-fault pattern, **41**-13
    single fault pattern, **41**-13
Stochastic Finite Element Analysis, **21**-1–33
Stochastic mechanics, research, **6**-1–11
    current methods, **6**-8–9
    deterministic input, **6**-7–8
    deterministic systems, **6**-2–4
    research needs, trends, **6**-9
    stochastic input, deterministic systems, **6**-5–7
Stochastic response, failure analysis, linear system
    damage, due to cyclic loading, probability of fatigue failure, **17**-16–17
    failure analysis, **17**-12–17
    fatigue failure, **17**-15–17
    first excursion failure, **17**-12–15
    nonstationary inputs, response to, **17**-11–12
    probabilistic analysis, **17**-10–17
    stationary inputs, response to, **17**-11
    stochastic response, **17**-11–12
Stochastic simulation, **20**-1–34
    bivariate transformation, **20**-4
    computing tail distribution probabilities, **20**-25–28
    covariance-based simulation models, **20**-12–15
    density decomposition, **20**-4–5
    dynamic Monte Carlo (DMC), **20**-22–25
    employing stochastic linear PDE solutions, **20**-28–30
    Gaussian vectors, **20**-506
    incorporating modeling uncertainties, **20**-30–31
    inverse probability transformation, **20**-3–4
    nonGaussian vectors, **20**-6–8
    one-dimensional random variables, **20**-3–5
    one-level hierarchial simulation models, **20**-11–16
    polynomial chaos series-based simulation models, **20**-15–16
    sampling acceptance-rejection, **20**-4
    sequential importance sampling (SIS), **20**-21–22
    simulation in high-dimensional stochastic spaces, **20**-21–31
    stochastic fields (or processes), **20**-8–21
    two-level hierarchical simulation models, **20**-16–21
    uniform random numbers, **20**-3
    vectors, with correlated components, **20**-508
Stochastic systems
    current methods, **6**-7–9
    deterministic input, **6**-7–8
    input, **6**-8–10
    nonlinear systems, **17**-17
    research needs, trends, **6**-7–9

Stochastic tools development, hybrid architectures, vehicles, **41**-3–4
Stochastic variability
    crack initiation, **28**-12–14
    crack propagation, **28**-17–21
Stoplight chart, expert knowledge reliability characterization, **13**-14
Stratified sampling, variance-reduction, **27**-6–8, **27**-16–17
Strength limit states, composite structures, materials, **44**-2–7, **44**-8–18
    analysis, **44**-8–9
    critical-component failure, **44**-15–16
    deterministic initial screening, **44**-14–15
    fast branch-and-bound method, **44**-14
    multiple failure sequences, **44**-6
    numerical examples, **44**-6–7, **44**-11–13, **44**-16–18
    ply-level limit states, **44**-3–4, **44**-9–10
    reliability, **44**-3–4
    shear deformation theory, **44**-8–9
    single failure sequence, **44**-4–6
    stiffness modification, progressive failure analysis, **44**-10–11
    system failure probability, **44**-4–6
    weakest-link model, **44**-15
Stress *vs.* strength plot, **26**-2
Structural reliability predictions, Bayesian modeling, updating, **22**-14–16
    asymptotic approximations, for large number of data, **22**-15–16
    computational issues, **22**-14–15
    formulation, as probability integral, **22**-14
Structural system modeling, reliability, **15**-2–10. *See also* Reliability
    formal representation of systems, **15**-6–10
    fundamental systems, **15**-2–4
    level N, **15**-4–5
    mechanism level, **15**-6
Structural system optimization, **24**-1–32
    aerodynamics, **24**-21–23
    aeroelasticity, **24**-21–23
    aircraft systems, **24**-23
    algorithms, **24**-12–15
    composite structures, **24**-21
    design sensitivities, **24**-9–11
    life-cycle cost-based, **24**-15–24
    life-cycle cost reliability-based optimization, **24**-15–16
    material cost-based, **24**-16–17
    multicriteria optimization, **24**-15–16
    nested optimization, **24**-12–14
    one-shot optimization methods, **24**-12–14
    optimization methods, **24**-12–15
    performance measure approach, **24**-7–8, **24**-11
    reliability index approach, **24**-5–7
    sensitivity analysis, **24**-8–11
    standard optimization problem, **24**-4–5
Subgrain level, mesodomain, mechanical component, **42**-5
Success and failure for problem area, in expert knowledge reliability characterization, **13**-13
Suppliers, emergence, decline of, in automotive design, **5**-12
System identification, **11**-17–21
    example, **11**-19–21
    optimal identification, **11**-17
    robustness, **11**-17–19
    uncertainty, **11**-17–19
System reliability, **15**-1–45. *See also* Reliability
    analysis, analytical methods, **38**-5
    assessment of probability of failure of parallel systems, **15**-19–20
    brittle elements, **15**-34–36
    Ditlevsen bounds, **15**-14–15
    ductile elements, **15**-32–34
    elastic plastic framed structure, **15**-36–38
    equally correlated elements, series systems with, **15**-15–16
    equivalent linear safety margins for parallel systems, **15**-22–24
    examples, **15**-32–43
    failure mechanisms, identification of, **15**-26–38
    formal representation of systems, **15**-6–10
    fundamental systems, **15**-2–4
    interface, **3**-6
    modeling of systems at level N, **15**-4–5
    modeling of systems at mechanism level, **15**-6
    optimal maintenance strategies, **15**-41–43
    parallel systems, reliability assessment of, **15**-18–26
    parallel systems with equally correlated elements, **15**-24–26
    parallel systems with unequally correlated elements, **15**-26
    probability of failure of series systems, **15**-12–13
    reliability bounds for parallel systems, **15**-20–22
    reliability bounds for series systems, **15**-13–15
    simple bounds, **15**-13–14
    steel-jacket offshore structure, **15**-40
    structural system modeling, **15**-2–10
    tubular joint, **15**-38–40
    unequally correlated elements, series systems with, **15**-17–18

T

Tail distribution probabilities, simulation in high-dimensional stochastic spaces, **20**-25–28
Tank roadarm durability, computer-aided engineering methodology, **36**-15–18
Taxonomies of uncertainty, **8**-6–18
Technology transition, mechanical design and, **4**-12–13
Theories of uncertainty, **8**-4–5, **10**-3
Three-dimensional analysis, composite structures, materials, **44**-8–13
    analysis, **44**-8–9
    numerical example (composite plate), **44**-11–13
    ply-level limit states, **44**-9–10
    shear deformation theory, **44**-8–9
    stiffness modification, progressive failure analysis, **44**-10–11
Time-dependent failure, weakest-link probabilistic failure, **43**-5–8
Time dependent processes, **7**-9–11
    distribution, process model, **7**-9–11
    process model, **7**-9–11

Time-dependent reliability assessment, **39**-20–35
   operational conditions, scenarios for, **39**-20–23
   sea states, scenarios for, **39**-20–23
Time-dependent structural damage, prediction of, **39**-12–19
   corrosion wastage models, **39**-13–15
   fatigue crack model, **39**-15–19
Time-variant reliability, **14**-19–21, **18**-1–15, **33**-1–28
   active load, frame subject to, **33**-11–12
   deformable system, **33**-3–5
   design values of loads, combination, **33**-12–15
   directional simulation, **18**-8–9
   downsizing dimension, reliability problem, **33**-22–23
   elastic-plastic frames, subject to combination of random loads, **33**-10–11
   ensembled upcrossing rate approach, **18**-9–10
   examples, **33**-25
   fast probability integration, **18**-7
   first passage probability, **18**-3–4
   Gaussian processes, **18**-7–8
   large mechanical systems, subjected to combination of diffusion-type Markov loads, by generalized Pontryagin equations, **33**-15–25
   linear limit state functions, **18**-7–8
   load combinations, **18**-11–13
   loads, as processes, **18**-2–3
   Markov loads, **33**-15–20
   mean out-crossing rate of column, stochastic loads, **14**-21
   Monte Carlo, **18**-8
   multiple loads: outcrossigs, **18**-3
   outcrossing rate estimation, **18**-5–6
   Pontryagin equations, **33**-20–22
   random loads, combination of, **33**-20
   redundant structures, subject to combinations of Markov-type loads, **33**-1–15
   snow load, **33**-18
   solution methods, **18**-10–11
   some remaining research problems, **18**-13–14
   strength (or barrier) uncertainty, **18**-6
   system life, system reliability, transition, **33**-23–25
   technological loads, **33**-19–20
   time-variant reliability estimation techniques, **18**-7–11
   unloading, frame subject to, **33**-11–12
   upcrossing rates estimation, **18**-4–5
   wind load, **33**-19
Tool sets in design environment, **3**-2
Traditional analysis, evidence theory, **10**-14–16
Trailer drawbar, computer-aided engineering methodology, **36**-3
Transformations, response surfaces for reliability assessment, **19**-10–11
Transient engine model, nondeterministic hybrid architectures, vehicles, **41**-9
Transition assessment, **3**-11–13
   design of experiments, **3**-13
   expert systems, **3**-13
   fuzzy theory, **3**-12
   interval arithmetic, **3**-12–13
   neural networks, **3**-11
   nondeterministic optimization, **3**-13
   response surface methods, **3**-13
Translation uncertainties, in automotive design, **5**-12
Transport aircraft wing optimization, **34**-18–21
Triangular conorm, **9**-14
Triangular norm, **9**-14
Trusses, adaptive, conventional, progressive, buckling, **35**-1–27
   adaptive structure, **35**-18–21
   buckling of columns, **35**-4
   conventional trusses, probabilistic progressive bucking of, **35**-4
   finite element model, **35**-2–3
   first buckled member, **35**-5–6
   initial eccentricity, probabilistic buckling, **35**-11–14
   intelligent structure, **35**-23–25
   mathematical models, validation of, **35**-4–5
   probabilistic model, **35**-4
   probabilistic truss end-node displacements, **35**-9–11
   second/third/fourth buckled members, **35**-7–8
   smart structure, **35**-21–22
Tubular joint, reliability of, **15**-38–40
Tunnel vulnerability assessment, reliability assessment, **32**-18–21
Turbine blade, probabilistic design, **30**-7–19
   blade frequency-probabilistic campbell diagram, **30**-12–13
   failure probability calculation, **30**-16–18
   field effects, **30**-19
   forced-response prediction, **30**-14–15
   geometry variation, **30**-10–12
   material capability, **30**-15–16
   mistuning, **30**-13–14
   predictions, updating of, **30**-18–19
Turbine engine, probabilistic, turbomachinery design, **30**-1–21
   blade frequency-probabilistic campbell diagram, **30**-12–13
   failure probability calculation, **30**-16–18
   fan, turbine blade probabilistic design, **30**-7–19
   field effects, **30**-19
   field-management risk assessments, **30**-7
   forced-response prediction, **30**-14–15
   future probabilistic efforts, **30**-7
   geometry variation, **30**-10–12
   material capability, **30**-15–16
   mistuning, **30**-13–14
   new parts, rotor design/fracture mechanics, probabilistic, **30**-5–7
   predictions, updating of, **30**-18–19
   rotor design/fracture mechanics, probabilistic, **30**-4–7
   traditional reliability engineering, **30**-2–4
Turbomachinery design, probabilistic gas turbine engine, **30**-1–21
   blade frequency-probabilistic Campbell diagram, **30**-12–13
   failure probability calculation, **30**-16–18
   fan, turbine blade probabilistic design, **30**-7–19
   field effects, **30**-19
   field-management risk assessments, **30**-7
   forced-response prediction, **30**-14–15

Index I-19

future probabilistic efforts, **30**-7
geometry variation, **30**-10–12
material capability, **30**-15–16
mistuning, **30**-13–14
new parts, rotor design/fracture mechanics, probabilistic, **30**-5–7
predictions, updating of, **30**-18–19
rotor design/fracture mechanics, probabilistic, **30**-4–7
traditional reliability engineering, **30**-2–4
Two-dimensional thin plate, stochastic finite element analysis, finite element formulations for random media, **21**-22–24

U

Ultimate hull girder strength models
   denting, local, effect of, **39**-11
   fatigue cracking, and local denting, effect of, **39**-11
Ultimate limit-state equations, hull girder, **39**-5–6
Uncertain load, opportunity from, **11**-9–11
Uncertainty. *See also under* specific type
   automotive design under, multiattribute, **37**-1–19
   body-door, **37**-5–8
   bounding in simulation models, **2**-6–7
   bumper-deflection requirement, probability of meeting, **37**-11
   bumper performance measures, preference function for, **37**-11
   bumper standardization, **37**-8–11
   bumper variety, **37**-10
   definitions, **2**-3, **8**-5–6
   description, and range of design variables, **37**-15
   engineering decision under, **1**-3
   final values, design variables, **37**-15
   finite element, vehicle side-impact example, **37**-14
   formal preference aggregation theory, **37**-4–5
   information-gap models of, **11**-28–29
   managing, **2**-5
   mathematical model validation, **37**-16
   modeling, **8**-15
   objectives, aggregation of, bumper, **37**-10
   performance measure, automotive body-door, **37**-6
   preference aggregation theory, bumper optimization using, **37**-9–11
   quality specification structure, **37**-2–5
   range of design variables, **37**-15
   reliability-based design optimization, **37**-4
   type of information, **8**-16–18
   types of, **2**-4–5, **8**-12–14
   vehicle-bumper interaction, design variables, **37**-9
   vehicle side-impact, **37**-13–16
Uncertainty-based information theory, **9**-1–3
Uncertainty characterization methods, in automotive design, **5**-10–11
Uncertainty in aeroelasticity, **29**-2–4
   identification, **29**-2–3
   mathematical models, **29**-3–4
   sources, **29**-3
Uncertainty in design decision making, **8**-1–20
   aleatory uncertainty, **8**-6, **8**-16–17
   automotive component, uncertainty in design of, **8**-5

automotive engineering, **8**-2
automotive example, **8**-1–20
causes of uncertainty, taxonomies according to, **8**-6–10
conflict, nonspecificity and, **8**-11
definition of uncertainty, **8**-5–6
epistemic uncertainty, **8**-17–18
extreme value distribution, **8**-15
first principles-based models of uncertainty, **8**-15
framing decision, **8**-13
human errors, **8**-10, **8**-18
information, types of, **8**-14–15
intervals, **8**-14
linguistic information, **8**-14
mathematical models, **8**-9
modeling uncertainty, **8**-15
models, **8**-15, **8**-18
nature of uncertainty, taxonomies according to, **8**-10–11
numerical solution of predictive models, **8**-18
outcome of action, predicting, **8**-13–14
outcomes of actions, evaluating payoff of, **8**-14
predicting response, **8**-4
Raleigh distribution, **8**-15
taxonomies of uncertainty, **8**-6–12
taxonomy, **8**-12–18
theories of uncertainty, **8**-4–5
type of information uncertainty, **8**-16–18
types of, **8**-9, **8**-12–14
uncertainty in models of uncertainty, **8**-13
vagueness, nonspecificity and, **8**-11
validation of mathematical models, **8**-5
Weibull distribution, **8**-15
Uncertainty measures, **2**-4–5
Uncertainty modeling
   historical account of, **2**-3
   interval methods, **12**-2–8
Uncertainty propagation, **1**-2–3
Uncertainty quantification, **2**-8–9
   generalized information theory, **9**-2
Uncertainty reduction, Bayesian modeling
   asymptotic approximation, information entropy, large number of data, **22**-10–11
   computational issues for finding optimal sensor configuration, **22**-12–13
   design optimal sensor configuration, **22**-11–12
   information entropy measure, parameter uncertainty, **22**-9–10
   optimal sensor configuration for selecting model class, **22**-13–14
   optimal sensor location, **22**-9–14
Unemployment levels, in automotive design, **5**-12
Unequally correlated elements, series systems with, **15**-17–18
Uniform distribution example plot, **26**-6
Uniform sampling, computer experiments, **38**-11–12
Uniform statistical distribution, **26**-5
Unknown-unknown errors, **3**-17
Upcrossing rates
   ensembled, time-variant reliability, time-variant reliability estimation techniques, **18**-9–10
   time-variant reliability, **18**-4–5

Updating, Bayesian, **22**-1–20
    asymptotic approximation, information entropy, large number of data, **22**-10–11
    asymptotic approximations, for large number of data, **22**-15–16
    computational issues, **22**-12–15
    damage detection, **22**-6–9
    design, **22**-11–12
    formulation, as probability integral, **22**-14
    information entropy measure, parameter uncertainty, **22**-9–10
    model class selection, **22**-5–6
    sensor configuration for selecting model class, **22**-13–14
    sensor location, **22**-9–14
    statistical modeling, updating, **22**-3–5
    structural reliability predictions, **22**-14–16
    symbols, **22**-19–20
Utility independence, determining, **23**-7
Utility methods, **23**-1–14
    design under uncertainty, **23**-12–13
    engineering applications, **23**-3
    engineering design example, **23**-2–3
    example, **23**-7–13
    expected utility, **23**-12–13
    independence, **23**-6–7
    individual attributes, utilities over, **23**-8–10
    lotteries, **23**-4–5
    mathematical background, **23**-4
    multiattribute utility, **23**-5–7, **23**-10–12
    notation, **23**-6
    utility independence, determining, **23**-7

V

Vagueness, nonspecificity, **8**-11
Validation errors, **3**-17
Validation of mathematical models, in automotive design, **5**-13
Variance-reduction techniques, **27**-1–19
    illustrative application, **27**-12–17
    importance sampling, **27**-8–15
    physical testing, **27**-4–6
    with physical testing, **27**-10–12
    practical issues, **27**-9, **27**-17
    rejection sampling, **27**-17
    robust engineering, uncertainty and, **8**-1–20
    stratified sampling, **27**-6–8, **27**-11–12, **27**-16–17
    virtual testing, using computer models, **27**-2–4
Vector extraction, feature, benefits of, **41**-19
Vectors, interval methods, **12**-6
Vehicle development process, **5**-2
    decision-analytic view, **5**-3
Vehicle engine prognostics health management, **41**-1–28
    ambiguity reduction, **41**-20
    automated diagnostics, **41**-19–27
    automated fault classification, **41**-24–25
    automatic feature extraction, **41**-22–23
    data uncertainty, incorporation of, **41**-20
    decision-level fusion, **41**-4, **41**-16–19
    decision-level fusion benefits, **41**-20
    fault basins of attraction, **41**-20
    fault-detection sensitivity, **41**-10–11
    feature mapping, **41**-20
    feature vector, engine behavior, **41**-6–8
    feature vector extraction, **41**-3
    feature vector extraction benefits, **41**-19
    functional variabilities, **41**-7
    in-operation conditions, predictions for, **41**-7
    mathematical model validation, **41**-4–6
    multiple-band-pass demodulation, **41**-19–27
    multiple-band-pass fault-severity index, **41**-26–27
    multiple fault detection, **41**-20
    multiple-fault pattern, stochastic fault diagnostics, **41**-13
    multivariate reliability analysis, **41**-3–4, **41**-12–16
    multivariate stochastic analysis, benefits of, **41**-20
    nondeterministic adaptive network-based fuzzy inference models, **41**-4–6
    partial-engine transient model, **41**-9–10
    prognostic output, **41**-20
    quasi-stationary engine model, **41**-8
    reliability-based reasoning, **41**-20
    robust operational performance, **41**-20
    single fault pattern, stochastic fault diagnostics, **41**-13
    sparse sensor array studies, **41**-8–10
    statistical variabilities, (random part) from functional variabilities, **41**-7
    stochastic fault diagnostics, aspects related to, **41**-13
    stochastic tools development, **41**-3–4
    transient engine model, **41**-9
VEPHM. *See* Vehicle engine prognostics health management
Vibrating system with uncertain dynamics
    example, **11**-15–17
    maneuvering, **11**-11–16
    model uncertainty, **11**-11–12
    performance optimization with best model, **11**-12–13
    robustness function, **11**-13–15
Virtual product development facilities, **2**-7–8
    safety assessment, **2**-8
    uncertainty quantification, **2**-8–9
Virtual testing, using computer models, **27**-2–4

W

Weakest-link probabilistic failure, **43**-1–10
    composite materials, applications to, **43**-8–9
    failure-location predictions, **43**-3–5
    static structures, theory, **43**-1–3
    time-dependent failure, extension to, **43**-5–8
Wei models, corrosion-fatigue, **28**-29–31
Weibull distribution, **8**-15, **26**-5
    example plots, **26**-4
Weibull models
    aircraft engine field support, **3**-8–9
    step-stress, accelerated life testing, **25**-20–22